The Translational Apparatus

Structure, Function, Regulation, Evolution

The Translational Apparatus

Structure, Function, Regulation, Evolution

Edited by

Knud H. Nierhaus
François Franceschi
Alap R. Subramanian

Max-Planck-Institut für Molekulare Genetik
Berlin, Germany

Volker A. Erdmann

Freie Universität Berlin
Berlin, Germany

and

Brigitte Wittmann-Liebold

Max-Delbrück-Centrum für Molekulare Medizin
Berlin, Germany

Springer Science+Business Media, LLC

Library of Congress Cataloging-in-Publication Data

The Translational apparatus : structure, function, regulation,
  evolution / edited by Knud H. Nierhaus ... [et al.].
       p.   cm.
     "Proceedings of an international conference on the translational
  apparatus, held October 31-November 5, 1992, in Berlin, Germany"-
  -Copr. p.
     Includes bibliographical references and index.
     ISBN 978-1-4613-6021-6   ISBN 978-1-4615-2407-6 (eBook)
     DOI 10.1007/978-1-4615-2407-6
     1. Genetic translation--Congresses. 2. Genetic translation-
  -Regulation--Congresses.   I. Nierhaus, Knud H.
  QH450.5.T723   1993
  574.87'3223--dc20   93-37131
                                                               CIP

Proceedings of an International Conference on The Translational Apparatus, held October 31–
November 5, 1992, in Berlin, Germany

ISBN 978-1-4613-6021-6

© 1993 Springer Science+Business Media New York
Originally published by Plenum Press, New York in 1993
Softcover reprint of the hardcover 1st edition 1993

PREFACE

The conference entitled "The Translational Apparatus" was held in Berlin from October 31 to November 5, 1992, *in honorem et memoriam* of H. G. Wittmann. The presentations of the rewarding, enjoyable and scientifically exciting week in Berlin were a fitting appreciation of this great man, and all participants and speakers contributed to the success of the meeting.

The second reason for holding the Berlin meeting was to convene a group of scientists for a state-of-the-art presentation on ribosomes and related subjects; the outcome is this book. We have broadened the spectrum of topics to cover activities preceding and following pure ribosomal functions, such as "Synthetases" and "Protein Sorting", respectively, and we took the liberty of ordering the contributions in a way which does not always reflect the order of speakers. We are grateful to the authors who patiently tolerated our organization and our regulations for manuscripts.

K.H. Nierhaus
F. Franceschi
A.R. Subramanian

CONTENTS

I. tRNA and Aminoacyl-tRNA Synthetases

II. rRNA and mRNA: Regulation, Processing, Assembly

III. Translational Initiation and Termination

IV. The Elongation Process: Sites, Factors, Nascent Chain

V. Accuracy in Translation

VI. Quaternary Structure,
Functional Centers and Domains in the Ribosome

VII. Translation in Cell Organelles

VIII. Post-Translation: Protein Sorting

IX. The Translational Apparatus and Evolution

THE CRYSTAL STRUCTURE OF SERYL-tRNA SYNTHETASE AND ITS COMPLEXES WITH ATP AND tRNA^Ser

S. Cusack, C. Berthet-Colominas, V. Biou, F. Borel,
M. Fujinaga, M. Hartlein, I. Krikliviy[2], N. Nassar,
S. Price, M.A. Tukalo[2], A.D. Yaremchuk[2] and R. Leberman

European Molecular Biology Laboratory Grenoble Outstation
c/o ILL, 156X, 38042 Grenoble Cedex, France

[2] Institute of Molecular Biology and Genetics
252627 Kiev, Ukraine

INTRODUCTION

Aminoacyl-tRNA synthetases charge specifically their cognate tRNAs with the correct amino acid in a two step reaction in which ATP is used to form the aminoacyl-adenylate intermediate. The high specificity of these enzymes, which conversely implies a strong discrimination against the majority of non-cognate tRNAs and amino acids, is crucial for the fidelity of translation of messenger RNA into protein. In recent years, genetic, biochemical and crystallographic studies have lead to spectacular advances in the understanding of the molecular basis of aminoacyl-tRNA synthetase function, in particular concerning the specific recognition of tRNA by synthetases. These studies have also lead to a definitive classification of the twenty synthetases into two quite different structural classes, representing two distinct solutions to the problem of aminoacylation. This discovery is clearly of major importance to the understanding of the evolution of the translation machinery even though it may be difficult, as with the genetic code, to explain how things came to be as they are.

In this paper, we will first introduce the notion of classes of aminoacyl-tRNA synthetases and then specialise to the structural studies on seryl-tRNA synthetase, the first class 2 synthetase to be structurally characterised, that have been carried out in our laboratory. A preliminary description of new crystallographic results on complexes of seryl-tRNA synthetase with ATP and with a cognate tRNA will be given.

THE CLASSIFICATION OF AMINOACYL-tRNA SYNTHETASES

The definitive classification of aminoacyl-tRNA synthetases into two distinct classes was given by Eriani *et al.* (1990) on the basis of primary sequence analysis, even though the existence of a second structural class was previously demonstrated by the crystal structure of the seryl-tRNA synthetase (Cusack *et al.* 1990). Table 1 summarise the members and the characteristics of each class.

TABLE 1. Characteristics of the two classes of aminoacyl-tRNA synthetases.

CLASS	I			II	
Aminoacylation site on terminal ribose	2'OH			3'OH (except Phe)	
Sequence motifs	..HIGH..			(1) ..GF/YXEVXTP.. (* only)	
	..KMSKS..			(2) ..FR/HXE/D.....R/HXXF..	
	..DWCISRQ..(^only)			(3) ..GXGXGXE/DR..	
Subclass	(a)	(b)	(c)	(a)	(b)
members	Leu^	Tyr	Glu	Ser*	Asp*
	Ile^	Trp	Gln	Thr*	Asn*
	Val^		Arg^	Pro*	Lys*
	Cys^			His*	(Gly)
	Met^			(Ala)	(Phe)
Fold of catalytic domain	Rossmann fold (Tyr, Met, Gln)			Anti-parallel fold (Ser, Asp)	

The ten class 1 synthetases all possess two 'signature' sequences which are usually referred to as the 'HIGH' motif (invariably close to the N-terminus) and the 'KMSKS' motif (where the bold type indicates the only absolutely conserved residues). Both these

TABLE 2. Domain structure of class 2 aminoacyl-tRNA synthetases from E. coli.
The conserved motifs in the catalytic domain are shown in black.
Homologous tRNA binding domains are shown with different shading.
Upon this basis, subclasses can be defined (Cusack et al. 1991).

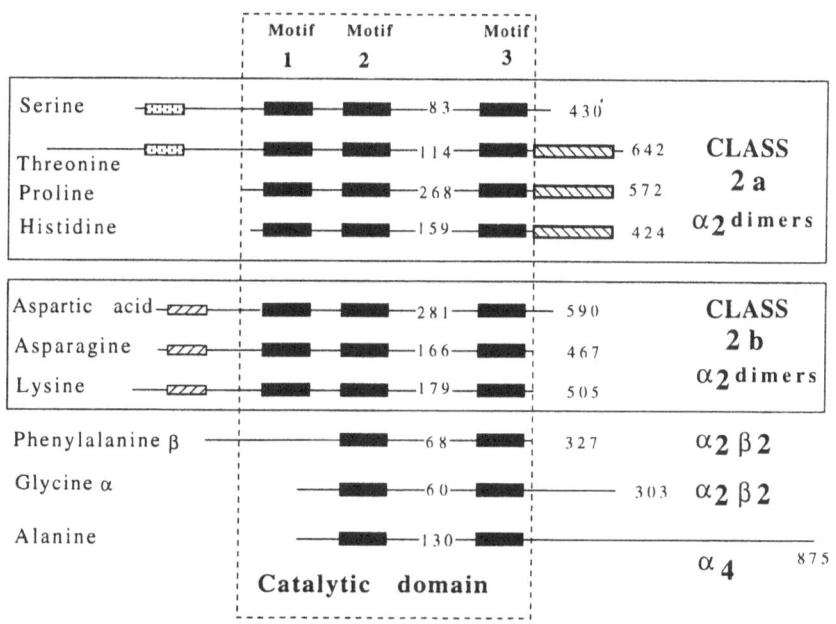

motifs occur in the active site, the former being involved with ATP binding (Brick *et al.* 1989, Rould *et al.* 1989) and the conserved lysine in the latter being implicated in stabilisation of the transition state during amino acid activation (Mechulam *et al.* 1991). A subgroup of six class 1 synthetases possess another motif 'DWCISRQ' (Eriani *et al.* 1991).

The ten class 2 synthetases also all possess two 'signature' sequences which are now referred to as motif 2 and motif 3 (Eriani *et al.* 1990). Motif 2 is described by the pattern 'FR/HXE/D.....R/HXXF' (see Table 3) and is involved in binding the ATP and the acceptor end of the tRNA (see below). Motif 3, 'GXGXGLER', contains an absolutely conserved arginine and it is interesting to speculate this residue plays the equivalent role to the conserved lysine in class 1 synthetases. A subgroup of seven class 2 synthetases also possess motif 1, 'GF/YXEVXTP'; all these enzymes are α_2 dimers and indeed this motif is an important part of the dimer interface; there is no evidence of this motif in the three other class 2 synthetases which all have a different oligomeric structure (Cusack *et al.* 1991).

The structural similarity between members within each class is in general restricted to the catalytic domain which is either based on a Rossmann fold for class 1 or an antiparallel fold for class 2 (see below). Functionally, the class distinction correlates with the initial site of aminoacylation on the tRNA (Eriani *et al.* 1990). These aspects have recently been reviewed by Moras (1992). Outside the catalytic domain, additional tRNA binding domains are much more variable as is clearly visible in the five known synthetase structures which all have different tRNA binding domains. However more extensive sequence homology exists between subclasses within each class (Table 1). In the case of class 2 synthetases these have been analysed in detail by Cusack *et al* (1991) the results of which are shown in Table 2.

SERYL-tRNA SYNTHETASE AND tRNA^{SER} IDENTITY

The serine system has a number of interesting features. Firstly, *E. coli* posseses five isoaccepting tRNA^{Ser}s in order to cope with the six codons for serine which are from two distinct codon classes. In addition the tRNA^{Selcys} (which has an opal stop anticodon) and a tRNA^{Ser} suppressor (which has an amber anticodon) are also specifically charged by seryl-tRNA synthetase. As a result there is no consistency in the anticodon bases of tRNA^{Ser}s. Another special feature is that with certain exceptions (notably in mitocondria) the tRNAs possess a long variable arm of 14-20 nucleotides rather than the usual 4 or 5. This feature is shared in procaryotes by tRNA^{Tyr} and tRNA^{Leu}.

What are the common features of tRNA^{Ser}s that are recognised by the synthetase? Not unsurprisingly it turns out that the anticodon is not recognised by the enzyme as was originally shown by Normanly *et al.* (1986,1992). This fact distinguishes serine from the majority of other systems where it has been shown that anticodon bases are essential to synthetase recognition (the other exceptions being leucine and alanine, Schulman 1991). The *in vivo* identity switch experiments of Normanly *et al.* (1992) have demonstrated that the discriminator base and bases from the first three pairs of the acceptor stem are part of the tRNA^{Ser} identity as is the base pair C11-G24. The latter is necessary to avoid competitive mischarging by leucyl- and glutamyl-tRNA synthetases and so maybe in fact be a negative determinant. Experiments by Himeno *et al* (1990) and also Normanly *et al.* (1992) have highlighted the importance of the long variable arm as an essential identity element. Other experiments by Schatz *et al.* (1991) using the phophorothiorate/iodine cleavage method have shown that certain phosphates in the acceptor stem, T-loop and variable arm are protected in the synthetase-tRNA complex. Clearly only a crystal structure of the complex can give a comprehensive picture of the recognition of seryl-tRNA synthetase for its cognate tRNAs, but of course one expects it to be consistent with the biochemical and mutagenesis results.

THE CRYSTAL STRUCTURE OF SERYL-tRNA SYNTHETASE

The seryl-tRNA synthetase from *E.coli* is a homodimer, each subunit having 430 residues and molecular weight 48414 daltons. The protein was cloned, sequenced and over-expressed and crystallised at EMBL in Grenoble as reviewed by Leberman *et al* (1990). The crystal structure refined at 2.5Å resolution was described by Cusack *et al.* (1990). A ribbon diagram of one-subunit of the enzyme is shown in Figure 1.

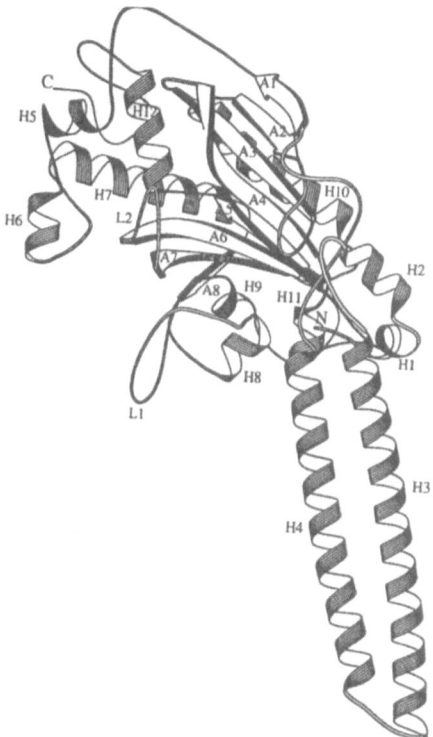

FIGURE 1. Ribbon diagram of one subunit of seryl-tRNA synthetase from *E. coli* showing the major secondary structure elements (Cusack *et al* 1991).

The structure of seryl-tRNA synthetase shows two remarkable features. Firstly the N-terminal domain (up to residue 100) is largely alpha-helical and includes a unique antiparallel coiled-coil (helices H3 and H4 in Figure 3) which stretches about 60Å out into the solution. This structure is stabilised by a inter-helical hydrophobic core containing in particular a ladder-like arrangement of leucine-leucine or leucine-isoleucine pairs (Cusack *et al.* 1990). The second unique feature of the structure is the C-terminal catalytic domain, which has a completely different tertiary fold from the catalytic domain of the three previously known class 1 enzymes, TyrRS (Brick *et al.*, 1989)), MetRS (Brunie *et al.*, 1990) and GlnRS (Rould *et al.* 1989). In seryl-tRNA synthetase the framework of the catalytic domain is a seven-stranded *antiparallel* β-sheet (sheet A in Figure 1), whereas in class 1 enzymes, the catalytic domain is basically a Rossmann nucleotide-binding fold with *parallel* β-strands and connecting helices (see Moras, 1992). The antiparallel-fold of seryl-tRNA synthetase is thus the first representative of a new ATP binding fold, which unlike the Rossmann-fold has not yet been found in other non-synthetase enzymes. The more recent structure of the aspartyl-tRNA synthetase from yeast (Ruff *et al.* 1991), shows that this fold is, as expected, conserved in another class 2 synthetase. Indeed it is possible to superpose 78 Cα positions between the catalytic domain of SerRS and AspRS with a root-mean-square deviation of 1.2Å, even though the sequence homology is virtually non-existant.

The crystal structure of seryl-tRNA synthetase from the extreme thermophile *Thermus thermophilus* has also now been determined and refined to 2.5Å resolution (Fujinaga *et al.* unpublished results). The crystals were first grown by Garber *et al.* (1990). The enzyme from two strains of *T. thermophilus* has been cloned and sequenced (Tukalo *et al.* unpublished results); both enzymes have 421 residues per subunit, but differ in six positions, and have an overall sequence identity with the *E. coli* enzyme of 37%.*T. thermophilus* seryl-tRNA synthetase is remarkably similar to that from *E. coli*, with only minor differences mainly in loop regions. However, as in this structure the dimer is not a crystallographic dimer, a significant asymmetry can be observed in the orientation of the helical arm of each subunit. This is almost certainly due to the intrinsic flexibility of the arms, which appear to be hinged at their base, allowing crystal packing forces to determine their orientation. An exactly similar result is observed in a new orthorhombic form of the seryl-tRNA synthetase from *E. coli* (Nassar and Cusack, unpublished results). In this crystal form the dimer is again asymmetric with different orientations of the arms, both of which are different from that observed in the original crystal form. The importance of the flexibility of the arm becomes apparent in the crystal structure of the complex with tRNA.

STRUCTURE OF THE SERYL-tRNA SYNTHETASE-ATP COMPLEX

Whereas it has never been possible to observe ATP and/or serine bound in the active site of the *E.coli* seryl-tRNA synthetase, probably due to crystal contacts subtly distorting the active site, ATP or ATP analogs can be readily soaked into and observed in crystals of seryl-tRNA synthetase from *T. thermophilus*. The structure of the ATP complex has been refined to 3Å resolution enabling the general mode of binding in the active site to be determined (Berthet *et al.* unpublished results). However the medium resolution of the data, the disorder of some of the active site residues and the partial hydrolysis of the β and γ phosphates do not permit a precise description of the interactions. Nevertheless it is clear that the ATP is bound in the part of the active site pocket formed by the class 2 conserved motifs 2 and 3 (Figure 2). Of the specific interactions, most are with conserved residues of motif 2 on either side of the flexible loop L2, which is presumed to bind specifically to the acceptor stem in class 2 synthetases (Table 3). The triphosphate is aligned parallel to the β-strand A5 from motif 3 (Figure 2). The adenine ring of the ATP stacks with the universally conserved phenylalanine (Phe-285 in the *E. coli* enzyme) and the N6 amino group of the adenine makes hydrogen bonds to both Glu-270 (which is either a glutamate or aspartate in all class 2 synthetases) and the main-chain carbonyl-oxygen of Met-284 (*E. coli* residues); these latter interactions might be responsible for the discrimination in favour of ATP and against GTP. Arg-268 (also nearly fully conserved in class 2 synthetases) interacts with the alpha-phosphate of ATP as does Lys-287 which is conserved only amongst the seryl-tRNA synthetases. It is interesting to note that the univerally conserved arginine from motif 3 (Arg-397 in the *E. coli* enzyme) makes no direct contacts with the ATP, suggesting a possible role for this residue in the catalytic process.

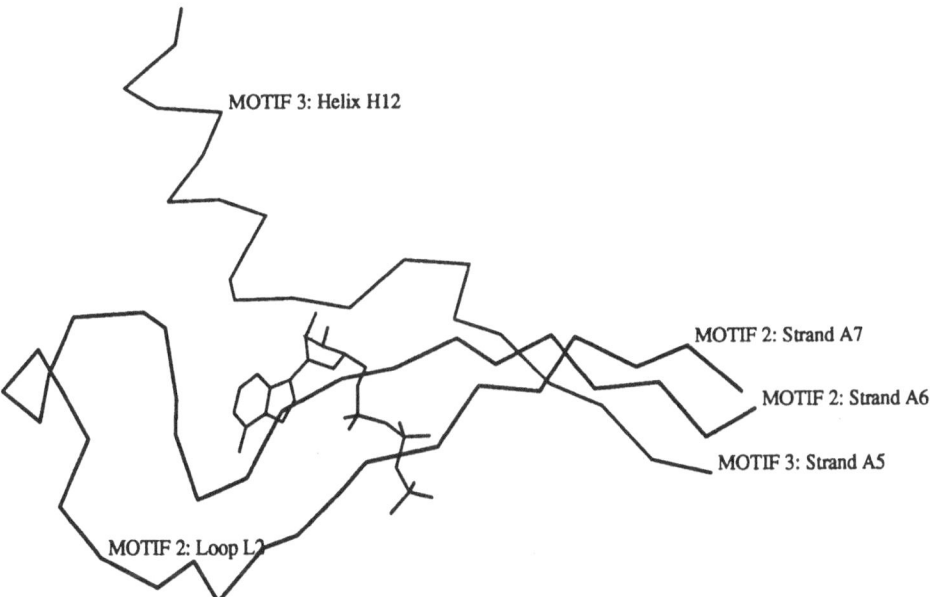

FIGURE 2. Orientation of the ATP with respect to the conserved class 2 motifs 2 and 3 in seryl-tRNA synthetase. (See also Table 3).

Table 3. Class 2 aminoacyl-tRNA synthetases: MOTIF 2 (* interacts with ATP)

```
           strand A7        LOOP L2             strand A6
           βββββββ                              βββββββ

                        * CLASS 2A                        *

SerEC  256  L P I K M T A H T P C F R S E A G S Y G R D T R G L I R M H Q F D K V E M V Q I V R P E D
SerTT  244  L P L R Y A G Y A P A F R S E A G S F G K D V R G L M R V H Q F H K V E Q Y V L T E A S L E A
SerCHF   ?  . . . I . H Y V G Y S s S C F R R E A G S H G K D A W G V F R V H A F E K I E Q F V I T E P E K
SerSC  267  L P I H Y V G Y S s S C F R R E A G S H G K D A W G V F R V H A F E K I E Q F V I T E P E K
ThrSC  437  L P W R V A D F G V I H R N E L S G A L S - - - G L T R V R R F Q Q D D A H I F C T H D
ThrHU  454  L P L R L A D F G G L H R N E L S G A L T - - - G L T R V R R F Q Q D D A H I F C A M E
ThrEC  351  L P L R M A E F G S C H R N E P S G S L H - - - G L M R V R G F T Q D D A H I F C T E E
ProDM 1340  L P I R L N Q W N N V V R W E F K Q P T - - - - P F L R T R E F L W Q E G H 2 F A D K E E
ProEC  128  L P L N F Y Q I Q T K F R D E V R P R F - - - - G V M R S R E F L M K D A Y S F H T S Q E
HisEC  101  Q E Q R L W Y I Q P M F R H E R P - . - - - - Q K G R Y R Q F H Q L G C E V F G - L Q
HisHU  145  T N I K R Y H I A K V Y R R D N P A M - - - T R G R Y R E F Y Q C D F D I A G N F D
HisSC  144  Q S I K R Y H I A K V Y R R D Q P A M - - - T K G R M R E F Y Q C D F D V A G T F E
AlaEC   63  T T S Q R C V R A G G K H N D L E N V G Y - - - T A R H H T F F E M L G N F S F G D Y F K H
AlaBM   70  V N T Q K C I R A G G K H N D L D D V G K - - - D V Y H H T F F E M N G N W S F G D Y F K K

                          CLASS 2B

AspEC  205  G F D R Y Y Q I V K C F R D E D - - - - - - L R A D R Q P E F T Q I D V E T S F M T A P Q
AspSC  313  D F E R V Y E I G P V F R A E N - - - - - - S N T H R H M T E F T G L D M E M A F E E H
AspHU  260  D F E K V F S I G P V F R A E D - - - - - - S N T H R H L T E F V G L D I E M A F N Y H
AsnEC  223  - L S K I Y T F G P T F R A E N - - - - - - S N T S R H L A E F W M L E P E V A F A N L N D
AsnSC  228  - L S R C W T L S P C F R A E K - - - - - - S D T P R H L S E F W M L E V E M C F V N S V N
LysEC  251  G F E R V F E I N R N F R N E G - - - - - - I S V R H N P E F T M M E L Y M A Y A D Y K D
LysSC  314  G L D R V Y E I G R Q F R N E G - - - - - - I D M T H N P E F T T C E F Y Q A Y A D V Y D
LysSCm 293  G L Q K V Y E I Q K V F R N E G - - - - - - I D S T H N A E F S T L E F Y E T Y M S M D D
PheEC  183  P P I R I I A P G R V Y R N D Y - - - - - - V D Q T H T P M F H Q M E G L I V D T N I S F
PheSC  347  K P T R L F S I D R V F R N E A - - - - - - V D A T H L A E F H Q V E G V L A D Y N I T L
GlyEC   58  Y V Q P S R R A T D G R Y G E N - - - - - - P N R L Q H Y Y Q F Q V V I K P S P D N I Q E L
```

EC: E. coli, TT: T. thermophilus, CHF: Chinese hamster fibroblasts, SC(m): S. cerevisiae(mitochond.), HU: human, DM: Drosophila, BM: B. mori

CRISTALLISATION OF SERYL-tRNA SYNTHETASE-tRNA^{SER} COMPLEXES

Crystals of the complex of seryl-tRNA synthetase with cognate tRNA^{Ser} have been obtained from both *T. thermophilus* and *E. coli* components.

The *E. coli* complex crystals

Orthorhombic crystals of the *E. coli* complex which diffract to 4Å resolution have been obtained from ammonium sulphate using the isoacceptor tRNA$_2$^{Ser} (S. Price *et al.* manuscript in preparation). The latter was obtained by overexpression *in vivo* from a synthetic gene (F. Borel *et al.* manuscript in preparation). To avoid plasmid instability, this required utilisation of pOU71, a runaway replication plasmid whose copy number is temperature inducible.

The crystal structure has been solved by a combination of molecular replacement (using the known *E. coli* synthetase structure), a mercury heavy atom derivative and solvent flattening. This resulted in a good low resolution electron density map of the complex in which the position of the complete tRNA backbone is clearly visible although it is difficult to build an accurate atomic model. A curious feature of these crystals is that the first two bases from the anticodon of tRNA$_2$^{Ser} (CGA) appear to make base-pairs with the equivalent bases from another tRNA related by a crystallographic 2-fold symmetry axis. Interestingly the complex with the suppressor tRNA^{Ser} which is identical to tRNA$_2$^{Ser} except for the having the anticodon CUA (in which the first two bases are not self-complimetary) does not crystallise.

The *T. thermophilus* complex crystals

Four crystal forms of this complex have been obtained all using ammonium sulphate as precipitant (Yaremchuk *et al.* 1992a, Yaremchuk *et al.* 1992b). Of these, two (denoted by Form III and Form IV) diffract to medium resolution. Using molecular replacement and heavy atom derivatives it has been shown that the crystallographic asymmetric unit of Form III crystals contains two synthetase dimers each with two tRNA molecules. On the other hand Form IV crystals, which diffract to 2.8Å resolution, *contain a single tRNA molecule bound to the synthetase dimer* (Yaremchuk *et al.* 1992b). This unexpected result is explained by the fact that the second tRNA site on the synthetase dimer is blocked by crystal contacts. Model building and refinement of this structure has been carried out with the aid of an atomic model of tRNA^{ser} kindly provided by Dr. E. Westhof (IMBC, Strasbourg). The current R-factor is about 21% at 2.9Å resolution (V. Biou and S. Cusack, unpublished results). Unfortunately two regions of the tRNA are absent in the electron density map, either due to disorder or degradation; these are the acceptor stem and the anticodon stem and loop.

Nevertherless all the interactions of the tRNA with the long helical arm of the synthetase and in particular the structure of the long variable arm of the tRNA are well defined. The structure of the complex into which the ATP analog AMPPCP has been soaked, has also been refined and confirms the position of the ATP described above for the native enzyme. An interesting observation concerning the Form IV crystals is that all crystal contacts are mediated by protein. This perhaps explains the relatively high diffraction quality and relative radiation insensitivity of the crystals compared to the other crystal forms of the complex.

THE STRUCTURE OF THE SERYL-tRNA SYNTHETASE-tRNA^{SER} COMPLEX

A ribbon diagram of the structure of the *T. thermophilus* complex derived from Form IV crystals is shown in Figure 3. From this structure and that of the *E. coli* complex at low resolution, the following points can be made about the recognition between synthetase and tRNA.

(a) The tRNA binds across the two subunits of the dimer. Only the acceptor end down to base-pair 5-68 is in contact with the active site domain of one subunit, the rest of the tRNA makes contact with the other subunit. This is the first time this has been observed in a synthetase-tRNA complex and demonstrates the absolute necessity for a dimeric molecule. Sequence homologies suggest strongly that the same will be true of other members of class 2a that is ThrRS, ProRS and possibly HisRS. On the otherhand, in the AspRS (and presumably the other members of class 2b) tRNA contact is almost exclusively with one subunit (Ruff *et al.* 1991).

(b) The anticodon loop is not in contact with the synthetase. This is demonstrated graphically by the base-pairing of the anticodon with that from a symmetry related tRNA molecule in the case of the *E. coli* complex and is consistent with the known features of tRNA^{Ser} identity (see above).

(c) Upon interaction with tRNA, the long helical arm of the synthetase is fixed in an orientation different from those previously observed in non-complexed synthetase structures. This orientation is essentially identical for the *E. coli* and *T. thermophilus* complexes. This supports the suggestion above that the arm is freely hinged and only takes up a fixed orientation when constrained by crystal contacts or by binding to tRNA.

(d) The helical arm of the synthetase passes between the long variable arm and the TΨC loop of the tRNA, contacts mainly being between polar residues from both helices and phosphate groups. Contacts on the variable arm extend from the second base-pair out to the fifth base-pair, indicating the importance of at least this length of the extra arm for synthetase recognition. In the Form IV structure, the end of the variable arm is not in contact with the synthetase and is disordered; this is consistent with the fact that longer variable arms (most notably in the tRNA^{Selcys}) can also be accomodated by the synthetase.

(e) The only base-specific contacts are made by Gln-45 (in the middle of the first long helix in the *T. thermophilus* synthetase) to the fourth base-pair of the long variable arm which in all known bacterial tRNA^{Ser}s is a G-C or C-G. In addition the end of the helical arm makes

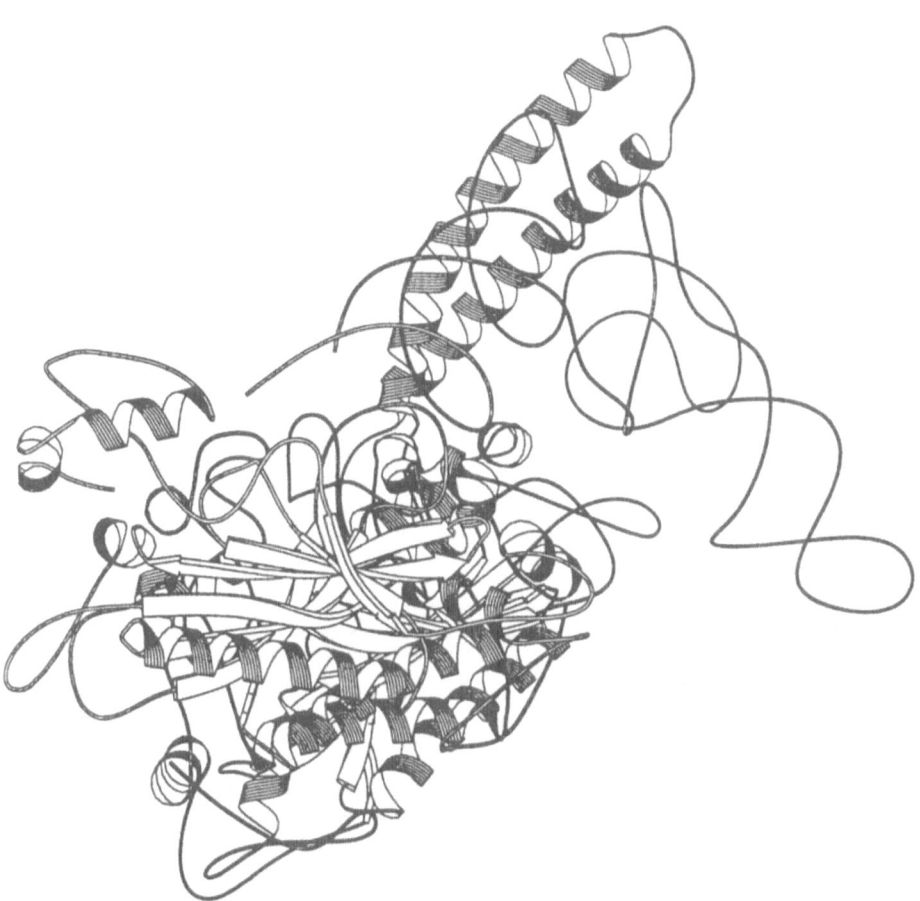

FIGURE 3. Ribbon diagram of the *T. thermophilus* seryl-tRNA synthetase-tRNASer complex as derived from Form IV crystals at 2.9Å resolution. Note that one of the helical arms of the synthetase is disordered in this structure.

a hydrophobic platform against which the base-pair G19-C56 packs. This interaction, although significant, is not specific, since this base-pair is conserved in all tRNAs and plays the important role of pining together the D-loop and T-loop.

CONCLUSION

The crystallographic structures of seryl-tRNA synthetase complexed with ATP and tRNA^{Ser} described here bring us closer to understanding structure-function relationships in this enzyme and in class 2 synthetases in general. The synthetase-tRNA complex enables many features of tRNA^{Ser} identity, previously suggested by genetic and biochemical experiments, to be understood in atomic detail, as well as providing for the first time the structure of a tRNA with a long extra arm. Due to the disorder of the tRNA in Form IV crystals, we have unfortunately no detailed information yet on the presumed interactions in the active site region of the motif 2 loop L2 with the identity elements in the acceptor stem of the tRNA^{Ser}. The other major area of remaining ignorance is in the exact site of binding of the serine. All crystallographic experiments to observe serine, seryl-adenylate or the inhibitor serine hydroxymate in the active site have so far failed. It is salutory to note that the amino acid binding site has so far only been determined crystallographically in one of the five crystal structures of synthetases known, the tyrosyl-tRNA synthetase (Brick *et al.* 1989).

REFERENCES

Brick, P. Bhat, T.N. and Blow, D.M. *J. Mol. Biol.* 1989, **208**:83-98.
Brunie, S., Zelwer, C. and Risler, J-L. *J. Mol. Biol.* 1990, **216**:411-424.
Cusack, S., Härtlein, M. and Leberman, R. *Nuc. Acids. Res.* 1991, **19**:3489-3498.
Cusack, S., Berthet-Colominas, C., Härtlein, M. Nassar, N. and Leberman, R. *Nature* 1990, **347**:249-255.
Eriani, G., Dirheimer, G. and Gangloff, J. Nuc. Acids. Res. 1991, 19:265-269.
Eriani, G., Delarue, M., Poch, O., Gangloff, J. and Moras, D. *Nature* 1990, **347**:203-206.
Garber, M. B., Yaremchuk, A.D., Tukalo, M.A., Egorova, S.P., Berthet-Colominas, C. and Leberman, R. *J. Mol. Biol.* 1990, **213**:631-632.
Himeno, H., Hasegawa, T., Ueda, T. , Watanabe, K. and Shimizu, M. *Nucl. Acids Res.* 1991 **18**:6815-6819.
Leberman, R., Härtlein, M. and Cusack, S. *Biochimica et Biophysica Acta* 1991, **1089**:287-298.
Mechulam, Y., Dardel, F., Le Corre, D., Blanquet, S. and Fayat, G. *J. Mol. Biol.* 1991, **217**:465-475.
Moras, D. *TIBS.* 1992, **17**:159-164.
Normanly J., Ollick T., and Abelson J. *Proc. Natl. Acad. Sci. USA* 1992, **89**:5680-5684.
Normanly, J., Ogden, R.C., Horvath, S.J. and Abelson, J. *Nature* 1986, **321**:213-219.
Rould, M.A., Perona, J.J., Söll, D. and Steitz, T.A. *Science* 1989, **246**:1135-1142.
Ruff, M., Krishnaswamy, S. Boeglin, M.m Poterszman, A., Mitschler, A., Podjarny, A., Rees, B., Thierry, J.C. and Moras, D. *Science* 1991, **252**:1682-1689.
Schulman, L. D. *Progress in Nucleic Acid Research and Molecular Biology* 1991, **41**:23-87.
Schatz, D., Leberman, R. and Eckstein, F. *P.N.A.S. USA* 1991, **88**:6132-6136.
Yaremchuk, A.D., Tukalo, M.A., Krikliviy, I., Mel'nik,V. N., Berthet-Colominas, C. and Cusack, S. and Leberman, R. *J. Mol. Biol.* 1992a, **224**: 519-522.
Yaremchuk, A., Tukalo, M.A., Krikliviy, I., Malchenko, N., Biou, V., Berthet-Colominas, C. and Cusack, S. *FEBS Letters* 1992b, **310**:157-161.

CHARGING OF RNA MICROHELICES AND DECODING GENETIC INFORMATION: EVALUATION OF FUNCTIONAL COUPLING BETWEEN DISTAL PARTS OF A tRNA STRUCTURE

Paul Schimmel

Department of Biology
Massachusetts Institute of Technology
Cambridge, Massachusetts 02139

INTRODUCTION

Aminoacyl tRNA synthetases are divided into two distinct and unrelated classes of ten enzyme each(Webster et al., 1984; Ludmerer and Schimmel, 1987; Schimmel, 1987; Eriani et al., 1990; Cusack et al., 1990, 1991; Moras, 1992). These enzymes interpret the genetic code through the aminoacylation reaction, whereby specific amino acids are joined to the 3'-ends of tRNAs. The tRNAs contain specific anticodon trinucleotide sequences which are located approximately 75 angstroms from the amino acid attachment site, so that the aminoacylation reaction per se establishes the trinucleotide-amino acid rules of the code.

The tRNA structure is comprised of two domains, which are schematically illustrated in Figure 1 (Kim, 1974; Robertus, 1974). One domain is comprised of the 12 base pair acceptor-TUC stem and loop. The second domain consists of the dihyrouridine-anticodon stem-biloop structure, which contains the anticodon. Although the anticodon specifies a specific amino acid in the language of the code, this structural element does not make contact with alanine tRNA synthetase(Park and Schimmel, 1988). In this instance, the connection between a specific amino acid and a trinucleotide sequence is made indirectly.

Instead, the enzyme interacts directly with a specific G3:U70 base pair in the tRNA acceptor stem and it is this interaction which is responsible for alanine acceptance(Hou and Schimmel, 1988; McClain and Foss, 1988; Francklyn and Schimmel, 1989). The significance of this interaction was demonstrated by the aminoacylation of truncated substrates comprised of RNA microhelices which reconstruct the seven base pair acceptor stem of tRNA^{Ala}(Francklyn and

Schimmel, 1989). Aminoacylation is dependent on the G3:U70 base pair--a property which is shared by the full tRNA and the microhelix. Transfer of this base pair into other tRNA or microhelix frameworks confers alanine acceptance(Hou and Schimmel, 1988; McClain and Foss, 1988; Hou and Schimmel, 1989a; Hou and Schimmel, 1989b; Francklyn and Schimmel, 1989; Francklyn et al., 1992; Shi et al., 1992).

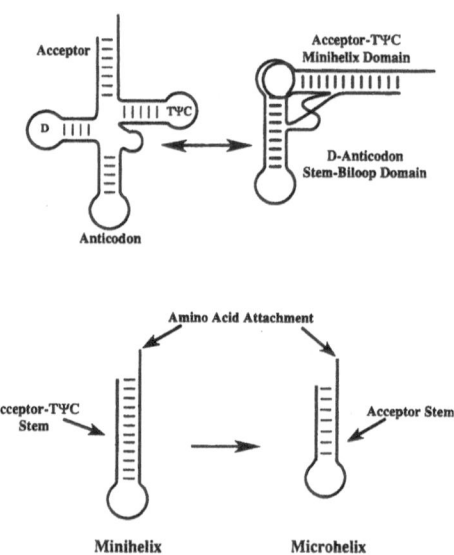

Figure 1. Schematic illustration of the organization of the tRNA structure into two domains and of the 12 bp minihelix and the 7 bp acceptor stem microhelix. Adapted in part from(Burbaum and Schimmel, 1991).

RNA microhelices based on the seven base pair acceptor stems of tRNAs have now been demonstrated to be sequence-specific substrates for aminoacylation by the class II alanine, glycine, and histidine tRNA synthetases(Francklyn and Schimmel, 1990; Francklyn et al., 1992) and the class I methionine tRNA synthetase(Martinis and Schimmel, 1992)(Figure 2). Aminoacylation of a 12 base pair minihelix based on the acceptor stem of tRNA^{Val} has also been reported(Frugier et al., 1992). Several more examples of sequence-specific aminoacylation of RNA microhelix substrates are expected. In these instances, signals for specific aminoacylation(in the form of atomic groups presented for synthetase recognition) are proximal to the amino acid attachment site at the 3'-end of the RNA substrate. With microhelix substrates, a set of rules that relates amino acids to specific nucleotides at specific positions can be formulated. These rules are distinct from those of the genetic code and yet the interpretation of the code is dependent upon them(Schimmel, 1991; Francklyn et al., 1992).

The sequence-specific aminoacylation of acceptor stem microhelices by methionine and valine tRNA synthetases is of particular interest because these enzymes are believed to interact directly with the anticodon, in addition to the acceptor stem(Schulman and Pelka, 1988; Frugier et al., 1992). That the anticodon is not required for aminoacylation in these instances suggests that alanine tRNA synthetase, which has no interaction with the anticodon, may be an extreme of the general case where acceptor stem nucleotides are sufficient for aminoacylation, but where interactions outside of the stem, such as at the anticodon, cooperate with acceptor stem interactions to give the maximal efficiency and specificity of aminoacylation. The general adaptation of acceptor stem sequences to charging with specific amino acids implies that motifs on synthetases that interact with acceptor stem nucleotides are within 10-20 angstroms of the catalytic site.

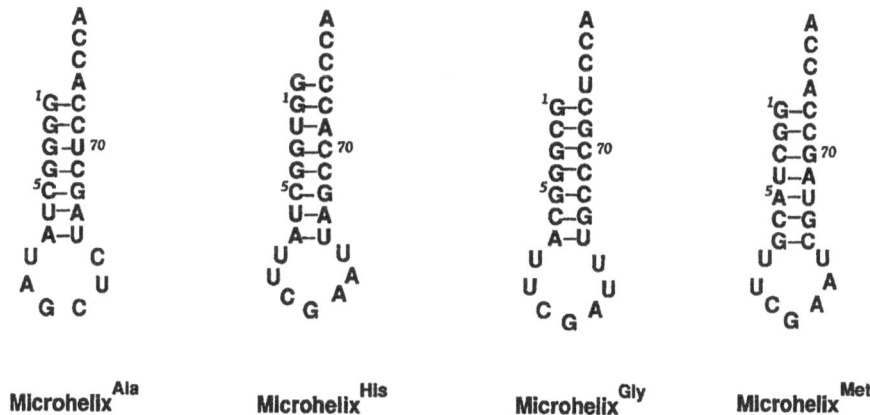

Figure 2. Microhelices based on the acceptor stems of specific tRNAs which are aminoacylated with the cognate amino acid.

Although acceptor stem interactions appear in general to have an essential role in determining aminoacylation specificity, the functional coupling between acceptor stem interactions with those at other places in the tRNA structure is not understood. To investigate the possible functional coupling of distal parts of the tRNA structure to acceptor stem interactions, two experimental systems have been used. In one example, the coupling of acceptor helix recognition to the anticodon-synthetase interaction of methionine tRNA synthetase was investigated, using a modified methionine tRNA synthetase in which the anticodon binding motif was disabled by an internal deletion(Kim and Schimmel, 1992). In the other example, the potential for functional coupling of distal parts of the tRNA^{Ala} structure was investigated, by selecting for intragenic mutations that compensate for a base pair substitution in the acceptor helix at the 3:70 position(Hou and Schimmel, 1992a).

PROBE FOR FUNCTIONAL COUPLING OF ACCEPTOR HELIX AND ANTICODON INTERACTIONS IN A CLASS I tRNA SYNTHETASE

Methionine tRNA synthetase is an alpha$_2$ dimer of identical polypeptides of 676 amino acids which can be truncated in the C-terminal region by approximately 125 amino acids to yield a monomeric N-terminal fragment with almost full

15

activity (Meinnel et al., 1990). The three dimensional structure of the enzyme without bound tRNA^{Met} has been solved(Brunie et al., 1990). As a typical class I enzyme, methionine tRNA synthetase has an N-terminal nucleotide binding fold of alternating beta strands and alpha helices which contains the site for adenylate synthesis and for docking of the 3'-end of the tRNA. This domain extends to approximately Ser366 and is followed by an alpha-helix-rich structure which encodes the polypeptide segments which interact with the tRNA^{Met} CAU anticodon. Trp461 is critical for the anticodon interaction(Ghosh et al., 1990; Meinnel et al., 1991)]. It is located on the surface, at the beginning of a flexible loop which starts after an extended alpha helix(Brunie et al., 1990). A Trp461-->Ala substitution lowers k_{cat}/Km for aminoacylation of tRNA^{Met} by a factor of 2×10^{-5}(Meinnel et al., 1991).

The approximately 25 amino acid peptide at the C-terminus of the truncated enzyme extends back to the N-terminal domain where it has a role in enzyme stability and activity(Meinnel et al., 1990). This segment links the C-terminal anticodon binding domain with the active site and the acceptor helix interactions in the N-terminal domain. To investigate the influence of the anticodon-binding determinant at Trp461 on the N-terminal catalytic site, in-frame deletions of the region of the gene encompassing this tryptophan were created. These deletion mutants were then investigated for enzyme stability, adenylate synthesis, and acceptor helix interactions, as measured by the aminoacylation of a methionine-specific microhelix(Kim and Schimmel, 1992).

These Trp461-encompassing deletions ranged from 4 to 87 amino acids. The largest stable deletions removed 11 amino acids. One of these removed Asp456 to Gln466 and the other deleted Tyr454 to Ala464. In both instances, the synthesis of methionyl adenylate was not significantly impaired by the deletion. For the Tyr454--Ala464 deletion mutant, kinetic parameters including Km's for methionine and ATP and k_{cat} were not altered beyond the range of experimental error(Kim and Schimmel, 1992). This result established the lack of functional linkage between the structure of the anticodon binding motif and aminoacyl adenylate synthesis.

The influence of the 11 amino acid deletion on acceptor helix contacts was investigated with the microhelix system. With the Tyr454--Ala464 deletion mutant, the rate of charging of tRNA^{Met} is severely reduced, as expected from the removal of a critical anticodon-binding interaction(Figure 3). In spite of this reduction in the rate of aminoacylation of the full tRNA, there is no effect of the deletion on the rate of sequence-specific charging of the methionine microhelix(Kim and Schimmel, 1992)(Figure 3). These results indicate a close functional integration of microhelix interactions with the site of adenylate synthesis, and suggest that microhelix interactions are functionally independent of the structure of the anticodon binding domain. The tight association between microhelix aminoacylation and adenylate synthesis reflects a protein design that has integrated specific recognition of a few nucleotides, near the site of amino acid attachment, with the catalytic center. This organization may reflect an early synthetase which was quite small.

DISTAL BASE PAIR SUBSTITUTION WHICH COMPENSATES FOR A MUTATION IN THE MAJOR IDENTITY DETERMINANT OF A CLASS II tRNA SYNTHETASE

An alternative approach to functional coupling between distal parts of a tRNA structure was investigated with the alanine tRNA synthetase-tRNA^{Ala}

complex. Although the acceptor stem G3:U70 base pair is essential for aminoacylation with alanine, contacts outside of the acceptor stem have been demonstrated by RNA footprint analysis. These contacts do not extend to the anticodon, but they do include the portion of the anticodon stem that is adjacent to the dihydrouridine stem(Park and Schimmel, 1988).

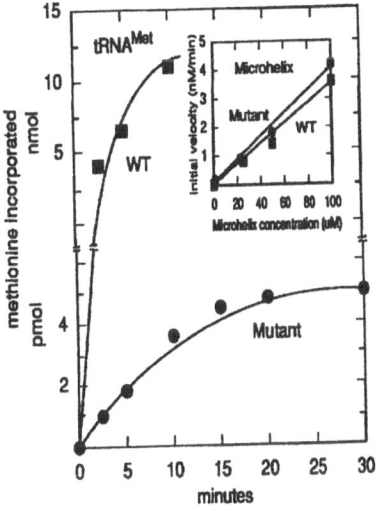

Figure 3. Aminoacylation of tRNA^{Met} and of the acceptor stem microhelix(inset) based on tRNA^{Met}(Figure 2), by wild-type and the Ala454-Tyr464 deletion mutant of methionine tRNA synthetase. The deletion mutant is severely impaired for aminoacylation of tRNA^{Met}, but not for charging of the microhelix(inset). The inset also shows that initial rate of aminoacylation of the microhelix is linear with enzyme concentration. Adapted in part from (Kim and Schimmel, 1992).

To investigate the potential for communication between critical acceptor stem contacts and those with other parts of the tRNA structure, the G3:U70 base pair was replaced with G3:C70(Hou and Schimmel, 1992a). This substitution eliminates alanine acceptance(Hou and Schimmel, 1988, Park et al., 1989, Francklyn and Schimmel, 1989). Mutations were then introduced to select for second-site intragenic mutations that compensated for the C3:C70-inactivating mutation. For this purpose, an *in vivo* screen with a mutagenized G3:C70 tRNA$^{Ala/CUA}$ amber suppressor(CUA anticodon) was used. The G3:C70 amber suppressor fails to suppress the alanine-requiring amber UAG234 codon(Murgola and Hijazi, 1983) in the alpha-subunit of tryptophan synthetase of *E. coli* strain *trpA*(UAG234)(Hou and Schimmel, 1988). Sequence degenerate oligonucleotides were used for the construction of a population of variant tRNAs with substitutions at 66 of 76 positions in the sequence of the structural gene for G3:C70 tRNA$^{Ala/CUA}$. The ten fixed positions in the G3:C70 tRNA$^{Ala/CUA}$ sequence were C70, the CCA trinucleotide at the 3'-end of all tRNAs, and six nucleotides in the anticodon loop from positions 33 to 38, including the CUA anticodon at positions 34-36. The remaining positions in the sequence had a frequency of substitution of 10%, to give on average 6.6 replacements per molecule(Hou and Schimmel, 1992b).

From a population of 10^5 transformants, approximately 250 were functional in protein synthesis *in vivo*, as judged by the suppression of an amber mutation in strain XAC-1(Hou and Schimmel, 1992b). This strain contains a lacI-Z fusion which can be suppressed by any of the twenty amino acids(Kleina et al., 1990). About 18% of these functional molecules differed from the starting molecule(G3:C70 tRNA$^{Ala/CUA}$) which is a weak suppressor on strain XAC-1 and which does not suppress the *trpA*(UAG234) amber allele(Hou and Schimmel, 1992a,b). Amongst these functional molecules, only one suppressed *trpA*(UAG234)(Hou and Schimmel, 1992a).

This suppressor has a C27:G43-->U:U substitution at the base of the anticodon stem(Figure 4). Determination of the amino acid at the position corresponding to the amber codon in a *fol*$_{am}$ gene product(cf. Normanly et al., 1986) confirmed the insertion of alanine(with about 30% tyrosine) by the U27:U43 amber suppressor. *In vitro* aminoacylation of U27:U43 tRNA$^{Ala/CUA}$ with alanine was also demonstrated, with an estimated k_{cat}/Km which is approximately 1.2% of that of wild-type tRNA^{Ala}(Hou and Schimmel, 1992a).

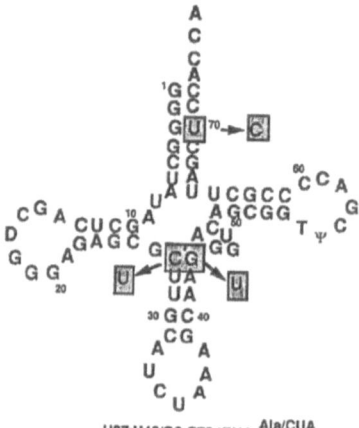

U27:U43/G3:C70 tRNA $^{Ala/CUA}$

Figure 4. The G3:C70 tRNA$^{Ala/CUA}$ amber suppressor which acquires alanine acceptance by virtue of the distal U27:U43 base pair substitution. Adapted from (Hou and Schimmel, 1992a).

The 27:43 substitution is adjacent to an enzyme-tRNA contact detected by RNA footprinting(Park and Schimmel, 1988). The ability of the U:U substitution at this position to confer alanine acceptance on a G3:C70-containing tRNA^{Ala} suggests functional communication between the acceptor and anticodon stems of G3:C70 tRNA$^{Ala/CUA}$. This base pair is located at the junction of the D- and anticodon stems, and may alter the coaxial stacking of these two helices(Figure 1). It is not likely that the placement of unpaired bases at this location *per se* creates a motif for alanine tRNA synthetase recognition, because substitution of a U27:C43 base combination did not confer alanine acceptance. A possible effect of the mismatch at 27:43, especially considering the flexibility it can introduce at the junction of two stems, is to alter the interaction at 3:70. In the wild-type acceptor helix, the enzyme makes a functional contact with the unpaired exocyclic 2-amino group of G3, which is located in the minor groove(Musier-Forsyth et al., 1991). In contrast to the wobble arrangement of the G3:U70 base pair, the G3:C70 base pair is in the Watson-Crick configuration and the 2-amino group is shifted into hydrogen bond formation with O2 of C70 and this difference is sufficient to eliminate alanine acceptance. Possibly the U27:U43 mismatch alters the tRNA structure in a way that the interaction at 3:70 is perturbed enough to afford a weak functional contact of the enzyme with the 2-amino group of G3 in the G3:C70 base pair.

CONCLUDING REMARKS

In the co-crystal of *E. coli* glutamine tRNA synthetase with $tRNA^{Gln}$, acceptor helix contacts are made with an insertion into the nucleotide binding fold(Rould et al., 1989). This insertion facilitates the docking of the 3'-end of $tRNA^{Gln}$ near the activated amino acid. An insertion into the adenylate synthesis domain of the class II alanine tRNA synthetase also provides acceptor helix contacts(Miller and Schimmel, 1992). Thus, the structures for acceptor helix interactions are intimately associated with the catalytic site for adenylate synthesis and aminoacylation. These structures for catalysis and acceptor helix interactions represent what may have been the primordial enzymes, which were too small to interact with RNA determinants more than a few nucleotides beyond the site for amino acid attachment.

Interactions with more distal parts of the tRNA L-shaped structure, outside of the acceptor-TΨC minihelix domain, may involve an additional domain which was recruited or developed for that purpose. These additional domains are evident in the crystal structures of the class I *E. coli* glutamine and methionine tRNA synthetases(Rould et al., 1989; Brunie, et al., 1990) and the class II yeast aspartate tRNA synthetase(Ruff et al., 1991). In each case, a separate anticodon-binding domain has been fused to the core enzyme which has the site for adenylate synthesis and acceptor helix contacts. Although glutamine and methionine tRNA synthetases belong to the same class by virtue of the structural relationship between their N-terminal nucleotide binding folds, their C-terminal anticodon-binding domains are unrelated. A beta-barrel structure interacts with the anticodon of $tRNA^{Gln}$(Rould et al., 1989) and an alpha-helix rich domain interacts with the anticodon of $tRNA^{Met}$(Brunie, et al., 1990). Thus, unlike the N-terminal domains, the C-terminal domains do not share a common evolutionary origin and may have been added later, after the development of the minimalist catalytic units which have the structural elements for acceptor helix interactions.

Anticodon nucleotides in $tRNA^{Gln}$(Jahn et al., 1991) and $tRNA^{Asp}$(Putz et al., 1991) contribute to Km and to k_{cat} for aminoacylation. The latter parameter is determined by the transition state free energy of activation within the enzyme-tRNA complex. A contribution to k_{cat} implies that interactions at the anticodon are directly coupled to the active site. The data presented above suggest that, for methionine tRNA synthetase, the structure of the anticodon domain per se does not influence acceptor helix interactions or the catalytic site. On the other hand, the isolation of a second site revertant of G3:C70 tRNA$^{\text{Ala/CUA}}$ with a distal base pair substitution(Hou and Schimmel, 1992a) suggests that functional communication between the two domains of tRNA is possible in principle. In this example, and possibly in the cases of tRNA^{Gln} and tRNA^{Asp}, functional communication between the two domains may require an intact tRNA structure through which a distal interaction can influence the transition state activation free energy. In point of fact, an isolated anticodon stem-loop hairpin helix has only a small stimulatory effect on the aminoacylation of a minihelix based on the acceptor stem of yeast tRNA^{Val}(Frugier et al., 1992). Continuity of the tRNA chain in this and other systems may be required for the full cooperation of distal anticodon interactions with the catalytic site and the associated acceptor helix interactions.

ACKNOWLEDGMENTS

This work was supported by National Institutes of Health grants GM15539 and GM23562. Figure 2 was provided by Dr. J-P. Shi and Figure 3 by Dr. S. Kim.

REFERENCES

Brunie, S., Zelwer, C. and Risler, J., 1990, *J. Mol. Biol.* 216:411-424.

Burbaum, J. J. and Schimmel, P., 1991, *J. Biol. Chem.* 266:16965-16968.

Cusack, S., Berthet-Colominas, C., Hartlein, M., Nassar, N. and Leberman, R., 1990, *Nature* 347:249-255.

Cusack, S., Hartlein, M. and Leberman, R., 1991, *Nucleic Acids Research* 19:3489-3498.

Eriani, G., Delarue, M., Poch, O., Gangloff, J. and Moras, D., 1990, *Nature* 347:203-206.

Francklyn, C., Musier-Forsyth, K. and Schimmel, P., 1992,

Francklyn, C. and Schimmel, P., 1989, *Nature* 337:478-481.

Francklyn, C. and Schimmel, P., 1990, *Chemical Reviews* 90:1327-1342.

Francklyn, C., Shi, J.-P. and Schimmel, P., 1992, *Science* 255:1121-1125.

Frugier, M., Florentz, C. and Giegé, R., 1992, *Proc. Natl. Acad. Sci. USA* 89:3990-3994.

Ghosh, G., Pelka, H. and Schulman, L. H., 1990, *Biochemistry* 29:2220-2225.

Hou, Y.-M. and Schimmel, P., 1988, *Nature* 333:140-145.

Hou, Y.-M. and Schimmel, P., 1989a, *Biochemistry* 28:4942-4947.

Hou, Y.-M. and Schimmel, P., 1989b, *Biochemistry* 28:6800-6804.

Hou, Y.-M. and Schimmel, P., 1992a, *Biochemistry* 31:10310-10314.

Hou, Y.-M. and Schimmel, P., 1992b, *Biochemistry* 31:4157-4160.

Jahn, M., Rogers, M. J. and Söll, D., 1991, *Nature* 352:258-260.

Kim, S. H., Suddath, F. L., Quigley, G. J., McPherson, A., Sussman, J. L., Wang, A. H. J., Seeman, N. C. and Rich, A., 1974, *Science* 185:435-440.

Kim, S. and Schimmel, P., 1992, *J. Biol. Chem.* 267:15563-15567.

Kleina, L. G., Masson, J.-M., Normanly, J., Abelson, J. and Miller, J. H., 1990, *J. Mol. Biol.* 213:705-715.

Ludmerer, S. W. and Schimmel, P., 1987, *J. Biol. Chem.* 262:10801-10806.

Martinis, S. A. and Schimmel, P., 1992, *Proc. Natl. Acad. Sci. USA* 89:65-69.

McClain, W. H. and Foss, K., 1988, *Science* 240:793-796.

Meinnel, T., Mechulam, Y., Dardel, F., Schmitter, J.-M., Hountondji, C., Brunie, S., Dessen, P., Fayat, G. and Blanquet, S., 1990, *Biochim.* 72:625-632.

Meinnel, T., Mechulam, Y., Le Corre, D., Panvert, M., Blanquet, S. and Fayat, G., 1991, *Proc. Natl. Acad. Sci. U. S. A.* 88:291-295.

Miller, W. T. and Schimmel, P., 1992, *Proc. Natl. Acad. Sci. USA* 89:2032-2035.

Moras, D., 1992, *Trends Biochem. Sci.* 17:159-164.

Murgola, E. J. and Hijazi, K. A., 1983, *Mol. Gen. Genet.* 191:132-137.

Musier-Forsyth, K., Usman, N., Scaringe, S., Doudna, J., Green, R. and Schimmel, P., 1991, *Science* 253:784-786.

Normanly, J., Ogden, R. C., Horvath, S. J. and Abelson, J., 1986, *Nature* 321:213-219.

Park, S. J., Hou, Y.-M. and Schimmel, P., 1989, *Biochemistry* 28:2740-2746.

Park, S. J. and Schimmel, P., 1988, *J. Biol. Chem.* 263:16527-16530.

Putz, J., Puglisi, J. D., Florentz, C. and Giegé, R., 1991, *Science* 252:1696-1699.

Robertus, J. D., Ladner, J. E., Finch, J. T., Rhodes, D., Brown, R. S., Clark, B. F. C. and Klug, A.,1974, *Nature* 250: 546-551.

Rould, M. A., Perona, J. J., Söll, D. and Steitz, T. A., 1989, *Science* 246:1135-1142.

Ruff, M., Krishnaswamy, S., Boeglin, M., Poterszman, A., Mitschler, A., Podjarny, A., Rees, B., Thierry, J. C. and Moras, D., 1991, *Science* 252:1682-1689.

Schimmel, P., 1987, *Ann. Rev. Biochem.* 56:125-158.

Schimmel, P., 1991, *The FASEB J.* 5:2180-2187.

Schulman, L. H. and Pelka, H., 1988, *Science* 242:765-768.

Shi, J.-P., Martinis, S. A. and Schimmel, P., 1992, *Biochemistry* 31:4931-4936.

Webster, T. A., Tsai, H., Kula, M., Mackie, G. A. and Schimmel, P., 1984, *Science* 226:1315-1317.

IDENTITY OF A PROKARYOTIC INITIATOR tRNA

Michael R. Dyson, Chan Ping Lee, Nripendranath Mandal,
Baik L. Seong, Umesh Varshney and Uttam L. RajBhandary

Department of Biology
Massachusetts Institute of Technology
Cambridge, Massachusetts, USA

Of the two classes of methionine tRNAs present in all organisms, the initiator is used exclusively for initiation of protein synthesis whereas the elongator is used for inserting methionine into internal peptide linkages (Kozak, 1983; Gold, 1988). To fulfill this special function, initiator tRNAs possess a number of properties which are different from those of elongator tRNAs. For eubacterial initiator tRNAs, these include: (i) formylation of the tRNA subsequent to aminoacylation, (ii) binding of the fMet-tRNA to the ribosomal P site, (iii) exclusion of the tRNA from the ribosomal A site and (iv) resistance of the fMet-tRNA to peptidyl-tRNA hydrolase (Figure 1).

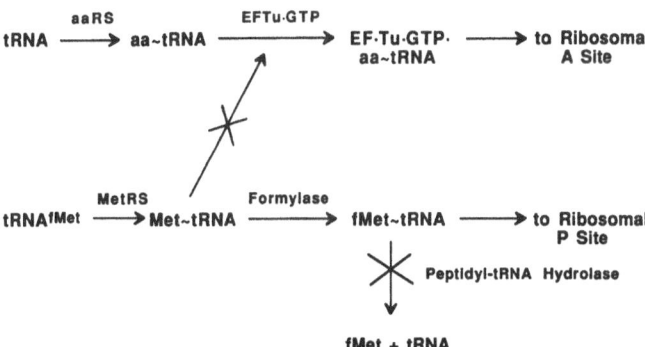

Figure 1. Differential utilization of elongator and initiator tRNAs in protein synthesis.

Along with these special properties, eubacterial initiator tRNAs also possess unique sequence and/or structural features that are not found in elongator tRNAs (Figure 2). These include: (i) absence of a Watson-Crick base pair between nucleotides 1 and 72 at the end of the acceptor stem, (ii) the presence of a sequence of three guanines and three cytosines at the bottom of the anticodon stem forming three consecutive G:C base pairs and (iii) the presence of a Purine 11: Pyrimidine 24 base pair in the dihydrouridine stem in contrast to a Pyrimidine 11: Purine 24 base pair in other tRNAs (Rich and RajBhandary, 1976). We are interested in identifying the sequence and/or structural features of E. coli initiator tRNA that are important

The Translational Apparatus, Edited by K.H. Nierhaus
et al., Plenum Press, New York, 1993

for specifying its distinctive properties and in defining the molecular basis of how the various proteins and/or the ribosome utilize these features.

Our approach is to introduce site specific mutations into the initiator tRNA gene and analyze the function of mutant tRNAs at each of the steps of protein synthesis *in vitro* and *in vivo*. For *in vitro* studies, we can isolate and purify the mutant tRNAs in large amounts. We have developed a method that allows us to analyze directly the effect of mutations on aminoacylation and formylation *in vivo* (Varshney et al., 1991a). In addition, we have developed methods for analyzing the function of mutant tRNAs in initiation and in elongation *in vivo* (Varshney and RajBhandary, 1990; Seong *et al.*, 1989). The combined results of *in vitro* and *in vivo* studies on mutant initiator tRNAs have allowed us to identify the main sequence and/or structural elements important for aminoacylation and for specifying each of the distinctive properties of *E. coli* initiator tRNA. These elements are clustered mostly in the acceptor stem and in the anticodon stem and loop. Of the three features unique to eubacterial initiator tRNAs, we have identified the role of two of them. We show that introduction of these two features into *E. coli* elongator methionine tRNA converts it into a tRNA that is active in initiation *in vivo*.

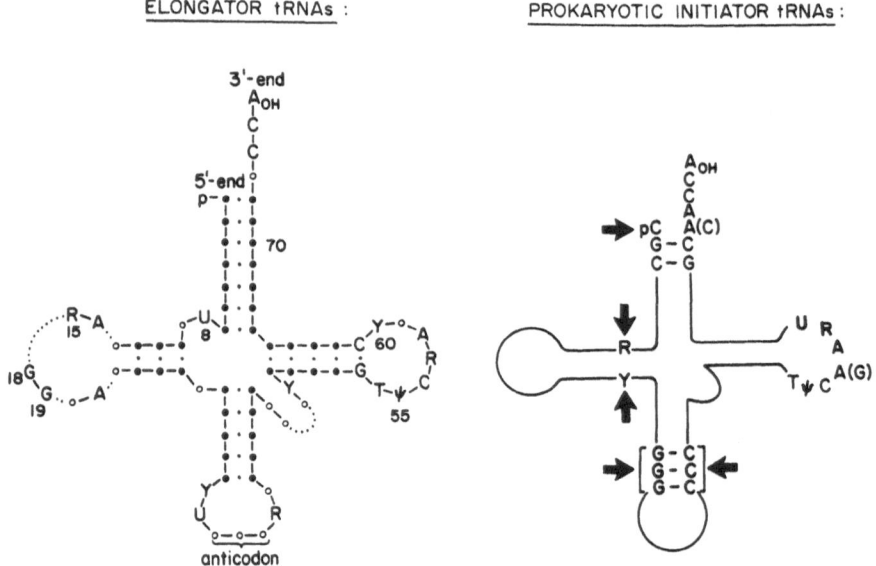

Figure 2. Unique features of prokaryotic initiator tRNAs (arrows).

ISOLATION OF MUTANT tRNAs

The *E.coli* K12 tRNA gene that we have mutagenized corresponds to tRNA$_2^{fMet}$ which differs from tRNA$_1^{fMet}$ only in having A at position 46 instead of 7-methyl G. The mutant tRNAs can be overproduced in *E. coli*. Because tRNA$_2^{fMet}$ is among the fastest migrating of the tRNA species in non-denaturing polyacrylamide gels, a single step of gel electrophoresis can separate wild type and mutant tRNA$_2^{fMet}$ from all other tRNAs (Seong and RajBhandary, 1987a). Recently, we have shown that *E. coli* B lacks the tRNA$_2^{fMet}$ species and that the *metY* locus of *E. coli* B encodes tRNA$_1^{fMet}$ instead of tRNA$_2^{fMet}$ (Mandal and RajBhandary, 1992). This has allowed us to express the mutant tRNA ^{fMet} genes in *E. coli* B and thereby obtain mutant tRNAs completely free of any wild type tRNA$_2^{fMet}$ (Figure 3). This in turn has greatly simplified interpretation of data on kinetic parameters in aminoacylation and formylation of mutant tRNAs.

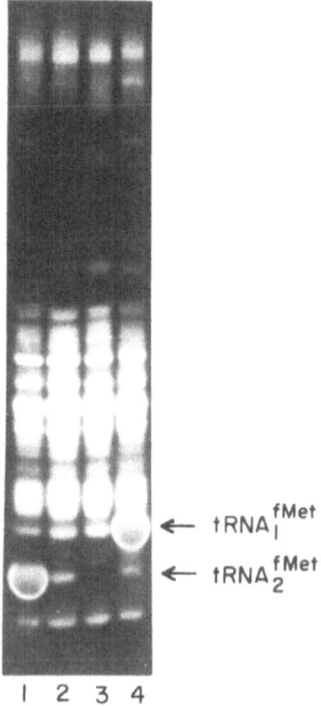

Figure 3. Separation on a native polyacrylamide gel of tRNA isolated from *E. coli* K12 (lane 2), *E. coli* B (lane 3) and *E. coli* K12 transformed with pUC19 carrying the *metY* locus of *E. coli* K12 (lane 1) or *E. coli* B (lane 4).

REQUIREMENTS FOR AMINOACYLATION

We showed that a gel electrophoretic procedure described previously (Ho and Kan, 1987) could be used to separate uncharged, aminoacylated and formylaminoacylated forms of the tRNA from each other (Varshney *et al.*, 1991a). This has allowed us to use Northern blot analysis of total tRNA isolated from cells under conditions that preserve the ester linkage between the amino acid and the tRNA, to assess directly the effect of mutations on aminoacylation and formylation *in vivo*. We have used this data along with measurements of kinetic parameters in aminoacylation *in vitro* to identify the sequences important for recognition of tRNA by Methionyl-tRNA synthetase (MetRS).

Of the number of mutants studied to date (Figure 4), those that affect significantly the kinetic parameters in aminoacylation lie in the anticodon sequence and the discriminator base 73. These results agree with the conclusions (Schulman, 1991; Schulman and Pelka, 1983) that the anticodon sequence is one of the crucial elements in the tRNA for MetRS recognition. Our results show that the nature of the discriminator base (Crothers et al., 1972) also affects aminoacylation. Mutation of A73 to C73 or G73 lowers Vmax/km by factors of about 30 (Table 1) and affects aminoacylation in vivo (Lee et al., 1992). Because A73 → U73 change has little effect and A73 and U73 have no functional groups in common, the detrimental effect of the C73 or G73 change is more likely from negative interactions and/or an altered structure than from loss of a direct contact between MetRS and the discriminator base.

Effect of the G73 mutation is strikingly exacerbated by an additional mutation of the neighboring nucleotide. Coupling of the G73 mutation with G72, which by itself

Table 1. Kinetic parameters in aminoacylation of tRNAs by E. coli MetRS.

tRNAs	V_{max} arbitrary units	$K_m{}^{app}$ (mM)	Relative $V_{max}/K_m{}^{app}$
tRNA$_2^{fMet}$	298.5	13.4	1.0
T73	293.3	32.1	2.4
C73	42.1	63.3	33.5
G73	49.9	65.1	29.1
G72	398.8	41.0	2.3
G72T73	232.6	30.8	2.9
G72C73	6.7	11.8	39.2
G72G73	1.5	31.3	465

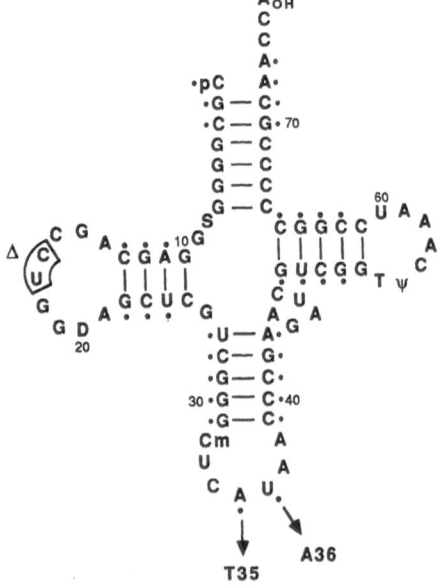

Figure 4. Sequence of E. coli tRNA$_2^{fMet}$ in cloverleaf form, sites of mutation are indicated by dots (substitution) or a triangle (deletion).

has a minor effect on MetRS recognition, produces a mutant tRNA (G72G73) that is an extremely poor substrate for MetRS *in vitro* (Table 1) and *in vivo* (Lee *et al.*, 1992). It is possible that the G72G73 mutant tRNA adopts a structure that cannot be easily altered to make the tRNA CCA end fit into the catalytic pocket of MetRS.

In general, the results of aminoacylation *in vivo* parallel the kinetic parameters for aminoacylation *in vitro*. An exception is the T1 mutant tRNA, which is a very good substrate for MetRS *in vitro* (Lee *et al.*, 1992) but is substantially uncharged *in vivo* (Figure 5). This is because the T1 mutant with a U1:A72 base pair is not only a good substrate for Met-tRNA transformylase, which converts it to fMet-tRNA, but is now also a substrate for peptidyl-tRNA hydrolase (Schulman and Pelka, 1975), which hydrolyzes the fMet-tRNA to formylmethionine and tRNA (Kossel and RajBhandary, 1968). The accumulation of uncharged T1 mutant tRNA suggests that peptidyl-tRNA hydrolase hydrolyzes fMet-tRNA almost as fast as MetRS aminoacylates the released tRNA. This conclusion is supported by the fact that all of the T1 mutant is present as fMet-tRNA in cells that are overproducing MetRS (Figure 5, lane 3) or in cells that have a temperature-sensitive mutation (Atherly and Menninger, 1972) in peptidyl-tRNA hydrolase (Figure 5, lanes 4 and 5).

Thus, an important role of the C1xA72 mismatch in E. coli initiator tRNA is to help maintain high steady state levels of fMet-tRNA in vivo by preventing its hydrolysis by peptidyl-tRNA hydrolase.

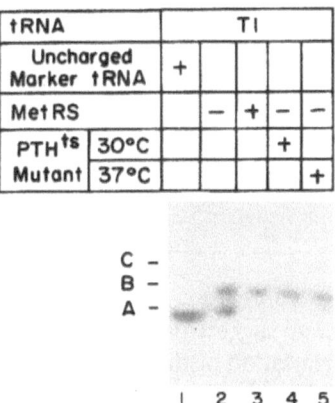

tRNA		T1			
Uncharged Marker tRNA	+				
MetRS		−	+	−	−
PTH ts Mutant	30°C				+	
	37°C					+

```
C  −
B  −
A  −
```
1   2   3   4   5

Figure 5. RNA blot analysis of tRNA₂^fMet from transformants carrying the T1 mutant tRNA gene. A, B, and C indicate locations of uncharged tRNA, fMet-tRNA, and Met-tRNA respectively. The T1 mutant was expressed in *E. coli* CA274 (lanes 2 and 3) and *E. coli* AA7852 [temperature-sensitive mutant of peptidyl-tRNA hydrolase, PTH ts] (lanes 4 and 5).

REQUIREMENTS FOR FORMYLATION

The sequence and/or structural elements in the tRNA important for formylation are clustered at the end of the acceptor stem (Lee *et al.*, 1991). The key determinants appear to be a mismatch (as in the wild type tRNA) or a weak base pair between nucleotides 1 and 72, a G:C base pair between nucleotides 2 and 71 and a C:G (or less preferably a G:C) base pair between nucleotides 3 and 70. Mutations in G4:C69 base pair affect formylation kinetics but have less of an effect than mutations at the above positions. Mutations at position 73 affect formylation minimally except for the G73 mutation (Guillon *et al.*, 1992b, Dyson and RajBhandary, unpublished). Because mutations of A73 to C or U have little or no effect (Dyson and RajBhandary, unpublished), the effect of G73 mutation is quite specific and is, therefore, likely due to negative interactions and/or an altered structure.

The nature of the mismatch between nucleotides 1 and 72 is less important for formylation than the fact that they do not form a strong Watson-Crick base pair (Lee et al., 1991, Lee et al., 1992, Guillon et al., 1992b). tRNAs carrying the wild-type CxA mismatch or the mutant CxC, AxA, or CxU mismatches are essentially fully formylated in vivo, whereas tRNAs carrying strong base pairs, such as C1:G72 or G1:C72, are not formylated to a significant extent. Interestingly, the severe effect on formylation of having a C1:G72 base pair can be compensated for by a secondary change of the neighboring base A to C or U (Table 2). The most likely explanation of this result is that the C1:G72 base pair, which is at the end of an RNA helix and may, therefore, have a tendency to "breathe," is stabilized by stacking (Wakao et al., 1989) of the neighboring base A73 on top of the C1:G72 base pair. Change of A73 to pyrimidine bases, such as C73 or U73, could destabilize the C1:G72 base pair due to loss of this stacking interaction (Sugimoto et al., 1987) and thereby lower the energetic cost of disrupting this base pair.

Table 2. Kinetic parameters in formylation of mutant tRNA by E. coli Met-tRNA transformylase.

tRNAs	Relative V	K_m^{app} (mM)	Relative V_{max}/K_m^{app}
tRNA$_2^{fMet}$	1.4	7.3	1.0
G72C73	0.76	14.6	3.7
G72T73	0.79	15.1	3.7
G72			495[*]

[*], Lee et al., 1991.

There is strong evidence to suggest that the base pair between nucleotides 1 and 72 must be disrupted for formylation of the tRNA. Among mutant tRNAs that have a base pair at this position, there is an excellent correlation between the strength of the base pair (Sugimoto et al., 1987) and how good a substrate the tRNA is for formylation in vitro (U1:A72 > A1:U72 > C1:G72 > G1:C72). The only mutant with a base pair at this position which is a good substrate for the formylating enzyme is the T1 mutant with a U1:A72 base pair. This is most likely because U1:A72 base pair at the end of an RNA helix is the weakest of the four possible base pairs and, therefore, most easily disrupted.

REQUIREMENTS FOR OVERALL ACTIVITY IN INITIATION

We have shown that mutants of initiator tRNA with an anticodon sequence change from CAU → CUA (T35A36 mutant) can initiate protein synthesis from a UAG codon (Varshney and RajBhandary, 1990). This result along with work of Schulman and coworkers (Chattapadhyay et al., 1990) showed that protein synthesis in E. coli could be initiated with codons unrelated to AUG and with amino acids other than methionine. More importantly, this finding provided us with an assay for *in vivo* function of mutant tRNAs in initiation. By coupling the CAU → CUA anticodon sequence change with mutations in the main body of the tRNA, we could determine how well they function in initiation *in vivo* by using a chloramphenicol acetyl transferase (CAT) gene carrying a AUG → UAG mutation in the initiation codon as reporter and measuring CAT activity in cell extracts. Since the T35A36 mutant tRNA with the CAU → CUA anticodon sequence change is aminoacylated with glutamine (Schulman and Pelka, 1985), initiation of CAT protein synthesis under these conditions occurs with formylglutamine. To minimize the contribution of effects of mutations in the main body of the tRNA on aminoacylation by glutaminyl-tRNA synthetase (GlnRS), CAT activity levels are measured also in extracts prepared from cells overproducing GlnRS.

Analysis of a large number of such mutants shows that the only mutations that affect activity of tRNAs in initiation lie in the acceptor stem and in the anticodon stem. In the acceptor stem, the effect of mutation depends upon the site of mutation and the nature of the mutation. There is an excellent correlation between formylation of tRNAs and activity in initiation (Figure 6). Mutations at the first, second and third base pairs in the acceptor stem which affect formylation greatly lower the activity of mutant tRNAs in initiation *in vivo*. Thus, formylation of initiator tRNA is important for initiation of protein synthesis in *E. coli*. Further evidence comes from recent work showing that disruption of the gene for Met-tRNA transformylase results in *E. coli* cells that grow extremely slowly (Guillon *et al.*, 1992a).

Correlation Between Formylation and Activity of Various Mutant tRNAs in Initiation

tRNA Mutant	% CAT Activity		% Formylation In Vivo
	-GlnRS	+GlnRS	
T35A36	100	118	>90
T1/T35A36	22	65	>90*
T1/T2:A71/T35A36	1	1	<5*
T1/T3:A70/T35A36	4	3	<5*
G3:C70/T35A36	76	64	>70*
T35A36/G72	1	1	<5*
T35A36/G72G73	0	0	ND*
T35A36/G72T73	7	22	>90*
T35A36/G72C73	5	44	>90*

N.D. not detectable

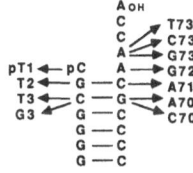

* Level of formylation in mutants without the T35A36 mutation

Figure 6. Correlation between formylation and activity of various mutant tRNAs in initiation.

In the anticodon stem, mutations that alter the 3 G:C base pairs lower activity of the mutant tRNAs in initiation (Mandal and RajBhandary, unpublished). A mutant with all three G:C base pairs changed has much reduced activity in initiation *in vivo* (Lee *et al.,* 1991) and *in vitro* (Seong and RajBhandary, 1987a). *In vitro* and *in vivo* studies have shown that mutations in the anticodon stem do not affect aminoacylation, formylation or binding of the fMet-tRNA to the initiation factor IF2. Mutations here specifically affect binding of the mutant tRNAs to the ribosomal P site (Seong and RajBhandary, 1987a). Thus, the 3 consecutive G:C base pairs conserved in the anticodon stem of initiator tRNAs are important for targeting the tRNA to the ribosomal P site. It is possible that this feature is used by the initiation factor IF3 to select the initiator tRNA from all other tRNAs at the P site (Hartz *et al.,* 1989).

REQUIREMENTS FOR EXCLUSION FROM THE RIBOSOMAL A SITE

The important structural feature that prevents the *E. coli* initiator tRNA from acting as an elongator is the absence of a Watson-Crick base pair between nucleotides 1 and 72 at the end of the acceptor stem. Single mutants that have either a U1:A72 or a C1:G72 base pair act as elongators *in vitro*, whereas a double mutant with a U1xG72 mismatch does not (Figure 7). The G72 mutant with a strong C1:G72 base pair is much more active than the T1 mutant with a weak U1:A72 base pair. This is also reflected in the relative affinity of the Met-tRNAs derived from these two mutants for EF-Tu.GTP (G72>T1>T1G72 = wild type tRNA). Thus the T1 and G72 mutant tRNAs are active in elongation because they bind EF-Tu.GTP in contrast to the wild type tRNA which does not (Schulman *et al.,* 1974, Fisher *et al.,* 1985, Seong and RajBhandary, 1987b).

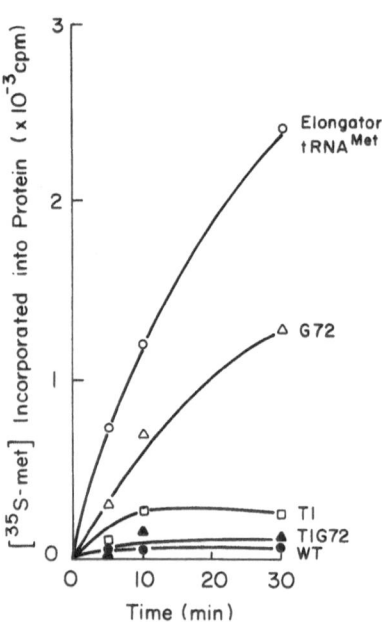

Figure 7. Incorporation *in vitro* of [^{35}S] Methionine from various mutant [^{35}S]Met-tRNAs in a MS2 RNA directed protein synthesis system.

By coupling mutations in the main body of the tRNA with a CAU → CUA anticodon sequence change which allows the tRNA to read UAG codons, we have analyzed a large number of mutant tRNAs for their ability to act as amber suppressors and, therefore, as elongators *in vivo* (Seong *et al.*, 1989). These studies confirm the results of *in vitro* analyses and show that virtually all the mutants that retain the C1xA72 mismatch are inactive in elongation *in vivo*. Exceptions are a G3:C70 mutant (Varshney *et al.*, 1991b) and the G73 mutant (Guillon *et al.*, 1992b, Dyson and RajBhandary, unpublished). The G3:C70 mutation generates six consecutive G:C base pairs in the acceptor stem with six Gs and six Cs in a row. It is possible that this mutant tRNA exists in alternate conformations, one of which can bind to EF-Tu.

MUTANTS OF INITIATOR tRNA THAT ARE ACTIVE BOTH IN INITIATION AND IN ELONGATION

Although the *E. coli* initiator tRNA appears to have been sequestered for use exclusively for initiation of protein synthesis, some of the mutant tRNAs that we have generated act as both initiators and as elongators (Varshney *et al.*, 1991b). The T1 mutant with a U1:A72 base pair is an example. Thus, while there may be selective advantages in sequestering a tRNA for use in initiation, activities of a tRNA in initiation and in elongation are not mutually exclusive. This may explain why many of the mitochondrial DNAs, with strong constraints on the genome size and on the number of tRNA genes, encode only one methionine tRNA (Attardi and Schatz, 1988) and makes plausible the notion that in such mitochondrial systems, the same methionine tRNA may act both as initiator (in the formylated form) and as elongator (in the unformylated form).

ROLE OF THE UNIQUE FEATURES OF E. COLI INITIATOR tRNA IN ITS FUNCTION

Of the features unique to *E. coli* and eubacterial initiator tRNAs, we have identified the role of two of them. (1) The GGG:CCC sequence in the anticodon stem is important for targeting the tRNA to the P site on the ribosome. (2) The lack of a Watson-Crick base pair between nucleotides 1 and 72 (CxA mismatch in the case of *E. coli* initiator tRNA) may be important for specifying three of its four distinctive properties. (i) The mismatch prevents the initiator tRNA from binding to EF-Tu.GTP and thereby from acting at the elongation step of protein synthesis. (ii) The mismatch maintains high steady state levels of fMet-tRNA *in vivo* by preventing hydrolysis of fMet-tRNA by peptidyl-tRNA hydrolase. (iii) The mismatch may also be important for formylation of the tRNA by Met-tRNA transformylase (see above). The absolute conservation of the mismatch between nucleotides 1 and 72 found in all eubacterial initiator tRNAs is understandable given the fact that this feature is utilized by three different proteins to distinguish this tRNA from other tRNAs. It is interesting to note that these three proteins recognize either aminoacyl-tRNA or acylaminoacyl-tRNA but not uncharged tRNA.

The role of the third feature, the Purine 11:Pyrimidine 24 base pair in the D stem, is not yet known. Mutation of A11:U24 to C11:G24 results in an approximately 7-fold increase in *Km* in the formylation reaction. However, a mutant with changes in 3 of the 4 base pairs in the D stem, including the same A11:U24 to C11:G24 change, behaves normally in formylation. When coupled to an anticodon sequence change from CAU → CUA, the mutant with 3 of the 4 base pair changes in the D stem is quite active in initiation *in vivo*. Thus the importance of the A11:U24 base pair for formylation depends upon the sequence context in and around the D stem (Lee *et al.*, 1991). The role of the A11:U24 base pair in eubacterial initiator tRNA structure and/or function could be more subtle than those of the other two unique features.

CONVERSION OF AN ELONGATOR tRNA TO AN INITIATOR tRNA

Having identified the key features important for the overall function of tRNAs in initiation, we have asked whether transplantation of these features into elongator tRNAs will allow them to act in initiation. We have introduced features important for formylation and for targeting the tRNA to the ribosomal P site onto *E. coli* methionine elongator tRNA. To allow assay for activity in initiation in *E. coli*, the mutant tRNAs also had a CAU → CUA anticodon sequence change. Introduction of a C1xA72 mismatch, which generates the minimal features necessary for formylation of the tRNA, produces a tRNA with extremely low, although measurable, activity in initiation (mutant M:i1, Table 3). However, further introduction of three consecutive G:C base pairs at the bottom of the anticodon stem produces a tRNA with significant activity in initiation (mutant M:i2, Table 3). The amount of CAT activity in extracts of cells carrying the mutant elongator tRNA is about 30% of that in cells carrying the *E. coli* initiator tRNA mutant with the corresponding anticodon sequence change (Varshney et al., 1993).

Table 3. Relative CAT activities (%) in extracts of *E. coli* CA274 transformants carrying the various mutant tRNA genes. The transformants also contained another plasmid which did not (-) or did (+) contain the gene for *E. coli* GlnRS or for *E. coli* MetRS.

tRNA Gene	-	+ GlnRS	+ MetRS
fM/T35A36[*]	100	118	263
M:i1	nd	nd	3
M:i2	.8	.8	81

[*]*E. coli* initiator $tRNA_2^{fMet}$ gene carrying the T35A36 mutation. CAT activity in cells containing this tRNA gene and not overproducing any of the aminoacyl-tRNA synthetases is fixed at 100%.

nd, not determined.

Acknowledgements

We thank our colleague Dr. C. Ming Chow for valuable suggestions on this manuscript and Annmarie McInnis for her usual care and patience in putting together this manuscript. This work was supported by grants GM 17151 from the National Institutes of Health and NP 114 from the American Cancer Society.

REFERENCES

Attardi, G., and Schatz, G., 1988, *Annu. Rev. Cell. Biol.* 4:289-333.

Atherly, A.G., and Menninger, J.R., 1972, *Nat. New Biol.* 240:245-246.

Chattapadhyay, R., Pelka, H., and Schulman, L.H., 1990, *Biochemistry* 29:4263-4268.

Crothers, D.M., Seno, T., and Soll, D., 1972, *Proc. Natl. Acad. Sci. USA* 69:3063-3067.

Fischer, W., Doi, T., Ikehara, M., Ohtsuka, E., and Sprinzl, M., 1985, *FEBS. Lett* 192:151-154.

Gold, L., 1988, *Annu. Rev. Biochem.* 57:199-233.

Guillon, J-M., Mechulam, Y., Schmitter, J-M., Blanquet, S., and Fayat, G., 1992a, *J. Bact.* 174:4294-4301.

Guillon, J-M., Meinnel, T., Mechulam, Y., Lazennec, C., Blanquet, S., and Fayat, G. 1992b, *J. Mol. Biol.* 224:359-367.

Hartz, D. McPheeters, D.S., and Gold, L., 1989, *Genes and Develop.* 3:1899-1912.

Ho, Y-H., and Kan, Y.W., 1987, *Proc. Natl. Acad. Sci. USA* 84:2185-2188.

Kossel, H. and RajBhandary, U.L., 1968, *J. Mol. Biol.* 35:539-560.

Kozak, M., 1983, *Microbiol. Rev.* 47, 1-45.

Lee, CP., Dyson, M.R., Mandal, N., Varshney, U., Bahramian, B. and RajBhandary, U.L., 1992, *Proc. Natl. Acad. Sci. USA* 89:9262-9266.

Lee, C. P., Seong, B. L., and RajBhandary, U.L., 1991, *J. Biol. Chem.* 266:18012-18017.

Mandal, N. and RajBhandary, U.L., 1992, *J. Bact.* 174:7827-7830.

Menninger, J.R., 1979, *J. Biol. Chem.* 253:6808-6813.

Rich, A., and RajBhandary, U.L., 1976, *Annu. Rev. Biochem.* 45:805-860.

Rould, M.A., Perona, J.J., Soll, D., and Steitz, T.A., 1989, *Science* 246:1135-1142.

Schulman, L.H., 1991, *Progr. Nucleic Acid Res. Mol. Biol.* 41:23-87.

Schulman, L.H., and Pelka, H., 1985, *Biochemistry* 24:7309-7314.

Schulman, L.H., and Pelka, H., 1975, *J. Biol. Chem.* 250:542-547.

Schulman, L.H., and Pelka, H., 1983, *Proc. Natl. Acad. Sci. USA* 80:6755-6759.

Schulman, L.H., Pelka, and Sundari, R.M., 1974, *J. Biol. Chem.* 249:7102-7110.

Seong, B.L., Lee, C.P., and RajBhandary, U.L., 1989, *J. Biol. Chem.* 246:6504-6508.

Seong, B.L., and RajBhandary, U.L., 1987a, *Proc. Natl. Acad. Sci. USA* 84:334-338.

Seong, B.L., and RajBhandary, U.L.. 1987b, *Proc. Natl. Acad. Sci. USA* 84:8859-8863.

Sugimoto, N., Kerzek, R., and Turner, D.H., 1987, *Biochemistry* 26:4554-4558.

Varshney, U., Lee, C.P., and RajBhandary, U.L., 1991a, *J. Biol. Chem.* 266:24712-24718.

Varshney, U., Lee, C.P., Seong, B.L., and RajBhandary, U.L., 1991b, *J. Biol. Chem.* 266:18018-18024.

Varshney, U., Lee, C.P., and RajBhandary, U.L., 1993, *Proc. Natl. Acad. Sci. USA* 90:2305-2309.

Varshney, U., and RajBhandary, U.L., 1990, *Proc. Natl. Acad. Sci. USA* 87:1586-1590.

Wakao, H., Romby, P., Westhof, E., Laalami, S., Grunberg-Manago, M., Ebel, J-P., Ehresman, C., and Ehresman, B., 1989, *J. Biol. Chem.* 264:20363-20371.

GENETIC SYSTEMS IN YEAST FOR ANALYSIS OF INITIATOR/ ELONGATOR tRNA SPECIFICITY

Anders S. Byström, Ulrich von Pawel-Rammingen, and Stefan U. Åström

Department of Microbiology
Umeå University
S-901 87 UMEÅ, Sweden

INTRODUCTION

Transfer RNAs are highly differentiated nucleic acids, which in the decoding of genetic information require unique properties. First, the tRNAs have the ability to recognize the correct codon by a codon- anticodon interaction. Second, each tRNA is charged by its cognate aminoacyl tRNA synthetase, which identifies the tRNA by different nucleotides within the structure (McClain and Foss, 1988; Normanly and Abelson, 1989; Schimmel, 1989). To study the function of a tRNA in protein synthesis, a general strategy is to generate mutations in the tRNA. To score their effects *in vivo*, the mutations are normally introduced in a suppressor tRNA (Murgola, 1985; Eggertsson and Söll; 1988). However, another *in vivo* system has been developed in which the rate of aminoacyl tRNA selection is measured in competion with a $tRNA^{Leu}_2$ frameshift suppressor (Curran and Yarus, 1989). The use of nonsense suppressor tRNAs has some disadvantages. In the translational process, a stop codon is a signal for the release factors to stop translation. Thus, a nonsense suppressor tRNA must compete with the release factors for the stop codon. A tRNA, such as methionine tRNAs, which has its synthetase recognition site within the anticodon (Schulman and Pelka, 1988; Cigan et al, 1988; Despons et al, 1992), would be greatly impaired in aminoacylation as a suppressor since the anticodon must be changed so that the tRNA would function as a nonsense suppressor. Thus, a phenotypic change due to a second site mutation in such a tRNA is not easily distinguished from a possible effect on aminoacylation. To overcome these problems, we have constructed *Saccharomyces cerevisiae* strains to measure the overall *in vivo* activity of a mutant $tRNA^{Met}_i$ or a mutant $tRNA^{Met}_m$, without using a suppressor tRNA.

The Translational Apparatus, Edited by K.H. Nierhaus
et al., Plenum Press, New York, 1993

Specific Features of Eukaryotic Initiator Methionine tRNAs

Initiator tRNAs only decode AUG start codons, whereas elongator methionine tRNAs decode the internal AUG codons of a mRNA (Housman *et al.*, 1970; Smith & Marcker, 1970.). Thus, the initiator tRNA recognizes the P-site on the ribosome. This discrimination in the translational process is believed to be associated with the highly conserved features found in initiator tRNAs but not in elongator tRNAs (Kozak, et al 1983; Spinzl et al, 1991; see Fig 1.). Most initiator tRNAs, independent of organism, have four conserved G-C base pairs in positions 3-70, 12-23, 30-40, and 31-39 (Calagan *et al.*,1980; Spinzl et al, 1991). Eukaryotic initiator tRNAs, compared to elongator tRNAs, have in positions 54-57 AU/ΨCG instead of TΨCG/A, an A instead of a pyrimidine at position 60, and an A-U base pair between position 1 and 72 in the acceptor stem, (Calagan *et al.*,1980; Kozak, 1983). Furthermore, loop I of eukaryotic initiator tRNAs consists of seven nucleotides in contrast to the eight- or nine- membered loop of elongator tRNAs (Calagan et al, 1980).

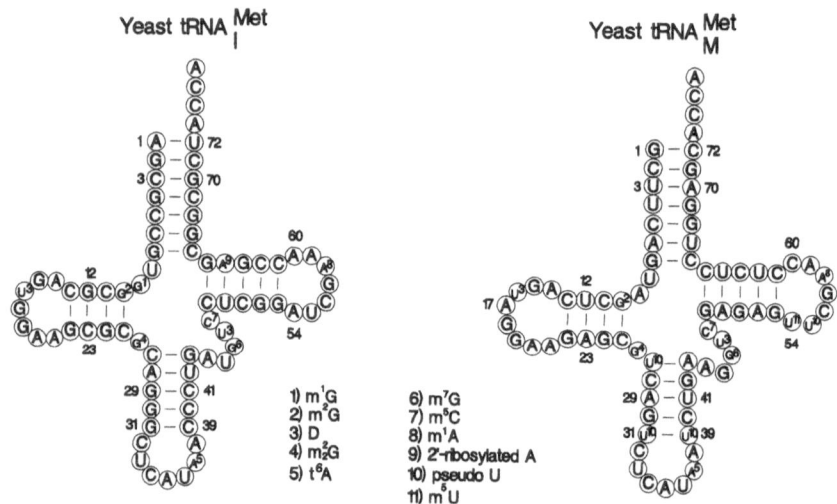

Figure 1. Schematic two-dimensional drawing of yeast tRNA$_i^{Met}$ and tRNA$_m^{Met}$, slightly modified from von Pawel-Rammingen et al, 1992. The sequences and numbering system were obtained from Sprinzl *et al.*, (1991). Nucleosides are circled and numbers within circles represent the following modifications: 1, 1-Methylguanosine (m^{1}G); 2, N^2 - Methylguanosine (m ^{2}G); 3, Dihydrouridine (D); 4, N^2,N^2 - Dimethylguanosine (m$_2^2$G); 5, N-(N-(9-ß-D-Ribofuranosylpurin-6-yl)carbamoyl)threonine (t^{6}A); 6, 7-Methylguanosine (m^{7}G); 7, 5-Methylcytidine (m^{5}C); 8, 1-Methyladenosine (m^{1}A); 9, 2'-1''-ß-(5''-phosphoryl)-ribosyl-adenosine; 10, Pseudouridine (Ψ); 11, 5-Methyluridine (m^{5}U). Base pair 3-70, 12-23, 30-40 and 31-39 are G-C base pairs almost conserved among all initiator tRNA species. Numbers outside circles represent positions of tRNA$_i^{Met}$ that have been mutagenized by oligodirected mutagenesis.

In addition, initiator tRNAs seem to have a unique structure in the anticodon loop, which was determined by S1-nuclease analysis (Wrede et al, 1979, Seong and RajBhandary, 1987). In *E. coli* met- tRNA^{fmet}, the mismatched first base pair of the aminoacyl stem, (C$_1$-A$_{72}$), is part of an identity element for the *E.coli* met-tRNA^{fmet} formyltransferase (Guillon *et al.*, 1992a; Lee *et al.*, 1991.). Formylation of met-tRNA^{fmet}

restricts its participation in translational elongation but improves it to function in translational initiation (Guillon et al, 1992a; Guillon et al, 1992b; Varshney et al; 1991; Sundari et al, 1976). Furthermore, the presence of three G-C base pairs in the anticodon stem directs the *E. coli* initiator tRNA to the ribosomal P-site (Seong & RajBhandary, 1987). On the other hand, the eukaryotic met-tRNA$_i^{Met}$ is not formylated (Smith and Marker, 1968). Instead, a discrimination by eIF2 between the side chains of the aminoacyl adducts of the tRNA$_i^{Met}$ have been suggested, as a misaminoacylated Ile-tRNA$_i^{Met}$ neither form ternary complex with eIF-2 and GTP or participates/ inhibits *in vitro* protein synthesis in rabbit reticulocytes (Wagner et al, 1984). In eukaryotes, the crystal structure of yeast tRNA$_i^{Met}$ has a unique substructure formed by a set of tertiary interactions between loop I and IV (Basavappa & Sigler, 1991). The importance of loop IV in the initiation of translation by yeast tRNA$_i^{Met}$ has been suggested from *in vitro* experiments in rabbit reticulocytes (Wagner *et al.*, 1989). In these experiments, the yeast met-tRNA$_i^{Met}$ participated only in the initiation of translation, whereas the *E. coli* met-tRNA^{fmet}, participated primarily in the elongation of translation. The most obvious difference between these two initiators is the lack of the highly conserved A$_{60}$ and A$_{54}$ in loop IV of the *E. coli* tRNA^{fmet}. A 2'-O-ribosylation modification has so far only been found at position 64 of initiator tRNAs from yeast and plants (Desgres *et al.*, 1989; Glasser et al, 1991). *In vitro*, the lack of this modification in the yeast initiator tRNA allows, in addition to initiate, the yeast initiator tRNA to elongate and to improve the formation of a ternary complex with GTP and the heterolougous *E. coli* EF-Tu (Kiesewetter *et al.*, 1990).

Number of *IMT* and *EMT* Genes in Various Yeast Strains

In *Saccharomyces cerevisiae* (bakers yeast), two isoaccepting species of methionine accepting tRNA, tRNA$_i^{Met}$ and tRNA$_m^{Met}$, are present (Gruhl & Feldman, 1976; Koiwai & Miyazaki, 1976). In contrast, brewers yeast has three isoaccepting methionine tRNA species, of which two represent the tRNA$_m^{Met}$ (Gruhl & Feldman, 1976; Koiwai & Miyazaki, 1976). The total number of Initiator Methionine tRNA (*IMT*) genes and Elongator Methionine tRNA (*EMT*) genes in *Saccharomyces cerevisiae* strain S288C (derivative of GRF88) was determined by genomic Southern blot analysis. Two sets of oligonucleotides, derived from the 5'- and 3'- region of the tRNA$_i^{Met}$ and tRNA$_m^{Met}$ RNA sequence, were used as probes (Simsek and RajBhandary, 1972, Gruhl & Feldman, 1976; Koiwai & Miyazaki, 1976). Four DNA fragments using the *IMT* oligonucleotides and five DNA fragments using the *EMT* oligonucleotides were obtained (Byström and Fink, 1989; Fig 2). For both the *IMT* oligonucleotides and the *EMT* oligonucleotides, fragments of identical size were obtained with both 5'- and 3'- oligonucleotides. The corresponding *IMT* and *EMT* genes were isolated from a YCp50 derived yeast DNA library (Rose et al, 1987) by colony hybridization using the *IMT* and *EMT* specific oligonucleotides (Byström and Fink, 1989; Åström et al, Data not shown). The potential positive clones were retested by Southern blot analysis. Clones, which hybridize strongly to the probes, were grouped based on restriction enzyme analysis. DNA sequence analysis of representatives from each group confirmed four *IMT* genes and five *EMT* genes, which contained identical tRNA$_i^{Met}$ and tRNA$_m^{Met}$ coding sequences but which contained different 5'- and 3'- flanking sequences (Byström and Fink, 1989; Åström et al, Data not shown). In strain S288C, the four *IMT* genes have been located on chromosomes V, X, XV, and XVI, whereas the locations of the *EMT* genes are unknown (Byström and Fink, 1989; Åström et al, Data not shown). However, in standard haploid laboratory yeast strains, the number of *IMT* genes may vary from four to five (strains GRF5, D273-10B, and A364A), and the number of *EMT* genes may vary from five to six (D273-10B) (Byström et al; 1989; Fig 2).

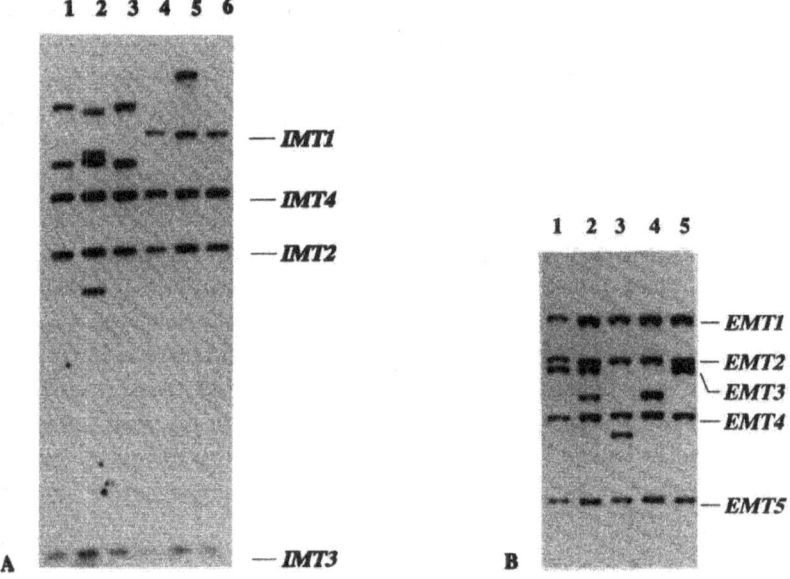

Figure 2A. Southern blot analysis of yeast chromosomal DNA. The four *IMT* gene signals of strain S288C are indicated. Lane 1, GRF5 (*Saccharomyces norbensis*); lane 2, C80-CG7 (*S. carsbergensis*); lane 3, D273-10B (Sherman strain); lane 4, FL100 (Lacroute strain); lane 5, A364A (Hartwell strain); lane 6, S288C. In all lanes genomic DNA was cut with the restriction enzymes *BamHI* and *HindIII*. Samples were electrophoresed on 1% agarose and blotted onto a nylon filter. An oligonucleotide homologous to position 9-28 of the IMT gene was used as a probe. Figure modified from Byström and Fink, 1989.
Figure 2B. Southern blot analysis of yeast chromosomal DNA. The five *EMT* gene signals of strain S288C are indicated. Lane 1, GRF5 (*Saccharomyces norbensis*); lane 2, D273-10B (Sherman strain); lane 3, A364A (Hartwell strain); lane 4, FL100 (Lacroute strain); lane 5, S288C. In all lanes genomic DNA was cut with the restriction enzymes *EcoRI* and *XhoI*. Samples were electrophoresed on 1% agarose and blotted onto a nylon filter. An oligonucleotide homologous to position 1-22 of tRNA$_m^{Met}$ was used as a probe.

Genes *IMT2, IMT3, and IMT4* of strain S288C are localized on the identical sized *BamHI -HindIII* fragments as in the other yeast strains, whereas the location of the *IMT1* gene varies (Fig 2a). The flanking sequences of the fifth gene in strain A364A differ from the four present in strain S288C (Byström and Fink, 1989; Venegas et al, 1982). *Saccharomyces carlsbergensis*, which is a diploid strain between *Saccharomyces cerevisiae* and another *Saccharomyces* species, show eight DNA fragments which hybridized the *IMT* oligonucleotides (Byström and Fink, 1989; Fig2A). The Southern hybridization patterns for the detection of *EMT* genes from various yeast strains are shown in Figure 2B. All strains tested showed five DNA fragments which hybridized the oligonucleotides, except strain D273-10B which had six. The DNA fragments corresponding to *EMT3* in strain S288C was missing in strains A364A and FL100 (Fig 2B). The fifth gene of these strains was located on a smaller sized DNA fragment and in strain FL100, the size of the fragment was similar to the sixth signal in strain D273-10B.

Construction of Yeast Strains

In separate strains with S288C background, each *IMT* or *EMT* gene was disrupted individually by replacing the wild-type copy with a mutant copy of the gene which contained a *TRP1* gene insertion (Byström and Fink, 1989; Åström et al, Data not shown). Since the genotype of the S288C derivative is *ura3-52, leu2-3, leu2-112, trp1Δ1*, the *TRP1* gene will serve as a genetic marker for the tRNA gene. Figure 3 outlines the procedure for the generation of a strain with null alleles of all four *IMT* genes such that growth of the strain is dependent on a plasmid derived *IMT* gene.

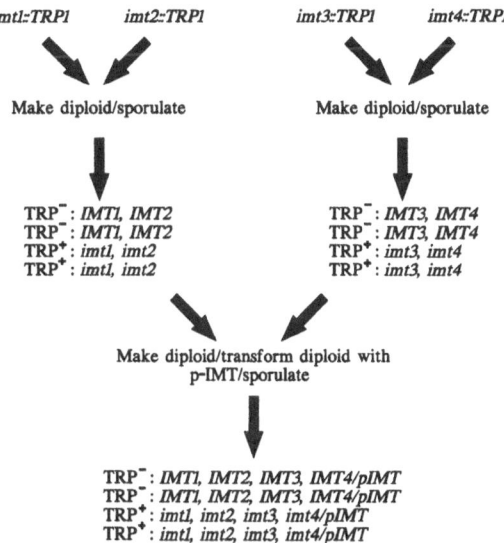

Figure 3. Strategy for construction of strains depleted of chromosomal *IMT* or *EMT* genes. The common genetic background of the strains are *ura3-52, leu2-3, leu2-112,trp1Δ 1*, mating type is not shown. A disrupted *imt* gene is marked by a *TRP1* gene, giving a TRP[+] phenotype. Diploids are formed between strains of opposite mating type, sporulated and subjected to tetrad analysis. In the tetrads the corresponding TRP[+] phenotype and *imt* genotype are shown. Yeast media and standard yeast genetic manipulations have been described by Rose *et al.* (1990). The selection of Ura[-] phenotypes with 5-fluoro-orotic acid was described by Boeke *et al.*, (1984).

The double mutant (*imt1::TRP1, imt2::TRP1*) was made by mating haploid strains of *imt1::TRP1* and *imt2::TRP1* with each other and the zygotes were isolated by micromanipulation and sporulated. Strains disrupted in genes *imt1* and *imt2* were identified in tetrads, which segregated the TRP phenotype as 2TRP[+]:2TRP[-]. The *IMT* genotype of these strains were confirmed by Southern blot analysis (Byström and Fink, 1989). Strains disrupted for both *imt3* and *imt4* were isolated in a similar manner. To obtain a strain with all four *imt* genes disrupted, a diploid formed between the two double mutants *imt1 imt2* and *imt3 imt4,* was transformed with a *URA3* based vector carrying a wild type *IMT* gene and sporulated (Fig 3).

Strains having all four *imt* genes disrupted by a *TRP1* gene were identified in tetrads, in which the TRP phenotype segregated as 2 TRP[+]:2TRP[-]. These tetrads were printed to plates containing 5-fluoro-orotic-acid (5-FOA), which selects against the *URA3* vector carrying the wild-type *IMT* gene, since the active *URA3* gene product converts 5-FOA to a toxic compound (Boeke et al, 1984). The TRP[+] progeny in these tetrads did not grow on 5-FOA plates and were confirmed by Southern blot analysis to have their chromosomal *IMT* genes disrupted (von Pawel-Rammingen et al, 1992). Strains in which the five elongator methionine tRNA genes (*EMT*) disrupted by a *TRP1* gene and for which the viability is dependent on a *URA3* yeast vector carrying an *EMT* gene were constructed in a similar way (Åström et al, Data not shown).

Features of *IMT* and *EMT* Assay Systems

In the tester strains, the chromosomal *IMT* and *EMT* genes are disrupted by a functional *TRP1* gene; the viability of the strain is maintained by a wild-type *IMT* or *EMT* gene on a *URA3* based plasmid (Fig 4 and Fig. 7).

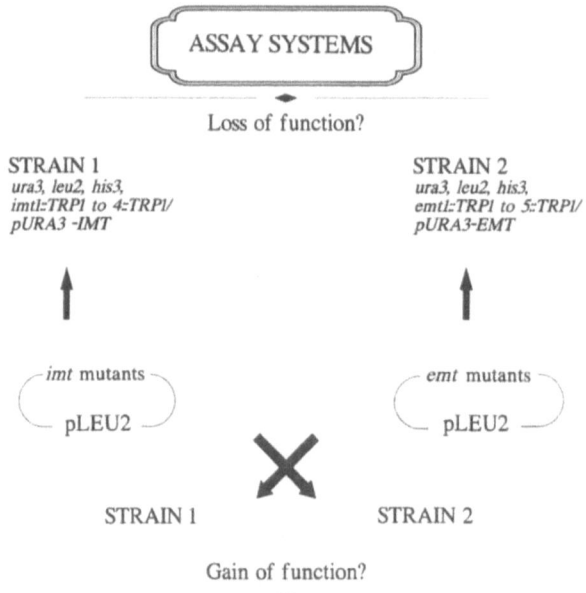

Figure 4. *In vivo* assay systems for mutated *imt* or *emt* genes. In separate strains of S288C genetic background, the four *IMT* or *EMT* genes have been disrupted by a functional *TRP1* gene. The viability of the strains are maintained by having a wild type *IMT* or *EMT* gene on a *URA3* based vector. By using a *LEU2* vector a second *IMT* or *EMT* gene, mutated or wild type, can be introduced into the strains. Thus, a strain that habors both a wild type gene (*URA3* vector) and a mutated gene (*LEU2* vector) can be obtained. If plating such a strain on medium supplemented with 5-FOA, the *URA3* vector will be selected against, since the active URA3 gene product converts 5-FOA to a toxic compound (Boeke et al, 1984). By this procedure the phenotype of the *LEU2* based mutant *imt* or *emt* gene will be uncovered. By having both systems, a mutant *imt* gene can be tested both for loss of function in the initiator depleted strain as well as gain of function as elongator in the elongator depleted strain. Like wise a mutant emt gene can be tested in both systems.

In addition to the *ura3* selectable marker, a *leu2* selectable marker is present in the strains. A second *IMT* or *EMT* gene, either mutated or wild-type, can be located on the *LEU2* (low- or high-copy) based plasmid and introduced into the strains. By plating the strain on 5-FOA containing media, the *URA3* wild-type derived *IMT* or *EMT* gene will be selected against and the phenotype of the *LEU2*-borne mutated *imt* or *emt* gene will be uncovered. This plasmid shuffling procedure was first described for the *CDC27* gene of *Saccharomyces cerevisiae* (Boeke et al, 1987). The shuffling procedure scores no growth if the mutation is lethal and growth if the mutant is viable. A viable mutant can also be distinguished by growth; i.e. normal growth or slow growth.

Since most of the signals for eukaryotic tRNA gene expression are located intragenically (Sharp et al, 1985; Geiduscheck and Tocchini-Valentini, 1988; Palmer and Folk, 1990; Gabrielsen and Sentenac, 1991), a mutation within the tRNA genes could affect the transcription or processing of the gene as well the participation of the mutant $tRNA_i^{Met}$ or mutant $tRNA_m^{Met}$ in the initiation or elongation of protein synthesis. To distinguish between these possibilities, total tRNA is prepared from the unshuffled viable strain containing the mutant tRNA gene of interest and a Northern blot is performed to determine whether the mutant tRNA transcript is synthesized and has the correct size (Fig 5). The hybridization conditions have been determined such, that a one nucleotide mismatch can be distinguished from the wild type tRNA (von Pawel-Rammingen et al, 1992).

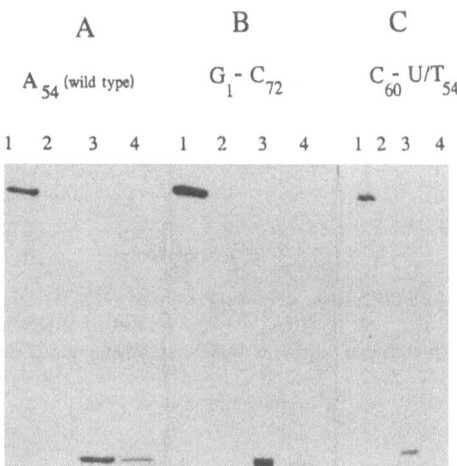

Figure 5. Northern blot analysis. Total tRNA preparations of unshuffled strains carrying, a wild type (panel A) or mutated *imt* gene (panels B and C) on a high copy vector. Positive controls in lanes 1 are digested plasmid DNA carrying the wild type *IMT* gene (A1) or the mutations G_1-C_{72} (B1) or $C_{60}-T_{54}$ (C1). Negative controls are digested plasmid DNA containing a wild type *IMT* gene in lanes B2 and C2, and a mutant *IMT* gene in lane A2. In lane A3, B3, and C3 total tRNA preparations from unshuffled strains carrying a wild type *IMT* gene (A3), or mutations G_1-C_{72} (B3) or $C_{60}-U/rT_{54}$ (C3) are present. The negative RNA controls in lanes B4 and C4 are total tRNA preparations of strains carrying a wild type IMT gene. In lane A4, total tRNA from a strain carrying a mutant *imt* gene is used as negative control. Three different gel runs are shown.

All mutant tRNAs showed the correct size and were synthesized to similar amounts as the wild type $tRNA_i^{Met}$. The only exceptions were initiator tRNAs containing single mutations U_{60} or U/rT_{54} which were probably destabilized due to an extra base pair in loop IV (von Pawel-Rammingen et al., 1992; see below).

In addition, the aminoacylation of the mutant tRNA must be investigated. This can be done by aminoacylating the crude tRNA preparation with S^{35} methionine and by analyzing the sample on NACS-20 column chromatography. This column easily separates the wild-type $tRNA_i^{Met}$ and $tRNA_m^{Met}$ from each other (Fig 6), and depending on the mutation, the mutant $tRNA_i^{Met}$ either separates from the corresponding wild-type tRNA or elutes with the wild type peak.

By utilizing both systems, a mutant *imt* tRNA can be assayed for a loss of activity in the initiator depleted strain, as well as a gain in elongator function in the elongator depleted strain. Similarly, a mutant *emt* tRNA can be assayed for a loss of activity in the elongator depleted strain and a gain in initiation function in the initiator depleted strain (Fig 4).

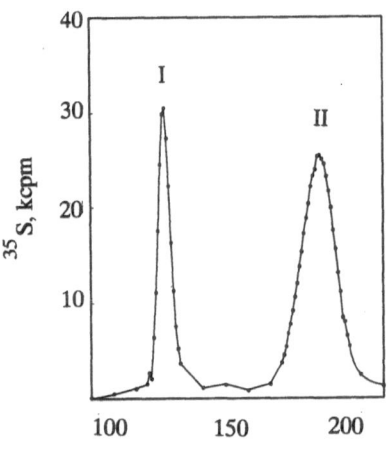

Fraction number

Figure 6. Separation on NACS-20 column chromatography of (^{35}S) Methionine aminoacylated tRNAs. Peak I represents $tRNA_i^{Met}$, and peak II $tRNA_m^{Met}$. The amount of $tRNA^{Met}$ species can be calculated from the total radioactivity in each peak of the NACS-20 chromatogram. Total tRNA was prepared from strain: L1937 (*Mata, ura3-52, leu2-3, leu2-112, trp1Δ1*), aminoacylated and separated on NACS-20 column as previously described by von Pawel-Rammingen et al, 1992.

Initiator Specific Features

By oligonucleotide directed mutagenesis, the nucleotides, conserved among eukaryotic initiators and believed to be important for initiator tRNA function, were mutated to the corresponding nucleotides of yeast $tRNA_i^{Met}$. In some positions, additional mutations were made. The effect of these mutant *imt* genes on growth, was investigated in the initiator depleted strain (von Pawel-Rammingen et al., 1992; Fig 7.).

Nucleoside changes in base pairs 3-70, 12-23, 31-39, 31-39/29-41 of the $tRNA_i^{Met}$, as well as loop I expansion at position 17 supported growth of the tester strain. Mutant U_{30}-A_{70} showed a slight effect on survival when present on a low-copy vector during plasmid shuffling, but not when present on a high-copy vector. In loop IV, the single mutations C_{54}, G_{54}, G_{60}, and C_{60} did not prevent cell growth. Single mutants U_{60} and U/rT_{54} were not viable, probably due to an extra base pair in loop IV, which affected the tRNA stability (von Pawel-Rammingen, et al 1992). The double mutants U_{60}-C_{54} and C_{60}-U/rT_{54} prevented this extra base pair. Mutation U_{60}-C_{54} behaved as the single C_{54} mutation, whereas C_{60}-U/rT_{54} blocked cell growth, even though the corresponding tRNA was synthesized and accepted methionine *in vitro*. Of the mutants in loop IV, G_{54}, C_{54}, and U_{60}-C_{54} had a reduced growth rate. Mutations C_{54} and U_{60}-C_{54} also showed a reduced viability, when present on low-copy vectors during plasmid shuffling, but not on high-copy vectors. Thus, in loop IV any nucleotide is allowed at position 60. In position 54, an A is optimal; a G or a C is tolerated, but a U/rT is detrimental (von Pawel-Rammingen et al, 1992). A U/rT in position 54 is a conserved feature of

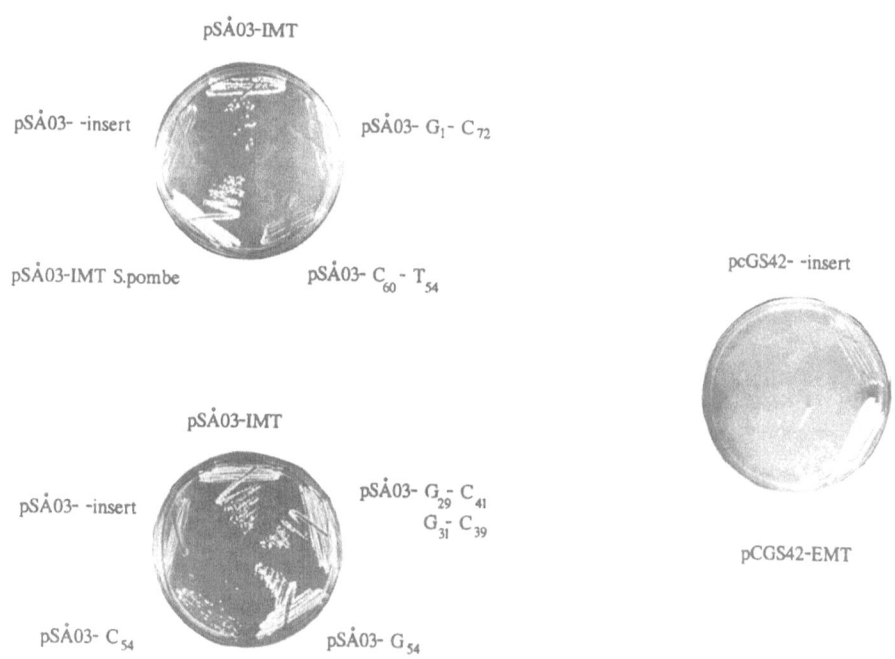

Figure 7. Growth of strains dependent on mutated *IMT* or *EMT* genes on 5-FOA plates. All strains are *ura3-52, leu2-3, leu2-112*,and *trp1Δ1*.The two plates to the left show strains depleted for *IMT* genes (*imt1* to *imt4*). The plate to the right shows strains depleted for *EMT* genes (*emt1* to *emt5*). These strains are also *his3ΔH3*. Growth is dependent on the low copy number plasmid pSÅ03, containing mutated or wild type *IMT* genes or on high copy number plasmid pCGS42 containing an *EMT* gene. The mutated positions of the tRNA$_i^{Met}$ are given; e.g. pSÅ03-C_{54} describes an *imt* gene in which the A in position 54 is changed to a C. The selection of Ura$^-$ phenotypes with 5-fluoro-orotic acid was described by Boeke *et al.*, (1984).

elongator tRNAs. Changing the A-U base pair 1-72, to the G-C of tRNA$_m^{Met}$ was deleterious for growth, although this tRNA was synthesized and accepted methionine *in vitro*. The *imt* mutants A_1-U_{72} and C_{60}-U/rT_{54}, act as elongator methionine tRNAs in the elongator depleted strain (Åström et al, Data not shown). Thus, position 54 and base pair 1-72 are important for tRNA$_i^{Met}$ identity per se and not for tRNA function in general.

Interestingly, position 1 of base pair 1-72, position 54, and position 64 of the tRNA$_i^{Met}$ (defined by Kiesewetter et al, 1990) are all located on the same surface of the tRNA tertiary structure (Bassavappa and Sigler, 1992; von Pawel-Rammingen et al; 1992). The striking difference between *E. coli* and yeast initiator tRNAs is the dependence of loop IV, which in yeast is not allowed to have a U/rT in position 54 (von Pawel-Rammingen et al, 1992). In *E. coli*, both the initiator and elongator tRNAs have a U/rT in position 54. The importance of loop IV of eukaryotic initiator tRNAs was first suggested from *in vitro* experiments by Wagner et al (1989). A second possible difference between yeast and *E. coli* initiator tRNAs is the dependence on G-C base pairs 'in the anticodon stem. In *E. coli*, the G-C base pairs are important for P-site recognition (Seong and RajBhandary, 1987). In yeast, a mutant, in which nearly the entire anticodon stem looks like tRNA$_m^{Met}$, supports growth to the same extent as wild- type tRNA$_i^{Met}$ (von Pawel-Rammingen et al, 1992; Fig 7).

SUMMARY AND OUTLOOK

We have constructed *in vivo* assays for the Initiator Methionine tRNA (*IMT*) and Elongator Methionine tRNA (*EMT*) genes (Fig 4). In these systems, a cell can be made absolutely dependent on a mutated tRNA$_i^{Met}$ or tRNA$_m^{Met}$ gene for growth. Thus, the systems measure the activity of a single mutated tRNA. We are presently using the systems for the investigation of the discriminatory elements between yeast tRNA$_i^{Met}$ and tRNA$_m^{Met}$ for participation in the initiation or elongation of protein synthesis and for pseudorevertant analysis of tRNAs defective in either process. Initiator tRNA depleted strains can also be used to investigate the activity of heterologous initiator tRNAs. Thus, a slow growing yeast strain with three out of four *IMT* genes disrupted (Byström and Fink, 1989), when supported with a plasmid in which the transcription of a human tRNA$_i^{Met}$ gene is driven by a yeast tRNA^{Arg} gene, restores growth (Francis and RajBhandary, 1990). Also, the *Schizosaccharomyces pombe* tRNA$_i^{Met}$ supports growth in a strain depleted of all four *IMT* genes (Fig 7). The systems can also be used to investigate other cellular processes, in which tRNA$_i^{Met}$ or tRNA$_m^{Met}$ participates. Thus, the yeast retrotransposon TY1 of *Saccharomyces cerevisiae* that transposes by a mechanism similar to retroviruses (Boeke and Sandmeyer, 1991), has by using the initiator depleted strain been shown to use tRNA$_i^{Met}$ as a primer for reverse transcriptase (Chapman et al, 1992).

Acknowledgements

We are grateful to G.R. Björk and D. Milton for critical reading of the manuscript. We would like to thank J. Kohli for providing us with plasmid pT3-B406 containing the serine methionine tRNA dimer of *S. pombe*. This work was supported by the Swedish Natural Science Research Council (project B-BU-4856-305) and the Swedish National Board for Technical Development (project 90-00270P) to A. S. Byström.

REFERENCES

Basavappa, R.,and Sigler, P.B., 1991, *EMBO J.* 10: 3105.
Boeke, J.D., Lacroute, F., and Fink, G.R., 1984. *Mol. Gen. Genet.* 197: 345.

Boeke, J.D., and Sandmeyer, S.B., 1991, p193. in "The Molecular and Cellular Biology of the yeast *Saccharomyces*. Genome Dynamics, Protein Synthesis, and Energetics, J.D. Broach, J.R. Pringle, E.W. Jones, eds., Cold Spring Harbor Laboratory Press, Cold Spring Habor, NY.

Boeke, J.D., Truehardt, J., Natsoulis, G., and Fink, G.R., 1987, *Methods in Enzymology* 154: 164.

Byström, A.S., and Fink, G.R., 1989. *Mol. Gen. Genet.* 216: 276.

Calagan, J.L., Pirtle, R.M., Pirtle, I.L., Kashdan, M.A., Vreman, H.J., and Dudock, D.S, 1980. *J. Biol. Chem.* 255: 9981.

Chapman, K.B., Byström, A.S., and Boeke, J.D., 1992, *Proc. Natl. Acad. Sci. USA* 89:3236.

Cigan, A.M., Feng, L., and Donahue, T.F., 1988, *Science* 242: 93.

Curran, J.F., and Yarus, M., 1989, *J. Mol. Biol.* 209:65.

Desgre`s, J., Keith, G., Kuo, K.C., and Gehrke, C.W., 1989. *Nucleic Acid Res.* 17: 865.

Despons, L., Senger, B., Fasiolo, F., and Walter, P., 1992, *J. Mol. Biol.* 225:897.

Eggertsson, G., and Söll, D., 1988, *Microbiol. Rev.* 52:354.

Francis, M.A., and RajBhandary, U.L., 1990, *Mol. Cell. Biol.* 10:4486.

Gabrielsen, O.S., and Sentenac, A., 1991, *TIBS* 16:412.

Geiduscheck, E.P., and Tocchini-Valentino, G.P., 1988, *Ann. Rev. Biochem.* 57: 873.

Guillon, J-M., Meinell, T., Mechulam, Y., Lazennec, C., Blanquet, S., and Fayat, G., 1992a. *J. Mol. Biol.* 224: 359.

Guillon, J-M., Mechulam, Y., Scmitter, J-M., Blanquet, S., and Fayat, G., 1992b, *J. Bacteriol.* 174:4294.

Glasser, A., Desgres, J., Heitzler, J., Gehrke, C. W., and Keith, G., 1991, *Nucleic Acid Res.* 19: 5199.

Gruhl, H., and Feldmann, H., 1976, *Eur. J. Biochem.* 68: 209.

Housman, D., Jacobs-Lorena, M., RajBhandary, U.L., and Lodish, H. F., 1970, *Nature (London)*, 227:913.

Kiesewetter, S., Ott, G., and Sprinzl, M., 1990, *Nucleic Acids Res.* 18: 4677.

Koiwai, O. and Miyazaki, M., 1976, *J. Biochem.* 80: 951.

Kozak, M., 1983, *Microbiol. Rev.* 47: 1.

Lee, C.P., Seong, B. L., and Rajbhandary, U.L., 1991, *J. Biol. Chem.* 266: 18012.

McClain, W.H., and Foss, K., 1988, *Science* 240: 793.

Murgola, E.J., 1985, *Annu. Rev. Genet.* 19:57.

Normanly, J., and Abelson, J., 1988, *Annu. Rev. Biochem.*. 58:1029.

Palmer, J.M., and Folk, W.R., 1990, *TIBS* 15: 300.

Rose, M.D., Novick, P., Thomas, J.H., Botstein, D., and Fink, G.R., 1987, *Gene* 60: 237.

Rose, M.D., Winston, F., and Hieter, P., 1990. *Methods in yeast genetics*. Cold Spring Harbor Laboratory Press, Cold Spring Harbor, NY.

Schimmel, P., 1989, *Biochemistry* 28:2747.

Schulman, L.H., and Pelka, H., 1988, *Science* 242:765.

Seong, B.L., and RajBhandary, U.L., 1987, *Proc. Natl. Acad. Sci. USA* 84: 334.

Sharp, S.J., Schaack, J., Cooley, L., Johnson Burke, D., and Söll, D., 1985, *Crit. Rev. Biochem.* 19: 107.

Simsek, M., and RajBhandary, U.L., 1972, *Biochem Biophys Res Commun* 49:508.

Smith, A.E., and Marcker, K.A., 1968, *J. Mol.Biol.* 38:241.

Smith, A.E. and Marcker, K.A., 1970. *Nature (London)* 226: 607.

Sprinzl, M., Dank, N., Nock, S., and Schön, A., 1991, *Nucleic Acids Res.* 19: Sequences Supplement 2127-2171.

Sundari, R.M., Stringer, E.A., Schulmann, L.M., and Maitra, U., 1976, *J. Biol. Chem.* 251: 3338.

Varshney, U., Lee, C.P., Seong, B.L., and RajBhandary, U.L., 1991, *J. Biol. Chem.* 266:18018.

Venegas, A., Gonzalez, E., Bull, P., and Valenzuela, P., 1982, *Nucleic Acids Res.* 10:1093.

von Pawel-Rammingen, U., Åström, S.U., and Byström, A.S., 1992. *Mol. Cell. Biol.* 12: 1432.

Wagner, T., Gross, M. and Sigler, P.B., 1984. *J. Biol. Chem.* 259: 4706.

Wagner, T., Rundquist, C., Gross, M. and Sigler, P.B., 1989. *J. Biol. Chem.* 264: 18506.

Wrede, P., Woo, N.H., and Rich, A., 1979, *Proc. Natl. Acad. Sci. USA* 76:3289.

SPECIFICITY IN RNA:PROTEIN INTERACTIONS; THE RECOGNITION OF ESCHERICHIA COLI GLUTAMINE tRNA

M. John Rogers, Ivana Weygand-Durašević, Etienne Schwob, Joyce M. Sherman, Kelley C. Rogers, H.-Ulrich Thomann, Lee A. Sylvers, Martina Jahn, Hachiro Inokuchi,[1] Eiko Ohtsuka,[2] and Dieter Söll

Department of Molecular Biophysics and Biochemistry
Yale University
New Haven, CT 06511 USA
[1]Department of Biophysics
 Kyoto University
 Kyoto, 606 Japan
[2]Faculty of Pharmaceutical Sciences
 Hokkaido University
 Sapporo, 060 Japan

INTRODUCTION

A major element ensuring the accuracy of translation of the genetic code is the recognition of tRNA by its cognate aminoacyl-tRNA synthetase. The tRNA molecule assumes many of the unusual elements of RNA structure (Delarue and Moras, 1989), and presumably this flexibility and variety in structure leads to the role of tRNA in other metabolic functions additional to protein synthesis (Söll, 1993). The application of a variety of genetic, biochemical and biophysical techniques have been used to elucidate factors determining the accuracy of the recognition by *Escherichia coli* glutaminyl-tRNA synthetase (GlnRS) for tRNA^{Gln}, and this system is probably one of the best understood tRNA:aminoacyl-tRNA synthetase systems (Söll, 1990), resulting in the first RNA-protein structure determined at the molecular level (Rould et al., 1989; Rould et al., 1991). This review will focus on our current understanding of this RNA:protein interaction.

A prokaryotic cell typically contains at least twenty aminoacyl-tRNA synthetases and about sixty tRNA species, and the sequences of aminoacyl-tRNA synthetases for each amino acid (Moras, 1992) and the tRNA gene sequences for all tRNA isoacceptors (Komine et al., 1990) are known for *E. coli*. Synthetases formally catalyze the same reaction; the aminoacylation of the 3'-terminal adenosine of the cognate tRNA with an amino acid (AA) in an ATP-dependent reaction:

$$AA + ATP + tRNA \rightleftharpoons AA\text{-}tRNA + AMP + PP_i$$

Aminoacyl-tRNA synthetases are remarkable for their diversity in sequence and oligomeric structure (Schimmel, 1989), particularly as the tertiary structure of the tRNA substrates they recognize is probably conserved to a large extent. There are fundamental differences between the synthetases in their presumed mode of ATP binding and in the site of aminoacylation of the terminal ribose of the tRNA (2'- or 3'-OH). The classification of aminoacyl-tRNA synthetases into two classes of ten enzymes each was originally based on sequence analysis (Eriani et al., 1990), and supported by biochemical data and the X-ray structure of Class I and Class II enzymes (Moras, 1992). However, it is also intriguing that Class II enzymes are predominantly specific for smaller amino acids (Carter, 1993). Although this may reflect the different architecture of the active site of Class II synthetases, this may also be indicative of an earlier, more restricted coding capacity. GlnRS possesses highly conserved structural motifs of a Rossmann fold-type of nucleotide binding site to bind ATP and to provide the presumed site of amino acid binding (Perona et al., manuscript in preparation). The fore-going and the fact that GlnRS aminoacylates the 2'-OH of the terminal adenosine show that GlnRS is a Class I enzyme (Englisch-Peters et al., 1991).

There are two factors operating in vivo to ensure the correct aminoacylation of tRNA; competition between synthetases for uncharged tRNA and the specificity of recognition of tRNA by the cognate synthetase. In addition, mechanisms of proofreading or editing of an incorrectly-charged tRNA act subsequently to the aminoacylation step (Freist, 1989; Jakubowski and Goldman, 1992). Competition between aminoacyl-tRNA synthetases and tRNA is a factor determining specificity (Yarus, 1972), as altering the ratio of either the tRNA or the synthetase may result in mischarging (a population of tRNA aminoacylated with the wrong amino acid). Therefore, the ratio of uncharged to charged tRNA is critical to determining overall accuracy. This was demonstrated for the glutamine system; when the in vivo level of GlnRS was elevated, mischarging with glutamine of the tRNA^{Tyr} amber suppressor resulted (Swanson et al., 1988). The effect was abolished by concomitant overproduction of tRNA^{Gln}, restoring the correct balance of tRNA to cognate synthetase. Mischarging of amber suppressor tRNA^{Tyr} was also abolished by elevating the level of tyrosyl-tRNA synthetase, thereby preventing uncomplexed tRNA^{Tyr} from being mischarged by GlnRS (Sherman et al., 1992a). Therefore, competition in vivo is a contributing factor in determining the accuracy of aminoacylation.

tRNA IDENTITY

The tRNA molecule contains a number of conserved nucleotides, and most tRNAs can be folded into the familiar cloverleaf structure. The informational content for specific recognition by a number of proteins and RNA of the translational apparatus is accommodated within the tRNA molecule. The limited number of nucleotides in tRNA that define the specificity of aminoacylation in vivo by the cognate synthetase are termed the identity elements (Normanly and Abelson, 1989). The identity elements may be examined by in vitro experiments and are referred to in this context as recognition elements (Schulman, 1991). The structure of a number of tRNAs have been established (Moras, 1992), although the structure of a tRNA with long variable loop has not been determined and the structure of tRNA^{Gln} uncomplexed with GlnRS is not known. From the structure of the tRNA^{Gln}:GlnRS complex (Rould et al., 1989; Rould et al., 1991), there are many RNA:protein interactions contributing to overall specificity. The tRNA also assumes a novel structure in complex with GlnRS, with deformation of the 3'-end of tRNA^{Gln}, unusual base-pairs extending the anticodon stem and disruption of the regular stacking of the bases of the anticodon. Therefore, the tRNA and protein are extensively linked in this system, leading to an intricate picture of glutamine identity.

In Vivo Identity

An in vivo approach to elucidating tRNA identity is based on the extensive suppressor tRNA genetics in *E. coli* (Eggertsson and Söll, 1988) or on an initiation assay based on mutating the anticodon of the initiator tRNA^{fMet} (Schulman, 1991). This is coupled with total tRNA gene synthesis or mutagenesis to construct mutated tRNAs, and purification of a reporter protein (usually dihydrofolate reductase; DHFR) from an *E. coli* strain containing the mutated tRNA. The amino acid(s) inserted by the mutated suppressor tRNA or the N-terminal (formylated) amino acid(s) by the mutated initiator tRNA is then determined by protein sequencing (Normanly and Abelson, 1989; Schulman, 1991).

The Discriminator Nucleotide (Position 73)

The nucleotide at position 73 in tRNA was termed the discriminator when a correlation was noted between the amino acid aminoacylated on a tRNA and the nucleotide at this position (Crothers et al., 1972). Genetic experiments conducted twenty years ago with mutants of the amber suppressor derived from tRNA^{Tyr} had identified the discriminator nucleotide as important for GlnRS recognition. This is an important nucleotide in the interaction of tRNA^{Gln} complexed with GlnRS, as G73 forms an intramolecular hydrogen bond to maintain the hairpin structure placing the 3'-terminal adenosine in the active site of GlnRS. A comprehensive study of discriminator base mutants of amber suppressor tRNAs derived from tRNA^{Gln} was undertaken to examine the role of this nucleotide in tRNA identity (Fig. 1): G73 was substituted with other nucleotides and the incorporation of amino acids in vivo by the DHFR assay was determined (Sherman et al., 1992b). All the discriminator mutants of the amber suppressor derived from tRNA^{Gln} still insert glutamine, suggesting that tRNA^{Gln} is "overdetermined" for glutamine identity and in vivo competition favors aminoacylation with glutamine. To test this hypothesis, the in vivo identity of discriminator mutants of the amber suppressor tRNA^{Gln} mutated at the first base pair (tRNA^{Gln} G1C72, Fig. 1) were tested. The rationale for this was, as the first base pair is disrupted in the tRNA^{Gln}:GlnRS structure, a stronger base pair would probably impede aminoacylation by GlnRS, leading to mischarging by other synthetases. This was indeed the case, as the G1C72A73 mutants and G1C72G73 mutants show minor levels of Tyr- and Ser-insertion respectively (Fig. 1). This study was extended for discriminator base mutants of amber suppressors derived from other tRNAs. The role of the discriminator base (G73) was unknown for tRNA^{Glu} as the amber suppressor (with an additional mutation A38 to improve suppressor efficiency) is mischarged with glutamine (Normanly and Abelson, 1989; and Fig. 1), and also the discriminator base (A73) in tRNA^{Tyr} was studied where its role as an identity element is established.

The results from the analysis of the discriminator mutants of tRNA^{Tyr} confirm the role of A73 as an identity element for TyrRS, as both the G73, A73 and U73 mutant tRNAs are mischarged with glutamine, and mischarging at low levels of tRNA^{Tyr}U73 by a number of amino acids indicates productive competition with other synthetases in vivo resulting in a tRNA with ambiguous identity (Fig. 1). The results

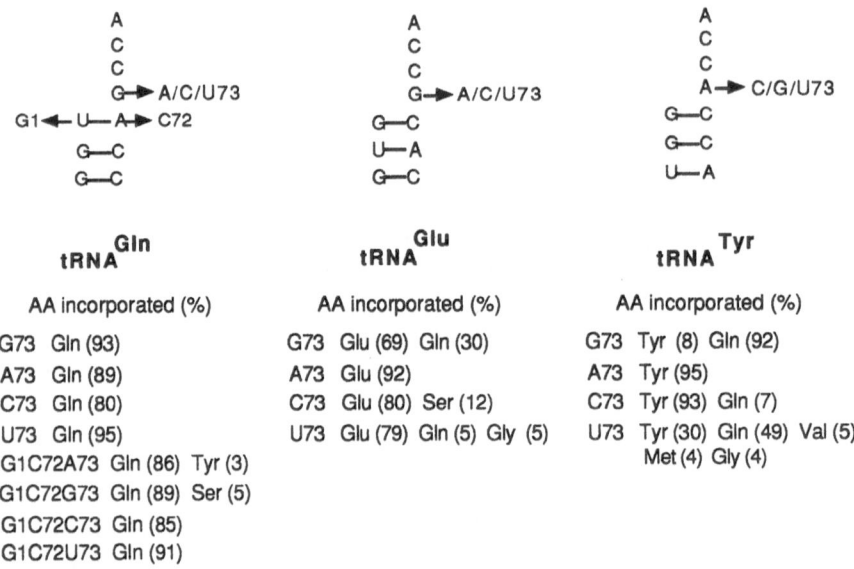

Fig. 1. Discriminator base mutants of tRNA^{Gln}, tRNA^{Glu} and tRNA^{Tyr}.
The acceptor stem of the amber suppressors derived from tRNA^{Gln}, tRNA^{Glu} and tRNA^{Tyr} are shown, indicating mutations made in the discriminator nucleotide (position 73) of each tRNA and the terminal base pair (1-72) of tRNA^{Gln} (Sherman et al., 1992b). The amino acid(s) inserted by these suppressors are indicated below, as determined by protein sequencing of DHFR.

from the discriminator mutants of tRNA^{Glu} are interesting, as mutating the wild-type G73 to A73 results in an efficient, glutamate-specific tRNA^{Glu} suppressor. Competition with GlnRS has then been abolished, and as the discriminator base is apparently not important for GluRS identity, only in vivo charging with the cognate enzyme is seen. Other amino acids are inserted at a low level by the C73 and U73 mutants of tRNA^{Glu}, which reflects the importance of, for example, U73 for glycine identity (Schulman, 1991). The discriminator base is strategically placed for recognition of the tRNA as it is the only non-conserved single-stranded nucleotide near the site of aminoacylation of tRNA, and these studies confirm that its role in vivo is complex, depending both on nucleotide context and synthetase competition.

The Anticodon

The amber suppressor derived from tRNA^{Trp} by a single point mutation (C→U35) in the anticodon was determined a number of years ago to be mischarged with glutamine in vivo and in vitro (Yaniv et al., 1974; Celis et al., 1976). This implicated the anticodon of tRNA, in particular U35, as being important for GlnRS recognition. The accurate recognition of the anticodon of tRNA^{Gln} is suggested from the structure of the tRNA^{Gln}:GlnRS complex (Rould et al., 1991), and confirmed by an in vitro analysis of tRNA^{Gln} mutants (Jahn et al., 1991). For example, the single mutation U→C35 in an in vitro transcript derived from tRNA^{Gln} reduces the specificity constant (k_{cat}/K_M) for aminoacylation by GlnRS by more than 10^5-fold. We were therefore interested in the amino acid(s) inserted by the opal suppressor of tRNA^{Gln} (su$^+$2 UGA) isolated by genetic selection (Inokuchi et al., 1990), as this has a C at position 35 (Fig. 2). The resulting suppressor also contained an additional mutation C→U70 in the acceptor stem which provides the G3-U70 base pair, shown previously to be important for alanine identity (Hou and Schimmel, 1988; McClain and Foss, 1988a).

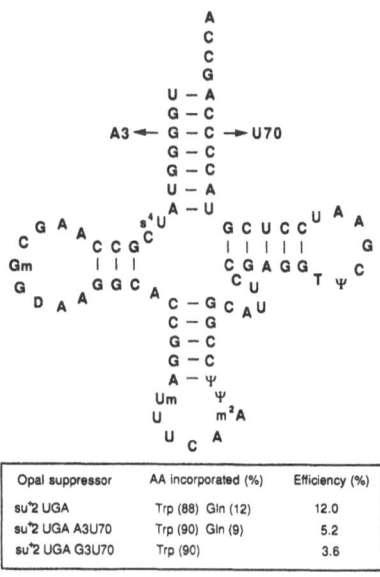

Opal suppressor	AA incorporated (%)	Efficiency (%)
su$^+$2 UGA	Trp (88) Gln (12)	12.0
su$^+$2 UGA A3U70	Trp (90) Gln (9)	5.2
su$^+$2 UGA G3U70	Trp (90)	3.6

Fig. 2. The opal suppressor su$^+$2 UGA derived from tRNA$_2^{Gln}$ by the anticodon mutations CUG→UCA.
Additional mutations of the 3-70 base pair are shown, with the amino acids incorporated determined by DHFR sequencing shown below with the efficiency of read-through (Rogers et al., 1992).

The opal suppressor of tRNA^{Gln} with mutations exclusively in the anticodon was made by in vitro mutagenesis (Fig. 2). Remarkably, the opal suppressor su$^+$2 UGA is mischarged efficiently with tryptophan as determined by the DHFR assay (Rogers et al., 1992; and Fig. 2) but also retains a small amount of glutamine-acceptance in vivo. Therefore, although in vitro the corresponding anticodon mutation results in a very poor substrate for GlnRS, the specificity of GlnRS in vivo is eased. Mutations of the 3-70 base pair reduce suppressor efficiency of the su$^+$2 UGA suppressor (Fig. 2), presumably by reducing aminoacylation with glutamine as the wild-type G3-C70 base pair is specifically recognized by GlnRS (Rould et al., 1989). The analysis of the opal suppressors derived from tRNA^{Gln}, together with in vitro aminoacylation by tryptophanyl-tRNA synthetase (TrpRS) (Rogers et al., 1992; Himeno et al., 1991) shows the importance of the anticodon, in particular C35, for TrpRS identity. In addition, Gln- and Trp- identities are interchangeable by substitutions in the anticodons of the tRNAs.

Anticodon recognition by GlnRS is currently being addressed by in vitro mutagenesis. For example, from the structure of the complex of tRNA^{Gln}:GlnRS, the specificity of recognition of position G36 of the anticodon of tRNA^{Gln} may be mediated by the side chain of R402 of GlnRS. To address this question, two mutants of GlnRS (Q402 and A402) were made at this position (U.Thomann, unpublished results). The ability of the mutant enzymes to aminoacylate cognate tRNA^{Gln} is severely reduced both in vivo and in vitro. The availability of amber and opal suppressor tRNAs derived from tRNA^{Gln} (with a G→A change at position 36) may reveal the basis of specific recognition of position 36 in vivo.

Acceptor stem recognition by GlnRS

The importance of the acceptor stem in accurate recognition of tRNA by GlnRS was first implicated from genetic experiments with the amber suppressor derived from tRNA^{Tyr} (Celis and Piper, 1982) containing mutations in the terminal base pairs of the acceptor stem. Additionally, the amber suppressor derived from tRNA$_1^{Ser}$ was converted to an efficient glutamine-accepting tRNA with additional mutations of the 1-72 and 3-70 base pairs (Rogers and Söll, 1988), indicating that discrimination between tRNA^{Gln} and tRNA^{Ser} is mediated by acceptor stem recognition. An in vivo approach has also been used to select for mutants of GlnRS that confer relaxed specificity for tRNA. This selection involves an amber mutation in the gene for ß-galactosidase (lacZ$_{1000}$) which is efficiently suppressed by glutamine, and not by most other amino acids, allowing detection of mischarging of amber suppressor tRNAs by GlnRS mutants. The lacZ$_{1000}$ selection was used to isolate mutants which mischarge the amber suppressor derived from tRNA^{Tyr} (Inokuchi et al., 1984; Rogers and Söll, 1990). The position of the mutated residues in GlnRS which confer relaxed specificity were implied from the tRNA^{Gln}:GlnRS structure as affecting the recognition of the acceptor stem of tRNA, in particular with the 3-70 base pair and the 3'-end of tRNA^{Gln} (Perona et al., 1989).

The genetic selection (Fig. 3) has been extended to select for mutants produced by in vitro saturation mutagenesis. A localized region of the protein was targeted from the structure as contacting the acceptor stem of tRNA (Weygand-Đurašević et al., 1993; vide infra). The substrate used for the genetic selection was the amber suppressor derived from tRNA$_1^{Ser}$, shown previously to be an efficient, serine-inserting amber suppressor not mischarged with glutamine unless acceptor stem mutations were superimposed on the tRNA (Rogers and Söll, 1988). This suppressor tRNA provides an ideal substrate for genetic selection of mutants of GlnRS with altered acceptor stem recognition. Two loop motifs of the acceptor binding domain of GlnRS were chosen for random mutagenesis; helix/loop E and loop 1 (Fig. 3). A mutation in helix E (GlnRS15; I→T129) was isolated by genetic selection from mischarging of the amber suppressor derived from tRNA^{Tyr} (Perona et al., 1989).

The subset of mutations which confer mischarging are centered around R130 and E131 in helix/loop E and I183, V184 and M185 of loop 1 (Weygand-Đurašević et al., 1993; Fig. 3). Crystallographic data shows that R130 contributes to the formation of a binding pocket for C74, and R133 interacts at several sites of the acceptor stem of tRNA^{Gln} (Rould et al., 1989); the phosphates of G73 and C74, and the ribose hydroxyl of A72. Since the GlnRS mutants with the strongest phenotypes contain mutations at positions 130 and 131, the wild type amino acid in these positions is likely to be critical for accurate recognition. Interestingly, a number of multiple mutations were isolated from the genetic selection, in addition to mutations (for example, R→P130) that are likely to introduce considerable structural changes in the positioning of helix/loop E. The R→P130, E→D131 double mutant enzyme was overexpressed and purified, and the in vitro properties of the mutant enzyme indicate that the kinetic parameters (k_{cat}/K_M) for cognate tRNA and for the non-cognate (mischarged) tRNA do not differ greatly from wild-type enzyme. Rather, the dissociation constant (K_d) for a non-cognate tRNA (tRNA^{Glu}) is 8-fold lower with the mutant enzyme. Therefore, the stability of non-cognate complex formation may be the basis for in vivo mischarging.

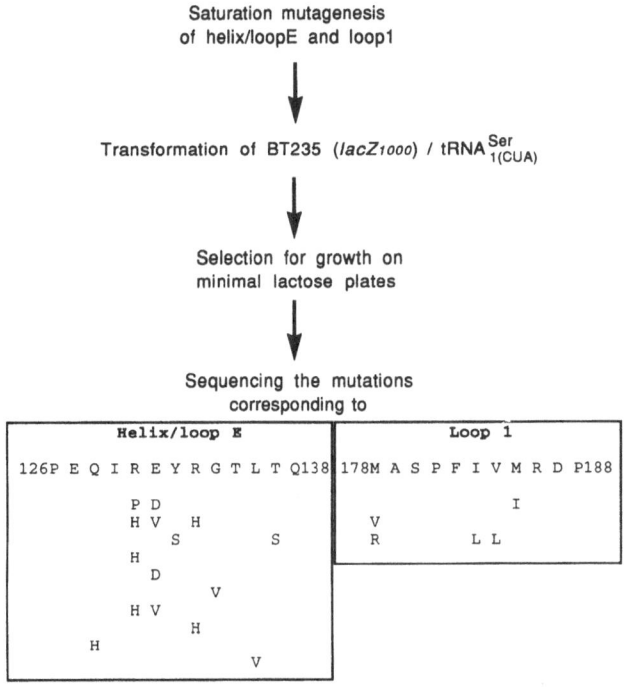

Saturation mutagenesis
of helix/loopE and loop1

Transformation of BT235 (*lacZ1000*) / tRNA$^{Ser}_{1(CUA)}$

Selection for growth on
minimal lactose plates

Sequencing the mutations
corresponding to

Helix/loop E	Loop 1
126P E Q I R E Y R G T L T Q138	178M A S P F I V M R D P188

Fig. 3. The selection for mutants of helix/loop E (amino acids 126-138) and loop 1 (amino acids 178-188) in GlnRS by mischarging of the amber suppressor derived from tRNA$^{Ser}_1$.
     The mutations responsible for the mischarging phenotype are shown below the wild-type amino acid (Weygand-Đurašević et al., 1993).

The residues corresponding to the DNA sequence targeted for mutagenesis are part of two finger-like structures penetrating the minor groove of the acceptor stem of tRNA^{Gln} (Fig. 4), and are probably important in forming a complementary surface for sequence-specific recognition of the conformation of the 3'-end of tRNA^{Gln}. The isolation of multiple mutations in many of the mischarging enzymes including mutations likely to effect considerable structural perturbation suggests that there is an sophisticated set of interactions changed by the altered conformation of the mutant enzymes. This change in a set of interactions is therefore necessary to detect mischarging of non-cognate tRNAs. The contribution of these regions (Fig. 4) is to reject non-cognate tRNAs through negative interactions with the acceptor stem of

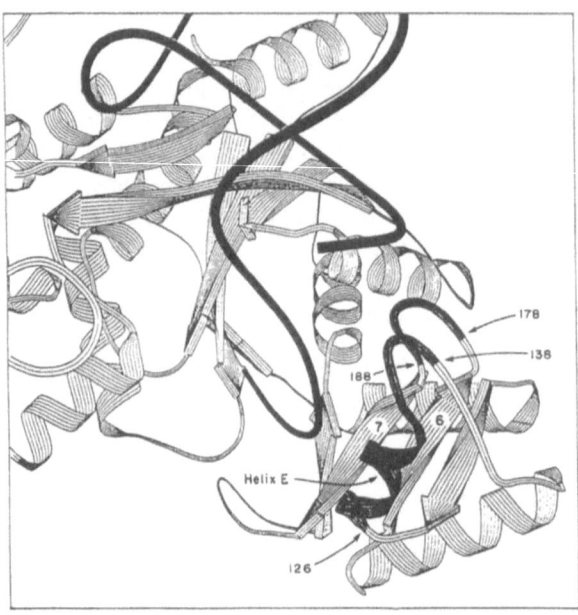

Fig. 4. The tRNA^{Gln}:GlnRS structure showing the acceptor binding domain and location of regions mutagenized interacting with the acceptor stem of tRNA^{Gln}. The two shaded regions (helix/loop E and loop 1) correspond to the DNA sequences mutagenized in GlnRS, and the acceptor stem of tRNA^{Gln} indicated in solid lines (Weygand-Đurašević et al., 1993; reprinted with permission).

these tRNAs, in addition to providing productive interactions with cognate tRNA.

As discussed above, the structure of the complex between tRNA^{Gln}:GlnRS shows a number of interactions with helix/loop E and loop 1 with tRNA (Rould et al., 1989). However, it is striking that I129 makes no direct contacts with tRNA, although the I→T129 mutation in GlnRS15 effects mischarging of the amber suppressor tRNA^{Tyr} (Perona et al., 1989). Additionally, the structure of the tRNA^{Gln}:GlnRS15 complex shows no additional changes in interactions with cognate tRNA (Arnez and Steitz, unpublished results). Clearly, to elucidate the structural basis of mischarging and selectivity among tRNA, the structure of the mischarged substrate used in the genetic selection with the mutated enzyme is needed. Also, the summation of conformational changes and distortions are important for the mischarging effected by the mutant GlnRS enzymes. For example, helix E is followed by L136, whose side chain interrupts the first base pair of tRNA^{Gln} (U1-A72) (Rould et al., 1989). Some of the mutations isolated in helix/loop E could lead to mis-positioning of L136, facilitating relaxed discrimination of tRNAs altered at the first base pair.

In Vitro Recognition

The recognition of tRNA in vitro can be evaluated from the kinetic parameters for the aminoacylation reaction. Essential to this approach is that tRNA substrates made by transcription with T7 RNA polymerase in vitro are usually good substrates for aminoacyl-tRNA synthetases (Sampson and Uhlenbeck, 1988). However, the lack of modified nucleosides in the RNA transcript may result in loss of specificity for aminoacylation, as modified bases may act as "anti-determinants" by blocking the recognition by non-cognate enzymes (Muramatsu et al., 1988; Perret et al., 1990). The kinetic analysis cannot usually take into account the effect of competition with other aminoacyl-tRNA synthetases, although an in vitro system can be devised to analyze the aminoacylation of a substrate with "ambiguous" identity (Sherman et al., 1992a).

The dissociation constant between a tRNA and its cognate aminoacyl-tRNA synthetase is typically in the 10^{-6}-10^{-7} M range, indicative of rapid tRNA turnover in aminoacylation with rather weak interactions (Schimmel and Söll, 1979). The usual analysis with a single synthetase gives an evaluation of the catalytic parameter k_{cat} and the Michaelis constant K_M. While the K_M is usually in the micromolar range for cognate tRNA, the K_M for non-cognate tRNAs may be within an order of magnitude. Therefore, discrimination by substrate binding alone is unlikely. Rather, the accuracy of aminoacylation is reflected in the specificity constant (k_{cat}/K_M), although correlation between in vivo identity and in vitro recognition is needed to build an overall picture of tRNA to include competition with other synthetases. In particular, the sensitivity of genetic assays coupled with the usual in vivo analysis at a plateau of aminoacylation may result in differences in the interpretation of in vivo and in vitro results.

The steady-state kinetic analysis of mutant tRNA substrates derived from tRNA^{Gln} shows that the acceptor stem and anticodon nucleotides play a major role in tRNA^{Gln} recognition (Jahn et al., 1991). In particular, single point mutations in the anticodon decrease the specificity constant by up to 10^5-fold and mutations in the acceptor stem, particularly at base pairs 2-71 and 3-70, decrease the specificity constant by 10^4-fold. The in vitro kinetic analysis generally supports the identity of tRNA^{Gln} determined from in vivo experiments. The molecular nature of some of these interactions were studied by inosine-substituted tRNAs made at positions where the 2-amino group of guanosines has been implicated as being important (Hayase et al., 1992). These inosine-substituted tRNAs, which then lack specific 2-amino groups, confirmed the role of the 2-amino group of G2, G3 and G10 in GlnRS recognition. With a combination of genetic, biochemical and biophysical techniques we feel that a complete set of identity/recognition elements has been described for tRNA^{Gln} (Fig. 5).

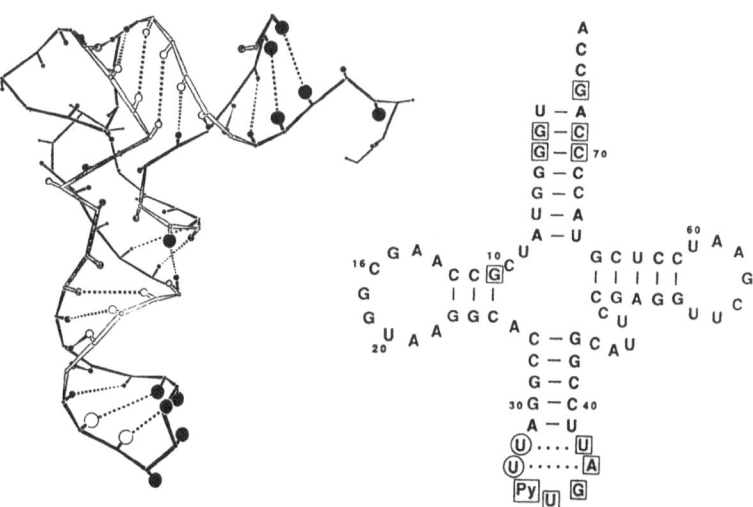

Fig. 5. The identity of tRNA^{Gln}. (Left): The tertiary structure of tRNA^{Gln}, when complexed with GlnRS, with the phosphate backbone and bases shown as spheres. The identity elements are highlighted. (Right): The secondary structure of unmodified tRNA$_2^{Gln}$ with the identity elements boxed and circled bases facilitating additional base-pairing important for identity (Hayase et al., 1992; reprinted by permission of Oxford University Press).

The location of the identity elements for tRNA^{Gln} are clustered in two regions; the acceptor stem and the anticodon (Fig. 5). The emerging picture of tRNA identity suggests that the acceptor stem and anticodon are major recognition sites in most tRNAs, although each system is idiosyncratic (Schulman, 1991; for example). Identity elements for other tRNAs include the unique G-1 position of tRNA^{His} (Himeno et al., 1989; Francklyn and Schimmel, 1990), the discriminator base (Sherman et al., 1992b; for example), the long variable arm of tRNA^{Tyr} and tRNA^{Ser} (Himeno et al., 1990), the variable pocket formed by the tertiary structure of tRNA^{Arg} (McClain and Foss, 1988b), and the modified nucleosides in the anticodon of tRNA^{Ile} (Muramatsu et al., 1988) and tRNA^{Glu} (K. Rogers et al., manuscript in preparation).

OUTLOOK

The GlnRS enzyme consists of a central catalytic domain with four additional domains (Rould et al., 1989). Are the other domains, responsible in part for tRNA selectivity, dispensable in vivo? Remarkably, genetic selection has been used to delete extensive regions of the acceptor binding domain and anticodon binding domain of the GlnRS enzyme (E. Schwob and D. Söll, manuscript in preparation). This then supports a minimal enzyme made up of the central catalytic core of Class I enzymes which is conserved structurally in evolution (Perona et al., 1991), with additional domains of Class I synthetases added for tRNA specificity. Intriguingly, since the mischarging enzymes isolated by genetic selection are relaxed generally in tRNA specificity, the elimination of negative interactions with non-cognate tRNAs may be the basis for this loss in specificity. Is it possible to isolate a GlnRS enzyme with super-specificity for cognate tRNA? The mischarging of the amber suppressor tRNA^{Tyr} by overproduction of the wild-type GlnRS enzyme suggests it is possible to isolate such a mutant by genetic selection.

The structure of the tRNA^{Gln}:GlnRS complex indicates a substantial region of the inside of the L-shaped tRNA molecule in contact with the protein (Rould et al., 1989; Rould et al., 1991). Included in these interactions are a number of tRNA backbone phosphate and 2'-OH groups interacting with protein. Because of a specific RNA context or conformation, these contribute to overall specificity and may be studied by exploiting current methods of RNA manipulation. Also, the contribution of negative elements on tRNA, preventing productive interactions with the identity elements, may be revealed by switching tRNA identity from one system to another. This is shown for anticodon switching between the Gln- and Trp-systems in vivo (Rogers et al., 1992), and for other systems in vitro (M. Frugier and K. Rogers, unpublished results).

The recognition of tRNA is a dynamic process, as the structure of the tRNA^{Gln}:GlnRS complex shows deformation of both the anticodon and acceptor stem of tRNA^{Gln} compared with the usual L-shaped tRNA (Rould et al., 1989; Rould et al., 1991). Is there a connection between the recognition of these parts of the tRNA molecule that signals the correct aminoacylation of the 3'-end of tRNA? The availability of amber and opal suppressor tRNAs, altered in both anticodon and acceptor stem recognition, may provide a genetic selection for compensating mutations in GlnRS. The location of these mutations may not be directly implicated in tRNA specificity but possibly important for transmitting the signal for correct anticodon recognition to the active site.

Is *E. coli* glutamine identity a special case in tRNA:synthetase interactions? In Gram-positive eubacteria and chloroplasts of higher plants, a GlnRS activity cannot be detected (Wilcox and Nirenberg, 1968; Schön et al., 1988). Rather, the glutamyl-tRNA synthetase (GluRS) in these systems is a naturally misacylating enzyme which charges both tRNA^{Glu} and tRNA^{Gln} with glutamate. The mischarged Glu-tRNA^{Gln} is then converted to Gln-tRNA^{Gln} by a specific amidotransferase (Schön et al., 1988). It is pertinent to note that the GluRS from *Bacilli* are apparently lethal if expressed in *E. coli* (Breton et al., 1990). Therefore, the comparison of the recognition of *E. coli* GlnRS with GluRS (and amidotransferase) from these species will be particularly interesting.

ACKNOWLEDGMENTS

The authors apologize for the incomplete reference list cited due to space considerations; where possible, the most comprehensive reviews have been included. We thank A. Lloyd for critical reading of the manuscript, and members of the laboratories for discussions and suggestions.

REFERENCES

Breton, R., D. Watson, M. Yaguchi, and J. Lapointe. 1990. *J. Biol. Chem.* **265**: 18248-18255.

Carter, C. W., Jr. 1993. *Annu. Rev. Biochem.* in press.

Celis, J. E., C. Coulondre, and J. Miller. 1976. *J. Mol. Biol.* **104**: 729-734.

Celis, J. E., and P. W. Piper. 1982. *Nucleic Acids Res.* **10**: r83-r91.

Crothers, D.M., T. Seno, and D. Söll. 1972. *Proc. Natl. Acad. Sci. USA* **69**: 3063-3067.

Delarue, M., and D. Moras. 1989. *In: Nucleic Acids and Molecular Biology* **3**: 182-196, F. Eckstein, and D. Lilley, ed., Springer-Verlag, New York.

Eggertsson, G., and D. Söll. 1988. *Microbiol. Rev.* **52**: 354-374.

Englisch-Peters, S., J. Conley, J. Plumbridge, C. Leptak, D. Söll, and M. J. Rogers. 1991. *Biochimie:* **73**: 1501-1508.

Eriani, G., M. Delarue, O. Poch, J. Gangloff, and D. Moras. 1990. *Nature* **247**: 203-206.

Francklyn, C., and P. Schimmel. 1990. *Proc. Natl. Acad. Sci. USA* **87**: 8655-8659.

Freist, W. 1989. *Biochemistry* **28**: 6787-6795.

Hayase, Y., M. Jahn, M. J. Rogers, L. A. Sylvers, M. Koizumi, H. Inoue, E. Ohtsuka, and D. Söll. 1992. *EMBO J.* **11**: 4159-4165.

Hou, Y.-M., and P. Schimmel. 1988. *Nature.* **333**: 140-145.

Himeno, H., T. Hasegawa, T. Ueda, K. Watanabe, K. Miura, and M. Shimizu. 1989. *Nucleic Acids Res.* **17**: 7855-7863.

Himeno, H., T. Hasegawa, T. Ueda, K. Watanabe, and M. Shimizu. 1990. *Nucleic Acids Res.* **17**: 6815-6819.

Himeno, H., T. Hasegawa, H. Asahara, K. Tamura, and M. Shimizu. 1991. *Nucleic Acids Res.* **19**: 6379-6382.

Inokuchi, H., P. Hoben, F. Yamao, H. Ozeki, and D. Söll. 1984. *Proc. Natl. Acad. Sci. USA* **81**: 5076-5080.

Inokuchi, H., K. Kondo, M. Yoshimura, and H. Ozeki. 1990. *Mol. Gen. Genet.* **223**: 433-437.

Jahn, M., M. J. Rogers, and D. Söll, D. 1991. *Nature* **352**: 258-260.

Jakubowski, H., and E. Goldman. 1992. *Microbiol. Rev.* **56**: 412-429.

Komine, Y., T. Adachi, H. Inokuchi, and H. Ozeki. 1990. *J. Mol. Biol.* **212**: 579-598.

McClain, W. H., and Foss, K. 1988a. *Science* **240**: 793-796.

McClain, W. H., and Foss, K. 1988b. *Science* **241**: 1804-1807.

Moras, D. 1992. *Trends Biochem. Sci.* **17**: 159-164.

Muramatsu, T., K. Nishikawa, F. Nemoto, Y. Kuchino, S. Nishimura, T. Miyazawa, and S. Yokoyama. 1988. *Nature* **336:** 179-181.

Normanly, J., and J. Abelson. 1989. *Annu. Rev. Biochem.* **58:** 1029-1049.

Perona, J. J., M. A. Rould, T. A. Steitz, J.-L. Risler, C. Zelwer, and S. Brunie. 1991. *Proc. Natl. Acad. Sci. USA* **88:** 2903-2907.

Perona, J. J., R. N. Swanson, M. A. Rould, T. A. Steitz, and D. Söll. 1989. *Science* **246:** 1152-1154.

Perret, V., A. Garcia, H. Grosjean, J.-P. Ebel, C. Florentz, and R. Giege. 1990. *Nature* **344:** 787-789.

Rogers, M. J., and D. Söll. 1988. *Proc. Natl. Acad. Sci. USA* **85:** 6627-6631.

Rogers, M. J., and D. Söll. 1990. *Prog. Nucl. Acid Res. Mol. Biol.* **39:** 185-208.

Rogers, M. J., T. Adachi, H. Inokuchi, and D. Söll. 1992. *Proc. Natl. Acad. Sci. USA* **89:** 3463-3467.

Rould, M. A., J. J. Perona, D. Söll, and T. A. Steitz. 1989. *Science* **246:** 1135-1142.

Rould, M. A., J. J. Perona, and T. A. Steitz. 1991. *Nature* **352:** 213-218.

Sampson, J. R., and O. C. Uhlenbeck. 1988. *Proc. Natl. Acad. Sci. USA* **85:** 1033-1037.

Schimmel, P. 1989. *Biochemistry* **28:** 2747-2759.

Schimmel, P. R., and D. Söll. 1979. *Annu. Rev. Biochem.* **48:** 601-648.

Schön, A., C. G. Kannangara, S. Gough, and D. Söll. 1988. *Nature* **331:** 187-190.

Schulman, L.H. 1991. *Prog. Nucl. Acid Res. Mol. Biol.* **41:** 23-87.

Sherman, J. M., M. J. Rogers, and D. Söll. 1992a. *Nucleic Acids Res.* **20:** 2847-2852.

Sherman, J. M., K. C. Rogers, M. J. Rogers, and D. Söll. 1992b. *J. Mol. Biol.* **228:** in press.

Söll, D. 1990. *Experientia* **46:** 1089-1096.

Söll, D. 1993. *In: The RNA World,* in press, J. Atkins, and R. Gesteland, ed., Cold Spring Harbor Laboratory Press, New York.

Swanson, R., P. Hoben, M. Sumner-Smith, H. Uemura, L. Watson, and D. Söll. 1988. *Science* **242:** 1548-1551.

Weygand-Ðurašević, I., E. Schwob, and D. Söll. 1993. *Proc. Natl. Acad. Sci. USA* in press.

Wilcox, M., and M. Nirenberg. 1968. *Proc. Natl. Acad. Sci. USA* **61:** 229-236.

Yaniv, M., W. R. Folk, P. Berg, and L. Soll. 1974. *J. Mol. Biol.* **86:** 245-260.

Yarus, M. 1972. *Nature New Biol.* **239:** 106-108.

CONFORMATIONAL CHANGE OF tRNA UPON INTERACTION OF THE IDENTITY-DETERMINANT SET WITH AMINOACYL-tRNA SYNTHETASE

Osamu Nureki,[1] Tatsuya Niimi,[1] Yutaka Muto,[1] Hideo Kanno,[1] Toshiyuki Kohno,[1] Tomonari Muramatsu,[1] Gota Kawai,[2] Tatsuo Miyazawa,[2] Richard Giegé,[3] Catherine Florentz,[3] and Shigeyuki Yokoyama[1]

[1]Department of Biophysics and Biochemistry, Faculty of Science, The University of Tokyo, Hong, Bunkyo-ku, Tokyo 113, Japan
[2]Faculty of Engineering, Yokohama National University, Hodogaya-ku, Yokohama 240, Japan
[3]UPR Structure des Macromolécules Biologiques et Mécanismes de Reconnaissance, Institut de Biologie Moléculaire et Cellulaire du CNRS, 67084 Strasbourg Cedex, France

INTRODUCTION

For correct interpretation of the genetic code, the twenty aminoacyl-tRNA synthetases are required to strictly recognize their cognate tRNA and amino acid. The amino acid specificity of tRNA depends on a set of a relatively small number of nucleotide residues (determinants for the identity of tRNA) such as the anticodon residues, the discriminator base at position 73, base pairs 1:72, 2:71, and 3:70 in the acceptor stem, base pair 10:25 in the D stem, and residue 20 (Shimura et al., 1972; Normanly et al., 1986; Hou and Schimmel, 1988; McClain and Foss, 1988ab; Muramatsu et al., 1988b; Schulman and Pelka, 1988; Normanly and Abelson, 1989; Schimmel, 1989; Himeno et al., 1990; Jahn et al., 1991; Pütz et al., 1991; Franklyn et al., 1992; Muramatsu et al., 1992; Schulman, 1991; Tamura et al., 1992). The three-dimensional structures of the complexes of *E. coli* tRNA^{Gln} and yeast tRNA^{Asp} with their cognate aminoacyl-tRNA synthetases, GlnRS and AspRS, respectively, have been determined by X-ray crystallography (Rould et al., 1989; Rould et al., 1991; Ruff et al., 1991). These synthetases recognize the identity determinants mainly in the two ends of the L-shaped structure of tRNA (the anticodon loop and the 5'/3' terminal region) (Jahn et al., 1991; Pütz et al., 1991) and concomitantly change the conformations of these regions (Rould et al., 1989; Rould et al., 1991; Ruff et al., 1991).

In *E. coli*, there are major and minor isoleucine tRNA species, tRNA$^{Ile}_1$ having the anticodon GAU and tRNA$^{Ile}_2$ having the anticodon LAU (L is a hypermodified cytidine, "lysidine") (Muramatsu et al., 1988a). For both of these isoleucine tRNA species, we have shown that the first letter of the anticodon is a major determinant for the tRNA identity (Muramatsu et al., 1988b; Muramatsu et al., 1992). Here we summarize our recent studies on the complete identity-determinant set of tRNA$^{Ile}_1$ and global and local conformational changes of the tRNA upon interaction with *E. coli* isoleucyl-tRNA synthetase (IleRS).

The Translational Apparatus, Edited by K.H. Nierhaus
et al., Plenum Press, New York, 1993

NMR ANALYSES OF SECONDARY AND TERTIARY BASE PAIRS OF *E. coli* tRNA$^{Ile}_1$ IN THE FREE AND IleRS-BOUND STATES

By ^{1}H and ^{1}H-^{15}N NMR spectroscopy (Redfield, 1978; Davis and Poulter, 1991), interaction of tRNA$^{Ile}_1$ with IleRS both from *E. coli* was analyzed (T. Niimi, Y. Muto, G. Kawai, M. Haruki, T. Kohno, T. Muramatsu, M. Takayanagi, T. Noguchi, N. Hayashi, K. Watanabe, T. Miyazawa, and S. Yokoyama, manuscript submitted). The imino proton resonances of free tRNA$^{Ile}_1$ were assigned to all the twenty secondary base pairs and five tertiary base pairs, indicating that tRNA$^{Ile}_1$ takes a typical L-shaped structure (Figure 1). In the case of tRNA$^{Ile}_1$ bound with IleRS, many of the imino proton resonances were observed still at the same chemical shifts as those of the free tRNA$^{Ile}_1$ but broadened because of the increase in the molecular mass. By contrast, the imino proton resonances of at least seven base pairs were appreciably shifted and/or significantly broadened, and those of at least four base pairs were missing, because the imino protons undergo rapid exchange with solvent water. This indicates disruption or breathing of these secondary and tertiary base pairs in the complex. It was indicated therefore that upon binding with IleRS drastic changes were induced in the conformations of three regions of tRNA$^{Ile}_1$ (Figure 1). First, two secondary base pairs were disrupted (or breathing) in the anticodon stem. Second, secondary base pairs including two G:U base pairs in the junction of the acceptor and T stems were distorted. Third, two tertiary base pairs were disrupted in the core region involving the T loop, the D arm and the extra loop. These drastic changes in the secondary and tertiary structures of tRNA$^{Ile}_1$ suggest that the recognition mechanism of tRNA^{Ile} by IleRS is more complicated than those found so far for other systems by X-ray crystallography (Rould et al., 1989; Rould et al., 1991; Ruff et al., 1991).

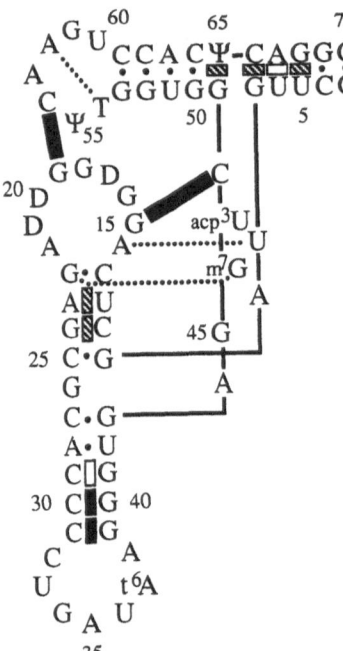

Figure 1. The nucleotide sequence of *E. coli* tRNA$^{Ile}_1$ arranged in the L-shaped structure. Upon binding with IleRS, imino proton resonances of secondary and tertiary base pairs indicated with closed bars were largely shifted or disappeared, indicating disruption of base pairs. Secondary base pair imino proton resonances indicated with shaded bars were appreciably broadened (with or without shift) and those indicated with open bars were appreciably shifted, indicating conformation changes and/or direct interactions with IleRS. Other observed secondary and tertiary base pairs are indicated with dots and dotted lines, respectively. Abbreviations: acp^{3}U, 3-(3-amino-3-carboxypropyl)uridine; D, dihydrouridine; m^{7}G, 7-methylguanosine; T, ribothymidine; t^{6}A, *N*-((9-β-D-ribofuranosylpurin-6-yl)carbamoyl)threonine; Ψ, pseudouridine.

ALKYLATION-MAPPING ANALYSES OF PHOSPHATE GROUPS OF *E. coli* tRNA$^{Ile}_1$ IN THE FREE AND IleRS-BOUND STATES

Then, we performed chemical footprinting (phosphate-ethylation mapping) analysis with *N*-nitroso-*N*-ethylurea (Romby et al., 1985; Theobald et al., 1988; Dietrich et al., 1990) of IleRS-bound tRNA$^{Ile}_1$ (Nureki et al., manuscript in preparation). Susceptibilities of phosphate groups of tRNA$^{Ile}_1$ to alkylation were compared between the free and IleRS-bound states. As for the anticodon arm, nearly all the phosphate groups in the loop and the 3'-strand of the stem were protected from alkylation by IleRS (Figure 2). The D stem was protected by IleRS at several phosphate groups on both strands around the U12•A23 base pair. These protected phosphate groups are located on "the front" of the duplex, while the back of the duplex is involved in the tertiary interaction with the U8-A9 stretch and the variable loop (Figure 2). In addition, phosphate groups at positions 70 and 72 in the acceptor stem are in contact with IleRS (Figure 2). In contrast, several phosphate groups are more susceptible to alkylation in the IleRS-bound state than in the free state. Protected phosphate groups 10, 19 and 59 are located at or adjacent to the sites of tertiary interactions which support the L-shape of tRNA (Figure 2), indicating a global change in the tertiary structure of tRNA$^{Ile}_1$ upon binding with IleRS. Further, the phosphate group of the t^{6}A residue at position 37 became susceptible to alkylation (Figure 2). This indicates that the conformation around the t^{6}A37 residue is significantly altered through interaction with IleRS. For the D loop or the TΨC arm, no phosphate protection was observed.

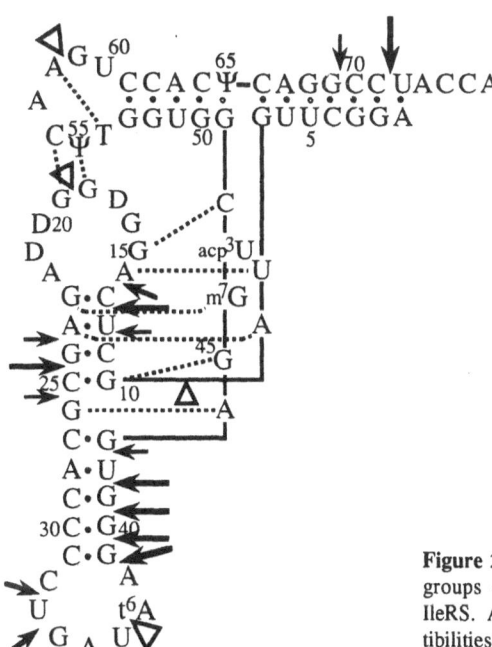

Figure 2. Changes in the susceptibilities of phosphate groups of tRNA$^{Ile}_1$ to alkylation upon binding with IleRS. Arrows indicate phosphate groups whose susceptibilities were decreased (the size of arrow represents the extent of decrease). On the other hand, open arrowheads indicate phosphate groups whose susceptibilities were increased upon binding.

AMINOACYLATION ANALYSES OF *E. coli* tRNA$^{Ile}_1$ VARIANTS

In vitro transcription with T7 RNA polymerase has been developed as a useful method for preparation of tRNA variants even though such T7 transcripts lack post-transcriptional modifications (Sampson and Uhlenbeck, 1988). We have found that the T7 transcript of *E. coli* tRNA$^{Ile}_1$ is aminoacylated with isoleucine specifically but much less efficiently than the fully-modified tRNA$^{Ile}_1$ (Muramatsu et al., 1992) primarily because of a 400-fold decrease in k_{cat}. It is possible that the t^{6}A37 residue adjacent to the anticodon is essentially involved in aminoacylation. However, the phosphate-alkylation mapping pattern in the IleRS-bound state was found to be essentially the same between the T7 transcript and the fully-modified tRNA$^{Ile}_1$, except only for the modified residues.

Thus, we could elucidate other unmodified members of the identity-determinant set of *E. coli* tRNA$^{Ile}_1$ by using the T7 transcription method (Figure 3); boxed bases/base pairs were changed to all the other possible three bases/base pairs for the anticodon and D arms, and to one or two bases/base pairs for the acceptor and T arms, and K_M and k_{cat} values for aminoacylation with IleRS of these variants were determined (Nureki et al., manuscript in preparation). All the three residues of the anticodon are essential for the aminoacylation. As for the anticodon arm, A38 and C29:G41 were found to contribute largely to the tRNA identity. Furthermore, U12:A23 in the D stem and C4:G69 in the acceptor stem were revealed to be important. These locations of identity determinants are characteristic of this tRNA$^{Ile}_1$. The discriminator base, A73, is also a strong determinant. All of these identity determinants are located at or near the sites of direct contact with IleRS as determined by the chemical footprinting analyses (Figure 2), and also consistent with the NMR results (Figure 1). It is also interesting that three invariant nucleotide residues, U33, C74, and A76 are important for aminoacylation.

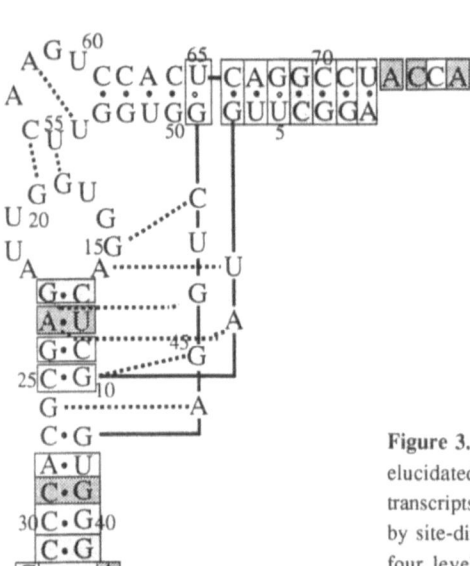

Figure 3. Identity determinants of *E. coli* RNA$^{Ile}_1$ as elucidated by aminoacylation analyses of *in vitro* T7 transcripts. Boxed residues or base pairs were examined by site-directed mutagenesis. Shading is performed at four levels 1 to 4 from the lightest (nothing) to the darkest, which indicate the decrease in k_{cat}/K_M to be negligible (level 1), not negligible but less than 10 fold (level 2), between 10 fold and 100 fold (level 3) and more than 100 fold (level 4).

In order to examine the role of the L-shaped structure, a series of variants defective in formation of one of the tertiary base pairs were prepared. As shown in Figure 4, it was found that U8:A14 and G22:G46 pairs are indispensable, but other tertiary base pairs are dispensable for the tRNA identity. This is consistent with the NMR results; the two indispensable tertiary base pairs were found to be retained, but observable members of the dispensable tertiary base pairs were indicated to be disrupted (or breathing) in the complex.

CONFORMATIONAL CHANGE OF *E. coli* tRNA$^{\text{Ile}}_1$ UPON INTERACTION OF THE IDENTITY-DETERMINANT SET WITH ISOLEUCYL-tRNA SYNTHETASE

Taken together, our NMR, chemical footprinting, and mutagenesis studies revealed the following mechanisms of the molecular recognition of *E. coli* tRNA$^{\text{Ile}}_1$ by IleRS. The anticodon loop and the 3'-strand of the anticodon stem are bound to IleRS, so that identity determinants, the anticodon (G34-A35-U36), t6A37, A38 and G41, in this stretch come into contact with IleRS. Concomitantly, the anticodon stem probably undergoes unwinding and base-pair disruption. On the other hand, IleRS binds to the D-stem duplex supported from the back by the two tertiary base pairs, and recognizes the base pair U12:A23 as an identity determinant. Further, the acceptor stem and the CCA terminus are bound to IleRS, where the base pair C4:G69 and the discriminator base A73 serve as identity determinants. Probably because three of the four arms are bound with IleRS for recognition of identity determinants located over these regions, a number of tertiary base pairs are disrupted and the secondary base pairs at the junction between the acceptor and TΨC stems are distorted.

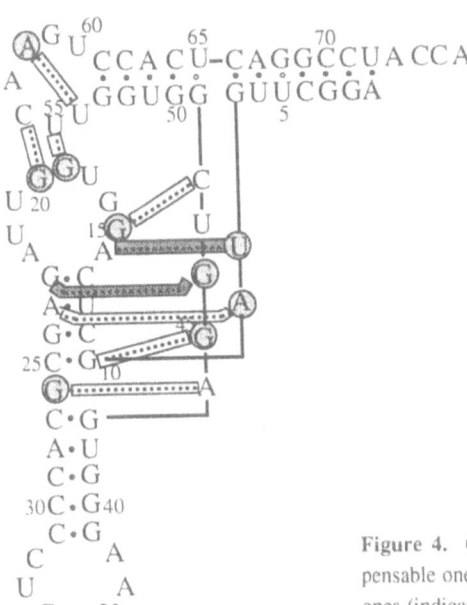

Figure 4. Classification of tertiary base pairs into indispensable ones (indicated with shaded bars) and dispensable ones (indicated with open bars) with regard to Ile-accepting activity, on the basis of site-directed mutagenesis analyses of residues indicated with shaded circles.

As described above, the TΨC arm of tRNA$^{Ile}_1$ has no identity determinant or exhibits no direct contact with IleRS, and the tertiary interaction involving this arm are dispensable for the identity. Thus, we deleted the TΨC arm of tRNA$^{Ile}_1$.and in turn inserted a sequence of U$_n$G (n=1-5) (Figure 5) (Nureki et al., manuscript in preparation). It has been found that mitochondrial tRNAs from Nematode worms *Caenorhabditis elegans* and *Ascaris suum* commonly lack the conventional TΨC arm (Wolstenholme et al., 1987); the sequence of U$_5$G was taken in the present study from the corresponding region of *Ascaris suum* tRNA^{Ile}. The *E. coli* tRNA$^{Ile}_1$ variant having U$_5$G remained an efficient substrate for IleRS; K_M and k_{cat} are 7.3 μM and 3.4×10^{-4} s^{-1}, respectively, for the U$_5$G variant and 8.2 μM and 40×10^{-4} s^{-1}, respectively, for the intact tRNA$^{Ile}_1$ (*in vitro* transcript). This confirms that the tightly-folded L-shaped structure of tRNA$^{Ile}_1$ is not essential for the specific interaction with IleRS. tRNA$^{Ile}_1$ variants having shorter loops (n = 1 - 4) exhibit lower

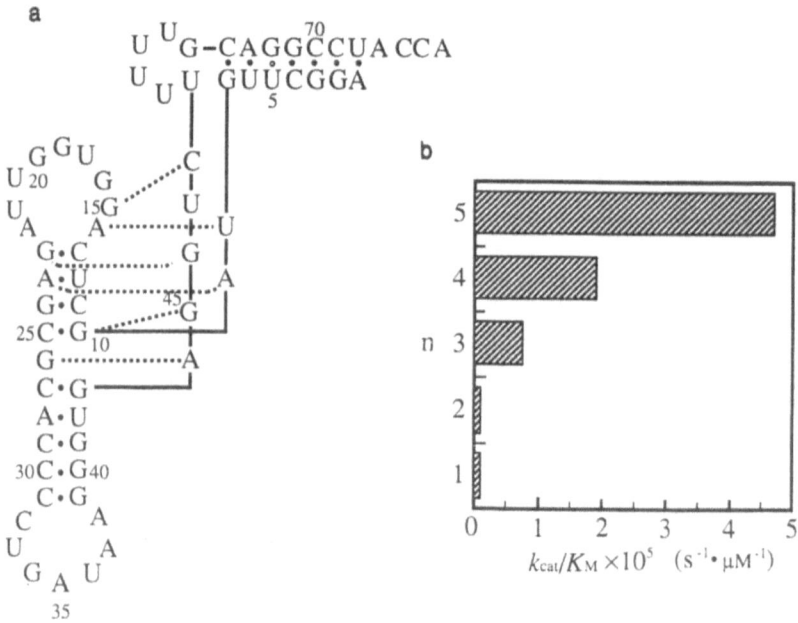

Figure 5. Deletion of the TΨC arm of *E. coli* tRNA$^{Ile}_1$. *a.* a variant of tRNA$^{Ile}_1$ having U$_5$G sequence in place of the TΨC arm. *b.* k_{cat}/K_M values of tRNA$^{Ile}_1$ variants having U$_n$G sequence (n = 1 - 5) in place of the TΨC arm.

amino acid-accepting activities than that of the U$_5$G variant (Figure 5b). Probably, the replacement of the TΨC arm by such a short loop drastically distorts conformations of other regions of tRNA, or impairs the inherent flexibility of the L-shaped structure, and therefore prevent the required mode of presentation of the identity elements towards IleRS.

MINIHELCES AND MICROHELICES OF *E. coli* tRNA$^{Ile}_1$

Minihelices and microhelices based on the acceptor-TΨC stem and the acceptor stem, respectively, have been found to be specifically aminoacylated in the Ala, His, Gly, Met and Val systems (Franklin and Schimmel, 1990a,b; Franklin et al., 1992; Frugier et al., 1992; Martinis and Schimmel, 1992; Shi et al., 1992). In the present study, we prepared the minihelix and microhelix of *E. coli* tRNA$^{Ile}_1$ and examined their Ile-accepting activities (Figure 6). The K_M values of the minihelix and microhelix of tRNA$^{Ile}_1$ are much the same as that of the *in vitro* transcript of the tRNA, while the k_{cat} values of these helices are much lower than that of the tRNA transcript (Figure 6). Both C4:G69→G4:C69 and A73→G73

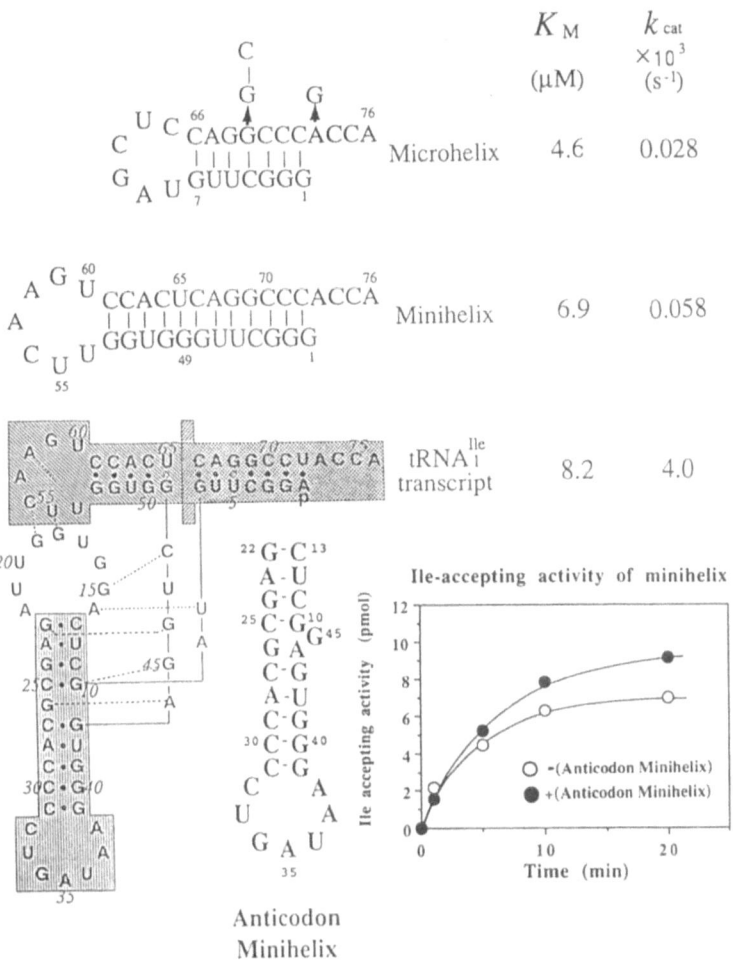

Figure 6. Structures of the Ile-accepting minihelix and microhelix and the anticodon minihelix derived from *E. coli* tRNA$^{Ile}_1$. The K_M and k_{cat} values of the Ile-accepting minihelix and microhelix are compared with those of the *in vitro* transcript of tRNA$^{Ile}_1$. Two variants of microhelix as having mutation(s) indicated with arrow were found to exhibit no Ile-accepting activity. Minihelices and microhelices were synthesized by *in vitro* transcription of double-stranded templates with T7 RNA polymerase and purified by electrophoresis on 20% polyacrylamide denaturing gel as described (Muramatsu et al., 1992). Aminoacylation assays of mini- and microhelices were performed at 37 °C in 100 mM Tris-HCl buffer (pH 7.5), 10 mM KCl, 20 mM MgCl$_2$, 2 mM ATP, 0.1 mM ^{14}C-isoleucine (11.7 MBq/µM), 10 µM *E. coli* IleRS, and various concentrations of mini- and microhelices (0.5-50 µM). *Inset*, time courses of isoleucylation of the minihelix (10 µM) in the absence and presence of the anticodon minihelix (64 µM). Taken from Nureki et al., in preparation.

mutations, which remarkably reduce the Ile-accepting activity of tRNA$^{Ile}_1$, were also found to impair the Ile-accepting activity of the microhelix, indicating that the aminoacylation of the mini- and microhelices is specific. The Ile-accepting activity of the minihelix was increased slightly but appreciably by addition of the anticodon minihelix of tRNA$^{Ile}_1$ (Figure 6, *inset*), which is quite similar to the case of yeast tRNA^{Val} (Frugier et al., 1992). Accordingly, for efficient aminoacylation of the acceptor stem of tRNA$^{Ile}_1$, the anticodon arm is required to be supported through the tertiary interactions.

Thus, the identity set of *E. coli* tRNA$^{Ile}_1$ and recognition mechanisms of the identity elements were elucidated at the level of base pair structures. These mechanisms are now being compared in detail with those in other systems, which may contributes to elucidation of general features of the molecular recognition of tRNAs by aminoacyl-tRNA synthetases.

ACKNOWLEDGEMENT: This work was supported in part by a Research Grant from the Human Frontier Science Program and Grants-in-Aid for Scientific Research on Priority Areas from the Ministry of Education, Science and Culture, Japan.

REFERENCES

Davis, R,D., and Poulter, C.D., 1991, *Biochemistry* 30:4223.

Dietrich, A., Romby, P., Marechal-Drouard, L., Guillemaut, P., and Giegé, R., 1990, *Nucl. Acids Res.* 18:2589.

Franklyn, C., and Schimmel, P., 1990a, *Nature* 337:478.

Franklyn, C., and Schimmel, P., 1990b, *Proc. Nat. Acad. Sci., U.S.A.* 87:8655.

Franklyn, C., Shi, J. P., and Schimmel, P., 1992, *Science* 255:1121.

Frugier, M., Florentz, C., and Giegé, R., 1992, *Proc. Natl. Acad. Sci.. U.S.A.* 89:3990.

Himeno, H., Hasegawa, T., Ueda, T., Watanabe, K., and Shimizu, M., 1990, *Nucleic Acids Res.* 18:6815.

Hou, Y.-M., and Schimmel, P., 1988, *Nature (London)* 333:140

Jahn, M., Rogers, M.J., and Söll, D., 1991, *Nature (London)* 352:258.

Martinis, S.A., and Schimmel, P., 1992, *Proc. Natl. Acad. Sci. U.S.A.* 89:65.

McClain, W.H., and Foss, K., 1988a, *Science* 240:793.

McClain, W.H., and Foss, K., 1988b, *J. Mol. Biol.* 202:697.

Muramatsu, T., Yokoyama, S., Horie, N., Matsuda, A., Ueda, T., Yamaizumi, Z., Kuchino, Y., Nishimura, S., and Miyazawa, T., 1988a, *J. Biol. Chem.* 263:9261.

Muramatsu, T., Nishikawa, K., Nemoto, F., Kuchino, Y., Nishimura, S., Miyazawa, T., and Yokoyama, S., 1988b, *Nature(London)* 336:179.

Muramatsu, T., Miyazawa, T., and Yokoyama, S., 1992, *Nuceosides Nuceotides* 11:719.

Nazarenko, I.A., Peterson, E.T., Zakharova, O.D., Lavrik, O.I.,and Uhlenbeck, O.C., 1992, *Nucl. Acids Res.* 20:475.

Normanly, J., and Abelson, J., 1989, *Annu. Rev. Biochem.* 58:1029.

Normanly. J., Ogden, R.C., Horvath, S.J., and Abelson, J., 1986, *Nature (London)* 321:213.

Pütz, J., Puglisi, J.D., Florentz, C., and Giegé, R., 1991, *Science* 252:1696.

Redfield, A.G, 1978, *Methods Enzymol.* 49:253.

Romby, P., Moras, D., Bergdoll, M., Domas, P., Vlassov, V.V., Westhof, E., Ebel, J.P., and Giegé, R., 1985, *J. Mol. Biol.* 184:455.

Rould, M.A., Perona, J.J., Söll, D., and Steitz, T.A., 1989, *Science* 246:1135.

Rould, M.A., Perona, J.J., and Steitz, T.A., 1991, *Nature (London)* 352:213.

Ruff, M., Krishnaswamy, S., Boeglin, M., Poterszman, A., Mitschler, A., Podjarny, A., Rees, B., Thierry, J.C., and Moras, D., 1991, *Science* 252:1682.

Sampson, J.R., and Uhlenbeck, O.C., 1988, *Proc. Natl. Acad. Sci. U.S.A.* 85:1033.

Schimmel, P., 1989, *Biochemistry* 28:2747.

Schulman, L.H., and Pelka, H., 1988, *Science* 242:765.

Schulman, L.H., 1991, *Prog. Nucleic Acids Res. Mol. Biol.* 41:23.

Shi, J.-P., Martinis, S.A., and Schimmel, P., 1992, *Biochemistry*, 31:4931.

Shimura, Y., Aono, A., Ozeki, H., Sarabahai, A., Lamfrom, H,. and Abelson, J., 1972, *FEBS Lett.* 22:144.

Tamura, K., Himeno, H., Asahara, H., Hasegawa, T., and Shimizu, M., 1992, *Nucleic Acids Res.* 20:2335.

Theobald, A., Springer, M., Grunberg-Manago, M., Ebel, J.P., Giegé R., 1988, *Eur. J. Biochem.* 175:511.

Wolstenholme, D.R., Macfarlane, J.L., Okimoto, R., Clary, D.O., and Wahleithner, J.A., 1987, *Proc. Natl. Acad. Sci. USA*, 84:1324.

FUNCTIONAL ASPECTS OF THREE MODIFIED NUCLEOSIDES, Ψ, ms^2io^6A, AND m^1G, PRESENT IN THE ANTICODON LOOP OF tRNA

Tord G. Hagervall[1], Birgitta Esberg[1], Ji-nong Li[1], Thérèse M. F. Tuohy[2], John F. Atkins[2], James F. Curran[3] and Glenn R. Björk[1]

1. Department of Microbiology, University of Umeå, S-901 87 Umeå, Sweden

2. Howard Hughes Medical Institute, and Department of Human Genetics, 6160 Eccles Genetics, Building 533, University of Utah, Salt Lake City, Utah 84112, USA

3. Department of Biology, Wake Forest Univesity, PO Box 7325, Winston-Salem, NC 27109 USA

INTRODUCTION

Transfer RNAs contain many modified nucleosides, which are derivatives of the four normal nucleosides. At present more than 75 different modified nucleosides are characterised (Edmonds *et al.*, 1991). The synthesis of the majority of the modified nucleosides is carried out on the preformed precursor tRNA except in two cases. Queuine and hypoxanthine are synthesised from guanine and adenine, respectively, and then incorporated into the tRNA through an exchange reaction. The synthesis of these 75 modified nucleosides is catalysed by enzymes, which are highly specific, not only for the nucleoside that they modify but also for the position of the target nucleoside in the tRNA. For example there are different enzymes catalysing the formation of Ψ in the anticodon stem and in the TΨC-loop (Singer *et al.*, 1972). The importance of tRNA modification is reflected by the fact that about 1% of the genetic information in *Escherichia coli* and *Salmonella typhimurium* is devoted to the synthesis of tRNA modifying enzymes, which is 4-fold more than that used for the synthesis of their substrate, tRNA (Björk, 1992). This paper will discuss the function of three modified nucleosides present in the anticodon region. Mutants defective in their synthesis have been used to study their role in cell physiology and in the decoding process.

The Translational Apparatus, Edited by K.H. Nierhaus
et al., Plenum Press, New York, 1993

RESULTS AND DISCUSSION

Presence of pseudouridine (Ψ), 2-methylthio-N^6-[4-hydroxyisopentenyl]adenosine (ms^2io^6A) and 1-methylguanosine (m^1G) in tRNAs from *Salmonella typhimurium*.

Pseudouridine, Ψ, is present in several positions in bacterial tRNA. A single enzyme, the tRNA(Ψ38,39,40)synthetase, catalyses the formation of this modified nucleoside in the anticodon loop (position 38) and in the anticodon stem (positions 39 and 40). About half of the tRNAs in *S. typhimurium* contain Ψ in these positions (Turnbough *et al.*, 1979) including tRNA^{His} (Singer *et al.*, 1972) and tRNA $^{Leu}_{1,2,3}$. Mutations in *hisT*, the structural gene for the tRNA(Ψ38,39,40)synthetase, causes deficiency in Ψ in these positions and because of this also derepression of several amino acid biosynthetic operons, for example the *his-* and *leu*-operons (Cortese *et al.*, 1974; Roth *et al.*, 1966; see below for further discussion).

The modified nucleoside ms^2io^6A is present in position 37 in tRNAs that read codons starting with U except in tRNA $^{Ser}_{I,V}$. (Grosjean *et al.*, 1985). The postulated pathway for the synthesis of ms^2io^6A and the genes involved (the genes *miaB/C* and *E* are tentative) are as follows:

$$A37 \xrightarrow{miaA} i^6A37 \xrightarrow{miaB} s^2i^6A37 \xrightarrow{miaC} ms^2i^6A37 \xrightarrow{miaE} ms^2io^6A37$$

[The *miaD* gene product may be involved in a demodification step or in the regulation of the synthesis of the *miaA* gene product (Connolly and Winkler, 1991)].

The last step, the hydroxylation reaction, occurs in *S. typhimurium* but not in *E. coli* (Buck *et al.*, 1982). The biosynthetic pathway has been postulated from precursor analyses of tRNA from methionine- or cysteine-starved *E. coli* cells (Agris *et al.*, 1975), and from mutants which are blocked in different steps in the pathway (Eisenberg *et al.*, 1979; Ericson and Björk, 1986). A mutation in the *miaA* gene results in the accumulation of an unmodified A37 instead of ms^2A37 in the tRNA. Thus, the MiaB enzyme seems to have a strict requirement for the isopentenyl group. This conclusion is also consistent with some early experiments on the biosynthesis of ms^2i^6A (reviewed in Hall, 1971). The starvation of *E. coli* (*rel, met, cys*) for methionine results in accumulation of a precursor to ms^2i^6A, which may be s^2i^6A37, whereas cysteine- or iron-starved *E. coli* accumulates i^6A (Agris *et al.*, 1975; Griffiths and Humphreys, 1978; Rosenberg and Gefter, 1969; Wettstein and Stent, 1968). Therefore it was concluded that the methyl-group of the ms^2- originates from AdoMet and the thiogroup from cysteine. Furthermore, the methylthiolation seems to require iron. Starvation for cysteine or iron in *S. typhimurium* also results in the accumulation of i^6A37 and only a small fraction (5%-12%) is in the hydroxylated form io^6A37 (Buck and Ames, 1984). A mutation likely to be in the *miaB* gene of *S. typhimurium* resulted in tRNA containing i^6A37 and no io^6A37 was present (data not shown). Thus, the hydroxylation reaction in *S. typhimurium* requires the ms^2-group and the methylthiolation reaction requires the isopentenyl group. Therefore, the modifying enzymes involved in the synthesis of ms^2io^6A act in a strictly sequential manner.

1-methylguanosine (m^1G) is present next to the 3′-end of the anticodon (position 37) in tRNAs that read CUN (leucine), CCN (proline) and CGG (arginine) codons from all organisms (Björk, 1984; Figure 1). This evolutionary conservation in the same subset of tRNAs suggests that the function of this modified nucleoside has one or more important functions. The enzyme tRNA(m^{1}G37)methyltransferase is encoded by the *trmD* gene, which is located in a four-cistron operon encoding the ribosomal protein S16 (*rpsP*), a 21kDa protein of unknown function, the tRNA(m^{1}G37)methyltransferase and the ribosomal protein L19 (*rplS*), in that order (Byström *et al.*, 1983).

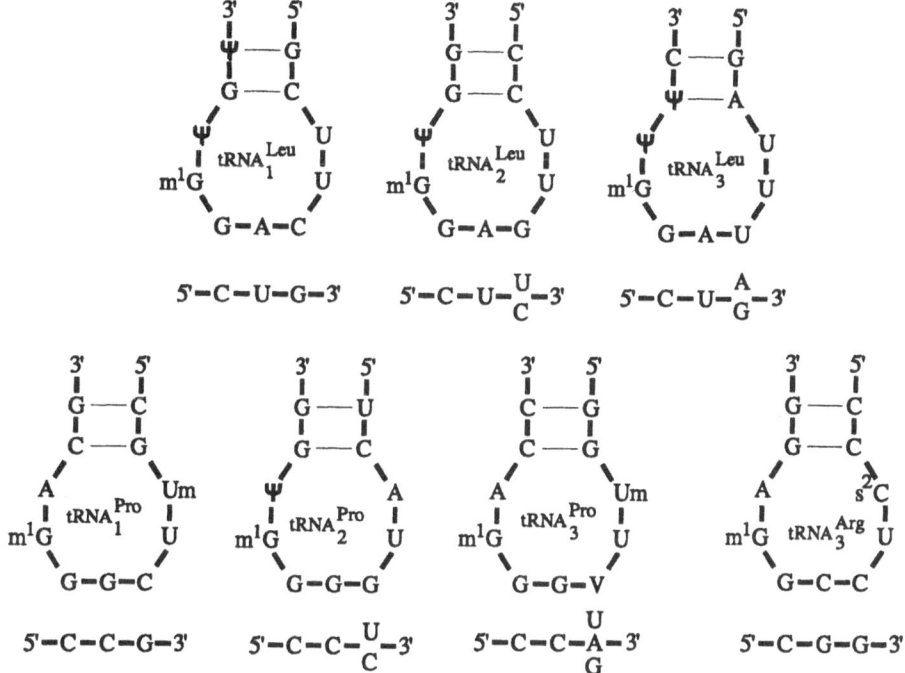

Figure 1. Sequences of the anticodon loops of tRNAs reading codons specific for leucine, proline, and arginine. The sequences are from *Salmonella typhimurium* in Sprinzl *et al.* (1991) with the exceptions below. The sequence of tRNA$_2^{Leu}$ is from *E. coli*. The sequence of tRNA$_2^{Pro}$ is from Sroga *et al.* (1992) and the sequence of tRNA$_3^{Leu}$ was deduced from the *E. coli* DNA sequence in Komine *et al.*, (1990). Abbreviations used: m^{1}G: 1-methylguanosine; V : uridine-5-oxyacetic acid; Ψ: pseudouridine; Um : 2'-*O*-methyluridine ; s^{2}C: 2-thiocytidine.

Lack of Ψ38,39,40, ms^{2}io^{6}A37, or m^{1}G37 reduces the growth rate and polypeptide step time and induces pleiotropic effects on cell physiology.

In *hisT1504*, *miaA1* or *trmD3* mutants of *S. typhimurium*, tRNA lacks Ψ38,39,40, ms^{2}io^{6}A37 or m^{1}G37, respectively. In glucose minimal medium the reductions in growth rate caused by these mutations are 16, 30 and 24%, respectively. The reduction in the polypeptide chain elongation rate is 23% in the *hisT1504* mutants whereas in both the *miaA1* and *trmD3* mutants the reduction is 31-32% (Hagervall *et al.*, 1990; Palmer *et al.*, 1983). However, lack of the ms^2-group of ms^{2}io^{6}A37 in the *miaB1* mutant does not influence either the growth rate or the polypeptide chain elongation rate (Hagervall *et al.*, 1990; data not shown).

These deficiencies in the tRNA modification also induce pleiotropic effects which can be traced back to the effect on transcriptional attenuation of several amino acid biosynthetic operons. For the *miaA1* mutant, these effects were attributed to a decrease in the polypeptide chain elongation rate (Ericson and Björk, 1986). The *hisT* mutants are derepressed for the *his*-operon (Roth *et al.*, 1966). This operon is preceded by a leader sequence containing seven histidine codons in a row (Barnes, 1978). The *hisT*-mediated derepression is caused by a slower step time at these histidine codons by a Ψ38,39-deficient tRNA[His] (Johnston *et al.*,1980). This is consistent with the observed reduced polypeptide chain elongation rate in *hisT* mutants (Palmer *et al.*, 1983). Similarly, the effect of *hisT* on other amino acid biosynthetic operons can be explained. The leader region of the leucine operon encodes a 28 amino acids long polypeptide containing four consecutive leucine codons (CUA-CUA-CUA-CUC)(Carter *et al.*, 1985). The tRNA$_2^{Leu}$ (CUC) and tRNA$_3^{Leu}$ (CUA) normally contain Ψ in the anticodon region (Figure 1). In a *hisT* mutant, these tRNAs are undermodified and contain U instead of Ψ. A mutation in the *hisT* gene leads to derepression of the *leu*-operon, which can be reconciled with a slower step time at the leucine codons. The tRNAs[Leu] decoding these CUA and CUC codons also contain m^{1}G37. However, in a *trmD3* mutant which contains G37 in place of m^{1}G37 in tRNA there was no effect on the expression of LeuB peptide (unpublished observation). Furthermore, the *hisT* mutant is resistant to the leucine analogue 5',5',5',-trifluoro-DL-leucine (Cortese *et al.*, 1974) whereas the *trmD3* mutant shows a slight sensitivity towards this analogue compared to the wild type control. Apparently, the slower step time of tRNA$_{2,3}^{Leu}$ lacking Ψ in the anticodon is enough to derepress the *leu*-operon whereas an m^{1}G37-deficiency of the same tRNA$_{2,3}^{Leu}$ leads to an unchanged or a slightly further repressed expression, implying an unaffected or even faster step time for these leucine tRNAs.

Ψ, ms^{2}io^{6}A37 and m^{1}G37 all affect the efficiency of translation but to different extent and by different mechanisms.

Pseudouridine is present in position 39, i. e. in the anticodon stem, of tRNA[Tyr]. A Ψ39-lacking derivative of an amber suppressor mutant of this tRNA has a two-fold reduced efficiency compared to the wild type. For six different codon contexts the reduction in amber suppression efficiency caused by the presence of U39 in place of Ψ39 is the same, suggesting that Ψ39 influences the efficiency but does not affect the apparent codon context sensitivity (Hagervall *et al.*, 1990). The amber suppressor derivative of tRNA[Gln] (*supE20*) contains both Ψ38 (in the anticodon loop) and Ψ39. The efficiency of this suppressor is decreased more than 10-fold in a *hisT* mutant (Bossi and Roth, 1980). Mistranslation of histidine codons by tRNA[Gln] is decreased if this noncognate tRNA has U38,39 rather than Ψ38,39 (Parker, 1982). However, mistranslation of asparagine codons by the noncognate tRNA[Lys], which normally has Ψ39, is not affected by a deficiency of this modification. Apparently, Ψ in the anticodon loop decreases the efficiency of the tRNA in a noncognate interaction whereas Ψ deficiency in the stem has only a minor, if any effect. Therefore, the impact of Ψ in the different positions 38 (anticodon loop) and 39,40 (anticodon stem) is not the same. Quantitatively, the Ψ modification seems to be more important when present in the loop (position 38) than in the stem (positions 39 and 40). This is reasonable, since Ψ and U form equally stable base pairs with A (Woese, 1967). Nevertheless, a minor but significant effect due to the lack of Ψ even in the stem (position 39) is still observed, suggesting a perturbation of the anticodon. Thus, depending on position in the tRNA, the decoding efficiency is diminished by Ψ-deficiency, but apparently to the same extent in different codon contexts.

The ms^2io^6A37 is present in most tRNAs that read codons starting with U and in all known amber suppressor tRNAs. We have combined the *miaA1* and *miaB1* mutations with the three amber suppressors *supF30* (tRNA^{Tyr}), *supD10* (tRNA^{Ser}) and *supJ60* (tRNA^{Leu}). Assays of suppression of amber codons in six different positions in the *lacI* mRNA shows that the efficiency of translation is decreased 99% by the *miaA1* and 85% by the *miaB1* mutations (Bouadloun *et al.*, 1986; Hagervall *et al.*, 1990). Furthermore, these mutations have been combined with a codon context mutation (Bossi and Roth, 1980) to investigate not only how *miaA1* and *miaB1* influence the efficiency of the tRNA but also how they respond to different nucleotides 3′ of the amber codon (Ericson and Björk, 1991). Two constructs in which MudK creates a translational fusion of *lacZ* to the 3'-proximal part of the *hisD* gene were used. Upstream of the fusion point they contain a UAG codon immediately followed by a C or an A. These constructs were transferred into strains containing the different suppressors and different alleles of the *miaA* or the *miaB* genes. The β-galactosidase measurements revealed that the relative ability to suppress UAG-A versus UAG-C at the same position in the *hisD* mRNA is 1.6 for the three fully modified suppressor tRNAs. These ratios are 5.2 - 5.7 for the three different suppressors in the *miaA1* strain (Ericson and Björk, 1991) whereas this ratio was about 1.9 - 2.7 in the *miaB1* strain (data not shown). Apparently, the ms^2io^6-modification not only increases the efficiency of the tRNA but also makes it less sensitive to the codon context. The ms^2-modification also improves the efficiency but not as much as the io^6-group does. Moreover, the codon context sensitivity of the tRNA is less affected by the ms^2-group than by the io^6-group.

The first step in the translation elongation cycle is the selection step of the binding of the ternary complex (EF-Tu-GTP-aatRNA) to the A-site on the mRNA-programmed ribosome. A competition assay for this aa-tRNA selection has been described (Curran and Yarus, 1989). The synthesis of release factor 2 (RF2) requires a +1 frameshift during the translation of its mRNA (Craigen *et al.*, 1985). If the *lacZ* gene is fused into the shifted frame (*i. e.* in +1 frame), the β-galactosidase activity becomes a measurement of the efficiency of frameshifting. The degree of frameshifting is dependent on the presence of a Shine-Dalgarno sequence upstream of the frameshifting site (Weiss *et al.*, 1988) and on the tRNA at the frameshifting point (Curran and Yarus, 1988). The primary selection at the sense codon just downstream of the frameshifting site will then influence the degree of frameshifting by the previous tRNA (Figure 2). The faster the selection of the cognate aa-tRNA for the test sense codon occurs, the smaller is the chance for frameshifting. Since the β-galactosidase activity is a direct measurement of the degree of frameshifting, the more efficiently the test codon is read, the lesser β-galactosidase activity is observed. The relative rates of selection at 29 different codons was determined using this assay system (Curran and Yarus, 1989). We have used this assay system to determine the influence of the Ψ, ms^2io^6A37 and m^1G37 in the aa-tRNA selection for several different tRNAs. We introduced plasmids with different sense codons into wild-type cells, *hisT1504*, *miaA1*, *miaB1* and *trmD3* mutants. Figure 2 shows the ratio of relative activities of β-galactosidase between mutant and wild-type cells. A ratio of about 1 indicates no difference between the undermodified tRNA and the fully modified tRNA at the selection step whereas a high ratio suggests that the mutant tRNA has a lower efficiency in the aa-tRNA selection step.

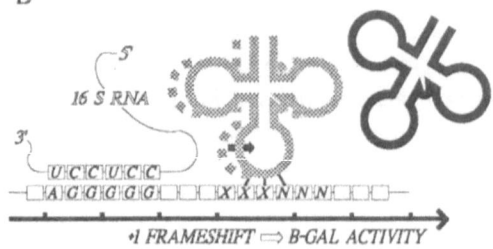

A   B

$-\boxed{A|G|G|G|G|G}\;\boxed{}\;\boxed{X|X|X|N|N|N}\;\boxed{}\;\rightarrow$ $-\boxed{A|G|G|G|G|G}\;\boxed{}\;\boxed{X|X|X|N|N|N}\;\boxed{}\;\rightarrow$

MAINTAINED FRAME ⟹ NO β-GAL ACTIVITY   *+1 FRAMESHIFT ⟹ β-GAL ACTIVITY*

C

	Test codon		
	CUA	*CCG*	*UUC*
hisT/hisT⁺	2.8	1.0	n.d.
trmD3/trmD⁺	1.4	8.5	n.d.
miaA/miaA⁺	n.d.	n.d.	0.9
miaB/miaB⁺	n.d.	n.d.	0.6

Figure 2. Comparison of aminoacyl-tRNA selection in the ribosomal A-site between the respective mutants and wild type cells using the competition assay system. A: Succesful binding to the A-site; no frameshifting. B: Unsuccesful binding to the A-site. The P-site tRNA shifts frame due to slow competition by the aa-tRNA. C: Ratios of β-galactosidase activities between the respective mutants and wild type cells. To correct for general effects on the translation elongation or effects on plasmid copy number, the β-galactosidase activities were divided with the β-galactosidase activities obtained from a construct with the lacZ gene in frame in the respective strains.

The Ψ-lacking tRNA$^{Leu}_1$ is less efficient in the selection at the CUG codon compared to fully modified tRNA$^{Leu}_1$ by this assay, while deficiency of m^{1}G37 has less effect at this codon. For tRNA^{Pro} (species 1 or 3) the selection at CCG is much reduced by m^{1}G37 deficiency but not at all by Ψ deficiency. This is to be expected since the potential proline tRNAs that read this codon do not normally have Ψ. Only a minor, if any effect on the selection of the phenylalanyl-tRNA was observed in the *miaA* and *miaB* mutants which are deficient in the biosynthesis of ms^{2}io^{6}A37. Thus, this hypermodified nucleoside that so dramatically influences the efficiency of translation if the whole translation cycle is measured (Bouadloun *et al.*, 1986; Ericson and Björk, 1991; Hagervall *et al.*, 1990), apparently has no effect on the primary selection step. This has also been observed *in vitro* (Diaz and Ehrenberg, 1992). Although all these three modified nucleosides increase polypeptide chain elongation rate, they apparently influence the aa-tRNA selection *in vivo* differently and in a tRNA dependent manner. Interestingly, whereas m^{1}G37 affects tRNA$^{Pro}_{1,3}$ it has only a minor effect on tRNA$^{Leu}_1$ and tRNA$^{Leu}_{2,3}$, if any (data not shown). The latter correlates with the inability to derepress the leucine operon. Our results imply that in the aa-tRNA selection step, the presence of m^{1}G37 has only a minor effect, if any on tRNA^{Leu}, but a significant effect on tRNA^{Pro}. Furthermore, our observed lack of effect of m^{1}G37 on the regulation of the leucine operon (see above), which would monitor the whole elongation cycle, would imply that the effect of m^{1}G, if any, may cause a faster step time. If so, the reduced polypeptide chain elongation rate in the *trmD3* mutant is largely due to the effect on tRNA^{Pro} (or tRNA^{Arg}) but to a lesser extent on tRNA^{Leu}. However, codon context may influence the step time of individual tRNAs, where an effect of m^{1}G37 on tRNA^{Leu} can not be excluded.

1-methylguanosine (m^{1}G) is important for the maintenance of the reading frame.

The *trmD3* mutant of *Salmonella typhimurium* lacks m^{1}G in tRNA at temperatures above 37°C (Björk *et al.*, 1989). When the mutant was characterised, it was shown that the deficiency of m^{1}G correlated with a +1 frameshifting activity at the sites CCCC and CCCU. In wild type cells, m^{1}G is present in position 37 in all three tRNA^{Pro} species. Therefore, it was proposed that in the *trmD3* mutant, one or several undermodified species of tRNA^{Pro} were responsible for the frameshifting activity that was observed at the above sites. A mechanism for the +1 frameshifting was proposed in which the G in position 37 would become part of a four base anticodon and when engaged in the codon pairing a quadruplet translocation would result (Figure 3A and B).   In the wild type tRNA, the

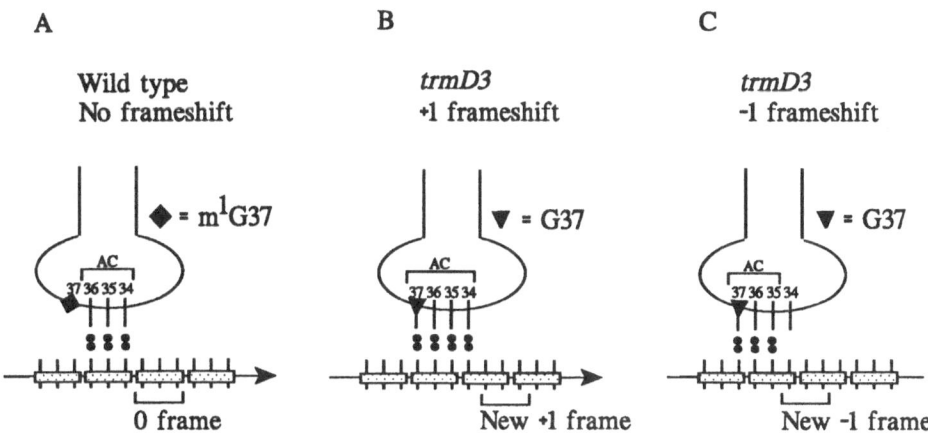

Figure 3. Mechanisms for *trmD3*-dependent frameshifting. A: Wild type. No frameshifting. B: *trmD3*. Proposed +1 frameshifting mechanism. C: *trmD3*. Proposed -1 frameshifting mechanism.

methyl group in m^{1}G37 would prevent this base pairing. In support of this idea, it has been shown that the presence of the methyl group prevents the formation of a poly-m^{1}G/poly-C helix *in vitro* (Newmark and Cantor, 1968). It has therefore been suggested that the function of m^{1}G is to maintain the correct reading frame by preventing the base pairing by G37 (Piecznik, 1980; Björk *et al.*, 1989). Björk *et al.* (1989) also proposed that a tRNA lacking m^{1}G might act as a -1 frameshift suppressor but this hypothesis was not tested (Figure 3A and C). In addition to the tRNA^{Pro} (CCN) species, m^{1}G is also found in position 37 in tRNA$^{Leu}_{1,2,3}$ (CUN) and tRNA$^{Arg}_3$ (CGG) (Figure 1), implying that *trmD3* may also act as a +1 frameshift suppressor at sites normally read by those tRNAs.

To test the above hypotheses, an assay system relying on a pBR322-derived plasmid, in which the *lacZ* gene is fused downstream of the *tac* promoter was used (Weiss *et al.*, 1987). In the 5'-proximal part of *lacZ*, different synthetic frameshift windows were inserted in such a way that β-galactosidase activity was dependent on either a +1 or a -1 frameshift (Figure 4). Thus, the activity of β-galactosidase was a direct measurement of the efficiency of frameshifting. NH$_2$-terminal sequencing of the hybrid translation product would reveal the amino acid inserted at the frameshifting site. It was found that *trmD3* was an efficient suppressor of all potential proline (cCC-N) frameshifting sites (The first c would be read by G37 and the capitalised bases, CC-N, would be read by the normal anticodon. The base after the hyphen ,N, is the first base of the next codon in the 0-frame) (Hagervall *et al.,* 1992). The β-galactosidase activities were 6-30 times higher in the *trmD3* mutant than in wild type cells. NH$_2$-terminal sequencing of the translation products from two of the constructs, cCC-U and cCC-A, revealed that a proline had been inserted at these sites, and moreover, a quadruplet translocation had occurred. Thus, a +1 frameshift was correlated with tRNA^{Pro} undermodified in m^{1}G, supporting the previously proposed model. Generally, the level of frameshifting at the leucine sites (cCU-N) was about 10-fold lower than at the proline sites (Hagervall *et al.,* 1992). However, at two of the sites, cCU-U and cCU-C, the frameshifting efficiency was 10-fold higher in the *trmD3* mutant than in the wild-type. Unfortunately, the low levels of β-galactosidase which were synthesised from the cCU-N (leucine) constructs did not allow NH$_2$-terminal sequencing and no conclusions about the mechanism can therefore be drawn.

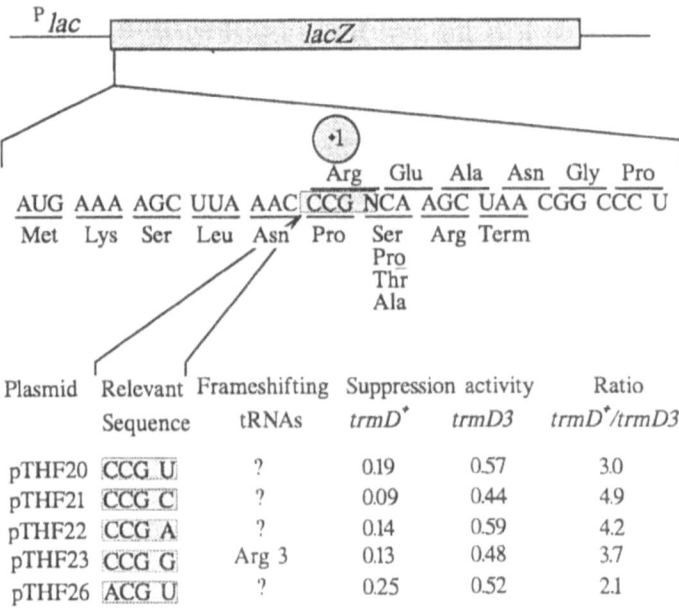

Figure 4. Frameshifting activities at cCG-N (arginine) sites in wild type and *trmD3* cells. The top of the figure shows the system used and the frameshifting window.

To test the possibility of frameshifting by tRNA^{Arg} species, plasmids pTHF20 - 23 (cCG-N) and plasmid pTHF26 (aCG-U) were introduced into the wild type and the *trmD3* mutant strains and β-galactosidase activities were determined (Figure 4). Only tRNA$^{Arg}_3$ (which reads CGG) contains m^{1}G and suppression would only be expected at the site cCG-G in plasmid pTHF23 since the model for *trmD3*-dependent frameshifting requires quadruplet base-pairing. Surprisingly, a low level suppression was detected at all arginine sites (cCG-N) and even at the site aCG-U in plasmid pTHF26. The latter site would not be expected to stimulate frameshifting if quadruplet base-pairing involving G37 is required. Thus, an alternative mechanism from that proposed by Björk *et al.* (1989) must be acting at the arginine sites.

The potential for a -1 frameshifting activity associated with *trmD3* was also tested using the assay system described above (Hagervall *et al.*, 1992). However, at the sites tested there were no significant difference between wild type and *trmD3* cells in suppression efficiency.

In some cases, +1 frameshifting has been shown not to require an interaction between the tRNA and the wobble base. Gaber and Culbertson (1984) isolated all possible base-substitution mutations of the wobble base of the yeast +1 frameshift suppressor *SUF16* and tested the specificity of suppression at GGGN frameshifting sites. Although possible interactions by the wobble base affect the efficiency of suppression, they are not required, and apparently the quadruplet translocation was primarily directed by the number of nucleotides in the loop. Likewise, the *S. typhimurium* suppressor *sufD*, an analog of *SUF16*, does not require a quadruplet interaction to shift to the +1 frame (Weiss *et al.*,1990). A mechanism for quadruplet translocation by tRNAs with abberant anticodon loops has been proposed (Curran and Yarus, 1987). The ability of these tRNAs to assume either of two stack conformations explain their frameshifting properties. A possibility of a four base pair interaction between the tRNA and the mRNA would stabilise the frameshifting stack conformation, thereby increasing the probability for quadruplet translocation. Such a mechanism could explain the low level suppression which was observed at the cCG-N (arginine) sites, although other alternatives certainly exist. A mechanism where the quadruplet translocation is directed by structural changes, secondarily caused by the lack of m^1G can not be excluded. Thus, it is possible that only one or two of the undermodified tRNA[Pro] species may be responsible for the frameshifting at cCC-N (Proline) and also for the low level suppression which was observed at the cCU-N (Leucine) and cCG-N (arginine) sites (Figure 4). In such a case, the only requirement for the undermodified tRNA[Pro] would be a doublet interaction with the two C's at the 5'-end of the frameshifting site although a triplet or quadruplet interaction would increase the efficiency. An alternative explanation for the low level suppression, that was observed at the cCU-N (leucine) and cCG-N (arginine) sites, would be that they are the result of several mechanistically different events although some of the suppressing activity may result from a +1 frameshift by tRNA[Leu] species and tRNA $^{Arg}_3$ (at cCG-G) according to the originally proposed model. The low level suppression at the aCG-U site could be explained by the *trmD3*- induced structural change of the frameshifting tRNA which was discussed above. However, any involvement of tRNA[Pro] species in the suppression of this site seems unlikely since only a single anticodon base interaction is possible. A more likely frameshifting agent at this site is the undermodified tRNA $^{Arg}_3$ where a doublet interaction is possible.

In summary, *trmD3* is an efficient +1 frameshift suppressor of all proline sites (cCC-N) (Hagervall *et al.*, 1992). The results obtained from the NH_2-terminal sequencing is in accordance with the model proposed by Björk *et al.* (1989) in which the G37 is involved in base-pairing. At the leucine (cCU-N) and arginine (cCG-N) sites, the frameshifting levels are much lower and the lack of protein sequence data does not allow any predictions about the mechanism. However, we favour a quadruplet interaction also at the cCU-N (leucine) sites. At the cCG-N (arginine) sites it is likely that an alternative mechanism is acting since sites not normally read by m^{1}G-containing tRNAs were suppressed, albeit with a very low efficiency. Finally, *trmD3* is not a -1 frameshift suppressor.

ACKNOWLEDGEMENTS

We are greatly indebted to the skilful technical assistance by Gunilla Jäger. This work was supported by the Swedish Cancer Society (project no. 680 to GRB) and The Swedish National Science Research Council (Project no. BBU 2930-105 to GRB). TGH was supported by a travel grant from The Swedish National Science Research Council (Project no. R-RA 9651-300). We thank Ray Gesteland for continued support in his laboratory, where part of TGH's work was performed.

REFERENCES

Agris, P.F., Armstrong, D. J., Schäfer, K. P., and Söll, D. (1975). Maturation of a hypermodified nucleoside in transfer RNA. *Nucl. Acids Res.* 2: 691-698.

Barnes, W. M. (1978). DNA sequence from the histidine operon control region: seven histidine codons in a row. *Proc. Natl. Acad. Sci.(USA)*, 75: 4281.

Björk, G. R. (1984). Transfer RNA modification in different organisms. *Chemica Scripta.* 26B: 91-95.

Björk G. R. (1992). The role of modified nucleosides in transfer RNA interactions. *in:* "Transfer RNA in protein synthesis". Hatfield, D. L., Lee, B. J., and, Pirtle R. M., eds, CRC press, Boca Raton, FL.

Björk, G. R., Wikström, P. M., and Byström, A. S. (1989). Prevention of translational frameshifting by the modified nucleoside 1-Methylguanosine. *Science* 244: 986-989.

Bossi, L., and Roth, J. R.. (1980). The influence of codon context on genetic code translation. *Nature* 286:123-127.

Bouadloun, F, Srichaiyo, T., Isaksson L. A., and Björk G. R. (1986). Influence of modification next to the anticodon in tRNA on codon context sensitivity of translational suppression and accuracy. *J. Bacteriol.* 166: 1022-1027.

Buck M., and Ames B. N. (1984). A modified nucleotide in tRNA as a possible regulator of aerobiosis: Synthesis of cis-2-methyl-thioribosylzeatin in tRNA of *Salmonella*. *Cell* 36: 523-531.

Buck, M., McCloskey, J. A., Basile, B., and Ames, B. N. (1982). *Cis*-2-methylthio-ribosylzeatin (ms^{2}io^{6}A) is present in transfer RNA of *Salmonella typhimurium,* but not *Escherichia coli. Nucl. Acids Res.* 10: 5649-5662.

Byström, A. S., Hjalmarsson, K. J., Wikström, P. M., and Björk, G. R. (1983). The nucleotide sequence of an *Escherichia coli* operon containing genes for the tRNA(m^{1}G)methyltransferase, the ribosomal proteins S16 and L19 and a 21-K polypeptide. *EMBO J.* 2: 899-905.

Carter, P. W., Weiss, D. L., Weith, H. L., and Calvo J. M. (1985). Mutations that convert the four leucine codons of the *Salmonella typhimurium leu* leader to four threonine codons. *J. Bacteriol.* 162: 943-949.

Connolly, D. M., and Winkler, M.E. (1991). Structure of *Escherichia coli* K-12 *miaA* and characterization of the mutator phenotype caused by *miaA* insertion mutations. *J. Bacteriol.* 173: 1711-1721.

Cortese, R., Kammen, H. O., Spengler, S. J., and Ames B. N. (1974). Biosynthesis of pseudouridine in transfer ribonucleic acid. *J. Biol. Chem.* 249: 1103-1108.

Craigen, W. J., Cook, R. G., Tate, W. P., and Caskey, C. T. (1985). *Proc. Natl. Acad. Sci. USA* 82: 3616-3620.

Curran, J. F., and Yarus, M. (1987). Reading frame selection and transfer RNA anticodon loop stacking. *Science* 238: 1545-1550.

Curran, J. F., and Yarus, M. (1988). Use of tRNA suppressors to probe the regulation of *Escherichia coli* release factor 2. *J. Mol. Biol.* 203: 75-83.

Curran, J. F., and Yarus, M. 1989. Rates of aminoacyl-tRNA selection at 29 sense codons *in vivo. J. Mol. Biol.* 209: 65-77.

Diaz, I., and Ehrenberg, M. (1992). ms^2i^6A deficiency enhances proofreading in translation. *J. Mol. Biol.* 222: 1161-1171.

Edmonds, C. G., Crain, P. F., Gupta, R., Hashizume, T., Hocart, C. H., Kowalak, J. A., Pomerantz, S. C.,. Stetter K. O, and McCloskey, J. A..(1991). Posttranscriptional modification of tRNA in thermophilic Archaea (Archaebacteria). *J Bacteriol.* 173: 3138-3148.

Eisenberg, S. P., Yarus, M., and, Soll, L. 1979. The effect of an *Escherichia coli* regulatory mutation on transfer RNA structure. *J. Mol. Biol.* 135: 111-126.

Ericson, J. U., and Björk, G. R. 1986. Pleiotropic effects induced by modification deficiency next to the anticodon of tRNA from *Salmonella typhimurium* LT2. *J. Bacteriol.* 166: 1013-1021.

Ericson, J. U., and Björk, G. R. (1991). tRNA anticodons with the modified nucleoside 2-methylthio-N^6-(4-hydroxyisopentenyl)adenosine distinguish between bases 3' of the codon. *J Mol. Biol.* 218: 509-516.

Gaber R. F. and, Culbertson M. R. (1984). Codon recognition during frameshift suppression in *Saccharomyces cerevisiae. Mol .Cell .Biol* 4: 2052-2061.

Griffiths, E., and, Humphreys, J. 1978. Alterations in tRNAs containing 2-methylthio-N^6-Δ^2-isopentenyl)-adenosine during growth of enteropathogenic *Escherichia coli* in the presence of iron-binding proteins. *Eur. J. Biochem.* 82: 503-513.

Grosjean , H., K. Nicoghosian, K., Haumont, E., Söll, D. and, Cedergren, R. 1985. Nucleotide sequences of two serine tRNAs with a GGA anticodon: The structure-function relationships in the serine family of *E. coli* tRNAs. *Nucl. Acid Res.* 13: 5697-5706.

Hagervall, T. G., Tuohy, T. M. F., Atkins J. F. and, Björk, G. R. (1992). Deficiency of 1-methylguanosine in tRNA from *Salmonella typhimurium* induces frameshifting by quadruplet translocation. *J. Mol Biol.* Submitted.

Hagervall, T. G., Ericson, J. U., Esberg, K. B., Ji-nong, L. and, Björk, G. R. 1990. Role tRNA modification in translation fidelity. *Biochem. Biophys.* Acta 1050: 263-266.

Hall, R. H. *The modified nucleosides in nucleic acids.* Columbia University Press. 1971. Johnston, H. M., Barnes, W. M., Chumley, F. G., Bossi, L., and Roth, J. R. Model for regulation of the histidine operon of *Salmonella. Proc. Natl. Acad. Sci. (USA).* 77: 508, 1980.

Komine Y., Adachi T., Inokuchi H., Ozeki H. (1990). Genomic organization and physical mapping of the transfer RNA genes in *Escherichia coli* K12. *J. Mol. Biol.* 212: 579-598.

Newmark R. A. & Cantor C. R. (1968). Nuclear magnetic resonance study of the interactions of guanosine and cytidine in dimethyl sulfoxide. *J. Am. Chem. Soc.* 90: 5010-5017.

Palmer, D. T., P. H. Blum, P. H., and S. W. Artz, S. W. 1983. Effects of the *hisT* mutation of *Salmonella typhimurium* on translation elongation rate. *J. Bacteriol.* 153: 357-363.

Parker, J. 1982. Specific mistranslation in *hisT* mutants of *Escherichia coli Mol Gen. Genet.* 187: 405-409.

Pieczenik G. (1980). Predicting coding function from nucleotide sequence or survival of "fitness" of tRNA. *Proc. Natl. Acad. Sci. USA.* 77: 3539-3543.

Rosenberg, A. H., and M. L. Gefter. 1969. An iron-dependent modification of several transfer RNA species in *Escherichia coli*. *J. Mol. Biol.* 46: 581-584.

Roth, J. R, D. N. Anton, and P. E. Hartman. 1966. Histidine regulatory mutants in *Salmonella typhimurium*. I. Isolation and general properties. *J. Mol. Biol.* 22: 305-323.

Singer, C., Smith G. R., Cortese C., and Ames B. N. (1972). Mutant tRNA ineffective in repression and lacking two pseudouridine modifications. *Nature (London) New Biol.* 238: 72.

Sprinzl M., Dank N., Nock S., Schön A. (1991). Compilation of tRNA sequences and sequences of tRNA genes. *Nucleic Acids Res.* 19: supplement, 2127-2171.

Sroga G. E., Nemoto F., Kuchino Y. and, Björk G. R. (1992). Insertion in the anticodon loop or base substitution (*sufC*) in the anticodon stem of tRNA $^{Pro}_2$ from *Salmonella typhimurium* induces suppression of frameshift mutations. *Nucl. Acids. Res.* 20: 3463-3469.

Turnbough Jr., C. L., Neill, R. J., Landsberg, R. and, Ames, B. N. (1979). Pseudouridylation of tRNAs and its role in regulation in *Salmonella typhimurium*. *J. Biol. Chem.* 254: 5111-5119.

Weiss R. B:, Dunn D. M., Atkins J. F., Gesteland R. F. (1987). Slippery runs, shifty stops, backward steps and forward hops: -2, -1, +1, +2, +5, and +6 ribosomal frameshifting. *Cold Spring Harbor Symp. Quant. Biol.* 52: 687-693.

Weiss, R. B., D. M. Dunn, J. E. Dahlberg, J. F. Atkins, and R. F. Gesteland. 1988. *EMBO J.* 7:1503-1507.

Weiss R. B., Dunn D. M., Atkins J. F., Gesteland R. F. (1990). Ribosomal frameshifting from -2 to +50 nucleotides. *Prog Nucl Acid Mol Biol* 39: 159-183.

Wettstein, F. O., and G. S. Stent. 1968. Physiologically induced changes in the property of phenylalanine tRNA in *Escherichia coli*. *J. Mol. Biol.* 38: 25-40.

Woese, C. R. 1967. *in:* "The genetic code. The molecular basis for genetic expression". Harper and Row, p. 134.

THE DETERMINATION OF POSTTRANSCRIPTIONAL MODIFICATION IN RNA

Jeffrey A. Kowalak and James A. McCloskey

Departments of Biochemistry and Medicinal Chemistry
University of Utah
Salt Lake City, Utah 84112 U.S.A

INTRODUCTION

Significant progress has been made in recent years in recognition and understanding of the diversity of functional roles played by posttranscriptional modification in RNA (e.g., Björk et al., 1987; Cunningham et al., 1990; Björk et al., 1992). The functional aspects of modification have been most extensively studied in the case of tRNA (Björk et al., 1987; Björk et al., 1992) in which approximately 80 different posttranscriptionally modified nucleosides have been identified (Sprinzl et al., 1989; Edmonds et al., 1991). Although at an earlier time the structural characterization of modified nucleosides was usually undertaken in conjunction with their recognition during sequence determination (McCloskey and Nishimura, 1977), advances in structural methods based on mass spectrometry (McCloskey, 1991) have more recently resulted in the discovery of a number of new nucleosides in unfractionated tRNA, for which sequence locations are not yet known. We presently summarize recent progress in the development of three mass spectrometry-based methods, using instrumentation now commercially available, for the determination of posttranscriptional modification in RNA, with recent examples of their applications.

ANALYSIS OF MODIFIED NUCLEOSIDES IN RNA BY DIRECTLY COMBINED LIQUID CHROMATOGRAPHY–MASS SPECTROMETRY (LC/MS)

The combination of high performance liquid chromatography and mass spectrometry in a single instrument (Fig. 1) provides the basis for a method for analysis of nucleosides in enzymatic hydrolysates of RNA that is significantly more effective than either technique alone (Edmonds et al., 1985). The resulting data consist of (*i*) relative HPLC retention times, which can be compared against standardized values for RNA nucleosides (Pomerantz and McCloskey, 1990), (*ii*), UV absorbance, recorded using dual wavelength or photodiode array detector, and (*iii*) mass spectra, usually consisting of ions representing the protonated molecule (MH^+), protonated base (BH_2^+, where B is the base fragment), and in the case of certain hypermodified nucleosides, side chain fragment ions (Pomerantz and McCloskey, 1990). Because mass (measured to within 1 Da) is a highly selective molecular parameter, nucleoside components in the enzymatic digest in general need not be chromatographically

The Translational Apparatus, Edited by K.H. Nierhaus
et al., Plenum Press, New York, 1993

resolved for detection and identification, in contrast to sole use of UV detection. Absolute sensitivity of detection is approximately 2-20 ng (10 pmoles) per component, for example permitting recognition of one modified nucleoside in a 50 μg digest of *E. coli* 16S rRNA (1542 nt), or one residue in 1 μg of isoaccepting tRNA (76 nt).

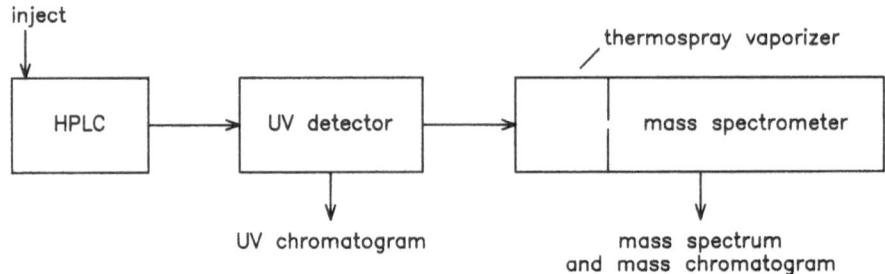

Figure 1. Instrument configuration for combined LC/MS (adapted from Pomerantz and McCloskey, 1990). A conventional HPLC chromatogram is obtained from the UV detector, or a chromatogram based on mass response is obtained from the mass spectrometer. Full mass spectra are recorded every 1.7 sec.

Detection and Identification of Nucleosides of Known Structure

In general, mass and relative retention time criteria for identification of nucleosides can be effectively applied for rapid screening of RNA digests for known components, in which nearly all nucleosides can be unambiguously identified in a single 40-minute experiment, plus data reduction time of about one hour. With rare exceptions, isomers are readily distinguished by retention times.

The LC/MS technique is of particular value for nucleosides whose UV absorbance is inadequate, or which are obscured by co-elution with other nucleosides or UV-absorbing impurities. A commonly encountered example is dihydrouridine, which exhibits virtually no UV absorbance and is therefore difficult to detect by conventional HPLC methods, but is readily detected on the basis of mass (2 Da higher than uridine) (Pomerantz and McCloskey, 1990). Fig. 2 demonstrates the presence of dihydrouridine in *E. coli* 23S rRNA, which had been isolated from 50S subunits and further purified by electrophoresis in a 1% agarose gel. Identification is made by detection of the MH^+ ion at the required elution time (Pomerantz and McCloskey, 1990). Dihydrouridine has not previously been reported in *E. coli* 23S rRNA, (Branlant et al., 1981), which we judge is in part a result of the difficulty of detection.

Recognition and Characterization of Nucleosides of Previously Unknown Structure

Structurally new or unexpected nucleosides can be most readily recognized in total hydrolysates of RNA by their molecular weights (simply derived from the MH^+ ion whose mass is 1 Da greater), compared with cataloged values of known RNA nucleosides. Recognition of ribose methylation is made from the mass spectrum, by difference in mass between the molecular ion and base ion (Pomerantz and McCloskey, 1990). This information, derived from a single 1.7 second mass spectrum scan, usually permits a partial structure of a new nucleoside to be assigned, directly from LC/MS analysis of the hydrolysate without necessity of isolation of the constituent in question. In favorable cases, complete structures can be assigned with sufficient certainty to warrant testing by chemical synthesis (e.g., Edmonds et al., 1987).

An example in which this method has been used effectively was in extensive mapping of tRNA posttranscriptional modification across the archaeal domain, as summarized in

Fig. 3. Nucleosides **3** through **9** were first discovered using LC/MS; structures **4 - 9** were assigned almost entirely on the basis of LC/MS data, and subsequently confirmed by chemical synthesis (see Edmonds et al., 1991, for leading references). In the case of more complex structures **2** (Gregson et al., 1992) and **3** (McCloskey et al., 1987), mass spectra from LC/MS provided initial structural data, and permitted the design of further experiments which were carried out following isolation of µg-level quantities of each nucleoside. Six nucleosides (**4 – 9**) are structurally novel in being modified both in the base and by methylation of ribose, and occur primarily in tRNA of extreme thermophiles.

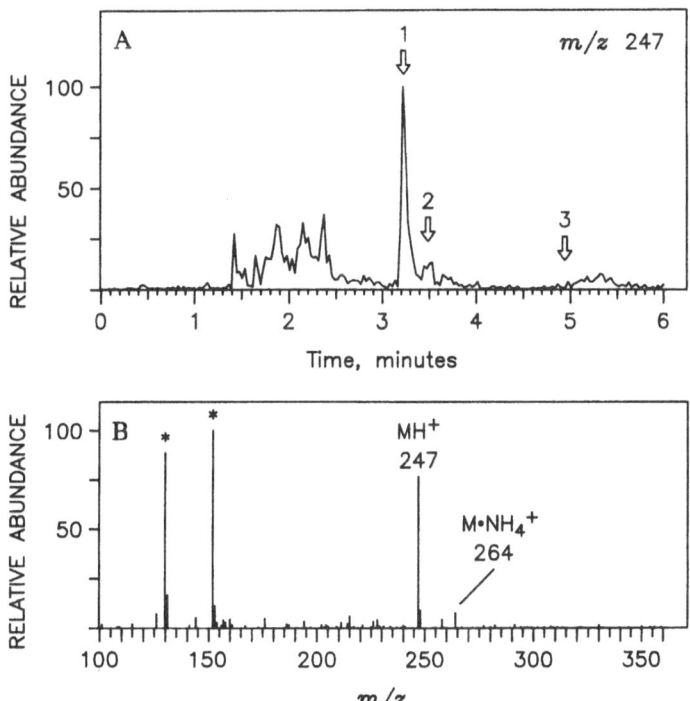

Figure 2. Identification of dihydrouridine (D) in a total enzymatic digest of *E. coli* 23S rRNA. A. HPLC analysis using m/z 247 detection channel, corresponding to the mass of protonated D. Arrows: 1, the required elution position of D under the standardized conditions employed (Pomerantz and McCloskey, 1990); 2 and 3, elution positions of pseudouridine and cytidine, respectively, determined from 254 nm UV detection (not shown). Pseudouridine and cytidine show no response in the m/z 247 detection channel because their molecular masses differ from that of D. B. Mass spectrum recorded at 3.22 min in panel A, showing protonated molecule and NH_4^+ adduct ions for dihydrouridine. Ions denoted by asterisks do not track with the profile of m/z 247 in panel A, and are due to background.

Ribose methylation (Kawai, et al., 1992a), in concert with certain forms of base modification (Horie et al., 1985) lends structural stability to individual nucleosides, contributing to overall stabilization of tRNA. Such stabilization is particularly notable in the case of N^4-acetyl-2′-*O*-methylcytidine (Kawai, et al., 1992b), which is common in the archaeal thermophiles (**6**, Fig. 3), and was unexpectedly discovered in 5S rRNA of *Pyrodictium occultum* (Bruenger et al., 1993, discussed in a later section). Overall, the rigorous characterization of 33 modified nucleosides in archaeal tRNA, compared with 18 known prior to screening by LC/MS, shows that a significant correlation exists between structural modification motifs in tRNA and the phylogenetic status of the organism, and that our knowledge of the diversity of posttranscriptional modification in tRNA, and hence of its function, is far from complete.

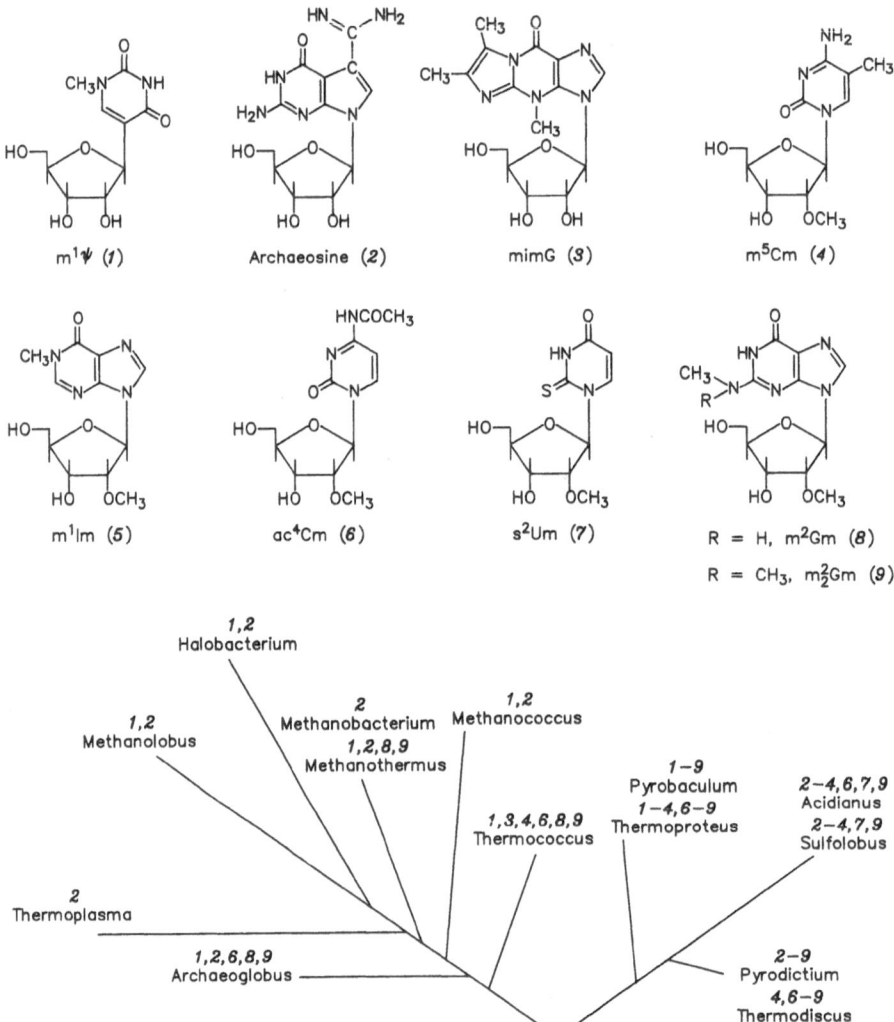

Figure 3. Structures of tRNA nucleosides (**1** - **9**) found to be unique to the archaeal (archaebacterial) phylogenetic domain (adapted from Edmonds et al., 1991), and their distribution across the archaeal tree based on 16S rRNA sequence comparisons (Woese, 1987). These and 24 additional modified nucleosides were mapped using LC/MS.

STRUCTURE DETERMINATION OF MODIFIED NUCLEOSIDES BY TANDEM MASS SPECTROMETRY

The ability to establish nucleoside structure by mass spectrometry rests on experiments in which gas-phase nucleoside ions dissociate into fragment ions, whose mass values can be correlated with substructural elements and thus used to derive the structure of the intact nucleoside. The dissociation step is crucial, because the resulting products can provide direct information concerning the nature and sites of attachment of subunits. In a tandem mass spectrometer, Fig. 4, ions of a single mass value are selected by the first mass spectrometer and subjected to collisional dissociation, with two important consequences. (*i*) The collision process often provides sufficient energy to produce structurally diagnostic dissociation reactions that are otherwise absent in a conventional single-stage mass spectrometer. (*ii*) All product ions formed and analyzed by the second mass spectrometer are derived from ions of known preselected mass, thus providing a clearer picture of the origins of substructure ions, and allowing a clean mass spectrum of one component to be acquired in the presence of other components so long as their mass values differ. The utility of this experimental technique is considerable in the case of complex nucleosides, for which alternate structural methods, such as NMR, may have sample size requirements several orders of magnitude higher (µg vs. ng range) and are thus less useful.

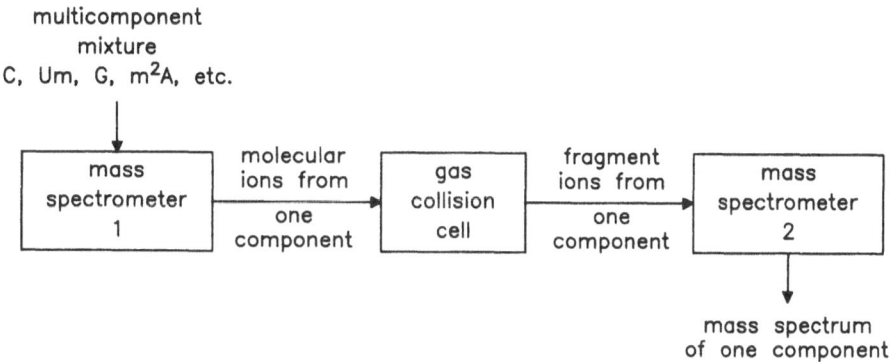

Figure 4. Instrument configuration for tandem mass spectrometry. Ions formed in the first mass spectrometer are selected within a ±0.3 Da range and transmitted to a collision cell, from which the resulting fragment ion products are measured as a mass spectrum in the second mass spectrometer.

This approach is demonstrated in two examples in which tandem mass spectrometry was used to establish key structural features in nucleosides of unknown structure isolated from tRNA. The adenosine derivative shown in Fig. 5 (as the chemically prepared trimethylsilyl derivative) was initially discovered using LC/MS, in unfractionated tRNA of ten organisms, including two thermophilic bacteria and six hyperthermophilic archaea (Reddy et al., 1992). The adenosine moiety and 3-hydroxynorvaline were observed as separate fragment ions in the mass spectrum acquired using LC/MS, but without direct evidence either of molecular weight, or of the nature of the linkage between the two major structural units. The fragment ion of *m/z* 510, used in conjunction with molecular weight and other features of the spectrum, established presence of the 28 Da carbonyl bridge as shown. The structure was confirmed by chemical synthesis, and thus shown to closely resemble nucleoside t⁶A (where threonine replaces 3-hydroxynorvaline), which occurs commonly at tRNA position 37 adjacent to the 3′-end of the anticodon.

Tandem mass spectrometry played a key role in characterization of archaeosine, a hypermodified nucleoside of unexpected structure (Gregson et al., 1992) which is distributed widely across the archaeal domain (see **2**, Fig. 3). Structure determination, carried out on µg-quantities of material, was confounded by initial data as summarized in Fig. 6, in which the elemental composition and number of exchangeable hydrogen atoms were established by mass spectrometry. These values were inconsistent with presence of a purine ring, even though G was specified by gene sequences for all cases in which both tRNA and gene sequences were known. This dilemma was resolved by tandem mass spectrometry of the chemically prepared permethyl derivative, which produced informative fragment ions not observed without the use of collisional activation. These spectra (Gregson et al., 1992) revealed presence of two exocyclic amino groups in the molecule, and showed an N-C-NH$_2$ sequence of atoms corresponding to the 2-aminopyrimidine portion of a purine ring. These data led to the pyrrolo[2,3-d]pyrimidine (7-deazaguanine)

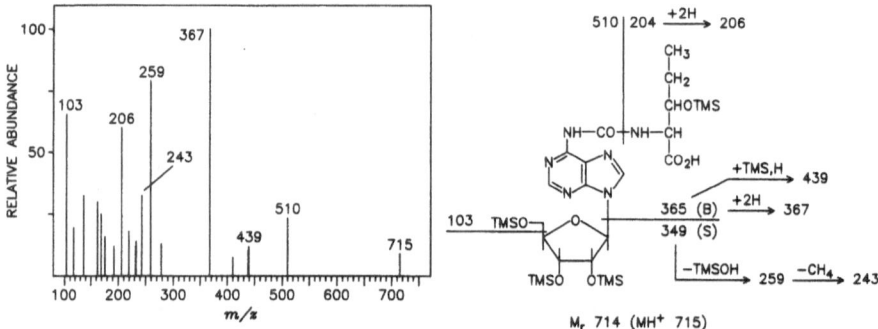

Figure 5. Mass spectrum of tRNA nucleoside hn^{6}A, used to establish molecular weight, presence of the 3-hydroxynorvaline-adenine carbonyl linkage, and other structural features. The spectrum was produced by tandem mass spectrometry, by collision-induced dissociation of the protonated trimethylsilyl (TMS) derivative (adapted from Reddy et al., 1992).

structure shown, which was then confirmed by chemical synthesis. Of phylogenetic significance, archaeosine is reported in isoaccepting tRNAs for 15 amino acids from archaea (Sprinzl et al., 1989), solely at position 15, a site never modified in all known

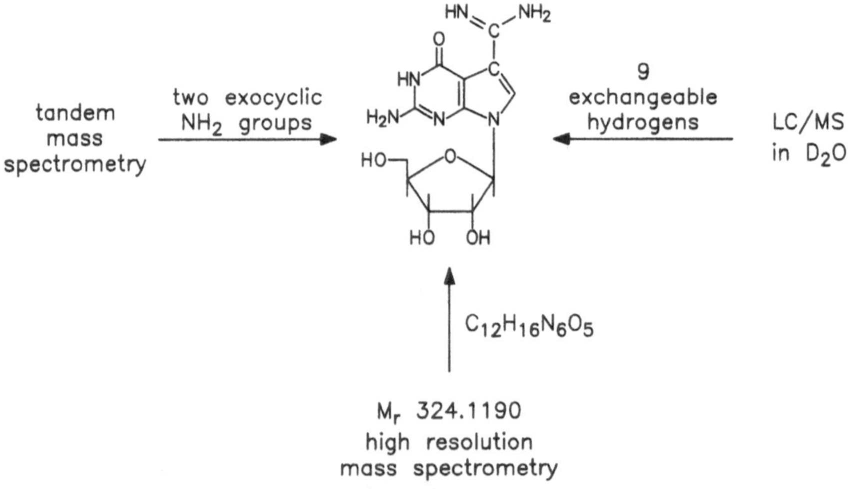

Figure 6. Central elements of the structure determination of the phylogenetically conserved tRNA nucleoside archaeosine (Gregson et al., 1992).

bacterial and eukaryotic tRNAs (Sprinzl et al., 1989). Archaeosine contains a charged formamidino side chain, not previously known in nucleic acids, while the base nucleus is the same as in queuosine (Q) (Kasai et al., 1975). Q is restricted to four isoacceptors (Tyr, His, Asp, Asn), at position 34 in the anticodon of tRNAs of bacteria and eukaryotes (Nishimura, 1983), but is absent in archaeal tRNA (Edmonds et al., 1991).

DETECTION AND LOCATION OF MODIFIED NUCLEOSIDES IN OLIGONUCLEOTIDES BY ELECTROSPRAY MASS SPECTROMETRY

Unprecedented progress has been made in recent years in development of new methods to produce gaseous ions from large biological molecules (Smith et al., 1990; Chait and Kent, 1992), and hence to acquire their mass spectra. In the case of polynucleotides (McCloskey and Crain, 1992; Nordhoff et al., 1992), these techniques open the door to new strategies for mapping of posttranscriptional modifications in rRNA, as outlined in Fig. 7 (Kowalak et al., 1992). Central to this approach is the recent finding that accurate molecular mass measurement of RNase T_1 fragments, using the technique of electrospray mass spectrometry, permits in nearly all cases the direct determination of oligonucleotide composition (Pomerantz et al., 1992). For example, using 50 pmoles of HPLC-isolated T_1-oligonucleotide, base composition can be *uniquely* assigned for all possible oligonucleotides at least through the 14-mer level by determination of mass to within ±0.01% (e.g., ±0.5 Da at 5 kDa), a measurement which can be carried out in 15 minutes. The first step (Fig. 7) is complete hydrolysis of the RNA to its nucleoside substituents, followed by LC/MS

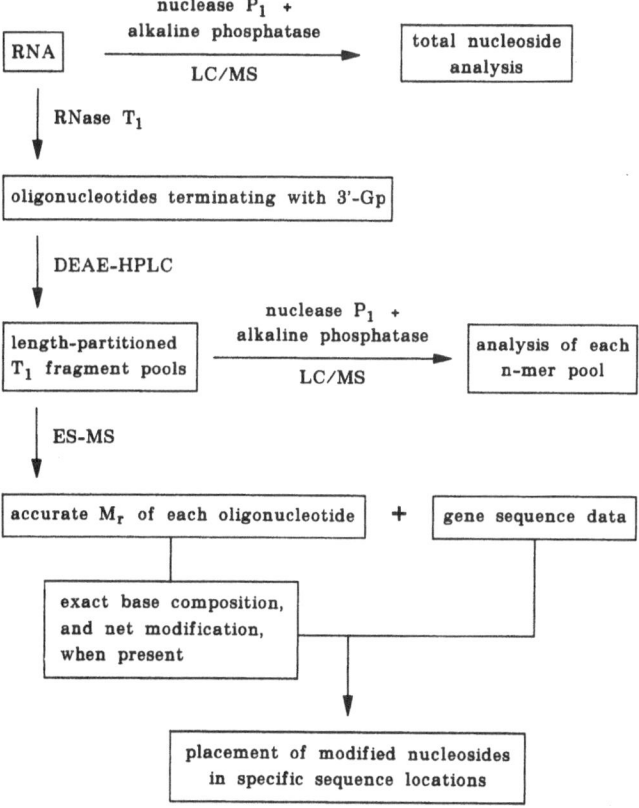

Figure 7. Protocol for determining sites of posttranscriptional modification in RNA by mass spectrometry.

analysis of the hydrolysate. This initial screen serves as a qualitative indicator of nucleoside content and can be used for recognition of unexpected or structurally new nucleosides. Next, the RNA is digested with RNase T_1 yielding a mixture of 3' Gp-terminated oligonucleotides of varying chain length. The T_1 fragments are then resolved primarily on the basis of chain length by DEAE anion exchange HPLC. An aliquot of each DEAE pool can be hydrolyzed to nucleosides and examined by LC/MS to identify modification-containing HPLC fractions. Oligonucleotide pools shown to contain modified nucleosides are examined by electrospray mass spectrometry to accurately determine the

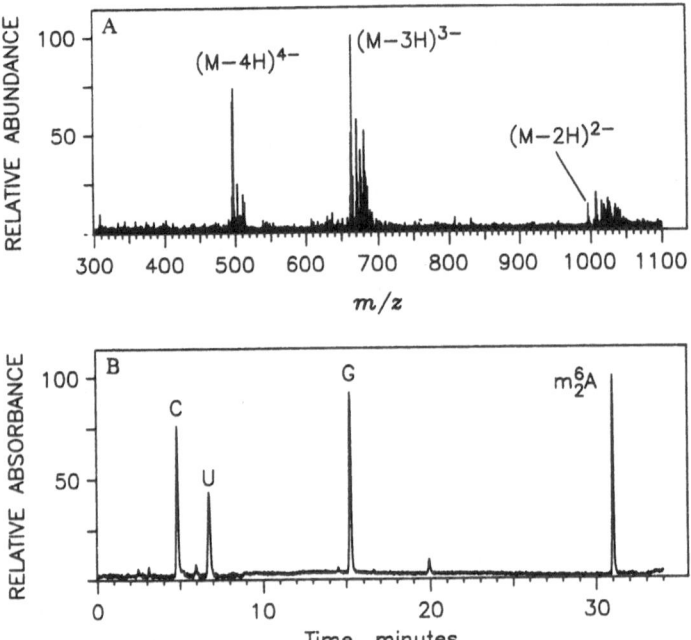

Figure 8. A. Electrospray mass spectrum of chromatographically isolated RNase T_1 oligonucleotide fragment from *E. coli* 16S rRNA. Deprotonated molecular ions: (-4H) m/z 497 not measured; (-3H) m/z 663.75; (-2H) m/z 996.15. The resulting molecular mass (M) of 1994.28 uniquely specifies the composition as C_2UA_2Gp + $4CH_2$, corresponding to the RNase T_1 fragment containing N^6,N^6-dimethyladenosine, $m_2^6Am_2^6ACCUGp$, as expected. B. HPLC chromatogram, with UV detection at 254 nm, from LC/MS analysis of RNase T_1 fragment, confirming the composition determined by mass measurement.

molecular mass of each oligonucleotide species. The experimentally determined M_r of each oligonucleotide and qualitative and quantitative knowledge of the nucleoside species contained in each pool are then used in conjunction with the corresponding gene sequence to determine the sequence location of modified nucleosides within the primary sequence of the RNA molecule. In less favorable cases the modification site will be restricted to a fixed RNase T_1 fragment, while in favorable cases a single sequence site will be identified.

An example of how the method is employed is given by data in Fig. 8, in which *E. coli* 16s rRNA is used as a model. RNase T_1 fragments were fractionated by DEAE HPLC; from one of these fractions LC/MS analysis revealed the presence of a m_2^6A-containing oligonucleotide. The electrospray mass spectrum presented in Fig. 8 corresponds to an oligonucleotide whose experimentally determined M_r of 1994.28 specifies a base composition of $C_2UA_2Gp + 4CH_2$ (M_r calc. 1994.29). The oligonucleotide was hydrolyzed using nuclease P_1–alkaline phosphatase and examined by LC/MS confirming the composition determined in the ES-MS experiment and further establishing the modification specifically as dimethylation of the N^6 position of adenosine. Other molecular combinations to accommodate four methyl groups, such as one trimethyladenosine and one monomethyladenosine, are excluded by data in Fig. 8B, which requires two dimethyladenosines. The experimentally determined base composition, used in conjunction with the *E. coli* 16s rRNA gene sequence, thus specifies the sequence and location as 1518-$m_2^6Am_2^6ACCUGp$-1523, as expected (Brosius et al., 1978).

This technique was recently employed in a study of 5S rRNA from *Pyrodictium occultum* (Bruenger et al., 1992), the most thermophilic organism presently known, with a optimal growth temperature of 105 °C (Stetter, 1982). LC/MS analysis of a nuclease P_1–alkaline phosphatase digest of the RNA revealed presence of N^4-acetylcytidine (ac^{4}C) and the dinucleotide ac^{4}CmpG, an interesting finding in view of the rarity of 5S rRNA modification (Wolters and Erdmann, 1988). Following cleavage by RNase T_1, an oligonucleotide fraction was examined by electrospray mass spectrometry, and found to contain two oligonucleotides, of M_r 2914.45 (M_a) and 2244.18 (M_b) (Fig. 9). The latter value uniquely specifies the nucleoside composition $C_2U_2A_2G$ (calc. 2244.35), and therefore represents 38-ACUCAUGp-44, predicted from the *P. occultum* 5S rRNA gene sequence (Kaine et al., 1989). The mass of the M_a permits no composition consistent with unmodified RNase T_1 products allowed by the gene sequence, but correlates with a nucleotide of composition $C_5A_3G + 56$ Da (calc. 2914.87), corresponding to nucleotides 28-36, dictated by the gene sequence. This mass value therefore reveals presence of the ac^{4}Cm residue found by LC/MS (acetyl = 42 Da, methyl = 14 Da net increase). Because, from LC/MS data, ac^{4}Cm must occur adjacent to the Gp-3′ oligonucleotide terminus, the position of C modification is established, leading to the RNA sequence 28-CAACACCac^{4}CmG>p-36. Using similar measurements, the ac^{4}C residue discovered by LC/MS was placed in one or both of the trinucleotides 45-UCG-47 and 80-CUG-82 (Bruenger et al., 1992).

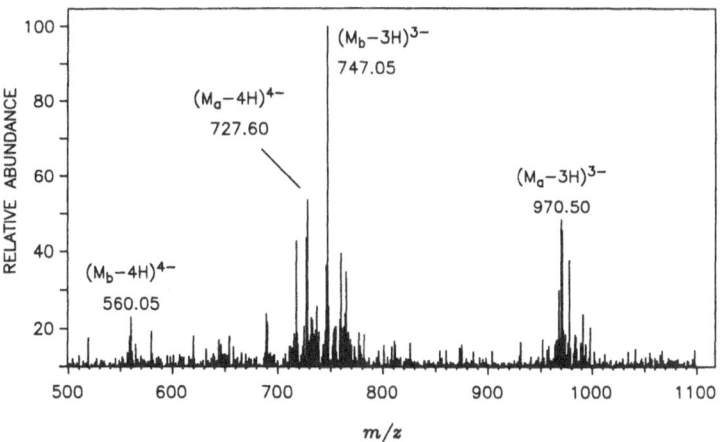

Figure 9. Electrospray ionization mass spectrum of RNase T_1 fragments from *Pyrodictium occultum* 5S rRNA, showing two oligonucleotides from which molecular masses are measured as M_a, 2914.45 (ac^{4}CmC$_4$A$_3$G>p) and M_b, 2244.18 ($C_2U_2A_2Gp$) (Bruenger et al., 1993).

Acknowledgements

This work was supported by grants GM21584 and GM29812 from the National Institute of General Medical Sciences. The authors are pleased to acknowledge the collaboration of P. F. Crain, R. Gupta, S. C. Pomerantz, D. M. Reddy, and K. O. Stetter in previously published portions of this work.

REFERENCES

Björk, G. R., Ericson, J. U., Gustafsson, C. E. D., Hagervall, T. G., Jönsson, Y. H., and Wikström, P. M., 1987, *Biochemistry* 56:263.

Björk, G. R., 1992, *in*: "Transfer RNA in Protein Synthesis, "Hatfield, D. L., Lee, B. J., and Pirtle, R. M., eds., CRC Press, Boca Raton, FL, p. 23.

Branlant, C., Krol, A., Machatt, M. A., Pouyet, J., and Ebel, J. P., 1981, *Nucleic Acids Res.* 9:4303.

Brosius, J., Palmer, M. L., Kennedy, P. J., and Noller, H. F., 1978, *Proc. Natl. Acad. Sci. USA* 75:4801.

Bruenger, E., Kowalak, J. A., Kuchino, Y., McCloskey, J. A., Mizushima, H., Stetter, K. O., and Crain, P. F., 1993, *FASEB Journal* 7:916.

Chait, B. T., and Kent. S. B. H., 1992, *Science* 257:1885.

Crain, P. F., Hashizume, T., Nelson, C. C., Pomerantz, S. C., and McCloskey, J. A., 1990, *in*: "Biological Mass Spectrometry," Burlingame, A. L., and McCloskey, J. A., eds., Elsevier, New York, p. 509.

Cunningham, P. R., Weitzmann, C. J., Nègre, D., Sinning, J. G., Frick, V., Nurse, K., and Ofengand, J., 1990, *in* "The Ribosome, Structure, Function and Evolution," W. E. Hill, A. Dahlberg, R. A. Garrett, P. B. Moore, D. Schlessinger, and J. R. Warner, eds., American Society for Microbiology, Washington, DC, p. 243.

Edmonds, C. G., Crain, P. F., Gupta, R., Hashizume, T., Hocart, C. H., Kowalak, J. A., Pomerantz, S. C., Stetter, K. O, and McCloskey, J. A., 1991, *J. Bacteriol.* 173:3138.

Edmonds, C. G., Crain, P. F., Hashizume, T., Gupta, R., Stetter, K. O., and McCloskey, J. A., 1987, *J. Chem. Soc., Chem. Commun.* 909.

Edmonds, C. G., Vestal, M. L., and McCloskey, J. A., 1985, *Nucleic Acids Res.* 13:8197.

Gregson, J. M., Crain, P. F., Edmonds, C. G., Gupta, R., Hashizume, T., Phillipson, D. W., and McCloskey, J. A., 1993, *J. Biol. Chem.* 268:10076.

Horie, N., Hara-Yokoyama, M., Yokoyama, S., Watanabe, K., Kuchino, Y., Nishimura, S., and Miyazawa, T., 1985, *Biochemistry* 24:5711.

Kaine, B. P., Schurke, C. M., and Stetter, K. O., 1989, *System. Appl. Microbiol.* 12:8.

Kasai, H., Ohashi, Z., Harada, F., Nishimura, S., Oppenheimer, N. J., Crain, P. F., Liehr, J. G., von Minden, D. L., and McCloskey, J. A., 1975, *Biochemistry* 14:4198.

Kawai, G., Yamamoto, Y., Kamimura, T., Masegi, T., Sekine, M., Hata, T., Iimori, T., Watanabe, T., Miyazawa, T., and Yokoyama, S., 1992a, *Biochemistry* 31:1040.

Kawai, G., Hashizume, T., Yasuda, M., Miyazawa, T., McCloskey, J. A., and Yokoyama, S., 1992b, *Nucleosides Nucleotides* 11:759.

Kowalak, J. A., Pomerantz, S. C., Crain, P. F., and McCloskey, J. A., 1993, *Nucleic Acids Res.*, in press.

McCloskey, J. A., 1991, *Accounts Chem. Res.* 24:81.

McCloskey, J. A., and Crain, P. F., 1992, *Int. J. Mass Spectrom. Ion Processes* 118/119:593.

McCloskey, J. A., Crain, P. F., Edmonds, C. G., Gupta, R., Hashizume, T., Phillipson, D. W., and Stetter, K. O., 1987, *Nucleic Acids Res.* 15:683.

McCloskey, J. A., and Nishimura, S., 1977, *Accounts Chem. Res.* 10:403.

Nishimura, S., 1983, *Prog. Nucleic Acid Res. Mol. Biol.* 28:49.

Nordhoff, E., Ingendoh, A., Cramer, R., Overberg, A., Stahl, B., Karas, M., Hillenkamp, F., and Crain, P. F., 1992, *Rapid Commun. Mass Spectrom.* 6:771.

Pomerantz, S. C., Kowalak, J. A., and McCloskey, J. A., 1993, *J. Am. Soc. Mass Spectrom.*, in press.

Pomerantz, S. C., and McCloskey, J. A., 1990, *Methods Enzymol.* 193:796.

Smith, R. D., Loo, J. A., Edmonds, C. G., Barinaga, C. J., and Udseth, H. R., 1990, *Anal. Chem.* 62:882.

Reddy, D. M., Crain, P. F., Edmonds, C. G., Gupta, R., Hashizume, T., Stetter, K. O., Widdell, F., and McCloskey, J. A., 1992, *Nucleic Acids Res.* 20:5607.

Sprinzl, M., Hartmann, T., Weber, J., Blank, J., and Zeidler, R., 1989, *Nucleic Acids Res. (Suppl)* 17:r1.

Stetter, K. O., 1982, *Nature* (London) 300:258.

Wolters, J., and Erdmann, V. A., 1988, *Nucleic Acids Res. (Suppl)* 14:r1.

RNA POLYMERASE I, THE NUCLEOLUS AND SYNTHESIS OF 35S rRNA IN THE YEAST <u>SACCHAROMYCES CEREVISIAE</u>

Masayasu Nomura[1], Yasuhisa Nogi[1,2], Ryoji Yano[1], Melanie Oakes[1], Daniel A. Keys[1], Loan Vu[1] and Jonathan A. Dodd[1]

[1]Department of Biological Chemistry
University of California
Irvine, CA 92717-1700

[2]Laboratory of Molecular Genetics
Keio University
Tokyo, Japan

INTRODUCTION

The synthesis of ribosomes is regulated in response to environmental changes, and this regulation is important in connection with the regulation of growth. This is true for both prokaryotic and eukaryotic organisms. In *E. coli*, the cellular concentration of ribosomes is roughly proportional to the growth rate except under slow growth conditions; this regulatory feature is called growth-rate-dependent control of ribosome synthesis. Extensive studies of this regulation in *E. coli* have shown that the regulation acts primarily on the synthesis of rRNA, and that growth-rate-dependent control of ribosomal protein (r-protein) synthesis is achieved indirectly through various autogenous (feedback) repression systems. In eukaryotes, although the synthesis of r-proteins may be regulated directly, and independently of rRNA synthesis, there is no doubt that the synthesis of rRNAs in the nucleolus is of primary importance in overall regulation of ribosome biosynthesis (see reviews, Warner, 1982; Sollner-Webb and Tower, 1986; Sollner-Webb and Mougey, 1991). A few years ago, our laboratory initiated studies on the synthesis of rRNA in the yeast *Saccharomyces cerevisiae* as a model eukaryotic organism, with the eventual goal of understanding molecular events involved in the regulation of ribosome biosynthesis.

There are several features of rRNA synthesis in eukaryotes that distinguish it from rRNA synthesis in prokaryotes. First, the genes for large rRNAs (called "35S rRNA genes" or "35S rDNA" in *S. cerevisiae*) are tandemly repeated. In the yeast *Saccharomyces cerevisiae*, a single 35S rRNA gene, together with a spacer containing the 5S RNA gene and an enhancer element, make up a single unit, and this unit is repeated approximately 120 times on chromosome XII. Second, in contrast to a single RNA polymerase in *E. coli*, eukaryotes contain three nuclear RNA polymerases (in addition to organelle RNA polymerases), RNA polymerase I, II and III, (Pol I, II, and III), and Pol I is exclusively utilized for the transcription of 35S rDNA. Third, eukaryotic rRNA transcription takes place in a cytologically visible structure, the nucleolus, where all the machinery required for rRNA transcription and processing (and perhaps ribosome assembly) is present and presumably highly organized for the purpose of efficiency and regulation. As described below our studies touch upon some of these unique features of the eukaryotic system.

The approaches we have taken are primarily genetic, but an *in vitro* rRNA transcription system has also been developed (Riggs and Nomura, 1990), anticipating the use of this system to directly study some regulatory features *in vitro*, and in addition, to study some of the mutationally altered components identified by genetic studies. Because our knowledge of the components involved in rRNA transcription was (and still is) very limited and because such knowledge is a prerequisite to regulation studies, genetic approaches were designed to identify these components. Two approaches have been used. One is the analysis of suppressors of defined temperature-sensitive (ts) Pol I subunit mutants, with the hope of identifying components interacting with the pertinent Pol I subunits. The second is the use of a *GAL7*-35SrDNA fusion system designed to identify components uniquely involved in 35S rRNA synthesis, as will be described below.

POL I SUBUNITS AND THEIR GENES IN S. CEREVISIAE

The three nuclear RNA polymerases in *S. cerevisiae* are perhaps the best characterized among eukaryotes (for reviews, Sentenac, 1985; Thuriaux and Sentenac, 1992). The subunit composition of *Saccharomyces cerevisiae* Pol I is given in Table 1. Pol I consists of 14 different subunits. The two largest subunits of Pol I, A190 and A135, like the two largest subunits of Pol II and Pol III, are homologous to *E. coli* RNA polymerase subunits ß' and ß, respectively. Two other subunits, AC40 and AC19, which are shared with Pol III, correspond to the α subunit of *E. coli* polymerase (Martindale, 1990; Dequard-Chablat *et al.*, 1991). Among the remaining 10 subunits, five are shared by Pol I, Pol II and Pol III, and the other five are unique to

Table 1. Subunits and genes of Pol I of *S. cerevisiae* and homologous *E. coli* RNA polymerase subunits

Subunit class	Subunit	Gene	Homologous E. coli subunit
I (core subunit)	A190	*RPA190*	ß'
	A135	*RPA135*	ß
	AC40	*RPC40*	α
	AC19	*RPC19*	α
II (subunits common for Pol I, II and III)	ABC27	*RPB5*	
	ABC23	*RPB6*	
	ABC14.5	*RPB8*	
	ABC10α	*RPC10*	
	ABC10ß	*RPB10*	
III (subunits unique to Pol I)	A49	*RPA49*	
	A43	*RPA43*	
	A34.5	*RPA34*	
	A14	(not cloned)	
	A12.2	*RPA12/RRN4*	

Pol I. Some of these conclusions were originally obtained by comparison of the sizes of subunits of the three purified polymerases and by the results of immunological studies using antibodies prepared against individual protein subunits (Huet *et al.*, 1982). However, cloning and sequencing of the genes for these protein subunits have given a more solid experimental basis for these conclusions (Woychik *et al.*, 1990, 1991; Treich *et al.*, 1992).

Regarding genes for Pol I subunits, work in our laboratory has contributed the cloning of the gene (*RPA135*) for the A135 subunit and the gene (*RPA12*) for the A12.2 subunit. *RPA135* was isolated as an allele-specific suppressor gene (*SRP3-1*) responsible for partial suppression of certain ts mutations in the A190 subunit (McCusker *et al.*, 1991; Yano and Nomura, 1991). Two mutations known to be suppressed by *SRP3-1* (*RPA135-1*) are in the putative zinc-binding domain of A190 located near the amino terminus of the protein (Wittekind *et al.*, 1988). The mutational alteration in A135 responsible for this suppression was found to be located within the putative zinc-binding domain of A135 located near its carboxy terminus (Yano and Nomura, 1991). These results, together with other considerations (Sawadogo and Sentenac, 1990), led to the suggestion that this putative zinc-binding domain of A135 is in physical proximity to and interacts with the putative zinc-binding domain of the A190 subunit.

The *RPA12* gene was isolated as the gene which complements mutations in *RRN4* (Nogi *et al.*, 1993). *RRN4* is one of 9 genes which are specifically involved in the synthesis of 35S rRNA and are defined by complementation analysis of *rrn* mutants (Nogi *et al.*, 1991a; see below and Table 2).

Except for the gene for A14, the genes for all other Pol I subunits in *S. cerevisiae* have now been cloned (Table 1; for a review, see Thuriaux and Sentenac, 1992). This will be very helpful for future study of Pol I structure and function, which in turn will provide a basis for understanding 35S rRNA transcription and its regulation.

SYNTHESIS OF LARGE rRNAs BY POL II IN POL I DELETION MUTANTS

We asked the question whether in yeast cells functional 35S precursor rRNA can be synthesized by Pol II from a fusion gene in which 35S rDNA is fused to a suitable Pol II promoter. We reasoned that if the answer were yes, we would be able to construct a system convenient for the isolation of mutants defective in 35S rRNA synthesis; this was the primary reason for asking this question. However, we realized that there is no guarantee of success in constructing a yeast strain synthesizing rRNA without intact Pol I. For example, it had been assumed without proof that the sole function of Pol I is the transcription of 35S rDNA. If this were not the case and there were other essential genes requiring Pol I for their transcription, Pol I-deficient yeast cells would not be viable even if the cells could synthesize 35S rRNA using Pol II. Fortunately, transcription of a plasmid-born fusion gene in which the 35S rRNA coding region is fused to the GAL7 promoter ("*GAL7*-35SrDNA"; see Fig. I) suppressed the lethality of Pol I mutations. Specifically, two mutations tested, a *rpa190* ts mutation (*rpa190-3*) and a *rpa135* deletion mutation (*rpa135::LEU2*), were suppressed (Nogi *et al.*, 1991b); i.e., these mutant strains carrying the fusion gene on a plasmid grew in the presence of galactose, but not in glucose. Furthermore, synthesis of large rRNAs in these strains growing in galactose was shown to be repressed by glucose (Nogi *et al.*, 1991b). These results demonstrate that the sole essential function of Pol I is, in fact, transcription of the rRNA genes. In addition, they also show that the tandemly repeated structure of the chromosomal rRNA genes is not required for the synthesis of rRNA and of ribosomes.

In the yeast *S. cerevisiae*, the nucleolus appears to be crescent-shaped, occupying a substantial fraction of the nucleus and having close contact with the nuclear envelope. The presence of several nucleolar-specific proteins has been demonstrated in this crescent-shaped structure; these components include the A190 subunit of Pol I (Clark *et al.*, 1990), SSB1 (Clark *et al.*, 1990), the yeast homologue of fibrillarin (also called NOP1; Henriquez *et al.*, 1990; Schimmang *et al.*, 1989), and GAR1 (Girard *et al.*, 1992). Proteins SSB1, fibrillarin and GAR1 are components of small nucleolar ribonucleoprotein particles and appear to participate in rRNA

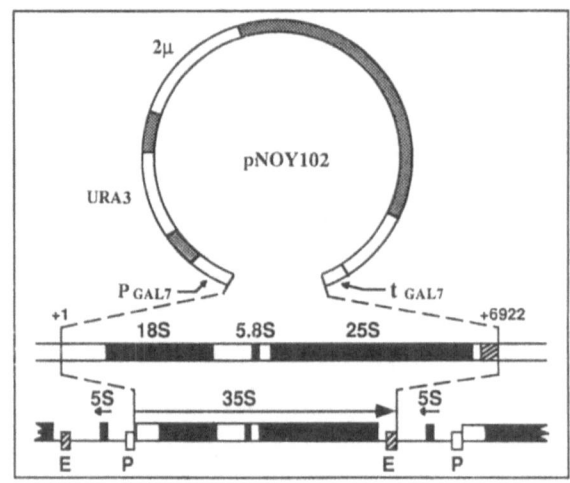

Figure 1. Structure of the plasmid (pNOY102), which carries the GAL7-35SrDNA fusion gene. The 18S, 5.8S and 25S rRNA coding regions are shown as black boxes, as is the 5S RNA-coding region. Boxes designated as P and E are the Pol I promoter and enhancer elements, respectively. Nucleotide sequence numbers +1 and +6922 are given to indicate the transcription start site and the HindIII site that defines the distal end of the enhancer element, respectively, and are fused to the GAL7 promoter and the GAL7 terminator, respectively (see Nogi *et al.*, 1991b)

processing (Clark *et al.*, 1990; Girard *et al.*, 1992; Tollervey *et al.*, 1991). U3 RNA, which is known to be complexed with fibrillarin (as well as other proteins), is the best studied small nucleolar RNA and its role in rRNA processing is firmly established (Kass *et al.*, 1990; Hughes and Ares, 1991). The presence of yeast U3 RNA in the nucleolus has also been demonstrated (Dvorkin *et al.*, 1991).

As described in the previous section, we have constructed Pol I deletion mutants of yeast which can grow in galactose media by synthesizing rRNA from the GAL7-35SrDNA fusion gene using Pol II. By immunofluorescence microscopy using antibodies against SSB1 and fibrillarin, we found that the intact crescent-shaped nucleolar structure is absent in these mutants. Instead, we observed the presence of several granules (called "mininucleolar bodies" or "MNBs") that were stained with these antibodies (M. Oakes, Y.

Nogi, M.W. Clark and M. Nomura, manuscript submitted). Thus, these mutant cells synthesize and process rRNA, produce functional ribosomes, and grow in the absence of an intact nucleolus, demonstrating that an intact crescent-shaped nucleolar structure is not absolutely required for rRNA processing, ribosome assembly or cell growth. MNBs, which contain nucleolar proteins required for rRNA processing, may function in rRNA processing in these Pol I deletion mutants. They may represent a constituent of the normal nucleolus, perhaps resembling prenucleolar bodies observed in higher eukaryotic cells during the late period of mitosis (for prenucleolar bodies, see De La Torre and Gimenez-Martin, 1982).

The absence of an intact nucleolus in growing Pol I deletion mutants demonstrates that the presence of Pol I is essential for the maintenance of an intact crescent-shaped nucleolar structure. Two possibilities can be considered. First, it may be the transcription of rDNA by Pol I or the presence of growing rRNA that is responsible for the maintenance of the nucleolar structure. Second, Pol I itself may play a role as a structural element for the maintenance of the nucleolar structure. To examine these possibilities, we studied nucleolar structures under several conditions that inhibit rRNA synthesis, and found that these conditions do not cause conversion of the nucleolus to MNBs (Oakes *et al.*, manuscript submitted). Thus, we favor the second possibility. Our studies on *SRP1*, to be described below, also suggest that the nucleolus is not a simple aggregate structure formed on rDNA and growing rRNA to which various proteins and ribonucleoproteins "self-assemble" by their affinity to rDNA and/or precursor rRNA. Rather, some nuclear proteins as well as nucleolar proteins seem to play structural roles in forming specific nucleolar structures, and Pol I might be a part of the nucleolar structure interacting with these other proteins.

ROLE OF SRP1 IN NUCLEOLAR STRUCTURE AND FUNCTION

As mentioned above, we have isolated mutants carrying suppressors of ts mutations in the zinc-binding domain of the A190 subunit of Pol I (McCusker *et al.*, 1991). Among many independently isolated mutants, most were mapped in (or very close to) one particular gene, *SRP1*, which turned out to be an interesting gene encoding a nuclear matrix/envelope protein. The *SRP1* suppressor (*SRP1-1*) was also shown to suppress ts mutations in the zinc binding domain, but not other ts mutations, of the Pol I A135 subunit (Yano *et al.*, 1992). The *SRP1* gene, originally as a mutant suppressor form, was cloned by its ability to suppress *rpa190* ts mutations and then sequenced (Yano *et al.*, 1992). The SRP1 protein, as deduced from the nucleotide sequence, has an interesting structure consisting of three domains: the central domain, which is composed of eight (degenerate) 42-amino acid contiguous tandem repeats, and the surrounding N-terminal and C-terminal domains, both of which contain clusters of acidic and basic amino acids and are highly hydrophilic. Both immunofluorescence microscopy using antibodies against the SRP1 protein and biochemical fractionation experiments indicated that SRP1 is a component of a larger macromolecular complex associated with the nuclear envelope/matrix (Yano *et al.*, 1992).

In addition to being a suppressor of certain specific Pol I ts mutations, a relationship between SRP1 protein and the nucleolus was revealed by SRP1 depletion experiments using a strain in which the production of SRP1 was controlled by the *GAL7* promoter (R. Yano, M. Oakes, L. Vu and M. Nomura, manuscript in preparation). Upon transfer of this strain from galactose to glucose media, striking morphological changes were observed as examined by DNA staining and by immunofluorescence staining using antibodies against nucleolar proteins. DNA threads, together with nucleolar proteins, first appeared as loops protruding from the main body of the nucleus ("unfolding" of the nucleolus), and then the loop structure disappeared and granules containing the nucleolar proteins, similar to the

MNBs mentioned above, were produced. Concomitant with the appearance of MNB-like granules, the RNA synthesis rate started to decrease. At the beginning, rRNA synthesis was affected somewhat preferentially, but eventually all the nuclear transcription activities were almost completely abolished. In addition, we observed that depletion of SRP1 ultimately led to inhibition of nuclear division or segregation. Thus, SRP1 appears to play an essential role in the maintenance of the nucleolar (and nuclear) structures. Perhaps SRP1 interacts with rDNA and/or Pol I and helps to keep very long rDNA regions, with many tandemly repeated rDNA genes, in a compact "folded" state. In this way, SRP1 may function as an important element in organizing the nucleolar structure.

It is an enigma that mutations in SRP1, which is localized mainly in the nuclear periphery, specifically suppress certain ts mutations of Pol I, which is localized mainly in the nucleolus; we do not know the exact mechanism of suppression. Nevertheless, since SRP1 is clearly a component of a large macromolecular complex associated with the nuclear matrix/envelope, and since Pol I is also, at least in part, associated with the nuclear matrix/envelope (Dickinson et al., 1990; Yano et al., 1992), we suggest that SRP1 interacts with Pol I either directly or indirectly through other components in the structure containing SRP1. As discussed in the previous section, there is evidence indicating that Pol I plays a structural role in the maintenance of the nucleolar structure. This is consistant with the presence of the proposed (direct or indirect) interaction between Pol I and SRP1. We speculate, as was done by Cook and his coworkers for mammalian Pol I systems, that association of Pol I with some nucleolar structure(s) (nuclear matrix/envelope) is important for its function and regulation. According to this model, the cessation of rRNA synthesis seen upon SRP1 depletion is a result of the disruption of the nucleolar structure, as was observed in the experiments, and the nucleolar structure is required for active transcription of rRNA genes by Pol I.

ISOLATION OF rrn MUTANTS WHICH ARE PREFERENTIALLY DEFECTIVE IN 35S rRNA SYNTHESIS

Although the subunit structure of S. cerevisiae Pol I is well defined and most of the genes for the subunits have been cloned and characterized, our knowledge is very limited regarding other components, such as transcription factors or structural components of the nucleolus, that might influence rRNA transcription. As mentioned above, suppressor analysis has been taken as one approach to remedy this situation. In addition, we have developed a more general method to isolate mutants that are primarily defective in 35S rRNA synthesis. This method utilizes the GAL7-35SrDNA fusion system described earlier. Chromosomal mutations affecting components specifically involved in 35S rRNA synthesis by Pol I can be suppressed by this hybrid gene in the presence of inducer (galactose). We looked for mutants whose growth depended on the presence of the plasmid expressing the fusion gene. Starting with an ade2 ade3 strain and using a plasmid carrying ADE3 in addition to the fusion gene, we were able to use a red/white colony color assay as the initial screen. We then further screened colonies that remained red with no white sectors (i.e., those which did not lose the plasmid on complex galactose plates) for galactose-dependent growth. Finally, galactose-dependent candidate mutants were directly tested for galactose-dependent synthesis of large rRNAs. In this way we have isolated many rrn (rRNA-synthesis defective) mutants and have identified at least 9 genes (RRN1 - RRN10; see Table 2 and its legend) by complementation analysis (Nogi et al., 1991a; other unpublished experiments). Using the rpa190 and rpa135 mutants available, two of the 9 rrn genes (RRN1 and RRN2) were shown to correspond to Pol I subunit genes (RPA190 and RPA135), justifying the strategy used for mutant isolation. To elucidate the nature of the remaining genes, two approaches have been used. One is a straight forward cloning of genes by complementation of mutational defects and subsequent analysis (including sequencing) of the cloned genes. The second approach is testing the ability of mutant extracts to carry out specific transcription of 35S rRNA genes *in vitro*.

Our *in vitro* extracts prepared from the wild type yeast strain can be separated into two fractions by phosphocellulose column chromatography. One fraction eluted with 300 mM salt ("PC300") contains Pol I (probably

Table 2. Summary of *rrn* mutants identified

Gene product	Number of indep. mutants isolated	Gene cloned & sequenced	Nature of gene
RRN1 (RPA190)	1	(yes)	(Pol I subunit A190)
RRN2 (RPA135)	6	(yes)	(Pol I subunit A135)
RRN3	5		
RRN4 (RPA12)	2	yes	Pol I subunit A12
RRN5	11	yes	Unknown
RRN6	5	yes	Transcription factor
RRN7	2	---	
RRN9	2	---	
RRN10	1	yes	Unknown

Mutants classified as rrn8 showed glucose inhibition of synthesis of all RNAs (Nogi *et al.*, 1991b), and subsequent experiments also showed that the RRN8 gene is probably not related to Pol I function, and hence the gene is omitted from this list.

together with some transcription factors) and the flow-through fraction, after further chromatography on DEAE-cellulose, ("D300") contains mostly transcription factors and little Pol I. Either fraction alone shows little or no specific transcriptional activity; the combination of the two fractions is required for activity. Some mutant extracts, such as extracts prepared from *rrn6* mutants, are inactive, but the addition of the factor fraction (D300) from the wild type make the mutant extracts transcriptionally active, suggesting that a certain transcription factor(s) is inactive in the mutant extracts due to mutational alteration. The component responsible for this apparent *in vitro* complementation of *rrn6* extracts was further fractionated by heparin agarose column chromatography (our unpublished experiments). Thus *RRN6* appears to encode a transcription factor.

The gene for RRN6 has also been cloned and its sequence has been determined (our unpublished experiments). The deduced amino acid sequence shows that the RRN6 protein is 894 amino acids long with a calculated molecular weight of 102,000 kDa. On the other hand, the component that complements mutant *rrn6* extracts *in vitro* (see above) appears to be 200,000 kDa or larger as judged from its behavior in gel filtration chromotography. Thus, the RRN6 protein may be complexed with other protein(s) (and/or with itself) to form a functional transcription factor. There is no recognizable similarity of RRN6 to the published sequence of UBFs from various higher eukaryotes. UBF is the only Pol I transcription factor whose sequence has been published [for reviews, see Reeder, 1990; Sollner-Webb and Mougey, 1991; it should be noted that a well characterized protein involved in Pol II function, TBP or TATA binding protein, has now been shown to be a component of SL1, another known Pol I transcription factor purified by Tjian and his coworkers (Comai *et al.*, 1992), but RRN6 is different from TBP, as expected]. Therefore, the RRN6 protein may represent a new transcription factor used by the yeast Pol I system.

Three other genes, *RRN4*, *RRN5* and *RRN10*, have also been cloned and sequenced [sequencing of *RRN5* and *RRN10* were carried out by E. Fantino and T. Nguyen in our laboratory, respectively]. As mentioned earlier, the protein encoded by *RRN4* was found to be the A12.2 subunit of Pol I and to be homologous to the B12.6 subunit of Pol II (Nogi *et al.*, 1993). The nature of the *RRN5* and *RRN10* gene products is currently being studied.

CONCLUDING REMARKS

In this article, we have reviewed our recent studies on the synthesis of rRNA in the yeast *S. cerevisiae*. Our genetic approaches, analysis of suppressors of Pol I ts mutations and the use of the *GAL7*-35SrDNA for new mutant isolation, have proved to be productive. These approaches have taken advantage of a unique feature of eukaryotes which is absent in prokaryotes, namely, the presence of Pol I as the enzyme dedicated solely to transcription of large rRNA genes. It should be noted that, despite extensive studies for many years, no clearly defined *E. coli* RNA polymerase mutants have been isolated that show defects specifically in the transcription of rRNA (or stable RNA) genes. By mutagenizing isolated genes for the Pol I-specific subunits, it should be easy to isolate conditionally lethal mutants as we have done with *RPA190* (Wittekind *et al.*, 1988) or with *RPA135* (our unpublished experiments). Suppressor analysis should then give information on components interacting with Pol I as our studies on the suppressors of *rpa190* ts have already demonstrated. Since the A135 subunit is a homologue of the ß subunit of the *E. coli* enzyme, suppressor analysis of well defined *rpa135* ts mutants may be of particular interest in this respect.

The construction of a Pol I deletion/*GAL7*-35SrDNA strain demonstrated that the only essential function of Pol I is to transcribe the larger rRNA genes. Starting in a wild-type background, the *GAL7*-35SrDNA system has proved to be effective in identifying new genes involved in rRNA synthesis. Since the distribution of independently isolated mutants into different genes as analyzed so far does not indicate saturation of possible *rrn* genes, continuation of the mutant isolation using this system is expected to lead to discovery of more genes that are specifically involved in 35S rRNA synthesis.

The Pol I-deletion/*GAL7*-35SrDNA system may also be useful to define regulatory systems acting on the the Pol I transcription machinery. Regulation acting uniquely on the Pol I machinery may be absent in strains synthesizing rRNA from the *GAL7*-35SrDNA fusion gene by Pol II. For example, the Pol I-deletion/*GAL7*-35SrDNA strains exhibit a very long time lag in the transition from the stationary phase to the exponentially growing phase, suggesting some defects in the regulation of rRNA synthesis in response to altered nutritional conditions or to the state of the cell cycle (our unpublished observations).

The Pol I-deletion/*GAL7*-35SrDNA system should also be useful for structure-function studies of ribosomes as well as studies of rRNA processing and ribosome assembly. In *E. coli*, mutagenesis of rRNA genes carried on a plasmid has been done extensively and the expression of the mutated genes has been examined against the background of ribosomes derived from chromosomal rRNA genes using various technical manipulations, e.g., by incorporating an antibiotic resistant markers in the plasmid-encoded rRNA genes and testing ribosome function in the presence of the antibiotics (e.g., see Triman *et al.*, 1989). In yeast, Planta and his coworkers devised a tagged rRNA gene system to study mutagenized rRNA genes with respect to ribosome assembly and function (Musters *et al.*, 1989). With the present system it should be simpler to carry out mutagenesis studies since all the large rRNAs in a cell are derived from the *GAL7*-35SrDNA fusion gene on a plasmid in this system.

Our work on *SRP1* strongly suggests the importance of the nucleolar structure for rRNA synthesis. It has been observed by previous investigators that rRNA genes are associated with an insoluble subnuclear fraction comprising the nuclear envelope/matrix (Davis *et al.*, 1983; Dickinson *et al.*, 1990; Jackson *et al.*, 1984; Keppel, 1986; Pardol and Vogelstein, 1980). In addition, most of the actively functioning Pol I was found to be bound to a similar insoluble nuclear envelope/matrix fraction (Dickinson *et al.*, 1990). Based on such observations, Cook and his coworkers proposed that Pol I (and other nuclear RNA polymerases) is bound to the nuclear envelope/matrix and the DNA template moves past the bound Pol I (and other nuclear RNA polymerases) (Dickinson *et al.*, 1990; for a review, see Cook, 1989). However, most of the previous experiments supporting this proposal were *in vitro* biochemical fractionation experiments and the possibility of artifactual interaction between Pol I (or rDNA) and insoluble materials in the nuclear envelope/matrix fraction has not been completely eliminated. No information is available on the chemical nature of the components proposed to interact with Pol I (or rDNA). Continuation of studies on SRP1 and further genetic analysis of the present yeast system will help clarify the nature of individual nucleolar (or nuclear) structural components interacting with rDNA or Pol I and dissect the complex structural organization of the nucleolus that must be important for rDNA transcription and its regulation.

ACKNOWLEDGEMENTS

The work described in this article was supported by grants from the National Institute of Health (R37GM35949) and from the National Science Foundation (DMB8904131). We thank Dr. J. Keener for reading the manuscript and S. Pfleiger for help in the preparation of the manuscript.

REFERENCES

Clark, M.W., Yip, M.L.R., Campbell, J., and Abelson, J., 1990, SSB-1 of the yeast *Saccharomyces cerevisiae* is a nucleolar-specific, silver binding protein that is associated with the snR10 and snR11 small nuclear RNAs, *J. Cell Biol.* 111:1741.

Comai, L., Tanese, N., and Tjian, R., 1992, The TATA-binding protein and associated factors are integral components of the RNA polymerase I transcription factor, SL1, *Cell* 68:965.

Cook, P.R., 1989, The nucleoskeleton and the topology of transcription, *Eur. J. Biochem.*, 185:487.

Davis, A.H., Reudelhuber, T.L., and Garrard, W.T., 1983, Variegated chromatin structures of mouse ribosomal RNA genes, *J. Mol. Biol.*, 167:133.

De La Torre, C., and Gimenez-Martin, G., 1982, The nucleolar cycle, in: "The Nucleolus," E.G. Jordan and C.A. Cullis, eds., Cambridge University Press, Cambridge.

Dequard-Chablat, M., Riva, M., Carles, C., and Sentenac, A., 1991, *RPC19*, the gene for a subunit common to yeast RNA polymerases A (I) and C (III), *J. Biol. Chem.* 266:15300.

Dickinson, P., Cook, P.R., and Jackson, D.A., 1990, Active RNA polymerase I is fixed within the nucleus of HeLa cells. *EMBO J.*, 9:2207.

Dvorkin, N., Clark, M.W., and Hamkalo, B.A., 1991, Ultrastructural localization of nucleic acid sequences in *Saccharomyces cerevisiae* nucleoli, *Chromosoma*, 100:519.

Girard, J.-P., Lehtonen, H., Caizergues-Ferrer, M., Amalric, F., Tollervey, D., and Lapeyre, B., 1992, GAR1 is an essential small nucleolar RNP protein required for pre-rRNA processing in yeast, *EMBO J.*, 11:673.

Henriquez, R., Blobel, G., and Aris, J.P., 1990, Isolation and sequencing of NOP1. A yeast gene encoding a nucleolar protein homologous to a human autoimmune antigen., *J. Biol. Chem.*, 265:2209.

Huet, J., Sentenac, A., and Fromageot, P., 1982, Spot-immunodetection of conserved determinants in eukaryotic RNA polymerases, *J. Biol. Chem.* 257:2613.

Hughes, J.M.X., and Ares, M., 1991, Depletion of U3 small nucleolar RNA inhibits cleavage in the 5' external transcribed spacer of yeast pre-ribosomal RNA and impairs formation of 18S ribosomal RNA, *EMBO J.*, 10:4231.

Jackson, D.A., Cook, P.R., and Patel, S.B., 1984, Attachment of repeated sequences to the nuclear cage, *Nucleic Acids Res.*, 12:6709.

Kass, S., Tyc, K., Steitz, J.A., and Sollner-Webb, B., 1990, The U3 small nucleolar ribonucleoprotein functions in the first step of preribosomal RNA processing, *Cell*, 60:897.

Keppel, F., 1986, Transcribed human ribosomal RNA genes are attached to the nuclear matrix, *J. Mol. Biol.*, 187:15.

Martindale, D.W., 1990, A conjugation-specific gene (cnjC) from *Tetrahymena* encodes a protein homologous to yeast RNA polymerase subunits (RPB3, RPC40) and similar to a portion of the prokaryotic RNA polymerase alpha subunit (rpoA), *Nucleic Acids Res.* 18:2953.

McCusker, J.H., Yamagishi, M., Kolb, J.M., and Nomura, M., 1991, Suppressor analysis of temperature-sensitive RNA polymerase I mutations in *Saccharomyces cerevisiae*: suppression of mutations in a zinc-binding motif by transposed mutant genes, *Mol. Cell. Biol.*, 11:746.

Musters, W., Venema, J., van der Linden, G., van Heerikhuizen, H., Klootwijk, J., and Planta, R.J., 1989, A system for the analysis of yeast ribosomal DNA mutations. *Mol. Cell. Biol.*, 9:551.

Nogi, Y., Yano, R., Dodd, J., Carles, C., and Nomura, M., 1993, Gene *RRN4* in *Saccharomyces cerevisiae* encodes the A12.2 subunit of RNA polymerase I, and is essential only at high temperatures, *Mol. Cell. Biol.*, in press.

Nogi, Y., Vu, L., and Nomura, M., 1991a, An approach for isolation of mutants defective in 35S ribosomal RNA synthesis in *Saccharomyces cerevisiae*, *Proc. Natl. Acad. Sci. USA*, 88:7026.

Nogi, Y., Yano, R., and Nomura, M., 1991b, Synthesis of large rRNAs by RNA polymerase II in mutants of *Saccharomyces cerevisiae* defective in RNA polymerase I, *Proc. Natl. Acad. Sci. USA*, 88:3962.

Pardoll, D.M., and Vogelstein, B., 1980, Sequence analysis of nuclear matrix associated DNA from rat liver, *Exp. Cell Res.*, 128:466.

Reeder, R.H., 1990, rRNA synthesis in the nucleolus, *Trends in Genetics*, 6:390.

Riggs, D.L., and Nomura, M., 1990, Specific transcription of *Saccharomyces cerevisiae* 35S rDNA by RNA polymerase I *in vitro*, *J. Biol. Chem.* 13:7596.

Sawadogo, M., and Sentenac, A., 1990, RNA polymerase B (II) and general transcription factors, *Annu. Rev. Biochem.*, 59:711.

Schimmang, T., Tollervey. D., Kern, H., Frank, R., and Hurt, E.C., 1989, A yeast nucleolar protein related to mammalian fibrillarin is associated with small nucleolar RNA and is essential for viability. *EMBO J.*, 8:4015.

Sentenac, A., 1985, Eukaryotic RNA polymerases, *Crit. Rev. Biochem.*, 18:31.

Sollner-Webb, B., and Mougey, E.B., 1991, News from the nucleolus: rRNA gene expression. *Trends in Biochem. Sci.*, 16:58-62.

Sollner-Webb, B., and Tower, J., 1986, Transcription of cloned eukaryotic ribosomal RNA genes, *Ann. Rev. Biochem.* 55:801.

Thuriaux, P., and Sentenac, A., 1992, Yeast nuclear RNA polymerases, *in*: "The Molecular and Cellular Biology of the Yeast *Saccharomyces*," J.R. Broach and J.R. Pringle, eds., Cold Spring Harbor Press, Cold Spring Harbor, in press.

Tollervey, D., Lehtonen, H., Carmo-Fonseca, M., and Hurt, E.C., 1991, The small nucleolar RNP protein NOP1 (fibrillarin) is required for pre-rRNA processing in yeast. *EMBO J.*, 10:573.

Treich, I., Carles, C., Riva, M., and Sentenac, A., 1992, RPC10 encodes a new mini subunit shared by yeast nuclear RNA polymerases, *Gene Expr.*, 2:31.

Triman, K., Becker, E., Dammel, C., Katz, J., Mori, H., Douthwaite, S., Yapijakis, C., Yoast, S., and Noller, H.F., 1989, Isolation of temperature-sensitive mutants of 16S rRNA in *Escherichia coli*, *J. Mol. Biol.*, 209:645-653.

Warner, J.R., 1982, The yeast ribosome: structure, function and synthesis, *in* "The Molecular Biology of the Yeast *Saccharomyces cerevisiae*," J. Strathern, E. Jones, and J.R. Broach, eds., Cold Spring Harbor Laboratory Press, Cold Spring Harbor.

Wittekind, M., Dodd, J., Vu, L., Kolb, J.M., Buhler, J.-M., Sentenac, A., and Nomura, M., 1988, Isolation and characterization of temperature-sensitive mutations in *RPA190*, the gene encoding the largest subunit of RNA polymerase I from *Saccharomyces cerevisiae*, *Mol. Cell. Biol.*, 8:3997.

Woychik, N.A., Lane, W.S., and Young, R.A., 1991, Yeast RNA polymerase II subunit RPB9 is essential for growth at temperature extremes, *J. Biol. Chem.*, 266:19053.

Woychik, N.A., Liao, S.-M., Kolodziej, P.A., and Young, R.A., 1990, Subunits shared by eukaryotic nuclear RNA polymerases, *Genes and Devel,*, 4:313.

Yano, R., Oakes, M., Yamagishi, M., Dodd, J.A., and Nomura, M., 1992, Cloning and characterization of *SRP1*, a suppressor of temperature-sensitive RNA polymerase I mutations, in *Saccharomyces cerevisiae*, *Mol. Cell. Biol.*, in press.

Yano, R., and Nomura, M., 1991, Suppressor analysis of temperature-sensitive mutations of the largest subunit of RNA polymerase I in *Saccharomyces cerevisiae*: a suppressor gene encodes the second-largest subunit of RNA polymerase I, *Mol. Cell. Biol.*, 11:754.

DNA ELEMENTS AND PROTEIN FACTORS INVOLVED IN THE REGULATION OF TRANSCRIPTION OF THE RIBOSOMAL RNA GENES IN YEAST

Rudi J. Planta and Tanja Kulkens

Department of Biochemistry and Molecular Biology
Vrije Universiteit
de Boelelaan 1083, 1081 HV Amsterdam, The Netherlands

INTRODUCTION

Regulation of transcription of the rRNA genes may be central to the intricate process of ribosome biosynthesis in response to environmental conditions. In *Saccharomyces cerevisiae* the rRNA genes are organized in a tandem array of ~150 units on chromosome XII (Petes, 1979; Warner, 1989). The genes encoding 17S, 5.8S and 26S rRNA are arranged in a pre-rRNA operon, which is transcribed by RNA polymerase I (Pol I) in the nucleolus (*cf.* Fig 1).

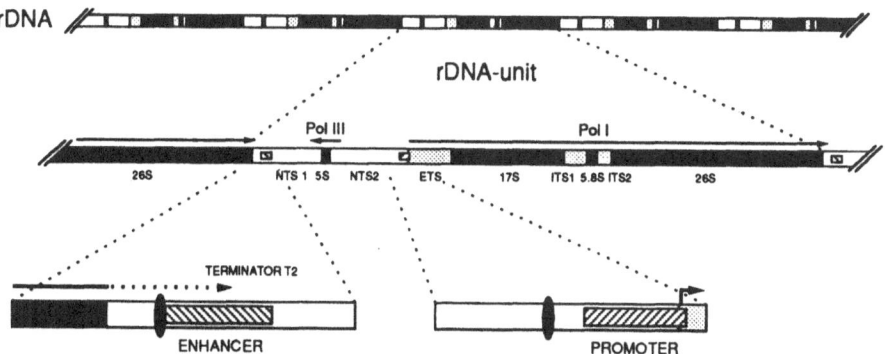

Figure 1. Organization of the rRNA genes of the yeast *Saccharomyces cerevisiae*. At the top, a part of the chromosomal tandem array of ~150 rDNA units is shown. One rDNA unit is enlarged. Regions encoding the rRNA genes are indicated by black bars. Shaded bars represent transcribed spacers and open bars represent Intergenic Spacers (ETS=External Transcribed Spacer, ITS=Internal Transcribed Spacer, NTS=Non-Transcribed Spacer). The Pol I and Pol III transcripts are indicated by arrows. At the bottom, the *cis*-acting elements involved in regulation of Pol I transcription are enlarged. The enhancer and promoter are indicated by striped boxes. The Pol I transcript proceeding to the main terminator within the enhancer is represented by a stippled arrow, which part is rapidly processed to form the mature end of the 26S rRNA. ●: binding sites for the protein RBP1/REB1.

Unlike other eukaryotes, the *S. cerevisiae* rDNA unit also contains a 5S rRNA gene, which is located within the spacer regions between the pre-rRNA operons and is transcribed by Pol III (Phillippsen et al., 1978). The *cis*-acting elements involved in yeast Pol I transcription have been extensively studied (reviewed by Raué and Planta, 1991). The boundaries and substructure of the Pol I promoter have been analyzed *in vivo* (Musters et al., 1989b) as well as *in vitro* (Kulkens et al., 1991). A region from −155 to +27 with respect to the Pol I transcription initiation site is required for accurate and efficient initiation of transcription. In addition a 170-190 bp enhancer element, responsible for a 15 to 30 fold increase in transcription has been identified in the spacer, ~2.2 kb upstream of the initiation site (Elion and Warner, 1984; 1986). The enhancer element also contains the main terminator, T2 (Kempers-Veenstra et al., 1986; Mestel et al., 1989; van der Sande et al., 1989). The location of the Pol I terminator within the enhancer has led us to propose a model that accounts for efficient recycling of Pol I molecules and/or transcription factors (Kempers-Veenstra et al., 1986). In this model, the so-called "ribomotor", the terminator/enhancer element is brought in the vicinity of the Pol I promoter by looping out the Pol I transcription unit. Pol I molecules which have terminated at T2 can immediately be passed on to the promoter by means of the enhancer. To induce the proposed association between promoters and enhancers and to stabilize such a structure (a) protein factor(s) may be involved. The ribosomal DNA binding protein RBP1/REB1, which binds the rDNA enhancer and also close to the Pol I promoter (Morrow et al., 1989; 1990; Kulkens et al., 1989) seems to be an attractive candidate for this function. In the loop structures that are supposed to be formed this way the promoter and enhancer of one and the same operon can be juxtaposed, or alternatively, promoter and enhancer of adjacent units can be brought together.

In this Chapter we will review the relevant data, which have contributed to our present knowledge of the regulation of rRNA synthesis in yeast.

INITIATION OF TRANSCRIPTION BY POL I IN YEAST

Pol I Promoter

The 5'-end of the yeast precursor rRNA transcript has been mapped. This rRNA molecule was shown to contain a 5'-triphosphate group, indicating that it indeed represents the genuine site of rDNA transcription initiation and has not yet undergone processing at its 5'-end (Klootwijk et al., 1979). The study of the yeast Pol I promoter started by the comparison of the sequences around the transcription initiation sites of four *Saccharomycetoideae*, which revealed significant homology around +1 and two additional short homologous elements at position -70 and -130 respectively (Verbeet et al., 1984). Sequence comparison between less related eukaryotic species, however, showed no homologies between the initiation regions.

The yeast Pol I promoter has been analyzed *in vivo* by transformation of yeast cells with (mutated) rDNA minigenes as well as *in vitro* using the same minigenes as templates (Musters et al., 1989b; Kulkens et al., 1991). Using a series of 5'- and 3'-deletion mutants, Musters et al. (1989b) showed that the promoter extends from −155 to +27, with 5'-deletions down to −134 and 3'-deletions up to −2 removing essential information. Analysis of the same deletion mutants in the Pol I *in vitro* transcription system provided similar results. Previously, using another type of minigene, not including the enhancer, Kempers-Veenstra et al. (1985) had shown that sequences downstream of position +15 are not required for correct initiation of transcription, while they mapped the 5'-border between positions −149 and −133. Analysis of the 5'-deletion mutants *in vitro* showed that weak transcriptional activity remains with deletions further downstream (Kulkens et al., 1991). A second significant decrease seems to occur between positions −91 and −76. However, transcriptional activity is still detectable with deletions down to −38. Transcription is no longer detectable with deletions to −26 and beyond, indicating that the 5'-border of the minimal promoter lies between these positions. The important role of the upstream part of the promoter is most pronounced under stringent *in vitro* conditions and *in vivo*. In addition, it was found that mutants with 3'-deletions up to –2 and –5, and therefore without the original start site, still allow a low level of transcription. From the length of these transcripts it was concluded that they start at a site corresponding to the original +1 site, suggesting that positioning of Pol I is directed by upstream sequences rather than by recognition of the primary sequence around the start site. A similar conclusion had been reached earlier for *Acanthamoeba* (Paule et al., 1991).

Comparison of the architecture of the yeast Pol I promoter with the generalized structure for a eukaryotic Pol I promoter (reviewed by Sollner-Webb and Mougey, 1991) reveals that yeast does not notably differ from higher eukaryotes. In general, a eukaryotic Pol I promoter is about 150 bp in size and consists of a minimal (core) promoter of ~35 bp upstream of the start site and an upstream element extending to about –150, which is stimulatory under stringent *in vitro* conditions and *in vivo*.

To investigate the internal organization of the yeast Pol I promoter Musters et al. (1989b) constructed linker scanning mutants, that traverse the Pol I promoter region and comprise between 5 and 12 clustered point mutations. Analysis of the transcription of these mutants was performed *in vivo* in transformed yeast cells (Musters et al., 1989b) as well as in the Pol I *in vitro* transcription system (Kulkens et al., 1991). The yeast Pol I promoter was found to consist of three domains, which is somewhat deviating from the two-domain-structure found in higher eukaryotes. Domain I (–28 to +8) in yeast may be equivalent to the core promoter element (CPE) found in other eukaryotes and domain III (–146 to –91) may be similar to the upstream control element (UCE). It is not known whether domain II (–76 to –51) must be regarded as belonging to a (bipartite) CPE or UCE.

Spacing and orientation between the promoter elements was found to be extremely important for efficient initiation of transcription. Musters et al. (1989b) showed that an insertion of 4 bp between domain I and II, which distorts the orientation of these domains by about half a DNA helix, drastically decreases transcription *in vivo*. Choe et al. (1992) tested a set of spacing mutants *in vitro*, which push apart and pull together the promoter domains. Changing the spacing at position –129/–102 by 5 bp drastically affects promoter activity. When spacing is increased or decreased 10 bp (one DNA helix turn), however, partial rescue of promoter activity is observed. These results suggest that the spacing between domain III and II is critical and that proteins binding to these domains must be positioned on the correct face of the DNA helix for maximal activity. Using a set of spacing mutants at position –49/–22 Choe et al. (1992) showed that correct spacing between domain II and I of the yeast promoter is also of critical importance. At this position the increase or decrease of spacing by 10 bp does not cause rescue of promoter activity.

Transcription Initiation Factors

In yeast not much is known about the transcription initiation factors yet. Our initial analyses of three crude fractions capable to reconstitute Pol I transcription *in vitro* (D.L. Riggs and T. Kulkens, unpublished results; *cf.* Kulkens, 1992) revealed characteristics similar to those of the transcription initiation factors UBF and SL1 described for other eukaryotes (Sollner-Webb and Mougey, 1991). The stimulatory activity found in one of the fractions is likely to be a yeast UBF-homologue, since this activity is in particular noticeable under stringent conditions and seems to be mediated via the upstream domain III of the promoter. This fraction may also contain a factor involved in the positioning of Pol I at +1. The DNase I footprinting pattern of another fraction is highly reminiscent of the footprints obtained with crude SL1-containing fractions in higher eukaryotes. Also, this fraction was shown to be essential for a minimal reconstituted system, and thus is likely to exert its action via the core promoter (domain I). Finally, the third fraction contains Pol I and maybe a factor which is involved in regulation of Pol I transcription in response to the growth rate of the yeast cells. A similar regulatory factor is found in other eukaryotes as well, although its mechanism of action is still unclear (Reeder, 1990; Sollner-Webb and Mougey, 1991).

TERMINATION OF TRANSCRIPTION

The study of the site of termination of transcription is always thwarted by the difficulty to distinguish 3'-end formation as a result of termination from 3'-ends created by processing of longer transcripts, in particular when processing occurs very soon. When a test assay is insufficient for detection of rapidly processed or unstable transcripts, it is inadequate to determine the genuine site of termination of transcription. Various assays therefore, have been used to obtain information on the most probable site of termination of Pol I transcription.

Veldman et al. (1980) showed that the 3'-end of the 37S precursor rRNA in yeast is 7 nucleotides longer than that of the mature 26S rRNA. Also, this 3'-end was shown to be produced via processing of a longer transcript, meaning that Pol I transcription continues further within the spacer region (Kempers-Veenstra et al., 1986). These processing events have been shown to occur also *in vitro* (Yip and Holland, 1989). Analysis of minigene transcription in a yeast mutant deficient in processing, showed transcripts with 3'-ends at +15 and +45 relative to the 3'-end of the 26S rRNA gene (Kempers-Veenstra et al., 1986). Using 5'-deletion analysis in minigenes Kempers-Veenstra et al. (1986) showed that removal of sequences from -149 to +18 required for the formation of the 3-ends of 26S rRNA and the precursor Pol I transcript, causes the appearance of transcripts ending at +50 and +210. The 3'-end at +210 remains unchanged upon transformation of the deletion mutants into the processing deficient strain. Deletion to +111 still shows the 3'-end at +210. A mutant with a deletion to +282 identified another candidate for the Pol I terminator at about +750. The 3'-end formating sites were called T0 (3'-ends of 37S precursor rRNA and mature 26S rRNA, generated by processing), T1 (+15 to +50), T2 (+210) and T3 (+750). Van der Sande et al. (1989) extended the termination studies by *in vitro* run-on transcription analysis and functional analysis *in vivo* of various parts of the intergenic spacer to distinguish processing from termination at these sites. T1 was found to behave as a processing site. T2 was shown to be an efficient, genuine Pol I terminator. T3 was identified to consist of two 3'-end generating sites, being T3A (+690) and T3B (+950). T3A and T3B may function as fail safe terminators, preventing Pol I molecules that did not recognize T2 from transcribing into the 5S rRNA gene. Johnson and Warner (1991) also studied termination of transcription by yeast Pol I. Using a different kind of minigene system they essentially confirmed the results obtained by van der Sande et al. (1989). The latter authors also showed that the sequences directly flanking T2 are conserved among different cloned yeast rDNA units, but that this conserved part is localized in a region of the intergenic spacer, which shows considerable sequence heterogeneity from one unit to another. However, T2 was mapped at precisely the same site in three different rDNA units. S1 mapping in T-rich regions on the other hand can lead to artefactual 3'-ends due to the AT-regions being sensitive to S1 nuclease. The stretch of T residues may be involved in influencing the efficiency of T2. The fact that the genuine Pol I terminator T2 is located within the enhancer element (see next section) has led to the proposition that within the chromosomal context the terminator/enhancer element is in the close vicinity of the Pol I promoter, thereby facilitating the immediate transfer of Pol I molecules from the terminator to the promoter.

REGULATION OF POL I TRANSCRIPTION

The regulation of transcription of the rRNA genes is an important parameter in the overall mechanism of regulation of ribosome biogenesis. Specific DNA elements in the (intergenic spacer of the) rDNA units as well as specific protein factors may play an important role in the control of rRNA synthesis. In the rDNA of other eukaryotes several

DNA elements have been shown to be able to stimulate Pol I transcription (Sollner-Webb and Tower, 1986). In the intergenic spacers of these eukaryotes duplications of the promoter are found, called spacer promoters, which are able to generate spacer transcripts and in addition can stimulate transcription starting at the genuine promoter. Furthermore, sometimes repetitive elements are present in the intergenic spacers, such as the 60/81-bp elements in *Xenopus* (*cf.* Reeder, 1990), which act as enhancers. In contrast to these eukaryotes, the spacer in *S. cerevisiae* rDNA has no repeated sequences and no promoter elements that can act *in vivo*. The spacer promoter, described by Swanson et al. (1985), which under centain conditions can promote Pol I transcription *in vitro*, is currently considered to be an artefact.

The sole stimulatory element found in yeast rDNA is a 170-190 bp enhancer element, identified by Elion and Warner (1984, 1986), located in the intergenic spacer, about 100 bp downstream of the rRNA operon and about 2.2 kb upstream of the next operon. This enhancer element can stimulate rRNA transcription about 20-30 fold, and (as described above) also contains the genuine Pol I terminator. In artificial minigene constructs the stimulating effect of the enhancer was found to be independent of its position or orientation, although the degree of stimulation is somewhat influenced by the sequence context. In this respect the yeast rDNA enhancer may be unique among the eukaryotic Pol I stimulating elements. Attempts to define specific subelements in this rather long enhancer, using either deletion analysis in minigenes (Mestel et al., 1989; Warner, 1989) or mutational analysis of enhancer sequences in templates transcribed by Pol I *in vitro* (Lue and Kornberg, 1990; Lorch et al., 1990) have not resulted in unambiguous conclusions. No one region of the 190-bp sequence is essential for full function.

The best defined part of the enhancer is the extreme 5'-region, which appears to be the binding site for an abundant and ubiquitous protein, designated REB1 (Morrow et al., 1989; 1990) or RBP1 (Kulkens et al., 1989). Another protein, called REB2 (Morrow et al., 1989) and later on identified to be ABF1, was found to bind about 20 bp downstream of the RBP1/REB1 binding site. However, the ABF1 binding site is not found in all rDNA units as cloned in different laboratories (Kulkens et al., 1989).

The RBP1/REB1 binding site in the 5'-end of the enhancer has an 8 bp core (CCGGGTAA) which is also present about 60 bp upstream of the 5'-end of the Pol I promoter, although in the opposite orientation (Kulkens et al., 1989; Morrow et al., 1989). This observation has brought us to the suggestion that the protein RBP1/REB1 may have an important function in the structural organization of the rDNA units in the nucleolus, for instance by anchoring consecutive enhancers and promoters to the nucleolar matrix in such a way that these elements are brought in close proximity to each other.

Analysis of the RBP1/REB1 binding sites by using episomal rDNA minigenes (Kulkens et al., 1989) or mini rDNA repeats integrated into a non-rDNA locus did not reveal a role for RBP1/REB1 in rRNA transcription (*cf.* Warner et al., 1990), although some results may suggest that RBP1/REB1 is involved in termination at a site different from T2 (Johnson and Warner, 1991). Also *in vitro* transcription analysis has suggested in some cases that RBP1/REB1 may function in transcription termination (Schultz and coworkers, unpublished results) but in this system the strong DNA-binding protein RBP1/REB1 may simply act as a road block for Pol I and thus induce artificial termination.

Obviously these experimental systems cannot account for all specific features of rDNA transcription in its natural locus. Specifically, the high degree of tandem repetition of the rDNA genes and their presence in a specialized area of the nucleus, the nucleolus, containing high concentrations of specific proteins important for rDNA transcription, cannot be mimicked by the systems exploited thusfar. We have sought a way to circumvent the abovementioned experimental problems and developed an experimental system that optimally minics the natural context of pol I transcription. To this end we used rDNA units carrying oligonucleotide tags in both their 17S and 26S rRNA genes (as developed by Musters et al., 1989a) for integration into the rDNA locus. The tags allowed for specific detection of only the transcripts from such a unit, and made it possible to study the effects of rDNA mutations on Pol I transcription within their natural chromosomal context. Tagged integrated rDNA units were shown to be transcribed with similar efficiency as normal, endogenous rDNA units (Kulkens et al., 1992).

Using this novel system we showed that the Pol I enhancer, initially identified using artificial minigenes, indeed functions as a stimulatory element when assayed within the rDNA locus. Furthermore, we found that enhancers exerted their function in two directions, mainly on their two most proximal rRNA operons. Deletion of the sequences between the enhancer and the Pol I promoter in the tagged, integrated unit indicated that this part of the intergenic spacer contains no other transcriptional regulatory elements for Pol I (Kulkens et al., 1992).

We also applied the system to study the function of the rDNA binding protein RBP1/ REB1 and showed that this protein is involved in efficient Pol I transcription within the chromosomal context. The protein is hypothesized to play a crucial role in keeping the chromosomal rDNA units in an optimal spatial configuration by anchoring consecutive enhancers and promoters to the nucleolar matrix in a highly ordered, linear fashion (see Figure 2), either directly (Figure 2A) or indirectly via some unknown protein (Figure 2B).

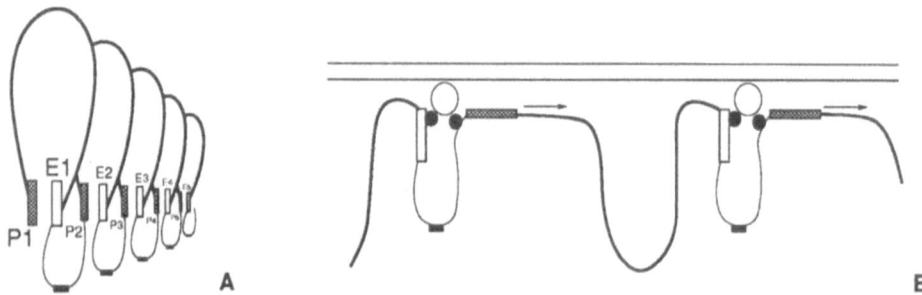

Figure 2. Model for the regulation of transcription by Pol I. (A) Five consecutive rDNA units are indicated. Large loops (thick black lines) represent Pol I operons, whereas the smaller loops (thin lines) represent the intergenic spacer, carrying the 5S rRNA gene (small black box). The Pol I promoter (P) and enhancer (E) are indicated by shaded and open boxes, respectively. (B) Matrix-attachment-model, in which the rDNA units are anchored to the nucle(ol)ar matrix via interaction of RBP1/REB1 (black circles) with some unknown (protein) component (white circles).

A similar matrix-attachment model has been proposed for Pol I transcription in HeLa cells by the group of Cook (Jackson and Cook, 1985; Cook, 1989; Dickinson et al., 1990). The DNA loops formed are thought to be organized in such a way as to locate all Pol I transcription units at one side and the 5S rRNA genes (transcribed by Pol III) at the other side. Terminating Pol I molecules can, in this model, by some as yet unknown mechanism be handed over to either the most proximal upstream or downstream promoter. It seems unlikely that just a high local concentration of Pol I molecules is responsible for the enhancing effect since this would result in additional activation of more distally located promoters. Overall regulation of rRNA transcription is thought to be exerted mainly at the level of the number of promoters and terminator/enhancer elements attached to the nucle(ol)ar matrix, in accordance with the electron microscopical finding that a given rDNA unit is either in a transcriptional on or off mode with little modulation at the level of the number of initiating Pol I molecules (Hamkalo, 1985).

When RBP1/REB1 binding to the enhancer in the tagged operon is abolished, this enhancer cannot be placed in its optimal position as in the natural context and thus is not able to function optimally. When the binding site near the promoter is mutated, the interaction of this promoter with both the upstream and downstream enhancer is affected, which explains the observed stronger negative effect of this mutation on transcription (Kulkens et al., 1992).

In the *in vivo* situation up- and down-regulation of Pol I transcription may very well be regulated by the number of rDNA transcription units organized in the proposed loop structure. This regulation is, however, not necessarily the effect of the direct action of RBP1/REB1 itself. In fact it is probably that regulation is exerted via another protein(s) whose state of post-translational modification reflects the metabolic state of the cell and, as a result, can or cannot couple RBP1/REB1 bound rDNA units to the nucleolar matrix. This proposal is instigated by the fact that the RBP1/REB1 protein is supposed to play a much more general role in the yeast cell than solely its involvement in the regulation of rDNA transcription. Binding sites for RBP1/REB1 have, for instance, also been identified in the upstream region of several Pol II transcribed genes, as well as in the centromere *CEN4* and the subtelomeric X and Y regions (Chasman et al., 1990; Wang et al., 1990). RBP1/REB1 appears to be identical to factor Y (Fedor et al., 1988), GRF2 (Chasman et al., 1990) and QBP (Brandl and Struhl, 1990). In some cases the protein stimulates transcription, whereas in others it acts as a repressor. According to Chasman et al. (1990) factor Y functions by influencing the chromatin structure and creating a nucleosome-free region surrounding its binding site in the promoter region. In view of all these possible functions it is not surprising that RBP1/REB1 is an essential protein for growth of the yeast cell (Ju et al., 1990). It has been shown that RBP1/REB1 is a phosphoprotein, suggesting that phosphorylation/dephosphorylation might be a mechanism by which its various (nucleolar) functions and/or DNA binding could be regulated (Morrow et al., 1990).

Unlike the rDNA in other eukaryotes the intergenic spacer in yeast rDNA contains the 5S rRNA gene, which is transcribed by Pol III (Phillippsen et al., 1978). Attachment of the rDNA units by RBP1/REB1 to the nucle(ol)ar matrix and the formation of a series of loops (*cf.* Figure 2), therefore, may be necessary also to arrange both transcription units in different compartments of the nucleus.

In summary, the results obained using our newly developed system for studying Pol I transcription within the chromosomal context (Kulkens et al., 1992) clearly show that in yeast the sole element involved in stimulation of Pol I transcription is the terminator/ enhancer element. This single yeast Pol I enhancer differs from the Pol I enhancer elements present in the rDNA of other organisms, in that it does not display a (partial) sequence identity with the gene promoter. Furthermore, the enhancer does contain the main Pol I termination site, which is an essential feature in models describing its mode of action. The intimate coupling of terminator and enhancer may imply that, in contrast to the Pol I enhancer elements present in the rDNA of other eukaryotes, the yeast Pol I enhancer will not function in the opposite orientation within the chromosomal context. This idea follows from the observation that the yeast Pol I terminator T2 is not functioning in the opposite orientation (van der Sande et al., 1989; Mestel et al., 1989).

Furthermore, our studies have provided convincing evidence that the protein RBP1/REB1 is an important structural factor, required for efficient Pol I transcription within the chromosomal context. This rDNA binding protein likely promotes the interplay of enhancer and promoter and transcription factors within the specific tandem array of rDNA units present in the nucleolus.

ACKNOWLEDGEMENTS

We thank Carine van der Sande, Karin Bergkamp-Steffens, Jean-Pierre Sibeyn and Albert Dekker for their contributions during the initial stages of the pORIS-project, and Harm van Heerikhuizen for helpful discussions. The work presented in this paper was continuously supported in part by the Netherlands Foundation for Chemical Research, with financial aid from the Netherlands Organization for Scientific Research (N.W.O.).

REFERENCES

Brandl, C.J., and Struhl, K., 1990, *Mol. Cell. Biol.*, 10:4256.

Chasman, D.I., Lue, N.F., Buchman, A.R., LaPointe, J.W., Lorch, Y., and Kornberg, R.D., 1990, *Genes Dev.*, 4:503.

Choe, S.Y., Schultz, M.C., and Reeder, R.H., 1992, *Nucleic Acids Res.*, 20:279.

Cook, P.R., 1989, *Eur. J. Biochem.*, 185:487.

Dickinson, P., Cook, P.R., and Jackson, D.A., 1990, *EMBO J.*, 9:2207.

Elion, E.A., and Warner, J.R., 1984, *Cell*, 39:663.

Elion, E.A., and Warner, J.R., 1986, *Mol. Cell. Biol.*, 6:2089.

Fedor, M.J., Lue, N.F., and Kornberg, R.D., 1988, *J. Mol. Biol.*, 204:109.

Hamkalo, B.A., 1985, *Trends in Genet.*, 9:255.

Jackson, D.A., and Cook, P.R., 1985, *EMBO J.*, 4:919.

Johnson, S.P., and Warner, J.R., 1991, *Mol. Cell. Biochem.*, 104:163.

Ju, Q., Morrow, B.E., and Warner, J.R., 1990, *Mol. Cell. Biol.*, 10:5226.

Kempers-Veenstra, A.E., Musters, W., Dekker, A.F., Klootwijk, J., and Planta, R.J., 1985, *Curr. Genet.*, 10:253.

Kempers-Veenstra, A.E., Oliemans, J., Offenberg, H., Dekker, A.F., Piper, P.W., Planta, R.J., and Klootwijk, J., 1986, *EMBO J.*, 5:2703.

Klootwijk, J., de Jonge, P., and Planta, R.J., 1979, *Nucleic Acids Res.*, 6:27.

Kulkens, T., van Heerikhuizen, H., Klootwijk, J., Oliemans, J., and Planta, R.J., 1989, *Curr. Genet.*, 16:351.

Kulkens, T., Riggs, D.L., Heck, J.D., Planta, R.J., and Nomura, M., 1991, *Nucleic Acids Res.*, 19:5363.

Kulkens, T., van der Sande, C.A.F.M., Dekker, A.F., van Heerikhuizen, H., and Planta, R.J., 1992, *EMBO J.*, 11 (in press).

Kulkens, T., 1992, *Ph.D. Thesis Free University Amsterdam*, FEBO, Enschede (NL).

Lorch, Y., Lue, N.F., and Kornberg, R.D., 1990, *Proc. Natl. Acad. Sci. USA*, 87:8202.

Lue, N.F., and Kornberg, R.D., 1990, *J. Biol. Chem.*, 265:18091.

Mestel, R., Yip, M., Holland, J.P., Wang, E., Kang, J., and Holland, M.J., 1989, *Mol. Cell. Biol.*, 9:1243.

Morrow, B.E., Johnson, S.P., and Warner, J.R., 1989, *J. Biol. Chem.*, 264:9061.

Morrow, B.E., Ju, Q., and Warner, J.R., 1990, *J. Biol. Chem.*, 265:20778.

Musters, W., Venema, J., van der Linden, G., van Heerikhuizen, H., Klootwijk, J., and Planta, R.J., 1989, *Mol. Cell. Biol.*, 9:551.

Musters, W., Knol, J., Maas, P., Dekker, A.F., van Heerikhuizen, H., and Planta, R.J., 1989, *Nucleic Acids Res.*, 17:9661.

Paule, M.R., Bateman, E., Hoffman, L., Iida, C., Imboden, M., Kubaska, W., Kownin, P., Li, H., Lofquist, A., Risi, P., Yang, Q., and Zwick, M., 1991, *Mol. Cell. Biochem.*, 104:119.

Petes, T.D., 1979, *Proc. Natl. Acad. Sci. USA*, 76:410.

Phillippsen, P., Thomas, M., Kramer, A., and Davis, R.W., 1978, *J. Mol. Biol.*, 123:387.

Raué, H.A., and Planta, R.J., 1991, *Progr. Nucleic. Acid Res. Mol. Biol.*, 41:89.

Reeder, R.H., 1990, *Trends in Genet.*, 6:390.

Sollner-Webb, B., and Tower, J., 1986, *Ann. Rev. Biochem.*, 55:801.

Sollner-Webb, B., and Mougey, E.B., 1991, *Trends Biochem. Sci.*, 16:58.

Swanson, M.E., Yip, M., and Holland, M.J., 1985, *J. Biol. Chem.*, 260:9905.

Van der Sande, C.A.F.M., Kulkens, T., Kramer, A.B., de Wijs, I.J., van Heerikhuizen, H., Klootwijk, J., and Planta, R.J., 1989, *Nucleic Acids Res.*, 17:9127.

Van der Sande, C.A.F.M., Kwa, M., van Nues, R.W., van Heerikhuizen, H., Raué, H.A., and Planta, R.J., 1992, *J. Mol. Biol.*, 223:899.

Veldman, G.M., Klootwijk, J., de Jonge, P., Leer, R.J., and Planta, R.J., 1980, *Nucleic Acids Res.*, 8:5179.

Verbeet, M.P., Klootwijk, J., van Heerikhuizen, H., Fontijn, R.D., Vreugdenhil, E., and Planta, R.J., 1984, *Nucleic Acids Res.*, 12:1137.

Wang, H., Nicholson, P.R., and Stillman, D.J., 1990, *Mol. Cell. Biol.*, 10:1743.

Warner, J.R., 1989, *Microbiol. Rev.*, 53:256.

Warner, J.R., Baronas-Lowell, D.M., Eng, F.J., Johnson, S.P., Ju, Q, and Morrow, B.E., 1990, *in:* "The Ribosome: Structure, Function and Evolution", W.E. Hill et al., eds., Am. Soc. Microbiol., Washington, Chapter 38:443.

Yip, M.T., and Holland, M.J., 1989, *J. Biol. Chem.*, 264:4045.

GENETIC APPROACHES TO THE STUDY OF EUKARYOTIC RIBOSOMES

Jonathan R. Warner, Josep Vilardell, Bernice E. Morrow, Qida D. Ju, Francis J. Eng, Mariana D. Dabeva, and Lefa E. Alksne

Department of Cell Biology
Albert Einstein College of Medicine
Bronx, NY 10461

PREFACE

The ribosome is an object of ancient beauty and function. As we seek to understand the ribosome and its place in the physiology of the cell, we must use every technique in our armamentarium. One of these is genetics. Amongst the eukaryotes, the yeast *Saccharomyces cerevisiae* has proved particularly useful for the application of genetics to the study of ribosome function and biosynthesis. This paper, which we dedicate to the memory of Prof. H.-G. Wittmann, describes our recent work on genetic approaches to the *S. cerevisiae* ribosome.

TRANSCRIPTION OF RIBOSOMAL RNA

The transcription of ribosomal RNA genes differs from that of most other genes in two respects: i) it employs a special enzyme, RNA Polymerase I; ii) there are roughly 140 identical copies of the gene in a single tandem array. To bring about the efficient transcription of these genes *S. cerevisiae* has evolved an enhancer element that differs from any described for the transcription of genes by RNA Polymerase II (Elion and Warner, 1984; Elion and Warner, 1986). It is found near the end of each transcription unit, where at least part of it is transcribed. It can enhance transcription not only of a downstream gene but also of an upstream gene, at a distance of more than six kb (Johnson and Warner, 1989). Because of its proximity to the termination of transcription, several models have been proposed describing how the enhancer might couple the termination of one transcription event with the initiation of another, on the same or an adjacent gene (Kempers-Veenstra et al., 1986; Elion and Warner, 1987; Johnson and Warner, 1989).

We have attempted to determine whether the presence of the enhancer leads inevitably to a stimulation of rRNA transcription, or whether it is involved in the regulation of transcription. An example of regulation is the 2-3 fold stimulation of rRNA transcription that occurs when cells are shifted from a nonfermentable carbon source, such as ethanol, to glucose (Kief and Warner, 1981). The effect of the enhancer on rRNA transcription in different carbon sources was assayed using a construct as shown in Figure 1, where two rRNA genes, most of whose sequences have been replaced by bacteriophage T7 sequences (Johnson and Warner, 1989), are immersed in the normal non-transcribed spacer sequences, with or without an enhancer.

The Translational Apparatus, Edited by K.H. Nierhaus
et al., Plenum Press, New York, 1993

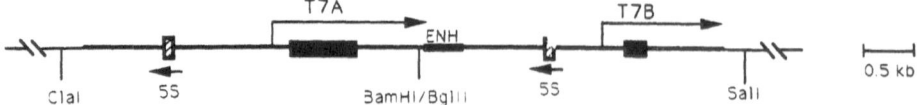

Figure 1. Diagram of the construct, inserted in the single copy vector YCP50, used for the experiment described in Table I.

Table I

	A gene		B gene	
	G/E	D	G/E	D
+ Enhancer	60	100	21	100
- Enhancer	14	13	10	11

The relative rate of transcription of the marked rRNA transcripts was assessed by quantitative Northern analysis of the RNA isolated from cells growing in log phase in nutrient broth containing as a C source either 2% glycerol and 2% ethanol (G/E) or 2% dextrose (D). The data are related to the value obtained with the enhancer in glucose, set arbitrarily to 100 (from Morrow, Johnson, and Warner in press).

Table I shows that in the absence of the enhancer, the transcription of the test genes is independent of the carbon source. On the other hand, in the presence of the enhancer, there is a substantial stimulation of transcription when cells are growing in glucose. This occurs very rapidly, reaching the steady state value in less than 30 minutes after addition of glucose (data not shown). It is interesting that the effect of the enhancer differs in the upstream and the downstream direction. For the upstream gene, the enhancer stimulates more than four fold in G/E and only an additional two-fold in glucose. For the downstream gene the effects are just the opposite.

Two paradigms have been used to explain the regulation of transcription (Reeder, 1989). The one most frequently invoked to explain the regulation of transcription of a gene by RNA polymerase II involves alteration in the receptivity of the gene for the polymerase, usually because of the binding of new or modified stimulatory factors to the controlling elements of the gene, or, less frequently, the removal of inhibitory factors. The other paradigm, for which there is some evidence in the transcription of rRNA in higher organisms, suggests that the receptivity of the gene is not altered; rather, the proportion of "activated" RNA Polymerase I molecules is altered (Tower and Sollner-Webb, 1987; Schnapp and Grummt, 1991). The view is that an active complex is permanently set up on the rRNA gene (Conconi et al., 1989), but that transcription is limited by the amount of polymerase. Stimulation of ribosomal RNA transcription is effected by increasing the number of "active" polymerase I molecules. Stated more simply, the first paradigm suggests an excess of "active" polymerase searching for an "active" gene; the second suggests a constant supply of "active" genes waiting for a limiting supply of "active" polymerase. In the latter view, the enhancer should play a role only in helping to establish "active" genes. Addition of a good carbon source should stimulate transcription to the same degree whether or not an enhancer is present. Table I shows, however, that this is not the case. Since stimulation of transcription is entirely dependent on the enhancer, we conclude that the enhancer plays a role in gene selection. Therefore, insofar as the enhancer is concerned, rRNA transcription in *S. cerevisiae* in response to carbon source is largely subject to the first paradigm; the enhancer increases the "activity" of its genes in response to glucose.

PROCESSING OF RIBOSOMAL RNA

A large number of steps are involved in the processing of the rRNA transcript to the three mature rRNA molecules found in the eukaryotic ribosome. These include endonuclease and exonuclease activity, methylation, pseudouridyation, and assembly with ribosomal proteins. Indeed, assembly with ribosomal proteins appears to be an essential element, upon which most of the subsequent processing steps depend (Warner amd Soeiro, 1967; Udem and Warner, 1972; Trapman et al., 1975). The ultimate aim is to identify the nucleases and other components that are responsible for individual steps and then to determine the basis for the specificity of the cleavage reactions. Two approaches have been used. We attempted the classical genetic approach, i.e, to identify mutants that accumulate an intermediate as a sign that the next step in the process is blocked. We screened more than 500 ts mutants for the accumulation of the 35S transcript, the 27S or 20S intermediates, or any aberrant processing products. This exercise was almost futile. We identified a single mutant, in a gene termed RRP2, in which a portion of the 5.8S RNA accumulated as a 7S' species, with an extra 148 nucleotides on the 5' end. We observed no mutant with substantial accumulation of intermediate species (Shuai and Warner, 1991). Lindahl et al. (1992) had the same experience, screening nearly 1000 ts mutants of another set, and finding only a single one in rRNA processing, another allele of RRP2. The lesson from this effort is one that has been learned many times before. Rarely is there an accumulation of an RNA molecule that is improperly treated; almost without exception they are rapidly degraded.

A more successful approach has been to identify molecules associated with the nucleolus, the site of rRNA processing (reviewed in Warner, 1990). A variety of methods have been employed to identify genes for the yeast version of fibrillarin (Tollervey et al., 1991), perhaps nucleolin (Lee et al., 1992), and other proteins (Girard et al., 1992), as well as at least three small nucleolar RNA's, U3 (Hughes and Ares, 1991), U14, and snR10 (Tollervey, 1987). Using the genetic methods available in *S. cerevisiae*, it has been possible to show that each of these is at least important, and in several cases essential, for the processing of rRNA. These frequently have a stronger effect on the processing steps leading to 18S RNA, leading to a deficiency of 40S subunits. This is consistent with the finding that while the 5' external spacer and the internal spacer are necessary for correct production of 18S RNA, only the short spacer between 25S and 5.8S RNA is essential for their proper maturation (van der Sande et al., 1992).

Yet, only the surface has been scratched. We know almost nothing about the nucleases, the modification enzymes, the substrate structure, the role of individual ribosomal proteins. Clearly, for a system this complex, genetics must eventually come to the rescue.

SPLICING IN THE REGULATION OF RIBOSOMAL PROTEINS SYNTHESIS

In the synthesis of ribosomes the cell faces a formidable problem of inventory control. With a few exceptions, each of the ribosomal proteins is needed in amounts equimolar with all the others. Limitation of almost any of the proteins will limit the production of ribosomes. Yet, as potent RNA-binding elements, ribosomal proteins are potentially hazardous to the health of the cell. Therefore cells must closely regulate the accumulation of ribosomal proteins on both a macro and a micro level. At the macro level, where the global synthesis of ribosomal proteins must be matched with that of rRNA, regulation occurs largely at the level of transcription, under the control of general transcription factors such as Rap1p and Abf1p, although other elements are likely to be involved as well (reviewed in Mager and Planta, 1991; Woolford and Warner, 1991).

At the micro level where the concentration of individual ribosomal proteins must be modulated, numerous mechanisms are employed (*ibid.*). In the case of one ribosomal protein, L32, we now have strong evidence that the protein itself is responsible for feedback regulation at the level of splicing. We found originally that the introduction into a cell of extra copies of the *RPL32* gene led to the accumulation of >50 fold excess of unspliced transcript (Warner et al., 1985). Further analysis showed that this was due to the overproduction of the protein itself, and not to saturation of the splicing apparatus (Dabeva et al., 1986). More recent experiments suggested that the regulation of splicing can be traced to a specific RNA structure into which the 5' exon of the *RPL32* transcript folds (Fig. 2) (Eng and Warner, 1991). The structure pairs nucleotides near the 5' end of the transcript with nucleotides at the 5' splice site, stabilized by an additional stem. Disruption of either base-paired stem abolishes the regulation of splicing *in vivo*. Evidence suggests that the nucleotides in the two bulges, but not those in the loop, also play a role in the regulation of splicing. Our hypothesis is that free L32 can bind to the structure shown in Figure 2, and prevent the association of U1 snRNA with the hidden 5' splice site.

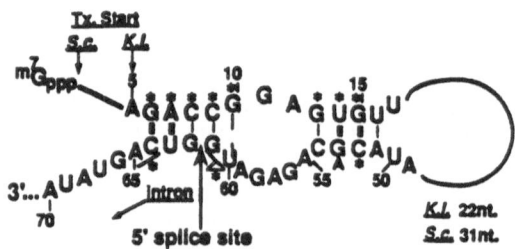

Figure 2. Proposed RNA secondary structure. Nucleotides drawn are only those within the conserved blocks of sequence at the 5' portion of both the *S. cerevisiae* and *K. lactis RPL32* transcripts. The region between the conserved blocks, which differs in length as indicated as well as sequence, is represented as a loop. Nucleotides are numbered according to their position in the *S. cerevisiae* transcript. The 5' splice site as well as transcription initiation sites are indicated. The pairing between G10 and U60 is drawn as uncertain. Asterisks denote nucleotides whose mutations disrupts regulation. C64A65 were changed together, to A64U65. Base pair interactions supported by compensatory mutations are represented by thick vertical bars.

In vitro Studies

Our studies have now been extended to an *in vitro* system. Although pure L32 appears to be too insoluble for use in biochemical analysis, we have constructed a protein in which L32 has been fused to the C terminal end of the *E. coli* maltose binding protein (MBP) (Guan et al., 1987). The fusion protein can be produced in large amounts in *E. coli*, and is readily purified on an amylose affinity column. We find that the MBP-L32 fusion protein binds effectively to a 75 nucleotide RNA molecule consisting of the 5' exon and 14 nucleotides of the intron of the *RPL32* gene, and somewhat less well to an RNA molecule consisting of the 5' exon, the entire intron, and a portion of the second exon. The relevance of this binding to the regulation of splicing is supported by the observation that the introduction of any of several mutations that abolish regulation of splicing *in vivo* also abolish the binding of the fusion protein *in vitro*. Using an *in vitro* splicing extract (Seraphin and Rosbash, 1989), we find that the splicing of an *RPL32* transcript is much less efficient than that of actin. Addition of the MBP-L32 fusion protein, but not of MBP itself, abolishes splicing of the *RPL32* transcript entirely, with little if any effect on the splicing of the actin transcript. Conversely, addition of the 75 nucleotide transcript as competitor stimulates *RPL32* splicing substantially. We surmise that the splicing extracts contain a certain amount of L32 that inhibits the splicing of the *RPL32* transcript; the addition of the 5' end of the transcript (the structure in Figure 2) may titrate out this L32, permitting more efficient splicing.

Is the Regulation of Splicing General?

The genes for ribosomal proteins of *S. cerevisiae* are unusual in that most carry an intron, while few other genes do so. It seems likely that during evolution *S. cerevisiae* has lost most of its introns (Fink, 1987). The retention of introns in ribosomal protein genes suggests that they provide some selective function. Yet, of the several genes that have been examined, only for *RPL32* is there an accumulation of unspliced transcript when the product is overproduced. There are now suggestions, however, that while *RPL32* is unusual in the accumulation of unspliced precursor in response to a regulatory signal, the splicing of transcripts of other genes may also be regulated, but with the consequence that the excess unspliced transcript is rapidly degraded. Thus, the mRNAs of the two genes coding for ribosomal protein 59, *CRY1* and *CRY2*, are normally found in a ratio of 10:1. When *CRY1* is deleted, however, the *CRY2* mRNA increases 8-fold (Woolford and Warner, 1991), due to sequences found within the intron (Li, Paulovich, and Woolford, pers. comm.). Furthermore, within the introns of the two genes for S13 (Mizuta et al., 1991; Vincent and Liebman, 1992) and those for L8 (Mizuta et al., 1992; Mizuta, pers. comm.) there has been an unusual sequence conservation, suggesting some selective pressure. We suggest that there may be many instances of regulated splicing among the ribosomal proteins, but that they are difficult to detect because the unspliced molecules are rapidly degraded. New assays must be developed to detect such regulation.

Regulation of Translation

An intriguing feature of the sequences regulating the splicing of the *RPL32* transcript is that they are nearly entirely within the 5' exon. Thus, the spliced mRNA is potentially also a target for binding by L32 (Figure 3), leading to an additional level of autoregulation, at translation. We asked if this were the case by fusing the 5' exon of *RPL32* to LacZ, with no intron, and expressing the gene in *S. cerevisiae*. The level of β-galactosidase activity in cells overexpressing L32 was only half that in control cells. Yet the level of mRNA in cells overexpressing L32 was twice that in control cells. Perhaps, as is the case with the unspliced transcript, the binding of L32 stabilizes the mRNA. In any case, it appears that excess L32 can inhibit translation of an mRNA carrying the *RPL32* 5' exon by as much as 75%.

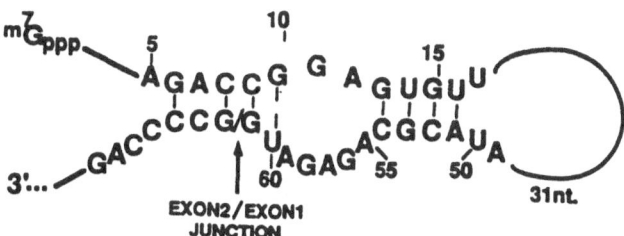

Figure 3. Potential structure of the 5' end of the spliced transcript of *RPL32* showing the similarity to the structure of the unspliced transcript (Figure 2).

Value of Related Organisms

An element of the genetic approach to the understanding of ribosome synthesis whose value can hardly be overestimated is the comparison of the sequences of distantly related organisms. In our case we have found *Kluyveromyces lactis* to be particularly useful. It is estimated to be removed from *S. cerevisiae* by 100 to 200 million years (Barns et al., 1991; Van de Peer et al., 1992). Non-essential elements have diverged substantially. But those elements for which there is strong selection can still be identified. Thus, the sequences responsible for the regulation of splicing of the *RPL32* transcript were first identified by comparing the non-translated regions of the genes from the two organisms (Eng and Warner, 1991) (Figure 2). Similarity of intron sequences between the *CRY2* gene from *S. cerevisiae* and the single *CRY* gene of *K. lactis* support the idea that there is regulation of splicing of this transcript (Li, Paulovich, and Woolford, pers. comm.). The sequences responsible for the binding of the Reb1p protein to the enhancer and promoter of the rRNA genes were identified by comparison of the amino acid sequences of the two proteins (Morrow et al., 1992).

RIBOSOME FUNCTION

Post-Translational Modification

It is possible, using the genetic approaches available in *S. cerevisiae*, to ask questions about the function of eukaryotic ribosomes that cannot be answered with mammalian systems. One element of potential importance is the role of post-translational modifications of ribosomal proteins. We studied the effect of phosphorylation of ribosomal protein S10, the yeast equivalent of S6 in mammals, whose phosphorylation has been connected circumstantially with many aspects of growth control (Erikson, 1991). By replacing the appropriate serine residues with alanine in both genes for S10, we eliminated the phosphorylation of the protein. Yet there was little, if any, effect either on the growth rate (Johnson and Warner, 1987), or on the spectrum of proteins synthesized (Johnson et al., 1988).

Similarly we have made use of a mutant strain in which the *NAT1* gene has been deleted. These cells are no longer able to N^{α}-acetylate proteins (Mullen et al., 1989). How prevalent is the N^{α}-acetylation of ribosomal proteins and what is its function? We find at least fourteen of the *S. cerevisiae* ribosomal proteins are N^{α}-acetylated, exclusively on serine residues (Takakura et al., 1992) (Figure 4). The N-terminal analysis possible with proteins isolated from the mutant strain provided the first sequence data for a number of these. Surprisingly, the lack of N^{α}-acetyl groups leads to relatively little effect on the growth rate of the cells (Mullen et al., 1989; Takakura et al., 1992) or on the spectrum of proteins synthesized by the ribosomes.

Deletion of Ribosomal Proteins

E. coli can survive without any one of several ribosomal proteins (Dabbs, 1986). *S. cerevisiae* can as well. At least five ribosomal proteins are dispensable, although sometimes with drastic effects on the growth rate of the cell (reviewed in Woolford and Warner, 1991). We have studied the case of L30. Deletion of both genes coding for L30 increases the doubling time by only about 30%; surprisingly, the presence of a truncated version of L30 is substantially more deleterious (Baronas-Lowell and Warner, 1990). Attempts to study the structure of the ribosomes lacking L30 have been frustrated by the extreme instability of the ribosomes *in vitro* (Lee, pers. comm.).

Accuracy in Translation

It is unlikely that any eukaryotic ribosome will receive such detailed study as has the *E. coli* ribosome over the past thirty years. Thus, it is important to know with what confidence we can extrapolate findings on the structure and function of the *E. coli* ribosome to the eukaryotic ribosome. Although some ribosomal proteins have retained identifiable similarities during the divergent evolution of the eubacteria and the eukaryotes, many have not. Some of the latter may be identified by analysis of the archaebacteria, which often occupy a middle ground (Wittmann-Liebold et al., 1990). One of these is S28 of *S. cerevisiae*, which is clearly related to S12 of archaebacteria, and thus to S12 of *E. coli*, although there are no identifiable homologies greater than 4 consecutive amino acids between the yeast and the eubacterial sequences (Alksne & Warner, in prep.). S12 is of particular interest

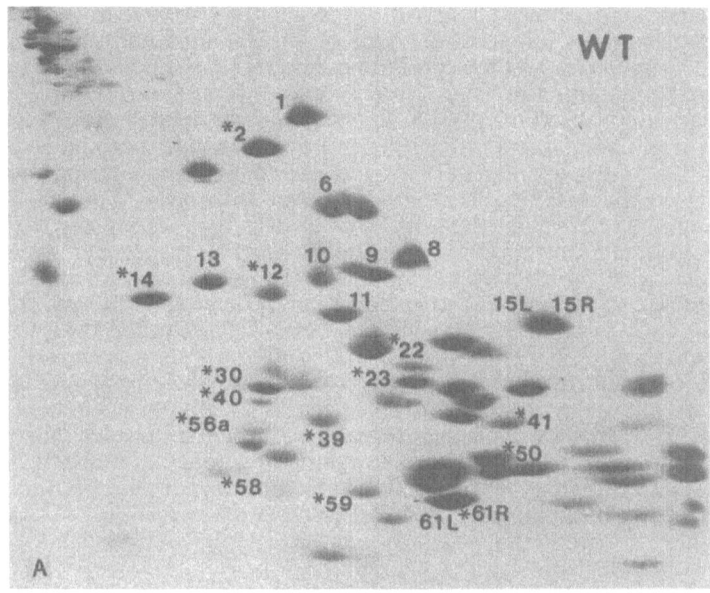

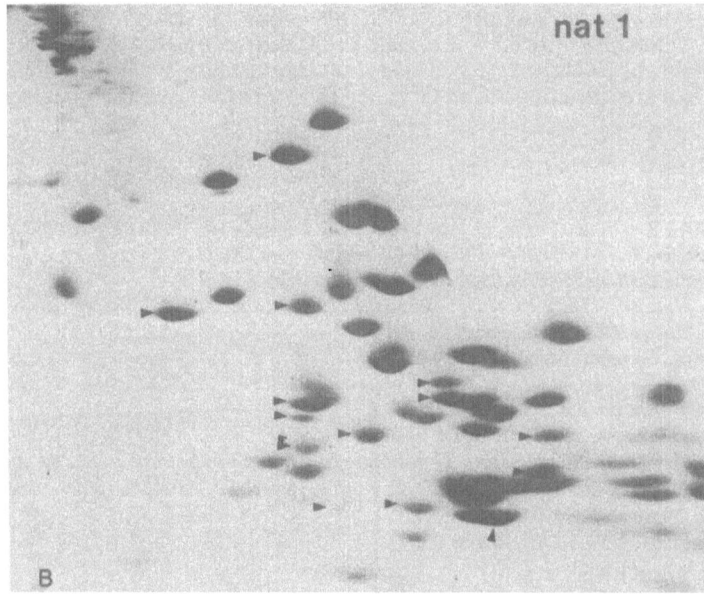

Figure 4. Ribosomal proteins of *NAT1* and *nat1* strains. Ribosomal proteins were prepared from strains carrying the wild type, *NAT1* and the *nat1::LEU2* alleles, and fractionated on two-dimensional polyacrylamide gels. Some of the proteins are identified according to the nomenclature orignally used with this gel system (Gorenstein & Warner, 1976). In two cases, rp15 and rp61, where it is now known that a single spot represents two proteins, have been indicated L and R. The *asterisks* denote proteins that, when isolated from the *nat1* strain, have a high mobility in the first dimension. The same proteins are indicated by *arrowheads* in the gel from the *nat1* strain. (Reprinted with permission from Takakura et al. (1992).)

because, together with S4 and S5, it appears to contribute to the optimal level of accuracy of the ribosome (Gorini, 1974). Mutants at K42 of S12 lead to increased accuracy, and streptomycin resistance. Mutants of S4 or S5 can lead to decreased accuracy and suppress those of S12. One of the few apparent regions of homology between S12 of *E. coli* and S28 of *S. cerevisiae* includes K42. In *S. cerevisiae*, this residue can be identified as K62. We have asked if mutations in this residue affect the accuracy of translation of the *S. cerevisiae* ribosome. The answer is yes! Mutation of K62 can render the cells more resistant to paromomycin, and can counteract the omnisuppressors *SUP44* and *SUP46*, both findings suggesting that the mutations have increased the accuracy of translation (Alksne, Anthony, Liebman, & Warner, in prep.). What is of particular interest is that *SUP44* and *SUP46* encode the yeast ribosomal proteins equivalent to S5 and S4, respectively of *E. coli* (All-Robyn et al., 1990; Mizuta et al., 1991; Vincent and Liebman, 1992). Thus the interactions of S4, S5, and S12 to establish an optimal level of accuracy has been maintained during the nearly two billion years of evolution that separates *E. coli* from *S. cerevisiae*.

We conclude from the data presented above and throughout this meeting that, just as has been the case for the ribosomal RNA's (Gutell et al., 1985), substantial divergence of the sequences of the ribosomal proteins during evolution masks a fundamental conservation of structure, and hence of function. We predict that the peptide structures that lead to the key functions of the ribosome, binding to mRNA, selection of amino-acyl tRNA, peptide bond formation, and translocation, will be highly conserved.

ACKNOWLEDGMENTS

This work was supported by grants from the NIH #GM25532 and the American Cancer Society #NP-793, and the Cancer Center #CA13330. L.A. was a Trainee on grant #GM07491. J.V. is a Fulbright Fellow from the Science Ministry of Spain. We are grateful to Ann Greaney for preparation of the manuscript.

REFERENCES

All-Robyn, J.A., Brown, N., Otaka, E., and Liebman, S.W., 1990, *Mol. Cell. Biol.* 10:6544.

Barns, S.M., Lane, D.J., Sogin, M.L., Bibeau, C., and Weisburg, W.G., 1991, *J. Bacteriol.* 173:2250.

Baronas-Lowell, D.M. and Warner, J.R., 1990, *Mol. Cell Biol.* 10:5235.

Conconi, A., Widmer, R.M., Koller, T., and Sogo, J.M., 1989, *Cell* 57:753.

Dabbs, E.R., 1986, *in:* "Structure, Function, and Genetics of Ribosomes," B. Hardesty and G. Kramer, eds., Springer-Verlag, New York.

Dabeva, M.D., Post Beittenmiller, M.A., and Warner, J.R., 1986, *Proc. Natl. Acad. Sci. USA* 83:5854.

Elion, E.A. and Warner, J.R., 1984, *Cell* 39:663.

Elion, E.A. and Warner, J.R., 1986, *Mol. Cell Biol.* 6:2089.

Elion, E.A. and Warner, J.R., 1987, *in:* "Transcriptional Control Mechanisms," D.K. Granner, G. Rosenfeld, and S. Chang, eds., Alan R. Liss, Inc., New York.

Eng, F.J. and Warner, J.R., 1991, *Cell* 65:797.

Erikson, R.L., 1991, *J. Biol. Chem.* 266:6007.

Fink, G.R., 1987, *Cell* 49:5.

Girard, J-P., Lehtonen, H., Caizergues-Ferrer, M., Amalric, F., Tollervey, D., and Lapeyre, B., 1992, *EMBO J.* 11:673.

Gorini, L., 1974, *in:* "Ribosomes," M. Nomura, A. Tissieres, and P. Lengyel, eds., Cold Spring Harbor Laboratory Press, New York.

Guan, C., Li, P., Riggs, P.D., and Inouye, H., 1987, *Gene* 67:21.

Gutell, R.R., Weiser, B., Woese, C.R., and Noller, H.F., 1985, *Prog. Nucl. Acid Res. & Molec. Biol.* 32:156.

Hughes, J.M.X. and Ares, M., Jr., 1991, *EMBO J.* 10:4231.

Johnson, S.P., McLaughlin, C., and Warner, J.R., 1988, *in:* "Genetics of Translation: New Approaches," M.F. Tuite, M. Picard, and M. Bolotin-Fukuhara, eds., Springer Verlag, New York.

Johnson, S.P. and Warner, J.R., 1987, *Mol. Cell Biol.* 7:1338.

Johnson, S.P. and Warner, J.R., 1989, *Mol. Cell Biol.* 9:4986.

Kempers-Veenstra, A.E., Oliemans, J., Offenberg, H., Dekker, A.F., Piper, P.W., Planta, R.J., and Klootwijk, J., 1986, *EMBO J.* 5:2703.

Kief, D.R. and Warner, J.R., 1981, *Mol. Cell Biol.* 1:1007.

Lee, W.-C., Zabetakis, D., and Melese, T., 1992, *Mol. Cell. Biol.* 12:3865.

Lindahl, L., Archer, R.H., and Zengel, J.M., 1992, *Nucleic Acids Res.* 20:295.

Mager, W.H. and Planta, R.J., 1991, *Mol. Cell Biochem.* 104:181.

Mizuta, K., Hashimoto, T., Suzuki, K., and Otaka, E., 1991, *Nucleic Acids Res.* 19:2603.

Mizuta, K., Hashimoto, T., and Otaka, E., 1992, *Nucleic Acids Res.* 20:1011.

Morrow, B.E., Ju, Q., and Warner, J.R., 1992, *Mol. Cell. Biol.*, in press.

Mullen, J.R., Kayne, P.S., Moerschell, R.P., Tsunasawa, S., Gribskov, M., Sherman, F., and Sternglanz, R., 1989, *EMBO J.* 8:2067.

Reeder, R.H., 1989, *Curr. Op. in Cell. Biol.* 1:466.

Schnapp, A. and Grummt, I., 1991, *J. Biol. Chem.* 266:24588.

Seraphin, B. and Rosbash, M., 1989, *Cell* 59:349.

Shuai, K. and Warner, J.R., 1991, *Nucleic. Acids. Res.* 19:5059.

Takakura, H., Tsunasawa, S., Miyagi, M., and Warner, J.R., 1992, *J. Biol. Chem.* 267:5442.

Tollervey, D., 1987, *EMBO J.* 6:4169.

Tollervey, D., Lehtonen, H., Carmo-Fonseca, M., and Hurt, E.C., 1991, *EMBO J.* 10:573.

Tower, J. and Sollner-Webb, B., 1987, *Cell* 50:873.

Trapman, J., Retel, J. and Planta, R.J., 1975, *Exp. Cell. Res.* 90:95.

Udem, S.A. and Warner, J.R., 1972, *J. Mol. Biol.* 65:227.

Van de Peer, Y., Hendriks, L., Goris, A., Neefs, J.-M., Vancanneyt, M., Kersters, K., Berny, J.-F., Hennebert, G.L., and De Wachter, R., 1992, *System. Appl. Microbiol.* 15:250.

van der Sande, C.A., Kulkens, T., Kramer, A.B., de Wijs, I.J., Van Heerikhuizen, H., Klootwijk, J., and Planta, R.J., 1989, *Nucleic. Acids. Res.* 17:9127.

van der Sande, C.A.F.M., Kiva, M., Van Nues, R.W., Van Heerikhuizen, H., Raue, H.A., 1992, *J. Mol. Biol.* 223:899.

Vincent, A. and Liebman, S.W., 1992, *Genetics* 132:375.

Warner, J.R. and Soeiro, R., 1967, *Proc. Natl. Acad. Sci. USA* 58:1984.

Warner, J.R., Mitra, G., Schwindinger, W.F., Studeny, M., and Fried, H.M., 1985, *Mol. Cell Biol.* 5:1512.

Warner, J.R., 1990, *Curr. Opin. Cell Biol.* 2:521.

Wittmann-Liebold, B., Kopke, A.K.E., Arndt, E., Kromer, W., Hatakeyama, T., and Wittmann, H.-G., 1990, *in:* "The Ribosome: Structure, Function, & Evolution," W.E. Hill, A. Dahlberg, R.A. Garrett, P.B. Moore, D. Schlessinger, and J.R. Warner, eds., ASM, Washington, D.C.

Woolford, J.L.,Jr. and Warner, J.R., 1991, *in:* "The Molecular and Cellular Biology of the Yeast Saccharomyces Genome Dynamics, Protein Synthesis, and Energetics," J.R. Broach, J.R. Pringle, and E.W. Jones, eds., Cold Spring Harbor Laboratory Press, New York.

REGULATION OF RIBOSOMAL RNA SYNTHESIS AND CONTROL OF RIBOSOME FORMATION IN *E. COLI*

Rolf Wagner, Günter Theißen[1], and Martin Zacharias[2]

Institut für Physikalische Biologie, Heinrich-Heine-
Universität Düsseldorf, D-4000 Düsseldorf 1
[1]present address: Max-Planck-Institut für Züchtungsforschung,
D-5000 Köln
[2]present address: Department of Chemistry, University of Houston,
Texas, U. S. A.

INTRODUCTION

The Importance of rRNA Synthesis for Ribosome Formation and Bacterial Growth

One of the long standing open questions in bacterial growth is to understand how bacteria manage the very efficient adaptation of their growth rates to changes in environmental conditions. The astonishing capability of rapid growth changes, in response to the nutritional conditions, is to a large part related to the fact that bacterial cells are able to adjust their protein synthesizing capacity precisely to the growth demand. One has to note, however, that the translational capacity is not increased by the speed of the translating ribosomes. Rather it is known that the rate of peptide bond formation does not exceed 20 amino acids per second. Therefore, the very fast doubling times of bacteria under optimal growth conditions can only be achieved by a rapid increase in the number of ribosomes. In fact, during exponential growth bacteria accumulate in very short time over 20.000 ribosomes per individual cell.

Ribosomes are multicomponent particles, composed of 3 RNA molecules and at least 52 different proteins. Therefore, coordination and balance in the synthesis of all components is a prerequisite for efficient ribosome formation.

The key molecules for the regulation of ribosome biogenesis in bacteria are the ribosomal RNAs (rRNAs). Unlike to the mRNA fraction, ribosomal RNA synthesis rates increase in proportion to the square of the cell growth rate (Gausing, 1977). Moreover, it is commonly accepted that the regulation of most, if not all, ribosomal proteins is determined by translational feedback in response to the available free ribosomal RNA (Jinks-Robertson et al., 1983; Gourse et al., 1985). Thus, the rate of synthesis of ribosomal RNA determines largely the rate of ribosome formation in the cell. Therefore, rRNA synthesis is of central importance for the establishment of conditions for rapid cell growth.

The general significance of the regulation of ribosomal RNA synthesis for the metabolism of the cell is possibly reflected by the wealth of different and complex control mechanisms underlying rRNA synthesis.

Despite intensive research on the regulation of rRNA expression in the past we are far from a complete understanding of all the regulatory events, and actually none of the principal molecular mechanisms is known in detail. Our present knowledge

The Translational Apparatus, Edited by K.H. Nierhaus
et al., Plenum Press, New York, 1993

has been summarized in a number of recent reviews (Nomura et al., 1984; Lamond, 1985; Lindahl and Zengel, 1986; Wagner, 1989), and although for some aspects detailed information is available, it should only be considered as a basis for further research. Based on the past investigations several regulatory mechanisms, both, transcriptional and post-transcriptional in nature can be distinguished. Here we report on the current status and recent progress we have made on some of the major control circuits presently known.

Location and Arrangement of rRNA Genes

There are 7 operons encoding ribosomal RNA transcription units in *E. coli* (*rrnA, B, C, D, E, G* and *H;* Lindahl and Zengel, 1986). Their locations on the circular *E. coli* chromosome have been mapped (Ellwood and Nomura, 1982), and the general arrangement of genes is well known. Each operon contains genes for 16S, 23S, 5S and one or several tRNAs. The structural genes are flanked by highly conserved leader and spacer regions which encode important regulatory signals as we shall explain in more detail below. Fig. 1 gives a schematic description of the representative arrangement of a rRNA operon (*rrnB*), where some of the known regulatory elements which are subject in this report are indicated.

For reasons of space, and because more information has been collected, we concentrate on the promoter and leader region preceding the 16S RNA, shown in enlarged presentation in the lower part of Fig. 1.

rrnB operon

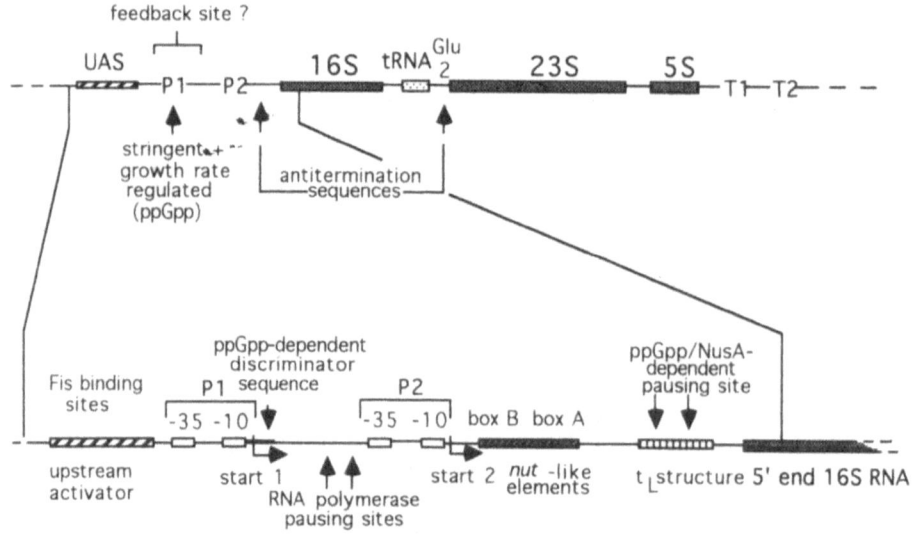

Figure 1. Schematic arrangement of the *rrnB* operon. The locations of some of the known regulatory signals are indicated. For details see text. The drawing is not to scale.

It should be noted here that in rapidly growing cells the rRNA gene dosage is not constant. Due to the clustering of rRNA transcription units around the origin of replication, with the transcription direction parallel to the bidirectional replication, an increase in gene dosage during rapid growth must be anticipated. There are some early reports where no significant consequences of changes in the gene dosage on rRNA expression have been observed (Morgan and Kaplan, 1976; Hill and Harnish, 1982; Ellwood and Nomura, 1980). Whether gene dosage effects or differences of the various chromosomal locations of the individual rRNA operons contribute to the very high rRNA synthesis rates or to a possible differential expression of rRNA operons is currently reinvestigated, and first positive results in this direction have been obtained (Condon et al., 1992).

TRANSCRIPTION ACTIVATION AND REGULATION OF RIBOSOMAL RNA PROMOTERS

Tandem Promoters

One of the characteristic features of all ribosomal RNA operons in *E. coli* is the arrangement of tandem promoters (P1 and P2 in Fig. 1). Both promoters are separated by about 120 base pairs (bp). As we shall see below, both promoters differ in the way they are regulated. There are some discrepancies about the contribution of the two promoters to the amount of transcripts in the cell. Both produce rRNA *in vivo* but originally it was believed that P1 is responsible for the very efficient transcription during rapid growth, while P2 was considered to produce a low basal level of rRNA. More recent observations seem to indicate, however, that P2 is an equally strong promoter *in vitro* and *in vivo*. Interestingly, transcription from promoter P1 seems to start generally with ATP (GTP, in case of the *rrnD* operon) while, unlike for most bacterial promoters, CTP is the starting nucleoside triphosphate for promoters P2.

Most of the regulatory phenomena known today are attributed to promoter P1. There is good evidence, however, that P2 is also under some kind of regulation. Studies on the *in vivo* strength of *rrnB* P2 fusion constructs under different physiological conditions revealed a two phase kinetics of expression, indicating that P2 is also sensitive to regulation (Lukacsovich et al., 1987; Csiszar et al., 1990).

Both tandem promoters are unusual in comparison to other strong *E. coli* promoters. Their rather moderate *in vitro* activity, for instance, stands in sharp contrast to the very high *in vivo* efficiency. It has also been demonstrated for the *rrnD* promoters, and is very likely true for all ribosomal promoters, that the open initiation complexes are exceptionally labile, and can be challenged with heparin or other competing promoters (Langert et al., 1992)

We have to confess that, despite intensive research, the physiological importance of the tandem arrangement of two promoters for rRNA expression is entirely unclear today.

Upstream Activating Sequences (UAS)

Ribosomal RNA promoters belong to the strongest of the cell despite the fact that their similarity to the consensus hexamer sequence (-10 and -35 regions) of prokaryotic promoters is rather weak (Hawley and McClure, 1983). Furthermore, they have a suboptimal 16 bp spacing between the -35 and -10 region, compared to the 17 bp spacing of other known strong promoters.

It is known for many highly transcribed genes that in addition to the promoter core elements sequences upstream to the -35 consensus hexamer are essential for the high transcriptional activity. These upstream activating sequences (UAS) extending roughly from position -40 to -150, relative to the transcription start site, are known to occur in front of all stable RNA (rRNA and tRNA) promoters. They do not show noticeable sequence conservation but are generally very rich in AT clusters, and, as a result of the helical phasing of the AT clusters, deviate from the linear B-form DNA (Lamond and Travers, 1983; Bauer et al., 1988; Vijgenboom et al., 1988). UAS regions are, therefore, characterized by a curved DNA backbone. This unusual DNA structure can readily be detected by a reduced electrophoretic mobility in polyacrylamide gels (Marini et al., 1982; Zacharias et al., 1990). The ribosomal RNA P1 promoters all contain such UAS regions and their presence activates transcription by a factor of 10 to 20. DNase I and hydroxyl radical footprints indicate extended RNA polymerase contacts to the promoter upstream region. Furthermore, using filter binding assays a lower association constant K_B was determined when the UAS region had been deleted (Gourse, 1988; Leirmo and Gourse, 1991).

To find out whether the curved DNA conformation of the UAS contributes to transcriptional activation *per se*, we have constructed a nested set of base change mutations spanning the UAS region from position -105 to -47, relative to the transcription start site of promoter P1. In another set of mutants we have additionally inserted oligonucleotides of variable length at position -47, thus shifting the UAS region further upstream, equivalent to half, one or two helical turns. Comparison of the gel electrophoretic mobilities of DNA fragments containing the different base changes in an approximately central position allowed us to detect changes in the sequence dependent DNA curvature, and to localize the center of bending of the P1 UAS region around position -90. Apparently, within the UAS region, sequences further upstream contribute stronger to the sequence dependent curvature. Base changes close to the promoter core, between positions -47 and -58, for instance, do not affect the gel electrophoretic mobility to a noticeable degree (Zacharias et al., 1992).

Transcription Activation by the Trans-activating Protein Fis

A number of *E. coli* proteins were claimed in the past to bind specifically to the UAS region of rRNA genes, and to act as positive regulators. However, their identity had been determined erroneously. It turned out that the small basic protein Fis (factor for inversion stimulation), known as a recombinational enhancer in a number of phage associated inversion reactions (Kahmann et al., 1985; Johnson and Simon, 1985), was actually the transcriptional activator under investigation (Nilsson et al., 1990; Ross et al.,. 1990; Zacharias et al., 1992). Using DNase I protection and gel retardation studies it was shown that Fis binds to three specific sites (sites I, II, III) within the UAS region which all contain the common consensus sequence known to be required for Fis binding (Ross et al., 1990; Hübner and Arber, 1989).

We used the above described UAS mutants to find out how *rrnB* P1 promoter activation due to Fis interaction is achieved. In some of the mutants the consensus motif for Fis interaction was altered in binding site I or II. First, we have determined Fis binding to the different UAS mutants by gel retardation assays. We could show that both, the primary DNA sequence and the DNA curvature are important parameters for stable Fis interaction. When we analyzed the effects of the extent and location of the DNA curvature, and the binding of Fis on the promoter activity of rRNA P1, it turned out that base changes affecting Fis binding sites, as well as sequence variants with no effect on Fis binding, can reduce promoter activity *in vivo* (Zacharias et al., 1992).

From these studies, performed in fis $^+$ and fis $^-$ background, one has to conclude that two mechanisms of activation are existing, one Fis-dependent and one independent on Fis. Both mechanisms seem to be non-additive and partly compensating. The signals for the Fis-independent mechanism are located between position -65 and -35, relative to the transcription start site, and stable curvature is obviously no prerequisite for Fis-independent activation.

Determination of the promoter activities of constructs where the Fis binding sites had been shifted systematically to a more upstream position by DNA insertions yielded the following results: Shifting the UAS region upstream by half a helical turn completely abolishes any Fis-dependent activation, while increasing the distance by a complete turn of the DNA helix restores activation partly. A further displacement of the Fis binding sites by two integral helical turns reduces the activity again completely. One has to conclude from these results that both, the angular orientation as well as the absolute distance of the Fis binding sites are crucial for the mechanism of Fis-dependent promoter activation. Presumably, RNA polymerase and Fis have to interact at the same face of the DNA helix in order to operate optimally (Zacharias et al., 1992).

It remains to be added that our studies on the mechanism of promoter activation by Fis, using abortive initiation assays, demonstrated that the activation step is due to an increase in the primary binding of RNA polymerase to the promoter (K_B), whereas the melting of the promoter (isomerisation rate k_f) which is mainly affected if superhelical or linear templates are compared, is not changed significantly (Zacharias et al., 1991).

The results of the studies on activation of rRNA promoter P1 due to the UAS region and the trans-activating protein Fis can be summarized as follows:

1. There are two mechanisms of rRNA P1 promoter activation due to the UAS region - one Fis-dependent and one independent on Fis.
2. Signals for the Fis-independent activation mechanism are located between positions -65 and the -35 hexamer sequence.
3. The sequence-dependent curvature of the UAS does very likely not contribute to the Fis-independent activation.
4. The Fis-dependent activation requires a precise distance of the Fis binding site as well as a correct angular orientation of Fis and the RNA polymerase bound to the promoter.
5. Prerequisite for stable Fis-DNA interaction is a defined base sequence (consensus site) as well as a certain DNA curvature (or flexibility).
6. The Fis-dependent activation mechanism almost certainly involves an increase in the RNA polymerase-promoter affinity, while changes in the rate of promoter melting (isomerisation) are not likely to be involved.

We like to mention that basically the same conclusions as summarized above were derived from an independent study by Newlands et al. (1992).

Are there Additional Trans-acting Factors Contributing to the Regulation of rRNA Transcription ?

There is growing evidence that the role of transcription factors in directing prokaryotic RNA polymerase has been underestimated in the past. Today many examples for gene-specific factors are known to operate during prokaryotic transcription. Some of these factors are specifically activated by phosphorylation/dephosphorylation events or small ligand binding (Kustu et al., 1991; Austin and Dixon, 1992; Sanders et al., 1992).

Recently, the identification of *E. coli* proteins with implications on rRNA expression has been reported. In one study the binding of (a)rather undefined protein(s) to the rRNA P2 downstream region was described. Binding of the protein(s) was discovered by mobility shift analysis. However, the identity or function of the protein fraction has not been solved so far (Csiszar et al., 1990).

In several different reports temperature sensitive mutations in the glycolytic fructose-1,6-diphosphate aldolase gene (FDA) were claimed to specifically repress stable RNA synthesis at the level of rRNA transcription initiation. The mechanism, however, is obscure so far (Singer et al., 1991a,b).

Recently we have discovered in our laboratory a factor with specific binding properties to the *rrnB* P1 promoter region (Zacharias et al., 1991). The protein fraction with the binding activity has been partly purified, and the binding domain at the rRNA promoter was analyzed by DNase I and hydroxyl radical footprinting. The results indicate that the protein binds to one site of the bent DNA, extending from position -90 to -15, relative to the transcription start site at promoter P1.

The identity and functional importance of this protein is under investigation in our laboratory and will be presented elsewhere.

Stringent and Growth Rate Regulation

Stringent and growth rate regulation are long known phenomena determining bacterial growth. They both affect stable RNA synthesis, and although they can be distinguished physiologically, they most likely share a common mechanism (Baracchini and Bremer, 1988). Stringent control is characterized by a rapid and coordinated inhibition of stable RNA synthesis with a concomitant increase in the concentration of the small effector molecule guanosine tetraphosphate (ppGpp) under conditions of amino acid deprivation (Gallant, 1979). Growth rate regulation, on the other hand, is defined by the fact that synthesis of stable RNA in exponentially growing bacteria is not constant but proportional to the square of the growth rate (Gausing, 1977; Nierlich, 1978). We and others have shown, that under different growth conditions the basal concentration of ppGpp is almost inverse proportional to the growth rate (Baracchini and Bremer, 1988; Zacharias et al., 1989). Therefore, guanosine tetraphosphate seems to be a common mediator for both types of control.

Virtually countless reports describing studies on the mechanism of stringent control and the function of ppGpp have accumulated in recent years. For almost every result presented a contradictory finding can be cited - a very unpleasant yet much too frequent situation in science! Consequently, unequivocal information and solid facts on how stringent control works and what function ppGpp plays in this elusive event are vanishingly few. The situation has not been changed basically by the studies we describe below, but at least, we hope that the evidence for some aspects of stringent control has been strengthened, and a common mechanistic explanation for stringent and growth rate control, with ppGpp as a central mediator, has become more likely.

There is little doubt today that both types of control are acting only at the P1 and not at the P2 promoters (Glaser et al., 1983; Sarmientos and Cashel, 1983; see, however, Kingston, 1983). Sequence comparison of many stringent regulated promoters has led to the proposal of a GC-rich discriminator sequence motif responsible for stringent control (Travers, 1980). Such a sequence is present upstream to the transcription start at all P1 promoters but lacking at the corresponding position of the P2 promoters, in accordance with the fact that transcription from P2 is not under stringent or growth rate control.

To find out the necessary or sufficient determinants for stringent or growth rate control we have constructed a series of different promoter variants with and without the GCGC discriminator motif at the corresponding position. In addition we have synthesized hybrid promoters where the upstream part to the discriminator motif was taken from a stringent regulated and the downstream part from an unregulated promoter, and *vice versa*. The promoters were fused to reporter genes, and their activity was determined *in vivo* under conditions of amino acid starvation and different growth rates (Zacharias et al., 1989; Zacharias et al., 1990b). The results allowed to draw the following conclusions :

1. The GCGC discriminator motif is a necessary but not sufficient requirement for both stringent and growth rate regulation.
2. Guanosine tetraphosphate (ppGpp) is the mediator for both types of control, indicating at least strong similarities in the two mechanisms.
3. The determinants for stringent and growth rate regulated promoters are not restricted to certain primary sequence motifs but involve the overall context of the complete promoter structure.

Using a bacterial strain that is unable to synthesize ppGpp Gourse and coworkers have obtained results contradicting the finding that growth rate regulation is related to the concentration of ppGpp in the cell (Gaal and Gourse, 1990). This report has recently been disputed (Hernandez and Bemer, 1992). Although the controversy is not completely settled, our results strongly support the finding of a functional role of ppGpp in the regulation of stable RNA expression. Moreover, the results make a partition model for RNA polymerase plausible (Ryals and Bremer, 1982; Little et al., 1983). According to this model different interconvertible RNA polymerases exist with high or low affinity for stringent controlled promoters. The interconversion of the two forms may be triggered directly or indirectly by ppGpp.

The small omega peptide, associated with RNA polymerase preparations, was shown to be the target for ppGpp dependent change in promoter affinity (Igarashi et al., 1989) - but see the introductory remarks - this finding too has now clearly been disputed (Gentry et al., 1991).

FUNCTION OF THE rRNA LEADER AND SPACER SEQUENCES IN TRANSCRIPTIONAL AND POST-TRANSCRIPTIONAL CONTROL

Processing Sites

To yield mature molecules the extra transcribed leader, spacer and trailer sequences of the rRNA primary transcripts are removed in a series of complex nucleolytic cleavage reactions. Some of these processing and maturation events take place before transcription is completed. The pathway leading to mature RNAs has been studied in several laboratories and the results are summarized in different reviews (Srivastva and Schlessinger, 1990; King et al., 1986; Gegenheimer and

Apirion, 1981). Several enzymes participating in the endonucleolytic cleavage reactions, like RNase III, RNase E and RNase P have been identified. However, not all the nucleases involved in the final maturation have been characterized in detail up to now. A summary showing the sites of cleavage and the enzymes involved is presented in Fig. 2.

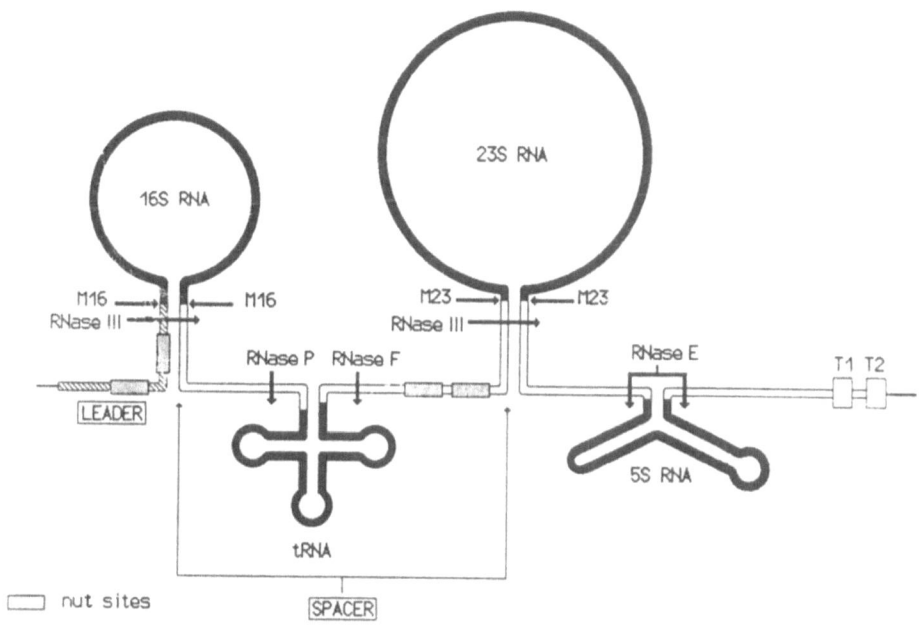

Figure 2. Schematic arrangement of rRNA processing sites. The sites of endonucleolytic cleavage and the corresponding enzymes are indicated by arrows. Structural genes are shown in black. The leader region is shaded, and *nut* site elements are shown as grey boxes. T1 and T2 indicate transcriptional terminators.

The signals for the processing reactions are not restricted to the extra transcribed sequences. For instance, correct RNase III cleavage can be obtained *in vitro* with pure isolated RNA but does depend on the presence and correct folding of certain sequences within the structural genes. Results obtained with mutants where the spacer tRNA gene had been removed or shifted distantly from the 3'end of 16S RNA demonstrated that structural genes are required for correct processing (Szymkowiak et al., 1988; Srivastava and Schlessinger, 1989). Furthermore, a number of maturation steps require pre-assembled particles, and final maturation is known to occur at the level of functional polysomes (Srivastava and Schlessinger, 1988). Obviously, the correct structure and assembly of ribonucleoprotein particles (RNPs) are required for maturation. Thus, quality control of the final particles may be a second important functional aspect of processing. Evidently, processing and assembly are tightly coupled interdependent events which occur in a co-ordinated and co-operative way (King et al., 1986).

Nut -like Elements and Antitermination

Ribosomal RNA leaders and spacers contain highly conserved sequence elements with strong similarity to the *nut L* and *nut R* sites of lambdoid phages (box A, B, C). These sequences are known to mediate N-dependent transcription antitermination during phage expression (Friedman et al., 1984; Li et al., 1984). Experimental evidence for suppression of premature transcription termination during rRNA transcription has been obtained in a number of studies. It was shown that polar mutations or transcription termination signals can effectively be suppressed when cloned downstream from the rRNA leader region (Holben and Morgan, 1984; Li et al., 1984; Berg et al., 1989; Albrechtsen et al., 1990). Antitermination in phage lambda is known to involve the phage encoded N protein as well as the trans-acting factors NusA, NusB, NusE (ribosomal protein S10 !) and NusG, which, together with the *nut* -sequences containing transcript, modify RNA polymerase to a termination resistant form (Barik et al., 1987; Horwitz et al., 1987; Mason and Greenblatt, 1991; Nodwell and Greenblatt, 1991; Li et al., 1992).

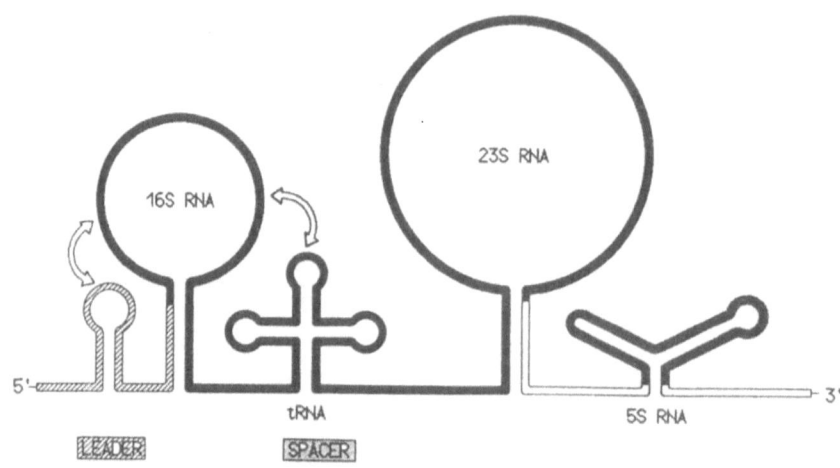

Figure 3. Model showing the proposed interaction of sequences flanking the structural genes for 16S and 23S RNA to facilitate structure formation. The arrows indicate putative direct or protein mediated interactions. Evidence for the tRNA-16S RNA interaction stems from Szymkowiak et al., (1988); Srivastava and Schlessinger, (1989). Support for leader-16S RNA interaction is summarized in last section of the manuscript, see also Mori et al., (1990).

Although antitermination systems in lambda and in case of rRNA operons share many cis- and trans-acting components, their mechanisms can not be completely identical. First, there is no protein equivalent to the phage encoded N protein known to exist in uninfected *E. coli* cells. Second, the order of *nut* -like elements boxA and boxB is different, and certain terminator structures are not suppressed to the same extend when rRNA leader or lambda *nut* site fusion constructs are compared. Finally, since additional functions like processing and maturation are encoded in the leader, and assembly to pre-ribosomal particles occurs already during transcription, any possible antitermination mechanism in rRNA transcription must be different from lambda, at least in some aspects. In line with this, a mutational study of the leader boxA sequence has revealed important post-transcriptional functions of this sequence element, implying that the mechanism responsible for the suppression of premature termination includes control of maturation and assembly (Theißen et al., 1990a).

RNA Polymerase Pausing sites

Pausing of RNA polymerase during transcription has since some time been recognized as an important mechanism controlling pro- and eukaryotic gene expression (Landick et al., 1987; Resnekov and Aloni, 1989; Platt, 1981; Yang et al., 1989). Pausing may be important to allow for correct RNA structure formation or to support specific RNA protein interactions necessary for the function of the growing transcript. There are several specific sites within the rRNA leader where RNA polymerase was shown to pause for a considerable time (Kingston and Chamberlin, 1981). Already in the early reports it was recognized that the length of some of the pauses was influenced by protein factors (NusA) and small molecular weight effector molecules (ppGpp). We have quantified several of the pauses known to occur during transcription of the rRNA leader in response to transcription factors or the template topology (Theißen et al., 1990b; Krohn et al., 1992). It turned out that some pauses in the early leader are dependent on the transcription factor NusA. In addition, high superhelical density of the template, known so far to be a prerequisite for efficient transcription initiation, was shown to enhance pausing strength at certain sites, and thereby offers a potential mechanism for the regulation of transcription elongation (Krohn et al., 1992).

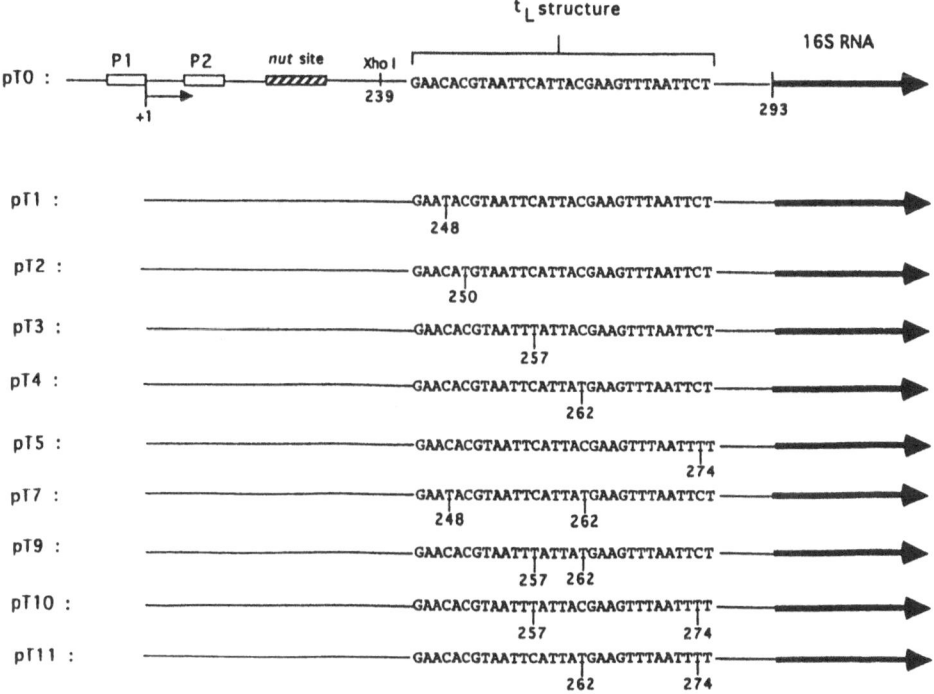

Figure 4. Schematic arrangement of the t_L transition mutations. The names of the mutant plasmids are given on the left. Base changes are indicated by numbers denoting the nucleotide position relative to the transcription start site at P1. pT0 harbors the genuine *rrnB* sequence with the exception of a single A to G base change creating a unique XhoI site. The promoters P1 and P2, the *nut* site sequence and the beginning of the 16S RNA gene are indicated.

One of the sites for which pausing is characteristic is termed t_L region (see Fig. 1). Our original supposition that the t_L structure might function as a discriminator, and facilitate antitermination has been discarded after it became evident that mutations in t_L did not affect the read-through potential of RNA polymerase but rather showed post-transcriptional effects (Theißen et al., 1990c). We now know (see below) that at least part of the t_L sequence is involved in the biogenesis of functional 30S subunits.

Function of the Leader in Structure Formation and Ribosome Biogenesis

The results outlined above already suggest that the rRNA leader must be involved in processes other than transcriptional control. In fact, there are some recent results which demonstrate that part of the leader sequence is involved in directing the formation of functionally active ribosomal particles. This activity can best be described by a chaperon-like function of the rRNA leader facilitating correct biogenesis of the 30S subunit. A similar function may reside in the spacer, and the processing defects triggered by removal or displacement of the spacer tRNA gene (Szymkowiak et al., 1988; Srivastava et al., 1989) can be taken as evidence in that direction. The involvement of leader and spacer sequences in chaperon-like support of ribosomal structure formation is schematically depicted in Fig. 3.

We like to outline several lines of evidence in support of this notion. First evidence for a chaperon-like action of the leader stems from studies performed with single and double base change mutations in the t_L region. They are based on the analysis of mutants which contain single or double C to T transitions within the t_L region (Theißen et al., 1990c). The site of mutations and the corresponding plasmid constructs are indicated schematically in Fig. 4. The mutations, when present on multicopy plasmids, confer changes in the growth phenotype of transformed cells ranging from slightly enhanced to severely retarded growth. We could show that ribosomes isolated from leader mutant strains showed functional defects. Isolated ribosomal subunits from slow growing mutants, for instance, had a strongly reduced capacity to associate to 70S particles. For one of the slow growing mutants a detailed structural analysis of the 30S subunits was performed. By S1 mapping and primer extension we could show that 3' as well as 5' maturation of the 16S RNA from the mutant 30S subunits was not significantly altered. However, several non-ribosomal proteins could be demonstrated to be associated with the free mutant 30S particle, compared to the corresponding wild type preparation. In line with this finding the mutant 30S particles showed a slightly enhanced mobility on low percentage polyacrylamide gels, indicative of a subtle structural alteration.

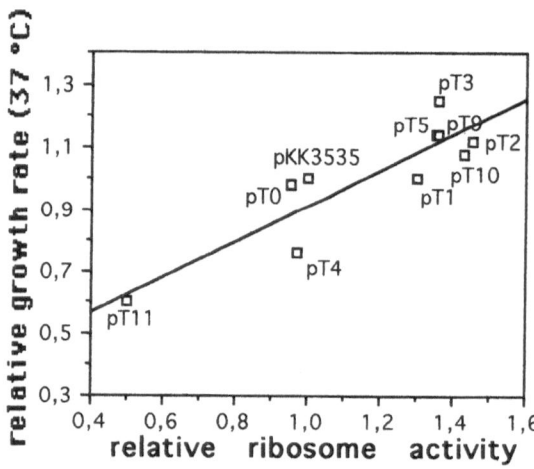

Figure 5. Correlation of the *in vitro* poly(U) dependent translational capacity of ribosomes from different leader mutant strains and the corresponding growth rates. Data were obtained from HB101 cells transformed with the plasmids shown in Fig. 4, and with ribosomes prepared from these cells, respectively. The data is normalized to 1 for the results obtained with plasmid pKK3535, which contains the unmodified *rrnB* operon.

Ribosomes isolated from various t_L mutants were tested for their *in vitro* translational activity. It turned out that the polypeptide synthesis efficiency is proportional to the growth rates noted for the different mutants (Fig. 5).

From these results we have to conclude that sequences within the leader region are responsible for the correct formation of 30S subunits, although these sequences are only transiently linked to the nascent pre-30S particles, and cleaved off during the normal processing events. We do not presently know if the leader sequence directly interacts with parts of the 16S RNA sequence in order to facilitate rRNA or pre-ribosome formation, or whether additional proteins are involved. Since the anti-termination factors NusA, B, E, and G are considered to interact with the rRNA leader *nut* site they are prime candidates for any possible protein mediated interaction (note that Nus E is identical with the ribosomal protein S10!).

A similar observation concerning the function of the rRNA leader was made during the study of rRNA mutants performed in Harry Noller's lab. There, teperature-sensitive revertants of a 16S RNA encoded spectinomycin resistance were isolated. For some of the mutants it could be shown by sequence analysis that single base changes are responsible for the phenotype. However, the remarkable point is again that these base changes are located outside the structural 16S RNA gene. They map within the t_L leader region! Since the mutant ribosomes are processed correctly, the altered leader sequence is physically separated from the mature ribosomes, yet the function of the resulting particles is clearly altered (Mori et al., 1990).

The mechanism of assisting correct folding and assembly by external rRNA sequences may not be restricted to bacteria. Analogous conclusions for the function of leader rRNA sequences were derived for yeast by Musters et al., (1990).

Acknowledgements

This article is dedicated to the memory of H. G. Wittmann who for many years has generously supported R. W. and the colleagues of his group. We like to thank all the people from our lab which do not appear as authors here, but have contributed in many ways to the results presented above. We are grateful to the Deutsche Forschungsgemeinschaft supporting our work over the last years.

REFERENCES

Albrechtsen, B., Ross, B. M., Squires, C. and Squires, C. L., 1984, *Nucleic Acids Res.* 19: 1845-1852.
Austin, S. and Dixon, R., 1992, *EMBO J.* 11: 2219-2228.
Baracchini, E. and Bremer, H., 1988, *J. Biol. Chem.* 263: 2597-2602.
Barik, S., Gosh, B., Whalen, W., Lazinski, D. and Das, A., 1987, *Cell* 50: 885-899.
Bauer, B. F., Kar, E. G., Elford, R. M. and Holmes, W. M., 1988, *Gene* 63: 123-134.
Berg, K. L., Squires, C. and Squires, C. L., 1989, *J. Mol. Biol.* 209:345-358.
Csiszar, K., Lukacsovich, T. and Venetianer, P., 1990, *Biochim. Biophys. Acta* 1050: 312-316.
Condon, C., Philips, J., Fu, Z-Y., Squires, C. and Squires, C. L., 1992, *EMBO J.* 11: 4175-4185.
Ellwood, M. and Nomura, M., 1980, *J. Bacteriol.* 143: 1077-1080.
Ellwood, M. and Nomura, M., 1982, *J. Bacteriol.* 149: 458-468.
Friedman, D. I., Olson, E. R., Georgopoulos, C., Tilly, K., Herskowitz, I. and Banuett, F., 1984, *Microbiol. Rev.* 48: 299-325.
Gaal, T. and Gourse, R. L., 1990, *Proc. Natl. Acad. Sci. USA* 87: 5533-5537.
Galant, J. A., 1979, *Annu. Rev. Genet.* 13: 393-415.
Gausing, K, 1977, *J. Mol. Biol.* 115: 335-354.
Gegenheimer, P. and Apirion, D., 1981, *Microbiol. Rev.* 45: 502-541.
Gentry, D., Xiao, H., Burgess, R. and Cashel, M., 1991, *J. Bacteriol.* 173: 3901-3903.
Glaser, G., Sarmientos, P. and Cashel, M., 1983, *Nature* 302: 74-76.
Gourse, R. L, 1988, *Nucleic Acids Res.* 16: 9789-9809.
Gourse, R. L, Takebe, Y. S., Sharrock, R. A. and Nomura, M., 1985, *Proc. Natl. Acad. Sci. USA* 82: 1069-1073.
Hawley, D. K. and McClure, W. R., 1983, *Nucleic Acids Res.* 11: 2237-2254.
Hernandez, V. J. and Bremer, H., 1992, personal communication
Hill, C. W. and Harnish, B. W., 1982, *J. Bacteriol.* 149: 449-457.
Holben, W. E. and Morgan, E. A., 1984, *Proc. Natl. Acad. Sci. USA* 81: 6789-6793.
Horwitz, R. J., Li, J. and Greenblatt, J., 1987, *Cell* 51: 631-641.
Hübner, P. and Arber, W., 1989, *EMBO J.* 8: 577-585.
Igarahsi, K., Fujita, N. and Ishihama, A., 1989, *Nucleic Acids Res.* 17: 8755-8765.
Jinks-Robertson, S., Gourse, R. L. and Nomura, M., 1983, *Cell* 33: 865-978.
Johnson, R. and Simon, M. I., 1985, *Cell* 41: 781-791.
Kahmann, R., Rudt, F., Koch, C. and Mertens, G., 1985, *Cell* 41: 771-780.
King, T. C., Sirdeskmukh, R. and Schlessinger, D., 1986, *Microbiol. Rev.* 50: 428-451.
Kingston, R. E., 1983, *Biochemistry* 22: 5249-5254.
Kingston, R. E. and Chamberlin, M. J., 1981, *Cell* 27: 523-531.

Krohn, M., Pardon, B. and Wagner, R., 1992, *Molecular Microbiol.* 6: 581-589.
Kustu, S., Nath, A. K. and Weiss, D. S., 1991, *Trends in Bochem. Sci.* 16: 397-402.
Lamond, A. I. 1985, *Trends in Biochem. Sci.* 6: 271-274.
Lamond, A. I. and Travers, A. A., 1983, *Nature*, 305: 248-250.
Landick, R., Carey, J. and Yanofsky, C., 1987, *Proc. Natl. Acad. Sci. USA* 84: 1507-1511.
Langert, W., Meuthen, M. and Müller, K., 1992, *J. Biol. Chem.* 266: 21608-21615.
Leirmo, S. and Gourse, R. L., 1991, *J. Mol. Biol.* 220: 555-568.
Li, S. C., Squires, C. L. and Squires, C., 1984, *Cell* 38: 851-860.
Li, J., Horwitz, R., McCracken, S. and Greenblatt, J., 1992, *J. Biol. Chem.* 267: 6012-6019.
Lindahl, L. and Zengel, J. M., 1986, *Annu. Rev. Genet.* 20: 297-326.
Little, R., Ryals, J. and Bremer, H., 1983, *J. Bacteriol.* 154: 787-792.
Lukacsovich, T., Gaal, T. and Venetianer, P., 1989, *Gene* 78: 189-194.
Marini, C. J., Levene, S. D., Crothers, D. M. and Englund, P. T., 1982, *Proc. Natl. Acad. Sci.USA* 79: 7664-7668.
Mason, S. W. and Greenblatt, J., 1991, *Genes & Development* 5: 1504-1512.
Morgen, E. A. and Kaplan, S., 1976, *Biochem. Biophys. Res. Commun.* 68: 969-974.
Mori, H., Dammel, C., Becker, E., Triman, K. and Noller, H. F.,1990, *Biochim. Biophys. Acta* 1050: 323-327.
Musters, W., Boon, K., van der Sande, C. A. F. M., van Heerikhuizen, H. and Planta, R. J., 1990, *EMBO J.* 9: 3989-3996.
Newlands, J. T., Josaitis, C. A., Ross, W. and Gourse, R. L., 1992, *Nucleic Acids Res.* 20: 719-726.
Nierlich, D. P., 1978, *Annu. Rev. Microbiol.* 32: 393-432.
Nilsson, L., Vanet, A., Vijgenboom, E. and Bosch, L.,1990, *EMBO J.* 9: 727-734.
Nodwell, J. R. and Greenblatt, J., 1991, *Genes & Development* 5: 2141-2151.
Nomura, M., Gourse, R. L. and Baughman, G., 1984, *Annu. Rev. Biochem.* 53: 75-117.
Platt, T., 1981, *Cell* 24: 10-23.
Resnekov, O. and Aloni, Y., 1989, *Proc. Natl. Acad. Sci. USA* 86: 12-16.
Ross, W., Thompson, J. F., Newlands, J. T. and Gourse, R. L., 1990, *EMBO J.* 9: 3733-3742.
Ryals, J. and Bremer, H., 1982, *J. Bacteriol.* 150: 168-179.
Sanders, D. A., Gillece-Castro, B. L., Burlingame, A. L. and Koshland, D. E., 1992, *J. Bacteriol.* 174: 5117-5122.
Sarmientos, P., Sylvester, J. E., Contente, S. and Cashel, M., 1983, *Cell* 32: 1337-1346.
Singer, M., Rossmiessl. P., Cali, B. M., Liebke, H. and Gross, C. A., 1991, *J. Bacteriol.* 173: 6242-6248.
Singer, M., Walter, W. A., Cali, B. M., Rouviere, P., Liebke, H., Gourse, R. L. and Gross, C. A., 1991, *J. Bacteriol.* 173: 6249-6257.
Srivastava, A. K. and Schlessinger, D., 1988, *Proc. Natl. Acad. Sci. USA* 85: 7144-7148.
Srivastava, A. K. and Schlessinger, D., 1989, *Nucleic Acids Res.* 17: 1649-1663.
Srivastava, A. K. and Schlessinger, D., 1990, *Annu. Rev. Microbiol.* 44: 105-129.
Szymkowiak, C., Reynolds, R. L., Chamberlin, M. J. and Wagner, R., 1988, *Nucleic Acids Res.* 16: 7885-7899.
Theißen, G., Behrens, S. E. and Wagner, R., 1990a, *Molecular Microbiol.* 4: 1667-1678.
Theißen, G., Pardon, B. and Wagner, R., 1990b, *Anal. Biochem.* 189: 254-261.
Theißen, G., Eberle, J., Zacharias, M., Tobias, L. and Wagner, R., 1990c, *Nucleic Acids Res.* 18: 3893-3901.
Travers, A. A., 1980, *J. Bacteriol.* 141: 937-976.
Vijgenboom, F., Nilsson, L. and Bosch, L., 1988, *Nucleic Acids Res.* 16: 10183-10197.
Wagner, R., 1989, *Life. Sci. Adv. Mol. Gen.* 8: 105-115.
Yang, X., Goliger, J. A. and Roberts, J. W., 1989, *J. Mol. Biol.* 210: 453-460.
Zacharias, M., Göringer, H. U. and Wagner, R., 1989, *EMBO J.* 8: 3357-3363.
Zacharias, M., Wagner, R. and Göringer, H. U., 1990a, *Nucleic Acids Res.* 18: 2827.
Zacharias, M., Göringer, H. U. and Wagner, R., 1990b, *Nucleic Acids Res.* 18: 6271-6275.
Zacharias, M., Theißen, G., Bradaczek, C. and Wagner, R., 1991, *Biochimie* 73: 699-712.
Zacharias, M., Göringer, H. U. and Wagner, R., 1992, *Biochemistry* 31: 2621-2628.

REGULATION OF THE ELEVEN GENE S10 RIBOSOMAL PROTEIN OPERON BY THE 50S SUBUNIT PROTEIN L4

Janice M. Zengel and Lasse Lindahl

Department of Biology
University of Rochester
Rochester, NY 14627

INTRODUCTION

The S10 operon of *Escherichia coli* encodes 11 r-proteins. Like most other r-protein operons, it is regulated by one of its own gene products, the 50S subunit protein L4 which is encoded by the third gene of the operon (Lindahl and Zengel, 1979; Yates and Nomura, 1980; Zengel et al., 1980). However, unlike these other operons, the autogenous regulation of the S10 operon is achieved by two different regulatory circuits, both activated by L4 and each contributing about 5-fold inhibition of the expression of the S10 operon (Freedman et al., 1987). The mechanism for one of these circuits is repression of transcription by induction of premature termination in the S10 leader about 30 bases upstream from the first gene of the operon (Zengel et al., 1980; Lindahl et al., 1983; Zengel and Lindahl, 1990a; Zengel and Lindahl, 1990b). This mechanism appears to be unique among the r-protein operons. The other circuit works by repressing translation of the mRNA (Yates and Nomura, 1980; Freedman et al., 1987). Such translation control, although varying in molecular detail, is common to all r-protein operons for which autogenous regulation has been studied (Lindahl and Zengel, 1986; Jinks-Robertson and Nomura, 1987), with the exception of the *trmD* operon (Wikström et al., 1988).

DOMAINS OF THE S10 mRNA LEADER

The organization of the 172 base long S10 leader has been analyzed by genetic studies (Freedman et al., 1987; Zengel and Lindahl, 1992), secondary structure mapping (Shen et al., 1988), and phylogenetic comparisons (Shen, 1991). Our current impression is that the leader is organized into functional modules, each of which is composed of one or more hairpins. The relationship between function and secondary structure is summarized in Figure 1. The three most promoter proximal hairpins, HA, HB and HC, are dispensable for both types of L4-mediated regulation and have no identified function in the expression of the S10 operon (J. M. Zengel, L. Cassidy, P. Manzanero and L. Lindahl, unpublished experiments). The module containing the largest hairpin, called HE, is required for both transcription and translation regulation by L4. The hairpin immediately upstream of this region, HD, is needed only for transcription regulation, while sequences downstream of HE are involved only in translation regulation. The modular nature of the leader is also evident from the finding that the HD and HE hairpins have separable functions in the transcription regulation (see below).

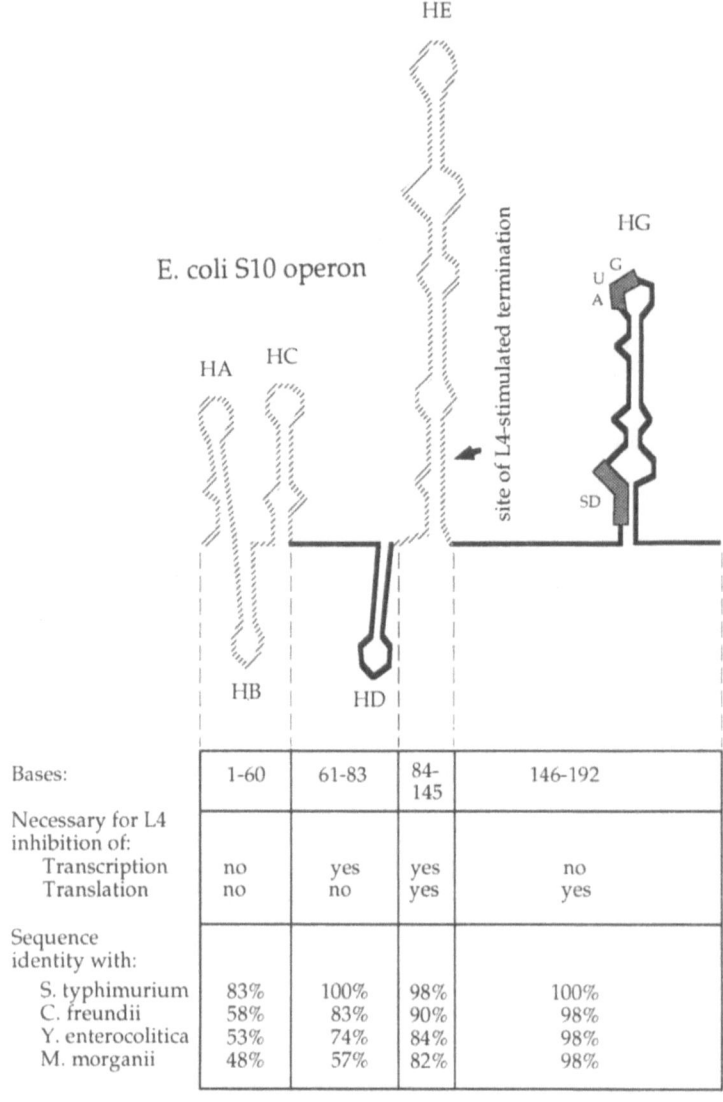

Bases:	1-60	61-83	84-145	146-192
Necessary for L4 inhibition of:				
Transcription	no	yes	yes	no
Translation	no	no	yes	yes
Sequence identity with:				
S. typhimurium	83%	100%	98%	100%
C. freundii	58%	83%	90%	98%
Y. enterocolitica	53%	74%	84%	98%
M. morganii	48%	57%	82%	98%

Figure 1. Domain structure of the leader of the S10 r-protein operon of *E. coli*.   The secondary structure of the leader is drawn according to Shen et al. (1988). Domains needed for L4-mediated regulation at the attenuator or at the S10 translation initiation site are indicated (Freedman et al., 1987; Zengel et al., 1990a; our unpublished experiments) . Data for the comparison of *E. coli* leader with the leaders of several other enterobacteria are from Shen (1991).

The proposed functional domains in the leader are supported by phylogenetic comparisons of the *E. coli* leader with the S10 leaders of four other enterobacteria: *Salmonella typhimurium, Citrobacter freundii, Yersinia enterocolitica*, and *Morganella morganii* (Shen, 1991). A summary of the comparison between *E. coli* and these other species is given in Figure 1. The promoter proximal half of the leader, containing the hairpins HA, HB and HC which are dispensable for L4 regulation, shows as little as 50% sequence identity between *E.coli* and the other bacteria. Nevertheless, these sequences are still compatible with the existence of three hairpins, although particularly in the cases of HA and HC there are substantial variations in hairpin size and structure. An even stronger case can be made for the conservation of hairpin HD, which is in a region of the leader that is essential for transcription (but not translation) regulation by L4. The primary sequence homology is fairly low, but sequences from all four enterobacteria fit the format of a six base pair stem and a five base loop. Indeed, several "compensatory changes" in the stem support the genetic evidence for the functional importance of the HD hairpin.

In contrast to the relatively weak homology in the upstream leader sequence, there is more than 80% identity in the 3' half of the leader containing the sequences required for both L4-mediated regulatory processes. The homology is especially striking for the sequence encompassing the S10 structural gene and the 25-30 bases upstream. In the HE hairpin, most of the changes involve bases predicted by our structure analysis to be non-base-paired (i.e., in an internal loop or bulge). No "compensatory changes" were observed in helical regions, but the few changes involving bases within helices were "neutral" substitutions (e.g., AU to GU) that probably preserve the overall secondary structure of the HE hairpin region.

L4 INTERACTION WITH RNA

The generic model for autogenous regulation of r-protein synthesis is based on the observation that most regulatory ribosomal proteins have been identified as primary rRNA binding proteins in ribosome reconstitution studies. Early studies showed, for example, that r-protein L4 can bind specifically to the 5' third of 23S rRNA (Spierer et al., 1975). Thus it is assumed that each regulatory ribosomal protein binds to a site in its own mRNA which is structurally homologous to its rRNA binding target (Nomura et al., 1980). Indeed several r-proteins have been demonstrated experimentally to bind directly to their mRNA (Johnsen et al., 1982; Deckman and Draper, 1985; Gregory et al., 1988; Philippe et al., 1990), even though the homology with the rRNA binding site is not always obvious (Tang and Draper, 1989). So far our attempts to bind L4 to the S10 mRNA have failed (P. Shen, J. M. Zengel and L. Lindahl, unpublished results). One possible reason for our failure to observe interaction between L4 and the S10 mRNA is that the L4 binding site might include protein determinants contributed by RNA polymerase, NusA or an initiating ribosome. Another possibility is that the nascent S10 leader RNA might undergo a structural change upon completion of its synthesis which "removes" the L4 binding site. So far, we have only used RNA containing the complete leader sequence, which has the structure shown in Figure 1; this RNA may have the "wrong" conformation for recognition by L4.

A hyphenated sequence homology between the S10 leader and domain II of 23S rRNA was previously suggested to be important for the regulation of the

S10 operon (Olins and Nomura, 1981), based on cross-linking of L4 to this region of 23S rRNA after UV irradiation of the intact 50S subunit (Maly et al., 1980). However, mutagenesis of the S10 leader failed to substantiate the importance of this sequence homology (Freedman et al., 1987; Zengel and Lindahl, 1990a; Zengel and Lindahl, 1990b). Furthermore, it was later found that L4 can also be cross-linked to domain I of 23S rRNA (Gulle et al., 1988). We therefore examined the issue of L4 binding to 23S rRNA by asking if specific fragments of 23S rRNA can compete for L4 in the *in vitro* transcription assay (see below) and thereby eliminate the effect of L4 on transcription elongation. Results from such experiments have demonstrated that domain I alone, but not domain II alone, can bind L4 (Zengel and Lindahl, 1991; J. M. Zengel and L. Lindahl, unpublished results). Thus, the L4 binding site is contained within domain I, not domain II, consistent with our genetic studies suggesting that the homologies between the domain II and the S10 leader most likely are fortuitous. Although our current model for L4 regulation of the S10 operon still assumes that the protein recognizes a site in the S10 leader that is very similar to its rRNA target, there are no obvious primary or secondary structure homologies between the S10 leader and domain I of 23S rRNA. Perhaps the recognition elements are too subtle, and require information about the tertiary structure of the two RNA molecules. We may also be looking at the "wrong" S10 leader structure if the nascent leader molecule transiently forms an alternative structure during its transcription. Hopefully, further refinements of the RNA structures will solve this problem. In any case, we believe that L4 must make some contact with the S10 leader, since this is the simplest way to account for the specificity of the L4-mediated regulation.

TRANSCRIPTION REGULATION

As already mentioned, L4 regulates transcription of the S10 operon by stimulating RNA polymerase to terminate prematurely within the S10 leader. Experiments with a purified transcription system (Zengel and Lindahl, 1990b) have shown that L4-mediated transcription regulation requires NusA, an auxiliary transcription factor originally identified because of its function in antitermination of transcription in bacteriophage lambda. NusA is necessary to provoke RNA polymerase to pause at the attenuator site (Zengel and Lindahl, 1992). In the absence of NusA, the enzyme does not pause and L4 has essentially no effect on the transcription. On the other hand, polymerases paused under the influence of NusA are inhibited strongly by L4 from returning to the elongation mode (Zengel and Lindahl, 1992). *In vivo* RNA synthesis measurements indicate that the augmented pause leads to true transcription termination, i.e. release of the RNA polymerase and transcript from the template. However, in the *in vitro* system relatively little release of polymerase and RNA is seen (such is frequently the case for transcription terminators analyzed in cell-free systems). Based on the *in vitro* experiments, we have proposed that transcription termination at the S10 attenuator is the result of a pathway of separable steps, each stimulated by separate proteins (Zengel and Lindahl, 1992). This tentative model is illustrated in Figure 2. In the first step, NusA induces RNA polymerase to pause at the attenuator. In the second step, L4 prevents the enzyme from resuming elongation. Finally, in a third step so far just inferred from comparing *in vivo* and *in vitro* transcription, the RNA polymerase and nascent RNA are released from the template.

The step-wise process illustrated in Figure 2 is compatible with the organization of the leader into the functional domains discussed above. That is, the HE region suffices for the NusA-mediated induction of a pause at the S10 attenuator, but HD is required for the L4-dependent stabilization of this paused transcription complex (Zengel and Lindahl, 1992). It is interesting that the L4 effect depends on a hairpin upstream of the attenuator structure, yet the protein stabilizes the paused complex even if it is not added to the reaction until RNA polymerase has already reached the pause/termination site. In other words, L4 does not need to be present while RNA polymerase is transcribing the HD and HE regions. The role of hairpin HD is not clear. It might represent the L4 binding site on the leader RNA. However, since HD is dispensable for L4-mediated repression of translation, this would imply that translation control involves a different leader binding site for L4. Preliminary studies of the HE hairpin has so far shown that the upper stem is required for

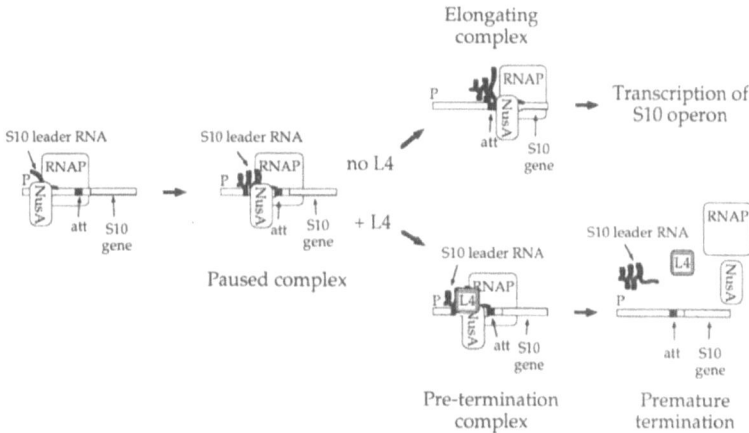

Figure 2. Model for L4-mediated transcription termination at the S10 attenuator (adapted from Zengel and Lindahl, 1992).

efficient pausing, even though this portion of the HE hairpin is about 15 bases upstream of the pause site (J. M. Zengel and L. Lindahl, preliminary experiments). Future studies will hopefully provide further details about the functions of the HD and HE hairpins in L4-mediated transcription termination.

TRANSLATION REGULATION

Our secondary structure probing experiments suggest that the S10 leader can assume several different conformations generated by gradually shortening the bottom of hairpin HE and simultaneously creating a new hairpin, HF, just upstream of the Shine-Dalagarno region (Shen et al., 1988). The two most extreme conformers are shown in Figure 3. We have proposed that the equilibrium between these structures could affect the translation efficiency of the S10 gene in the following way. The sequestering of the Shine-Dalgarno region in hairpin HG in both structures suggests that in neither case can ribosomes bind directly to the S10 translation initiation region. Rather, we speculate that the single-stranded region between HE and HG in the "translatable" structure shown in Figure 3 serves as an "entry-site" where the ribosome makes its first, perhaps weak, association with the mRNA (Shen et al., 1988). Once the ribosome is associated with the message, we propose that it can more efficiently gain access to the Shine-Dalgarno region, perhaps by contributing directly to the melting of basepairs in the HG hairpin. Thus, the S10 mRNA is translatable when this single-stranded entry-site is available. Conversely, if the S10 leader assumes the conformation in which hairpin HF forms (the "untranslatable" structure in Figure 3) the entry-site is sequestered. This hypothesis would then provide a working model for translation repression by L4 if we assume that the r-protein somehow favors the untranslatable conformation.

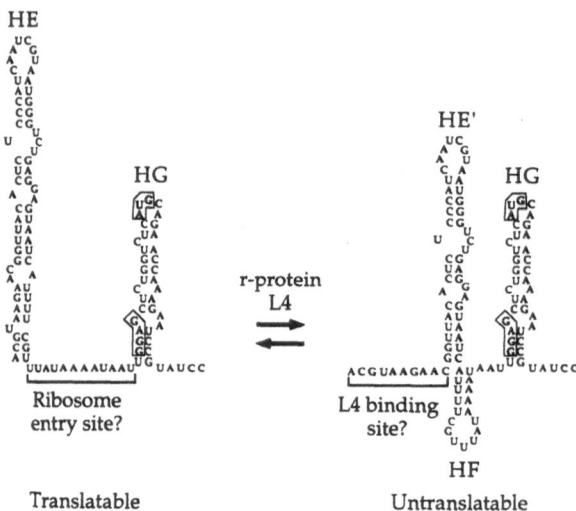

Figure 3. Model for the L4-mediated repression of translation of the S10 operon. Only nucleotides involved in hairpins HE, HF and HG are shown.

Consistent with this working model, an eight base deletion in the region designated as the ribosome entry site reduces the translation efficiency 10-fold and eliminates L4-mediated translation repression (J. M. Zengel and L. Lindahl, unpublished experiments). However, this model is clearly an oversimplification of the parameters involved in regulating translation of the S10 operon. For example, it cannot account for our observation that translation regulation is affected by several mutations in the upper stem-loop of hairpin HE, well upstream of the ribosome entry-site and the Shine-Dalgarno sequence (Freedman et al., 1987). In that connection, it is interesting to note that the *in vivo* translation-control phenotype of these hairpin mutations more or less correlates with the effect on NusA-dependent pausing observed in the *in vitro* transcription system (our unpublished experiments). Therefore, it is tempting to speculate that L4 can mediate its effect on translation of the S10 gene only if it is bound to the mRNA during a pause in transcription. For example, the L4 binding site may involve a region of the S10 leader on the ascending side of the HE hairpin, still single-stranded during the NusA-dependent pause. Once bound, the protein might prevent the lower stem of HE from forming upon completion of the leader synthesis and thereby favor formation of hairpin HF, i.e. the untranslatable form. This hypothesis could also explain why we have failed to observe L4 binding to the leader, since this single-stranded target is sequestered by the lower stem of HE present in the full length mRNAs used in our binding experiments.

PERSPECTIVES

Fairly detailed molecular models for regulating several r-protein operons in *E. coli* have emerged in recent years. Nevertheless, there are many aspects of the regulation which are not understood. Furthermore, there are important new problems to be addressed in the S10 operon and other r-protein operons. Among these is the mapping of functional domains in the regulatory proteins themselves. For example, are the same or different domains used for autogenous regulation and for incorporation and/or function in a ribosome? Another interesting question is how the regulatory mechanisms evolved. The sequences of many ribosomal protein genes in many bacterial species are now available. It is evident that many of the r-protein genes are maintained in the same order, even though some genes have been deleted or inserted during evolution. However, there are striking differences in the positions of promoter and terminators. As a result, some transcription units contain more than one of the genes corresponding to the *E. coli* regulatory r-proteins, while other transcription units have none of the *E. coli* regulatory genes. Yet there is evidence that at least some other bacterial species are using regulatory mechanisms similar to the autogenous control found in *E. coli* (Grundy and Henkin, 1991). Thus, it will be interesting to compare the structural details of the regulatory units in different bacterial species. This could provide clues to our understanding of how the regulatory proteins were recruited as regulators. It might also help determine if autogenous regulation of r-protein synthesis was "invented" more than once in evolution, or if all systems of autogenous regulation evolved from one master system.

REFERENCES

Deckman, I. C., and Draper, D. E., 1985, Specific interaction between ribosomal protein S4 and the α operon messenger RNA. *Biochemistry* 24:7860.

Freedman, L. P., Zengel, J. M., Archer, R. H., and Lindahl, L., 1987, Autogenous control of the S10 ribosomal protein operon of *Escherichia coli*: genetic dissection of transcriptional and post-transcriptional regulation. *Proc. Natl. Acad. Sci. USA* 84:6516.

Gregory, R. R., Cahill, P. B. F., Thurlow, D. L., and Zimmermann, R. A., 1988, Interaction of *Escherichia coli* ribosomal protein S8 with its binding sites in ribosomal RNA and messenger RNA. *J. Mol. Biol.* 204:295.

Grundy, F. J., and Henkin, T. M., 1991, The *rpsD* gene, encoding ribosomal protein S4, is autogenously regulated in *Bacillus subtilis*. *J. Bacteriol.* 173:4595.

Gulle, H., Hoppe, E., Osswald, M., Greuer, B., Brimacombe, R., and Stöffler, G., 1988, RNA-protein cross-linking in *E. coli* 50S ribosomal subunits: determination of sites on 23S RNA that are cross-linked to proteins L2, L4, L24 and L27 by treatment with 2-iminothiolane. *Nucleic Acids Res.* 16:815.

Jinks-Robertson, S., and Nomura, M., 1987, Ribosomes and tRNA, *in*: "*Escherichia coli* and *Salmonella typhimurium*: Cellular and Molecular Biology," F. C. Neidhardt, J. L. Ingraham, K. B. Low, M. Schaechter and E. Umbarger, ed., American Society for Microbiology, Washington DC.

Johnsen, M., Christensen, T., Dennis, P. P., and Fiil, N. P., 1982, Autogenous control: ribosomal protein L10-L12 complex binds to the leader sequence of its mRNA. *EMBO J.* 1:999.

Lindahl, L., Archer, R., and Zengel, J. M., 1983, Transcription of the S10 ribosomal protein operon is regulated by an attenuator in the leader. *Cell* 33:241.

Lindahl, L., and Zengel, J., 1986, Ribosomal genes in *Escherichia coli*. *Ann. Rev. Genet.* 20:297.

Lindahl, L., and Zengel, J. M., 1979, Operon-specific regulation of ribosomal protein synthesis in *Escherichia coli*. *Proc. Natl. Acad. Sci. USA* 76:6542.

Maly, R., Rinke, J., Sweib, C., and Brimacombe, R., 1980, Precise localization of the site of cross-linking between L4 and 23S ribonucleic acid induced by mild ultraviolet irradiation of 50S ribosomal subunits. *Biochemistry* 19:4179.

Nomura, M., Yates, J. L., Dean, D., and Post, L. E., 1980, Feedback regulation of ribosomal protein gene expression in *Escherichia coli*: Structural homology between ribosomal RNA and ribosomal protein mRNA. *Proc. Natl. Acad. Sci. USA* 77:7084.

Olins, P. O., and Nomura, M., 1981, Regulation of the S10 ribosomal protein operon in *E. coli*: nucleotide sequence at the start of the operon. *Cell* 26:205.

Philippe, C. C., Portier, C., Mougel, M., Grunberg-Manago, M., Ebel, J. P., Ehresmann, B., and Ehresmann, C., 1990, Target site of *Escherichia coli* ribosomal protein S15 on its messenger RNA. Conformation and interaction with the protein. *J. Mol. Biol.* 211:415.

Shen, P. (1991). RNA secondary structure and the regulation of the S10 ribosomal protein operon in *Escherichia coli*. Ph.D. Thesis, The University of Rochester.

Shen, P., Zengel, J. M., and Lindahl, L., 1988, Secondary structure of the leader transcript from the *Escherichia coli* S10 ribosomal protein operon. *Nucleic Acids Res.* 16:8905.

Spierer, P., Zimmermann, R. A., and Mackie, G. A., 1975, RNA-protein interactions in the ribosome: binding of 50S subunit proteins to 5' and 3' terminal fragments of the 23S rRNA. *Eur. J. Biochem.* 52:459.

Tang, C., and Draper, D. E., 1989, Unusual mRNA pseudoknot structure is recognized by a protein translational repressor. *Cell* 57:531.

Wikström, P. M., Byström, A. S., and Björk, G. R., 1988, Non-autogenous control of ribosomal protein synthesis from the *trmD* operon in *Escherichia coli*. *J. Mol. Biol.* 203:141.

Yates, J. L., and Nomura, M., 1980, *E. coli* ribosomal protein L4 is a feedback regulatory protein. *Cell* 21:517.

Zengel, J. M., and Lindahl, L., 1990a, *Escherichia coli* ribosomal protein L4 stimulates transcription termination at a specific site in the leader of the S10 operon independent of L4-mediated inhibition of translation. *J. Mol. Biol* 213:67.

Zengel, J. M., and Lindahl, L., 1990b, Ribosomal protein L4 stimulates *in vitro* termination of transcription at a NusA-dependent terminator in the S10 operon leader. *Proc. Natl. Acad. Sci. USA* 87:2675.

Zengel, J. M., and Lindahl, L., 1991, Ribosomal protein L4 of *Escherichia coli*: *In vitro* analysis of L4-mediated attenuation control. *Biochimie* 73:719.

Zengel, J. M., and Lindahl, L., 1992, Ribosomal protein L4 and transcription factor NusA have separable roles in mediating termination of transcription within the leader of the S10 operon of *E. coli*. *Genes and Dev.* in press

Zengel, J. M., Mueckl, D., and Lindahl, L., 1980, Protein L4 of the *E. coli* ribosome regulates an eleven gene r-protein operon. *Cell* 21:523.

FIS-DEPENDENT *TRANS*-ACTIVATION OF STABLE RNA OPERONS AND BACTERIAL GROWTH

Leendert Bosch[1], Lars Nilsson[2], Erik Vijgenboom[1], and Hans Verbeek[1]

[1]Department of Biochemistry, Leiden University, The Netherlands
[2]Department of Cell Biology, Stockholm University, Sweden

INTRODUCTION

A characteristic feature of *Escherichia coli* is its rapid response to environmental signals. These bacterial cells, growing relatively slowly in a poor medium, promptly speed up their growth upon addition of nutrients to the medium. This rather sudden increase of cell division is attended by a drastic metabolic change. In order to meet the demands of accelerated growth the bacterium has to synthesize more protein per unit of time. Since ribosomes generally move along the messenger RNA at a maximal rate (K. Gausing, 1979) the cell increases its output of protein synthesis by making more ribosomes and tRNA (S. Jinks-Robertson and M. Nomura, (1987). The concentration of ribosomes and tRNA in cells growing exponentially under steady state conditions thus appears to be proportional to the growth rate, a phenomenon now well-known as growth rate-dependent control (Dennis and Bremer, 1974; Gausing, 1974; Gausing, 1979).

The primary event in the expansion of the translational apparatus is the synthesis of ribosomal RNA and tRNA molecules, which occurs by transcription activation of the genes specifying these molecules. The formation of additional ribosomal proteins is not regulated at the gene level and is a secondary event which will not concern us here. Each *E. coli* cell harbours some fifty to sixty different tRNA genes and seven rRNA operons, each comprising three different rRNA genes. Both the tRNA and the rRNA operons are scattered over the entire bacterial genome. All these genes have to be activated in a highly coordinated fashion (Nomura *et al.*, 1984).

Two basic mechanisms have been described to regulate the expression of stable RNA operons, one acting by repression, the other by stimulation of transcription initiation. The repressive control system has been ascribed to feedback inhibition of rRNA and tRNA synthesis by ribosomes not participating in translational movement along the mRNA (ribosome feedback control (Cole *et al.* (1987); Jinks-Robertson *et al.* (1983); Jinks-Robertson *et al.* (1987) Lindahl and Zengel, (1982); Lindahl and Zengel, (1986) and/or inhibition by ppGpp (Cashel and Rudd, 1987; Ryals *et al.*, 1982). These regulatory systems are not fully understood, a number of mechanistic details are unknown, some being object of debate and controversy. In general derepression of the repressive control system results in transcription activation of stable RNA genes. Some years ago, when studying a transcription unit of four tRNA genes and the *tufB* gene (Figure 1A), we found that tRNA and rRNA genes can also be regulated by a positively acting process, thus by gene activation *strictu sensu* (Vijgenboom *et al.*, 1988; Nilsson *et al.*, 1990; Bosch *et al.*, 1990). We presented evidence that this and a number of other stable RNA operons, share a common *trans*-activating protein that binds to *cis*-activating regions upstream of the promoters of these operons. This indicated that more stable RNA operons are regulated by this *trans*-activation system.

In this paper we discuss the regulation of the *thrU(tufB)* operon in relation to cellular growth and present a model for the mechanism of the *trans*-activation.

FIS-DEPENDENT *TRANS*-ACTIVATION OF THE *THRU(TUFB)* OPERON UNDER VARYING GROWTH CONDITIONS

In order to study the transcription regulation of this operon the *tufB* gene was fused to a promoterless *galK* gene, thus putting the expression of this reporter gene under the control of the tRNA operon (Figure 1B). This plasmid-borne construct was introduced into *galK* defective bacterial cells, and the amount of galactokinase activity per femtomole of plasmid per hour was taken as a measure of the transcriptional activity of the operon. Deletion of increasing parts of the DNA starting at position -500 with respect to the transcription start point (arrow in Figure 1) revealed that the sequence extending from position -134 to approximately -60 activates transcription (van Delft *et al.* 1987; Vijgenboom *et al.*, 1988). It was called upstream activator sequence (UAS).

Upstream activator sequences are also found upstream of other tRNA operons and rRNA operons such as illustrated in Figure 2 (Lamond and Travers, 1983; Travers, 1984; Gourse *et al.*, 1986; Bauer *et al.*, 1988; Verbeek *et al.*, 1990). The UAS upstream of the P1 promoter of the *rrnB* operon was described and characterized for the first time by the group of Nomura *et al.* (Gourse *et al.*, 1986). At that time the activator activity was ascribed to the DNA sequence as such and no role of an activating protein was invoked in rRNA operon expression. Evidence obtained in our laboratory suggested that the UAS of the *thrU(tufB)* operon is the target of an activating protein. In fact we isolated a protein that not only binds specifically to this UAS but also to the other upstream sequences illustrated in Figure 2. Subsequently we demonstrated both *in vitro* and *in vivo* that this protein activates the *thrU* operon (Nilsson *et al.*, 1990 and 1992). Our finding that one and the same protein binds specifically to the UAS of each of the operons depicted in Figure 2 was in line with a coordinated activation of these and possibly more tRNA and rRNA operons by a common activating protein.

The nucleotide sequence of the upstream DNA is rich in AT and TA base pairs, indicative of an unusual physical conformation, such as kinking or bending of the DNA helix (Bossi and Smith, 1984; Drew and Travers, 1984; Gourse *et al.*, 1986). Bending is also indicated by the retardation of the electrophoretic migration of DNA fragments derived from the UAS through a gel. *In vitro* binding of the protein to a DNA fragment containing the UAS results in the formation of three protein/DNA complexes (not shown). Their electrophoretic migration in the gel is retarded with respect to that of free DNA, most likely as a result of enhanced DNA bending caused by the binding of the *trans*-activating protein to the UAS. Footprinting using DNase I identifies two regions on the UAS as the binding sites of the protein: one delimited by positions -48 and -81, the other by -118 and -131 (Figure 1C). After deletion of the -118 to -131 protein binding site, transcription drops considerably and drops further when both binding regions are deleted. The correlation between the drop in transcription with that of protein-DNA complex formation strongly suggests that complex formation or bending is a requisite for gene activation (Vijgenboom *et al.*, 1988; Verbeek *et al.*, 1992).

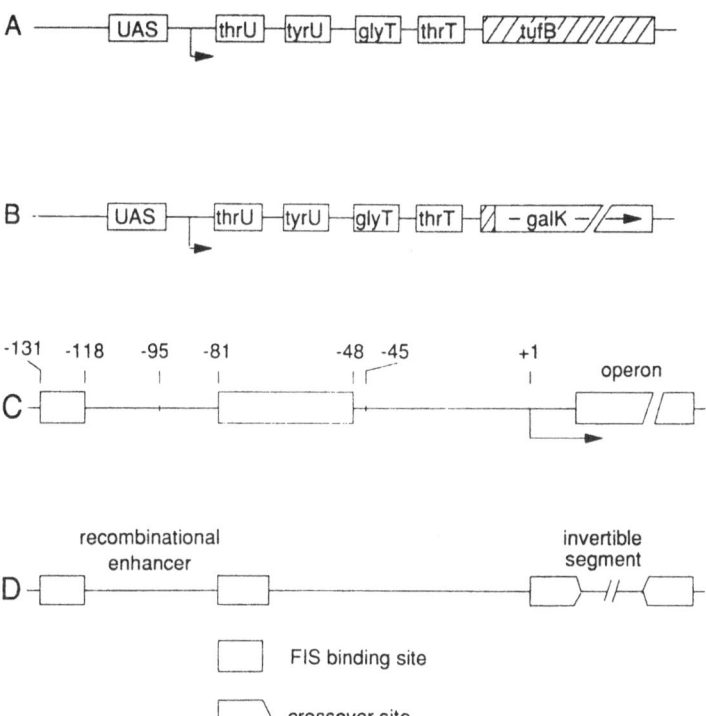

Figure 1. The *thrU(tufB)* operon of *E. coli* (Figures A, B and C) and the invertible DNA segment of certain bacterial viruses (Figure D). A. *TufB* encodes the polypeptide chain elongation factor EF-Tu. It is preceded by four different tRNA genes. B. Translation stop codons in three reading frames between the junction point and *galK* prevent translational readthrough from the cloned *tufB* fragment into *galK*. C. FIS binding regions of the UAS. D. The invertible DNA segment and the recombinational enhancer.

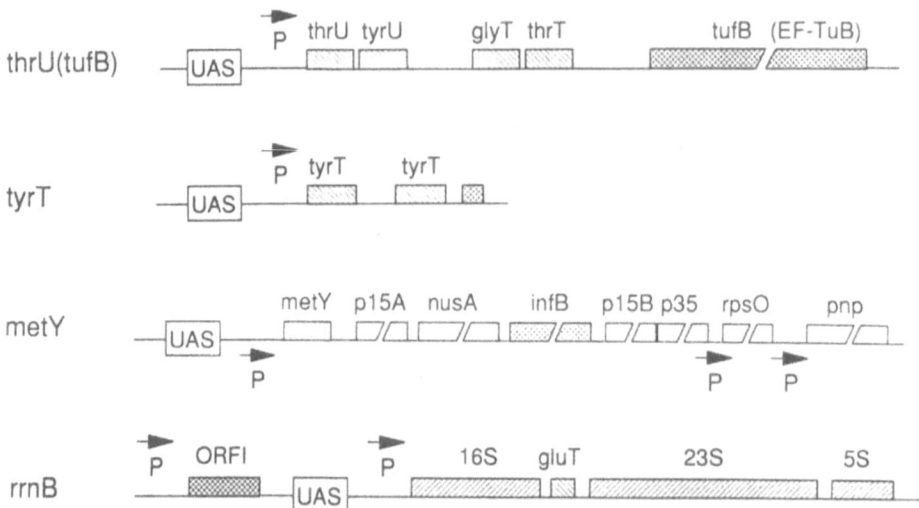

Figure 2. Four stable RNA operons and their respective upstream activator sequences.

Studies in our laboratory (Nilsson *et al.*, 1990; Nilsson *et al.* 1992) and in that of Gourse and coworkers concerning the *rrnB* operon (Ross *et al.*, 1990), revealed the nature of this *trans*-activating protein. As mentioned above, this protein causes bending of DNA. A number of such proteins are known, one of them called FIS appeared to be quite similar to the protein isolated by us. Up till recently, this heat stable protein was only known to play a role in the replication of certain bacterial viruses. These viruses switch the orientation of a certain DNA segment of their genome (Figure 1D). By doing this, they express a different set of tail fiber genes thus altering the host range. The site-specific event that is responsible for the inversion takes place between two inverted repeated recombination sites and is catalyzed by a virus encoded enzyme. The bacterial protein FIS (Factor for Inversion Stimulation) stimulates the inversion by binding to the recombinational enhancer. FIS also plays a role in other recombination systems such as that of the phages lambda and Mu and in transpositions (for a review see S.E. Finkel and R.C. Johnson, 1992).

For some time it was not clear why the bacterium produces the protein FIS, considering its role in virus metabolism. It did not seem to contribute to the welfare of the bacterium itself. Figure 3 shows, however, that the bacterium benefits very much from having this protein. In this experiment the expression of the tRNA operon is studied during a normal bacterial growth cycle. Immediately after initiation of growth a large increase of transcriptional activity takes place. This increase is due to UAS-dependent activation: deletion of the UAS abolishes this increase almost entirely (Figure 3A). This activation is much less in cells unable to make the protein FIS (*fis* cells, Figure 3B). In the early log phase of the growth cycle *fis* cells do show a rise of transcriptional activity but it is much less pronounced than in wild-type cells. In the absence of FIS derepression of the repressive control system occurs. Interestingly, an effect of the UAS is seen in *fis* cells, which becomes apparent by the ratio between UAS-dependent and UAS-independent transcriptional activity. This ratio remains constant throughout the entire growth cycle. In wild-type cells this ratio increases in the early log phase, reaches a maximum and decreases thereafter. A ratio larger than one in *fis* cells reflects a conformational *cis* effect of the UAS which is induced by the nucleotide sequence and which is not affected by environmental signals. Apparently, *cis*- and *trans*-activation of the operon can be observed *in vivo* under appropriate conditions.

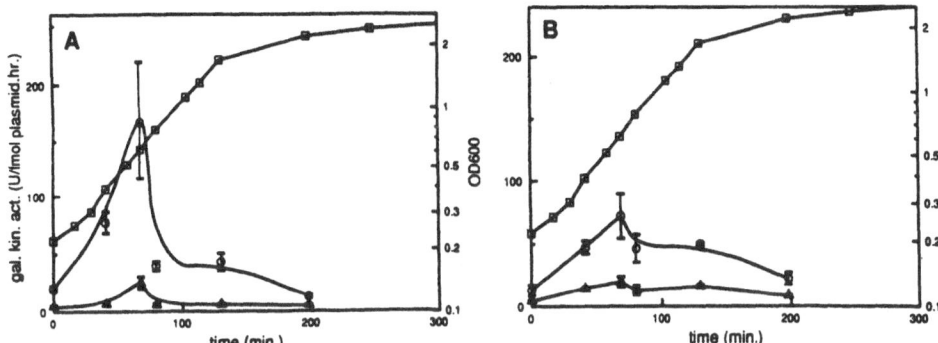

Figure 3. Effects of FIS and UAS on *thrU(tufB)* expression during a growth cycle. Wild type (A) and *fis* cells (B) carrying the *thrU(tufB'):galK* operon fusion were diluted to an OD600 of 0.2 in LB medium and grown at 37°C. Galactokinase activities are expressed in units/fmol plasmid.hr. o - o UAS⁺; Δ - Δ UAS⁻; ◻ - ◻ OD₆₀₀.

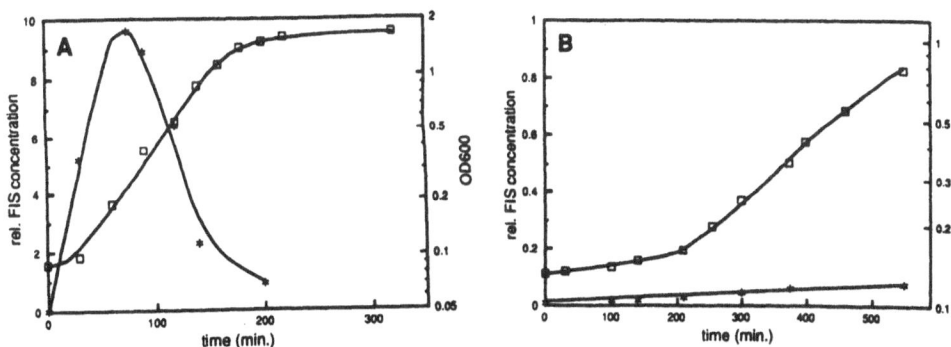

Figure 4. Cellular FIS concentration during a growth cycle. Wild type cells were grown in LB medium plus 1% glucose (A) or in minimal medium (1% succinate; B). The FIS concentration was determined by Western blotting. * - * relative FIS concentration, ◻ - ◻ OD₆₀₀.

Concomitant with the increase in transcriptional activity (Figure 3A), the cellular FIS concentration of wild type cells rises (Figure 4A). Cells growing slowly in a poor medium do not show such a rise of FIS (Figure 4B), nor an increased operon transcription in early log phase (not shown). After addition of nutrients to the medium these cells promptly respond with an increase both of transcriptional activity (Figure 5A) and of the FIS concentration (Figure 5B). Enhanced expression of the *thrU(tufB)* operon is greatly due to a rise in transcription activation, since the ratio of UAS-dependent and UAS-independent galactokinase activities increases approximately fourfold over a period of 2 hr after the nutritional upshift. The rise of FIS occurs within a 1-hr period, whereafter it declines rather steeply and levels off at a relatively elevated level. We conclude that FIS-dependent *trans*-activation acts as a sensor of environmental signals. *fis* cells also sense these signals but the rise in galactokinase activity is much less pronounced than that of wild-type cells and the ratio of UAS-dependent and UAS-independent galactokinase activities remains constant (Figure 5C). Most likely they respond with derepression of the repressive control system.

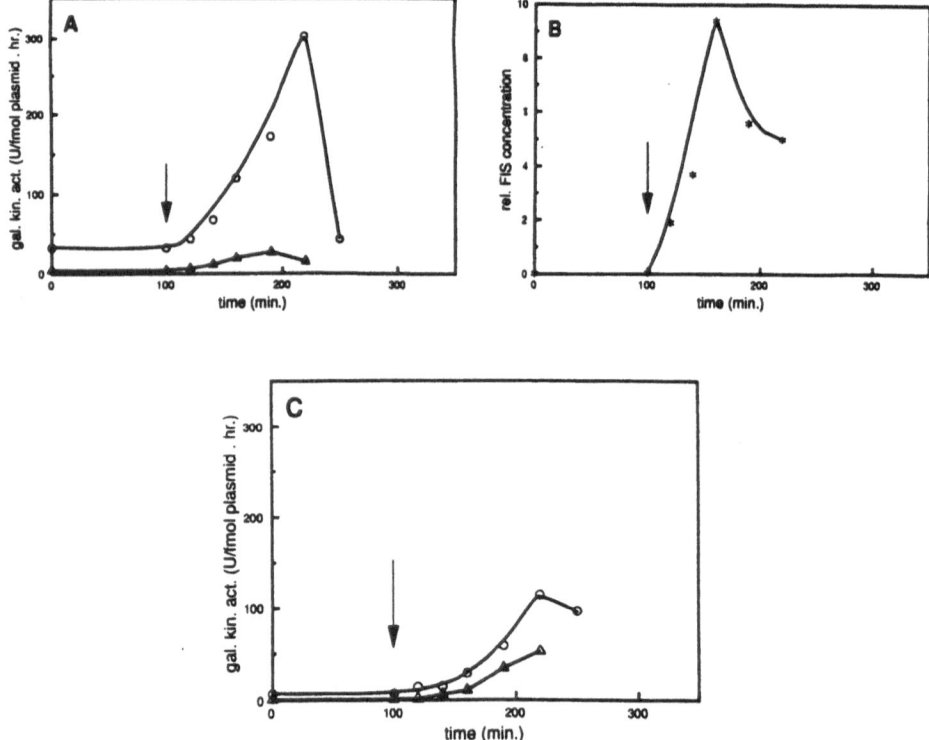

Figure 5. Expression of the *thrU(tufB)* operon after a shift (arrow) from minimal medium to LB medium plus 1% glucose. (A) UAS-dependent (o --- o) and UAS-independent galactokinase (Δ --- Δ) activities in wild-type cells. (B) Relative FIS concentrations (* — *) in wild-type cells. (C) UAS-dependent (o --- o) and UAS-independent (Δ --- Δ) galactokinase activities in *fis* cells.

As mentioned above the cellular concentration of ribosomes during balanced growth of bacterial cultures varies in a linear fashion with the growth rate. Gourse *et al.* (1986) showed that this growth rate-dependent control of the *rrnB* operon is still observed when the UAS upstream the P1 promoter is deleted. This indicates that cells incapable of *trans*-activation regulate ribosome synthesis through the repressive control system. This does not mean, however, that *trans*-activation does not play any role in growth rate-dependent control of wild-type ribosome synthesis, as has been suggested. Figure 6 shows that both cells transformed with the plasmid carrying the *thrU(tufB'):galK* fusion (UAS⁺ cells) and cells transformed with the plasmid deletion derivative lacking the UAS (UAS⁻ cells) show a linear relationship between the galactokinase activities and the growth rates, but the increase of these activities in the range of growth rates studied is 4.6 times for UAS⁺ and 2.6 times for UAS⁻ cells. The conclusion, therefore, is that growth rate-dependent regulation of stable RNA synthesis is governed by the *trans*-activating as well as by the repressive control system. The operation of positive and negative control systems may allow fine tuning of tRNA and rRNA synthesis (see also below). Furthermore FIS enables the cell to achieve very rapid growth as becomes apparent when the growth of wild-type and *fis* cells is compared in very rich media (Nilsson *et al.*, 1992). This may be a significant advantage when cells grow under the conditions prevailing in nature. Such an advantage has been demonstrated in chemostat experiments, showing that wild-type cells outcompete *fis* cells under appropriate conditions (Nilsson *et al.*, in press).

REGULATION OF FIS SYNTHESIS AND ITS RELATION TO GROWTH

The amount of FIS protein in the *E. coli* cell is subject to drastic changes during a normal growth cycle (Figure 4 and Thompson *et al.*, 1987). Rapid but transient increases in FIS levels occur in exponentially growing cells when exposed to a nutrient upshift (Figure 6).

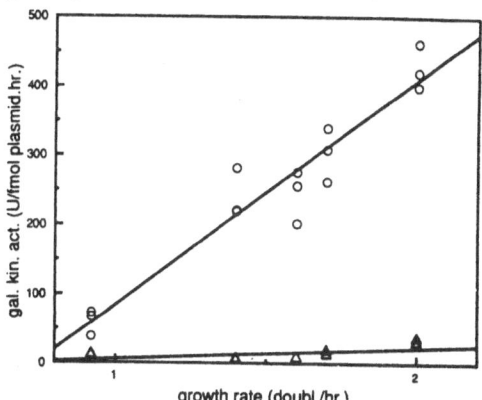

Figure 6. Relationship between the expression of the *thrU(tufB)* operon and the growth rates of exponentially growing cells. Wild-type cells, transformed with the plasmid carrying the *thrU(tufB'):galK* operon fusion (o—o) or with the plasmid deletion derivative lacking the UAS (Δ — Δ), were grown overnight in LB medium and diluted to an OD$_{600}$ of 0.1 in different media. Growth rates were determined during the early log phase and galactokinase activities were measured at an OD$_{600}$ of 0.4.

How do these cells sense the nutritional quality of the medium and respond with altering the level of FIS and of *trans*-activation? Recently (Ninnemann *et al.*, 1992) showed that *fis*mRNA undergoes similar changes as the protein. This suggested that FIS synthesis is regulated at the transcriptional level. Interestingly, the authors found that the growth phase regulation of the *fis* promoter depends on the presence of a GC motif downstream of the -10 region. The promoter is strongly repressed by amino acid starvation in a *Rel*A-dependent manner, an effect which is linked to the presence of the GC motif. The conclusion is that the transcription of the *fis* gene is controlled by ppGpp. Furthermore, mutational analysis using promoter-*lac*Z fusions revealed that FIS has a negative effect on *fis* transcription. This effect is directly linked to the presence of two FIS binding sites, overlapping the RNA polymerase binding site responsible for transcription initiation. These and other data show that the *fis* promoter is autoregulated by FIS.

The findings offer a ready explanation for the rapid changes of stable RNA operon expression in response to environmental signals. During outgrowth of wild-type stationary cells in fresh medium the ppGpp concentration rapidly drops which will result in derepression of the repressive control system and in enhanced transcription initiation of both stable RNA operons and the *fis* gene. The rising level of FIS will then cause further elevation of stable RNA synthesis, thus amplifying the effect of the environmental signal. By contrast *fis* cells, although responding to the lowering of the ppGpp level with enhancement of stable RNA operon expression, do not show effect amplification of the nutritional signal.

As yet we do not know the molecular mechanism through which cells accelerate growth upon expansion of the translational apparatus. Nonetheless it is conceivable that wild type cells are better equipped to meet the demands of accelerated growth than *fis* cells. As mentioned above, exponentially growing wild-type cells achieve high growth rates in rich media (*e.g.* LB medium plus 1% glucose), their doubling times exceeding those of *fis* cells. Chemostat experiments showed that they are able to compete out their mutant counterparts under appropiate conditions (Nilsson *et al.*, in press).

Autoregulation of *fis* transcription will cause a decline of FIS in wild-type cells when all FIS-binding sites on the UAS of the stable RNA operons are occupied and *trans*-activation of these operons has reached its maximum. Ribosome feedback and/or a rise of ppGpp (due to excessive ribosome synthesis) may accentuate this decline (*cf.* Figure 3A). Stringent conditions will lead to simultaneous repression of transcription initiation at the *fis* and stable RNA promoters, thus leading to amplification of the shut off signal. As pointed out by the authors "such a regulatory circuit is ideally suited for a rapid and precise adaptation to physiological changes".

A MODEL FOR THE MECHANISM OF FIS-DEPENDENT *TRANS*-ACTIVATION

The enzyme RNA polymerase is predominantly bound to one face of the DNA helix (U. Siebenlist *et al.*, 1980). Insertion of short DNA fragments at position -45 (see Figure 1C) alters the spatial orientation of the UAS bend with respect to the RNA polymerase. Insertions comprising less than a full helix turn or a deletion of 4 base pairs reduce transcription initiation in wild type cells. Insertions of one or two complete helix turns partially restore transcription initiation. These results (not shown here; *cf.* Verbeek *et al.*, 1991) are best explained by an abnormal conformation of the UAS DNA helix, such as DNA bending. They suggest very strongly that bending of the UAS DNA helix is involved in activation of the operon.

Electrophoretic analysis of circularly permuted UAS DNA fragments located a bending centre around position -95 ± 15 bp (with respect to the transcription start site), in between the two FIS-binding regions (*cf.* Figure 1C). FIS binding to these fragments, resulting in the formation of three FIS/DNA complexes and in increased bending of the UAS, does not alter the location of the bending center nor the number of complexes (Verbeek *et al.*, 1991).

Occupation by FIS of both FIS binding regions is required for optimal *trans*-activation. This has been concluded from deletion studies (Verbeek *et al.*, 1992) shown in Figure 7. A deletion extending from -176 to -88, eliminating the upstream FIS-binding region at -131 to -118 and most of the intervening sequence between the two FIS-binding regions, abolishes FIS-dependent *trans*-activation. FIS-independent *cis*-activation is also reduced, but not completely. This deletion apparently removes DNA determinants for both FIS-dependent and FIS-independent transcription activation. DNA fragments that have undergone this deletion form only one FIS/DNA complex. *Cis*-activation of transcription drops further by extending the deletion to -56, eliminating the entire UAS. This can only be observed in FIS-producing cells, since derepression of the repressive control in *fis* cells counteracting this drop, is observed when the deletion is extended from -81 to -56 (*cf.* Figure 7).

Specific helical positioning of the two binding regions is also essential. Deletion of one base pair or the insertion of three base pairs in the bending centre destroys *trans*-activation, despite the fact that the formation of three FIS/DNA complexes remains possible. Apparently, for optimal *trans*-activation the FIS molecules must be bound to both binding regions in a very specific spatial orientation. This suggests that the FIS molecules bound at these regions have to interact with other macromolecules for proper *trans*-activation and raises the question what is the nature of these other macromolecules. First, it may be assumed that the FIS molecules interact with each other, bringing the FIS-binding regions in close proximity. This, however, would require very strong bending of the intervening DNA double helix, which comprises approximately 40 bp, *i.e.* about 4 complete helix turns. Sterically this does not seem quite feasible. A more likely explanation is FIS interaction with the RNA polymerase (RNAP). A model visualizing the latter possibility is presented in Figure 8. A prominent feature of it is that the DNA with the two FIS binding sites each occupied by FIS, wraps around the RNAP. RNAP/DNA interaction beyond the classical boundary at position -45 (Siebenlist *et al.*, 1980), was first observed by Travers and coworkers during their studies of the tRNA operon *tyrT* (*cf.* Figure 2 and Travers *et al.*, 1983; Travers, 1984; Drew and Travers, 1984). They showed that RNAP protects the UAS of the *tyrT* operon against DNaseI up to position -65 and the UAS of a mutant *tyrT* promoter to at least -130. These results, obtained in the absence of FIS, led them to propose that more than one RNAP binds to the UAS of stable RNA promoters, which might activate transcription. In our view the results of Figure 7, showing a continuous decrease of FIS-independent *cis*-activation of the *thrU(tufB)* operon upon progressive deletion of the UAS, reflects RNAP/DNA interaction over large parts of the UAS as depicted in Figure 8. It may be expected that the interaction around and upstream of the bending center will be disrupted by insertions or deletions of a few base pairs at the bending centre. Such manipulations have been shown indeed to reduce FIS-independent *cis*-activation (Verbeek *et al.*, 1992). FIS-dependent *trans*-activation is strongly reduced by these manipulations. These findings are in line with wrapping of the UAS around the RNAP in the presence and absence of FIS. Since the target sites of FIS on the RNAP should be identical or nearly identical, the two α subunits are reasonable candidates. More important in favour of this assumption are the investigations by Igarashi and Ishihama (1991) and Igarashi *et al.* (1991), showing that the α subunit is essential for the communication between activator proteins and RNAP on certain promoters. According to these authors the C-terminal region of α carries a contact site for some transcription activators that bind, like FIS, DNA sites located upstream of the basic promoter elements. The topography of the RNAP with a 15 nm long axis of the RNAP ellipsoid section (Meisenberger *et al.*, 1981; Heumann *et al.*, 1988) may allow FIS molecules, separated by an intervening sequence of about four DNA helical turns, to interact with the α subunits. The bending of the UAS DNA, wrapped around the RNAP in the absence of FIS, will increase in the presence of FIS but this will not alter the bending centre of the UAS. UAS/RNAP interaction may be less stable or transient in the absence of FIS. FIS/RNAP interaction will stabilize the RNAP/DNA interaction over a considerable part of the UAS and thus may trigger a conformational change of the RNAP, favouring unwinding of the DNA double helix at the transcription start site, thereby converting the closed promoter complex to an open complex. The dynamic interaction between RNAP and DNA in the absence of FIS most likely leads less frequently to effective opening of the closed promoter complex. This frequency is further reduced by elimination of the UAS preventing DNA wrapping around the RNAP.

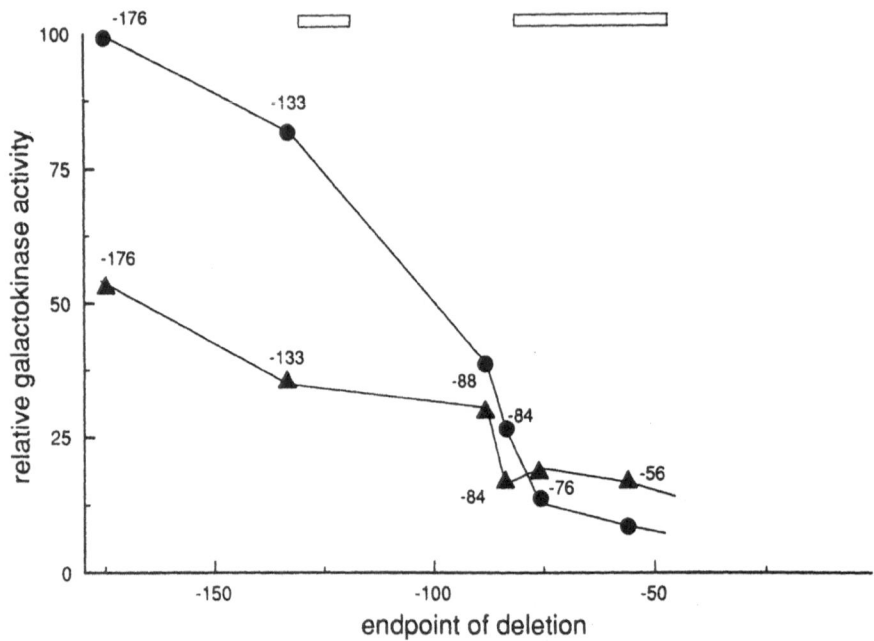

Figure 7. Promoter upstream deletions affecting FIS-dependent (o --- o) and FIS-independent (Δ --- Δ) activation of the *thrU(tufB)* operon. Galactokinase activities of cells transformed with plasmid deletion derivatives harbouring the *thrU(tufB'):galK* fusion operon are expressed as percentage of the activity measured with an intact UAS. Bars indicate the two FIS-binding regions (-131 to -118 and -81 to -48).

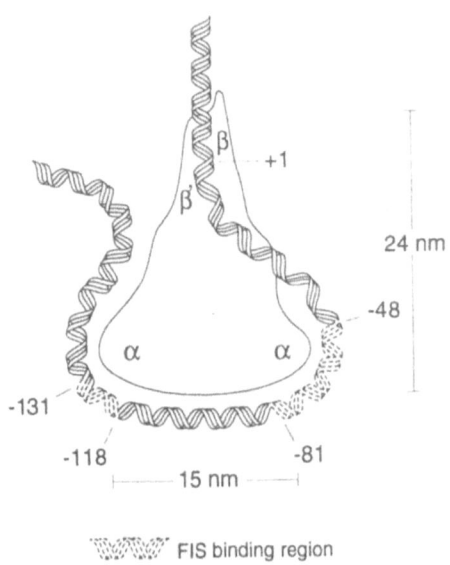

Figure 8. Model of the FIS/RNAP interaction at the *thrU(tufB)* promoter and its UAS.

Trans-activation of stable RNA operons and DNA inversion have some common features. The two FIS binding sites on the recombinational enhancer are separated by a well defined intervening sequence that allows for a conformational change of the DNA upon interaction with FIS. This change is manifested by DNA bending, which is essential for recombinational enhancer activity. FIS binding to the enhancer DNA does not alter the bending centre. Correct helical positioning of the two FIS binding sites on the enhancer DNA is essential for activity. Insertions other than an integral number of helical turns between the two sites destroy the function, although FIS binding to the DNA is not abolished. FIS mutations have been described that affect recombination but do not affect FIS/DNA binding. Models of enhancer action have been proposed (for a review see Verbeek *et al.*, 1992). According to these models, the FIS-enhancer complex and the recombinase-crossover sites complexes have first to interact before recombination can take place. This predicted interaction between FIS and recombinase could be necessary in order to align the crossover sites properly in a synaptic complex. In essence this model is similar to the one we propose for *trans*-activation.

PERSPECTIVES

The X-ray structure of FIS, determined by Kostrewa *et al.* (1991) and Yuan *et al.* (1991), revealed that the protein is composed of four α helices intertwined to form a globular dimer with two protruding helix-turn-helix motifs. The two recognition helices are separated by only 25 A, which implies that DNA has to bend for tight binding of FIS to two major DNA grooves. So far no FIS-DNA co-crystals have been described. When they become available more details about the DNA-protein interaction may be expected.

Cloning of the *fis* gene has enabled isolation of FIS in amounts sufficient for high resolution proton NMR studies.

Consensus FIS-binding sequences can be recognized upstream of all rRNA and 13 tRNA operons (Verbeek *et al.*, 1990) It will be of interest to see whether varying affinity of FIS for these sequences permits differential expression of tRNA genes.

ACKNOWLEDGEMENTS

The investigations were supported in part by the Commission of the European Communities, Biotechnology Action Programme (BAP), Directorate-General Science, Research and Development, Brussels. We thank Drs. B. Kraal and C.W.A. Pley for helpful discussions.

REFERENCES

Bauer, B.F., Kas, E.G., Elford, R.M., and Holmes, W.M., 1988, *Gene* 63:123 (1988).

Bosch, L., Nilsson, L, Vijgenboom, E., and Verbeek, H., 1990, *Biochim. Biophys. Acta* 1050:293.

Cashel, M., and Rudd, K.E., 1987, in: "*Escherichia coli* and *Salmonella typhimurium*, Cellular and Molecular Biology", Vol. 2, F.C. Neidhardt, J.L. Ingraham, K. Brooks Low, B. Magasanik, and M. Schaechter, eds., Am. Soc. Microbiol., Washington, D.C.

Cole, J.R., Olsson, C.R., Hershey, J.W.B., Grunberg-Manago, M., and Nomura, M., *J. Mol. Biol.* 198:383.

Dennis, P.P., and Bremer, H., 1974, *J. Mol. Biol.* 84:407.

Drew, H.R., and Travers, A.A., 1984, *Cell* 37:491.

Feng, J.-A, Yuan, H.S., Finkel, S.E., Johnson, R.C., Kaezor-Grzeskowick, M., and Dickerson, R., 1992, *in*: "Structure and Function, Vol. 2: Proteins", R.H. Sarma, and M.H. Sarma, eds, Adenine Press, Schenectady, N.Y.

Finkel, S.E., and Johnson, R.C., 1993, *Molec. Microbiol.*, in press.

Gausing, K., 1974, *Mol. Gen. Genet.* 129:61.

Gausing, K. 1977, *J. Mol. Biol.* 115:335.

Gausing, K., 1979, *in*: "Ribosomes, Structure, Function and Genetics', G. Chambliss, G.R. Craven, J. Davies, L. Kahan and M. Nomura, eds., University Park Press, Baltimore.

Gourse, R.L, de Boer, H.A., and Nomura, M., 1986, *Cell* 44:197.

Heumann, H., Lederer, H., Baer, G., May, R.P. Kjems, J.K., and Crespi, H.L., 1988, *J. Mol. Biol.* 201:115.

Igarashi, K., and Ishihama, A., 1991, *Cell* 65:1015.

Igarashi, K, Hanamura, A., Makino, K., Aiba, H., Mizuno, T., Nakata, T., and Ishihama, A., 1991, *Proc. Natl. Acad. Sci. USA* 88:8958.

Jinks-Robertson, S., R.L. Gourse and M. Nomura, 1983, *Cell* 33:865.

Jinks-Robertson, S., and Nomura, M., 1987, *in*: "*Escherichia coli* and *Salmonella typhimurium*, Cellular and Molecular Biology", Vol. 2, F.C. Neidhardt, J.L. Ingraham, K. Brooks Low, B. Magasanik and M. Schaechter, eds., Am. Soc. Microbiol., Washington, D.C.

Kostrewa, D., Granzin, J., Koch, C., Choe, H.-W., Raghunathan, S., Wolf, W., Labahn, J., Kahmann, R., and Saenger, W., 1991, *Nature* 349:178.

Lamond, A.I., and Travers, A.A., 1983, *Nature*, 305:248.

Lamond, A.I., 1985, *Trends Biochem. Sci.* 10:271.

Lindahl, L., and Zengel, J.M., 1982, *Adv. Genet.* 21:53.

Lindahl, L., and Zengel, J.M., 1986, *Annu. Rev. Genet.* 20:297.

Meisenberger, O., Heumann, H., and Pilz, I., 1981, *FEBS Lett.* 123:22.

Nilsson, L., Vanet, A., Vijgenboom, E., and Bosch, 1990, *EMBO J.* 9:727.

Nilsson, L., Verbeek, H., Vijgenboom, E., van Drunen, C., Vanet, A., and Bosch, L., 1992, *J. Bacteriol.* 174:921.

Nilsson, L., Verbeek, H., Hoffmann, U., Haupt, M., and Bosch, L., 1993, *FEMS Lett.*, in press.

Ninnemann, O., Koch, C., and Kahmann, R., 1992, *EMBO J.* 11:1075.

Nomura, M., Gourse, R., and Baughman, G., 1984, *Ann. Rev. Biochem.* 53:75.

Ross, W., Thompson, J.F., Newland, J.T., and Gourse, R.L., 1990, *EMBO J.* 9:3733.

Ryals, J., Little, R., and Bremer, H., 1982, *J. Bacteriol.* 151:1261.

Siebenlist, U., R. B. Simpson, R.B., and Gilbert, W., 1980, *Cell* 20:269.

Thompson, J.F., Moitoso de Vargas, L., Koch, C., Kahmann R., and Landy, A, 1987, *Cell* 50:901.

Travers, A.A., Lamond, A.I., Mace, H.A.F., and Berman, M.L., 1983, *Cell* 35:265.

Travers, A.A., 1984, *Nucleic Acids Res.* 12:2605.

van Delft, J.H.M., Marion, B., Schmidt, D.S., and Bosch, L., 1987, *Nucleic Acids Res.* 15:9515 (1987).

Verbeek, H., Nilsson, L., Baliko, G., and Bosch, L., 1990, *Biochim. Biophys. Acta* 1050:302.

Verbeek, H., Nilsson, L., and Bosch, L., 1991, *Biochimie* 73:713.

Verbeek, H., Nilsson, L., and Bosch, L., 1992, *Nucleic Acids Res.* 20:4077.

Vijgenboom, E., Nilsson, L., and Bosch, L., 1988, *Nucleic Acids Res.* 16:10183.

RIBOSOMAL RNA PROCESSING IN *SACCHAROMYCES CEREVISIAE*

Rob W. van Nues, Jaap Venema, Rudi J. Planta, and Hendrik A. Raué

Department of Biochemistry and Molecular Biology
Vrije Universiteit
de Boelelaan 1083, 1081 HV Amsterdam, The Netherlands

INTRODUCTION

Ribosome biogenesis in eukaryotic cells requires an intricate interplay between a large number of molecules. In the course of this process some 80 ribosomal proteins (r–proteins) have to assemble in an ordered fashion with the four rRNA molecules, which themselves are formed by stepwise maturation of primary transcripts produced by two different RNA polymerases. rRNA processing and r–protein assembly occur concomitantly and are interdependent. Furthermore, a growing number of non-ribosomal components, both proteins and ribonucleoprotein (RNP) particles, is being identified, that are required in *trans* for the correct and efficient formation of eukaryotic ribosomes. Because of its accessibility to genetic and physiological manipulation, the yeast *Saccharomyces cerevisiae* has become one of the most popular organisms for studying eukaryotic ribosome biogenesis. In this Chapter we will discuss recent progress in the identification of both *cis*-acting elements and *trans*-acting factors involved in the formation of the mature 17S, 26S and 5.8S rRNA species in yeast.

THE rRNA PROCESSING PATHWAY

Analogous to other eukaryotic cells the 17S, 5.8S and 26S rRNAs of *S. cerevisiae* are transcribed by RNA polymerase I as part of a single 37S precursor transcript that is processed into the mature species via an ordered series of steps as outlined in Figure 1. This processing involves the removal of two external (5' ETS and 3' ETS) and two internal (ITS1 and ITS2) transcribed spacers from the precursor rRNA. The maturation largely takes place in the nucleolus, where the precursor assembles with both ribosomal proteins and non-ribosomal components into a 90S ribonucleoprotein particle and is concomitantly modified by methylation and pseudouridylation. Most of the 3' ETS is rapidly removed by an endonucleolytic cleavage, and transcripts extending to the actual transcription termination site are not detected in yeast cells under normal conditions. Only seven of the 210 nucleotides of the 3' ETS remain in the largest observable 37S pre-rRNA (reviewed in Klootwijk and Planta, 1989; Raué and Planta, 1991).

Until recently, the 5' ETS (699 nt) was assumed to be removed in one step by endo-nucleolytic cleavage at A1 producing the mature 5' end of the 17S rRNA molecule (Veinot-Debrot et al., 1988; Klootwijk and Planta, 1989). However, Hughes and Ares (1991) have

The Translational Apparatus, Edited by K.H. Nierhaus
et al., Plenum Press, New York, 1993

reported evidence for an additional cleavage event, called A0, occurring 89/90 nucleotides upstream from the mature 5' terminus. The cleavage at A0 can precede the one at A1 because at least a portion of the 32S precursor hybridizes with a probe complementary to the extreme 3' end of the 5' ETS (Hughes and Ares, 1991). On the other hand, the same probe failed to detect the 20S precursor, demonstrating that cleavage at A2 follows the cuts at A0 and A1 in accordance with the results of 5'–end mapping on the 20S species (Klootwijk and Planta, 1989). The failure to detect the 32S intermediate in earlier experiments (Veinot-Debrot et al., 1988) can be explained by the use of a probe complementary to sequences upstream from A0. Thus, 5' ETS removal in yeast appears to be a multi-step process as is the case in mammalian cells (Kass et al., 1987).

After removal of the 5' ETS and most of the 3' ETS, cleavage at A2 within ITS1 splits the 90S particle into a 43S and a 66S RNP, containing the 20S and 29S$_A$ precursor rRNAs, respectively. The 43S particle is exported from the nucleolus into the cytosol, where the remaining part of ITS1 is removed to generate mature 17S rRNA (reviewed by Raué and Planta, 1991). Accumulation of discrete 5'–terminal ITS1 fragments as long as 209 nucleotides in a yeast strain that is deficient in a 5'→3' exonuclease confirmed that formation of the mature 3' end of 17S rRNA is due to an endoribonucleolytic cleavage at A3 (Stevens et al., 1991) followed by rapid degradation of the excised ITS1 sequences.

The 153 ITS1 nucleotides at the 5' end of 29S$_A$ pre-rRNA are removed in the nucleolus, thereby creating the 5' end of 5.8S (B1). Ten percent of the 5.8S rRNA population consists of molecules having a 5'-terminal extension of 6-7 nucleotides (Klootwijk and Planta, 1989). The level of the shorter form drops dramatically in *rat1–1* temperature-sensitive mutants at the non-permissive temperature. The *RAT1* gene has been identified to encode a protein involved

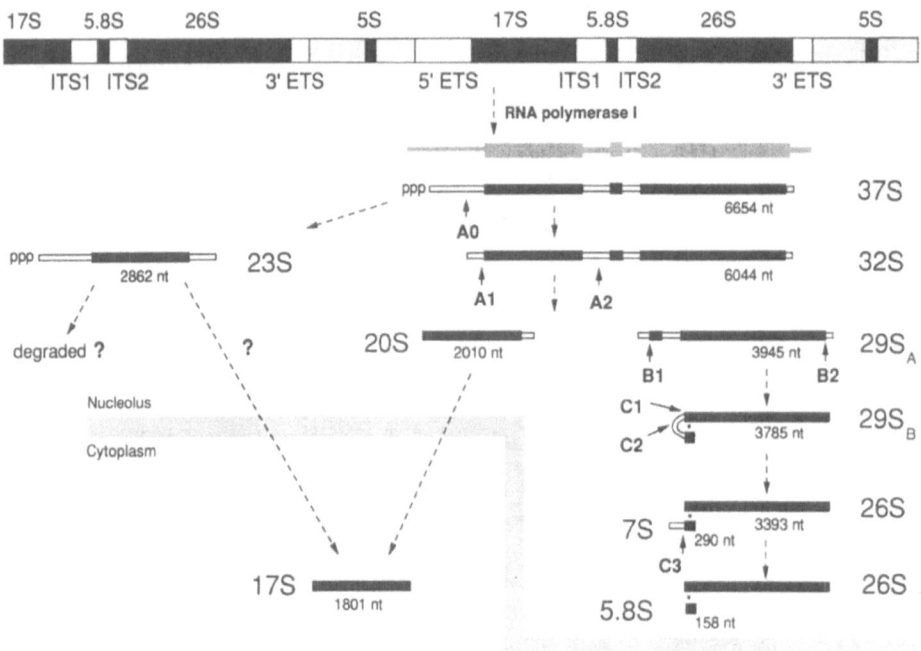

Figure 1. The processing pathway of the yeast 37S ribosomal RNA precursor. The top line represents the structure of two of the 150-200 tandemly repeated rDNA units of *S. cerevisiae*. Regions encoding mature rRNAs are shown in black; transcribed spacers and non-transcribed sequences are depicted in white and grey, respectively. The primary RNA polymerase I transcript, which is normally undetectable, is drawn as a grey silhouette. Mature and spacer sequences in the detectable precursors are again shown in black (as are the final mature rRNAs) and white, repectively. The abnormal 23S molecule is generated by initial cleavage at B1.

in the transport of mRNA across the nuclear envelope, but this protein also shows $5' \rightarrow 3'$ exonuclease activity (Amberg et al., 1992). Therefore, the mature end of 5.8S rRNA may be formed by a two-step reaction: an endonucleotic attack 6-7 nucleotides upstream from the 5' terminus of 5.8S rRNA in most cases followed by exonucleolytic removal of the extra nucleotides. However, it cannot be excluded that all of the ITS1 sequences are removed from the $29S_A$ molecule by an exonuclease starting further upstream and pausing at the 5' end of the extended form of 5.8S rRNA. The finding that processing at B1 still takes place when endonucleolytic cleavage at A2 is blocked (see below) argues in favour of the two-step mechanism.

Almost coincidently with B1, cleavage at B2 separates the remnant of the 3' ETS from the $29S_A$ pre-rRNA. The resulting $29S_B$ molecule is converted into mature 26S rRNA and the 7S precursor of 5.8S rRNA by virtually simultaneous endonucleolytic attacks at C1 and C2, although the 7S molecule remains associated with 26S rRNA by non-covalent interactions. The cut at C2 leaves 132-135 nt of ITS2 attached to the 3' end of 5.8S rRNA. Finally, cleavage at C3 produces mature 5.8S rRNA and the resulting 60S subunit is also transported across the nuclear envelope into the cytoplasm to take up its role in mRNA translation.

TRANS-ACTING FACTORS

Since processing of yeast pre-rRNA occurs at the level of RNP particles, ribosomal proteins are obvious candidates for *trans*-acting factors. The requirement for particular r–proteins in specific steps of the processing pathway has been observed for both the small and large subunits. On the one hand, lack of the non-essential small subunit protein S37 (Raué and Planta, 1991) affects the efficient formation of mature 40S subunits by retarding the conversion of the 20S precursor into 17S rRNA (Finley et al., 1989). On the other hand, perturbing the nuclear import of the large subunit protein L29 causes a defect in the processing of the 29S precursors (Underwood & Fried, 1990). A similar phenotype is observed when a mutant carrying a cold-sensitive mutation in a gene for another large subunit protein, *i.e.* L16, is grown at the non-permissive temperature (Moritz et al., 1991). This suggests that the correct three-dimensional rRNA structure required for processing depends on the ordered assembly with ribosomal proteins.

Buoyant density measurements of yeast ribosomal precursor particles indicated a protein content exceeding that of mature ribosomes, which suggested the presence of non-ribosomal components possibly involved in pre-rRNA processing and assembly (reviewed by Klootwijk and Planta, 1989). The involvement of a growing number of non-ribosomal *trans*-acting factors in rRNA processing has since indeed been demonstrated by genetic methods. With the exception of small nucleolar RNPs (see below), these factors have not yet been characterized biochemically and almost nothing is known concerning their precise function. It should be noted, however, that nearly all of the genetic defects disturb the biogenesis of only one of the two ribosomal subunits. The first such *trans*-acting factor reported was the product of the *RRP1* gene, inactivation of which blocked processing of the 29S precursors in strain ts351 (Andrew et al., 1976). Carter and Cannon (1980) showed that an uncharacterized genetic defect in yeast strain CLP–8 affects the formation of 40S subunits. Both the transport of 20S pre-rRNA across the nuclear envelope and its cytoplasmic maturation are slowed down. Sachs and Davis (1989; 1990) have found that mutation of a gene encoding a putative RNA helicase causes a cold-sensitive phenotype characterized by a reduction in the formation of 60S subunits.

A cold-sensitive phenotype consisting of a severe defect in the whole pre-rRNA processing pathway is caused by disruption of the *NSR1* gene. Cold-shock of the mutant strain completely blocks processing of the 37S precursor at A0, A1 and A2 whereas some cleavage at B1 still occurs (Kondo et al., 1992; Kondo and Inouye, 1992). At 30°C the same mutant severely underproduced 40S subunits due to a defect in 17S rRNA synthesis (Lee et al., 1992; Kondo and Inouye, 1992). Again the exact function of the NSR1 protein in ribosome maturation and assembly is unknown. Notwithstanding some structural similarity and sequence identity with the metazoan nucleolin, NSR1 cannot be replaced by this protein in yeast (Kondo and Inouye 1992; Lee et al., 1992). Interestingly, the gene was originally identified to encode a protein that recognizes nuclear localization signals, a property suggesting a role in nuclear import of r–proteins (Lee et al., 1991; 1992)

Three nucleolar proteins, called NOP1, GAR1 and SSB1, have been shown to play important roles in pre-rRNA processing in yeast (Tollervey et al., 1991; Girard et al., 1992; Clark et al., 1990). Each of these proteins is associated with small nucleolar RNAs (snoRNAs) and it is the resulting snoRNPs that constitute the actual *trans*-acting factors. Using genetic and biochemical approaches, direct evidence has now been obtained for involvement in pre-rRNA processing of four of the at least eleven yeast snoRNPs, namely U3 (snR17), snR10, U14 (snR128) and snR30 (Hughes and Ares, 1991; Beltrame and Tollervey; 1992; Tollervey, 1987; Li et al., 1990; Morrissey and Tollervey, 1992). NOP1, the yeast homolog of metazoan fibrillarin, is a component of all the known snoRNPs (Tollervey et al., 1991; Jansen et al., 1991). GAR1 associates with snR10 and snR30 as well as some of the other snoRNAs (Girard et al., 1992) while SSB1 binds strongly to snR10 and weakly with other snoRNAs (Clark et al., 1990). These observations are based on immunoprecipation experiments which are not completely informative on account of possible epitope masking under the conditions used (Clark et al., 1990)

Remarkably, interfering with the function of any one of these snoRNPs, either by mutation or depletion of one of the proteins or snoRNAs, results in a very similar phenotype in all cases: a block or, in the case of the non-essential snR10 RNP, severe reduction in the formation of mature 17S rRNA because of loss of the cleavages at A0, A1 and A2 (Hughes and Ares, 1991; Li et al., 1990; Tollervey 1987; Morrissey and Tollervey 1992; Tollervey et al., 1991; Girard et al., 1992). Instead of the normal 20S precursor an abnormal 23S molecule accumulates. This aberrant pre-rRNA species probably results from cleavage of the 37S ribosomal RNA precursor at B1, because it contains ETS and ITS1 sequences (*cf.* Figure 1), but is not detected by ITS2-specific probes (Tollervey, 1987). It may function as an intermediate in an alternative processing route, taken when cleavage at A0, A1 or A2 is prevented, since some mature 17S rRNA could still be detected in these experiments. The fact that this route does not function as efficiently as the major pathway might be explained by instability of the 5' end of the 37S precursor leading to its rapid degradation when one of the essential snoRNPs is absent (Tollervey et al., 1991). On the other hand, the possibility that the 23S species is not a precursor of mature 17S rRNA but only represents a relatively stable degradation intermediate cannot be ruled out. The genes studied were either non-essential (*i.e.* *snR10*) or placed under control of the UAS$_{GAL}$ to allow their expression to be down-regulated by growth on glucose. Depletion via repression of the *GAL* promoter is known to be incomplete and even at late time points after the shift to glucose-based medium the phenotype cannot be assumed to be equivalent to that of a null mutation (Girard et al., 1992). Therefore, the small amount of 17S rRNA seen in cells lacking snR10 or depleted of U3, U14 or snR30 snoRNP may still have been synthesized via the normal processing route.

In contrast to 17S rRNA synthesis, formation of 26S rRNA is hardly affected in snoRNP-depleted cells because alternative cleavage of the 37S precursor at B1 produces 27S$_B$ pre-rRNA. Apparently, further processing of this precursor to mature 26S and 5.8S rRNA does not require functional U3, snR10, U14 or snR30 snoRNP. It is likely that the list of snoRNPs involved in yeast pre-rRNA processing will be extended because the majority of the snoRNA species has been shown to associate with pre-rRNAs. However, most if not all of the additional snoRNAs are encoded by non-essential genes (Tollervey, 1987; Zagorski et al., 1988; Parker et al., 1988) demonstrating that unlike U3, U14 and snR30 they are not vital for the production of the mature rRNA species.

Recently, a temperature-sensitive mutant of yeast (*rrp2*) has been shown to lack cleavage at B1 at the restrictive temperature. Instead, cleavage at the 3' end of 5.8S rRNA (C3) occurred, followed by processing at A2, as was concluded from the accumulation of a 24S molecule hybridizing to ITS1 and 5.8S probes, and of an abnormal 7S intermediate that contains sequences derived from 5.8S rRNA as well as the 3' end of ITS1. Neither species is recognized by oligonucleotides complementary to ITS2 sequences. The aberrant 7S rRNA can be incorporated into 60S subunits and is also found in the polysomal fraction (Shuai and Warner, 1991; Lindahl et al., 1992).

Almost all of the nucleases involved in yeast rRNA processing remain to be identified. Apart from the above mentioned *RAT1* gene product, so far only the endonuclease encoded by the *RNA82* gene has been implicated in rRNA maturation, because mutation in this gene abolishes the cleavage within the 3' ETS that removes all but 7 nt of this region (Kempers-Veenstra et al., 1986). Interestingly, the same mutation blocks 3' processing of pre-5S rRNA.

CIS-ACTING ELEMENTS

In yeast, *trans*-acting factors and their function in ribosome biogenesis can be studied in various ways, e.g. by construction and analysis of conditional expression mutants. Identification of *cis*-acting elements within the ribosomal RNA precursors, however, requires a faithful *in vitro* processing system, which as yet is not available, or a technique to carry out mutational analysis *in vivo*. In the latter case one has to solve the problem how to detect the products derived from the mutant rDNA unit against a massive background of transcripts produced by the approximately 200 wild-type chromosomal copies. The "tagged ribosome" (pORCS) system developed in our laboratory offers such a solution. It is based upon a centromeric plasmid carrying one complete rDNA unit in which the genes for 17S and 26S rRNA contain small unique oligonucleotide insertions. These insertional tags do not interfere with processing and accumulation of the mature rRNAs in the 40S and 60S subunits of actively translating ribosomes. They allow specific detection of the high-molecular-weight (pre-)rRNA species originating from the plasmid-encoded rDNA unit (Musters et al., 1989; 1990*a*; 1990*b*). Using this system we found that relatively large deletions in the 5' ETS and both internal spacers are detrimental, demonstrating that these spacers are necessary for normal formation of either 17S or 26S rRNA and that they contain elements controlling their own removal (Musters et al., 1989; 1990*a*; 1990*b*; Van der Sande et al., 1992). Furthermore, it was shown that in some cases transcribed spacers from other yeast species can function in the *S. cerevisiae* context, despite poor conservation of primary structure (Van der Sande et al., 1992; our unpublished results). Therefore, it is likely that the *cis*-acting elements that direct pre-rRNA processing are organized in conserved higher-order structures.

5' External Transcribed Spacer

As discussed above, removal of the 5' ETS by cleavage at A0 and A1 requires participation of the U3 snoRNP. The site of interaction of U3 with the yeast 37S pre-rRNA has been traced to this spacer by RNA-RNA cross-linking experiments using psoralen (Beltrame and Tollervey, 1992). Two cross-linking sites were identified *in vivo* at positions +470 and +655 downstream of the transcription initiation site respectively. The first one is located within a 10 nt long region (positions +470-479) that displays perfect complementarity to nt 39-48 of U3 snoRNA. Deletion analysis of the 5' ETS using a variant of the "tagged ribosome system" demonstrated that the region complementary to U3 is essential for 17S rRNA formation but dispensable for 26S rRNA synthesis (Beltrame and Tollervey, 1992). This is in agreement with previous results from our laboratory which showed that deletion of various portions of the 5' ETS, all of which included the U3 complementarity, caused the same processing defect (Musters et al., 1990*a*; 1990*b*). Deletion of the region that contains the second *in vivo* cross-linking site (nt +644-661) had no detectable effect on synthesis of either 17S or 26S rRNA. No complementarity between U3 snoRNA and this region of the 5' ETS is apparent (Beltrame and Tollervey, 1992).

Depletion of yeast cells for snR10, snR30 or U14 snoRNA results in a phenotype very similar to that described above for U3-depleted cells. As far as the first two snoRNPs are concerned no information indicating their site of interaction with the pre-rRNA is available. Yeast U14 snoRNA, however, contains two 13 nt long sequences that are complementary to different regions in domain I of mature 17S rRNA. This complementarity has been conserved between yeast and mouse (Li et al., 1990). Formation of 17S rRNA in yeast cells was severely reduced when the first of these two U14 sequences was mutated (Jarmolowski et al., 1990) suggesting that base pairing between this sequence and its complement immediately downstream of helix 6 in domain I of the 17S rRNA plays an important role in pre-rRNA processing. Mutations in the second U14 sequence, which is complementary to helix 14 of 17S rRNA, had a much less severe impact upon 17S rRNA synthesis (Jarmolowski et al., 1990), although the corresponding region of mouse U14 snoRNA is exclusively responsible for base pairing with 18S rRNA *in vitro* (Shanab and Maxwell, 1992). Interestingly, we have found that structural modifications of helix 6 also cause a severe reduction in the amount of 17S rRNA produced by the mutant rDNA unit (unpublished data). Possibly, the structurally altered forms of helix 6 interfere with the association of U14, even though this helix is part of one of the eukaryotic expansion segments (*i.e.* V1) that show considerable structural variation from one organism to another (Raué et al., 1988). These results, together with earlier data from our laboratory on the effect of a structural alteration in the V3 region of 17S rRNA (Musters et al., 1990a), demonstrate that correct and efficient removal of the transcribed spacers also depends upon elements within mature portions of the precursor.

Internal Transcribed Spacer 1

The first internal transcribed spacer, separating the 17S rRNA and 5.8S rRNA coding sequences by 362 bases, is removed from the ribosomal RNA precursor via cleavages at A2, A3 and B1 (Figure 1). Processing at A2 is carried out by an endoribonuclease leaving 209 nucleotides of ITS1 attached to the 3' terminus of the 17S rRNA sequence, while generating the 5' end of the $29S_A$ intermediate 153 nt upstream from the mature 5' end of 5.8S rRNA (Klootwijk and Planta, 1989). It is not yet clear, whether the loss of cleavage at A2 upon snoRNP-depletion is due to a direct participation of snoRNPs in this processing step. Alternatively, cleavage at A2 might depend on prior removal of the 5' ETS, which is blocked in snoRNP-depleted cells. Recently, however, snR30 was reported to be directly involved in recognition of A2 (Morrissey and Tollervey, 1992). The cleavage at A2 is essential for the formation of a ribosomal RNA precursor that efficiently matures into 17S rRNA, since deletion of 160 nucleotides surrounding A2 strongly interfered with synthesis of 40S subunits containing tagged 17S rRNA. This deletion led to an accumulation of tagged 32S pre-rRNA, suggesting that processing at A0 or A1 can proceed independently from that at A2. The same

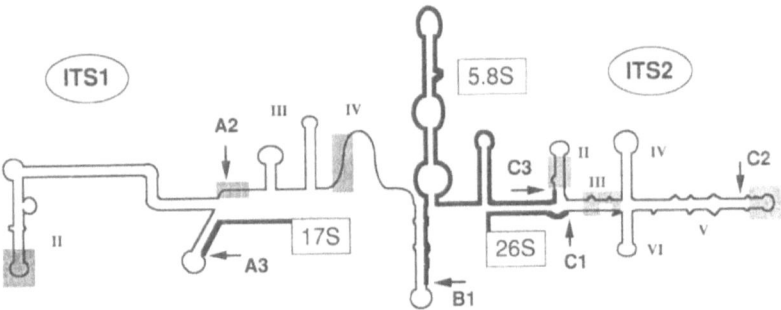

Figure 2. Schematical representation of the secondary structure of ITS1, 5.8S and ITS2 in 37S pre-rRNA (Yeh and Lee, 1991; plus our modifications for ITS1 on account of phylogenetic data). Sequences of mature 17S, 5.8S and 26S rRNAs are symbolized by bold lines, the transcribed spacers by thin ones. The processing sites are indicated and the shaded boxes symbolize conserved regions. Domains of ITS1 as mentioned in the text are indicated. The helices of ITS2 are numbered according to Yeh and Lee (1990).

deletion did not disturb the formation of tagged 26S rRNA (Musters et al., 1990a; 1990b). This result, together with the data concerning the effect of snoRNP depletion discussed in the previous section, suggests that cleavage at A2 is a key step for the separation of the small and large subunit rRNAs along the normal processing pathway.

Using the secondary-structure model of ITS1 as well as phylogenetic comparison of corresponding regions from other yeast species we set out to identify *cis*-acting elements in ITS1 involved in pre-rRNA processing in more detail. We determined the nucleotide sequence of ITS1 of additional members of the subfamily *Saccharomycetoidae*, namely *Torulaspora delbrueckii* (formerly called *Saccharomyces rosei*), which is closely related to *S. cerevisiae*, *Kluyveromyces lactis*, *K. marxianus* and *Hansenula wingei*, a more distantly related yeast (Kreger-Van Rij, 1973). Although primary sequence comparison reveals only a low degree of homology, all four ITS1 regions can be folded into a secondary structure comparable to that derived for *S. cerevisiae* ITS1. In addition a number of primary structural elements appears to be highly conserved, *i.e.* the apical end of domain II and two single stranded regions, one surrounding A2 in domain III, the other located in domain IV (Figures 2 and 3A).

Replacement studies revealed that, despite the considerable sequence differences with their *S. cerevisiae* counterpart, both the *K. lactis* and *T. delbrueckii* ITS1 are fully functional when inserted into the *S. cerevisiae* rDNA unit. As shown in Figure 3A for the *K. lactis* ITS1

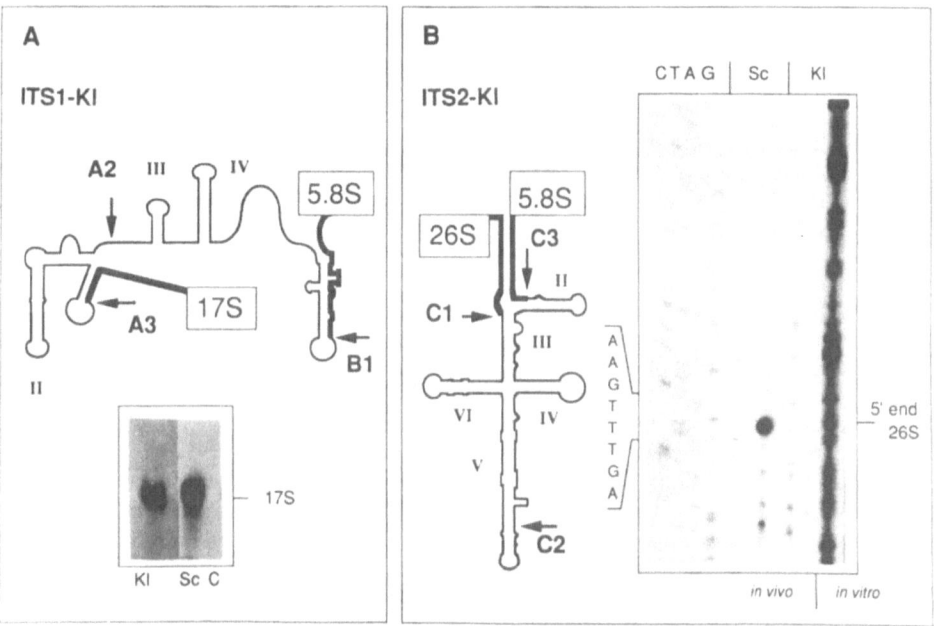

Figure 3. Replacement studies of internal transcribed spacers. (A) ITS1 of *S. cerevisiae* (362 bp) was exchanged for its counterpart from *K. lactis* (226 bp, ITS1-Kl) in a tagged rDNA unit using conserved restriction sites in the adjacent rRNA genes. The restriction sites were chosen in such a way that no mutations in mature sequences were introduced. The secondary structure of the *K. lactis* spacer is represented. Highly conserved regions are boxed. The lower panel displays Northern analysis of RNA isolated from untransformed cells (C), cells transformed with either the intact *S. cerevisiae* tagged rDNA unit (Sc) or with a tagged rDNA unit containing the *K. lactis* spacer (Kl). Hybridization to the tag in 17S rRNA is shown. (B) ITS2 of *S. cerevisiae* (234 bp) was replaced by its *K. lactis* counterpart (246 bp, ITS2-Kl; symbols and mutagenesis as in (A)). The photograph shows primer extension analysis of *in vivo* produced rRNA using a primer complementary to the tag in 26S rRNA. *In vitro* synthesized RNA was used as a control for artificial reverse transcription stops (See Musters et al. 1989; 1990a for procedures).

(lane Kl), formation of 17S rRNA is not detectably affected by the spacer replacement. The major difference between the *S. cerevisiae* and *K. lactis* ITS1 regions is the much smaller size of domain II in the latter (*cf.* Figures 2 and 3A). A similar size difference exists between domains II of *S. cerevisiae* and *T. delbrueckii* ITS1 (not shown). Thus, a large part of domain II is apparently dispensable for correct and efficient pre-rRNA processing. This even seems to include the highly conserved apical end of this domain. In preliminary experiments we did not detect any significant disturbance in the formation of both tagged rRNAs when this conserved region was almost completely deleted from the *S. cerevisiae* ITS1.

We have also individually exchanged the 5'- or 3'-terminal halves (on either side of position 220 downstream from the 3' end of the 17S rRNA sequence) of *S. cerevisiae* ITS1 with the corresponding regions of *T. delbrueckii* ITS1. In both cases normal accumulation of both 17S and 26S rRNA was maintained indicating the absence of essential higher-order interaction between the two spacer portions. This is in striking contrast to the situation in ITS2, where long-distance interactions appear to be functionally important (see next section).

An interesting aspect of the ITS1 secondary-structure model is the proposed base pairing between the 3'-terminal sequence and the 5' region of 5.8S rRNA in the 37S precursor (Figure 2; Yeh and Lee, 1991). In mature 60S subunits the same region of 5.8S rRNA binds non-covalently to sequences in domain I of 26S rRNA (Yeh and Lee, 1991; Raué et al., 1988). Therefore, the organization in 37S pre-rRNA might reflect the way in which processing of the $29S_B$ molecule is regulated. Only after the separation from the $29S_A$ precursor via B1, the 5' end of the 5.8S rRNA coding region would become available to interact with 26S rRNA sequences. This might lead to a conformational change that exposes ITS2 to an attack by endoribonucleases.

Internal Transcribed Spacer 2

In contrast to processing of ITS1, the correct and efficient removal of ITS2 does depend upon higher-order structure involving widely-spaced portions of this transcribed spacer. Relatively large deletions affecting the 5' part or the middle of ITS2 completely prevented formation of mature 26S rRNA by blocking the processing steps subsequent to B1. A similar deletion in the 3' part still allowed some correctly processed 26S rRNA to be synthesized but at a severely reduced level (Musters et al. 1990*a*; 1990*b*; Van der Sande et al., 1992). Each of these mutations causes substantial disruption of the secondary structure of ITS2 (Figure 2), which was deduced from chemical and enzymatic probing experiments (Yeh and Lee, 1990; 1991) and further supported by phylogenetic comparison (Van der Sande et al., 1992; our unpublished data). More subtle mutational analysis demonstrated that disruption of the base pairing in the middle part of helix V lacking significant primary structure conservation has little or no effect. In contrast, a similar alteration of the secondary and primary structure of a conserved element in the lower portion of this helix, around C2, severely affected the efficiency, though not the accuracy, of 26S rRNA formation (Van der Sande et al., 1992). Because all these mutations block further processing of the $29S_B$ precursor, their primary effect seems to be to prevent the cleavage at C2. Thus, this cleavage appears to be the key step in the formation of both mature and 26S and 5.8S rRNA, analogous to the central position of cleavage at A2 in 17S rRNA synthesis.

Replacement experiments show that the ability of heterologous or chimeric yeast ITS2 regions to function in the *S. cerevisiae* context is much more limited than in the case of ITS1. While ITS1 from both *T. delbrueckii* and *K. lactis* can functionally replace its *S. cerevisiae* counterpart, in the case of ITS2 such a functional replacement is possible only for the more closely related *T. delbrueckii* species. Apparently, while the *K. lactis* ITS1 is compatible with the *S. cerevisiae* processing machinery, the ITS2 is not, since its presence in the 37S precursor blocks formation of mature tagged 26S rRNA. As shown in Figure 3B, compared to the large amount of correctly formed 26S rRNA in case of *S. cerevisiae* (lane Sc), hardly any mature 26S transcripts were derived from the rDNA unit containing the ITS2 of *K. lactis* (ITS2-Kl, lane Kl). The *K. lactis* ITS2 can be folded into a secondary structure very similar

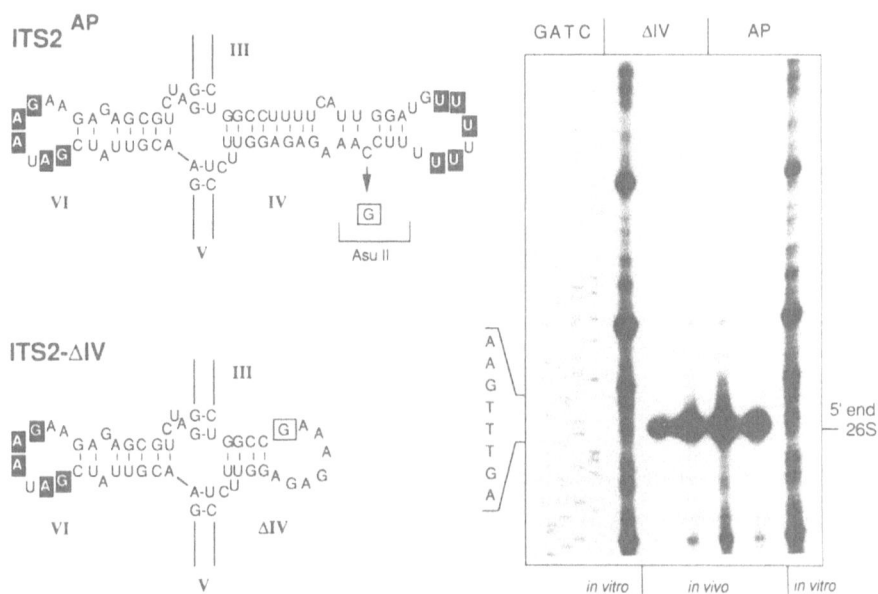

Figure 4. Functional analysis of a specific secondary-structure element in ITS2 (helix IV). **(A)** Replacement of the Hpa I - Asu II fragment of pORCS-ITS2^AP (ITS2^AP, Van der Sande et al., 1992) by synthetic oligonucleotides keeping the secondary structure of domain III intact, but decreasing the length of helix IV and altering the nucleotides in the loop (ITS2-ΔIV). A putative interaction between the helices IV and VI via the nucleotides shown in reversed contrast is thereby disturbed. **(B)** Primer extension by reverse transcriptase from the 26S rRNA tag was performed on rRNA isolated from cells transformed with either pORCS-ITS2^AP(AP) or pORCS-ITS2-ΔIV (ΔIV) as described in the legend of Figure 3.

to that of its *S. cerevisae* and *T. delbrueckii* counterparts. Furthermore it also contains all major sequence elements conserved between the latter two spacers. Consequently, its failure to function in the heterologous context must be due to relatively minor deviations in primary and/or secondary structure outside the conserved elements.

The high sensitivity of ITS2 to relatively minor structural alterations is further underlined by the observation that replacement of only the 3' half of this spacer by the corresponding region of the *T. delbrueckii* ITS2 disrupts $29S_B$ processing, even though the entire *T. delbrueckii* ITS2 is fully functional in the *S. cerevisiae* context (Van der Sande et al., 1992). The structure of the chimeric spacer is closely similar to that of either of its two intact parents. However, the chimera has lost the potential for a tertiary interaction between the loops closing helices IV and VI (*cf.* Figure 4A), which is conserved in all yeast ITS2 regions sequenced to date (Van der Sande et al., 1992). In order to determine whether this putative tertiary interaction is important for processing we altered the structure of helix IV resulting in a shorter stem and a different sequence of the loop (Figure 4; ITS2-ΔIV), while leaving the secondary structure of the rest of the *S. cerevisiae* ITS2 intact. Although the mutant ITS2 would be unable to form tertiary base pairs between helices IV and VI, it did not prevent efficient formation of correctly processed 26S rRNA. This is demonstrated by the primer extension analysis shown in Figure 4B, in which total RNA isolated from cells with the mutated ITS2 (lane ΔIV) is compared to RNA of cells with a functional ITS2 (lane AP). Furthermore, introducing the same mutation into the *S. cerevisiae/T. delbrueckii* ITS2 chimera does restore the complementarity between the two loops but not its function in

A   **B**

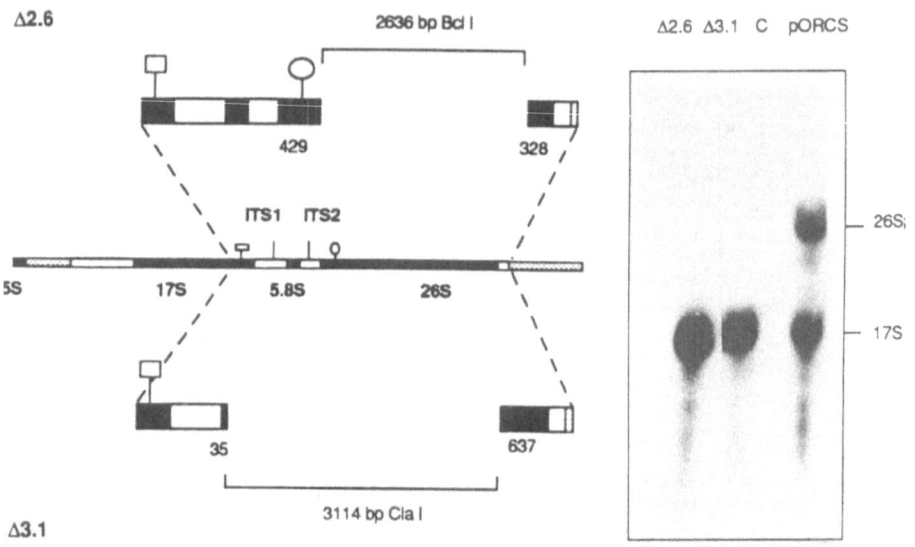

Figure 5. Functional analysis of large deletions in 26S rRNA. (**A**) Deletions of 2.6 kb (Δ2.6) and 3.1 kb (Δ3.1) were introduced into the tagged rDNA unit by means of digestion with Bcl I and Cla I, respectively. The lengths (in bp) of the deleted segments and the remaining sequences coding for mature rRNA are indicated. (**B**) Northern analysis of RNA isolated from untransformed host cells (C), cells transformed with the intact rDNA unit carrying both tags (pORCS), and transformants containing either mutant rDNA unit (Δ2.6, Δ3.1), was performed according to Musters et al. (1989; 1990a). Probes recognizing the tags were used simultaneously.

processing. We conclude, therefore, that tertiary interaction between helices IV an VI plays no role in ITS2 processing.

Interestingly, we found that complete deletion of ITS2 prevents formation of mature 26S rRNA (Musters et al., 1990a; 1990b), but not of 17S rRNA (Van der Sande et al., 1992). The fusion of the 5.8S and 26S rRNA sequences resulting from this deletion would have a virtually identical structure to that of the base-paired individual molecules, normally present in the 60S subunit. Therefore, we propose that ITS2 plays an important role at some stage of subunit assembly subsequent to the formation of the 29S$_B$ precursor.

THE SPLIT MATURATION OF THE 37S PRECURSOR

From the data discussed in the previous sections it appears that the two portions of the primary pre-rRNA transcript that give rise to the small and large subunit rRNAs mature independently. Mutations in either 5' ETS or ITS1 perturb formation of only 17S rRNA, whereas the effect of structural alterations in ITS2 is limited to disturbance of 26S rRNA synthesis. Similarly, functional inactivation of individual *trans*-acting factors including r–proteins affects processing of either one or the other part of the pre-rRNA, but not both simultaneously. Even though the initial processing steps are carried out on the 90S RNP particle containing the full-length pre-rRNA transcript, it seems that this particle is already organized in two separate domains. Correct and efficient processing/assembly of the two different ribosomal subunits apparently does not require cross-talk between these domains.

As far as maturation of the large subunit is concerned evidence for this conclusion comes from experiments carried out in our laboratory in which we showed that removal of most of the 5' ETS, nearly half of ITS1, or almost two-thirds of the 17S rRNA gene from an rDNA unit does not affect the production of functional 60S subunits from that same unit (Musters et al., 1989; 1990a; 1990b). The experiment shown in Figure 5 demonstrates that the reverse is also true. As shown in panel A, we introduced two extensive deletions into the genes coding for the large subunit rRNAs, the largest of which leaves just 35 nt of the 5'-terminal and 637 nt of the 3'-terminal sequence of the genes encoding 5.8S and 26S rRNA, respectively. Northern analysis of RNA isolated from cells transformed with these mutant rDNA units (Figure 5B) showed that neither deletion interferes detectably with formation of tagged 17S rRNA. Thus, it appears that the organization of the three eukaryotic rRNA genes into one operon is not a prerequisite for the correct processing of pre-rRNA molecules to the mature species and their incorporation into ribosomal subunits.

In conclusion it can be stated that in the past few years progress has been made in the identification of both *cis*-acting elements and *trans*-acting factors that participate in the complex process by which eukaryotic pre-rRNA matures and is assembled into functional ribosomal subunits. However, the inventory is certainly not complete and may eventually reach a complexity comparable to or exceeding that of the pre-mRNA splicing apparatus. The challenge for the immediate future lies in establishing how this multitude of components acts together to ensure the correct and efficient conversion of the primary pre-rRNA transcript into the mature species coupled to the ordered assembly of ribosomal proteins. *In vitro* systems in which individual steps in this process can be mimicked should be an important aid in detailed identification of specific *cis*-acting elements and their interaction with individual *trans*-acting factors. The development of such systems for mouse (Kass et al., 1987) has established a basis for preparation of similar extracts from yeast cells. Together with its potential for genetic modification this should make yeast an even more valuable tool to elucidate the molecular mechanisms by which eukaryotic ribosome biogenesis is driven.

ACKNOWLEDGEMENTS

We express our gratitude to Erwin Mollee and Caroline Sluiter for their assistance in performing mutational analysis of ITS1 and ITS2. We also thank Jan Boesten for synthesizing the oligonucleotides used in our experiments. This work was supported by the Netherlands Foundation for Chemical Research, with financial aid from the Netherlands Organization for Scientific Research.

REFERENCES

Amberg, D.C., Goldstein, A.L., and Cole, C.N., 1992, *Genes Dev.*, 6:1173.
Andrew, C., Hopper, A.K., and Hall, B.D., 1976, *Molec. Gen. Genet.*, 144:29.
Beltrame, M., and Tollervey, D., 1992, *EMBO J.* 11:1531.
Carter, C.J., and Cannon, M., 1980, *J. Mol. Biol.*, 143:179.
Clark, M.W., Yip, M.L.R., Campbell, J., and Abelson, J., 1990, *J. Cell Biol.*, 111:1741.
Finley, D., Bartel, B., and Varshavsky, A., 1989, *Nature*, 338:394.
Girard, J-P., Lehtonen, H., Caizergues-Ferrer, M., Amalric, F., Tollervey, D., and Lapeyre, B., 1992, *EMBO J.*, 11:673.
Hughes, J.M.X., and Ares, M., 1991, *EMBO J.*, 10:4231.
Jansen, R.P., Hurt, E.C., Kern, H., Lehtonen, H., Carmo-Fonseca, M., Lapeyre, B., and Tollervey, D., 1991, *J. Cell Biol.*, 113:715.
Jarmolowski, A., Zagorski, J., Li, H.V., and Fournier, M.J., 1990, *EMBO J.*, 9:4503.
Kass, S., Craig, N., and Sollner-Webb, B., 1987, *Mol. Cell. Biol.*, 7:2891.
Kempers-Veenstra, A.E., Oliemans, J., Offenberg, H., Dekker, A.F., Piper, P.W., Planta, R.J., and Klootwijk, J., 1986, *EMBO J.*, 5:2703.
Klootwijk, J., and Planta, R.J., 1989, *Meth. in Enzymology*, 180:96.
Kondo, K., and Inouye, M., 1992, *J. Biol. Chem.*, 267:16252.
Kondo, K., Kowalski, L.R.Z., and Inouye, M., 1992, *J. Biol. Chem.*, 267:16259.

Kreger-Van Rij, N.J.W., 1973, *in* "The Fungi," G.C. Aisworth, F.S. Sparrow, and A.S. Sussman, eds., Academic press, London.

Lee, W-C., Xue, Z., and Mélèse, T., 1991, *J. Cell Biol.*,113:1.

Lee, W-C., Zabetakis, D., and Mélèse, T., 1992, *Mol. Cell. Biol.*, 12:3865.

Li, H.V., Zagorski, J., and Fournier, M.J., 1990, *Mol. Cell. Biol.*, 10:1145.

Lindahl, L., Archer, R.H., and Zengel, J.M., 1992, *Nucl. Acids Res.*, 20:295.

Moritz, M., Pulaski, B.A., and Woolford, J.L., 1991, *Mol. Cell. Biol.*, 11:5681.

Morrissey, J.P., and Tollervey, D., 1992, *Yeast*, 8:S211.

Musters, W., Venema, J., Van der Linden, G., Van Heerikhuizen, H., Klootwijk, J., and Planta, R.J., 1989, *Mol. Cell. Biol.*, 9:551.

Musters, W., Boon, K., Van der Sande, C.A.F.M., Van Heerikhuizen, H., and Planta, R.J., 1990a, *EMBO J.*, 9:3989.

Musters, W., Planta, R.J., Van Heerikhuizen, H., and Raué, H.A., 1990b, *in*:"The Ribosome: Structure, Function and Evolution," W.E. Hill, A.E. Dahlberg, R.A. Garrett, P.B. Moore, D. Schlessinger and J.R. Warner, eds, Amer. Soc. Microbiol., Washington, DC., Chapter 37:435.

Parker, R., Simmons, T., Shuster, E.O., Siliciano, P.G., and Guthrie, C., 1988, *Mol. Cell. Biol.*, 8:3150.

Raué, H.A., Klootwijk, J., and Musters, W., 1988, *Prog. Biophys. Molec. Biol.*, 51:77.

Raué, H.A., and Planta, R.J., 1991, *Progr. Nucl. Acid Res. Mol. Biol.* 41:89.

Sachs, A.B., and Davis, R.W., 1989, *Cell*, 58:857.

Sachs, A.B., and Davis, R.W., 1990, *Science*, 247:1077.

Shanab, G.M., and Maxwell, E.S., 1992, *Eur. J. Biochem.*, 206:391.

Shuai, K., and Warner, J.R., 1991, *Nucl. Acids Res.*, 19:5059.

Stevens, A., Hsu, C.L., Isham, K.R., and Larimer, F.W., 1991, *J. Bact.*, 173:7024.

Tollervey, D., 1987, *EMBO J.*, 6:4169.

Tollervey, D., Lehtonen, H., Carmo-Fonseca, M., and Hurt, E.C., 1991, *EMBO J.*, 10:573.

Underwood, M.R., and Fried, H.M., 1990, *EMBO J.*, 9:91.

Van der Sande, C.A.F.M., Kwa, M., Van Nues, R.W., Van Heerikhuizen, H., Raué, H.A., and Planta, R.J., 1992, *J. Mol. Biol.*, 223:899.

Veinot-Debrot, L.M., Singer, R.A., and Johnston, G.C., 1988, *J. Mol. Biol.*, 199:107.

Yeh, L-C.C., and Lee, J.C., 1990, *J. Mol. Biol.*, 211:699.

Yeh, L-C.C., Thweatt, R., and Lee, J.C., 1990, *Biochemistry*, 29:5911.

Yeh, L-C.C., and Lee, J.C., 1991, *J. Mol. Biol.*, 217:649.

Zagorski, J., Tollervey, D., and Fournier, M., 1988, *Mol. Cell. Biol.*, 8:3282.

ANALYSIS OF MUTATIONS IN THE 23S rRNA

Urmas Saarma,[1] Birgit T. U. Lewicki,[2] Tõnu Margus,[3] Sulo Nigul,[1] and
Jaanus Remme[3]

[1]Estonian Biocentre, Jakobi st. 2 Tartu, EE2400, Estonia
[2]Max-Planck-Inst. f. Molekulare Genetik, Ihnestr. 73, D-1000 Berlin 33
 Germany
[3]Inst. Chem. Phys. & Biophys., Jakobi st. 2 Tartu, EE2400, Estonia

INTRODUCTION

Several experimental approaches have been employed to study the structure and function of ribosomes (reviewed in Dahlberg, 1989, and Leclerc and Brakier-Gingras, 1990). One such approach is that of site-directed mutagenesis of the rRNA, most commonly in the rRNA of the bacterium *Escherichia coli*. This organism has seven rRNA operons and thus separation of the host encoded rRNA from that of the mutated plasmid-encoded rRNA is necessary for further biochemical studies. The problem can be overcome by using *in vitro* synthesized rRNA for reconstitution of the subunit. This has been sucessfully achieved for total reconstitution of functional 30S particles. In contrast, transcribed and unmodified 23S rRNA cannot form active particles in reconstitution assays. Therefore, 23S rRNA should be expressed *in vivo* followed by isolation of the ribosomes. It is possible to discriminate between activities of mutant and wild-type ribsomes by using a second mutation in the gene of 23S rRNA. One such mutation is the A to T transversion corresponding to the position 1067 of 23S rRNA which confers resistance to thiostrepton during cell-free translation (Thompson *et al.*, 1988). Thiostrepton has a high affinity to wild-type ribosomes. U1067 ribosomes bind the drug with a reduced affinity (Thompson and Cundliffe, 1991). Therefore, A to U substitution at position 1067 can be used for selective inactivation of wild-type ribosomes and for physical separation of mutant ribosomes. This experimental approach has been used to analyze functional properties of mutations corresponding to positions G2505 and G2583 of 23S rRNA (Saarma and Remme, 1992). The most interesting finding was that mutations at position 2583 increased the translational accuracy. These results, together with those of mutations at positions G2505, G2583, G2607, and G2608 *in vitro* and G2505 and G2583 *in vivo* will be summarized here.

Furthermore, data concerning the formation of 50S subunits are presented. The assembly of ribosomes *in vivo* is strongly coupled with the transcription of the ribosomal RNA and includes the rRNA processing and rRNA modification. All attempts to show this process *in vivo* and to mimic it *in vitro* have to date failed. As a first approach the properties

The Translational Apparatus, Edited by K.H. Nierhaus
et al., Plenum Press, New York, 1993

of ribosomes whose rRNA was derived either by T7 RNA polymerase or *E. coli* RNA polymerase were compared with regard to both the *in vivo* transcription of the plasmid-encoded *rrn*B operon and ribosomal assembly. To distinguish between plasmid- and chromosomal-derived rRNA the above described A to T transversion at the position 1067 in the plasmid-encoded 23S rRNA gene was used. The relative amount of the plasmid-derived ribosome fraction in the mixed population was determined by sequencing the 1067 region of the 23S rRNA by means of reverse transcriptase, whereas the activity of the mutant particles could be measured in a poly(U) dependent poly(Phe) synthesis system in the presence of thiostrepton. It was found that the 50S particles containing rRNA transcribed by the viral T7 RNA polymerase were practically inactive in contrast to ribosomes synthesized by the host transcriptase. The significance of these results is discussed.

ANALYSIS OF 50S ASSEMBLY

To study the influence of the RNA polymerase species on the *in vivo* transcription of the *rrn*B operon and the ribosomal assembly, two series of plasmids were constructed. All of them carry the A to T transversion at the 1067 position in the 23S rRNA gene. In the first set the ribosomal RNA genes are under the control of a T7 late promoter instead of the natural P1P2-tandem promoters. These constructs are designated pT7... and were transformed into *E. coli* BL21(DE3). This strain contains a chromosomal copy of the gene for the T7 RNA polymerase under control of the IPTG-inducible *lac*UV5 promoter. The *rrn*B operon on pT7-1 is complete, whereas on pT7-2 the spacer following the P2 promoter and the 16S rRNA gene was excised, except for 39 bases at the 3' end. The construct pT7-3 furthermore lacks in addition the 5' half of the spacer between the 16S and 23S rRNA genes including the 5' half of the spacer tRNA^{Glu2}. The second series is called ptac-1, ptac-2 and ptac-3 and contains the identical gene arrangement than the first series but here the rRNA genes are under the control of the *tac* promoter instead of the T7 promoter. The second series of constructs were transformed into *E. coli* JM109 where the transcription from the *tac* promoter by the *E. coli* RNA polymerase is induced by the addition of IPTG.

An effect on the cell growth could be observed only in the presence of IPTG with *E. coli* cells carrying the pT7 constructs. Here the cell growth stopped about 45 min after induction whereas the growth was not disturbed upon induction of the T7 RNA polymerase encoded in the chromosome in the absence of any plasmid. In contrast, the growth of the host with the ptac plasmid was hardly affected by induction (Lewicki *et al.*, 1993) . These results were not surprising since the T7 late promoter is very strong and highly specific for its own RNA polymerase resulting in a large overproduction of rRNA. It is known that a large excess of rRNA over ribosomal proteins increases the formation of inactive particles which could be one of the reasons for the premature inhibition of growth of the cells containing the pT7 constructs. This overproduction of rRNA in the presence of T7 RNA polymerase was confirmed when the cells containing pT7-3 grew with $^{32}PO_4$ either under steady state conditions (the radiolabled orthophosphate was present for at least 4 generations) or pulse labelling conditions (the label was added 15 min prior to the corresponding harvesting point). Evidently 85 % of the rRNA overproduced by the T7 RNA polymerase was degraded whereas only about 15 % of the excess 23S rRNA was stably incorporated into ribosomal particles.

To determine the functional activity of the mutant ribosomes, the distribution of ribosomal particles and their content of plasmid-born rRNA was determined by analysis of sucrose gradients (Figure 1). The particles corresponding to the 70S and 50S fraction were isolated, and the 23S rRNA sequenced around nucleotide 1067. The amounts of chromosomal and plasmid-derived rRNA were determined and the relative amounts of plasmid-born 23S rRNA calculated, as percentages of the total 23S rRNA below the 70S and 50S peaks, respectively.

As it is shown in Figure 1 (left side) all three pT7 containing cell lysates have a normal 70S peak but show overproduction of 50S subunits and the presence of precursor particles. The plasmid-coded rRNA is unequally distributed between the 70S and 50S fraction

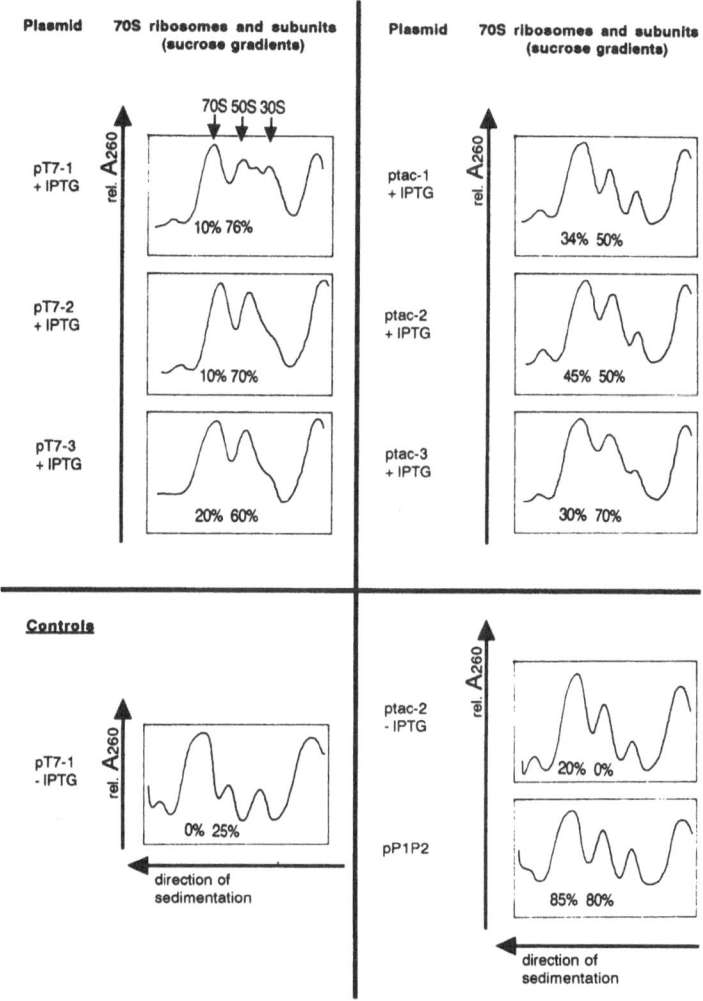

Figure 1. Sucrose gradient analysis of cell lysates with various plasmid constructs. The 23S rRNA derived from 50S and 70S fractions, respectively, were sequenced around the 1067 region and the relative amount of plasmid-born rRNA was determined as described elsewhere (Lewicki *et al.*, 1993). The percentage values below the peaks give the relative amounts of mutant (plasmid-born) 23S rRNA as compared to total 23S rRNA derived from plasmids plus chromosome.

containing 10 - 20 % and 70 ± 10 % rRNA synthesized by the viral transcriptase, respectively. Obviously, no free exchange exists between the free pool of 50S subunits carrying plasmid-encoded rRNA and the 50S subunits in the 70S ribosomes which represents the fraction active in protein synthesis. These results leads to the conclusion that most of the particles which contain rRNA transcribed by T7 RNA polymerase seem to be inactive. A different picture was evident when the plasmid-encoded rRNA was transcribed by the *E. coli* RNA polymerase. At least the presence of pT7-1 containing the entire *rrn*B operon resulted in an indistinguishable sucrose gradient pattern in the presence and absence of the inducer and a much more equal distribution of plasmid-derived rRNA between the 50S and 70S fraction (Figure 1, right side). These findings agreed with the relative degrees of thiostrepton resistance observed with the lysates in poly(Phe) synthesizing systems. The lysates with the pT7 plasmids showed low activity in the presence of thiostrepton (Table 1), whereas about 3-fold higher activity was found with all ptac constructs.

Table 1. Poly(Phe)-synthesis activity of 70S ribosomes in the presence of thiostrepton (3 µM).

plasmid	induction	poly(Phe) synthesis		relative activity in the presence of thiostrepton
		- thio	+ thio	
		[cpm]		[%]
no	-	4100	45	1
pT7-1	no	1800	25	1.4
	yes[1]	6700	500	7
pT7-2	no	2350	33	1
	yes	4700	300	6
pT7-3	no	2100	40	2
	yes	3200	180	5.6
ptac-1	no	5451	113	2
	yes	7450	1363	18
ptac-2	no	9617	586	6
	yes	8550	1577	18
ptac-3	no	8207	254	3
	yes	7496	1349	18

[1] In the case of induction IPTG was added 60 min before the harvesting point.

It was considered whether the inactivity of the large excess of 50S particles produced in the presence of pT7-3 was caused by functionally inactive 23S rRNA or, alternatively, by an accumulation of assembly-defective particles containing normal 23S rRNA. To test these alternatives the rRNA from the corresponding inactive 50S particles was isolated and a total reconstitution in the presence of wild-type TP50 was performed. A relative activity in the presence of thiostrepton was found which was the same as that of reconstituted particles containing ptac derived 23S rRNA. It follows that the pT7-3 derived rRNA in native particles is functional, and that the inactivity of the corresponding particles is caused by a defective assembly process. It cannot be excluded that one or more of the rRNA modifications are lacking since it was shown that modified nucleotides are important for a correct assembly process in the case of 16S rRNA (Cunningham *et al.*, 1991). A second reason could be the T7 RNA polymerase itself whose transcription elongation rate is about five times faster compared to that of the host polymerase (Chamberlin & Ring, 1973). This higher velocity of the transcriptase might cause premature folding of the nascent rRNA chain and/or premature association of ribosomal proteins.

To analyze the question whether the assembly defect is a cause of the high expression level of the rRNA or is due to the T7 RNA polymerase itself, the transcription initiation rate of the viral enzyme was decreased by using very low amounts of IPTG (0.03 mM). This condition reduces the overproduction of rRNA and should lead to an increased production

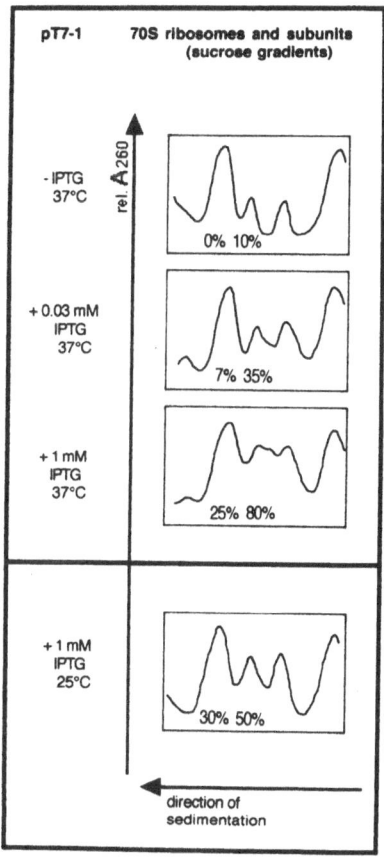

Figure 2. Sucrose gradient analysis of lysates prepared from cells grown in the presence of different IPTG concentrations and temperatures, and the relative amounts of plasmid-born 23S rRNA within the 70S and 50S fractions. For further details see legend of Fig. 1.

of active particles. Interestingly, the same unequal distribution of plasmid-encoded rRNA between the 70S and 50S fraction was observed (Figure 2, top) whereas the absolute amount of rRNA produced by T7 RNA polymerase was lowered. The exchange and therefore the activity of 50S particles containing pT7-1 derived rRNA between 50S and 70S pool is not increased indicating that the assembly defects are independent of the expression levels of the rRNA genes on the pT7-1 plasmid and therefore are caused by the T7 RNA polymerase itself. However, when the growth temperature was lowered to 25°C the rRNA analysis showed a distribution of plasmid-derived rRNA in the 50S and 70S fraction similar to that from the ptac constructs (Figure 2, bottom). Note, that both experiments lead to very similar sucrose gradient patterns with only very little amount of overproduced 50S and 30S particles. Under conditions in which the transcription initiation rate was lowered (0.03 mM IPTG) the distribution of rRNA, synthesized by T7 RNA polymerase, between the 50S and 70S pool was unequal, whereas by lowering the transcription elongation rate of the viral enzyme (at 25°C) this distribution was much more equal indicating the increased production of active 50S particles.

EFFECT OF MUTATIONS ON THE FUNCTION OF RIBOSOMES

The mutant rRNA genes were cloned under control of an inducible promoter (lambda P_L in pNO2680, Gourse *et al.*, 1983). Conditional expression was necessary because the expression of mutant 23S rRNA had a deleterious effect on the bacterial growth. The mutations at positions G2505, G2583, G2607, and G2608 did not prevent 23S rRNA from beeing processed and assembled into 50S subunits, although the expression of 23S rRNA containing C2583 and U 2583 was accompanied by the appearance of incomplete particles (25S). Mutant rRNA formed 40 - 50 % of the total 70S ribosome population indicating that mutant 50S subunits were able to associate with 30S subunits.

The ability of mutant ribosomes to participate in cell-free translation was tested during poly(U) directed poly(Phe) synthesis. In the presence of 5-fold molar excess of thiostrepton over ribosomes, poly(U) translation on the wild-type ribosomes was inhibited 97 - 99 %. The ribosomes derived from pNO1067U (a mixed population containing 50% mutant ribosomes) were inhibited by 50 %, whereas the ribosomes encoded by the double mutant plasmid were inhibited by 54 - 71 %. This indicates that the double mutant ribosomes were able to translate poly(U). Thereby, in the presence of the drug only mutant ribosomes were active. Thiostrepton resistance of the double mutant ribosomes depended on the Mg^{2+} concentration. The effect of the magnesium ion concentration on the translational acitivity of mutant ribosomes was analyzed during poly(U) directed cell-free protein synthesis in the presence of thiostrepton. Neither thiostrepton nor the U1067 mutation had an effect on the magnesium dependence of *in vitro* translation as compared to the wild-type ribosomes. The results indicated that the nucleotide substitutions at different positions led to similar alterations of the response by the mutant ribosomes to the Mg^{2+} concentration. All mutants exhibited a lower acitivity at 10 mM $MgCl_2$ and a higher acitivity at 25 mM $MgCl_2$ as compared to the single mutant U1067 or wild-type ribosomes. G to A transitions corresponding to positions G2583 and G2608 had the smallest effect, as expected. Mutant ribosomes containing G to C transversions at positions G2505, G2583, G2607, and G2608 exhibited lower acitivity at low Mg^{2+} concentration as compared to the corresponding G to U transversions. Taking into account the known influence of magnesium concentration on the ribosome, the results suggested that mutations affected tRNA binding to the ribosome.

The sensitivity of mutant ribosomes to antibiotics, which are known to interfere ribosomal peptidyltransferase activity, was tested in two different systems. The first system was poly(U) directed translation according to Rheinberger and Nierhaus, 1990. The poly(U) translation is less sensitive to inhibitors as compared to the natural mRNA translation (Pestka, 1977). Wild-type ribosomes were inactivated by thiostrepton as previously described. By comparing the results of poly(U) translation on wild-type and U1067 ribosomes with respect to the sensitivity to chloramphenicol and erythromycin, it was evident that the U1067 mutation had no influence on the sensitivity to peptidyltransferase inhibitors. On the other hand, the sensitivity of the mixed U1067 ribosome population exhibited no difference in the presence or absence of thiostrepton. Thus, the translocation inhibitor thiostrepton does not effect the sensitivity of ribosomes to the peptidyltransferase inhibitors (chloramphenicol and erythromycin). Inhibition of poly(U) directed translation on the mutant ribosomes was tested in the presence of thiostrepton. The concentrations of the antibiotics corresponding to the 50 % inhibition level are indicated in Table 1. The results obtained with erythromycin were similar for all the mutants within the experimental error. By contrast, inhibition of cell-free translation by chloramphenicol was effected on the U2583 and C2583 double mutant ribosomes in the reprocucible manner indicating that mutations at this position conferred hypersensitivity of poly(U) translation to chloramphenicol.

The second system to assay the sensitivity of the mutant ribosomes to antibiotics was based on the ability of thiostrepton to block translocation on the wild-type ribosomes. NacPhe-tRNA^{Phe} was bound to the ribosomal A site following preoccupation of the P site by deacylated tRNA. After EF-G promoted translocation, peptidyl-tRNA becomes competent to puromycin. When this reaction is carried out in the presence of thiostrepton, translocation occurs only on the ribosomes containing U1067. Consequently, this route allows analysis of the peptidyltransferase reaction catalyzed by mutant ribosomes using a mixed ribosome population. Chloramphenicol and lincomycin inhibited the transfer of Nac-Phe to puromycin, whereas erythromycin had no effect on the peptide bond formation, which has also been shown in earlier reports (Pestka, 1977). The concentration of drugs corresponding to 50 % inhibition of peptidyltransferase reaction on the wild-type and mutant ribosomes is indicated in Table 1. The inhibition pattern of chloramphenicol was similar indicating that mutations in 23S rRNA had no effect during this particular assay. By contrast, mutations C2505, U2583, and C2583 led to an increased sensitivity of peptide bond formation by lincomycin. It is known that both, chloramphenicol and lincomycin, inhibit the acceptor substrate binding (Pestka, 1977) but different effects of mutations on the sensitivity of these two drugs show that inhibition is caused by a different way of action. In addition, both, erythromycin and chloramphenicol, have been shown to indure the release of oligopeptidyl-tRNA during cell-free translation (Pestka, 1977; Rheinberger and Nierhaus, 1990). Hypersensitity of mutants U2583 and C2583 to chloramphenicol during poly(U) translation and to lincomycin during puromycin reaction can be caused by a reduced binding affinity of the 3' end of tRNA to the ribosomal peptidyltransferase centre.

The influence of mutations in 23S rRNA on the fidelity of translation was analyzed during poly(U) directed translation system in the presence and absence of thiostrepton. Single mutant U1067 ribosomes exhibited the same Leu misincorporation level as wild-type ribosomes both in the presence and absence of thiostrepton. Consequently, A1067 to U transversion and the presence of thiostrepton had no effect on the translational accuracy. The comparison of the missense error frequencies in the presence and absence of the drug showed that only the double mutants at position G2583 exhibited significant difference. Data on the Leu misincorporation in the presence of thiostrepton is presented in Table 2. It is interesting to note that the effect on the translational accuracy correlates with the sensitivity to chloramphenicol during cell-free translation. One possible way to explain this phenomenon is to assume that binding of the aa-tRNA is weaker (or alternatively the transfer reaction is slower) on the C or U2583 mutant. In this way, ribosomes would have more time for aa-tRNA dissociation during proofreading. Accordingly, the peptidyltransferase reaction must be limiting and an irreversible step for aa-tRNA selection. This model predicts that mutant ribosomes need more GTP's to hydrolyze for aa-tRNA binding.

The effect of mutations in 23S rRNA on the translational accuracy was tested *in vivo*. Plasmids containing mutant rRNA genes were introduced into cells with nonsense mutants in ß-galactosidase. In this system the ribosomes must translate through a stop codon in order to synthesize functional ß-galactosidase. Changes in the level of enzyme synthesis, which are due to the presence of mutant ribosomes, were detected as described in Prescott and Kornau, 1992. The strains used were suppressor-free and therefore the level of enzyme synthesis reflected the mistranslation at nonsense codons (Petrullo *et al.*, 1983). The activity of ß-galactosidase was normalized to the level of ß-lactamase activity synthesized at the same point in time. The ß-lactamase gene contains no internal in-frame stop codon. Therefore, the amount of this enzyme synthesized is a reflection of overall *in vivo* translation activity. The ratio of ß-galactosidase activity relative to that of ß-lactamase represents the level of mistranslation to translational acitvity.

The relative level of mistranslation at UAA nonsense condon was apparently unaffected by the presence of mutation U1067a. The double mutants caused the efficiency of mistranslation to be decreased in the following order: U1067>C2505>A2583>U2583>C2583. Thus these results are in good agreement with the results obtained *in vitro*. On the other hand an increased level of translational readthrough

Table 2. Functional effects of mutations in 23S rRNA during poly(U) directed cell-free translation and peptidyltransferase reaction (PTR). Assays were performed as described in Saarma and Remme, 1992.

Ribosomes	Poly(U) translation				PTR	
	Thio[1] resist	Inhibition[2]		Leu/Leu+Phe)[3]	Inhibition[4]	
		Cm (mM)	Ery (mM)		Cm (µM)	Lin (mM)
wild-type	0.04	>3	4	0.0122	4	0.35
U1067	0.52	>3	4	0.0123	4	0.35
U1067/C2505	0.32	2.5	5	0.0188	4	0.07
U1067/A2583	0.43	>3	4	0.0101	4	0.15
U1067/U2583	0.35	1.0	5	0.0057	4	0.10
U1067/C2583	0.29	2.0	5	0.0039	4	0.10
U1067/C2607	0.39	>3	6	0.0120	ND	ND
U1067/A2608	0.46	>3	4	0.0121	ND	ND
U1067/U2608	0.41	>3	8	0.0122	ND	ND
U1067/C2608	0.38	>3	8	0.0121	ND	ND

[1] Thiostrepton resistance is strongly dependent upon Mg^{2+} concentration. The data correspond to 12 mM $MgCl_2$ and represent the ratio of cpms ± thiostrepton (1 µM).
[2] Inhibition was detected in the presence of thiostrepton on the mutant ribosomes and in the absence of drug on the wild-type ribosomes. Poly(U) translation was inhibited by 50 % at the indicated concentrations; Cm = chloramphenicol, Ery = erythromycin.
[3] Leu misincorporation was measured in the presence of thiostrepton on the mutant ribosomes and in the absence of drug on the wild-type ribosomes.
[4] PTR was inhibited by 50 % with either both wild-type and mutant ribosomes (Cm) or with mutant ribosomes (Lin) at the indicated concentrations; Lin = lincomycin.
ND - not determined.

was observed at UAG stop codon due to the presence of U1067 ribosomes. This effect was suppressed by the second mutation at position G2583. The UAG specific readthrough on the U1067 ribosomes can be explained by affecting RF-1 and RF-2 binding in a different way. Therefore, mutations at position G2583 reduce the level of mistranslation at both UAA and UAG nonsense codons.

Taking into account that mutant ribosomes form about 40 - 60 % of the total ribosome population (Saarma and Remme, 1992), the effects of mutations in 23S rRNA on the translational accuracy should be higher than those observed in these experiments. A comparison of the results obtained *in vivo* and *in vitro* show that the effect of mutations at position 2583 of 23S rRNA on the translational accuracy are in agreement in both systems. Surprisingly, C2505 had a small effect *in vivo* but not *in vitro*.

The A1067 to U transversion is known to reduce the affinity of thiostrepton to ribosomes 100 times (Thompson and Cundliffe, 1991) allowing the physical separation of mutant ribosomes as it was proposed (Thompson et al., 1988). Several solid supports were tested in respect to binding of ribosomes to the immobilized thiostrepton. It was found that the acid-hydrolyzed Sephadex can be used as a support for affinity purification of U1067 ribosomes. Evidently polysaccharide chains are long enough spacers to avoid steric hindrances in ribosome-thiostrepton interaction. Physical separation of mutant ribosomes would be important for the analysis of effects of mutations in 23S rRNA concerning the structure and function of the ribosome. Direct measurements of the binding affinities of aminoacyl-oligonucleotides to the peptidyltransferase centre of the mutant ribosomes would answer to the question whether the mutations in the 23S rRNA effect the binding of substrates to the peptidyltransferase centre.

CONCLUSION

Thiostrepton is known to inhibit both elongation factor dependent allosteric transition and nonenzymatic translocation of the wild-type ribosomes (Hausner et al., 1988). The A to U transversion at position 1067 of 23S rRNA has been shown to confer resistance to thiostrepton during cell-free translation on the E. coli ribosomes (Thompson et al., 1988). U1067 has proven to be a useful mutation to discriminate between the activities of chromosomal and plasmid encoded ribosomes.

In this study the thiostrepton resistance was used first to study the 50S assembly process. It was shown that 50S particles containing transcripts synthesized by the T7 RNA polymerase at 37°C are inactive: 1) The corresponding 50S particles accumulated in the 50S pool and could not enter the 70S pool *in vivo* indicative of their inactivity. 2) *In vitro* the plasmid-derived ribosomes were not active in poly(Phe) synthesis. 3) The pT7-1 derived transcripts accumulate in the 50S fraction even when the transcription initiation rate was lowered which caused low expression levels of the plasmid rRNA genes. 4) The 50S subunits with pT7-1 rRNA could partially enter the 70S fraction when the transcription elongation rate was decreased by incubation at 25°C. It follows that it is neither the transcription providing defective rRNA nor the overproduction of ribosomal RNA but rather the assembly of defective 50S subunits which caused the inactivity of the 50S particles *in vivo* when the rRNA is synthesized with the T7 RNA polymerase. A likely explanation is that the coupling beween rRNA transcription and ribosomal assembly ("assembly gradient", for review see Nierhaus, 1991) is a prerequisite for the formation of active particles and can be disturbed in the presence of T7 RNA polymerase. The viral enzyme is five times faster than the host transcriptase and so the finely tuned interaction between rRNA folding and protein binding might be uncoupled.

Further, the thiostrepton resistance was used as a second mutation for the functional characterization of single point mutations at the peptidyltransferase region of E. coli 23S rRNA. It was found that mutations corresponding to positions G2505 and G2583 affected ribosomal sensitivity following exposure to the inhibitors of peptide bond formation, whereas mutations at G2607 and G2608 had no effect. An increase in translational accuracy was attributed to the mutations at position 2583. The hyperaccurate phenotype of these mutations was confirmed *in vivo*. In addition, it was found that the U1067 mutation had an influence on the UAA directed termination event. The nucleotide 1067 is situated at a region which might be related to the factor dependent GTPase activity. Therefore, mutations at this position might affect factor directed reaction on the ribosomes and are possibly not suitable for the study of mutations which interfere with factor-dependent reactions. On the other hand, U1067 had no influence on the ribosomal peptidyltransferase centre and can indeed be used as a second mutation to study the role of 23S rRNA in the peptide bond formation.

Acknowledgements

K. H. Nierhaus from Max-Planck Institute of Molecular Genetics and R. Villems from Estonian Biocentre are acknowledged for support and discussions. We would like to thank C. Prescott for advise and help in experiments *in vivo*. We thank A. Liiv and T. Tenson for assistance and C. Prescott and Mrs. K. Ustav for correcting the manuscript.

REFERENCES

Chamberlin, M., and Ring, J.; 1973; *J. Biol. Chem.* 248:2235

Cunningham, P. R., Richard, R. B., Weitzmann, C. J., Nurse, K., and Ofengand, J.; 1991; *Biochimie* 73:789

Dahlberg, A.; 1989; *Cell,* 57:525

Gourse, R. L., Stark, M. J. R., and Dahlberg, A.; 1983; *Cell,* 32:1347

Hausner, T.-P., Geigenmüller, U., and Nierhaus, K. J.; 1988; *J. Biol. Chem.* 263: 13103

Krzyzosiak, W., Denman, R., Nurse, K., Hellmann, M., Boublik, M., Gehrke, C. W., Agris, P. F., and Ofengand, J.; 1987; *Biochemistry* 26:2353

Leclerc, D. and Brakier-Gingras, L.; 1990; *Biochem. Cell Biol.* 68:169

Lewicki, B. T. U., Margus, T., Remme, J., and Nierhaus, K. H.; 1993; *J. Mol. Biol.*, in press.

Melançon, P., Gravel, M., Boileau, G., and Brakier-Gingras, L.; 1987; *Biochem. Cell Biol.* 65:1022

Nierhaus, K. H.; 1991; *Biochimie* 73:739

Pestka, S.; 1977; In: "Molecular Mechanisms of Protein Biosynthesis", Weisbach, H., Pestka, S. eds. Academic Press. New-York, San-Francisco, London

Petrullo, L. A., Gallagher, P. J., and Elseviers, D.; 1983; *Mol. Gen. Genet.* 190:289

Prescott, C. D., and Kornau, H.-C.; 1992; *Nucl. Acids Res.* 20:1567

Rheinberger, H.-J., and Nierhaus, K. H.; 1990; *Eur. J. Biochem.* 193:643

Saarma, U., and Remme, J.; 1992; *Nucl. Acids Res.* 20:3147

Thompson, J., and Cundliffe, E.; 1991; *Biochimie* 73:1131

Thompson, J., Cundliffe, E., and Dahlberg, A.; 1988; *J. Mol. Biol.* 203:457

EXTRINSIC FACTORS IN RIBOSOME ASSEMBLY

Jean-Hervé Alix

Institut de Biologie Physico-Chimique
13, rue Pierre et Marie Curie
F-75005 Paris

Total reconstitution, *in vitro*, of ACTIVE procaryotic 30s and 50s ribosomal subunits by self-assembly does not exclude the participation of extrinsic (non ribosomal) factors involved in ribosomal assembly *in vivo*, as already pointed out by Nierhaus (1991). The facts available at present in support of this idea are : the essentially different characteristics between *in vitro* and *in vivo* ribosome assembly, the occurence of post-translational modifications affecting several ribosomal proteins, the discovery of the *rim* genetic loci, and of other extraribosomal genetic determinants revealed as extragenic suppressors of mutations blocking ribosome assembly, and finally the possible role of the chaperone DnaK in ribosome biogenesis.

DIFFERENCES BETWEEN *IN VITRO* AND *IN VIVO* RIBOSOME ASSEMBLY

In vitro reconstitution of *E. coli* ribosomal subunits from their r-RNA and protein components is performed under ionic, temperature and time conditions which are far away from physiological conditions.
This is not astonishing since molecular components to be assembled (r-RNAs and ribosomal proteins) are not the same *in vitro* and *in vivo* : in the first case they are mature rRNA molecules (fully processed, possessing their trimmed 5' and 3' ends and all their post-transcriptional modifications (Björk, 1987) as well as ribosomal proteins harboring their post-translational modifications (see next paragraph) , whereas *in vivo* they are unprocessed, unmodified, nascent molecules (and perhaps unfolded in the case of ribosomal proteins). These differences have led to the proposal of "an assembly gradient", *i.e.* the coupling of rRNA synthesis and ribosomal assembly (Nierhaus, 1991).

The Translational Apparatus, Edited by K.H. Nierhaus
et al., Plenum Press, New York, 1993

POST-TRANSLATIONAL MODIFICATIONS TO RIBOSOMAL PROTEINS

Several *E. coli* ribosomal proteins are affected by post-translational modifications (Table 1) and, although very little research on their role in ribosome function or assembly has been performed so far, these modifications could play a role during the integration of ribosomal proteins into ribosomal particles. An indication of a role of this type exists in the case of the METHYLATION of *E. coli* ribosomal protein L3 (an assembly-initiator protein of the large ribosomal subunit ; Nowotny and Nierhaus, 1982) since a defect in the genetic locus (*prmB*) which governs it leads to a slight cold-sensitive growth phenotype and defective ribosomal pattern (Lhoest and Colson, 1981). On the other hand, inactivation by gene disruption of the *prmA* genetic locus, encoding the L11-methyltransferase, does not produce any apparent modification of cell growth characteristics(Vanet and Alix, 1992) and therefore argues against a role of the post-translational methylation of L11 in ribosome assembly.
In yeast, it has been suggested that covalent, transient association between ubiquitin and some ribosomal proteins promotes their incorporation into nascent ribosomes, indicating a chaperone function for ubiquitin (Finley *et al.*, 1989).

RIM LOCI

Isolation of cold-sensitive mutants of *E.coli* defective in 50s ribosomal subunit assembly permitted detection of genetic determinants, named *rim*A, *rim*B, *rim*C and *rim*D, the products of which are involved in ribosomal assembly. As they do not correspond to any of the structural genes of ribosomal components (with the exception of *rim*A which most likely is an allele of the *rpm*H gene, coding for ribosomal protein L34 ; Hansen *et al,* 1982) they probably encode extraribosomal assembly factors (Bryant and Sypherd, 1974 ; Sypherd *et al.*, 1974 ; Bryant *et al.*, 1974) although no precise biochemical function has yet been attributed to them (the mutant allele *rim*B is dominant ; Grogan and Cronan, 1984). Two other genetic determinants (*rbaA* and *rbaB*) involved in ribosome assembly have been detected in a similar way (Nashimoto *et al.*, 1985).

EXTRAGENIC SUPPRESSORS OF MUTATIONS AFFECTING RIBOSOME ASSEMBLY

The selection of extragenic suppressors (second-site revertants) of a conditionally lethal mutant (for example a temperature-sensitive mutant) is an invaluable genetic tool to search for loci encoding molecular components which are FUNCTIONALLY related to the component afflicted by the conditionally lethal mutation. These compensatory mutations are easily selected as temperature-resistant pseudorevertants and can be classified on the basis of their map locations on the *E. coli* chromosome. This strategy has been applied with great success to an *E.coli* thermosensitive mutant bearing a missense mutation (*rplX* 19) in the structural gene *rplX* encoding ribosomal protein L24, which leads to impaired assembly of 50s ribosomal subunits at 42°C (Cabezón *et al.*, 1977 ; Marvaldi *et al.*, 1979 ; Nishi and Schnier, 1986 ; Nishi *at al.*, 1987).

Table 1. *E. Coli* genetic determinants, mapping outside the known ribosome structural gene clusters, possibly involved in ribosome assembly. *rim*A and *prm*A are not included (see text).

Genetic loci	Chromosomal map position (min.)	Phenotype or biochemical function	References
rimB	37	cold-sensitivity; ribosomal assembly defect	Bryant and Sypherd 1974; Sypherd *et al*, 1974; Bryant *et al*, 1974; Nashimoto *et al*, 1985
rimC	26		
rimD	87		
rbaA (era)	55		
rimE	72	thermosensitivity alteration in 50S subunit proteins	Kushner *et al*, 1977
rimH	14	thermolethality; ribosome maturation	Johnson *et al*, 1976
rimF (res)	1	ribosomal restriction	Garvin and Gorini, 1975.
rimG(ramB)	1	alteration of ribosomal protein S4	Zimmermann *et al*, 1973
rimI	99	post-translational acetylation of S18	Yoshikawa *et al*, 1987
rimJ	24	post-translational acetylation of S5	
rimK	19	post-translational addition of glutamic acid residues to S6	Kang *et al*, 1989
rimL	34	post-translational acetylation of L12 converting it into L7	Tanaka *et al*, 1989
fusB	14	pleiotropic effects on ribosomes and protein S6	Isaksson and Takata, 1978; Takata and Isaksson,1978
prmB	50	post-translational methylation of L3	Lhoest and Colson, 1981
genes unknown		post-translational methylations affecting S11,L7/L12,L16 and L33 ribosomal proteins	review, Alix, 1988
srmB (rhlA) *(rbaB)*		Extragenic suppressors of the mutation *rplX*19-ts (L24)	Nishi *et al*, 1988 Kalman *et al*, 1991 Nashimoto, 1992
zu	71		Vanet and Nishi, (see text)
zup	5		
rit	87	thermoresistance of the 50S subunit	Ono and Kuwano, 1978
dnaK	1	protein folding and assembly	present work

Since on the one hand L24 is an initiator protein for the *in vitro* assembly of the 50S subunit (Nowotny and Nierhaus, 1982) and since on the other hand its binding region near the 5' end of 23s rRNA is well characterised (there are in fact two binding sites, site A from nucleotide 260 to 280 and site B from nucleotide 480 to 510 ; Egebjerg *et al.*, 1987), this mutant (*rplX19*-ts) having an altered L24 ribosomal protein is particulary suitable for this type of study (Schnier and Nishi, 1988).

A mutation, in a plasmid-coded gene for 23S RNA, which suppresses the lesion in strain *rplX19*-ts has, in fact, been isolated, and shown to be a C to U transition at residue 33 of the 23S RNA (Nishi and Schnier, 1986) *i.e.* adjacent to binding site B in the putative secondary structure of this RNA (Egebjerg *et al.*, 1987). There is thus very good agreement concerning the L24-23s rRNA interaction, between these genetic results, studies of deletions within the 5' region of the 23s RNA gene which block ribosome assembly (Skinner *et al.*, 1985) and studies on protection of 23s RNA against ribonuclease and chemical attack by protein L24 (Egebjerg *et al.*, 1987).

Moreover several other extragenic chromosomal suppressor mutants have been isolated from the *rplX19*-ts mutant. One has an altered protease LA encoded by the *lon* gene (Nishi and Schnier, 1988), suggesting a participation of this heat-shock gene product in the turnover of ribosomal protein L24. In two others the suppressor mutations, *zu* and *zup*, are also extragenic because the electrophoretically altered L24 mutant proteins are still present and they have been mapped at minutes 5 and 71 of the *E. coli* chromosome (outside ribosomal protein genes). Gene cloning of their corresponding wild-type alleles is in progress (Vanet and Nishi, unpublished results).

Finally, phenotypic suppression of a conditionally lethal mutation can also be obtained by overexpression of a wild-type gene (cloned on a multicopy plasmid) the product of which interacts with, or compensates for, the defective component in the mutant grown under non-permissive conditions. An *E. coli* gene, *srmB*, that can suppress the effect of the *rplX19*-ts mutation when expressed at high copy number was identified by this means (Nishi *et al.*, 1988, 1989). The deduced aminoacid sequence of SrmB includes an aspartate-glutamate-alanine-aspartate (D-E-A-D) motif found in several RNA helicases, RNA binding proteins and RNA-dependent ATPases (Linder *et al*, 1989 ; Wassarman and Steitz, 1991 ; Schmid and Linder, 1992) and, moreover the SrmB protein possesses an RNA-dependent ATPase activity suggesting that it may stabilize an unstable ribosome assembly product by some specific alterations of rRNA secondary structure (unwinding). Comparable results recently obtained in *S. cerevisiae* (the discovery of a putative RNA helicase SPB4 involved in the formation of the 60S ribosomal subunit) support this idea (Sachs and Davis, 1990). This opens a promising way of finding other extraribosomal gene products playing a role in ribosome assembly since *spb* mutants in *S. cerevisiae* as well as thermoresistant pseudorevertants in *E. coli* can be positively selected, and in this manner a multicopy plasmid bearing an RNA helicase-like gene(harboring a D-E-A-D box motif and named *deaD*) has been isolated as a suppressor of a temperature-sensitive ribosomal protein S2 mutation (*rpsB*) in *E. coli*, but there is no cross-suppression of the *rplX* and *rpsB* mutations by the overexpressed *deaD* or *srmB* genes, which thus have different functions in the cell (Toone *et al*, 1991).

Another approach is to follow the trail of D-E-A-D, *i.e.* to screen an *E. coli* genomic library with a probe encoding a D-E-A-D- box motif and this has recently revealed five RNA helicase-like (*rhl*) D-E-A-D box genes (including *srmB* renamed *rhlA* and *deaD* renamed *rhlD*) at widely separated chromosomal locations (Iggo *et al*, 1990 ; Kalman *et al*, 1991). Further studies will show if *rhl* genes have functions related or not to the biosynthesis of ribosomal RNA.

A POSSIBLE ROLE OF THE CHAPERONE DNA K IN RIBOSOME ASSEMBLY

Evidence is accumulating that protein folding and assembly, long considered to be spontaneous processes, are actively catalysed *in vivo* by "chaperones" (Langer *et al.*, 1992) and we have recently undertaken a study (Alix and Guérin, 1992) of the possible role in ribosome assembly of the most abundant chaperone in *E. coli*, the product of the heat-shock gene *dnaK* (LaRossa and Van Dyk, 1991).

Labelling with ^{3}H-uridine of ribosomes synthesized by an *E. coli* strain bearing a thermosensitive allele *dnaK756* one hour after the transfer of the bacterial culture from 30°C to the non-permissive temperature (45°C) revealed accumulation of ribosomal particles with approximate sedimentation coefficients of 45S, 35S and 25S, along with normal 50S and 30S ribosomal subunits (Fig. 1).
Several arguments lead to the conclusion that they represent ribosomal precursor particles :

1) Such a ribosomal pattern is not observed when labelling is made at 30°C in the same strain (panel 1 of Fig.3), or at 45°C in a control strain bearing the wild-type *dnaK* allele : in both cases more than 95% of the ribosomal subunits display normal sedimentation coefficients and association properties in the presence of 10 mM Mg^{++}. The ribosomal pattern shown in Fig. 1 is therefore clearly dependent on the expression of the *dnaK756*-ts allele at the non-permissive temperature.

2) These ribosomal particles are not unstable, nor are they artefacts formed during cell extraction or sucrose gradient sedimentation because they can be isolated from a preparative sucrose gradient, identical to that of Fig.1A and rerun on analytical sucrose gradients prepared in ribosomal subunit dissociation conditions. As can be seen in Fig. 2, they display unchanged sedimentation coefficients. Moreover the type of rRNA they contain was shown to be 23S for the 45S and 35S particles and 16S for the 25S ribosomal particles. This confirms that 45S and 35S particles are related to 50S ribosomal subunits and 25S particles to 30S subunits. Therefore the *dnaK756*-ts allele perturbs the assembly of both ribosomal subunits at the non-permissive temperature.

3) These ribosomal particles have sedimentation coefficients close to those of the natural precursor particles detected by pulse labelling under normal growth conditions (Hayes and Hayes, 1971; Nierhaus et al., 1973 ; Turco et al., 1974 ; Lindahl, 1975).

4) Ribosomes isolated from the dnaK756-ts mutant labelled with ^{3}H-uridine at 30°C and then transferred to 45°C for one hour in a non radioactive growth medium are composed of structurally and functionally normal 30S and 50S subunits (Fig. 3), demonstrating that it is NOT THE THERMOSTABILITY of ribosomal subunits once assembled but their ASSEMBLY itself which is affected in this mutant at 45°C.

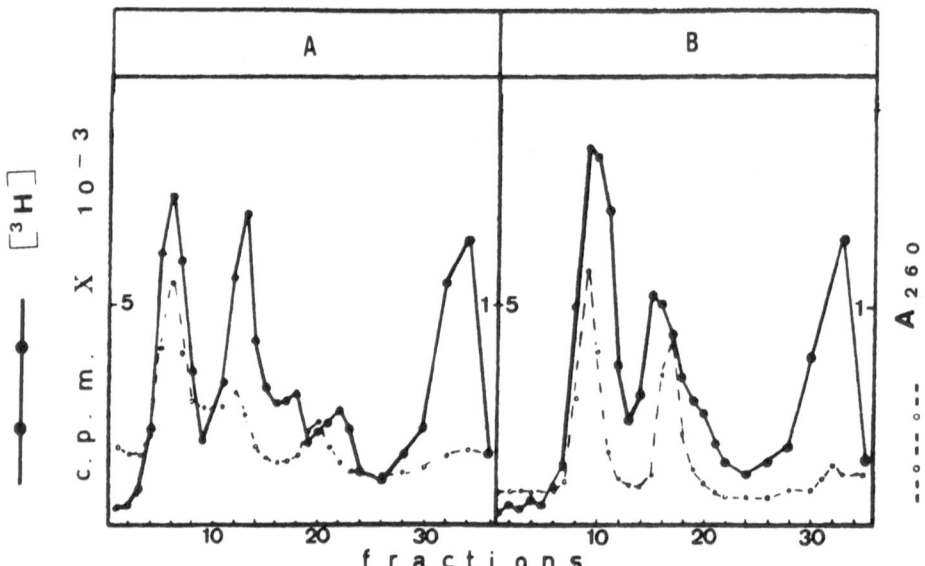

FIGURE 1. Sedimentation profiles on 10% to 30% linear sucrose gradients of ribosomes and their subunits prepared from the dnaK756-ts mutant grown at 30°C in minimal medium supplemented with 0.2 % glucose, 1µg/ml thiamine, 0.01% threonine, 0.2% casaminoacids, and , one hour after transfer from 30°C to 45°C labelled with ^{3}H-uridine (1µCi/ml ; 3µM) for a further hour. Ionic conditions in the gradient in panel A are those that permit association of ribosomal subunits (10mM magnesium acetate) whereas those in panel B dissociate ribosomes (10mM magnesium acetate, 400mM sodium chloride). Sedimentation is from right to left. Positions of the reference 70S, 50S and 30S ribosomal particles from E. coli wild-type were monitored at 260nm (o--- ---o).

5) Transformation of precursor particles synthesized and labelled with ^{3}H-uridine in the dnaK756-ts mutant at 45°C into normal ribosomal subunits is observed when bacteria are transferred to the same growth medium containing 0.1 mM non radioactive uridine instead of ^{3}H-uridine and incubated at the permissive temperature (30°C) for 15 min. to one hour.

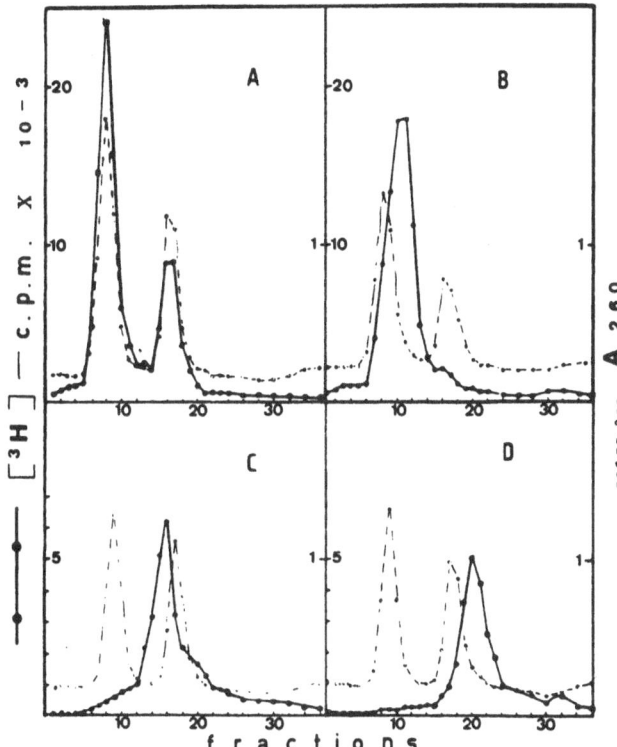

FIGURE 2. Sedimentation of ^{3}H-uridine labelled ribosomal particles synthesized by the dnaK756-ts mutant at 45°C, on 10% to 30% sucrose gradients containing 10mM magnesium acetate and 400mM sodium chloride. Large ribosomal subunits found unassociated (50S-45S) in conditions that permit association of ribosomal subunits (panel A of Fig.1) were collected and rerun here (panel B). 35S and 25S ribosomal particles, and as a control, ribosomes sedimenting at 70S were also collected from the same preparative sucrose gradient and rerun here (panels C, D and A respectively). Sedimentation is from right to left. Positions of the reference 70S, 50S and 30S ribosomal particles from *E. coli* wild-type were monitored at 260nm (o--- ---o).

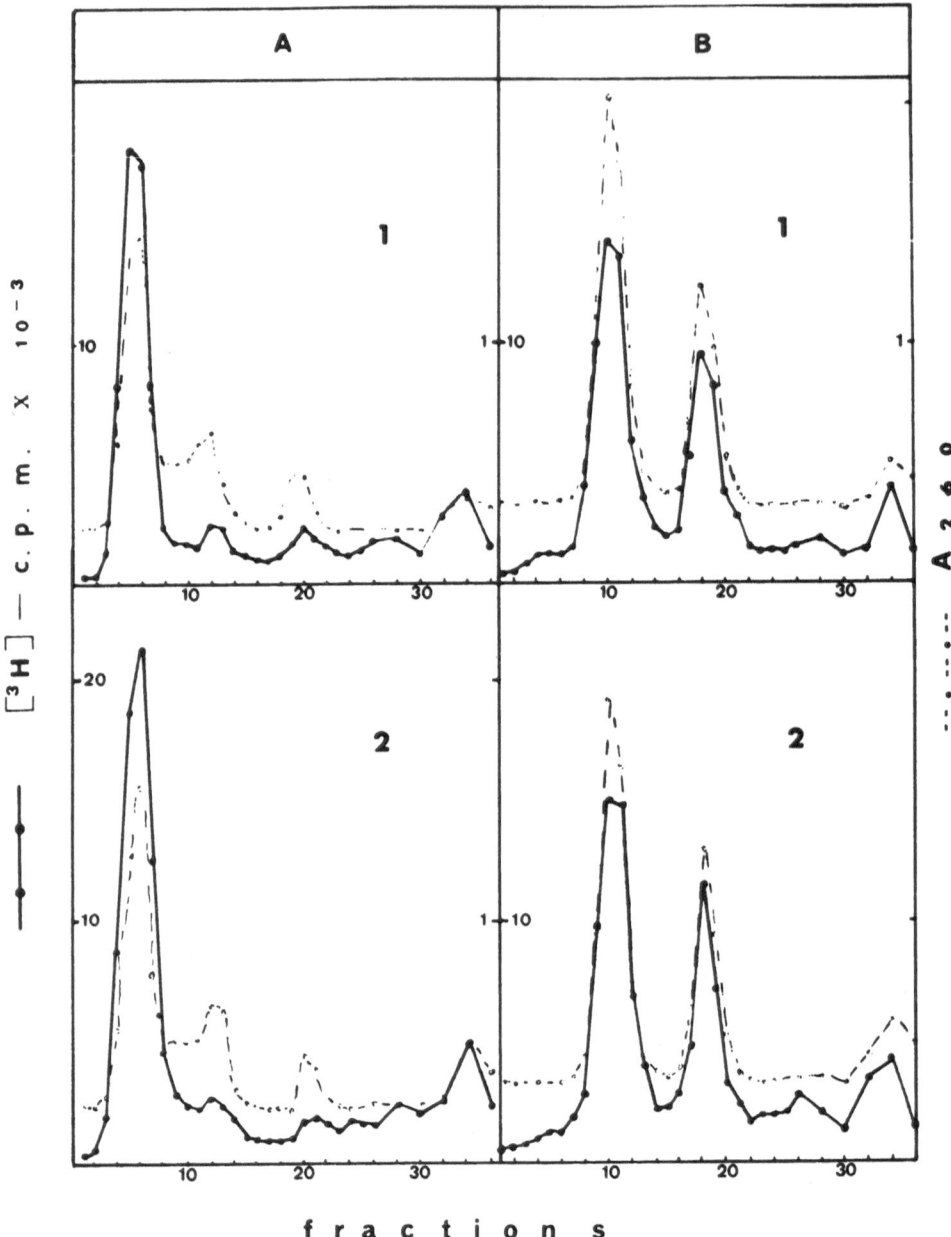

FIGURE 3. Sedimentation profiles of ribosomes and their subunits prepared from the
dnaK756-ts mutant labelled with ^{3}H-uridine at 30°C (labelling conditions are the
same as in Fig.1, except temperature) and chased for one hour in non radioactive
growth medium at 45°C (panel 2), or at 30°C as control (panel 1) . Ionic conditions in
panels 1A and 2A are those that permit association of ribosomal subunits, whereas
those in panels 1B and 2B dissociate ribosomes. Sedimentation is from right to left.
Positions of the reference 70S, 50S and 30S ribosomal particles from E. coli wild-type
were monitored at 260 nm (o--- ---o).

The conclusion of this study is therefore that the *dnaK756*-ts allele affects ASSEMBLY of BOTH ribosomal subunits, which suggests that the chaperone DnaK is implicated in ribosome assembly. It remains to be shown whether this effect is direct or indirect. This is under current investigation, but the following observations can already be made :

a) The fact that the *dnaK756*-ts allele affects assembly of both ribosomal subunits is in contrast with the effect on ribosome assembly of another *dnaK* mutation (the newly isolated *dnaK25*-ts allele; Gaitanaris *et al.*, 1990 and 1992) which causes accumulation of 45S but not of 35S and 25S ribosomal precursor particles at the non-permissive temperature. This argues against an indirect effect of the *dnaK*-ts mutations on ribosome assembly, which would not lead to two different types of ribosome assembly defects, one characteristic of the *dnaK756* mutation and the other one of the *dnaK25* mutation.

b) The triggering of the stringent response in the *relA*[+], *dnaK756*-ts mutant at the non-permissive temperature (Itikawa *et al.*, 1986) is not the cause of the ribosomal assembly defects observed here, because we found no such ribosomal assembly defects in a *relA*[+], thermosensitive valyl-tRNA-synthetase (*val* S7) mutant labelled with ^{3}H-uridine at 34°C, in spite of the induction of the stringent and heat-shock responses known to occur at this semi-permissive temperature (Grossmann *et al.*, 1985).

c) A constitutive synthesis of the heat-shock proteins which takes place in the *dnaK756*-ts mutant due to the failure to turn off the heat-shock response could also cause the ribosomal assembly defects, but this uncontrolled heat-shock response is observed even at 30°C (Tilly *et al.*, 1983) and therefore ribosome assembly defects would also have been seen at this temperature, which is not the case (panel 1 of Fig.3).Nevertheless studies on ribosome assembly in a *rpoH*, *dnaK756*-ts double mutant (devoid of the continuous synthesis of heat-shock proteins) will produce a definitive answer.

d) Another possibility is that, without active DnaK chaperone, ribosomal proteins are normally synthesized, but cannot be integrated into ribosomal subunits either because they are not in a proper (unfolded) conformation or because they are extensively degraded. This would explain the presence of ^{3}H-labelled normal 30S and 50S subunits synthesized in the mutant *dnaK756*-ts à 45°C (Fig.1) which would result from assembly of rRNA synthesized at the non-permissive temperature with preexisting ribosomal proteins, as well as accumulation of precursor particles resulting from imbalance between rRNA in excess and available ribosomal proteins. Examples are known in which excess synthesis of rRNA relative to ribosomal proteins results in failure of ribosome assembly (see Discussion in Dodd *et al.*, 1991). If this were so, it would indicate the necessity of the DnaK chaperone for the protection of ribosomal proteins or for their acquisition of a correct folded state prior to ribosome assembly.

The above points once clarified, it will also be worthwhile to examine the participation of other chaperones or heat-shock gene products in the ribosome assembly process, in particular that of

ClpB. The heat-shock gene *clp*B immediately precedes the *rrn*G operon on the *E. coli* chromosome, and Squires and Squires (1992) noted that at high temperatures, physical proximity of the two genes could ensure local concentration of ClpB for the protection of ribosomes being assembled from the *rrn*G operon.

Acknowledgments: I thank M.F. Guérin for expert technical assistance, M. Gottesman for the gift of the *dnaK*-ts mutants, M. Grunberg-Manago in whose laboratory this work was done, D. Hayes for reading the manuscript, and University Paris 7 and the Centre National de la Recherche Scientifique (URA 1139) for financial support.

REFERENCES

Alix, J.H., 1988, Post-translational methylations of ribosomal proteins, *in* : "Advances in Post-translational Modifications of Proteins and Aging. Advances in Experimental Medicine and Biology", vol.231. V. Zappia, P. Galletti, R. Porta, F. Wold, eds. Plenum Press, pp. 371-385.

Alix, J.H. and Guérin, M.F., 1992, submitted to *Proc. Natl. Acad. Sci. USA*.

Bryant, R.E. and Sypherd, P.S., 1974, *J. Bact.* 117: 1082-1092.

Bryant, R.E., Fujisawa, T. and Sypherd P.S., 1974, *Biochemistry* 13, 2110-2114

Björk, G .R., 1987, Modification of stable RNA, *in* : "*Escherichia coli* and *Salmonella typhimurium* : Cellular and Molecular Biology". F.C. Neidhardt, J.L. Ingraham, K.B. Low, B. Magasanik, M. Schaechter and H.E. Umbarger, eds. American Society for Microbiology, pp. 719-731.

Cabezón, T., Herzog, A., Petre, J., Yaguchi, M. and Bollen, A., 1977, *J. Mol. Biol.* 116 : 361-374.

Dodd, J., Kolb, J.M. and Nomura, M. 1991, *Biochimie* 73 : 757-767

Egebjerg, J., Leffers, H., Christensen, A., Andersen, H. and Garrett, R.A., 1987, *J. Mol. Biol.* 196 : 125-136.

Finley, D., Bartel, B. and Varshavsky, A., 1989, *Nature* 338 : 394-401.

Gaitanaris, G.A., Papavassiliou, A.G., Rubock, P. Silverstein, S.J. and Gottesman, M.E., 1990, *Cell* 61 : 1013-1020 and 1992, *Cell* 70 : 714.

Garvin, R.T. and Gorini, L., 1975, *Mol. Gen. Genet.* 137 : 73-78.

Grogan, D.W. and Cronan, J.E., 1984, *Mol. Gen. Genet.* 196 : 367-372.

Grossman, A.D., Taylor, W.E., Burton, Z.F., Burgess, R.R. and Gross, C.A., 1985, *J. Mol. Biol.* 186 : 357-365.

Hansen, F.G., Hansen, E.B., Atlung, T., 1982, *EMBO J.* 1 : 1043-1048.

Hayes, F. and Hayes, D.H., 1971, *Biochimie* 53 : 369-382.

Iggo, R., Picksley, S., Southgate, J., McPheat, J. and Lane, D.P., 1990, *Nucleic Acids Res.* 18 : 5413-5417.

Isaksson, L.A. and Takata, R., 1978, *Mol. Gen. Genet.* 161 : 9-14.

Itikawa, H., Fujita, H., Wada M., 1986, *J. Biochem.* 99 : 1719-1724.

Johnson, S.C., Watson, N. and Apirion, D., 1976, *Mol. Gen. Genet.* 147 : 29-37.

Kalman, M., Murphy, H. and Cashel, M., 1991, *The New Biologist* 3 : 886-895.

Kang, W.K., Icho, T. Isono, S., Kitakawa, M. and Isono, K., 1989, *Mol. Gen. Genet.* 217 : 281-288.

Kushner, S.R., Maples, V.F. and Champney, W.S., 1977, *Proc. Natl. Acad. Sci.,* USA 74 : 467-471.

Langer, T. Lu, C., Echols, H., Flanagan, J., Hayer, M.K. and Hartl, F.U., 1992, *Nature* 356 : 683-689.

LaRossa, R.A. and Van Dyk, T.K. 1991, *Mol. Microbiol.* 5 : 529-534.

Lhoest, J. and Colson, C., 1981, *Eur. J. Biochem.* 121 : 33-37.

Lindahl, L. 1975, *J. Mol. Biol.* 92 : 15-37

Linder, P., Lasko, P.F., Ashburner, M., Leroy, P., Nielsen, P.J., Nishi, K., Schnier, J. and Slonimski, P.P., 1989, *Nature* 337 : 121-122.

Marvaldi, J., Pichon, J. and Marchis-Mouren, G., 1979, *Mol. Gen. Genet.* 171 : 317-325.

Nashimoto, H., Miura, A., Saito, H. and Uchida, H., 1985, *Mol. Gen. Genet.* 199 : 381-387.

Nashimoto, H., Communication at the Conference "The Translational Apparatus", Berlin, 1992.

Nierhaus, K.H., Bordasch, K. and Homann, H.E., 1973, *J. Mol. Biol.* 74: 587-597.

Nierhaus, K.H., 1991, *Biochimie* 73 : 739-755.

Nishi, K., and Schnier, J., 1986, *EMBO J.* 5 : 1373-1376.

Nishi, K., Müller, M. and Schnier, J., 1987, *J. Bact.* 169, 4854-4856.

Nishi, K. and Schnier, J., 1988, *Mol. Gen. Genet.* 212 : 177-181.

Nishi, K., Morel-Deville, F., Hershey, J.W.B., Leighton, T. and Schnier, J., 1988, *Nature* 336 : 496-498 and1989, *Nature* 340 : 246.

Nowotny, V. and Nierhaus, K.H., 1982, *Proc. Natl. Acad. Sci.* USA 79 : 7238-7242.

Ono, M. and Kuwano, M., 1978, *J. Bact.* 134 : 677-679.

Sachs, A.B. and Davis, R.W., 1990, *Science* 247 : 1077-1079.

Schmid, S.R. and Linder, P., 1992, *Mol. Microbiology* 6 : 283-291.

Schnier, J. and Nishi, K., 1988, *Methods Enzymol.* 164 : 706-709.

Skinner, R.H., Stark, M.J.R. and Dahlberg, A.E., 1985, *EMBO J.* 4 : 1605-1608.

Squires, C. and Squires, C.L., 1992, *J. Bact.* 174 : 1081-1085.

Sypherd, P.S., Bryant, R., Dimmitt, K. and Fujisawa, T., 1974, *J. Supramolecular structure* 2 : 166-177.

Takata, R. and Isaksson, L.A., 1978, *Mol. Gen. Genet.* 161 : 15-21.

Tanaka, S., Matsushita, Y., Yoshikawa, A. and Isono, K., 1989, *Mol. Gen. Genet.* 217 : 289-293.

Tilly, K., McKittrick, N., Zylicz, M. and Georgopoulos, C., 1983, *Cell* 34 : 641-646.

Toone, W.M., Rudd, K.E.,Friesen, J.D.,1991, *J. Bact.*173 : 3291-3302.

Turco, E., Altruda, F., Ponzetto, A. and Mangiarotti, G., 1974, *Biochemistry* 13 : 4752-4757.

Vanet, A. and Alix, J.H. Abstract n°105 presented at the Conference 'The Translational Apparatus", Berlin, 1992.

Wassarman, D.A., and Steitz, J.A., 1991, *Nature* 349 : 463-464.

Zimmermann, R.A., Ikeya, Y., and Sparling, P.F, 1973, *Proc. Natl. Acad. Sci.* USA 70 : 71-75.

Yoshikawa ,A., Isono, S., Sheback, A. and Isono, K., 1987, *Mol. Gen. Genet.* 209 : 481-488.

NON-RIBOSOMAL PROTEINS AFFECTING THE ASSEMBLY OF RIBOSOMES IN ESCHERICHIA COLI

Hiroko Nashimoto

Department of BioSciences
Teikyo University
1-1, Toyosatodai
Utsunomiya 320, Japan

INTRODUCTION

We have previously described isolation of suppressors to a temperature-sensitive mutation of ribosomal protein S12; one such suppressor mutation was found to be an alteration of the RNaseIII gene on the *E. coli* chromosome (Nashimoto, et al., 1985; Nashimoto & Uchida, 1985). In these studies, a series of cold-sensitive mutations defective in the assembly of ribosomes at 20°C was isolated by localized mutagenesis of a region near the rnc gene on the *E. coli* chromosome. The genes defined by the cold sensitive mutations were named rbaA and rbaB, for ribosome assembly. Although both of these genes map close to the rnc gene on the *E. coli* chromosome, at least rbaB was found to be definitely different from the rnc gene. Both of the rba genes accumulated premature ribosomal subunits during growth at 20°C. Otherwise, they were not characterized further at that time. Now, they were mapped more precisely, the genes cloned, and their nucleotide sequence determined.

The results showed that the rbaA gene is identical to era of *E. coli*, which in turn has strong similarity to the yeast RAS gene and encodes a GTP-binding protein (Ahnn, J. et al., (1986). A search by computer showed that our translation of the rbaB sequences was found to be identical to the amino acid sequence of the product of the *E. coli* srmB gene which, if present in high gene dosage, is able to suppress a temperature-sensitive mutation in the gene encoding the ribosomal protein L24 (Nishi, K., et al., 1988, 1989; Swiss-Prot Acc.#:

The Translational Apparatus, Edited by K.H. Nierhaus
et al., Plenum Press, New York, 1993

P21507). The srmB protein has been biochemically examined and shown to exhibit RNA-dependent ATPase activity, but the site of the srmB gene on the *E. coli* chromosome remained unknown. Although there are several differences between the nucleotide sequences of the srmB gene as published (EMBL Acc.#: X14152) and the sequence of rbaB that we have determined, identity of the gene products strongly suggested the identity of the srmB gene with the rbaB gene. Our cold-sensitive mutants in the rbaB gene showed delayed processing of the precursor rRNA at 20°C, and accumulation of the "precursor 16S rRNA" was remarkable (Nashimoto, et al., 1985). In the present communication, we will describe also that disruption of the wild-type rbaB gene by introduction of a Kan^r gene block (Winans, et al., 1985) converted the wild-type cell into a cold-sensitive one, but they grew normally at 37°C. Function of the rbaB gene, therefore, is dispensable for growth at higher temperature.

The D-E-A-D gene family of RNA helicases has been identified in bacteria, yeast, fly, mouse and man (Linder, et al., 1989), of which the rbaB gene is obviously a member. More recently, a multiple of the D-E-A-D box genes have been described in *E. coli* (Iggo, et al., 1990; Toone, et al., 1991). Obvious question to be asked is to inquire if the loss of these D-E-A-D genes also affect the assembly of *E. coli* ribosomes. Preliminary attempts were made, therefore, to disrupt the genes by introduction of gene blocks.

MATERIALS AND METHODS

Bacterial and Plasmid Strains

Lambda phage clones containing segements of the *E. coli* chromosome (Kohara, et al., 1987) were obtained from Y. Kohara.

Table 1. Bacterial strains and plasmids.

Strain name	Relevant charcters	References / Source
HN343	F$^-$arg nadB purL supE44	P1(malA$^+$)/W3110 into PA3306(CGSC#4537)
HN1001	purL$^+$ cold-sensitive rbaA1001	HN343; Nashimoto
HN1002	purL$^+$nadB$^+$cold-sensitive rbaB1002	HN343; Nashimoto
HN1003	purL$^+$ nadB$^+$cold-sensitive rbaA1003	HN343; Nashimoto
HN1007	purL$^+$nadB$^+$cold-sensitive rbaB1007	HN343
HN1009	purL$^+$nadB$^+$cold-sensitive rbaB1009	HN343
HN1011	purL$^+$ cold-sensitive rbaA1011	HN343
HN1016	purL$^+$nadB$^+$cold-sensitive rbaB1016	HN343
HN1017	purL$^+$nadB$^+$cold-sensitive rbaB1017	HN343
JC7623	F$^-$supE44 recB21 recC22 recF143 sbc15	GSRC (Japan)
Km gene block	Kan^r GenBlock^{TM} EcoRI	Pharmacia

Accession Numbers: The following accession numbers were registered at EMBL or Swiss-Prot Databanks:

Genes	EMBL Accession #	Swiss-Prot #
dbpA	X52647	P21693
deaD	M63288	P23304
rnc-era	X02946, M14658	P06616
rbaB	D13169	
srmB	X14152	P21507

Cold-Sensitivity: Cold-sensitivity was defined as inability to form colonies on nutrient agar after incubation for 48 hrs at 20°C. After 96 hrs of incubation at 20°C, they form tiny colonies which are barely countable. On the other hand, cold-insensitive bacteria formed tiny colonies after incubation for 48 hrs at 20°C.

Gene Disruption: The site-directed disruption of a gene was done by the method described by Winans, et al (1985). Namely, the Kanamycin resistance (Km^r) determinant (1.5 kb Kan^r GenBlock^{TM}; Pharmacia, Product number 27-4897) was inserted into the gene to be disrupted with or without partial deletion of the gene. The linear DNA fragment containing the disrupted gene was isolated and used to transform strain JC7623 (Oishi & Cosloy, 1972; Wackernagel, 1973) to Km^r. Proper insertion of the gene block was confirmed by restriction mapping. P1 lysates were prepared on these transformant, and used to transduce the disrupted gene to other strain of *E. coli.*

DNA Sequencing: A.L.F. DNA Sequencer (Pharmacia) was used mostly to determine nucleotide sequences of DNA fragments. However, some of DNA sequence analysis was performed by the dideoxynucleotide-chain termination sequencing method with sequencing kits obtained from TAKARA (Japan). For DNA fragments cloned in pUC118 or pUC119 plasmids, deletions of various lengths were isolated by TAKARA Kilo-sequence deletion kit which utilizes ExoIII digestion. Fluorescent primer oligonucleotides were prepared by a Gene Assembler Special (Pharmacia).

DNA-Directed Expression of Genes *in vitro*: *In vitro* DNA-directed transcription-translation kit for procaryotic DNA was purchased from Amersham. Molecular weights of the products synthesized were determined by electrophoresis in SDS acrylamide gels.

RESULTS

Mapping and Identification of rbaA and rbaB Mutations

The original cold-sensitive mutants of *E. coli* were isolated by localized mutagenesis of the purL - nadB region of the *E. coli* chromosome, and classified by complementation with λpurLnadB transducing phages containing various deletions. Thus, the rbaA mutation was complemented by λpurLnadBΔ52 but not by λpurLnadBΔ5, whereas rbaB mutation was complemented by λpurLnadBΔ5, but not by λpurLnadBΔ52 (Nashimoto, et al., 1985). The observation suggested that rbaA mutations are within the DNA fragment "c", where the rnc gene is located.

Some subclonings of the fragment "c" into pBR322 plasmid were also described (Nashimoto & Uchida, 1985). We have presently found that DNA fragments 50 and 50-12 were unable to complement the cold-sensitivity of rbaA mutant cells. Since each of the fragment "c" and pBR322 contained one NruI restriction site, digestion by NruI of pBR322 DNA containing the fragment "c" at its EcoRI site followed by ligation yielded a 6.5 kb Amp^{r}Tet^s plasmid. The plasmid was double digested with EcoRI and NruI, and a 2.7 kb fragment was produced, and the fragment was named Δ2.7w (Fig. 1). The fragment was recloned into pUC118 or pUC119. The Δ2.7w fragment contained genes for rnc, era and a part of recO (Ahnn, et al., 1986; Morrison, et al., 1989; Takiff, et al, 1989). The fragment complemented the cold-sensitivity of rbaA mutant bacteria. Corresponding 2.7 kb DNA fragments were isolated from each of rbaA mutant bacteria, HN1001, HN1003 and HN1011, and cloned into the EcoRI site of pUC118 or pUC119. These clones were checked for the presence of the rnc DNA by colony hybridization using HincII-ClaI DNA fragment (Fig. 1) isolated from the rnc gene as the probe. Mutant genomic clones could not complement the cold-sensitivity of corresponding rbaA mutants. However, corresponding clone containing the wild-type allele isolated from the parental strain complemented the cold-sensitivity of the mutant bacteria. Nucleotide sequences of approximately 2.1 kb DNA flanked by NruI and HincII and containing complete rnc and era genes and a proximal half of the recO gene (Fig. 1) was determined for each clone isolated from 2.7 rbaA mutants as well as from the wild-type bacteria. For both of the HN1001 and HN1011 mutants, a cytosine at 155 counting from the start of the era gene was converted to thymine (Thr to Ile), and for HN1003, a cytosine at 623 was converted to thymine (Ala to Val). No other difference from the wild-type genes was detected for rnc, era or the proximal half of recO. Therefore, it was concluded that the cold-sensitivity of the ribosomal assembly in rbaA mutant bacteria was due to amino acid changes in the era gene.

λpurLnadBΔ5 transducing fragment was found to contain a DNA region of about 4.5 kb long, bracketted by an EcoR1 and a PvuII sites (Fig. 2). Since the region was found to harbor the rbaB gene as shown by complementation of various cold-sensitive mutations isolated previously, the DNA fragment containing the region was cloned into pUC118 as well as into pUC119, and the nucleotide sequence of the whole insert was determined.

Various deletions were isolated by utilizing TAKARA Kilobase deletion kit which were utilized to sequence the region. Fluorescent primer oligonucleotides (mostly 25-mers) needed were synthesized by Gene Assembler (Pharmacia). The results of DNA sequencing

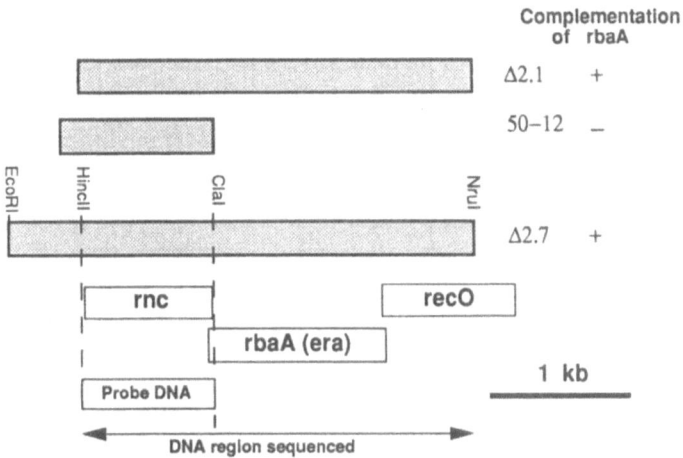

Fig. 1. Mapping the rbaA mutations. The nucleotide sequences as indicated were determined for each of the genomic DNA fragment isolated from rbaA mutant bacteria and cloned into pUC118 plasmid. The Figure shows three known genes, and HincII-ClaI fragment used as the probe to isolate the fragments.

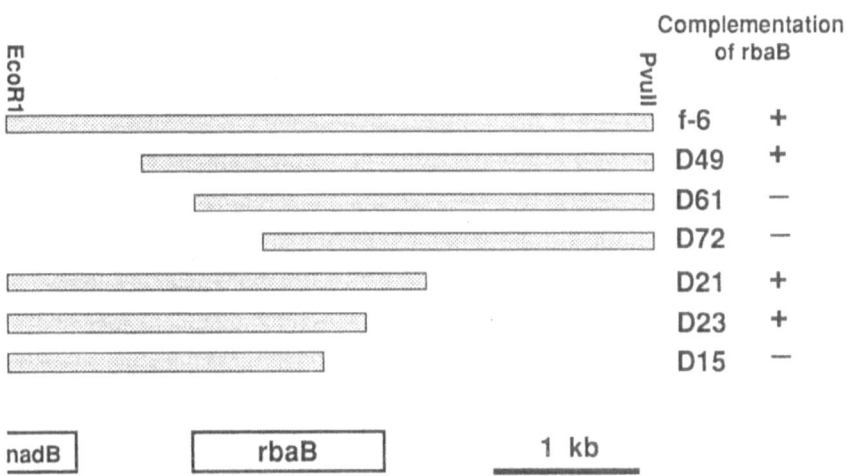

Fig. 2. Deletion mapping of the rbaB gene. A 4.5 kb EcoRI-PvuII fragment named f-6 of λpurLnadBΔ5 containing a part of nadB gene and the rbaB gene was cloned into EcoRI-SmaI site of pUC118 (pUC119). Various deletion derivatives of the DNA fragment were prepared by ExoIII digestion using TAKARA kilo-sequence deletion kit, and their sizes were determined by DNA sequencing. Complementations were carried out by testing cold-sensitivity of rbaB bacteria harboring plasmids containing the deleted fragments.

indicated that the region has only one open reading frame of significant size as shown in Fig. 2. The molecular weight of the putative gene product as estimated from the amino acids composition was calculated to be 49,911. Molecular weights of gene products encoded on pUC118 DNA containing the 4.5 Kb DNA fragement bracketted by EcoRI-PvuII (designated as f-6 in Fig. 2) were determined using the DNA-directed *in vitro* transcription-translation kit for procaryotes (Amersham). The results presented in Fig. 3 indicated that a 50K protein is the only product directed by the f-6 DNA (Fig. 3).

Genomic DNA was isolated from each of the rbaB mutants, digested with EcoRI and PvuII, and separated by electrophoresis on agarose gels to fractionate EcoRI-PvuII DNA fragements of 4.5 kb. They were cloned into pUC118, transformed into JM109 (TAKARA, competent cells) and Amp^r Xgal/white cells were selected. Colonies containing the mutant rbaB gene were detected by colony hybridization with approximately 1 kb NruI-NruI fragment of the rbaB gene. Using 25-mer primers, nucleotide sequences of mutant rbaB genes were determined. rbaA and rbaB mutations identified in the present study were listed in the following :

Table 2. Nucleotide alterations of rbaA and rbaB mutations.

Mutant strains	location (from start)	wild-type allele	mutant allele
rbaA:			
HN1001	#155	C (Thr)	T (Ile)
HN1003	#623	C (Ala)	T (Val)
HN1011	#155	C (Thr)	T (Ile)
rbaB:			
HN1007	#1036	G (Gly)	A (Ser)
HN1009	#415	C (Gln)	T (Stop)
HN1016	#624	G (Val)	A (Val)
and	#850	C(Gln)	T (Stop)
HN1017	#415	C (Gln)	T (Stop)

Disruption of the rba Genes

Since we now know nucleotide sequences of the rba genes, disruption of these genes by application of the site-directed disruption utilizing recBCsbc mutant of *E. coli*

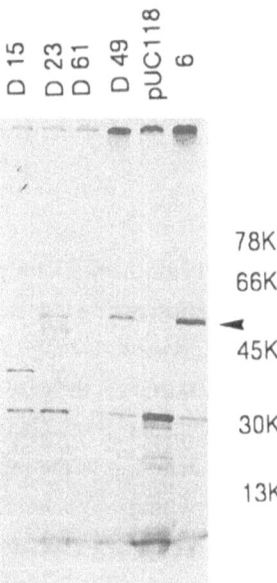

D 15 D 23 D 61 D 49 pUC118 6

78K

66K

◄

45K

30K

13K

Fig. 3. *In vitro* transcription-translation of the 4.5 Kb EcoRI-PvuII DNA fragment and deletions cloned in pUC118. Structures of cloned fragments were described in Fig. 2. The product of the rbaB gene is indicated by an arrow. D23 produces inactive gene product by deletion of 36 amino acids from the carboxyl terminus , and fusion to the proximal part of the lac gene of the vector DNA.

(Winans, et al., 1985; Wackernagel, 1973) was tried. 1.5 kb Kan^r GenBlock^{TM} EcoRI (Pharmacia) was inserted at single PstI site of the rbaA gene on the 4.5 kb EcoRI fragment "c" (Nashimoto, et al., 1985), and the disrupted DNA fragment was used to attempt transformation of JC7623, selecting for kanamycin resistance with the DNA. However, transformants were not obtained by incubation at 37°C .

The rbaB gene contains two NruI sites within the gene at 258 and 1263 counting from the start of the gene. Therefore, three types of disruptions were prepared; the first having the Kan^r GenBlock introduced at 258th nucleotide, the second having the block at 1263rd sites on the rbaB gene, and the third deleting and substituting the middle portion of the rbaB gene from 258 to 1263 with the GenBlock. Kan^r GenBlock was therefore introduced into the rbaB gene present in the 4.6 kb DNA which is cloned in the pUC118 vector. After introduction of the GenBlock, circular DNA was digested with BamHI and EcoRI to make 4.6 kb linear DNA. Each type of disrupted gene fragments were transformed into JC7623 by selecting for kanamycin resistance at 37°C. Km^r transformants were mostly cold-sensitive, irrespective of sites where GenBlock was introduced within the rbaB gene. Cellular DNA was isolated from cold-sensitive transformants, and digested with EcoRI. Southern hybridizations were performed using 1kb long NruI-NruI DNA fragment of the rbaB gene, as well as the Kan^{r}GeneBlock DNA, indicating successful insertion of the GenBlock. The disruption showed that the rbaB gene is dispensable at least for growth at 37°C. P1 phage lysates were prepared on Km^r JC7623, and used to transduce HN343 to Km^r. Km^r transductants were mostly cold-sensitive, but about one half of them were nadB$^+$.

Disruption of the dbpA Gene

Iggo, et al. (1990) described identification of a gene called dbpA for dead box protein at 29.8 min on the *E. coli* chromosome by low stringency screening of an *E. coli* genomic library with a p68 cDNA of *S. pombe*. Biological function of the gene, however, remained unknown. Since nucleotide sequence of the gene and its vicinity was determined, it was possible to disrupt the gene with the Kan^r GenBlock, by the same procedure employed to disrupt the rba genes. According to the report by Iggo, et al., (1990), the 1296 bp long dbpA gene is located within a 3.5 kb BamHI fragment, of which the nucleotide sequence of 1668 bp long containing dbpA has been determined. The dbpA gene has two NruI sites located at 1023 and 1073. The 3.5 kb BamHI DNA fragment was isolated from a λ transducing phage containing the region of the *E. coli* genome (Kohara clone # 262, Kohara, et al., 1987) and the fragment was cloned into pUC118. Kan^r GenBlock was then inserted substituting 1023-1073 fragment of the dbpA gene. The circullar DNA was linearized by digestion with BamHI, and JC7623 cells were transformed with the linear DNA. Kanamycin resistant transductants were obtained, and they were mostly cold-sensitive.

DISCUSSION

We have previously reported suppressors of temperature-sensitive mutations in a ribosomal protein gene, rpsL(S12), of *Escherichia coli* K12, and one of them was identified as an alteration of RNaseIII (Nashimoto, et al., 1985; Nashimoto & Uchida, 1985). During the study, we have also described a series of cold-sensitive mutations of *Escherichia coli* which defined two new genes named rbaA and rbaB in the vicinity of rnc, the gene for RNaseIII on the chromosome of *E. coli*. Mutant cells were defective in ribosomal assembly at 20°C (Nashimoto, et al., 1985). In one of the rbaB mutants, HN1002, the assembly of 50S particles was severely reduced at 20 °C with accumulation of 32S particles, reminding us of the behavior of mutant Sad19 cells (Guthrie et al., 1969). In the rbaA mutant named HN1003, the 30S ribosomal particles contained mostly RNA species which migrated even more slowly than the "precursor" 16S RNA (Nashimoto, et al., 1985). Since we know the chromosomal locations of rbaA and rbaB genes, and that their mutant alleles are recessive to the wild type allele, we have mapped and cloned them in detail and determined the nucleotide sequences.

Results of nucleotide sequence determination indicated clearly that rbaA is the same as era, which is located adjacent and downstream to rnc (Ahnn, et al., 1986; Takiff, et al., 1989). Takiff, et al. (1989) presented a negative experiment to disrupt the era gene using antibiotic cassettes and concluded that era is essential for *E. coli* growth. We have also attempted to disrupt era by an introduction of Kan^r GenBlock which was not successful. Nevertheless, we could determine nucleotide alterations in three different rbaA mutants, two of which were repetitive mutations at the same locus altered in the same way. If era were essential, rbaA mutations should be those which do not alter the indispensable function assigned to the era gene. Our determination of mutational alterations in the rbaA gene identified that cold-sensitive mutations we have isolated were substitution of amino acids (Table 2).

On the other hand, the function of the rbaB is dispensable for growth of the cell, in the sense that the gene can be disrupted by the Kan^r GenBlock, even allowing substantial part of the gene be substituted with the GenBlock. Furthermore, analyses of mutant genes disclosed frequent alteration of an amino acid to stop codon (Table 2). Although function of the rbaB gene is dispensable for growth of the cells at 37°C, growth at low-temperature were impaired (cold-sensitive) exhibiting defective assembly of ribosomes.

The present determination of DNA sequences strongly suggested that the rbaB gene is identical to the srmB gene (Nishi, et al., 1988, 1989). The srmB is a member of the so-called D-E-A-D gene family defined on the basis of the sequence homology (Linder, et al., 1988), which are thought to play roles in mRNA translation and ribosome assembly. More recently, deaD and dbpA genes have been described as new members of the D-E-A-D family in E. coli (Iggo, et la., 1990; Toone, et al., 1991). Although overexpression of srmB or deaD proteins from high copy number plasmids were shown to be able to suppress defects in cells harboring temperature-sensitive mutations in ribosomal proteins, their normal functions remained unknown (Nishi, et al., 1988; Toone, et al., 1991). As for the dbpA gene, the phenotype of dpbA mutants was unknown (Iggo, et al., 1990). We now know that disruption of the gene was possible, and the disruption made the cell cold-sensitive. We have also disrupted the deaD gene using the Kan^r GenBlock, but the disrupants grew at 20°C (data not shown). Although rbaB, dbpA and deaD genes are all classified as members of the D-E-A-D family, biological role of the deaD gene may be different from others. In fact, numbers of amino acids constituting proteins dictated by rbaB and dbpA genes are 444 and 432, respectively, those of the deaD gene contains 571 amino acids.

The reconstitution of active ribosomes of *Escherichia coli in vitro* proceeds at high-temperature involving unimolecular structural rearrangement of the intermediate RI particles (Traub and Nomura, 1969). However, *in vivo* assembly of ribosomes obviously proceeds at much lower temperature. Therefore, participation of non-ribosomal cellular components in *in vivo* processes of ribosomal assembly is an obvious possibility. Since ribosomal components of *Escherichia coli* and their reconstitution *in vitro* have completely been described, a genetic approach will be the first step to identify cellular components or functions affecting the assembly and control of ribosomes *in vivo* (Cole, et al., 1987). We have here many possible candidates which are mutationally characterized and which may function and act directly or indirectly in the assembly of ribosomes *in vivo*. Specific clones and sequence information will be useful to plan further experiments to clarify biological importance of these genes.

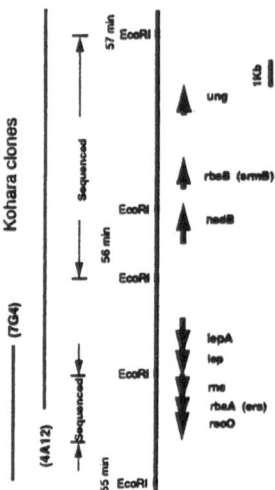

Fig. 4. Organization of the rbaA-rbaB region of *Eschericia coli* chromosome. The region corresponds mostly to the 55-57min region of the *E. coli* K12 map, which is covered by Kohara clones 7G4 and 4A12 (Kohara, et al., 1987). Approximate locations of known genes are shown. DNA sequences as indicated were determined in the present study .

Acknowledgements

The author thanks Drs. Hideo Ikeda and Hisao Uchida for their support in this project. This research was supported in part by grants-in-aid from the Ministry of Education, Science and Culture, Japan.

REFERENCES

Ahnn, J., March, P. E., Takiff, H. E., and Inouye, M., 1986, A GTP-binding protein of *Escherichia coli* has homology to yeast RAS proteins, *Proc. Natl. Acad. Sci. USA*, 83: 8848.

Cole, J. R., Olsson, C. L., Hershet, J. W. B., Grunberg-Manago, M., and Nomura, M., 1987, Feedback regulation of rRNA synthesis in *Escherichia coli*. Requirement for initiation factor IF2. 1987, *J. Mol. Biol.* 198: 383.

Guthrie, C., Nashimoto, H., and Nomura, M., 1969, Structure and function of *E. coli* ribosomes, VIII. Cold-sensitive mutants defective in ribosome assembly. *Proc. Natl. Acad. Sci., USA.,* 63: 384.

Iggo, R., Picksley, S., Southgate, J., McPheat, J., and Lane, D. P., 1990, Identification of a putative RNA helicase in *E. coli, Nucleic acids Res.* 18: 5413.

Kohara, Y., Akiyama, K., and Isono, K., 1987, The physical map of the whole *E. coli* chromosome: Application of a new strategy for rapid analysis and sorting of a large genomic library. *Cell,* 50: 495.

Linder, P., Lasko, P. F., Ashburner, M., Leroy, P., Nielsen, P. J., Nishi, K., Schnier, J., and Slonimski, P. P.,1989, Birth of the D-E-A-D box, *Nature,* 337: 121.

Morrison, P. T., Lovett, S. T., Gilson, L. E., and Kolodner, R., 1989, Molecular analysis of the *Escherichia coli recO* gene. *J. Bacteriol.,* 171: 3641.

Nashimoto, H., and Uchida, H., 1985, DNA sequencing of the *Escherichia coli* ribonuclease III gene and its mutations., *Mol. Gen. Genet.,* 201: 25.

Nashimoto, H., Miura, A., Saito, H., and Uchida, H.,1985,Suppressors of temperature-sensitive mutations in a ribosomal protein gene, *rps* L (S12), of *Escherichia coli* K12, *Mol. Gen. Genet.,* 199: 381.

Nishi, K., Morel-Deville, F., Hershey, J. W. B., Leighton, T., and Schnier, J., 1988, An elf-4A-like protein is a suppressor of an *Escherichia coli* mutant defective in 50S ribosomal subunit assembly, *Nature,* 336: 496.

Nishi, K., Morel-Deville, F., Hershey, J. W. B., Leighton, T., and Schnier, J., 1989, Corrigendum. *Nature,* 340: 246.

Oishi, M., and Cosloy, S. D., 1972, The genetic and biochemical basis of the transformability of *Escherichia coli* K12, *Biochem. Biophys. Res. Comm.* 49: 1568.

Sachs, A. B., and Davis, R. W., 1990, Translation initiation and ribosomal biogenesis: Involvement of a putative rRNA helicase and RPL46. Science, 247, 1077.

Takiff, H. E., Su-Min Chen, and Court, D. L., 1989, Genetic analysis of the rnc operon of *Escherichia coli, J. Bacteriol.,* 171:2581.

Toone, W. M., Rudd, K. E., and Friesen, J. D.., 1991, *dea*D, a new *Escherichia coli* gene encoding a presumed ATP-dependent RNA helicase, can suppress a mutation in *rps*B, the gene encoding ribosomal protein S2, *J. Bacteriol.* 173:3291.

Traub, P., and Nomura, M., 1969, Structure and function of *E. coli* ribosomes. VI. Mechanism of assembly of 30S ribosomal particles *in vitro, J. Mol. Biol.* 40:391.

Wackernagel, W., 1973, Genetic transformation in *E. coli*: The inhibitory role of the *rec*BC DNase, *Biochem. Biophys. Res. Comm.* 51: 306.

Winans, S.C., Elledge, S. J., Krueger, J. H., and Walker, G. C., 1985, Site-directed insertion and deletion mutagenesis with cloned fragments in *Escherichia coli, J. Bacteriol.,* 161:1219.

MECHANISMS OF TRANSLATIONAL INITIATION AND REPRESSION IN PROKARYOTES

David E. Draper

Department of Chemistry
The Johns Hopkins University
Baltimore, Maryland 21218

INTRODUCTION

Perhaps the first example of gene regulation at the translational level was the observation that the MS2 coat protein represses synthesis of the replicase (Nathans *et al.*, 1969). Many other examples of translational regulation by protein repressors have been found since. Especially noteworthy is the fact that the translation of most ribosomal proteins in *E. coli*, which accounts for a substantial fraction of the total protein synthesis, are autogenously regulated (Nomura *et al.*, 1984).

The mechanism by which translational repression takes place has seemed obvious: it is well-known that mRNA secondary structure within the ribosome binding site supresses translation, and a protein bound to the mRNA could also prevent initiating ribosomes from finding the mRNA initiation site. This is a simple competition, and whether the ribosomes or the repressor protein binds more tightly determines which wins. A problem with this view of repression is that the ribosome initiation complex is extremely stable, while many repressor proteins bind mRNA with only modest affinities. How can a repressor ever hope to compete? It is the object of this review to show that repressors can and do use several subtle strategies besides brute force to slow initiation. I will begin by reviewing and extending a framework for understanding two basic kinds of repression mechanisms, and then discuss several repression systems for which there are the most quantitative data in the literature.

TRANSLATIONAL INITIATION: QUANTITATIVE CONSIDERATIONS

In initiation of translation, 30S subunits and $tRNA_f^{met}$ bind to mRNA in random order to form a "pre-ternary" complex, followed by an isomerization of this complex to the initiation complex. The scheme is usually written as an initial equilibrium (K_{30S}) followed by an irreversible step (k_2) (Gualerzi & Pon, 1990), as suggested in Figure 1A. Kinetic schemes of this sort are common and have been extensively studied. For instance, in transcriptional initiation there is a first equilibrium binding of RNA polymerase to the promoter, followed by an irreversible kinetic step which involves the transition to an open

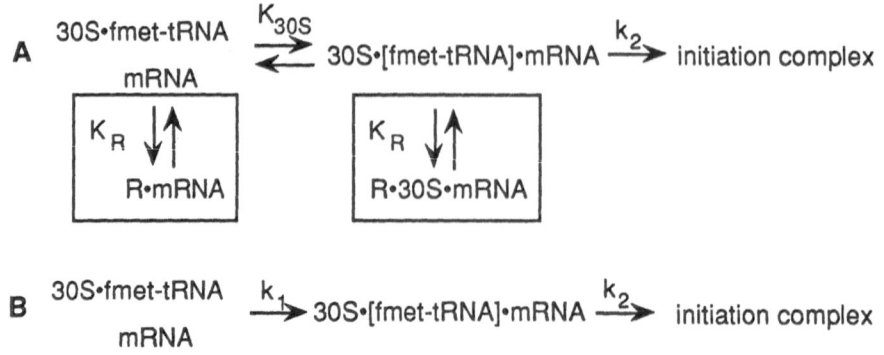

Figure 1. Two related schemes for initiation of translation. **A**, subunits and mRNA equilibrate rapidly to form a pre-initiation complex (K_{30S}), followed by a slow isomerization step (k_2). fmet-tRNA is bound in the pre-initiation complex, but not paired correctly with the initiation codon. Two points at which a repressor could bind the mRNA are illustrated in the boxes. **B**, dissociation of the pre-ternary complex is slow compared to the isomerization step, and initiation takes place in two consecutive irreversible steps. For simplicity the tRNA is shown pre-bound to the 30S subunit; the order of binding to the mRNA could be reversed.

complex. Even though more steps have now been resolved in transcriptional initiation, the simple two step description has been a very powerful first approximation for describing and rationalizing the effects of promoter sequence and repressor or enhancer proteins on transcriptional efficiency (McClure, 1985).

Unfortunately, detailed and reliable measurements of the equilibrium and rate constants for the simple scheme shown in Figure 1A have not been made for any single translational initiation site. Instead we must rely on scattered measurements performed on different RNAs under varying conditions, which are considered below.

Ribosome - mRNA Affinity

The binding of 30S subunits to RNA has been measured by several different methods. Non-specific interactions with poly(U) or poly(A) have modest affinities on the order of $10^6 - 10^7$ M^{-1}, as measured by a sucrose gradient method (Draper & von Hippel, 1979) or by the fluorescence of modified polymers (Draper & Gold, 1980). The same sucrose gradient method yields a binding constant of ~2×10^7 M^{-1} for 30S subunit binding to T4 gene 32 mRNA (unpublished observations), and a similar estimate has been made for a short RNA containing a Shine-Dalgarno sequence (Calogero *et al.*, 1988), though these workers did not allow for dissociation of the complex during sedimentation and may have underestimated the affinity. In the latter study, deletion of the Shine-Dalgarno sequence weakened binding by at least 10 fold. Lastly, a "toeprint" of 30S subunits bound to T4 gene 32 mRNA has been observed with MMLV reverse transcriptase; while the experiment has not been quantitated, appearance of a signal at a subunit concentration of 0.02 μM is consistent with a binding constant on the order of 10^7 M^{-1} (Hartz *et al.*, 1991).

Binding of tRNA$_f^{met}$ to 30S subunits probably does not increase the affinity of the subunits for mRNA by more than 10 fold. Ellis & Conway (1984), using a crude *in vitro* system to study R17 and Qβ coat protein translation, deduced from kinetic studies that tRNA$_f^{met}$ bound to the subunits first, and that the Michaelis constant for mRNA dissociation from this 30S-tRNA$_f^{met}$ complex was $0.3 - 0.7 \times 10^{-7}$ M. This suggests a binding affinity of the same order of magnitude as the direct measurements of subunit - mRNA affinities.

Isomerization Rates

An efficiently translated mRNA is initiated *in vivo* about every 3 sec (Kennell & Riezman, 1977), which sets a lower limit for any isomerization step of ~0.3 sec^{-1}. A direct measure of k_2 has been made utilizing tRNA labeled with a fluorescent tag, and gives a value of ~0.1 sec^{-1} (Wintermeyer & Gualerzi, 1983). This step is probably rate-limiting for translation. It is possible that isomerization depends on the mRNA sequence, and is slower for inefficiently translated mRNAs.

Does a Pre-Equilibrium Scheme Apply?

The kinetic studies of Ellis & Conway (1984) showed that the binding of a 30S-tRNA$_f^{met}$ complex to mRNA comes to equilibrium, in accordance with the way Figure 1A is drawn. Their conclusion may not apply *in vivo*, since they were using appearance of the protein itself to monitor the kinetics of initiation, and not the formation of the initiation complex. Elongation in their system, at 0.1 - 0.2 amino acids per second, was two orders of magnitude slower than *in vivo* value of ~17 (Kjeldgaard & Gausing, 1974), which may have slowed initiation sufficiently to allow binding steps to come to equilibrium.

Unfortunately, direct kinetic measurements of 30S - mRNA association are lacking. If binding is diffusion controlled, the association rate should be on the order of 10^7 M^{-1}sec^{-1}, and, given the estimate of the equilibrium binding constant as 10^7 - 10^8 M^{-1}, dissociation rates must be on the order of 0.1 to 1 sec^{-1}. This simple calculation suggests that k_{-1} is as slow or slower than k_2, and the first step cannot come to equilibrium. If this is the case, the scheme is better thought of as two sequential steps, as shown in Figure 1B. However, these rate constant estimates can easily be off by an order of magnitude in either direction. It is well known that DNA binding proteins can undergo facilitated diffusion by first binding non-specifically to the DNA, and then diffusing along the DNA to the target site (von Hippel & Berg, 1989). This accelerates the association rates by two or three orders of magnitude over the diffusion-controlled limit. Much smaller rate enhancements are likely for ribosomes initiating *in vivo*, since the relatively short leader sequences, as well as the coupling of transcription and translation, must limit the average "landing site" available to the ribosome. If the ribosome is able to bind the mRNA within (for instance) ±10 nucleotides of the initiation site, and then rapidly find the Shine-Dalgarno sequence, an order of magnitude rate enhancement is feasible. In support of this model, prokaryotic ribosomes have been shown to diffuse along an mRNA for distances of 40 nucleotides (Adhin & van Duin, 1990). (There seems to be some "sloppiness" in the ribosomal machinery, since it tolerates some variability in the position of the Shine-Dalgarno sequence relative to the initiation codon. This may allow for a few nucleotides variability in the initial position of ribosome contact with mRNA, even in the absence of sliding.)

Another mechanism which might increase the association rate dramatically *in vivo* is an entry site upstream of the ribosome binding site. As a ribosome clears from the initiation site, another ribosome, already non-specifically bound to an upstream "entry" site, could move into position. A small cooperativity in 30S subunit binding to polymers has been detected (Draper & Gold, 1980), which could make this a very efficient process orders of magnitude faster than any diffusion-controlled association. This mechanism would not be detected by *in vitro* studies of initiation complex formation, since it would accelerate the binding of the second and subsequent ribosomes to translate the mRNA, and not affect the rate at which the first ribosome finds the initiation site.

It is also possible that the steric requirements of fitting the mRNA into the ribosome binding site and aligning the Shine-Dalgarno sequence with the 16S rRNA is the rate-limiting step for subunit association, and much slower than diffusion controlled. Different mRNA sequences may achieve different balances in the relative rates of the binding and isomerization steps, and require different strategies for regulation by repressors.

REPRESSION MECHANISMS

In a previous discussion of translational repression, two kinds of repression mechanisms were defined (Draper, 1987). "Displacement" mechanisms are those in which binding of repressor prevents ribosomes from binding the mRNA; i.e., the repressor affects the initial binding step. "Entrapment" mechanisms involve repressor binding to the ribosome-mRNA complex and affecting the isomerization rate k_2. These two mechanisms are illustrated by the two boxes in Figure 1A. The two kinds of mechanisms are operationally defined by whether mRNA binding by the repressor and ribosome are mutually exclusive (displacement) or compatible (entrapment).

Equations have been derived to predict the effects of repressor concentration on translation rate, given the pre-equilibrium scheme of Figure 1A (Draper, 1987). They are

$$\text{(displacement)} \quad \text{relative rate} = \frac{1 + K_{30S}[30S]}{1 + K_{30S}[30S] + K_R[R]} \tag{1}$$

$$\text{(entrapment)} \quad \text{relative rate} = \frac{1}{1 + K_R[R]} \tag{2}$$

where the relative translation rate is the rate of translation in the presence of a given repressor concentration divided by the rate in the absence of repressor. This is the reciprocal of the "repression ratio" frequently measured experimentally. [30S] and [R] are the concentrations of free 30S subunits and repressor protein, respectively. Note that the repression levels in the entrapment mechanism are independent of the ribosome concentration; this is because there is no competition between the two. A protein with a modest mRNA affinity should be able to effectively repress translation by an entrapment mechanism without having to accumulate to very high concentrations.

If the consecutive kinetic steps of Figure 1B apply, the concentration of initiation complex never comes to equilibrium and steady-state kinetics have to be used to determine the effect of repressor. The resulting equation for the displacement reaction is

$$\text{relative rate} = \frac{1 + \left(\dfrac{k_1}{k_2}\right)[30S]}{1 + \left(\dfrac{k_1}{k_2}\right)[30S] + K_R[R]} \tag{3}$$

and the corresponding equation for the entrapment reaction is

$$\text{relative rate} = \frac{1 + \left(\dfrac{k_1}{k_2}\right)[30S]}{1 + \left(\dfrac{k_1}{k_2}\right)[30S](1 + K_R[R])} \tag{4}$$

If $(k_1/k_2)[30S] \gg 1$, this latter equation simplifies and is identical to the pre-equilibrium binding case (equation 2).

Although one might expect a small repressor to equilibrate rapidly with the mRNA, as drawn in Figure 1, it is possible that the kinetics of repressor association are comparable to those of ribosome binding. In that case the competition will be entirely a kinetic race between the repressor and ribosome for reaching the mRNA, and equilibrium considerations will be irrelevant.

mRNA SECONDARY STRUCTURE AS A TRANSLATIONAL REPRESSOR

A displacement type mechanism also describes the way secondary structure of the mRNA itself can repress translation by sequestering the ribosome binding site. The stability of the structure is the analog of the repressor binding potential, and the term $K_R[R]$ in equations (1) and (3) is simply replaced by the equilibrium stability of the structure, K_f. A classic paper by de Smit & van Duin (1990) measured the *in vivo* translation rates of the MS2 coat protein in a series of mRNAs with increasingly stable hairpins containing the Shine-Dalgarno sequence. This experiment is the exact analog of measuring translational repression at increasing levels of repressor.

When de Smit & van Duin plotted relative translation rate as a function of K_f, the curve was exactly of the form predicted by either equations (1) or (3) (see Figure 2 for an example of the predicted curve). To fit the data to the pre-equilibrium model (1), it was necessary to assume $K_{30S}[30S] \approx 2 \times 10^4$. An estimate of 8.5 µM for the *in vivo* concentration of 30S subunits then required that $K_{30S} \approx 2.4 \times 10^9$ M^{-1}, one or two orders of magnitude too large. Since the time that paper was written, the rules for predicting hairpin stabilities have been revised to reflect new data on hairpin loops (Jaeger *et al.*, 1989) and G-U pairs (He *et al.*, 1991). Using these newer values, the stabilities of the hairpins average about 2 kcal more negative, which requires that $K_{30S}[30S] \approx 10^6$ and predicts that K_{30S} is $>10^{11}$ M^{-1}. This is much too large to be consistent with the binding constant estimates discussed above. It appears that initiating ribosomes are much more effective at denaturing mRNA secondary structure than expected on the basis of their binding affinities.

An alternative to consider is that coat protein translation is kinetically controlled. In equation 3, a fit to the de Smit & van Duin data demands that $(k_1/k_2)[30S] = 10^5$ to 10^6, allowing for the uncertainty in the helix stability measurements. If k_2 is ~ 0.1 and $[30S] \approx 10$ µM, $k_1 = 10^9$ to 10^{10} $M^{-1}sec^{-1}$. This is faster than diffusion controlled by a factor of 10^2 to 10^3. A factor of 10^3 enhancement from "sliding" is probably unlikely, but the "entry site" mechanism discussed above might be able to accelerate the association kinetics by a large enough factor.

T4 regA PROTEIN

The product of the regA gene of T4 phage regulates the translation of a number of genes by binding to a short sequence within the ribosome binding site. The protein has been purified and its binding to polynucleotides studied *in vitro*. It specifically recognizes an RNA dodecamer containing the genetically-defined regulatory site, and single mutations within this site decrease the binding affinity by as much as 100 fold (Webster & Spicer, 1990). The binding constant at 37° is 2.5×10^6 M^{-1}. *In vitro* repression studies have also been carried out at 37°, and show that half-maximal repression takes place at 0.6 µM regA protein, i.e. when $K_R[R] \approx 1.5$ (Adari & Spicer, 1986). Although the concentration of ribosomes in the coupled transcription - translation system is not precisely determined, it is on the order of 1 µM to give $K_{30S}[30S] \gtrsim 10$. The simple equilibrium displacement mechanism of repression predicts less than 15% repression under these conditions, while the entrapment model predicts about the 50% level actually observed. If the de Smit & van Duin (1990) measurement of the *in vivo* potential of translating ribosomes is applicable in this case, and $K_{30S}[30S]$ or $(k_1/k_2)[30S]$ is $\sim 10^5$, then the difference between the two models is more extreme.

Although this might seem a clear case of the entrapment mechanism, two experimental observations have been interpreted as evidence for displacement. First, the regions protected from RNAse by regA protein in the gene 44 and rIIB mRNAs include either the initiation codon or the Shine-Dalgarno sequence (Webster & Spicer, 1990). This suggests a direct competition of the regA protein for the ribosome recognition features. Second, competition between the regA protein and ribosomes for mRNA binding has (apparently) been observed in a "toeprint" assay (Winter et al., 1987).

What are we to make of the apparent paradox that regA protein competes with ribosomes in the in vitro toeprint assay even though it should not have sufficient binding affinity? I believe the toeprint assay has been misinterpreted. The assay looks for a truncated reverse transcriptase product to appear when a ternary $tRNA_f^{met}$ -30S subunit-mRNA initiation complex is formed on an mRNA. In the absence of initiation factors, this complex forms irreversibly, with a lifetime of hours, and cannot be displaced by the AMV transcriptase (Spedding et al., 1992). Quantitation of the intensity of the truncated transcript relative to the full length transcript is a good measure of the fraction of mRNAs bound in initiation complexes. Ribosome-mRNA and most repressor protein - mRNA complexes, which do not have such long lifetimes as the initiation complex, are not detected by the AMV transcriptase, even though they may have substantial binding affinities. The fact that a toeprint signal disappears upon adding the repressor protein can therefore mean either that the ribosome has been displaced from the mRNA by the repressor, or that binding of $tRNA_f^{met}$ has been prevented, such that only the shorter-lived binary ribosome-mRNA complex is present. Thus the toeprint experiment does not distinguish between displacement and entrapment mechanisms.

Exactly this kind of result has been obtained with the S15 repression system, in which addition of the repressor causes the AMV transcriptase toeprint to disappear. But when MMLV transcriptase, which is sensitive to much shorter-lived complexes, is used, a new toeprint appears which clearly indicates the existence of a ternary S15-30S-mRNA complex (C. Ehresmann, personal communication).

An entrapment mechanism for regA requires that the protein remain bound to the ribosome binding site when the 30S subunit binds. While the protein is certainly sensitive to the RNA sequence, it is not at all clear what the protein actually recognizes: there is no consensus sequence (Webster et al., 1989), and NMR studies of a dodecamer containing the site and a single base mutant show that both are stacked single strands with no unusual structure (Szewczak et al., 1991). While a regA-30S-mRNA complex may seem unusual, there is no reason at present to discount its possibility, and an entrapment mechanism for the regA protein is presently the most plausible explanation for the available data.

T4 GENE 32 PROTEIN

The T4 gene 32 product (p32) binds cooperatively to single stranded nucleic acids, and functions as part of the T4 DNA replication apparatus. Titrations of an in vitro translation system with gene 32 protein showed that it represses its own translation (Lemaire et al., 1978). A plot of repression as a function of p32 concentration is very sharp, suggesting that cooperative binding of at least several p32 monomers to the mRNA is required. The specificity of p32 for its own mRNA is attributed to the presence of a structure, probably a pseudoknot, which is about 40 nt upstream of the initiation codon and has an unusually high affinity for p32. Nuclease protection studies show that the the region around this structure becomes protected at ~ 2 μM protein, while protection of nucleotide downstream require progressively increasing concentrations of protein (McPheeters et al., 1988). p32 also reduces the intensity of a toeprint signal on the mRNA, again with a sharp, cooperative concentration dependence; higher concentrations of protein are required to see the same inhibition if the potential pseudoknot structure is removed from the 5' terminus. The cooperative binding of p32 to the mRNA, nucleated upstream of the ribosome initiation site and then extended into it at higher p32 concentrations, is proposed to prevent ribosome binding and thus inhibit translation (McPheeters *et al.*, 1988).

The addition of a single p32 protein to a cluster of cooperatively bound proteins has an equilibrium constant on the order of 10^7 M^{-1} for poly(A) at physiological salt and temperature (Kowalczykowski *et al.*, 1981). About 3 µM protein is needed to begin to see repression in the *in vitro* system, for a $K_R[R]$ of 30. This compares favorably with the estimated $K_{30S}[30S] \approx 10$ at the concentration of ribosomes used in these assays, but is too small if the larger value derived from the de Smit & van Duin (1990) measurements is used. From the slope of the plot of toeprint signal vs. log[p32], the number of p32 proteins which must be cooperatively bound to inhibit ribosome binding can be estimated to be 4 - 5. This is a minimum number, since an unknown concentration of other mRNA is present that undoubtedly binds some p32 and reduces the slope. Probably five bound proteins are needed to reach the edge of the ribosome binding site from the pseudoknot. If the ribosome must displace two bound p32 (a total of 6-7 p32 bound), then the binding affinity is quite adequate to directly compete with ribosomes for mRNA binding.

The one discrepancy among these sets of data is in comparing the nuclease protection experiment with toeprint assays. 4-5 µM p32 are needed for 50% maximal protection of sequences within the ribosome binding site, while toeprint assays done under the same conditions with the same mRNA show that only 2 µM protein decreases the toeprint intensity by 50%. The latter assay measures the kinetics of intiation complex formation, which appear to be particularly fast with the p32 mRNA (Spedding *et al.*, 1992). A tempting speculation is that the ribosome takes advantage of the unusually large unstructured region around the ribosome binding site (McPheeters *et al.*, 1988) to "slide" into place, and that protein binding in the upstream region alone is sufficient to slow the kinetics significantly.

R17 BINDING VS. REPRESSION

An extensive study of the R17 coat protein binding to its target site in the replicase gene ribosome initiation site was done by Carey and Uhlenbeck (1983). Under physiological conditions, they estimate the equilibrium binding affinity to be 3×10^5 M^{-1}. The protein accumulates to 5 - 10 µM in infected cells, which means that substantial repression of the replicase gene takes place when $K_R[R] \approx 1.5$ to 3. The repressor binding site is a simple hairpin which base pairs the Shine-Dalgarno sequence and part of the initiation codon. There is no way ribosomes and repressor can bind to the mRNA simultaneously, and only a displacement mechanism is feasible.

At first glance it would appear that $K_R[R]$ is much too small to support significant repression by a displacement type of mechanism. However, binding of repressor adds to the stability of a hairpin, and it is the overall stability of this complex which is important. The degree of repression will be the ratio of synthesis rates in the presence of hairpin alone and hairpin plus repressor,

$$\text{relative rate} = \frac{1 + K_f + K_{30S}[30S]}{1 + K_f K_R[R] + K_{30S}[30S]} \tag{5}$$

The stability of the target hairpin has been experimentally measured, and is -7.7 kcal/mol at 37° (Groebe & Uhlenbeck, 1989). K_f is therefore 2.7×10^5, in the same range as $K_{30S}[30S]$ (or $(k_1/k_2[30S])$) as measured by de Smit & van Duin (see above). The hairpin by itself should reduce the translation rate by only a factor of two, but the additional stability conferred by protein binding should start to inhibit translation more substantially. This is

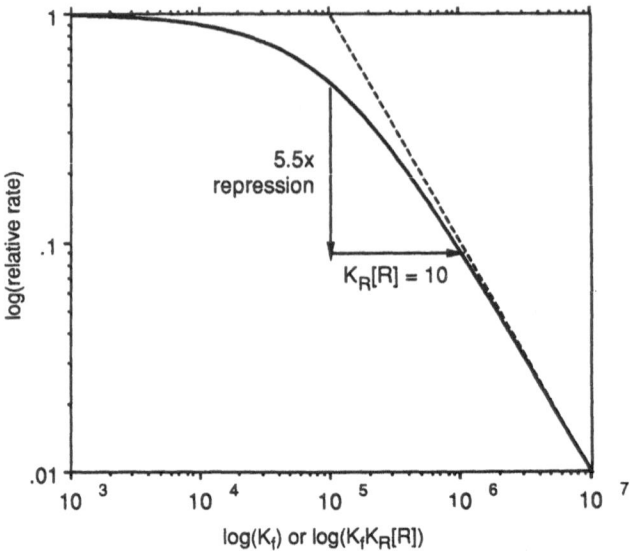

Figure 2. Predicted levels of translational repression as a function of either K_f or the product $K_f K_R[R]$, using equation 1 of the text and $K_{30S}[30S] = 10^5$. Arrows indicate the additional repression expected if a hairpin with $K_f = 10^5$ is bound by a protein with $K_R[R] = 10$. Dashed line indicates the limiting slope obtained at high levels of repression.

illustrated in the graph in Figure 2. For $K_R[R] = 10$, for example, translation is repressed by over five fold. The modest $K_R[R]$ estimated for the R17 coat protein *in vivo* is therefore close to the range needed for substantial repression. In effect, the R17 hairpin gives the repressor a "boost" to get into the range needed for repression. A prediction of this calculation is that a small destabilization of the hairpin will decrease repression substantially, even if the repressor binding constant remains unchanged.

S4 REPRESSION OF THE α OPERON

One of the most highly structured ribosome binding sites yet described is in the α operon of ribosomal proteins. A very stable hairpin with a 12 base pair stem and 29 nucleotide loop is located just upstream of the S13 ribosome binding site, and this loop pairs with sequences following the initiation codon to make three helical segments. The result is a complex pseudoknot extending from -77 to +32 relative to the initiation codon (Tang & Draper, 1989). The structure has been deduced by assaying compensatory base pair changes for their ability to bind the S4 repressor protein. Recent thermodynamic studies show that the pseudoknot is a stable structure; the downstream helical segments melt cooperatively with $\Delta G_{37°} \approx -7.4$ kcal/mol (Spedding *et al.*, 1992).

Comparison of the S4 affinities of mutants measured *in vitro* with the repression levels seen *in vivo* suggested that repression takes place by an indirect, allosteric mechanism (Tang & Draper, 1990). It was observed that some mutations reduce repression severely, but have little effect on S4 binding affinity. This can happen if the ribosome and S4 bind to separate sites on the mRNA, but S4 binding prevents translation (either by displacement or entrapment). S4, in this model, is an allosteric effector of translation. The model predicts

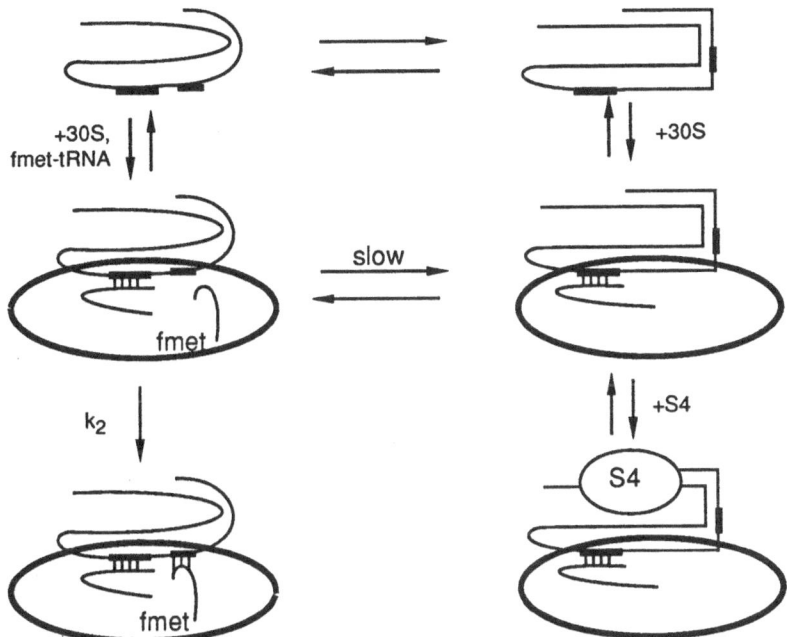

Figure 3. Translational repression in the α operon. The mRNA leader can adopt either of two conformations, both of which bind 30S subunits (schematically represented by Shine-Dalgarno pairing of the 16S mRNA 3' terminus with the mRNA). One conformation (left) is able to bind fmet-tRNA and form an initiation complex rapidly (k_2), while the other (right) binds S4 protein and is unable to initate translation. Under *in vitro* conditions, in the absence of initiation factors, interconversion of the two 30S-mRNA complexes is slow.

that the mRNA can exist in two conformations, one supporting translation and the other binding S4.

Evidence for such a conformational switch comes from studies of the rate of initiation complex formation with the pseudoknot, using the toeprint assay. At temperatures >45° all of the mRNA reacts rapidly with 30S subunits and $tRNA_f^{met}$ to form initiation complexes, while at 37° only half of the mRNAs react rapidly. Below ~32° the rapidly reacting fraction disappears entirely (Spedding *et al.*, 1992). This suggests that the mRNA undergoes a temperature-dependent switch between two different conformations. Further support for this comes from similar studies on the association of S4 and 30S subunits with the mRNA (G. Spedding & D.E.D., ms. submitted). It appears that subunits and S4 do not compete for binding to the mRNA, but bind independently; this rules out a displacement type of mechanism. The appearance of S4-30S-mRNA complexes again has slow and fast components, but in this case low temperature favors the complex. The net result is that at temperatures near 37°, 30S subunits bind to all of the mRNA molecules, but roughly half of the 30S-mRNA complexes react rapidly with $tRNA_f^{met}$ to form initiation complexes, and the other half reacts rapidly with S4 to form a "dead-end" complex unable to bind $tRNA_f^{met}$.

The scheme is illustrated in Figure 3. Interconversion of the "active" and "inactive" conformations of the mRNA is relatively fast (Spedding *et al.*, 1992), but binding of the 30S subunits slows down the isomerization to a time scale of hours. We doubt these slow kinetics hold *in vivo*. The experiments supporting this scheme have been carried out in the absence of initiation factors, which are known to drastically affect the kinetics of initiation complex formation. [The ternary $tRNA_f^{met}$-30S-mRNA complex, for instance, has a half-life of hours in the absence of factors but exchanges on a much more rapid time scale in the presence of IF3 (Risuleo *et al.*, 1976).] This point remains to be confirmed for the α mRNA.

There are two unexpected features of this translational repression scheme. First is the participation of an RNA conformational switch. Although the R17 replicase hairpin can also be considered a conformational switch, with "open" and "closed" states affecting translation, the α operon switch is not a simple denaturation of the pseudoknot, and probably involves some tertiary structure (T. Gluick and D.E.D., unpublished observations). The second unusual aspect is the fact that repression occurs by an entrapment; this is the first system for which this mechanism has been directly observed.

SUMMARY

Extrapolations from *in vitro* to *in vivo* conditions have large uncertainties, of course, and some of the discussions presented here are therefore quite speculative. But in each case, a quantitative consideration of how translational repression works *in vivo* has suggested new aspects of translational initiation and repression which can be tested by direct experiment. Repression by the gene 32 protein seems to be a straightforward competition between ribosomes and the protein, but there is a hint that protein binding upstream of the initiation site could alter the ribosome binding kinetics. The R17 coat protein is another case of direct competition between ribosome and repressor, though the potentially crucial role of the hairpin stability in adjusting the repression level has not been realized before. The regA protein is unlikely to displace ribosomes from mRNA, and the possibility that this protein traps the ribosome on the mRNA needs to be re-examined. Lastly, a comparsion of *in vivo* and *in vitro* results for the S4 protein suggested an allosteric, entrapment type of mechanism, which has since received experimental support.

It is worth noting that little attention has been paid to the crucial issue of what steps are rate limiting for translational initiation *in vivo*. Some mRNAs may be limited entirely by kinetics, while others may involve a pre-equilibrium binding step. This certainly makes a difference for the way translation is regulated, and for any attempts to increase *in vivo* expression of a gene. It is hoped that this review and discussion of translation will stimulate experiments to test a wider range of translational initiation and repression mechanisms than previously envisioned.

REFERENCES

Adari, H.Y. and Spicer, E.K., 1986, *Proteins: Structure, Function, and Genetics* 1:116.

Adhin, M.R. and van Duin, J., 1990, *J. Mol. Biol.* 213:811.

Calogero, R.A., Pon, C.L., Canonaco, M.A. and Gualerzi, C.O., 1988, *Proc. Natl. Acad. Sci. USA* 85:6427.

Carey, J. and Uhlenbeck, O.C., 1983, *Biochemistry* 22:2610.

de Smit, M.H. and van Duin, J., 1990, *Proc. Natl. Acad. Sci. USA* 87:7668.

Draper, D.E., 1987, *in*" Translational Regulation of Gene Expression," Ilan, J., ed., Plenum Press, New York.

Draper, D.E. and Gold, L., 1980, *Biochemistry* 19:1774.

Draper, D.E. and von Hippel, P.H., 1979, *Biochemistry* 18:753.

Ellis, S. and Conway, T.W., 1984, *J. Biol. Chem.* 259:7607.

Groebe, D.R. and Uhlenbeck, O.C., 1989, *Biochemistry* 28:742.

Gualerzi, C. and Pon, C.L., 1990, *Biochemistry* 29:5881.

Hartz, D., McPheeters, D.S. and Gold, L., 1991, *J. Mol. Biol.* 218:99.

He, L., Kierzek, R., SantaLucia, J., Walter, A.E. and Turner, D.H., 1991, *Biochemistry* 30:11124.

Jaeger, J.A., Turner, D.H. and Zuker, M., 1989, *Proc. Natl. Acad. Sci. USA* 86:7706.

Kennell, D. and Riezman, H., 1977, *J. Mol. Biol.* 114:1.

Kjeldgaard, N.O. andGausing, K., 1974, *in* "Ribosomes," Nomura, M., Tiessières, A. & Lengyel, P., ed., Cold Spring Harbor Laboratory, Cold Spring Harbor, New York.

Kowalczykowski, S.C., Lonberg, N., Newport, J.W. and von Hippel, P.H., *J. Mol. Biol.* 145:75.

Lemaire, G., Gold, L. andYarus, M., 1978, *J. Mol. Biol.* 126:73.

McClure, W.R., 1985, *Ann. Rev. Biochem.* 54:171.

McPheeters, D.S., Stormo, G.D. and Gold, L., 1988, *J. Mol. Biol.* 201:517.

Nathans, D., Oeschger, M.P., Polmar, S.K. andEggen, K., 1969, *J. Mol. Biol.* 39:279.

Nomura, M., Gourse, R. and Baughman, G., 1984, *Ann. Rev. Biochem.* 53:75.

Risuleo, G., Gualerzi, C. and Pon, C., 1976, *Eur. J. Biochem.* 67:603.

Spedding, G. and Draper, D.E., *ms. submitted.*

Spedding, G., Gluick, T.C. and Draper, D.E., 1992, *J. Mol. Biol.* in press.

Szewczak, A.A., Webster, K.R., Spicer, E.K. and Moore, P.B., 1991, *J. Biol. Chem.* 266:17832.

Tang, C.K. and Draper, D.E., 1989, *Cell* 57:531.

Tang, C.K. and Draper, D.E., 1990, *Biochemistry* 29:4434.

von Hippel, P.H. and Berg, O.G., 1989, *J. Biol. Chem.* 264:675.

Webster, K.R., Adari, H.Y. andSpicer, E.K., 1989, *Nucleic Acids Res* 17:10047.

Webster, K.R. and Spicer, E.K., 1990, *J. Biol. Chem.* 265:19007.

Winter, R.B., Morrissey, L., Gauss, P., Gold, L., Hsu, T. and Karam, J., 1987, *Proc. Natl. Acad. Sci. USA* 84:7822.

Wintermeyer, W. and Gualerzi, C., 1983, *Biochemistry* 22:690.

THE CENTRAL PSEUDOKNOT CONNECTING THE THREE MAJOR DOMAINS IN 16S rRNA IS REQUIRED FOR TRANSLATIONAL INITIATION

Marcel F. Brink, Martin Ph. Verbeet and Herman A. de Boer

Department of Biochemistry, Gorlaeus Laboratories
University of Leiden
Einsteinweg 55, 2333CC Leiden
The Netherlands

INTRODUCTION

Detailed models have been proposed for the structure of the 16S rRNA molecule of *E.coli* and its folding in the 30S subunit (Expert-Bezancon and Wollenzien, 1985; Brimacombe et al., 1988; Stern et al., 1988; Hubbard and Hearst, 1991). In these models, the 5'-domain of the 16S rRNA molecule mainly constitutes the body of the 30S subunit, the central domain is part of the platform and the 3'-major domain is located within the head of this subunit.

The formation of a translationally active ribosomal particle is likely to be dependent on the correct conformation of the rRNA molecules. By probing the structure of 16S rRNA within translationally active 30S subunits, Moazed and Noller (1989) showed that the 16S rRNA does not have a rigid conformation, but that structural changes occur within this molecule during the various stages of the translation process. Thus, it is not surprising that local disturbances in the highly complex structure of this molecule can have a dramatic effect on the activity of the small ribosomal subunit. This is illustrated by the fact that, mutations introduced in phylogenetically conserved rRNA regions negatively affect the translational activity of the ribosome (Dahlberg, 1989).

Mutations in the rRNA molecules may have an effect on the translational activity of the ribosome if they impair mRNA- or tRNA-binding, the association of ribosomal proteins, or the interaction with translational initiation or elongation factors. Also, mutations in rRNA regions which are directly involved in reactions such as peptide bond formation or tRNA translocation, will inhibit the translation process. Ofengand et al. (1988) developed a system for the *in vitro* formation of ribosomes containing 16S rRNA with any desired mutation. In this *in vitro* system, the effect of rRNA mutations on subunit assembly, tRNA-binding or peptide bond formation can be assessed.

The Translational Apparatus, Edited by K.H. Nierhaus
et al., Plenum Press, New York, 1993

The construction of a plasmid containing the entire rrnB operon (Brosius et al., 1981) allows for the introduction and study of rRNA mutations *in vivo*. The effect of such mutations on the translational activity of the ribosome has been assessed by monitoring the growth rate of cells. Thus, rRNA mutations could be classified as dominant if they resulted in a decrease in growth rate or cessation of growth (Gourse et al., 1982), whereas mutations were classified as recessive if they did not affect the growth rate.

Dominant mutations may exert their effect if the mutant ribosomes still can bind to mRNAs, but fail to carry out the elongation. Thus, they will also hinder the chromosomally encoded wildtype ribosomes in translating the mRNAs, resulting in a decrease in growth rate. Since ribosomes with such dominant mutations are expected to be toxic to the cell, it is difficult, if not impossible, to isolate and study this interesting class of mutant ribosomes *in vivo*. Therefore, a rRNA-operon was placed under the control of an inducible promoter (Gourse et al., 1985; Steen et al., 1986) allowing for control over the accumulation of toxic ribosomes. Upon induction of the synthesis of such ribosomes, their effect on the ribosomal activity can be assessed by monitoring the growth rate.

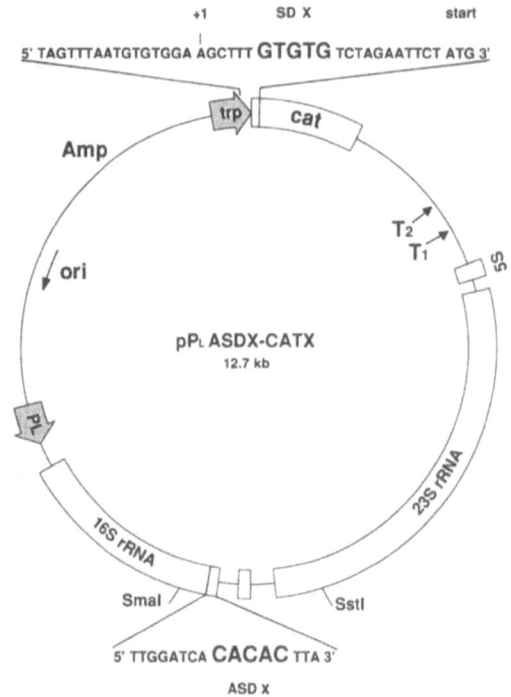

Figure 1. **Structure of the plasmid encoding the specialized ribosome system.** The sequences of the 5'-end of the cat-gene harboring the modified Shine-Dalgarno sequence (SDX: 5'GTGTG3') and those at the 3'-end of the 16S rRNA gene containing the complementary anti-Shine-Dalgarno sequence (ASDX: 5'CACAC3') are shown. The 16S rRNA gene also harbors a C to U change at position 1192, confering resistance to the antibiotic spectinomycin. The cat-gene is under control of a constitutive trp-promoter, whereas the rRNA-operon rrnB is driven by a thermo-inducible lambda P_L-promoter. Negative control cells contain pASDXΔSma-Sst. This plasmid is identical to the specialized ribosome system, except for a deletion in the rrnB operon between the SmaI restriction site at position 1383 of 16S rRNA and the unique SstI site in 23S rRNA (restriction sites are indicated). Therefore, upon induction of the lambda P_L-promoter, cells containing pASDXΔSma-Sst do not accumulate specialized ribosomes.

Obviously, recessive mutations which do not impair the translational activity will not affect the growth rate. However, mutations which affect the translational activity by impairing ribosome synthesis or inhibiting the formation of a 30S initiation complex will not affect the growth rate either, as endogenous wildtype ribosomes encoded by the seven, virtually identical chromosomal operons, will maintain growth. In order to determine whether recessive mutations do affect the ribosomal activity *in vivo*, Sigmund et al. (1988) introduced a C to U change at position 1192 of a plasmid borne 16S rRNA gene, thus confering resistance to the antibiotic spectinomycin. Consequently, in the presence of spectinomycin, cell growth is only supported by plasmid derived ribosomes and thus the effect of recessive rRNA mutations can be assessed, by monitoring the growth rate.

Since monitoring growth rate is a rather indirect way of determining the effect of mutations on ribosomal activity, we use the specialized ribosome system (Hui and de Boer, 1987 and Figure 1). Specialized ribosomes do not translate any of the endogenous mRNAs, but only one particular exogenous mRNA species. Due to the altered anti-Shine-Dalgarno (ASD) sequence ($5'CACAC3'$) at the 3'-end of 16S rRNA, ribosomes containing this rRNA molecule are "specialized", as they only translate the single mRNA species with the modified Shine-Dalgarno sequence (SD: $5'GTGTG\ 3'$) that is complementary to the altered ASD sequence. In addition, a C to U change at position 1192 has rendered specialized ribosomes resistant to spectinomycin. In the system that we use in our laboratory, the mRNA species encodes chloramphenicol acetyl transferase (CAT). The suitability of this system for studying the effect of recessive mutations in 16S rRNA at the translational level is based on the fact that the specialized ribosomes only translate the CAT-mRNA species, while the chromosomally encoded wildtype ribosomes lacking this complementary ASD sequence are unable to translate the CAT-mRNA. Thus, rather than monitoring a change in growth rate, the effect of mutations on ribosomal activity can be assessed by studying the translation of CAT-mRNA. Moreover, as specialized ribosomes do not translate the endogenous mRNAs, dominant mutations in this non-essential subpopulation of ribosomes will not impair growth. In this way, any desired rRNA mutation can be studied under normal physiological conditions i.e. a growing cell.

MUTAGENESIS AT THE CENTRAL PSEUDOKNOT IN 16S rRNA

We have studied the function of one of the three postulated RNA pseudoknot structures in 16S rRNA in the translation process (Pleij et al., 1985; Gutell et al., 1986; Woese and Gutell, 1989). Such RNA pseudoknots are formed when several residues in the loop of a hairpin structure form standard Watson-Crick basepairs with a complementary sequence outside this loop (Pleij et al., 1985). Consequently, a pseudoknot consists of two helices: the helix forming the stem and the helix formed by basepaired residues in the loop of the hairpin. These pseudoknots may be of functional importance as they occur in phylogenetically conserved rRNA regions (Powers and Noller, 1991).

One of the pseudoknots, located at the center of the secondary structure of 16S rRNA, is formed when residues $U_{17}C_{18}A_{19}$ in the loop of the 5'-terminal hairpin structure basepair with residues $U_{916}G_{917}A_{918}$ near the center of the rRNA molecule (Pleij et al., 1985). Stern et al. (1988) have proposed that this central pseudoknot is coaxially stacked with the helix formed by the rRNA regions 27-37 and 547-556, thus forming a higher order structure of three coaxially stacked helices which together are defined as the pseudoknot helix (see Figure 2). This pseudoknot helix forms a central core element in the 30S ribosomal subunit from which the three major structural domains of 16S rRNA emerge. In order to study the function of the central pseudoknot in the translation process, we have disrupted part of this this structure by introducing mutations in the helix formed by residues $U_{17}C_{18}A_{19}$ and $U_{916}G_{917}A_{918}$ (helix II).

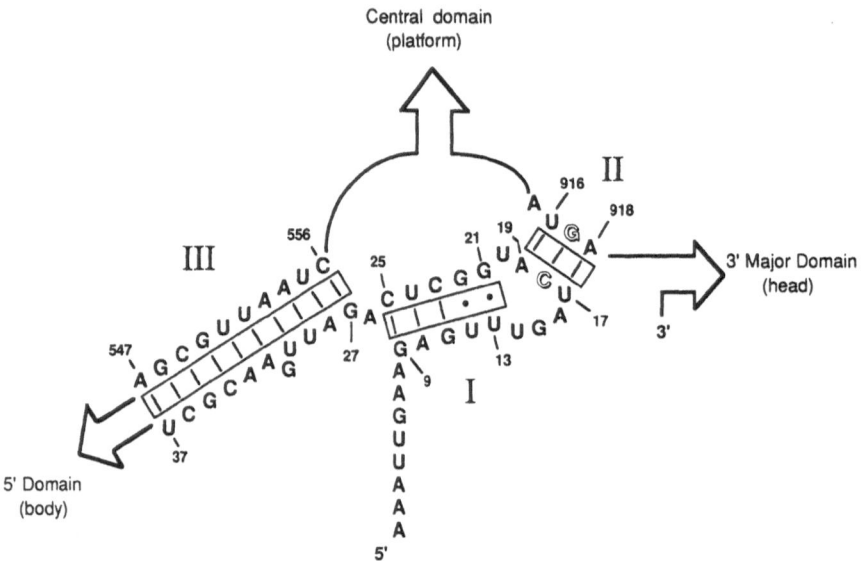

Figure 2. Schematic diagram of the pseudoknot helix in 16S rRNA. The pseudoknot helix is formed when the central pseudoknot which consists of helix I (nucleotides 9-13/21-25) and helix II (nucleotides 17-19/916-918) is coaxially stacked with helix III (nucleotides 27-37/547-556). The three helices are indicated with shaded boxes and the arrows indicate the relative orientations of the three major domains of 16S rRNA as they protrude from the pseudoknot helix. The C_{18}-G_{917} basepair that is subjected to mutational analysis, is presented with open letters.

AN INTACT PSEUDOKNOT STRUCTURE IS REQUIRED FOR RIBOSOMAL ACTIVITY

The C-residue at position 18 of 16S rRNA of the specialized ribosomes was substituted with either an A-, G- or U-residue. Thus the mutants A_{18}, G_{18} and U_{18} were made. Upon induction of the synthesis of specialized ribosomes, samples were taken at 30 min intervals and the synthesis of CAT was determined by measuring the CAT-activity in the cell lysates. Figure 3A shows that, after induction of the synthesis of specialized ribosomes with the wildtype pseudoknot, the CAT-activity increases linearly, while for the A_{18}-mutant the CAT-activity in samples taken two hours after induction is merely 5% of the wildtype activity. The CAT-activity for the G_{18}- and U_{18}-mutant is reduced to 10% and 25% of the wildtype level, respectively.

To answer the question whether any mutation that restores helix II would also result in recovery of ribosomal activity, substitutions at position 917 were introduced that are complementary to the mutation at position 18 of 16S rRNA. Thus, the mutants $A_{18}U_{917}$, $G_{18}C_{917}$ and $U_{18}A_{917}$ were made. Figure 3B shows that the CAT-activity in cells harboring specialized ribosomes with such mutations is completely

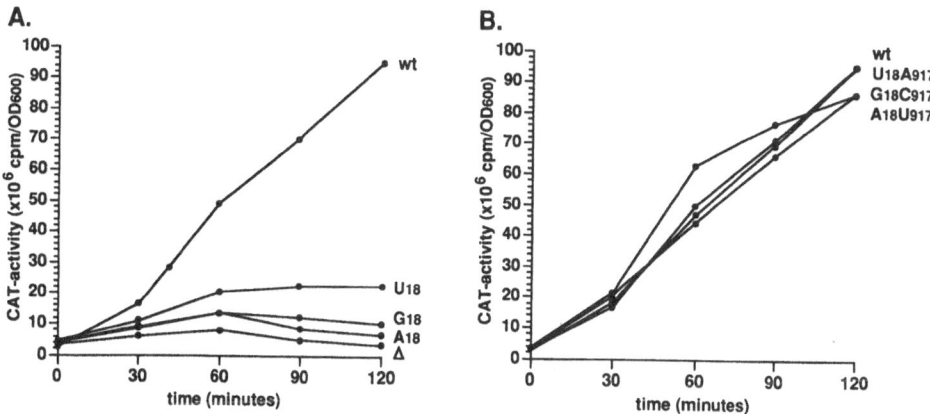

Figure 3. Ribosomal activity of specialized ribosomes with a wildtype or mutated central pseudoknot. The ribosomal activity of the specialized ribosomes was assessed by measuring the CAT-activity in cell lysates. Upon temperature induction of the synthesis of specialized ribosomes (t=0), samples were taken at 30 min intervals. The CAT-activity was determined by measurement of the amount of [^{3}H]di-acetyl-chloramphenicol formed. (A) The CAT-activity in cells harboring specialized ribosomes with the wildtype pseudoknot (wt) compared with cells harboring the pseudoknot mutants A_{18}, G_{18} and U_{18} or cells harboring the negative control plasmid (see Figure 1). (B) The CAT-activity in cells harboring specialized ribosomes with the wildtype pseudoknot (wt) compared with cells harboring the pseudoknot mutants $A_{18}U_{917}$, $G_{18}C_{917}$ and $U_{18}A_{917}$.

restored to wildtype levels. These results demonstrate that the central pseudoknot indeed exists and that its restoration, regardless the nature of the introduced Watson-Crick basepair, results in full recovery of the ribosomal activity.

The observed reduction in CAT-activity in cells harboring mutants A_{18}, G_{18} or U_{18} is caused by a reduction in the synthetic rate of CAT or is the result of reduced translational fidelity. A reduction in fidelity could have caused premature termination of the translation of the CAT-mRNA or may have resulted in a higher ratio of erroneously incorporated amino acids. Thus, the ribosomes with disruptive mutations in the central pseudoknot either may have synthesized a truncated form of CAT or, due to the presence of misincorporated amino acids, an unstable or enzymatically inactive form of CAT. The sizes of CAT synthesized by specialized ribosomes with the wildtype or mutated pseudoknot were determined by SDS/polyacrylamide gel electrophoresis. Figure 4 shows the protein profile in [^{35}S]-L-methionine labeled cells harboring mutants A_{18}, G_{18} or U_{18} (lanes 3, 5 and 7, respectively). Clearly, as compared to CAT synthesized by specialized ribosomes with the wildtype pseudoknot (lane 2), the size of CAT synthesized by these disruptive mutants (lanes 3 to 8) is the same. As no truncated forms of CAT are detected, we conclude that the ribosomes with the disruptive mutations do not prematurely terminate the translation of the CAT-mRNA. This experiment also shows that the quantity of CAT synthesized by the disruptive mutants corresponds with their measured CAT-activity (Figure 3). This means that the reduction in CAT-activity is not caused by a gross increase in translational errors leading to protein instability or loss of enzymatic activity, but rather by a reduced rate of CAT-synthesis. The size and quantity of CAT synthesized by mutants having a central pseudoknot with a different basepair (lanes 4, 6 and 8), is similar to the size and quantity of CAT synthesized in cells containing specialized ribosomes with the wildtype pseudoknot. This is in agreement with the restored CAT-activity observed for these pseudoknot mutants.

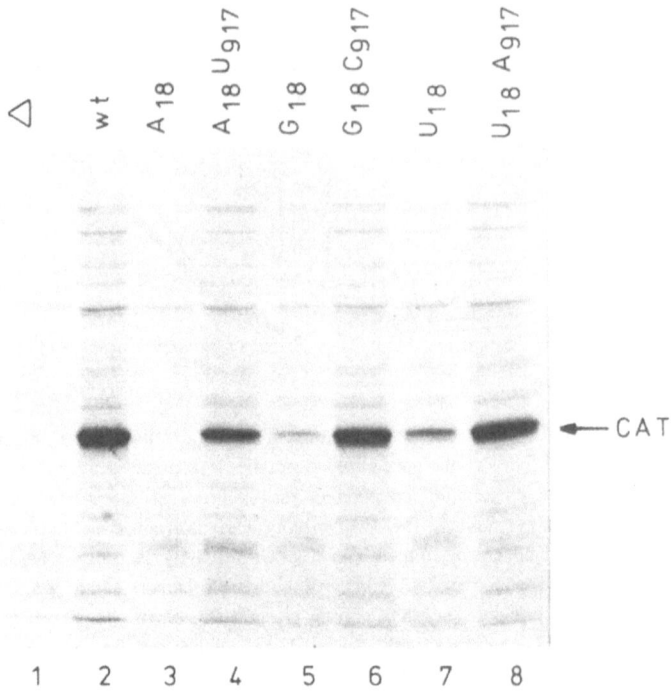

Figure 4. Protein synthesis by specialized ribosomes with a wildtype or mutated central pseudoknot. Cells were grown for 3 h in M9-medium at 30°C and synthesis of specialized ribosomes was induced at 42°C for 2 h. Protein synthesis by wildtype ribosomes was blocked by addition of spectinomycin (0.5 mg/ml) and after 15 min *de novo* synthesized proteins were metabolically labeled by addition of [³⁵S]-L-methionine (760 μCi/μmol). Proteins were separated by SDS/polyacrylamide gel electrophoresis (12.5%). Lane 1 shows the proteins synthesized in cells harboring the negative control plasmid (see Figure 1). Lane 2 shows the proteins synthesized by specialized ribosomes with the wildtype pseudoknot and lanes 3 to 8 show the proteins synthesized by specialized ribosomes with indicated mutations.

MUTATIONS IN THE CENTRAL PSEUDOKNOT DO NOT INTERFERE WITH THE ASSEMBLY OF 16S rRNA INTO RIBOSOMAL PARTICLES

The observed reduction in CAT-synthesis can either be caused by the reduced translational activity of the specialized ribosomes with mutations A_{18}, G_{18} or U_{18}, but may also be due to the defective assembly of the mutant 16S rRNA into ribosomal particles, which results in a lower number of specialized ribosomes capable of translating the CAT-mRNA. To answer the question whether the disruptive mutations interfere with the formation of specialized ribosomes, we determined the relative levels of such ribosomes in the cell. The cells were harvested two hours after temperature induction and the ribosomes were obtained by high speed centrifugation. From this ribosomal pellet, total rRNA was isolated and the relative levels of chromosomally encoded wildtype 16S rRNA and plasmid encoded specialized 16S rRNA were determined using the primer extension method (Sigmund et al., 1988).

As shown in Figure 5A, a [³²P]-end-labeled oligonucleotide complementary to the region 1194-1210 of 16S rRNA is extended in the presence of ddGTP using reverse transcriptase. When annealing to chromosomally encoded wildtype 16S rRNA, termination will occur at position C_{1192}, resulting in a 19-mer. When annealing to plasmid encoded specialized 16S rRNA, termination will not occur at position 1192 due to the presence of an U-residue. Instead, termination will occur at the next C-residue which is at position 1172 and a 39-mer will be made. Since total rRNA is isolated from the ribosomal pellet which contains both wildtype and specialized ribosomes and because the oligonucleotide recognizes the same sequence in wildtype and in specialized 16S rRNA, the relative intensities of the end-labeled 19-mer and 39-mer should correlate directly with the relative levels of wildtype and specialized ribosomes, respectively, present in the ribosomal pellet.

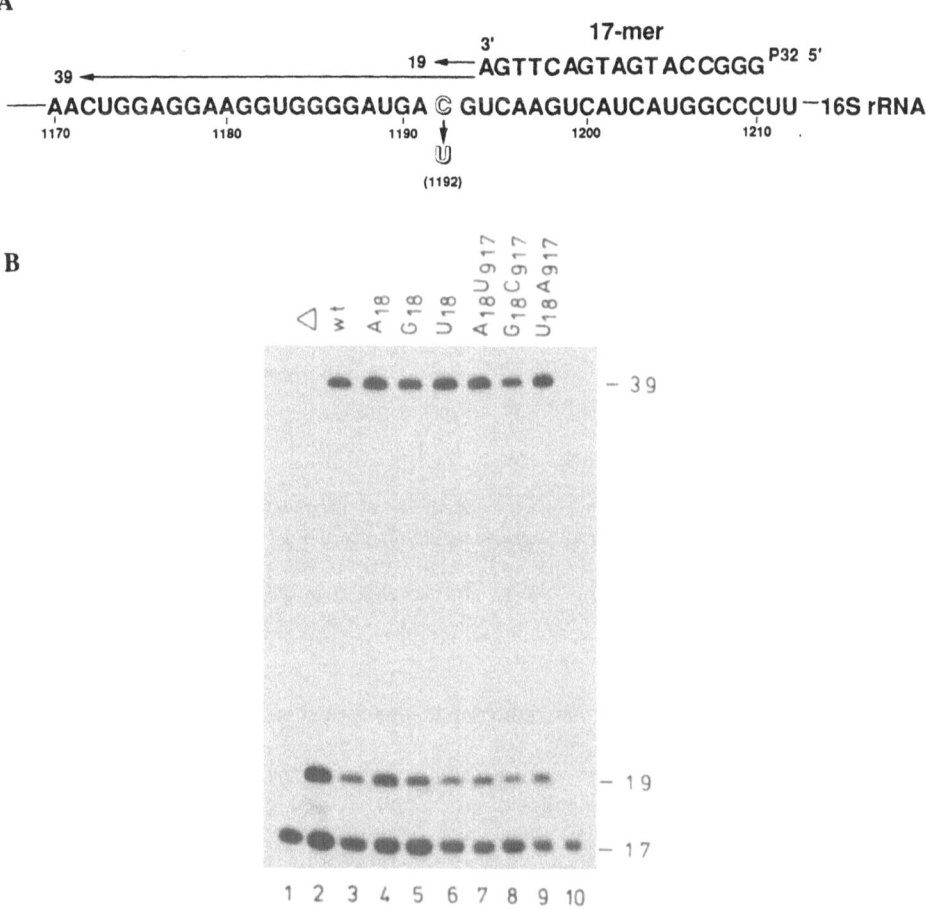

Figure 5. Determination of the assembly into ribosomal particles of 16S with a wildtype or mutated central pseudoknot, using the primer extension method. (A) Sequence of the primer and the sizes of expected extension products. The reaction was done in the presence of ddGTP using a [^{32}P]-end-labeled-oligonucleotide complementary to region 1194-1210. The C-residue at position 1192 which distinguishes wildtype 16S rRNA from specialized 16S rRNA with U_{1192} is presented with an open letter. The sizes of the expected extension products are indicated. (B) Extension products synthesized using total rRNA from cells harboring specialized ribosomes with a wildtype or mutated central pseudoknot. Lanes 1 and 10 show the position of the unextended primer and lane 2 shows the extension products synthesized using total rRNA from cells harboring the negative control plasmid (see Figure 1). Lane 3 shows the extension products synthesized using total rRNA from cells harboring specialized ribosomes with the wildtype pseudoknot and lanes 4 to 9 show the extension product when using total rRNA from cells harboring specialized ribosomes with the indicated mutations. The sizes of the unextended and extended primer are shown.

Figure 5B shows the results of such a primer extension experiment. The presence of the 39-mer in lanes 4 to 9, indicates that ribosomal particles with mutations in central pseudoknot of 16S rRNA do accumulate. To determine the relative levels of these particles, the radioactivity present in the 19-mer and 39-mer was counted. As shown in Table 1 (left column), the relative levels of specialized ribosomes with the indicated mutations in the central pseudoknot range between 50%-60%. These relative levels are similar to that of specialized ribosomes with the wildtype pseudoknot. Thus, we conclude that the mutations in the central pseudoknot do not interfere with the assembly of the rRNA into ribosomal particles which are isolated by high speed centrifugation.

Table 1. Quantification by primer extension of the relative levels (%) of specialized ribosomes with a wildtype or mutated central pseudoknot present in whole cell lysates or the 30S, 70S, disome or trisome fraction of polysome profiles[a].

mutant	whole cell[b]	ribosomal fraction			
		30S	70S	disomes	trisomes
wildtype	54	51	41	34	41
A_{18}	53	53	9	7	9
G_{18}	50	54	6	3	4
U_{18}	59	52	24	19	22
$A_{18}U_{917}$	57	54	33	32	35
$G_{18}C_{917}$	56	59	41	33	36
$U_{18}A_{917}$	59	59	37	37	37

a) Relative levels were determined by direct measurement of the radioactivity (cpm) present in the extension products (i.e. 19- and 39-mers). Relative levels of specialized ribosomes (%) were calculated as: [(cpm in 39-mer)/(cpm in 19-mer + cpm in 39-mer)] x 100%
b) Relative levels of specialized ribosomes (% of total) present in the high speed pellet from cell lysates.

It should be kept in mind that, inherent to this isolation procedure, the ribosomal pellet not only contains mature 30S, 50S and 70S ribosomal particles, but it can also harbor those ribosomal particles that do not contain all the ribosomal proteins, but which are large enough to be pelleted. Therefore, it can not be excluded that the mutations in the pseudoknot may impair the binding of ribosomal proteins, or even later stages in ribosome biogenesis such as the processing at the 5'- and 3'-regions of the precursor-rRNA.

MUTATIONS IN THE CENTRAL PSEUDOKNOT DO NOT INTERFERE WITH PROCESSING AT THE 5'-REGION OF PRECURSOR-rRNA

Correct processing of precursor-rRNA to mature 16S rRNA is essential for the formation of translationally active 30S ribosomal subunits (Wireman and Sypherd, 1974; Nomura and Held, 1974). Several endonucleases are involved in the maturation process leading to mature 16S rRNA. They do not seem to process the precursor in a fixed sequence of events and they can act both on naked rRNA as well as on precursor-rRNA which is already assembled into 30S subunits (Srivastava and Schlessinger, 1988).

To assess whether mutations in the central pseudoknot interfere with correct processing, we determined the 5'-end of 16S rRNA using the primer extension method. As shown in Figure 6A, a [³²P]-end-labeled oligonucleotide which is complementary to region 19-35 of 16S rRNA is extended in the presence of ddGTP. When annealing to chromosomally encoded wildtype 16S rRNA, termination will

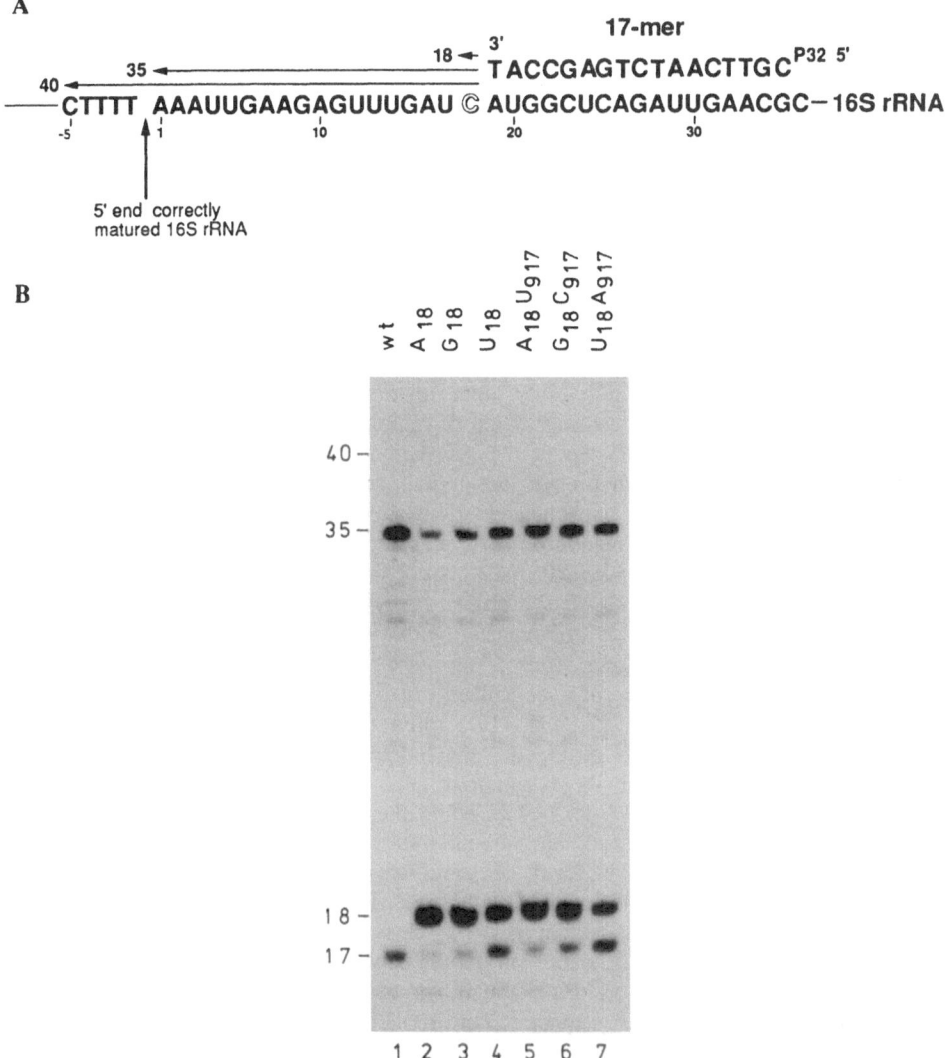

Figure 6. Determination of the 5'-end of 16S rRNA of specialized ribosomes with a mutated central pseudoknot, using primer extension. (A) Sequence of the primer and sizes of expected extension products. The reaction was done in the presence of ddGTP using a [³²P]-end-labeled oligonucleotide complementary to region 19-35. The C-residue at position 18 which distinguishes wildtype 16S rRNA from specialized 16S rRNA with A₁₈, G₁₈ or U₁₈ is presented with an open letter. The 5'-end of correctly matured 16S rRNA and the sizes of the expected extension products are indicated. (B) Extension products synthesized using total rRNA from cells harboring specialized ribosomes with a wildtype or mutated central pseudoknot. Lane 1 shows the extension products synthesized in the presence of dNTPs using total rRNA from cells harboring specialized ribosomes with a wildtype pseudoknot. Lanes 2 to 7 show the extension products synthesized using total rRNA from cells harboring specialized ribosomes with the indicated mutations at the central pseudoknot. The sizes of the unextended and extended primer are shown.

occur at position C_{18}, resulting in an 18-mer. When annealing to plasmid encoded specialized 16S rRNA with a mutated central pseudoknot, termination will not occur at position 18, due to the presence of an A-, G or U-residue. Instead, a 35-mer run-off transcript will be made if the 16S rRNA containing these mutations is correctly processed. If this were not the case, termination will occur at the next C-residue which is 5 nucleotides upstream from the proper cleavage site (see Figure 6A), resulting in a 40-mer.

Figure 6B shows the results of such a primer extension experiment. When the size of the run-off transcript synthesized on 16S rRNA lacking mutations in the pseudoknot (lane 1) is compared with the sizes of the run-off transcripts synthesized on rRNA with a mutated central pseudoknot (lanes 2 to 7), it appears that a 35-mer is present in all cases. A 40-mer is never detected, which proves that the 5'-region of the 16S rRNAs with mutations in the central pseudoknot is correctly processed.

DISRUPTION OF THE CENTRAL PSEUDOKNOT INTERFERES WITH THE FORMATION OF POLYRIBOSOMES

Having excluded the possibility that disruptive mutations in the central pseudoknot interfere with processing at the 5'-region of the 16S rRNA, we infer that such mutations must affect a step in the translation process. In order to find out at which stage in the translation process the mutant ribosomes are impaired, polysome profiles of cells harboring specialized ribosomes with disruptive or restoring mutations in the central pseudoknot were prepared and the relative levels of the mutant 16S rRNAs in the 30S, 70S, disome or trisome fraction were determined, using the primer extension method. Table 1 shows that, in the 30S fraction, the relative levels of specialized ribosomes with a wildtype or mutated central pseudoknot are similar, ranging from 50 to 59%. In the 70S, disome or trisome fraction, the relative level of specialized ribosomes with a wildtype central pseudoknot ranges between 34% and 41%, while the relative levels for specialized ribosomes with mutations A_{18} or G_{18} are much lower, ranging from 7% to 9% and 3% to 6%, respectively. In these fractions, the relative level of the ribosomes containing U_{18} is between 19% and 24%, while the relative levels of ribosomes with restoring mutations in the pseudoknot, in all three cases, are between 32% to 41%. As compared to the relative level of specialized ribosomes with the wildtype pseudoknot present in the 70S, disome or trisome fraction, the moderate reduction in the relative level for the U_{18}-mutant is in agreement with the reduced translational activity for this mutant. Also, the similarity between the relative levels for the specialized ribosomes with restoring mutations in the pseudoknot and specialized ribosomes with the wildtype sequence corresponds with their similar translational activities. Considering these results, we conclude that, upon disruption of the central pseudoknot in 16S rRNA, 30S ribosomal subunits are unable to form 70S ribosomes (or polyribosomal complexes).

DISCUSSION

The Formation of the Central Pseudoknot Structure is Required for Translational Initiation

We have demonstrated that disruption of the central pseudoknot by substituting the C-residue at position 18 of 16S rRNA results in a dramatic decrease in translational activity. This substitution does not have a deleterious effect on the assembly of the mutant 16S rRNA into ribosomal particles. Moreover, disruption of this pseudoknot does not affect processing at the 5'-region of precursor-rRNA. The relative levels of specialized ribosomes with these disruptive mutations, present in the various fractions of the polysome profiles, show that the mutant 30S subunits are not capable of forming a 70S ribosome (or polyribosomal complexes).

The translational activity of these pseudoknot mutants was reduced to merely 5% of the wildtype activity for the A_{18}-mutant, to 10% for G_{18}-mutant and to 25% for the U_{18}-mutant, respectively. In the latter case, the potential formation of an U_{18}-G_{917} "wobble" basepair, could maintain the pseudoknot structure, which may explain the moderate effect on the translational activity of the introduction of U_{18}. The calculated free energy (ΔG_0; Freier et al., 1986) of helix II with the wildtype sequence is -4.1 kcal/mole, while the ΔG_0 for helix II with a center U-G basepair is -1.2 kcal/mole and +0.8 kcal/mole with the mutations A_{18} or G_{18}.

Although the moderate reduction in translational activity of ribosomes with U_{18} in the pseudoknot may suggest that there is a correlation between the stability of helix II and the translational activity of the ribosome, we did not see any differences between the translational activities of ribosomes having a restored pseudoknot structure with a different set of basepairs. In vitro studies on short RNA oligonucleotides capable of forming a stable pseudoknot structure (Puglisi et al., 1988) showed that the most stable one of the two composing helices determines the overall stability of a pseudoknot. In the central pseudoknot, the free energy of helix I (Figure 1) is -6.4 kcal/mole, compared to -4.1, -2.0 or -1.2 kcal/mole for helix II with a center G-C, A-U or U-G basepair, respectively. Thus, the overall stability of the central pseudoknot should be dependent on the stability of helix I. Although this could account for the similar translational activities of mutants with a center A-U or G-C basepair in helix II, it does not explain why the translational activity is reduced to 25% of the wildtype activity, when helix II contains a center U-G basepair.

Since the translational activity of specialized ribosomes with a central pseudoknot with different basepairs is completely restored to the wildtype level, we conclude that, the postulated central pseudoknot of 16S rRNA indeed exists and that its formation, regardless of its composing basepairs, is essential for the formation of functionally active 30S subunits. As 30S subunits, containing 16S rRNA with disruptive mutations in the central pseudoknot, do not form 70S ribosomes, we speculate that this pseudoknot is required for the formation of a 30S initiation complex or the association with the 50S subunit to form a 70S initiation complex. Consequently, these mutant 30S subunits will not pass the stage of translational inititiation.

REFERENCES

Brimacombe, R., Atmadja, J., Stiege, W. and Schüler, D. (1988) *J. Mol. Biol.*, **199**, 115-136.

Brosius, J., Ullrich, A., Raker, M.A., Gray, A., Dull, T.J., Gutell, R.R. and Noller, H.F. (1981) *Plasmid*, **6**, 112-118.

Dahlberg A.E. (1989) *Cell*, **57**, 525-529.

Freier, S.M., Kierzek, R., Jaeger, J.A., Sugimoto, N., Caruthers, M.H., Neilson, T. and Turner, H. (1986) *Proc. Natl. Acad. Sci. USA*, **83**, 9373-9377.

Expert-Bezancon, A. and Wollenzien, P. (1985) *J. Mol. Biol.*, **184**, 53-66.

Gutell, R., Noller, H.F. and Woese C.R. (1986) *EMBO J.*, **5**, 1111-1113.

Gourse, R.L., Stark, M.J. and Dahlberg, A.E. (1982) *J. Mol. Biol.*, **159**, 397-416.

Gourse, R.L., Takebe, Y., Sharrock, R. & Nomura, M. (1985) *Proc. Natl. Acad. Sci. USA*, **82**, 1069-1073.

Hogan, J.J. and Noller, H.F. (1978) *Biochemistry*, **17**, 587-593.

Hubbard, J.M. and Hearst, J.E. (1991) *J. Mol. Biol.*, **221**, 889-907.

Hui, A.S. and de Boer, H.A. (1987) *Proc. Natl. Acad. Sci. USA*, **84**, 4762-4766.

Moazed, D., Stern, S. and Noller H.F. (1987) *J. Mol. Biol.*, **187**, 399-416.

Moazed, D. and Noller, H.F. (1989) *Nature*, **342**, 142-148.

Neefs, J.M., Van de Peer, Y., Hendriks, L. and de Wachter, R. (1990) *Nucleic Acids Res.*, **18**, supplement 2237-2317.

Nomura, M. and Held, W.A. (1974) in *Ribosomes* (M. Nomura, A. Tissières and P. Lengyel, eds.) Cold Spring Harbor Laboratory, Cold Spring Harbor, NY, 193-223.

Ofengand, J., Denman, R., Negre, D., Krzyzosiak, W., Nurse, K. & Colgan, J. (1988) in *Structure & Expression* (R.H. Sarma and M.H. Sarma, eds.), vol.1, Adenine Press, 209-228 .

Pleij, C.W.A., Rietveld, K. and Bosch, L. (1985) *Nucleic Acids Res.*, **13**, 1717-1731.

Pleij, C.W.A. (1990) *Trends Biochem. Sci.*, **14**, 143-147.

Powers, T. and Noller, H.F. (1991) *EMBO J.*, **10**, 2203-2214.

Puglisi, J.D., Wyatt, J.R. and Tinoco, I. Jr. (1988) *Nature*, **331**, 283-286.

Sigmund, C.D., Ettayebi, M. and Morgan, E.A. (1984) *Nucleic Acids Res.*, **12**, 4653-4663.

Sigmund, C.D., Ettayebi, M., Borden, A. and Morgan, E.A. (1988) *Methods Enzymol.*, **164**, 673-690.

Srivastava, A.K. and Schlessinger, D. (1988) *Proc. Natl. Acad. Sci. USA*, **74**, 7144-7148.

Steen, R., Jemiolo, D.K., Skinner, R.H., Dunn, J.J. and Dahlberg, A.E. (1986) in *Prog. Nucl. Acids Res. Mol. Biol.*, vol.33, 1-18.

Stern, S., Weiser, B. and Noller, H.F. (1988) *J. Mol. Biol.*, **204**, 447-481.

Wireman, J.W. and Sypherd, P.S. (1974) *Nature*, **247**, 552-554.

Woese, C.R. and Gutell, R.R. (1989) *Proc. Natl. Acad. Sci. USA*, **86**, 3119-3122.

Wollenzien, P.L. and Cantor, C.R. (1982) *J.Mol.Biol.*, **159**, 151-166.

RNA HELICASE ACTIVITY IN TRANSLATION INITIATION IN EUKARYOTES

Arnim Pause and Nahum Sonenberg

Department of Biochemistry and McGill Cancer Center
McGill University
Montreal, Quebec
Canada, H3G 1Y6

INTRODUCTION

Eukaryotic translation initiation is significantly different than in prokaryotes in several respects. 1) A cap structure (m^7GpppX, where X is any nucleotide) that is present at the 5' end of most eukaryotic mRNAs enhances translational efficiency. 2) mRNA binding to ribosome is dependent on ATP hydrolysis in eukaryotes, but not in prokaryotes. 3) A Shine-Dalgarno sequence plays a critical role in the selection of the initiator codon in prokaryotes, but not in eukaryotes. These differences in the mechanisms of translation are reflected in the considerable larger number of initiation factors in eukaryotes as compared to prokaryotes.

Cap-stimulated mRNA binding to the small ribosomal subunit requires at least three initiation factors; eIF-4A, eIF-4B and eIF-4F, in addition to the hydrolysis of ATP (reviewed in Edery et al. 1987, Sonenberg, 1988; Rhoads, 1988). eIF-4F is a multi-subunit complex consisting of three major polypeptides of 24-, 50-, and 220-kilodaltons (kDa) (Tahara et al., 1981; Grifo et al., 1983; Edery et al., 1983). The 24 kDa polypeptide is the cap binding subunit which also exists in a free form, termed eIF-4E (Sonenberg et al., 1979). The 50 kDa polypeptide is an eIF-4A variant. Two eIF-4A species (eIF-4AI and eIF-4AII) were reported in mice and rabbits and are differentially expressed in various tissues (Nielsen et al., 1985; Nielsen and Trachsel, 1989). One of these forms, eIF-4AII, is preferentially associated with the eIF-4F complex (Conroy et al., 1990). eIF-4A is an essential protein in yeast, and its ablation results in arrest of protein synthesis (Blum et al., 1989). The 220 kDa component of eIF-4F (p220) has not been well characterized, except that its structural integrity is required for eIF-4F activity (Etchison et al., 1984). eIF-4B is an 80 kDa polypeptide that contains an RNA recognition motif 7(RRM; Milburn et al., 1990).

It was proposed that cap function is mediated by initiation factors that interact with the cap structure and subsequently use the energy generated from ATP hydrolysis to unwind 5' proximal mRNA secondary structure, thereby facilitating 40S subunit attachment (Sonenberg, 1981; Sonenberg et al., 1981). Numerous lines of evidence support the model that eIF-4A, eIF-4B and eIF-4F are responsible for unwinding of mRNA secondary structure (reviewed in Edery et al., 1987; Sonenberg, 1988). The first indication was provided by Ray et al. (1985), who showed that eIF-4F, and to a lesser extent eIF-4A, induces conformational changes in mRNA structure as monitored by a nuclease sensitivity assay. eIF-4B, which by itself possesses no unwinding activity, stimulated eIF-4A melting activity. eIF-4A has been characterized as an ATP dependent single-stranded RNA binding protein, whose ATPase activity is stimulated by the presence of single-stranded RNA (Grifo et al., 1984; Abramson et al., 1987). The involvement of an RNA unwinding activity during translation initiation is consistent with the inhibitory effect of excessive mRNA 5' proximal secondary structure on translation (Pelletier and Sonenberg, 1985). Subsequently, Rozen et al. demonstrated, by using a direct assay, helicase activity for eIF-4A and eIF-4F in the presence of eIF-4B (Rozen et al, 1990).

eIF-4A is a member of a rapidly growing protein family of putative RNA helicases, termed the DEAD protein family (Linder et al., 1989). This family includes about 30 proteins in a wide range of organisms from bacteria to humans (Schmid and Linder, 1991b). These proteins are involved in diverse cellular functions such as translation initiation, RNA splicing, ribosome assembly, spermatogenesis, embryogenesis, cell growth and division (Wassarman and Steitz, 1991). The DEAD protein family is characterized by 8 highly conserved amino acid regions, one of which is the DEAD region (Schmid and Linder, 1991b). Two members of the DEAD family, eIF-4A and p68, a mammalian protein found in the nuclei of dividing cells, have been shown to exhibit ATP dependent RNA helicase activity (Rozen et al., 1990; Hirling et al., 1989). It is likely that all the DEAD proteins share the same mechanism for RNA unwinding, although they are involved in different processes in the cell. The functional regions for the ATPase and RNA helicase activities of the DEAD family proteins have not been determined. Comparison of the DEAD proteins show a common core region represented by eIF-4A, that is flanked in other proteins by N- and C-terminal extensions. These extensions share little sequence homology and are probably required for specialized functions of the individual proteins. Since eIF-4A is the best characterized member of the DEAD protein family, it is an excellent prototype protein for the analysis of the mode of action of RNA helicases. Here, we describe the characteristics of mammalian eIF-4A as an RNA helicase and its structure-function relationship.

CHARACTERIZATION OF RNA HELICASE ACTIVITY OF INITIATION FACTORS

RNA helicase activity was demonstrated for the combination of eIF-4F and eIF-4B and for the combination of eIF-4A and eIF-4B proteins purified from rabbit reticulocytes (Lawson et al., 1989; Rozen et al., 1990). These factors have been traditionally purified from the ribosomal salt wash, which contains other RNA binding proteins, that may also possess RNA unwinding activity. Preparations of eIF-4A were about 95% pure, whereas preparations of eIF-4B were 80-90% pure, and some contained a contaminating RNA unwinding activity (Rozen et al., 1990; Anthony and Merrick, personal communication). Unwinding of duplex RNA requires a large excess of eIF-4A and eIF-4B (Rozen et al., 1990). Therefore, although highly unlikely, it was difficult to formally rule out the possibility that the observed RNA helicase activity of eIF-4A and eIF-4B was due to an unrelated contaminating protein. To address this problem, we expressed the mouse eIF-4A and human eIF-4B cDNAs in *E. coli* and purified the factors to apparent homogeneity for use in the RNA unwinding assay. An RNA substrate was used which contains a 10 bp duplex region and 30 nucleotide long single stranded 3' ends (Figure 1). The duplex RNA was stable at 37°C during a 30 min incubation period (lane 1). The faster migrating monomer was generated after heat denaturation (lane 2). Incubation of RNA and eIF-4A in the presence of ATP caused minimal unwinding (varying from 0 - 5% in different experiments; lane 3). Incubation with higher amounts (10µg) of eIF-4A alone caused approximately 20% unwinding (data not shown). eIF-4A and eIF-4B incubated with RNA in the absence of ATP also caused very minimal unwinding (varying between 0 - 2% in different experiments; lane 4). However, when eIF-4A and eIF-4B were incubated in the presence of ATP, quantitative RNA unwinding activity was observed and showed an eIF-4A dose dependent response (lanes 5-7).

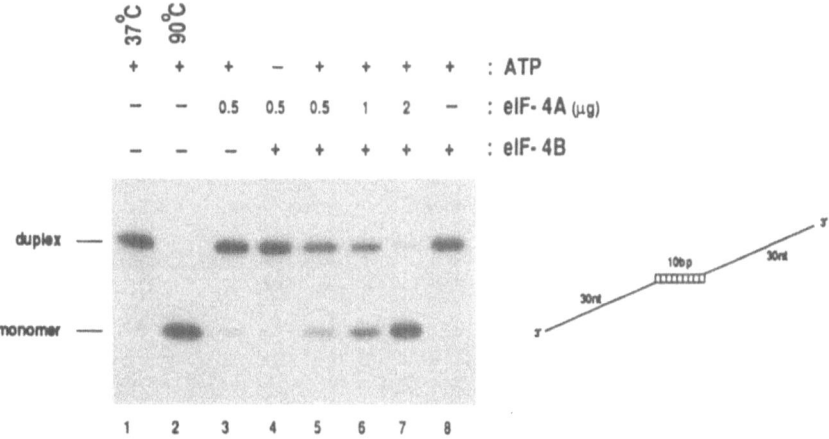

Figure 1. Recombinant eIF-4A and eIF-4B are active in RNA unwinding. The RNA substrate, shown schematically on the right, contains a double stranded region of 10 bp ($\Delta G = -35$ kCal/mole; alternating G-C nucleotides) with 3' terminal extensions of single stranded tails of 30 nucleotides. Indicated amounts of eIF-4A and 0.3 µg of eIF-4B were incubated with 0.4 ng of radiolabelled duplex RNA for 30 min at 37°C. Measurement of unwinding was performed by separating the duplex RNA from the single stranded, monomer RNA on a SDS-15% polyacrylamide gel followed by autoradiography. Lane 1: Duplex RNA incubated under unwinding conditions without protein for 30 min at 37°C. Lane 2: Duplex RNA incubated under the same conditions for 5 min at 90°C.

Thus, eIF-4B is essential for efficient unwinding of the duplex RNA. eIF-4B alone, in the presence of ATP, did not cause any unwinding (lane 8). These results substantiate the conclusions derived from the experiments using rabbit reticulocyte factors (Rozen et al., 1990), and rule out the possibility that the helicase activity of eIF-4A in combination with eIF-4B is due to contaminating proteins. Also, the results presented below rule out a helicase contaminant of bacterial origin, since mutations in eIF-4A abrogate helicase activity.

BIDIRECTIONAL UNWINDING

The previous results indicate that eIF-4A in combination with eIF-4B could unwind duplex RNA in a bidirectional manner. To further substantiate this conclusion, we wanted to demonstrate bidirectional unwinding on a single molecule. To this end, we prepared a trimer RNA that contained duplexes at each end of a central single-stranded region of 30 nucleotides (Fig.2A). The free energy of each duplex was -35 kCal/mol (similar to the duplex used above [Fig.1]). Dissociation of the short RNAs could be initiated only from the middle single-stranded region, since the ends are flush double stranded. In this experiment the only RNA labeled to high specific activity was the long RNA (62 nucleotides), and dissociation was followed by a shift in the migration of this RNA. Figure 2B shows the helicase activity of eIF-4A plus eIF-4B, or eIF-4F plus eIF-4B on this RNA. Lanes 1 to 7 show the positions of the different possible hybrids that can be formed between the three different RNAs. The migration of the free 62-nucleotide RNA is shown in lane 1, and the migration of the duplexes containing the 62-nucleotide RNA and either of the shorter RNAs is shown in lanes 2 and 4.

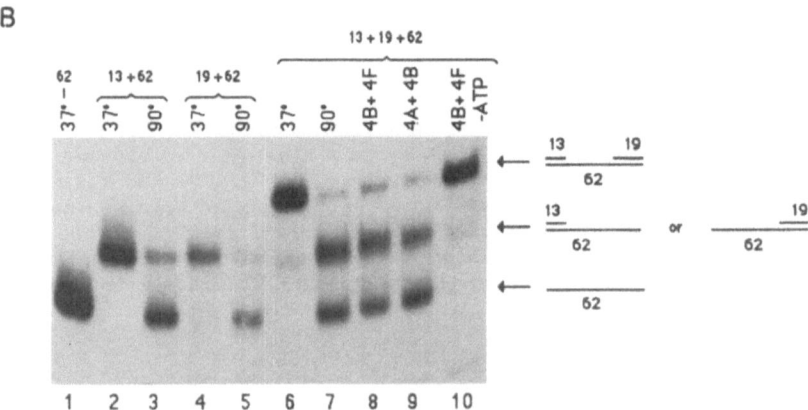

Figure 2. Bidirectional Unwinding from a trimer RNA.

(A) Nucleotide sequence of RNA templates and predicted base pairing to generate a trimer RNA.

(B) Transcription of the RNA templates was done as described (Rozen et al, 1990). Uncapped duplex or trimer RNA templates were generated by mixing a 62 nt RNA (lane 1) with either a 13nt RNA (lanes 2,3) or a 19nt RNA (lanes 4,5) or both (lanes 6-10) in 0.4M NaCl, 3mM Mg(OAc)$_2$, 20mM Tris-Hcl, 7.5 and 1mM DTT, heating at 90°C for 5 min and slow cooling to 40°C. Unwinding of the trimer RNA was performed with 2μg of eIF-4B plus 2μg of eIF-4F (lanes 8 and 10), or 2μg eIF-4B plus 6μg eIF-4A (lane 9) in the presence (lanes 8 and 9) or absence (lane 10) of 1.3 mM ATP and 0.6 mM Mg(OAc)$_2$ for 2 hours at 37°C.

Heating the duplexes at 90°C resulted in the release of most of the 62-nucleotide RNA (lanes 3 and 5). When all three RNAs were hybridized together, a trimer was formed (lane 6); upon heating, this trimer dissociated into the free 62-nucleotide RNA and a dimer RNA (lane 7; it is possible that heating did not cause the complete dissociation of the trimer RNA components, because of excess of the short RNAs that resulted in partial rehybridization; these conditions were different from those used for a formation of the duplexes described above). Incubation of the trimer RNA with eIF-4B plus eIF-4F or eIF-4B plus eIF-4A yielded a similar pattern to that obtained upon heating of the RNA to 90°C (Compare lanes 8 and 9 with lane 7). Importantly, the factor-mediated dissociation is completely dependent on ATP (compare lane 10 with lane 9). The appearance of the 62-nucleotide RNA in the presence of protein factors in an ATP-dependent manner is evidence for the dissociation of both short RNAs from the 62-nucleotide RNA and strongly indicates that the helicase can function in a bidirectional fashion. Furthermore, the helicase is not limited to binding single-stranded ends, but can access the RNA by binding to an internal region devoid of secondary structure.

OLIGONUCLEOTIDE DIRECTED MUTAGENESIS IN HIGHLY CONSERVED REGIONS OF EIF-4A

Proteins of the DEAD family contain 8 highly conserved regions (Fig. 3), and are thought to exhibit ATPase and RNA unwinding activity, as such activities have been demonstrated for eIF-4A and p68 (Rozen et al, 1990; Hirling et al, 1989). To investigate the contribution of the conserved amino acid regions to ATP binding, ATP hydrolysis and RNA unwinding, we performed site-directed mutagenesis of the conserved regions in eIF-4A. The DEAD family shares eight conserved amino acid regions (Fig. 3). The most N-terminal region (AXXXXGKT), part of the ATPase A motif (Walker et al., 1982), is present in many nucleotide binding proteins, and according to X-ray crystallography forms a phosphate binding loop (P-loop; Fry et al., 1986; Saraste et al., 1990; Story and Steitz, 1992). Alanine 76, preceding the GKT region (AXXXXGKT), is found in the majority of DEAD proteins. To determine the significance of alanine at position 76, it was mutated to serine, glycine or valine (S, G, VXXXXGKT). Valine was chosen as it abolished the function of yeast eIF-4A, TIF1, and another ATPase, STE6, *in vivo* (Schmid and Linder, 1991a; Berkower and Michaelis, 1991). We have previously mutated lysine 82 in the ATPase A motif (Rozen et al., 1990), which is invariant in all nucleotide binding proteins, and binds directly to the β and γ phosphates of the bound nucleotide (Fry et al., 1986; Story and Steitz, 1992). Replacement of lysine 82 with asparagine (AXXXXGNT) abolished ATP binding to eIF-4A (Rozen et al., 1989).

The second motif we investigated was suggested to be a variant of the Walker ATPase B motif (Linder et al, 1989). X-ray crystallography showed that the first aspartate residue of the DEAD region is in close proximity to the ATPase A motif, and binds Mg^{2+} through a water molecule (Pai et al., 1990; Story and Steitz, 1992). The Mg^{2+} is complexed to the β and γ phosphates of the nucleotide. We replaced the N-proximal aspartate (aspartate 182) in the DEAD region with either glutamate or asparagine (EEAD and NEAD), to assess whether the size or the charge is important for function. We replaced the highly conserved glutamate at position 183 by glutamine (DQAD). We also mutated the C-proximal aspartate in the DEAD region (aspartate 185) to histidine (DEAH). Replacement of aspartate 182 by glutamate

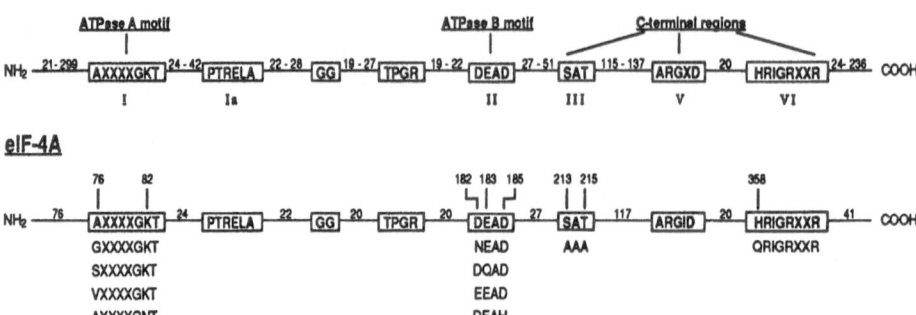

Figure 3. Schematic representation of highly conserved amino acid regions in the DEAD family and the eIF-4A protein (Schmid and Linder 1991b). The conserved regions are in bold and boxed and the numbers between the boxes indicate the distance in amino acid residues. The roman numbers below the boxes indicate the nomenclature of the regions after Hodgman (1988). Mutations introduced into the eIF-4A protein are shown below the corresponding boxed motif of the eIF-4A. The numbers above the amino acids in the eIF-4A indicate the position of the residue in the eIF-4A protein. X indicates any amino acid.

(EEAD) and aspartate 185 by histidine (DEAH) in yeast eIF-4A led to the loss of activity *in vivo* (Schmid and Linder, 1991a). The most conserved amino acid regions in the C-terminus of the DEAD proteins are the SAT and the HRIGRXXR regions. No function can yet be attributed to these regions, but it is likely that they constitute important motifs for the protein function. We mutated serine 213 and threonine 215 of the SAT region to alanine (AAA), and histidine 358 of the HRIGRXXR region to glutamine (QRIGR). Wild type and mutant eIF-4A proteins as well as the eIF-4B protein were expressed in E. coli and purified to apparent homogeneity. The mutations did not affect protein stability or levels of expression.

COMPARISON OF THE BIOCHEMICAL ACTIVITIES OF THE EIF-4A MUTANTS

The results from the ATP cross-linking, ATPase and RNA unwinding assays are summarized in Table I. Conservative mutations in the ATPase A motif (SXXXXGKT and GXXXXGKT) did not severely affect the biochemical activities of eIF-4A. However, nonconservative mutations like the VXXXXGKT and AXXXXGNT mutations abolished ATP binding and subsequently ATP hydrolysis and RNA unwinding activities. Mutations in the ATPase B motif (DEAD) had drastic effects on RNA helicase activity. Mutants NEAD and DQAD were still capable of binding ATP, but lost the hydrolyzing and unwinding activities. For the conservative EEAD mutant ATP binding and ATPase activity was comparable to wildtype but RNA helicase activity was abolished. For the mutants DEAH and AAA ATPase activity was enhanced or reduced (QRIGR mutant), but these mutants showed no, or drastically reduced, unwinding activity. These mutations (EEAD, DEAH, AAA, QRIGR) resulted in uncoupling of the ATPase and the RNA helicase activity of the eIF-4A protein. They are still capable of binding to RNA in an ATP dependent manner, but are incapable of unwinding the duplex RNA, because the functional link between the two activities is curtailed by the point mutations.

In summary, the conserved eIF-4A regions investigated by us can be categorized into three functional groups: a) the ATPase A motif (AXXXXGKT) is involved in ATP binding, b) the DEAD region is involved in the ATP hydrolysis reaction and c) the SAT region plays a critical role in the RNA unwinding reaction. Another conclusion is that the DEAD region couples ATP hydrolysis to the RNA unwinding activity. Blum et al, (1992), performed a mutational analysis on yeast eIF-4A. It was shown that only mutants which are capable of hydrolyzing ATP and unwinding a duplex RNA were viable *in vivo* and could restore translation in translation extracts *in vitro*. These findings demonstrate that the RNA helicase activity of eIF-4A is an essential activity for translation initiation in eukaryotes.

Table I. Biochemical activities of eIF- 4A mutants[a]

	eIF- 4A	X- linking [%]	ATPase [%]	Unwinding [%]
	WT	100	100	100
ATPase A motif (AXXXXGKT)	SXXXXGKT	220	110	110
	GXXXXGKT	250	160	50
	VXXXXGKT	7	0	0
	AXXXXGNT	2	0	0
ATPase B motif (DEAD)	NEAD	150	0	0
	DQAD	60	0	0
	EEAD	110	80	0
	DEAH	410	340	10
C - terminal regions (SAT, HRIGR)	AAA	340	240	0
	QRIGRXXR	60	30	0

[a] ATP cross linking, ATPase and RNA helicase assays were performed and quantitated as described (Pause and Sonenberg, 1992).

REFERENCES

Abramson,R.D., Dever,T.E., Lawson,T.G., Ray,B.K., Thach,R.E. and Merrick,W.C. 1987, *J. Biol. Chem.* 262:3826-3832.

Berkower,C. and Michaelis,S. 1991, *EMBO J.* 10:3777-3785.

Blum,S., Muller,M., Schmid,S.R., Linder,P. and Trachsel,H. 1989, *Proc.Natl. Acad. Sci.* USA. 86:6043-6046.

Blum, S., Schmid, S.R., Pause, A. Buser, P. Linder, P. Sonenberg, N. and Trachsel, H. 1992. *Proc. Natl. Acad. Sci.* USA. 89:7664-7668.

Bourne,H.R., Sanders,D.A. and McCormick,F. 1991, *Nature.* 349:117-127.

Company,M., Arenas,J. and Abelson,J. 1991, *Nature*, 349:487-493.

Conroy,S.C., Dever,T.E., Owens,C.L. and Merrick,W.C. 1990, *Arch. Biochem. Biophys.*, 28:363-371.

Edery,I., Humberlin,M., Darveau,A., Lee,K.A.W., Milburn. S., Hershey,J.W.B., Trachsel,H. and Sonenberg,N. 1983, *J. Biol. Chem.*, 258:11398-11403.

Edery, I., Pelletier, J., and Sonenberg, N. 1987, in *"Translational Regulation of Gene Expression* (Ilan, J.ed). p.335, Plenum Publishing Co. New York.

Fry,D.C., Kuby,S.A. and Mildvan,A.S. 1986, *Proc. Natl. Acad. Sci.* USA.,83:907-911.

Grifo,J.A., Tahara,S.M., Shatkin,A.J., and Merrick,W.C. 1983, *J. Biol. Chem.*,258:5804-5810.

Grifo,J.A., Abramson,R.D., Satler,C.A. and Merrick,W.C. 1984, *J. Biol. Chem.*,259:8648-8654.

Hirling,H., Scheffner,M., Restle,T. and Stahl,H. 1989, *Nature*, 339:562-564.

Hodgman,T.C. 1989 *Nature*, 333: 22-23.

Lawson,T.G., Lee,K.A., Maimone,M.M., Abramson,R.D., Dever,T.E.,Merrick,W.C. and Thach,R.E. 1989, *Biochemistry*, 28:4729-4734.

Linder, P., Lasko,P.F., Ashburner,M., Leroy,P., Nielsen, P.J.,Nishi,K.,Schnier,J and Slonimsky,P.P. 1989 *Nature*,337:121-122.

Milburn,S.C., Pelletier,J., Sonenberg,N. and Hershey,J.W.B. 1988 *Arch. Biochem. Biophys.*, 264: 348-350.

Milburn,S.C., Hershey, J.W.B., Davies,M.V., Kelleher,K. and Kaufman,R. 1990, *EMBO J.*, 9: 2783-2790.

Nielsen, P.J, McMaster,G.K. and Trachsel,H. 1985, *Nucleic Acids Res.*,13:6867-6880.

Nielsen,P.J. and Trachsel,H. 1989, *EMBO J.*, 7: 2097-2105.

Pai,E.F., Krengel,U., Petsko,G.A., Gody,R.S., Katsch,W. and Wittinghofer, A. 1990, *EMBO J.*, 9: 2351-2359.

Pause, A. and Sonenberg, N. 1992, *EMBO J.* 11: 2643-2654.

Ray,B.K., Lawson,T.G., Kramer,J.C., Cladaras,M.H., Grifo,J.A.,Abramson,R.D., Merrick,W.C. and Thach,R.E. 1985 *J. Biol. Chem.*, 260: 7651-7658.

Rhoads, R.E. 1988. *Trends Biochem. Sci.*13:52-56.

Rozen,F., Pelletier,J., Trachsel,H. and Sonenberg,N. 1989 *Mol. Cell. Biol.*,9: 4061-4063.

Rozen,F., Edery,I., Meerovitch,K., Dever,T.E., Merrick,W.C. and Sonenberg,N. 1990 *Mol. Cell. Biol.*, 10: 1134-1144.

Saraste,M., Sibbald,P.R. and Wittinghofer,A. 1990, *Trends Biochem. Sci.* 15: 430-434.

Schmid,S.R. and Linder,P. 1991a, *Mol. Cell. Biol.*, 11: 3463-3471.

Schmid,S.R. and Linder,P. 1991b, *Molec. Microbiol.*, 6: 283-292.

Sonenberg,N., Rupprecht, K.M., Hecht, S.M. and Shatkin, A.J. 1979, *Proc. Natl. Acad. Sci.* USA, 76:4345-4349.

Sonenberg,N. 1988, *Prog. Nucleic Acid Res. Mol. Biol.* 35: 173-207.

Story,R.S. and Steitz,T.A. 1992 *Nature*, 355: 374-376.

Tahara, S.M., Morgan, M.A., and Shatkin, A.J. 1981, *J. Biol. Chem.* 256:7691-7694.

Walker,J.E., Saraste,M., Runswick,M.J. and Gay,N.J. 1982, *EMBO J.* 1: 945-951.

Wassarman,D.A. and Steitz,J.A. 1991, *Nature,* 349: 463-464.

TRANSLATION INITIATION BY INTERNAL RIBOSOME BINDING OF EUKARYOTIC mRNA MOLECULES

Chang-You Chen, Dennis G. Macejak, Soo-Kyung Oh, and Peter Sarnow

Molecular Biology Program, and Department of Biochemistry, Biophysics
and Genetics
University of Colorado Health Sciences Center
4200 East Ninth Avenue
Denver, CO 80262

INTRODUCTION

A unique feature of all eukaryotic mRNA molecules is the presence of a m^7GpppN (where N can be any nucleotide) "cap" structure at the 5' end of the RNAs (Shatkin, 1976; Banerjee, 1980). This cap structure interacts with eukaryotic initiation factor (eIF) 4F, a three subunit protein complex composed of eIF-4E, eIF4A and p220 (Tahara et al. 1981; Edery et al., 1983; Grifo et al., 1983). It is thought that binding of eIF-4F to the 5'- terminal cap-structure facilitates the recruitment of the 40S ribosomal subunit to the 5' end of the mRNA (Ray et al., 1985; Rozen et al., 1990).Subsequently, the 40S ribosomal subunit, carrying factors eIF2-GTP, eIF3 and the initiator tRNA (tRNA^{met}i), is postulated to move in a 5' to 3' direction along the mRNA until an appropriate AUG codon is encountered, which is then used to start protein synthesis. This model is known as the "scanning mechanism" for translation initiation (Kozak, 1989a). Evidence supporting this model includes the inability of circular RNAs to bind eukaryotic ribosomes (Kozak, 1979; Konarska et al., 1981), suggesting that the 40S subunit can enter the mRNA only at its free 5' end. Furthermore, 40S subunits can be trapped both upstream of the AUG start codon if ATP is depleted (Kozak, 1980) and upstream of stable RNA hairpin structures (Kozak, 1989b). Thus, the scanning mechanism can explain translation initiation of most vertebrate mRNAs containing short 5' noncoding regions (5'NCRs), harboring one or only a few AUG codons.

Certain mRNAs, notably encoding proto-oncogenes and regulatory genes, contain long 5'NCRs with multiple AUG codons (Kozak, 1991).Thus, it is predicted that the translation initiation rate of these mRNAs is quite low because of RNA structures and AUG codons present in the 5'NCRs regions; alternatively, other translation initiation mechanisms may operate to ensure efficient translation. Two such mechanisms have been under intense scrutiny.

The first mechanism, the "reinitiation mechanism" (a modification of the scanning model), operates on the *GCN4* mRNA in the budding yeast *Saccharomyces cerevisiae* (Hinnebusch, 1988). The 5' NCR of *GCN4* mRNA contains four open reading frames. The fifth open reading frame is used to produce the *GCN4* protein when amino acids are limited in the cell (Hinnebusch, 1988). When amino acids are readily available, 40S subunits resume scanning after the termination codon of the first open reading frame is reached and reinitiate at the fourth open reading frame; following translation of the fourth open reading frame, however, no reinitiation occurs at the fifth open reading frame. The proposed mechanism is that under amino acid starvation, the phosphorylation of eIF2 is induced and thus of eIF2-GTP is formed more slowly. As a consequence, 40S subunits scanning downstream from the first open reading frame are not yet competent to initiate translation at the fourth open reading frame. Instead, most 40S subunits acquire eIF2-GTP only while scanning from the fourth to the fifth AUG codon, resulting in the formation of 80S ribosomes at the fifth AUG codon, the start codon for *GCN4* (Dever et al., 1992).

Although a reinitiation mechanism could in principle be used to translate mRNAs with long 5' noncoding regions and multiple AUG codons, it is hard to imagine how such mRNAs could be translated efficiently in cells under normal growth conditions. An explanation of this dilemma has been provided by virologists. It is well-known that the 5' NCRs of picornaviruses are very long, 600 to 1200 nucleotides in length, and are burdened with many AUG codons (Toyoda et al., 1986). Moreover, during many picornaviral infections the p220 component of eIF-4F is cleaved, thereby inhibiting the translation of capped cellular mRNAs while the translation of the naturally uncapped picornaviral mRNAs is greatly stimulated (Etchison et al., 1982). What is the mechanism of the cap-independent translation of picornaviral mRNAs in infected cells at a time when cap-dependent translation of cellular mRNAs is inhibited? Pelletier and Sonenberg (1988) and Jang and colleagues (1989) have shown that picornaviral mRNAs are translated by an "internal ribosome binding" mechanism, in which the long 5'NCRs, with multiple AUG codons, of the viral RNAs do not inhibit translational initiation.

Because viruses usually subvert preexisting host macromolecular mechanisms to their own advantage, we have investigated whether an internal ribosome binding mechanism may be used to initiate the translation of any cellular mRNAs as well. We will summarize here some of our recent results on the translation by internal ribosome binding of two cellular mRNAs, encoding the immunoglobulin heavy chain binding protein (BiP) of humans and the *Antennapedia* gene product of *Drosophila melanogaster*. In addition, we will outline a method for the production of large quantities of circular RNA molecules which can be used to study RNA sequences that mediate translation initiation by internal ribosome binding.

TRANSLATION OF A CELLULAR mRNA ENCODING AN 80KD PROTEIN (BiP) IS STIMULATED IN POLIOVIRUS-INFECTED CELLS WHEN CAP-DEPENDENT TRANSLATION IS INHIBITED

Initially, we constructed mutations in an infectious poliovirus cDNA to study the

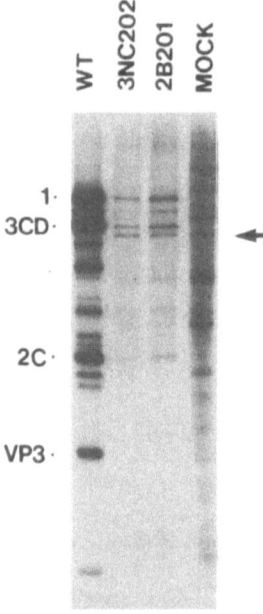

Figure 1. Synthesis of a cellular 80kd protein in cells infected with different poliovirus mutants. ^{35}S-methionine-labeled proteins from infected and mock-infected cells were analyzed by SDS-polyacrylamide gel electrophoresis. An autoradiograph is shown. The arrow denotes the 80kd protein.

structure and function of the viral genome. Several poliovirus mutants could be isolated after transfection of mutated cDNAs into mammalian cells. Two such mutants, 3NC202 and 2B201, harbored mutations in the 3'NCR and the coding region, respectively. To assess the effects of these mutations on translation of the viral RNAs, human HeLa cells were infected with wild-type and mutant polioviruses and labeled with ^{35}S-methionine. Cytoplasmic extracts were prepared and analyzed in SDS-containing polyacrylamide gels.

Figure 1 shows that, as expected, cap-dependent translation of cellular mRNAs was inhibited in cells infected with wild-type poliovirus (WT) and only the viral genome was translated at this time, three hours after infection, producing known viral proteins. Translation of most cellular mRNAs was also inhibited in mutant-infected cells, although lower amounts of viral proteins were synthesized. We noted the synthesis of an 80kd protein , denoted by the arrow in Figure 1, in mutant-infected cells but not in wild-type-infected cells. Interestingly, a protein of similar size was synthesized in uninfected cells as well. Comparison of proteolytic peptide maps generated by digestion of the 80kd protein isolated from mutant-infected and

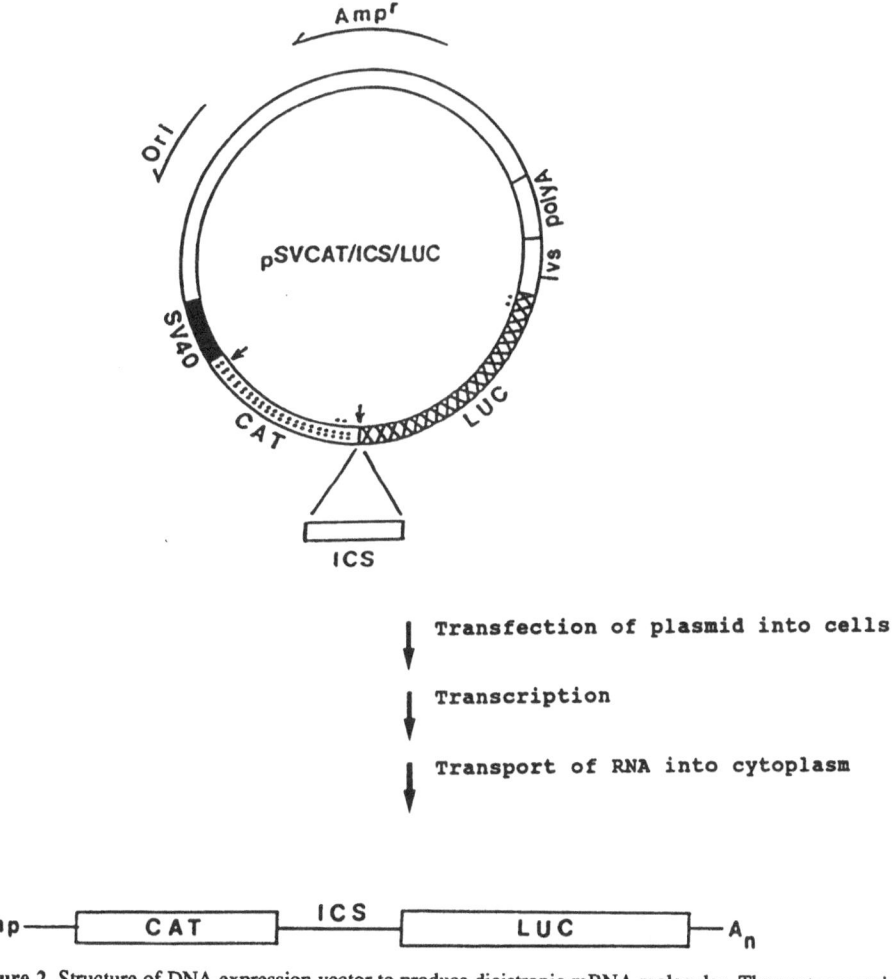

Figure 2. Structure of DNA expression vector to produce dicistronic mRNA molecules. The vector contains promoter and enhancer elements from SV40, followed by the coding region for CAT, an ICS (intercistronic spacer), and a second cistron encoding LUC. Reprinted with permission from OH et al. (1992).

uninfected cells revealed a high degree of identity   (Sarnow, 1989). Subsequently, this protein was identified as immunoglobulin heavy chain binding protein (BiP), also known as glucose-regulated protein 78 (Lee et al., 1981).

THE 5'NONCODING REGION OF BIP MEDIATES TRANSLATION BY AN INTERNAL RIBOSOME BINDING MECHANISM

The steady-state levels of BiP mRNA were the same in infected and in uninfected cells (Sarnow, 1989), suggesting that the synthesis of BiP in mutant poliovirus-infected cells occurred by the same mechanism as the synthesis of viral proteins. To test directly whether BiP mRNA could be translated by an internal ribosome binding mechanism, we constructed vectors that could be used to direct the synthesis of dicistronic mRNAs in mammalian cells (Pelletier and Sonenberg, 1988; Jang et al., 1989). Figure 2 depicts pSVCAT/ICS/LUC, a vector that should express capped, dicistronic 5'CAT-ics-LUC3' mRNAs in transfected mammalian cells. The first open reading frame, encoding chloramphenicol acetyltransferase (CAT), should be translated by a cap-dependent initiation mechanism, while the second cistron, encoding luciferase (LUC), should be translated only if preceded by intercistronic spacer (ICS) sequences that can mediate internal ribosome binding.

We constructed several dicistronic vectors that contained, in the ics, either the 5'NCR of poliovirus as a positive control, *Drosophila* sequences ED as a negative control or the 5'NCR of BiP. Upon transfection into HeLa cells, all three vectors produced dicistronic mRNAs that directed the synthesis of comparable amounts of CAT protein (Macejak and Sarnow, 1991). However, LUC protein was only detected in cells expressing mRNAs containing either the 5'NCRs of poliovirus or BiP (Macejak and Sarnow, 1991).This indicated that the 5'NCR of BiP was sufficient to direct the synthesis, by internal ribosome binding, of the second LUC cistron in this dicistronic construct.

To exclude the possibility that LUC was produced from dicistronic 5'CAT-BiP-LUC3' mRNAs by a reinitiation mechanism in which the 40S subunits resumed scanning after the termination codon of the first cistron, CAT, we compared the translation of dicistronic mRNAs containing or lacking a stable RNA hairpin structure in the 5'NCR of CAT (Figure 3). A reinitiation mechanism predicts that translation of the CAT cistron should be required for the LUC cistron to be translated. Internal ribosome binding-mediated translation of LUC, however, should be independent of CAT translation. A stable RNA hairpin present in the 5'NCR of CAT should inhibit the translation of the CAT cistron (Fig.3,

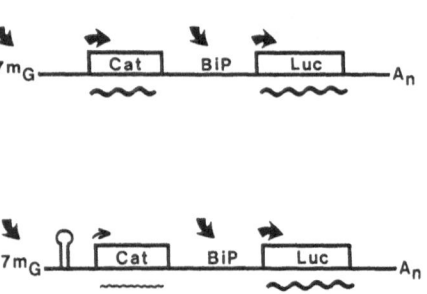

Figure 3. Diagram of dicistronic mRNAs lacking (top) or containing (bottom) a stable RNA hairpin structure in the 5'NCR of the first CAT cistron. The arrows indicate entry and movement of ribosomal subunits on the mRNAs. Reprinted with permission from Macejak and Sarnow (1991).

Pelletier and Sonenberg, 1985), allowing us to test whether LUC translation was coupled to CAT translation, or not. Figure 4 shows that introduction of a stable RNA hairpin structure into the 5'NCR of CAT inhibited, as expected, the synthesis of the CAT cistron in this dicistronic mRNA. However, the synthesis of the second cistron, LUC, was unaffected, indicating that the 5'NCR of BiP could mediate internal ribosome binding in this dicistronic mRNA molecule.

Inspection of the 5'NCR of human BiP, 220 nucleotides in length, revealed no AUG codons upstream of the initiator AUG (Ting and Lee, 1988). Thus, if it had not been for the observed enhanced translation of BiP mRNA in poliovirus-infected cells, its 5'NCR would not have been suspected to function as an internal ribosome entry site (IRES). It is likely that the BiP mRNA can be translated by a cap-dependent scanning mechanism (Kozak, 1989a) as well. If this is the case, the BiP mRNA can be translated by two different translation initiation mechanism. Is internal ribosome binding the preferred initiation mechanism at a certain time in the cell cycle or during particular growth conditions? To gain insight into these questions, we have searched for additional mRNAs that can mediate internal ribosome binding.

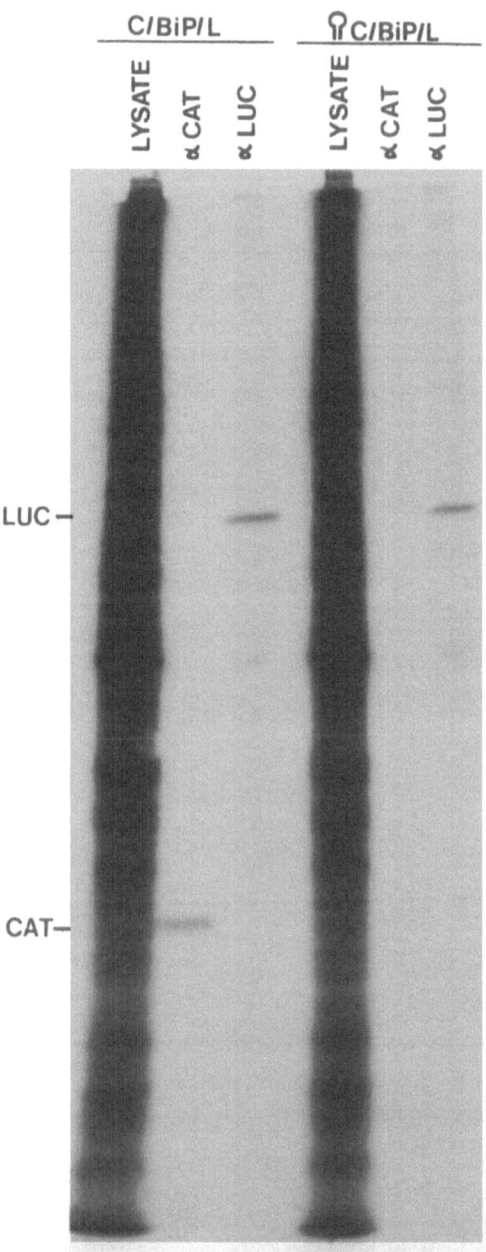

Figure 4. Synthesis of CAT and LUC in cells expressing dicistronic mRNAs without (three left lanes) or with (three right lanes) a stable RNA hairpin structure in the 5'NCR of the first cistron, CAT. Whole cell lysates and immunoprecipitations of extracts with α-CAT and α-LUC antibodies are shown. Reprinted with permission from Macejak and Sarnow (1991).

MANY *DROSOPHILA* mRNAs CONTAIN COMPLEX 5' NONCODING REGIONS

We noted that many *Drosophila* mRNAs contain very long 5'NCRs with many AUG codons. Specifically, 42% of mRNAs from *Drosophila melanogaster* contain one or more AUG codons upstream of the initiator AUG codon (Cavener and Cavener, 1992). Furthermore, the average length of *Drosophila* 5'NCRs is 248 nucleotides (Cavener and Cavener, 1992), considerably longer than the 40- to 80-nucleotide 5'NCRs of most vertebrate mRNAs. Curiously, vertebrate mRNAs with longer 5'NCRs often encode genes involved in growth control (Kozak, 1991). In this context, it is noteworthy that most *Drosophila* genes have been identified on the basis of mutant phenotypes, suggesting that their gene products have regulatory functions. The features of several 5'NCRs of *Drosophila* mRNAs of developmentally important regulatory genes are summarized in Table 1.

Table 1. Structural features of 5' noncoding regions of *Drosophila* mRNAs.[a]

Embryonic mRNA	length (nt)	Upstream AUGs	Consensus AUGs[b]	Stop codons	References
Abdominal-B (pH189)	2800	31	13	108	DeLorenzi et al.,1988
Abdominal-B (pH200)	494	1	0	22	DeLorenzi et al.,1988
Antennapedia P1	1512	8	2	78	Laughon et al., 1986 Stroeher et al., 1986
Antennapedia P2	1727	15	6	72	Laughon et al., 1986 Stroeher et al., 1986
bicoid	171	2	1	5	Berleth et al.,1988
caudal (maternal)	275	3	2	11	Mlodzik et al., 1987
caudal (zygotic)	461	4	2	21	Mlodzik et al., 1987
Deformed	490	4	0	23	Regulski et al.,1987
E74 A	1891	17	4	95	Burtis et al.,1990
E74 B	794	6	4	34	Burtis et al.,1990
fushi tarazu	120	1	0	3	Laughon, Scott, 1984
hunchback (3.2 kb transcript)	511	1	0	13	Tautz et al., 1987
Krüppel	185	2	2	9	Rosenberg et al,1986
labial	239	0	0	11	Diederich et al.,1989 Mlodzik et al., 1988
nanos	262	1	0	11	Wang and Lehmann,1991
Notch	799	7	1	21	Kidd et al., 1986
Sex combs reduced	626	5	1	23	Lemotte et al.,1989
sevenless	824	10	4	22	Bowtell et al.,
terminus	155	0	0	6	Baldarelli et al.,1988
Ultrabithorax	965	2	0	47	Kornfeld et al.,1989

[a]The 5' end of each of these mRNAs has been determined experimentally.

[b] *Drosophila* consensus sequence for translation initiation is $^{C}/_{A}AAA^{A}/_{C}AUGN$ (Cavener, 1987).

Reprinted with permission from OH et al. (1992).

Again, the translation of these mRNAs should be very inefficient according to both the cap-dependent scanning and reinitiation mechanisms. Thus, we have begun to test whether some of these mRNAs contain IRES elements in their 5'NCRs. We have found that mRNAs encoding the homeotic *Antennapedia* (see below) and *Ultrabithorax* (not shown) genes harbor sequence elements that can function as internal ribosome entry sites in dicistronic mRNAs in transfected into *Drosophila* cells.

A CONSERVED EXON OF THE HOMEOTIC *ANTENNAPEDIA* GENE ENCODES AN IRES ELEMENT IN THE 5' NONCODING REGION OF THE RNA TRANSCRIPT

The *Antennapedia (Antp)* gene of *Drosophila melanogaster* contains two promoters, P1 and P2, producing two transcripts that differ in their respective 5'NCRs (Laughon et al., 1986; Schneuwly et al., 1986; Stroeher et al., 1986). Figure 5 shows the P1-derived 5'NCR, 1512 nucleotide in length, comprising exons A, B, D and part of E, and the P2-derived 5'NCR, 1727 nucleotide in length, containing exons C, D and part of E. If a scanning mechanism were used to produce the *Antp* protein from the P2 mRNA, for example, the 43S ternary complex would have to scan 1730 nucleotides of the 5'NCR, bypassing 15 AUG codons, six of which having a consensus sequence for translation initiation in *Drosophila* (see Table 1). Alternatively, an internal ribosome binding

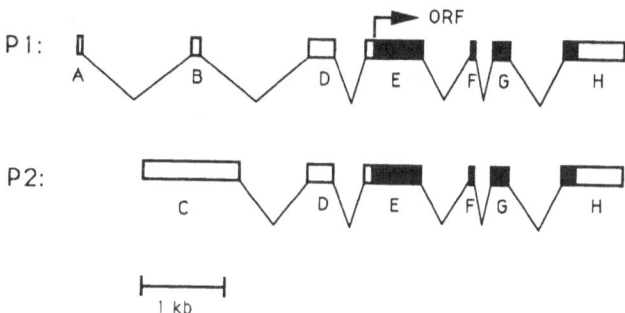

Figure 5. Organization of noncoding (open rectangles) and coding (filled rectangles) exons of *Antennapedia* transcripts from P1 and P2 promoters.

mechanism could be used to initiate translation of the P2 mRNA more efficiently. To test this second hypothesis, dicistronic mRNAs with various parts of the *Antp* P2 5'NCR inserted between the CAT and LUC cistrons (see Figure 2), were produced in vitro by T7 RNA polymerase, and transfected directly into cultured *Drosophila* cells. The fate of the transfected RNAs was monitored and quantified by nuclease protection assays (shown in OH et al., 1992). Next, the production of CAT and LUC proteins in extracts from transfected cells was examined. Table 2 shows that each transfected, capped RNA produced a comparable amount of active CAT protein. In addition, capped dicistronic mRNAs containing the entire 5'NCR

Table 2. Translation of dicistronic mRNAs after direct transfection of RNA molecules into cultured Drosophila melanogaster SL2 cells.

RNA	% chloramphenicol conversion/10^6 cells	LUC light units/10^6 cells
CDE	68±2	16,000±2,000
DE	67±3	7,000±1,000
D	49±5	3,500±500
E	49±2	240±200
C	30±2	300±100
ED	60±3	240±40
DE+AUG	66±3	170±60
uncapped CDE	9	13,600

All dicistronic mRNAs contained a m⁷GpppG-cap structure, except "uncapped CDE". Reprinted with permission from OH et al. (1992).

235

of *Antp* P2 (exons C, D and E) mediated the synthesis of active luciferase protein. Importantly, the same <u>uncapped</u> dicistronic mRNA did not direct the synthesis of CAT, yet the amount of LUC synthesized, was approximately the same as from the capped mRNA. This indicated that the translation of the second cistron was independent of the translation of the first cistron, and suggested that an internal ribosome entry site was provided by the 5'NCR of *Antp* P2. Exons DE were sufficient to mediate translation of luciferase, while exon C sequences or the reverse complement of DE sequences (ED) were not (Table 2). Interestingly, insertion into the DE sequences of an AUG codon that was in-frame with the luciferase initiator AUG abolished the production of active luciferase protein (Table 2). However, analysis by Western blotting revealed that an apparently non-functional luciferase protein was synthesized, containing extra amino acids at the N-terminus (not shown). This indicates that ribosomes could enter internally to exon DE and subsequently use the inserted AUG codon for initiation of protein synthesis. Further analysis revealed that sequences present in exon D alone were sufficient to function as an IRES (Table 2); these sequences are present in transcripts from both P1 and P2 promoters (Figure 5). Inspection of sequences in exon D revealed a 55-nucleotide RNA element that is highly conserved among *Drosophila* species whose common ancestors lived 60 million years ago (Hooper at al., 1992). Preliminary experiments have shown that this 55-nucleotide RNA element alone can indeed function as an IRES (OH, unpublished data).

Translation Initiation by Internal Ribosome Binding in *Drosophila*

Is there a reason to think that internal initiation may be used in the fly as well as in the cultured cells we have studied? During early stages of *Drosophila* development, the embryo is a single-cell syncytium that contains many rapidly dividing nuclei. It is expected that many maternal and early zygotic mRNAs are translated at that time. However, it is known that cap-dependent translation is severely impaired in mammalian cells undergoing mitosis, because of the underphosphorylation of the cap-binding protein eIF-4E (Bonneau and Sonenberg, 1987; Huang and Schneider, 1991). If this is the same in the *Drosophila* embryo, it is not clear how early embryonic mRNAs such as *bicoid* (Berleth et al., 1988) and *nanos* (Wang and Lehmann, 1991) could be translated efficiently. A cap-independent translation mechanism such as internal ribosome binding could provide a means by which mRNAs are translated in the absence of functional cap-binding protein complexes.

PRODUCTION OF CIRCULAR RNAs AS TOOLS FOR STUDYING INTERNAL RIBOSOME BINDING

It has been shown that viral (reviewed in Jackson et al., 1990) and cellular (Macejak and Sarnow, 1991; OH et al., 1992) IRES elements can mediate internal ribosome binding. However, it is not clear whether ribosomal subunits can bind directly to IRES sequences or whether non-ribosomal factors interact with the IRES to bind ribosomal subunits or to transfer them directly to the initiator AUG codons. Because IRES-containing dicistronic RNAs may bind ribosomal subunits both at the 5' end and internally, it is difficult to study the two ribosome entry sites independently. However, it was shown that small circular RNAs, 25 to 110 nucleotides in length, were unable to bind eukaryotic ribosomes in a cell-free system, although they were able to bind prokaryotic ribosomes (Kozak, 1979; Konarska et al., 1981). From this it was concluded that eukaryotic ribosomes could only bind to RNAs via their free 5' ends. In contrast, one would predict that IRES elements should function in RNAs without free 5' ends. Circular IRES-containing RNAs should, therefore, associate with ribosomes which could then direct the synthesis of a defined protein product.

To test this idea, we developed a method to produce large quantities of circular RNAs, up to 1000 nucleotides in length. We used a modification of a method originally described by Moore and Sharp (1992), outlined in Figure 6. Briefly, plasmids vectors containing the promoter for T7 RNA polymerase were linearized and used as templates in in vitro transcription reactions containing 500μM each of CTP, ATP, UTP and guanosine nucleoside and only 50μM of GTP.Thus, RNA molecules were produced that contained hydroxyl groups at their 5' termini. After addition of a radiolabeled phosphate to the 5' ends of the RNAs by T4 polynucleotide kinase, the RNAs were annealed to oligodeoxynucleotides containing sequences complementary to the 5' and 3' end of the RNAs. This DNA "splint"

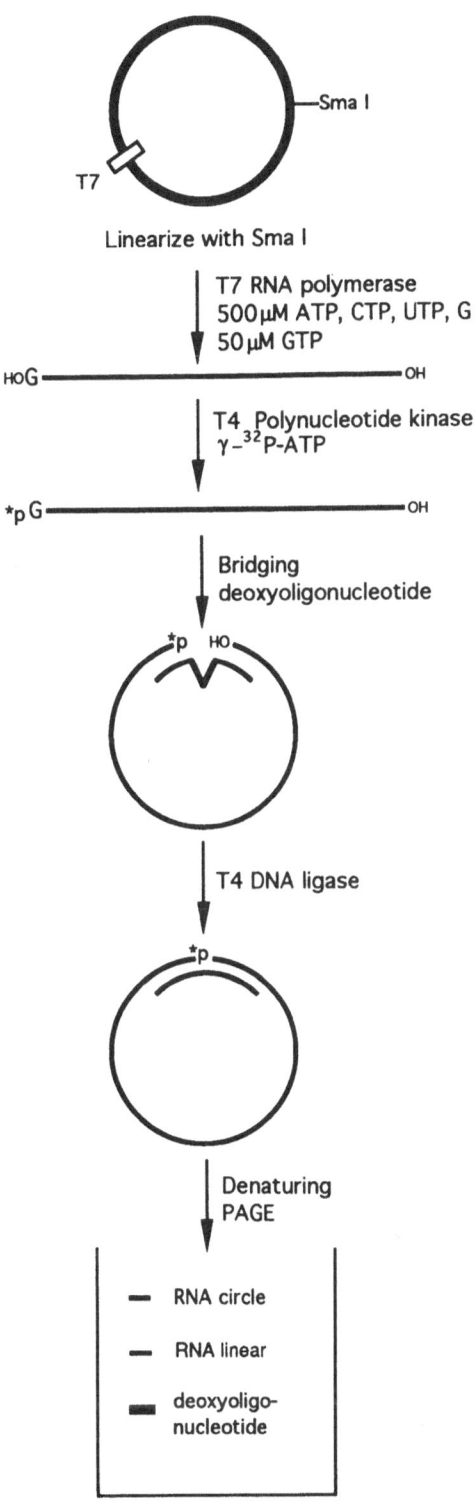

Figure 6. Modified Moore and Sharp method (Moore and Sharp, 1992) for the production of circular RNAs.

should result in the formation of circular DNA-RNA hybrids (Figure 6). The RNAs can then be covalently closed by the addition of T4 DNA ligase (Moore and Sharp, 1992), which has been shown to be able to efficiently ligate oligoribonucleotides in the presence of complementary deoxyribonucleotides (Kleppe et al., 1970) and has a much lower K_m for polynucleotides than RNA ligase (Romaniuk and Uhlenbeck, 1983). Subsequently, the circular RNA species were separated in urea-containing polyacrylamide gels from the oligodeoxynucleotides, linear and putative dimeric RNA species.

An example of an RNA circularization experiment is shown in Figure 7a. A 450-nucleotide transcript was synthesized in vitro by T7 RNA polymerase and ^{32}P-labeled at its 5' end (lane 1). After annealing to complementary (lane 5) or non-complementary (lane 2) oligodeoxynucleotides and incubation with T4 DNA ligase, a discrete RNA species was detected that migrated more slowly than the linear 450-nucleotide fragment. Incubation of the RNA/non-complementary oligodeoxynucleotide hybrids with RNA ligase (lane 3) or calf intestine phosphatase (lane 4) did not result in the formation of the slow-moving RNA species. To ensure that the slow-moving RNA species represented RNA circles and not dimer-length RNA molecules, we extracted this species from the denaturing gels and treated it with various phosphatases and nucleases. The results from such analyses clearly revealed

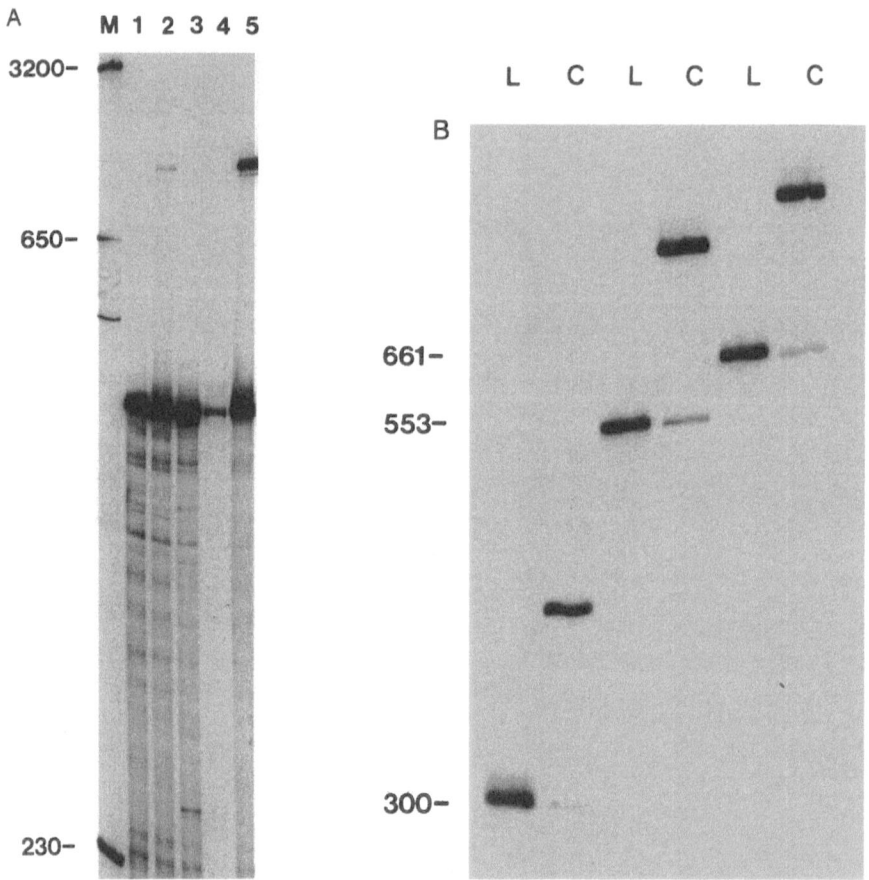

Figure 7. Analysis of circular RNAs in urea-containing polyacrylamide gels. a. Production of 450-nucleotide circular RNAs. Lane 1: 450-nt linear RNA; lane 2: 450-nt linear RNA plus non-complementary oligodeoxynucleotide incubated with T4 DNA ligase; lane 3: 450-nt linear RNA plus non-complementary oligodeoxynucleotide incubated with T4 RNA ligase; lane 4: same reaction as in lane 3 plus additional incubation with phosphatase; lane 5: 450-nt linear RNA plus complementary oligodeoxynucleotide incubated with T4 DNA ligase. An autoradiograph is shown. b. Production of RNA circles 300, 553 and 661 nucleotides in length, in pairs of lanes from left to right. An autoradiograph displaying the linear (L) and circular (C) RNA species is shown.

238

that the slow-moving RNA species represented RNA circles; for example, the first product of partial digestion with any endonuclease co-migrated with the 450-nucleotide linear monomer (data not shown). So far, we have constructed RNA circles of various lengths. This is illustrated in Figure 7b, where RNA circles ranging from 300 to 661 nucleotides in length are displayed.

In preliminary experiments, we have incubated circular RNAs containing various IRES elements with eukaryotic ribosomes in the presence of the elongation inhibitor sparsomycin and followed the fate of the circles after sedimentation in sucrose gradients. We have observed that IRES-containing circles can be found in a fraction sedimenting at 80S, indicating that an intact ribosome was attached to the circle. If this is true, these circles should be translated to produce unique proteins of predicted length and amino acid composition.

Thus, the RNA circles should aid in studying the mechanism with which IRES elements mediate translation initiation. Lastly, RNA circles should provide valuable tools for identifying further RNA sequences that can function as IRES elements.

Acknowledgements

We would like to thank Karla Kirkegaard for stimulating discussions and critical reading of this manuscript. This work was supported in part by a grant from the Lucille P. Markey Charitable Trust, The Council for Tobacco Research, U.S.A., and grants AI 25105 and AG 07347 from the National Institutes of Health. P.S. acknowledges the receipt of a Faculty Research Award from the American Cancer Society.

REFERENCES

Baldarelli, R.M., Mahoney, P.A., Salas, F., Gustavson, E., Boyer, P.D., Chang, M.-F., Roark, M., and Lengyel, J.,1988, *Dev. Biol.* 125:85-95.

Banerjee, A.K., 1980, *Microbiol. Rev.* 44:175-205.

Berleth, T., Burri, M., Thoma, G., Bopp, D., Richstein, S., Frigerio, G., Noll, M., and Nüsslein-Volhard, C.,1988, *EMBO J.* 7:1749-1756.

Bonneau, A.-M., and Sonenberg, N., 1987., *J. Biol. Chem.* 262:11134-11139.

Bowtell, D.D.L., Simon, M.A., and Rubin, G.M., 1988,*Genes & Dev.* 2:620-634.

Burtis, K.C., Thummel, C.S., Jones, W.C., Karim, F.D., and Hogness, D.,1990, *Cell* 61: 85-99.

Cavener, D.R., 1987, *Nucleic Acids Res.* 15:1353-1361.

Cavener, D.R., and Cavener, B.A., 1992, *in:* "Atlas of Drosophila Genes," G. Maroni, ed., in press.

DeLorenzi, M., Ali, N., Saari, G., Henry, C., Wilcox, M., and Bienz, M., 1988, *EMBO J.* 7:3223-3231.

Dever, T.E., Feng, L., Wek, R.C., Cigan, A.M., Donahue, T.F., and Hinnebusch, A.G., 1992, *Cell* 68:585-596.

Diederich, R.J., Merrill, V. K.L., Pultz, M.A., and Kaufman, T.C., 1989, *Genes. & Dev.* 3:339-414.

Edery,I., Humbelin, M., Darveau, A., Lee, K.A.W., Milburn, S., Hershey, J.W.B., Trachsel, H., and Sonenberg, N., 1983, *J. Biol. Chem.* 256:11398-11403.

Etchison, D., Milburn, S.C., Edery, I., Sonenberg, N., and Hershey, J.W.B., 1982, *J. Biol. Chem.* 257:14806-14810.

Grifo, J.A., Tahara, S.M., Morgan, M.A., Shatkin, A,J., and Merrick, W.C., 1983, *J. Biol. Chem.* 258:5804-5810.

Hinnebusch, A.G., 1988, *Microbiol. Rev.* 52:248-273.

Hooper, J.E., Perez-Alonso, M., Bermingham, J.R., Prout, M., Rockline, B.A., Wagenbach, M., Edstrom, J.-E., De Frutos, R., and Scott, M.P., 1992, *Genetics*, in press.

Huang, J., and Schneider, R.J.,1991, *Cell* 65:271-280.

Jackson, R.J., Howell, M.T., and Kaminski, A., 1990, *Trends biochem. Sci.* 15:477-483.

Jang, S.K., Davies, M.V., Kaufman, R.J., and Wimmer, E., 1989, *J. Virol.* 63:1651-1660.

Kidd, S., Kelly, M.R., and Young, M.W., 1986, *Mol. Cell. Biol.* 6:3094-3108.

Kleppe, K., van de Sande, J.H., and Khorana, H.G. 1970. *Proc. Natl. Acad. Sci. USA* 67:68-72.

Konarska, M., Filipowicz, W., Domdey, H., and Gross, H.J., 1981, *Eur. J. Biochem.* 114:221-227.

Kornfeld, K., Saint, R.B., Beachy, P.A., Harte, H.J., Peattie, D.A., and Hogness,D.S.,1989, *Genes & Dev.* 3:243-258.

Kozak, M., 1979, *Nature* 280:82-85.

Kozak, M., 1980, *Cell* 22:459-467.

Kozak, M., 1989a, *J. Cell. Biol.* 108:229-241.

Kozak, M., 1989b, *Mol. Cell. Biol.* 9:5134-5142.

Kozak, M., 1991, *J. Cell. Biol.* 115:887-903.

Laughon, A., and Scott, M.P.,1984, *Nature* 310: 25-31.

Laughon, A., Boulet, A.M., Bermingham, J.R., Laymon, R.A., and Scott, M.P.,1986, *Mol. Cell.Biol.* 6:4676-4689.

Lee, A.S., Delegeane, A., and Scharff, D., 1981, *Proc. Natl. Acad. Sci. USA* 78:4922-4925.

Lemotte, P.K., Kuroiwa, A., Fessler, L.I., and Gehring, W.J.,1989, *EMBO J.* 8:219-227.

Macejak, D.G., and Sarnow, P., 1991, *Nature* 353:90-94.

Mlodzik, M., and Gehring, W.J.,1987, *Cell* 48:465-478.

Mlodzik, M., Fjose, A., and Gehring, W.J.,1988, *EMBO J.* 7: 2569-2578.

Moore, M.J., and Sharp, P.A., 1992, *Science* 256:992-997.

OH, S.-K., Scott, M.P., and Sarnow, P., 1992, *Genes & Dev.* 6:1643-1653.

Pelletier, J., and Sonenberg, N., 1985, *Cell* 40:515-526.

Pelletier, J., and Sonenberg, N., 1988, *Nature* 334:320-325.

Ray, B.K., Lawson, T.G., Kramer, J.C., Cladaras, M.H., Grifo, J.A., Abramson, R.D., Merrick, W.C., and Thach, R.E., 1985, *J. Biol. Chem.* 260:7651-7658.

Regulski, M., McGinnis N., Chadwick, R., and McGinnis, W.,1987, *EMBO J.* 6:767-777.

Romaniuk, P.J., and Uhlenbeck, O.C.1983. *Methods Enzymol.* 100:52.

Rosenberg, U.B., Schröder, C., Preiss, A., Kienlin, A., Côte, S., Riede, I., and Jäckle, H.,1986, *Nature* 319:336-339.

Rozen, F., Edery, I., Meerovitch, K., Dever, T.E., Merrick, W.C., and Sonenberg, N., 1990, *Mol. Cell. Biol.* 10:1134-1144.

Sarnow, P., 1989, *Proc. Natl. Acad. Sci. USA* 86:5795-5799.

Schneuwly, S., Kuroiwa, A., Baumgartner, P., and Gehring, W.J.,1986, *EMBO J.* 5:733-739.

Shatkin, A.J., 1976, *Cell* 9:654-653.

Stroeher, V.L., Jorgensen, E.M., and Garber, R.L.,1986, *Mol. Cell. Biol.* 6:4667-4675.

Tahara, S.M., Morgan, M.A., and Shatkin, A.J., 1981, *J. Biol. Chem.* 256:7691-7694.

Tautz, D., Lehmann, R., Schnürch, H., Schuh, R., Seifert, E., Kienlin, A., Jones, K., and Jäckle, H.,1987, *Nature* 27:383-389.

Ting, J., and Lee, A.S., 1988, *DNA* 7:275-286.

Toyoda, H., Kohara, M., Kataoka, Y., Suganuma, T., Omata, T., Imura, N., and Nomoto, A., 1984, *J. Mol. Biol.* 174:561-585.

Wang, C., and Lehmann R.,1991, *Cell* 66:637-647.

NOVEL STRUCTURAL AND FUNCTIONAL ASPECTS OF
TRANSLATIONAL INITIATION FACTOR IF2

Roberto Spurio,[1] Manuela Severini,[1,2] Anna La Teana,[1,2]
Maria A. Canonaco,[2] Roman T. Pawlik,[2] Claudio O. Gualerzi,[1]
and Cynthia L. Pon[1]

[1]Dipartimento di Biologia MCA
 Università di Camerino
 62032 Camerino (MC) Italy
[2]Max-Planck-Institut für Molekulare Genetik
 Ihnestraße 73
 W-1000 Berlin 33, Germany

IF2 is the largest of the proteins involved in translation interacting directly with the ribosome and, along with the elongation factors EF-Tu and EF-G, belongs to the growing family of the GTP/GDP binding proteins which are involved in a large number of cellular regulatory functions. From the structural point of view, the best characterized examples of this group of proteins are H-ras p21 oncogene protein and the above-mentioned EF-Tu for which a refined crystal structure is available at 1.35 Å (Pai et al., 1990) and at 2.6 Å (Clark et al., 1990) resolution, respectively. The primary structure of IF2 revealed large sequence homologies with EF-Tu; however, these are restricted to the GTP/GDP binding motifs (Sacerdot et al., 1984) situated in the middle and in the N-terminal portion of the molecule of IF2 and EF-Tu, respectively.

Function of IF2 in Protein Synthesis

IF2 is an essential protein in *Escherichia coli* since reduction of the expression of its structural gene below 50% results in increasingly severe reduction of the growth rate (Cole et al., 1987). The function of IF2 which has been recognized for a long time and probably accounts for the main (but likely not the only) activity of the factor in vivo is the stimulation of the binding of fMet-tRNA to ribosomes during formation of the 30S and 70S initiation complexes. The available evidence indicates that IF2 binds to 30S ribosomal subunits in the presence of GTP and, probably through an effect on the ribosomal subunit, accelerates the formation of codon-anticodon interaction at the P-site and slows down the dissociation of the ribosome-bound aminoacyl-tRNA. Thus, overall, IF2 favors the formation over the dissociation of the 30S initiation complexes. Furthermore, since in this activity IF2 makes use of its specific and fairly strong affinity for the initiator tRNA, which is recognized through its blocked α-amino group, the function of IF2 at this stage is also that of enhancing the selectivity of the initiation process by favoring the binding of the initiator tRNA over spurious non-initiation aminoacyl-tRNA binding events (for a review see Gualerzi et al., 1986; Gualerzi and Pon, 1990). In a subsequent step, IF2 stimulates the joining of the 50S subunit with the 30S initiation complex (Godefroy-Colburn et al., 1975) in the process which leads to the formation of a 70S initiation complex and, in turn, stimulates IF2 to hydrolyze GTP to GDP and Pi. Formation of the first peptide bond between fMet and the amino acid encoded by the triplet present at the A-site either precedes or follows the release of IF2 from the 70S ribosome.

From this short outline, it appears clear that IF2 is involved in at least four functional interactions (with GTP, 30S, 50S and fMet-tRNA) and it has been shown that the factor can indeed bind each of these ligands, albeit with different affinities (Petersen et al., 1979; Weiel and Hershey, 1982; Pon et al., 1985). Furthermore, IF2 is involved in a ribosome-dependent GTPase activity and, as it will be shown later, the catalytic center has also been localized within the factor molecule; many questions concerning the functional significance of this activity still remain unanswered, however.

Unfortunately, due to its large size and to persistent failures in obtaining workable crystals of IF2, neither NMR spectroscopy nor X-ray crystallography could be applied to this molecule. Consequently, the only relevant information on the 3D structure of this factor is that concerning the architecture of its GTP binding site which, in light of the above-mentioned extensive sequence homology, can be assumed to be nearly identical to that determined for EF-Tu (see Fig. 2B). Thus, amongst the several open questions concerning IF2, the most pressing concern the structural basis sustaining its function, the nature of its active sites and a more precise definition of possible additional activities carried out by this protein. In fact, if the role of IF2 were only that of promoting the binding of fMet-tRNA to the ribosomal P-site, a function homologous to that displayed by EF-Tu in the binding of elongator aminoacyl-tRNAs to the ribosomal A-site, it would be difficult to understand why a protein with twice the size of EF-Tu should be required for this function. Indeed, several lines of evidence suggest that the function of IF2 in initiation is more complex than that originally thought; in fact, it has been proposed that IF2 promotes an adjustment (sliding) of the mRNA on the 30S subunit in preparation for the binding of initiator tRNA to the P-site (Canonaco et al., 1989) and here we present some evidence indicating that IF2 also plays a role in a codon-dependent and GTPase-dependent adjustment of the initiator tRNA in the P-site (i.e. in a site where the formyl-methionine can react, under the auspices of the ribosomal peptidyltransferase, with the incoming aminoacyl-tRNA or with puromycin).

Finally, the observation that the N-terminal 25-30% of the four IF2 molecules known so far hardly share any homology with each other and are not directly required for any translational function (Gualerzi and Pon, 1990) seems to strengthen the idea that IF2 may also play additional roles in cellular functions other than translation, such as transcription (Travers et al., 1980) or protein secretion (Shiba et al., 1986).

Interaction of IF2 with fMet-tRNA

The formation of a complex between IF2 and fMet-tRNA was recognized at an early stage (e.g. see Rudland et al., 1971); taking advantage of the selective protection from spontaneous hydrolysis of fMet-tRNA in the presence of IF2, Petersen et al. (1979) determined the apparent association constant of this complex (10^6 M^{-1} under physiological conditions) and, in contrast to what occurs in the interaction of EF-Tu with aminoacyl-tRNAs, found it to be unaffected by GTP. An additional difference with EF-Tu is that most likely IF2 does not function as an aminoacyl-tRNA carrier but makes use of its affinity for fMet-tRNA only at the ribosomal level (Gualerzi and Wintermeyer, 1986; Canonaco et al., 1986).

The sites of the initiator tRNA in contact with IF2 or structurally affected by interaction with the factor have been studied by protection from chemical modification and enzymatic digestion (Petersen et al., 1981; Wakao et al., 1989). They were found primarily within the 3'-half of the cloverleaf mainly in the T loop and the minor groove of the T stem.

For what concerns the site of IF2 interacting with initiator tRNA, only recently some information has been obtained, and this will be discussed below.

Purification of fMet-tRNA by Affinity Chromatography on Matrix-bound IF2

The selective affinity of IF2 for fMet-tRNA and the greater stability of thermophilic IF2 and of its complex with the initiator tRNA have been exploited to develop an affinity chromatography procedure which allows the one-step purification of fMet-tRNA from a mixture of tRNAs or aminoacyl-tRNAs (Canonaco et al., in prep.). In Fig. 1, we present the elution profile of fMet-tRNA^{Metf} and Met-tRNA^{Metf} from a matrix-bound IF2 column. As seen in the figure, the formylated form of initiator tRNA is eluted at 2 M NH$_4$Cl, while the non-formylated form is eluted from the column at low ionic strength. Elongation aminoacyl-tRNAs are eluted at the same position as Met-tRNA; upon blocking the α-amino group, as in NAcPhe-tRNA, however, the elution profile of these non-initiator tRNAs becomes similar to that of fMet-tRNA. This behavior provides further evidence for the importance of a blocked α-amino group for the recognition by IF2. In our laboratory, we found this purification procedure particularly useful to deplete selectively fMet-tRNA from postribosomal supernatants and for the purification of the minute amounts of fMet-[^{32}P]tRNA obtainable from extracts of [^{32}P]labelled cells.

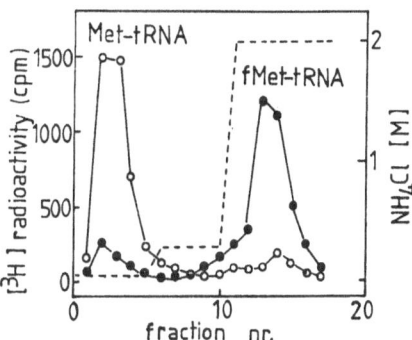

Figure 1. Chromatographic elution of fMet-tRNA^{Metf} and Met-tRNA^{Metf} from a Bio-Gel P300 column coupled with *Bacillus stearothermophilus* IF2 by reaction with glutaraldehyde. The experimental details will be given in a forthcoming paper (Canonaco et al., in prep.). (o) Met-tRNA; (•) fMet-tRNA; (----) NH$_4$Cl concentration.

Ligand Protection Experiments and Identification of Natural Domains of *Bacillus stearothermophilus* IF2

In an earlier work, GTP and GDP were found to protect *E. coli* IF2 against digestion with three different proteases (Pon et al., 1985). These experiments had also shown that the partial proteolysis of IF2 with the same enzymes yielded discrete fragments of similar size suggesting the existence in this protein of structurally compact domains and the possible occurrence of ligand-triggered conformational changes. The existence of these conformational changes has also been suggested by small angle neutron diffraction studies carried out on IF2 in the presence and absence of GTP (Cannistraro et al., 1987).

Work along these lines was further pursued as soon as large amounts of the more stable factor from *Bacillus stearothermophilus* became available (Brombach et al., 1986) and limited digestion with trypsin of purified IF2 in the presence and absence of GTP (Severini et al., 1990) and of fMet-tRNA (Severini et al., 1992) allowed us to identify the main sites of cleavage as well as the peptide bonds protected by these two ligands (Fig. 2A).

As seen from the figure, the proteolysis proceeds with cleavage at positions A,B,C and D (i.e. after K146, K154, R519 and R308, respectively), while the N-terminal portion of the molecule is degraded rapidly yielding relatively small peptides. Thus, the proteolysis generates a 24.5 kD C-terminal peptide which remains absolutely resistant to further digestion with trypsin and also with other proteases and a 40 kD central peptide which is cleaved at D to yield two smaller fragments (17kD and 23kD) displaying intermediate stability. The cleavage at this point is prevented, however, by the presence of either GTP or fMet-tRNA. The

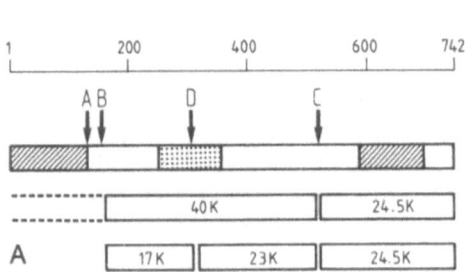 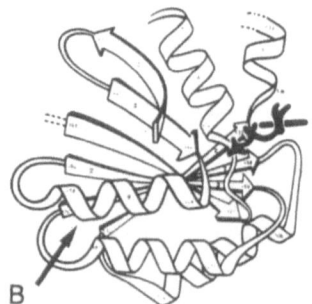

Figure 2. (A) Schematic representation of *B. stearothermophilus* IF2 indicating the main trypsin cleavage sites within the molecule, the sequential order of their appearance (in alphabetical order) and the sizes of the main digestion products. The dotted area indicates the GTP binding domain, the striped area on the left indicates the region of the molecule deleted in IF2ΔN, the striped area on the right indicates the Cla-core of the C-domain which is hyperproduced by pXP401Cla and which has been deleted in IF2ΔCla. Further details are given in the text. The figure is taken from Severini et al. (1990). (B) Model of the 3D structure of *E. coli* IF2 G-domain based on the crystal structure of EF-Tu. The arrow indicates the site of cleavage D (see Fig. 2A) which is strongly protected by GTP or by fMet-tRNA. (Modified from Cenatiempo et al., 1987).

initiator tRNA also weakly protects IF2 from cleavage at two of the other above-mentioned sites (A and B). Since the fMet-tRNA binding site of IF2 is contained within the carboxyl terminal domain of the protein (see below), the protection provided by the initiator tRNA should be attributed to ligand-induced conformational changes; another explanation could be that the protected sites are located close to the factor-bound fMet-tRNA in the tertiary structure of the protein. The cleavage site (R308) protected by GTP, on the other hand, appears to be located at a considerable distance from the bound GTP (Fig. 2B) so that it is likely that the protection is due to a local conformational change of IF2 induced by this ligand. The limited proteolysis experiments indicated that, at least from the structural point of view, IF2 is constituted by three domains: a very labile N-terminal domain, a very stable C-terminal domain and a central domain (G-domain) which remains structurally stable only in the presence of GTP or GDP.

Genetic Manipulations of *Bacillus stearothermophilus infB* and Hyperproduction of IF2 Domains and Fragments

To determine whether specific functions could be attributed to the above-mentioned domains and to have the possibility to perform more detailed structural analyses, we manipulated *B. stearothermophilus infB* to obtain expression vectors capable of hyperproducing these natural domains of IF2.

B. stearothermophilus IF2 was overproduced in *E. coli* K12ΔH1Δtrp cells harboring pIM401 (Brombach et al., 1986) that had been constructed by placing the *B. stearothermophilus infB* gene behind the lambda pL promoter in pPlc2833. *E. coli* IF2 was likewise overproduced in the same strain harboring pXP101. Construction of the expression vector pXP401G for hyperproduction of the G-domain was carried out as described (Gualerzi et al., 1991). Construction of the vector expressing the C-domain (pXP401C) was carried out by excising from pIM401 the 700bp PstI-PstI DNA fragment corresponding to the 3'-end of *infB* and cloning it into the PstI site in the polylinker of pTZ19R. A clone with the insert in the appropriate orientation was digested with EcoRI and HindIII to obtain a 740 bp fragment which was then placed into pEV1 (Crowl et al., 1985) behind the lambda pL promoter. Upon induction, this expression vector overproduces a polypeptide corresponding to the C-fragment with the following additional amino acids encoded by nucleotides from the plasmid polylinkers included in the N-terminal portion of the C-domain:

```
5'TTAATATGAATGAATTCGAGCTCGGTACCCGGGGATCCTCTAGAGTCGACCTGCAGGAGCAG-
       M   N   E   F   E   L   G   T   R   G   S   S   R   V   D   L   Q   E   Q -
```

Construction of deletion mutant pIM401ΔCla was obtained straigthforwardly by removal from pIM401 of the ~0.4 kb ClaI-ClaI fragment located in the distal portion of *infB*. Following religation, one obtains a vector which hyperproduces IF2 molecules (IF2ΔCla) carrying a large deletion (134 residues) in the central portion of the C-domain. In turn, the deleted peptide was hyperproduced from pXP401Cla which was constructed by inserting the ClaI-ClaI fragment into the ClaI site of pEV1 and then adding a SMURFT linker (Pharmacia) at the HindIII site of pEV1 after ascertaining the appropriate orientation of the inserted fragment. The polypeptide (Cla-core of the C-domain) hyperproduced from this construct contained the additional amino acids (indicated by lower case letters) at the ends:

```
5'-AATTAATATGAATGAATTCGGATCCATCGATGTC--GAAATCGATAAGCTCTAGCTAG-3'
       m   n   e   f   g   s   I   D   V --   E   I   D   k   l   *
```

Functional Properties of the IF2 Domains

The above-mentioned C-domain, G-domain, Cla-core of the C-domain as well as IF2 molecules with either an extensive N-terminal deletion (IF2ΔN) or a large deletion (134 aa residues) corresponding to the Cla-core of the C-domain (IF2ΔCla) (Fig. 2A) were hyperproduced in *E. coli*, purified and tested for their activity in various partial reactions of the translation initiation pathway.

The results show that IF2ΔN displays near-normal activity in all partial reactions of initiation. This finding is in agreement with the notion that a similarly truncated form of *E. coli* IF2 can sustain bacterial growth in *infB* null mutations when supplied in excess *in trans* (Laalami et al., 1991). In some experiments our IF2ΔN mutant showed, in comparison with the intact molecule, a reduced retention by DNA cellulose suggesting that the N-terminal peptide may favor or strengthen an interaction of IF2 with nucleic acids.

The G-domain was found to be totally inactive in stimulating protein synthesis and in the ribosomal binding of fMet-tRNA but displayed substantial activity in several partial reactions. Thus, the G-domain was shown to bind GTP retaining a fairly compact configuration resistant to proteolytic cleavage in the presence of this ligand. In addition to an active GTP/GDP binding site, the G-domain retains an intact site for the interaction with the 50S ribosomal subunit and at least part of the site for the interaction with the 30S ribosomal subunit. Interaction of the G-domain with ribosomes in the presence of GTP is also capable of activating the GTPase activity of IF2. The GTPase activity of the G-domain and that catalyzed by the native molecule were shown to have the same Vmax and nearly the same Km for the substrate (Fig. 3A).

In the absence of ribosomes, the GTPase activity of IF2 is negligible, under normal conditions. As is the case of EF-Tu in the presence of kirromycin, and of EF-G in the presence of aliphatic alcohols, however, also IF2 can be activated to hydrolyze GTP in the absence of ribosomes provided that 20% ethanol is added (Fig. 3B). This finding demonstrates that, as with the other ribosomal GTP-binding proteins, the ribosome functions as an effector of the IF2 GTPase activity whose catalytic center is located within the factor (Severini et al., 1991). The question then arises as to whether both subunits are required to stimulate the GTPase activity and, if this is the case, what is their function in triggering this effect. Earlier studies had implicated the 50S subunit assigning a fundamental role to proteins L7/L12 in the activation of the GTPase of IF2 (as well as of EF-Tu and EF-G) but the presence of 30S had always been found to be necessary (Parmeggiani and Sander, 1981).

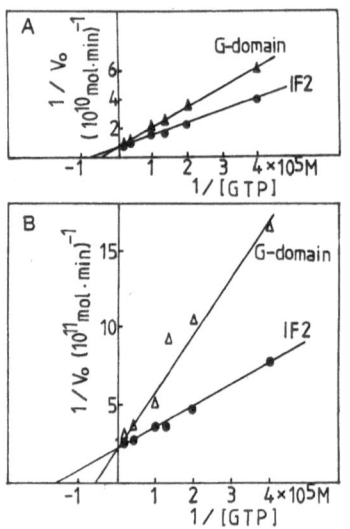

Figure 3. Lineweaver-Burk plot of the initial velocity of IF2- and G-domain-dependent GTPase. (A) ribosome-dependent. (B) ribosome-independent in the presence of 20% ethanol. The figures A and B are taken from Gualerzi et al. (1991) and Severini et al. (1991), respectively, where further experimental details can be found.

In the experiment shown in Fig. 4, we took advantage of the greater stability of the interactions between IF2 and ribosomal subunits derived from a thermophile and determined the GTPase activity of IF2 in the presence of the subunits taken individually or in combination. The results demonstrate that, although maximum stimulation of the IF2 GTPase is obtained in the presence of both subunits, substantial activity can also be obtained in the presence of 50S subunits alone, while negligible activity over the background represented by the ribosome-independent GTPase activity of IF2 is observed in the presence of the 30S alone. This finding indicates that the IF2-50S interaction is both necessary and sufficient to trigger the activation of the catalytic center and suggests that the function of the 30S subunit is either that of holding the factor in a favorable orientation or, more likely, that of stabilizing the weak interaction of IF2 with the large subunit. It should be noted, in this connection, that studies carried out with *E. coli* IF2 have shown that the affinity of IF2 for the 70S monomer is approximately one order of magnitude greater than that displayed for the 50S subunit (Pon et al., 1985).

Like native IF2, also the G-domain of the factor can be activated by alcohol to hydrolyze GTP in the absence of ribosomes. Also in this case, the reaction catalyzed by the G-domain and that catalyzed by the native factor were shown to have similar kinetic parameters (Fig. 3B). This finding demonstrates that not only the GTP binding site but also the GTPase catalytic center of IF2 is localized in its G-domain.

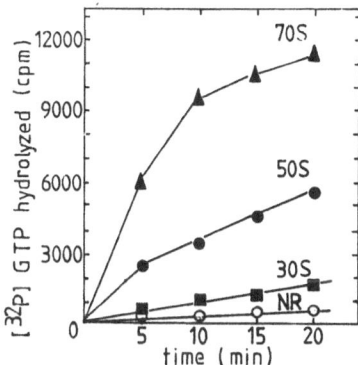

Figure 4. Time course of IF2-dependent GTP hydrolysis in the presence of 70S ribosomes (Δ), 50S subunits (•), 30S subunits (■), no ribosomes (o). The reaction conditions are essentially the same as reported in Gualerzi et al. (1991).

Unlike the G-domain, which proved to be completely inactive in this function, the C-domain was found to bind specifically fMet-tRNA and to protect it from Tris-base catalyzed hydrolysis albeit with a somewhat (5-10-fold) reduced efficiency when compared to the native molecule (see Fig. 6); this suggests that some additional bonds stabilizing the interaction might be lost upon molecular dissection of the factor.

In spite of its total lack of affinity for the ribosomes, the C-domain was also found to provide a weak stimulation of the binding of fMet-tRNA or NAcPhe-tRNA to 30S and 70S ribosomes programmed with random poly(AUG) or poly(U), respectively. If the consequences of the interaction with the C-domain on the conformation of fMet-tRNA are similar to those brought about by the interaction with native IF2 (Wakao et al., 1989), this stimulation could be attributed to the acquisition of a more favorable conformation of the anticodon stem and loop of the initiator tRNA. No stimulation of fMet-tRNA binding was observed in the presence of MS2 RNA or other mRNAs, however, and the fMet-tRNA bound in the presence of the C-domain to 70S ribosomes programmed with poly(AUG) was not reactive with puromycin. These findings underline a requirement for native IF2 in binding the initiator tRNA in response to natural mRNAs and for its subsequent placement in the ribosomal P-site (see below).

Narrowing Down the fMet-tRNA Binding Site of IF2

In full agreement with the conclusion that the C-domain contains the fMet-tRNA binding site of IF2, deletion of the Cla-core of the C-domain yielded an IF2 molecule which proved to be not only totally inactive in the stimulation of initiator tRNA binding to ribosomes (Fig. 5A), but also in the formation of a binary complex with fMet-tRNA, at least as judged from its incapacity to protect it from hydrolysis (Fig. 5B). Conversely, the deleted peptide (i.e. the Cla-core of the C-domain) displayed substantial activity in protecting the initiator tRNA (Fig. 5C). In light of the low solubility of this fragment, its activity in binding initiator tRNA might also be somewhat underestimated from the results shown in the figure. Taken together, these findings suggest that all or most of the structural elements constituting the fMet-tRNA binding site of IF2 are localized within the ~14 kD peptide constituting the Cla-core of the C-domain. This premise is further supported by the analysis of a C-domain mutant displaying a reduced activity in fMet-tRNA binding (see below).

In the case of EF-Tu, very little is known concerning the site where the aminoacyl-tRNA is bound within the EF-Tu-GTP-aminoacyl-tRNA ternary complex. The available evidence indicates, however, that two domains of the factor (e.g. the GTP binding domain I and domain II) are implicated in the interaction since several residues belonging to either domain display protection from chemical modification (i.e. His66 and 118 and Lys2, 4, 237, 248, 263 and 282) or an enhanced reactivity (i.e. Lys208 and 390). Similarly, Lys 208 and 237 were crosslinked to a 3'-oxidized tRNA in the presence of kirromycin or ribosomes (Jonak et al., 1984; Antonsson and Lebermann, 1984; van Noort et al., 1989). With the exception of Lys390, no residue belonging to domain III, which probably acts as a hinge between domain I and II, was found to be affected by aminoacyl-tRNA.

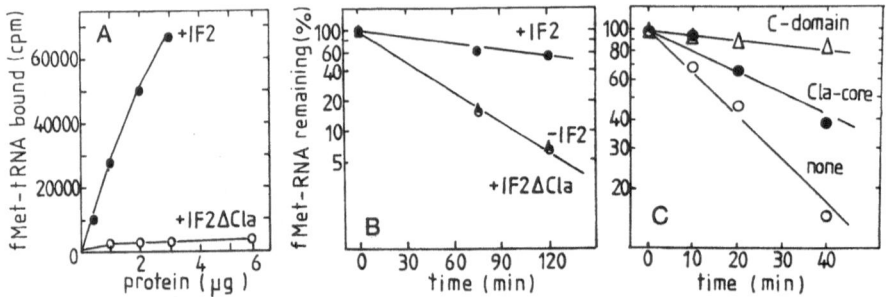

Figure 5. Lack of activity of IF2ΔCla in (A) stimulation of the ribosomal binding of fMet-tRNA in the presence of random poly(AUG) and (B) protection of fMet-tRNA from spontaneous hydrolysis. (C) Activity of the C-domain and of the Cla-core of the C-domain in the protection of fMet-tRNA from spontaneous hydrolysis. Further details will be given elsewhere.

In the case of IF2, preliminary chemical modification experiments carried out by Dr. T. Suryanarayama had suggested the involvement of a His residue in the interaction between the C-domain and fMet-tRNA. In light of the above-mentioned role played by histidine in the EF-Tu-aminoacyl-tRNA interaction, this finding suggested the possible existence of an homology between IF2 and EF-Tu in their mechanism of interaction with aminoacyl-tRNA. To test this hypothesis, we replaced the two histidine residues present in the C-domain of IF2 and tested the resulting mutants for their capacity to interact with fMet-tRNA. Thus, oligonucleotide-directed mutagenesis was performed according to the altered sites mutagenesis system (Promega) to generate four mutants [pXP401C(H575→D), pXP401C(H575→Y), pXP401C(H621→D), pXP401C(H621→Y)]. After screening for the desired DNA sequences, the four mutant proteins were overproduced as described for pXP401C, purified and tested

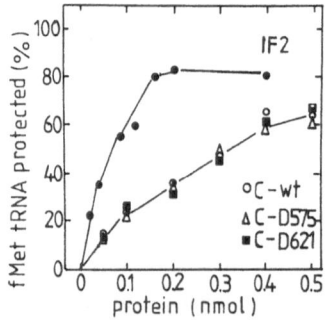

Figure 6. Protection of fMet-tRNA from spontaneous hydrolysis by increasing amounts of IF2, wild type C-domain and two histidine mutants of the C-domain generated by site-directed mutagenesis as described in the text. The experimental conditions are as described in Gualerzi et al., 1991.

in the fMet-tRNA binding assay. The results obtained with the two H→D mutants are presented in Fig. 6; they are basically identical to those obtained with the H→Y mutants. As seen from the figure, none of the mutants displayed any reduction of activity when compared to wild type C-domain indicating that neither one of the two histidines is directly involved in the interaction with fMet-tRNA.

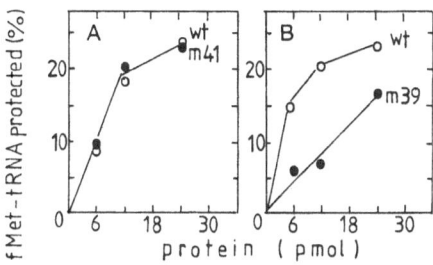

Figure 7. Protection of fMet-tRNA from spontaneous hydrolysis by increasing amounts of wild type C-domain and by (A) mutant 41 and (B) mutant 39 of the C-domain. See text for further details.

An alternative approach towards identifying the amino acid residues involved in the interaction with initiator tRNA was to subject the DNA fragment encoding the C-domain to random mutagenesis with hydroxylamine and to screen for clones hyperproducing proteins with phenotypic alterations of their fMet-tRNA binding capacity. This approach yielded a large number of isolates, most of them displaying wild type fMet-tRNA binding capacity (e.g. mutant 41 in Fig. 7A). In at least one case (mutant 39 in Fig. 7B), however, we found a protein with a clearly reduced capacity to interact with initiator tRNA. DNA sequence determination of the corresponding clone revealed that two G→A transitions had occurred in this mutant resulting in the substitutions of Ala594→Thr and Val601→Ile. Even though the second replacement is conservative by nature and is not expected to have particularly drastic effects, we do not yet know whether or not both amino acid changes are necessary to produce the defective phenotype.

In Fig. 8, we present a comparison of the amino acid sequence of the C-domain of *B. stearothermophilus* with those of three other known corresponding sequences. The figure highlights the homologies and the two arrows mark the limits of the Cla-core; as seen in the figure, the two non-essential histidine residues (encircled) and the two residues changed in mutant 39 (enclosed in a triangle) are within this Cla-core.

A Function of the IF2 GTPase in Protein Synthesis

To determine the biological function of the ribosome-dependent GTPase activity of IF2 is an important goal of our research. At least three not mutually exclusive functions can be envisaged for this catalytic activity: the adjustment of fMet-tRNA in the P-site, the ejection of IF2 from the ribosome and the participation in a kinetic proofreading mechanism aimed at increasing the fidelity of initiation. Concerning the first function, clear evidence that 70S-bound fMet-tRNA is not puromycin-reactive if GTP is replaced by GMP-PCP was obtained many years ago (Thach and Thach, 1971; Mazumder 1973), but this was generally attributed to a failure of the system to release IF2 in the absence of GTP hydrolysis. The persistent presence of IF2 on the ribosome would hinder the peptidyl-transferase activity as well as inhibit the binding of the EF-Tu-aminoacyl-tRNA-GTP ternary complex, alledgedly by use of overlapping ribosomal binding sites by the two factors (Benne et al., 1973). More recently, however, it has been shown that IF2 binds to the 70S ribosomes with approximately the same affinity ($Ka \cong 3 \times 10^6$ M^{-1}) in the presence of either GDP or GTP or in the absence of both nucleotides. This finding indicates that a "GDP conformation" of IF2 incompatible with the binding to 70S does not exist. Consequently, the ejection of IF2 may simply result from a reduction of its affinity for the ribosome caused by subunit association. In fact, this process may loosen the IF2-30S interaction both directly and indirectly, through the ejection of IF1 and IF3 which contribute a great deal to the stability of the IF2-30S complex. Binding of IF2-GTP to free 30S subunits to form a complex having a greater association constant would also contribute to shifting the equilibrium towards the dissociation of the 70S-IF2 complexes (Weiel and Hershey, 1982; Pon et al., 1985).

In light of that discussed above, the possibility that the adjustment of the initiator tRNA in the P-site may require *per se* IF2-dependent GTP hydrolysis seems to be a reasonable hypothesis; this was tested by looking at the effect of GTP on the formation of fMet-puromycin and of the dipeptide fMet-Phe directed by random poly(AUG) or a natural mRNA. For these experiments, we used ribosomal subunits and IF2 from both *E. coli* and *B. stearothermophilus* in the two homologous systems as well in a series of heterologous combinations. Preliminarily, it was shown that almost 100% of the fMet-tRNA bound to the 30S subunit programmed with either mRNA in the presence of all factors becomes puromycin-reactive upon addition of 50S subunits. Our results also showed that the puromycin reaction is almost completely dependent on the presence of IF2 at low (5-7 mM) $[Mg^{2+}]$ and remains strongly dependent on the factor even at much higher (18 mM) $[Mg^{2+}]$. Furthermore, it has been shown that the binding of fMet-tRNA to *B. stearothermophilus* 30S subunits is virtually unaffected by temperature changes between 10°C and 40°C in the complete system as well as under conditions lacking GTP, mRNA or both GTP and mRNA (Fig. 9A); the puromycin reactivity of the bound fMet-tRNA, on the other hand, is strongly influenced by

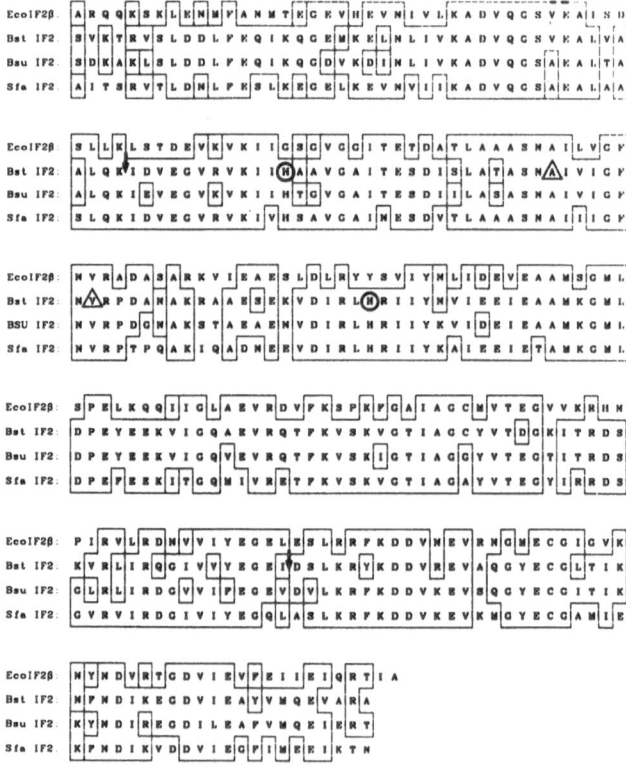

Figure 8. Sequence comparison of the C-domains of *E. coli* (Sacerdot et al., 1984), *B. stearothermophilus* (Brombach et al., 1986), *B. subtilis* (Shazand et al., 1990) and *Streptococcus faecium* (Friedrich et al., 1988) IF2. Identical residues are boxed. For the other symbols, see text.

variations of these parameters since it displays strong temperature-dependence and does not occur at all if both GTP and mRNA are simultaneously omitted. Compared to the complete system, fMet-puromycin formation occurs to a substantial (yet much lower) extent in the absence of GTP or of mRNA (Fig. 9B). The latter activity is characteristic of the *B. stearothermophilus* system since in that of *E. coli*, fMet-puromycin formation is almost completely abolished in the absence of the template. By use of heterologous mixtures and of *E. coli / B. stearothermophilus* chimeric 70S ribosomes it has been shown that the thermophilic ribosomes (primarily the large subunit) are responsible for this activity (La Teana et al., manuscript in preparation).

The kinetics of fMet-puromycin formation has been measured at various temperatures in the complete system as well as in the absence of GTP or of mRNA or in the presence of GDP in place of GTP. The results indicate that while fMet-puromycin formation is much faster in the presence than in the absence of GTP, the reactions have nearly the same Arrhenius activation energy. On the contrary, a much higher activation energy is required in the absence of mRNA or in the presence of GDP. Finally, the comparison of the Arrhenius plots obtained with homologous and heterologous *E. coli / B. stearothermophilus* systems, display a characteristic break at a critical temperature which depends on the type of ribosome and not on the type of IF2 (La Teana et al., manuscript in preparation). These results, together with those described above concerning the mRNA-independent fMet-puromycin formation, clearly suggest that the ribosome plays an active role, besides that of providing the peptidyl-transferase activity, in the adjustment of fMet-tRNA in the P-site. As to the mechanism of the adjustment and the role played by GTP in this process, our data clearly indicate that the omission of GTP slows down the process in otherwise complete systems. The presence of GTP, on the other hand, becomes essential in placing fMet-tRNA in the P-site when the anticodon stem of the tRNA is not properly oriented by the codon-anticodon interaction. Finally, even though we have not directly determined the fate of IF2 under the various experimental conditions, the differences in the puromycin reactivity of fMet-tRNA and the differences in the extent to which the GTPase activity is required to obtain the transpeptidation are difficult to reconcile with the idea that the adjustment of the initiator tRNA in the P-site might be controlled by the ejection of IF2 from the ribosome.

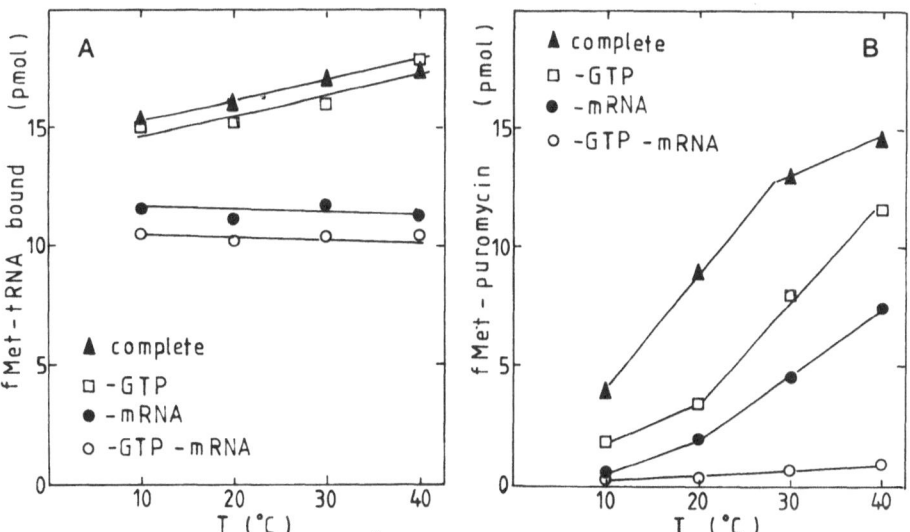

Figure 9. Temperature dependence of the (A) poly(AUG)-dependent binding of fMet-tRNA to 30S ribosomal subunits and (B) fMet-puromycin formation upon addition of 50S subunits in *B. stearothermophilus* systems, either complete or with the indicated omissions (Further details will be given in La Teana et al., manuscript in preparation).

ACKNOWLEDGMENTS

This work was supported in part by grants from the Italian MURST and National Research Council (CNR) Target Projects on Biotechnology and Bio-instrumentation and Genetic Engineering.

REFERENCES

Antonsson, B. and Lebermann, R., 1984, *Eur. J. Biochem.* 141:483.

Benne, R., Ebes, F. and Voorma, H.O., 1973, *Eur. J. Biochem.* 38:265.

Brombach, M., Gualerzi, C.O., Nakamura, Y. and Pon, C.L. 1986, *Mol. Gen. Genet.* 205:97.

Cannistraro, S., Petrillo, C., Sacchetti, F., Calogero, R. and Gualerzi, C.O., 1987, *Chem. Phys. Lett.* 139:116.

Canonaco, M.A., Calogero, R.A. and Gualerzi, C.O., 1986, *FEBS Lett.* 207:198.

Canonaco, M.A., Gualerzi, C.O. and Pon, C.L., 1989, *Eur. J. Biochem.* 182:501.

Cenatiempo, Y., Deville, F., Dondon, J., Grunberg-Manago, M., Sacerdot, C., Hershey, J.W.B., Hansen, H.F., Petersen, H.U., Clark, B.F.C., Kjeldgaard, M., la Cour, T.F.M., Mortensen, K.K, and Nyborg, J., 1987, *Biochemistry* 26:5070.

Clark, B.F.C., Kjeldgaard, M., la Cour, T.F.M., Thirup, S. and Nyborg, J., 1990, *Biochim. Biophys. Acta* 1050:203.

Cole, J.R., Olsson, C.L., Hershey, J.W.B., Grunberg-Manago, M. and Nomura, M., 1987, *J. Mol. Biol.* 198:383.

Crowl, R., Seamans, C., Lomedico, P. and McAndrew, S., 1985, *Gene* 38:31.

Friedrich, K., Brombach, M. and Pon, C.L., 1988, *Mol. Gen. Genet.* 214:595.

Godefroy-Colburn, T., Wolfe, A.D., Dondon, J., Grunberg-Manago, M., Dessen, P. and Pantaloni, D., 1975, *J. Mol. Biol.* 94:461.

Gualerzi, C.O., Pon, C.L., Pawlik, R.T., Canonaco, M.A., Paci, M. and Wintermeyer, W., 1986, in "Structure, Function, and Genetics of Ribosomes," B. Hardesty and G. Kramer, ed., Springer-Verlag, New York.

Gualerzi, C.O. and Pon, C.L., 1990, *Biochemistry* 29:5881

Gualerzi, C.O. and Wintermeyer, W., 1986, *FEBS Lett.* 202:1.

Gualerzi, C.O., Severini, M., Spurio, R., La Teana, A. and Pon, C.L., 1991, *J. Biol. Chem.* 266:16356.

Jonak, J., Petersen, T.E., Meloun, B. and Rychlik, I., 1984, *Eur. J. Biochem.* 144:295.

Laalami, S., Putzer, H., Plumbridge, J.A. and Grunberg-Manago, M., 1991, *J. Mol. Biol.* 220:335.

Mazumder, R., 1973, *Proc. Nat. Acad. Sci. USA* 70:1939.

Pai, E.F., Krengel, U., Petsko, G.A., Goody, R.S. Kabsch, W. and Wittinghofer, A., 1990, *EMBO J.* 9:2351.

Parmeggiani, A. and Sander, G., 1981, *Mol. Cell Biochem.* 35:129

Petersen, H.U., Kruse, T.A., Worm-Leonhard, H., Siboska, G.E., Clark, B.F,C., Boutorin, A.S., Remy, P., Ebel, J.P., Dondon, J. and Grunberg-Manago, M., 1981, *FEBS Lett.* 128:161.

Petersen, H.U., Røll, T., Grunberg-Manago, M. and Clark, B.F.C., 1979, *Biochem. Biophys. Res. Comm.* 91:1068.

Pon, C.L., Paci, M., Pawlik, R.T. and Gualerzi, C.O., 1985, *J. Biol. Chem.* 260:8918.

Rudland, P.S., Whybrow, W.A. and Clark, B.F.C., 1971, *Nature New Biology* 231:76.

Sacerdot, C., Dessen, P., Hershey, J.W.B., Plumbridge, J.A. and Grunberg-Manago, M., 1984, *Proc. Nat. Acad. Sci. U.S.A.* 81:7787.

Severini, M., Choli, T., La Teana, A. and Gualerzi, C.O., 1990, *FEBS Lett.* 276:14.

Severini, M., Choli, T., La Teana, A. and Gualerzi, C.O., 1992, *FEBS Lett.* 297:226.

Severini, M., Spurio, R., La Teana, A., Pon, C.L. and Gualerzi, C.O., 1991, *J. Biol. Chem.* 266:22800.

Shazand, K., Tucker, J., Chiang, R., Stansmore, K., Sperling-Petersen, H.U., Grunberg-Manago, M., Rabinowitz, J.C. and Leighton, T., 1990, *J. Bacteriol.* 172:2675.

Shiba, K., Ito, K., Nakamura, Y., Dondon, J. and Grunberg-Manago, M., 1986, *EMBO J.* 5:3001.

Thach, S.S. and Thach, R.E., 1971, *Proc. Nat. Acad. Sci. USA* 68:1791.

Travers, A.A., Debenham, P.G. and Pongs, O., 1980, *Biochemistry* 19:1651

van Noort, J.M., Kraal, B., Bosch, L., la Cour, T.F.M., Nyborg, J. and Clark, B.F.C., 1989, *Proc. Nat. Acad. Sci. USA* 81:3969.

Wakao, H., Romby, P., Westhop, E., Laalami, S., Grunberg-Manago, M., Ebel, J.P., Ehresmann, C. and Ehresmann, B. 1989, *J. Biol. Chem.* 264:20363.

Weiel, J. and Hershey, J.W.B., 1982, *J. Biol. Chem.* 257:1215

TRANSLATIONAL STOP SIGNALS: EVOLUTION, DECODING FOR PROTEIN SYNTHESIS AND RECODING FOR ALTERNATIVE EVENTS

Warren P. Tate, Frances M.Adamski, Chris M. Brown, Mark E. Dalphin, Jason P. Gray, Jules A. Horsfield, Kim K. McCaughan, John.G. Moffat, Robert J. Powell, Kirsten M. Timms, and Clive N. A. Trotman

Department of Biochemistry and Centre for Gene Research
University of Otago
Dunedin, New Zealand

INTRODUCTION

Termination or translational stopping involves a close relationship between the ribosome, the mRNA, and the polypeptide chain release factors. The discovery of an array of alternative events occurring at stop codons in the mRNA has focussed attention on how the decoding mechanism discriminates between simple 'stop' signals and alternative events outside the normal constraints of the genetic code (Tate and Brown, 1992). This latter phenomenon has been recently called 'recoding' and the signals 'recoding signals' (Gesteland et al., 1992). Is the normal role of the stop codon merely overriden by the recoding signals or does the stop component of such a signal contribute to a finely tuned regulation at sites where the alternative events occur? These questions demand a re-examination of how a stop signal recognition mechanism might have evolved, first for its function in translational stopping, and second for a possible role in gene regulation.

EVOLUTION OF A STOP SIGNAL RECOGNITION MECHANISM

Examination of the genetic codes used by organisms and organelles today suggests that codons were once either assigned as sense for amino acids, or were true nonsense codons with no specific recognition function (Lehman and Jukes, 1988; Osawa et al., 1992). Remnants of these nonsense codons are seen in the AT-rich mycoplasma (Oba et al., 1991) and in GC-rich *Microccocus luteus* (Kano et al., 1991) where certain codons specify neither sense nor stop. One might visualize an early stage in the evolution of the termination of protein synthesis as being a simple dissociation of the protein from the proto-ribosome when a nonsense codon appeared in the appropriate site. In the further evolution of triplet codes certain codons have evolved as specific stop signals (for example, UAA, UAG and UGA of today's universal code) but clearly they had the capacity along with sense codons to further change their roles. In the codes used by mitochondria there is clear evidence of this having occurred; for example a number of species use UGA as a sense codon for tryptophan, while in a few cases the sense codons for arginine are now used as specific stop signals. These concepts are summarised in Figure 1.

The Translational Apparatus, Edited by K.H. Nierhaus
et al., Plenum Press, New York, 1993

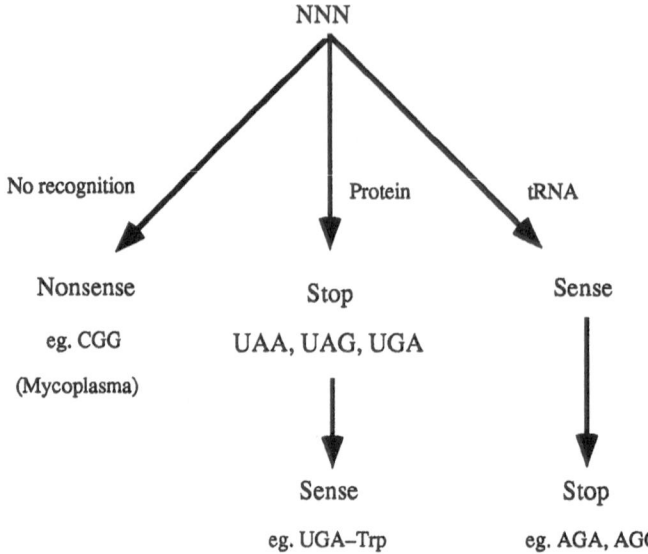

Figure 1. Evolution of signals for 'sense', and for 'stop', from unassigned codons.
NNN refers to any unassigned codon, and the arrows suggests possible routes by which sense, stop or unassigned codons seen today might have arisen. Alternative routes such as the use of UGA for sense first and then a change to stop are also possible.

Indeed if the protoribosome were made of RNA, as is suggested from the recent demonstration that rRNA can catalyse the formation of peptide bonds (Noller et al., 1992), then the evolution of a specific mechanism for termination of protein synthesis involving protein factors was probably a relatively late event. It is now well documented that prokaryotes using the universal code have two such protein factors, each of which recognises two of the three stop codons, whereas eukaryotes have only one factor which can recognise all three.

What is the situation in organisms or organelles that depart from the so-called universal code? We have demonstrated that rat mitochondria have the factor that recognises UAA and UAG only, the equivalent of the bacterial RF-1 (Lee et al., 1987), and they have no factor which recognises the UGA codon, now used for tryptophan. It is interesting that although UAG is an unused codon in rat mitochondria, the mtRF1 factor has retained the ability to recognise this codon as a specific stop signal (at least on *Escherichia coli* ribosomes). The yeast, *Saccharomyces cerevisiae* has two putative mitochondrial release factors (Powell and Tate unpublished). One appears to be equivalent to the mitochondrial RF1 (mtRF1), and its gene has also been serendipitously isolated from a yeast defective in mRNA splicing (Pel et al., 1992). This protein however, also recognises UGA, even though this codon specifies tryptophan. In this manner the specificity of the factor for stop codons is more like a eukaryotic factor, but it has all of the specific characteristics and requirements of a prokaryotic factor and none of a typical eukaryotic factor. This adds to our confidence that we have isolated a yeast mitochondrial RF and not the yeast cytoplasmic factor. The second factor detected in yeast mitochondria albeit with low activity has specificity equivalent to the bacterial RF-2, recognising UAA and UGA. It remains a puzzle why an organelle which now uses UGA as a sense codon and therefore has no specific need for the factor has retained it. These data are summarised in Table 1. The simple *in vitro* assay used to determine codon specificities can detect activity with the triplet codons at high concentrations in the prokaryotic and organelle cases but not the eukaryotic cases where there is an absolute requirement for a minimum of a tetranucleotide.

Table 1. Examples of classes of protein release factors and their determined codon specificities

Release factor	Organism/Organelle	Codon Specificity
RF1	*E. coli, B. subtilis, T. thermophilis*	UAA, UAG
RF2	*E. coli, B. subtilis,*	UAA, UGA
eRF	rabbit, *Artemia*	UAAN, UAGN,UGAN[1]
mtRF1	rat mitochondria	UAA, UAG[2]
mtRF1	*Saccharomyces* mitochondria	UAA, UAG[2], UGA[2]
mtRF2?		UAA,UGA[2]

[1]Release factor requires tetranucleotides for activity.
[2]These codons do not encode stop in mRNAs from these organelles

The isolation and characterisation of a putative eukaryotic release factor cDNA clone (i.e. the rabbit eRF) has given paradoxical information. The sequence showed no homology to the bacterial factors but rather to bacterial tryptophanyl-tRNA synthetases (WRS) (Lee et al., 1990). This raised the intriguing possibility that the protein mediating the specific stop mechanism in eukaryotes might have a dual function in protein synthesis: termination and aminoacylation. From an evolutionary perspective this implied the radical possibility that specific protein-mediated stopping mechanisms might have evolved following the split between eukaryotes and prokaryotes.

The weight of evidence, however now suggest that the proteins eRF and WRS might be different proteins but closely associated in a complex. This may have physiological importance. We have used two different purification schemes to isolate apparently homogeneous eRF (Dalphin and Tate, unpublished results); half of the protein from such an eRF preparation can be removed from solution without significant loss of eRF activity, using the original monoclonal antibody that isolated the putative eRF rabbit clone, and an antimouse IgG linked to Sepharose. We have used the rabbit cDNA to isolate a number of human cDNA clones, and all of these show sequence identity to the human WRS. Purified WRS protein in contrast has been obtained by a different scheme and has no apparent eRF activity (Frolova et al submitted). After expression of the protein from the rabbit cDNA clone, however, we have demonstrated induction of WRS activity. Furthermore human cells treated with interferon induce a protein with WRS activity but with no apparent eRF activity (Rubin et al., 1991).

An unresolved question remains however, in that Lee et al. (1990) were able to measure eRF activity when the cDNA was expressed as protein, although a successful assay required ethanol. There is also indirect evidence for the stimulation of apparent eRF function in reticulocyte lysates following translation of the RNA derived from the human WRS gene (Buwitt et al., 1991). One possible resolution to the paradox would be if the WRS protein were amplifying the function of eRF already present in low amounts on the ribosome added to the assay, by acting as a stabilizing factor. Although the weight of the data lean towards eRF and WRS being distinct polypeptides, the question will not be resolved until a separate gene has been isolated, expressed and shown to have eRF activity and not WRS activity.

THE SIGNAL FOR TERMINATION

With the evolution of protein-directed mechanisms for terminating protein synthesis the process becomes inherently more sophisticated than a simple dropping-off when a nonsense codon is encountered. When the release factor forms a complex with the ribosome and the mRNA it has the potential to make contact with many more than three nucleotides in the mRNA. Thus there is the potential for nucleotides on either side of a triplet stop codon to superimpose additional instructions, particularly in the sequence following which is unconstrained noncoding sequence. The discovery of alternative events at sites of termination codons, such as suppression or recoding signals, suggests there might be a heirarchy of signals of varying strengths. It has long been known that stop codons differ in their suppressibility depending on their context, but distinguishing whether this is a result of context on suppression, on termination, or on both is not a trivial problem.

We approached the question of the nature of the stop signal initially by utilising the burgeoning nucleic acid data bases to analyse the sequences around stop codons of many genes. This followed indications in the literature of context bias from much smaller studies (Kohli and Grosjean, 1981). There is a preferred subset of stop signals used by genes in a number of different organisms, with a significant bias in the nucleotide following the conventional stop codon, and some other less defined biases. For example, in *Escherichia coli* the termination sequences,UAA(U/G), are strongly favoured, particularly in a subset of genes which are highly expressed, whereas in *Saccharomyces cerevisiae* the preferred signals are UAAR (where R=A or G), and although in humans the distinctions are not quite as marked there is a preference for codons followed by a purine, as illustrated in Table 2 (Brown et al., 1990a & 1990b; Tate & Brown, 1992). Organisms with high GC content, such as *Pseudomonas* Spp. and *Streptomyces* Spp., prefer UGAG to UAAU. Thus AT or GC pressure has evidently influenced the evolution of the preferred signal in any particular organism.

Table 2. The preferred stop signals in genes of several organisms

Organism	Signal	Occurrence (%)	
Saccharomyces		Total	Highly Expressed
	UAAG	18	50
	UAAA	20	31
Human			
	UGAG	20	24
	UAAG	8	16
	UAAA	10	15
	UAGG	7	13
Escherichia coli			
	UAAU	28	56
	UAAG	15	32

The occurrence of the preferred stop signals in genes are shown as a percentage of the total signals, and as a percentage of the signals found in highly expressed genes

A model in which 4 bases are important recognition determinants in a stop signal has a number of consequences. Firstly 12 alternative signals can be derived from the three conventional stop codons. This can allow for a range of signal strengths which might have functional significance particularly at sites where there are other recoding signals. Indeed, Pedersen and Curran (1991), found that the selection of RF-1 at UAGN signals in *E.coli* is influenced by the 4th base according to the predictions of our hypothesis; UAGU was the best and UAGA the poorest. It is intriguing that infreqently occurring stop signals at natural termination sites are often found at sites of recoding, where the failure to terminate efficiently is well documented. For example UGAC is found at the recoding site in formate dehydrogenase in *E. coli* where selenocysteine is incorporated (Bock et al., 1991), and at the frameshift site in the RF-2 gene where it fails to signal stop in 30-50% of ribosomal passages (Craigen and Caskey 1986; Donly et al., 1990). We have shown that this stop signal can be strengthened in *trans* by increasing the concentration of the RF-2, which decodes the UGAC signal, or weakened by providing a disabled release factor to compete with the native factor for decoding the signal (Donly et al., 1990). Now, attempts are underway to replace the UGAC with all of the other possible signals and determine whether the competitive ability of the signal at the recoding site can be influenced in *cis*. The RF-2 frameshift site has the potential of an ideal test system to test the strengths of stop signals since distinct recoding signals favouring frameshifting are separated from the UGAC (Weiss et al., 1988).

DECODING OF STOP SIGNALS FOR TERMINATION

Models for the recognition of stop signals have fallen into two classes; those which presume that there is direct physical recognition and decoding of the nucleotides in the mRNA by the protein factor, and those which assume the interaction of the factor with the ribosome is an allosteric consequence of mRNA and rRNA decoding (base pairing). Conceivable scenarios for mechanisms of RF-signal interaction range widely in the relative contributions of the RF and ribosome. At one extreme is a "tRNA analogue model" where RF binding is determined by a specific "protein anticodon", with the rRNA merely providing a scaffold for the recognition event. At the other extreme is a model where the stop signal in the mRNA is base-paired to the rRNA, and provides only one of the many determinants within the RF binding site. In this case there is no necessity for the signal to make actual contact with the RF, subsequent to it contributing to the correct conformation of the binding site.

It should be noted that several scenarios are still possible, since, while there is much indirect evidence, there is as yet only one definitive study which provides evidence for direct recognition of the stop signal by the factor. We have used a defined mRNA with a thio-U as the first nucleotide of the stop signal to form a termination complex with the ribosome and the release factor. After irradiation of the complex at a wavelength that promotes zero length crosslinking from the thio-U residue, we found evidence for a crosslink between the factor and the mRNA, the yield of which was influenced by conditions which affected the stability of the termination complex (Tate et al., 1990). However, further evidence from alternative approaches are needed before models of codon recognition which exclude direct factor stop signal contact can be unequivocally eliminated.

One concern was how the factor could recognise the signal when it seemed to be physically distant from it. We had shown by immunoelectron microscopy that that the primary ribosomal binding site of RF-2 was on the side opposite the decoding site at the interface between the two subunits, although it was not clear how far the factor penetrated into the cavern between the subunits and hence across to the decoding site (Kastner et al., 1990). When the factor was crosslinked in a functional complex to the isolated small subunit however, there were epitopes of the factor on both sides of the subunit, implying the factor can extend across the subunit (Kastner and Tate, unpublished). More recent crosslinking studies from thio-U residues at different positions in mRNA are inconsistent with some features of current models of the three- dimensional arrangement of the 16S rRNA and 30S proteins, and the orientation of the substrates in the decoding site (Brimacombe 1992; Dontsova et al., 1992). It now seems that the decoding site is more towards the centre of the subunit than previously thought, placing it nearer to the major binding site of the release factor, and that helices 18 (530 loop), 34, and the single stranded region between helices 28 and 44 of the 16S rRNA (the 1400 region), must be in close proximity (Brimacombe 1992).

These findings may be crucial to an understanding of the mechanism of codon recognition by the release factor. In the termination complex of the release factor, the ribosome, and mRNA, the major crosslink from the thio-U in the stop codon was to A 1408, part of the single stranded region between helix 28 and 44 of the 16S rRNA (Figure 2A , B) (Tate et al., 1990).When this U was displaced to the position following the stop codon it crosslinked to A1396. Furthermore, an mRNA containing UAAU crosslinked to both 1408 and 1396 (Brown and Tate unpublished-Fig2). Nucleotide, 1408, is protected by A site tRNA, and it is likely that A site decoding occurs near here. The unpaired nucleotides, 1406-1408, UmCA, have the potential to form 2 or 3 base pairs with the three stop codons, UAA, UGA and UAG, and therefore form a base-paired template stacked on helix 44 for factor recognition (see Figure 2). The fourth position crosslinking to 1396 could be explained by the 1400 region forming a single stranded A helix thus placing 1396 near to the fourth position of the stop codon (Figure 2B). It is not clear at this stage how the less specific 4th base of a termination signal would be accommodated. The answer may be that the factor recognises a number of nucleotides in and around the decoding site, perhaps some in a base paired structure and some unpaired. Recognition of the preferred U or G as the 4th base in the stop signal by an RF protein could involve a common structural element in these bases, perhaps the adjacent carbonyl group and ring nitrogen. A base pairing model was proposed by Murgola et al.(1988) involving helix 34 and nucleotides 1199-1204 which could accommodate UGA, but it was a puzzle that this appeared so distant from the decoding site (but not from the RF binding site) in the models of the rRNA (Brimacombe et al., 1988; Stern et al., 1988). The new cross link information placing helix 34 close to the decoding site indicated in Figure 2B(Dontsova et al., 1992), together with experiments showing that perturbations of helix 34 can suppress all three termination codons (Prescott et al., 1991, Prescott and Kornau 1992) indicate, not unexpectedly, that the conformation surrounding the decoding site is important in maintaining fidelity of codon recognition and RF binding. A change in the 530 loop equivalent (at the end of helix 18) in yeast mitochondrial ribosomes (Shen and Fox, 1989) causing UAA suppression was not easily explained before these new data which suggest that this feature is also in the vicinity of the decoding site. Together the picture emerging for codon recognition is of a release factor binding between the subunits controlled by contacts with the termination signal in the mRNA, and with adjacent decoding site rRNA (1440 region, 530 loop, helix 34) and ribosomal proteins (S3, S4, S5, S7, S10, L7/L12, L11).

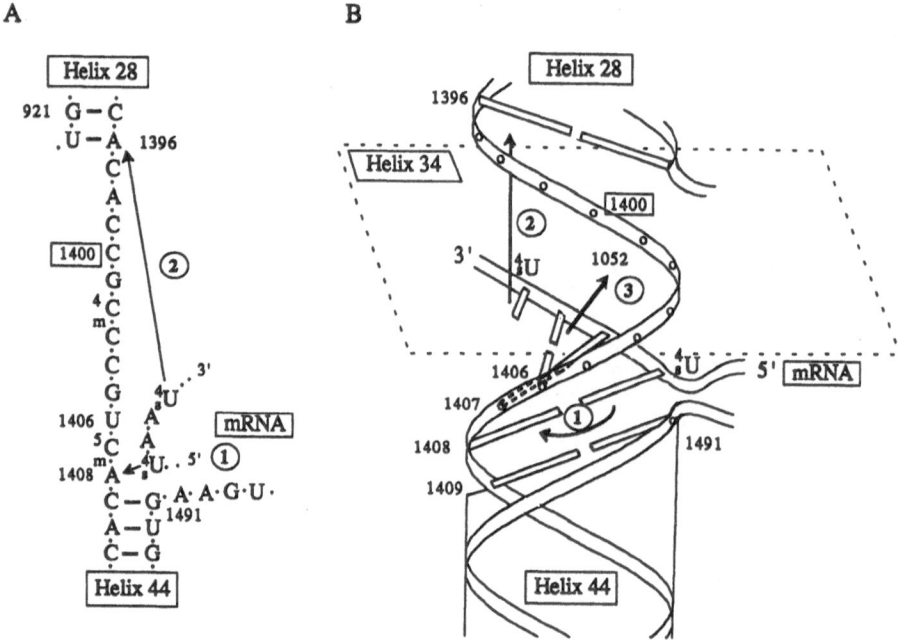

Figure 2. Crosslinking from stop signals to the 16SrRNA
A. Secondary structural model of part of helix 44 and the 1400 region between this helix and helix 28. The 16S rRNA is drawn as Brimacombe (1992). Two crosslinks (arrows) are shown from thio U (4sU) residues in an artificial mRNA, from the first position of the stop signal to A1408 of the 16SrRNA (Tate et al 1990)(1) and from the fourth position to ~1396 (2).
B. A possible tertiary structural model of this region . The base of helix 44 is shown as an A helix , the unpaired 1400 region extends up toward helix 28 as a loose A helical structure. Such looping of the 1396-1408 region would bring A1396 close to the 4th position of the stop signal as indicated by crosslink 2. Potential base pairing between the stop codon and 1406-1408 is shown, these base pairs are shown stacking on helix 44. Also shown is a crosslink (3) between a thio-U in the 3' position of an A site codon to 1052 of helix 34 (dotted), this crosslink indicates that !!.. A site codon is near helix 34, and suggests that the stop signal is also near helix 34.

THE FUNCTIONAL SITES ON THE RELEASE FACTOR

The release factor makes contact with the ribosome at the decoding site, and somehow also affects the peptidyl transferase centre to change it from a bond forming to a hydrolysis mode. This implies there are domains of structure within the release factor molecules with defined function, for example a codon recognition domain and a peptidyl-tRNA hydrolysis domain. The striking pattern of conservation and divergence of regions of the RF genes has been the strongest indicator of the arrangement of potential functional sites in the protein, but until now models for the functions and structural arrangement of particular parts of the RF primary sequence have been limited more to one-dimensional speculation (Mikuni et al., 1991; Pel et al., 1992). Our goal is to correlate the different functions of RF to defined parts of the sequence, and to probe the interactions between regions of the primary structure in the three dimensional molecule.

We started our analysis with the weight of biochemical studies suggesting that there might be two major domains within the release factor each making contact with opposite sides of the ribosome (Moffat et al., 1991). Examination of the primary sequences of the bacterial factors, and proteolytic digestion of the proteins suggested a highly exposed interdomain peptide turn around amino acid position 200 of RF-1. Therefore a chimeric gene between *Escherichia coli* release factors 1 and 2 (RFX-1 in Figure 3A) was created at the site to determine whether the codon recognition domain was within one of these apparent structural domains. Disappointingly the expressed protein had no activity indicating that the original model had been too simplistic. It is now clear that the amino terminal two-thirds of the RF molecules are more resistant to proteolytic digestion and may contribute to the globular core of the factors, whereas the carboxyl terminal one-third, which contains a region highly conserved among the release factors, has highly hydrophilic and protease-sensitive elements. This region may contain important surface features of the factors involved in recognising RNA or protein determinants at the decoding or hydrolysis sites.

A further chimeric gene with the link between RF-1 and RF-2 occurring within the highly conserved region was created (RFX-2, Figure 3A) and the protein expressed from this construct had partial activity for peptidyl-tRNA hydrolysis but no codon recognition activity. A hybrid of the two chimeras (RFX-3) which contained only 10 RF-2-specific residues within the highly conserved region had normal RF-1 codon recognition but impaired peptidyl-tRNA hydrolysis. Disruption of the amino terminus of RF-2 by fusion to a fragment of TrpE (Mr 37 000) had a similar phenotype. From these two constructs it would appear the highly conserved region in the RF molecules and a region near the amino terminus might be important for peptidyl-tRNA hydrolysis. What is more both regions have to be from the same factor since RFX-1, with the amino terminal half of RF-1 and the carboxy terminal half of RF-2, had no activity. Studies with these chimeras have enabled us to propose a model then suggesting that the hydrolysis domain seems to require an intact amino terminus and the highly conserved region towards the carboxy terminus. Similarly since the combination of RF-1 amino terminal amino acid sequence with the RF-2 carboxy terminal sequence does not result in functional codon recognition, this domain may also require at least two separate regions of the peptide chain, one towards the middle of the molecule and one at the carboxy terminus. These ideas are summarised in Figure 3B. Further refinement is now possible by site-directed mutagenesis of targeted residues within these regions.

EFFICIENCY OF DECODING STOP SIGNALS

The proposal that the termination signal is at least 4 bases, with additional bases 3' to the signal possibly contributing as weak recognition determinants, has consequences for the efficiency by which particular signals might be decoded by release factor. It is significant that bacteria use a subset of the possible stop signals for their highly expressed genes (Table 2), which include those genes essential for translation. Such signals might be decoded more efficiently by the release factor. Curran and Petersen, (1991) provide evidence that at least the UAGN set of stop signals are selected at different rates by RF-1 for decoding. The efficiency of decoding a particular stop signal can also be increased by increasing the concentration of a release factor in a transcription/translation system *in vitro* (Donly et al.,1990).

When might an organism want to increase the efficiency of decoding its stop signals? It is well established in bacteria that at fast growth rates the translational rate must increase (Bremer and Dennis 1987). Translational termination has been determined to be a slow step at least on a eukaryotic ribosome (Wolin and Walter 1988). Therefore termination must also become more efficient as growth rate increases, otherwise it may limit a desired increase in translational rate. This could be achieved by an increase in efficiency of decoding the stop

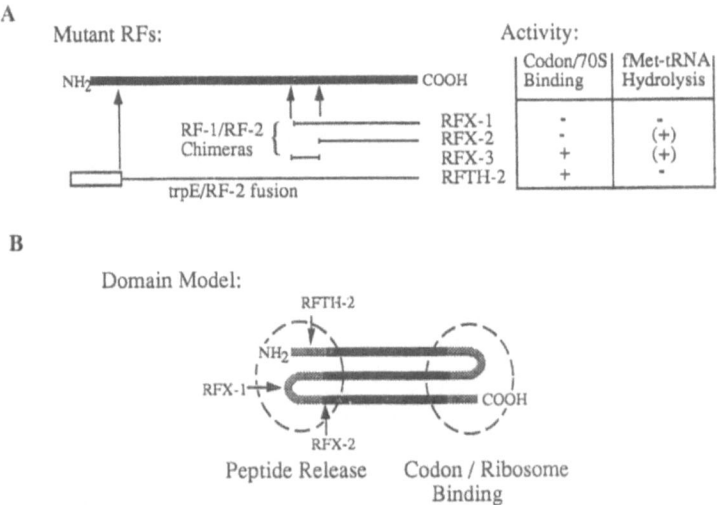

Figure 3. Separation of the codon recognition and peptide release activities of *E.coli* RF-1 and RF-2.
A. Disrupted RF genes were produced by replacing parts of the RF1 sequence with the homologous fragment of the RF2 gene as indicated by the thin horizontal lines (RFX-1 is RF1 amino acids 1--206, then RF2 223-365; RFX-2 is RF1 amino acids 1-238, then RF-2 255-365, RFX-3 is RF1 1-206, RF-2 223-255, then RF-1 238-360), and by fusion of the RF2 gene to the 3' end of a *trpE* gene fragment encoding a protein of Mr ~70 000 (RFTH-2).The fusion and chimeric proteins were overexpressed in *E.coli*, purified and assayed for activity in the partial reactions of *in vitro* termination as shown in the inset Table.
B Idealised representation of the functional and structural organisation of the release factor sequence, based on the observed partial activities and the inability of parts of the RF-1 and RF-2 sequences to complement each other for specific functions. This model suggests two possible domains and functional sites, but does not exclude other sites as being important for function, and it does not predict secondary or tertiary structure.

signals, but the same end would be achieved by increasing the concentration of the factors, or by changing their specific activity, or by a combination of these.

We have now shown that the number of release factor molecules per cell increases in a co-ordinated manner with the translational apparatus, as bacterial growth rate increases. Co-ordinated expression of the release factors with the translational apparatus will ensure that translational stopping rather than increased suppression or frameshifting will occur as the growth rate increases. It has been shown that if release factor activity is decreased then suppression increases in a context specific manner (Kopelwitz et al., 1992). We have found that there was no change in the specific activities of the two release factors at the different growth rates; nor does one factor predominate at the faster growths, implying that neither of the two release factors was used preferentially to decode the UAAU or UAAG signals of the highly expressed genes (Adamski et al., unpublished).

One unexpected finding of these studies was that although the specific activities of the two factors do not change with growth rate, the specific activity of RF-2 was more than 3 fold lower than that of RF-1 in the *in vitro* termination reaction measuring codon-dependent release of model peptide from the ribosome. This assay reflects the catalytic activity of the release factors. Such a finding was pertinent to studies we have undertaken with a number of clones carrying plasmids encoding RF-2. In this case there was a clear cut correlation between the degree of expression and the specific activity of the release factor. A fall in specific activity of over 10 fold was observed as the expression of the factor, and thereby its concentration, increased in the cell until it made up several percent of the total cell protein. This was not simply due to the expression of inactive RF-2 since total activity was increasing, but not at the same rate as RF-2 protein. Moreover the specific activity of this expressed factor in a stoichiometric ribosome binding assay seems to be unaltered.

These findings have led us to postulate that there might be a cellular mechanism for activating RF-2, and at any one time in the cell only a fraction of the RF-2 molecules may be able to efficiently catalyse the termination reaction. Such an 'activation system' becomes saturated as expression of the factor off the plasmids is increased. A preliminary search for a putative 'activator' has suggested that RF-2 is phosphorylated, although as yet we have been unable to correlate phosphorylation status with function.

Prior to these studies observation on the regulation of release factor amounts or activity have all revolved around RF-2. It has the translational frameshift site in its mRNA, a novel and exciting new dimension to regulation mechanisms when it was discovered in 1985 (Craigen et al., 1985), which can regulate the amount of the RF-2 protein present in the cell. We now have preliminary evidence as described above that the activity of the synthesised protein may also be regulated by a post translational mechanism. This is a puzzle because it can potentially blunt the elegance and fine-tuned precision of the translational frameshift control mechanism. RF-2 may also be regulated in the cell at the transcriptional level since the upstream sequences of the gene show a stringent discriminator. No experiments have yet determined the importance of the stringent response, however.

What is exciting is our recent observation that as RF-2 expression is increased the activity of RF-1 also increases, and this reflects new synthesis of RF-1 protein. From a scan of available sequence, RF-1, lacks a distinct stringent discriminator in its putative promoter region. It also lacks a translational frameshift site, does not display a changed specific activity on expression and is apparently not phosphorylated. It may be regulated by RF-2 at some pretranslational step so the activities of the two factors are co-ordinated in the cell. Our current understanding of regulation affecting release factors is shown in Figure 4. For some steps such as the translational frameshift of RF-2 extensive studies have been carried out, whereas for the others for most of the other steps the evidence is more preliminary.

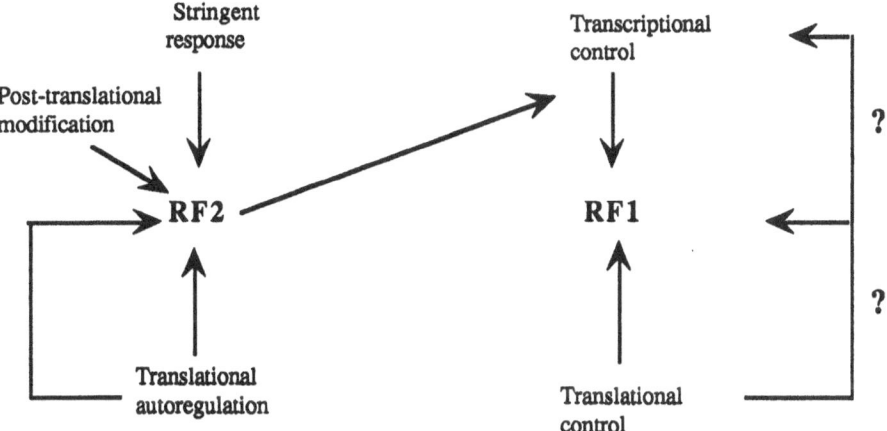

Figure 4. Possible points of control of RF-1 and RF-2 expression *in vivo*

ACKNOWLEDGEMENTS

This article is dedicated to the late Professor H.G. Wittmann, his vision, and his belief in the potential of young scientists around the globe, especially those from its southern-most university.

REFERENCES

Bremer, H., and Dennis, P.P., 1987 in "*Escherichia coli* and *Salmonella typhimurium*." F. C. Neidhart, ed., American Society for Microbiology. Washington DC.

Bock, A., Forchhammer, K., Heider, J., Leinfelder, W., Sawers, G., Veprek, B., and Zinoni, F., 1991 Mol. Microbiol. 5:515.

Brimacombe, R. 1988 Biochemistry 27:4207.

Brimacombe, R. 1992 Biochimie 74: 319-326.

Brown, C. M., Stockwell, P. A., Trotman, C. N. A.,and Tate, W. P., 1990a Nucleic Acids Res. 18:2079.

Brown, C. M., Stockwell, P. A., Trotman, C. N. A., and Tate, W. P., 1990b Nucleic Acids Res. 18:6339.

Buwitt, U., Flohr, T., and Bottger, E.C., 1992 EMBO J. 11:489.

Craigen, W. J., and Caskey, C.T., 1986 Nature 322:273.

Craigen, W.J., Cook, R.G., Tate, W.P., and Caskey , C.T., 1985 Proc. Natl. Acad. Sci. U.S.A . 82:3616.

Donly, B. C., Edgar, C. D., Adamski, F. M., and Tate, W. P., 1990 Nucleic Acids Res. 18:6517.

Dontsova, O., Dokudodovskaya, S., Kopylov, A., Bogdanov, A., Rinke-Appel J., Junke, N., and Brimacombe, R., 1991 EMBO J. 10:2613.

Gesteland, R.F., Weiss, R.B, and Atkins, J.F., 1992 Science 257:1640.

Kastner, B., Trotman, C., and Tate, W. P., 1990 J. Mol. Biol. 212:241.

Kohli, J., and Grosjean, H., 1981 Mol. Gen. Genet. 182:430.

Kopelowitz, J., Hampe, C., Goldman, R., Reches, M., and Engelberg-Kulka , H., 1992 J. Mol. Biol. 225:261.

Lee, C. C., Craigen, W. J., Muzny, D. M., Harlow, E., and Caskey, C. T., 1990 Proc. Natl. Acad. Sci. U.S.A 8: 3508.

Lee, C. C., Timms, K. M., Trotman, C. N., and Tate, W. P., 1987 J. Biol. Chem. 262:3548.

Lehman, N., and Jukes, T. H., 1988 J. Theor. Biol. 135:203.

Moffat, J. G., Timms, K. M., Trotman, C. N. A., and Tate, W. P., 1991 Biochimie 73:1113.

Murgola, E. J., Hijazi, K. A., Goringer, H. U., and Dahlberg, A. E., 1988 Proc. Natl. Acad. Sci. U. S. A. 85:4162.

Mikuni, O., Kawakami, K., and Nakamura, Y., 1991 Biochimie 73:1509.

Noller, H.F., Hoffarth, H., and Zimniak, L., 1992 Science 256:1416.

Oba, T., Andachi, Y., Muto, A.,and Osawa, S., 1991 Proc. Natl. Acad. Sci. U. S. A. 88:921.

Osawa, S., Jukes,. T.H., Watanabe, K., Muto, A., 1992 Microbiol. Rev. 56:229.

Pedersen, W. T., and Curran, J. F. 1991 J. Mol. Biol. 219:231.

Prescott, C.D., Krabben, L., and Nierhaus, K.H., 1991 Nucleic Acids Res. 19:5281.

Prescott, C.D., and Kornau, H-C., 1992 Nucleic Acids Res. 20:1567.

Pel, H.J., Rep, M., and Grivell, L.A., 1992 Nucleic Acids Res. 20:4423.

Rubin, B. Y., Anderson, S. L., Xing, L., Powell, R. J., and Tate, W. P., 1991 J. Biol. Chem. 266:24245.

Shen, Z., and Fox , T.D., 1989 Nucleic Acids Res. 17:4535.

Stern, S., Powers, T., Changchien, L.-M., and Noller, H. F., 1989 Science 244:783.

Tate, W., Greuer, B., & Brimacombe, R., 1990 Nucleic Acids Res. 18:6537.

Tate, W.P., and Brown, C.M., 1992 Biochemistry 31:2443.

Weiss, R.B., Dunn, D.M., Dahlberg, A.L., Atkins, J.F., and Gesteland, R.F., 1988 EMBO J. 7:1503.

Wolin, S. L., & Walter, P., 1988 EMBO J. 7:3559.

Yano, R., & Yura, T., 1989 J. Bacteriol. 171:1712.

THE ALLOSTERIC THREE-SITE MODEL AND THE MECHANISM OF ACTION OF BOTH ELONGATION FACTORS EF-Tu AND EF-G

Knud H. Nierhaus[1], Reinald Adlung[1], Thomas-Peter Hausner[1], Susanne Schilling-Bartetzko[1], Tomasz Twardowski[1,2], and Francisco Triana[1]

[1]Max-Planck-Institut für Molekulare Genetik, Abteilung Wittmann, Ihnestr. 73, Berlin, Germany
[2]Institute of Bioorganic Chemistry, Polish Academy of Sciences, Noskowskiego 12, Poznan, Poland

INTRODUCTION

It is now well established that ribosomes in all living cells contain three binding sites for tRNA: the A site (A for aminoacyl-tRNA), where the decoding takes place; the P site (P for peptidyl-tRNA), and the E site (E for exit) which is specific for deacylated tRNA. An analysis of the functional importance of the third site, the E site, has revealed an allosteric linkage between the first (A) and the third site (E) and led to the allosteric three-site model (for reviews see Nierhaus, 1990; and Rheinberger et al., 1990). After a brief outline of the two essential features of the allosteric three site model and its importance, the implications for the role of both elongation factors and for the importance of the α-sarcin stem-loop structure of 23S rRNA in these activities will be surveyed here.

THE TWO ESSENTIAL FEATURES OF THE ALLOSTERIC THREE-SITE MODEL

The first feature concerns an allosteric linkage between A site and E site, the first and the third site. The coupling between A and E sites is governed by negative cooperativities, i.e. if the A site is occupied the E site has a low affinity and thus cannot bind a tRNA and *vice versa*, if the E site is occupied the A site has a low affinity for tRNA (Rheinberger and Nierhaus, 1986a; Gnirke et al., 1989). This relationship has four important consequences:

1) The ribosome oscillates between two main states during the elongation cycle: The pre-translocational state with A and P sites of high affinities and the E site of low affinity, and the post-translocational state with P and E site of high affinities and the A site of low affinity.

2) Since the high-affinity sites are occupied with tRNAs the ribosome always carries two tRNAs during the elongation cycle.

3) Only the two main states are seperated by high activation-energy barriers. For this reason both states can be formed *in vitro* with a homogenity of at least 75%. Additional states therefore must be considered as "substates", e.g. the states after binding of a ternary complex to the A site, after EF-Tu dependent GTPase activity and after peptide-bond formation are substates of the pre-translocational state (Fig. 1, Schilling-Bartetzko et al., 1992a).

4) The central role of both elongation factors is the reduction of the activation-energy barrier, thus enormously accelerating the rate of protein synthesis. EF-Tu catalyses the transition from the post-translocational state to the pre-translocational state, EF-G the pre→post transition (Fig. 1, Schilling-Bartetzko et al., 1992a).

Allosteric three-site model

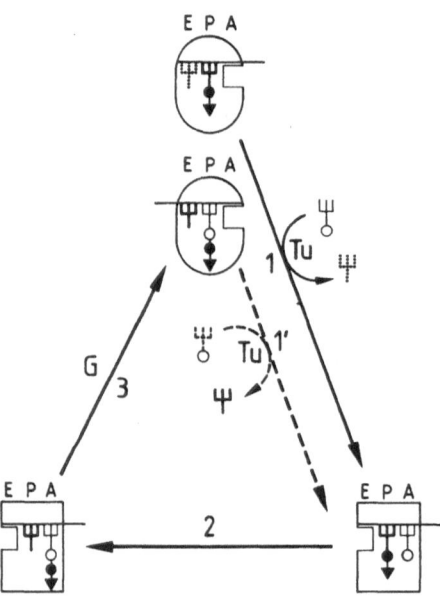

Fig. 1. *Ribosomal elongation cycle depicted in the frame of the allosteric three-site model. The three basic reactions are shown. Reaction 1: A-site binding, causing the allosteric transition from the post-translocational state (round ribosome, A site with low affinity) to the pre-translocational state (rectangular ribosome, E site with low affinity). Reaction 2: Peptidyltransfer. Reaction 3: Translocation, causing the allosteric transition from the pre- to the post-translocational state.*

The allosteric interplay between A and E site has been demonstrated in *E. coli* systems with homopolymeric mRNA (poly(U), Rheinberger and Nierhaus, 1986) and a heteropolymeric mRNA which displays different codons at the three sites (Gnirke et al., 1989), and even *in vivo* evidence supporting the allosteric linkage has been presented (Remme et al., 1989). Thus in prokaryotes this feature seems to be well documented. Since the allosteric linkage has been observed in archaebacterial systems (Saruyama and Nierhaus,

1986) and in eukaryotic systems as well (Triana et al., this volume), it seems to be a universal feature.

The second feature was also unexpected: Both tRNAs on the ribosome, regardless as to whether they are in A and P sites or in P and E sites, simultaneously undergo codon-anticodon interaction. This has been shown in experiments where a deacylated tRNA was directly bound to the E site (poly(U): Rheinberger et al., 1986; heteropolymeric mRNA: Gnirke et al., 1989) and with chasing experiments. An example of the latter is demonstrated in Fig. 2. The P site of non-programmed ribosomes has been occupied with [^{14}C]tRNA^{Asp}.

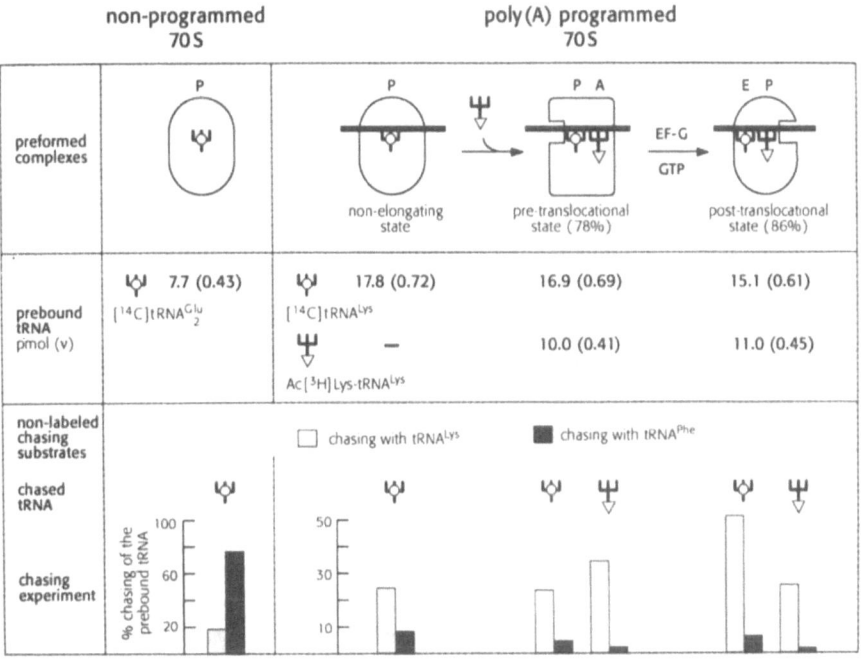

Fig 2. *Testing whether or not codon-anticodon interaction exists at A, P and E sites during elongation. Non-programmed ribosomes: 7.7 pmol [^{14}C]tRNA$_2$^{Glu} (specific activity 110 cpm / pmol) was bound to 18 pmol 70S ribosomes. For chasing, 108 pmol non-labeled tRNA^{Phe} or tRNA^{Lys} was added and the residual binding assessed. Data were taken from Gnirke et al., 1989. Poly(A) programmed ribosomes: 15.1 to 17.8 pmol of [^{14}C]tRNA^{Lys} (60 cpm per pmol) were bound to 24 pmol 70S in the three subsequently established states shown, before the chasing substrates were added. The corresponding numbers for the binding of Ac[^{3}H]Lys-tRNA^{Lys} (50 cpm / pmol) were 10 and 11 pmol in the pre- and post-translocational states, respectively. The numbers in parenthesis give the number of tRNAs bound per ribosome. According to the puromycin reaction, 78 % of the AcLys-tRNA was found at the A site in the pre-translocational state and 86 % at the P site in the post-translocational state after Ef-G dependent translocation. For chasing, 120 pmol of non-labeled tRNA^{Phe} (non-cognate) or non-labeled tRNA^{Lys} (cognate) was added. Data are taken from Rheinberger and Nierhaus, 1986b.*

When unlabelled tRNA^{Phe} or tRNA^{Lys} is added in excess as chasing substrate, tRNA^{Phe} is very efficient in chasing, in contrast to tRNA^{Lys}. It follows that tRNA^{Phe} has a high intrinsic affinity and tRNA^{Lys} a low intrinsic affinity for ribosomes in the absence of codon-anticodon interaction. However, if [^{14}C]tRNA^{Lys} is bound to poly(A)-programmed ribosomes, allowing codon-anticodon interaction at the P site, the chasing efficiencies are reversed. The cognate tRNA^{Lys} is now much more efficient as chasing substrate than the non-cognate tRNA^{Phe}. The latter relationship is therefore indicative of codon-anticodon interaction (high / low chasing efficiencies with tRNA^{Lys} / tRNA^{Phe}, respectively), the first one for its absence (low / high with tRNA^{Lys} / tRNA^{Phe}). When an Ac[^{3}H]Lys-tRNA is bound to the A site and a corresponding chasing experiment is performed, the chasing pattern clearly indicates the presence of codon-anticodon interaction for both tRNAs, and the same result is found after translocating the tRNAs from A and P to P and E sites, respectively. The conclusion is that both tRNAs present in either state do undergo codon-anticodon interaction simultaneously.

IMPORTANCE OF THE ALLOSTERIC THREE-SITE MODEL

Both main features of the allosteric three site model outlined in the preceding section seem to be of pivotal importance for the translation process. Concerning the adjacent codon-anticodon interactions throughout the elongation cycle, we assume that this feature is responsible for maintaining the reading frame. It is clear that maintainance of the reading frame is a prerequisite for the translation of the message. A frameshift is normally a rare event, only one frameshift being observed in more than 6000 elongation cycles (Jørgensen and Kurland, 1991). The hypothesis is that a ribosome allows just a six base-pair connection between mRNA and tRNA in a structurally well defined "window", i.e. two adjacent codon-anticodon interactions are possible before and after translocation. It is expected that anything that disturbs the formation of two adjacent codon-anticodon interactions will cause an increase in frameshifting. This hypothesis has still to be examined experimentally .

The importance of the other features, the allosteric coupling of A and E sites in the sense of a negative cooperativity, has already been analysed (Geigenmüller and Nierhaus, 1990; for review see Nierhaus, 1990). It turned out that the allosteric interplay solves a serious problem of molecular recognition during the decoding process. 41 tRNA species with different anticodons compete for the codon displayed in the A site of a ribosome (Fig. 3). The problem is caused by the fact that the free energy of A-site binding $\Delta G°$ is not very different for the 41 different corresponding ternary complexes. E.g., the elongation factor EF-Tu, an integral component of all tenary complexes, increases the affinity by up to two orders of magnitude over that of the mere aminoacyl-tRNA (Schilling-Bartetzko et al., 1992b). The EF-Tu dependent fraction of the free energy of binding is identical for all 41 ternary complexes and is thus non-discriminatory. All tRNA-ribosome interactions outside the anticodon region are also non-discriminatory but are essential for a precise fitting of the aminoacyl residue (about 75Å distant from the anticodon) into the peptidyltransferase center, i.e. they are a prerequisite for peptide-bond formation. Only codon-anticodon interaction represents the discriminatory fraction of the free energy of the binding, and less than 2% of the mass of a ternary complex makes the anticodon.

The fact that on the one hand we have a large free energy of binding, only a small fraction of which can be used for selection, and that, on the other hand, 41 very similar complexes are competing for the same codon, is a serious problem for the selection process.To exploit the discrimination power equilibrium must be reached, and with a large negative $\Delta G°$ equilibrium will only be achieved after a relatively long period. (For a binding reaction A+B $\underset{k_{-1}}{\overset{k_1}{\rightleftharpoons}}$ C is $\Delta G° = -RT\ln K_a$, K_a is the affinity constant, $K_a = k_1/k_{-1}$. If we assume that k_1 is the same for all ternary complexes (since diffusion controlled), k_{-1} must be slow with a large negative $\Delta G°$, i.e. equilibrium will be reached after a relatively long period).Therefore, even a non-cognate tRNA having an anticodon dissimilar to the cognate one should significantly interfere with the selection process and, before equilibrium has been achieved, should eventually be incorporated into the nascent peptide chain.

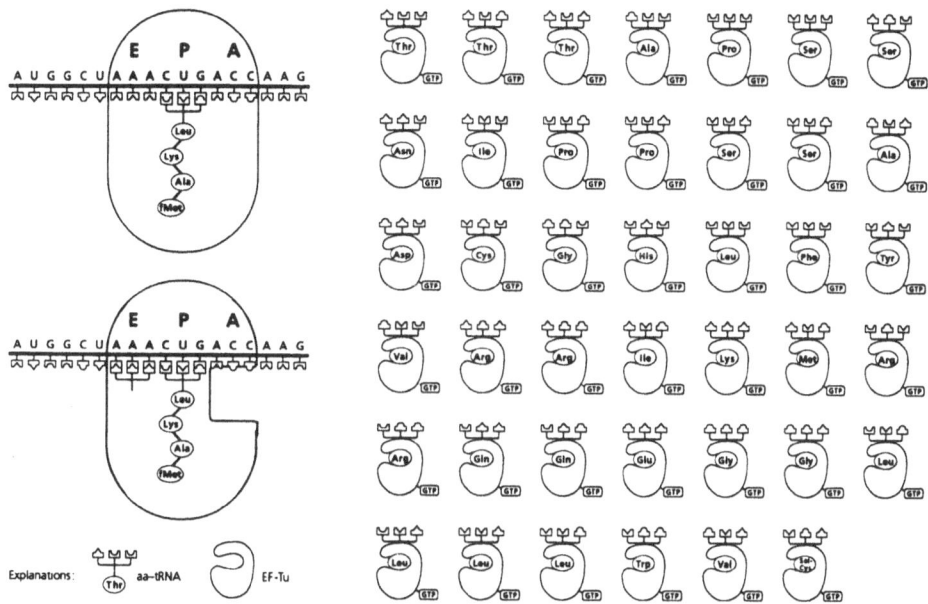

Fig 3. *41 different ternary complexes are competing for the A site. Left side above: When the E site is free the A site has a high affinity for ternary complexes outside the encoding center where codon-anticodon interaction takes place; any of the 41 ternary complexes will therefore interfere with the selection process. Left side below: When the E site is occupied with the cognate deacylated tRNA the A site has a low affinity, abolishing contacts outside the decoding center but allowing codon-anticodon interaction. For further explanation see text.*

Precisely this could be shown (Geigenmüller and Nierhaus, 1990): when the E site was not occupied (leaving the A site in a high-affinity state), poly(U) programmed ribosomes incorporated significant amounts of Asp residues, although the codon of tRNA^{Asp} (GAU/C) is dissimilar to that displayed at the A site (UUU). However, when the E site was charged with tRNA^{Phe} (inducing a low affinity A site), no incorporation of Asp residues could be observed whereas the incorporation of correct Phe residues was not hampered. The interpretation was that the low-affinity state abolishes the non-discriminatory interactions but does not affect or might even improve the discriminatory ones, i.e. codon-anticodon interaction. Interestingly, addition of near-cognate tRNA^{Leu_2} (codon CUU/C) could not trigger the transition to the post-translocational state (A site with low affinity), indicating that codon-anticodon interaction at the E site is the signal for the "flip" into the post-translocational state.

There is another interesting point in the E site/A site interplay. A consequence of the allosteric coupling of these two sites is that A-site occupation must occur in at least two steps. In the first one codon-anticodon interaction takes place (probably a fast reaction), a correct interaction triggers (or allows) the allosteric transition to the pre-translocational state (high-affinity A site, low-affinity E site). Since the second step obviously requires a gross conformational change it is probably a slow reaction. A sequence of two reactions, the second of which is very much slower than the first one, means that the first one practically runs at equilibrium even under steady-state conditions, thus exploiting the discrimination potential of codon-anticodon interaction.

THE ELONGATION FACTORS AND THE α-SARCIN STEM-LOOP STRUCTURE OF 23S rRNA

As mentioned already, the important tasks of the elongation factors are 1) to reduce the activation energy barrier between the pre- and post-translocational states, thus triggering an allosteric transition, and 2) to trap the newly established state after the allosteric transition, *viz.* EF-Tu the pre-translocation state and EF-G the post-translocational state (Fig. 1). Elongation factors might execute additional functions, e.g. it is assumed that EF-Tu plays a role in ribosomal proofreading activities. However, it is questionable whether or not a proofreading step is involved at all in the ribosomal selection process (for discussion see Nierhaus, 1990).

The site of the first ribosomal contact for the elongation factors and thus probably involved in reduction of the activation energy barrier seems to be the α-sarcin stem-loop structure of 23S-type rRNA. This structure comprises helix 95 (*E. coli* numbering) and a loop with 15 nucleotides. The loop contains a sequence of 12 nucleotides, which is the longest universally conserved sequence known in rRNAs (Fig. 4A). Exceptions to the complete conservation are observed only in animal mitochondria, where one or two replacements can be found. Protein synthesis in this organelle from animals has a number of deficiencies (e.g., "cripple" tRNAs as tRNA^{Ser}; loss of the codon UCU, which allows the number of tRNAs required for protein synthesis to be reduced below the minimal number of 23, (for review see Osawa et al., 1992)) and works with low efficiency. As Endo and Wool have shown, the cleavage of one phosphodiester bond in the loop after G2661 (*E. coli* numbering) by the RNase α-sarcin blocks protein synthesis completely (for review see Wool, 1984), whereas up to 10 cuts statistically scattered over the rRNAs have no significant effect.

The conclusion that the α-sarcin stem-loop structure might be the first essential contact site for both elongation factors was deduced from the observation that the single cleavage after G2661 exclusively impaired the binding of either elongation factor, whereas the intrinsic ribosomal functions such as the formation of 70S ribosomes from subunits, tRNA binding to A, P and E sites, peptidyltransfer and EF-G independent (spontaneous) translocation were not affected at all. It follows that the cleavage after G2661 does not induce a gross conformational change of the ribosome, but rather a local disturbance which prevents factor binding. Thus, the specifity of the α-sarcin effect is an indication that the α-sarcin stem-loop structure is an essential component of the binding site of both elongation factors. Furthermore, binding of either elongation factor blocks modification of nucleotides within the loop by some chemical reagents. EF-G protects in addition certain nucleotides in the loop around A1067 (Moazed et al., 1988), which is in accord with the notion that both factors do not bind to one and the same binding site but rather to overlapping binding sites (Richter, 1973). The overlapping region contains the α-sarcin stem-loop structure, and the common function of both factors is the reduction of the activation energy barrier between the two main states of the elongation ribosome. We assume that a contact with the α-sarcin stem-loop structure reduces the activation energy barrier, thus triggering a conformational change of the ribosome. Antisense DNA complementary to the α-sarcin stem-loop structure provided evidence for this hypothesis. We prepared a number of oligo-DNAs complementary to the loop (group 1, Fig. 4) and observed moderate binding (0.1 to 0.6 per ribosome) without notably affecting ribosomal functions (poly(Phe) synthesis, tRNA binding) as has been described already by White et al. (1988). The same was true when the oligo-DNA covered the loop plus the 5'-strand of the 95-helix (group 2). However, strong and specific binding was observed when the antisense DNA was complementary to the loop and 3'-strand of the duplex (group 3).

A

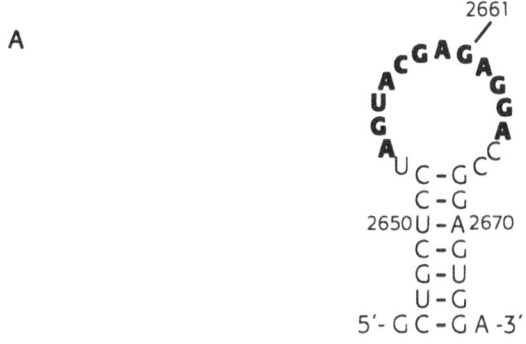

B

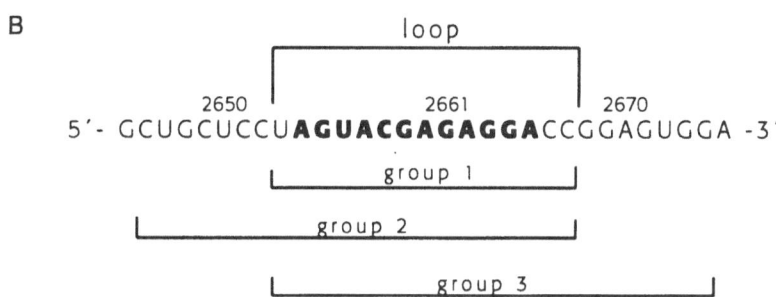

Fig 4. *A, the α-sarcin stem-loop structure (helix 95). B, the same structure unfolded, groups 1 - 3 indicate three families of antisense DNA complementary to the indicated region of the stem-loop structure. From Twardowski and Nierhaus., 1993.*

All ribosomal functions tested were nullified, and a systematic analysis revealed a dissociation of the tightly coupled ribosomes, causing the multiple inactivations observed upon hybridization of these antisense DNAs, which triggered a conformational change of the 50S subunits. These subunits showed impaired activities even after removal of the oligo-DNA. The conclusion was that binding of the stem-loop structure and melting of the stem generate a "functional dead end" in the large subunit by a conformational change (Twardowski and Nierhaus, 1993). The elongation factors may also use binding and melting as a trigger for elongation factor dependent transitions but preventing disasterous effects probably via the second step of the binding reaction, i. e. trapping the newly established state after the allosteric transition.

If the availability of the 3'-strand of the duplex is also required for the function of an elongation factor, the factor has to melt the α-sarcin stem in order to trigger the allosteric transition in one or the other direction. In this case, the stem should not be of maximal stability. Therefore, we compared the stability of the α-sarcin stem-loop structure with that of a stem-loop motif (helix 81), the secondary structure of which was conserved but not the primary sequence. The comparison revealed that the free energy change per base-pair formation was significantly lower throughout all kingdoms/domains in the case of the α-sarcin stem-loop structure (Fig. 5). The only exceptions found were three thermophilic archaebacteria (*Thermoproteus tenax*, *Thermophilum pendens*, *Desulfurococcus mobilis*, $\Delta G°37$ bp = -1.36 ± 0.06 kcal/mol), where the thermophily might require an increased stability for appropriate function (Nierhaus and Triana, 1993).

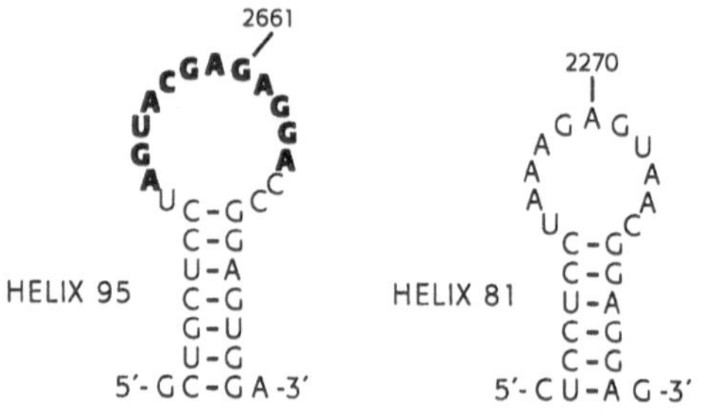

domain/ organelle	no. of sequences	$\Delta G°_{37,bp}$ (kcal/mol)	$\Delta G°_{37,bp}$ (kcal/mol)
Bacteria	16	-0.68±0.21	-1.37±0.09
Archea (no thermophiles)	6	-0.63±0.23	-1.23±0.18
Eucarya	8	-0.60±0.22	-1.00±0.08
Plastids	9	-0.58±0.24	-1.40±0.08
Mitochon- dria	13	-0.28±0.10	-1.09±0.09

Fig5. *The structural weakness of helix 95 is universally conserved as compared to helix 81 (E. coli numbering). Adapted from Nierhaus and Triana, 1993.*

It follows that the α-sarcin stem-loop structure is characterized by two universally conserved features: 1) The primary sequence of the dodecamer in the loop, and 2) the structural weakness of the stem.

HYPOTHESIS FOR THE MECHANISM OF THE RIBOSOMAL ELONGATION FACTORS

The various sets of information discussed in the preceding section are combined in a hypothesis of the mechanism of elongation factors in Fig 8. Both factors are described in a symmetric fashion; we consider for example EF-G. This factor binds in the GTP conformation loosely to the pre-translocational state and contacts the α-sarcin stem-loop structure, where the universally conserved dodecamer sequence is important for specific binding of the factor (first step of factor binding). The factor melts duplex 95 and thus triggers the allosteric transition to the post-translocational state. The post-translocational state is precisely adjusted by tight binding of EF-G still in its GTP conformation (second step of factor binding). In the post-translocational state a ribosomal structure induces the GTP dependent GTPase activity. GTP is cleaved, EF-G adopts its GDP conformation, in which the factor has only a low affinity for the ribosome, and therefore dissociates. The stem of duplex 95 reforms, which erects the activation energy barrier again, thus trapping the post-translocational state.   With EF-Tu the corresponding sequence of events might occur, but now starting with the post-translocational state. The essential difference is that the contact

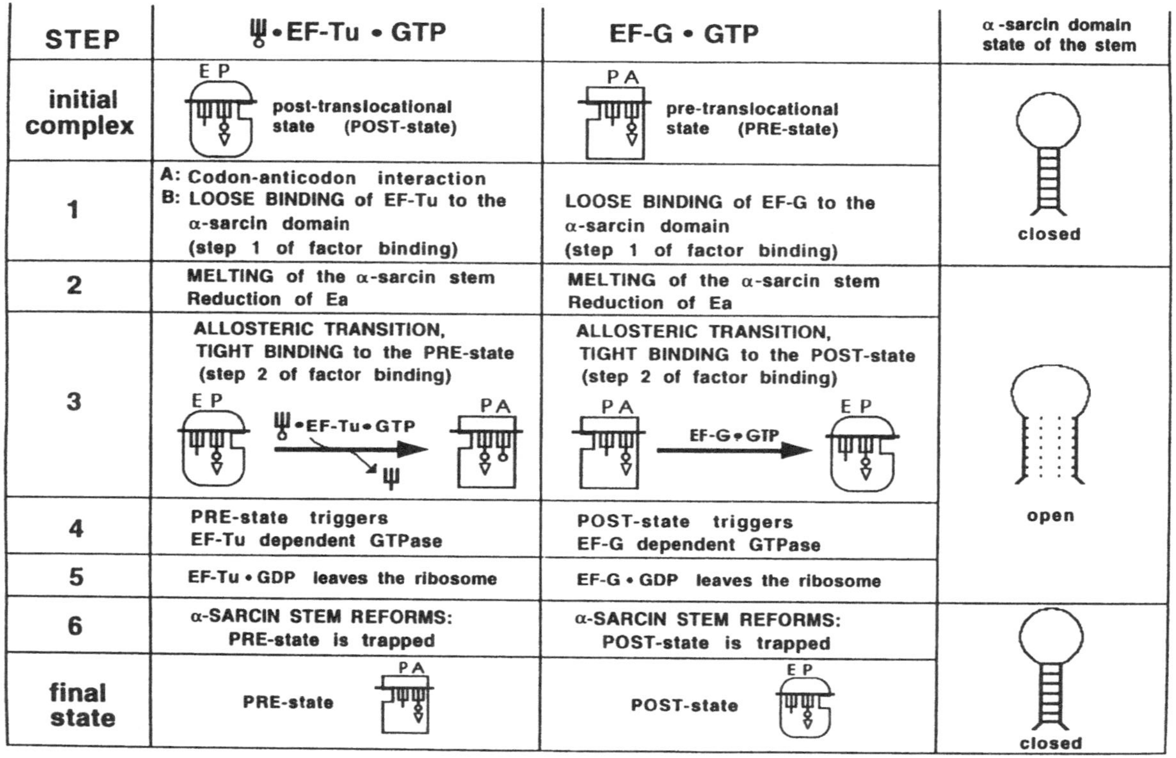

STEP	Ψ·EF-Tu·GTP	EF-G·GTP	α-sarcin domain state of the stem
initial complex	E P — post-translational state (POST-state)	P A — pre-translocational state (PRE-state)	closed
1	A: Codon-anticodon interaction B: LOOSE BINDING of EF-Tu to the α-sarcin domain (step 1 of factor binding)	LOOSE BINDING of EF-G to the α-sarcin domain (step 1 of factor binding)	
2	MELTING of the α-sarcin stem Reduction of Ea	MELTING of the α-sarcin stem Reduction of Ea	open
3	ALLOSTERIC TRANSITION, TIGHT BINDING to the PRE-state (step 2 of factor binding) E P — Ψ·EF-Tu·GTP — P A	ALLOSTERIC TRANSITION, TIGHT BINDING to the POST-state (step 2 of factor binding) P A — EF-G·GTP — E P	
4	PRE-state triggers EF-Tu dependent GTPase	POST-state triggers EF-G dependent GTPase	
5	EF-Tu·GDP leaves the ribosome	EF-G·GDP leaves the ribosome	
6	α-SARCIN STEM REFORMS: PRE-state is trapped	α-SARCIN STEM REFORMS: POST-state is trapped	closed
final state	P A — PRE-state	E P — POST-state	

Fig 6. *Hypothesis of the mechanisms of elongation factors.*

271

to the α-sarcin stem- loop structure has to be preceded by codon-anticodon interaction, whereby an appropriate fit is a prerequisite for initiating the factor-induced allosteric transition.

This hypothesis provides a frame which integrates various pieces of evidence concerning the mechanism of action of the elongation factors. Many details have still to be worked out, and nothing is known about the underlying ribosomal mechanism.

ACKNOWLEDGEMENTS

We thank Richard Brimacombe, Beatrix Röhrdanz, Jörg-Uwe Bittner, Ralf Jünemann, Christian Spahn and Martin Rühl for help and discussion.

REFERENCES

Geigenmüller, U., and Nierhaus, K.H., 1990, *EMBO J.* 9: 4527.
Gnirke, A., Geigenmüller, U., Rheinberger, H.-J., and Nierhaus, K.H., 1989, *J. Biol. Chem.* 264: 7291.
Hausner, T.-P., Atmadja, J., and Nierhaus, K.H., 1987, *Biochimie* 69: 911.
Jørgensen, F., and Kurland, C.G., 1991, *J. Mol. Biol.* 215: 511.
Moazed, D., Robertson, J.M., and Noller, H.F., 1988, *Nature* 334: 362.
Nierhaus, K.H., 1990, *Biochemistry* 29: 4997.
Nierhaus, K. H., and Triana, F., 1993, in "Translational Regulation of Gene Expression II", J. Ilan, ed., Plenum Publishing Corporation, New York, in press.
Osawa, S., Jukes, T.H., Watanabe, K., and Muto, A., 1992, *Microbiol. Rev.* 56: 229.
Remme, J., Margus, T., Villems, R., and Nierhaus, K.H., 1989, *Eur. J. Biochem.* 183:281.
Rheinberger, H.-J., Geigenmüller, U., Gnirke, A., Hausner, T.-P., Remme, J., Saruyama, H., and Nierhaus, K.H., 1990, *in* "The Ribosome: Structure, Function, and Evolution," W.E. Hill, A. Dahlberg, R.A. Garrett, P.-B. Moore, D. Schlessinger, and J.R. Warner,eds., American Society for Microbiology, Washington D.C., p. 318.
Rheinberger, H.-J., and Nierhaus, K.H., 1986a, *J. Biol. Chem.* 261: 9133.
Rheinberger, H.-J., and Nierhaus, K.H., 1986b, *FEBS Lett.* 204: 97
Rheinberger, H.-J., Sternbach, H., and Nierhaus, K.H., 1986, *J. Biol. Chem.* 261: 9140.
Richter, D., 1973, *J. Biol. Chem.* 248: 2853.
Saruyama, H., and Nierhaus, K.H., 1986, Mol. Gen. Genet. 204: 221.
Schilling-Bartetzko, S., Bartetzko, A., and Nierhaus, K.H., 1992a, *J. Biol. Chem.* 267: 4703.
Schilling-Bartetzko, S., Franceschi, F., Sternbach, H., and Nierhaus, K.H., 1992b, *J. Biol. Chem.* 267: 4693.
Triana, F., Nierhaus, K.H., Ziehler, J., and Chakraburtty, K., this volume.
Twardowski, T., and Nierhaus, K.H., 1993, submitted.
White, G. A., Wood, T., and Hill, W.E., 1988, *Nucleic Acids Res.* 16: 10817.
Wool, I.G., 1984, *Trends in Biochem. Sci.* 9: 14.

ELONGATION FACTOR Tu FROM <u>THERMUS THERMOPHILUS</u>, STRUCTURE, DOMAINS AND INTERACTIONS

Mathias Sprinzl[1] and Rolf Hilgenfeld[2]

[1] Laboratorium für Biochemie
    Universität Bayreuth, Germany
[2] Protein Crystallography
    Hoechst AG, 6230 Frankfurt 80, Germany

INTRODUCTION

Elongation factor Tu (EF-Tu) promotes the binding of aminoacyl-tRNA (aa-tRNA) to ribosomes during polypeptide chain elongation. In this process, EF-Tu interacts sequentially with GTP, aa-tRNA, ribosomes, GDP and EF-Ts (Miller and Weissbach, 1977). EF-Tu belongs to a class of regulatory GTPases which includes translational factors, signal transducing proteins (Kaziro et al., 1991), and proteins of the *ras* gene family (Grand and Owen, 1991). The GTPases function by switching between a GTP-bound "signal on" active conformation and a GDP-bound "signal off" latent conformation. Within this regulatory cycle EF-Tu controls the rate and fidelity of protein biosynthesis (Thompson and Karim, 1982). Bacterial EF-Tu consists of three domains (Kjeldgaard and Nyborg, 1992 and references therein). The GTPase activity is located in domain I. Less information is available about the function of domains II and III. These smaller structural units, as compared to domain I, were identified as a part of the tRNA binding surface (Kinzy et al., 1992), as the site interacting with EF-Ts (Peter et al., 1990a) and as a binding site for the antibiotic kirromycin (van Noort et al., 1984).

The three-dimensional structure of proteolytically cleaved *E. coli* EF-Tu·GDP missing 14 amino acid residues in the domain I was determined by Jurnak (1985) and La Cour et al. (1985) at 2.7 Å and 2.9 Å resolution, respectively. Recently the refined crystal structure of this EF-Tu·GDP complex was refined to 2.6 Å resolution. In the new model, in addition to domain I, the fold of domains II and III is described (Kjeldgaard and Nyborg, 1992). A high resolution X-ray structure analysis is also available for another GTPase, the H-*ras* p21, complexed

The Translational Apparatus, Edited by K.H. Nierhaus
et al., Plenum Press, New York, 1993

either with GDP (deVos et al., 1988) or analogues of GTP (Pai et al., 1989; Schlichting et al., 1990). Comparison of the "active" GTP-conformation with the "latent" GDP-structure of this protein might be instructive for the analysis of similar conformational changes in other structurally related GTPases (Jurnak et al., 1990). In contrast to GTP/GDP-binding translation factors and signal transducing heterotrimeric G-proteins, the H-*ras* p21 consists of a single GTP/GDP binding domain.

Proteins from thermophilic bacteria are resistant to heat denaturation and are generally more stable as the related proteins from mesophiles. For this reason they were frequently used as models to study the molecular bases of thermal stability. Recently several proteins and macromolecular complexes were isolated from bacteria of species *Thermus* and crystallized. Crystals diffracting to high resolution were obtained for *T. aquaticus* EF-Tu·GDP (Lippmann et al., 1992), *T. thermophilus* phenylalanyl-tRNA synthetase (Chernaya et al., 1987), threonyl-tRNA synthetase (Garber et al., 1990), glutamyl-tRNA synthetase (Nureki et al., 1992), methionyl-tRNA synthetase (Nureki et al., 1991), seryl-tRNA synthetase (Garber et al., 1990) and isopropylmalatdehydrogenase (Imada et al., 1991). Crystallization of ribosomes and ribosomal subunits from *Thermus thermophilus* was also reported (Trakhanov et al., 1989; Hansen et al., 1990). Crystallization of the *T. thermophilus* EF-Tu·GppNHp complex and its structure determination to 1.45 Å resolution was recently accomplished in our laboratories (Reshetnikova et al., 1991; Reshetnikova et al., 1992; Berchtold et al., in preparation).

This communication summarizes the recent biochemical and structural work on intact EF-Tu·GppNHp complex from *T. thermophilus* performed in our laboratories. The purpose of this research is to obtain information about structure and function-related dynamics of EF-Tu and structurally related signal transducing GTPases.

Genes and Sequences of <u>Thermus thermophilus</u> EF-Tu

Elongation factor Tu is coded in *T. thermophilus* as in *E. coli* by two genes located in unlinked operones. *Tuf* A is coexpressed with ribosomal proteins S7, S12 and the elongation factor G (Seidler et al., 1987; Kushiro et al., 1987). The *tuf* B gene (Satoh et al., 1991) is a part of an operon containing a cluster of four tRNA genes upstream of the EF-Tu gene. These code for tRNA^{Thr1}, tRNA^{Tyr}, tRNA^{Gly} and tRNA^{Thr2} (Weißhaar et al., 1990). In the two *tuf* genes there are 10 differences in the nucleotide sequence. Six mutations are silent. In four cases the nucleotide exchanges lead to amino acid replacement in EF-Tu (Fig. 1). The sequence analysis of the EF-Tu isolated from mid-log *T. thermophilus* cells grown at 65 °C indicated a preferential expression of the *tuf* B gene product (65% of the overall EF-Tu). The following amino acids in *tuf* A product are replaced in EF-Tu coded by *tuf* B: valine 6 for isoleucine, tyrosine 33 for phenylalanine , alanine 35 for threonine and finally arginine 264 for lysine (Satoh et al., 1991). The portions where the changes occur are not conserved among different bacterial EF-Tus (Fig. 1) and the replacing amino acids have structurally related side chains. Recently the

T. thermophilus tuf A was successfully overproduced in *E. coli* (Ahmadian et al., 1991) and a protein with a homogeneous EF-Tu(A) sequence was isolated. An *in vitro* comparison of biochemical properties and thermostability of pure EF-Tu(A) with 1:2 mixture of EF-Tu(A):EF-Tu(B) isolated from *T. thermophilus* cells did not reveal any differences in functional properties or stability. Thus the structural and functional consequences of these replacements are not yet understood.

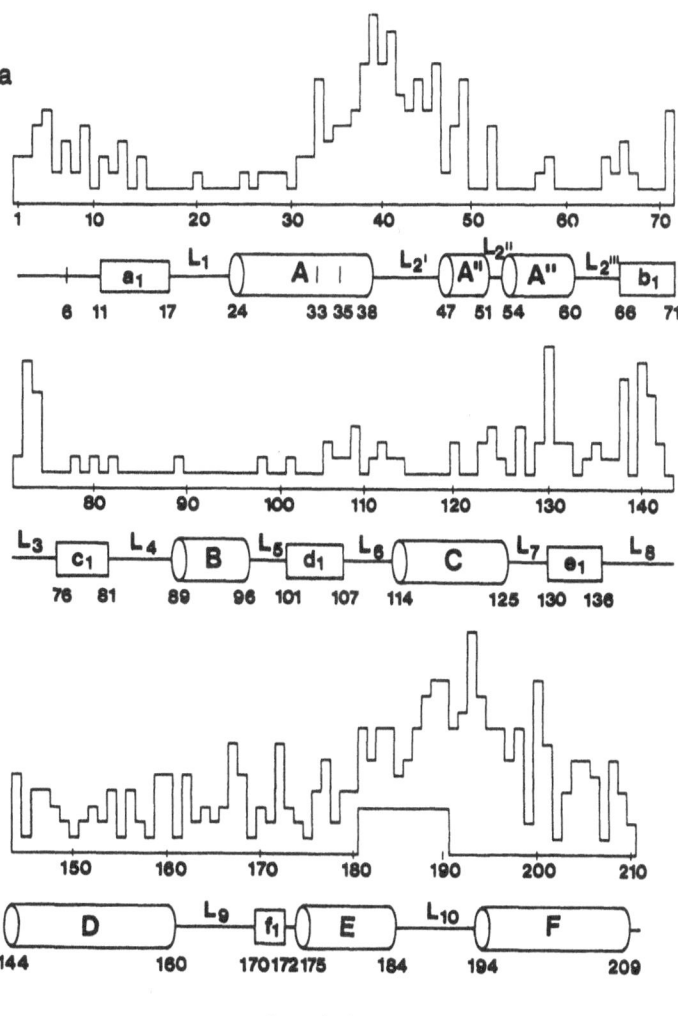

domain I

Figure 1. Amino acid variability in particular sequence positions in domains I (a), II (b) and III (c) of bacterial elongation factors Tu. Sequences of 36 elongation factors Tu (except those from archaebacteria) were evaluated. The numbering and secondary structure elements are related to the 3-D structure of *T. thermophilus* EF-Tu•GppNHp. The height of the bars corresponds to the occurrence of different amino acids in particular positions. The minimal value represents one amino acid for an invariable position. Missing line between amino acids 181-190 indicates deletions. Loops in domain I are indicated by L, ß-strands by small letters and indices indicating the particular domain. α-helices occur only in domain I and are indicated by capital letters.

b

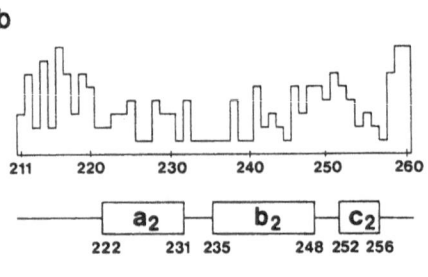

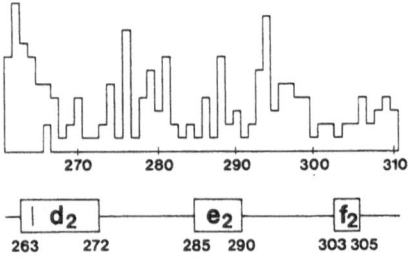

domain II

c

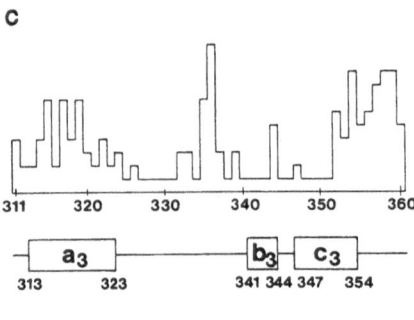

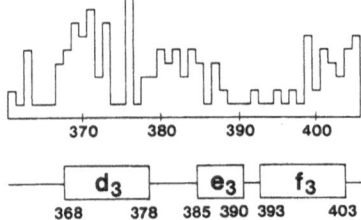

domain III

Figure 1b (above) and 1c.

Several genes of bacterial elongation factors Tu and eukaryotic elongation factors 1 were sequenced and the sequences used for phylogenetic studies (Ludwig et al., 1991). The three dimensional structures of *E. coli* EF-Tu·GDP and *T. thermophilus* EF-Tu·GppNHp now known allow an alignment of sequences

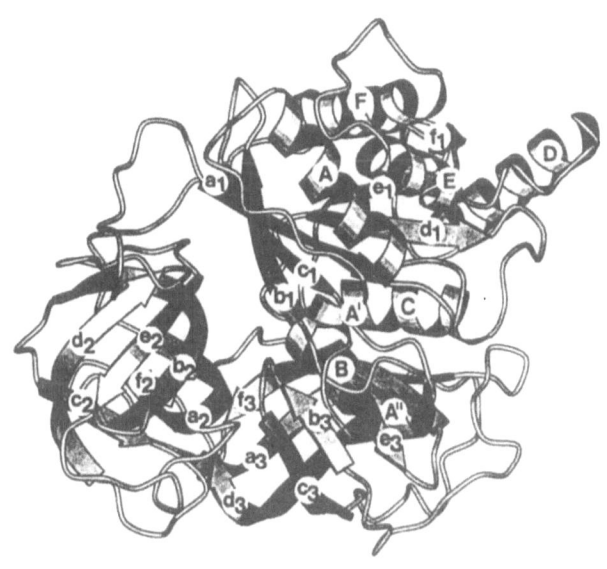

Figure 2. Schematic model of *T. thermophilus* EF-Tu·GppNHp. The secondary structure elements are defined by symbols used in Fig. 1.

according to the function of particular domains. In Fig. 1 the variability of amino acids in the sequences of EF-Tus is shown. The alignment was performed on the basis of the 3 dimensional model of *T. thermophilus* EF-Tu·GppNHp (Berchtold et al., in preparation). The number of invariant amino acids in bacterial EF-Tus goes far beyond the consensus sequences 18GXXXXGKS/T25, T62, 81DXXG84 and 136NKXD139 (numbering according to *T. thermophilus* EF-Tu) known as GTP/GDP interacting sites present in all GTPases (Dever et al., 1987). In domains I of bacterial EF-Tus the sequences forming ß-strands b_1, c_1, d_1 and α-helices B, C are remarkably well conserved. Extremely high variability is in the second half of helix A and the first half of the region connecting helix A with ß-strand b_1. However, after residue 50 this region is composed of conserved amino acid residues. Loops L_3 and L_{10} are again very variable. Interestingly, the loop L_3 and the variable part of L_2 and L_{10} are in close vicinity in the 3-D structure (Fig. 2) and may play a role as an interacting site in a species-specific process, e.g. ribosomal binding. This would explain the variability of sequences of this part of EF-Tu as compared to invariant sequences such as loops L_1 and L_4, or helix B, which have a function in all EF-Tus as "switch-", GTPase-, or subunit-interacting regions.

Stability of T. thermophilus EF-Tu

A remarkable property of T. thermophilus EF-Tu as compared to EF-Tu from E. coli is its thermostability. The molecular basis of the temperature resistance is not yet understood and cannot be deduced from the comparison of protein sequences. As can easily be observed by studying the temperature denaturation profiles shown in Fig. 3, the bound nucleotide is very important for stabilization of E. coli EF-Tu. The nucleotide-free EF-Tu from T. thermophilus is reasonably stable even at high temperatures. This is an advantage for many biochemical experiments since it allows a preparation of EF-Tu complexes with stoichiometric amounts of GTP, GDP, nucleotide analogues and labeled nucleotides for affinity labeling, NMR studies and crystallization experiments.

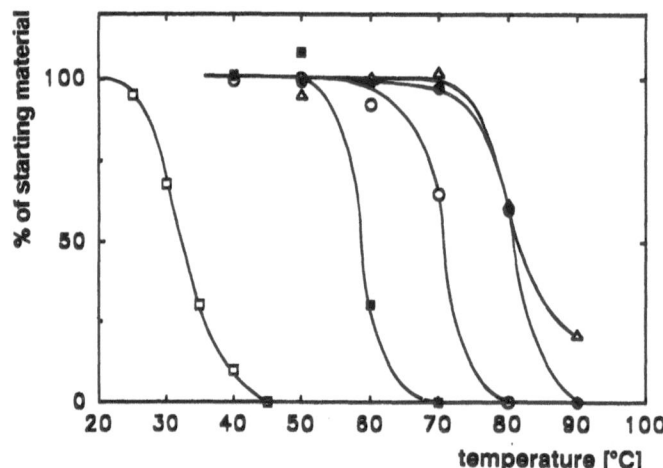

Figure 3. Thermostability of EF-Tu·GDP from E. coli (■) and Thermus thermophilus (•), nucleotide-free E. coli (□) and Thermus thermophilus (○) EF-Tu and domain II/III (Δ) at different temperatures. Protein samples were incubated for 5 min in 50 mM sodium borate pM 7.5, 50 mM KCl and 10 mM $MgCl_2$. The samples were centrifuged and the soluble proteins determined either by GDP binding test or Biorad protein assay.

As demonstrated in Fig. 3, the denaturation midpoint of nucleotide-free E. coli EF-Tu is at 32 °C under given conditions. Binding of GDP increases the denaturation midpoint by 25 °. In comparison the nucleotide-free T. thermophilus EF-Tu is denatured to 50% at 72 °C, and the addition of GDP shifts this temperature only by 10 °C to 82 °C. Interesting results were provided by the temperature denaturation profiles of isolated domains of the thermostable protein. The domain II/III consisting of amino acid residues 209 - 405, which can be prepared by limited proteolytic cleavage of nucleotide-free *T. thermophilus* EF-Tu (Peter et al., 1990a), has the same stability as the intact molecule (Fig. 3). Recently the isolated GTP-binding domain I of *T. thermophilus* EF-Tu (residues 1 - 210) was prepared by expression of the corresponding *tuf* A part in *E. coli* (Kreutzer et al., in preparation). Despite the fact that its sequence is derived from a thermophile, this protein is remarkably sensitive to elevated temperatures and denatures quickly, like *E. coli* EF-Tu, at temperatures above 35 °C (Fig. 4). The experiments in Fig. 4 demonstrate a further typical property of thermostable enzymes, namely that the high thermostability of these proteins is often associated by the loss of activity at low temperatures. Since in the nucleotide exchange reaction tested in these experiments, the dissociation of the non-radioactive GDP is the rate-limiting step, the slow exchange rate at low temperatures must be due to a conformation of the GTP binding domain in the intact *T. thermophilus* EF-Tu preventing the release of the nucleotide. Thus the temperature-stable, rigid structure of domains II/III is important for stabilization of the structure of domain I as well as the regulation of its function as GTPase.

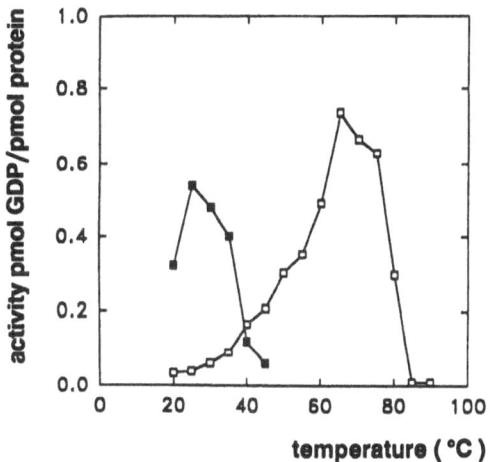

Figure 4. Temperature dependence of GDP exchange activity of intact *T. thermophilus* EF-Tu (□) and of the isolated domain I (■) of this protein. Samples were incubated at given temperatures for 10 min with the labeled GDP and the protein-bound GDP was determined.

Domain I and the effector region of EF-Tu

The loop L_2 connecting the helix A with antiparallel ß-sheets b and c in H-*ras* p21 was named the "effector loop" (deVos et al., 1988). It is involved in GTP/GDP binding, participates in the GTPase reaction and is expected to interact with the until now unknown "effector" of p21. As effector, a protein or protein complex is assigned to the signal from which the active form of GTP/GDP binding protein is transmitted under simultaneous induction of GTPase. Consequently the effector region of EF-Tu, which includes two short helices (Fig. 2) should have the following properties: a) the ability to interact with GTP, b) the ability to interact with ribosomes and the co-effector, aminoacyl-tRNA and c) participation in the GTPase reaction. Effector loops or effector domains of different lengths were identified in all known GTPases. In *T. thermophilus* EF-Tu this region would correspond to amino acid residues 41 - 62 (Fig. 1).

As mentioned above (Fig. 1) the residues 40 - 51 in the effector region of elongation factors Tu are hypervariable, and at least part of the ribosomal binding site of domain I is harbored here. This is supported by the observation that modifications in this area of the effector region influence the ribosome-dependent GTPase of the elongation factors (Peter et al., 1990b). The second part of the effector loop in *T. thermophilus* EF-Tu consists of predominantly conserved amino acids residues. This is consistent with the function as a part of the guanosine nucleotide interaction- and GTPase-site. As shown in Fig. 5, which was derived from the 3-D structure of *T. thermophilus* EF-Tu·GppNHp (Berchtold et al., in preparation), tyrosine 47 and threonine 62 of the *T. thermophilus* EF-Tu effector region participate in binding the α- and γ-phosphate of GTP, respectively. The side-chain of aspartic acid 51 binds via a water molecule to the Mg^{2+} ion, which in turn is coordinated to the γ- and ß-phosphates of GTP. No information in this regard is available from the X-ray structure analysis of *E. coli* EF-Tu·GDP since this part of the effector region (residues 44 - 58) is missing in the crystallized protein.

From Fig. 5 it is obvious that the effector region (residues 40 - 61) will change its conformation by exchange of GDP for GTP and by GTP hydrolysis. Indeed, we found that interaction of nucleotide exchange factor EF-Ts with EF-Tu significantly changes the accessibility of arginine 59 for proteolytic cleavage (Schirmer et al., 1991). This experiment indicates the effect of EF-Ts on the conformation of the effector region and suggests how EF-Ts accelerates the rate of GDP dissociation from EF-Tu and facilitates GTP binding. At least part of the mechanism leading to this exchange involves a change in the structure of the effector region. However, this is probably not achieved by a direct interaction of domain I with EF-Ts but rather by a long-range effect in which the binding of EF-Ts to domain II/III is transmitted to domain I. The possibility of such structural communication between domains of EF-Tu was demonstrated by several experiments. It could be shown that the accessibility of peptide bonds in the connecting region between domain I and II (residue 208) or within domain II (residue 275) is dependent on the occupation of the nucleotide binding site in domain I (Peter et al., 1990b). Pronounced stabilization of the domain I against temperature denaturation by covalently linked domain II/III polypeptide also demonstrates the mutual interaction of the individual domains (Fig. 3 and 4).

Cleavage of the effector region either in position 52 or 59 selectively influences the GTPase activity of EF-Tu. Whereas the ribosome-independent GTPase remains essentially unchanged after cleavage at positions 52 or 59, these modifications completely cancel the ability of ribosomes to stimulate the EF-Tu-

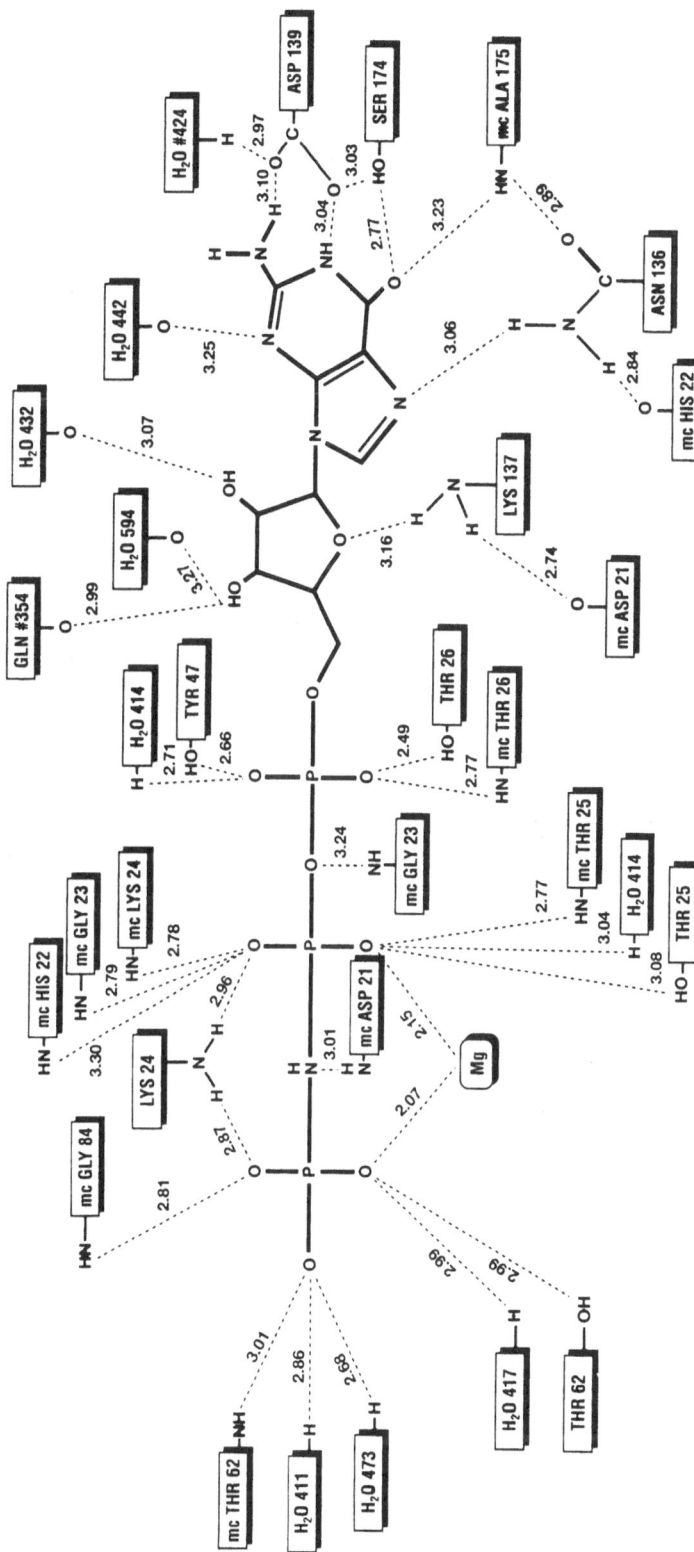

Figure 5. Aminoacyl residues of *T. thermophilus* EF-Tu involved in interaction with GppNHp.

dependent GTPase. The cleavage of the effector loop in the region 52 - 59 uncouples its ribosomal binding region (40 - 51) from the part (52 - 62) which is involved in GTP binding and GTPase reaction.

Aminoacyl-tRNA, together with the ribosome, can be considered as a co-effector for EF-Tu·GTP. Therefore the question arises whether EF-Tu, which is modified in the effector region, is still able to form a ternary complex with aminoacyl-tRNA and GTP. Masuda et al. (1985) reported that the EF-Tu missing 14 amino acid residues from the loop L_2 is inactive in aminoacyl-tRNA binding, whereas Ott et al. (1989) observed an interaction between aminoacyl-tRNA and this strongly modified protein. Since different assays were used for the detection of ternary complexes in the two investigations, the results are not contradictory but offer evidence for the location of the effector loop near the 3'-end of aminoacyl-tRNA. The tRNA binding site on domain I of EF-Tu is therefore located somewhere near amino acid residues 60 - 83 (as shown by affinity labeling and EF-Tu modification) and is modulated by the structure of the effector region and phosphate binding loop L_4. The observation that methylation of lysine 56 in *E. coli* EF-Tu mediates the aminoacyl-tRNA induced GTPase activity without affecting the binding of the aminoacyl-tRNA to EF-Tu·GTP is in support of the above model (van Noort et al., 1986).

Domain II/III of EF-Tu

In contrast to domain I, the isolated domain II/III does not bind GTP or GDP (Peter et al., 1990a). Interaction between aminoacyl-tRNA and domain II/III could not be detected either. This result is unexpected since both bacterial EF-Tu (Antonsson et al., 1984) and eukaryotic elongation factor 1 α (Kinzy et al., 1992) were successfully crosslinked to aminoacyl-tRNA. In both reports the domain II and regions near the aminoacyl-stem of the tRNA were the predominant crosslinking regions. The lack of interaction with the isolated domain II/III is therefore probably related to the absence of a domain I mediated conformation of domain II/III.

Removal of GTP from intact EF-Tu does not prevent the binding of aminoacyl-tRNA. The dissociation constant for the nucleotide-free factor is only slightly higher than that for EF-Tu·GTP (Peter et al., 1990a). This suggests that the lack of interaction of isolated domain II/III with aminoacyl-tRNA is due to the absence of a long-range effect by which domain I, even in the nucleotide-free form, codetermines the structure of domain II/III. Experimental evidence with EF-Tu mutants or protease cleavage of EF-Tu supports this interpretation. Thus, the binding of aminoacyl-tRNA to EF-Tu requires both the GTP/GDP-binding domain I and domain II/III. The mutual orientation of domains I and II of EF-Tu necessary for aminoacyl-tRNA binding is not dependent on the presence of GTP in domain I. On the other hand GTP hydrolysis and GDP formation induce a conformation of EF-Tu which excludes the binding of aminoacyl-tRNA.

EF-Tu interacts with the nucleotide exchange factor EF-Ts in its functional cycle. This factor resembles the function of a receptor molecule in the case of signal transducing G-proteins, where binding of the receptor increases the rate of

GDP dissociation. This binding event therefore influences the structure of the domain to which the nucleotide is bound. It can be demonstrated by polyacrylamide gel electrophoresis that *T. thermophilus* EF-Ts forms a complex with the isolated domain II/III (Peter et al., 1990a). It is known that the intact *T. thermophilus* EF-Tu forms a very stable complex with EF-Ts which can be isolated by gel permeation chromatography (Arai et al., 1978). This was, however, not the case for the domain II/III·EF-Ts complex, indicating that the stabilizing effect of domain I is necessary for the tight interaction of EF-Ts with the domain II/III of the intact EF-Tu. The binding of EF-Ts to domain II/III is in agreement with the previously reported observation that EF-Ts has no significant effect on the isolated domain I of *E. coli* EF-Tu (Parmeggiani et al., 1987). The ability of EF-Ts to interact with the domain II/III indicates that the receptor binding function of EF-Tu is probably localized here.

ACKNOWLEDGEMENT

We thank H. Berchtold, H. Faulhammer, N. Grillenbeck, R. Kreutzer, G. Ott, C. Reiser and N. Schirmer for cooperation in the still unpublished work described here, M. Lenk for preparation of the figures and H. Hohmann and M. Daniel for typing the manuscript. This work was supported by the Deutsche Forschungsgemeinschaft, SFB 213, Project D5.

REFERENCES

Ahmadian, M.R., Kreutzer, R. and Sprinzl, M. (1991) Biochemie 73, 1037-1043.
Antonsson, B. and Leberman, R. (1984) Eur. J. Biochem. 141, 483-487.
Arai, K.-I., Arai, N., Nakamura, S., Oshima, T. and Kaziro, Y. (1978) Eur. J. Biochem. 92, 521-531.
Chernaya, M.M., Korolev, S.V., Reshetnikova, L.S. and Safro, M.G. (1987) J. Mol. Biol. 198, 555-556.
de Vos, A.M., Tong, L., Milburn, M.V., Matias, P.M., Jancarik, J., Noguchi, S., Nishimura, S., Miura, K., Otsuka, E., and Kim, S.-H. (1988) Science 239, 888-893.
Dever, T.E., Glynias, M.J. and Merrick, W.C. (1987) Proc. Natl. Acad. Sci. USA 84, 1814-1818.
Garber, M.B., Yaremchuk, A.D., Tukalo, M.A., Egorova, S.P., Fomenkova, N.P. and Nikonov, S.V. (1990) J. Mol. Biol. 214, 819-820.
Grand, R.J.A. and Owen, D. (1991) Biochem. J. 279, 609-631.
Hansen, H.A.S., Volkmann, N., Piefke, J., Glotz, C., Weinstein, S., Makowski, I., Meyer, S., Wittman, H.G. and Yonath, A. (1990) Biochim. Biophys. Acta 1050, 1-7.
Imada, K., Sato, M., Tanaka, N., Katsube, Y., Matsuura, Y. and Oshima, T. (1991) J. Mol. Biol. 222, 725-738.
Jurnak, F. (1985) Science 230, 32-36.
Jurnak, F., Heffron, S., Schick, B. and Delaria, K. (1990) Biochim. Biophys. Acta 1050, 209-214.
Kaziro, Y., Itoh, H., Kozasa, T., Nakafuka, M. and Satoh, T. (1991) Ann. Rev. Biochem. 60, 349-400.
Kinzy, T.G., Freeman, J.P., Johnson, A.E. and Merrick, N.C. (1992) J. Biol. Chem. 267, 1623-1632.
Kjeldgaard, M. and Nyborg, J. (1992) J. Mol. Biol. 223, 721-742.
Kushiro, A., Shimizu, M. and Tomita, K.-I. (1987) Eur. J. Biochem. 170, 93-98.
La Cour, T.F.M., Nyborg, J., Thirup, S. and Clark, B.F.C. (1985) EMBO J. 4, 2385-2388.
Lippmann, C., Betzel, C., Dauter, Z., Wilson, K. and Erdmann, V.A. (1988) FEBS Lett. 240, 139-142.
Ludwig, W., Wallner, G., Tesch, A. and Klink, F. (1991) FEMS Microbiol. Lett. 78, 139-144.
Masuda, E., Louie, A. and Jurnak, F. (1985) J. Biol.Chem. 260, 8702-8705.
Miller, D.L. and Weissbach, H. (1977) in "Molecular Mechanisms of Protein Biosynthesis," Weissbach, H. and Pestka, S. eds. Academic Press, New York, 323-373.

Nureki, O., Muramatsu, T., Suzuki, K., Kohda, D., Matsuzawa, H., Ohta, T., Miyazawa, T. and Yokoyama, S. (1991) J. Mol. Biol. 226, 3268-3277.

Nureki, O., Suzuki, K., Hara-Yokoyama, M., Kohno, T., Matsuzawa, H., Ohta, T., Shimizu, T., Morikawa, K., Miyazawa, T. and Yokoyama, S. (1992) Eur. J. Biochem. 204, 465-472.

Ott, G., Faulhammer, H.G. and Sprinzl, M. (1989) Eur. J. Biochem. 184, 345-352.

Pai, E.F., Kabsch, W., Krengel, U., Holmes, K.C., John, J. and Wittinghofer, A. (1989) Nature 341, 209-214.

Parmeggiani, A., Swart, G.W.M., Mortensen, K.K., Jensen, M., Clark, B.F.C., Dente, L. and Cortese, R. (1987) Proc. Natl. Acad. Sci. USA 84, 3141-3145.

Peter, M. E., Reiser, C.O.A., Schirmer, N.K., Kiefhaber, T., Ott, G., Grillenbeck, N.W. and Sprinzl, M. (1990a) Nucleic Acids Res. 18, 6889-6893.

Peter, M.E., Schirmer, N.K., Reiser, C.O.A. and Sprinzl, M. (1990b) Biochemistry 29, 2876-2884.

Reshetnikova, L.S., Reiser, C.O.A., Schirmer, N.K., Berchtold, H., Storm, R., Hilgenfeld, R. and Sprinzl, M. (1991) J. Mol. Biol. 221, 375-377.

Reshetnikova, L.S., Schirmer, N.K., Reiser, C.O.A., Berchtold, H., Storm, R., Hilgenfeld, R. and Sprinzl, M. (1992) J. Crystal Growth 122, 360-365.

Satoh, M., Tanaka, T., Kushiro, A., Hakoshima, T. and Tomita, K.-I. (1991) FEBS Lett. 288, 98-100.

Schirmer, N.K., Reiser, C.O.A. and Sprinzl, M. (1991) Eur. J. Biochem. 200, 295-300.

Schlichting, I., Almo, S.C., Rapp, G., Wilson, K., Petratos, K., Lentfer, A., Wittinghofer, A., Kabsch, W., Pai, E.F., Petsko, G.A. and Goody, R.S. (1990) Nature 345, 309-315.

Seidler, L., Peter, M., Meissner, F. and Sprinzl, M. (1987) Nucleic Acids Res. 15, 9263-9277.

Thompson, R.C. and Karim, A.M. (1982) Proc. Natl. Acad. Sci. USA 79, 4922-4926.

Trakhanov, S., Yusupov, M., Shirokov, V., Garber, M., Mitschler, A., Ruff, M., Tierry, J.-C. and Moras, D.J. (1989) J. Mol. Biol. 209, 327-328.

van Noort, J.M., Kraal, B., Bosch, L., la Cour, T.F.M., Nyborg, J. and Clark, B.F.C. (1984) Proc. Natl. Acad. Sci. USA 81, 3969-3972.

van Noort, J.M., Kraal, B., Sinjorgo, K.M.C., Persoon, N.L.M., Johanns, E.S.D. and Bosch, L. (1986) Eur. J. Biochem. 160, 551-561.

Weißhaar, M., Ahmadian, R., Sprinzl, M., Satoh, M., Kushiro, A. and Tomita, K. (1990) Nucleic Acids Res. 18, 1902.

TOPOGRAPHY OF THE TERNARY EF-Tu/GTP/Phe-tRNA[Phe] COMPLEX AS STUDIED BY CROSSLINKING AND LIMITED PROTEOLYSIS

Joseph Reinbolt, Marie-Hélène Metz, Pascale Romby
Bernard Ehresmann and Chantal Ehresmann

UPR "SMBMR" N° 9002 CNRS
IBMC, 15, rue René Descartes
F-67084 Strasbourg Cedex

INTRODUCTION

The prokaryotic elongation factor EF-Tu is a monomeric protein of Mr 43200 found in the cytoplasm and constitues 5-10% of the cell proteins. Its function is to recognize, to transport and to position correctly the aminoacyl-tRNA onto the ribosome through the correct codon-anticodon base-pairing. During the polypeptide chain elongation cycle, EF-Tu undergoes series of conformational adaptations allowing the protein to recognize in a coordinate manner different substrates including GDP, GTP, aminoacyl-tRNA, elongation factor EF-Ts and the 70S ribosome. Hydrolysis of GTP in GDP and interaction with the 50S subunit leads to the release of EF-Tu in the form of a binary EF-Tu/GDP complex exhibiting a low affinity for tRNA and ribosome. Reactivation of the factor requires a GDP/GTP exchange which is promoted by a transient interaction with EF-Ts. To improve our understanding of the molecular mechanism of polypeptide elongation, a detailed knowledge of the specificity of the ternary complex formation and of the conformation of the protein and tRNA molecules within the complex is required. Numerous investigations of the EF-Tu/GTP/aminoacyl-tRNA ternary complex have been carried out. The formation of the ternary complex has been extensively studied by kinetic studies[32,26,25] and its conformation has been approached by small X-ray[31], neutron scattering[2,30] and NMR studies[14]. It has also been demonstrated that the ternary complex formation is tRNA sequence dependent. Comparative study of the binding of the initiator and elongator Met-tRNA complex indicates that the base pair including the first nucleotide in the tRNA is one of the essential requirement for the ternary complex formation[11]. In order to determine the regions of the tRNA molecule in close contact with EF-Tu within the ternary complex and to study the conformational changes of the tRNA induced by the protein, several approaches have been used : footprinting experiments, either by chemical probing[3,33,7] or by nuclease digestion[15,4,40,42,13] and crosslinking experiments[20]. Some of these results suggest a model of interaction in which the L-shaped tRNA lies parallel onto the protein as already proposed by Kabsch et al.[19]. This model provides large surface contacts as supported by the high binding constants[26,25] and involvement of the tRNA acceptor-arm and the T-arm are strongly suggested in the interaction with EF-Tu. Several amino acid residues of EF-Tu were found protected against selective chemical modifications by the tRNA binding : e.g. histidines 66 and 118 become protected by the aminoacyl-tRNA from photooxidation[17,18] and lysines 2, 4, 237, 248, 263 and 282 at their ε-amino group from labelling with ethylacetimidate[1]. Furthermore, the binding of L-1-tosylamido-2-phenylethylchloro-methylketone (TPCK) to cystein 81 prevents the interaction of EF-Tu with the aminoacyl-tRNA[16]. Specific crosslinks between EF-Tu and the aminoacyl-tRNA have been generated within the ternary complex : one takes place through histidine 66 and the amino acid residue of N-bromoacetyllysyl-tRNA[8], the other through lysine 208 and lysine 237 with the 3' oxidized tRNA in the presence of kirromycin[38,37].

The use of *trans*-dichlorodiammine platinum (II) (*trans*-DDP) as a reversible crosslinking agent has been developed in our laboratory and has been applied to promote RNA-protein crosslinks in the ribosome[36], in aminoacyl-tRNA synthetase/tRNA complexes[36] and to determine the 16S rRNA region crosslinked to initiation factor 3 within specific IF3-30S complex[10,29]. As previously described[36], *trans*-DDP has a square planar geometry where the two chlorines span a 7Å long distance. Both chlorines can easily be substituted by stronger nucleophilic groups. In ribonucleoprotein complexes binding positions are specific : on the RNA side primarily to position N7 of guanines and to a lesser extent to N1 of adenines and N3 of cytosines. On the protein side, at neutral pH, it binds to the sulfur atom of cysteines and methionines and to the unprotonated imidazole ring of histidines[36]. As to the coordination bonds induced between platinum and the acceptors, these crosslinks can either be kept stable under certain solvent conditions or can be reversed by the addition of stronger nucleophilic groups.

Previously, we carried out two distinct studies using *trans*-DDP to induce specific crosslinks within the ternary EF-Tu/GTP/Phe-tRNA^Phe complex at a sufficient yield to isolate the coordinated complex. Wikman et al.[41] determined two tRNA regions crosslinked to EF-Tu : a major one (nucleotides 58 to 65) located in the 3'-part of the T-stem and a minor one (nucleotides 31 to 42) including the anticodon arm. More recently, we isolated the crosslinked peptides/tRNA complex and we characterized two regions of EF-Tu crosslinked to Phe-tRNA[Phe] : a major one (75%) including residues 56 to 68 and a minor one (25%) containing residues 118 to 124[28].

This paper reviews the two previous studies and presents additional results obtained by a methodology designed to correlate tRNA and EF-Tu crosslinked regions. A short peptide/oligonucleotide crosslinked complex has been characterized. We also report the tRNA induced protection of amino acid residues of EF-Tu against protease digestion. These results are discussed in the light of the latest knowledge of the topography of EF-Tu/tRNA complex.

RESULTS

1)   Specificity of the crosslinking reaction by *trans*-DDP

After platination of the ternary EF-Tu/GTP/[14C]Phe-tRNA[Phe] complex, the non crosslinked species were dissociated in the presence of 0.5M NaCl. After elimination of the salt, a second platination procedure was performed by addition of a large excess of [14C]Phe-tRNA[Phe] in order to crosslink the totality of the free crosslinking sites. The recovery of a 1:1 EF-Tu/[14C]Phe-tRNA[Phe] complex indicates that the tRNA was crosslinked at its specific sites. In a previous paper[28] we indicated that an average of 35 to 40% of complex is crosslinked.

2)   Identification of crosslinking sites in the EF-Tu/GTP/tRNA ternary complex

Three distinct procedures were employed for the identification of crosslinking sites.

2.1 Characterization of crosslinking sites within tRNA

Using the procedure previously described[41], two crosslinking sites within the tRNA have been unambiguously identified. A RNase T1 octamer (nucleotides 58-65) containing a part of the T-loop and the T-stem was found crosslinked at a high and stable extent. Two subfragments encompassing nucleotides 58-63 and 61-65 resulting from non enzymatic phosphodiester cleavage were also found. A second tRNA region (nucleotides 25-45) was also found crosslinked to EF-Tu at a weaker and more variable extent than the octamer 58-65. This region spreads over the anticodon stem and loop.

2.2 Characterization of crosslinking sites within EF-Tu

The procedure described previously[28] led to the isolation of two Staphylo-coccus aureus V8 protease fragments. The major moiety (75%) includes peptides S1a (residues 56-68) and S2a (residues 64-68). The minor moiety (25%) contains peptide Sb (residues 118-124). This result is confirmed by identification of two large crosslinked peptides Ta (residues 45-74) and Tb (residues 117-154) obtained after 6h of tryptic digestion.

2.3 Attachment sites between EF-Tu and tRNA^{Phe}

We developed a procedure allowing to isolate the peptide(s)/oligonucleotide(s) rapidly and in sufficient quantity to be sequenced : EF-Tu/GTP was first platinated before interaction with tRNA leading to the formation of the crosslinked ternary complex, the non-crosslinked one was dissociated by 0.45 M NaCl. Both the crosslinked and free tRNA were directly digested by RNase T1. The resulting crosslinked EF-Tu/oligonucleotides complexes were fractionated from free oligonucleotides by filtration on glass fiber filters. Then, the crosslinked EF-Tu/oligonucleotide(s) were digested by Staphylococcus aureus V8 protease and peptide(s)/oligonucleotide(s) separated on HPLC. Figure 1 shows the

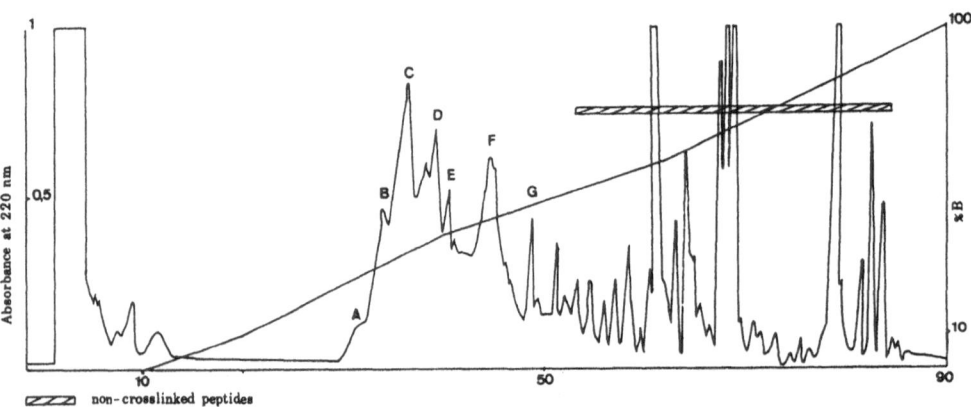

Figure 1 . Separation of peptide/oligonucleotide complexes on HPLC. Reverse-phase HPLC on a Delta-Pak TMC18 column Waters-Millipore, 15 x 0.2 cm with a HPLC apparatus Waters-Millipore. Flow-rate 0.25 ml/min at 25°C. Time of chromatography : 90 min. Solvent A : 0.1% trifluoroacetic acid ; solvent B : acetonitrile. The gradient is indicated by the line. Monitoring at 220 nm.

elution pattern obtained. Comparison with Staphylococcus aureus V8 protease digest of free EF-Tu allowed to locate non-crosslinked peptides (dashed box). This was confirmed by direct automated Edman degradation. Fractions A to G were submitted to automatic sequencing by the Edman degradation procedure and to oligonucleotide sequence analysis. We found that fractions A, C, D, E, F and G contain exclusively free oligonucleotides unspecifically retained on glass fiber filters. Only fraction B contains one unique peptide identified with a fragment corresponding to residues 65 to 70, resulting from the cleavage of the peptide bond : Thr64-Ser65. Table 1 shows the stepwise degradation of this peptide. PTH His66 was not found, strongly suggesting that His66 is involved in the crosslinking site. This result would be consistent with the trans-DDP binding specificity.

Table 1. Automatic Edman degradation of the peptide/oligonucleotide complex

Cycle	Residue	Amino acid
1	65	Ser
2	66	---
3	67	Val
4	68	Glu
5	69	Tyr
6	70	Asp

In order to characterize RNA fragment(s) crosslinked to this peptide fragment, fraction B was submitted tó the crosslinking reversion procedure and 5'-end labelling. The recovered oligonucleotides were fractionated by two successive cycles of gel electrophoresis on 15% and 20% polyacrylamide ⁻8 M urea. After autoradiography, three major bands could be detected. The corresponding RNA fragments were excised and eluted from the gel. It became evident that a RNase T1 fragment (nucleotides 62 to 76) containing a part of the T-loop, the T-stem and the 3'acceptor end was found crosslinked at a high yield. Two subfragments were also identified : nucleotides 64 to 76 and nucleotides 64-75 most probably resulting from unspecific phosphodiester cleavages. Since the crosslinks were reversed before sequencing, the platinated residues could not be directly identified. However the T1 RNase resistant G65 and G71 could be regarded as potential platination sites.

In order to insure that no contaminating RNA or EF-Tu fragments are coeluted with the crosslinked peptide/oligonucleotide complexes, two control experiments were achieved. In the first control non-platinated ternary complex was codigested by RNase T1 and Staphylococcus aureus V8 protease using the same procedure as for platinated complex. The sample was then fractionated by HPLC and analysed as previously described : no detectable amount of both components, peptide or oligonucleotide, could be detected at the position of peak B. In the second control, platinated EF-Tu was submitted to digestion steps by RNase T1 and Staphylococcus aureus V8 protease. The final digest was filtered as previously described and fractionated by HPLC. Peak B is not present in this chromatogram (results not shown).

3. Protection of some residues of EF-Tu/GTP by tRNA within the crosslinked EF-Tu/GTP/tRNA complex against proteolytic digestion

In order to determine the relative positions of the two macromolecules EF-Tu and tRNA within the ternary complex, limited endoprotease digestions of the crosslinked ternary EF-Tu/GTP/tRNA complex were achieved. A time limited Staphylococcus aureus V8 protease digestion of the crosslinked ternary complex was carried out by incubation in the dark during 6 h at 37°C (enzyme/substrate ratio : 1/20 by weight). A large crosslinked peptide was characterized (residues 41-70) containing residues Asp47, Asp50, Glu54, Glu55 and Glu68. In an experimental control of free EF-Tu/GDP, all these residues are cleavage sites for the Staphylococcus aureus V8 protease under the same limited conditions. A similar result was obtained by time limited tryptic digestion of the crosslinked ternary complex EF-Tu/GTP/tRNA by incubation in the dark during 6 h at 37°C (enzyme/substrate ratio : 1/20 by weight). Again, we isolated a large peptide stretching from residues 45 to 74 (Table 2) and covering about the same region as the peptide generated by Staphylococcus aureus V8 digestion overlapping region 41-70. In the crosslinked ternary complex, the peptidyl bonds with Lys56 and Arg68 escape tryptic digestion, quite the contrary to free EF-Tu/GDP. Interestingly, the large peptides obtained by Staphylococcus aureus V8 as well as by trypsine digestion, are consistent with a protection by tRNA of the region encompassing residues 41 to 74 against enzymatic digestion.

Table 2. Peptide 45-74

```
45                      Me              66              74
                         |
A F D Q I D N A P E E K A R G I T I N T S H V E Y D T P T R
```

DISCUSSION

Altogether our results bring strong evidences that the region containing residues 45 to 74 from prokaryotic elongation factor EF-Tu might be associated with tRNA binding. In the refined structure of elongation factor EF-Tu from *E.coli* by Kjeldgaard and Nyborg[23] this peptidic fragment corresponds to the so called effector loop, which is thought to be involved in the tRNA-dependent GTP hydrolysis[39,35]. It is remarkable to point out that region 45 to 74 of prokaryotic EF-Tu is highly conserved in all bacterial species studied so far (Table 3). Moreover our results strongly suggest that peptide 64-68 is linked by *trans*-DDP to nucleotides 58-65 within the tRNA, and most probably through the amino acid His 66 and the nucleotide G65 or G71. In this arrangement, the T-stem of tRNA would be in close contact with EF-Tu. As above mentioned (see Table 3) His66 is an amino acid remarkably conserved in all bacterial species studied so far, as well as the region 45 to 74, which belongs to the domain I of EF-Tu within the model proposed by Kjeldgaard and Nyborg[23]. His66 was already found to be protected by aminoacyl-tRNA and crosslinked to the chemically modified amino acid of the aminoacyl-tRNA[8]. This would also fit with the location of the 3'-end of the tRNA near the GDP-binding site[5]. In this arrangement, the 3'-strand of the acceptor- and T-stem are in contact with EF-Tu but not the T-loop. This orientation is also consistent with the footprinting experiments[5]. However, this localization of the 3'-end of the tRNA is not in agreement with the crosslinking of the 3'-oxidized tRNA to lysines 208 and 237 in the presence of kirromycin[38,37]. Furthermore our findings that the peptidic fragment 45 to 74 from EF-Tu appears protected by the tRNA against enzymatic digestion fit very well with the tridimensional model of EF-Tu/GTP/tRNA complex from *Thermus thermophilus* proposed by Sprinzl, where the T-arm of tRNA interacts with a EF-Tu region encompassing approximately the peptidic fragment 45 to 74 we characterized as protected by tRNA (see Hilgenfeld et al., this review).

Table 3. Sequence comparison of the tryptic peptide 45-74 from *E.coli* (1) EF-Tu crosslinked to the tRNA with EF-Tu sequences from *Euglena gracilis* chloroplast (2), *Saccharomyces cerevisiae* mitochondria (3) and *Thermus thermophilus* (4). Homologous areas are boxed.

```
     45                                                                    74

(1)  A F D Q -│I D│N│A P E E K A R G I T I N T│S│H V E Y D T│P T│R│
(2)  K R Y E D│I D│S│A P E E K A R G I T I N T│A│H V E Y E T│K N│R│
(3)  L D Y A A│I D│K│A P E E R A R G I T I│S│T│A│H V E Y E T│A K│R│
(4)  K D Y G D│I D│K│A P E E R A R G I T I N T│A│H V E Y E T│A K│R│
```

It must also be stressed that our results do not bring evidence for a strong interaction between the anticodon arm and EF-Tu in the ternary complex. However multiple orientations of the anticodon arm with respect to EF-Tu remain possible, especially if one would conciliate these results with the protection of the lysines in domain II[17,18]. Otherwise, footprinting data, small X-ray and small-angle scattering studies[31,2] suggest that the anticodon loop is protruding from the protein.

More recently Kinzy et al.[22] suggested a structural homology between both elongation factors, prokaryotic EF-Tu and eukaryotic elongation factor 1α. They proposed a possible ribosome binding site on the "upper" part of domain I of the elongation factor, this contact site allowing the positioning of the anticodon and the GTP near the ribosome. Indeed the EF-Tu/GTP/tRNA ternary complex seems to undergo a conformational change to form a quaternary structure with the ribosome itself, where domain I of elongation factor is playing the most important functional role. Moreover, Ehrenberg et al.[9] postulated that the ternary complex itself is extended to two EF-Tu molecules, one molecule of EF-Tu participating in A-site binding and the another one catalyzing the release of a deacylated tRNA from the E site of the ribosome (for a review see (9)).

REFERENCES

1. Antonsson, B. and Leberman, R. (1984) *Eur. J. Biochem.* 141, 483-487.
2. Antonsson, B., Leberman, R., Jacrot, B. and Zaccaï, G. (1986) *Biochemistry* 25, 3655-3659.
3. Bertram, S. and Wagner, R. (1982) *Biochemistry Internatl.* 4, 117-126.
4. Boutorin, A.S., Clark, B.F.C., Ebel, J.P., Kruse, T.A., Petersen, H.U., Remy, P. and Vassilenko, S. (1981) *J. Mol. Biol.* 152, 593-608.
5. Clark, B.F.C., La Cour, T.F.M., Nielsen, K.M., Nyborg, J., Petersen, H.U., Siboska, G.E. and Wikman, F.P. (1984) *In : Gene Expression, A. Benzon Symposium 19 (Clark, B.F.C. and Petersen, H.U. eds)* pp. 127-148, Munksgaard, Copenhagen.
6. Dirheimer, G. and Ebel, J.P. (1967) *Bull. Soc. Chim. Biol.* 49, 1679-1687.
7. Douthwaite, S., Garrett, R.A. and Wagner, R. (1983) *Eur. J. Biochem.* 131, 261-269.
8. Duffy, K.L., Gerber, L., Johnson, A.E. and Miller, D.L. (1981) *Biochemistry* 20, 4663-4666.
9. Ehrenberg, M., Rojas, A.M., Diaz, I., Bilgin, N., Weiser, J., Claesens, F. and Kurland, C.G. (1990). *In : "The Ribosome : structure, function and evolution (Hill, W.E. et al., eds)* pp. 373-379, American Society for Microbiology, Washington , USA.
10. Ehresmann, C., Moine, H., Mougel, M., Dondon, J., Grunberg-Manago, M., Ebel, J.P. and Ehresmann, B. (1986) *Nucleic Acids Res.* 14, 4803-4821.
11. Fischer, W., Doi, T., Ikehara, M., Ohtsuka, E. and Sprinzl, M. (1985) *FEBS Lett.* 192, 151-154.
12. Hachmann, J., Miller, D.L. and Weissbach, H. (1971) *Arch. Biochem. Biophys.* 147, 457-466.
13. Hansen, P.K., Wikman, F., Clark, B.F.C., Hersheu, J.W.B. and Petersen, H.U. (1986) *Biochimie* 68, 697-703.
14. Hilbers, C.W., Heershap, A., Walters, J.A. and Haasnoot, C.A. (1983) *"Nucleic Acids : The Vectors of Life" (Pullman B. and Jortner, J., eds.)* vol. 16, pp. 427-441. Reidel Publishing Company, London.
15. Jekowsky, E., Schimmel, P.R. and Miller, D.L. (1977) *J. Mol. Biol.* 114, 451-458.
16. Jonak, J., Petersen, T.E., Clark, B.F.C. and Rychlik, I. (1982) *FEBS Lett.* 150, 485-488.
17. Jonak, J., Petersen, T.E., Meloun, B. and Rychlik, I. (1984) *Eur. J. Biochem.* 144, 295-303.
18. Jonak, J., and Rychlik, I. (1987) *Biochim. Biophys. Acta* 908, 97-102.

19. Kabsch, W., Gast, W.H., Schulz, G.E. and Leberman, R. (1977) *J. Mol. Biol.* 117, 999-1012.

20. Kao, T.H., Miller, D.L., Abo, M. and Ofengand, J. (1983) *J. Mol. Biol.* 166, 383-405.

21. Kern, D., Giegé, R., Robbe-Saul, S., Boulanger, Y. and Ebel, J.P. (1975) *Biochimie* 57, 1167-1176.

22. Kinzy, T.G., Freeman, J.P., Johnson, A.E. and Merrick, W.C. (1992) *J. Biol. Chem.* 267, 1623-1632.

23. Kjeldgaard, M. and Nyborg, J. (1992) *J. Biol. Chem.* 223, 721-742.

24. Leberman, R., Antonsson, B., Giovanelli, R., Guariguata, R., Schumann, R. and Wittinghofer, A. (1980) *Anal. Biochem.* 104, 29-36.

25. Louie, A. and Jurnak, F. (1985) *Biochemistry* 24, 6433-6439.

26. Louie, A., Ribeiro, N.S., Reid, B.R. and Jurnak, F. (1984) *J. Biol. Chem.* 259, 5010-5016.

27. Maxam, A.M., and Gilbert, W. (1977) *Proc. Natl. Acad. Sci., USA* 74, 560-564.

28. Metz-Boutigue, M.H., Reinbolt, J., Ebel, J.P., Ehresmann, C. and Ehresmann, B. (1989) *FEBS Lett.* 245, 194-200.

29. Moine, H., Bienaimé, C., Mougel, M., Reinbolt, J., Ebel, J.P., Ehresmann, C. and Ehresmann, B. (1988). *FEBS Lett.* 228, 1-6.

30. Osterberg, R., Elias, P., Kjems, J. and Bauer, R. (1986) *J. Biomol. Str. Dyn.* 3, 1111-1120.

31. Osterberg, R., Sjöberg, B., Ligaarden, R. and Elias, P. (1981) *Eur. J. Biochem.* 117, 155-159.

32. Pingoud, A., Urbanke, C., Krauss, G., Peters, F. and Maass, G. (1977) *Eur. J. Biochem.* 78, 403-409.

33. Riehl, N., Giegé, R., Ebel, J.P. and Ehresmann, B. (1983) *FEBS Lett.* 154, 42-46.

34. Silberklang, M., Gillum, A.M. and RajBhandary, U.L. (1977) *Nucleic Acids Res.* 4, 4091-4108.

35. Toledo, H. and Jerez, C.A. (1989) *FEBS Lett.* 252, 37-41.

36. Tukalo, M.A., Kubler, M.D., Kern, D., Mougel, M., Ehresmann, C., Ebel, J.P., Ehresmann, B. and Giegé, R. (1987) *Biochemistry* 26, 5200-5208.

37. Van Noort, J.M., Kraal, B. and Bosch, L. (1988) *Proc. Natl. Acad. Sci.,* USA 82, 3212-3216.

38. Van Noort, J.M., Kraal, B.., Bosch, L., La Cour, T.F.M., Nyborg, J. and Clark, B.F.C. (1988) *Proc. Natl. Acad. Sci.,* USA 82, 3969-3972.

39. Van Noort, J.M., Kraal, B., Sinjorgo, K.M., Persoon, N.L.M., Johanns, E.S.D. and Bosch, L. (1986) *Eur. J. Biochem.,* 160, 557-561.

40. Weygand-Duracevic, I., Kruse, T.A.. and Clark, B.F.C. (1981) *Eur. J. Biochem.* 116, 59-65.

41. Wikmann, F.P., Romby, P., Metz, M.H., Reinbolt, J., Clark, B.F.C., Ebel, J.P., Ehresmann, C. and Ehresmann, B. (1987) *Nucleic Acids Res.* 15, 5787-5801.
42. Wikman, F.P., Siboska, G.E., Petersen, H.U. and Clark, B.F.C. (1982) *Embo J.* 1, 1095-1100.

SPECIFIC FUNCTIONS OF ELONGATION FACTOR Tu, A MOLECULAR SWITCH IN PROTEIN BIOSYNTHESIS, AS STUDIED BY SITE-DIRECTED MUTAGENESIS

Albert Weijland, Kim Harmark[*], Pieter H. Anborgh, and Andrea Parmeggiani

Structure Diverse d'Interventions n° 61840 du Centre National de la
Recherche Scientifique, Laboratoire de Biochimie, Ecole Polytechnique
F-91128 Palaiseau Cedex, France

INTRODUCTION

The elongation factor Tu (EF-Tu) is historically the first GTP-binding protein that was studied with respect to the relation between structure and function (for refs, see Weijland et al., 1992). It is therefore not surprising that it has widely been used as a model to investigate the common properties of the diverse families of these proteins that act as molecular switches in fundamental cellular pathways (Bourne et al., 1990; 1991). EF-Tu is a key component for protein biosynthesis and cell growth. As all GTP-binding proteins, it cycles between a GTP- and a GDP-bound state; in the GTP-induced state, it forms a complex with aa-tRNA that binds to the A-site of the mRNA-programmed ribosome. The interactions with aa-tRNA and ribosome are disrupted by the hydrolysis of the bound GTP that causes the release of EF-Tu•GDP, because its affinity for aa-tRNA is lower than that of EF-Tu•GTP by five orders of magnitude. The EF-Tu•GDP complex is several hundreds of times tighter than the EF-Tu•GTP complex and is characterized by a slow dissociation rate. Therefore the intervention of elongation factor Ts (EF-Ts), a GDP dissociation stimulator of EF-Tu, is crucial for a rapid regeneration of the active EF-Tu•GTP complex, as is required for the physiological rate of protein biosynthesis (10-20 amino acid residues incorporated•sec^{-1}). This canonical scheme of the EF-Tu cycle presents some controversial aspects: recently it has been proposed that two molecules of EF-Tu•GTP, instead of one, interact with one molecule of aa-tRNA, thus forming a pentameric complex ; two EF-Tu-bound GTP molecules should be hydrolyzed per each amino acid incorporated (Ehrenberg et al., 1990; Bilgin et al., 1992). In line with this, genetic and biochemical observations indicate the existence of intermolecular interactions between EF-Tu molecules, though so far this phenomenon has only been described for specific EF-Tu mutants (Vijgenboom and Bosch, 1989; Anborgh et al., 1991).

In the past years, we have tried to clarify some aspects of the structure-function relationships of EF-Tu from *E. coli*, carrying out site-directed mutagenesis of its encoding *tufA* gene. The positions to be mutated were chosen on the basis of the 3-D structure of EF-Tu and the structural analogy with other GTP binding proteins, in particular the mammalian protein c-H

The Translational Apparatus, Edited by K.H. Nierhaus
et al., Plenum Press, New York, 1993

ras p21. X-ray diffraction studies have shown that the EF-Tu molecule consists of 3 domains: N-terminal (N-) domain, middle (M-) domain and C-terminal (C-) domain (for refs, see Kjeldgaard and Nyborg, 1992; Figure 1). In EF-Tu the three consensus sequence elements that characterize the GTP-binding proteins concern loops 1, 4 and 8, three of the four loops that define the guanine nucleotide binding pocket. The first (G18HVDHGKT25) and the second (D80CPG83H) motifs are involved in the interaction with the phosphoryl groups, and the third one (N135KCD138) with the guanine base. To overcome the limits imposed by the present 3-D model EF-Tu$_{E.coli}$, which was derived from a nicked EF-Tu•GDP complex   lacking   residues

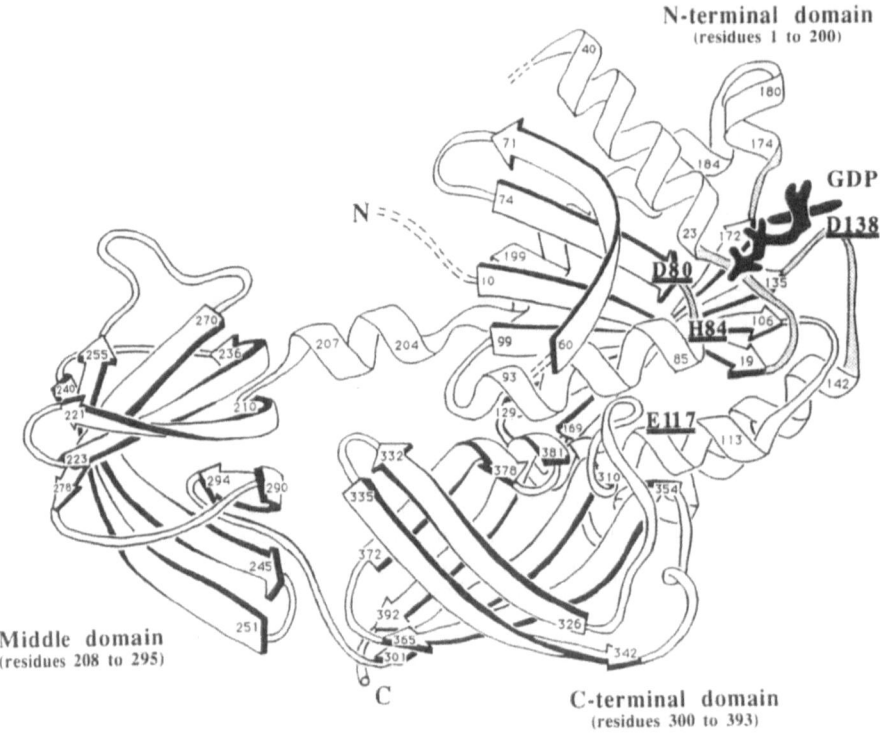

Figure 1. Three-dimensional model of a nicked EF-Tu•GDP complex lacking residues 45-58. The four loops defining the GDP-binding pocket are shaded. The positions of residues Asp80, His84, Glu117 and Asp138 in the N-terminal domain is indicated. Derived from Clark et al. (1990) and Kjeldgaard and Nyborg (1992).

45-58 and refined to medium resolution (2.6 Å), we had to take into account the 3-D models of protein p21•GppNp (for references, see Wittinghofer and Pai, 1991) and p21•GDP (for references, see Privé et al., 1992) elucidated at higher resolution (1.4 and 2.2 Å, respectively). This was justified by the observation that the secondary elements of the N-terminal domains of EF-Tu and p21 have the same topology and that 85 % of the main-chain coordinates of both models are superposable, despite a mere 17 % identity in terms of primary structure (Wittinghofer and Pai, 1991). Moreover, a growing number of experimental evidence indicates that in GTP-binding proteins residues in corresponding positions, even when nonconserved, can mediate similar functions. Therefore, function-structure relationships derived from one of these proteins can be applied to the members of the other families.

Several technical difficulties for an efficient expression and overproduction of mutated EF-Tus had to be overcome. EF-Tu is the most abundant protein in *E. coli* (5-10 % of total cellular protein) and is encoded by two nearly identical unlinked *tuf* genes (for references, see

Weijland et al., 1992). The overproduced mutated EF-Tu, if partially active, can compete with the chromosomal wild-type factor whose presence is required to ensure the viability of the cell. This competition may give rise to toxic effects. As an example overproduction of EF-TuN135D, mutated in a residue important for the architecture of the GTP-binding pocket, blocks the growth of the host *E. coli* cells. The normal growth is reestablished if the additional mutation Asp138→Asn, that fully abolishes the binding of guanine nucleotides, is introduced (Gümüsel et al., 1990).

The separation of plasmid-encoded mutated EF-Tu from the chromosome-borne EF-Tu, has already been accomplished several years ago with the aid of kirromycin, the first identified of the three antibiotics (kirromycin, pulvomycin and GE2270 A, for references see Parmeggiani and Swart, 1985; Anborgh and Parmeggiani, 1991) that inhibit protein biosynthesis by specifically affecting EF-Tu. Kirromycin, that inhibits the release of EF-Tu·GDP from the mRNA-programmed ribosome, competes with EF-Ts for binding to EF-Tu and changes the electrophoretic behaviour of EF-Tu. The last two properties were utilized for the separation of kirromycin-sensitive EF-Tu mutants from a chromosomal kirromycin-resistant EF-Tu (Swart et al., 1987; Anborgh et al., 1991). Recently, the production of engineered EF-Tu fused to glutathione-*S*-transferase (Knudsen et al., 1992; P.H. Anborgh, R.H. Cool, R. te Biesebeke, W. Bruns, and A. Parmeggiani, unpublished results) or containing a tag of His residues (Boon et al., 1992) has been successfully used for the rapid purification of mutated EF-Tu factors.

In this article we deal with the present state of knowledge of the structure-function relationships concerning the interaction with GTP and GDP and its intrinsic catalytic activity. In this context, a mutated EF-Tu with altered substrate specificity was utilized to investigate the energy consumption specifically depending on the EF-Tu interaction with aa-tRNA and the mRNA-programmed ribosome during the elongation cycle.

FUNCTIONS OF THE EF-Tu DOMAINS

Our first concern was to engineer the isolated N-terminal domain (including the first 202 N-terminal residues) that was named G domain and that contains the binding site for GDP and GTP (see Figure 1). However, the affiniy for GDP ($K_d = 2.1$ μM) approached that for GTP (K_d = 1 nM and 0.6 μM, respectively, at 0 °C) (see Table 1). The G domain could express an intrinsic GTPase activity, that was higher than the intrinsic GTPase of EF-Tu, as expected from the higher affinity for GDP of EF-Tu, that creates a thermodynamically unfavorable situation hindering the regeneration of GTP and thus the catalytic activity. The activities of the G domain were not affected by any of the ligands of the intact molecule, except for a small stimulation of its GTPase activity by the 70 S ribosome. These interesting properties of the G domain suggested to us its use as a suitable tool to study the structure-function relationships of the basic activities of EF-Tu (the binding of the nucleotide and the catalytic activity), avoiding the constraints imposed by the middle- and C-terminal domains. Recent experiments (P.H. Anborgh, unpublished results) have shown that already the removal of either the middle domain or of the C-terminal domain is sufficient to abolish almost entirely the ability of EF-Tu to distinguish between GTP and GDP. In both cases the K_d values for GTP and GDP lie in the μM range as for the G domain (Table 2). The action of the EF-Tu effectors is nevertheless differentially affected in these truncated molecules. For instance the dissociation rate of the GTP complex of the EF-TuΔM-domain was retarded by kirromycin, a typical effect of this antibiotic on the intact EF-Tu, even though the extent of the inhibitory effect was decreased. In contrast to this, no effect at all could be observed on EF-TuΔC-domain, that in turn was influenced by GE 2270 A. The EF-TuΔM-domain could form a stable complex with aa-tRNA somewhat more efficiently than the EF-TuΔC-domain but much less efficiently than the intact EF-Tu. The precise characterization of the activities of these two deleted EF-Tu molecules with respect to the

Table 1. Dissociation constants of EF-Tu, its isolated N-terminal domain (G domain), GST-EF-TuΔM and GST-EF-TuΔC for GDP and GTP

	K_d (GDP) μM	K_d (GTP) μM
EF-Tu	0.001	0.6
GST-EF-Tu	0.002	0.9
G-domain	2.1	3.6
GST-EF-TuΔM	3.0	10.0
GST-EF-TuΔC	2.1	6.0

These constants were determined at 0 °C.

interaction with the different ligands of EF-Tu is currently in progress. In Table 2, we resume the requirement of the EF-Tu-domains for its basic activities (GTPase and nucleotide binding) and for the formation of a stable complex between EF-Tu•GTP and aa-tRNA or the antibiotics kirromycin and GE2270 A, as obtained from results of our laboratory (Parmeggiani et al., 1987; P.H Anborgh, unpublished results).

Table 2. Requirement for N-, M - and C-domain for EF-Tu functions

	N-domain	M-domain	C-domain
Binding of:			
GDP or GTP	•		
kirromycin	•		•
GE2270 A	•	•	
aa-tRNA	•	•	
	•		•
Discrimination between GDP and GTP	•	•	•
Intrinsic GTPase activity	•		

EFFECT OF SUBSTITUTIONS His84→Gly AND Asp80→Asn IN THE SECOND CONSENSUS MOTIF OF THE G DOMAIN

Determination of the stereochemical course of the GTP hydrolysis, using chiral analogs of GTP, showed that the cleavage takes place in a single step, in line transfer of the γ-phosphate to a water molecule (Eccleston and Webb, 1982). The reaction proceeds with inversion of the configuration at the transferred phosphorus, implying the absence of a phospho-enzyme intermediate. The 3-D model of EF-Tu and analogy with other nucleotide binding proteins, such as DNAse 1 already suggested a few years ago that the second consensus sequence element (D80CPGH84) could be involved in the catalytic activity. As structural background for this reaction, Swart (1987) proposed a mechanism of a general acid-base catalysis, in which the side-chain carboxyl of Glu117 accepts a proton from the side-chain-nitrogen of His84 that activates a water molecule acting as nucleophile on the γ-phosphate. Although in the 3-D model the carboxyl group of the side-chain of Glu117 and the amino group of the side-chain of His84- are in hydrogen bond distance, this mechanism of general acid-base catalysis was not supported by the substitution Glu117→Gln, that did not essentially affect the GTPase activity of the G domain of EF-Tu (Table 3, Harmark et al., 1990). Nonetheless the possibility of a basic catalysis mechanism carried out by His84 alone, remained open. His84 is a conserved residue in elongation and initiation factors, that is situated at the edge of the second consensus sequence element. In the 3-D model of EF-Tu•GDP, its side-chain points away from the phosphoryl groups, the secondary amino group lying 11.4 Å from the next oxygen of the β-phosphoryl group (J. Nyborg, personal communication). Though this orientation allows neither a direct nor an indirect interaction between the side-chain of His84 and the substrate, the possibility of pronounced 3-D differences between the enzyme conformation of the GDP-bound state and the transition state, suggested us to test the functional consequences of the substitution of His84 by Gly (Cool and Parmeggiani, 1991). As a first important point, this mutation revealed little, if any, alterations of the nucleotide binding parameters, whereas the GTPase activity was strongly reduced to about 5 % of the wild-type activity, implying an effect on the catalysis (Table 3). The determination of the free energy of activation showed that in EF-TuH84G an uncharged hydrogen bond was disrupted, very likely involving an interaction with a water molecule. Two

Table 3. Dissociation constants and k_{cat} of wild-type G domain and three mutated G domains

G domain	K_d(GDP) μM	K_d(GTP) μM	k_{cat} $s^{-1} \cdot 10^3$
wild-type	2.1	3.6	0.06-0.12[a]
D80N	2.3	92	3.65[b]
H84G	2.5	4.1	0.078[c]
E117Q	2.7	2.3	0.13

[a] Depending on the experimental conditions. [b] In the presence of 35 % glycerol; under the same experimental conditions the k_{cat} of the wild-type G domain was 0.06 s⁻¹•10³. For details, see Harmark et al. (1992). [c] At 2 M KCl. For details, see Cool and Parmeggiani (1991).

possibilities arose from these results. The side-chain of His84 could activate a water molecule, that in turn acts as nucleophile on the γ-phosphate. Alternatively the side-chain of His84 could interact via a water molecule with an oxygen of the γ-phosphate contributing to the stabilization of the transition state of GTP. Against the former possibility, that was supported by the 3-D situation of the homologous residue (Gln61) in the p21•GppNp model (Pai et al., 1990; Krengel et al., 1990), was the presence of a significant residual GTPase activity that was exposed to the same cationic and pH-dependent constraints as the activity of the wild-type G domain (Cool and Parmeggiani, 1991). His84 can therefore, in strict sense, not be considered a catalytic residue. In our opinion the second mechanism, in which His84 contributes to stabilize the transition state of the substrate together with other residues, is more probable. A marked stabilization of the transition-state configuration of the substrate by the enzyme would lower the energy barrier of the hydrolysis via the binding energies of the interacting residues, thus facilitating the γ-phosphate cleavage. This mechanism can represent a major catalytic factor, as described for the hydrolysis of ATP by aminoacyl-tRNA synthetase (Fersht et al., 1988), removing the need for a strong nucleophile, such as a specifically activated water molecule.

In EF-Tu there are a considerable number of other interactions that can favor the catalysis. The γ-phosphate is polarized by the complexed Mg^{2+} ion, the side-chain of Lys24, probably also by Arg 58 and Thr61. To examine the role of the substrate-coordinated Mg^{2+} in EF-Tu GTPase, we substituted Asp80 (with Asn). Asp80 is the first residue of the second consensus sequence element and is a conserved in all GTPases. The 3-D model of EF-Tu•GDP as well as the analogy with the p21 model, strongly suggests that Asp80 is a component of a chelation ring such as Asp80-H_2O-Mg^{2+}-Thr25-Asp80. The functional effects of this substitution supported a key role of Asp80 in in the overall stabilization of the G domain, in the coordination of the Mg^{2+}•GTP complex and in the catalytic reaction. In fact, its substitution markedly destabilized the activity of the G domain, weakened the binding of GTP, and as the most striking result induced a strong increase in the intrinsic catalytic activity of the G domain, by almost two orders of magnitude (Table 3). The GTPase of G domainD80N was characterized by such a pronounced instability that under normal conditions it vanished within 5-10 min. This handicap was overcome by the addition of 30-40% glycerol that induced a linear multiple-round GTPase activity for several hours. That glycerol did not influence directly the catalysis was confirmed by the observation that under the same conditions the wild-type G domain was only little stimulated; moreover, the initial GTPase in the absence of glycerol corresponded closely to the turnover linear activity obtained in its presence (Harmark et al, 1992). Whereas the binding affinity for GDP of EF-TuD80N was about the same as that of the wild-type G domain, its affinity for GTP was reduced by almost 50 times. Most important, EDTA could mimic the effect of the mutation; in its presence the K_d of the wild-type G domain•GTP complex strongly increased, resembling that of the mutated complex that was essentially unaffected by EDTA. Interestingly, the GDP complexes of both mutated and wild-type G domain were virtually unaffected by EDTA, showing that the effect of magnesium concerned primarily the GTP complex. Taken together these observations indicate that the γ-phosphate cleavage involves a relieve of the constraints imposed by Asp80 on the Mg^{2+}•GTP coordination. Metal ions can favor the phosphate transfer by direct and indirect mechanism (Cooperman, 1982). A direct mechanism has been found to operate in enzymatic systems (Sigel et al., 1984). In this case a nucleophilic attack by a Mg^{2+}-bound hydroxide should take place in an intramolecular fashion, implying a penta-coordinate transition state of the γ-phosphate. A mechanism involving the substrate-coordinated magnesium could also be evoked to explain the dramatic stimulatory effect of the mRNA-programmed-ribosome on the EF-Tu-dependent GTPase activity, by five orders of magnitude. A picture may arise in which positive charges of EF-Tu residues such as Arg58, situated on the loop corresponding to the effector loop in p21 and thus possibly involved in the interaction with the ribosome, could shift the substrate-coordinated magnesium ion, triggering the burst-like cleavage of GTP, via direct or indirect mechanisms. But at the present state of our knowledge of the 3-D model of EF-Tu, the role of magnesium in the EF-Tu catalysis can only be speculative. It is expected that the 3-D model of the GppNp complex of EF-Tu from *Thermus thermophilus* (Reshetnikova et al., 1991) will contribute in an essential way to understand the mechanism of the EF-Tu GTPase.

ISOLATION AND PROPERTIES OF A MUTANT EF-Tu WITH ALTERED SUBSTRATE SPECIFICITY

The possibility of disposing of an EF-Tu with an altered substrate specificity was of great importance to define the precise role of the factor in the overall elongation process with respect to energy consumption. In elongation, the action of EF-Tu is associated with that of elongation factor G (EF-G) which is also a GTPase and supports the translocation of the polypetidyl-tRNA chain from the ribosomal A-site to the P site after the release of EF-Tu•GDP and peptide bond formation. Consequently, in the system utilizing wild-type EF-Tu, the precise determination of the contribution by these two factors to the dissipation of energy requires the introduction of correction factors that are difficult to estimate. This is particularly true considering that only a small portion of ribosomes (approx. 10 %, Wagner et al. 1982) is fully active in polypeptide elongation *in vitro*. It is noteworthy that EF-G can express a high GTPase activity uncoupled from polypeptide chain synthesis by interacting with ribosomes that are deficient in peptidyl transferase activity. Moreover it is difficult to eliminate completely the ribosome-bound GTPase activities that are not specifically involved in the basic reactions of the elongation cycle. In 1987, Hwang and Miller, using a maxi cell system reported that EF-TuD138N was able to bind XTP instead of GTP in the presence of aa-tRNA and ribosomes. This result was predicted by the 3-D model of EF-Tu in which the exocyclic amino group of the base forms a hydrogen bond with the side-chain carboxyl group of Asp138. These observations paved the way for a new approach of investigating the precise stoichiometry of the energy requirement of EF-Tu in the elongation cycle. The EF-Tu-dependent GTPase has since long been correlated with the proof-reading mechanism selecting the proper codon-antidocon interaction (Hopfield, 1974; Ninio, 1975; Ruusala et al., 1982; Thompson et al., 1981). A reinvestigation of these problems has become particularly actual considering the new model of Ehrenberg and Kurland (1990) proposing a pentameric complex between two molecules of EF-Tu•GTP and one of aa-tRNA, results which other authors have failed to confirm (Bensch et al., 1991).

To this purpose the substitution Asp138→Asn was introduced into the *tufA* gene cloned on pEMBL9⁺ and the mutated gene expressed from pCP40 under control of λPL promotor (Parlato and Parmeggiani, 1988). After transfer of the mutated gene into a pTTQ18 vector, under control of the *tac* promotor, EF-TuD138N was produced in *E. coli* and purified to homogeneity, completely free from the chromosome-borne kirromycin-resistant EF-Tu (A. Weijland and A. Parmeggiani, in preparation). To our fortune, the possibility to use EF-TuD138N as a suitable substitute of wild-type EF-Tu was supported by the observation that EF-TuD138N displayed an activity in poly(U)-directed poly(Phe) synthesis, that was virtually the same as that of wild-type EF-Tu (Figure 2).

This feature was complemented by another important finding: in the absence of EF-Tu the complete system used for polypeptide synthesis *in vitro* was practically devoid of any XTPase activity, while a strong background EF-Tu-independent GTPase (in the most purified system mainly dependent on EF-G-GTPase uncoupled from protein biosynthesis) was present. The characteristics of this system convinced us that the XTP-dependent EF-TuD138N could have been a proper tool to analyze the EF-Tu-mediated energy consumption during the elongation cycle.

The unequivocal answer obtained with this EF-Tu mutant was that two nucleotide triphosphates were hydrolyzed by EF-Tu for each amino acid incorporated into the polypeptide

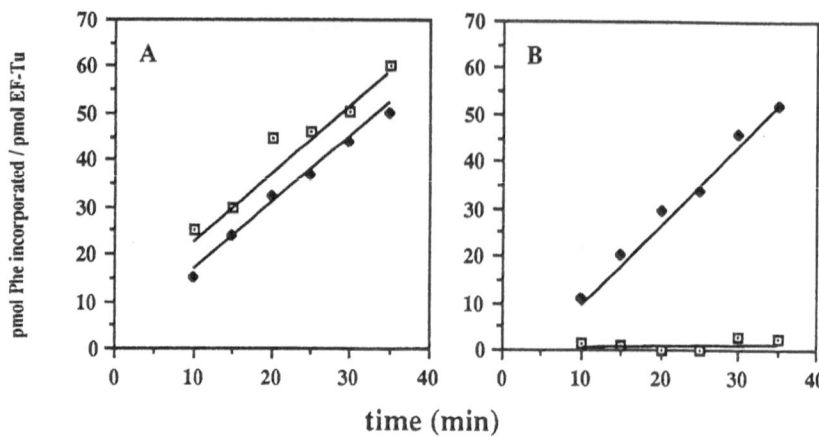

Figure 2. Dependence on XTP of the poly(Phe) synthesis sustained by EF-TuD138N. The reaction mixture (400 μl) contained 50 mM Tris-HCl, pH 7.6, 70 mM NH₄Cl, 7 mM MgCl₂, 7 mM 2-mercaptoethanol, 0.025 μM wild-type EF-Tu (A) or EF-TuD138N (B), 0.2 μM EF-G, 0.2 μM EF-Ts, plus XTP (filled symbols), minus XTP (open symbols), 1 mM ATP, 100 μM GTP, 130 μg poly(U), 0.5 μM ribosomes, 6 μM (^{14}C)Phe (0.05 Ci•mmol^{-1}, Amersham), 4 μM tRNA^{Phe} (70 % pure) and saturating amounts of phenylalanyl-tRNA synthetase. The reaction was carried out at 30 °C and triggered by adding Phe-tRNA^{Phe}. At the times indicated, a sample (50 μl) was withdrawn and the poly(Phe) incorporated measured by precipitation in hot trichloroacetic acid as reported (Swart et al., 1987). The blanks obtained with the complete system minus EF-Tu were subtracted. Highly purified NH₄Cl-washed ribosomes, EF-Tu, EF-Ts, EF-G and phenylalanyl-tRNA synthetase were prepared as described (Swart et al., 1987).

chain (Weijland and Parmeggiani, 1993). This stoichiometry was obtained with poly(U)-programmed ribosomes in diverse experimental conditions, such as by varying the substrate concentration, the EF-Tu:ribosome ratio, the rate of poly(Phe) synthesis or the temperature (between 20 and 40 °C) at which the reaction was carried out. The same 2:1 stoichiometry was also obtained in the absence of EF-G, that is by measuring the XTP hydrolyzed per Phe-tRNA^{Phe} enzymatically bound to the poly(U)•ribosome complex (Table 4). These results raise a number of questions. The most obvious ones are: why should two γ-phosphates be split per each cognate aa-tRNA properly positioned on the ribosome? Is this stoichiometry correlated to the proof-reading process? Is it mediated by the formation of a pentameric complex between two

Table 4. Stoichiometry between XTP hydrolyzed by EF-TuD138N and phenylalanine incorporated into a polypeptide chain or enzymatically bound to the ribosomal A-site.

		pmol XTP hydrolyzed/ pmol of Phe incorporated	pmol XTP hydrolyzed/ pmol of Phe bound
Poly(Phe) translation system with EF-TuD138N	plus EF-G	2.0	
	minus EF-G		2.0

molecules of EF-Tu•GTP and aa-tRNA? At present it is not yet possible to answer these questions in a satisfactory way. EF-TuD138N is also involved in the aa-tRNA selection on poly(U)-programmed ribosomes. When we use aa-tRNAs with different affinity for poly(U), the ratio between XTP hydrolyzed and amino acid bound to the ribosome changes as expected from the proof-reading mechanism (Ruusala et al., 1984; Thompson, 1988). In fact, in the presence of quantities of XTP limiting the regeneration of the complex between EF-TuD138N•XTP and aa-tRNA the amount of XTP hydrolyzed increases with increasing the affinity of the codon-anticodon interaction (Weijland and Parmeggiani, 1993). This suggests that the extent of the codon-anticodon interaction matching the productive interaction between the ribosome and EF- Tu directly influences the hydrolysis of GTP. A proof-reading mechanism with two molecules of GTP hydrolyzed sequencially would have the ability to increase the accuracy of the system, according to Hopfield (1974) and Ninio (1975). Additional experiments, also involving a reexamination of the EF-Tu•GTP•aa-tRNA interaction and possibly utilizing biochemical and genetic techniques in a coordinated fashion, will be required to solve these questions in a less hypothetical way.

ACKNOWLEDGEMENTS

This work was carried out in the framework of grant BAP-0066-F (CD) from the Biotechnology Action Programme of the Commission of the European Community and was supported by the "Association pour la Recherche contre le Cancer" and the "Institut National de la Santé et de la Recherche Médicale". K.H. was recipient of a fellowship from the Carlsberg Foundation.

FOOTNOTES

* Present address: Russell Grimvade School of Biochemistry, University of Melbourne, Parkville, Victoria 3053, Australia.

REFERENCES

Anborgh, P.H., and Parmeggiani, A., 1991, *EMBO J.* 10:779.
Anborgh, P.H., Swart, G.W.M., and Parmeggiani, A., 1991, *FEBS Lett.* 292:232.
Bensch, K., Pieper, U., Ott, G., Schirmer, N., Sprinzl, M., and Pingoud, A., 1991, *Biochimie* 73:1045.
Bilgin, N., Claesens, F., Pahverk, H., and Ehrenberg, M., 1992, *J. Mol. Biol.* 224:1011.
Boon, K., Vijgenboom, E., Madsen, L., Talens, J., Kraal, B., and Bosch, L., 1992, *Eur J. Biochem.* 210:177.
Bourne, H.R., Sanders, D.A., and McCormick, F., 1990, *Nature* 348:125.
Bourne, H.R., Sanders, D.A., and McCormick, F,. 1991, *Nature* 349:117.
Cool, R.H., and Parmeggiani, A., 1991, *Biochemistry* 30:362.
Cooperman, B.S., 1982, *Methods in Enzymology* 87:526.
Eccleston, J.F. and Webb, M.R., 1982, *J. Biol. Chem.* 257:5046.
Ehrenberg, M., Rojas, A-M., Weiser, J., and C.G. Kurland, 1990, *J. Mol. Biol.* 211:739.
Fersht, A.R., Knill-Jones, J.W., Bedouelle, H., and Winter, G., 1988, *Biochemistry* 27:1585.
Gümüsel, F., Cool, R.H Weijland, A., Anborgh, P.H., and Parmeggiani, A., 1990, *Biochim. Biophys. Acta* 1050:215.
Harmark, K., Cool, R.H., Clark, B.F.C., and Parmeggiani, A., 1990, *Eur. J. Biochem.* 194:731.

Harmark, K., Anborgh, P.H., Merola, M., Clark, B.F.C., and Parmeggiani, A., 1992, *Biochemistry* 31:7367.

Hopfield, J.J., 1974, *Proc. Natl. Acad. Sci. USA* 71:4135.

Hwang, Y.-W., and Miller, D., 1985, *J. Biol. Chem.* 260:11496.

Kjeldgaard, M., and Nyborg, J., 1992, *J. Mol. Biol.* 223:721.

Knudsen, C.R., Clark, B.F.C., Degn, B., and Wiborg, O., 1992, *Biochem. Internat.* 28:353.

Krengel, U., Schlichting, I., Scherer, A., Schumann, R., Frech, M., John, J., Pai, E.F., and Wittinghofer, A., 1990, *Cell* 62:539.

Ninio, J., 1975, *Biochimie* 57:587.

Pai, E.F., Krengel, U., Petsko, G.A., Goody, R.S., Kabsch, W. and Wittinghofer, A., 1990, *EMBO J.* 9:2351.

Parlato, G., and Parmeggiani, A., 1988, *Italian J. Biochem.* 37:353.

Parmeggiani, A., and Swart, G.W.M., 1985, *Annu. Rev. Microbiol.* 39:557.

Privé, G.G., Milburn, M.V., Tong, L., De Vos, A.M., Yamaizumi, Z., Nishimura, S., and Kim, S.-H., 1992, *Proc. Natl. Acad. Sci. USA* 89:3649.

Reshetnikova, L.S., Reiser, C.O.A., Schirmer, N.K., Berchtold, H., Storm, R., Hilgenfeld, R., and Sprinzl, M., 1991, *J. Mol. Biol.* 221:375.

Ruusala, T., Ehrenberg, M., and Kurland, C., 1982, *EMBO J.* 6:741.

Ruusala, T., and Kurland, C.G., 1984, *Mol. Gen. Genet.* 198:100.

Sigel, H., Hofstetter, F., Martin, R.B., Milburn, R.M., Scheller-Krattiger, V., and Scheller, K. H., 1984, *J. Am. Chem. Soc.* 106:7935.

Swart, G.W.M., 1987, Ph. D. Thesis, Leiden University.

Swart, G.W.M., Parmeggiani, A., Kraal, B., and Bosch, L., 1987, *Biochemistry* 26:2047.

Thompson, R.C., Dix, D.B., Gerson, R.B., Karim, A.M., 1981, *J. Biol. Chem.* 256:81.

Thompson, R.C., 1988, *Trends Biochem. Sci.* 13:91.

Vijgenboom, E., and Bosch, L., 1989, *J. Biol. Chem.* 264:13012.

Wagner, E.G.E., Jelenc, P.C., Ehrenberg, M., and Kurland C., 1982, *Eur. J. Biochem.* 122:193.

Weijland, A., Harmark, K., Cool, R.H., Anborgh, P.H., and Parmeggiani, A., 1992, *Mol. Microbiol.* 6:683.

Weijland, A., and Parmeggiani, A., 1993, *Science* 259:1311.

Wittinghofer, A., and Pai, E.F., 1991, *Trends Biochem. Sci.* 16:382.

EF-Tu STOICHIOMETRIES IN CODE TRANSLATION

Måns Ehrenberg, Neşe Bilgin, and Judith Scoble

Department of Molecular Biology
BMC, Box 590
S-751 24 Uppsala
Sweden

BACKGROUND

According to early observations the number of GTP's hydrolyzed on EF-Tu per peptide bond is close to one [Gordon, 1969; Miller and Weissbach, 1977; Thompson and Stone, 1977]. These measurements were made in partial translation systems lacking EF-G, and it took more than a decade before the corresponding experiments could be made in a full translation system [Ruusala et al, 1982a]. In the latter system, which had been optimized for rate as well as accuracy [Jelenc and Kurland, 1979; Wagner et al, 1982; Ehrenberg et al, 1990ab], the quantitative evaluation of the experiments depended on external estimates of the concentration of active EF-Tu as well as of the dissociaton rate of GDP from the factor [Ruusala et al, 1982a; Ehrenberg and Kurland, 1988]. They found that one GTP is hydrolyzed per peptide bond in accord with previous observations. A more direct assay was designed by Kakniashvili et al [1986], and they also concluded that one molecule of GTP is dissipated in EF-Tu function per cognate peptide bond. These latter results were ambiguous, however, since the rate of exchange between labelled and unlabelled GTP in the ternary complex did not seem to be controlled properly, as explained by Rojas [1988] and Ehrenberg et al [1990a].

At this time the estimate of the GTP to peptide bond stoichiometry in translation seemed like an established fact, and it was not until the development of new and technically superior assays, suitable for quench-flow studies [Bilgin et al, 1992], that new experimental

The Translational Apparatus, Edited by K.H. Nierhaus
et al., Plenum Press, New York, 1993

results made a revision of previous views necessary . With these improved methods we found, much to our surprise, that there are two GTP's hydrolyzed per peptide bond in EF-Tu function in poly(Phe) synthesis [Ehrenberg et al, 1990abc].

One important feature of our new experiments is that they encompass a number of internal controls that make estimates of the GTP to peptide bond stoichiometry particularly reliable. Since our controversial discovery has been remarkably ignored in the literature and since it may turn out to be fundamental for our understanding of translation in prokaryotes, the basic experiment will be recapitulated here.

Ternary complex is prepared with ^{3}H-GTP in two different ways. In the first the complex is made with an excess of Phe-tRNA^{Phe} and with guanine nucleotides only from EF-Tu and from ^{3}H-labelled GDP. After preincubation more than 98% of the GDP is converted to GTP by an energy pump, and the major fraction of GTP molecules in the mix is in ternary complex. In the second case we add a precisely determined amount of unlabelled GDP to the mixture, which increases the total amount of guanine nucleotides about threefold in relation to the first case [Ehrenberg et al, 1990ab].

The incubation starts by addition of preinitiated ribosomes, EF-G and a huge excess of unlabelled GTP to either of the ternary complex mixes. In the subsequent burst of poly(Phe)-synthesis the preexisting ternary complex is consumed in less than five seconds. In this time all EF-Tu·GTP in the ternary complex is changed to free EF-Tu·GDP.

There is no exchange between labelled and unlabelled GTP in the ternary complex during this rapid burst, which is followed by a slow steady state phase. In the latter almost all EF-Tu is in complex with GDP, since dissociation of GDP from EF-Tu is very slow in the absence of EF-Ts [Ruusala et al, 1982b]. As soon as a ^{3}H-labelled GDP leaves EF-Tu it is rapidly converted to ^{3}H-GTP by the energy regeneration system. The EF-Tu molecule is soon in complex with an unlabelled GTP drawn from the large pool of non radioactive GTP in the incubation mix, and then it rapidly turns to unlabelled EF-Tu·GDP.

Fig. 1b shows the natural logarithm of the ratio between ^{3}H- labelled GDP and total labelled guanine nucleotide as a function of time. The upper curve is from case one and the lower from case two with extra GDP added. The exponents of the intercepts at the abscissa give the fractions of GDP that are in complex with EF-Tu at zero incubation time in each case, and the ratio between these two fractions can be used to calculate the total amount of guanine nucleotide in each factor mix [Ehrenberg et al,1990ab]. When the fraction of guanine nucleotide that is on EF-Tu as well as the total amount of guanine nucleotide in each factor mix are determined, precise estimates of how much GTP that has been hydrolyzed on EF-Tu during the burst can be obtained [Ehrenberg et al, 1990ab].

From Fig.1 we get for this particular experiment that 823 pmoles of GTP were consumed in the initial phase. The slope of the lines gives the dissociation rate of GDP from EF-Tu, which in this case is 0.010 s^{-1}.

The extent of poly(Phe)-synthesis is shown in Fig.1a as a function of time. The intercept at the abscissa gives the amount of poly(Phe)-synthesis during the burst, and it is in this case 420 pmoles of poly(Phe). The number of GTP's hydrolyzed per peptide bond (f_c) is therefore 1.96. The slope of the straight line is determined by the amount of active EF-Tu in the experiment (Tu_0), the dissociation rate of GDP from EF-Tu (k_d) and by the GTP-hydrolysis to peptide bond formation stoichiometry (f_c) according to [Ruusala et al, 1982a; Ehrenberg et al, 1990a]

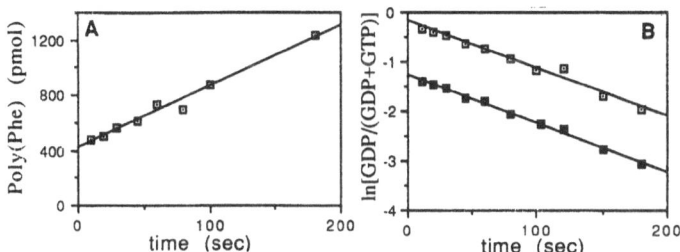

Figure 1. (A): Amount of poly(Phe) synthesized at different time points from incubation start. The intercept is 420 pmol Phe and the slope is 4.43 pmol Phe s^{-1}. (B): Log plots of the ratio r between ^{3}H-GDP and ^{3}H-GDP + ^{3}H-GTP at different time points from incubation start. The upper curve intercepts the y-axis at -0.18 ($r_1=0.835$) and the lower at -1.260 ($r_2=0.284$). The lower curve results from addition of 1908 pmol extra GDP to the factor mix. The total amount of guanosine nucleotide in the factor mix is 986 and the amount of GTP-hydrolysis in burst is 823 pmol (=active EF-Tu). The release rate of GDP from EF-Tu is 0.010 s^{-1} from both curves. f_c is 823/420=1.96 (from burst phase). f_c=823·0.010/4.43=1.86 (from steady state).

$$slope = \frac{Tu_0 \cdot k_d}{f_c}$$

Estimates for Tu_0 and k_d are obtained from the data in Fig.1b. From these and from the slope in Fig.1a, we calculate f_c to be 1.86 in the steady state phase of the experiment. The results are very straightforward, but their interpretation requires further considerations.

THE EXCESS DISSIPATION OF GTP IN EF-Tu FUNCTION IS NEITHER DUE TO AGGRESIVE PROOFREADING NOR TO INACTIVE RIBOSOMES

About thirty percent of the ribosomes are active in elongation in our experiments [Ehrenberg et al, 1990abc, Bilgin et al, 1992]. It is therefore possible *a priori* that the result in Fig. 1 can be accounted for by the existence of an idling reaction taking place when ternary complexes encounter inactive ribosomes and which leads to GTP hydrolysis without peptide bond formation.

Control experiments that were performed in the absence of EF-G show, however, that this type of idling is not strong enough to account for an increase in f_c from one to two [Ehrenberg et al, 1990a; Bilgin et al, 1992]. Comparatively strong idling reactions have been observed for hyperaccurate ribosomes with amino acid changes in protein S12, but not for wild type ribosomes [Bilgin et al, 1992].

At this point one cannot exclude the existence of ternary complex idling at inactive ribosomes that is somehow induced by EF-G, but models with such complicated behaviour are not plausible. A definitive refutation of explanations based on the interference of inactive ribosomes is discussed in the next section.

Another option that could account for why f_c is two and not one is that wild type ribosomes may be very aggressive proofreaders, so that one cognate aminoacyl-tRNA is discarded in proofreading after GTP-hydrolysis on EF-Tu for every aminoacyl-tRNA that participates in peptidyl transfer. If this were true one would predict that streptomycin (Sm), an antibiotic that turns off ribosomal proofreading of non cognate tRNAs almost completely [Ruusala and Kurland, 1984; Bilgin and Ehrenberg, in preparation], would reduce f_c from two to a value close to one [see Ehrenberg et al, 1990 for a dicussion of this].

In contrast, experiments [Ehrenberg et al, 1990abc] show that f_c is two also in the presence of Sm and that f_c is unchanged by such alterations in the ribosomal protein S4 that reduce proofreading [Andersson and Kurland, 1983]. These observations make it unlikely that aggresive proofreading is the explanation for the high f_c -values, and this tentative conclusion is confirmed by more direct experiments, which will be described in the next section.

Yet another *a priori* explanation for the anomalous f_c -values is that free EF-Tu·GTP may interact with active ribosomes and that this may lead to hydrolysis of GTP without concomitant peptide bond formation. This explanation is, however, excluded by the combination of two types of observations. The first is that all EF-Tu in our assays is active in aminoacyl-tRNA binding (more than 95%) as seen in non denaturing gel systems as well as by gel filtration columns [N. Bilgin, in preparation]. The second is that f_c remains near two also when the concentration of aminoacyl-tRNA is titrated to such high values that the concentration of free EF-Tu becomes negligible.

85% OF THE COGNATE AMINOACYL-tRNAS GO DIRECTLY FROM TERNARY COMPLEX TO PEPTIDYL TRANSFER

The experiments so far indicate that the excess hydrolysis of GTP is due to active ribosomes, and we will now consider two different main models that can explain the phenomenon.

The first model involves a revision of our notions about the ternary complex itself. It is thus conceivable that the ternary complex consists of two EF-Tu·GTP's bound to one aminoacyl-tRNA. When, by hypothesis, this extended ternary complex enters the A-site, both GTP molecules are rapidly hydrolyzed, and two EF-Tu·GDP's leave the ribosome for every aminoacyl-tRNA that is delivered to the A-site [Ehrenberg et al, 1990abc].

Direct experimental support for an extended ternary complex has been given by Tapio et al [1990] and Ehrenberg et al [1990abc], but their results were recently questioned by Bensch et al [1992].

Before discussing experiments that directly address the stoichiometry of the ternary complex, there is one important experimental consequence of the putative extended ternary complex which may be contrasted with the consequences of another possible explanation for why f_c is two.

This second model is that two "conventional" ternary complexes may operate in a sequence. One could, for instance, imagine that a ternary complex enters the ribosome and that, subsequently, there is hydrolysis of one GTP. Then some vital function requires the action of a second ternary complex and this leads to hydrolysis of a second GTP.

In this model one aminoacyl-tRNA goes either from the first or from the second ternary complex to the A-site, and one aminoacyl-tRNA becomes free in solution for every elongation cycle of the ribosome. This behaviour, where one tRNA goes to solution for every one that participates in peptidyl-transfer, is in sharp contrast with the first case, where essentially every tRNA that goes from the ternary complex receives a nascent polypeptide.

To distinguish between these two hypotheses, assays are required that score the probability that an aminocyl-tRNA that leaves the ternary complex in a ribosome catalyzed reaction goes directly to peptidyl-transfer.

If this probability is close to one, then the second hypothesis is refuted. If, in contrast, the probability is less than a half then this contradicts the first hypothesis. The experiments that settled this crucial question were designed as follows [J. Scoble, in preparation].

Aminoacyl-tRNA can leave the ternary complex along two pathways. The first involves a spontaneous dissociaton of tRNA from EF-Tu, and the second is a route that is catalyzed by ribosomes.

The two options may be summarized in a scheme like

$$Tern.complex + Ribosome \xrightarrow{k_{Ribosome}} C \xrightarrow{peptidyltransfer}$$

$$k_{spont} \downarrow \qquad\qquad proofreading \downarrow$$

$$unreacted\ aa\text{-}tRNA \qquad unreacted\ aa\text{-}RNA$$

The fate of an aminoacyl-tRNA in ternary complex is either spontaneous dissociaton with a rate constant k_{spont}, or interaction with ribosomes which occurs with a rate constant $k_{Ribosome}$.

Now, given that the aminoacyl-tRNA is released from ternary complex over the ribosome dependent pathway, it will participate in peptidyl-transfer with probability P_{pep}, or remain as aminoacyl-tRNA with the probability P_{diss}. In the latter scenario all pathways that lead from ribosomes to the free state of the tRNA are included.

If the first hypothesis above is correct and if the dissociation of tRNA from ribosomes in proofreading occurs with a low frequency, then P_{diss} is near zero and P_{pep} is close to one. If the second hypothesis is true, then P_{diss} must be larger than a half.

The experiment that measures P_{diss} and P_{pep} must fulfill three stringent requirements. The first is that $k_{Ribosome}$ is much larger than k_{spont} in the scheme above, so that aminoacyl-tRNA only leaves the ternary complex via ribosomes and that, accordingly, the spontaneous pathway can be disregarded. This condition is controlled by titrating with ribosomes until $k_{Ribosome}$ (proportional to the ribosome concentration) is much larger than k_{spont}.

The second requirement is that an aminoacyl-tRNA that has left the ternary complex on the "ribosome route" and has gone to its free state is not given a second chance to enter a ribosome and participate in peptidyl transfer. Such rebinding to a ribosome may occur if there is free EF-Tu·GTP, which can catalyze a second association of the aminoacyl-tRNA to the A-site.

This condition is fulfilled by a chase as follows. The amino acid of the aminoacyl-tRNA in the original ternary complex is radioactively labelled. The incubation starts by addition of preinitiated ribosomes together with a vast excess of the same cognate aminoacyl-tRNA which, however, carries an unlabelled amino acid. Since the experiment is designed in such a way that the labelled ternary complex interacts much faster with ribosomes than it dissociates spontaneously, the chase is effective only for those labelled tRNAs that have gone to the free state after interaction with a ribosome. The chase thus makes sure that an aminoacyl-tRNA only has one chance to participate in peptidyl-transfer. At the same time the chase does not interfere with the first ternary complex-ribosome interaction.

The third condition is that all aminoacyl-tRNAs that are present when the incubation starts are in ternary complex. This requirement comes down to having enough EF-Tu in the factor mix, and it is controlled by titrating with EF-Tu.

The experiment thus consists of three titrations, where each controls one of the requirements that must be fulfilled to obtain a proper measurement of P_{diss} and P_{pep}: A ribosome titration, a titration with the unlabelled chasing aminoacyl-tRNA in the ribosome mix, and an EF-Tu titration.

After a short incubation (typically 15s) the reaction is stopped. HPLC-technique is used to measure the amount of aminoacyl-tRNA that did not participate in peptidyl-transfer, and we also measured the amount of poly(Phe) synthesis from hot TCA precipitation counted on glass filters. From this experiment we were able to conclude that as much as 85% of all aminoacyl-tRNA that leaves ternary complex along the "ribosome route" go directly to peptidyl-transfer (P_{pep}=0.85) and as little as 15% remains as aminoacyl-tRNA after the incubation (P_{diss}=0.15).

This result is compatible with the first hypothesis above which implies that amino acyl-tRNA enters the ribosome in an extended ternary complex. It is not, however, in accordance with the second hypothesis, which predicts that one aminoacyl-tRNA should become free in solution for every one that goes to peptidyltransfer.

The result is powerful also in other respects. First, it excludes strictly the possibility that f_c is two due to aggressive proofreading of cognate tRNAs by wild type ribosomes. If this were the case, then P_{diss} must be larger than 0.5 and the value of 0.15 for P_{diss} that we measure here is by far too small to be consistent with such an explanation.

The result also excludes the possibility that there is an idling reaction where ternary complexes are dissipated by the interference of inactive ribosomes, since also in this case the prediction would be that P_{diss} must be larger than 0.5.

One question that remains unsettled is the contribution of proofreading to the 15% of aminoacyl-tRNAs that do not go directly to peptidyl transfer. The figure of 15% thus includes, apart from proofreading, all aminoacyl-tRNAs that are delivered from ternary complex to inactive ribosomes, where they may stay or from which they may dissociate. It also includes such aminoacyl-tRNAs that for some reason are inactive in translation. To identify the proofreading flow itself thus requires further work, which is beyond the scope of the present investigation. The only thing that presently can be said is that the proofreading losses of cognate aminoacyl-tRNA from the ribosome are less than 15%.

The results of this section have considerably strengthened the idea that we may indeed have to revise our previous notion about the ternary complex. They furthermore strongly indicate that the anomalous stoichiometry of two GTP's per peptide bond is a phenomenon that may be relevant for our understanding of translation, not an artifact due to some unidentified side reaction.

The next section is devoted to experiments directly focussed on the stoichiometry of the ternary complex itself.

THE COMPOSITION OF THE TERNARY COMPLEX DEPENDS ON TEMPERATURE

Confirmation of the hypothesis of the extended ternary complex [Ehrenberg et al, 1990abc] requires a direct demonstration of two molecules of EF-Tu associated with one molecule of aminoacyl-tRNA in a pentameric complex. One intricate problem here is that there at the moment is no known feature of translation that seems to require such an expensive device. The fact that aminoacyl-tRNA enters the ribosome in complex with EF-Tu·GTP and that its delivery to the A-site is accompanied by hydrolysis of one molecule of GTP may be rationalized as a requirement to achieve high accuracy by proofreading [Kurland and Ehrenberg, 1987, Ehrenberg et al, 1990c]. Dissipation of two molecules of GTP may in principle push the accuracy to much higher levels than can be achieved from one GTP [Ehrenberg and Blomberg, 1980]. This requires, however, rather special assumptions concerning sophisticated multiple proofreading mechanisms and so far we lack experimental evidence for such schemes in *E. coli* translation. The notion of the extended ternary complex is thus not only controversial [Bensch et al, 1991], but it also suffers from the fact that it at the present time is not rationalizing or explaining well known features of prokaryotic translation. The notion of the extended ternary complex is therefore counter intuitive, and this naturally raises the demands one puts on experiments that are designed to test its existence.

Our first assays to investigate the affinity between aminoacyl-tRNA and EF-Tu as well as the stoichiometry of their complex [Tapio et al, 1990; Ehrenberg et al, 1990a] were designed to suit the conditions of our optimized *in vitro* experiments for translation [Jelenc and Kurland, 1979; Wagner et al, 1982; Ehrenberg et al, 1990c]. This means, in particular, that they should work well at $37^{\circ}C$ in polymix buffer, where the ternary complex is comparatively unstable [Tapio et al, 1990; Ehrenberg et al, 1990a].

Another ambition was that the assays should be easy to interpret and evaluate. We wanted, in particular, to avoid that the concentration of aminoacyl-tRNA is changing with time during the incubation, which is a drawback in the protection experiments suggested by Pingoud et al [1977], as pointed out by Abrahamson et al [1985].

The solution to these experimental problems was to design an aminoacyl-tRNA protection assay with some new features. Our assay thus contains a small amount of Phe-synthetase, that continuously recharges aminoacyl-tRNAs as they spontaneously deacylate. The deacylation rate is accurately measured via the recharging of tRNA with a second label [Tapio et al, 1990; Ehrenberg et al, 1990a].

This assay showed that under our experimental conditions the association constant for the binding of Phe-tRNA^{Phe} to EF-Tu·GTP is about 10^7 M^{-1}. It also gave the surprising result that there are two molecules of EF-Tu bound to one tRNA molecule [Tapio et al, 1990; Ehrenberg et al, 1990abc].

In order to subject this result to independent tests we developed other techniques to measure the stoichiometry of the ternary complex. Among these were non denaturing gel methods, initiated in collaboration with T. Gluick [N. Bilgin, T.Gluick and M. Ehrenberg, Manuscript], as well as FPLC columns [N. Bilgin, in preparation]. To our disappointment it was very hard to get consistent results with these methods. Titration experiments evaluated with non denaturing gels gave 2:1 stoichiometry for the ternary complex, but other experiments, based on FPLC columns, indicated that the conventional 1:1 stoichiometry between aminoacyl-tRNA and EF-Tu may be correct, after all.

In the latter experiments we used a new type of gel filtration column from Pharmacia (Superdex 75), which completey resolved ternary complex from free EF-Tu·GTP and aminoacyl-tRNA. The amino acid (Phe or Leu) on tRNA was labelled either with ^3H or ^{14}C. EF-Tu was prepared from bacteria grown either with ^3H-Leu or ^{35}S-Met in the medium, and the ternary complex peak was evaluated with respect to the tRNA to EF-Tu stoichiometry point by point from double label experiments. One combination was to have ^3H-labelled tRNA in complex with ^{35}S-labelled EF-Tu and in the other case we used ^{14}C-labelled tRNA together with ^3H-labelled EF-Tu [N. Bilgin, in preparation]. For both combinations we obtained results close to a 1:1 stoichiometry between EF-Tu and aminoacyl-tRNA. The gel as well as the column experiments were both run at low temperature (about $4^{\circ}C$), and attempts to apply these methods to high temperature conditions ($37^{\circ}C$), were not succesful since the stability of the complex decreases drastically with increasing temperature.

The column experiments with double labels appeared convincing and they left us with two options. The first was that our original stoichiometry experiments, based on titrations, were wrong for unknown reasons. The second option was that temperature may make a significant difference not only for the binding constant between EF-Tu and aminoacyl-tRNA , but in fact also for the stoichiometry itself. Given that the binding constant changes with orders of magnitude when the temperature is varied from $4^{o}C$ to $37^{o}C$, a switch in stoichiometry certainly cannot be excluded *a priori.* .

While we were brooding over these experimental results, our hypothesis about the extended ternary complex was questioned in a publication by Bensch et al [1991]. They presented experiments that seemed to confirm the traditional view that the ternary complex consists of one EF-Tu and one aminoacyl-tRNA. They used fluorescence methods as well as assays based on the observation that aminocyl-tRNA is protected from RNAse A attacks when it is in complex with EF-Tu·GTP [Tanada et al, 1981; Louie and Jurnak, 1985].

All the experiments by Bensch et al [1991] were made at low temperature. The interpretation of their experiments, in terms of a stoichiometry for the ternary complex, depended on the assumption that only 60% of their EF-Tu was active in tRNA binding, and this had only weak experimental support. The results that they showed were, furthermore, obtained from a heterologous system with Phe-tRNA^{Phe} from yeast and EF-Tu from *E. coli.*. Their conclusions were, in short, not based on very convincing data. More important, however, is that their results were beside the point with respect to our conclusions, since they rested on the tacit assumption that the ternary complex is always the same irrespective of experimental conditions.

It was now clear that to get further we had to find reliable assays to analyze the ternary complex that were more direct than our previous experiments [Ehrenberg et al, 1990abc; Tapio et al, 1990] and that could be used in polymix buffer at $37^{o}C$. If the ternary complex is changing personality with temperature, our high temperature conditions are certainly more relevant to the situation *in vivo* than the conditions of our columns and gels or the conditions used by Bensch et al [1991]. In fact, a series of experiments from our laboratory, where the growth behaviour of bacterial mutants could be succesfully interpreted, had clearly demonstrated the relevance of our *in vitro* translation system [Andersson et al, 1986ab; Andersson and Kurland, 1983; Bohman et al, 1984; Ruusala et al, 1984; Diaz et al, 1986; Diaz and Ehrenberg, 1991; Bilgin et al, 1992; Tubulekas and Hughes, 1992].

The relevant experimental task was therefore to confirm or to reject the idea that the two GTP's per peptide bond that we observe in translation experiments at $37^{o}C$ can be associated with an extended ternary complex as claimed previously [Ehrenberg et al,1990abc]. To do this we must investigate the stoichiometry under the same conditions as used for the f_c-measurements described above, and not under radically different buffer conditions and at low temperatures, which happen to be convenient for nuclease protection assays as well as for column and gel runs.

As a first step in this direction we adjusted the nuclease protection assay [Tanada et al,

1981; Louie and Jurnak, 1985; Bensch et al, 1991] to high concentration conditions, where the aminoacyl-tRNA was kept close to 10^{-5} M, as *in vivo*. Stoichiometric titrations with EF-Tu at 0°C gave about one molecule of EF-Tu per aminoacyl-tRNA in the ternary complex, in accordance with our double label column experiments. Under low concentration conditions, the same as used by Bensch et al [1991], we consistently obtained 2:1 stoichiometry with this method [N. Bilgin, in preparation]. Modification of this method to experiments at 37°C required the solution of two technical problems. The first is that the rate constant for dissociation of Phe-tRNA^{Phe} from EF-Tu·GTP is as fast as 0.2 s^{-1} in polymix buffer at 37°C . This means that the amount of tRNA protected by EF-Tu must be estimated from extrapolations back to zero time from curves that describe how much aminoacyl-tRNA remains intact at different time points after incubation starts.

The second problem is that such extrapolations may be ambiguous, because the concentration of free EF-Tu·GTP is changing with time. Thus, if the experiment is designed so that there is aminoacyl-tRNA in excess at time zero, then there is essentially no free EF-Tu·GTP when the incubation starts. As time goes on, however, more and more aminoacyl-tRNA is digested by the nuclease and this leads to an increase in the pool of free EF-Tu·GTP. If now the amount of RNAse A in the assay is insufficient, then a tRNA molecule that leaves the ternary complex may rebind to EF-Tu·GTP instead of being digested. Since the amount of EF-Tu·GTP increases with time the rate of disappearence of Phe-tRNA^{Phe} will tend to slow down with increasing incubation times [N. Bilgin, in preparation]. Extrapolations back to zero time will under such conditions lead to systematic underestimates of the amount of aminoacyl-tRNA that can be protected by a certain amount of EF-Tu·GTP, and thus to an artificially high stoichiometry for the ternary complex.

This potential source of errors was eliminated by titrating with RNAse A to such high values that all extrapolations back to zero time became independent of the nuclease concentration [N, Bilgin, in preparation].

When these technical problems had been removed, our experiments showed a remarkable variation of the stoichiometry with temperature. In this high concentration range of Phe-tRNA^{Phe} about one molecule of aminoacyl-tRNA was protected by one EF-Tu·GTP at temperatures between 0 and 20 °C. Above this temperature the stoichiometry started to shift, so that at 37 °C and above close to two molecules of EF-Tu were required for the protection of one tRNA [N. Bilgin, in preparation].

This result is important in several ways and it contains an obvious lesson.

If, as our experiments indicate, the stoichiometry really changes with temperature then the inconsistencies between our different data sets have been removed.

We have then, furthermore, obtained with a different method a confirmation of our original experimental results on the stoichiometry of the ternary complex [Tapio et al, 1990; Ehrenberg et al, 1990abc].

The obvious lesson, finally, is that our temperature result in a dramatic way illustrates the potential danger of choosing conditions for *in vitro* experiments that one knows in advance must deviate radically from the *in vivo* situation that one wants to explain. The new element here, in comparison with our previous discussions concerning this point [Kurland and Ehrenberg, 1984; Ehrenberg et al,1990a], is that we now have results that clearly demonstrate how the *in vitro* conditions may influence not only the quantitative but also the qualitative behaviour of the translation system.

ACKNOWLEDGMENTS

This work was supported by the Swedish Natural Science Research Council.

REFERENCES

Abrahamson, J.K., Laue, T.M., Miller, D.L. and Johnson, A.E., 1985, *Biochemistry*. 24:692.
Andersson, D.I., and Kurland, C.G., 1983, *Mol. Gen. Genet.* 187:378.
Andersson, D.I., Andersson, S.G.E., and Kurland, C.G., 1986a, *Biochimie*. 68:705.
Andersson, D.I., van Verseveld, H., Stouthamer, A.H., and Kurland, C.G., 1986b, *Arch. Microbiol.* 144:96.
Bensch, K., Pieper, U., Ott, G., Schirmer, N., Sprinzl, M., and Pingoud, A., 1991, *Biochimie* 73:1045.
Bilgin, N., Claesens, F., Pahverk, H., and Ehrenberg, M., 1992, *J. Mol. Biol.* 224:1011.
Bohman, K., Ruusala, T., Jelenc, P.C., and Kurland, C.G., 1984, *Mol. Gen. Genet.* 198:90.
Diaz, I., Ehrenberg, M., and Kurland, C.G., 1986, *Mol. Gen. Genet.* 202:207.
Diaz, I., and Ehrenberg, M., 1991, *J. Mol. Biol.* 222:1161.
Ehrenberg, M., and Blomberg, C., 1980, *Biophys. J.* 31:333.
Ehrenberg, M., and Kurland, C.G., 1988, *Meth. in Enzymol.* 164:611.
Ehrenberg, M., Rojas, A.-M., Weiser, J., and Kurland, C.G., 1990a, *J. Mol. Biol.* 211:739.
Ehrenberg, M., Bilgin, N., and Kurland, C.G., 1990b, *Ribosomes and Protein Synthesis.*The practical approach series. IRL press, Oxford and Washington.
Ehrenberg, M., Rojas, A.-M., Diaz, I., Bilgin, N., Weiser, J., Claesens, F., and Kurland, C.G., 1990c, In: *"The Ribosome: Structure, Function, and Evolution"*, W.F. Hill, A. Dahlberg, R.A. Garret, P.B. Moore, D.Schlessinger, and J.R. Warner, eds., Am. Soc. for Microbiol., Washington D.C.
Gordon, J., 1967, *Proc. Nat. Acad. Sci. U.S.A.* 58:1574.
Jelenc, P.C., and Kurland, C.G., 1979, *Proc. Nat. Acad. Sci. U.S.A.* 76:3174.
Kakniashvili, D.G., Smailov, S.K., and Gavrilova, L.P., 1986, *FEBS Letters.* 196:103.
Kurland, C.G., and Ehrenberg, M., 1987, *Annu. Rev. Biophys. Biophys. Chem.* 16:291.
Louie, A., and Jurnak, F., 1985, *Biochemistry*. 24:6433.
Miller, D.L., and Weissbach, H., 1977, in *"Molecular Mechanisms of Protein Synthesis"*, H. Weissbach and S. Pestka, eds, Academic Press, New York.
Pingoud, A., Urbanke, C., Krauss, G., Peters, F., and Maas, G., 1977. *Eur. J. Biochem.* 78:403.
Rojas, A.-M., 1988, *Ph.D. Thesis.* Uppsala University, Uppsala, Sweden.
Ruusala, T., Ehrenberg, M., and Kurland, C.G., 1982a, *EMBO J.* 1:741.
Ruusala, T., Ehrenberg, M., and Kurland, C.G., 1982b, *EMBO J.* 1:75.
Ruusala, T., and Kurland, C.G., 1984, *Mol. Gen. Genet.* 198:100.
Ruusala, T., Andersson, D., Ehrenberg, M., and Kurland, C.G., 1984, *EMBO J.* 3:2575.
Tanada, S., Kawakami, M., Yoneda, T., and Takemura, S., 1981, *J. Biochem.* 89:1565.
Tapio, S., Bilgin, N., and Ehrenberg, M., 1990, *Eur. J. Biochem.* 188:347.
Thompson, R.C., and Stone, P.J., 1977, *Proc. Nat. Acad. Sci. U.S.A.* 74:198.
Tubulekas, I., and Hughes, D., 1992, *Mol. Microbiol.* In Press.
Wagner, E.G.H., Jelenc, P.C., Ehrenberg, M., and Kurland, C.G., 1982, *Eur. J. Biochem.* 122:193.

KINETIC FLUORESCENCE STUDY ON EF-Tu-DEPENDENT BINDING OF Phe-tRNA^{Phe} TO THE RIBOSOMAL A SITE

Marina V. Rodnina, Rainer Fricke,
and Wolfgang Wintermeyer

Institut für Molekularbiologie
Universität Witten/Herdecke
D-5810 Witten, Germany

INTRODUCTION

The elongation cycle of ribosomal protein synthesis begins with the binding of the substrate, aminoacyl-tRNA, to the A site of the ribosome. The reaction is a quite complex process which involves an enzyme, elongation factor Tu (EF-Tu), and GTP. It is the complex of aminoacyl-tRNA with EF-Tu and GTP ("ternary complex") which rapidly associates with the A site (Miller and Weissbach, 1977; Kaziro, 1978). Subsequently, the recognition of the codon takes place. The formation of the codon-anticodon complex provides the signal which triggers the hydrolysis of GTP. The latter enables the dissociation of EF-Tu·GDP from the ribosome which in turn allows the aminoacyl-tRNA to accommodate in the A site such as to be positioned correctly for peptidyl transfer. There is good evidence supporting the view that the complexity of the A-site binding reaction, at least in part, is due to the necessity to optimize speed and accuracy of aminoacyl-tRNA selection by the inclusion of proofreading during the process (Thompson and Dix, 1982; Thompson and Karim, 1982; Ehrenberg and Kurland, 1984; Ruusala et al., 1984).

The sequence of events during A-site binding depicted above primarily has been deduced from steady-state biochemical data. Data from pre-steady-state kinetic experiments, which are required to formulate a detailed mechanism, are rather scarce (Dix et al., 1990; Eccleston et al., 1985; Thompson et al., 1986; Bilgin et al., 1992). Furthermore, on the molecular level the process is complicated in that the interacting species undergo conformational changes which have not yet been characterized

The Translational Apparatus, Edited by K.H. Nierhaus
et al., Plenum Press, New York, 1993

sufficiently. Of those, the transition of EF-Tu from the GTP to the GDP form, which switches the affinity of the factor for binding to the ribosome (and to aminoacyl-tRNA) from high to low, is established best. Also, a change of the ribosome from a post- to a pretranslocation state, brought about by the EF-Tu-dependent occupancy of the A site (Möller and Maassen, 1986; Hausner et al., 1987), as well as changes of the conformation of charged tRNA, taking place upon A-site binding (Robertson and Wintermeyer, 1981; Bertram et al., 1983) have been reported.

The present article contributes to two aspects of A-site binding, the kinetics of the process and the conformational changes of the interacting molecules, in particular of the aminoacyl-tRNA. The experimental results have been obtained by applying the fluorescence stopped-flow technique utilizing the fluorescence signal of two different fluorophores located in the tRNA moiety of the ternary complex EF-Tu·GTP·Phe-tRNA^{Phe}. The fluorescent labels used were either the Y base (wyebutine) naturally present at position 37 next to the anticodon of yeast tRNA^{Phe} or proflavin chemically incorporated into yeast tRNA^{Phe} at positions 16 or 17 in the D loop (tRNA^{Phe}(Prf16/17)). It has been shown previously that both labels sensitively report conformational changes of the labeled parts of the tRNA molecule (Paulsen et al., 1982; Robertson and Wintermeyer, 1981). Complementary information regarding the status of EF-Tu is provided by the fluorescent derivative of GTP, mant-GTP.

BINDING OF THE TERNARY COMPLEX TO THE A SITE
GIVES RISE TO THREE DISTINCT KINETIC STEPS

The complexes of either Phe-tRNA^{Phe}(Prf16/17) or Phe-tRNA^{Phe} with EF-Tu and GTP or GTP analogs were formed in 2.5-fold excess of EF-Tu and purified by gel filtration on Superdex 75. During the chromatography, the EF-Tu·GTP·Phe-tRNA^{Phe} complex was completely purified from nonbound GTP and Phe-tRNA^{Phe} (tRNA^{Phe}), and partially from excess EF-Tu·GTP (not shown). The stability of the isolated complexes was checked by both RNAse A digestion and reaminoacylation. Less than 10% of Phe-tRNA^{Phe} was released from the complex during the time required for a stopped-flow experiment (about 30 min at 20°C).

To measure the kinetics of A-site interaction, the purified complex EF-Tu·GTP·Phe-tRNA^{Phe}(Prf16/17) and poly(U)-programmed ribosomes (three-fold excess), which in the P site carried either AcPhe-tRNA^{Phe} or deacylated tRNA^{Phe}, were rapidly mixed in a fluorescence stopped-flow apparatus. When the proflavin fluorescence is monitored, two steps are distinguished (Figure 1A): a fast fluorescence increase ($k_{app1} = 18$ s^{-1}) is followed by a decrease ($k_{app2} = 8$ s^{-1}). Practically the same picture is observed with either AcPhe-tRNA^{Phe} or deacylated tRNA^{Phe} in the P site.

When the Y base is monitored, a biphasic increase of the fluorescence is observed (Figure 1B); the two reactions are characterized by apparent rate constants of 16 s^{-1} and 4 s^{-1}. The latter step (step 3), which is not reported by the proflavin label, is not observed when the P site is occupied with deacylated tRNA^{Phe} instead of AcPhe-tRNA^{Phe} (Figure 1B).

To check whether the occupancy of the A site influences the Y base fluorescence of P site-bound tRNA^{Phe} or AcPhe-tRNA^{Phe}, experiments were carried out with the non-

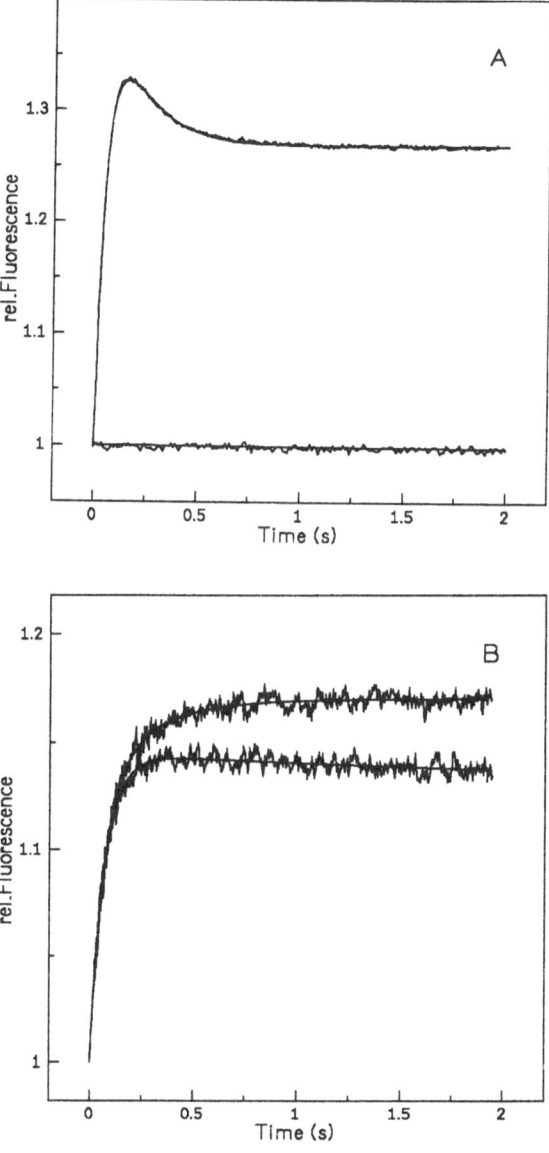

Figure 1. Time course of ternary complex binding to the A site. EF-Tu·GTP·Phe-tRNA^{Phe}(Prf16/17) (final concentration 0.1 μM), isolated by gel chromatography, was rapidly mixed with poly(U)-programmed ribosomes (0.3 μM) carrying in the P site either AcPhe-tRNA^{Phe} (A, B) or tRNA^{Phe} (B) at 20°C; buffer: 25 mM Tris·HCl, pH 7.5, 50 mM NH$_4$Cl, 10 mM MgCl$_2$, 0.5 mM EDTA, 1 mM DTE. (A) Proflavin fluorescence (excitation at 436 nm, emission filter KV 500), relative to the initial fluorescence of the ternary complex. The smooth line represents the function fitted with the sum of two exponential terms: k_{app1}= 18 s^{-1}, A_1= 49%, k_{app2}= 6 s^{-1}, A_2= -22%. The control (bottom curve) shows mixing of the ternary complex with buffer. (B) Y base fluorescence (excitation 312 nm, emission filter KV 408). Upper curve, AcPhe-tRNA^{Phe} in the P site; two-exponential fit: k_{app1}= 15 s^{-1}, A_1= 11%, k_{app3}= 4 s^{-1}, A_3= 4%. Lower curve, tRNA^{Phe} in the P site; two-exponential fit: k_{app1}= 14 s^{-1}, A_1= 13%, k_{app4}= 0.2 s^{-1}, A_4= -2%. The reproducibility of the rate constants is about 15%, the one of the amplitudes about 20% throughout. The displayed curves are an average of four to eight measurements.

fluorescent ternary complex EF-Tu·GTP·Phe-tRNA^{Phe}(E. coli). When this complex is bound to the A site, P site-bound tRNA^{Phe} shows no effect, while AcPhe-tRNA^{Phe} exhibits a small fluorescence increase (about 1%) with an apparent rate constant of about 4 s^{-1} (not shown). This indicates that the same step is reported by the Y base in both A and P sites, albeit with rather different amplitudes. The fact that step 3 is observed only when AcPhe-tRNA^{Phe} is in the P site strongly suggests that this effect is related to the formation of the peptide bond.

The kinetic picture is further simplified when GTP in the ternary complex with Phe-tRNA^{Phe}(Prf16/17) is replaced with the non-hydrolyzable analog, GMPPNP. In the stopped-flow experiment analogous to the one of Figure 1A, only the fast fluorescence increase (step 1) is observed, while step 2 (and step 3) is absent (not shown). Thus, step 2 is probably due to GTP hydrolysis and/or the subsequent dissociation of EF-Tu·GDP. As described below, it is the latter reaction which gives rise to the fluorescence decrease of step 2.

Lowering the Mg^{2+} concentration from 10 mM to 5 mM considerably reduces the amplitude of the fast fluorescence increase of step 1 without affecting either the amplitude of the second step or the rates of both steps significantly. The addition of spermidine, or the use of polymix buffer (Jelenc and Kurland, 1979), increases the amplitude of the first step and the rates of both steps, although the kinetic picture is not changed qualitatively.

k_{app1} and k_{app2} are temperature dependent to a somewhat different extent, both reaching about 50 s^{-1} at 37°C. (k_{app3} has not been measured at different temperatures.) All three steps are concentration dependent, reaching 40 s^{-1}, 20 s^{-1}, and 10 s^{-1}, respectively, at high ribosome concentration and 20°C (not shown). Thus, at optimum conditions with respect to concentration, ionic composition of the buffer, and temperature, A-site binding will be rather rapid, with k_{app1} and k_{app2} approaching 100 s^{-1}. To keep both the rates and the material consumption in a manageable range, the following experiments were performed at conditions of intermediate rates.

THE FLUORESCENT GTP DERIVATIVE, mant-GTP, REPORTS TWO STEPS OF TERNARY COMPLEX INTERACTION WITH THE A SITE

The fluorescent GTP derivative, 3'-O-(N-methylanthraniloyl)-2'-deoxyguanosine triphosphate (mant-GTP), has been shown to be readily hydrolyzed by ras p21 (Rensland et al., 1991). It binds to EF-Tu with a 30% fluorescence increase (data not shown), similar to mant-GDP (Eccleston et al.,1989). The binding of Phe-tRNA^{Phe} to the EF-Tu·mant-GTP complex does not change the fluorescence further. The ternary complex is as stable as the GTP complex and can be purified from excess mant-GTP by gel chromatography, as mentioned above.

Upon interaction of the complex EF-Tu·mant-GTP·Phe-tRNA^{Phe} with poly(U)-programmed P site-blocked ribosomes, the fluorescence of mant-GTP changes in a biphasic manner (Figure 2). The amplitudes of the two steps nearly cancel each other. This strongly suggests that the two steps reflect, respectively, the binding of the ternary complex to the A site and the dissociation of EF-Tu·mant-GDP following the hydrolysis of mant-GTP. Strikingly, the apparent rate constants of the two steps, 17 s^{-1} and 9 s^{-1},

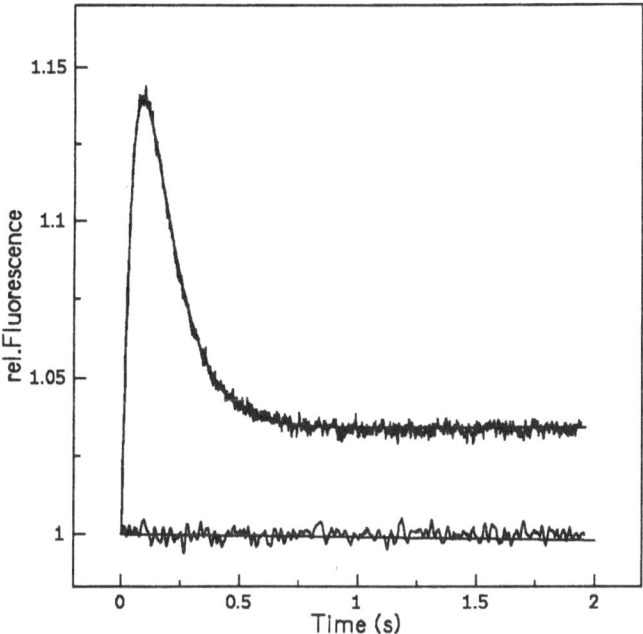

Figure 2. Fluorescence changes of mant-GTP during A-site binding of ternary complex. The experiment was carried out with EF-Tu·mant-GTP·Phe-tRNA^{Phe} and poly(U)-programmed ribosomes with AcPhe-tRNA^{Phe} in the P site as in Figure 1. Excitation at 365 nm, emission filter KV 408. Parameters of the two-exponential fit: $k_{app1} = 17\ s^{-1}$, $A_1 = 62\%$, $k_{app2} = 9\ s^{-1}$, $A_2 = -48\%$.

within the experimental error are the same as k_{app1} and k_{app2} found with the label in the tRNA (Figure 1A). Therefore, we conclude that the two major biochemical events in A-site binding, the association of the ternary complex with, and the dissociation of EF-Tu·GDP from, the ribosome are accompanied by conformational changes of the tRNA.

INITIAL BINDING OF THE TERNARY COMPLEX AND CODON-ANTICODON RECOGNITION

In order to identify the codon-dependent step in the sequence of A-site binding, stopped-flow experiments were performed with poly(A)-programmed ribosomes carrying tRNA^{Lys} in the P site and EF-Tu·GTP·Phe-tRNA^{Phe}(Prf16/17) (Figure 3). The change of proflavin fluorescence indicates that an interaction takes place which, however, qualitatively differs from cognate binding. There is a fast fluorescence increase ($k_{app1} = 25\ s^{-1}$) the amplitude of which is about 40% of that observed with poly(U)-programmed ribosomes, the decrease (k_{app2} in poly(U) experiments) is no longer seen, and there is a further slow increase of the signal ($k_{app} = 3\ s^{-1}$). The biochemical analysis shows that, upon rapid filtration without prior dilution, a small amount (about 15%) of the non-cognate complex is retained on millipore filters; however, the complex appears quite unstable and readily dissociates upon dilution shortly before filtration.

The formation of the noncognate A-site complex is very sensitive to the ionic conditions: the amplitude of the fluorescence change decreases two to three-fold upon increasing the concentration of monovalent ions from 75 to 165 mM, and the effect becomes very small when the Mg^{2+} concentration is decreased from 10 mM to 5 mM (Figure 3). In contrast, the effect of lowering the Mg^{2+} concentration on cognate complex formation is confined to a decrease of the amplitude of the initial binding step.

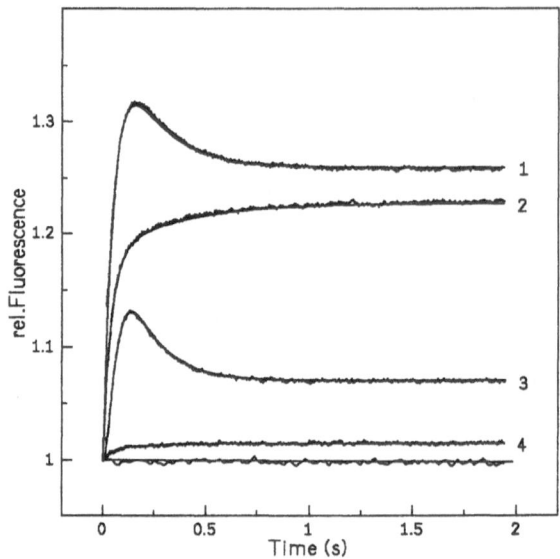

Figure 3. Influence of codon-anticodon interaction on A-site binding of EF-Tu·GTP·Phe-tRNA^{Phe}(Prf16/17). The time course of A-site binding was monitored by proflavin fluorescence. The stopped-flow experiment was performed either at 10 mM Mg^{2+} (cf. Figure 1) (1, 2) or at 5 mM Mg^{2+} in the same buffer (3, 4). The ribosomes were either programmed with poly(U) and had tRNA^{Phe} in the P site (1, 3), or with poly(A) and tRNA^{Lys} in the P site (2, 4). Parameters of the two-exponential fits:

(1) $k_{app1}= 19$ s^{-1}, $A_1= 44\%$, $k_{app2}= 6$ s^{-1}, $A_2= -18\%$;

(2) $k_{app1}= 26$ s^{-1}, $A_1= 18\%$, $k_{app2}= 3$ s^{-1}, $A_2= 5\%$;

(3) $k_{app1}= 17$ s^{-1}, $A_1= 25\%$, $k_{app2}= 7$ s^{-1}, $A_2= -15\%$;

(4) $k_{app1}= 25$ s^{-1}, $A_1= 1\%$, $k_{app2}= 4$ s^{-1}, $A_2= 1\%$.

The Y base fluorescence does not change upon binding of EF-Tu·GTP·Phe-tRNA^{Phe} to poly(A)-programmed ribosomes (not shown). This suggests that, in the poly(A) system, the anticodon of Phe-tRNA^{Phe} does not even reach the decoding site in the A site. Thus, the fast fluorescence increase of proflavin in the D loop, observed in the poly(A) system, probably reflects a codon-independent step which is common for both poly(U) and poly(A)-programmed ribosomes and preceeds the codon recognition step. The latter then shows up as a further increase of the fluorescence during the formation of the cognate complex (Figure 3). In this case, the two reactions are not observed as separate steps since their rate constants are too close to be resolved.

The rate constants of codon-independent binding of EF-Tu·GTP·Phe-tRNA^{Phe} were determined from the dependence of the apparent rate constant, k_{app1}, on the concentration of poly(A)-programmed ribosomes (up to 1 μM). From the linear titration curve (not shown), k_1 and k_{-1} were estimated to be $4 \cdot 10^7$ $M^{-1}s^{-1}$ and 20 s^{-1}, respectively ($20°C$, 10 mM Mg^{2+}).

A-SITE OCCUPANCY DOES NOT INHIBIT INITIAL BINDING OF TERNARY COMPLEX

To characterize further the initial binding state of the ternary complex, kinetic experiments were performed with ribosomes which carried AcPhe-tRNA^{Phe} both in the P site and in the A site. Upon interaction with EF-Tu·GTP·Phe-tRNA^{Phe}(Prf16/17), a biphasic fluorescence increase is observed (Figure 4), which shows that there is binding of the ternary complex to A site-blocked ribosomes. The similarity of the picture with the one seen with mis-programmed ribosomes (Figure 3) indicates that only initial binding takes place, while further rearrangements of the initial complex are blocked. Also the results of the nitrocellulose filtration assay are similar for the two complexes, in that only a small amount of Phe-tRNA^{Phe} is bound to the ribosome in a labile fashion.

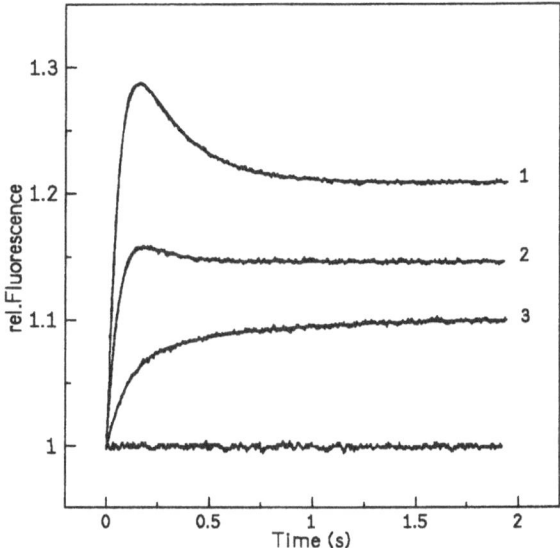

Figure 4. Effect on ternary complex binding of blocking the A site with AcPhe-tRNA^{Phe}. The kinetic experiments were performed with EF-Tu·GTP·Phe-tRNA^{Phe}(Prf16/17) and poly(U)-programmed ribosomes with AcPhe-tRNA^{Phe} in the P site (cf. Figure 1A) with the A site free (1), partially blocked (2), or fully blocked (3) with AcPhe-tRNA^{Phe} present in excess over ribosomes (up to five-fold). According to the filtration assay, the fully blocked ribosomes contained 1.97 pmol of AcPhe-tRNA^{Phe} per pmol of ribosomes before and after the kinetic experiment. Parameters of the two-exponential fit of curve 3: $k_{app1} = 9$ s^{-1}, $A_1 = 7\%$, $k_{app2} = 2$ s^{-1}, $A_2 = 3\%$.

The conclusion from this result is that the binding sites of the aminoacyl-tRNA in the A site and of the ternary complex in the initial binding state do not overlap. Thus, the initial binding state, as defined by the present results, differs from the previously described recognition state, or A/T state (Moazed and Noller, 1989), in which the anticodon region of the incoming ternary complex is positioned in the decoding region of the A site. In the initial binding state, the ternary complex seems to be in contact with both subunits, but neither the acceptor domain nor the anticodon region of the aminoacyl-tRNA are in the positions of the final A site-bound state. It is to be noted, though, that establishing the initial binding state involves a conformational change of the tRNA reported by a fluorescence change. This indicates some specificity and distinguishes the initial binding state from the first encounter complex the formation of which is not seen in our experiments.

CONTRIBUTION OF THE RIBOSOMAL SUBUNITS TO A-SITE BINDING

The first codon-independent interaction of the ternary complex with the ribosomes, which is reflected in the fast fluorescence increase of Phe-tRNA^{Phe}(Prf16/17), might be expected to take place on the 50S subunit, since it has been shown that GTP hydrolysis by EF-Tu is stimulated by 50S subunits alone (Sander et al., 1980), and the binding site of EF-Tu has been localized at the base of the L7/L12 stalk (Spirin & Vasiliev, 1989). On the other hand, 30S subunits also have been shown to bind the ternary complex (Langer et al.,1984; Langer and Lake, 1986; Moazed and Noller, 1990).

In order to determine the contribution to the initial binding of each subunit, the binding of the ternary complex EF-Tu·GTP·Phe-tRNA^{Phe}(Prf16/17) was studied in stopped-flow experiments with isolated 50S subunits as well as poly(U)-programmed 30S subunits. In both cases, the proflavin fluorescence increases only very little (about 1%) with k_{app} around 15 s^{-1} (data not shown). The Y base fluorescence shows no effect with either subunit. This suggests that the interaction of the ternary complex with the isolated subunits does not induce the conformational transitions of the tRNA observed with 70S ribosomes.

The interaction of the ternary complex with programmed 30S subunits according to the biochemical data leads to a stable complex (not shown). Thus, the failure to observe any effect on Y base fluorescence may indicate that the structure of the anticodon loop in the complex with isolated 30S subunits differs from the one on 70S ribosomes.

CONCLUSIONS

Several steps in the reaction sequence of A-site binding are distinguished and cha-racterized by the present kinetic results. The codon-independent initial binding of EF-Tu·GTP·Phe-tRNA^{Phe} to the ribosome is accompanied by a conformational change in the D loop of the tRNA. Both ribosomal subunits are involved in the initial binding, indicating that the site is located on the interface side of both subunits. The binding site does not overlap with the site occupied by aminoacyl-tRNA accommodated in the A site. The step following the initial binding, i.e., codon-anticodon recognition, involves con-formational rearrangements in both anticodon and D loops. Subsequently, GTP is hydrolysed and EF-Tu·GDP dissociates from the ribosome, which is again reflected in a conformational change of the D loop region of the tRNA accompanying the accomodation in the A site. Finally, a conformational change takes place in the anticodon region of the A site-bound aminoacyl-tRNA which reflects the formation of the peptide bond.

The overall picture of the mechanism of EF-Tu-dependent binding of aminoacyl-tRNA to the A site appears as a series of conformational transitions which are induced upon each other by the interacting macromolecules. The present kinetic data also provide some idea about the timing of those changes in the tRNA molecule. On this basis, more detailed studies of the individual steps, e.g., with respect to the branching of the sequence in proofreading, will be performed in future.

ACKNOWLEDGEMENTS

We thank N. Bilgin, Uppsala, for sharing with us the experience in purifying the ternary complex, and R. Goody, Heidelberg, for a gift of mant-GTP. The work described in this article has been supported by the Deutsche Forschungsgemeinschaft (Wi 626/2-4).

REFERENCES

Bertram, S., Göringer, U., and Wagner, R., 1983, *Nucl. Acids Res.* 11:575.

Bilgin, N., Claesens, F., Pahverk, H., and Ehrenberg, M., 1992, *J. Mol. Biol.* 224:1011.

Dix, D.B., Thomas, L.K., and Thompson, R.C., 1990, *in*: "The Ribosome. Structure, Function, and Evolution," W.E. Hill, A. Dahlberg, R,A, Garrett, P.B. Moore, D. Schlessinger, and J.R. Warner, eds., American Society for Microbiology, Washington, D.C.

Eccleston, J.F., Dix, D.B., and Thompson, R.C., 1985, *J. Biol. Chem.* 260:16237.

Eccleston, J.F., Kanagasabai, T.F., Molloy, D.P., Neal, S.E., and Webb, M.R., 1989, *in*: "The Guanine Nucleotide Binding Proteins," L. Bosch, B. Kraal, and A. Parmeggiani, eds., Plenum Publishing Corp., New York

Ehrenberg, M., and Kurland, C.G., 1984, Q. Rev. Biophys. 17:45.

Hausner, T.-P., Atmadja, J., and Nierhaus, K.H., 1987, *Biochimie* 69:911.

Jelenc, P.C., and Kurland, C.G., 1979, *Proc. Natl. Acad. Sci. USA* 76:3174.

Kaziro, Y., 1978, *Biochim. Biophys. Acta* 505:95.

Langer, J.A., and Lake, J.A., 1986, *J. Mol. Biol.* 187:617.

Langer, J.A., Jurnak, F., and Lake, J.A., 1984, *Biochemistry* 23:6171.

Miller, D.L., and Weissbach, H. 1977, *in*: "Molecular Mechanisms of Protein Biosynthesis", H. Weissbach and S. Pestka, eds., Academic Press, New York.

Moazed, D., and Noller, H.F., 1989, *Nature* 342:142.

Moazed, D., and Noller, H.F., 1990, *J. Mol. Biol.* 211:135.

Möller, W., and Maassen, J.A., 1986, *in*: "Structure, Function, and Genetics of Ribosomes," B. Hardesty and G. Kramer, eds., Springer-Verlag, New York

Paulsen, H., Robertson, J.M., and Wintermeyer, W., 1982, *Nucl. Acids Res.* 10:2651.

Rensland, H., Lautwein, A., Wittinghofer, A., and Goody, R.S., 1991, *Biochemistry* 30:11181.

Robertson, J.M., and Wintermeyer, W., 1981, *J. Mol. Biol.* 151:57.

Robertson, J.M., and Wintermeyer, W., 1987, *J. Mol. Biol.*. 197:525.

Ruusala, T., Andersson, D., Ehrenberg, M., and Kurland, C.G., *EMBO J.* 3:2575.

Sander, G., Ivell., R., Crechet, J.B., and Parmegiani, 1980, Biochemistry 19:165.

Spirin, A.S., and Vasiliev, V.D., 1989, *Biol. Cell* 66:215.

Thompson, R.C., and Dix, D.B., 1982, *J. Biol. Chem.* 257:6677.

Thompson, R.C., and Karim, A.M., 1982, *Proc. Natl Acad. Sci. USA* 79:4922.

Thompson, R.C., Dix, D.B., and Karim, A.M., 1986, *J. Biol. Chem.* 261:4868.

DEFINING THE FUNCTION OF EF-3, A UNIQUE ELONGATION FACTOR IN LOW FUNGI

Francisco J. Triana,[1] Knud H. Nierhaus,[1] Jeffrey Ziehler,[2] and Kalpana Chakraburtty[2]

[1]Max-Planck-Institut für Molekulare Genetik, Abteilung Wittmann, Berlin, Germany
[2]Medical College of Wisconsin, Department of Biochemistry, Milwaukee, Wisconsin, USA

INTRODUCTION

The discovery of a protein factor that in addition to elongation factor 1 (EF-1) and elongation factor 2 (EF-2) is absolutely required for translation in the yeast *Saccharomyces cerevisiae*, opened a series of puzzling questions regarding the elongation phase of the protein biosynthesis. Elongation factor 3 (EF-3), originally detected in cell free extracts as a particularly active ribosome-dependent GTPase (Skogerson and Wakatama, 1976), has been later shown to be a ribosome-dependent ATPase that can also accept GTP and ITP as substrates and to a lesser extent pyrimidine nucleotides (Dasmahapatra and Chakraburtty, 1981; Uritani and Miyazaki, 1988a; Kamath and Chakraburtty, 1989). This wide substrate specificity constitute one principal difference from most other translational factors which show a stringent specificity for GTP (Moldave, 1985).

Fungal Ribosomes Require EF-3 for Translation

The functional requirement of yeast ribosomes for EF-3 was first shown in experiments with yeast and rat liver ribosomes and elongation factors (Skogerson and Engelhardt, 1977; Kamath and Chakraburtty, 1986) and later with heterologous combinations of ribosomal subunits from yeast and wheat germ (Chakraburtty and Kamath, 1988). In both cases, EF-1α and EF-2 are able to perform their basic functions in heterologous systems regardless of the origin of the ribosomes, while EF-3 shows effect only in the homologous system with the yeast ribosomes. The experiments with the ribosomal subunits indicated that the requirement of EF-3 in polypeptide synthesis is dependent upon the source of the 40S subunits (Kamath and Chakraburtty, 1989). Since the ribosomes from higher eukaryotes and plants tested so far (liver, reticulocytes, brine shrimp and wheat germ) function in the absence of this factor (Chakraburtty and Kamath, 1988), it appears that the requirement for EF-3 is a unique property of the fungal translational machinery as was originally proposed by Skogerson (1979).

EF-3 was purified to homogeneity by Dasmahapatra and Chakraburtty (1981) and it was demonstrated to consist of a single polypeptide chain with a molecular weight of 125,000. Antibodies raised against the purified factor showed no cross reaction with EF-1α or EF-2 and exhibited a potent inhibitory effect in poly(U) directed polyphenylalanine synthesis that could be overcome by the addition of an excess of EF-3. In contrast, ribosomes from wheat germ or rabbit reticulocytes were not affected by the antibodies to EF-3 (Kamath and Chakraburtty, 1989). A monoclonal antibody independently obtained by Hutchinson et al. (1984) showed also that the factor is essential for the translation of poly(U) as well as natural mRNA and indicated that EF-3 is associated with polysomes but not with ribosomal subunits and it is required for every cycle in the elongation phase.

The structural gene encoding EF-3 (YEF3) have been cloned by immunoscreening of a yeast genomic library (Sandbaken et al., 1990a ,b; Qin et al., 1990). The DNA sequence analysis revealed an open reading frame of 1044 codons and the deduced amino acid sequence indicate that EF-3 should be a soluble protein and is also in close agreement with the previously reported molecular weight (Dasmahapatra and Chakraburtty, 1981). The codon usage corresponds to the characteristic of yeast genes expressed at moderately high level (Sharp et al., 1990). The analysis of the deduced primary sequence of EF-3 reveals a number of interesting features. The carboxyl terminal end is highly positively charged and contains three basic polylysine blocks. This end of the molecule may be involved in the interaction with nucleic acids and/or with the ribosome. The amino-terminal end shows homology to valyl-tRNA synthetase from *B. stearothermophilus*. The significance of this observation is not clear at this time.

The availability of specific antibodies made possible the search for analogs of EF-3 in cell free extracts from different sources. The immunoblot analyses revealed that all the fungi extracts tested (*Candida, Cryptococcus, Aspergillus*, and *Neurospora*) contain a single polypeptide cross reacting with anti-EF-3 antibody, while extracts from mammals (liver and reticulocytes) or plants (wheat germ) did not cross react (Chakraburtty and Kamath, 1988). It appears that a physical analog of yeast EF-3 is absent in mammals and in lower eukaryotes other than fungi. A recent report, where the presence of a protein with nucleotide hydrolase activity in the ribosomes isolated from *Tetrahymena pyriformis* was implied (Miyazaki and Kagiyama, 1990), and an effect of anti-EF-3 antibodies on polyphenylalanine synthesis in the same system (Miyazaki et al., 1988) suggest the existence of ribosome bound EF-3 analogs in non-fungal eukaryotic systems. However, a comparative data base analysis of the deduced amino acid sequence of EF-3 with that of known ribosomal proteins have not revealed the existence of an EF-3 like protein in higher eukaryotes (Chakraburtty, unpublished data). It will be interesting to compare the structural properties of EF-3 with that of the ribosome-associated ATPases.

The gene coding for the EF-3 analogs in *Candida albicans* and in *Pneumcystis carinii* have been recently isolated and sequenced (DiDomenico et al, 1992; Ypma-wong et al., 1992). The deduced primary sequence of EF-3 from these two species show over 75% identity and more than 88% similarity. The functional identity of EF-3 among different fungi was shown by the experimental demonstration that *Candida* EF-3 gene can supplement the disrupted EF-3 gene in *S. cerevisiae* (DiDomenico et al., 1992)

EF-3 is a Ribosome Dependent ATPase

The most notable structural feature in EF-3 sequence is the presence of an internal repeat domain of approximately 200 amino acids which contains two bipartite nucleotide-binding sites. One of this motifs, GX_4GKS/T, is characteristic of many ATP-binding proteins and of all GTP-binding proteins (Dever et al., 1987). The second duplicated element is found only in ATP binding proteins (Walker et al., 1982) and has been referred as the ATP-Binding Cassette (ABC; Higgins et al., 1986), often found in proteins associated with transport. These proteins bind and hydrolyse ATP ("traffic ATPases"; Ames et al., 1992). This structural data correlate well with biochemical studies which strongly indicate that the nucleotidase activity associated with the factor is indeed an intrinsic property of EF-3 and not the result of a EF-3 mediated stimulation of a ribosomal ATPase (Miyazaki et al., 1988; Kamath and Chakraburtty, 1989). Furthermore, the inhibitory effect of vanadate, a potent inhibitor of a number of ATPases (Macara, 1980), suggests that a phosphoenzyme intermediate may be involved in the mechanism of hydrolysis (Miyazaki et al., 1988; Chakraburtty, unpublished observation)

The nucleolytic activity of EF-3 is a strict requirement for its function (Kamath and Chakraburtty, 1989; Uritani and Miyazaki, 1988b). All observed effects of the factor are dependent on the presence of purine nucleotides which cannot be replaced by non-hydrolyzable analogs. This hydrolytic activity is stimulated by two orders of magnitude by yeast ribosomes which in turn are devoid of intrinsic nucleotide hydrolase activity. Interestingly, ribosomes from non-fungal eukaryotic sources hydrolyze ATP and GTP at significant rates in the absence of added protein factors (Chakraburtty, unpublished observation; El'skaya et al., submitted).

EF-3 Facilitates the Binding of Aminoacyl-tRNA to Programmed Ribosomes

Early attempts to define the function of EF-3 gave not clear results. The partial reactions of the elongation cycle, namely EF-1α-dependent aminoacyl-tRNA binding, peptide bond formation and EF-2-dependent translocation as well as the nucleotide exchange reactions of either EF-1α or EF-2 appeared to be not significantly affected by the presence of the factor (Skogerson and Engelhardt, 1977; Dasmahapatra and Chakraburtty, 1981). More detailed examination of the aminoacyl-tRNA binding step revealed that EF-3 acts at this level by stimulating the binding process when EF-1α is present in catalytical amounts, the previous determinations were done always in the presence of saturating amounts of EF-1α, possibly masking the action of EF-3 (Chakraburtty and Kamath, 1988; Kamath and Chakraburtty, 1989; Uritani and Miyazaki, 1988b).

Two additional functional aspects of EF-3 action have been determined. The ribosome dependent nucleotide hydrolyitic activity of EF-3 has been shown to be stimulated by deacylated tRNA (Sandbaken and Chakraburtty, manuscript in preparation). A recent analysis shows a facilitated release of radiolabeled EF-1α from the ribosome in the presence of EF-3 and ATP (Yang and Chakraburtty, unpublished results).

These findings, together with the requirement for poly(U) directed polyphenylalanine synthesis or endogenous mRNA translation constitute the major known effects of EF-3 in the protein-biosynthesis machinery (for review see Chakraburtty, 1992).

The allosteric three-site model proposed for *Escherichia coli* ribosomes (for review see Nierhaus, 1990) provides a new frame in which the mechanism of action of the elongation factors can be studied. According to this model, the ribosome oscillates between two main conformational states during the elongation cycle, the pre-translocational state, in which the ribosomal P and A sites are occupied with tRNA, and the post-translocational state with P and E sites occupied (the E site accepting only deacylated tRNA). The occupation of E and A sites, respectively, is governed by mutually negative cooperativity (Gnirke et al. ,1989), and the function of the elongation factors is proposed to be the reduction of the activation energy between the two conformational states of the ribosome (EF-Tu facilitates the transition post-pre and EF-G the transition pre-post; Schilling-Bartetzko et al., 1992a; Nierhaus et al., this volume). This model has been demonstrated to be valid for archaebacterial systems (Saruyama and Nierhaus, 1986) and the reported tRNA binding capacity of rat liver ribosomes indicate that three sites also exist in higher eukaryotes (Rodnina et al., 1988). In order to use the potential of this model in the investigation of the function of EF-3, the tRNA binding capabilities of 80S ribosomes from yeast were examined.

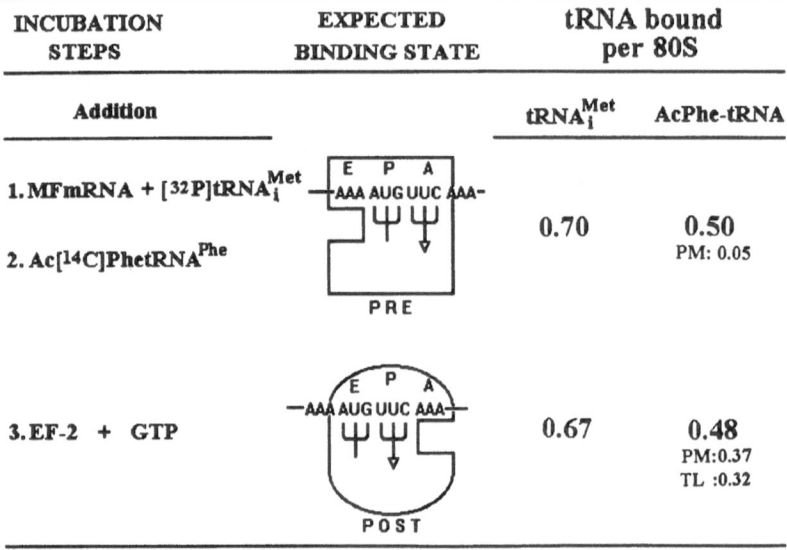

Figure 1. Cotranslocation of Ac[14C]Phe-tRNAPhe ( ) and [32P]tRNAiMet ( ). A pre-translocational complex (PRE) was formed by sequential addition of [32P]tRNAiMet and Ac[14C]Phe-tRNAPhe to 80S ribosomes from yeast programmed with a heteropolymeric mRNA that contains two unique codons for Met and Phe: GGGAAAAGAAAAGAAAAGAAA **AUG UUC** AAAAGAAAAGAAAAGAAAAU(MFmRNA). The binding of both tRNAs and the puromycin reaction were measured before and after an EF-2-dependent translocation step (POST=post-translocational state). The binding is expressed in mols of tRNA/mol 80S (Triana et al., manuscript in preparation).

THE "E" SITE IS PRESENT IN 80S RIBOSOMES FROM YEAST

Under optimal conditions for tRNA binding (Dasmahapatra et al., 1981) poly(U) programmed ribosomes from yeast bind at saturation three times more deacylated tRNAPhe than N-Acetyl-Phe-tRNAPhe (AcPhe-tRNA), while non-programmed 80S bind AcPhe-tRNA poorly and deacylated tRNAPhe to the same level as programmed 80S AcPhe-tRNA (Triana et al., unpublished observation). These data suggest the existence of three tRNA binding sites on 80S ribosomes from yeast as has been reported for *E. coli* (Rheinberger et al., 1981) and for rat liver (Rodnina et al., 1988), and also suggest the validity of the exclusion principle for AcPhe-tRNA binding proposed for the *E. coli* system (Rheinberger et al., 1981; Geigenmüller et al., 1986).

Additional evidence for the third site appears when translocation experiments are performed with a heteropolymeric messenger RNA containing two unique codons for Met and Phe (MFmRNA; see Fig.1). The results presented in Fig.1 show that the deacylated tRNA remains firmly bound to the post-translocational complex, in other words, the "E" site is functionally demonstrated for 80S ribosomes of *Saccharomyces cerevisiae*.

In order to check a possible effect of EF-3 in the translocation reaction, experiments were performed in the presence of varying amounts of EF-3 and with limiting amounts of EF-2. As could be predicted according to early observations (Skogerson, 1976; Dasmahapatra and Chakraburtty, 1981) and in contrast to a more recently claimed effect of the factor in the translocation reaction (Miyazaki et al, 1988), neither the extent and rate of the translocation of AcPhe-tRNA nor the binding of deacylated tRNA to the P or E sites were significantly affected by the presence of EF-3 (Triana et al., unpublished results).

EF-3 FUNCTIONS IN THE ALLOSTERIC TRANSITION FROM THE POST-TRANSLOCATIONAL STATE TO THE PRE-TRANSLOCATIONAL STATE

Next we tested a possible interplay between the different tRNA-binding sites. A new model heteropolymeric messenger containing three unique codons, the MVFmRNA (see Fig. 2), was designed and synthesised using phosphoramidate chemistry (Scaringe et al., 1990). This messenger allowed to establish a post-translocational state with Valyl-tRNA^{Val} located in the P site and deacylated tRNA$_i^{Met}$ in the E site. In the final step, the kinetics of EF-1α·GTP·Phe-tRNA^{Phe} binding were followed in the presence or absence of EF-3. As can be seen in Fig. 2, in the absence of EF-3, both the binding of ternary complex and the release of deacylated tRNA are restricted to a minimal value while in the presence of the factor both binding to the A site and release from the E site increase three to four-fold approaching the same level of binding for the Valyl-tRNA^{Val} located in the P site. It is notable that the binding of ternary complex (A site) and the release of deacylated tRNA$_i^{Met}$ (E site) proceed with the same rate, i.e., for each ternary complex bound to the A site a deacylated tRNA is released from the E site. It follows that a negative cooperativity linkage between the A and the E sites first observed in *E. coli* ribosomes (Rheinberger and Nierhaus, 1986a; Gnirke et al., 1989) is also valid in the yeast system.

The striking effect of EF-3 in the specific facilitation of ternary complex binding to the A site of a post-translocational complex in contrast to the poor effect observed on Ac-acyl-tRNA binding indicate that this factor functions synergistically with EF-1α facilitating the transition from the post-translocational state to the pre-translocational state during the elongation cycle. This conclusion is additionally supported by control experiments in which the binding of a ternary complex to ribosomes programmed with MVFmRNA with or without prebound tRNA in the P site but leaving the E site free was tested (see Fig.3). In both cases the addition of EF-3 has very little effect indicating that the primary action of EF-3 is directed to ribosomes carrying tRNAs at P and E sites (post-translocational state). In other words, a strong effect of EF-3 is only observed when the E site is occupied. These findings specify the formerly observed A-site effects of EF-3 in poly(U) systems (Kamath and Chakraburtty, 1989; Uritani and Miyazaki, 1988b).

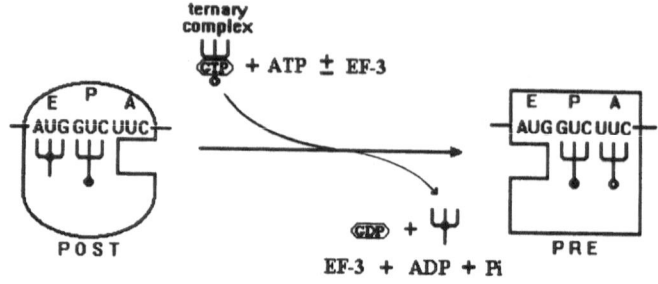

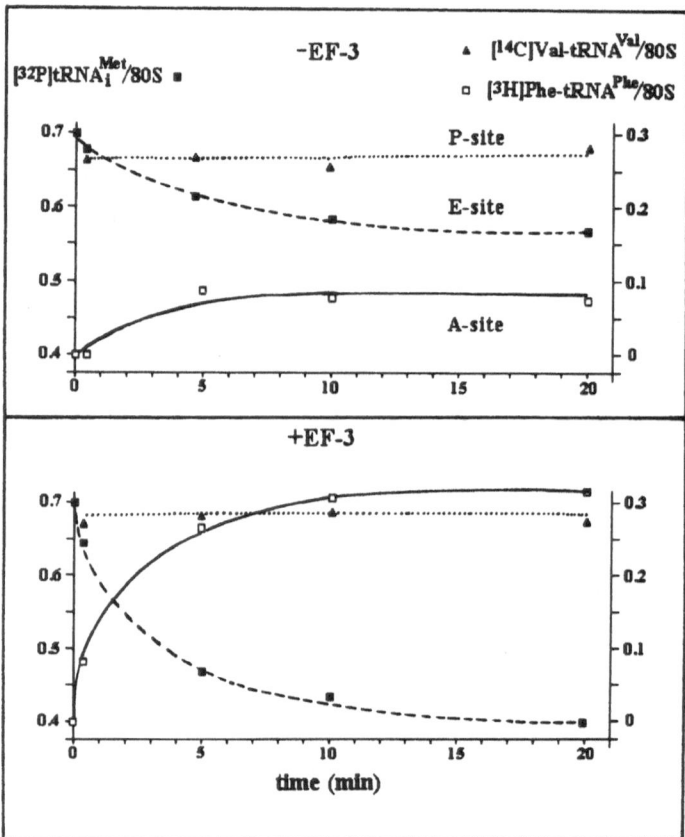

Figure 2. Kinetics of ternary complex binding to the post-translocational complex of 80S ribosomes programmed with MVFmRNA (containing three unique codons for Met, Val and Phe): CCCCCCCCCCCCCCCCCC **AUG GUC UUC** CCCCCCCCCCCCCCCC. The post-translocational complex (POST) was obtained after an EF-2 dependent translocation of an initial complex formed by sequential addition of [^{32}P]tRNA$_i^{Met}$ and [^{14}C]Val-tRNA^{Val}·EF-1α·GTP and EF-2 dependent translocation step. The tRNA binding was determined at the indicated times after addition of [^{3}H]Phe-tRNA^{Phe}·EF-1α·GTP in the presence or the absence of EF-3. The binding is expressed in mols of tRNA/mol 80S (Triana et al., manuscript in preparation).

Since the E site occupation appears to be a crucial factor and a prerequisite for the experimental determination of EF-3 activity, a closer look to this aspect of tRNA binding to programmed 80S ribosomes became imperative.

EF-3 IS THE KEY FOR THE E SITE:
THE FACTOR IMPROVES THE ACCESSIBILITY TO THIS SITE

One issue largely discussed about the quality of tRNA binding to the E site in *E. coli* ribosomes has been the question of codon-anticodon interaction. Experiments of direct binding to the site or chasing of an already bound tRNA gave a clear answer since only the

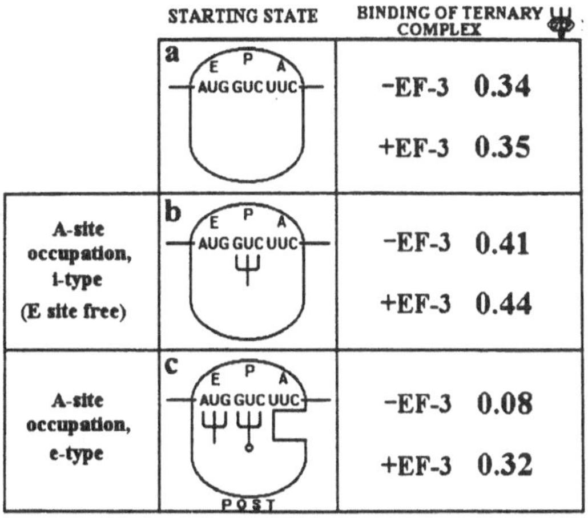

Figure 3. Effect of EF-3 in the binding of ternary complex ($[^3H]$Phe-tRNA^{Phe}·EF-1α·GTP)to three different ribosomal complexes: a) 80S programmed with MVFmRNA; b) 80S programmed with MVFmRNA carrying a deacylated tRNA^{Val} in the P-site (A-site occupation of the i-type; "i" for initiation-like); c) post-translocational complex formed as indicated in figure 2 (A-site occupation of the e-type; "e" for elongation-like). The binding is expressed in mols of tRNA/mol 80S (Triana et al., manuscript in preparation).

cognate tRNA was able to bind or chase with maximal efficiency (Rheinberger et al., 1986; Rheinberger and Nierhaus, 1986b). Despite the discussion about the contribution of this interaction to the general affinity of the E site (Lill and Wintermeyer, 1987), all groups investigating the problem have found that the binding of cognate tRNA is favored over near-cognate or non-cognate tRNAs (Wintermeyer et al, 1990; Rheinberger et al., 1990; Rheinberger, 1991). The different ionic conditions used by different groups for binding or chasing experiments could at least in part explain the reports of "weak" E site codon-anticodon interaction since this interaction becomes more important when the ionic environment is more physiological (Lill and Wintermeyer, 1987; Schilling-Bartezko et al., 1992b).

Under nearly physiological conditions (Mg^{2+} 3-6 mM plus polyamines) the binding to the E site appears to be restricted to the cognate tRNA (Gnirke et al, 1989) and a cognate tRNA at the E site has been shown to be of principal importance for the accuracy of tRNA selection at the A site (Geigenmüller and Nierhaus, 1990). According to this criterion the conditions selected for the tRNA binding experiments presented in this article, based on a previous optimisation for binding of aminoacyl-tRNA (Dasmahapatra and Chakraburtty, 1981), are in the appropiate range since the optimal concentration of Mg^{2+} was reduced to 5 mM in the presence of 2 mM spermidine, the major polyamine in yeast cells (White-Tabor and Tabor, 1985).

The first attempt to evaluate the codon-anticodon interaction in the E site of yeast ribosomes gave surprising results. After establishing a post-translocational state with MFmRNA (see Fig.1) programmed ribosomes containing $[^{32}P]tRNA_i^{Met}$ in the E site and $Ac[^{14}C]Phe$-tRNA in the P site, a five-fold excess of several non radioactive tRNAs was added for chasing in separate incubations. On line with the expected results in case of positive codon-anticodon interaction in the E site, the chasing effect was maximal in the presence of the cognate tRNA. However, quantitatively, the chasing (~12%) was five times less than that reported in the *E. coli* system (Rheinberger and Nierhaus, 1986b). The near cognate tRNAs ($tRNA^{Val}$ or $tRNA^{Ile}$) could chase up to 6% and the chasing with non-cognate $tRNA^{Thr}$ was indistinguishable from the control incubation. Since this indication of tight binding (very slow off rate) could be a special property of the initiator tRNA, the experiment was repeated using the MVFmRNA as template and radioactive $tRNA^{Val}$ at the E site as the substrate to chase. The chasing with cognate (cold $tRNA^{Val}$) and near cognate tRNA ($tRNA_i^{Met}$) were improved (22% and 11% respectively) indicating that indeed the initiator tRNA could have an unusually tight binding. Nevertheless, the amount of tRNA chased was still low if compared with the *E. coli* standards (Rheinberger and Nierhaus, 1986b). It follows that the deacylated tRNA is firmly and tightly bound to the E site of post-translocational ribosomes.

Taking into account that the effect of EF-3 in ternary-complex binding to the A site is linked to the release of deacylated tRNA from the E site (see Fig. 2), it was of interest to examine if EF-3 has an effect on the process of chasing. Data presented in Fig. 4 show the results of chasing experiments performed using MFmRNA and MVFmRNA as templates and with ten fold excess of cold tRNA for chasing. The assay system using either of the mRNAs has the advantage of being symmetrical with respect to cognate and near-cognate tRNAs when the messenger changes, allowing a direct qualitative estimation of the influence of codon-anticodon interaction in the efficiency of chasing. It is striking that the ratio of the amounts of tRNA chased with cognate and near-cognate tRNA, is nearly the same with both mRNAs, indicating that regardless which tRNA species is bound to the E site and whether or not EF-3 is present, the codon-anticodon interaction is an important feature of the binding to the E site. When EF-3 is not present, the absolute chasing, in general low, appears to differ according to the species of tRNA bound to the E site. In presence of EF-3 the chasing is enhanced to approximately the same value (~4-fold increase in the case of MFmRNA and ~2-fold for MVFmRNA), indicating that the quality of the E-site interaction is changed to a more uniform state in which the "off rate" of the tRNA bound is increased to a similar final level, yet preserving the codon-anticodon interaction as a major determinant of binding, and probably without severe changes in the general affinity of the site for the cognate tRNA. This last affirmation implies that the "on rate" to the E site should be also enhanced by EF-3. The results in translocation, tRNA saturation, as well as control experiments of direct binding to the E site (data not shown) give support to this notion.

Since all the previously known effects of EF-3 depend on the hydrolysis of a purine nucleotide (Kamath and Chakraburtty, 1989), the dependence on nucleotide hydrolysis was

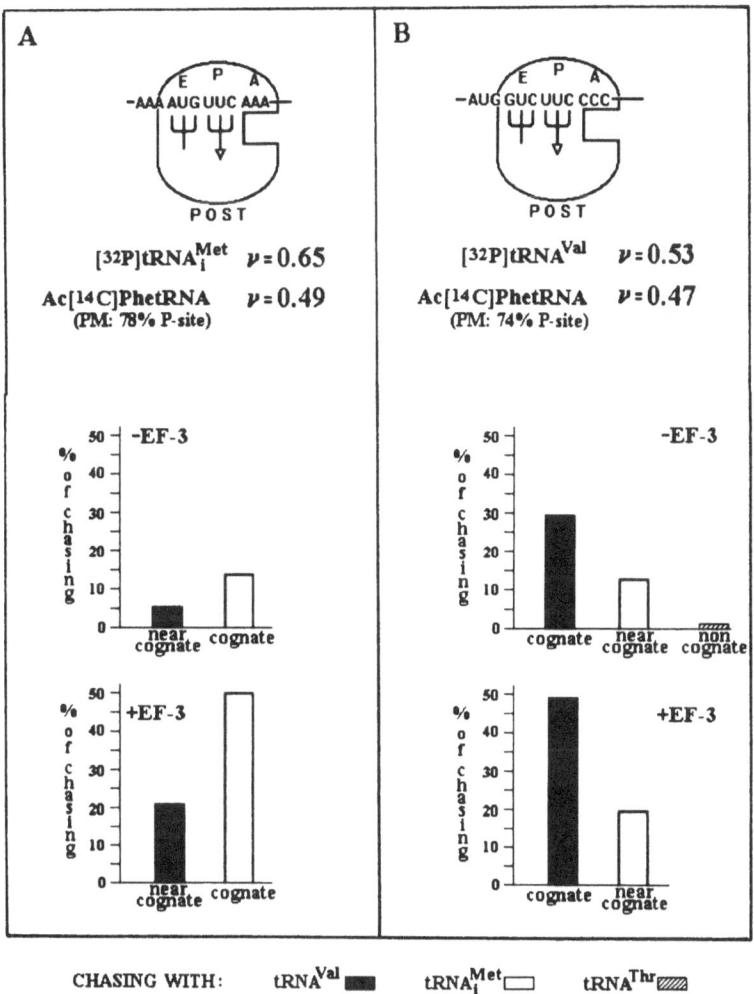

Figure 4. Chasing of the tRNA bound to the E-site. Post-translocational complexes formed as indicated in Fig. 1 using MFmRNA (A) or MVFmRNA(B) carrying the indicated amounts of Ac[14C]Phe-tRNA^Phe in the P site and either [32P]tRNA$_i$^Met (A) or [32P]tRNA^Val (B) in the E site were incubated with a ten-fold excess of the indicated non-radioactive tRNAs in the presence or the absence of EF-3. The occupation number "ν" is given in mols of tRNA/mol 80S. The percentage of chasing is calculated taking the initial binding of [32P]tRNA as 100% (Triana et al., manuscript in preparation).

also investigated for the new effect in the chasing of the E site bound tRNA. The results presented in Fig. 5 show that the EF-3 induced effects on the E site require hydrolysis of ATP since these effects are lost in the absence of this nucleotide or when it is replaced by the non-hydrolysable analog AMPPNP. In spite of the wide substrate specificity showed in vitro by EF-3, the requirement for ATP should reflect the actual in vivo situation since this species is the major NTP in the cell and also the preferred substrate in vitro (Uritani and Miyazaki, 1988a; Kamath and Chakraburtty, 1989).

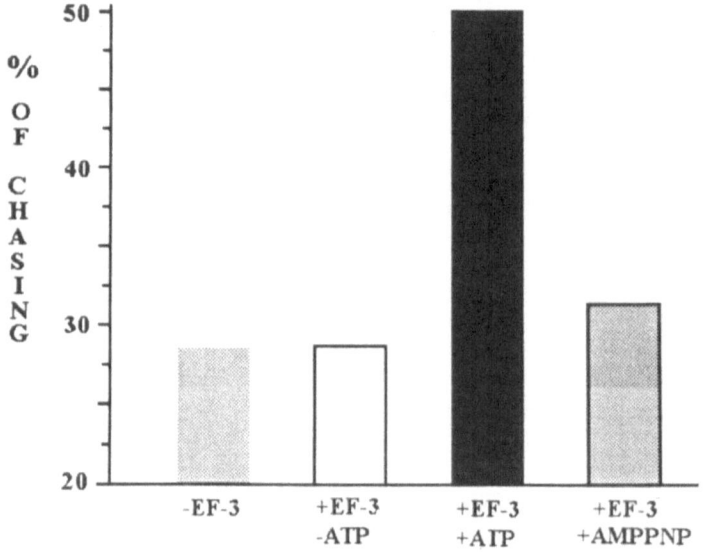

Figure 5. Chasing with cognate tRNA. The chasing was performed as described in Fig. 4B using a ten-fold excess of cold tRNA^{Val}.

CONCLUSIONS

The demonstration of two strong effects of EF-3, first stimulation of ternary-complex binding to the ribosomes in the post-translocational state, and second, facilitation of the chasing of the E site bound deacylated tRNA, suggests the following:

During the elongation cycle EF-3 functions with a strict requirement for ATP hydrolysis by changing the conformation of the ribosome in a way that facilitates the release of deacylated tRNA from the E site when a cognate ternary complex binds to the A site. This effect promotes the allosteric transition from the post- to the pre-translocational state.

The observation that yeast strains overproducing EF-3 grow slower than the wild type strains and are more sensitive to antibiotics that cause mistranslation (Sandbaken et al,

1990a) is in line with the proposed action of EF-3. In presence of abnormally high amounts of the factor, the release of deacylated tRNA from the E site could occur too early or in a way non-coordinated with the selection of the ternary complex at the A site increasing the error frequency. This thesis also implies that the E-site interaction plays a role in the accuracy of selection at the A site in yeast ribosomes, similar to that demonstrated for *E. coli* ribosomes (Geigenmüller and Nierhaus, 1990).

The action of EF-3 can be incorporated to the allosteric three-site model for the elongation cycle as indicated in figure 6. The need for an additional elongation factor for the transition post-pre is a difference with the well characterised situation in *E. coli* that should not be overlooked as an exotic property of the fungi since it could have a more general meaning for eukaryotic systems.

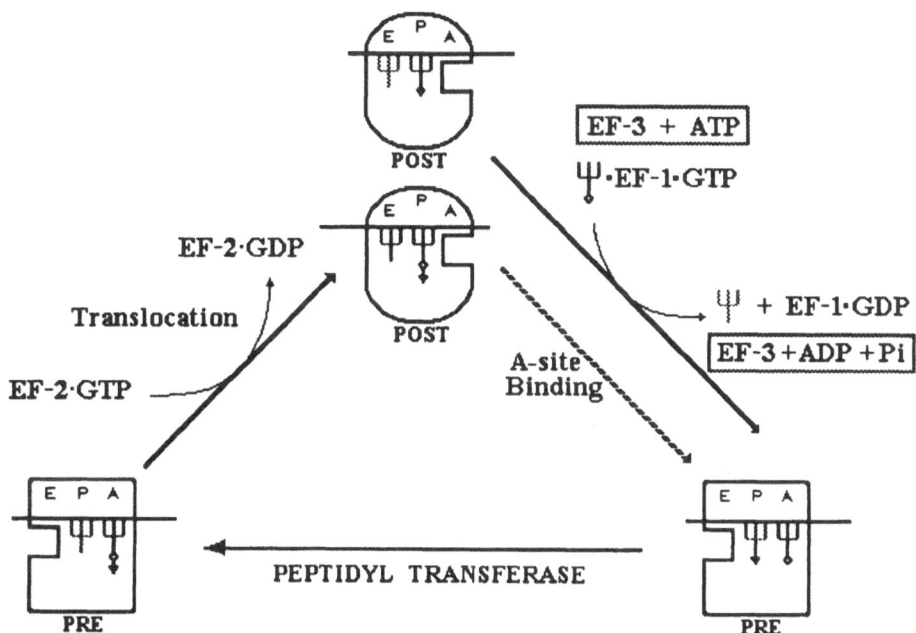

Figure 6. Allosteric three-site model for the elongation cycle in *Saccharomyces cerevisiae*.

The mysterious "built-in" ATPase and GTPase activities found in all the eukaryotic ribosomes tested from non-fungal sources (Miyazaki et al, 1988; Kamath and Chakraburtty, 1989, El'skaya et al., submitted) could play a role similar to EF-3. The experimental strategy defined in this work offers a simple way to test this idea.

ACKNOWLEDGEMENTS

We are grateful to Doris Finkelmeier and Doris-Anna Limbers for technical assistance.

REFERENCES

Ames, G.F.L., Mimura, C.S., Holbrook, S.R., and Shymala, V., 1992, *Adv. Enzymol.* 65: 1

Chakraburtty, K., 1992, *in*: "New Approaches to Anti-fungal Agents", P. Fernendes ed., Burkhauser publication. p. 114

Chakraburtty, K., and Kamath, A., 1988, *Int. J. Biochem.* 20: 581.

Dasmahapatra, B., and Chakraburtty, K., 1981, *J. Biol. Chem.* 256: 9999.

Dasmahapatra, B., Skogerson, L., and Chakraburtty, K., 1981, *J. Biol. Chem.* 256: 10005.

Dever, T.E., Glynias, M.J., and Merrick, W.C., 1987, *Proc. Natl. Acad. Sci. USA* 84:1814.

Di Domenico, B.J., Lupisella, J., Sandbaken, M., and Chakraburtty, K., 1992, *Yeast*, 8: 337

Geigenmüller, U., and Nierhaus, K.H., 1990, *EMBO J.* 9: 4527.

Geigenmüller, U., Hausner, T.-P., and Nierhaus, K.H., 1986, *Eur. J. Biochem.* 161: 715.

Gnirke, A., Geigenmüller, U., Rheinberger, H.-J., Nierhaus, K.H., 1989, *J. Biol. Chem.* 264: 7291.

Higgins, C.F., Hiles, I.D., Salmond, G.P.C., Gill, D.R., Downie, J.A., Holland, I.B., Gray, L. Buckel, S.D., Bell, A.W., and Hermodson, M.A., 1986, *Nature* 323: 448

Hutchinson, J.S., Feinberg, B., Rotwell, T.C., and Moldave, K., 1984, *Biochemistry* 23: 3055.

Kamath, A., and Chakraburtty, K., 1986, *J. Biol. Chem.* 261: 12593.

Kamath, A., and Chakraburtty, K., 1989, *J. Biol. Chem.* 264: 15423.

Lill, R., and Wintermeyer, W., 1987, *J. Mol. Biol.* 196: 137.

Macara, I.G., 1980, *TIBS* 5: 92.

Miyazaki, M., and Kagiyama, H., 1990, *J. Biochem* (Tokio) 108: 1001.

Miyazaki, M.,Uritani, M., and Kagiyama, H, 1988, *J. Biochem.* (Tokio) 104: 445.

Moldave, K., 1985, Annu. Rev. Biochem. 54: 1109.

Nierhaus, K.H., 1990, Biochemistry, 29: 4997.

Nierhaus, K.H., Adlung, R., Hausner, T.-P., Schilling-Bartezko, S., Twardowsky, T., and Triana, F., this volume.

Qin, S., Xie, A., Christina, M., Banato, M., and McLaughlin, C., 1990, *J. Biol. Chem.* 265: 1903.

Rheinberger, H.-J., 1991, *Biochimie*, 73: 1067.

Rheinberger, H.-J., and Nierhaus, K.H., 1986a, *Eur. J. Biol. Chem.* 261: 9133.

Rheinberger, H.-J., and Nierhaus, K.H., 1986b, *FEBS Lett.* 204: 97.

Rheinberger, H.-J., Geigenmüller, U., Gnirke, A., Hausner, T.-P., Remme, J., Saruyama, H., and Nierhaus, K.H., 1990, *in*: "The Ribosome: Structure, Function, and Evolution", W.E. Hill, A. Dahlberg, R.A., Garrett, P.-B. Moore, D. Schlessinger, and J.R. Warner, eds., American Society for Microbiology, Washington D.C., p. 318

Rheinberger, H.-J., Sternbach, H. and Nierhaus, K.H., 1981, *Proc. Natl. Acad. Sci. USA* 76: 5310.

Rheinberger, H.-J., Sternbach, H. and Nierhaus, K.H., 1986, *J. Biol. Chem.* 261: 9140

Rodnina, M.V., El`skaya, A.V., Semenkov, Y.P., and Kirillov, S.V.,1988, *FEBS Lett.* 231: 71

Sandbaken, M., Lupisella, J.A., DiDomenico B. and Chakraburtty, K.,1990a, *J. Biol. Chem.* 265: 15838.

Sandbaken, M., Lupisella, J. A., DiDomenico B. and Chakraburtty, K., 1990b, *Biochim. Biophys. Acta* 1050: 923.

Saruyama, H., and Nierhaus, K.H., 1986, *Mol. Gen. Genet.* 204: 221

Scaringe, S.A., Franklyn, C., and Usman, N., 1990, *Nucleic Acids Res.* 15: 4403.

Schilling-Bartetzko, S., Bartetzko, A., and Nierhaus, K.H., 1992a, *J. Biol. Chem.* 267: 4703.

Schilling-Bartetzko, S., Franceschi, F., Sternbach, H., and Nierhaus, K.H., 1992b, *J. Biol. Chem.* 267: 4693.

Sharp, P., Cowe, E., Higgins, D.G., Shields, D. C., Wolfe, K. H., and Wright, F., 1988, *Nucleic Acids Res.* 16: 8287

Skogerson, L., 1979, *Methods Enzymol.* 60: 676.

Skogerson, L., and Engelhardt, D., 1977, *J. Biol. Chem.* 252: 1471.

Skogerson, L., and Wakatama, E., 1976, *Proc. Nat. Acad. Sci. USA* 37: 73.

Uritani, M., and Miyazaki, M., 1988a, *J. Biochem.* (Tokio) 103: 522.

Uritani, M., and Miyazaki, M., 1988b, *J. Biochem.* (Tokio) 104: 118.

Walker J. E., Saraste, M., Runswick, M.J. and Gay, N.J., 1982, *Embo J.* 1: 945.

White-Tabor, C., and Tabor, H., 1985, *Microbiol. Rew.* 49: 81.

Wintermeyer, W., Lill, R., and Robertson J.M., 1990, *in*: "The Ribosome: Structure, Function, and Evolution", W.E. Hill, A. Dahlberg, R.A., Garrett, P.-B. Moore, D. Schlessinger, and J.R. Warner, eds., American Society for Microbiology, Washington D.C., p. 348

Ypma-Wong, M., Fonzi, W. and Sypherd, P.S., 1992, (Submitted)

ANTIBIOTIC AND PROTEIN INTERACTIONS WITH THE GTPase
AND PEPTIDYL TRANSFERASE REGIONS IN 23S rRNA

Stephen Douthwaite, Birte Vester, Claus Aagaard,
and Gunnar Rosendahl

Department of Molecular Biology
Odense University
DK-5230 Odense M
Denmark

INTRODUCTION

Specific regions of rRNAs have been linked to events in protein synthesis, paving the way for studies on how the functions of these regions are defined by their structures. To this end, footprinting has been an invaluable technique for investigating rRNA structure and its interaction with r-proteins, antibiotics and other ligands. Results from footprinting have been used in conjunction with phylogenetic data to build structural models that can be tested by molecular genetics. Mutations that break and remake putative rRNA secondary and tertiary interactions, or perturb the sites of protein and antibiotic interaction, generally have an effect on the phenotype resulting in, for example, temperature sensitivity, slow growth or antibiotic resistance.

The application of footprinting and mutagenesis has recently been expanded by an allele-specific priming technique. Briefly, mutant rRNAs are expressed from high copy number plasmids in *Escherichia coli*. As the chromosomal rRNA operons remain unaltered, cells contain a heterogeneous ribosome population. Unique hybridization sites, that enable specific primer extension analysis of mutagenised rRNA regions, have therefore been introduced in the plasmid-coded rRNAs. Here, we have studied the structure of the 23S rRNA peptidyl transferase and GTPase regions and their interactions with proteins and antibiotics. Two unique priming sites (Figure 1) were used to determine the effects on structure and ligand interaction caused by mutations in the GTPase and peptidyl transferase regions. These effects are correlated with phenotypic changes observed in cells harboring the mutant rRNAs.

THE PEPTIDYL TRANSFERASE REGION

Evidence has recently been presented suggesting that rRNA directly catalyses peptide bond formation (Noller et al, 1992). Figure 2 summarises experimental data that has linked peptidyl transferase activity to an internal loop formed at the junction of five helices in domain V of the 23S rRNA secondary structure. Chemical probing and phylogenetic data suggest that the peptidyl transferase loop is folded into a conserved tertiary conformation that determines its function in protein synthesis. A range of antibiotics, including macrolides and lincomycins, inhibit protein synthesis by recognising and interacting with essential structural elements within the peptidyl transferase region.

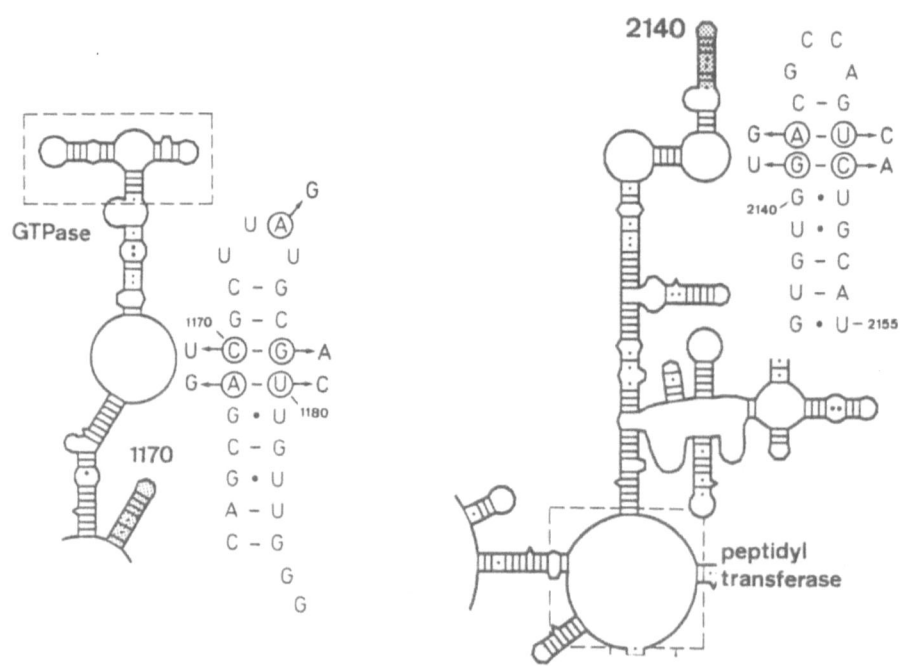

Figure 1. Unique hybridization sites for allele-specific priming and analysis of the GTPase and peptidyl transferase regions in plasmid-coded 23S rRNA. The sites are positioned at a suitable distance 3' to these regions. The base changes preserve the secondary structures of the priming sites and additionally introduce new restriction enzyme sites to aid DNA analysis and cloning (Aagaard et al, 1991).

Antibiotic interactions

The inhibitory characteristics of an antibiotic are determined by its ability to gain access to its target site on the ribosome, its binding affinity at that site, and the manner in which it perturbs the rRNA structure. Lincomycin and clindamycin (7-chloro-7-deoxylincomycin) block peptide bond formation (Gale et al, 1981; Vázquez, 1979). Of the two, clindamycin is the more potent inhibitor of Gram negative bacteria. This is at least in part due to the higher lipid solubility of clindamycin that enables it to permeate the outer membrane of these bacteria. The affinity of the two drugs for the ribosome, determined by footprinting, is approximately the same (Douthwaite, 1992b). The protection afforded by both drugs is limited to the 23S rRNA peptidyl transferase loop, but their interaction here is subtly different. Both drugs strongly protect bases A2058, A2451 and

G2505, and weakly protect G2061 from chemical modification. The modification patterns differ in that A2059 is additionally protected by clindamycin but not by lincomycin (Douthwaite, 1992b), which leaves open the possibility that there is a difference in the function of the antibiotics.

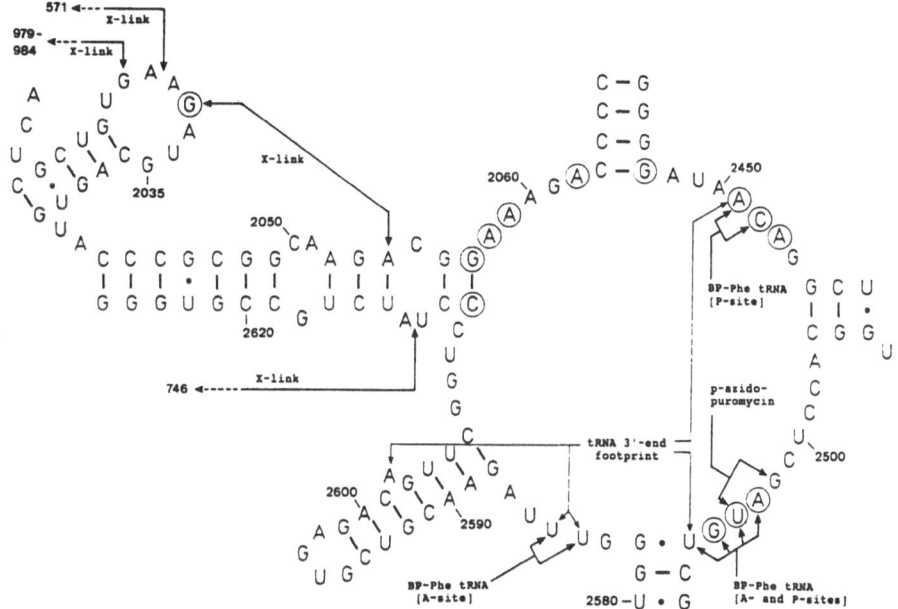

Figure 2. Secondary structure of the 23S rRNA peptidyl transferase loop and adjacent sequences (Noller, 1984). Arrows indicate cross-linked bases (Döring et al, 1991), and the positions photoaffinity labeled by benzophenone-derivatized (BP-) Phe-tRNA (Barta et al, 1990) and p-azidopuromycin (Cooperman et al, 1990). Sites where the 3'-end of tRNA has been footprinted (Moazed and Noller, 1991) are shown. The positions at which mutations confer drug resistance are encircled (for references see Douthwaite, 1992a).

Point mutations within and adjacent to the peptidyl transferase loop confer a range of altered drug tolerances (reviewed in Douthwaite, 1992a; Garrett and Rodriguez-Fonseca, 1992). Mutations that open either side of the 2057G-2611C base pair which borders the loop result in erythromycin resistance (Ettayebi et al, 1985; Sor and Fukuhara, 1984). When mutagenised from G→A, the base at position 2057 apparently remains stacked on the end of its helix (Figure 2) while the two neighbouring bases 2058A and 2059A become more reactive to chemical reagents. The erythromycin resistant 2058G (Vester and Garrett, 1987) and 2058U mutants (Sigmund et al, 1984) in this loop region also takes on a more open conformation (S.D. and C.A., unpublished).

In wild-type 23S rRNA, erythromycin interacts with the peptidyl transferase loop protecting 2058A and 2059A from chemical modification (Moazed and Noller, 1987), and these bases are approximately 50% protected by a drug concentration of 10^{-8} M (Douthwaite et al, 1989). In the 2057A mutant, the same two bases are protected although 20 times the level of drug is required to obtain comparable protection. The 2058U and 2058G mutants require 10^{-5} and 10^{-4} M erythromycin, respectively, for this degree of protection at position 2059A (S.D. and C.A., unpublished). The most straightforward interpretation of these observations is that erythromycin recognises and binds to a specific conformation of

341

nucleotides centered at 2058A and 2059A which constitute the drug target site. In this model, resistance to the drug is conferred by specific alterations in the conformation of the putative target site, with the highest resistance involving a change in the chemical groups presented at position 2058. Drug footprinting (Douthwaite, 1992b; Egebjerg and Garrett, 1991; Moazed and Noller, 1987), and multiple drug resistances conferred by mutations and *in vivo* methylation (Cundliffe, 1990) in this region, suggest that other drugs also recognise and bind the same or an overlapping target site.

The function of the peptidyl transferase loop is influenced by the structure of the adjacent hairpin loop around position 2032 and a stem-loop in domain II of the RNA (Douthwaite, 1992a). Position 2032 has been shown by cross-linking to be in close proximity to the peptidyl transferase loop (Döring et al, 1991). Mutagenesis of 2032G→A confers resistance to lincomycins and chloramphenicol, and additionally, hypersensitivity to erythromycin (Cseplö et al, 1988; Douthwaite, 1992a). The 2032A mutation alone has no detectable effect on the conformation of the peptidyl transferase loop, whereas combination of the 2032A mutation in the same 23S rRNA molecule as 2058U leads to an opening of the structure here with unwinding at the end of the helical stem at 2063C-2447G (S.D. and C.A., unpublished). This is accompanied by slow cell growth and drug sensitivity. The peptidyl transferase region is a finely tuned mechanism involving positions 2032 and 2058; combined changes at both these positions cannot be accommodated in a functional structure.

Methyl transferase interaction with the peptidyl transferase loop

Actinomycetes that produce the antibiotic types discussed here protect their own ribosomes by methylating the rRNA (Cundliffe, 1990). Erythromycin is synthesised by *Saccharopolyspora erythrea*, and this organism produces a methyl transferase (the *erm*E gene product) that modifies 23S rRNA at a single site, converting adenosine 2058 to N6, N6-dimethyladenosine. This modification lies within the center of the drug target site described above, and confers resistance to erythromycin and to other macrolides, lincosamides, and streptogramin B type antibiotics (MLS resistance) (Cundliffe, 1990). Methyl transferases offer great potential for studying functional protein-RNA interactions: both components are amenable to study by genetic and biochemical approaches; and a functional interaction between the two components leads to a stable end-product, a modified base, that can be readily assayed *in vitro* by primer extension, or *in vivo* by growth on antibiotics.

The *erm*E gene has been isolated (Bibb et al, 1985) and brought under control of the *lac* promoter in an *E. coli* "runaway" R1-derivative plasmid (B.V.and S.D., unpublished). This plasmid-type is single copy at 37°C, but on incubation of cells at 42°C, it replicates in an uncontrolled manner amplifying the copy number approximately one thousand fold (Larsen et al, 1984). This plasmid system is compatible with the pBR322-derivative plasmids containing the *rrn*B operon coding for 23S rRNA with new priming sites (Aagaard et al, 1991). Methyl transferase expression in this system can be induced by derepressing the *lac* promoter or by increasing the plasmid copy number, and confers high resistance to clindamycin and erythromycin.

Dimethylation of 23S rRNA by the methyl transferase causes reverse transcriptase to stop at position 2058. The degree of methylation of plasmid-coded mutant rRNAs, analysed by primer extension from the unique site at 2140, has been estimated relative to the methylation of chromosome-coded wild-type 23S rRNA which serves as an internal standard. Mutagenesis of position 2058 completely abolishes methylation, while changing position 2057 lowers the amount of methylation. The latter mutant rRNA is a poorer substrate for the enzyme probably because of the more open conformation of its peptidyl transferase loop.

The ErmE methyl transferase has been partially purified and functions *in vitro* in the presence of the methyl-group donor (S-adenosyl methionine) and 23S rRNA. *In vitro* transcription from the T7 promoter of progressively shorter rRNA fragments shows that all the necessary recognition and binding signals are contained within 23S rRNA nucleotides 2000 to 2630 (domain V) (B.V. and S.D., unpublished). We are presently establishing the minimal RNA structure that is specifically recognised and methylated by the enzyme.

THE GTPase REGION

The region between nucleotides 1050 and 1105 in domain II of 23S rRNA has been associated with factor-dependent GTPase activity (Figure 3). The antibiotics thiostrepton and micrococcin interact with this region within the binding site of r-protein L11 affecting elongation factor G-dependent GTP hydrolysis (Gale et al, 1981; Cundliffe and Thompson, 1981). Binding and inhibition by these drugs is dependent on the presence of L11, which in turn binds co-operatively with, and adjacent to, the $L10(L12)_4$ pentameric complex (Egebjerg et al, 1990, and refs therein). This region has been analysed by chemical and enzymatic probes in naked 23S rRNA, in ribosomes, and also in reconstituted complexes with the r-proteins (Egebjerg et al, 1990) and with antibiotics (Egebjerg et al, 1989). These results together with phylogenetic data formed the basis for building a model of the tertiary

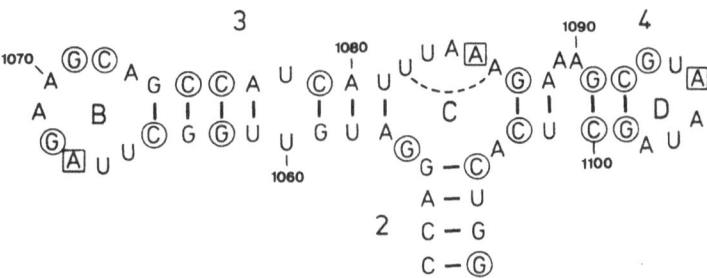

Figure 3. Secondary structure of the r-protein L11 binding site in *E. coli* 23S rRNA. The broken line in loop C indicates the base pair interaction between 1082U and 1086A which determines the relative orientation of helices 3 and 4 in the tertiary structure (Egebjerg et al, 1990; Figure 4). The encircled bases are the positions of single base bisulfate mutations (G→A and C→U). The boxed adenosines have been changed to the other three possible bases by oligonucleotide-directed mutagenesis. L11 binding to naked, intact 23S rRNA is reduced by the changes at 1062, 1068, 1071, 1072, 1085, 1087, 1091, 1092, 1093, 1099, 1100, 1102 and 1104. Thiostrepton and micrococcin binding to ribosomes is reduced by the changes at 1067 and 1095 (G.R. and S.D., unpublished).

folding of this rRNA region (Egeberg et al, 1990; Figure 4). One of the main features of this model is that the orientation of helices 3 and 4 is constrained by a Watson-Crick base pair interaction between positions 1082 and 1086, and a base triplet interaction with 1056G, bringing loops B and D in close proximity to each other. The 1082-1086 interaction has subsequently been confirmed by site-directed mutagenesis (Ryan and Draper, 1991).

Ribosomal protein interactions

Bisulfate-induced mutations and specific oligonucleotide-directed mutations have been engineered in this region of plasmid-coded 23S rRNA (Figure 3). The 23S rRNA contains a prior mutation (2058A→G) for monitoring rRNA function by growth on antibiotics and rRNA incorporated into polysomes (Aagaard et al, 1991). Structure and interactions at the mutant GTPase regions were determined by allele-specific priming from the 1170 site.

A number of mutations falling within the L11 binding site reduce the binding of this protein to naked, intact 23S rRNA (Figure 3). A notable example is the 1092C→U change which abolishes L11 binding to naked 23S rRNA. However, mutant ribosomes isolated from

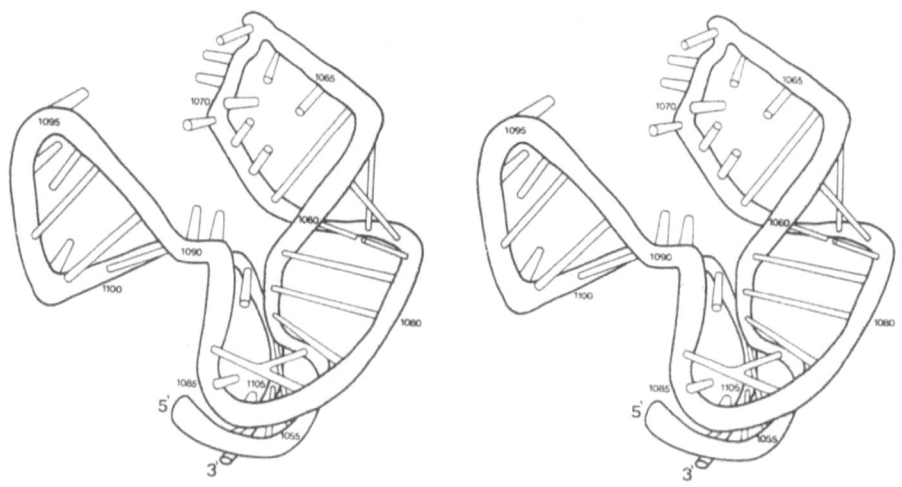

Figure 4. Model for the tertiary folding of the L11 binding site; the model is based on structural probing and phylogenetic data (Egebjerg et al, 1990). (a) The hydroxyl radical footprint of L11 binding to naked, wild-type 23S rRNA: the protected riboses are indicated by the shaded areas on the RNA backbone. (b) Riboses protected from hydroxyl radicals in L11-minus ribosomes that have been reconstituted with L11 (G.R. and S.D., unpublished).

cells grown at 37°C contain a stoichiometric amount of L11. *In vitro* binding studies indicate that the L10(L12)₄ complex and possibly other r-proteins are involved in the assembly of L11 into the 50S subunit. The 1092U mutant also confers temperature sensitivity - mutant ribosomes are active in protein synthesis at 37°C but are nonfunctional at 42°C. The temperature sensitive phenotype is relieved by the compensatory base change 1099G→A, but this does not re-establish binding between the naked rRNA and L11. All the point mutations indicated in Figure 3 that reduce or abolish L11 binding to naked 23S rRNA *in vitro*, assemble L11 in ribosomes *in vivo* in a 1:1 stoichiometry.

In a damage-selection study, binding of L11 to an RNA fragment implicated numerous bases in the interaction (Karaoglu and Thurlow, 1991); whereas a mutagenesis study on a similar fragment indicated that few bases were directly involved in L11 binding (Ryan et al, 1991). L11 has been shown to interact with its equivalent binding site in eucaryotic rRNA (Raué et al, 1988), and the *E. coli* L11 binding site is functionally active when substituted in yeast rRNA (Musters et al, 1991). Our data indicate that base-L11 interactions are possibly important in the initial recognition and binding of the protein to its site in naked rRNA, however, when L11 has found its site either on naked 23S rRNA or via co-operative effects with other r-proteins, its interaction is determined by the secondary and tertiary structure of the rRNA. This interaction is probably with the backbone sugar groups (Ryan et al, 1991), and we have therefore investigated these contacts by hydroxyl radical footprinting. Riboses protected by L11 are within the binding site previously defined by chemical and ribonuclease footprinting (Egebjerg, et al, 1990). The smaller size and the difference in reactivity of the hydroxyl radical probe enables sharper definition of the main L11 contact site, which lies within the minor groove of helix 3 (Figure 4a). The $L10(L12)_4$ complex makes far fewer ribose contacts and these are limited to helix 2 and the internal loop below this helix (Figure 3). Possibly the L10 moiety of the complex relies more on specific base interactions at its binding site. Binding of L11 together with the pentameric complex leads to protection of additional riboses in loop C.

To investigate whether L11 interacts differently with the rRNA backbone in completely assembled ribosomes, we have footprinted the binding site in L11-minus ribosomes (Stöffler et al, 1980) before and after reconstitution with L11. The footprint covers the same region as in the free complex and extends into loop C (Figure 4b), reflecting co-operative interactions with the other r-proteins.

Interactions of thiostrepton and micrococcin

The chemical and ribonuclease footprints of thiostrepton and micrococcin on 23S rRNA within 50S subunits are very similar showing altered reactivities at numerous nucleotides within the L11 binding region (Egebjerg et al, 1989). The footprints differ at the N1 position of 1067A which is protected by thiostrepton but rendered more reactive by micrococcin, and this correlates with the effects of the drugs on GTP hydrolysis, which is inhibited by thiostrepton but stimulated by micrococcin (Cundliffe and Thompson, 1981).

Despite the relatively large size of these two antibiotics, the extensive footprinting patterns are presumed to result from a limited number of direct drug contacts. These contacts could tighten the structure of the rRNA region or strengthen the L11-23S rRNA interaction resulting in the extensive protection (Egebjerg et al, 1989). This seems to be borne out by mutagenesis of this region in 23S rRNA. The drug-ribosome interactions, as estimated by allele-specific footprinting, are not appreciably altered by the single substitutions shown in Figure 3, with the exception of changes at 1067A and 1095A. Nucleotide 1067A is the site of 2'-O methylation in the thiostrepton-producing bacterium *Streptomyces azureus* (Cundliffe, 1990), the ribosomes from which are resistant to both drugs; and transversion mutations at 1067A confer drug resistance in *E. coli* ribosomes (Thompson et al, 1988). In the tertiary structure model, 1067A in loop B is in close proximity to 1095A in loop D (Figure 4). Hydroxyl radical footprinting shows that the antibiotics protect riboses in loop B, and thiostrepton shows minor protection effects in loop D. All the drug-induced effects on naked rRNA and in ribosomes are dependent on the prior binding of L11. The data presently indicate that thiostrepton and micrococcin slot into the folded rRNA structure bridging the gap between loops B and D by interacting with 1067A and 1095A.

ALLELE-SPECIFIC PRIMING IN OTHER rRNA REGIONS

Other unique priming sites that have been introduced around positions 1370 and 2800 in plasmid-coded 23S rRNA provide a specific means of probing the structure of mutagenised helix 1200-1250 and α-sarcin regions (Aagaard et al, 1991). Incorporation of the 2058G resistance marker shows that none of the priming sites adversely affect 23S rRNA function. Similar priming site constructions in 16S rRNA can be tested with a spectinomycin resistant marker (Sigmund et al, 1984), and specific priming of plasmid-coded 16S rRNA has recently been used to study a functional pseudoknot in the 530 region (Powers and Noller, 1991). The method is therefore versatile and can be employed to correlate structure and function in most regions of *E. coli* rRNA.

REFERENCES

Aagaard, C., Rosendahl, G., Dam, M., Powers, T., and Douthwaite, S., 1991, Biochimie 73:1439.

Barta, A., Kuechler, E., and Steiner, G., 1990, pp 358-365, in: "The Ribosome: Structure Function and Evolution," W.E. Hill, A.E. Dahlberg, R.A. Garrett, P.B. Moore, D. Schlessinger, J.R. Warner, ed., American Society for Microbiology, Washington, D.C.

Bibb, M.J., Janssen, G.R., and Ward, J.M., 1985, Gene 38:215.

Cooperman, B.S., Weitzmann, C.J., and Fernández, C.L., 1990, pp 491-501, in: "The Ribosome: Structure Function and Evolution," W.E. Hill, A.E. Dahlberg, R.A. Garrett, P.B. Moore, D. Schlessinger, J.R. Warner, ed., American Society for Microbiology, Washington, D.C.

Cseplö, A., Etzold, T., Schell, J., and Schreier, P.H., 1988, Mol. Gen. Genet. 214:295.

Cundliffe, E., 1990, pp479-490, in: "The Ribosome: Structure Function and Evolution," W.E. Hill, A.E. Dahlberg, R.A. Garrett, P.B. Moore, D. Schlessinger, J.R. Warner, ed., American Society for Microbiology, Washington, D.C.

Cundliffe, E., and Thompson, J., 1981, Eur. J. Biochem., 118:47.

Douthwaite, S., 1992a, J. Bacteriol. 174:1333.

Douthwaite, S., 1992b, Nucl. Acids Res. 20:4717.

Douthwaite, S., Powers, T., Lee, J.Y., and Noller, H.F., 1989, J. Mol. Biol. 209:655.

Döring, T., Greuer, B., and Brimacombe, R., 1991, Nucl. Acids Res., 19:3517.

Egebjerg, J., Douthwaite, S., and Garrett, R.A., 1989, EMBO J. 8:607.

Egebjerg, J., Douthwaite, S.R., Liljas, A., and Garrett, R.A., 1990, J. Mol. Biol. 213:275.

Egebjerg, J., and Garrett, R.A., 1991, Biochimie 73:1145.

Ettayebi, M., Prasad, S., and Morgan, E.A., 1985, J Bacteriol. 162:551.

Gale, E.F., Cundliffe, E., Reynolds, P.E., Richmond, M.H., and Waring, M.J. (1981) The Molecular Basis of Antibiotic Action, John Wiley and Sons, London.

Garrett, R.A., and Rodriguez-Fonseca, C., 1992, in: "Ribosomal RNA: Structure, Evolution, Processing and Function in Protein Synthesis," R.A. Zimmermann and A.E. Dahlberg, ed., CRC Press.

Karaoglu, D., and Thurlow, D.L., 1991, Nucl. Acids Res. 19:5293.

Larsen, J.E.L., Gerdes, K., Light, J., and Molin, S., 1984, Gene 28:45.

Moazed, D., and Noller, H.F., 1987, Biochimie, 69:879.

Moazed, D., and Noller, H.F., 1991, Proc. Natl. Acad. Sci. USA. 88:3725.

Musters, W., Gonçalves, P.M., Boon, K., Raué, H.A., van Heerikhuizen, H., and Planta, R., 1991, Proc. Natl. Acad. Sci. USA. 88:1469.

Noller, H.F., 1984, Ann. Rev. Biochem. 53:119.

Noller, H.F., Hoffarth, V:, and Zimniak, L., 1992, Science 256:1416.

Powers, T., and Noller, H.F., 1991, EMBO J. 10:2203.

Raué, H.A., Klootwijk, J., and Musters, W., 1988, Proc. Biophys. Mol. Biol. 51:77.

Ryan, P.C., and Draper, D.E., 1991, Proc. Natl. Acad. Sci. USA. 88:6308.

Ryan, P.C., Lu, M., and Draper, D.E., 1991, J. Mol. Biol. 221:1257.

Sigmund, C.D., Ettayebi, M., and Morgan, E.A., 1984, Nucl. Acids Res. 12:4653.

Sor, F., and Fukuhara, H., 1984, Nucl. Acids Res. 12:8313.

Stöffler, G., Cundliffe, E., Stöffler-Meilicke, M., and Dabbs, E.R., 1980, J. Biol. Chem. 255:10517.

Thompson, J., Cundliffe, E., and Dahlberg, A.E., 1988, J. Mol. Biol. 203:457.

Vázquez, D. (1979) Inhibitors of Protein Synthesis. Springer-Verlag, Berlin.

Vester, B., Garrett, R.A., 1987, Biochimie 69:891.

EXTENSION AND FOLDING OF NASCENT PEPTIDES ON RIBOSOMES

Boyd Hardesty, O.W. Odom, Wieslaw Kudlicki, and Gisela Kramer

Department of Chemistry & Biochemistry
The University of Texas at Austin
Austin, Texas 78712

THE PEPTIDYL TRANSFERASE REACTION

The ribosomal synthesis of a peptide bond takes place by transfer of the peptidyl ester of peptidyl-tRNA to the amino acid amino group of an incoming aminoacyl-tRNA. Attempts to isolate an acyl-ribosome intermediate of the kind found for many enzyme-catalyzed hydrolyses or transpeptidation reactions have been unsuccessful. This has led many investigators to speculate that transpeptidation by the ribosome is brought about by an appropriate spatial orientation and alignment of the aminoacyl-tRNA and peptidyl-tRNA without the catalytic involvement of special nucleophilic groups of the large ribosomal subunit as discussed by Spirin (1986). The range of covalent derivatives other than peptides or amides that can be formed: esters (Fahnestock et al., 1970), thioesters (Gooch and Hawtrey, 1975), thioamides (Victorova et al., 1976), phosphinoamides (Tarussova et al., 1981) support this hypothesis with the implication that the peptidyl transferase reaction itself is of the S_N2 type with nucleophilic substitutions through a tetrahedral intermediate. However, these considerations only serve to emphasize the importance of understanding how the 3' ends of two tRNAs are brought precisely into reactive proximity to facilitate the reaction that by many measures is the most evolutionarily conserved and fundamental process of life, the reaction system by which genetic information encoded in nucleic acid is translated into protein.

RIBOSOMAL SITES

Early studies prompted Watson (1964) to propose the classical two-site model in which two tRNA binding sites were envisioned, the P or peptidyl site and the A or acceptor ribosomal site. Peptide bond formation was thought to involve the physical transfer of the nascent peptide from the P site tRNA to the amino acid of tRNA in the A site. Although intuitively attractive, problems soon arose. The conceptual requirements and data indicated that there must be at least three functionally distinct sites to which tRNA can be bound (Noll, 1966; Hardesty et al., 1969): the entry, acceptor, and donor sites as we called them. Evidence was presented that three tRNAs could be bound simultaneously to the same ribosome (Rheinberger and Nierhaus, 1980; Lill et al., 1984) although the functional significance of such binding was not clear (Spirin, 1985).

Using fluorescence techniques we found that the center portion and amino acid stem of tRNA which had been bound to ribosomes as a peptidyl-tRNA analogue moved relative to the ribosomes as the peptidyl transferase reaction took place and proposed the displacement model to account for these observations as well as the existence of three sites (Hardesty et al., 1986). The central feature of this model is that the nascent peptide stays in one position, the peptidyl transferase center, relative to the ribosomes, as the chemical reaction of peptide transfer to an incoming aminoacyl-tRNA takes place. The requisite movement is accomplished by movement of the tRNA rather than the peptide. That the nascent peptide does not move more than several Å, whereas the amino acid stem of the tRNA moves more than 20 Å, was subsequently demonstrated directly (Odom et al., 1990). It appears that movement of the tRNA relative to the ribosome as it is deacylated during the peptidyl transferase reaction is associated with a conformational change in the tRNA (Odom and Hardesty, 1987). It was also demonstrated unequivocally that one molecule each of deacylated tRNA, N-acetyl Phe-tRNA (a peptidyl tRNA analogue) and puromycin (an analogue of aminoacyl-tRNA) could be bound simultaneously to the same ribosome (Odom et al., 1991). Other results favor the interpretation that the anticodons of only two tRNAs interact simultaneously with mRNA on the small ribosomal subunit under conditions of efficient peptide synthesis (Moazed and Noller, 1989) and prompted Noller to propose the hybrid state model with three tRNA binding sites on the large ribosomal subunit and two on the small subunit (Noller et al., 1990).

RNA IN THE PEPTIDYL TRANSFERASE CENTER

The experiments showing that the position of the peptide relative to the ribosome did not change appreciably during peptide transfer were carried out with a peptidyl-tRNA analogue in which a fluorophore was covalently attached to the amino acid amino group of aminoacyl-tRNA. For these experiments, tRNA^{Phe} was enzymatically aminoacylated, isolated, and then mercaptoacetylated by reaction with the succinimide ester of dithiodiglycolic acid (the disulfide of mercaptoacetic acid) followed by reduction of the disulfide. The resulting mercaptoacetylated Phe-tRNA was then labeled by reaction of the sulfhydryl group with either CPM (3-(4-maleimidophenyl)-7-diethylamino-4-methyl coumarin) or IAEDANS (5-[(2-iodoacetylaminoethyl)amino]napthalene-1-sulfonic acid) to yield probe-S-AcPhe-tRNA. When bound to ribosomes the acylamino acid moiety of the tRNA could be transferred to puromycin or could form the amino terminus of a nascent peptide (Picking et al., 1991a). The results demonstrate that the acyl-aa-tRNA derivative can be bound into the peptidyl transferase center and can function as a peptidyl-tRNA analogue in the peptidyl transferase reaction. Fluorescence from the probe was used to characterize the peptidyl transferase center and the environment of the N-terminus of the nascent peptide as it is extended from the center.

Changes in the emission spectrum and fluorescence quantum yield that occur upon binding of the CPM or AEDANS derivatives to the ribosome indicate that the probes are held in a relatively hydrophobic environment. Fluorescence anisotropy increased to nearly the theoretical maximum, 0.40, indicating that the probe and presumably the amino acid to which it was attached are held rigidly in the peptidyl transferase center in a way that allows very little movement of the probe independent of the ribosome (Odom et al., 1991). Also, upon binding to the ribosome, the coumarin probe became inaccessible to anticoumarin IgG (Picking et al., 1992a). Considered together, these results suggest that the acyl amino acid or nascent peptide of the tRNA in the peptidyl transferase center is held within or between structured elements of the large ribosomal subunit.

The accessibility of the CPM probe was investigated further using methyl viologen, MV^{2+} (1,1'-dimethyl-4,4'-bipyridinium dichloride), as a quenching agent. Initially we assumed that quenching would decrease when the CPM-S-AcPhe-tRNA was bound to the ribosome due to shielding from MV^{2+} in the solvent phase. However, more than a two-fold increase in quenching was observed (Picking et al., 1992b). A detailed analysis of this result using fluorescence lifetime measurements indicated that the observed increase in quenching is due entirely to static quenching by MV^{2+} that is bound in the peptidyl transferase center within a few Å of the probe rather than dynamic quenching due to collision of MV^{2+} in the solvent phase with the probe (Picking et al., 1992b). By analogy with the well characterized quenching by MV^{2+} of fluorescence from ethidium bromide intercalated into double-stranded DNA (Atherton, 1988) with which binding of MV^{2+} appears to result from direct electrostatic interaction of the bipyridine with the ionized phosphate residues along the backbone of the nucleic acid, it appears that MV^{2+} binds to RNA in the ribosomes. Since MV^{2+} quenching of coumarin fluorescence is due to electron transfer over a very short distance, it follows that the MV^{2+} is bound to RNA in the peptidyl transferase center in the immediate vicinity of the nascent peptide. Recently, Noller and his coworkers (Noller et al., 1992) reported that 50S ribosomal subunits from *Thermus aquaticus* retained their ability to form a peptide between fMet-tRNA and puromycin after the subunits had been incubated with proteinase K and extracted with phenol to remove up to 99% of the ribosomal proteins, thus indicating that the peptidyl transferase reaction is mediated by RNA. The MV^{2+} quenching effects appear to be entirely consistent with this conclusion.

THE CONFORMATION OF NASCENT HOMOPOLYMERIC PEPTIDES

The conformation of nascent polyphenylalanine and polylysine chains synthesized with poly(U) and poly(A), respectively, was investigated by incorporating a fluorescently derivatized amino acid residue into the N-terminus of the nascent peptides, as indicated above. The Lys-tRNA derivative initially appeared to be held rigidly in a relatively hydrophobic environment within the peptidyl transferase center as described above for the derivative of Phe-tRNA. However, striking differences in fluorescence were noted as the polyphenylalanine and polylysine peptides were extended. With polyphenylalanine, fluorescence anisotropy remained high and the probe appeared to be in an increasingly hydrophobic environment, as judged by the quantum yield and emission spectrum (Picking et al., 1991a). Considered together with other data the results appear to indicate that polyphenylalanine was formed as an insoluble mass located at or near the peptidyl transferase center. It should be noted that polyphenylalanine is strikingly insoluble in nearly all aqueous and organic solvent systems. The conclusion that polyphenylalanine is formed as an insoluble mass was reinforced by results with the antibiotic erythromycin (Odom et al., 1991). This antibiotic blocks the extension of most nascent peptides beyond the di- or tri-peptide stage but does not inhibit the synthesis of the first peptide bond. Also, inexplicably, it does not inhibit the poly(U)-directed synthesis of polyphenylalanine. There is evidence indicating that it is bound to the 50S ribosomal subunit at a point that is very near to the peptidyl transferase center. The erythromycin binding site appears to be so close to the acyl-amino acid of aminoacyl-tRNA in the peptidyl transferase center that bound erythromycin perturbs the local environment of the fluorescent probe. Also, aminoacyl-tRNA in the peptidyl transferase center has a similar effect on fluorescence from coumarin covalently linked to bound erythromycin (Odom et al., 1991).

Erythromycin does not bind to ribosomes that bear nascent peptides formed with mRNA or RNA homopolymers, including polyphenylalanine formed with poly(U). However, if fluorescently labeled erythromycin was bound to the ribosomes before polyphenylalanine synthesis was carried out, subsequent peptide synthesis resulted in increasing hydrophobicity of the environment of the probe and the bound erythromycin could no longer be readily exchanged with unlabeled erythromycin (Odom et al., 1991). The bound erythromycin appears to be buried under or within an insoluble mass of nascent polyphenylalanine.

Exactly where erythromycin is bound to the 50S subunit and how it blocks the extension of most peptides but not of polyphenylalanine remain intriguing questions. Presumably, erythromycin binds to a site near the peptidyl transferase center that is occupied by the nascent peptide after it is extended to a length of three or four amino acid residues or more. This might be the entrance to a channel (Ryabova et al., 1988) or tunnel

(Yonath et al., 1987) through which the nascent peptide is extended to reach the point at which it leaves the surface of the ribosome in the "exit domain" (Bernabeu and Lake, 1980).

The situation with polylysine is strikingly different from that described above for polyphenylalanine (Picking et al., 1991a). The environment of the probe becomes more hydrophilic and fluorescence anisotropy drops rapidly to near the levels observed for free probe-Lys-tRNA as the polylysine chains are extended to a length of three or four residues. Fluorescence parameters do not change significantly as the nascent polylysine peptides are extended beyond this length. It should be noted that polylysine in aqueous solution exists primarily as a random coil at neutral pH (Hartman et al., 1974).

The physical properties of both polyphenylalanine and polylysine are atypical in ways that may compromise their utility for study of nascent peptides on ribosomes. It was deemed desirable to use homopolymeric peptides of other amino acids. This was accomplished by constructing synthetic tRNAs in which the natural anticodon-amino acid relations were altered. Nucleotide sequences corresponding to tRNA^{Ser} (Picking et al., 1991b) and elongator tRNA^{Ala} as well as tRNA^{Ala} species with a primary sequence corresponding to initiator tRNA^{Met} (Picking et al., 1991c) were cloned behind a T7 promoter then transcribed by the viral polymerase. Each of these tRNA species had an AAA anticodon, was enzymatically aminoacylated, and functioned in peptide synthesis with poly(U). The initiator tRNA^{Ala} could be enzymatically bound to ribosomes at low Mg^{2+} concentrations with IF-2/GTP and would generate the amino terminus of nascent peptides but would not function in peptide elongation. Polyalanine or polyserine peptides containing more than 100 residues could be formed from the respective elongator tRNA$_{AAA}$ with poly(U). Peptides were initiated with coumarin-alanine or coumarin-serine, then the environment and accessibility of the probe were monitored by changes in fluorescence as a function of the length of the nascent peptides as they were elongated. The anisotropy of fluorescence from the N-terminus of nascent polyalanine remained high at about 0.37 as the peptides were extended to 30-40 residues, then gradually declined as the peptides were extended to lengths in the order of 60-70 residues. Anisotropy declined more rapidly as the peptides were extended to lengths greater than 70-80 residues (Picking et al., 1992a). Polyserine anisotropy declined considerably faster than that of polyalanine peptides of the same average length, apparently reflecting differences in the secondary structure of the two types of nascent peptides. The pattern for both peptides was strikingly different from that for polyphenylalanine or polylysine. The differences in the anisotropy data for polyalanine and polyserine probably reflect the propensity of the two types of polypeptides to form α helices. They suggest that the secondary structure of nascent polyalanine is primarily α-helical to peptide lengths of at least 70 residues.

Experiments were carried out to determine the accessibility of N-terminal coumarin on nascent polyalanine or polyserine peptides to anticoumarin IgG or Fab fragments derived from it. Binding of either IgG or Fab to N-terminal coumarin on the nascent peptides caused large changes in fluorescence quantum yield, emission spectrum, and anisotropy. For CPM-SAc-Ala-tRNA free in solution, antibody binding caused more than a two-fold increase in quantum yield and a shift in the emission maximum from 480 nm to 460 nm, apparently reflecting the hydrophobic nature of the coumarin binding site in the antibody. During peptide synthesis the N-terminal coumarin of nascent chains became accessible to anti-coumarin IgG at a nascent peptide length of about 70 alanine residues or 60 serine residues, and to Fab fragments of the same IgG at an average peptide length of about 55 alanine and 45 serine residues, respectively. The differences in accessibility with the polyalanine and polyserine peptides apparently reflect differences in secondary structure whereas the differences between IgG and its Fab fragment probably reflect differences in their molecular mass. Sensitivity of the nascent chains to proteinase K (M_r 27,000) was also monitored. The N-terminal coumarin was relased from the nascent peptide at an average length of about 40 alanine residues and 25-30 serine residues, respectively. Work from other laboratories carried out with different ribosomes, different nascent peptides and different proteases indicates that the nascent peptides of proteins are hydrolyzed by proteolysis at a somewhat variable length generally in the order of 30-35 amino acid residues (Malkin and Rich, 1967; Blobel and Sabatini, 1970; Smith et al., 1978; Ryabova et al., 1988). It should be noted that polyalanine peptides have a high tendency to form α helices in aqueous solution at neutral pH (Chou and Fasman, 1974); α-helix formation is somewhat less with polyserine but the tendency of both polyamino acids to form α helices would increase appreciably if an α-helical core existed within the peptide (Picking et al., 1992a; Creighton, 1983). Spirin and Lim (1986) proposed that an α-helix is formed by the peptidyl transferase reaction as the nascent peptide is formed.

Our results described above indicate a distinct difference in the accessibility of the N-terminus of the nascent polyalanine and polyserine chains based on the molecular size of the assay protein, i.e., IgG vs. Fab or proteinase K, and that the N-terminus of the polyalanine peptides remain protected to a somewhat longer length than nascent polyserine.

FOLDING OF NASCENT MS2 COAT PROTEIN

One of the most important questions of biochemistry and molecular biology is how the native tertiary structure of proteins, both globular and fibrous, is formed (Richards, 1991). Chantrenne (1961) appears to have been the first to clearly enunciate the vectorial folding hypothesis, the proposition that folding of nascent peptides, possibly into specific domains or regions of the native protein, may occur as the nascent peptides are formed from their N-terminal to C-terminal ends on ribosomes. The role the ribosome itself may play in this process, beyond peptide bond formation and addition of individual amino acid to the nascent peptide, is largely unknown. The homopolymeric peptides considered above do not fold into the complex tertiary structure of a typical globular protein. MS2 coat protein was chosen for initial experiments in which the same general strategy described above was used to monitor the extension of nascent homopolymeric peptides (Picking et al., 1992c). This protein contains 130 amino acids including the N-terminal methionine (Jou et al., 1972) and is folded into a well-defined globular-like structure containing seven antiparallel β-strand and two short α-helices (Valegard et al., 1990). To label the N-terminus of the nascent protein, a coumarin derivative was covalently attached to the methionine amino group of Met-tRNA$_f$, which was then bound to ribosomes in the presence of MS2 RNA. Synthesis of specific lengths of nascent MS2 coat protein was accomplished by limiting tRNA or amino acids in the polymerization reaction mixture or by truncating the MS2 RNA with ribonuclease H after hybridization with an antisense DNA oligomer. Fluorescence anisotropy was used as a measure of probe mobility as a function of nascent peptide length. The resulting anisotropy profile was most similar to that for polyalanine rather than polyserine, polylysine, or polyphenylalanine. Anisotropy remained very high (near 0.38) until a peptide length of 30-40 residues was reached followed by a slow decline to values near those for full-length MS2 coat proteins that had been released from the ribosome.

MV^{2+} quenching of fluorescence from N-terminal coumarin-methionine was used to characterize the environment of the probe as the nascent peptides were extended. Stern-Volmer quenching constants were determined in reaction mixtures in which translation had been stopped to give nascent peptides of specific lengths. Measurements were made for peptides bound to the ribosomes or after they had been released from the ribosome by reaction with puromycin. Similar experiments were carried out with nascent polyalanine peptides and the results are shown in Figure 1. As described above, quenching for nascent polyalanine on ribosomes declined to the value for puromycin-released chains at a chain length of about 60 residues. However, although quenching for the ribosome-bound MS2 protein initially declined rapidly as the chains were extended, values remained well above those for the corresponding puromycin-released material to lengths of about 125 residues. This result appears to indicate that MV^{2+} quenching of fluorescence from the N-terminal coumarin is subject to influence by the ribosome as long as even nearly full-length peptides remain bound to the ribosomes. This is a strikingly different result than was observed for nascent polyalanine.

Experiments similar to those described above were carried out with anticoumarin IgG and Fab derived from it with the results shown in Figure 2. Both IgG and Fab interact readily with N-terminal coumarin of all lengths of puromycin-released MS2 peptides. Coumarin becomes accessible to Fab at an average length of about 45 residues. This is approximately the length at which N-terminal coumarin on nascent polyalanine reacts with the Fab. However, in contrast to the results with nascent polyalanine with which N-terminal coumarin became accessible at an average chain length of about 60 residues, the N-terminus of all lengths of ribosome-bound nascent MS2 peptides are largely shielded from IgG. The different accessibility for Fab and IgG is consistent with the hypothesis that the N-terminal coumarin of nearly full-length peptides is shielded from IgG by the

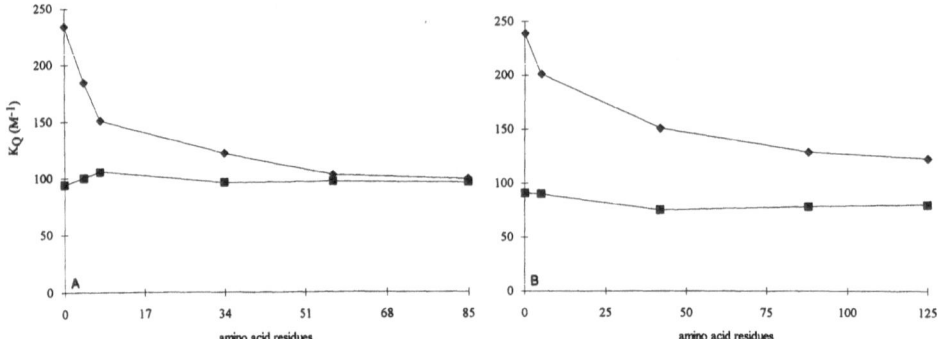

Figure 1. MV^{2+} quenching of coumarin at the N-terminus of nascent polyalanine (A) and MS2 coat protein (B). Polyalanine was synthesized using polyuridylic acid and a synthetic tRNA^{Ala} after prebinding CPM-SAcAla-tRNA (Picking et al., 1992a). Chain lengths of nascent MS2 coat protein were limited as described (Kolb et al., 1987; Ryabova et al., 1988). In all cases ribosomes bearing the fluorescent nascent peptides were isolated by Sephacryl S300 chromatography and divided into 2 fractions. Quenching by MV^{2+} was measured directly on one fraction (♦). The second fraction was incubated with 1 mM puromycin prior to measuring MV^{2+} quenching (■). The Stern-Volmer quenching constant, K_Q, was calculated for each sample and plotted versus peptide length.

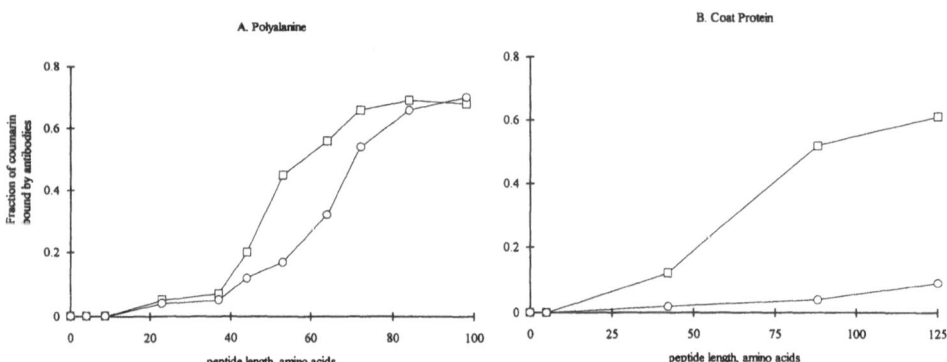

Figure 2. Accessibility of coumarin on the N-terminus of nascent polyalanine (A) and MS2 coat protein (B) to anticoumarin IgG and Fab derived from it. Various lengths of nascent polyalanine on MS2 coat protein were obtained as indicated for Figure 1. In all cases ribosomes bearing nascent peptides were isolated by gel filtration to remove unbound material and were then incubated with an excess of rabbit anti-coumarin antibodies (O) or Fab fragments prepared from these antibodies (□). The fraction of the amino-terminal CPM bound by antibody or Fab was calculated from the increase in fluorescence at 450 nm before and after release of the chains by puromycin.

ribosome in a conformation or position in which they are accessible only to the smaller Fab molecule.

Both MV^{2+} quenching and the antibody accessibility experiments suggest that the nascent MS2 peptides of about 125 residues are in a folded conformation while they are bound to the ribosome. This conclusion was tested directly by non-radiative energy transfer between coumarin covalently attached to cysteine at position 47, the first cysteine residue in the MS2 sequence, and fluorescein on the N-terminal methionine. The N-terminal fluorescein-methionine was incorporated into the peptide from Met-tRNA$_f$ that had been derivatized as described above. Coumarin-cysteine was incorporated from Cys-tRNA that had been derivatized by direct reaction with CPM. With ribosome-bound nascent peptides of about 125 amino acid residues energy transfer of more than 75% was determined. The same amount of energy transfer was observed after the nascent peptides were released from the ribosomes by puromycin and with full-length coat protein that had been released from the ribosome by termination factors at the UAA-UAG termination sequence for the coat protein. The observed level of energy transfer appears to be a minimum value in that a number of factors that might reduce energy transfer exist in the experiments. A high level of energy transfer, probably greater than 95%, would be predicted from the distance between the N-terminus and cysteine[47] in the crystal structure of the native protein in the phage (Valegard et al., 1990). The relatively high level of energy transfer that was observed is consistent with a distance between the probes that is shorter than anticipated if the nascent peptides were in the form of an α-helix or random coil. A corollary of this conclusion is that the nascent peptides are folded into some type of compact tertiary structure while they are bound to the ribosomes. The relation of this pre-native state to the conformation of folded globular proteins is not known.

SYNTHESIS AND FOLDING OF ENZYMES

Enzymatic activity provides the strongest indication that a protein is in its native conformation. This critical test has been used to evaluate folding of proteins formed *in vitro*. We have developed a fractionated cell-free system from *E. coli* that carries out coupled transcription and translation from nonlinearized plasmids (Kudlicki et al., 1992). Thus far we have used the system to synthesize *in vitro* enzymatically active dihydrofolate reductase, chloramphenicol acetyl transferase, and rhodanese. The latter enzyme is particularly interesting as a mammalian protein (M_r 33,000) that is encoded by a nuclear gene but is transported from the cytoplasm into the mitochondrial matrix. Rhodanese has been used as a model enzyme to study renaturation after its denaturation in 6 M urea. The enzyme refolds spontaneously to a very limited extent into its native conformation. Recovery of enzymatic activity from the denatured state is aided by the chaperonins GroEL and GroES in an ATP-dependent reaction (Mendoza et al., 1991). Our recent results demonstrate a requirement for GroEL and GroES during synthesis to form enzymatically active rhodanese in the cell-free *E. coli* system (Tsalkova et al., 1992). The results strongly suggest that the chaperonins must interact with incomplete nascent peptides while they are bound to the ribosome as peptidyl-tRNA. How this is accomplished is under investigation.

DISCUSSION

All of our results suggest that nascent polyalanine chains bound to ribosomes as peptidyl-tRNA exist primarily as α-helices at least to lengths of less than 65 to 75 residues, the length at which an N-terminal probe becomes accessible to anticoumarin IgG and escapes the influence of the ribosomes. This corresponds to a linear distance along the α-helix of about 105 Å which is appreciably less than the 140 Å estimated by Bernabeu and Lake (1980) to be the distance between the presumptive site of peptide bond formation at the base of the central protuberance and the point at which the nascent peptide leaves the surface of the ribosome in the exit domain. The difference may be accounted for by error or incorrect assumptions inherent in either determination. Due caution is warranted in interpreting the absolute distance values. In any event, it is clear that N-terminal coumarin is shielded from anti-coumarin IgG on longer nascent polyalanine peptides than from Fab fragments derived from the IgG. The N-terminal probe on nascent polyalanine becomes accessible to the corresponding Fab fragment at a peptide chain length of about 40-45 residues, i.e., about 65 Å if the nascent peptide was in an α-helical conformation. Thus, there appears to be a domain or cavity within the 50S subunit into which Fab can penetrate about 40 Å further than IgG. The results from the MV^{2+} quenching experiments as well as changes in fluorescence quantum yield and anisotropy that occur during elongation of polyalanine are in general agreement with this conclusion.

The results with MS2 coat protein strongly indicate that it exists in a folded conformation while it is bound to ribosomes as peptidyl-tRNA. The accessibility of N-terminal coumarin to IgG and its Fab provides a direct indication that the nascent peptides are shielded from large proteins in the solvent phase as long as they are attached as peptidyl-tRNA to the ribosome. Considered together, we believe these results indicate that nascent peptides may fold as they are formed on a ribosome in accord with the vectorial folding hypothesis. Further, the results suggest that this folding may take place in a cavity or domain of the large ribosomal subunit, a "folding domain," in which the nascent peptide may be transiently shielded while folding progresses towards the conformation of the native protein.

ACKNOWLEDGMENTS

We gratefully acknowledge the assistance of Amy Whitworth in preparing the typescript. This work was supported by grants to Boyd Hardesty from the National Science Foundation and the Foundation for Research.

REFERENCES

Atherton, S., 1988, *in* "Light in Biology and Medicine," Vol. 1, R.H. Douglas, J. Moan, and F. Dall-Acqua, eds., Plenum Press, New York, pp. 71-84.

Bernabeu, C., and Lake, J.A., 1980, *Proc. Natl. Acad. Sci. USA* 79:3111-3115.

Blobel, G., and Sabatini, D., 1970, *J.Cell Biol.* 45:130-145.

Chantrenne, H., 1961, *in* "Modern Trends in Physiological Science," P. Alexander and Z. Bacq, eds., Pergamon Press, p. 122.

Chou, P.Y., and Fasman, G.D., 1974, *Biochemistry* 13:211-222.

Creighton, T.E., 1983, "Proteins: Structures and Molecular Properties," W.H. Freeman and Co., New York.

Fahnestock, S., Neumann, H., Shashoua, V., and Rich A., 1970, *Biochemistry* 9:2477-2483.

Gooch, J., and Hawtrey, A.O., 1975, *Biochem. J.* 149:209-220.

Hardesty, B., Culp, W., and McKeehan, W., 1969, *Cold Spring Harbor Symp. Quant. Biol.* 34:331-345.

Hardesty, B., Odom, O.W., and Deng, H.-Y., 1986, *in* "Structure, Function, and Genetics of Ribosomes," B. Hardesty and G. Kramer, eds., Springer-Verlag, New York, pp. 495-508.

Hartman, R., Schwaner, R.C., and Hermans, J., Jr., 1974, *J. Mol. Biol.* 90:415-429.

Jou, W.M., Haegeman, G., Ysebaert, M., and Fiers, W., 1972, *Nature (London)* 237:82-88.

Kolb, V.A., Kommer, A.A., and Spirin, A.S., 1987, *Dokl. Akad. Nauk. SSR.* 296:1497-1501.

Kudlicki, W., Kramer, G., and Hardesty, B., 1992, *Analytical Biochemistry* 206, in press.

Lill, R., Robertson, J.M., and Wintermeyer, W., 1984, *Biochemistry* 23:6710-6717.

Malkin, L.I., and Rich, A., 1967, *J.Mol.Biol.* 26:329-346.

Mendoza, J.A., Rogers, E., Lorimer, G.H., and Horowitz, P., 1991, *J. Biol. Chem.* 266: 13044-13049.

Moazed, D., and Noller, H.F., 1989, *Nature* 342:142-148.

Noll, H., 1966, *Science* 151: 1241-1245.

Noller, H.F., Hoffarth, V., and Zimniak, L., 1992, *Science* 256:1416-1419.

Noller, H.F., Moazed, D., Stern, S., Powers, T., Allen, P.N., Robertson, J.M., Weiser, B., and Triman, K., 1990, *in* "The Ribosome: Structure, Function, & Evolution," W.E. Hill, A. Dahlberg, R.A. Garret, P.B. Moore, D. Schlessinger, and J.R. Warner, eds., American Society for Microbiology, Washington, D.C., pp. 73-92.

Odom, O.W., and Hardesty, B., 1987, *Biochimie* 69:925-938.

Odom, O.W., Picking, W.D., and Hardesty, B., 1990, *Biochemistry* 29:10734-10744.

Odom, O.W., Picking, W.D., Tsalkova, T., and Hardesty, B., 1991, *Eur. J. Biochem.* 198:713-722.

Picking, W.D., Kolb, V.A., Odom, O.W., Picking, W.L., Spirin, A.S., and Hardesty, B., 1992c, submitted.

Picking, W.D., Odom, O.W., and Hardesty, B., 1992b, *Biochemistry*, in press.

Picking, W.D., Odom, O.W., Tsalkova, T., Serdyuk, I., and Hardesty, B., 1991a, *J. Biol. Chem.* 266:1534-1542.

Picking, W.D., Picking, W.L., Odom, O.W., and Hardesty, B., 1992a, *Biochemistry* 31:2368-2375.

Picking, W.L., Picking, W.D., and Hardesty, B., 1991b, *Biochimie* 73:1101-1108.

Picking, W.L., Picking, W.D., Ma, C., and Hardesty, B., 1991c, *Nucl.Acid.Res.* 19:5749-5754.

Rheinberger, H.-J., and Nierhaus, K.H., 1980, *Biochem. Int.* 1:297-303.

Richards, F.M., 1991, *Scientific American* 264:54-63.

Ryabova, L.A., Selivanova, O.M., Baranov, V.I., Vasiliev, V.D., and Spirin, A.S., 1988, *FEBS Letters* 226:255-260.

Smith, W.P., Tai, P.-C., and Davis, B.D., 1978, *Proc.Natl.Acad.Sci.U.S.A.* 75:5922-5925.

Spirin, A.S., 1985, *Prog. Nucl. Acid. Res. Mol. Biol.* 32:75-114.

Spirin, A.S., 1986, "Ribosome Structure and Protein Biosynthesis," Benjamin/Cummings Publishing Company, Menlo Park.

Spirin, A.S., and Lim, V.I., 1986, *in* "Structure, Function, and Genetics of Ribosomes," B. Hardesty and G. Kramer, eds., Springer-Verlag, New York, pp. 556-572.

Tarussova, N.B., Jacovleva, G.M., Victorova, L.S., Kukhanova, M.K., and Khomutov, R.M., 1981, *FEBS Letters* 130:85-87.

Tsalkova, T., Zardeneta, G., Kudlicki, W., Kramer, G., Horowitz, P., and Hardesty, B., 1992, submitted.

Valegard, K., Liljas, L., Fridborg, K., and Unge, I., 1990, *Nature (London)* 345:36-41.

Victorova, L.S., Kotusov, L.S., Azhayev, A.V., Krayevsky, A.A., Kukhanova, M.K., and Gottikh, B.P., 1976, *FEBS Letters* 68:215-218.

Watson, J., 1964, *Bull. Soc. Chim. Biol.* 46:1399-1425.

Yonath, A., Leonard, K.R., and Wittmann, H.G., 1987, *Science* 236:813-816.

PROTEIN SYNTHESIS AND SECRETION
AS SEEN BY THE NASCENT PROTEIN CHAIN

Arthur E. Johnson, Kathleen S. Crowley, Steven K. Shore, and Gregory D. Reinhart

Department of Chemistry and Biochemistry
University of Oklahoma
Norman, Oklahoma, 73019

INTRODUCTION

One of the features that distinguishes the ribosome from most other enzymes is the need to retain the product of each transpeptidation reaction, specifically the nascent or growing polypeptide chain, until the mRNA-dependent polymerization of amino acids is terminated by a stop codon. Since many proteins are more than 1000 amino acids in length, the space occupied by the nascent chain attached to the ribosome-bound peptidyl-tRNA can be substantial. The ribosome must therefore be designed to minimize the interference between the growing nascent chain and the molecular traffic associated with decoding and protein chain elongation (tRNA and elongation factors). The probable solution to this structural issue was provided by Malkin and Rich (1967) and by Blobel and Sabatini (1970), who found that the ribosome protected the C-terminal 40 or so amino acids of the nascent chain from proteolytic digestion. This suggested that the nascent chain was not exposed to the cytoplasm near the peptidyltransferase center, but instead left the ribosome far from its active site. This model was later supported by immunoelectron microscopy data that detected nascent chain folding outside the ribosome at the base of the large ribosomal subunit (Bernabeu and Lake, 1982) and by x-ray diffraction data that revealed a region of low electron density in the large ribosomal subunit that extended approximately from its base to the peptidyltransferase center located near the base of the central protuberance (Yonath et al., 1987). The resultant model, that the nascent chain moves through a tunnel in the ribosome and exits near the base of the large subunit, was later questioned by others based on their electron microscopy data. Ryabova et al. (1988) concluded that the nascent chain becomes exposed to the cytoplasm near the peptidyltransferase center, while Wagenknecht et al. (1989) identified a potential nascent chain exit site on the back of the large subunit near the base of the central protuberance. Each of these groups suggest that the nascent chain then moves along a groove in the surface of the ribosome, protected from proteolytic digestion, until reaching the base of the large subunit where nascent chain folding is localized. Thus, the path taken by the nascent chain as it moves from the peptidyltransferase center to the site of folding remains unidentified and in dispute.

The Translational Apparatus, Edited by K.H. Nierhaus
et al., Plenum Press, New York, 1993

Even more controversial has been the nature of the pathway followed by nascent secretory proteins as they pass through the membrane of the endoplasmic reticulum (ER) in eukaryotic cells. Ribosomes synthesizing proteins destined for secretion, for localization within intracellular organelles such as the Golgi, or for integration into membranes are directed to the ER membrane via a signal sequence, typically found at the N-terminal end of the nascent chain (for references, see Walter and Lingappa, 1986; Sanders and Schekman, 1992; Nunnari and Walter, 1992). When the signal sequence emerges from the ribosome, the signal recognition particle (SRP) binds to the signal sequence via a direct interaction between the signal sequence and the 54 kDa protein subunit of SRP. Translation is blocked or slowed in the resulting ribosome•nascent chain•SRP complex while it diffuses to the ER membrane where the SRP receptor or docking protein is located. A GTP-dependent interaction between the SRP and the SRP receptor then results in the binding of the ribosome to the ER membrane, the release of SRP and SRP receptor from the ribosome, the resumption of protein chain elongation, and the initiation of nascent chain translocation across, or integration into, the ER membrane. As noted elsewhere in this book, many of the molecular details of the targeting process have now been identified. However, the molecular mechanisms that accomplish translocation are essentially unknown. In fact, two very different models for translocation have been proposed and are still viable, despite the fact that they are mutually exclusive.

Our lack of knowledge about the translocation process is demonstrated by the fact that the location and interactions of the signal sequence are unknown from the time that it is released from the SRP on the cytoplasmic side of the ER membrane to the time at which the signal sequence is cleaved from the nascent chain by a signal peptidase on the lumenal side of the ER membrane. Because each signal sequence contains a stretch of nonpolar amino acids, it has been proposed that the translocation process involves the spontaneous insertion of the signal sequence into the nonpolar core of the phospholipid bilayer, followed by the movement of the nascent chain itself into the membrane and ultimately into the lumen of the ER (von Heijne and Blomberg, 1979; Engelman and Steitz, 1981). This view is supported by the observation that isolated signal peptides and analogues can insert spontaneously into the nonpolar interior of a phospholipid bilayer (Killian et al., 1990; McKnight et al., 1991; and references therein). According to these models, it is unnecessary for the signal sequence and nascent chain to be near or interact with an ER membrane protein. An opposing view, first detailed by Blobel and Dobberstein (1975), argues that protein components of the ER form an aqueous pore in the membrane through which the nascent secretory protein is transported into the lumen. Evidence supporting this model includes (i) an increase in the conductivity of the ER membrane when nascent chains are released from membrane-bound ribosomes by puromycin (Simon and Blobel, 1991), presumably because the loss of the nascent chain exposes an aqueous channel capable of passing ions, and (ii) the opening of ion-conducting channels in the plasma membrane of *E. coli* (and by extrapolation in eukaryotes) by isolated signal peptides, presumably by the binding of the signal sequence to a channel protein that opens a pore in the membrane that transports the secretory protein (Simon and Blobel, 1992). The fact that neither one of these two totally different models of translocation, nor any of the various modifications proposed over the years, has been eliminated on the basis of experimental data only proves the inadequacy of past approaches in examining nascent chain interactions and environment at the ER membrane. These approaches failed for two primary reasons: (i) an inability to locate probes selectively at the site of interest, in this case in the nascent chain (the substrate of the process), and (ii) the unavailability of a method to create a homogeneous sample of nascent chains.

THE OPTIMAL APPROACH: PROBES IN THE NASCENT CHAIN ITSELF

The most direct method of investigating the interactions and environment of the nascent chain and signal sequence at the ER membrane is to position probes at specific sites in the nascent chain itself. Then one can be assured of "seeing what the nascent chain sees." However, since the probes cannot be chemically attached solely and specifically to nascent chains when ribosomes and ER membrane proteins are present, one must incorporate probes into the nascent chain during its synthesis on the ribosome by using analogues of aminoacyl-tRNA. The viability of this approach was first demonstrated by Johnson et al. (1976), who devised a method for selectively modifying the side chain amino group of the lysine in Lys-tRNA and then showed that the N^ϵ-modified lysine was incorporated into complete globin chains in an *in vitro* translation system. Lys-tRNA analogues have since been used successfully to examine nascent secretory and membrane proteins in ribosomes and membranes, as is discussed below. In addition, Lys-tRNA analogues have been used to probe the structure of individual components of the translation and translocation machinery, specifically the ribosome (Johnson and Cantor, 1980), elongation factor Tu (Johnson et al., 1978; Duffy et al., 1981; Guerrier-Takada et al., 1981), elongation factor 1α (Johnson and Slobin, 1980; Kinzy et al., 1992), and SRP (Krieg et al., 1986; Kurzchalia et al., 1986; Zopf et al., 1990; High and Dobberstein, 1991).

It is also important to emphasize another major advantage of this approach. Since probes will be incorporated into nascent polypeptides only by active ribosomes and will be targeted to the ER membrane only by active SRP and ER membrane proteins, a probe in a nascent chain that is bound to membranes must be associated and engaged with functionally active components of both the translation and translocation systems. Thus, by synthesizing the labeled nascent chain *in situ*, we take advantage of the biochemistry of the system to ensure that we are examining only functional and fully-assembled complexes and intermediates of translation, translocation, or integration.

During the past twenty years, we have attached a variety of probes to Lys-tRNA, including fluorescent dyes, photoreagents, spin labels, antigenic determinants, and chemically reactive groups. In each case, the modified Lys-tRNAs were functionally active. Even fluorescein-labeled lysines were incorporated into complete globin chains in a reticulocyte lysate, despite the fact that the dye contains four 6-carbon aromatic rings. In principle, and so far in practice, this approach can be used with any probe that can be esterified with N-hydroxysuccinimide. In order to examine the environment of the nascent chain pathway in the ribosome and the ER membrane, we have utilized both photoreactive probes and fluorescent probes because the information provided by the two types of probes is complementary: the former shows what, if anything, is located close enough to the nascent chain to crosslink to it, while the latter monitors directly the polarity of the immediate environment of the nascent chain and its accessibility to the cytoplasm.

Until recently, it was impossible to obtain a homogeneous sample containing only a single defined intermediate of translation, translocation, or integration. The primary problem had been the difficulty in synchronizing either the initiation or the cessation of translation so that all nascent chains had the same length. To circumvent this, we decided several years ago to use truncated mRNAs, synthesized by *in vitro* RNA transcription of a plasmid that had been cleaved by a restriction endonuclease at a specific site in the coding region of the protein of interest (Krieg et al., 1989). Since truncated mRNAs lack a stop codon, normal termination of translation does not occur. Instead, the ribosome translates to the end of the mRNA and then stops, leaving a nascent chain of defined length (dictated by the length of the truncated mRNA) bound to the ribosome as a peptidyl-tRNA. When SRP and microsomal membranes are included in the translation mixture, a specific translocation intermediate is formed in which the ribosome is bound to the ER membrane, and the signal sequence and nascent chain of defined length are functionally engaged at the translocon which has been defined by Walter and Lingappa (1986) to be the site of translocation or integration in the ER membrane. This approach therefore allows one to create a homogeneous sample of fully-assembled, functional complexes that are specific intermediates in the these processes.

A SPECIFIC SET OF ER MEMBRANE PROTEINS ARE ADJACENT TO NASCENT SECRETORY PROTEINS DURING TRANSLOCATION OR NASCENT MEMBRANE PROTEINS DURING INTEGRATION

The two extreme models for secretory protein translocation across the ER membrane differ greatly since one predicts that nascent chains translocate across the ER membrane without the assistance of ER proteins, while the other predicts that the nascent chain moves through a pore and hence will be adjacent to ER membrane proteins during translocation. To distinguish between these two hypotheses, we incorporated photoreactive probes into a nascent preprolactin secretory protein using N^ϵ-(5-azido-2-nitrobenzoyl)-Lys-tRNA (ϵANB-Lys-tRNA) and trapped the ribosomal complex at an intermediate stage in translocation by using a truncated mRNA (Krieg et al., 1989). By varying the length of the truncated mRNA and hence nascent chain, we were able to position photoreactive probes at several different points along the nascent chain pathway through the ER membrane. In addition, we devised a method to form "staged" samples, in which photoreactive probes were incorporated only into lysine positions in the mature prolactin sequence and were prevented from being incorporated into the two lysine positions in the preprolactin signal sequence (Krieg et al., 1989). After photolysis, both staged and unstaged nascent chains were found crosslinked to a discrete set of ER proteins, with the primary target being a 39 kDa transmembrane glycoprotein of the ER that we termed mp39 (Krieg et al., 1989). Since mp39 was crosslinked to probes positioned at different locations along the nascent chain pathway through the membrane, mp39 appears to act as a guide or pore protein during translocation (Krieg et al., 1989). Thus, the results of this photocrosslinking study support the model of Blobel and Dobberstein, since a discrete set of ER membrane proteins are located adjacent to nascent chains as they are translocated across the ER membrane.

Using photoreactive Lys-tRNA analogues and SRP-arrested translation intermediates, Wiedmann et al. (1987) were the first to report the crosslinking of a nascent secretory protein to an ER protein. They named the photocrosslinking target, a 34 kDa ER glycoprotein, the signal sequence receptor or SSR (later SSRα), though there were no functional data to support this apellation. Their subsequent experiments (Wiedmann et al., 1989) confirmed the results reported by Krieg et al. (1989), who had concluded that mp39 did not function as a signal sequence receptor. Prehn et al. (1990) purified a 34 kDa ER glycoprotein and identified it as SSRα. However, recent reconstitution studies have shown that the protein identified as SSRα is not essential for protein translocation (Migliaccio et al., 1992), and a re-examination of earlier results has shown that SSRα photocrosslinks preferentially to long, rather than short, nascent chains (Görlich et al., 1992). Neither of these observations is consistent with SSRα acting as a signal sequence receptor when nascent chains are first bound to the ER membrane. Thus, there is no current evidence that the signal sequence in a membrane-bound ribosomal complex binds to a specific ER membrane protein(s) during the initial stages of translocation, only evidence that the signal sequence is *adjacent* to a discrete set of ER membrane proteins.

The recent success in reconstituting translocation activity from partially purified ER membrane proteins has provided a means to evaluate the functional importance of those ER proteins that have been photocrosslinked to the nascent chain. Using this technique, translocation activity was shown to be unaffected by the removal of SSRα (Migliaccio et al., 1992), but to require the presence of a different ER glycoprotein termed TRAM (Görlich et al., 1992). Since TRAM is a multispanning membrane protein and is photocrosslinked to nascent chains with a higher efficiency than is SSRα (Görlich et al., 1992), it appears that TRAM comprises part of a translocation tunnel through the ER membrane for the nascent secretory chain. SSRα and mp39 have recently been shown to be different proteins because mp39 contains at least one cysteine residue (Tober, Andrews, and Johnson, unpublished data), while SSRα contains no Cys (Prehn et al., 1990). It therefore seems likely that mp39 and TRAM are the same protein.

Since the SRP-dependent targeting of ribosomes to the ER membrane is the same for both nascent secretory proteins and nascent membrane proteins (for references, see Thrift et al., 1991), one would expect translocation and integration to be initiated at the same sites on the ER membrane. However, at some point subsequent to targeting, the processing pathways of the secretory proteins and the membrane proteins must diverge. Since it is possible that the long stretch of nonpolar amino acids that constitute a transmembrane or stop-transfer sequence may insert spontaneously into the membrane, without any assistance from an ER protein, we sought to determine whether any ER proteins were adjacent to a nascent membrane protein during its integration into the ER membrane. Photoreactive εANB-Lys residues were therefore incorporated into a nascent membrane protein immediately adjacent to an IgM transmembrane sequence using truncated mRNAs in the presence of microsomal membranes and SRP. After photolysis, these nascent chains were found crosslinked to several ER membrane proteins, including mp39 and a non-glycosylated protein that is 1-2 kDa larger than mp39 (Thrift et al., 1991). The same or very similar ER proteins were also found photocrosslinked to the nascent chains of different membrane proteins (High et al., 1991). Interestingly, we found that the transmembrane sequence of a single-spanning membrane protein remains in the translocon until the termination of translation, even after the cytoplasmic tail of the protein had been lengthened by 141 amino acids (nearly 500 Å of fully extended polypeptide) beyond the stop-transfer sequence (Tober, Andrews, Walter, and Johnson, unpublished data; Thrift et al., 1991). Thus, secretory proteins and membrane proteins are located adjacent to ER membrane proteins during translocation and integration, respectively, and the two types of nascent proteins appear to use the same translocon.

THE SIGNAL SEQUENCE IS HELD IN AN AQUEOUS COMPARTMENT DURING THE EARLY STAGES OF PROTEIN TRANSLOCATION AT THE ER MEMBRANE

At the conclusion of the targeting process, the location of the signal sequence is unknown. The two extreme possibilities, suggested by the models noted above, are that the signal sequence is buried in the hydrophobic core of the phospholipid bilayer or that the signal sequence is sequestered inside an aqueous channel through the membrane (Figure 1). Since the primary difference between these two possibilities is the polarity of the environment of the signal sequence, the best way to evaluate which model is correct is to incorporate a probe in the signal sequence that is sensitive to the polarity of its environment and then determine directly whether the signal sequence is in a hydrophobic or hydrophilic environment.

The optimal technique to use in addressing this issue is fluorescence spectroscopy, both

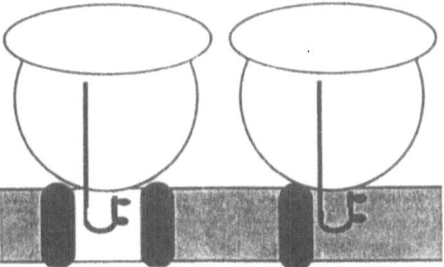

Figure 1. Signal sequence environment at the ER membrane. The two extreme possibilities, located in an aqueous pore (shown in white) or in the nonpolar core of the bilayer (shown in gray), are shown for a pPL$_{64}$ nascent chain. The dark ovals in the membrane represent the ER membrane proteins that have been shown by photocrosslinking to be adjacent to the nascent chain as it passes through the membrane. Although no interactions are shown between these ER membrane proteins and the signal sequence and/or nascent chain, such interactions may occur. The small black ovals represent the approximate locations of the two lysines in the signal sequence, 4 and 9 amino acids from the N-terminal end of the nascent chain (Sasavage et al., 1982).

because the emission properties of many fluorescent dyes are strongly dependent upon solvent polarity and because the sensitivity of this technique permits one to make meaningful measurements at probe concentrations below 5 nM. The latter reason is particularly important in this case because the availability of material is limited: the probes must be incorporated into the nascent chains by *in vitro* translation and then targeted successfully to the ER membrane vesicles in larger than radioisotope-detectable quantities. Thus, Lys-tRNA was selectively modified at the ε-amino group of lysine with 6-(7-nitrobenz-2-oxa-1,3-diazol-4-yl)aminohexanoic acid (NBD) as described (Johnson et al., 1976; Johnson and Cantor, 1980; Krieg et al., 1986) and then purified to yield εNBD-Lys-tRNA (Crowley et al., submitted).

Despite the large size of the fluorophore, the modification of the Lys-tRNA did not alter its activity. In a wheat germ translation system programmed with full-length preprolactin, 60% of the added εNBD-Lys-tRNA incorporated its fluorescent lysine into preprolactin. Furthermore, both the rate and the total extent of incorporation were the same for modified and unmodified lysines using either truncated or full-length preprolactin mRNAs. Similarly, when full-length preprolactin mRNA was translated in the absence or presence of SRP and microsomal membranes, the extent of translocation was unaffected by the presence of the NBD dye. To demonstrate that the NBD remained covalently attached to the nascent chain during translation and translocation, newly-synthesized polypeptides were immunoprecipitated with an antiserum that binds specifically to the NBD dye. Since equivalent amounts of preprolactin and prolactin were immunoprecipitated by antibodies to NBD and by antibodies to prolactin, the fluorescent probes were not lost or degraded during translation or translocation. Thus, fluorescent-labeled lysines in the signal sequence or elsewhere in the nascent chain do not interfere with translation, with SRP-dependent targeting to the ER membrane, with the translocation of secretory proteins across the membrane, or with signal sequence cleavage. In short, fluorescent nascent chains are synthesized and processed normally; the probes do not interfere with function.

The suitability of the NBD dye as a probe of polarity was demonstrated by the 4-fold increase in its emission intensity, the 40 nm decrease in its wavelength of maximum emission, and the increase in its fluorescence lifetime from 1.4 nsec to 10.1 nsec when the dye was shifted from an aqueous solvent to chloroform, a nonpolar solvent. To assess whether NBD was an acceptable reporter group with membranes, we examined NBD dyes covalently attached either to cholesterol (NBD-cholesterol) or to the end of a 12-carbon fatty acyl chain of phosphatidylcholine (NBD-PC) because each probe is positioned within the nonpolar region of the bilayer, although the latter dye appears to loop back toward the surface of the membrane (Chattopadhyay and London, 1987). We found that the fluorescence lifetimes of these probes were 7-8 nsec (Table 1), and these long lifetimes indicate that the NBD prefers to remain within the hydrophobic core of the membrane rather than be exposed to an aqueous environment, even if its position within the core is anomalous. Since the molar fraction of dyes in a polar or nonpolar environment in a heterogeneous sample can be determined directly from the phase and modulation data used to obtain fluorescence lifetimes, we chose to monitor the fluorescence lifetime of NBD in order to assess its environment in a sample.

When a truncated mRNA that coded for the N-terminal 64 residues of preprolactin (pPL_{64}) was translated in a wheat germ system in the presence of εNBD-Lys-tRNA, NBD dyes were incorporated only into the signal sequence of the nascent chains, at positions 4 and 9 (Fig. 1). If the translation was carried out in the presence of SRP and washed microsomal membranes (EKRM), the fluorescent-labeled nascent chains were targeted to the ER membrane and ribosome•NBD-pPL_{64}•EKRM complexes were formed. After the free and membrane-bound ribosomal complexes had been purified extensively to remove residual unincorporated εNBD-Lys, the samples were examined spectroscopically. The phase and modulation data were analyzed assuming one, two, or three species with either a discrete lifetime or a distribution of lifetimes. The data were fit best by assuming a distribution of short lifetimes and a discrete long lifetime component. In the few cases where a better χ^2 value was obtained by assuming three

Table 1. Fluorescence lifetimes of NBD probes in various complexes

Sample	τ_1 (ns)	W_1 (ns)	f_1	τ_2 (ns)	f_2	χ^2
NBD-cholesterol / PC[1]	8.5		1.00			16.3
NBD-PC / PC[2]	7.3		1.00			1.5
εNBD-lysine	1.4		1.00			1.1
NBD-pPL$_{64}$•ribosome	0.4	2.5	0.97	7.5	0.03	1.7
NBD-pPL$_{64}$•ribosome•EKRM	0.7	2.7	0.94	8.7	0.06	15.3
NBD-pPL$_{86}$•ribosome	1.0	0.8	0.96	8.4	0.04	7.2
NBD-pPL$_{86}$•ribosome•EKRM	0.9	2.9	0.94	9.1	0.06	23.6
NBD-pPL-ssK$_{56}$•ribosome	0.2	2.3	0.98	7.7	0.02	3.1
NBD-pPL-ssK$_{56}$•ribosome•EKRM	0.5	2.5	0.98	7.5	0.02	10.0

Samples were prepared and purified as described in Crowley et al. (submitted). Fluorescence data were fit to a two-component model in which the first component conformed to a lorentzian distribution (Alcala et al., 1987) about a central lifetime value (τ_1) with the indicated width (W_1) and mole fraction of the species (f_1). The second component was a discrete exponential decay with indicated lifetime (τ_2) and mole fraction (f_2). χ^2 values were calculated as described previously (Jameson et al., 1984).
[1]22-(N-(7-nitrobenz-2-oxa-1,3-diazo-4-yl)amino)-23,24-bisnor-5-cholen-3b-ol in small unilamellar vesicles of dioleoyl-sn-glycero-3-phosphocholine at a molar ratio of 1:100, respectively.
[2]1-palmitoyl-2-(12-((7-nitro-2-1, 3-benzoxadiazol-4-yl)amino)dodecanoyl)-sn-glycero-3-phosphocholine in small unilamellar vesicles of dioleoyl-sn-glycero-3-phosphocholine at a molar ratio of 1:100, respectively.

lifetime components, the improvement in χ^2 was not large enough to warrant discarding the simpler two-state model. Thus, the lifetime data for both the free and membrane-bound ribosome•pPL$_{64}$ complexes indicated that NBD dyes are located in two separate environments in the samples. However, in each case, nearly all (> 94%) of the NBD dyes had short fluorescence lifetimes and hence were in an aqueous environment (Table 1). More important, there was no significant change in the fluorescence lifetime distribution of the NBD dyes incorporated into the preprolactin signal sequence upon its binding to the ER membrane. Therefore nearly all of the NBD dyes are located in an aqueous environment in the NBD-pPL$_{64}$•ribosome•membrane complex.

An 86-residue-long preprolactin nascent chain was also examined to determine if a longer nascent chain would allow the signal sequence to move into the hydrophobic core of the bilayer. However, as shown in Table 1, the binding of the NBD-pPL$_{86}$•ribosome complexes to the membrane did not alter the fraction of NBD dyes with a short fluorescence lifetime and hence did not increase the number of signal sequence probes in a nonpolar environment. Since the NBD dyes in the membrane-bound pPL$_{64}$ and pPL$_{86}$ complexes did not have a long lifetime similar to that of NBD-PC, the signal sequence probes were not embedded in the nonpolar core of the ER membrane. Instead, these data show that the signal sequence of a nascent secretory protein in eukaryotes is in an aqueous environment after the translocation process has been initiated and the ribosome•nascent chain is functionally engaged at the translocon in the ER membrane, at least for nascent chains less than 87 amino acids in length.

In fact, since the NBD dyes prefer a nonpolar environment to a polar one (cf. NBD-PC), it appears that the NBD dyes at the translocon are prevented from partitioning into the nonpolar core of the ER membrane. This result is particularly striking because the NBD dye is covalently attached to the polypeptide backbone via a long (17 Å when fully extended) tether that is largely hydrocarbon. Hence, fluorescent probes in the nascent chain would appear to have both the inclination and the range to bury themselves in the hydrophobic core of the bilayer if possible. Since this did not happen, we conclude that the signal sequence and its probes in the complex were prevented from becoming embedded in the hydrophobic region of the membrane. This in turn suggests that ER membrane proteins prevent a direct interaction between the hydrophobic core of the signal sequence and the hydrophobic core of the ER membrane. This could be accomplished either by an ER membrane protein binding to the signal sequence and actively keeping the NBD probes in an aqueous milieu or by ER proteins forming a proteinaceous pore that passively prevents the signal sequence from being exposed to the nonpolar core of the membrane. Our current data do not distinguish between these two possibilities. The most likely candidates for the ER membrane proteins that prevent signal sequence exposure to the hydrophobic interior of the membrane are those that have been photocrosslinked to the signal sequence and nascent chain (reviewed in Sanders and Schekman, 1992; Nunnari and Walter, 1992).

THE RIBOSOME AND ER MEMBRANE FORM A TIGHT JUNCTION, CREATING AN AQUEOUS COMPARTMENT FOR THE SIGNAL SEQUENCE AND NASCENT CHAIN THAT IS SEALED OFF FROM THE CYTOPLASM

Since small molecules and ions do not pass freely across the ER membrane, the transport of secretory proteins through the membrane must occur in a manner that maintains the impermeability of the bilayer to small molecular species. The extent to which the ribosome contributes to maintaining a permeability barrier is unknown. In fact, the nature and extent of the interaction of the ribosome with the ER membrane is largely unknown, even though some ER membrane proteins have been shown to bind or crosslink to the ribosome (reviewed in Sanders and Schekman, 1992; Nunnari and Walter, 1992). Is the signal sequence exposed to the cytoplasm in a membrane-bound ribosomal complex? Is the junction between the ribosome and the ER membrane sealed and impermeable to small ions? Does the ribosome make contact with ER proteins around the entire circumference of the putative aqueous pore so as to seal it off from the cytoplasm?

Collisions between fluorescent dyes and certain solutes, such as iodide ions, result in a quenching of fluorescence because the excited state energy of a dye is lost when it contacts a quencher. Thus, the exposure of a particular dye to the solvent and the dissolved iodide ions can be determined by measuring emission intensity as a function of quencher concentration. For steady-state collisional quenching of fluorescence, a linear plot is obtained when data are analyzed according to the Stern-Volmer law:

$$(F_0/F) - 1 = K_{sv} \, [I^-] \, ,$$

where K_{sv} is the Stern-Volmer quenching constant, F_0 is the emission intensity in the absence of iodide ions, and F is the emission intensity in the presence of iodide ions at concentration $[I^-]$. K_{sv} is equal to $k_q \tau_0$, where k_q is the bimolecular quenching constant and τ_0 is the fluorescence lifetime in the absence of quencher. Since the lifetimes of the NBD dyes in our samples do not differ significantly (Table 1), the relative magnitudes of the bimolecular quenching constants in different samples can be estimated by comparing their K_{sv} values.

Thus, iodide ion quenching experiments provide a means to determine directly the extent to which the nascent chain and signal sequence in membrane-bound complexes are exposed to the cytoplasm. Both free εNBD-lysine and purified NBD-pPL$_{64}$ have large K_{sv} and k_q values, consistent with the NBD dyes being exposed to the solvent and hence to iodide ions (Table 2). Similarly, the NBD dyes in NBD-pPL$_{64}$·ribosome complexes are exposed to the solvent (cytoplasm) since the K_{sv} for this complex is more than half of that of the free nascent chain, even though the ribosome is likely to reduce the rate of quenching by sterically blocking some of the iodide ion access routes to the NBD dyes. However, in contrast, iodide ions quench the NBD dyes in the pPL$_{64}$·ribosome·ER membrane complex very poorly, if at all. Since the KI was added (actually, mixtures of KCl and KI were added to four parallel samples so that the

Table 2. Collisional quenching of NBD-containing complexes

Sample	K_{SV} (M^{-1})
εNBD-Lysine	11.0
NBD-pPL$_{64}$	4.3
NBD-pPL$_{37}$•ribosome	3.1
NBD-pPL$_{64}$•ribosome	2.3
NBD-pPL$_{64}$•ribosome•ER membrane	0.3
NBD-pPL$_{86}$	3.7
NBD-pPL$_{86}$•ribosome	3.1
NBD-pPL$_{86}$•ribosome•ER membrane	0.6
NBD-pPL-ssK$_{47}$•ribosome	1.5
NBD-pPL-ssK$_{47}$•ribosome•ER membrane	0.2
NBD-pPL-ssK$_{56}$•ribosome	2.4
NBD-pPL-ssK$_{56}$•ribosome•ER membrane	− 0.2

Samples were prepared and purified as described (Crowley et al., submitted). Stern-Volmer constants for individual experiments were determined by linear least squares analysis. The K_{SV} values shown above are the average of 2 or more independent determinations.

ionic strength was maintained at a constant level) to the cytoplasmic side of the microsomal vesicles, the almost total lack of iodide ion quenching indicates that the NBD probes in the signal sequence are not exposed to the cytoplasm. Instead, the ribosome apparently binds to the ER membrane so as to form a tightly sealed junction that will not pass small ions and molecules. This presumably involves a tight association of the ribosome with ER membrane proteins, but may include ribosome-phospholipid interactions. The most likely candidates for the ER proteins involved in forming this junction are those that have been photocrosslinked to the nascent chain (see above). Whatever the specifics of the ribosome-membrane association, small molecule exchange is prevented between the cytoplasm and the compartment formed by the ribosome and ER membrane. Thus, at early stages of the translocation process, the signal sequence and nascent chain are located in an aqueous compartment that is sealed off from the cytoplasm

THE NASCENT CHAIN PASSES THROUGH AN AQUEOUS TUNNEL IN THE RIBOSOME, NOT ALONG A DEEP GROOVE ON ITS SURFACE

The nature of the nascent chain pathway in or on the ribosome can be examined using the same approaches that have been used to investigate the nascent chain environment at the ER membrane. In order to focus solely on the nascent chain pathway in the ribosome, we have either used short nascent chains of preprolactin in which the probes in the signal sequence have not yet emerged from the ribosome, or a preprolactin derivative, generously provided by Dr. David Andrews, that lacks 8 amino acids in the signal sequence, including the two lysines. Despite the removal of a portion of the signal sequence, the targeting and translocation of this modified preprolactin, that we have termed pPL-ssK, is unimpaired (Andrews et al., 1992). Translation of the truncated mRNA for pPL–ssK$_{56}$ yields a ribosome•nascent chain complex that contains only a single lysine in the nascent chain, located about 3/4 of the way through the ribosome-protected region and 29 amino acids away from the peptidyltransferase center. Analysis of the lifetime data for the ribosome-bound NBD-pPL-ssK$_{56}$ sample demonstrated that the NBD dyes (98%) had a short lifetime and hence were in an aqueous medium (Table 1).

In addition, the pPL_{86} nascent chain contains two lysines that are positioned inside the ribosome, 9 and 15 residues from the peptidyltransferase center, in addition to the two lysines in the signal sequence. Since the measured NBD lifetimes for pPL_{86}•ribosome complexes (Table 1) are not distinguishably different from those of pPL_{64}•ribosome complexes, which do not have any probes inside the ribosome, the NBD dyes in the ribosomal nascent chain pathway appear to be in an environment that is equivalent to that experienced by dyes in the signal sequence located outside the ribosome. Thus, the nascent chain moves along an aqueous pathway in the ribosome.

Collisional quenching experiments demonstrate that nascent chains bound to free ribosomes are accessible to ions dissolved in the aqueous cytoplasm. NBD dyes incorporated into pPL_{37} were positioned 29 and 34 residues from the peptidyltransferase center, while probes in NBD-pPL-ssK$_{47}$ and NBD-pPL-ssK$_{56}$ were located 20 and 29 residues, respectively, from the peptidyltransferase center. The high K_{SV} values observed with each of these complexes and with the NBD-pPL$_{86}$•ribosome complex indicate that probes located along the entire length of the nascent chain pathway are in an aqueous environment and are accessible to iodide ions dissolved in the cytoplasmic space (Table 2).

Although the nascent chain probes inside the ribosome were accessible to iodide ions, they were not accessible for binding to antibodies. For example, the addition of an excess of NBD-specific antibodies to NBD-pPL$_{37}$•ribosome complexes did not reduce the NBD emission intensity significantly, though the binding of the antibodies to εNBD-lysine reduces its emission intensity by 80%. Thus, the NBD probes in the ribosomal nascent chain pathway were not exposed sufficiently for association with antibodies, despite the fact that the NBD moieties were tethered to the polypeptide chain via a linker arm that spans 17 Å between the lysine α-carbon and the dye when fully extended. These data therefore suggest that the nascent chain is not exposed on the surface of the ribosome, but either lies in a deep (> 20 Å) groove on the surface of the ribosome or passes through a tunnel in the ribosome. Similarly, Picking et al. (1992) concluded, using samples with a distribution of nascent chain lengths, that antibodies have limited access to the N-terminus of nascent chains less than 50 residues in length.

To determine whether the nascent chain pathway in the ribosome is a groove or a tunnel, we examined the accessibility of membrane-bound nascent chain•ribosome complexes to iodide ions. Since the collisional quenching observed with free ribosomal complexes is reduced to zero or nearly zero for membrane-bound ribosomal complexes (compare the K_{SV} values in Table 2 ± EKRM), the binding of the ribosomal complex to the ER membrane blocks iodide ion access to the nascent chain in the ribosome. If the nascent chain pathway were an aqueous groove on the surface of the ribosome, then one would expect to observe collisional quenching of the fluorescent dyes because the iodide ions would have access to the probes in the groove whether or not the ribosome was bound to the membrane. Since essentially no quenching was observed for any of the probes in the membrane-bound complexes, the nascent chain must not be exposed to the cytoplasm in membrane-bound ribosomes. The nascent chain can be sealed off from the cytoplasm only if the nascent chain is surrounded on all sides by the ribosome or, in other words, passes through a tunnel in the ribosome.

Besides separating the peptidyltransferase center from the region where nascent chain folding occurs, the nascent chain tunnel in the ribosome provides a means to control nascent chain movement. This is particularly important for secretory proteins, since they must be directed across the ER membrane. As the nascent chain occupies and ultimately fills the restricted space in the ribosome tunnel, any extension of the nascent polypeptide at the peptidyltransferase end of the tunnel must be accompanied by the movement of polypeptide out of the tunnel. Thus, during co-translational translocation, the tunnel will direct the nascent chain towards the ER membrane and lumen by passive diffusion because there is no alternative path to follow. Movement across the bilayer, probably through an aqueous pore, is then likely to be ensured by the folding of the nascent chain in the ER lumen and/or by glycosylation and/or by nascent chain binding to lumenal proteins in a process that may be energy-dependent. The tunnel, unlike a groove, also prevents the nascent chain from diffusing into the cytoplasm and thereby interfering with the translocation process by folding prematurely and/or interacting with cytoplasmic components. The tight seal between the ribosome and the ER membrane also prevents nascent secretory chain movement into the cytoplasm, as well as creating a permeability barrier that blocks the exchange of small molecules and ions between the cytoplasm and the ER lumen. Thus, the ribosome tunnel constitutes a topographical feature that dictates the co-translational movement of the nascent chain into and through the membrane, whether or not this movement is facilitated and made irreversible by interactions within the membrane or by folding, interactions, or modifications within the ER lumen.

Our data therefore strongly indicate that the nascent chain passes through a tunnel in the ribosome, and hence are consistent with a tunnel in the large ribosomal subunit as proposed by Yonath et al. (1987). The opposing view, that the nascent chain exits the ribosome near the peptidyltransferase center and then travels along a groove on the surface of the ribosome (Ryabova et al., 1988), is not supported by our iodide ion quenching data. Also, in contrast to the data reported by Ryabova et al. (1988), neither Picking et al. (1992) nor we (Shore, 1991) have observed antibody binding to fluorescent-labeled nascent chains. Furthermore, we have found that anti-fluorescein antibodies, which quench fluorescein emission upon binding to the dye, do not bind to εFlSAc-Lys-tRNA when it is bound in either the A or the P site of the ribosome. Our quenching data also do not support the model of Wagenknecht et al. (1989) because an exit site at the base of the central protuberance on the back side of the ribosome would appear to be accessible to iodide ions in membrane-bound ribosomes. In view of these discrepancies, it is worth noting that that the fluorescence data reported here were obtained with samples that (i) were formed by functional ribosomes and membranes, (ii) were not exposed to denaturing conditions, and (iii) were assayed after the spectral measurements to assess the biochemical integrity of the samples. In addition, each of the fluorescent nascent chains in the sample contributed to the observed signal. In contrast, a selection of individual ribosomal complexes were observed by the electron microscope, and no data were presented to indicate what fraction of the ribosome-associated tRNA in the samples was bound in the A or P site instead of adsorbed to the surface of the ribosome. It is possible that the antigenic determinants observed on the surface of some ribosomes by Ryabova et al. originated from peptidyl-tRNAs that were not bound to functional tRNA binding sites on the ribosome.

PROSPECTUS

Very few of the molecular mechanisms that mediate translocation and integration at the ER membrane have been elucidated, and a myriad of questions about the interactions and environment of the signal sequence and nascent chain at the ER membrane remain to be answered. These include: Does the signal sequence enter the nonpolar core of the membrane when the nascent chain is longer than 86 amino acids? Does the nascent secretory protein pass through an aqueous pore in the ER membrane? Is the nascent chain exposed to phospholipid molecules at any point during translocation or integration? Does the stop-transfer sequence bind to an ER membrane protein during its integration into the membrane? Is the tight ribosome-membrane junction maintained during the integration of nascent membrane proteins? Since the stop-transfer of a single-spanning membrane protein remains in the translocon until the termination of translation, is the cytoplasmic domain of the nascent protein directed into the cytoplasm through the ribosome-membrane junction while the polypeptide is completed, and if so, is the permeability barrier maintained? Because of the success of our approach in resolving some long-standing controversial issues in translation and translocation, as noted above, it seems likely that this approach will continue to play a major role in future examinations of (i) the molecular mechanisms that mediate secretory protein translocation and membrane protein integration, and (ii) the topography and interactions of the membrane-bound multicomponent complexes that accomplish these processes. In particular, the incorporation of fluorescent probes into specific sites in the nascent chain to create defined, intact translocation and integration intermediates will enable one to detect specific interactions and to quantify and characterize the affinities, conformational changes, and kinetics associated with those interactions. Such studies will benefit greatly from the recent advances in the reconstitution of translocation-active membranes from partially-purified components of the ER membrane (Nicchitta and Blobel, 1980; Migliaccio et al., 1992; Görlich et al., 1992).

REFERENCES

Alcala, J. R., Gratton, E., and Prendergast, F. G., 1987, *Biophys. J.* 51:587.
Andrews, D. W., Young, J. C., Mirels, L. F., and Czarnota, G. J., 1992, *J. Biol. Chem.* 267:7761.
Bernabeu, C., and Lake, J. A., 1982, *Proc. Natl. Acad. Sci. USA* 79:3111.
Bernabeu, C., Tobin, E. M., Fowler, A., Zabin, I., and Lake, J. A., 1983, *J. Biol. Chem.* 96:1471.
Blobel, G., and Dobberstein, B., 1975, *J. Cell Biol.* 67:835.
Blobel, G., and Sabatini, D., 1970, *J. Cell Biol.* 45:130.
Chattopadhyay, A., and London, E., 1987, *Biochemistry* 26:39.
Crowley, K. S., Reinhart, G. D., and Johnson, A. E., 1992, Submitted.
Duffy, L. K., Gerber, L., Johnson, A. E., and Miller, D. L., 1981, *Biochemistry* 20:4663.
Engelman, D. M., and Steitz, T. A., 1981, *Cell* 23:411.
Görlich, D., Hartman, E., Prehn, S., and Rapoport, T. A., 1992, *Nature* 357:47.
Guerrier-Takada, C., Johnson, A. E., Miller, D. L., and Cole, P. E., 1981, *J. Biol. Chem.* 256:5840.
High, S., and Dobberstein, B., 1991, *J. Cell Biol.* 113:229.
High, S., Görlich, D., Wiedmann, M., Rapaport, T. A., and Dobberstein, B., 1991, *J. Cell Biol.* 113:35.
Jameson, D. M., Gratton, E., and Hall, R. D., 1984, *Appl. Spec. Rev.* 20:55.
Johnson, A. E., and Cantor, C. R., 1980, *J. Mol. Biol.* 138:273.
Johnson, A. E., Miller, D. L., and Cantor, C. R., 1978, *Proc. Natl. Acad. Sci. USA* 75:3075.
Johnson, A. E., and Slobin, L. I., 1980, *Nucleic Acids Res.* 8:4185.
Johnson, A. E., Woodward, W. R., Herbert, E., and Menninger, J. R., 1976, *Biochemistry* 15:569.
Kellaris, K. V., Bowen, S., and Gilmore, R., 1991, *J. Cell. Biol.* 114:21.
Killian, J. A., Keller, R. C. A., Struyvé, M., de Kroon, A. I. P. M., Tommassen, J. and de Kruijff, B., 1990, *Biochemistry* 29:8131.
Kinzy, T. G., Freeman, J. P., Johnson, A. E., and Merrick, W. C., 1992, *J. Biol. Chem.* 267:1623.
Krieg, U. C., Johnson, A. E., and Walter, P., 1989, *J. Cell. Biol.* 109:2033.
Krieg, U. C., Walter, P., and Johnson, A. E., 1986, *Proc. Natl. Acad. Sci. USA* 83:8604.
Kurzchalia, T. V., Wiedmann, M., Girshovich, A. S., Bochkareva, E. S., Bielka, H., and Rapoport, T. A., 1986, *Nature* 320:634.
Malkin, L. I., and Rich, A., 1967, *J. Mol. Biol.* 26:329.
McKnight, C. J., Rafalski, M., and Gierasch, L. M., 1991, *Biochemistry*, 30:6241.
Migliaccio, G., Nicchitta, C. V., and Blobel, G., 1992, *J. Cell. Biol.* 117:15.
Müsch, A., Weidmann, M., and Rapoport, T. A., 1992, *Cell* 69:343.
Nicchitta, C. V., and Blobel, G., 1990, *Cell* 60: 259.
Nunnari, J., and Walter, P., 1992, *Curr. Opin. Cell Biol.* 4:573.
Picking, W. D., Picking, W. L., Odom, O. W., and Hardesty, B., 1992, *Biochemistry* 31:2368.
Prehn, S., Herz, J., Hartmann, E., Kurzchalia, T. V., Frank, R., Roemisch, K., Dobberstein, B., and Rapoport, T. A., 1990, *Eur. J. Biochem.* 188:439.
Ryabova, L. A., Selivanova, O. M., Baranov, V. I., Vasiliev, V. D., and Spirin, A. S., 1988, *FEBS Lett.* 226:255.
Sanders, S. L., and Schekman, R., 1992, *J. Biol. Chem.* 267:13791.
Sanders, S. L., Whitfield, K. M., Vogel, J. P., Rose, M. D., and Schekman, R., 1992, *Cell* 69:353.
Simon, S. M., and Blobel, G., 1991, *Cell* 65:371.
Simon, S. M., and Blobel, G., 1992, *Cell* 69:677.
Shore, S. K., 1991, Ph. D. Dissertation, University of Oklahoma, Norman, OK.
Thrift, R. N., Andrews, D. W., Walter, P., and Johnson, A. E., 1991, *J. Cell. Biol.* 112:809.
von Heijne, G., and Blomberg, C., 1979, *Eur. J. Biochem.* 97:175.
Wagenknecht, T., Carazo, J. M., Radermacher, M., and Frank, J., 1989, *Biophys. J.* 55:455.
Walter, P., and Lingappa, V. R., 1986, *Annu. Rev. Cell Biol.* 2:499.
Wiedmann, M., Goerlich, D., Hartmann, E., Kurzchalia, T. V., and Rapoport, T. A., 1989, *FEBS Lett.* 257:263.
Wiedmann, M., Kurzchalia, T. V., Hartmann, E., and Rapoport, T. A., 1987, *Nature* 328:830.
Yonath, A., Leonard, K. R., and Wittmann, H. G., 1987, *Science* 236:813.
Zopf, D., Bernstein, H. D., Johnson, A. E., and Walter, P., 1990, *EMBO J.* 9:4511.

MUTANTS OF tRNA, RIBOSOMES AND mRNA AFFECTING FRAMESHIFTING, HOPPING OR STOP CODON READ-THROUGH

J.F. Atkins[1], K. Herbst[1,2], M. O'Connor[1,3], T.M.F. Tuohy[1], R.B. Weiss[1], N.M. Wills[1,2] and R.F. Gesteland[1,2].

[1]Department of Human Genetics and [2]Howard Hughes Medical Institute, 6160 Eccles Genetics Bldg., University of Utah, Salt Lake City, Utah 84112 [3]Section of Biochemistry, Brown University, Providence, Rhode Island 02912

INTRODUCTION

Programmed non-standard translation elongation events, frameshifting, hopping and read-through, are important in the decoding of a minority of mRNAs from diverse sources. In this chapter we provide an overview of some of the approaches we are taking to investigate this "recoding" (Gesteland et al., 1992). Also included is an overview of some current studies on mutants which cause frameshifting and read-through, at sites where it does not normally occur i.e., non-programmed events.

RIBOSOMAL PROTEIN MUTANT WHICH RESTORES HIGH-LEVEL HOPPING

Fifty nucleotides separate the first 46 codons from the last 114 codons in the mature mRNA of bacteriophage T4 gene 60 (Huang et al., 1988; Weiss et al., 1990). Ribosomes hop over this 50-nucleotide coding gap at high efficiency (Weiss et al., 1990; Dayhuff et al., 1992). One of the several necessary requirements for this hop is an mRNA stem-loop structure which contains the 5' end of the coding gap (Weiss et al., 1990). Herbst et al. (unpublished) have now isolated a mutant of Escherichia coli which restores high efficiency hopping to variants of gene 60 mRNA in which the stem-loop is altered. The mutation mapped to rplI, the gene coding for protein L9 of the large ribosomal subunit. In the mutant the codon for serine$_{93}$ was altered (Herbst et al., unpublished). At least some non-programmed hopping events were also significantly affected by the mutant.

The Translational Apparatus, Edited by K.H. Nierhaus
et al., Plenum Press, New York, 1993

MUTANTS OF THE mRNA CONTEXT AND STRUCTURE FOR READTHROUGH OF THE MURINE LEUKEMIA VIRUS *gag* GENE TERMINATOR

The *gag* and *pol* genes of Mo-MuLV and a minority of other retroviruses are separated by an in-frame UAG codon. Approximately 5% of ribosomes read through this stop codon and synthesize a gag-pol fusion polypeptide that is the sole source of *pol* gene products. N-terminal sequencing of the protease product (Yoshinaka *et al.*, 1985) showed that its coding region spanned the UAG codon and that glutamine is encoded by the UAG. The "leakiness" of this UAG, in contrast to most UAG codons, is due to a signal programmed in the mRNA that acts as a "stimulator" for read-through. The stimulator in this case is a pseudoknot located 8 nucleotides 3' to the UAG codon (Wills *et al.*, 1991; Feng *et al.*, 1992; see also Felsenstein, K.M. and Goff, S.P., 1992). Disruption of basepairing in either of the two stems of the pseudoknot greatly diminishes or abolishes readthrough. Regeneration of the structure via compensatory mutations restores read-through.

Although the pseudoknot is essential for read-through, the 8-nucleotide purine-rich sequence "spacer" between the UAG and the start of the pseudoknot is also important (Honigman *et al.*, 1991; Feng *et al.*, 1992). Alteration of several spacer nucleotides has a deleterious effect on read-through (Wills *et al.*, unpublished). The length of the spacer region is also important. Decreasing the spacer by 3 nucleotides or increasing the spacer by 3 nucleotides reduces read-through, though caution is required in distinguishing a spacing requirement from sequence requirements.

Unlike MuLV, in the majority of retroviruses the *pol* gene is in the -1 frame in relation to the *gag* gene and frameshifting is required to generate the gag-pol fusion protein. The stimulator for this frameshifting, at least in Mouse Mammary Tumor Virus (MMTV), is a pseudoknot (Chamorro *et al.*, 1992), as had been first found for frameshifting in the Coronavirus, Infectious Bronchitis Virus (Brierley *et al.*, 1989). The spacing between pseudoknots which promote frameshifting and their shifty sites fall within a narrow range of 3-9 nucleotides. Interestingly, a similar spacing is found between pseudoknots which stimulate read-through and their stop-codons.

As the pseudoknot required for MuLV stop codon read-through appears grossly similar to its counterpart in MMTV required for frameshifting, we replaced the MuLV pseudoknot with the pseudoknot sequence from the *gag-pro* shift site of MMTV. The MuLV-MMTV hybrid showed 10% of the MuLV read-thorough activity (Wills *et al.*, 1991), indicating that stem structures alone are not sufficient. We have focused our attention on the 18 nucleotides in the second loop of the MuLV pseudoknot.

A series of 3-nucleotide deletions has defined a region of importance at the 5' end of loop 2. In contrast, any, or all, of the remaining nucleotides in loop 2 can be deleted with no effect (Wills *et al.*, unpublished). Single nucleotide replacements within the critical codons implicate particular bases in those codons as specific determinants for read-through. These results are perhaps surprising in light of the absence of sequence requirements in loop 2 of the IBV pseudoknot (Brierley *et al.*, 1991). Perhaps this indicates a specific interaction of these nucleotides with rRNA, ribosomal protein(s) or release factor to effect read-through or some additional RNA structure.

MUTANTS OF THE 5'C OF THE CONSERVED CCA AT THE 3' END OF tRNAs

It has been proposed that the movement of tRNA from the A to P to E sites during translation involves the dissolution and formation of pairing between the CCA tRNA tail of tRNA and successive series of nucleotides in the large subunit rRNA (Moazed and Noller, 1989). In a selection for mutants of translation components which cause frameshifting at discrete places where it is not programmed to occur O'Connor *et al.*, (unpublished) have isolated mutants of *E. coli* tRNA$_1^{Val}$ which have an altered the CCA terminus. These mutants also cause stop codon read-through. The available data can be interpreted by a model in which a non-standard interaction between the 3' three nucleotides of the tRNA and 23S rRNA results in an altered conformation of 23S rRNA in a neighboring ribosomal site, the A site, with consequences for decreased discrimination.

FRAMESHIFTING CAUSED BY A tRNA WITH 10 EXTRA BASES

In collaboration with Z. Li and M. Deutscher, we are investigating the processing of an unusual mutant tRNA$_2^{Arg}$. This mutant is a frameshift mutant suppressor which reads the 4-base sequence CCGU as a single codon (Tuohy *et al.*, 1992). It is unusual in that in addition to a predicted 9-nucleotide anticodon loop, the DNA sequence of the cloned tRNA gene predicts a 10-nucleotide duplication of the 3' part of the TFC arm. The sequence of the 10 bases is 5'AATCCTCCCG3'. There are a number of possibilities for the structure of the mature tRNA; it may be that the 10-nucleotide duplication is precisely excised from the precursor tRNA. It is also possible that the precursor tRNA is processed in a different way resulting in the deletion of the last 10 nucleotides of the mature tRNA requiring the mature tRNA to refold in such a way as to approximate the structure of a conventional tRNA, but with two mismatches in the acceptor stem, and a possible turnover of CCC3' to CCA3'. The third possibility under consideration is that the extra 10 nucleotides are not processed out of the mature form, but that they remain, in a unconventionally folded tRNA. This tRNA may not only function in normal translation (if it does), but can function as a frameshift suppressor.

ACKNOWLEDGEMENTS

We thank Linda Nichols for dedicated technical assistance. This work was supported by the Howard Hughes Medical Institute and by NIH grant #GM48152.

REFERENCES

Brierley, I., Digard, P., and Inglis, S.C., 1989, Characterization of an efficient coronavirus ribosomal frameshifting signal: requirement for an RNA pseudoknot, *Cell* 57:537.

Brierley, I., Rolley, N.J., Jenner, A.J., and Inglis, S.C., 1991, Mutational analysis of the RNA pseudoknot component of a coronavirus ribosomal frameshifting signal, *J. Mol. Biol.* 220:889.

Chamorro, M., Parkin, N., and Varmus, H.E., 1992, An RNA pseudoknot and an optimal heptameric shift site are required for highly efficient ribosomal frameshifting on a retroviral messenger RNA, *Proc. Natl. Acad. Sci. USA* 89:713.

Dayhuff, T.J., Gesteland, R.F., and Atkins, J.F., 1992, Electrophoresis, autoradiography and electroblotting of peptides: T4 gene *60* hopping, *BioTechniques* 13:499.

Felsenstein, K.M., and Goff, S.P., 1992, Mutational analysis of the *gag-pol* junction of Moloney murine leukemia virus: requirements for expression of the *gag-pol* fusion protein, *J. Virol.* 66:6601.

Feng, Y.-X., Yuan, H., and Rein, A., and Levin, J.G., 1992, Bipartite signal for readthrough suppression in murine leukemia virus mRNA: an eight-nucleotide purine-rich sequence immediately downstream of the *gag* termination codon followed by an RNA pseudoknot, *J. Virol.* 66:5127.

Gesteland, R.F., Weiss, R.B., and Atkins, J.F., 1992, Recoding: Reprogrammed genetic decoding, *Science* 257:1640.

Huang, W.M., Ao, S.Z., Casjens, S., Orlandi, R., Zeikus, R., Weiss, R., Winge, D., and Fang, M., 1988, A persistent untranslated sequence within bacteriophage T4 DNA topoisomerase gene *60*, *Science* 239:1005.

Honigman, A., Wolf, D., Yash, S., Falk, H., and Panet, A., 1991, *cis* Acting RNA sequences control the *gag-pol* translation readthrough in murine leukemia virus, *Virol.* 183:313.

Moazed, D., and Noller, H.F., 1989, Intermediate states in the movement of transfer RNA in the ribosome, *Nature* 342:142.

Tuohy, T.M.F., Thompson, S., Gesteland, R.F., and Atkins, J.F., 1992, Seven, eight and nine-membered anticodon loop mutants of tRNA$_2^{Arg}$ which cause +1 frameshifting. Tolerance of DHU arm and other secondary mutations, *J. Mol. Biol.* **228**:1042.

Weiss, R.B., Huang, W.-M., and Dunn, D.M., 1990, A nascent peptide is required for ribosomal bypass of the coding gap in bacteriophage T4 gene *60*, *Cell* 62:117.

Wills, N.M., Gesteland, R.F., and Atkins, J.F., 1991, Evidence that a downstream pseudoknot is required for translational read-through of the Moloney murine leukemia virus *gag* stop codon, *Proc. Natl. Acad. Sci. USA* 88:6991.

Yoshinaka, Y., Katoh, I., Copeland, T.D., and Oroszlan, S., 1985, Murine leukemia virus protease is encoded by the *gag-pol* gene and is synthesized through suppression of an amber termination codon, *Proc. Natl. Acad. Sci. USA* 82:1618.

USE OF RIBOSOMAL ACCURACY MUTANTS TO PROBE MECHANISMS OF PROGRAMMED TRANSLATIONAL FRAMESHIFTS IN *Escherichia coli*

John Sipley[1] and Emanuel Goldman[2]

Department of Microbiology & Molecular Genetics
New Jersey Medical School-UMDNJ
185 South Orange Avenue
Newark, NJ 07103, USA

INTRODUCTION

In recent years, a number of "programmed" translational frameshifts have been identified, in which ribosomes, to varying extents, shift their reading frame at specific sites while translating a message. This has been found in diverse species including eucaryotic viruses and bacteria, among others (reviewed in Atkins et al., 1990).

Bacterial systems offer the capability of manipulating the level of accuracy of the translation process (Gorini, 1971; reviewed in Parker, 1989). Earlier studies (Atkins et al., 1972; Weiss and Gallant, 1983) showed that translational accuracy did affect at least some low frequency frameshift errors of translation the same way that it affected other mistakes of translation, i.e., increased accuracy decreased frameshifts and other errors, while decreased accuracy increased these errors. However, some higher frequency frameshifts were reported to be unaffected by increased translational accuracy conditions (Parker and Precup, 1986; Tsuchihashi, 1991).

We have systematically manipulated the level of accuracy of translation in *Escherichia coli* in order to ask the question as to what relationship, if any, does translational accuracy have to two different programmed translational frameshifts. In one case, with a -1 shift (phage T7 gene *10*), translational accuracy changes had no effect (Sipley et al., 1991a). In another case, with a +1 shift (a modified RF2 system), increased translational accuracy significantly *increased* the frameshift (Sipley and Goldman, 1993). These results, in conjunction with those of several other groups, lead to a coherent model for programmed frameshifting, in which at least two different mechanisms can be distinguished: +1 (rightward) frameshifts characterized by vacant ribosomal A sites; -1

---

[1] Present address: Biology Dept., Brookhaven National Laboratory, Upton, NY 11973, USA

[2] Corresponding author

(leftward) frameshifts characterized by filled ribosomal A sites. Both mechanisms have in common the need for ribosomal pausing (facilitated by a "stimulator"), which can be accomplished by several different routes; but it is the occupancy of the A site which appears to determine whether the shift will respond to ribosomal accuracy and tRNA availability.

MANIPULATIONS OF TRANSLATIONAL ACCURACY

The conditions varying translational accuracy which we have employed are outlined in Table 1. A set of isogenic strains derived from *Escherichia coli* strain XAc was obtained from Dr. Leif Isaksson, Stockholm, Sweden. These strains contained different ribosomal fidelity alleles, and have been extensively tested with respect to their fidelity characteristics (Olsson and Isaksson, 1979; Andersson and Kurland, 1983; Petrullo et al., 1983; Bohman et al., 1984; Ruusala and Kurland, 1984; Faxén et al., 1988; Tapio and Isaksson, 1988).

Table 1. Manipulation of translational accuracy.

Conditions	Relative Translational Accuracy	Ribosomal Protein Affected	Strain XAc Ribosomal Fidelity Allele
wild-type (wt)	+ +		
wt + streptomycin	+		
Streptomycin-resistant (Sm^r)	+ + +	S12	rpsL 224 rpsL 282
Sm^r + streptomycin	+ +		
Streptomycin-pseudodependent (Sm^p)	+ + + +	S12	rpsL 1204
Sm^p + streptomycin	+ +(+)		
Ribosomal ambiguity (ram)	+	S4	rpsD 14

These conditions were used to test whether we could affect the frequency of the phage T7 gene *10* programmed frameshift (Dunn and Studier, 1983; Condron et al., 1991a; Sipley et al., 1991b), and also a modified version of the *E. coli* RF2 programmed frameshift (Craigen and Caskey, 1986; Weiss et al., 1988; Curran and Yarus, 1988).

THE PHAGE T7 GENE *10* FRAMESHIFT

This -1 shift normally occurs at a 10-15% frequency. The site of the shift is a run of 4 U's spanning codons 340-341 of the 344 residue 10A protein (excluding the processed N-terminal methionine), leading to an additional 53 codons to yield the 10B protein. The precise nature of the ribosome pause element (stimulator) is not entirely clear, but seems to involve long-range RNA structures; the shift itself seems to be similar to retroviral "simultaneous slip" shifts (Condron et al., 1991b).

Phage T7 infections were carried out in varying conditions of translational accuracy as outlined in Table 1. Synthesis of the 10A and 10B proteins was monitored by pulse labeling with [^{35}S]methionine during the course of the infection, followed by separation of proteins by SDS-polyacrylamide gel electrophoresis. Frameshifting was determined by quantitation of the ratio of the two gene *10* products.

Figure 1 panel A presents the phage specific proteins produced late in infection of cells containing wild-type ribosomes, "hyperaccurate" streptomycin pseudodependent ribo-

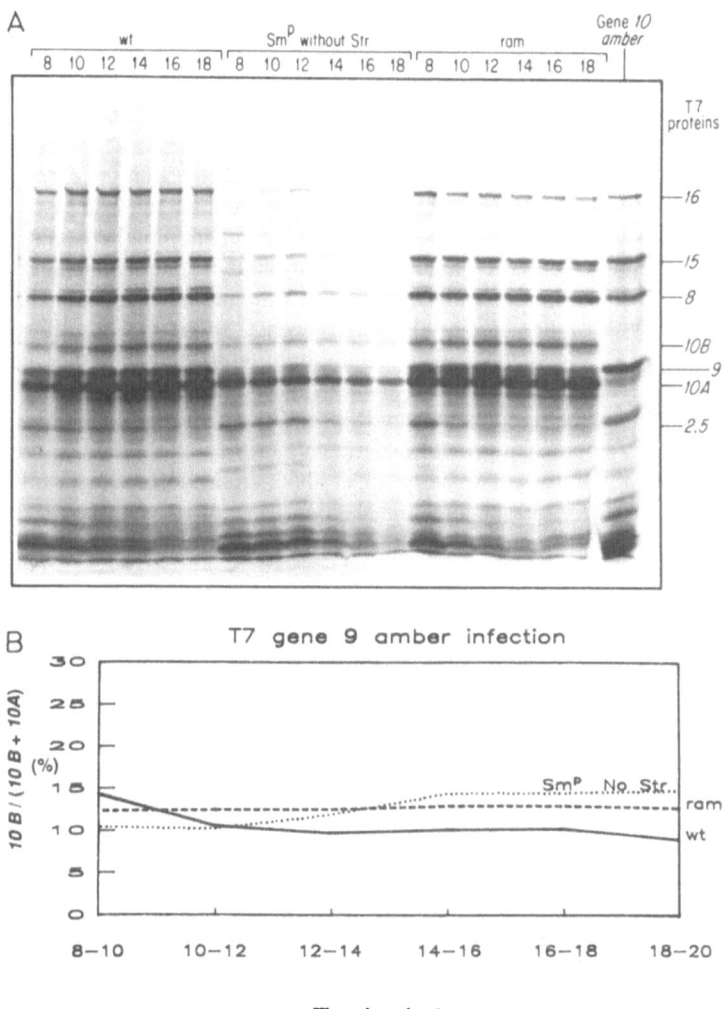

Figure 1. Time course of late protein synthesis during T7 gene *9* amber phage infection of cells with wt ribosomes (wt), streptomycin-pseudodependent ribosomes without streptomycin (Sm^P without Str), and ram ribosomes (ram). A. Cultures were each infected with T7 *9am17*; after 8 min, the cultures were pulse labeled with [^{35}S]methionine for 2 min intervals, extracted, and analyzed by SDS-polyacrylamide gel electrophoresis. Lane numberings indicate the time of infection in minutes. T7 *9am17* appeared to be somewhat leaky in these experiments. Protein products from T7 gene *10* amber phage (T7 *10am13*) infected wt strain *XAc* were run in the outside lane of the gel to help identify the precise location of the 10A and 10B proteins. B. Quantitation of the frameshift ratio. Protein bands corresponding to 10A and 10B were cut out from the gel and quantitated. The range of actual counts/min recovered from the gel slices varied from approximately 50,000 to 384,000 for 10A, and 9,000 to 55,000 for 10B. Reprinted, with permission, from Sipley et al., 1991a.

somes without streptomycin ("Sm^P without Str"), and "inaccurate" ram ribosomes. The outer lane ("gene *10 amber*"), which lacks the 10A and 10B proteins, serves as an aid to help identify the individual 10A and 10B bands in the other lanes (the minor band immediately above 10B is a host band which is labeled due to incomplete shutoff of host protein synthesis). As indicated, the gene *9* protein is the major band migrating between the 10A and 10B proteins under these conditions of electrophoresis. In an effort to facilitate quantitation of 10B relative to 10A, a T7 gene *9* amber mutant was utilized to minimize incorporation of label into protein migrating in the 10A region of the gel. However, the same results were also obtained with wild-type T7 (Sipley, 1992). Overall, the amount and kinetics of synthesis of the late T7 proteins are similar in the wt and ram strains; the levels of 10B and 10A expression also appear proportional. However, in the Sm^P strain, total incorporation of label is noticeably reduced, and synthesis of T7 late proteins appears delayed; in some experiments, proteins characteristic of earlier timepoints are still apparent at later timepoints (Sipley, 1992). Quantitation of the frameshift ratio expressed as [10B/(10B + 10A)] x 100, of each lane of the gel is presented in the corresponding graph (Figure 1 panel B). No significant difference in the amount of frameshifting among the three strains was apparent. Similar results were also obtained under conditions of reduced translational accuracy caused by addition of exogenous streptomycin (Sipley, 1992).

The lack of effect of altering ribosomal fidelity on the programmed gene *10* frameshift contrasts with earlier studies of low-frequency frameshifts as translational errors (Atkins et al., 1972; Weiss and Gallant, 1983), but is consistent with other studies of high-frequency shifts (Parker and Precup, 1986; Tsuchihashi, 1991). Unexpectedly, any changes in fidelity conditions (both increased and decreased accuracy) led to inhibition of phage T7 morphogenesis in single-step growth experiments (Sipley et al., 1991a). Taken by themselves, these results don't tell us very much about frameshifts *per se*, although they do have implications for understanding translational fidelity as well as T7 morphogenesis. However, these results become more significant for understanding frameshifting when seen in the context of the next set of experiments.

THE MODIFIED *E. coli* RF2 FRAMESHIFT

The RF2 frameshift is viewed as a kind of autoregulatory phenomenon (Craigen and Caskey, 1986). An in-frame UGA triplet at the 26th codon is the site of the shift. When sufficient RF2 is available, ribosomes will terminate at the UGA stop codon, mediated by this factor. When levels of RF2 are insufficient, ribosomes will shift to the +1 frame at this site with at least a 50% efficiency, and continue to translate the message to make more RF2. A stimulator of this shift has been identified as a translated upstream Shine & Dalgarno sequence, which facilitates the ribosome realignment (Weiss et al., 1988). The UGA codon can be replaced by many sense codons, and various levels of frameshifting will continue to occur (Curran and Yarus, 1989), although not to the same extent as when UGA is at the site. Curran and Yarus (1989) exploited this system by fusing an out-of-frame β-galactosidase reporter gene downstream of the shift site, and examining frameshift frequencies for 29 sense codons replacing the UGA. Replacement of UGA with a sense codon eliminates one of the competing variables in the RF2 frameshift, in that absolute levels of of RF2 *in vivo* no longer affect the shift in these RF2/*lacZ* fusion constructs. One of the highest frameshift frequencies obtained (11%) was with the relatively low-usage UGG codon (which is the sole codon for Trp), and this could be decreased 3-fold by adding a plasmid overproducing tRNA^{Trp}. One of the lowest frequencies obtained (~2%) was with the high-usage CUG (Leu) codon.

Several mutant *rpsL* (S12) alleles were tested along with wild type in an isogenic background, as outlined in Table 1. Each of these isogenic strains was transformed with either of three homologous plasmids (obtained from James Curran, Wake Forest Univ., N.C.): a) pJC27, a plasmid containing an in-frame pseudowild-type *lacZ* allele (Curran and Yarus, 1988); b) pUGG, a plasmid containing a 23 nucleotide sequence from the RF2 frameshift region inserted near the 5' end of the β-galactosidase gene in pJC27. These 23 nucleotides include the upstream Shine & Dalgarno region, and have substituted a TGG Trp codon for the original TGA in the RF2 gene. In order for β-galactosidase to be synthesized, the ribosomes must shift to the right one nucleotide into the +1 reading frame; c) pCUG, same as b, above, except that the original TGA of the RF2 gene has been substituted with a CTG Leu codon (Curran and Yarus, 1989).

The inserted sequences, placed in the zero frame near the 5' end of the coding region for β-galactosidase, are: CTT AGG GGG TAT CTT (TGG or CTG) CTA CG (Curran and Yarus, 1988). The shift occurs when the tRNA$_2^{Leu}$ which has just read the CUU codon preceding the frameshift site pairs with a codon in the new frame (UUU for the TGG construct, UUC for the CTG construct), via a first position wobble (Curran and Yarus, 1989).

Cells were grown to early log phase, and IPTG was added to induce transcription of the β-galactosidase genes in the various constructs. One hour after induction, the amount of β-galactosidase produced was measured. Table 2 shows the results with wt and several increased accuracy ribosomal mutants. The actual units measured are shown for the in-frame constructs, and the frameshift frequencies, i.e., the relative amounts of β-galactosidase synthesized in each strain, are indicated for the two out-of-frame constructs.

Table 2. Effect of ribosomal accuracy alleles on the modified RF2 frameshift.[1]

ribosomal S12 allele	β-gal units for in-frame construct	frameshift frequency (%), relative β-gal units, codon at frameshift site	
		CUG	UGG
wt	5764	1.6	13.3
Sm^r (*rpsL282*)	3309	2.5	28.9
Sm^r (*rpsL224*)	2232	3.0	46.5
Sm^p (*rpsL1204*)	1666	1.8	51.0

[1] XAc cells containing the various S12 (*rpsL*) alleles and modified RF2-reporter fusion plasmids indicated above were grown to 1 x 10^8 cells/ml in minimal medium with appropriate supplements. IPTG was added; after 1 hour, β-galactosidase was assayed as in Curran and Yarus (1989). Shown in this table and in Table 3 are the means of 3-5 experiments; the standard error of the mean was 15% or less. Reprinted from Sipley and Goldman, 1993.

With CUG at the frameshift site, wt ribosomes allowed less than a 2% shift, and there was little effect of changing the accuracy of the ribosome. With UGG at the frameshift site, wt ribosomes allowed an ~13% shift, in good agreement with data of Curran and Yarus (1989). However, the increased accuracy alleles showed a considerable increase in shifting, to a maximum of 51% for the most restrictive pseudodependent host (Sm^p). Similar results were obtained for different lengths of induction time (Sipley, 1992). The amount of β-galactosidase made from the in-frame construct decreased in the more accurate strains, which is not unexpected since ribosomes with increased accuracy are known to be slower in translation (Bohman et al., 1984; Ruusala et al., 1984; Andersson et al., 1986), a result we have also obtained in our studies of phage T7 infection of the Sm^p strain (Sipley et al., 1991a; also, see above). There appeared to be a proportionality

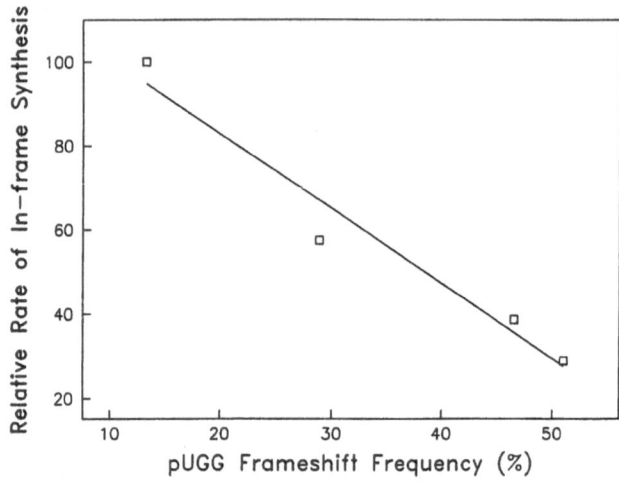

Figure 2. Slower ribosomes increase the frameshift. Frameshift frequencies (%) of the UGG frameshift in Table 2 are plotted against the "relative rate of in-frame synthesis," which was derived from the actual β-galactosidase units measured for the in-frame construct in Table 2, normalizing to a value of 100 for the wt ribosomal allele. Plot obtained by linear regression; the coefficient of determination (r^2) was 0.96. Reprinted from Sipley and Goldman, 1993.

between the extent of reduction of synthesis for the in-frame reporter (as judged by amount of β-galactosidase made in one hour), and the extent of UGG frameshift for the out-of-frame reporter (Figure 2), as if the slower the ribosomes, the greater the shift.

Curran and Yarus (1989) hypothesized that the slower the tRNA selection for the codon at the frameshift site, the greater the shift. In support of this, they found a 3-fold reduction in the UGG frameshift when a plasmid overproducing cognate $tRNA^{Trp}$ was introduced to the system. We have followed this approach, and introduced several additional plasmids containing specific tRNA genes into our strains. In Table 3, wt ribosomes showed a 2-fold reduction in the UGG shift when $tRNA^{Trp}$ was overproduced, similar to Curran and Yarus (1989). The $tRNA^{Trp}$ gene in this plasmid is also under IPTG control, and by 1 hour after induction, has approximately an order of magnitude increase in $tRNA^{Trp}$ (Rojiani et al., 1989). When the $tRNA^{Trp}$ plasmid was introduced into the hyperaccurate Sm^p strain, the frameshift was reduced from 51% to ~7% (Table 3). The $tRNA^{Trp}$ plasmid had no effect on the CUG frameshift in either host.

Introduction of excess cognate tRNA for the CUG codon, $tRNA_1^{Leu}$, caused at least a five-fold reduction in the CUG shift in both wt and Sm^p hosts. In wt, this $tRNA_1^{Leu}$ plasmid had no effect on the UGG shift, and in Sm^p, the reduction of the UGG shift was only about 2-fold. It should be noted that some of the tRNA constitutively overproduced from this plasmid is hypomodified, and leads to a general reduction in protein synthesis rates (Wahab et al., 1992). Introduction of excess $tRNA_2^{Leu}$ into wt cells was of interest because this is the slippery tRNA which shifts from the CUU codon preceding the shift site into the new frame. However, no significant change in the extent of shifting was observed. Similarly, introduction of excess $tRNA^{Phe}$, which is expected to be unrelated to this shift mechanism, had no effect on the extent of shifting. This tRNA is overproduced and charged by an order of magnitude (unpublished data).

All strains containing plasmids leading to overproduced tRNAs showed reductions, generally about 2-fold, in the amount of β-galactosidase made for the in-frame reporter pJC27 (Table 3). This result is not specific for overproduced tRNAs, however; a similar reduction was also obtained with just the vector alone (pBR322), without any tRNA genes

Table 3. Effect of overproduced tRNA species on the modified RF2 frameshift.[1]

tRNA species	β-gal units for in-frame construct	frameshift frequency (%), relative β-gal units, codon at frameshift site	
		CUG	UGG
A. wt ribosomes			
none	5764	1.6	13.3
tRNA^{Trp}	3235	2.6	7.0
tRNA$_1^{Leu}$ *(reads CUG)*	2488	0.3	12.0
tRNA$_2^{Leu}$ *(reads CUC/U)*	2005	3.1	10.1
tRNA^{Phe}	2358	3.3	11.3
pBR322 *(vector for tRNAs)*	2127	2.9	12.6
B. Sm^P ribosomes			
none	1666	1.8	51.0
tRNA^{Trp}	1454	1.4	7.4
tRNA$_1^{Leu}$ *(reads CUG)*	1030	<0.1	21.6

[1] Performed as in Table 2, except that cells contained additional plasmids with tRNA genes, as indicated. Reprinted from Sipley and Goldman, 1993.

inserted (Table 3). More likely, the extra plasmid simply competes for the biosynthetic capacity of the cell, hence the reduction in extent of synthesis.

Strain *Xac*, deleted for the *lac* operon, does not contain *lac* repressor. Since the modified RF2 system is under control of the *lac*UV5 promoter, male versions of the *Xac* strains had been made for these experiments, containing an F' factor with the *lacI*^{Q} gene. Unfortunately, the male version of our ram strain was not amenable as a host for the modified RF2 plasmids and could not be grown without consistently obtaining an unacceptable level of revertants. Therefore, exogenous addition of sublethal amounts of streptomycin to the male version of strain *Xac* was our only condition of decreased ribosomal fidelity. There was no effect of exogenous streptomycin on the frequency of the modified RF2 frameshifts in cells with wt ribosomes. However, Sm^P and Sm^r cells showed up to 2-fold reductions in the UGG shift upon addition of streptomycin (Sipley, 1992).

Curran and Yarus (1989) had found a relatively small contribution from differences in mRNA levels to the extent of frameshifted product synthesized. There were a number of additional reasons to presume that varying levels of mRNA could not explain the differences in levels of β-galactosidase obtained in our experiments. For example, the possibility that more mRNA accounted for the increased shift in Sm^P cells was minimized by the tRNA^{Trp} reversal, the low level of shift with CUG at the frameshift site, and by always comparing to the standard of the in-frame reporter. As a more direct way to determine the extent to which variable mRNA levels might or might not contribute to the different levels of β-galactosidase obtained, analysis of β-galactosidase mRNA was carried out by slot-blot and Northern hybridizations. These experiments indicated that differences in mRNA levels were too small to account for the different levels of β-galactosidase obtained with wt ribosomes, in agreement with Curran and Yarus (1989), and also that induced mRNA levels on a per cell basis were not significantly different in Sm^P cells compared to wt (Sipley, 1992). A Northern analysis shows that there are no major differences in mRNA levels in Sm^p cells with the modified RF2-reporter fusion plasmids in the presence or absence of excess tRNA^{Trp} (Figure 3). Relative mRNA levels in cells harboring pJC27, pUGG, or pCUG were estimated by liquid scintillation counting of radioactivity from the hybridized probe in Fig. 3 (Sipley, 1992). This yielded very similar results in the Sm^p strain to those reported by Curran and Yarus (1989) for wt cells. Since the UGG frameshift frequency in Sm^p cells drops from 51% to ~7% in the presence of excess tRNA^{Trp} (Table 3), these mRNA analyses confirm that this drop cannot be accounted for by differences in mRNA levels, and hence must be due to differences in translation.

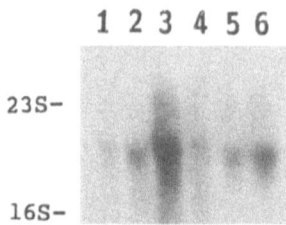

Figure 3. Northern analysis of β-galactosidase mRNA in Sm^p cells with and without excess tRNA^{Trp}. Sm^p cells containing the modified RF2-reporter fusion plasmids either with (lanes 1-3) or without (lanes 4-6) a plasmid overproducing tRNA^{Trp} were grown and induced as in Table 3. Lanes 1, 4: pCUG. Lanes 2, 5: pUGG. Lanes 3, 6: pJC27 (in-frame). Cells (3 ml) were pelleted and resuspended in 0.1 ml of MOPS-cracking buffer (0.02 M MOPS, pH 7.0, 8 mM Na acetate, 2 mM Na$_2$EDTA, 1% SDS, 1% β-mercaptoethanol, 10% glycerol, 0.05% bromocresylgreen, and 0.05% xylenecyanole FF), boiled in capped tubes for 2 min, immediately quick-frozen on dry-ice, thawed in the presence of formaldehyde (final concentration ~6%), mixed, incubated for 15 min at 65°C, and placed on ice. Condensates were centrifuged briefly, and aliquots (equivalent to the same number of cells for each sample) were subjected to electrophoresis on a 1.5% agarose gel containing ~6% formaldehyde in a buffer of 0.02 M MOPS, pH 7.0, 8 mM Na acetate, 1 mM EDTA (Sambrook et al., 1989). Migration of 23S and 16S rRNAs is indicated; these were visualized by cutting away and staining parallel lanes overnight in 0.5 μg/ml ethidium bromide, 0.1 M ammonium acetate, and destaining for 1 h in 0.1 M ammonium acetate. The gel was blotted onto nitrocellulose and hybridized (Sambrook et al., 1989) at 48°C with a ^{32}P-labeled probe specific to the RF2-reporter fusions. XAc-derived cells which did not contain the RF2-reporter fusion plasmids did not show any hybridization to the probe (Sipley, 1992). Following a procedure from James Curran, our probe was an end-labeled oligonucleotide (17 bases), synthesized on a DNA synthesizer, with sequence CGATAATTTCACCGCCG, which is specific for the lacZ gene. Reprinted from Sipley and Goldman, 1993.

CONCLUSIONS

Our results can be rationalized as follows. Ribosome pause time at the frameshift site determines the frequency of the shift; this pause time can be affected by cognate tRNA availability; the longer the pause, the greater the shift frequency. Increased accuracy alleles increase shifting by slowing the ribosomes (Bohman et al., 1984; Ruusala et al., 1984; Andersson et al., 1986; Sipley et al., 1991a). The increased accuracy phenotype of rpsL mutants has been attributed to increased translational proofreading (Ruusala et al., 1984). Our results suggest that hyperaccurate rpsL mutants are also slower in initial tRNA selection as well, since a vacant A site is needed for the frameshift. Earlier in vitro experiments had also suggested slower initial tRNA selection by rpsL mutants (Yates, 1979).

Overproduction of tRNA cognate to the frameshift site codon reduces the shift by decreasing the time the A site is vacant, keeping the ribosomes in frame, thereby reducing production of the out-of-frame reporter. To our surprise, overproduction of even $tRNA_1^{Leu}$, cognate to the high-usage CUG codon, significantly reduced the shift occurring in that construct. Since this species is already the most abundant isoacceptor in *E. coli*, it would not be expected that this tRNA species is limiting for translation; however, the cognate codon is also the most frequently used codon in *E. coli* as well (Zhang et al., 1991), raising the possibility that excess demand could make even an abundant tRNA limiting. As alternate possible considerations for the reduced shift with excess $tRNA_1^{Leu}$, perhaps this tRNA species is competing with the slippery $tRNA_2^{Leu}$ at the preceding codon, thus reducing the shift. There is evidence both *in vivo* (Holmes et al., 1977) and *in vitro* (Goldman et al., 1979) for $tRNA^{Leu}$ species substituting for one another depending on conditions. This is not likely to be the entire explanation, however, since there is no effect of $tRNA_1^{Leu}$ on the UGG shift with wild type ribosomes, and only about a 2-fold effect with Sm^P ribosomes (Table 3). Alternately, uncharged and/or hypomodified $tRNA_1^{Leu}$ might stall the ribosome at the CUG codon, keeping the A site filled and preventing the shift. There is evidence that excess uncharged tRNA inhibits translation (Rojiani et al., 1990; Goldman and Jakubowski, 1990), and the overproduced $tRNA_1^{Leu}$ is only about 40% charged (Wahab et al., 1992). It should also be noted that the frameshift assay monitors the outcome of a competition between frameshifting and normal translation by the cognate tRNA. In this view, it might be expected that an increased concentration of any cognate charged tRNA would decrease frameshifting, even if the tRNA is not normally "limiting" for translation. Whatever the explanation for the effect of $tRNA_1^{Leu}$, our data provide strong support for the model of ribosome pause time governing frameshift frequencies, and for cognate tRNA availability influencing the process.

The results from both the T7 gene *10* and modified RF2 systems indicate that ribosomal fidelity alleles do not ordinarily monitor the reading frame *per se*, rather just the accuracy of triplet decoding. The fact that slower ribosomes with increased accuracy lead to an increase in the modified RF2 shift suggests slower tRNA selection by ribosomes under increased accuracy conditions, since a vacant A site is required for this shift. In this context, it is understandable that accuracy manipulations have no effect on the -1 T7 gene *10* shift, and others of this class, since the A site is filled prior to the shift. The results with both of these two different kinds of programmed frameshifts imply that these high-frequency shifts should not be considered as errors in translation, but are in fact normal responses of the translational apparatus to specific, evolved signals in the message.

ACKNOWLEDGEMENTS

This work was supported by grants from the National Institutes of Health (GM27711 and 2 SO7 RR05393) and from the Foundation of UMDNJ (#9-93).

REFERENCES

Andersson, D.I., and Kurland, C.G., 1983, Ram ribosomes are defective proofreaders. *Mol. Gen. Genet.* 191:378-381.

Andersson, D.I., van Verseveld, H.W., Stouthamer, A.D., and Kurland, C.G., 1986, Suboptimal growth with hyper-accurate ribosomes, *Arch. Microbiol.* 144:96-101.

Atkins, J.F., Elseviers, D., and Gorini, L., 1972, Low activity of β-galactosidase in frameshift mutants of *Escherichia coli*, *Proc. Natl. Acad. Sci. USA* 69:1192-1195.

Atkins, J.F., Weiss, R.B., and Gesteland, R.F., 1990, Ribosome gymnastics - degree of difficulty 9.5, style 10.0, *Cell* 62:413-423.

Bohman, K., Ruusala, T., Jelenc, P.C., and Kurland, C.G., 1984, Kinetic impairment of restrictive streptomycin-resistant ribosomes, *Mol. Gen. Genet.* 198:90-99.

Condron, B.G., Atkins, J.F., and Gesteland, R.F., 1991a, Frameshifting in gene *10* of bacteriophage T7, *J. Bacteriol.* 173:6998-7003.

Condron, B.G., Gesteland, R.F., and Atkins, J.F., 1991b, An analysis of sequences stimulating frameshifting in the decoding of gene *10* of bacteriophage T7, *Nucl. Acids Res.* 19:5607-5612.

Craigen, W.J., and Caskey, C.T., 1986, Expression of peptide chain release factor 2 requires high-efficiency frameshift, *Nature (London)* 322:273-275.

Curran, J.F., and Yarus, M., 1988, Use of tRNA suppressors to probe regulation of *Escherichia coli* release factor 2, *J. Mol. Biol.* 203:75-83.

Curran, J.F., and Yarus, M., 1989, Rates of aa-tRNA selection at 29 sense codons *in vivo*, *J. Mol. Biol.* 209:65-77.

Dunn, J.J., and Studier, F.W., 1983, Complete nucleotide sequence of bacteriophage T7 DNA and the locations of T7 genetic elements, *J. Mol. Biol.* 166:477-535.

Faxén, M., Kirsebom, L.A., and Isaksson, L.A., 1988, Is efficiency of suppressor tRNAs controlled at the level of ribosomal proofreading in vivo? *J. Bacteriol.* 170:3756-3760.

Goldman, E., Holmes, W.M., and Hatfield, G.W., 1979, Specificity of codon recognition by *Escherichia coli* tRNA^{Leu} isoaccepting species determined by protein synthesis *in vitro* directed by phage RNA, *J. Mol. Biol.* 129:567-585.

Goldman, E., and Jakubowski, H., 1990, Uncharged tRNA, protein synthesis, and the bacterial stringent response, *Molecular Microbiology* 4:2035-2040.

Gorini, L., 1971, Ribosomal discrimination of tRNAs, *Nature New Biol.* 234:261-264.

Holmes, W.M., Goldman, E., Miner, T.A., and Hatfield, G.W., 1977, Differential utilization of leucyl-tRNAs by *Escherichia coli*, *Proc. Natl. Acad. Sci. USA* 74:1393-1397.

Olsson, M.O., and Isaksson, L.A., 1979, Analysis of *rpsD* mutations in *Escherichia coli*. I. Comparison of mutants with various alterations in ribosomal protein S4, *Mol. Gen. Genet.* 169:251-257.

Parker, J., 1989, Errors and alternatives in reading the universal genetic code, *Microbiol. Rev.* 53:273-298.

Parker, J., and Precup, J., 1986, Mistranslation during phenylalanine starvation, *Mol. Gen. Genet.* 204:70-74.

Petrullo, L.A., Gallagher, P.J., and Elseviers, D., 1983, The role of 2-methylthio-N6-isopentenyladenosine in readthrough and suppression of nonsense codons in *E. coli*, *Mol. Gen. Genet.* 190:289-294.

Rojiani, M.V., Jakubowski, H., and Goldman, E., 1989, Effect of variation of charged and uncharged tRNA^{Trp} levels on ppGpp synthesis in *Escherichia coli*, *J. Bacteriol.* 171:6493-6502.

Rojiani, M.V., Jakubowski, H., and Goldman, E., 1990, Relationship between protein synthesis and concentrations of charged and uncharged tRNA^{Trp} in *E. coli*, *Proc. Natl. Acad. Sci. USA* 87:1511-1515.

Ruusala, T., and Kurland, C.G., 1984, Streptomycin preferentially perturbs ribosomal proofreading, *Mol. Gen. Genet.* 198:100-104.

Ruusala, T., Andersson, D., Ehrenberg, M., and Kurland, C.G., 1984, Hyper-accurate ribosomes inhibit growth, *EMBO J.* 3:2575-2580.

Sambrook, J., Fritsch, E.F., and Maniatis, T., 1989, "Molecular Cloning: A Laboratory Manual," 2nd Edition, Cold Spring Harbor Laboratory Press, NY, chapters 7, 9, and 10.

Sipley, J., 1992, "Effects of Manipulating Ribosomal Fidelity on Two Programmed Translational Frameshifts and Bacteriophage T7 Morphogenesis in *Escherichia coli*," Ph.D. Thesis, UMDNJ, Newark, NJ

Sipley, J., Dunn, J., and Goldman, E., 1991a, Bacteriophage T7 morphogenesis and gene *10* frameshifting in *Escherichia coli* showing different degrees of ribosomal fidelity, *Mol. Gen. Genet.* 230:376-384.

Sipley, J., Stassi, D., Dunn, J., and Goldman, E., 1991b, Analysis of bacteriophage T7 gene *10A* and frameshifted *10B* proteins, *Gene Expression* 1:127-136.

Sipley, J., and Goldman, E., 1993, Increased ribosomal accuracy *increases* a programmed translational frameshift in *Escherichia coli*, *Proc. Natl. Acad. Sci. USA* 90:in press.

Tapio, S., and Isaksson, L.A., 1988, Antagonistic effects of mutant elongation factor Tu and ribosomal protein S12 on control of translational accuracy, suppression and cellular growth, *Biochimie* 70:273-281.

Tsuchihashi, Z., 1991, Translational frameshifting in the *Escherichia coli dnaX* gene *in vitro*, *Nucl. Acids Res.* 19:2457-2462.

Wahab, S.Z., Rowley, K.O., and Holmes, W.M., 1992, Effects of tRNA$^{Leu}_1$ overproduction in *Escherichia coli*, *Molecular Microbiology* 6:in press.

Weiss, R., and Gallant, J., 1983, Mechanism of ribosome frameshifting during translation of the genetic code, *Nature (London)* 302:389-393.

Weiss, R.B., Dunn, D.M., Dahlberg, A.E., Atkins, J.F., and Gesteland, R.F., 1988, Reading frame switch caused by base pair formation between the 3' end of 16S rRNA and the mRNA during elongation of protein synthesis in *Escherichia coli*, *EMBO J.* 7:1503-1507.

Yates, J.L., 1979, Role of ribosomal protein S12 in discrimination of aminoacyl tRNA, *J. Biol. Chem.* 254:11550-11554.

Zhang, S., Zubay, G., and Goldman, E., 1991, Low-usage codons in *Escherichia coli*, yeast, fruit fly and primates, *Gene* 105:61-72.

23S rRNA TUNES RIBOSOMAL ACCURACY

Neşe Bilgin and Måns Ehrenberg

Department of Molecular Biology
University of Uppsala, BMC, Box 590
S-751 24 Uppsala, Sweden

INTRODUCTION

New techniques to study RNA have lead to a rapid accumulation of data indicating that ribosomal RNA (rRNA) is part of every functional site on the ribosome. The emphasis is now on the functional role played by rRNA (Reviewed by Noller, 1991), rather than by ribosomal proteins.

We now know that base changes in 16S rRNA suppress nonsense mutations (Murgola et al, 1988, Shen and Fox, 1989), frameshift mutations (Weiss-Brumer and Hüttenhofer, 1989) and cause increased translational misreading (Allen and Noller, 1991). There is also experimental evidence that rRNA is the target for universal inhibitors of translation (e.g. Noller, 1991), and earlier experiments have demonstrated factor free aminoacyl-tRNA binding (Gavrilova et al., 1976) and translocation (Gavrilova and Spirin, 1974). These findings, together with recent experiments that indicate the existence of protein free peptidyl-transfer (Noller et al., 1992), have lead to a revival of the idea (Crick, 1968) that the primary ribosomes were made solely of RNA and that they became co-functional with proteins only later in evolution (e.g. Noller, 1991).

One concern of the present review is how base changes in a particular region of 23S rRNA affect accuracy. However, our results cannot be understood without bringing ribosomal proteins into the picture. It must not be forgotten that ribosomal proteins and

The Translational Apparatus, Edited by K.H. Nierhaus
et al., Plenum Press, New York, 1993

translational factors are also involved in important reactions on the ribosome (Kaziro, 1978; Spirin, 1985). Reactions that occur under artificial conditions without proteins are normally many orders of magnitude slower than when proteins are present. It is furthermore, far from settled that there ever was a protein free ribosome.

To understand the relation between efficient ribosome function and optimal growth of living cells, it is required that we study the "now living ribosome", with the intricate and poorly understood interplay between its protein and RNA parts. One challenge is to identify the contributions from each of these components to specific functions of the ribosome. Another, and more demanding one, is to understand how these various parts of the ribosome interact with each other.

Mutations and universal inhibitors of protein synthesis that alter ribosome function have been very useful for such studies. One early example is Gorini's *rpsL* mutants, altered in ribosomal protein S12 (Gorini et al., 1966; Ozaki et al., 1969). These strains have hyperaccurate ribosomes, and this shows that accuracy tuning in translation depends on components other than codons and anticodons (Gorini, 1971; Ninio, 1974). Gorini's findings lead Ninio (1974) to reformulate the accuracy problem, hitherto described in terms of structure, in a kinetic frame of reference. This conceptual breakthrough was followed by experimental studies of structural properties of tRNA that lead to efficient and accurate translation (Yarus, 1982; Murgola, 1985). Fidelity mutants of tRNA (Murgola, 1985) and EF-Tu (Hughes et al., 1987) have been isolated. Recently, 16S rRNA mutations that cause ambiguity in code reading have been reported (Allen and Noller, 1991).

We have recently found that a 2661G to C transversion in 23S rRNA enhances the accuracy of code translation (N. Bilgin and M. Ehrenberg, in preparation). We have also found that the effects of this transversion depends strongly on the *rpsL* background as well as on the presence of Sm. Our experiments emphasize further that there is a structural interplay between proteins and RNA in translation.

We have interpreted our results assuming that ribosomes achieve their high accuracy through a proofreading mechanism (Hopfield, 1974; Ninio, 1975). This model has considerable experimental support (Thompson and Stone, 1977; Ruusala et al., 1982; Kakniashvili et al., 1986) but was recently questioned by Nierhaus (1990). In the next section, we inspect his arguments, and discuss the recent findings by Moazed and Noller (1989) that lead to the hybrid-site model. This model is highly relevant for how protein S12 and the 2660 region of 23S rRNA may influence initial selection and proofreading.

In the following section new experiments (Bilgin et al.,1992) concerning hyperaccurate ribosomes altered in S12 are summarized. We emphasize such aspects that are relevant for the functional interplay between the 2660-region on 23S rRNA and ribosomal protein S12. Then, in the final section, we describe how the 2661G to C transversion affects accuracy in different contexts and present estimates of the maximal possible selectivity (the d-value, Kurland and Ehrenberg, 1984; 1987) for the initial selection step on *E. coli* ribosomes.

IS THERE A ROLE FOR PROOFREADING IN *E. COLI* TRANSLATION?

The proofreading function of *E. coli* ribosomes was recently discussed in two reviews (Nierhaus, 1990; Noller, 1991). In the first Nierhaus (1990) deals with negative cooperativity between A-site binding of aminoacyl-tRNA and E- site binding of deacylated tRNA. He argues that proofreading may not exist in *E. coli* translation, and that the negative cooperativity between A- and E-site is a better solution to the accuracy problem. The second review (Noller, 1991) contains an interesting discussion about how proofreading may take place on the ribosome. Both reviews bring together the notion of proofreading and structural aspects of the ribosome.

Nierhaus' (1990) first argument against the existence of proofreading in translation is that the single step selectivity (the d-value) is so high that repeated selection is not necessary. The d-value in poly(U) translation, when cognate Phe-tRNA^{Phe} competes with Leu-tRNA$^{Leu}_2$ has previously been estimated to be 10^3 or larger (Thompson and Dix, 1982). If the ribosome selects tRNA in a single step and if the accuracy (A) is close to its maximum (d=10^3), then translation is exceedingly inefficient (Kurland and Ehrenberg, 1984; 1987) and when A approaches d, the rate of translation approaches zero.

The d-value of 10^3 obtained by Thompson and Dix (1982), was calculated for tRNA$^{Leu}_2$, and must be smaller for tRNA$^{Leu}_4$ which is considerably more error-prone in poly(U)-translation (Ruusala et al., 1982) . The translation error that we measure for tRNA$^{Leu}_4$, is about $2.6 \cdot 10^{-4}$ (A=$3.8 \cdot 10^3$) for wild type and about $3.5 \cdot 10^{-5}$ (A=$2.9 \cdot 10^4$) for SmP ribosomes, and both these error levels are out of reach for a single step selection mechanism with d near 10^3. Therefore Nierhaus' first argument seems to support rather than to contradict proofreading.

Nierhaus' (1990) second argument is that "not one mechanism has been proposed up to now that satisfactorily explains a repeated melting and rejoining of codon-anticodon interactions". Since this "melting and rejoining" does not occur in the original descriptions of proofreading (Hopfield, 1974; Ninio, 1975; Kurland, 1978; Ehrenberg and Blomberg, 1980) nor in succeeding experimental papers or reviews dealing with the subject (Thompson and Stone, 1977; Ruusala et al,, 1982; Kurland and Ehrenberg, 1984; 1987) the argument appears to be beside the point.

The experiments that show that proofreading exists in translation are instead that the number of GTP's hydrolyzed in EF-Tu function per non-cognate peptide bond (f_w) is always much larger than the corresponding number for cognate peptide bond (f_c) (Thompson and Stone, 1977; Ruusala et al., 1982; Kakniashvili et al., 1986). Knowledge of f_w and f_c can also be used to estimate how much proofreading (F) contributes to the overall accuracy (A): A=$I \cdot f_w/f_c$=$I \cdot F$, where I is the accuracy in the first selection step (Kurland and Ehrenberg, 1984;1987).To disprove the existence of proofreading in translation it must be demonstrated that the large values of f_w which have been observed by several groups (Thompson and Stone, 1977; Ruusala et al., 1982; Kakniashvili et al., 1986) are due to a side reaction unrelated to the accuracy of ribosome. It should be noted that the interpretation that the high f_W-values áre caused by a proofreading of noncognate tRNAs and not by a side reaction has survived several critical tests (Ruusala et al., 1982; Ehrenberg et al., 1990; Kakniashvili et al, 1986). Therefore, we will here describe the effects of S12 changes and of the 2661G to C substitution in 23S rRNA on ribosomal accuracy using a model containing proofreading (N. Bilgin and M. Ehrenberg, in preparation).

Noller's (1991) review deals with the recently proposed hybrid sites for tRNA in translation (Moazed and Noller, 1989). Ironically, the experimental results of Moazed and Noller (1989) and their interpretation is somehow a vindication of the long struggle of Rheinberger and Nierhaus to obtain recognition for their discovery of the E-site (Rheinberger et al., 1981). The hybrid site model is highly relevant for the interpretation of our data on ribosome function. When aminoacyl-tRNA enters the A-site in ternary complex with EF-Tu and GTP (Kaziro, 1978; Ehrenberg et al, this volume), then, according to the hybrid site concept, the anticodon part makes contact with 16S rRNA and there is, at this time, no direct interaction between tRNA and 23S rRNA. It is here, in this A/T configuration, that initial selection occurs. The ternary complex, cognate or non-cognate, may either dissociate from the ribosome or be transformed by GTP hydrolysis to EF-Tu·GDP which rapidly leaves the ribosome, and to aminoacyl-tRNA positioned in the proofreading site. In this site tRNA has its anticodon end bound to mRNA as well as to 16S rRNA, while the CCA-end is free and without a detectable contact with the ribosome (Noller, 1991). Aminoacyl-tRNA may now be discarded from the ribosome or, alternatively, its CCA-end may dock to 23S rRNA. When docking occurs, selection is over and aminoacyl-tRNA becomes stably bound to the A/A-site and peptidyl transfer can occur. These features of the hybrid site model are attractive in several ways.

Firstly, they give to EF-Tu a prominent role in tuning the accuracy of the initial selection step, in accord with results from experiments on EF-Tu mutants (Tubulekas and Hughes, 1992).

Secondly, it solves a problem previously discussed by Kurland and Ehrenberg (1984; 1987) which can be referred to as the accuracy-processivity dilemma as follows. When aminoacyl-tRNA in ternary complex interacts with the ribosome or when free aminoacyl-tRNA is checked in the proofreading step, then a high accuracy requires comparatively low affinity between tRNA and the ribosome. However, after peptidyl-transfer high processivity (Jörgensen and Kurland, 1990) requires a high affinity between ribosome and tRNA. In Noller's (1991) model there is a switch between low affinity, as required for selection, and high affinity, as required for processivity, which occurs in a natural way when the CCA-end of tRNA docks to 23S rRNA. This feature of the hybrid site model may eliminate the accuracy-processivity dilemma.

Thirdly, Noller's scheme illustrates clearly how proofreading may work. In the initial selection the ribosome chooses between cognate and non-cognate ternary complexes. In the second selection step the choice is between cognate and non-cognate tRNAs, after EF-Tu·GDP has left the ribosome. There is, accordingly, no "melting and rejoining" of mRNA-anticodon complexes (c.f. Nierhaus, 1990). Proofreading requires only that there must be a net outflow of aminoacyl-tRNA over the proofreading discard branch. There can not, in other words, be a significant influx of free aminoacyl-tRNA to the ribosome's proofreading site. This means that aminoacyl-tRNA must enter the ribosome only as ternary complex and not as a free molecule. This apparent break of the law of detailed balance is ultimately accomplished by a shift of GTP and H_2O high above equilibrium with GDP and phosphate, *in vivo* as well as *in vitro* (Kurland, 1978). This GTP to GDP displacement from equilibrium is also driving the ternary complex out of equilibrium with free aminoacyl-tRNA, GDP and phosphate.

Forthly, the docking of the CCA-end of aminoacyl-tRNA to 23S rRNA can be viewed as a rotation of the tRNA-molecule, so that one part (the anticodon end) is fixed and one part (the CCA-end) moves. If one assumes, which seems quite reasonable, that this rotation occurs in a diffusion like movement, and if the tRNA molecule can dissociate from the ribosome at every position of the path, then this may be a clever realization of a multiple proofreading process (Ehrenberg and Blomberg, 1980). With such a mechanism built into the translation machinery a given accuracy can be obtained with extremely small kinetic losses and this is in contrast to the initial selection step. There is, furthermore, virtually no limit to the accuracy in tRNA selection even for moderate d-values (c.f. Kurland and Ehrenberg, 1987).

Such multiple proofreading schemes have been described before (e.g. Ehrenberg and Blomberg, 1980), and there have been speculations that such devices may play a role in translation (Blomberg et al., 1980). At that time two major objections to such schemes were made. The first is that such multiple step proofreading is very time consuming and that it therefore will slow down peptide elongation rate. The other objection is that the construction of multiple steps that proofread tRNA will add to the complexity of the ribosome and therefore to the cell's already expensive investment in ribosomal proteins (c.f. Ehrenberg and Kurland, 1984).

It seems, however, that Noller's (1991) model has removed both these objections. The tRNA may, for all that we know, complete its rotation in less than a couple of microseconds, and a site allowing such movement probably does not have to be very complicated and expensive to build. The experimental tests for such a hypothesis are not altogether trivial, since the mechanism will look like a single step in most types of assays (M. Ehrenberg, in preparation).

HYPERACCURATE RIBOSOMES CHANGED IN S12

Streptomycin resistance (SmR) is often associated with changes in the ribosomal protein S12 in the small ribosomal subunit (Gorini et al., 1966; Ozaki et al., 1969). A subclass of SmR ribosomes is hyperaccurate in code translation *in vivo* (Gorini, 1971) as well as *in vitro* (Ruusala et al., 1984; Bohman et al., 1984). *In vitro* measurements show that Sm increases missense errors by as much as three orders of magnitude for the wild type ribosome

(Ruusala and Kurland, 1984). Although Sm affects both proofreading and initial selection , the drug's ability to drastically reduce proofreading of non-cognate tRNAs was emphasized by Ruusala and Kurland (1984), and Sm has ever since been associated with a perturbation of the proofreading part of tRNA selection (e.g. Andersson et al., 1986; Noller ,1991).

Hyperaccurate SmR ribosomes inevitably have reduced k_{cat}/K_M-values for their interaction with ternary complexes. Early experiments indicated that they are aggressive proofreaders (high f_C values) of cognate tRNAs (Ruusala et al., 1984; Andersson et al., 1986), and it was argued that the reduction in their efficiency to interact with cognate ternary complexes was due to these enhanced f_C-levels (Andersson et al., 1986). From these observations a simple picture emerged, which can be described as follows: The drug Sm turns off ribosomal proofreading and Sm resistance occurs by such changes in S12 that in the presence of Sm restore proofreading and accuracy. In the absence of Sm, these ribosomes are "overcompensated" in their proofreading step so that they are both hyperaccurate and aggressive proofreaders of cognate tRNAs.

Recent error measurements *in vitro* combined with determinations of proofreading parameters for cognate (f_C) and non-cognate (f_W) aminoacyl-tRNAs, show that the effects of Sm on accuracy are evenly distributed between initial selection and proofreading for wild type ribosomes (N. Bilgin, in preparation; next section). Thus Sm is not only affecting the proofreading function of ribosomes, and the binding of Sm to rRNA can accordingly not reveal the location of a specific proofreading region (c.f. Noller, 1991).

Quench-flow studies on hyperaccurate ribosomes in a partial translation system lacking EF-G have revealed new features of such changes in S12 that give Sm resistance and hyperaccuracy. These results have forced us to revise some of our previous views concerning the phenotype of these ribosomes (Bilgin et al., 1992). While the efficiency of ternary complex interaction for these hyperaccurate variants is the same as in full translation systems, there is no significant increase in the cognate proofreading parameter f_c in the partial system (Bilgin et al, 1992). Thus, the efficiency loss due to the S12 change is here not due to an enhanced discard probability of cognate tRNAs in proofreading. Instead, it is the efficiency of the initial selection step where the ternary complex itself is accepted or rejected that is impaired. Consistent with this we find that the rate of GTP-hydrolysis on EF-Tu when the ternary complex sits in the A-site is significantly reduced for all hyperaccurate ribosomes.

The apparent discrepancy between experiments performed in assays without EF-G and steady-state measurements including EF-G is probably removed by another set of observations. It turns out that all hyperaccurate ribosomes that we have investigated, are associated with a strong, ternary complex dependent GTP idling. This reaction depends on correct codon-anticodon interaction. It is very small for wild-type ribosomes, and very large for the least efficient S12-altered ribosomes. It probably interferes with our steady state measurements and leads to artificially high f_c-values. The alternative explanation that the proofreading properties of the ribosome change drastically by the addition of EF-G appears less likely (Bilgin et al, 1992).

These new results concerning the kinetic properties of hyperaccurate ribosomes are important for how we interpret the effects of the 2661G to C transversion, which is described in the next section.

23S rRNA TUNES RIBOSOMAL ACCURACY

Recently a G to C substitution was constructed at position 2661 of the 23S rRNA, and introduced into the rrnB operon on the high copy number plasmid pKK3535 (Tapprich and Dahlberg, 1990). The 2660 region is highly conserved, and one may have expected a significant change in the growth characteristics of bacteria transformed by this mutated plasmid (pKK2661C). However, Tapprich and Dahlberg found growth inhibition caused by the 2661G to C transversion only in such *rpsL* backgrounds that are associated with streptomycin resistance (SmR) and enhanced accuracy in translation. For a slightly restrictive SmR variant there was a significant reduction in growth rate (Tapprich and Dahlberg, 1990). Attempts to transform highly restrictive S12-altered variants with pKK2661C lead to cell death, unless these were combined with a third, tufA (Aa), mutation (Tapio and Isaksson, 1991).

Ribosomes isolated from the slightly restrictive *rpsL* strain appeared to have a very high affinity for EF-Tu, indicating that there may be synergistic effects between alterations in the ribosomal protein S12 and the 2660-region of the 23S rRNA. Tapprich and Dahlberg (1990) speculated that these doubly altered ribosomes may have an increased affinity for ternary complexes, and that such an enhanced stability may explain their anomalous EF-Tu binding as well as how they cause growth inhibition or cell death. An alternative explanation was offered by Noller (1991), who suggested that the rate of GTP-hydrolysis on ternary complex may be slowed down by the combined effect of changes in S12 and the G to C substitution at position 2661. When the latter interpretation was made it had little support from experimental data, but our findings that hyperaccurate ribosomes altered in S12 always have significantly reduced rates of GTP-hydrolysis (Bilgin et al., 1992; above) support Noller's proposition. If the 2661G to C transversion by itself tends to slow down the rate of GTP-hydrolysis on EF-Tu, this may not be so dangerous when S12 is wild type. When S12 is altered, however, so that the rate of this kinetic step is already impaired, then the combined effect of those two mutations may become significant.

One important property of hyperaccurate ribosomes altered in S12 is the reduction in k_{cat}/K_M-value for their interaction with ternary complex (Ruusala et al., 1984; Diaz et al., 1986; Bilgin et al., 1992). Another important feature is that addition of Sm to these ribosomes restores their behaviour to almost wild type. We reasoned that if Sm restores the normal function of restrictive S12 mutants, the drug may also make restrictive S12/2661C double mutants viable. Indeed, with Sm in the medium, we were able to overcome the lethal effect of the 2661C mutation in a strongly restrictive S12 background and transform a hyperaccurate *rpsL* strain, FJU122 which is pseudodependent on Sm (SmP) (Zengel et al., 1977;Jörgensen and Kurland, 1992) with the pKK2661C plasmid (Bilgin and Ehrenberg, in preparation). We anticipated that detailed studies of the combined effects of this extreme S12 phenotype and the 2661G to C transversion would reveal interesting features of how rRNA and ribosomal proteins jointly define the functional properties of the ribosome. Accordingly we purified ribosomes from this transformed strain grown in the presence of Sm so that we could investigate four ribosome combinations: w.t./w.t., w.t./2661C, SmP/w.t. and SmP/2661C. These ribosomes were, furthermore, analyzed both in the presence and absence of Sm.

Our first discovery was that the efficiency of the ternary complex interaction with the ribosome (k_{cat}/K_M) is reduced due to the 2661C mutation, whether or not these ribosomes contain a restrictive S12 change. Secondly, we found that this reduction of k_{cat}/K_M, normally associated with increased accuracy (Kurland and Ehrenberg, 1984) was correlated with a reduction in the error frequency also for this w.t./2661C combination. This is the first direct evidence that a single base change in 23S rRNA can modulate translational accuracy. An interesting question was now whether this reduction in missense errors was caused by initial selection or proofreading. If the interaction between EF-Tu and the 2660-region is weakened, one would expect enhanced accuracy in the initial selection according to the hybrid site model. Alternatively, there may be a perturbation of the tRNA-rRNA contact. In this case the hybrid site model would predict a change in proofreading instead, since the 23S rRNA-tRNA interaction is not established until the initial selection is over (Noller, 1991).

Comparisons of the proofreading parameters f_c and f_w for wild type ribosomes with or without the 2661C substitution showed that the RNA sequence change only affects the accuracy in the initial selection of ternary complex (Bilgin and Ehrenberg, in preparation). Our result is thus compatible with the idea that the 2661G to C transversion perturbes the interaction between the protein part of the ternary complex and the ribosome, but other models cannot be excluded.

One primary concern of our study was to explain why the SmP/2661C combination is lethal in the absence of Sm. The succesful transformation of a restrictive SmP strain by the pKK2661C plasmid in the presence of Sm showed that we had discovered a new type of SmD. This suggested that ribosomes from the SmP/2661C strain may in fact be functionally similar to SmD ribosomes perturbed in S12 (Ruusala et al., 1984). One dramatic change associated with such SmD ribosomes is their significantly reduced rate of GTP hydrolysis on EF-Tu when the ternary complex is in the A-site (Bilgin et al, 1992). Our experiments showed, indeed, that the rate of GTP-hydrolysis for SmP/2661C ribosomes is considerably slower than for the SmP/w.t. variant. This is in accord with Noller's (1991) prediction, but there are other effects of the 2661G to C transversion in this S12 background. The increase in the average peptide bond formation time (c.f. Bilgin et al., 1992) by the 2661G to C transversion is so large that it cannot be explained by the decreased rate of GTP hydtrolysis only. We concluded that one or several steps after GTP hydrolysis leading to and including peptidyl transfer, must also be slowed down by the 2661C mutation. This means that either the rate of EF-Tu·GDP release from the ribosome or the rate of peptidyl transfer is impaired.

If the rate of release of EFTu·GDP from SmR/2661C (as SmP/2661C) ribosomes is significantly slowed by the 2661G to C change, this may explain the apparently high affinity between EF-Tu and the SmR/2661C combination as observed by Tapprich and Dahlberg (1990). This explanation has strong support in the dramatic increase that we observe for the time that EF-Tu spends to complete one round of its cycle for the SmP/2661C ribosomes. This slow cycling of EF-Tu is perfectly compatible with the idea that the rate of release of EF-Tu·GDP is very slow for the SmP/2661C ribosome variant. We therefore suggest that it is not an increased affinity between the ternary complex and the ribosome, but rather of EF-Tu·GDP and the ribosome, that explains the EF-Tu-ribosome complexes observed by Tapprich and Dahlberg (1990). Both the k_{cat}/K_M effect and the reduction in EF-Tu cycle rate are reversed to near wild type levels by the addition of Sm, and this, we think, explains the Sm dependence of these ribosomes.

The higher accuracy for w.t./2661C ribosomes in relation to w.t./w.t. ribosomes, due to enhanced initial selectivity, is accompanied by a reduction in k_{cat}/K_M for the interaction between these ribosomes and cognate ternary complex as mentioned above. Assuming that the d-value, i.e. the maximum possible discrimination in the first selection step, is not changed by the RNA alteration one can calculate its value from this type of experiments. We obtain a d-value of about 400 in this case (Bilgin and Ehrenberg, in preparation), which compares well with the earlier estimate of d=1000 by Thompson and Dix (1982) obtained for a different system and at lower temperature.

The SmP/w.t. ribosome is also more accurate than the w.t./w.t. variant and has a lower k_{cat}/K_M-value for its cognate ternary complex interaction. A significant part of the accuracy increase is also in this case due to the initial selection, and a calculation similar to the previous one gives a d-value close to 400. If these d-estimates are true, it means that wild-type ribosomes under our *in vitro* conditions only use about 20% of the maximum selectivity of the initial discrimination step (I=60, d=400). It also means that the reduction in k_{cat}/K_M due to losses of cognate ternary complex from the A-site is smaller than 20%. These calculations are indirect. They assume, first, that proofreading does exist and that the d-value itself is not perturbed by either the rRNA, or the protein change. The fact that the results are so similar indicates however, that the procedure may be justified (M. Ehrenberg, in preparation).

One remarkable outcome of the error experiments is that introduction of the 2661G to C transversion in SmP ribosomes in the presence of Sm leads to an error increase instead of a decrease. This error increase is, furthermore, due to a reduction of the accuracy in the proofreading step while the initial selectivity is unchanged (Bilgin and Ehrenberg, in preparation).

This clearly shows that the effects of the 2661G to C change depend qualitatively on the ribosomal context. It is therefore unlikely that the effects of the 2660 region on ribosome function can be understood without thorough investigations of its subtle interplay with ribosomal proteins and antibiotics. Further support to the notion of an intricate interplay between the 2660 region and S12 comes from the observation that when the 2661G to C change is introduced in SmP ribosomes, the error drops. It decreases, however, in this case by enhanced proofreading rather then by an increased initial selection.

To summarize, our results show that the 2661G to C change enhances the accuracy in the initial selection step in one context (wild type). In other contexts (SmP) the same base change lowers (+Sm) or enhances (-Sm) the accuracy in the proofreading step.

ACKNOWLEDGMENTS

This work is supported by Swedish National Science Research Council.

REFERENCES

Allen, P.N., and Noller, H.F., 1991, *Cell* 66:141.

Andersson, D.I., Andersson, S.G.E., and Kurland, C.G., 1986, *Biochimie* 68:705.

Bilgin, N., Claesens, F., Pahverk, H., and Ehrenberg, M., 1992, *J. Mol. Biol.* 224:1011.

Blomberg, C., Ehrenberg, M., and Kurland, C.G., 1980, *Quart. Rev. of Biophys.* 13:231.

Bohman, K., Ruusala, T., Welenc, P.C. and Kurland, C.G., 1984, *Mol. Gen. Genet.* 198:90.

Crick, F.H.C., 1968, *J. Mol. Biol.* 38:367.

Diaz, I., Ehrenberg, M., and Kurland, C.G., 1986, *Mol. Gen. Genet.* 202:207.

Ehrenberg, M., and Blomberg, C., 1980, *Biophys. J.* 31:333.

Ehrenberg, M., and Kurland, C.G., 1984, *Quart. Rev. of Biophys.* 17:4582.

Ehrenberg, M., Rojas, A.-M., Weiser, J., and Kurland, C.G., 1990, *J. Mol. Biol.* 211:739.

Gavrilova, L.P., Kostiashkina, O.E., Koteliansky, V.E., Rutkevitch, N.M., and Spirin, A.S., 1976, *J. Mol. Biol.* 101:537-552.

Gavrilova, P., Spirin, A.S., 1974, *Methods. Enzymol.* 30:452.

Gorini, L., 1971, *Nature New Biol.* 234: 261.

Gorini, L., Jacoby, G.A., Breckenridge, L., 1966, *Cold Spring Harbor Symp. Quant. Biol.* 26:173.

Hopfield, J.J., 1974, *Proc. Natl. Acad. Sci. USA* 71:4135.

Hughes, D., Atkins, J.F., and Thompson, S., 1987, *EMBO J.* 6:4235.

Jörgensen, F., and Kurland, C.G., 1990, *J. Mol. Biol.* 215:511.

Kakniashvili, D.G., Smailov, S.K., and Gavrilova, L.P., 1986, *FEBS Letters* 196: 103

Kaziro, Y., 1978, *Biochim. Biophys. Acta* 505:95.

Kurland, C.G., 1978, *Biophys. J.* 22:373.

Kurland, C.G., and Ehrenberg, M., 1984, *Prog. Nucleic Acids Res. Mol. Biol.* 31:191.

Kurland, C.G., and Ehrenberg, M., 1987, *Annu. Rev. Biophys. Biophys. Chem.* 16:291.

Moazed, D., and Noller, H.F., 1989, *Nature* 342:142.

Murgola, E.J., 1985, *Ann. Rev. Genet.* 19:57.

Murgola, E.J., Hijazi, K.A., Göringer, H.U., and Dahlberg, A.E., 1988, *Proc. Natl. Acad. Sci.USA* 85:4162.

Nierhaus, K.H., 1990, *Biochemistry* 29:4997.

Ninio, J., 1974, *J. Mol. Biol.* 84:297.

Ninio, J., 1975, *Biochimie* 57:587.

Noller, H.F., 1991, *Annu. Rev. Biochem.* 60:191.

Noller, H.F., Hoffarth, V., Zimniak, L., 1992, *Science* 256:1416.

Ozaki, M., Mizushima, S., Nomura, M., 1969, *Nature* 222:333

Rheinberger, H.J., Sternbach, H., and Nierhaus, K.H., 1981, *Proc. Natl. Acad. Sci. USA* 76:5310-5314.

Ruusala, T. and Kurland, C.G., 1984, *Mol. Gen. Genet.* 198:100.

Ruusala, T., Andersson, D., Ehrenberg, M., and Kurland, C.G., 1984, *EMBO J.* 3:2575.

Ruusala, T., Ehrenberg, M., and Kurland, C.G., 1982, *EMBO J.* 1:741.

Shen, Z., and Fox., T.D., 1989, *Nucl. Acids Res.* 17:4535.

Spirin, A.S., 1985, *Prog. Nucleic Acid Res. Mol. Biol.* 32:75.

Tapio, S., and Isaksson, L.A., 1991, *Eur. J. Biochem.* 202:981.

Tapprich, W.E., and Dahlberg, A.E., 1990, EMBO J. 9:2649.

Thompson, R.C., and Dix, D.B., 1982, *J. Biol. Chem.* 257:6677.

Tubulekas, I., and Hughes, D., 1992, Manuscript.

Thompson, R.C., and Stone, P.J., 1977, *Proc. Natl. Acad. Sci. USA* 74:198.

Weiss-Brummer, B., and Hüttenhofer, A., 1989, *Mol. Gen. Genet.* 217:362.

Zengel, J.M., Young, R., Dennis, P.P., and Nomura, M. (1977). *J. Bacteriol.* 129:1320.

TOWARDS ATOMIC RESOLUTION OF PROKARYOTIC RIBOSOMES:

CRYSTALLOGRAPHIC, GENETIC AND BIOCHEMICAL STUDIES

François Franceschi#, Shulamith Weinstein*, Ute Evers°, Evelyn Arndt#, Werner Jahn^, Harly A.S. Hansen°, Klaus von Böhlen°, Ziva Berkovitch-Yellin*°, Miriam Eisenstein*, Ilana Agmon*, Jesper Thygesen°, Niels Volkmann°, Heike Bartels°, Frank Schlünzen°, Anat Zaytzev-Bashan*, Ruth Sharon*, Inna Levin*, Alex Dribin*, Irit Sagi*, Theodora Choli-Papadopoulou+, Paraskevi Tsiboli+, Gitay Kryger*, William S. Bennett° and Ada Yonath°

#Max-Planck-Inst. for Mol. Genetics, Berlin,FRG; *Dept. Struct.Biol., Weizmann Inst., Rehovot,Israel; °Max-Planck-Lab. for Ribosomal Structure, Hamburg, FRG; ^Max-Planck-Inst. for Medical Res., Heidelberg, FRG; +Aristotelian Uni., Chem. School, Thessaloniki, Greece

1. INTRODUCTION

The studies reported here were initiated and inspired by the late Prof. H.G. Wittmann. From the early stages of this project, when it was widely believed that even the initial steps in determining the molecular structure of ribosomes are impossible, until his last days, Prof. Wittmann was actively involved in the experimental design and in the actual studies. We have no doubt that without his motivation, optimism, guidance and support, this project would not have reached its current stage.

In this chapter we describe the progress since the last "Ribosome Meeting" (summarized by Hill et al., 1990). Of particular significance are: the growth of improved ribosomal crystals, which diffract to almost atomic resolution; the collection and the evaluation of X-ray and neutron crystallographic data of high quality; the combination of metallo organic biochemistry, genetic manipulations and functional studies, which led to quantitative labeling of ribosomal crystals with an improved form of a multi-heavy-atom cluster (undecagold); the crystallization of thermophilic ribosomes trapped in defined functional states; the quantitative binding of the undecagold cluster to tRNA^{phe} in a fashion which does not hamper its recognition by its synthetase and its binding to ribosomal particles; the preparation and crystallization of core ribosomal particles lacking one or a few selected proteins; the progress in amino acid sequencing of the halophilic and thermophilic ribosomal proteins; the genetic insertion of -SH reactive groups to the surface of the halophilic ribosome; the hybridization of cDNA oligomers, which are complementary to exposed single strands of rRNA of halophilic and thermophilic ribosomes; the purification and characterization of a ribonucleoprotein complex from halophilic ribosomes; the growth of highly ordered two-dimensional arrays of ribosomal particles, which may be investigated by electron microscopy without staining; and new suggestions for alternative assignments of functional sites in the electron-microscopical reconstructed models.

2. CRYSTALLOGRAPHY

I. Crystals of Ribosomal Particles Diffract to 2.9 Å Resolution

The most striking recent achievement is the growth of crystals of 50S subunits from *Haloarcula marismortui*, which have a low mosaic spread and improved mechanic strength, yielding usable data to almost atomic resolution (Fig. 1). This high internal order was most unexpected crystallographically and in view of the apparent sequence heterogeneity, in the two genes encoding the rRNA of this subunit (Mylvaganam and Dennis, 1992). These crystals grow in solutions mimicking the halophilic intracellular environment in the presence of a few mM Cd^{++} (von Böhlen et al., 1991). The exceptional effect of the Cd^{++} is noteworthy, as crystals grown under the same conditions but without it diffracted nominally to 4.5 Å, but due to their high mosaicity and tendency for fragmentation, their X-ray crystallographic data were useful to only 7-9 Å resolution.
Considerable difficulties were encountered in efforts to crystallize halophilic 70S ribosomes, most probably due to their marked tendency to dissociate. To avoid dissociation the halophilic 70S ribosomes were kept at 1.9 M ammonium sulfate. Initial indications for the growth of microcrystals of the halophilic 70S ribosomes were obtained by investigating crystallization mixtures in the electron microscope. In addition, crystals of reasonable size were recently grown from halophilic 30S subunits.

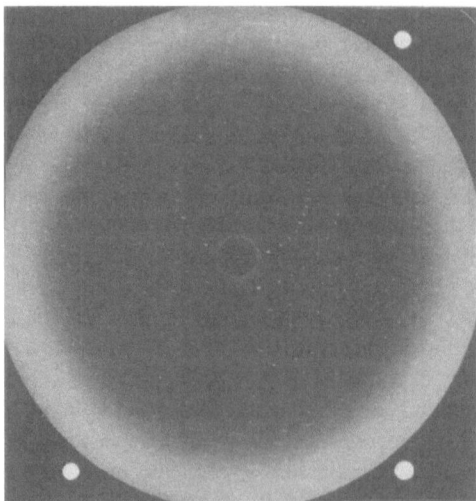

Figure 1. A rotation photograph of a crystal of 50S subunits from *H. marismortui* grown as described in von Böhlen et al., 1991. The pattern was obtained at 90 K at Station F1/CHESS, operating at 5.3 Gev and 50 mA. Crystal to film distance=220 mm, collimator=0.1 mm; wave length = 0.9091 Å.

II. Specific Concerns in Data Collection and Evaluation

Due to the weak diffracting power and the large unit cells of the crystalline ribosomal particles (TABLE I), virtually all the crystallographic studies have to be performed with intense synchrotron radiation. To eliminate the extreme radiation damage of these crystals, data are collected at cryogenic temperatures (about -180°C) from shock frozen crystals. Under these conditions, the ribosomal crystals diffract with no observable decay for periods longer than needed for collecting a full set of data (typically a few days), and the irradiated crystals can be stored for months. To facilitate cryogenic data collection we designed experimental procedures which accommodate the special features of the ribosomal crystals, i.e. fragility, sensitivity, thin edges, etc. (Berkovitch-Yellin et al., 1993). Data are collected using the screenless rotation method. The weak diffraction power of ribosomal crystals dictates extremely long exposure times (8-24 min./deg.), resulting in high background levels, which limit the maximum oscillation range. On the other hand, due to the mosaicity, which even for our best crystals is still significant (0.3-1.0°), large rotation ranges are desirable for obtaining a sufficient number of fully recorded reflections, required for internal scaling. A third factor influencing the data collection strategy is the extremely steep falloff of intensities as a function of resolution. Thus, under conditions optimized for collecting the high resolution reflections, which are the weakest, the low-resolution reflections are overexposed, even when using Imaging Plates, which have a large dynamic range. These contradictory requirements dictate data collection in several resolution shells, a procedure which became feasible only after introducing data collection at cryogenic temperature.

TABLE I. PARAMETERS OF RIBOSOMAL CRYSTALS

Source	Grown From*	Cell Dimensions (Å)	Resolution (Å)
70S T.t.	MPD	524x524x306; $P4_12_12$	20
70S T.t. complexed +	MPD	524x524x306; $P4_12_12$	15
30S T.t	MPD	407x407x170; $P42_12$	7.3
50S H.m.°	PEG	210x300x581; $C222_1$	2.9
50S T.t.	AS	495x495x196; $P4_12_12$	15
50S B.st.^	A	360x680x920; $P2_12_12$	18
50S B.st.^#	PEG	300x547x384; 114°; C2	11

B.st=*Bacillus stearothermophilus*; T.t.= *Thermus thermophilus*; H.m.=*Haloarcula marismortui*

* crystals were grown by vapor diffusion in hanging drops from solutions containing methyl-pentane-diol (MPD), polyethyleneglycol (PEG), ammonium sulfate (AS) or low molecular weight alcohols (A). For exact conditions see (Berkovitch-Yellin et al., 1992).

Reported is the "Useful resolution": the limit at which a significant amount (above 40%) of data could be properly evaluated. The "potential resolution" is the highest Bragg spacing at which sharp diffraction spots are consistently observed.

+ A complex of 70S ribosomes, two molecules of phetRNA[phe] and an oligomer of 35 uridines.

° Same form and parameters for crystals of 50S lacking protein HmaL11.

^ Same form and parameters for crystals of 50S lacking protein BL11 and for modified particles with an undecagold-cluster.

Same form and parameters for crystals of a complex of 50S subunits, with one tRNA molecule and a segment (18-20 mers) of a nascent polypeptide chain.

The high resolution diffraction data sets contain hundreds of thousands reflections. For example, 1,434,786 reflections were measured for the shell 10-3.5 Å from crystals of the halophilic 50S subunits, yielding 145,249 unique intensities. It was found that the evaluation of the ribosomal data is still not a routine task, although some special computational tools have been developed specifically for this purpose. Despite these difficulties, the evaluated data are of quality comparable to that obtained from crystalline proteins of average size. Thus, for above 50% completeness, the typical values for R-merge (I) at relatively high resolutions (6 Å) are in the range of 5-10% (Berkovitch-Yellin et al., 1993).

3. PHASING STRATEGIES

The phasing of X-ray amplitudes which enables calculation of electron density maps remains the most difficult part in crystal structure determination. The classical methods for phasing are MIR and SIR (Multiple and Single Isomorphous Replacement) which require the specific binding of heavy atoms at limited number of sites within the unit cell. Both methods are based on differences in intensities of the reflections of native and derivatized crystals, thus are dependent on close to ideal isomorphism. Due to the large size and the internal flexibility of ribosomal particles, achieving ideal isomorphism is doubtful. In some instances there is an apparent isomorphism. In others, a significant nonisomorphism is evident even between native crystals from the same batch. Currently it is not clear whether the variability in unit cell dimensions is an inherent property or induced by the cooling. Whether apparent or real, the lack of isomorphism complicates phasing by difference methods. However, successive exposures at different wave lengths of crystals derivatized by compounds with a significant anomalous scattering component may facilitate phasing by MAD (Multi-wavelength Anomalous Diffraction). Attempts in this direction are underway.

I. Preparation of Heavy Atom Derivatives: an Improved Gold Cluster

Heavy atom derivatives are obtained by introducing electron-dense compounds to the crystalline lattice at distinct locations. Conventional derivatization is achieved either by soaking the crystals in solutions of, or by co-crystallization with electron-dense compounds. For proteins of average size, useful isomorphous derivatives consist of one or a few heavy-metal atoms. For ribosomal crystals much heavier compounds are required. To reach the required electron-density and to achieve quantitative and site specific derivatization, we developed a monofunctional gold cluster (GC, m.w.=6200 Da), with a core of 11 gold atoms linked directly to each other (Weinstein et al., 1989). Simulation studies have indicated the potential phasing power of this cluster.

Maleimido derivative   Iodoacetamide derivative

Figure 2. The new short arms of the gold cluster

The first targets for binding the cluster to the ribosomes were exposed free sulfhydryls. To enable the binding of the gold cluster, we attached to it a short chemically reactive arm of about 3 to 4 Å in length (Fig. 2) with either a maleimido or iodoacetyl moieties at its end. The latter was designed to avoid the chirality introduced by the reaction with the double bond of the maleimido moiety. The improved cluster was quantitatively bound to base 47 of tRNAphe of E. coli (see 5.I) and to natural or genetically engineered sulfhydryl groups of isolated ribosomal proteins, which should subsequently be incorporated into core particles lacking them, in a fashion similar to that reported previously (Weinstein et al., 1989, 1992; Yonath et al., 1990).

II. Crystals of Halophilic Ribosomal Cores, Lacking Selected Proteins

Four ribosomal proteins were selectively detached from the 50S subunits from *H. marismortui* by dioxane. This approach was chosen due to the resistance of archaebacteria, to most antibiotics used for mutant selection (Amils et al., 1990). All the removed proteins were fully reconstituted. One of them, HmaL11, binds reagents specific to -SH moieties, but it was found that once its cysteine is modified, the protein could not be incorporated into core particles. This way we obtained 50S subunits, lacking protein HmaL11 which crystallize under the conditions used for the native subunits, and diffract to 10 Å resolution. We also crystallized 50S particles from which proteins HmaL1, HmaL11 and HmaL12 were removed by chemical methods.

The crystals of the depleted particles may be useful for locating the sites of the missing proteins at low resolution, thus providing an anchor for initial interpretation (see 6).

III. Inserting exposed sulfhydryls into the surface of the halophilic ribosomes

```
                11                              33
    1 AGTIEVLVPG G EANPGPPLG PELGPTPVDV QA V VQEINDQ TAAFDGTEVP VTVKYDDDGS
                C                              C

   61 FEIEVGVPPT AELIKDEAGF ETGSGEPQED FVADLSVDQV KQIAEQKHPD LLSYDLTNAA

                127                142
  121 KEVVGT C TSL GVTIEGENPR E F KERIDAGE YDDVFAAEAQ A
                S                  C
```

mutated HmaL11 protein	(wt)	(1)	(2)	(3)	(4)
reconstitution into minus L11 ribosomes	+	+	+	+	+
binding of NEM to the isolated protein	+	-	+	+	+
reconstitution of NEM-labeled protein into minus L11 ribosomes	-	-	-	-	+

Figure 3. Genetic insertions of cysteines into protein HmaL11. Amino acid sequence of protein HmaL11. Four different mutant proteins were produced, namely (1) HmaL11mut11, (2)HmaL11mut11/127, (3) HmaL11mut11/127/142 and (4) HmaL11mut11/33/127. The binding of NEM to the different proteins and its reconstitution into ribosomes lacking L11 was tested.

As the halophilic r-proteins HmaL1 HmaL11 and HmaL12, can be removed from the core particles and as their genes have been cloned (Arndt and Weigel, 1990), the insertion of -SH groups into the surface of the ribosome became feasible. Several mutants were produced by oligonucleotides directed in-vitro mutagenesis. The cys codon of the wild type HmaL11 was exchanged to a ser codon, and codons specific for cys were inserted into different positions. A similar procedure was applied to HmaL1, except for the necessity to eliminate natural cysteine, since HmaL1 does not contain any. The mutated genes were cloned into the vector (pet11d) and the mutated proteins were overexpressed in E. coli. These mutated, overexpressed proteins could be reconstituted into ribosomal cores lacking them. Screening of the suitability of the mutated proteins for binding heavy atom clusters, is in progress, and indications for the incorporation of protein HmaL11 in which amino acid 34 is a cysteine, into core 50S particles, have been obtained (Fig. 3).

For further insertions of exposed cysteines in other halophilic r-proteins we exploit the advanced stage of genetic sequencing (see 6.III), the determination of conditions for reconstitution of halophilic ribosomes (Sanchez et al., 1990), the mapping of the surface of the halophilic ribosome by limited proteolysis (Kruft and Wittmann-Liebold, 1991) and by the determination of exposed -SH groups as a function of the ionic strength (Weinstein et al., 1989; Sagi and Weinstein, to be published).

IV. Other Phasing Methods

The molecular replacement method is based on the search for best positioning of a model of the investigated compound in the crystallographic unit cell. The low-resolution models of 70S and 50S ribosomal particles from *B. stearothermophilus* (Sec. 4 and in Yonath et al., 1987; Arad et al., 1987; Berkovitch-Yellin et al., 1990), are being employed for these searches (Eisenstein et al., 1991), together with X-ray crystallographic data collected from three-dimensional crystals of ribosomal particles from *T. thermophilus* and *H. marismortui*. The use of reconstructed images of ribosomal particles of one bacterium with data obtained from crystals of ribosomes from another one, is based on the assumption that at the resolution limits of the reconstructions the gross structural features of prokaryotic ribosomes are similar. Parallel phasing attempts, exploiting maximum entropy, gas condensation and other computational methods for phasing, have also yielded encouraging results (N. Volkmann, M. Roth, E. Pebay-Peroula, S. Subbiah, M. Eisenstein, D. Rabinovitch, Z. Berkovitch-Yellin, A. Zaytzev-Bashan, unpublished).

4. ELECTRON MICROSCOPY AND IMAGE RECONSTRUCTION

In 1987 we reconstructed models for the 50S subunit and 70S ribosomes at 28-47 Å, using electron micrographs of tilt series of two-dimensional sheets (Yonath et al., 1987: Arad et al., 1987). Despite their low resolution, we detected in these models several key features, most of which are associated with internal vacant spaces or partially filled hollows, not observed earlier in prokaryotic ribosomes. The significant similarities in specific features in the reconstructed models were used to assess their reliability and to tentatively assign biological functions to some structural features.

I. The Original Interpretation of the Reconstructed Models

The existence of an internal tunnel in large ribosomal subunits was suggested more than two decades ago and further substantiated recently, as a result of several biochemical

402

experiments which showed that the ribosome masks the newly synthesized protein chains (for review see Yonath and Wittmann, 1989; Yonath et al., 1990; Berkovitch-Yellin et al., 1990). So far this tunnel was revealed only by diffraction studies. It was first observed as a narrow elongated region of low density in images reconstructed at very low resolution from 80S ribosomes of chick-embryos (Milligan and Unwin, 1986). As our image reconstruction studies were performed at somewhat higher resolution (28 Å), they resulted in a more precise description, albeit still suffering from the inherent shortcomings of electron microscopy at ambient temperature (Yonath et al., 1987; Berkovitch-Yellin et al., 1992). An almost identical model, including an internal tunnel, was recently reconstructed from two-dimensional sheets of 50S particles from *T. thermophilus*, obtained on lipid mono-layers, and investigated at cryo-temperature (Yoshinori Fujiyoshi, personal communication). In addition, a low density region was observed in features seen in the density map obtained from three-dimensional crystals of halophilic 50S subunits at 30 Å resolution, using neutron diffraction data, phased by direct methods (Eisenstein et al., 1991).

Image reconstructions using diffraction data from several tilt series of two-dimensional sheets of 70S ribosomes gave rise to a few models, sharing several common features (Yonath and Wittmann, 1989, Yonath et al., 1990; Berkovitch-Yellin et al., 1990). The two most deviating ones differ mainly in their bulkiness (called below "thick" and "thin"). However both contain a few rather long tunnels. The most striking similarity between the two models is in their internal void, which accounts for about 20% of their volume. Steric considerations showed that this void is spacious enough to accommodate up to three tRNA molecules together with other non-ribosomal components participating in protein biosynthesis. Therefore, it is conceivable that this void is the location for the various enzymatic activities of protein biosynthesis.

As we had no reason to prefer one reconstruction on the other, we first focused on the "thin" one, since it is more detailed and therefore more restrictive. Interestingly, our original assignments in the "thin" model fit well also the "thick" one. Part of the 70S particle was tentatively identified as the large subunit, based on a visual and computational fit of the overall shapes as well as on the directionality of the longest tunnel. A distinct region rich in rRNA was revealed (by uranyl-acetate staining) within the part of the 70S ribosome left for the 30S subunit (Fig. 4 and in Yonath and Wittmann, 1989, Berkovitch-Yellin et al., 1990; Yonath et al., 1990). Similarly, a region of a high stain-density was detected by electron microscopical investigations of isolated 30S subunits (Oakes et al., 1990). As it is known that the ribosome masks stretches of 30-40 nucleotides of mRNA molecules (Kang and Cantor, 1985), we assigned as the mRNA path, a groove within this region compatible with the biochemical findings.

In a model-building experiment, a molecule of tRNA was placed in the intersubunit space with its anticodon close to the tentative mRNA binding site and its CCA-terminus positioned so that the peptidyl group could extend into the tunnel (Fig. 4). In this orientation the tRNA molecule may form a variety of interactions with the walls of the intersubunit space. It should be mentioned that at the current resolution the two crystallographically determined conformations of tRNA (for review see Moras, 1989) can be placed in the intersubunit space in a very similar manner.

II. Alternative Interpretations of the Reconstructed Models

Although we do not doubt the feasibility of our original interpretation, we have been stimulated to consider tentative assignments in a model-building exercise. Recently a model for the 70S ribosome from *E. coli* was reconstructed by averaging the electron microscopical shapes of a large number of single particles embedded in vitrified ice (Frank et al., 1991).

Despite the different technique and the use of ribosomes from a different source, this reconstruction led to a model almost identical in shape to our "thick" 70S model. Since this new reconstruction is based on investigation of unstained samples, the contrast between regions rich in rRNA and the rest of the particle could be observed. The assignments based on this information (Frank et al. 1991) are consistent with our suggestion that the intersubunit void provides the site of the peptidyl transferase reaction (Arad et al, 1987; Yonath and Wittmann, 1989; Berkovitch-Yellin et al., 1990; Yonath et al., 1990; Hansen et al., 1990).

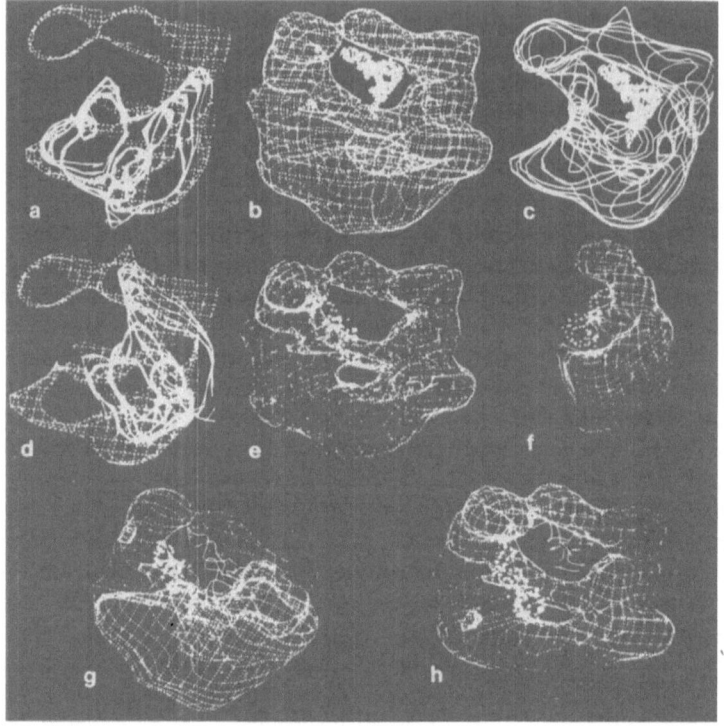

Figure 4. Functional assignments in reconstructed 70S ribosomes.

Top panel: The original assignments, based on the best fit in the direction of the long tunnel and in the shape of the 50S subunit and the part assigned to it in the "thin" reconstructed model of the 70S ribosome; (a) A slice of 50 Å in depth of the superposition of the reconstructed models of the "thin" 70S ribosome (shown as a net) and the 50S subunit (in full lines); (b) and (c) tRNA molecules, model built into the intersubunit free space in the "thick" and the "thin" models of 70S, respectively, with its CCA end pointing into the tunnel.

Middle Panel: The alternative assignments, based on the best fit in the direction of the shorter tunnel and on the distribution of mass within the ribosome (Frank et al, 1991); (d) A slice of 50 Å in depth of the superposition of the reconstructed models of the "thin" 70S ribosome (shown as a net) and the 50S subunit (in full lines); (e and f) Two orthogonal views of the reconstructed "thick" model of the 70S ribosome. A tRNA molecule is "Model built" into the intersubunit free space, with its CCA end pointing into the tunnel.

Bottom Panel: (g) and (h) Two views of the "thick" 70S ribosome. The intersubunit free space of each model contains two model-built tRNA molecules, each representing a different assignment. That placed according to the original interpretation is shown as a continuous line. The second, placed according to the alternative interpretation, is shown as a chain of plus (+) signs.

To assess the compatibility of the recent assignment with our original one, we attempted to fit our reconstructed model of the 50S subunits into the "thin" and the "thick" 70S models. We found that positioning the large subunit in the "thin" 70S model at any location other than our original one, leaves hardly any density for accommodating the small subunit, whereas the "thick" model possesses sufficient density for the 30S subunit in almost any orientation of the 50S model. However, even the best alignment between the 50S and the "thick" ribosome, based on the direction of the shorter tunnel, suffers from rather poor fit between the shapes of the 50S and the part assigned to it on the 70S ribosome, leaving several extensive regions of unmatched density. Remarkably, the density assigned to the 30S subunit contains a groove, in a region poor in rRNA with dimensions suitable to host a stretch of 30-40 nucleotides of mRNA. Like the groove described in our original interpretation (Arad et al, 1987; Yonath and Wittmann, 1989; Berkovitch-Yellin et al., 1990), the newly assigned one is positioned so that a molecule of tRNA can bridge between it and the entrance to the tunnel in the 50S subunit (Fig. 4) while interacting with the walls of the intersubunit space. Interestingly, a third interpretation, which utilizes the intersubunit void for the peptidyl transferase was recently suggested (Lim et al., 1992).

The fact that our original suggestion to assign the intersubunit void as the center of the biosynthetic process became widely accepted is most satisfying, although it is clear that neither of the reconstructions is of sufficient resolution to provide an unambiguous assignment. Perhaps the main lesson from these model building exercises is that the intersubunit void is sufficiently large, so that its ability to accommodate the components of protein synthesis in a sterically reasonable arrangement does not critically depend on the assignment of the small and the large subunits. It is evident that a more detailed interpretation may be feasible only after the maturation of further diffraction studies at higher resolution. Attempts in this direction are currently underway. High quality ordered arrays of ribosomal particles, which diffract to 12-15 Å resolution and can be studied unstained in vitrified ice, have been recently obtained (W. Chiu, M. Schmidt, T. L. Guan, T. Arad, A. Yonath, J. Piefke and F. Franceschi, to be published).

5. FUNCTIONAL EXPERIMENTS INFLUENCING STRUCTURAL STUDIES

Despite the uncertainties in our assignments, the detection of internal features in the ribosome stimulated further functional and structural investigations. As a result it was shown that ribosomes mask nascent natural proteins more efficiently than artificial homo-polypeptides (Evers and Gewitz, 1989; Kolb et al., 1990) and that the latter may choose an exit path slightly different than that of naturally occurring proteins (Hardesty et al., 1990). It is conceivable that a common feature located at the amino termini of natural proteins (e.g. fMet for prokaryotes) has a role in guiding the nascent protein chain into the tunnel. A failure to enter the tunnel may influence the biosynthetic machinery at early stages and lead to the termination of the process. This hypothesis may explain why usually only 40-60% of well prepared ribosomes are active in in vitro production of relatively long polypeptides, although almost all ribosomes bind quantitatively mRNA and tRNA (Rheinberger and Nierhaus, 1990) and why short and long chains of newly synthesized polylysine or polyphenylalanine migrate in different modes (Hardesty et al., 1990).

I. Crystals of complexes of ribosomal particles in defined functional states

(a) **Crystals of complexes of 50S subunits:** To investigate the chemical properties of the exit path of nascent polypeptide chains, we obtained small three-dimensional crystals and two-dimensional sheets of reasonable-size from 50S subunits of either *H. marismortui* or *B. stearothermophilus*, with a short polyphenylalanine or polylysine (8-18 amino acids) and one molecule of their cognate tRNA (Gewitz et al., 1988; Müssig et al., 1989). Such complexes may be used for illuminating the exit path of the nascent protein, as well as for derivatization, providing the possibility to label the nascent chain with heavy atoms (see 6).

(b) **Crystals of complexes of 70S ribosomes:** We assumed that the intersubunit space contributes to the motion of the ribosome, allowing for the dynamics involved in the biosynthetic process. At the same time it contributes to the poor internal order of the crystals of 70S ribosomes. To minimize the flexibility and to increase the homogeneity of the crystallized material, complexes composed of 70S ribosomes from *T. thermophilus*, with one or two Phe-tRNA^{phe} molecules and chains of about 9 or 35 uridyl residues were crystallized (Yusupova et al., 1991; Hansen et al., 1990). In this complexes the intersubunit space may have a limited motional freedom, as a result of trapping the ribosomes into relatively rigid conformations and the mRNA is of a length which may fit into the groove in the 30S subunit, so that no long stretches of it project into the solvent.

Crystallographic data collected from the crystals of the complex with oligo(U)-35-mer showed that indeed the crystals of the complex are superior to those of isolated ribosomes, despite its non optimal composition (e.g. using a homopolynucleotide rather than mRNA chains of a designed sequence). We observed a dramatic improvement in the reproducibility in crystal growth and in the internal order of the crystals. Thus, the crystals of the complex diffract to a resolution of 15 Å, compared with about 20 Å, obtained for the best crystals of 70S ribosomes (Hansen et al., 1990; Trackhanov et al., 1989; Berkovitch-Yellin et al., 1991). It is conceivable that better crystals may be obtained by using mRNA chains of defined lengths and sequences. Therefore we synthesized mRNA chains of a length which can be accommodated within the intersubunit space and a sequence suitable for productive binding to thermophilic ribosomes, with elements promoting initiation of protein biosynthesis. As little is known about the functional aspects of thermophilic ribosomes, we are currently engaged in their characterization. Attempts to produce and crystallize similar complexes of halophilic 70S ribosomes are underway, and the purification of several halophilic tRNA molecules from *H. marismortui* has been initiated (Safro et al., 1992).

(c) **Quantitative labeling of tRNA:** Since tRNA is an essential part of the crystalline functional complexes, it may be exploited as a carrier of heavy-atom clusters. Base 47 of *E. coli* tRNA^{phe} is a naturally modified uridine nucleoside (ACP3U), containing an exposed reactive primary amino group. Iminothiolane was used for converting this amino group into a reactive sulfhydryl, which was used, in turn, for binding the gold cluster with its short or long arms. The resulting modified tRNA (tRNA^{phe}-GC) was visualized by dark field scanning transmission electron microscopy at cryogenic temperature. Using radioactive tRNA^{phe}-GC (Boeckh and Wittmann, 1991), it was established that the modified tRNA molecule can be aminoacylated by its cognate synthetase and binds to 70S ribosomes and to 30S subunits from *T. thermophilus* with the same stoichiometry found for native tRNA^{phe}, in the presence or the absence of poly(U) (Weinstein et al., 1992).

II. DNA Oligomers, Complementary to Exposed rRNA

Exposed single-strand rRNA segments which may be complemented by DNA oligomers (e.g., Weller and Hill, 1991) are currently being located on halophilic and thermophilic ribosomes. A variety of DNA oligomers, targeting naturally exposed rRNA regions as well as those which become exposed by removing selected ribosomal proteins, have been synthesized. These may be used for derivatization, as we prepared them with a thiol group at the 5' end and bound the gold cluster to it (J.Späthe, S.Hottenträger, B.Wittmann-Liebold and F.Franceschi, unpublished). The regions of rRNA which were targeted by us are: bases 1125-1158 of 23S RNA of the mutant lacking protein L11 from *B. stearothermophilus*, which, in the wild type are masked by protein L11; the last 14 nucleotides from the 3' end of 16S RNA (the vicinity of the "Shine Dalgarno" position) of *T. thermophilus*; bases 2646-2667, the "α-sarcin binding site" on 23S RNA of *T. thermophilus* and *H. marismortui* and bases 1422-1432, the "thiostrepton binding site", of the latter.

6. TOWARDS THE INTERPRETATION OF CRYSTALLOGRAPHIC MAPS

Assuming successful phasing and the availability of the sequences, significant difficulties in the interpretation of the electron density maps of the ribosomal particles are expected. Our ability to insert reactive sites (e.g., -SH) at desired locations on selected ribosomal components provides means for inserting powerful markers, as the localization of the heavy atom in the electron density map should not only facilitate phasing, but also provide us with flags and markers for chain tracing.

We expect to interpret the electron density map in an iterative fashion. We shall first focus on the determination of low resolution (20-30 Å) envelopes. At this stage it should be possible to localize the heavy atom clusters and since their binding sites on the ribosome are known, also to localize the components to which they are bound. The subsequent medium resolution (8-20 Å) electron density map is likely to reveal several internal features, such as regions rich in rRNA. For the interpretation of conformational elements of the rRNA we hope to benefit from the extensive non-crystallographic information about the relative locations of several double helical segments and loop-out regions.

Additional markers should be provided from comparisons of three-dimensional images reconstructed from unstained two-dimensional sheets by comparing images of native, mutated, depleted and chemically modified two-dimensional sheets (see above, 4.II). At this stage we may also be able to distinguish between less dense RNA regions and proteins. A large volume of data concerning the proximities of rRNA to specific ribosomal proteins is also available. In addition, we plan to take advantage of the available information concerning the relative positioning of centers of mass of the ribosomal proteins (obtained by neutron scattering combined with triangulation) as well as of the locations of surface ribosomal proteins (obtained by immuno-electron-microscopy).

I. Localization of Specific Sites and Functional Centers

It is conceivable that once the envelope of the ribosome has been elucidated, for the determination of the higher resolution structure, more conventional heavy atom derivatives, containing 4-6 heavy metal ions will be useful. These may also provide a tool for the identification of specific functional elements, such as following the path of the emerging nascent protein by attaching heavy atoms to its cysteins. Furthermore, the non-ribosomal components of the crystallized functional ribosomal complexes (tRNA and mRNA) may be used not only for improving crystal quality, but also as carriers of the heavy atom clusters. The crystallographic localization of these compounds should also shed light on their mode of interaction with the ribosome.

II. Sequencing the Ribosomal Proteins from *H. marismortui* and *T. thermophilus*

Protein chemical methods coupled with cloning techniques were used for sequencing the ribosomal proteins from *H. marismortui*. So far the primary structure of 52 ribosomal proteins have been established (Arndt et al., 1991; Scholzen and Arndt, 1991; Krömer and Arndt, 1991; Scholzen and Arndt, 1992). In these studies advantage has been taken of the conservation of the cluster arrangement of a large number of ribosomal genes.

Comparison of the primary structures of ribosomal proteins offers a promising tool for establishing phylogenetic relationships between the archaea, eucarya and eubacteria (Fig. 4). These comparisons may also provide valuable information about important functional sites, since regions crucial for assembly or for the translational process are expected to be conserved throughout evolution. Partial results show that the halophilic r-proteins are in general more similar to their eukaryotic than to the eubacterial counterparts, although some proteins are exclusively related to their eubacterial homologs. Interestingly, some of the halophilic r-proteins do not show similarities to any other proteins which have been sequenced so far.

Fifteen proteins of the 30S subunits from *T. thermophilus* have been partially sequenced and found to be of a high homology but different electrophoretic mobilities than those of their counterparts form *T. aquaticus, B. stearothermophilus* and *E. coli*. As the preferred nomenclature relates to sequence homologies, the original numbers assigned to some 30S r-proteins had to be reassigned (e.g. TthS14 is located on two-dimensional gels in the position of E. coli S21).

A low molecular weight protein containing 26 amino acids, of which 13 are basic and only one acidic, was detected in the 30S subunits. Interestingly, a homologous protein was detected in ribosomes from spinach chloroplast (Schmidt et al., 1992), but not in any eubacterial ribosome (Choli-Papadopoulou et al., 1992).

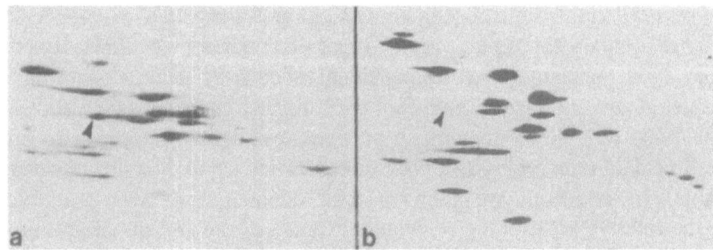

Figure 5. Two-dimensional gel electrophoresis of two samples of r-protein extracted from 50S subunits from *H. marismortui*. (a) Total proteins of the 50S subunits (the arrow points at protein HmaL1). (b) The fraction left after the separation of the HmaL1-23S RNA complex (the arrow points at the position of HmaL1).

III. The Isolation of in-situ Ribonucleoprotein Complexes

The knowledge of the accurate atomic structures of individual ribosomal components may be of instrumental assistance in the determination of the structure of ribosomes using the molecular replacement method. Therefore we purified a ribonucleoprotein complex with a defined composition, high stability and relatively low molecular weight, which may retain its native conformation.

After establishing a quantitative procedure for separating selected proteins from the ribosomal core, an internal complex composed of 23S RNA and protein L1 from *H.*

marismortui was isolated and characterized (Fig. 5). The rRNA fragments protected from RNase A digestion by protein HmaL1 show two regions of a high homology to the corresponding region of E. coli rRNA, despite the evolutionary distance. Stable heterologous complexes could be formed between the halophilic 23S RNA chain and protein L1 from E. coli, and between the rRNA of E. coli and the halophilic protein (Evers et al., submitted). Interestingly, although these complexes contain a non-halophilic component, they form under halophilic conditions. Protein L1 is one of the ribosomal proteins which act as translational repressors by binding to their own mRNA. In this capacity, there are considerable differences between the two bacteria, although their genes are clustered in a similar fashion. The production of the halophilic complex is currently being scaled up, aiming at its crystallization.

7. CONCLUDING REMARKS

The results reviewed in this chapter demonstrate that crystallographic studies on intact, modified, complexed and mutated ribosomal particles are well underway. From the early stages of this work, it has been clear that a straight-forward application of conventional concepts and techniques of macromolecular crystallography would not be adequate. Therefore we have devised an approach that combines the exploitation of the extensive information available on the genetic, functional and chemical properties of ribosomes for a rational design of innovative protocols in crystallization and in specific derivatization with multi heavy-atom clusters, together with the extension of the existing techniques of X-ray crystallography, image reconstruction, electron microscopy, neutron diffraction and cryogenics. We do expect that our efforts will lead to an understanding of the many functions of the ribosome at the molecular level.

Acknowledgments

We thank H.S. Gewitz, T. Arad, I. Dunkel, Y. Halfon, C. Glotz, G. Thoms, S. Meyer, J. Piefke, J. Müssig, M. Laschever, B. Romberg, R. Hasenbank, A. Bruhnsen, C. Paulke, B. Donzelmann, S. Hottenträger, G. Idan and B. Dressler for excellent technical assistance and advice, and the staff of EMBL/DESY, CHESS, SSRL and KEK/PF for providing us with X-ray diffraction facilities and of ILL for facilitating neutron diffraction experiments. This work was supported by research grants from BMFT(MPBO 180), NIH (GM34360), DFG (Yo-11/1-2), DARA (50QV 86061), NCRD (334190) and the Kimmelman Center for Biomolecular Structure and Assembly. A.Y. holds the Martin S. Kimmel Professorial chair.

REFERENCES

Amils R., Ramirez L., Sanz J. L., Martin I., Pisabarro A.G., Sanchez E., and Ureña D.,1990, In Hill et al., 1990:645

Arad T., Piefke J., Weinstein S., Gewitz H.S., Yonath A. and Wittmann H.G., 1987, Biochimie, 69:1001

Arndt E. and Francheschi F. 1992, Abs. "The Translational Apparatus" Berlin, P. 86

Arndt E. and Weigel C., 1990, Nucleid Acid Res., 18:1285

Arndt E., Scholzen T., Krömer W., Hatakayama, T. and Kimura, M. 1991, Biochemie, 73:657

Boeckh T. and Wittmann H.G., 1991, Biochem. Biophys. Acta, 1075:50

Berkovitch-Yellin Z., Wittmann H.G. and Yonath A., 1990, Acta Cryst. B46:637

Berkovitch-Yellin Z., Hansen H., Bennett W.S., Sharon R., von Böhlen K., Volkmann N., Piefke J., Yonath A. and Wittmann H.G., 1991, J. Crystal Growth, 110:208

Berkovitch-Yellin Z., Bennett W.S. and Yonath A., 1992, CRC Rev. Biochem. & Mol. Biol. 27:403

Berkovitch-Yellin Z., Hansen H.A.S., Weinstein S., Eisenstein M., von Böhlen K., Agmon I., Evers U., Thygesen J., Volkmann N., Bartels H., Schlünzen F., Zaytzev-Bashan A., Sharon, R., Levine I., Dribin A., Kryger G., Bennett W.S., Franceschi F. and Yonath A., 1993, in "Synchrotron Radiation and Mol. Biol.", (N. Sakabe, Ed.) Oxford U. Press,

von Böhlen, K., Makowski I., Hansen H.A.S, Bartels H., Berkovitch-Yellin Z, Zaytzev-Bashan A., Meyer S., Paulke C., Franceschi F. and Yonath A., 1991, J. Mol. Biol. 222:11

Choli-Papadopoulou T., Wittmann-Liebold B. and Yoanth, A. Abs. "The Translational Apparatus" Berlin, 1992, P. 94

Eisensten M., Sharon R., Berkovitch-Yellin Z., Gewitz H.S., Weinstein S., Pebay-Peyroula E., Roth M. and Yonath A., 1991, Biochemie, 73:897

Evers U. and Gewitz H.S., 1989, Biochem. Internat., 19:1031

Frank J., Penczek P., Grassucci, R., and Srivastava, S., 1991, J. Cell Biology, 15:579

Gewitz H.S., Glotz C., Piefke J., Yonath A. and Wittmann H.G., 1988, Biochimie, 70:645

Hansen H.A.S, Volkmann N., Piefke J., Glotz C., Weinstein S., Makowski I., Meyer S., Wittmann H.G. and Yonath A., 1990, Bioche. Biophys. Acta 1050:1

Hardesty B., Picking W.D. and Odom O.W., 1990, Bioche. Biophys. Acta, 1050:197

Hill E.W., Dahlberg A. Garrett R.A., Moore P.B., Schlesinger D. and Warner J.R., 1990, ASM, Washington, USA, "The Ribosome: Structure, Function and Evolution"

Kang C. and Cantor C.R., 1985, J. Mol. Biol., 210:659

Kolb V.A., Kommer A., Spirin A.S., 1990, Workshop on Translation, Leiden, p.84a.

Koepke A.K.E., Paulke C., Gewitz H.S., 1990, J. Biol. Chem. 265:6436

Krömer W.J. and Arndt E., 1991, J. Biol. Chem. 266:24573

Kruft V. and Wittmann-Liebold B., 1991, Biochemistry, 30:11781

Lim V., Venclovas C., Spirin A., Brimacombe R., Mitchell P and Müller F., 1992, Nuc. Acid Res. 20:2627

Milligan R.A. and Unwin P.N.T., 1986, Nature, 319:693

Mylvaganam S. and Dennis P.P., 1992, Abs. "The Translational Apparatus" Berlin, P. 232

Moras D., 1989, Nucleic Acids. Springer Verlag. Berlin Heidelberg and NY. 1.

Müssig J., Makowski I., von Böhlen K., Hansen H., Bartels K.S., Wittmann H.G. and Yonath A., 1989, J. Mol. Biol. 205:619

Oakes M., Scheiman A., Atha T., Shakweiler G. and Lake J., 1990, In Hill et al., 1990:180

Rheinberger H.J., Nierhaus K.H., 1990, Europ J. of Biochem. 193:643

Safro M, Reshetnikova L., Goldschmidt-Raisin S. and Franceschi, F., 1992, Abst. "The Translational Apparatus" Berlin, P. 136

Sanchez M.E., Urena D., Amils R. and Londei P., Biochemistry, 1990, 29:9256

Schmidt J., Weglöhner W., Giese K., Schröder W. and Subramanian A.,1992, Abst. "The Translational Apparatus" Berlin, P. 225

Scholzen T. and Arndt E.,1991, Mol. Gen. Genet.228:70

Scholzen T. and Arndt E., 1992, J. Biol. Chem. 267:12123

Trackhanov S.D., Yusupov M.M., Shirokov V.A., Garber M.B., Mitschler A., Ruff M., Thierry J.C., Moras D., 1989, J. Mol. Biol. 209:327

Weller J., and Hill W.E., Biochemie, 1991, 73:971

Weinstein S., Jahn W., Hansen H.A.S., Wittmann H.G. and Yonath A., 1989, J. Biol. Chem., 264:19138

Weinstein S., Jahn W., Laschever M., Arad T., Tichelaar W., Haider M., Glotz C., Boeckh T., Berkovitch-Yellin Z., Franceschi F.and Yonath A., 1992, J. Cryst. Growth, 122:286

Yonath A., Leonard K.R. and Wittmann H.G., 1987, Science, 236:813

Yonath A. and Wittmann H.G., 1989, TIBS, 14:329

Yonath A., Bennett W., Weinstein S. and Wittmann H.G., 1990, in Hill et al., 1990:134

Yusupova G., Yusupov M., Spirin A., Ebel J.P., Moras D., Ehresmann C., Ehresmann B., 1991, FEBS Letters, 290:69

FUNCTIONAL SITE DETERMINATIONS IN THREE DIMENSIONS ON EUKARYOTIC AND EUBACTERIAL RIBOSOMES

Adriana Verschoor, Suman Srivastava, Michael Radermacher,
Joachim Frank,
Robert R. Traut[1],
Marina Stöffler-Meilicke[2], and
Dohn Glitz[3]

Wadsworth Center for Laboratories and Research,
New York State Dept. of Health, Albany, NY 12201

[1]Dept. of Biological Chemistry, School of Medicine,
University of California, Davis, CA 95616

[2]Institut für Klinische und Experimentelle Virologie,
Freie Universität Berlin, Berlin, Germany

[3]Dept. of Biological Chemistry, UCLA School of Medicine,
University of California, Los Angeles, CA 90024

INTRODUCTION

Immuno electron microscopy has been a powerful tool in efforts to map exposures of individual r-proteins and functional sites on the ribosome, in that it has provided direct visual identification of the contact between an antibody and its epitope, as seen in an electron micrograph of a single-particle preparation of ribosomes. However, the information has been only two-dimensional: the site is seen on a projection image of the particle. If it can be identified on several different projections -- representing different orientations of the ribosome -- some three-dimensional (3D) information can be deduced, but the localization can be only approximate until 3D reconstruction techniques are applied to the problem.

3D functional site determination proceeds in two steps. First of all, the structure of the unlabeled ribosome must be determined, with greater accuracy than the models derived from visual synthesis of the various projections of the ribosomal particle. Secondly, the labelled ribosome must be reconstructed with the same technique, so that the precise location of the ligand can be determined in a difference map.

The Translational Apparatus, Edited by K.H. Nierhaus
et al., Plenum Press, New York, 1993

Through the use of image processing techniques pioneered by our group, we have achieved 3D reconstructions of the eubacterial 70S ribosome (Frank et al., 1991) and its 50S large subunit (Radermacher et al., 1987a,b), and the eukaryotic 80S ribosome (Verschoor and Frank, 1990) and its 40S small subunit (Verschoor et al., 1989). The random conical reconstruction technique (Radermacher et al., 1986, 1987a, 1992), also termed the single-exposure conical reconstruction technique (SECReT), was first applied to the 50S subunit (Radermacher et al., 1987b). This method has allowed us to combine information from hundreds of particles of a ribosomal specimen, with known angles relating the projections, and with minimized radiation damage. The most fundamental requirement is that the sample contain a sufficient number of particles having identical structure.

Our reconstruction technique is based on the fact that most specimens of macromolecules show particles in one or more preferred orientations relative to the supporting carbon foil, and random orientations within the specimen plane. If such a specimen is tilted by a large angle, typically 50° to 60°, then the images of the single particles that can be seen in the micrograph form a conical tilt series, in which the cone angle is the angle by which the specimen is tilted, with random azimuthal angles describing the in-plane orientations of the particles. From such a conical tilt series the three-dimensional structure can be calculated. Because only a single image of each particle is needed, the specimen is exposed to a minimal electron dose. The main restriction of the method is that the particles must display some degree of preference for certain orientations on the grid: the more marked the preference, the fewer micrographs need be recorded and analyzed. When the particles lack such preferences, the method cannot easily be applied.

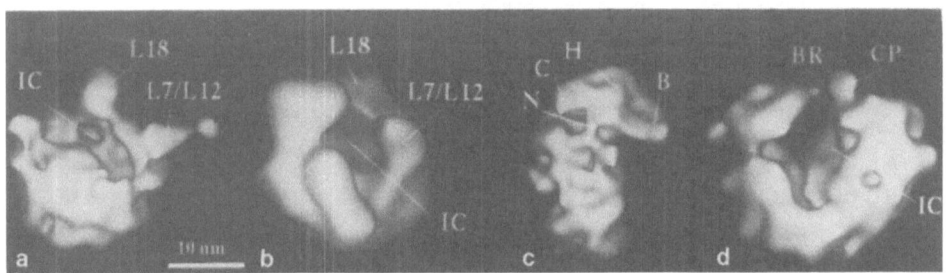

Fig. 1. The 3D structures of ribosomal particles determined using the random conical reconstruction method.
(a) The 50S large subunit from the *E. coli* ribosome. This reconstruction was obtained from a negatively stained specimen (Radermacher et al., 1987b); resolution 3.0 nm. Major features are the L7/L12 stalk (L7/L12), the interface canyon (IC) and the site of anti-L18 binding on the central protuberance (L18). In contrast to earlier models, the subunit exhibits a convex front surface.
(b) The 70S ribosome from *E. coli*, reconstructed from a frozen hydrated preparation (Frank et al., 1991), resolution 4.0 nm. The major features of the 50S subunit can be seen (labelled as in (a)); the 30S subunit lies in front.
(c) The 40S small subunit from rabbit reticulocyte ribosomes, reconstructed from a negatively stained specimen (Verschoor et al., 1989), resolution 3.8 nm. Labelled are: (H), the head; (N), the neck; (C), the crest; and (B), the beak.
(d) The 80S ribosome from rabbit reticulocytes (Verschoor and Frank, 1990), reconstructed from a negatively stained specimen; resolution 3.7 nm. Marked are the bridge (BR) between the small and large subunits, and the central protuberance (CP) and interface canyon (IC) of the large (60S) subunit. All resolution values have been determined with the 45° phase residual criterion.

For this reason we have not yet succeeded in reconstructing the eubacterial 30S subunit or the eukaryotic 60S subunit by this approach.

An older method of structure determination in electron microscopy is reconstruction from a *single axis tilt series* (Hoppe et al., 1974). In this, a series of images of a specimen is recorded, typically starting at -60° tilt and continuing in small increments up to a maximum tilt angle of +60°. Such a tilt series provides a series of views from each particle in the field, and allows each particle to be reconstructed separately. In some cases the reconstructions of several particles can be averaged (Knauer et al., 1983). This method has obvious drawbacks, because it does not allow the electron dose to be minimized. However, it is the method of last resort when the requirements of the random conical technique are not met. In fact, all volumes that are used as references or controls for our ligand binding studies have been reconstructed with the random conical method (Fig. 1).

LIGAND BINDING SITES

Once we have the 3D structure of the ribosomal particle, we are in a position to undertake the determination of functional sites or epitopes *in three dimensions* on the structure, through what we term *3D immuno electron microscopy*. In this method we reconstruct a ribosomal particle complexed with a bound ligand. This ligand may be either a natural ligand involved in the translational process (e.g., tRNAs, mRNA, initiation factors) or an antibody raised against a specific ribosomal component. To effect the quantitative determination of the ligand binding site, we subtract the control reconstruction of the unliganded structure from the reconstruction of the ribosome-ligand complex. As long as gross conformational changes do not accompany ligand binding, the main difference between the two volumes will obviously be the mass attributable to the ligand. The ribosomal particle should be essentially identical in both volumes, so that subtraction will 'cancel it out'. Thus, the difference volume contains a representation of the reconstructed ligand only, in the 3D geometric framework defined by the reconstruction of the complex.

Several problems exist in the practical implementation of the method. First, the attachment of the ligand to the ribosome introduces variability among the particles, in one of the following respects. If the antibody is an Fab, its unbound end is free to assume a range of positions: the ligand appears in a large number of similar but not identical orientations among all of the individual complexes averaged into the final result. Only the common information, the contact between the antibody and its epitope on the ribosome, is fixed, and thus strongly reinforced. Accordingly, in a reconstruction combining images of hundreds of complexes, a blurring of the Fab is expected. However, at the site where the information is constant -- the antibody-epitope contact -- all images will reinforce one another. In the difference volume, this point will show as a 'hot spot' of maximal difference between the control and the reconstruction of the complex.

If the antibody used is an IgG, we have an even greater variability of the ligand-bound complexes. We expected that we would be restricted to the single-axis reconstruction method for experiments using IgG's to link dimers of ribosomal particles: Since no two dimers will be identical in terms of antibody-linker angles or relative positions of the two particles, and since the two particles may overlap in the tilt projection, each dimer must be treated as a unique particle, and the images cannot be combined in the random conical scheme.

One final problem has proved to be serious, hindering our efforts to apply the technique to a number of specimens. We have had difficulties with stabilization of antibody binding in several experiments using different antibodies. We had hoped that Fab experiments would produce plentiful antibody-ribosome complexes, amenable to the random-conical technique. Unfortunately, to date, only one antibody, an IgG raised against protein L9 (Nag et al., 1991), has shown suitable binding to produce a sufficient number of complexes for this method to be attempted.

When only a low yield of complexes is obtained, we must resort to the use of the 'fallback' reconstruction technique, collection of single-axis tilt series, and reconstruction of *individual* ribosome-antibody complexes. The resulting 3D reconstruction has poorer resolution, and it also depicts only a single ribosome, rather than a statistically well defined average.

The most interesting ligands are natural ligands, components of the translational process such as non-ribosomal RNAs or protein factors. If a complex of such a ligand with the ribosome or ribosomal subunit can be stabilized sufficiently for electron microscopy to be performed, we can directly determine the functional site represented by the contact between the ligand and the ribosome. An example for such a natural ligand complex is the *native 40S subunit* of the mammalian ribosome.

NATIVE 40S SUBUNIT

By reconstructing the so-called native 40S subunit, we have localized the binding site of eukaryotic initiation factor 3, or eIF-3. The native subunit is a complex of the 40S subunit with this large (Mw = 650,000) protein factor, which has as one of its principal known functions the capability of preventing association of the 40S and 60S subunits into the monomeric ribosome (Lutsch et al., 1986). The 40S subunit lacking bound eIF-3 is termed the *derived subunit*; we had previously calculated a reconstruction of this structure, from a specimen of rabbit reticulocyte ribosomes (Verschoor et al., 1989).

The native subunit is isolated at a low (e.g., 25-75 mM) KCl concentration (Lutsch et al., 1986). At higher KCl concentrations, the factor dissociates, giving rise to the derived 40S subunit lacking non-ribosomal components. After reconstructing the native subunit using the random conical technique (Srivastava et al., 1992; Fig. 2), we were able to refine the analysis in two respects: to distinguish two morphologies for eIF-3 in our particle population through a multivariate statistical analysis, and to perform a difference calculation with the 3D reconstruction of the derived subunit. Through this difference calculation, the factor could be investigated in detail, free of obscuration by the 40S subunit (A. Verschoor, S. Srivastava, and J. Frank, in preparation).

40S - ANTI-S6 COMPLEX

The 40S subunit should be a good candidate for the random conical reconstruction method, due to its preference for assuming either of two lateral views with high frequency.

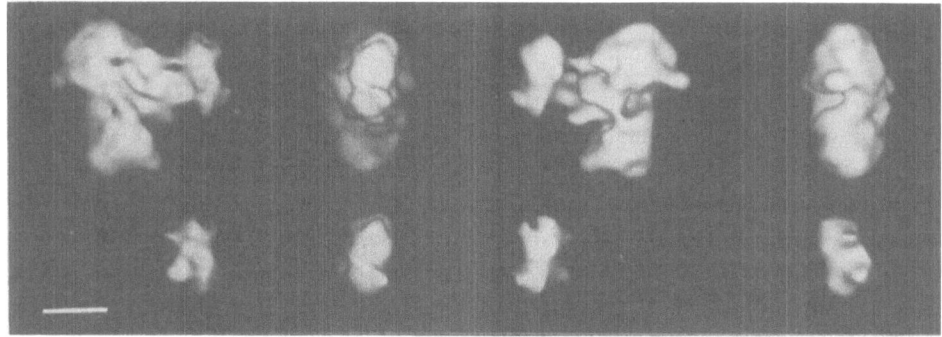

Fig. 2. 3D localization of binding site of eukaryotic initiation factor eIF-3 on the 40S subunit from rabbit reticulocyte ribosomes.

Top row: Orthogonal views of reconstruction of 'native' 40S subunit, shown in surface representation. First panel shows the 40S subunit in its strongly preferred lateral-view orientation.

Bottom row: Corresponding views of the initiation factor, eIF-3, shown in surface representation. To obtain these views of the factor alone, a 3D reconstruction of the 'derived' 40S subunit lacking non-ribosomal components (Verschoor et al., 1989) was subtracted from the 3D reconstruction of the native subunit shown in the top row. The details of the two strands that attach the factor to the 'back lobes' of the 40S subunit can be studied, as can the dumbbell-shaped morphology of the factor. Scale bar, 10 nm.

Thus far, however, it has proved less satisfactory for ligand studies (with the notable exception of eIF-3), because of the general difficulties in preparing and stabilizing eukaryotic specimens, especially with bound ligands. We have been attempting to localize the peptide in protein S6 which may be multiply phosphorylated (e.g., Heinze et al., 1988), through analysis of binding of monospecific antibodies raised against the corresponding synthetic peptide (unpublished results of R.R. Traut, J.R. Etchison, A. Verschoor, S. Srivastava, and J. Frank).

Several repetitions of the binding experiments produced inconsistent results, as have many studies involving S6 (e.g., Lutsch et al., 1983). A first experiment produced a limited number of 40S dimers linked by IgG's, but the antibodies appeared to bind to several spatially distinct sites on the 40S subunit. Although reconstructions of several individual dimers were calculated from single axis tilt series, the resolution and signal to noise ratio were rather low, in addition to the problem of the apparent multiple binding sites.

Several attempts to use Fab's were made, to obtain single particles with bound antibodies (as opposed to IgG-linked dimers); we hoped that this would enable us to apply the preferred random conical reconstruction method. Due apparently to poor stabilization by crosslinking, however, yields of clearly identifiable complexes were low, thus making random conical reconstruction impractical. Thus, we again had to resort to the single axis tilt scheme for these monomeric complexes. Several reconstructions showed antibody binding to the so-called head region of the 40S subunit. Attempts to improve the reconstructions using the restoration method known as POCS (Projection Onto Convex

Sets; Carazo and Carrascosa, 1987) were partially successful in compensating for the loss of information due to the limited range of tilting in the electron microscope.

Two reconstructions of individual 40S particles are shown. One, in a quasisymmetrical view (Fig. 3a,b) strongly resembling the eubacterial 30S subunit, has the Fab apparently bound to the middle portion of head; the other, in the classical lateral view (Fig. 3c,d), has the Fab bound to the back of the head. Currently the localization of the binding site is ambiguous, principally because of the multiple binding site problem. We hope to repeat the experiment and obtain a greater yield of bound complexes; with suitable stabilization through crosslinking we may be able to determine more precise information about the true location of the epitope of this antibody.

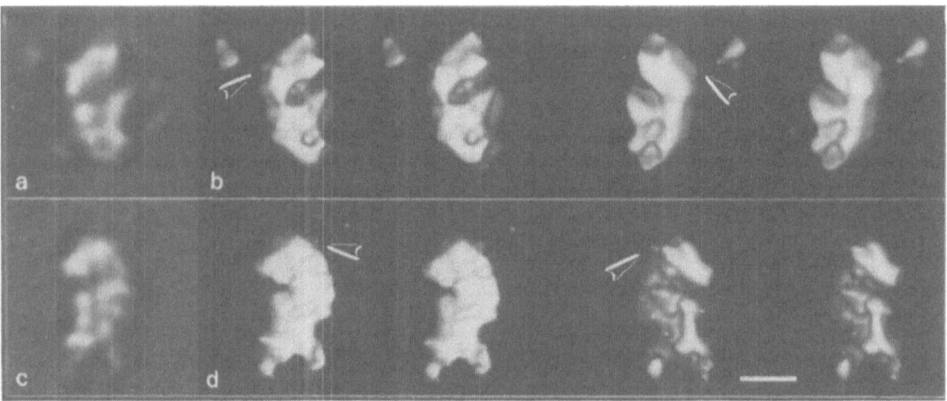

Fig. 3. Two 3D reconstructions of 40S subunit complexed with anti-S6 antibody raised against a synthetic peptide corresponding to the peptide that may be multiply phosphorylated *in vivo*.

In (a) and (b), a reconstructed 40S subunit is seen in the quasisymmetric orientation in which it strongly resembles the eubacterial (30S) small subunit.

Panels (c) and (d) show a second reconstructed complex, in which the particle is seen in its lateral-view orientation.

The surface views (b,d) of the reconstructions do not preserve the entire antibody (marked by arrow) as an intact linear feature contacting the ribosome, at the density threshold used for this representation. The continuity can, however, be traced in computed projections (a,c) of the reconstructions. Scale bar, 10 nm.

50S - ANTI-L18 COMPLEX

Several antibody studies on the eubacterial (*E. coli*) ribosome have also been undertaken in our laboratory. We have localized the antigenic determinant for an antibody raised against protein L18 of the 50S subunit (Stöffler-Meilicke and Stöffler, 1988) to the L1-facing aspect of the central protuberance of the 50S subunit (Fig. 4). This experiment utilized the same single-axis tilt geometry as the eukaryotic S6 study, but a greater overall tilting range allowed improved resolution due to a reduction in the artefacts. In this study we were able to reconstruct several individual IgG-linked dimers of ribosomal particles, and to attempt a 3D averaging of several of the individual ligand-bound 50S reconstructions (Srivastava et al. 1992).

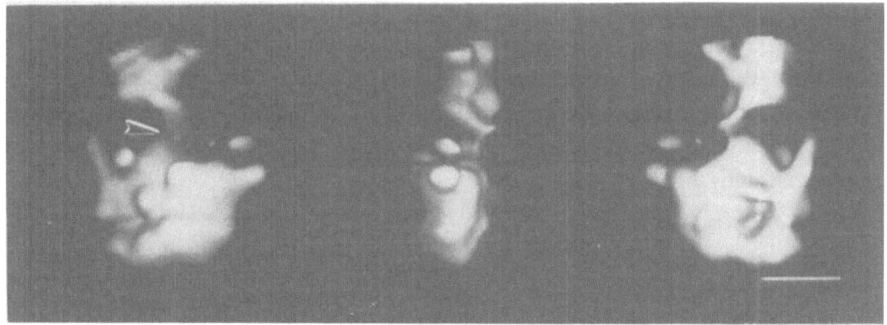

Fig.4. Surface representation showing approximately half of a reconstruction of a 50S-IgG-50S dimer. Three orthogonal views, depicting one 50S subunit, with a bound anti-L18 antibody, are shown. With the 50S subunit in crown orientation, we see the binding site (arrow) on the central protuberance, on the side towards the L1 shoulder. The elaborate structure attached to the central protuberance of the 50S subunit represents the linking IgG, plus a portion of the second 50S subunit of the dimer (masked off at top of frame). Scale bar, 10 nm.

50S-ANTI-L9 COMPLEX

In an ongoing study we are attempting to localize the epitope of an antibody to protein L9 of the 50S subunit from the *E.coli* ribosome. Although the antibody is an IgG, dimer formation does not occur (Fig. 5). Thus, we intend to apply the random conical method to this data set.

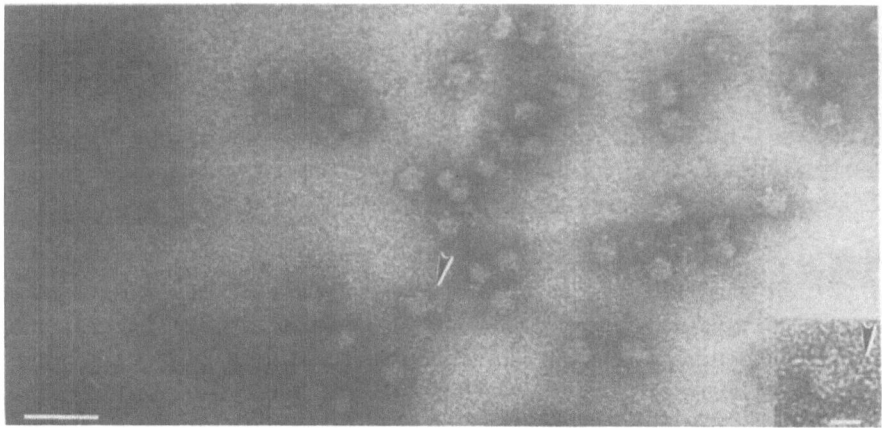

Fig. 5. Portion of electron micrograph showing 50S - anti-L9 complexes. Arrow indicates a single ribosome - IgG complex. Scale bar, 50 nm. Inset shows enlarged image of one complex. Arrow indicates position of bound IgG. Scale bar, 10 nm.

30S - T-RNA COMPLEX

An ongoing subject of investigation in our laboratory is the determination of the binding of tRNA's to both eubacterial and eukaryotic specimens. In a previous study

(Wagenknecht et al., 1988) we had attempted to directly localize in two dimensions the P site on the 30S subunit through photocrosslinking of a valyl-tRNA to C1400 of the 16S rRNA. However, as we have noted, the 30S subunit does not assume preferred orientations, which makes 2D averaging, as well as 3D reconstruction, particularly difficult. Nevertheless, a statistically significant peak was detected in the platform lip region in that study.

More recently, we have achieved preliminary 3D reconstructions of several individual 30S - tRNA complexes by the single-axis reconstruction technique. In a reconstruction of a 'Lake-view' particle (Fig. 6), we see an avidin-linked biotinylated tRNA making contact with the 30S subunit at the lip of the platform. The actual binding site is difficult to deduce at present, since we lack a control structure (3D reconstruction of uncomplexed 30S subunit). It is likely that the anticodon loop is binding within the 'cup' of the 30S platform structure, rather than on its rim, but the actual site can only be determined at higher resolution. Efforts are underway to refine the reconstruction to this end.

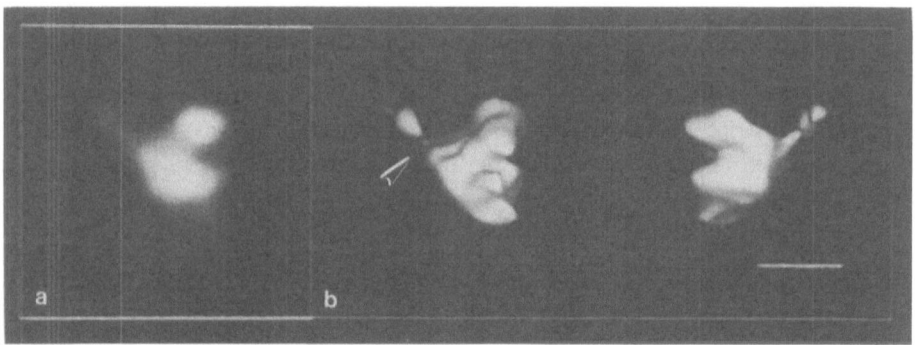

Fig.6. Reconstruction of 30S-tRNA complex. 30S subunit is seen in Lake-view orientation. (a) Projection through 3D volume; (b) surface representations of two opposite sides of the reconstructed complex. Scale bar, 10 nm.

CONCLUSIONS

The foregoing survey of some of the ligand projects underway in our laboratory suggests the potential of our techniques to directly demonstrate the binding site in three dimensions on the ribosomal particle. It is clear that we are able to obtain a certain amount of information even from the limited-resolution single-axis tilt studies (currently not better than 5 -6 nm). In fact, we believe that the resolution of these can be improved, if we make efforts to optimize several aspects of the method -- notably the technique for alignment of the set of tilt images. Our application of POCS to these single-axis reconstructions has already improved their ability to provide quantitative site information in the third dimension. Again, we hope to improve our antibody preparations to the point where we can apply the random conical method, SECReT, to the specimens, in order to obtain the highest possible resolution, and thus the most precise localizations. As we refine and complete these studies, we will for the first time be in a position to label quantitatively derived binding sites on our quantitatively derived models. As we begin to understand the spatial relationships among functional sites, we will develop a greater capability of understanding numerous aspects of the process of translation.

ACKNOWLEDGMENTS

This work was supported in part by grant 1R01 GM 29169 from the National Institutes of Health.

REFERENCES

Carazo,J-M. and Carrascosa,J.L., 1987, *J. Microsc.* 145:23.

Frank,J., Penczek,P., Grassucci,R., and Srivastava,S., 1991 *J. Cell Biol.* 115:597.

Heinze,H., Arnold,H.H., Fischer,D., and Kruppa,J., 1988, *J. Biol. Chem.* 263:4139.

Hoppe,W., Gaßmann,J., Hunsmann,N., Schramm,H.J., and Sturm,M., 1974, *Hoppe-Seyler's Z. Phys. Chem.* 355:1483

Knauer V., Hegerl R., Hoppe W., 1983, *J. Mol. Biol.* 163:409

Lutsch,G., Bielka,H., Enzmann,G., and Noll,F., 1983, *Biomed. Biochim. Acta* 42:705.

Lutsch,G., Benndorf,R., Westermann,P., Bommer,U.A., and Bielka,H., 1986, *Eur. J. Cell Biol.* 40:257.

Nag,B., Akella,S.S., Cann,P.A., Tewari,D.S., Glitz,D.G., and Traut,R.R., 1991, *J. Biol. Chem.* 266:22129.

Radermacher,M., Wagenknecht,T., Verschoor,A., and Frank,J., 1986, *J. Microsc.* 141:RP1.

Radermacher,M., Wagenknecht,T., Verschoor,A., and Frank,J., 1987a, *J. Microsc.* 146:113.

Radermacher,M., Wagenknecht,T., Verschoor,A., and Frank,J., 1987b, *EMBO J.* 6:1107.

Radermacher,M., Srivastava,S., and Frank,J., 1992, Proc. 10th Eur. Congr. El. Micr. Vol. III, 19-20

Srivastava,S., Radermacher,M., and Frank,J., 1992a, Proc. 10th Eur. Congr. El. Micr. Vol. I, 421-422

Srivastava,S., Verschoor,A., and Frank,J., 1992b, *J. Mol. Biol.* 226:301.

Stöffler-Meilicke,M. and Stöffler,G., 1988, *Meth. Enzymol.* 164:503.

Verschoor,A. and Frank,J., 1990, *J. Mol. Biol.* 214:737.

Verschoor,A., Zhang,N.-Y., Wagenknecht,T., Obrig,T., Radermacher,M., and Frank,J., 1989, *J. Mol. Biol.* 209:115.

Wagenknecht,T., Frank,J., Boublik,M., Nurse,K., and Ofengand,J., 1988, *J. Mol. Biol.* 203:753.

MESSENGER RNA PATH THROUGH THE PROCARYOTIC RIBOSOME

Alexey A.Bogdanov, Olga A.Dontsova and Richard Brimacombe*

A.N.Belozersky Institute of Physico-Chemical
Biology and Department of Chemistry
Moscow State University
Moscow 119899, Russia
*Max-Planck-Institut für Molekulare Genetik
Ihnestrasse 73, D-1000 Berlin 33, Germany

INTRODUCTION

The major function of the ribosome is to translate mRNA. However, until very recently our knowledge about the shape, conformation and topography of mRNA on the ribosome has been very limited. It is still not understood how a template polynucleotide moves through the ribosome and what ribosomal components are responsible for this fundamental process. Moreover, these problems were obviously off the "mainstream" of ribosomology. Typically, only a few chapters of the previous monograph on the ribosome (the so-called the "blue" Ribosome-Book) were partially devoted to mRNA-ribosome interactions and none of them discussed the topography and spatial organization of mRNA on the ribosome.

Although the interaction of mRNA with two cognate tRNAs is one of the major events in protein synthesis, the mRNA binding site of a ribosome is not limited to its decoding region. It has been known for many years that from 30 (Steitz, 1975) to 50 (Kang and Cantor, 1985) nucleotide long mRNA segments are in physical contact with the small ribosomal subunit in a *E.coli* 70S ribosome initiation complex. It was also suggested that apart from the well-characterized Shine-Dalgarno (SD) sequences, procaryotic mRNAs contain several ribosome binding regions (both upstream and downstream of the initiation codon and the SD sequence) required for their translation (reviewed by Shatsky et al., 1991). Therefore, the mRNA-binding center of the ribosome occupies a substantial part of its surface and a considerable number of rRNA segments and ribosomal proteins must be involved in its formation.

Many experimental approaches have been used to elucidate the conformation and topography of mRNA on the ribosome. Most commonly these studies have led to rather conflicting conclusions about the organization of the ribosomal mRNA binding site (summarized by Brimacombe, 1992). The results of some of these approaches are discussed in subsequent sections of this review. However, the focus of this paper will be on the identification of the components of the mRNA binding center of *E.coli* ribosomes by means of cross-linking to mRNA modified with photoaffinity probes at specific sites. We will demonstrate that data recently obtained in our studies can be used as a solid background for a description of the mRNA path through the ribosome.

CROSS-LINKING OF mRNA TO 16S rRNA AND PROTEINS OF *E.coli* RIBOSOMES

Earlier Studies

The affinity labeling approach is one of the most powerful and widely used methods for identification of the mRNA neighborhood on the ribosome (see for review Cooperman, 1988). Because of the lack of a convenient procedure for identification of the cross-linked sites on the rRNA, the ribosomal proteins were the major object of investigation until recently. The most popular models of mRNA were homogeneous poly- and oligonucleotides modified with different chemical and photochemical probes (Towbin et al., 1975, Fiser et al., 1975). In some cases the initiator AUG codon was attached to the 5'-end of a synthetic oligonucleotide (Vladimirov et al., 1990). Almost all proteins of the small ribosomal subunit and up to ten proteins of the large subunit were found to be cross-linked to a template in the presence of the corresponding tRNA. An attempt to use long polycistronic mRNA isolated from bacteriophages with a direct UV irradiation approach also led to the cross-linking of twelve 30S ribosomal proteins to mRNA (Broude et al., 1983). A binary complex of mRNA with ribosomes was used in this work.

There are several evident explanations of these results. Firstly, the long templates were not properly fixed in the mRNA binding sites, due to the absence of cognate tRNA (as in the case of phage RNA) or nucleotide sequences required for the formation of initiation complexes. Secondly, it has been shown that oligonucleotides have more than one binding site on *E.coli* ribosomes (Bakin et al., 1991) and they can be cross-linked to proteins located outside of the mRNA binding center. Thirdly, the reagents used in many studies had a rather large length and (what is more important) needed a very long incubation time for reaction to occur. Being attached to a flexible template, these reagents could label ribosomal components within a large sphere of rotation.

Cross-Linking of Short cro-mRNA-Like Templates

Design of mRNA and the Strategy of Analysis. To overcome the technical problems described in the previous section and to get reliable data on the composition of the mRNA path through the ribosome we used a new type of mRNA carrying a photochemical probe and developed a new strategy for the identification of cross-linked ribosomal components. This work represents a collaboration between laboratories in Moscow State University and the Max-Planck-Institut für Molekulare Genetik in Berlin.

As the first step a family of short mRNAs was constructed on the basis of the 5'-end part of the phage Cro-mRNA gene (Skripkin et al.,1986). These mRNAs contained a long SD sequence and were shown to form very stable complexes with both 30S subunits and 70S ribosomes in the presence of initiator tRNA with the AUG codon located exactly at the ribosomal P-site (Balakin et al,1990, 1992). It was also shown that mRNAs of this type carrying conventional photoreactive reagents (a diaziril group, for example) at their 5'-ends cross-linked to a reasonable set of 30S subunit proteins (Dontsova et al.,1990, 1992a).

4-thio-uridine, an analogue of uridine, was used as a cross-linking reagent which has some advantages in comparison with other photoaffinity reagents. Thio-UTP can be easily synthesized and used in the T7 RNA-polymerase reaction instead of UTP (Stade et al., 1989). Thio-U has a similar (although not identical) conformation to uridine and, in contrast to many other cross-linking reagents, its incorporation into RNA chains should not significantly disturb their macromolecular structure. Upon UV-irradiation under mild conditions (λ > 300nm) thio-U forms "zero-length" cross-links with both RNA and protein constituents. Since 70S ribosome activity is retained after irradiation the cross-linking reaction does not damage the ribosomal components. In comparison with 5-azido-U, another analogue of uridine producing "zero-length" cross-links, the yield of reaction with thio-U is about an order of magnitude higher.

The sequences of mRNA used in our study were designed so as to have a thio-U residue at definite positions downstream from the initiator AUG codon, and to cover all positions from +4 to +16 (A in the AUG codon is considered as +1). The segments upstream from the AUG codon were the same in each mRNA species and contained U residues at positions -1,-3 and -4 (Dontsova et al., 1992).

Sequence	Position
GGG<u>AAGGAGG</u>UUGU<u>AUG</u>UACCAACGCAAAGGACAG	+4
GGG<u>AAGGAGG</u>UUGU<u>AUG</u>GUACAACGGAAAGGACAG	+5
GGG<u>AAGGAGG</u>UUGU<u>AUG</u>GAUCAACGGAACUGCCAG	+6, +16
GGG<u>AAGGAGG</u>UUGU<u>AUG</u>GAAUACCGGAAAGGACAG	+7
GGG<u>AAGGAGG</u>UUGU<u>AUG</u>GAACUACGGUAACGACAG	+8, +13
GGG<u>AAGGAGG</u>UUGU<u>AUG</u>GAAACUCGCAAACGACAG	+9
GGG<u>AAGGAGG</u>UUGU<u>AUG</u>GAAAGCUAACGCGACAG	+10
GGG<u>AAGGAGG</u>UUGU<u>AUG</u>GAACAAGUGAAACGACAG	+11
GGG<u>AAGGAGG</u>UUGU<u>AUG</u>GAACAACGUGAACGACAG	+12
GGG<u>AAGGAGG</u>UUGU<u>AUG</u>GAACAACGCAU	+14
GGG<u>AAGGAGG</u>UUGU<u>AUG</u>GAACAACGCAAUCGACAG	+15

The identification of the mRNA-rRNA cross-links is based on the site-specific hydrolysis of 16S rRNA with RNase H (Bogdanov et al., 1988). At the first stage the purified cross-linked complexes of [^{32}P]-mRNA with 16S rRNA were digested with RNase H in the presence of pairs of decadeoxyribonucleotides complementary to 16S rRNA regions separated by 100-200 nucleotides (Rinke-Appel et al., 1991). From an analysis of the mobility of the [^{32}P]-labelled rRNA bands on gel electrophoresis, the location of the cross-links can be roughly estimated. An example of such an analysis is shown in Fig.1A. One can perform a more precise localization of the cross-linking site using other pairs of oligodeoxyribonucleotides in the RNase H hydrolysis to narrow down the modified rRNA region to a 20-40 nucleotide long segment.

The last stage of the localization of the cross-linked site on 16S rRNA is the primer extension analysis, which has been successfully used for the identification of modified bases in large RNAs (Moazed et al., 1986). It must be emphasized that the interpretation of data obtained by this approach is not an easy task. The reverse transcriptase used for the primer extension will stop not only at the modified base but also at any fortuitous chain break. It is difficult to discriminate such stops even when control samples are run in parallel. The separation of cross-linked and non-cross-linked fragments of rRNA is necessary to increase the yield of the specific reverse transcriptase stops. This is particularly important when thio-U is used as the cross-linking reagent since the enzyme does not show a strong stop at each thio-U generated cross-link. In fact, one can only be convinced of the reliability of the identification of a cross-linked nucleotide if the reverse transcriptase stops occures within the short region of 16S rRNA detected by independent RNase H analysis.

In cases where the mRNA contains several uridines and is statistically modified with thio-U, the question arises as to which U residue was cross-linked to 16S rRNA in each particular case. To be able to answer this question the mRNA primary structure was

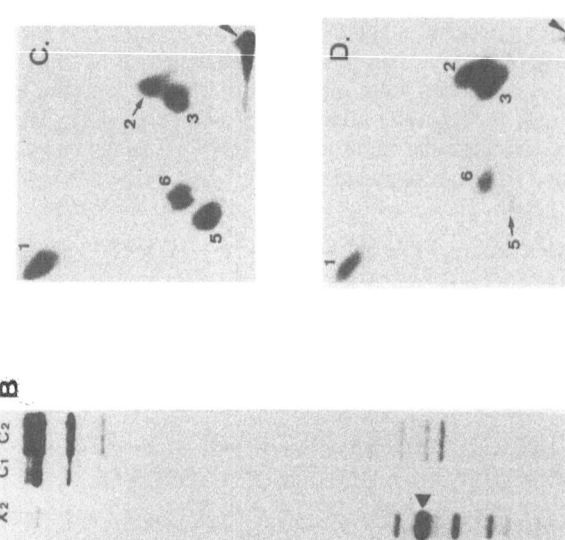

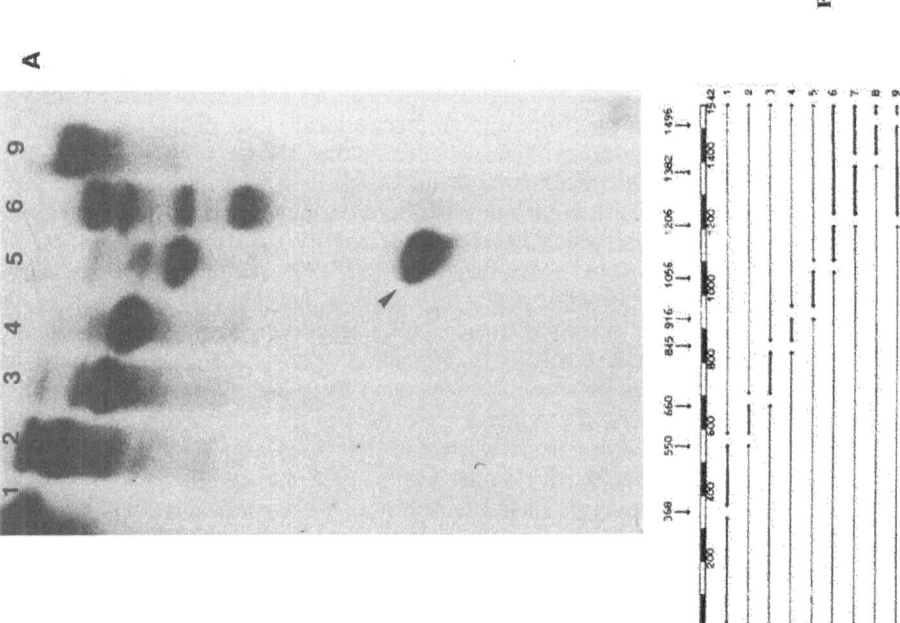

Figure 1A. Analysis of cross-links from "+6" mRNA. A: RNase H analysis of "+6" mRNA in the 70S initiation complex according to the scheme shown below the gel. B: Primer extension localization of the +6 cross-link. C: T1 fingerprint of the "+6" mRNA. D: T1 fingerprint of purified cross-link from mRNA "+6" to the 1050 region of 16S RNA.

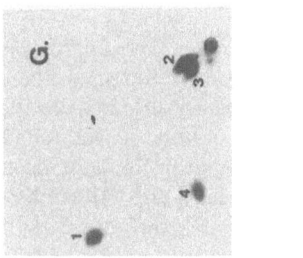

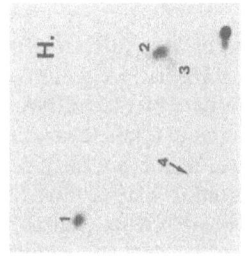

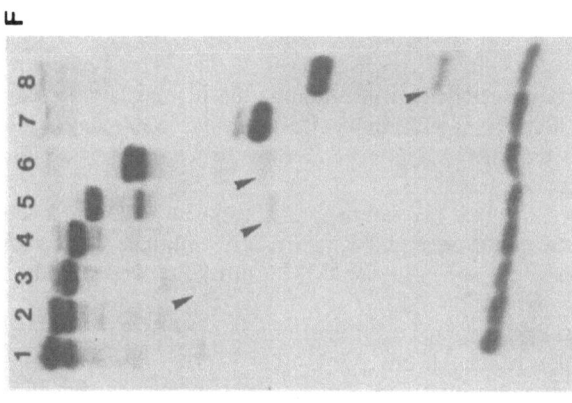

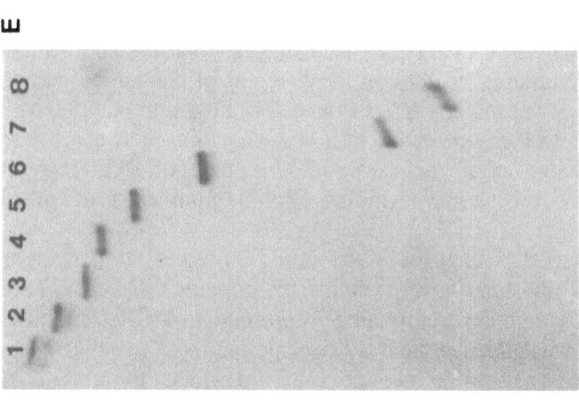

Figure 1B. Analysis of cross-links from "+9" mRNA. E: RNase H analysis of cross-links in a pre-translocation complex with the "+9" mRNA. F: RNase H analysis of cross-links of "+9" mRNA in the post-translocation complex according to the scheme below the gel (see Fig. 1A). G: T1 fingerprint of "+9" mRNA. H: T1 fingerprint of the "+9" mRNA cross-linked to the 1050 region of 16S RNA after translocation.

425

designed so that each U residue belonged to a specific mRNA fragment produced by RNase T1 hydrolysis. Thus, the cross-linked oligonucleotide is simply missing from the mRNA fingerprint (Dontsova et al., 1991). An example of such an analysis is presented in Fig. 1B.

The Composition of the mRNA Binding Center. The approaches described above allowed us to define 16S rRNA regions and ribosomal proteins which are involved in the organization of the part of the mRNA binding center proximal to the P- and A-sites of the *E.coli* 30S subunit (Dontsova et al., 1991, 1992). It was found that in the 70S initiation complex (i.e. in the presence of P-site bound tRNA) the upstream region of the mRNA near the initiator AUG codon (positions -1, -3 and -4) can be cross-linked to the 3'-end region of 16S rRNA as well as to proteins S7, S18 and S21. The two last proteins allso cross-link to the initiator codon itself. The first nucleotides of the second (position +4) and the third (position +7) codons in mRNA cross-link to the 16S rRNA region 1409-1450 and to the nucleotide 1395, respectively. On the other hand, the mRNA cross-links with good yield from position +6 to U1052 in 16S rRNA. As far as other nucleotide residues of these two codons are concerned, they can also form cross-links to position 1395, although only with a very low yield. Thus, the A-site region of the mRNA binding site of the ribosome is formed with 16S rRNA segments which belong to rather distant elements of its secondary structure. The nucleotide residues of the fourth codon of mRNA participating in the initiation of translation cross-link to 16S rRNA at position 532 and are therefore located near its "530" loop. No cross-links between 16S rRNA and the mRNA region +13 to +16 were observed. However, this region can be cross-linked to protein S3.

It should be noted that the involvement of the A-site codon in codon-anticodon interactions after addition of the second cognate tRNA eliminates the cross-links from positions +4 and +6, but does not influence the +7 cross-link and slightly increases the yield of the cross-link from position +11.

The cross-linking data obtained with the initiation complexes are completely born out by translocation experiments: all cross-links found in ternary complexes for downstream mRNA regions were also observed after one step of translocation. For example, a thio-U residue at mRNA position +9 formed the cross-link with position 1052 after translocation to position +6, and so on. On the other hand, several new interesting features were found when the movement of mRNA through the ribosome was followed by the cross-linking technique.

After +9 to +6 translocation of thio-U, a cross-link to the 1210 region of 16S RNA was observed in addition to U1052. This suggests that the conformation of the A-site part of the mRNA binding center can be different in elongation complexes (with a tRNA presumably located at the E-site) as opposed to initiation complexes. Further, one translocation step leads to the appearance of a new cross-link between the region located upstream from the initiator codon and the 660-700 segment of 16S rRNA. An analogous cross-link was observed for binary (minus tRNA) complexes of mRNA with the ribosome.

The most remarkable changes were found however in protein cross-linking. Whereas in binary 70S ribosome-mRNA complexes proteins S18 and S21 were the major targets, in the initiation complexes protein S7 appeared to be also modified. When the second cognate tRNA was added protein S7 became the most strongly modified protein, but after one step of translocation the yield of cross-links to proteins S18 and S21 increased again (Dokudovskaja et al., in preparation).

Cross-Linking of Other Natural-Like mRNA and tRNA to Ribosome. The data obtained with cro-mRNA-like templates correlate well results of other cross-linking studies. For example, Tate and co-workers (1990) found a cross-link between A1408 in *E.coli* 16S rRNA and position +4 in an mRNA analogue with a termination codon. The cross-link between A532 and mRNA position +11, which may reflect an important mRNA-ribosome contact (see below), was observed when the ACC codon of an mRNA

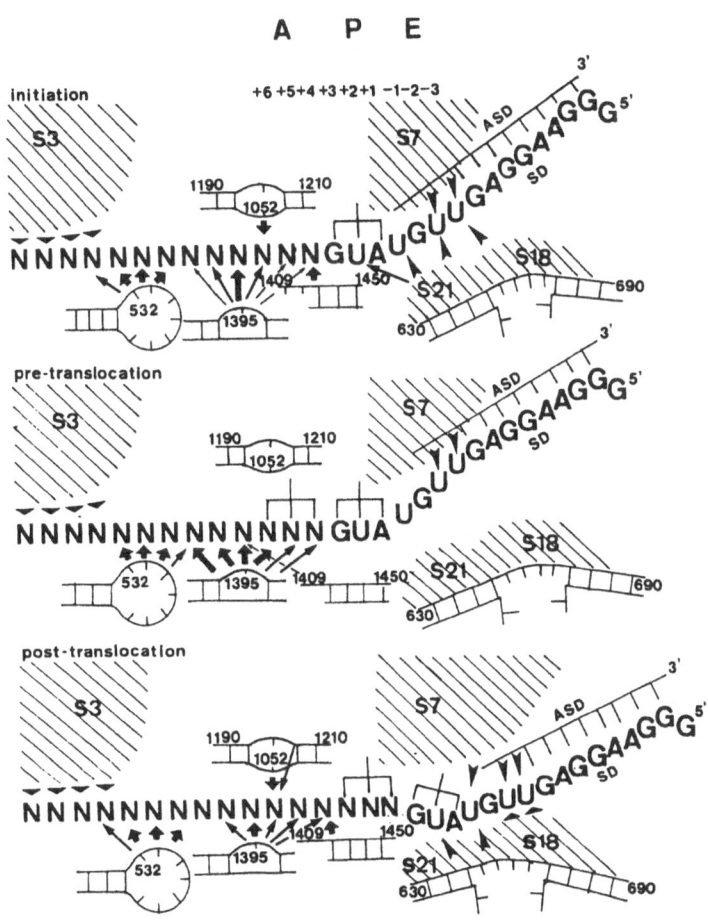

Figure 2. Diagram of cross-links of ribosomal components to mRNA in the initiation, pre-translocation and post-translocation complexes.

was fixed at the P-site by means of tRNA (Rinke-Appel et al., 1991). All the other 16S rRNA cross-linking sites described in the previous section were also revealed with synthetic mRNA analogues containing different spacer regions between the SD sequence and the initiator codon (Dontsova et al., 1992).

On the other hand, Wollenzien and co-workers (Wollenzien et al, 1991); Bhangu et al., 1992) found numerous cross-links of thio-U-containing mRNA to all domains of 16S rRNA. Only one cross-link at position 532 was in agreement with our findings. The reason for these discrepancies is not yet clear. The location of these cross-links on the mRNA was not determined in these studies and one can assume that some of them involve mRNA positions downstream from position +16. It must be emphasized however that the primer extension analysis was the only approach used for the evaluation of the cross-linking sites in this latter study.

Two thoroughly documented "zero-length" cross-links between the anticodon loop of P-site bound tRNA and the ribosome have been obtained. In the first case Prince and co-workers (1982) have shown that 5-carboxymethoxyuridine at position 34 of *E.coli* tRNA forms a cyclobutane dimer with C1400 of 16S rRNA when irradiated with UV-light. In the second case Sylvers and co-workers (1992) demonstrated that 2-azido-A, also placed in the anticodon loop of tRNA, is cross-linked to protein S7 both from the P- and A-sites. It is clear that these data are in good correlation with our conclusions about the composition of the mRNA binding region at 30S subunit decoding center.

Comparison of Cross-Linking Data with the Results of the Other Methods

Chemical Probing. Moazed and Noller (1986) discovered two sites on the *E.coli* 16S rRNA which could be protected by tRNA from chemical modification only in the presence of poly(U). One site is part of the 530 loop and the second site is located at positions A1408, A1492 and A1493 which are close together in the secondary structure of 16S rRNA. By comparison with the cross-liking data described above one can now conclude that these sites are directly involved in the organization of the 30S mRNA binding center.

Not many attempts have been undertaken to study the accessibility of mRNA nucleotides in functional complexes with the ribosome. It was found (Balakin et al., 1990) that nucleotide bases in the SD sequence and the initiator codon of a Cro-mRNA-like template are protected from chemical modification both in the binary mRNA-ribosome complex in the ternary complex with initiator tRNA. At the same time, the bases in the second (A-site codon) were slightly accessible and the bases at mRNA positions from +9 to +24 were strongly modified in the initiation complex. It was concluded that mRNA has a single stranded conformation when it passes through the ribosome. Also, it suggests that mRNA bases at positions +10, +11 and +12 are not involved in Watson-Crick base paring with the 530 loop nucleotides as proposed by Trifonov (1992) for some theoretically found mRNA motifs. One can assume that this part of the mRNA interacts with the ribosome via its ribose-phosphate backbone. It has also been suggested that mRNA bound to the ribosome retains some elements of its free state secondary structure in downstream regions quite distant from the decoding center. These elements of secondary structure can participate in modulation of ribosome activity (Heider et al., 1992).

Site-Specific Mutagenesis of rRNA. Our recent knowledge on the functional role of the rRNA in protein biosynthesis is based mainly on site-directed mutagenesis (see Dahlberg, 1989, and Tapprich et al., 1990, for reviews). It is noteworthy that numerous *E.coli* 16S rRNA mutations which affect the ribosomal functions coincide with cross-linking sites discovered in our study. For example, mutations in the 530 loop cause resistance of ribosomes to streptomycin (Melancon et al., 1988) or even lethality of *E.coli* cells (Pavers and Noller, 1991). Prescott and co-workers (1991) characterized several mutations in the region which includes our cross-linking sites at positons 1052 and 1210. The mutations affected termination of translation in different way and thus correlate very well with the positioning of these sites in the A-site part of the mRNA binding center.

In addition of course many mutations are known for the region surrounding the cross-linking position 1395 (see, for example Denman et al., 1989). Some them affect the initiation of translation and some of them are lethal, which confirms the localization of this position of the 16S rRNA directly in the decoding center.

Toeprint Analysis. It has been found that reverse transcriptase, primed at a distant part of mRNA, stops strictly at position +16 when it reaches the site of contact between the mRNA and the ribosome in the initiation complex (Hartz et al., 1988). Since the spatial arrangement of the reverse trancriptase-mRNA-ribosome complex is unknown, it is not easy to answer the question which of mRNA-ribosome contacts (if any) discovered by us are responsible for the enzyme stop. However, it was shown recently (Hartz et al., 1991) that in the binary mRNA-ribosome complex the reverse transcriptase stops at 4-5 bases downstream from the SD-anti-SD duplex. This suggests that the contact between mRNA (positions +11 and +12) and the 532 region of 16S rRNA could be responsible for the reverse transcriptase stop when initiation complexes are analyzed by the toeprinting technique.

Immune Electron Microscopy. The results obtained by the IEM technique and their relation to mRNA topography on the ribosome were recently discussed in detail by Shatsky et al. (1991). As far as mRNA cross-linking data are concerned, the use of IEM allows one to localize at least two of them on the morphological models of the ribosome with a high accuracy. These are the 3'-end of 16S rRNA (Shatsky et al., 1979; Olson and Glitz, 1979), which is involved in the SD-anti-SD interactions, and the 7-methyl-G residue (Trempe et al, 1982), which is a part of the 530 loop. They are localized on opposite sites of the junction of the head and the body of the small ribosomal subunit at the level of the groove which separates these parts of the 30S subunit. The cross-linking data allow us therefore unambiguously estimate the orientation of the mRNA on the ribosome in the initiation complex (Fig.3). If one looks at the 50S side of the 30S subunit the mRNA region upstream from the initiator codon is located on the right side of the groove and its downstream region on the left side. Therefore during translocation the mRNA moves between the ribosomal subunits from left to right. A similar conclusion concerning the orientation of mRNA on the ribosome has been made by McKuskie-Olson and co-workers (1988), who studied complexes of 30S subunits with template oligodeoxyribonucleotides by IEM.

The mRNA part between the SD sequence and positions +10 to +12 also has to be "dipped" into the 30S subunit groove. Indeed, in accordance with recent data on the distribution of 16S rRNA regions within of the small ribosomal subunit (see, for example, discussion in Brimacombe et al., 1988), those 16S rRNA parts which can be simultaneously cross-linked to codons at the A- and P-sites are located in the one case (the 1052 region) in the head and in the other case (the 1400 region) in the body of the 30S particle. (Fig.3). As was pointed out by Shatsky and co-workers (Shatsky et al., 1991), the placement of the decoding site into the 30S major groove does not contradict the IEM data if a reliable model for the 30S subunit structure is used.

The general shape of a synthetic template on the ribosome was also established by IEM (Evstafieva et al., 1983). Complexes of 30S subunits and 70S ribosomes with 30-55 nucleotide long poly(U) molecule modified at their 5'- or 3'-ends by dinitrophenyl haptens have been studied. Since the ends of poly U appeared to be in close proximity on the solvent side of the 30S subunit it was concluded that the message forms a "U"-turn when it passes through the ribosome. For this reason the shape of the mRNA binding center in Fig.3 is shown as a "necklace" surrounding the neck of the 30S particle as it was originally proposed by Shatsky et al. (1991). The loop shape of the mRNA on the ribosome was confirmed by singlet-singlet energy transfer measurements using poly U (Bakin et al.; 1991) an synthetic messengers carring fluorophores at their 5'- or (and) 3'-ends.

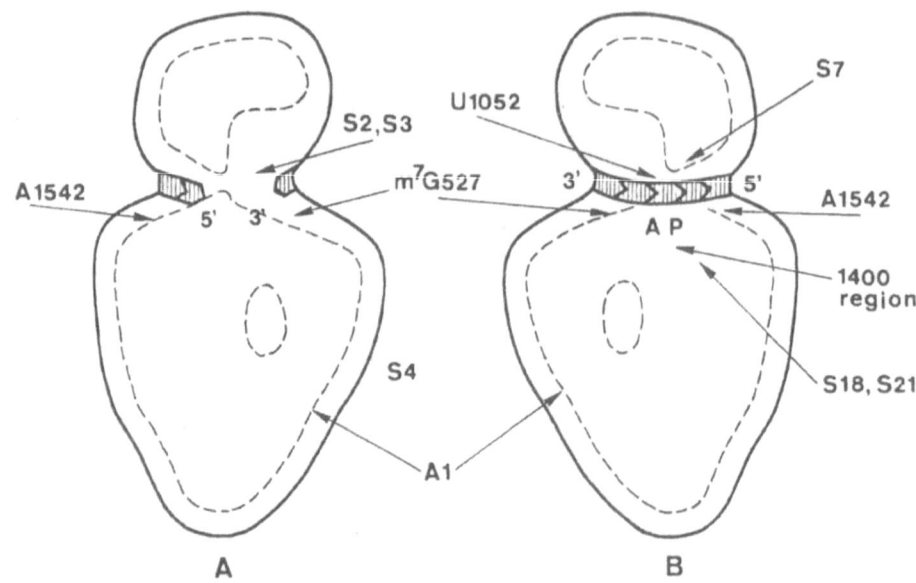

Figure 3. Possible arrangement of the mRNA path on the 30S subunit. The sketch of the 30S particle with the internal distribution of 16S RNA (shown with dotted line) deduced from the 70S ribosome model of Frank et al. (1991). The mRNA binding center is presented as a necklace with arrows showing the direction of mRNA movement.

Finally it should be mentioned that the locations of proteins S7, S18 and S21 on the 30S subunit established by IEM (Stöffler-Melicke and Stöffler; 1987) are also in good agreement with our cross-linking data (Fig.3).

CONCLUDING REMARKS

The new strategy for the identification of the cross-linked components of the ribosome with mRNA allows us to describe the composition of the mRNA-binding center in many details. It opens up unique possibilities for studing alterations in the structure of ribosomes during the translation cycle. At the same time the data recently obtained in our studies allow us to discuss the topography of the mRNA pathway through the ribosome with higher confidence.

Much more structural information is needed however to build a three-dimensional model of the organization of this important functional center of the ribosome. Future studies will attempt to reveal the functional role of every constituent of the mRNA binding center which now remains obscure.

Acknowledgements

We thank S. Dokudovskaya, I. Lavrik, J. Bittner and K. Rosen for the manuscript preparation and Dr.V. Tishkov for helpful discussion.

This work was supported by Grants from Max-Planck-Institut für Molekulare Genetik and Russian State Programme "Bioengineerings. Protein Biosynthesis".

REFERENCES

Bakin, A.V., Borisova, O.F., Shatsky, I.N., and Bogdanov, A.A., *J.Mol.Biol.* 221:441.

Balakin, A.G., Skripkin, E.A., Shatsky, I.N,. and Bogdanov, A.A., 1990, *Biochim.Biophys.Acta* 1050:119.

Balakin, A.G., Skripkin, E.A., Shatsky, I.N., and Bogdanov, A.A., 1992, *Nucl.Acids Res.* 20:563.

Bhangu, R., Wollenzein, P., 1992, *Biochemistry* 31:5937.

Bogdanov, A.A., Chichkova, N.V., Kopylov, A.M., Mankin, A.S., and Skripkin, E.A., 1988, *Methods in Enzymol.* 164:440.

Brimacombe, R., 1992, *Biochimie* 74:319.

Brimacombe, R., Atmadja,J., Stiege, W., and Schüler, D., 1988, *J.Mol.Biol.* 199:115.

Broude, N.E., Kussova, K.S., Medvedeva, N.I., and Budovsky, E.I., 1983, *Eur.J.Biochem.* 12:139.

Cooperman, B.S., 1988, *Methods in Enzymol.* 164:341.

Dahlberg, A.E., 1989, *Cell* 57:525.

Denman, R., Nerge, D., Cunningham, P.R., Nurse, K., Colgan, J., Weitzmann, C., and Ofengand, J., 1989, *Biochemistry* 28:1012.

Dontsova, O.A., Bogdanova, S.L., Rosen, K.V., Skrjabin, G.A., Skripkin, E.A., Kopylov, A.M., and Bogdanov, A.A., 1990, *Dokl.Acad.Sci.USSR* 133:730.

Dontsova, O.,Dokudovskaya, S., Kopylov, A.,Bogdanov, A., Rinke-Appel, J.,Junke, N., and Brimacombe, R., 1992b, EMBO J. 11:3105.

Dontsova, O., Kopylov,A., and Brimacombe,R., 1991, *EMBO J.* 10:2613.

Dontsova, O.A., Rosen, K.V., Bogdanova, S.L., Skripkin, E.A., Kopylov, A.M., and Bogdanov, A.A., 1992a, *Biochimie* 74:363.

Evstafieva, A.., Shatsky, I., Bogdanov, A., Semenkov., V. and Vasiliev, V.D., 1983, *EMBO J.* 2:799.

Fiser, I., Margaritella, P., Kuechler, E., 1975, *FEBS Lett* 52:282.

Frank, J., Penczek, P., Grassucci, R. and Srivastava, S., 1991, *J.Cell.Biol.* 115:597.

Hartz., D., McPheeters, D.S., Traut, R., and Gold, L., 1988, *Methods in Enzymol.* 164:419.

Heider, J., Baron,C ., and Bock, A., 1992, *EMBO J.* 11:3759.

Kang, C. and Cantor, C.R., 1985, *J.Mol.Biol.* 181:241.

McKuskie-Olson, H., Lasater, L.S., Cann,P.A., and Glitz, D.G., 1988, *J.Biol.Chem.* 263:15196.

Melancon, P., Lemieux, C., and Brakier-Gingras, L., 1988, *Nucl.Asids Res.* 16:9631.

Moazed, D., and Noller, H.F., 1986, *Cell* 47:985.

Moazed,D., Stern, S., and Noller, H.F., 1986, *J.Mol.Biol.* 187:399

Olson, H.M., and Glitz, D.G., 1979, *Proc.Natl.Acad.Sci.USA* 76:3769.

Prince, J.B., Taylor, B.H., Thurlow, D.L., Ofengand, J., and Zimmermann, R.A., 1982, *Proc.Natl.Acad.Sci.USA* 79:5450

Rinke-Appel, J., Junke, N., Stade, K., and Brimacombe, R., 1991, *EMBO J.* 10:2195.

Skripkin, E.A., Bogdanova, S.L., Kopylov, A.M., and Bogdanov, A.A., 1986, *Dokl.Acad.Sci.USSR* 287:237.

Shatsky, I.N., Bakin, A.V., Bogdanov, A.A., and Vasiliev, V.D., 1991, *Biochimie* 73:937.

Shatsky, I.N., Mochalova, L.V., Kojouharova, M.S., Bogdanov, A.A., and Vasiliev, V.D., 1979, *J.Mol.Biol.* 133:501.

Steitz, J.A.,1975, in: RNA Phages, N.D. Zinder, ed., *Cold Spring Harbor Laboratory*, New York.

Stoffler-Melicke, M., and Stoffler, G., 1987, *Biochimie* 69:1040.

Sylvers, L.A., Kopylov, A.M., Wower, J., Hixson, S.S. and Zimmermann, R.A. 1992, *Biochimie* 74:381.

Tapprich, W.E., Goringer, H.U.,De Stasio, E., Prescott, C., and Dahlberg, A. 1990, *in: The Ribosome. Strucure, Function and Evolution*, W.E.Hill, A. Dahlberg, R.A. Garrett, P.B. Moore, D. Schlessinger ,and J. Warner, eds., ASM, Washington,D.C.

Tate, W., Greuer, B., and Brimacombe, R.,1990, *Nucl. Acids Res.* 18:6537.

Towbin, N, Elson, D., 1978, *Nucl.Acids Res.* 5:3389.

Trempe, M.R., Ohgi, K., Glitz, D.G., 1982, *J.Biol.Chem.* 257:9822.

Trifonov, E.N., 1992, *Biochimie* 74:357.

Vladimirov, S., Babkina, G., Veniaminiva, A., Gimautdinova, O., Zenkova, M., Karpova, G., 1990, *Biochem.Biophys.Acta* 1048:284.

Wollenzien, P., Expert-Bezancon, A., Favre, A., 1991, *Biochemistry* 30:1788.

MAPPING THE FUNCTIONAL CENTRE OF THE
ESCHERICHIA COLI RIBOSOME

Richard Brimacombe, Thomas Döring, Barbara Greuer,
Nicole Jünke, Philip Mitchell, Florian Müller,
Monika Osswald, Jutta Rinke-Appel and Katrin Stade

Max-Planck-Institut für Molekulare Genetik
Abteilung Wittmann, Ihnestrasse 73
1000 Berlin 33, Germany

INTRODUCTION

The principal functional components which are attached to the ribosome during the process of polypeptide chain elongation are the mRNA, two tRNA molecules (either at the A- and P-sites, or the P- and E-sites), and the nascent protein. When the two tRNAs are present at the A- and P-sites, they are tightly constrained by the concomitant requirements (i) that their respective CCA 3'-termini must be close together at the peptidyl transferase centre, in order to allow peptide bond formation to occur, and (ii) that their respective anti-codons must also be close, to enable base-pairing to take place with the appropriate adjacent codons on the mRNA. It is known from fluorescence measurements (Johnson et al, 1982; Paulsen et al, 1983) that in this situation the angle between the planes of the L-shaped tRNA molecules must be relatively small, and there are thus two basically different possible configurations for the two tRNAs; in one the angle between the tRNA planes is approximately 90° (the so-called 'R' configuration (Rich, 1974; Lim et al, 1992)) and in the other it is approximately 270° (the so-called 'S' configuration (Sundaralingam et al, 1975; Lim et al, 1992)). A tRNA molecule at the E-site is not subject to the same constraints, since - having lost its peptidyl residue - the CCA terminus of this tRNA need no longer be close to the peptidyl transferase centre. The anticodon loop of the E-site bound tRNA on the other hand either still undergoes codon-anticodon interaction (Rheinberger et al, 1986), or is at least still fairly close to its mRNA codon (Paulsen and Wintermeyer, 1986).

It follows from these considerations that the mRNA, tRNAs and growing peptide chain form an interconnected topographical unit in the functioning ribosome, and the question which we want to address is how this unit is arranged in relation to the various

The Translational Apparatus, Edited by K.H. Nierhaus
et al., Plenum Press, New York, 1993

ribosomal components, with particular emphasis on the ribosomal RNA. The most direct experimental approach for the investigation of this problem is the use of 'site-directed cross-linking'. Here, a photoreactive label is introduced at a specific position in the ligand of interest, the modified ligand is bound to the ribosome under standard conditions, and - after photoactivation of the label - the cross-linked ribosomal components are analysed. A number of data points have already been established in this way, such as the well-known cross-link between the anticodon loop of P-site bound tRNA and position C-1400 of the 16S RNA (Prince et al, 1982), or the cross-links from affinity analogues of the amino acid residue in aminoacyl tRNA (at the A- or the P-site) to the peptidyl transferase region of 23S RNA (Steiner et al, 1988). We have extended this approach by undertaking a systematic site-directed cross-linking study of the functional components - i.e. the mRNA, tRNA at all three ribosomal sites, and the growing peptide chain - with a view to mapping the environment of the active centre of the ribosome in   detail. By combining the site-directed cross-linking information with other cross-linking data (in particular intra-RNA cross-links within the 16S or 23S RNA, or cross-links between the 16S and 23S molecules) we expect to be able to build up a topographical model of this active centre, to which progressively more and more elements of the ribosomal RNA structures can be added.

This mode-building strategy is substantially different from that used in developing previous three-dimensional models (Brimacombe et al, 1988; Stern et al, 1988; Mitchell et al, 1990) for the ribosomal RNA, which relied on RNA-protein cross-linking (Brimacombe et al, 1988; Mitchell et al, 1990) or footprinting (Stern et al, 1988) data to 'connect' the various structural elements of the RNA to the known arrangement of the ribosomal proteins (e.g. Capel et al, 1988). The latter approach has the inherent disadvantage that the ribosomal proteins are of unknown shape and dimensions, and that the cross-link or footprint sites are to undefined points on the surface of the protein concerned. The resolution of these older models is thus correspondingly low, and it has already been shown (Dontsova et al, 1992) that the models for the 16S RNA (Brimacombe et al, 1988; Stern et al, 1988) are not reconcilable with the newest site-directed cross-linking data. We believe that the new strategy will not only serve to resolve these discrepancies, but should also ultimately lead to definitive answers to other structure-function related questions, such as whether the nascent peptide exits through a tunnel in the 50S subunit (Yonath et al, 1987) or rather along a channel on the surface of the subunit (Ryabova et al, 1988), and whether the tRNAs lie in the 'R' or the 'S' configuration mentioned above (Lim et al, 1992).

THE SITE-DIRECTED CROSS-LINKING METHOD

For site-directed cross-linking of mRNA, suitable ^{32}P-labelled mRNA analogues are prepared from DNA templates by T7 transcription (Lowary et al, 1986), using 4-thio-UTP instead of UTP so as to incorporate thio-uridine residues into the sequence. In our earlier experiments (Stade et al, 1989; Rinke-Appel et al, 1991) each synthetic mRNA analogue contained only a single thio-uridine residue, which was then substituted with p-azido-phenacyl bromide, thus introducing an aromatic azide group as the photo-reactive moiety. More recently (Dontsova et al, 1991; 1992), we have used mRNA analogues containing several thio-uridine residues at selected positions, and these are photo-activated directly (without substitution) by irradiation at wavelengths above 300 nm (Hajnsdorf et al, 1989; Tate et al, 1990). Each mRNA molecule contains an AUG initiator codon as well as a Shine-Dalgarno sequence, and is bound to the ribosome in the presence of either tRNA^{fMet} alone or tRNA^{fMet} together with an appropriate aminoacyl tRNA cognate to the adjacent A-site codon in the mRNA sequence (Dontsova et al, 1992). Either 70S 'tight couples' or 'initiation complexes' (cf. Dontsova et al, 1991) are used.

After photo-activation, the cross-linked 70S complex is dissociated first into 30S and 50S subunits, and then into RNA and protein fractions, by a series of sucrose gradients. Cross-linking is exclusively to the 30S subunit, and the cross-link sites on the 16S RNA are determined (Rinke-Appel et al, 1991; Dontsova et al, 1992) by a combination of ribonuclease H digestions (in the presence of pairs of oligodeoxynucleotides complementary to appropriate regions of the 16S RNA), followed by primer extension analysis with reverse transcriptase (Moazed et al, 1986). The individual thio-uridine residues involved in each cross-link are identified by simple ribonuclease T_1 fingerprinting of the isolated mRNA-16S RNA fragment complexes obtained by ribonuclease H digestion. These experiments have led to the characterization of a highly specific series of cross-links to the 16S RNA from several defined positions in the mRNA, which have been observed with a number of different mRNA sequences and under a variety of different conditions (Dontsova et al, 1992); the cross-links are furthermore entirely dependent on the presence of appropriate cognate tRNA molecules, and the data are described below in the section entitled 'The Decoding Region'.

For the corresponding site-directed cross-linking programme with tRNA we make use of the naturally-occurring modified bases (Sprinzl et al, 1989) in order to attach a photoreactive group (cf. Podkowinski and Gornicki, 1991). The most useful modified bases for this purpose are either the so-called 'X'-base - 3-(3-amino-3-carboxypropyl) uridine, which contains an aliphatic primary amine group - and the various thio-bases; the X-base can be substituted by an activated ester derivative, whereas the thio-bases react with bromo-ketones (e.g. Ofengand et al, 1988). As photoreactive groups either aromatic azides (cf. Chen et al, 1985; Brimacombe et al, 1993) or, more recently, aromatic tri-fluoromethyl diazirine derivatives (cf. Bochkariov and Kogon, 1992) are suitable, the latter reagents being activated at wavelengths above 300 nm. The X-base and the thio-bases occur in a variety of 'strategically useful' positions in tRNA, for example tRNA^{Phe} from E. coli has an X-base at position 47, and lupin tRNA^{Met} has the same base at position 20a (Podkowinski and Gornicki, 1991); both these positions are in the 'elbow' region of the tRNA molecule. In the case of the thio-bases, a number of E. coli tRNA species have a 4-thio-uridine at position 8 (also in the elbow region), whereas tRNA^{Arg-I} has thio-cytidine at position 32 and tRNA^{Lys} has a modified 2-thio-uridine at position 34, both of these being in the anticodon loop.

The derivatized tRNA molecules can of course be bound to the ribosome either in the A-, P- or E-sites under appropriate conditions, so that the amount of data potentially available from this approach is enormous. The specificity of a particular cross-link for one of the three tRNA binding sites is primarily controlled by observing its absence in corresponding experiments with the tRNA bound to either of the other two sites (see also below). The same chemistry can furthermore be applied in the case of site-directed cross-linking with the growing peptide chain. Here the α-amino group of the N-terminal amino acid can be derivatized, or the ε-amino group of a lysine residue (e.g. Bochkariov and Kogon, 1992) at an internal position; again, either aromatic azide or trifluoromethyl diazirine derivatives can be used, coupled to an appropriate activated ester. In all cases, after photo-activation of the 70S ribosomal complex containing the derivatized tRNA or peptide analogue, the cross-linked ribosomal RNA is isolated and the sites of cross-linking are analysed as described above for the site-directed cross-linking of mRNA. A number of such analyses are currently in progress, but we confine our report here to three examples, (i) cross-links from the thio-cytidine residue at position 32 of P-site bound tRNA$^{\text{Arg-I}}$, derivatized with trifluoromethyl-(3-bromacetyl)-aminophenyl diazirine (cf. Bochkariov and Kogon, 1992; Brimacombe et al, 1993), to the 16S RNA, (ii) a cross-link from the X-base at position 47 of P-site bound tRNA^{Phe}, derivatized with 4-azido-hippuryl N-hydroxysuccinimide ester (Brimacombe et al, 1993), to the 23S RNA, and (iii) cross-links from the aminoacyl residue of P- or A-site bound Phe-tRNA^{Phe} (cf. Steiner et al, 1988; Brimacombe et al, 1993), also derivatized with 4-azido-hippuryl N-hydroxysuccinimide ester, to the 23S RNA. The detailed experimental procedures for these analyses will be presented elsewhere (in preparation), the results being included in the following section and Figure 1.

THE DECODING REGION

Figure 1 shows a sketch of two tRNA molecules at the ribosomal A- and P-sites in the 'R' configuration of Lim et al (1992), together with a segment of mRNA. The tRNA-mRNA complex is surrounded by the secondary structural elements of 16S RNA (Brimacombe, 1991) and 23S RNA (Brimacombe et al, 1990) which have been identified by the site-directed cross-linking experiments outlined above. Other relevant intra- or inter-RNA cross-linking data are also included. (There is at present not sufficient cross-linking data concerning the E-site tRNA to merit its inclusion in the Figure).

The volume occupied by the anticodon loops of the A- and P-site tRNA molecules, together with the two cognate codons on the mRNA is of the order of 40Å x 40Å x 30Å, and since this corresponds to a substantial portion of the 30S subunit it is more appropriate to refer to it as the 'decoding region' rather than the 'decoding site'. The data in Figure 1 which involve the decoding region are cross-links 'a' to 'j' in the 16S RNA. Cross-links 'a' and 'a″' are from the trifluoromethyl diazirine derivatized thiocytidine residue at position 32 of tRNA$^{\text{Arg-I}}$ in the P-site, as mentioned in the previous section. Cross-links (using a different photoreactive derivative) from this position to the 3'-half of the 16S RNA have been described by other authors (Chen et al, 1985), but the precise sites of cross-linking were not localized. Cross-link 'a' to the modified nucleotides in the loop-end of helix 31 of the 16S RNA has already been reported (Brimacombe et al, 1993), and we have since

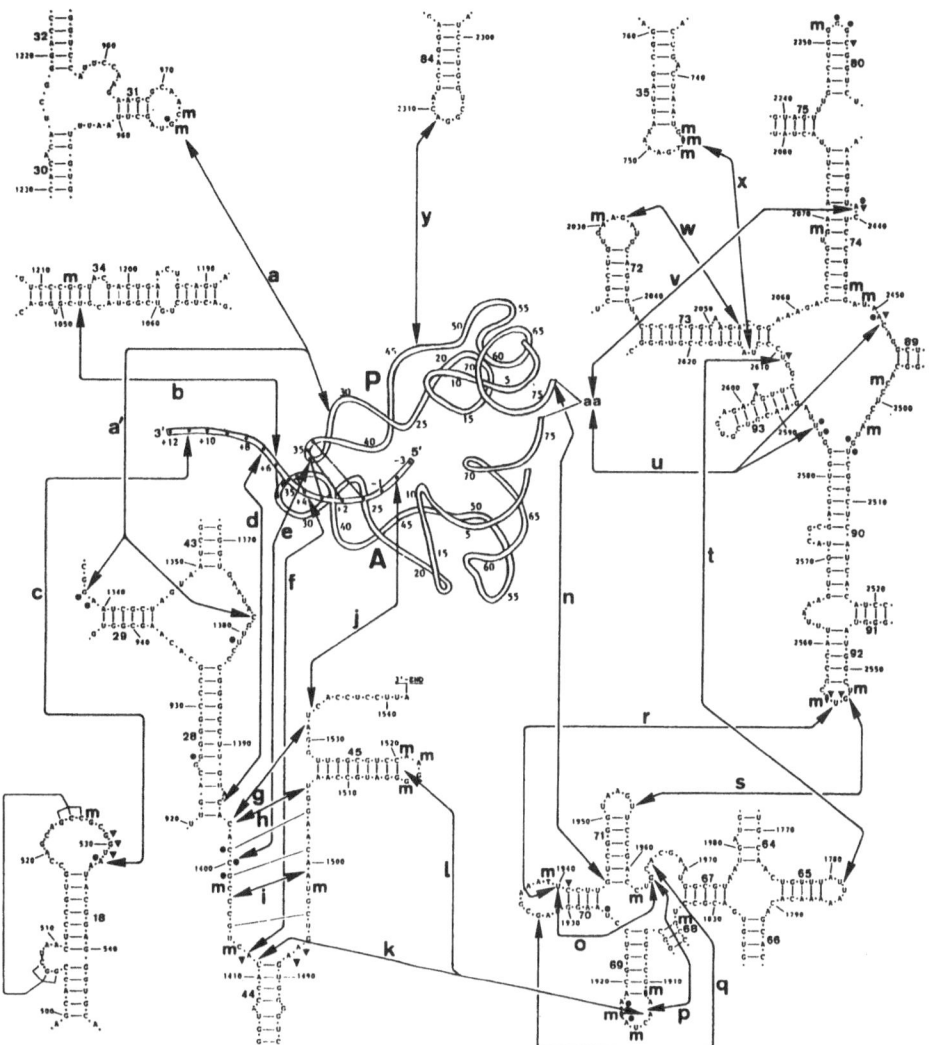

Figure 1. Cross-linking data to 16S and 23S RNA in the vicinity of the functional centre of the ribosome. The tRNA-mRNA complex is shown in the centre of the Figure, with the P- and A-site tRNAs in the 'R' configuration (Lim et al, 1992); 'aa' denotes the aminoacyl moiety. The mRNA segment shown runs from position -3 to +12, and its 5'- and 3'-ends are marked. Secondary structural elements of the 16S RNA (on the left) or 23S RNA (on the right) are arranged around the tRNA-mRNA complex, and are connected to it by arrowed lines representing the individual cross-links. Each of the latter is marked with a letter ('a' to 'y' in a counterclockwise direction), and the data sources are described in the text. Footprint sites for tRNA (Moazed and Noller, 1989; 1990) are indicated by filled triangles (A-site) and circles (P-site), respectively, and 'm' denotes a modified nucleotide.

characterized the two concomitantly formed cross-links 'a″' to positions 1338 and 1381 in the neighbourhood of helices 28 and 29. The finding of several simultaneous cross-links from a single site on the tRNA is a common phenomenon with the highly reactive diazirine derivatives, and in this instance it is noteworthy that all three positions are identical with or very close to P-site tRNA footprints on the 16S RNA, as observed by Moazed and Noller (1990). The cross-link from position 34 of the P-site bound tRNA to position 1400 of the 16S RNA (Prince et al, 1982) - cross-link 'e' in Figure 1 - also coincides with P-site tRNA footprint positions. However, a note of caution needs to be repeated here (cf. Brimacombe et al, 1993), since even a minor contribution from tRNA binding to the E- or A-sites could be misleading in a site-directed cross-linking study (if the tRNA in the minor binding site is able to cross-link with substantially higher efficiency); the P-site location of the tRNA in cross-links 'a' and 'a″' can thus only be unequivocally demonstrated when the corresponding cross-link patterns from the E- and A-sites have been determined, and these experiments are still in progress.

The other site-directed cross-links in the decoding region involve the mRNA (Dontsova et al, 1992). Cross-link 'b' is from position +6 on the mRNA (the A-residue of the AUG initiator codon is defined as +1) to nucleotide 1052 in helix 34, which is located in the 'head' of the 30S subunit (Brimacombe et al, 1988; Stern et al, 1988). The latter helix also contains a modified nucleotide at position 1207 (Carbon et al, 1979), a site (nucleotide 1054) which has been implicated in the termination event by site-directed mutagenesis (Murgola et al, 1988), and sites of footprinting (Moazed and Noller, 1987) or resistance (Sigmund et al, 1984) to the antibiotic spectinomycin. Cross-link 'c' is from position +11 on the mRNA to nucleotide 532 in the loop-end of helix 18 in the 'body' of the 30S subunit; this loop-end similarly contains a modified nucleotide, footprint sites primarily to A-site tRNA (Moazed and Noller, 1990), a site (nucleotide 517) implicated in termination (Shen and Fox, 1989), and a number of sites where resistance to streptomycin has been observed (Powers and Noller, 1991, and references therein). Cross-links 'd' and 'f' are from positions +7 and +4 on the mRNA to nucleotides 1395 (Dontsova et al, 1992) and 1408 (Tate et al, 1990), respectively, on the 16S RNA, whereas cross-link 'j' is from the 'upstream' part of the mRNA to a site (not yet precisely identified) close to the extreme 3'-terminus of the 16S molecule (Stade et al, 1989; Dontsova et al, 1991). These positions are interconnected by the ultraviolet induced intra-RNA cross-links 'g', 'h' and 'i' (Döring et al, 1992), as well as by the phylogenetically conserved tertiary interactions that have been identified in this region (Haselman et al, 1989).

The elements of the tRNAs and mRNA in or close to the decoding region form a tightly constrained topographical unit, and cross-links 'a' to 'j' provide corresponding constraints on the three-dimensional arrangement of those elements of the 16S RNA that are involved. As has already been discussed (Brimacombe et al, 1993), these cross-links serve to bring every one of the modified nucleotides in the 16S RNA into a cluster at the decoding region, and together with the new cross-links 'a″' almost all of the tRNA footprint sites reported by Moazed and Noller (1990) now also belong to this cluster, (the exceptions being the P-site footprints at positions 693 and 794-795). Furthermore, the sites mentioned above that have been implicated in the termination process (at positions 1054 and 517) can now clearly be placed immediately 'above' and 'below' the A-site codon-anticodon interaction region (Figure 1), respectively in the head and the body of the 30S subunit; the sites of mutation causing resistance to spectinomycin (in helix 34) and streptomycin (in helix 18) in these two thus neighboured regions are 'bridged' by ribosomal protein S5, which also contains sites of resistance to both these antibiotics, lying in separate domains of its crystal structure (Ramakrishnan and White, 1992). Protein S5 is located - as would be expected from this argument - at the junction between the head and body of the 30S subunit (Oakes et al, 1990; Stöffler-Meilicke and Stöffler, 1990).

THE PEPTIDYL TRANSFERASE REGION

A similar type of topographical clustering is observed in the peptidyl transferase region of the 23S RNA, as evidenced by cross-links 'n' to 'x' on the right-hand side of Figure 1. The peptidyl transferase centre (Vester and Garrett, 1988) consists primarily of the 'ring' enclosed by helices 73, 74, 89, 90 and 93, and the cross-link sites found by Steiner et al (1988) from a benzophenone derivatized phenylalanine residue in Phe-tRNA^{Phe} at the P- or A-site (cross-links 'u' in Figure 1) lie directly within this ring; the ring contains a number of modified nucleotides (Branlant et al, 1981; Smith et al, 1992), as well as several tRNA footprint sites (Moazed and Noller, 1989). On the other hand, our own similar experiments using phenylalanine derivatized with 4-azido-hippuryl N-hydroxysuccinimide ester (Brimacombe et al, 1993) in Phe-tRNA^{Phe} identified a site just outside the peptidyl transferase ring at position 2439 in helix 74 (cross-link 'v'); the cross-link was observed from both P- and A-site bound aminoacyl tRNA, and it is noteworthy that also this position (2439) directly neighbours a modified nucleotide and is identical with a tRNA footprint site.

Wower et al (1989) identified a cross-link from the 3'-terminal residue of P-site bound tRNA to nucleotide 1945 in helix 71 of the 23S RNA (cross-link 'n'). This is to a different domain of the 23S molecule, but - as has been previously discussed (Brimacombe et al, 1993) - it is clear from Figure 1 that the latter domain is topographically connected to the peptidyl transferase area by the ultraviolet induced intra-RNA cross-links 'r', 's' and 't' (Mitchell et al, 1990). Other intra-RNA cross-links ('o', 'p' and 'q' (Döring et al, 1991)) indicate that the 23S RNA region comprising helices 67 to 71 is compactly folded, and again a number of modified nucleotides and tRNA footprint sites are located in the immediate vicinity.

Other cross-linking data serve to bring yet more regions of the 23S RNA into the neighbourhood of the peptidyl transferase centre. Thus, cross-link 'x' (Figure 1) links the highly-modified loop-end of helix 35 (which is nearly 2000 bases away in the primary sequence) to helix 73, and the loop-end of helix 72 is also connected to the latter by cross-link 'w'. However, the loop-end of helix 72 itself is involved in other intra-RNA cross-links (Döring et al, 1991) to the remote helices 25 and 40 (not shown in Figure 1). Furthermore, protein L27 - which was found by Wower et al (1989) to be cross-linked to the 3'-terminal nucleotide of tRNA (and which we have also found to be cross-linked to the aminoacyl residue; our unpublished data) - has a number of direct cross-link sites to the 23S RNA in the region of helices 81, 83 and 84 (Brimacombe, 1991). These latter helices thus indirectly form part of the peptidyl transferase cluster.

THE 'ELBOW' REGIONS OF tRNA, AND THE INTERFACE BRIDGE

As just mentioned, protein L27 has been cross-linked to helix 84 of the 23S RNA (at the base of the stem-loop structure), and - similarly - protein L5 has been cross-linked to the loop-end of this helix (Brimacombe, 1991); immuno electron microscopy studies place these proteins at the base (L27) and the tip (L5) of the central protuberance of the 50S subunit (Stöffler-Meilicke and Stöffler, 1990). In our site-directed cross-linking studies with tRNA, we have now identified a cross-link from the X-base (modified with 4-azido-hippuryl N-hydroxysuccinimide ester) at position 47 in the elbow region of tRNA^{Phe} at the P-site to nucleotide 2309 of the 23S RNA (cross-link 'y' in Figure 1). This site is also at the loop-end of helix 84, and thus provides a link from the elbow region of the P-site tRNA to the top of the central protuberance of the 50S subunit and protein L5, and hence - via protein L27 - to the complex cluster of 23S RNA elements in the peptidyl transferase region as discussed in the previous section. In contrast to the case of cross-links 'a' and 'a''' (see above), cross-link 'y' can be clearly attributed to the P-site, since this cross-link was not observed in corresponding experiments with A- or E-site bound tRNA (our unpublished data).

Since the decoding region is in the 30S subunit and the peptidyl transferase centre in the 50S, it follows that the elbow regions of the P- and A-site tRNAs must *a priori* lie in the subunit interface area, and in fact we have recently been able to analyse cross-links between the 16S and 23S RNA (Mitchell et al, 1992) which provide a direct connection between the decoding and peptidyl transfer regions. These cross-links ('k' and 'l' in Figure 1) link either the loop-end of helix 45 or the base of helix 44 in the 16S RNA to the loop-end of helix 69 in the 23S RNA, and - yet again - involve RNA regions rich in modified nucleotides and tRNA footprint sites. Cross-links 'k' and 'l' correspond well with the 'interface bridge' observed by Frank et al (1991) in their three-dimensional reconstruction of the 70S ribosome from electron micrographs in ice, and the next (and final) section describes how the tRNA-mRNA complex - and the cross-linking data of Figure 1 - can be fitted to the Frank model.

LOCATING THE FUNCTIONAL COMPLEX IN THE 70S RIBOSOME

Figure 2 shows a sketch of the 70S ribosome, adapted from the reconstruction of Frank et al (1991), together with the tRNA-mRNA complex (to scale) in the same orientation as that of Figure 1. It can be seen that the complex fits neatly into the interface cavity, with the decoding region located at the head-body junction of the 30S subunit, and the peptidyl transferase centre (as indicated by the position of the aminoacyl residue) lying at the base of the central protuberance of the 50S subunit. In accordance with the data just described, position 47 of the P-site tRNA is close to the tip of the central protuberance, and the A-site tRNA lies roughly along the interface 'bridge' connecting the decoding and peptidyl transferase areas. Our model-building studies - which are still far from complete - indicate that all of the 16S and 23S RNA elements shown in Figure 1 can be arranged via the cross-linking data in a coherent structure which fits closely to the geometry of the interface cavity as illustrated.

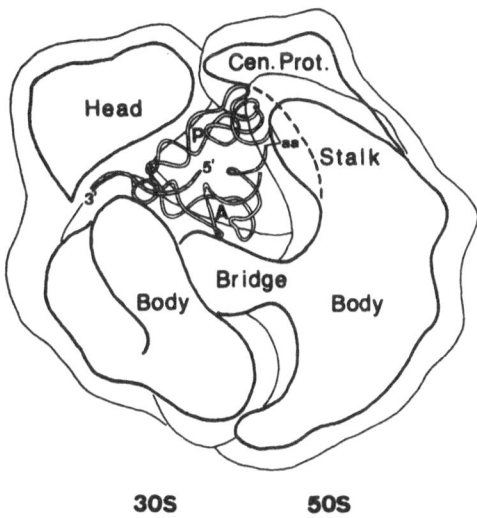

30S   **50S**

Figure 2. Incorporation of the tRNA-mRNA complex into the electron microscopic model of Frank et al (1991). The tRNA-mRNA complex is in the same orientation as that of Figure 1, and lies deep within the interface cavity so that the outgoing 5'-side of the mRNA (if prolonged) would pass out downwards through the plane of the paper, between the P-site tRNA and the head of the 30S subunit. 'Cen. prot.' is an abbreviation for central protuberance.

The model of Frank et al (1991) was reconstructed from micrographs of ribosomes in amorphous ice, and shows important differences to previous electron microscopically derived models (e.g. Oakes et al, 1990; Stöffler-Meilicke and Stöffler, 1990), where the ribosomes or their subunits were subjected to negative staining. The most obvious difference is the 'bridge', a feature which has not been reported before, (although it is interesting to note that the reconstruction from stained particles made by Yonath et al (1990) shows a remarkable similarity to that of Frank et al (1991); however, Yonath et al (1990) interpreted the plane of the subunit interface as being 'horizontal', rather than 'vertical' as in Figure 2). Just as important is the observation that in the Frank model the upright 'platform' feature as seen by Oakes et al (1990) is rather a flattish horizontal 'ledge' (not visible in Figure 2), and it is this difference which enables the two tRNAs to be fitted into the interface cavity in the manner shown in Figure 2. In this arrangement the tRNAs are in the 'R' configuration (Lim et al, 1992) as opposed to the 'S' configuration preferred by Wower et al (1989), (see Lim et al (1992) for discussion).

As noted in the Introduction, we believe that the further application of the site-directed cross-linking approach should lead to a definitive answer to the question of which tRNA configuration ('R' or 'S') is the correct one. At the same time, the geometry of the interface cavity (Figure 2) suggests that the site-directed cross-linking data, combined with other cross-linking data or secondary structural constraints, should enable a considerable proportion of the 16S and 23S RNA molecules to be incorporated directly or indirectly into a model involving at least parts of all the major features (head, body, protuberances, etc.) of the 30S and 50S subunits. We expect that it will then be possible to extrapolate this structure, with the help of the RNA-protein cross-linking and footprinting data used to construct the earlier models (Brimacombe et al, 1988; Stern et al, 1988; Mitchell et al, 1990), towards a model of the complete ribosomal RNA.

REFERENCES

Bochkariov, D., and Kogon, A., 1992, *Analyt. Biochem.* 204:90.

Branlant, C., Krol, A., Machatt, M.A., Pouyet, J., and Ebel, J.P., 1981, *Nucleic Acids Res.* 9:4303.

Brimacombe, R., Atmadja, J., Stiege, W., and Schüler, D., 1988, *J. Mol. Biol.* 199:115.

Brimacombe, R., Greuer, B., Mitchell, P., Osswald, M., Rinke-Appel, J., Schüler, D., and Stade, K. 1990, *in* "The Ribosome; Structure, Function and Evolution", W.E. Hill, A.E. Dahlberg, R.A. Garrett, P.B. Moore, D. Schlessinger, and J.R. Warner, eds., ASM Press, Washington DC, p.93.

Brimacombe, R., 1991, *Biochimie* 73:927.

Brimacombe, R., Mitchell, P., Osswald, M., Stade, K., and Bochkariov, D., 1993, *FASEB J.*, in press.

Capel, M.S., Kjeldgaard, M., Engelman, D.M., and Moore, P.B., 1988, *J. Mol. Biol.* 200:65.

Carbon, P., Ehresmann, C., Ehresmann, B., and Ebel, J.P., 1979, *Eur. J. Biochem.* 100:399.

Chen, J.K., Franke, L.A., Hixson, S.S., and Zimmermann, R.A., 1985, *Biochemistry* 24:4777.

Dontsova, O., Kopylov, A., and Brimacombe, R., 1991, *EMBO J.* 10:2613.

Dontsova, O., Dokudovskaya, S., Kopylov, A., Bogdanov, A., Rinke-Appel, J., Jünke, N., and Brimacombe, R., 1992, *EMBO J.* 11:3105.

Döring, T., Greuer, B., and Brimacombe, R., 1991, *Nucleic Acids Res.* 19:3517.

Döring, T., Greuer, B., and Brimacombe, R., 1992, *Nucleic Acids Res.* 20:1593.

Frank, J., Penczek, P., Grassucci, R., and Srivastava, S., 1991, *J. Cell Biol.* 115:597.

Hajnsdorf, E., Favre, A., and Expert-Bezançon, A., 1989, *Nucleic Acids Res.* 17:1475.

Haselman, T., Camp, D.G., and Fox, G.E., 1989, *Nucleic Acids Res.* 17:2215.

Johnson, A.E., Adkins, H.J., Matthews, E.A., and Cantor, C.R., 1982, *J. Mol. Biol.* 156:113.

Lim, V., Venclovas, C., Spirin, A., Brimacombe, R., Mitchell, P., and Müller, F., 1992, *Nucleic Acids Res.* 20:2627.

Lowary, P., Sampson, J., Milligan, J., Groebe, D., and Uhlenbeck, O.C., 1986, *in* "Structure and Dynamics of RNA", P.H. Van Knippenberg, and C.W. Hilbers, eds., NATO ASI Series Vol. 110, Plenum Press, New York, p.69.

Mitchell, P., Osswald, M., Schüler, D., and Brimacombe, R., 1990, *Nucleic Acids Res.* 18:4325.

Mitchell, P., Osswald, M., and Brimacombe, R., 1992, *Biochemistry* 31:3004.

Moazed, D., Stern, S., and Noller, H.F., 1986, *J. Mol. Biol.* 187:399.

Moazed, D., and Noller, H.F., 1987, *Nature* 327:389.

Moazed, D., and Noller, H.F., 1989, *Cell* 57:585.

Moazed, D., and Noller, H.F., 1990, *J. Mol. Biol.* 211:135.

Murgola, E.J., Hijazi, K.A., Göringer, U., and Dahlberg, A.E., 1988, *Proc. Natl. Acad. Sci. USA* 85:4162.

Oakes, M., Scheinman, A., Atha, T., Shankweiler, G., and Lake, J.A., 1990, *in* "The Ribosome; Structure, Function and Evolution", W.E. Hill, A.E. Dahlberg, R.A. Garrett, P.B. Moore, D. Schlessinger, and J.R. Warner, eds., ASM Press, Washington DC, p. 180.

Ofengand, J., Denman, R., Nurse, K., Liebman, A., Malarek, D., Focella, A., and Zenchoff, G., 1988, *Methods Enzymol.* 164:372.

Paulsen, H., Robertson, H.J., and Wintermeyer, W., 1983, *J. Mol. Biol.* 167:411.

Paulsen, H., and Wintermeyer, W., 1986, *Biochemistry* 25:2749.

Podkowinski, J., and Gornicki, P., 1991, *Nucleic Acids Res.* 19:801.

Powers, T., and Noller, H.F., 1991, *EMBO J.* 10:2203.

Prince, J.B., Taylor, B.H., Thurlow, D.L., Ofengand, J., and Zimmermann, R.A., 1982, *Proc. Natl. Acad. Sci. USA* 79:5450.

Ramakrishnan, V., and White, S.W., 1992, *Nature* 358:768.

Rheinberger, H.J., Sternbach, H., and Nierhaus, K.H., 1986, *J. Biol. Chem.* 261:9140.

Rich, A., 1974, *in* "Ribosomes", M. Nomura, A. Tissières, and P. Lengyel, eds., Cold Spring Harbor Press, New York, p.871.

Rinke-Appel, J., Jünke, N., Stade, K., and Brimacombe, R., 1991, *EMBO J.* 10:2195.

Ryabova, L.A., Selivanova, O.M., Baranov, V.I., Vasiliev, V.D., and Spirin, A.S., 1988, *FEBS Lett.* 226:255.

Shen, Z., and Fox, T.D, 1989, *Nucleic Acids Res.* 17:4535.

Sigmund, C.D., Ettayebi, M., and Morgan, E.A., 1984, *Nucleic Acids Res.* 12:4653.

Smith, J.E., Cooperman, B.S., and Mitchell, P., 1992, *Biochemistry* 31:10825.

Sprinzl, M., Hartman, T., Weber, J., Blank, J., and Zeidler, R., 1989, *Nucleic Acids Res.* 17:r1.

Stade, K., Rinke-Appel, J., and Brimacombe, R., 1989, *Nucleic Acids Res.* 17:9889.

Steiner, G., Kuechler, E., and Barta, A., 1988, *EMBO J.* 7:3949.

Stern, S., Weiser, B., and Noller, H.F., 1988, *J. Mol. Biol.* 204:447.

Stöffler-Meilicke, M., and Stöffler, G., 1990, *in* "The Ribosome; Structure, Function and Evolution", W.E. Hill, A.E. Dahlberg, R.A. Garrett, P.B. Moore, D. Schlessinger, and J.R. Warner, eds., ASM press, Washington DC, p. 123.

Sundaralingam, M., Brennan, T., Yathindra, N., and Ichikawa, T., 1975, *in* "Structure and Conformation of Nucleic Acids and Protein-Nucleic Acid Interactions", M. Sundaralingam, and S.T. Rao, eds., University Park Press, Baltimore, p.101.

Tate, W., Greuer, B., and Brimacombe, R., 1990; *Nucleic Acids Res.* 18:6537.

Vester, B., and Garrett, R.A., 1988, *EMBO J.* 7:3577.

Wower, J., Hixson, S.S., and Zimmermann, R.A., 1989, *Proc. Natl. Acad. Sci. USA* 86:5232.

Yonath, A., Leonard, K.R., and Wittmann, H.G., 1987, *Science* 236:813.

Yonath, A., Bennett, W., Weinstein, S., and Wittmann, H.G., 1990, *in* "The Ribosome; Structure, Function and Evolution", W.E. Hill, A.E. Dahlberg, R.A. Garrett, P.B. Moore, D. Schlessinger, and J.R. Warner, eds., ASM Press, Washington DC, p.134.

THE ARRANGEMENT OF tRNA IN THE RIBOSOME

Alexander S. Spirin,[1] Valery I. Lim,[1]
and Richard Brimacombe[2]

[1]Institute of Protein Research, Russian Academy of
 Sciences, Pushchino, Moscow Region, 142292, Russia, and
[2]Max-Planck-Institut für Molekulare Genetik, Abteilung
 Wittmann, Ihnestrasse 73, 1000 Berlin 33, Germany

INTRODUCTION

The main reaction performed by the translating ribosome is *trans-peptidation*. Two substrates participating in the reaction are peptidyl-tRNA and aminoacyl-tRNA:

$$\text{Pept(n)-tRNA}' + \text{Aa-tRNA}'' \longrightarrow \text{tRNA}'_{OH} + \text{Pept(n+1)-tRNA}''.$$

The substrate-binding sites of the ribosome for the two substrates, peptidyl-tRNA and aminoacyl-tRNA, are called the *P* and *A sites*, respectively. This is the principal definition of the two sites.

The question is where the A and P sites are located on the ribosome. We intentionally confine ourselves to the consideration of just the A and P sites as two principal positions of the reacting substrates of the ribosome, at the same time being aware of the existence of a series of intermediate states of the tRNAs on the way of codon-dependent binding, during translocation and on leaving the ribosome, including the E-site state.

BASIC POSTULATES AND FACTS

The following obvious postulates must be put at the basis of any considerations concerning the tRNA positions and movements in the translating ribosome (Rich, 1974; Sundaralingam et al., 1975; Spirin, 1985): (1) The anticodons of the two tRNAs, one in the A site and the other in the P site, must be drawn together, in order to provide their interactions with neighbour codons along mRNA. (2) The acceptor ends of the two tRNAs also must be in close proximity, in order to provide the transpeptidation reaction. The central cores ("elbows") of the two L-shaped tRNAs may be drawn apart. Thus, the two tRNAs form a tRNA pair that can be considered as a unit in the search of its position on the ribosome.

The interpretations of electron-microscopic images of ribosomes and their subunits made by various research groups have evolved to the point where there is a virtual consensus of agreement between different models.

The Translational Apparatus, Edited by K.H. Nierhaus
et al., Plenum Press, New York, 1993

Recently the most objective and detailed description of the ribosome structure has been provided by the image processing (Van Heel and Stöffler-Meilicke, 1985) and three-dimensional image reconstruction techniques (Frank et al., 1988, 1990, 1991), and these results are taken here as a structural basis for considering the problem of the tRNA position on the ribosome. The main features of the refined model are the very constricted neck of the small subunit, and the "interface canyon" of the large subunit, with the large area between the L7/L12 stalk and the central protuberance being exposed (non-covered by the small subunit). Schematic contour representation of the so-called overlap projection of the ribosome (70S) is given in Figure 1.

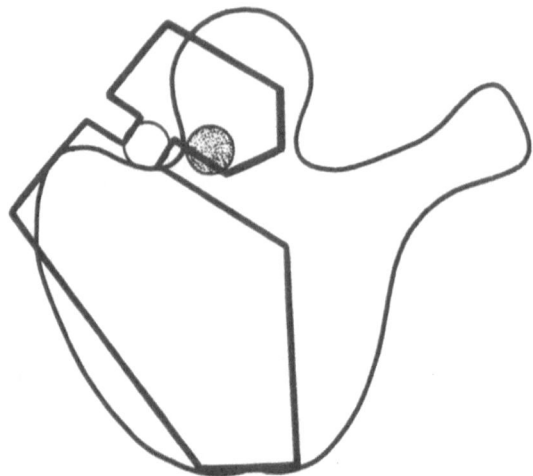

Figure 1. Schematic contours of two ribosomal subunits (30S and 50S) in the 70S ribosome "overlap" projection, as viewed from the external surface of the small (30S) subunit. The open circle designates the position of the tRNA anticodons (AC) on the small (30S) subunit, and the shaded circle is the position of the peptidyl transferase center (PT) on the large (50S) subunit. It is seen that the anticodon region is hanging over the peptidyl transferase thus implying the axis of translocation (see text) to be not far declined from perpendicular to the interface.

Immuno-electron microscopy, foot-printing studies, and chemical cross-linking experiments give strong evidence for localization of the tRNA anticodons and the tRNA acceptor ends, as well as elongation factors on the ribosome (see the review by Lim et al., 1992). The basic facts are as follows:
(1) The anticodons of the tRNAs and codons of mRNA are at the 30S subunit of the translating ribosome, in the cleft separating the head on one side and the "platform" and the body on the other, i.e., at the neck (see Figure 1, open circle).
(2) The acceptor ends of the tRNAs (the peptidyl transferase center) are at the 50S subunit, in the interface canyon under the central protuberance (see Figure 1, shaded circle).
(3) The elongation factors, EF-Tu and EF-G, have their binding site in the region of the L7/L12 stalk and its base at the 50S subunit. When the ternary complex (Aa-tRNA·EF-Tu·GTP) is bound with its cognate codon and with the factor-binding site, the position of the tRNA seems to be close to that of tRNA in the A site: the foot-prints on 16S RNA are identical, though those on 23S RNA are not yet realized (Noller et al., 1990).

MUTUAL ARRANGEMENT OF THE TWO tRNAs

For simplicity we designate the two sides of the L-shaped tRNA as the *D side* (the D-loop is on this side) and the *T side* (it is marked by the TΨ-loop). When aminoacyl-tRNA and peptidyl-tRNA are present in the ribosome, the axis connecting the region of the anticodons of the two tRNAs and their acceptor ends will be considered as the *axis of translocation*. The dihedral angle between the planes of the tRNAs is symbolized as ω. We shall assume that $\omega = 0°$ when the tRNA planes are oriented in parallel, with the T side of the A-site tRNA facing the D side of the P-site tRNA. The detailed stereochemical analysis of the plausible mutual orientations of the two tRNA molecules has been recently reported by us (Lim et al., 1992). It is based on the stereochemistry for the anticodon loops and for the rest of the tRNA molecules as determined by crystallographic studies (Kim et al., 1974; Robertus et al., 1974; Moras et al., 1980), and we have made no serious distortions of this crystal structure.

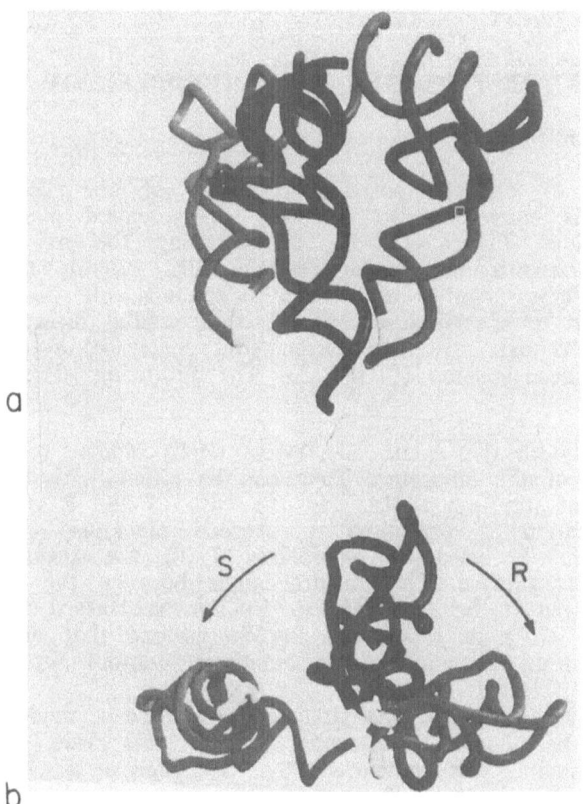

Figure 2. Mutual orientation of tRNA molecules. The ribose-phosphate backbones of the tRNA molecules are depicted; the A-site tRNA darker, the P-site tRNA lighter, with the anticodon regions in white. The A-site tRNA and the right-hand P-site tRNA are in the R-orientation. The A-site tRNA and the left-hand P-site tRNA are in the S-orientation.

a: Projection with the anticodons down and the A-site tRNA elbow facing the viewer. The axis of translocation is vertical.

b: Projection with the anticodons facing the viewer and the A-site tRNA elbow up. The axis of translocation is perpendicular to the figure plane.

Two principally different orientations can be considered (Figure 2). The first is the approximately perpendicular orientation with $\omega = 90°$ ($\pm 45°$) first proposed by Rich (1974) and analyzed in detail by Spirin and Lim (1986); we refer to it as the *R type orientation*. Here the T side of the A-site tRNA faces the D side of the P-site tRNA, and translocation will proceed *clockwise* if viewed from anticodons along the translocation axis. The parallel orientation with $\omega = 0°$ and up to $+45°$ (which is the extreme R type) is physically impossible because the maximal distance between neighbouring nucleosides of two adjacent codons (9 Å for a fully extended conformation) is shorter than the distance between parallel anticodons in this orientation (10 to 20 Å). The second is also a roughly perpendicular orientation, but $\omega = 270°$ ($\pm 45°$); it was proposed and analyzed by Sundaralingam et al. (1975) and so it is designated as the *S type orientation*. In this case the D side of the A-site tRNA faces the T side of the P-site tRNA, and the translocational movement will be *counterclockwise*. Versions of the same S type orientation were also discussed by McDonald and Rein (1987), Prabahakaran and Harvey (1989) and Nagano et al. (1991). The parallel orientation with $\omega = 360°$ (or up to $-45°$) is stereochemically possible and can be considered just as the extreme S orientation. A roughly coplanar arrangement with $\omega = 180°$ ($\pm 45°$) can be assigned as the intermediate *R/S type orientation*.

POSITION OF THE TWO tRNAs-mRNA COMPLEX ON THE RIBOSOME

Classification of Models

In order to classify possible positions of the tRNA pair on the ribosome, it is convenient to consider the so-called overlap projection of the 70S particle (Figure 1). If viewed along the axis of translocation, from the codon-anticodon region on the 30S subunit (the neck) to the peptidyl transferase center on the 50S subunit, all possible positions of the tRNAs can be classified according to the radial directions of the tRNA planes from the axis of translocation (the center of rotation) towards the periphery of the ribosome (Figure 3). The following classes of models can be discussed.

Class I: Both the A-site and P-site tRNA elbows (corners) are on the left (L1) side of the ribosome. They can be either with the R or with the S orientation of the two tRNAs. In the case of the R orientation (subclass I R) translocation is presumed to proceed clockwise in this projection, whereas in the S orientation (subclass I S) the translocational rotation will be counterclockwise. The dihedral angle between the planes of the two tRNAs is allowed to be from 45° to 135° in subclass I R and from 0° to 135° (360° to 225°) in subclass I S. We assume that all positions of the tRNA pair within the left-hand hemicircle contour in Figure 3, upper panel, belong to class I.
Lake (1977, 1980) was the first who proposed a model for the position of tRNAs on the ribosome; the model was of this class, of subclass I R. I S subclass model was advanced by Ofengand et al. (1986) and more recently by Wower et al. (1989).

Class II: Both the A-site and P-site tRNA elbows are on the right (L7/L12) side of the ribosome. Again, models of this class can be either with the R or with the S orientations of the two tRNAs. The corresponding II R and II S subclass models will be characterized by the same directions of translocational rotation and the same limits of the dihedral angles as in the case of the R and S orientations, respectively, in the class I models. Class II is assumed to include all positions of the tRNA pair within the right-hand hemicircle in Figure 3, middle panel.

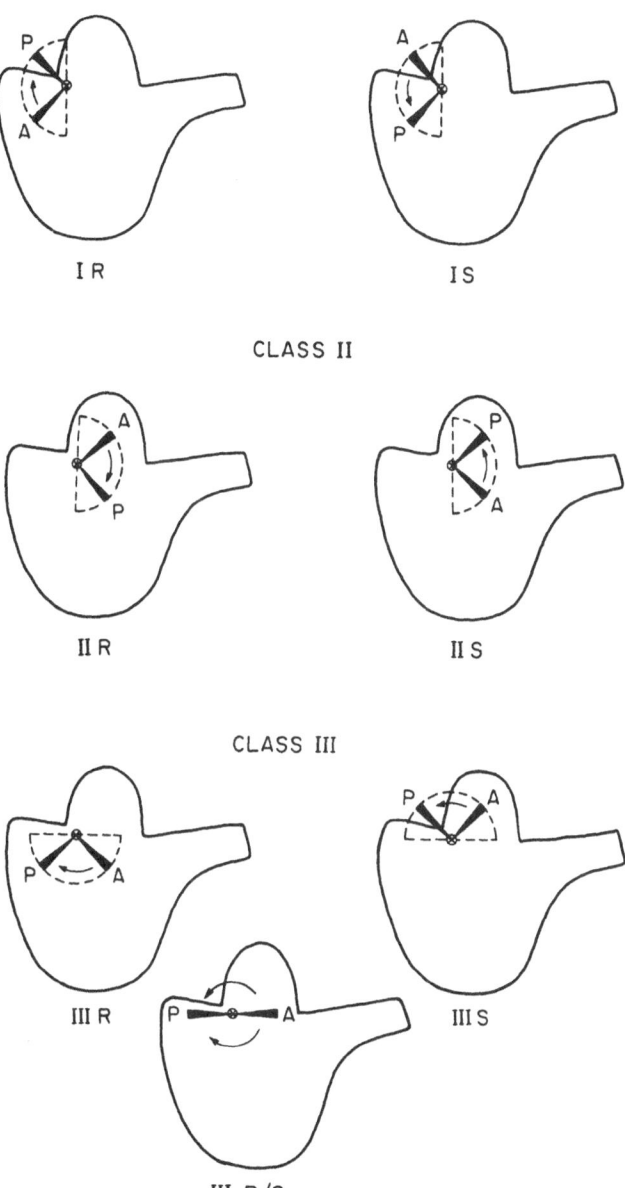

Figure 3. Contours of the large (50S) ribosomal subunit as viewed from the interface, with the positions of the tRNA pair. The axis of translocation marked by crossed circle is perpendicular to the figure plane (this plane does not necessarily coincide with the interface plane). The two tRNAs are symbolized by hands, the narrow ends being the overlapped anticodon and acceptor regions and the thick ends representing the central cores ("elbows"). A and P are positions of the A-site and P-site tRNAs, respectively, and the arrow is the direction of translocational rotation of the A-site tRNA. The hemicircle outlines the area allowed for occupation by the tRNAs.

The model of this class, the II R subclass, was proposed by Spirin (1983) and supported by Wagenknecht et al. (1989). Though given in more general terms, the model of Gold (1988) seems also to fit class II and rather subclass II R.

Class III: The elbow of the A-site tRNA is on the right (L7/L12) side whereas that of the P-site tRNA is on the left (L1) side, within the limits of the right and the left hemicircles, respectively. Three subclasses can be distinguished: III R, III S, and III R/S.

The model recently proposed and advocated by us (Figure 5,a in Lim et al., 1992; see also Figure 8,a in Dontsova et al., 1992) belongs to the subclass III R. The models earlier proposed by Noller et al. (1990) and Wower and Zimmermann (1991) seem to be a case of the III S subclass.

Formally, **Class IV** models also could be considered where the elbow of the A-site tRNA is on the left (L1) site whereas that of the P-site tRNA is on the right (L7/L12). Models of this class, however, strongly contradict the experimental data available, and first of all they are not consistent with the position of EF-Tu on the right side. Nobody seems to have proposed a model of this class. We shall not discuss this unlikely theoretical case.

Testing the Models

The models classified and presented schematically in Figure 3 can be tested by experimental facts and the knowledge available.

(1) The position of EF-Tu on the right side of the overlap projection of the ribosome (Girshovich et al., 1986) presumes that the A site is to be located somewhere in the vicinity of the L7/L12 stalk base, at least at this side of the ribosome. Only in this case serious perturbations of tRNA contacts with 16S RNA can be avoided during the EF-Tu release and A site occupation (Noller et al., 1990). This strong test rejects the class I models.

(2) The effect of the P-site tRNA anticodon loop on the A-site occupancy implies a steric contact, specifically of the 5'-side of the P-site anticodon loop, with the A-site codon-anticodon duplex (Smith and Yarus, 1989). This is the case of the R series models (Lim et al., 1992). The effect is hardly consistent with all models of the S series, namely I S, II S, and III S, as the parts to be in contact are drawn apart at this tRNA orientation (see Figure 2).

(3) The III R/S subclass models can be rejected by the tRNA-tRNA distance measurements made by the energy transfer technique (Johnson et al., 1982; Paulsen et al., 1983): the distance between the central parts of two tRNAs in the co-planar orientation would be significantly longer than 30 or 35 Å measured between the A-site and P-site tRNAs.

(4) The experiments on chemical cross-linking of the A-site and P-site tRNAs with the ribosome could be very promising, but unfortunately the results (reviewed by Wower and Zimmermann, 1991) show a distribution of agreements and disagreements with any class of the models. Sometimes there are mutual contradictions even in the results of the same authors (see the discussion in Lim et al., 1992). At the same time, the results of cross-linking of mRNA codon or near-codon regions with the ribosome (Stade et al., 1989; Rinke-Appel et al., 1991; Dontsova et al., 1991, 1992) give some useful information. The cross-link of the near-codon "outgoing" (5') mRNA section with protein S7 contradicts the models I R, I S and III S. The cross-links of the A-site codon with protein S5 and helix 34 contradict the model III S. In more detail these results are discussed elsewhere (Lim et al., 1992; Dontsova et al., 1992).

The summary of testing the models reviewed in Figure 3 is given in Table 1.

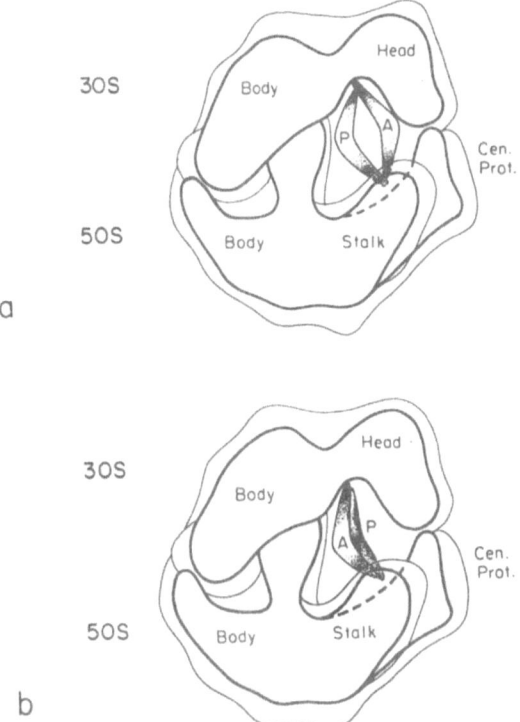

Figure 4. The tRNA pair superimposed on the 70S ribosome model of Frank et al. (1991). The 30S subunit is up, and the L7/L12 stalk is towards the reader. The A-site and P-site tRNAs are depicted in the R-orientation.
a: Class II model (II R).
b: Class III model (III R).

Table 1. Testing the models proposed for tRNA positions on the ribosome by experimental data available

Models:	Experimental data groups:				
	EF-Tu position	P-site tRNA effect	Energy transfer: tRNA-tRNA	mRNA codon-30S 5'-side - protein S7	head cross-links: 3'-side - protein S5, helix 34
I R	−	+	+	−	+
I S	−	−	+	−	+
II R	+	+	+	+	+
II S	+	−	+	+	+
III R	+	+	+	+	+
III S	+	−	+	−	−
III R/S	+	±	−	±	+

CONCLUSION

Thus, just two model subclasses have stood all the tests, namely II R and III R. The corresponding two models are shown schematically superimposed on the 70S model of Frank et al. (1991) in Figure 4. One of them (Figure 4,a) is a slight modification of the model proposed by Spirin (1983): both the tRNAs occupy the cavity of the 70S ribosome at the base of the L7/L12 stalk with the elbows facing the L7/L12 stalk side (i.e. the viewer), the anticodons protruding into the 30S subunit neck and the acceptor ends reaching the groove under the 50S central protuberance. The second one (Figure 4,b) has been recently proposed by us (Lim et al., 1992): the tRNA pair is turned along the translocation axis in such a way (by about 90°) that the tRNA cores are mainly in the interface canyon, with the same positions of their anticodons and acceptor ends as in the previous model; the elbows are directed towards the 50S body.

We believe that only these two principal positions of the tRNA pair (the A-site and the P-site) on the ribosome are consistent with the bulk of experimental data available. We hope that more detailed model constructions and further experimental work may derive benefit from the classification proposed and the selection made.

REFERENCES

Dontsova, O., Kopylov, A. and Brimacombe, R., 1991, The location of messenger-RNA in the ribosomal 30S initiation complex. Site-directed cross-linking of messenger-RNA analogs carrying several photo-reactive labels simultaneously on either side of the AUG start codon, *EMBO J.* 10:2613.

Dontsova, O., Dokudovskaya, S., Kopylov, A., Bogdanov, A., Rinke-Appel, J., Jünke, N., and Brimacombe, R., 1992, Three widely separated positions in the 16S RNA lie in or close to the ribosomal decoding region; a site-directed cross-linking study with mRNA analogues, *EMBO J.* 11:3105.

Frank, J., Radermacher, M., Wagenknecht, T., and Verschoor, A., 1988, Studying ribosome structure by electron microscopy and computer-image processing, *Methods Enzymol.* 164:3.

Frank, J., Verschoor, A., Radermacher, M., and Wagenknecht, T., 1990, Morphologies of eubacterial and eucaryotic ribosomes as determined by three-dimensional electron microscopy, in: "The Ribosome. Structure, Function, and Evolution", W.E.Hill, A.E.Dahlberg, R.A.Garrett, P.B.Moore, D.Schlessinger, and J.Warner, eds., ASM Press, Washington, D.C., p.107.

Frank, J., Penczek, P., Grassucci, R., and Srivastava, S., 1991, Three-dimensional reconstruction of the 70S *Escherichia coli* ribosome in ice: the distribution of ribosomal RNA, *J. Cell Biol.* 115:597.

Girshovich, A.S., Bochkareva, E.S., and Vasiliev, V.D., 1986, Localization of elongation factor Tu on the ribosome, *FEBS Lett.* 197:192.

Gold, L., 1988, Posttranscriptional regulatory mechanisms in *Escherichia coli*, *Annu. Rev. Biochem.* 57:199.

Johnson, A.E., Adkins, H.J., Matthews, E.A., and Cantor, C.R., 1982, Distance moved by transfer RNA during translocation from the A site to the P site on the ribosome, *J. Mol. Biol.* 156:113.

Kim, S.H., Suddath, F.L., Quigley, G.J., McPherson, A., Sussman, J.L., Wang, A.H.J., Seeman, N.C., and Rich, A., 1974, Three-dimensional tertiary structure of yeast phenylalanine transfer RNA, *Science* 185:435.

Lake, J.A., 1977, Aminoacyl-tRNA-binding at the recognition site is the first step of the elongation cycle of protein synthesis, *Proc. Natl. Acad. Sci. USA* 74:1903.

Lake, J., 1980, Ribosome structure and functional sites, in: "Ribosomes. Structure, Function, and Genetics", G.Chambliss, G.R.Craven, J.Davies, K.Davis, L.Kahan, and M.Nomura, eds., University Park Press, Baltimore, p.207.

Lim, V., Venclovas, C., Spirin, A., Brimacombe, R., Mitchell, P., and Müller, F., 1992, How are tRNAs and mRNA arranged in the ribosome? An attempt to correlate the stereochemistry of the tRNA-mRNA interaction with constraints imposed by the ribosomal topography, *Nucleic Acids Res.* 20:2627.

McDonald, J.J., and Rein, R., 1987, A stereochemical model of the transpeptidation complex, *J. Biomolec. Struct. Dynamics* 4:729.

Moras, D., Comarmond, M.B., Fischer, J., Weiss, R., Thierry, J.C., Ebel, J.P., and Giegé, R., 1980, Crystal-structure of yeast transfer RNA-Asp, *Nature* 288:669.

Nagano, K., Takagi, H., and Harel, M., 1991, The side-by-side model of transfer-RNA molecules allowing the alpha-helical conformation of the nascent polypeptide during the ribosomal transpeptidation, *Biochimie* 73:947.

Noller, H.F., Moazed, D., Stern, S., Powers, T., Allen, P.N., Robertson, J.M., Weiser, B., and Triman, K., 1990, Structure of rRNA and its functional interactions in translation, in: "The Ribosome. Structure, Function, and Evolution", W.E.Hill, A.E.Dahlberg, R.A.Garrett, P.B.Moore, D.Schlessinger, and J.R.Warner, eds., ASM Press, Washington, D.C., p.73.

Ofengand, J., Ciesiolka, J., Denman, R., and Nurse, K., 1986, Structural and functional interactions of the tRNA-ribosome complex, in: "Structure, Function, and Genetics of Ribosomes", B.Hardesty, and G.Kramer, eds., Springer-Verlag, New York, p.473.

Paulsen, H., Robertson, H.J., and Wintermeyer, W., 1983, Topological arrangement of two transfer RNAs on the ribosome. Fluorescence energy transfer measurements between A and P site-bound tRNA^Phe, *J. Mol. Biol.* 167:411.

Prabahakaran, M., and Harvey, S.C., 1989, Models for two tRNAs bound to successive codons on mRNA on the ribosome, *J. Biomolec. Struct. Dynamics* 7:167.

Rich, A., 1974, How transfer RNA may move inside the ribosome, in: "Ribosomes", M.Nomura, A.Tissières, and P.Lengyel, eds., Cold Spring Harbor Press, New York, p.871.

Rinke-Appel, J., Jünke, N., Stade, K., and Brimacombe, R., 1991, The path of messenger-RNA through the *Escherichia coli* ribosome. Site-directed cross-linking of messenger-RNA analogs carrying a photo-reactive label at various points 3' to the decoding site, *EMBO J.* 10:2195.

Robertus, J.D., Ladner, J.E., Finch, J.T., Rhodes, D., Brown, R.S., Clark, B.F.C., and Klug, A., 1974, Structure of yeast phenylalanine tRNA at 3 Å resolution, *Nature* 250:546.

Smith, D., and Yarus, M., 1989, tRNA-tRNA interactions within cellular ribosomes, *Proc. Natl. Acad. Sci. USA* 86:4397.

Spirin, A.S., 1983, Location of tRNA on the ribosome, *FEBS Lett.* 156:217.

Spirin, A.S., 1985, Ribosomal translocation: facts and models, *Prog. Nucleic Acid Res. Mol. Biol.* 32:75.

Spirin, A.S., and Lim, V., 1986, Stereochemical analysis of ribosomal transpeptidation, translocation, and nascent peptide folding, *in*: "Structure, Function, and Genetics of Ribosomes", B.Hardesty, and G.Kramer, eds., Springer-Verlag, New York, p.556.

Stade, K., Rinke-Appel, J., and Brimacombe, R., 1989, Site-directed cross-linking of mRNA analogues to the *Escherichia coli* ribosome; identification of 30S ribosomal components that can be cross-linked to the mRNA at various points 5' with respect to the decoding site, *Nucleic Acids Res.* 17:9889.

Sundaralingam, M., Brennan, T., Yathindra, N., and Ichikawa, T., 1975, Stereochemistry of messenger RNA (codon)-transfer RNA (anticodon) interaction on the ribosome during peptide bond formation, *in*: "Structure and Conformation of Nucleic Acids and Protein-Nucleic Acid Interactions", M.Sundaralingam, and S.T.Rao, eds., University Park Press, Baltimore, p.101.

Van Heel, M., and Stöffler-Meilicke, M., 1985, Characteristic views of *E.coli* and *B.stearothermophilus* 30S ribosomal subunits in the electron microscope, *EMBO J.* 4:2389.

Wagenknecht, T., Carazo, J.M., Radermacher, M., and Frank, J., 1989, Three-dimensional reconstruction of the ribosome from *Escherichia coli*, *Biophys. J.* 55:455.

Wower, J., and Zimmermann, R.A., 1991, A consonant model of the transfer RNA-ribosome complex during the elongation cycle of translation, *Biochimie* 73:961.

Wower, J., Hixson, S.S., and Zimmermann, R.A., 1989, Labeling the peptidyl-transferase center of the *Escherichia coli* ribosome with photoreactive tRNA[Phe] derivatives containing azidoadenosine at the 3' end of the acceptor arm: A model of the tRNA-ribosome complex, *Proc. Natl. Acad. Sci. USA* 86:5232.

A MODEL OF THE tRNA BINDING SITES ON THE *ESCHERICHIA COLI* RIBOSOME

Jacek Wower[1], Lee A. Sylvers[1,3], Kirill V. Rosen[1], Stephen S. Hixson[2] and
Robert A. Zimmermann[1]

[1]Department of Biochemistry and Molecular Biology
[2]Department of Chemistry
[3]Program in Molecular and Cellular Biology
 University of Massachusetts
 Amherst, Massachusetts 01003 (USA)

INTRODUCTION

In the course of the elongation cycle of translation, three different functional forms of tRNA bind to mRNA-programmed ribosomes. Two of them, aminoacyl-tRNA and peptidyl-tRNA, participate in the formation of the peptide bond while accommodated in the A (acceptor) and P (peptidyl) sites. A third form, deacylated tRNA, a product of the transpeptidation reaction, is transferred from the P site to the E (exit) site during translocation before it leaves the ribosome. Since binding of tRNA to the ribosome is essential for protein synthesis, numerous studies have been designed to delineate the topography of the ribosomal A, P and E sites (for review see Cooperman, 1980; Ofengand et al., 1986; Cooperman, 1987).

While a variety of techniques have contributed to our present knowledge about the spatial organization of tRNA binding sites on the ribosome, tRNA-ribosome cross-linking has undoubtedly been the most informative. This chapter will first review recent progress in cross-linking methods applicable to the study of tRNA-ribosome complexes. Cross-linking data pertinent to the topography of tRNA binding sites on the *E. coli* ribosome will then be presented. Finally, a model for the arrangement of the A, P and E site-bound tRNAs on the ribosome will be discussed.

NEW REAGENTS FOR CROSS-LINKING tRNA TO THE RIBOSOME

Although a host of chemically and photochemically reactive reagents are available, photochemical approaches for cross-linking tRNA to the ribosome have predominated in the past few years (Cooperman, 1987). Direct UV irradiation is the simplest method for inducing the formation of covalent, intermolecular bonds (Budowsky and Abdurashidova, 1989). The usefulness of this method is limited by the low and quite similar photoreactivity of the four nucleotide bases commonly found in RNA. Moreover, the cross-linking of unmodified tRNA to ribosomes, usually brought about by irradiation at 254 nm (far-UV), leads to the formation of spurious intraribosomal cross-links that inactivate the ribosomal particles. The adverse effects of far-UV radiation on ribosomes may be avoided by using tRNA probes carrying photoreactive moieties that are activated by near-UV light (300-380 nm). A variety of such probes have been prepared by derivatizing tRNA molecules with photolabile

The Translational Apparatus, Edited by K.H. Nierhaus
et al., Plenum Press, New York, 1993

reagents. The main drawback of this method is that the site of cross-linking is necessarily at some distance, often 10-20Å, from the actual tRNA binding site.

A number of cross-linking studies have exploited the unusually high photoreactivity of certain modified bases such as 5-carboxymethoxyuridine (cmo⁵U) or 5-methoxyuridine (mo⁵U) at position 34 in tRNA$_1^{Val}$ (Prince et al., 1982; Ehresmann et al, 1984), wyosine (Y) at position 37 in tRNA^{Phe} (Matzke et al., 1980) and 4-thiouridine (s⁴U) which occurs at position 8 in a variety of different tRNAs (Ofengand et al., 1974). Because of the scarcity of tRNAs with naturally photoreactive residues, we have explored the utility of introducing photoreactive nucleosides into tRNA molecules for the purpose of cross-linking them to ribosomes. Our first choice was 8-azidoadenosine (8N$_3$A), probably the most common photolabile nucleoside in use today (Haley, 1983). When tRNA containing this nucleoside was bound to the ribosome and irradiated with near UV light, it readily cross-linked to ribosomal components in its immediate neighborhood (Wower et al., 1988). A similar result was obtained when 2-azidoadenosine (2N$_3$A) was incorporated into the tRNA structure (Sylvers et al., 1989; Wower et al., 1989). Of the two derivatives, 2N$_3$A is better suited for tRNA-ribosome cross-linking experiments as replacement of the 3'-terminal A76 with 8N$_3$A renders tRNA inactive in the aminoacylation reaction (Wower et al., 1988). The observation that tRNA containing 2N$_3$A at its 3' terminus can be efficiently aminoacylated, whereas that bearing 8N$_3$A cannot, may be explained by the fact that the former assumes the normal *anti* conformation while the latter prefers the *syn* conformation (Czarnecki, 1984). Most recently, we have synthesized 2,6-diazido-9-(ß-D-ribofuranosyl)purine (2,6diN$_3$R). Incorporation of this highly photoreactive nucleoside into the 3' terminus of tRNA does not affect its amino acid acceptance or its ability to interact with the ribosome, and the corresponding tRNA derivative has been cross-linked to the P site in good yield (our unpublished results).

The choice of nucleosides for constructing tRNA probes for use in photoaffinity studies need not be limited to 2N$_3$A, 8N$_3$A and 2,6diN$_3$R. Figure 1 depicts several other photoreactive nucleoside derivatives which may be useful for modifying tRNA: 5-bromouridine (BrU), 4-thiouridine (s⁴U), 5-azidouridine (5N$_3$U), 2-azidoinosine (2N$_3$I), 8-azidoinosine (8N$_3$I), 8-azidoguanosine (8N$_3$G), and 6-thioguanosine (s⁶G). It has been shown that nucleic acids containing BrU or s⁴U are much more sensitive to UV irradiation than unmodified polynucleotides. Since both photoreactive pyrimidines absorb light at longer wavelengths than ordinary nucleotides, they can be selectively photolyzed with near-UV light. A new and very photoreactive analog of uridine, 5-azidouridine, has recently been synthesized and shown to be over 1000 times more photoreactive than BrU (Evans et al., 1986). While the utility of 5N$_3$U for nucleic acid cross-linking has been demonstrated in the case of mRNA-ribosome complexes (Dontsova et al., 1992a), it has not yet been incorporated into tRNA. 2N$_3$I and 8N$_3$I which can be prepared by deamination of 2N$_3$A and 8N$_3$A (J. W., unpublished), may be useful for replacing either adenosine or guanosine. However, 8N$_3$I might be less suited for introduction into tRNA than 2N$_3$I since, like 8N$_3$A, it has a preference for the *syn* conformation. The same applies to 8N$_3$G, despite its usefulness in studies of GTP-binding proteins (Haley, 1983).

CONSTRUCTION OF PHOTOREACTIVE tRNA DERIVATIVES

An obvious way to incorporate photoreactive nucleotides into tRNA is by *in vitro* transcription of tRNA genes using a DNA-dependent RNA polymerase in the presence of a photoreactive nucleoside triphosphate. Conditions can be worked out to yield an incorporation of approximately one reactive base per tRNA, randomly distributed as to position. While several photoreactive nucleotides, including 5N$_3$UTP (Evans et al., 1986), s⁴UTP and BrUTP (Tanner et al., 1988), and s⁶GTP (J. W., unpublished) have been demonstrated to serve as substrates for T7 RNA polymerase, others, such as the azidopurine derivatives, do not (Knoll et al., 1992; J. W., unpublished). Aside from this limitation, the use of randomly labeled tRNA transcripts also entails identification of the cross-linked base(s). In many situations it is therefore desirable to employ a tRNA derivative in which the photoreactive nucleoside is substituted at a single position in the polynucleotide chain.

We have used "recombinant RNA" technology (reviewed in Cedergren and Grosjean, 1987) to prepare a number of tRNA molecules specifically substituted with photoreactive bases (Wower et al., 1988; Sylvers et al., 1992). This approach, in simple terms, involves the "cutting" and "pasting" of

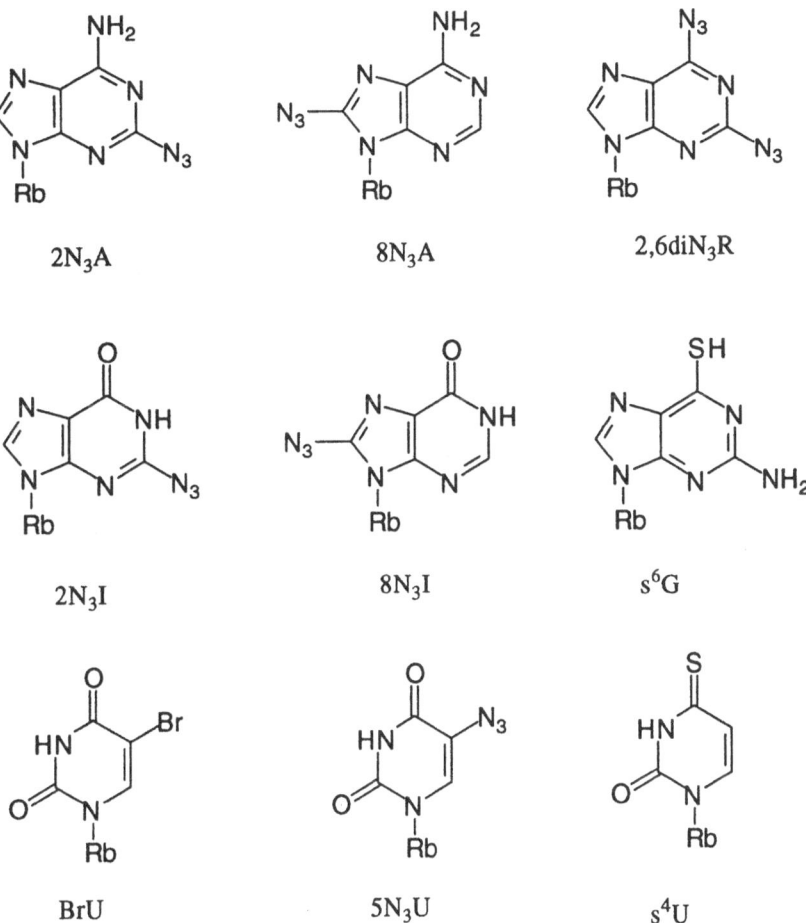

Figure 1. Photoreactive nucleosides. Rb - ribose, see text for further description.

RNA fragments to produce the desired RNA molecule. The "cutting" is carried out with controlled nucleolytic and/or chemical cleavage of tRNA, while the "pasting" entails the introduction of specific mono- and/or oligoribonucleotides, and the reconstruction of the full-length polynucleotide chain, with T4 RNA ligase. The "cut and paste" technique has been used frequently in the past to generate tRNAs with altered anticodons or T loop sequences (Cedergren and Grosjean, 1987). However, this approach has been used to construct photoreactive tRNA derivatives only recently. In the past few years, for example, we have created tRNA derivatives substituted within the 3'-terminal -ACCA sequence (Wower et. al, 1988, 1989, 1993), the anticodon loop (Sylvers et al., 1992; Wower et al., 1993), and the anticodon arm (Wower et al., 1991) with various azidonucleotides, including $2N_3A$, $8N_3A$, and $2,6diN_3R$.

With the advent of site-specific cleavage techniques based on the use of chimeric oligonucleotides and RNase H (Hayase et al., 1990, 1992), it is now possible to incorporate a photoreactive nucleotide virtually anywhere in a tRNA molecule. We are currently using this approach to place photoreactive nucleosides at specific positions within the D and T loops of tRNA^{Phe}. This technique can be further embellished by using synthetic RNAs and *in vitro* transcripts containing photoreactive bases. Such engineered oligoribonucleotides would open virtually limitless horizons for the creation of novel photoreactive RNA molecules.

IDENTIFICATION OF CROSS-LINKED COMPONENTS

In a typical cross-linking experiment, the tRNA derivatives, containing ^{32}P adjacent to the photoreactive nucleotide, are bound to 70S ribosomes in the presence of mRNA and irradiated with light of the appropriate wavelength. The covalent tRNA-ribosome complexes are then separated into 30S- and 50S-subunit fractions by sucrose-gradient centrifugation at low Mg^{++} concentration and each fraction is further dissociated into its protein and RNA constituents by a second round of sucrose-gradient centrifugation in the presence of sodium dodecyl sulfate (SDS) and LiCl. As the labeled proteins are still attached to intact tRNA at this stage, they are prepared for subsequent analysis by digestion with T1 and/or pancreatic RNase which leaves only the radioactively labeled oligonucleotides linked to the proteins.

In the past, cross-linked proteins were generally characterized by one- and two-dimensional polyacrylamide gel electrophoresis. One-dimensional gel analysis in the presence of SDS provides a reasonably accurate estimate of the molecular weight of the labeled protein as long as the size of the attached oligonucleotide is known. In most of the commonly used two-dimensional gel systems, however, the influence of covalently bound oligonucleotides on the mobility of the complex is less predictable, often leading to greater than expected shifts in protein position and, in some cases, errors in protein identification. Thus while the identities of the larger proteins can be determined with some confidence, those with molecular weights below 18,000 daltons must always be checked by an alternative method as they can be strongly affected by the presence of bound oligonucleotides.

The two most reliable methods for identifying ribosomal proteins in radioactively labeled protein-oligonucleotide complexes entail the use of specific antibodies or HPLC. In one recently developed immunological technique, protein identities can in most cases be determined using only a few hundred cpm of radioactively labeled complex (Gulle et al., 1988). In the reverse-phase HPLC system of Cooperman and co-workers, practically all of the *E. coli* ribosomal proteins can be resolved by appropriate adjustments to the elution regime (Cooperman et al., 1988). Moreover, the protein retention times are quite insensitive to the presence of covalently bound oligonucleotides (B.S. Cooperman, personal communication). We have used this system to advantage in identifying proteins cross-linked to tRNA$_f^{Met}$ randomly labeled with s^{4}U (Rosen et al., 1993).

Methods for determining the sites at which tRNA is cross-linked to the 16S and 23S rRNAs have improved notably over the past decade. Some of our earlier successes were achieved by what can at best be described as a "brute force" approach. Thus, characterization of the cross-link betweem cmo^{5}U34 within the anticodon of P site-bound tRNA^{Val} and nucleotide C1400 of the 16S rRNA depended in large measure upon the photoreversibility of the cross-link and the fortuitous labeling of only one of the two cross-linked T1 oligonucleotides by RNA ligase (Prince et al., 1982). Now a far more systematic methodology has been elaborated for pinpointing cross-links in the large rRNAs, due mainly to the work of Noller, Brimacombe and their colleagues (Barta et al., 1984; Stern et al., 1988; Mitchell et al., 1990; Dontsova et al., 1992b). In this approach, complementary oligodeoxyribonucleotides 10 to 15 residues in length are first annealed to the rRNA, either singly or in pairs. The heteroduplex segments are then cleaved with RNase H and the products separated by denaturing polyacrylamide gel electrophoresis. If the cross-link is bracketed by the two cleavage sites, or by a single cleavage site and the 5' or 3' end of the rRNA, then the radioactive label will be associated with a fragment of predictable length. In this way, the entire 16S or 23S rRNA molecule is "scanned" for cross-links at roughly 200-base intervals. Once the cross-linking site has been narrowed down, the precise site of attachment is determined by synthesizing a cDNA from the relevant portion of the rRNA template with reverse transcriptase, and scoring the appearance of new termination sites in the DNA transcript (Barta et al., 1984; Dontsova et al., 1992b). We have used RNase H cleavage to delimit the sequence within the 16S rRNA that is cross-linked by [2N$_3$A37]tRNA^{Phe} in the E site (Wower et al., 1993), and the reverse transcription procedure to confirm the site of tRNA^{Val} attachment to 16S rRNA in the P site (J. W., unpublished; see also Denman et al., 1988).

When the photoreactive base has been incorporated into tRNA by transcription *in vitro* with T7 RNA polymerase, it is also necessary to ascertain the position of the cross-linked nucleotide(s) in the tRNA molecule. An effective method for obtaining this information for protein-rRNA complexes was

Table 1. tRNA-Ribosome Cross-Links

Ribosomal Site	tRNA	Position in tRNA	Probe	Labeled component	Ref.
P	tRNA^{Phe} Yeast	A76 A73	3-4Å (azide)	L15,L16,L27 G1945/23S RNA L27	(1)
P	tRNA^{Phe} Yeast	A76 A73	3-4Å (azide)	L27 L2, L27	(2)
P	tRNA^{Phe} E. coli	acp^{3}U47	14-20Å (300nm)	S19,L5,L27	(3)
P	tRNA^{Phe} Yeast	G43	3-4Å (azide)	S13,S19	(4)
P	tRNA$_f^{Met}$ E. coli	U47 U17/17.1	3-4Å s^{4}U	S19 L1,L27	(5)
P	tRNA$_m^{Met}$ Yeast	t^{6}A37	17Å 24Å	S7,L1,L27,L33 S7	(6)
P	tRNA^{Phe} Yeast	Y37	3-4Å (azide)	S7	(7)
P	tRNA$_f^{Val}$ E. coli	cmo^{5}U34	3-4Å (300nm)	C1400/16S RNA	(8)
P	tRNA$_f^{Arg}$ E. coli	s^{2}C32	10-12Å TBAPD	956-986/16S RNA	(9)
P	tRNA$_m^{Met}$ Lupine	acp^{3}U20.1	10-11Å (300nm)	S7,L1,L33	(10)
P	tRNA$_m^{Met}$ Lupine	acp^{3}U20.1	18-19Å (300nm)	S5,S7,L1,L5,L33 1302-1398/16S RNA 2281-2348/23S RNA	(10)
P	tRNA^{Phe} E. coli	U60 U45 A21	2-4Å (254nm)	S9 S7 S5	(11)
P$_t$	tRNA^{Phe} E. coli	U60 C56 C17	2-4Å (254nm)	S9 L27 L2	(12)
P$_b$	tRNA^{Phe} E. coli	U60 C56 G44 C17	2-4Å (254nm)	S9 L27 L5 L2	(12)
A$_b$	tRNA^{Phe} Yeast	U59 A26 G18 A9	2-4Å (254nm)	L2 S7 L27 S10	(12)
A	tRNA^{Phe} Yeast	A76	3-4Å (azide)	L27	(13)
A	tRNA^{Phe} Yeast	Y37	3-4Å (azide)	S7	(7)
A	tRNA^{Phe} E. coli	s^{4}U8	9Å	S19	(14)
E	tRNA^{Phe} E. coli	A76	3-4Å (azide)	L33	(15)
E	tRNA^{Phe} Yeast	Y37	3-4Å (azide)	S11 1533-1543/16S RNA	(15)

References:

1 Wower et al., 1989
2 Wower et al., 1988
3 Ofengand et al., 1986
4 Wower et al., 1990
5 Rosen et al., 1993

6 Podkowinski & Gornicki, 1989
7 Sylvers et al., 1992
8 Prince et al., 1982
9 Brimacombe et al., 1993
10 Podkowinski & Gornicki, 1991

11 Abdurashidova et al., 1989
12 Abdurashidova et al., 1990
13 Sylvers et al., unpublished
14 Lin et al., 1984
15 Wower et al., 1993

worked out in the laborstory of Budowsky and co-workers (Abdurashidova et al.,1990). Here, the tRNA is labeled at either the 5' or 3' end and the complex subjected to limited alkaline hydrolysis. Oligonucleotides remaining attached to the protein are then removed by phenol extraction. When the soluble fraction is subjected to denaturing polyacrylamide gel electrophoresis, a normal "ladder" is observed consisting of bands extending from the labeled end to the cross-linked base; above that point no further bands appear as the corresponding oligonucleotides were attached to the protein and removed in the previous step. The position of the cross-linked base can thus be read directly from the gel by comparison with a set of tRNA sequencing lanes. This technique permitted us to determine that 30S-subunit protein S19 was cross-linked to U47 of tRNA$_f^{Met}$ (Rosen et al., 1993).

MODELING tRNA BINDING SITES ON THE RIBOSOME

Our first perspective on the arrangement of protein and RNA in the *E. coli* ribosome came from immune electron microscopy, a technique in which antibodies specific for individual proteins or for modified positions in the rRNA are visualized after attachment to their ribosomal substrates (Stöffler and Stöffler-Meilicke, 1986, 1990; Oakes et al., 1986, 1989).Over the years, the results of these studies have been well substantiated by protein-protein cross-linking, protein-rRNA cross-linking and a variety of other data (Traut et al., 1986; Walleczek et al., 1988). A second, totally independent, view of protein organization within the ribosome has been provided by neutron scattering experiments from which a matrix of distances between the centers of mass of all proteins in the 30S subunit (Capel et al., 1987) and a substantial fraction of those in the 50S subunit (May et al., 1992) have been derived. The neutron scattering map of the *E.coli* 30S subunit was subsequently used as a scaffold for modeling the three-dimensional structure of 16S rRNA within the small ribosomal particle (Brimacombe et al.,1988; Stern et al., 1988). Stern and co-workers made extensive use of modification-protection experiments to predict the portions of the rRNA in proximity to the 21 30S subunit proteins. Brimacombe's efforts, on the other hand, relied mainly on protein-rRNA cross-linking to define the pathway of the rRNA through the subunit, subject to constraints imposed by long-range RNA-RNA cross-links. Although differing in the emphasis placed upon various sets of experimental evidence, the two model-building approaches are similar in concept and both are limited by the precision of the neutron scattering map and, in particular, the approximation of protein shapes as spheres.

Our model of tRNA binding sites on the ribosome, illustrated in Figure 2, is based upon the ribosomal subunit images deduced from immune electron microscopy. The reasons for this choice include the fact that most of the tRNA-ribosome cross-links elucidated thus far are to proteins or rRNA segments already mapped by this procedure, that similar data are available for both 30S and 50S subunits, and that the tRNA positions have a clear relationship to well-characterized morphological features of the ribosomal particles. Nonetheless, as in the case of the 30S subunit models based on neutron scattering, our model is only as accurate as the immune electron microscopy data. The arrangement of the tRNAs in Figure 2 is predicated mainly on the tRNA-ribosome cross-linking results compiled in Table 1, in which the tRNA sites were well-defined, the identities of the labeled ribosomal components were determined unambiguously, and the cross-linking procedures yielded chemical bonds of known lengths.

Figure 2 portrays the A, P and E sites on both ribosomal subunits as well as on the 70S ribosome. Likely positions for the peptidyl transferase center, the factor-binding site and the decoding domain are indicated. The disposition of the tRNAs is consistent with the orientation of the mRNA defined by Olson et al. (1988). As required for protein synthesis, the 3' termini of the A- and P-site tRNAs meet at the peptidyl transferase center, while the two anticodons are juxtaposed at the decoding site. It is of some interest that photoreactive bases placed at or near the 3' end of tRNA in either the P or the A sites label mainly L27 (Wower et al., 1988, 1989; L.A.S. and J. W., unpublished). This protein may, therefore, play a role in positioning the aminoacyl arms of the tRNAs during peptide bond formation. Cross-linking of protein S7 from position 37 of tRNA in both the A and P sites is also in accord with the requirement that the two anticodons be in close proximity during mRNA decoding (Sylvers et al., 1992). We have recently placed the E site adjacent to the P site on the basis of cross-links from the 3' end of tRNA to L33 on the 50S subunit, and from the anticodon loop to S11 and the 3'-terminal portion of 16S rRNA in the 30S subunit (Wower et al., 1993). Thus, whether or not the E-site tRNA participates in

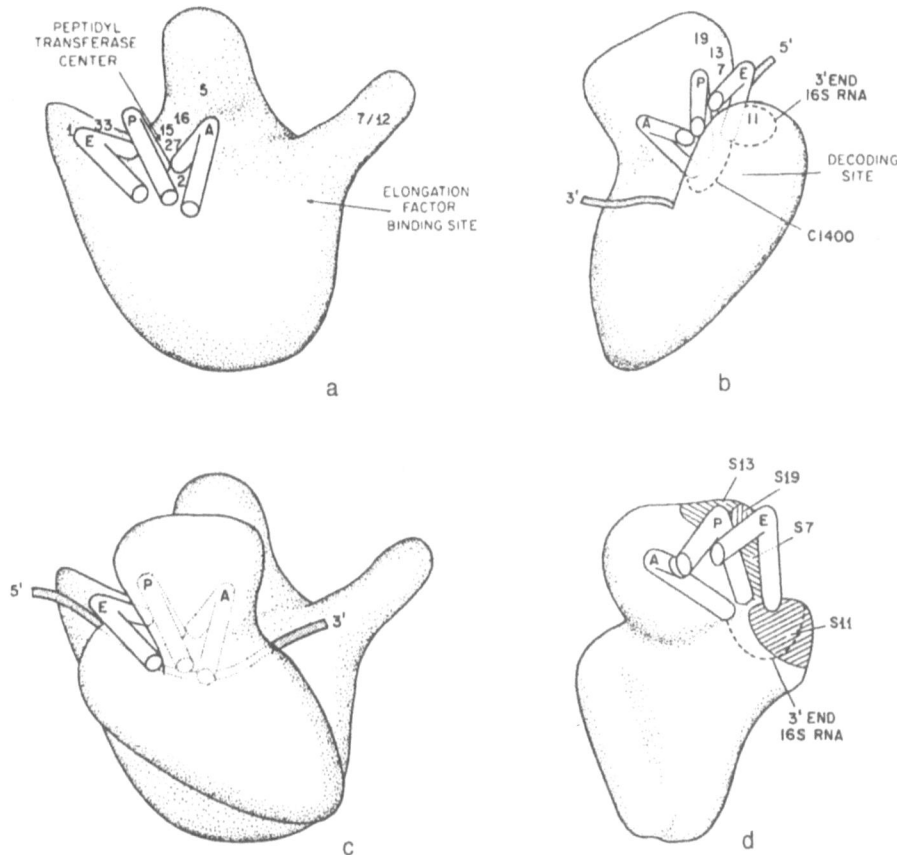

Figure 2. Model of the arrangement of tRNAs on the *E. coli* ribosome. (a) 50S subunit, (b), 30S subunit, and (c) 70S ribosome according to Oakes et al. (1986). (d) 30S subunit according to Stöffler and Stöffler-Meilicke (1986). A, P, and E designate tRNAs bound to aminoacyl, peptidyl, and exit sites, respectively. Numbers indicate the positions of ribosomal proteins derived by immune electron microscopy. The mRNA, represented as a shaded ribbon, is oriented according to Olson et al. (1988). A model similar to that in (c) has been proposed by Noller and co-workers (Noller et al., 1990).

codon-anticodon interaction, its anticodon is clearly in contact with the 30S subunit. While the Lake and Stöffler representations of the 50S subunit are very similar and, for the most part, can be used interchangeably, we have employed the Lake model of the 30S subunit more extensively than that of Stöffler because it includes the position of C1400 and thereby an approximate location of the decoding domain. However, as there is some controversy regarding the depth—or even the existence—of the cleft between the platform and body of the 30S subunit, we show that the cross-linking results can also be accommodated quite well by a 30S-subunit model in which the cleft is much less pronounced (Fig. 1d).

Although the protein, and in certain cases rRNA, neighborhoods of the acceptor end and the anticodon loop of tRNA are now reasonably well defined for all three tRNA binding sites, it is vital to delineate the ribosomal components that adjoin the tRNA helices and, in particular, the "elbow" or "central fold" of tRNA. Owing to new evidence, we can now locate several positions within the central portion of P site-bound tRNA. A few years ago, we reported a short-range cross-link between $2N_3A$ at position 43, near the junction of the anticodon stem and the variable loop, with 30S-subunit protein S13 (Wower et al., 1990). Using *in vitro* transcripts of E. *coli* tRNA$_f^{Met}$ randomly substituted with s^{4}U, we have recently labeled S19 from position 47, near the junction of the variable loop with the T stem (Rosen et al., 1993). S19 has also been cross-linked from position 47 of *E. coli* tRNA^{Phe}, although in this case the probe length was over 15 Å (Ofengand et al., 1986). We note that positions 43 and 47 are situated on the "T-loop" face of the tRNA molecule at a distance of approximately 12 Å from one another. The order of components on the surface of the 30S subunit that are in close proximity to this face of the tRNA is therefore C1400, S7, S13 and S19, running from the anticodon to the central fold. We have also determined that 50S-subunit proteins L1 and L27 can be labeled from position 17 within the D loop of s^4U-labeled tRNA$_f^{Met}$ (Rosen et al., 1993). As this portion of the D loop is on the tRNA face opposite to positions 43 and 47, it appears that the P site-bound tRNA may be sandwiched between the two subunits. In general agreement with our results, Podkowinski and Gornicki (1991) have shown that lupine tRNA$_m^{Met}$ derivatized at position 20.1 of the D loop with 10- to 19-Å probes can cross-link to 50S-subunit proteins L1 and L33, as well as to 30S-subunit protein S7 (see Table 1 and Figure 2).

Based on an analysis of the stereochemistry of codon-anticodon interaction at the A and P sites, and mRNA-ribosome cross-linking data, Brimacombe, Spirin and colleagues concluded that the anticodon/D-stem helix of the tRNA projects outward from the decoding domain of the 30S subunit and that the acceptor/T-stem helix lies parallel to the interface surface of the 50S subunit (Lim et al., 1992). Almost 20 years ago, the laboratories of Rich and Sundaralingam adavanced contrasting hypotheses about the stereochemistry of codon-anticodon interaction (Rich, 1974; Sundaralingam et al., 1975). While our model implicitly juxtaposes the A- and P-site tRNAs in a manner consistent with the Sundaralingam or "S" configuration, that of Lim et al. (1992) orients the tRNAs according the the Rich or "R" configuration. We note, however, that both the Sundaralingam and Rich proposals assume that the three-dimensional structure of the anticodon loop of ribosome-bound tRNA is identical to that derived from the X-ray structure of free tRNA (Quigley et al., 1975). As should be apparent, our cross-linking results shed no light on the stereochemistry of the codon-anticodon interaction. Nonetheless, as described above, the short-range cross-links that have been formed between bases within, or aligned with, the anticodon/D-stem helix to proteins S7, S13 and S19 strongly suggest that at least one face of this helix is in contact with the 30S subunit from the anticodon to the central fold of the tRNA molecule.

In the past, we have pointed out that tRNA can be used as a "molecular yardstick" for defining ribosome topography if sufficient short-range tRNA-ribosome cross-links could be established and characterized (Wower et al., 1990). Underlying this notion is the assumption that the tRNA maintains its three-dimentional crystal structure during its interaction with the ribosome. However, the anticodon, D, T, and variable loops are potentially flexible, as is the 3' end of the acceptor arm. The anticodon loop and the 3' terminus have in fact been shown to undergo significant deformations upon interaction with aminoacyl-tRNA synthetase (Rould et. al., 1989, 1991). It follows that tRNA may not be a very useful metric device for delineating the fine structure of the ribosome, especially in the vicinity of the loops and the acceptor end. Nonetheless, the notion of tRNA as a "molecular yardstick" is likely to hold up rather well at lower resolution since the overall size and shape of tRNA, unlike mRNA, is constrained in a rather rigid form by the lengths of the component helices and the web of tertiary base-base interactions that stabilize the central fold. For example, our results show that protein S7 and nucleotide C1400 of the 16S rRNA are much closer to one another than implied by immune electron microscopy, since both are linked from the anticodon loop (Sylvers et al., 1992). In a similar vein, S13, cross-linked from position 43, and S19, cross-linked from position 47, can be no more than 30 and 40 Å, respectively, from the anticodon loop. Additional cross-linking results will undoubtedly permit us to further improve our conception of the way in which the components for the ribosome are organized.

ACKNOWLEDGEMENT

We are most grateful to Mareile Fenner for preparing the ribosome illustrations and to Stefan Gross for contributing his skills in desktop publishing. Work in the authors' laboratory is supported by grant GM22807 from the National Institutes of Health.

REFERENCES

Abdurashidova, G.G., Tsvetkova, E.A., and Budowsky, E.I., 1989, *FEBS Lett.* 243:299.
Abdurashidova, G.G., Tsvetkova, E.A., and Budowky, E.I., 1990, *FEBS Lett.* 269:398.
Barta, A., Steiner, G., Brosius, J., Noller, H.F., and Kuechler, E., 1984, *Proc. Natl. Acad. Sci. USA* 81:3607.

Brimacombe, R., Atmadja, J., Stiege, W., and Schüler, D., 1988, *J. Mol. Biol.* 199:115.

Brimacombe, R., Mitchell, P., Osswald, M., Stade, K., and Bochkariov, D., 1993, *FASEB J.*, submitted.

Budowsky, E.I. and Abdurashidova, G.G., 1989, *Progr. Nucleic Acid Res. Molec. Biol.* 37: 1.

Capel, M.S., Engelman, D.M., Freeborn, B.R., Kjeldgaard, M., Langer, J.A. Ramakrishnan, V., Schindler, D.G., Schneider, D.K., Schoenborn, B.P., Sillers, I.Y., Yabuki, S., and Moore, P.B., 1987, *Science* (Washington, D.C.) 238:1403.

Cedergren, R., and Grosjean, H., 1987, *Biochem. Cell Biol.* 65:677.

Cooperman, B.S., 1980, in "Ribosomes: Structure, Function and Genetics," University Park Press, Baltimore, Chambliss, G., Craven, G.R., Davies, J., Davis, K., Kahan, L., and Nomura, M., eds., pp. 531-554.

Cooperman, B.S., 1987, *Pharmacol. Ther.* 34:271.

Cooperman, B.S., 1988, *Methods in Enzymol.* 164:341.

Czarnecki, J.J., 1984, *Biochim. Biophys. Acta* 800:41.

Denman, R., Colgan, J., Nurse, K., and Ofengand, J., 1988, *Nucleic Acids Res.* 16:165.

Dontsova, O., Rosen, K.V., Bogdanova, S.L., Skripkin, E.A., Kopylov, A.M., and Bogdanov, A.A., 1992a, *Biochimie*, 74:363.

Dontsova, O., Dokudovskaya, S., Kopylov, A., Bogdanov, A., Rinke-Appel, J., and Brimacombe, R., 1992b, *EMBO J.* 11:3105.

Ehresmann, C., Ehresmann, B., Millon, R., Ebel, J.-P, Nurse, K., and Ofengand, J., 1984, *Biochemistry* 23:429.

Evans, R.K., Johnson, J.D., and Haley, B.E., 1986, *Proc. Natl. Acad. Sci. USA* 83:5382.

Gulle, H., Hoppe, E., Osswald, M., Greuer, B., Brimacombe, R., and Stöffler, G., 1988, *Nucleic Acids Res.* 16:815.

Haley, B.E., 1983, *Fed. Proc., Fed. Am. Soc. Exp. Biol.* 42:2831.

Hayase, Y., Inuoue, H., and Ohtsuka, E., 1990, *Biochemistry* 29:8793.

Hayase, Y., Jahn, M., Rogers, M.J., Sylvers, L.A., Koizumi, M., Inuoue, H., Ohtsuka, E., and Söll, D., 1992, *EMBO J.* 11:4159.

Knoll, D.A., Woody, R.W., and Woody, A.Y.M., 1992, *Biochim. Biophys. Acta* 1121:252.

Lim, V., Venclovas, C., Spirin, A., Brimacombe, R., Mitchel, P., and Müller, F., 1992, *Nucleic Acids Res.* 20:2627.

Lin, F. L., Kahan, L., and Ofengand, J., 1984, *J. Mol. Biol.*, 172:77.

Matzke, A.J., Barta, A., and Kuechler, E., 1980, *Proc. Natl. Acad. Sci. USA* 77:5110.

May, R.P., Novotny, V., Novotny, P., Voss, H., and Nierhaus, K.H., 1992, *EMBO J.*, 11:373

Mitchell, P., Osswald, M., Schueler, D., and Brimacombe, R.,1990, *Nucleic Acids Res.* 18:4325.

Noller, H.F., Moazed, D., Stern, S., Powers, T., Allen, P.N., Robertson, J.M., Weiser, B., and Triman, K., 1990, in "The Ribosome," American Society for Microbiology, Washington, D.C., Hill, W.E., Dahlberg, A., Garrett, R.A., Moore, P.B., Schlessinger, D., and Warner, J.R., eds. pp. 73-92.

Oakes, M., Henderson, E., Scheiman, A., Clark, M., Lake, J.A., 1986, in "Structure, Function and Genetics of Ribosomes," Hardesty, B., and Kramer, G., eds., Springer, New York, pp. 47-67.

Ofengand, J., Delaney, P., and Bierbaum, J., 1974, *Methods Enzymol.* 29:673.

Ofengand, J., 1980, in "Ribosomes: Structure, Function, and Genetics," University Park Press, Baltimore, Chambliss, G., Craven, G.R., Davies, J., Davis, K., Kahan, L., and Nomura M., eds. pp. 497-529

Ofengand, J., Ciesiolka, J., Denman, R., and Nurse, K., 1986, in "Structure, Function and Genetics of Ribosomes,"Springer, New York, Hardesty, B., and Kramer, G. eds., pp. 473-494.

Olson, H.M., Lasater, L.S., Cann, P.A., and Glitz, D.G., 1988, *J. Biol. Chem.* 263:15196.

Podkowinski, J. and Gornicki, P., 1989, *Nucleic Acids Res.* 17:8767.

Podkowinski, J. and Gornicki, P., 1991, *Nucleic Acids Res.* 19:801.

Prince, J.B., Taylor, B.H., Thurlow, D.L., Ofengand, J., and Zimmermann, R.A., 1982, *Proc. Natl. Acad. Sci. USA* 79:5450.

Quigley, G.J., Seeman, N.C., Wang, A.H., Suddath, F.L., and Rich, A., 1975, *Nucleic Acids Res.* 2:2329.

Rich, A. , 1974, in "Ribosomes," Cold Spring Harbor Laboratory, Cold Spring Harbor, NY, Nomura, M., Tissieres, A., and Lengyel, P., eds., pp. 871-885.

Rosen, K.V., Wower, J., and Zimmermann, R.A., 1993, *Biochemistry*, submitted.

Rould, M.A., Perona, J.J., Söll, D., and Steitz, T.A., 1989, *Science* (Washington, D.C.) 246:1135.

Rould, M.A., Perona, J.J., and Steitz, T.A., 1991, *Nature* 352:213.

Stern, S., Weiser, B., and Noller, H.F., 1988, *J. Mol. Biol.* 204:447.

Stöffler, G. and Stöffler-Meilicke, M., 1986, in "Structure, Function, and Genetics of Ribosomes," Springer, New York, Hardesty, B., and Kramer, G., eds., pp. 28-46.

Stöffler-Meilicke, M., and Stöffler, G.,1990, in "The Ribosome," American Society for Microbiology, Washington, D.C., Hill, W.E., Dahlberg, A., Garrett, R.A., Moore, P.B., Schlessinger, D., and Warner, J.R., eds., pp. 123-133.

Sundaralingam, M., Brennan, T., Yathindra, N., and Ichikawa, T., 1975, in "Structure and Conformation of Nucleic Acids and Protein-Nucleic Acid Interactions," University Park Press, Baltimore, Sundaralingam, M., and Rao, S.T., eds., pp. 101-115.

Sylvers, L.A., Wower, J., Hixson, S.S., and Zimmermann, R.A., 1989, *FEBS Lett.* 245:9.

Sylvers, L.A., Kopylov, A.M., Wower, J., Hixson, S.S., and Zimmermann, R.A., 1992, *Biochimie* 74:381.

Tanner, N.K., Hanna, M.M., and Abelson, J., 1988, *Biochemistry* 27:8852.

Traut, R.R., Tewari, D.S., Sommer, A., Gavino, G.R., Olson, H.M., and Glitz, D.G., 1986, in "Structure, Function, and Genetics of Ribosomes," Springer, New York, Hardesty, B., and Kramer, G., eds., pp. 286-308.

Walleczek, J., Schüler, D., Stöffler-Meilicke, M., Brimacombe, R., and Stöffler, G., 1988, *EMBO J.* 7:3571.

Wower, J., Hixson, S.S., and Zimmermann, R.A., 1988, *Biochemistry* 27:8114.

Wower, J., Hixson, S.S., and Zimmermann, R.A., 1989, *Proc. Natl. Acad. Sci. USA* 86:5232.

Wower, J., Malloy, T.A., IV, Hixson, S.S., and Zimmermann, R.A., 1990, *Biochim. Biophys. Acta.* 1050:38.

Wower, J., Scheffer, P., Sylvers, L.A., Wintermeyer, W., and Zimmermann, R.A., 1993, *EMBO J.*, in press.

PHOTOLABILE OLIGODEOXYRIBONUCLEOTIDE PROBES OF
E. coli RIBOSOME STRUCTURE

Barry S. Cooperman, Parimi Muralikrishna, and Rebecca W. Alexander

Department of Chemistry
University of Pennsylvania
Philadelphia, PA 19104-6323
USA

INTRODUCTION

The extensive attention that has been directed recently toward the structure and function of ribosomal RNA (rRNA) has led to the identification of sites within rRNA that are directly linked to specific aspects of ribosomal function. One important approach has been the use of oligoDNAs that are complementary to single-stranded rRNA sequences to probe the structure of rRNA. This approach was first introduced by Bogdanov and his co-workers (Skripkin et al., 1979; Mankin et al., 1981) to probe both isolated rRNA and rRNA within ribosomal subunits. More recently, Hill and his co-workers (1990) have demonstrated that single-stranded regions of rRNA can form stable complexes with their complementary oligodeoxyribonucleotides, as evidenced by both filter binding assays using a ^{32}P-labeled probe and by the demonstration that treatment of the complex with RNase H cleaves rRNA at the appropriate position. By measuring the effects of added probes on specific ribosomal functions, they obtain evidence regarding the functional importance of a targeted rRNA sequences. A further application involves the attachment of electron microscopic markers to complementary oligoDNAs and the visualization of the complexes of such oligoDNAs bound to ribosome subunits, thus providing a three-dimensional localization of the targeted rRNA sequence. This approach has been called DNA hybridization electron microscopy (Olson et al., 1988; Lasater et al., 1989, 1990; Oakes et al., 1990a).

In this article we describe the use of radioactive, photolabile, complementary oligoDNA probes as photoaffinity labels to identify ribosomal components that neighbor functionally important single-stranded sequences of rRNA.

GENERAL METHODS OF PROCEDURE

Synthesis of Radioactive, Photolabile, OligoDNA Probes

We synthesize oligoDNA probes using standard phosphoramidite chemistry on a Milligen Biosearch Cyclone automated synthesizer. Thus far, we have incorporated photolability into these probes at both the 5' and 3' termini although incorporation of photolability at an internal position of the probe is also possible.

5'-Derivatives have been made as follows. Prior to deblocking, oligoDNAs are coupled at the 5'-terminus with O-[N-(4-monomethoxytrityl)-6-aminohexyl] O -(2-cyanoethyl N,N diisopropylphosphoramidite) according to the protocol and with reagents obtained from Glen

The Translational Apparatus, Edited by K.H. Nierhaus
et al., Plenum Press, New York, 1993

Research (Sterling, VA). Following deblocking and RP-HPLC purification, the 5'-amino-derivatized oligoDNA is derivatized with N-hydroxysuccinimidyl 4-azidobenzoate (HSAB), yielding photolabile oligoDNA derivatives of the type shown in Figure 1A. In these derivatives, with n =6, the maximal distance between the photogenerated nitrene and the 5'-terminal complementary base is 24Å. This distance can be varied between 19Å and 31Å (n=2 to n=12) with commercially available reagents. The photolabile oligoDNA is then made radioactive by coupling at the 3'-end with [α-^{32}P]ddATP in a reaction catalyzed by terminal deoxynucleotidyl transferase (Yousaf et al., 1984).

3'-Derivatives have been made through the application of a Glen Research product called 3'-Amino-modifier C7 CPG (Controlled pore glass) which places a 1^0 amine into the 3'-terminus and is designed for use in automated synthesizers.This reagent introduces a branched chain containing a protected primary amine into the normal reagent used to initiate oligoDNA synthesis. Following oligoDNA synthesis and deprotection, the amino group is available for derivatization. Condensation of this amino group with HSAB yields photolabile oligoDNA derivatives of the type shown in Figure 1B. In this derivative the maximal distance between the photogenerated nitrene and the 3'-terminal complementary base is 21 Å. Photolabile oligoDNAs of this type are made radioactive by phosphorylating at the 5'-end with [γ-^{32}P] ATP and polynucleotide kinase (Sambrook et al., 1989).

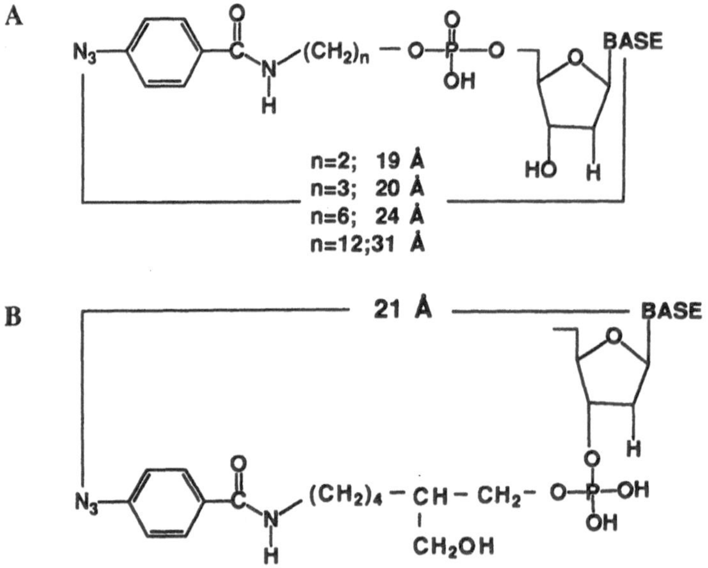

Figure 1. Photolabile derivatives of oligoDNA. A. Derivatization at the 5'-end. B. Derivatization at the 3'-end.

In addition to these syntheses, several variations on the basic theme are now readily available due to the work of a number of other groups that have been concerned with introducing photolability and radioactivity into oligoDNA probes. An excellent review summarizing this work has recently appeared (Goodchild, 1990).

Evidence for Site-Specific Binding and Photoincorporation

We and others (see Hill et al., 1990) typically measure the noncovalent binding of the radioactive, photolabile oligoDNA probe to the ribosome (or subunit) by Millipore filtration and demonstrate the site-specificity of such interaction by determining the sizes of the products of RNase H digestion of the probe-ribosome complex. However, our results and those of others (for details see Muralikrishna and Cooperman, 1991) make it clear that even when addition of the oligoDNA probe induces only a single site for RNase H cleavage, it is still possible for the oligoDNA probe to bind noncovalently to ribosomes at sites different from the target site, in ways not requiring heteroduplex formation. Such binding is of great concern for our photoincorporation studies, since it raises the possibility of photoincorporation into ribosomal components that do not neighbor the target site.

We use two control experiments to identify those ribosomal components that are labeled from the target site. The first asks whether added complementary oligoDNA reduces photoincorporation of the photolabile, radioactive oligoDNA. Such reduction indicates that photoincorporation is occurring from a specific site within the ribosome for which both oligoDNAs compete. The second asks what effect adding mismatched oligoDNA (MM-oligoDNA), in which two of the bases in the middle of the complementary oligoDNA are deliberately mismatched vis-a-vis the target site, has on photoincorporation of the photolabile, radioactive oligoDNA. Here the reasoning is that with two mismatches in the middle of the sequence, MM-oligoDNA would compete poorly for binding to the target site, but with most of the bases conserved (e.g., 7 for a 9-mer probe), could compete for binding to sites other than the target site. Therefore, lack of reduction in the photoincorporation of the photolabile, radioactive oligoDNA by added MM-oligoDNA coupled with reduction in such photoincorporation by added complementary oligo-DNA provides a strong indication that photoincorporation is occurring from the target site.

Identification of Cross-Linked Sites Within rRNA

Overall incorporation of photolabile oligoDNA probe into rRNA is measured by subjecting the photolyzed probe:subunit complex to sucrose gradient centrifugation in an SDS-urea buffer. Under these conditions, non-covalent binding is virtually abolished, permitting this method to serve as a useful screen for optimizing photoincorporation.

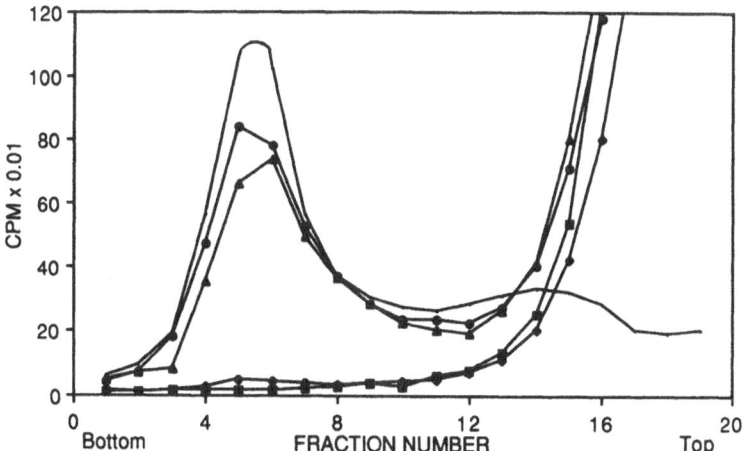

Figure 2. Photoincorporation of [^{32}P]-photolabile 2497-2505 probe into 23S rRNA isolated from labeled 50S subunits as determined on sucrose gradients in the presence of urea and SDS. (■), no photolysis; (●), with photolysis; (♦) with photolysis in presence of unmodified 2505-2497 probe; (▲) with photolysis in presence of mismatched 2505-2497 probe. The solid line is absorbance at 260 nm. (From Muralikrishna and Cooperman, 1991)

A typical gradient is shown in Figure 2 for the labeling of 23S rRNA on photolysis of the 2497-2505 probe complex with 50S subunits. These results show that labeling of 23S rRNA is completely light-dependent, and is site-specific, since it is fully inhibited by adding excess unmodified complementary 2505-2497 oligoDNA, but is virtually unaffected by the addition of mismatched oligoDNA. These results are typical for all of the oligoDNA probes examined so far. That is, covalent incorporation into rRNA is occurring predominantly or exclusively from the target site.

We employ a two step approach to identify the sites of photo-induced covalent incorporation of the photolabile oligoDNA probes into rRNA, in which the first step localizes a site of rRNA labeling to a limited region of rRNA (Brimacombe et al., 1990a) and the second step identifies the exact labeled base or bases (Barta et al., 1984). The first step is carried out using pairs of oligoDNAs complementary to different rRNA sequences that are simultaneously hybridized to a labeled rRNA. The resulting complex is digested with RNase H, and the radioactivity in the excised piece of RNA is determined, following PAGE separation, by autoradiography. With a set of oligoDNAs in hand that are placed along the entire length of both 16S rRNA and 23S rRNA, it is possible to quickly localize the site of labeling to within 100 bases. Interestingly, probe-labeled rRNA is a substrate for RNase H even in the absence of added oligoDNA. This is because the specific heteroduplex between photolyzed, covalently incorporated oligoDNA probe and rRNA is reformed under the conditions used for RNase H digestion and provides a site for RNase H cleavage. Because this first step allows quantification of the extent of labeling, it is particularly useful for the analysis of the results of controls for the photoaffinity labeling experiments, testing for example the effects of light fluence, of added native ligand as a competitor, and of the concentration of the photoaffinity label.

In the second step labeled rRNA is hybridized with a single-stranded oligoDNA that is complementary to a region of rRNA to the 3'-side of the region of rRNA found to be labeled in the first step. This heteroduplex is then used as a substrate for reverse transcriptase, and the exact position (or positions) of labeling is (are) identified, using a classical sequencing gel, as the position(s) following that (those) for which a halt or pause is observed. Typical results are displayed for the 2497-2505 probe in Figure 3.

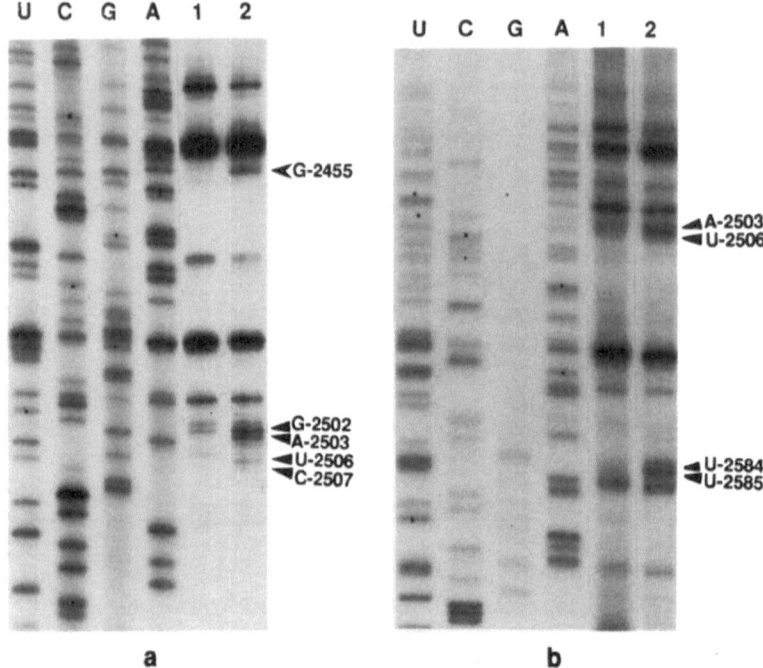

Figure 3. Autoradiograms showing reverse transcriptase analysis of 23S rRNA labeled with photolabile 2497-2505 probe. Primers used are 2576-2560 in part a and 2697-2681 in part b (from Muralikrishna and Cooperman, 1991).

Identification of Proteins Cross-Linked to Photolabile OligoDNA Probe

We identify proteins labeled by the radioactive, photolabile complementary oligoDNAs by the now classical methods of SDS-PAGE, RP-HPLC, and specific immunochemical analyses.

Each of these methods has its own advantages and drawbacks. Used in conjunction with autoradiographic analysis (with ^{32}P-labeled oligoDNA probes) one-dimensional SDS-PAGE permits rapid analysis of incorporation into proteins. Because many samples can be analyzed simultaneously, it permits facile optimization of experimental conditions (with respect to variables such as concentration and time of irradiation) as well as an assessment of various control experiments. The major drawback is that because many ribosomal proteins have similar molecular weights, identification of labeled protein is usually not possible except for the heavier proteins (> 20 kDa) which are generally fully resolved when analyzed as part of TP30 (total protein from 30S subunit) or TP50 (total protein from 50S subunit). Two-dimensional PAGE analysis is more highly resolving, and is useful for identification purposes, but it is not as rapid to perform and changes in migration on protein modification can introduce ambiguity into the identification process.

RP-HPLC analysis (Cooperman et al., 1988) well complements SDS-PAGE analysis. By providing another dimension of protein analysis, it aids in the identification of labeled protein. The major difficulty with RP-HPLC analysis used alone is that here too changes in elution on protein modification can introduce ambiguity into the identification process.

In principle, immunochemical analysis provides an unequivocal way of identifying a labeled ribosomal protein. However, for some proteins there is no available antibody, and some of the antibodies give assays with high background values, indicating some cross-reactivity of the antibody preparations with more than one ribosomal protein. Because of these problems, we consider immunochemical identification as a verification method, with PAGE and RP-HPLC analyses as the primary methods. To minimize background problems, the samples subjected to immunochemical analysis are first purified, either by PAGE (using gel elution) or by RP-HPLC. These samples contain the labeled protein of interest, and other proteins that either comigrate or coelute with the labeled protein. Immunochemical analyses are performed with antibodies to proteins migrating or eluting in the general vicinity of the labeled protein.

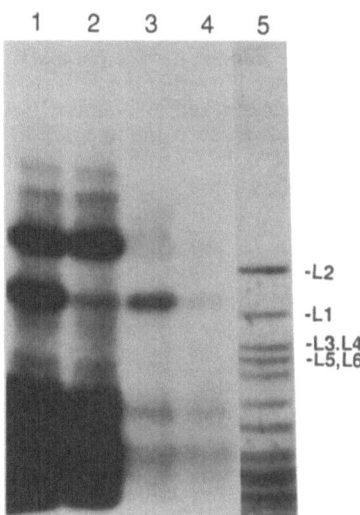

Figure 4. Autoradiogram of SDS-PAGE analysis of TP50 extracted from 50S subunits labeled with [^{32}P]-photolabile 2497-2505 probe. Lane 1, acetic acid soluble fraction in the absence of added cDNA 2497-2505; lane 2, acetic acid soluble fraction in the presence of added cDNA 2497-2505; lane 3, acetic acid insolube fraction in the absence of added cDNA 2497-2505; lane 4, acetic acid insolube fraction in the presence of added cDNA 2497-2505. (From Muralikrishna and Cooperman, 1991).

Two practical points are worth noting with respect to SDS-PAGE analysis of ^{32}P-labeled proteins to which oligoDNAs remain attached. First, although acetic acid extraction is a very efficient way of removing proteins from ribosomal subunits (Hardy et al., 1969; Jaynes et al., 1978) some protein containing oligoDNA is left behind in the acetic acid-insoluble (i.e., RNA) fraction. Moreover, we consistently find that the proportion left behind is greatest for protein labeled from the target site (Figure 4).

We speculate that this partitioning reflects an incomplete denaturation of the specifically formed heteroduplex between the photolyzed oligoDNA probe and rRNA. By adjustment of the acetic acid percentage used in extraction of proteins from subunits, it is possible to prepare acetic acid-insoluble fractions that give SDS-PAGE autoradiographs that show target site-labeled proteins rather clearly, even in the presence of a large amount of nonspecific labeling of r-proteins. This capability considerably facilitates analysis of our result. Second it appears that a protein to which oligoDNA is covalently attached migrates on SDS-PAGE with an apparent molecular weight equal to the sum of its weight and that of the attached probe. For example, L3 labeled with the 2497-2505 probe migrates as a 26 kDa band, equal to the weight of L3 (22.2 kDa) and the attached oligoDNA, and this phenomenon appears to be general for the labeled proteins we have identified.

From Figure 4 it is clear that much of the photoincorporation of the 2497-2505 probe into 50S protein is not site-specific in nature, emphasizing the importance of the control experiments discussed above for identifying proteins labeled from the target site. Non-specific photoincorporation into ribosomal protein is a general phenomenon for all of the probes we have studied thus far, and stands in marked contrast to the high degree of site-specific photoincorporation of these same probes into rRNA noted above. This result supports the notion that the nonspecific binding of complementary oligoDNAs to ribosomes is due to interaction with ribosomal proteins.

RESULTS OF SPECIFIC APPLICATIONS

Ribosomal Components Neighboring 23S rRNA Nucleotides 2497-2505 Within the Peptidyl Transferase Center

The 23S rRNA sequence 2497-2505 falls within the central loop of domain V. The evidence placing it at the peptidyl transferase center is summarized in Cooperman et al. (1990). It contains two nucleotides that on mutation confer chloramphenicol resistance on the ribosome. It also contains incorporation sites for photoaffinity labels for puromycin and for the 3'-end of a charged tRNA bound in either the A or the P site. Chemical modification of nucleotides within this sequence is blocked by bound tRNAs and by chloramphenicol. Most pertinent for our work is the observation that the oligoDNA 5'-CATCGAGGT-3' that is complementary to nucleotides 2505-2497 competes with the binding of both deacylated tRNA and chloramphenicol to the ribosome.

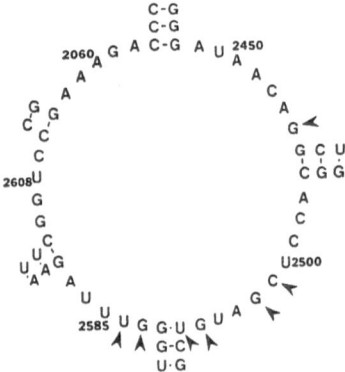

Figure 5. Sites of labeling by the photolabile 2497-2505 probe within the secondary structure of the central loop of domain V of 23S rRNA. (from Muralikrishna and Cooperman, 1991)

We have identified the ribosomal components labeled specifically from the target site by both a 5'-photolabile derivative (Muralikrishna and Cooperman, 1991) and a 3'-derivative (Muralikrishna and Cooperman, unpublished results) of the 2505-2497 oligo DNA probe. On photolysis in the presence of 50S ribosomes, the 5'-photolabile derivative site-specifically incorporates into protein L3, as identified both by SDS-PAGE (Figure 4) and immunologically, and into three separate 23S rRNA regions: nucleotides 2454; 2501, 2502, 2505, 2506; and 2583, 2584. While photoincorporation into the proximal nucleotides between 2501 and 2506 is to be expected and provides no new information about ribosome structure, photincorporation of probe into nucleotides 2454, 2583, and 2584 is a clear demonstration that each of these nucleotides is within 24 Å (the distance between G2505 and the photogenerated nitrene) of G2505, and is an important confirmation of the notion that these nucleotides are brought close together within the central loop of domain V (Figure 5).

Photoincorporation of the 5'-derivative into L3 represents the first direct physical evidence placing L3 at or near the peptidyl transferase center, although it is important to emphasize that L3 is one of only three proteins (L2 and L4 are the others) that might be essential for peptidyl transferase activity (Tate et al., 1987). Further, in 50S assembly experiments, the presence of L3 facilitates the uptake of protein L23 (Herold and Nierhaus, 1987), which has been placed at the peptidyl transferase center because of its photoaffinity labeling by puromycin as well as by a photolabile puromycin derivative (Cooperman et al., 1990). L3 has been shown to bind to nucleotides 2634-2638 and 2770-2785 by footprinting experiments (Leffers et al., 1988) and to cross-link near nucleotide 2832 (Brimacombe et al., 1990b) (Figure 6), thus providing evidence for the proximity of these nucleotides to 2497-2505.

Unfortunately, the 3'-photolabile 2497-2505 probe showed site-specific photoincorporation only into protein L3 and nucleotide 2497, providing no additional information regarding 50S components neighboring the 2497-2505 sequence.

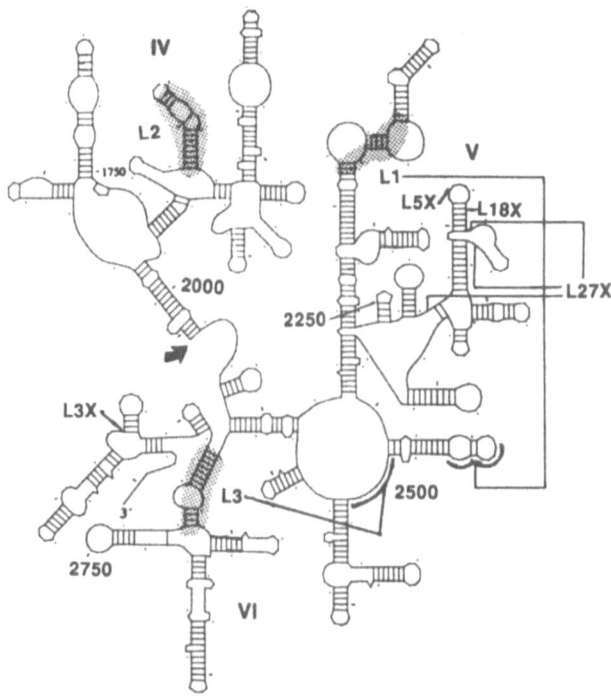

Figure 6. Crosslinking of photolabile 23S rRNA probes 2497-2505 and 2475-2483 considered in the light of other known protein: 23S rRNA interactions in domains IV-VI. Shaded areas refer to interaction sites defined by footprinting experiments (Egebjerg et al., 1991). Proteins followed by X are placed to indicate positions of crosslinking to 23S rRNA (Brimacombe et al., 1990c).

Ribosomal Components Neighboring 23S rRNA Nucleotides 2475-2483 Near the Peptidyl Transferase Center

The 23S rRNA sequence 2475-2483 forms part of a small loop which has been crossliked to protein L6 and which is connected by a long stem to the central loop of domain V (Figure 6). The 3'-derivative probe for 2475-2483 site-specifically photoincorporates into 23S rRNA only at the proximal positions 2469, 2470, and 2471, but also photoincorporates site-specifically into protein L1 (Muralikrishna and Cooperman, in preparation), which protects nucleotides in the 2120-2128 and 2160-2173 from chemical modification, thus providing evidence for the proximity of these nucleotides to 2475-2483 (Figure 6).

Ribosomal Components Neighboring 16S rRNA Nucleotides 1397-1405 Within the mRNA Decoding Site

The 16S rRNA sequence 1397-1405 is highly conserved evolutionarily. Nucleotides C1400 and G1401 have been shown by photoaffinity labeling and mutagenesis studies to be important for binding of the anticodon loop of tRNA to the P-site (Prince et al., 1982; Cunningham et al., 1992). Poly (U)-dependent tRNA^{Phe} competes with the complementary oligoDNA for binding to 1397-1405 (Hill et al., 1990). This sequence was located at the level of the neck, near the cleft and most likely on the head of the small subunit by DNA-hybridization electron microscopy (Oakes et al., 1986). 30S subunits as isolated are inactive in protein synthesis but may be activated by briefly heating at 37°C in the presence 10 mM Mg^{2+} (Zamir et al., 1971). Weller and Hill (1992) have demonstrated that the sequence 1397-1404 is essentially equally available to oligoDNA probe binding in both states.

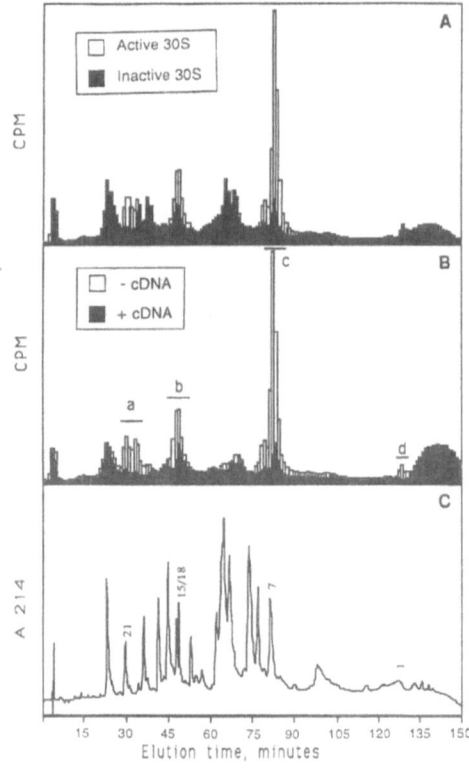

Figure 7. HPLC analysis of TP30 extracted from 30S subunits labeled with [^{32}P]-photolabile 1397-1405 probe. Panel A: radioactivity incorporated into active vs. inactive 30S subunits. Panel B: radioactivity incorporated into active subunits in the absence and presence of underivatized cDNA for 1397-1405. Panel C: Absorbance at 214 nm.

Consistent with this observation, the 5'-derivative photolabile probe for 1397-1405 binds non-covalently to both active and inactive 30S subunits to the same extent. On photolysis of the complex with active 30S subunits, the probe site-specifically photoincorporates into 16S rRNA nucleotides A1394, A1396, G1405, C1407, A1408, A1492, and A1493 as well as into proteins S1, S7, S18, and S21. Photoincorporation into inactive 30S subunits proceeds to a much lower extent, as shown for protein incorporation in Figure 7. This may indicate that the inactive to active transition in 30S subunits may be accompanied by a closing up of the ribosome structure in the vicinity of the target site. The data presented in Figure 7 also show that S7 is by far the most highly labeled protein in active subunits.

These results are generally consistent with what is known from other studies regarding the ribosomal neighborhood of 16S nucleotides 1397-1405. Thus, the accepted 2° structure model for 16S rRNA shows base pairing between nucleotides 1409-1413 and 1488-1491, placing 1492 and 1493 in the vicinity of nucleotide 1405. Moreover, Döring et al. (1992) have recently demonstrated direct u.v.-induced crosslinks between nucleotides 1402/1405 and 1498/1504 and footprinting experiments implicate nucleotides 1408, 1492 and 1493 at the A-site for tRNA (Moazed and Noller, 1990). Furthermore, protein S7 has been shown by cross-linking and footprinting experiments to interact with nucleotides 1360-1382 (Brimacombe et al., 1990c; Ehresmann et al., 1990; Noller et al., 1990), proteins S7, S18, and S21 cross-link to the 3'-terminus of 16S rRNA (Brimacombe et al., 1990c) and the labeling of protein S1 by mRNA affinity labels (as reviewed in Cooperman, 1987) link this protein to the anticodon site.

Ribosomal Components Neighboring 16S rRNA Nucleotides 518-526: An Unexpected Result

Nucleotides 518-526 are part of the 518-533 loop structure that has been implicated in several aspects of ribosome function. Complementary oligoDNA probes to parts of this sequence competitively inhibit poly (U) binding to the 30S subunit (Hill et al., 1990) supporting the proposal of Trifonov (1987) that the loop may form part of the mRNA binding site. This sequence was localized at the level of the neck, between the platform and head on the cytoplasmic side of the 30S subunit by DNA hybridization electron microscopy (Oakes et al., 1990b; Lasater et al., 1990). It also appears to be at or near the binding sites for elongation factors Tu (Langer and Lake, 1986) and G (Girshovich et al., 1974) , as shown by immunoelectron microscopy. Furthermore, an aminoacyl-tRNA.EF-Tu.GTP ternary complex bound in the A site protects bases G529 and G530 against chemical modification (Moazed and Noller, 1990), an A-->C transversion at 523 causes resistance to streptomycin (Melançon et al., 1988), and substitutions at G530 suggest it is involved in an allosteric fashion with A-site tRNA binding (Powers and Noller, 1990).

Our experiments (Alexander, Muralikrishna and Cooperman, in preparation) provide clear evidence that the 5'-derivative probe for 518-526 site-specifically photoincorporates into 16S rRNA only at the proximal positions C526, G527, and C528, but also site-specifically photoincorporates into proteins S7, S12, and, more tentatively (see below), into protein S4. Evidence that all three proteins are labeled by photolabile oligoDNA probe bound to the same (target) site comes from a detailed study measuring the decrease in covalent incorporation into each protein by a fixed amount of photolabile oligoDNA probe, as a function of increasing concentration of underivatized 518-526 probe, which competes for the target site. In these studies, covalent incorporation was determined by HPLC analysis. The results (Figure 8), when fitted to Michaelis-Menten saturation curves, gave essentially identical dissociation constants for the competitive binding of the underivatized probe, whether measured by inhibition of incorporation into S12 (K_d 2.7 nM) or S7 (K_d 2.5 nM). The data for S4 was less conclusive, due to nonspecific labeling of S3, which is not well resolved from S4 in the chromatogram employed. Nevertheless, the data for S4 photoincorporation may be adequately fit with a K_d value of 2.5 nM.

That S12 and S4 are specifically labeled by the 5'-derivative probe for 518-526 is fully in accord with footprinting experiments showing that both of these proteins strongly interact with 16S nucleotides within or near to the 530 loop (Noller et al., 1990). The labeling of S7, however, is unexpected. First, there is no evidence for S7 interaction with the 530 loop. More strikingly, labeling of S7 and either S4 or S12 from the same site, even taking into account the 24 Å between the photogenerated nitrene and the 5'-terminal complementary

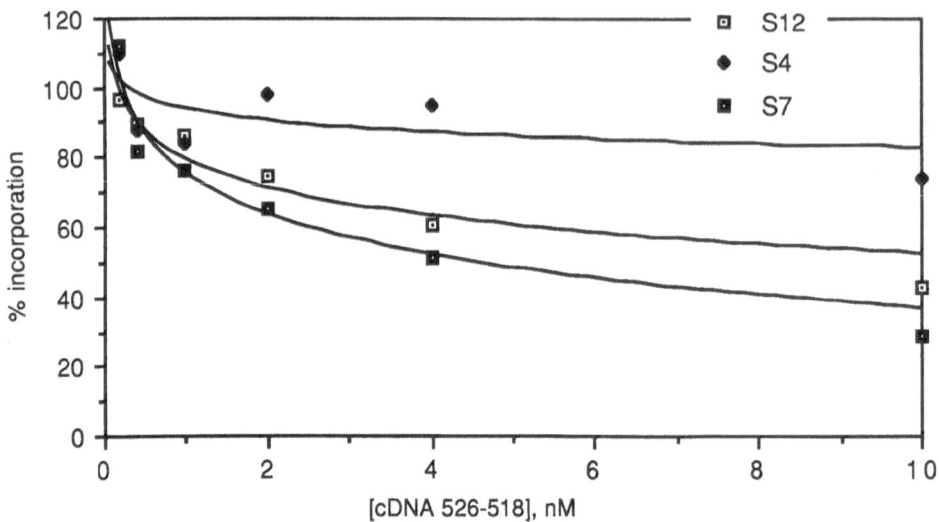

Figure 8. Photoincorporation of [^{32}P]-photolabile 518-526 probe into specifically labeled 30S proteins as a function of added underivatized 518-526 cDNA. (K$_d$ values for inhibition of photoincorporation into S7, S12, and S4, determined by curve fitting, were 2.5 nM, 2.7 nM, and 2.5 nM, respectively.)

base, appears to be inconsistent with the interprotein center of mass distances within the 30S subunit determined by neutron diffraction (Capel et al., 1988) which are: S4-S12, 40 Å; S7-S12, 112 Å; S4-S7, 113 Å. It is possible that S4 and S12 are labeled from a site different from that responsible for S7 labeling, and that the two sites just happen to have the same K$_d$ for the underivatized probe, but we find this highly unlikely since only one of the two putative sites could be complementary to the added probe. A very elongated conformation for S7 within the 30S subunit could provide an explanation for our results, but, this possibility aside, our results raise the question of whether some distances determined in the neutron diffraction study are in error. We also note that the fact that S7 is labeled by photolabile probes complementary to both nucleotide sequences 518-526 and 1397-1405 indicates some proximity of the 530 loop and the 1400 region, which is consistent with recent results of Rinke-Appel et al. (1991) showing simultaneous cross-linking of a photolabile mRNA probe to both of these 16S rRNA sites.

CONCLUSIONS AND FUTURE PROSPECTS

The results presented in this paper provide a clear illustration of the utility of photolabile oligoDNAs for identifying ribosomal components that neighbor targeted, single-stranded regions of rRNA. As rRNA regions that are most interesting from a functional point of view are likely to be in contact with solution, the approach we have demonstrated has wide potential application for determining the local structure of functional sites. An obvious limitation is that not all such sites within intact subunits bind from solution the relatively long oligoDNAs (8-9mers) that we have employed so far. One simple extension of our work will be to explore the site-specific binding of photolabile oligoDNA probes of shorter length, much as Hill and his co-workers (Hill et al., 1990; Weller and Hill, 1992) have done for underivatized cDNAs. It also should be possible to bind photolabile oligoDNAs to rRNA at different stages of subunit reconstitution at sites that are unavailable in the intact subunit. At a minimum, studies with such complexes should provide at least partial information about the ribosomal components neighboring a targeted sequence. In favorable cases, it may be possible to complete reconstitution following probe binding, thus allowing experiments to be performed on the intact particle.

REFERENCES

Barta, A., Steiner, G., Brosius, J., Noller, H.F., and Kuechler, E., 1984, *Proc. Natl. Acad. Sci. USA* 81:3607.

Brimacombe, R., Greuer, B., Gulle, H., Kosack, M., Mitchell, P., Osswald, M., Stade, K., and Stiege, W., 1990a, *in* "Ribosomes and Protein Synthesis, a Practical Approach," Spedding, G., ed., p. 131, Oxford UP, Oxford.

Brimacombe, R., Gornicki, P., Greuer,B., Mitchell, P., Osswald, M.,Rinke-Appel, J., Schuüler, D., and Stade, K. , 1990b, *Biochim. Biophys. Acta* 1050:8.

Brimacombe, R., Greuer, B., Mitchell, P., Osswald, M.,Rinke-Appel, J., Schuüler, D., and Stade, K.,1990c, *in* "The Ribosome: Structure, Function, and Evolution," Hill, W.E., et al., eds., p.93, American Society of Microbiology, Washington.

Capel, M., Kjeldgaard, M., Engelman, D.M., and Moore, P.B.,1988, *J. Mol. Biol.* 200:65.

Cooperman, B.S., 1987, *Pharmac. Ther.* 34:271.

Cooperman, B.S., Weitzmann, C.J., and Buck, M.A., 1988, *Methods Enzymol.* 164:523.

Cooperman, B.S., Weitzmann, C.J., and Fernandez, C.L., 1990, *in* "The Ribosome: Structure, Function, and Evolution," Hill, W.E., et al., eds., p.491, American Society of Microbiology, Washington.

Cunningham, P.R., Nurse, K., Weitzmann, C.J., Negre, D., and Ofengand, J., 1992,*Biochemistry* 31:7629.

Döring, T., Greuer, B., and Brimacombe, R., 1992, *Nucleic Acids Res.* 20:1593.

Egebjerg, J., Christiansen, J., and Garrett, R.A.,1991, *J. Moi. Biol.* 222:251

Ehresmann, B., Ehresmann, C., Romby, P., Mougel, M., Baudin, F., Westhof, E., and Ebel, J.P., 1990, *in* "The Ribosome: Structure, Function, and Evolution," Hill, W.E., et al., eds., p. 148, American Society of Microbiology, Washington.

Girshovich, A.S., Bochkareva, E.S., Kramarova, V.A., and Ovchinnikova, Y.A., 1974, *FEBS Lett.* 42:213.

Goodchild, J., 1990, *Bioconjugate Chem.* 1:165.

Hardy, S.J.S., Kurland, C.G., Voynow, P., and Mora, G., 1969, *Biochemistry* 8:2897.

Herold, M., and Nierhaus, K.H., 1987, *J. Biol. Chem.* 262:8826

Hill, W.E., Weller, J., Gluick, T., Merryman, C., Marconi, R.T., Tassanakajohn, A., and Tapprich, W.E., 1990, *in* "The Ribosome: Structure, Function, and Evolution," Hill, W.E., et al., eds., p. 253, American Society of Microbiology, Washington.

Jaynes, E.N. Jr., Grant, P.G., Giangrande, G., Wieder, R., and Cooperman, B.S., 1978, *Biochemistry* 17:561.

Langer, J.A., and Lake, J.A., 1986, *J. Mol. Biol.* 187:617.

Lasater, L.S., Cann, P.A., and Glitz, D.G., 1989, *J. Biol. Chem.* 264:21798.

Lasater, L.S., Montesano-Roditis, L., Cann, P.A., and Glitz, D.G., 1990, *Nucleic Acids Res.* 18:477.

Leffers, H., Egebjerg, J., Andersen, A., Christensen, T., and Garrett, R.A., 1988, *J. Mol. Biol.* 204:507.

Mankin, A.S., et al., 1981, *FEBS Lett.* 131:253.

Melançon, P., Lemieux, C., and Brakier-Gingras, L., 1988, *Nucleic Acids Res.* 16:9631.

Moazed, D., and Noller, H.F., 1990, *J. Mol. Biol.* 211:135.

Muralikrishna, P., and Cooperman, B.S., 1991, *Biochemistry* 30:5421

Noller, H., Moazed, D., Stern, S., Powers, T., Allen, P.N., Robertson, J.M., Weiser, B., and Triman, K., 1990, *in* "The Ribosome: Structure, Function, and Evolution," Hill, W.E., et al., eds., p. 73, American Society of Microbiology, Washington.

Oakes, M.I., Scheinman, A., Atha, T., Shankweiler, and Lake, J.A., 1990a, *in* "The Ribosome: Structure, Function, and Evolution," Hill, W.E., et al., eds., p. 180, American Society of Microbiology, Washington.

Oakes, M.I., Kahan, L., and Lake, J.A., 1990b, *J. Mol. Biol.* 211:907.

Oakes, M.I., Clark, M.W., Henderson, E., and Lake, J.A., 1986, *PNAS (USA)* 83:275.

Olson, H.M., Lasater, L.S., Cann, P.A., and Glitz, D.G., 1988, *J. Biol. Chem.* 263:15196.

Powers, T., and Noller, H.F., 1990, *PNAS (USA)*87:1042.

Prince, J.B., Taylor, B.H., Thurlow, D.L., Ofengand, J., and Zimmermann, R.A., 1982, *Proc. Natl. Acad. Sci. USA* 79:5450.

Rinke-Appel, J., Jünke, N., Stade, K., and Brimacombe, R., 1991, *EMBO J.* 10:2195.

Sambrook, J., Maniatis, T., and Fritsch, E.F., 1989, *in* "Molecular Cloning: A Laboratory Manual," 2d ed., Cold Spring Harbor Laboratory, Cold Spring Harbor, NY.

Skripkin, E.A., Kopylov, A.M., Bogdanov, A.A., Vinogradov, S.V., and Berlin, Yu.A., 1979, *Mol. Biol. Rep.* 5:221.

Tate, W.P., Sumpter, V.G., Trotman, C.N.A., Herold, M. and Nierhaus, K.H., 1987, *Eur. J. Biochem.* 165:403

Trifonov, E.N., 1987, *J. Mol. Biol.* 194:643.

Weller, J.W., and Hill, W.E., 1992, *Biochemistry* 31:2748.

Yousaf, S.I., Carroll, A.R., and Clarke, B.E., 1984, *Gene* 27:309.

Zamir, A., Miskin, R., and Elson, D., 1971, *J. Mol. Biol.* 60:347.

THE SIMPLICITY BEHIND THE ELUCIDATION
OF COMPLEX STRUCTURE IN RIBOSOMAL RNA

Robin Ray Gutell

MCD Biology
Campus Box 347
University of Colorado
Boulder, Colorado
80309-0347 USA
email: rgutell@boulder.colorado.edu

ABSTRACT

The elucidation of 16S and 23S rRNA Higher–Order Structure has been addressed by Comparative Sequence Methods for more than a decade. During these years our comparative methods have evolved as the number of complete 16S and 23S rRNA sequences have increased significantly, resulting in the maturation of the higher–order structure models for 16S and 23S rRNA.

With over 1000 16S (and 16S–like) and 200 23S (and 23S–like) sequences at this time, we have strong comparative evidence for the vast majority of all secondary structure base pairings, and are thus quite confident of the majority of the proposed *Escherichia coli* 16S and 23S rRNA secondary structure. Within the past few years additional rRNA Higher–Order structure constraints have been elucidated; constraints that reveal various RNA structural forms, including lone canonical pairings, pseudoknots, non–canonical pairings, tetra loops, canonical and non–canonical pairings that together forms a parallel (*vs.* the usual antiparallel) stranded structural element, and suggestive evidence for coaxial stacking of adjacent helices.

At this time we question what additional RNA structural constraints can be deciphered with comparative structure methods. To answer such questions, the rRNA sequence collection will need to continue to grow in both number and diversity, and our comparative structure algorithms need to evolve to a more sophisticated level. In an effort to establish the limits for structural similarity, we need to address how different two higher–order structures can be and still be considered analogous.

Introductory Statements

Since the first complete 16S (Brosius *et al.* 1978) and 23S (Brosius *et al.* 1980) rRNA sequences were determined, comparative analysis of these molecules has progressed in a variety of ways. Maybe foremost for the majority (especially for this audience) is the resulting higher–order structures, which ribosome–ologists utilize to map and/or design their experiments onto. While there is a wealth of information that can and should be elucidated from the sequences that make up the 16S and 23S rRNA datasets, this article will focus on the most obvious and probably experimentally meaningful structural features, namely secondary structure helices, tertiary interactions, and a few interesting examples of other comparatively derived structural constraints. And since much has already been written on comparatively derived rRNA structure [and most recently for an upcoming book on ribosomal RNA (Gutell *et al.* 1993), this article will only briefly touch on some of the emerging RNA structural features and new structural possibilities that have been uncovered within the past year or so, leaving the interested reader to investigate elsewhere for a more encompassing perspective of the details of the comparatively derived rRNA structures.

Comparative Structure Analysis

What is the basis of this method? What might we expect to decipher?

This method is rooted in the simple concept that similar or analogous three–dimensional structure can be composed of different primary structures, or in other words many different primary structures can fold into the same isomorphic 3–dimensional structure. Thus natural selection can maintain and act on the higher–order structures of RNA while the primary structure is free to change, although constrained in its divergence. The ribosomal RNA is an ideal molecule to apply such methodology to due to its structural and functional role in the ribosome, and the ribosome's position in protein synthesis and the evolution of the cell (Woese 1980).

Underlying this method are a number of key questions that cannot be answered *a priori*.

> How much and what types of variance can be tolerated in higher–order structure before these structures are not considered isomorphic?

> How much overall similarity should we expect to find for any RNA molecule? How much over-all variance should we expect to find for any RNA molecule (*i.e.*, tRNA: type I *vs.* type II)?

> To what extent can these methods identify general folding patterns and to what extent can these methods identify and distinguish subtle and detailed RNA structure (*i.e.*, elucidate the general-ized three–dimensional structure for tRNAs; elucidate the detailed features recognized by each of the aminoacyl synthetases for their cognate tRNA)?

> Will all RNA structural motifs be identified with such methods, or will only a subset of these structural elements be amenable to such methods, *i.e.*, should we expect secondary and all tertia-ry interactions to be equally decipherable?

> And lastly, should we anticipate the same overall and/or detailed structural (and even biological) constraints within phylogenetically related *vs.* distant structures?

A quick glimpse at the progression of our rRNA structure models

Although these questions are not (yet) answerable, the comparative analysis of the rRNAs (and all RNAs for that matter) has advanced in stages, in part so the results from each stage with their underlying assumptions can be evaluated before moving on to the next stage, and in part due to the significant increase in the number of sequences, development of the underlying correlation analysis algorithms, and the fact that we believe there is more structural detail to be found at the completion of each stage.

This analysis started with basic assumptions that were congruent with principles elucidated with exper-imental methods. Initially the comparative structure searched for the helices that compose the overall secondary structure [For 16S rRNA: (Woese *et al.* 1980, Stiegler *et al.* 1980, Zwieb *et al.* 1981), and 23S rRNA: (Noller *et al.* 1981, Glotz *et al.* 1981, and Branlant *et al.* 1981) rRNAs]. These methods specifically searched for canonical base pairings (*ie.* A–U and G–C) arranged contiguously and in an antiparallel orientation. These structures were tested and evaluated with each new rRNA sequence, resulting in numerous refinements in the secondary structures [Many references not noted here].

The specific search for helical elements gave way to a more generalized, non–structure based method. This method (Gutell *et al.* 1985) transformed the pattern of nucleotides at each column in the sequence alignment to a number pattern, which was based on the pattern of conservation and variance at every position in the molecule. Similar number patterns were subsequently grouped and analyzed, resulting in refinements in the secondary structure, and several proposed tertiary interactions (Gutell *et al.* 1985, 1986). Equally significant this simple algorithm uncovered a few basic principles of RNA structure, namely canonical pairings, and contiguous and antiparallel arrangement of such pairings [It should be noted that this method searched for columns (nucleotide positions) with similar patterns of variation or covariance, regardless of the nucleotide and pairing types. It so happened that the underlying pairs at covarying positions were canonical. Further, covariance between any two positions was independent of the pattern of change at flanking positions. Also resulting was the finding that the majority of the covarying positions were flanked by other correlating positions, and arranged in a helical fashion]. However, this method did have its faults, missing some correlating pairs when the pattern of variation at the two positions of interest were not identical.

Several quanitative methods have been developed, in part to address this issue while developing more powerful correlation analysis methods (Olsen 1983, Haselman *et al*. 1988, Chiu and Kolodziejczak 1991, Gutell *et al*. 1992b). Preliminary analysis of the 16S and 23S rRNA is most encouraging; the previously proposed structures are by and large substantiated while newer tantalizing correlations are beginning to emerge (Gutell *et al*. 1992b, unpublished analysis). More work remains to be done before these weaker, but statistically significant correlations can be appreciated in light of the known structures.

Ribosomal RNA Database

A large, diverse, and trustworthy collection of 16S and 23S rRNA sequences is essential for these comparative analysis efforts to accurately and thoroughly decipher structural information. The first complete 16S and 23S rRNAs sequences were determined in 1978 and 1980 respectively for *Escherichia coli* (Brosius *et al*. 1978, 1980). Over the past 15 years, the number of 16S and 23S sequences has increased significantly; The number of complete (or nearly so) 16S and 16S–like rRNA sequences has surpassed 1000 while the number of complete (or nearly so) 23S rRNA sequences is approaching 250. While the first few 16S and 23S rRNA sequences were primarily determined to facilitate studies on rRNA structure, almost all of the sequences determined since have been completed to address the study of phylogenetic relationships [The interested reader is encouraged to see Woese (1987) for a comprehensive review of the phylogenetic ramifications resulting from the analysis of rRNA sequences].

The rate at which these sequences are being determined continues to increase as sequencing methodology improves and parallels the widespread interest in rRNA for phylogenetic purposes. Of the 1000 plus 16S and 16S–like rRNA sequences, nearly 60% are from bacteria, (approximately 95% of these are *(eu)bacterial*, 5% are *archaea*), 30% are *eucarya*, with the organelles making up the remaining 10%. (70% of these are mitochondrial, 30% are plastid). The 23S and 23S–like rRNA sequence collection, with only 250 or so sequences, is comparable to the 16S rRNA collection in phylogenetic breadth. 50% of these sequences are organellar, of which 75% is mitochondrial, 25% plastid. Of the remaining 50%, 25% are *eucarya*, 25% are bacterial, and of these 65% are *(eu)bacterial*.

A Look at the Constraints Acting on 16S and 23S rRNA

Current Version of the 16S and 23S rRNA Higher–Order Structures

Over the past 10 years, comparative analysis of the 16S and 23S rRNA datasets has established and refined the *Escherichia coli* 16S and 23S rRNA higher–order structure models, as displayed in Figures 1 and 2. These models are by and large equivalent to those recently presented (Gutell *et al*. 1993), with only minor revisions. At this time, we are quite confident in the secondary structure component of these models, for almost all of these pairings contain at least two compensatory base changes with a minimal number of exceptions. Several tertiary interactions have been proposed within the past few years, although we anticipate more being elucidated as the sequence database grows and our correlation analysis algorithms mature. Of those found, essentially all of them contain a minimum of two compensatory changes, and thus we are quite confident of the accuracy of those proposed. [A detailed analysis of each proposed pair, revealing the number of coordinated changes that have occured throughout the phylogenetic tree, and the nature of the allowed pairs within the various phylogenetic assemblages will be presented elsewhere (Analysis in progress, R.Gutell)].

The comparatively derived structures for the 16S (and –like) and 23S (and –like) have been elaborated on in numerous articles, the most recent (Gutell *et al*. 1993) discusses many of the new refinements in the 16S and 23S rRNA models, while emphasizing some of the more interesting RNA structural motifs. For the purposes of this article, I will only touch on those themes raised elsewhere, focusing more attention on a few interactions that have only recently been elucidated and/or tentatively proposed interactions that now have additional comparative evidence.

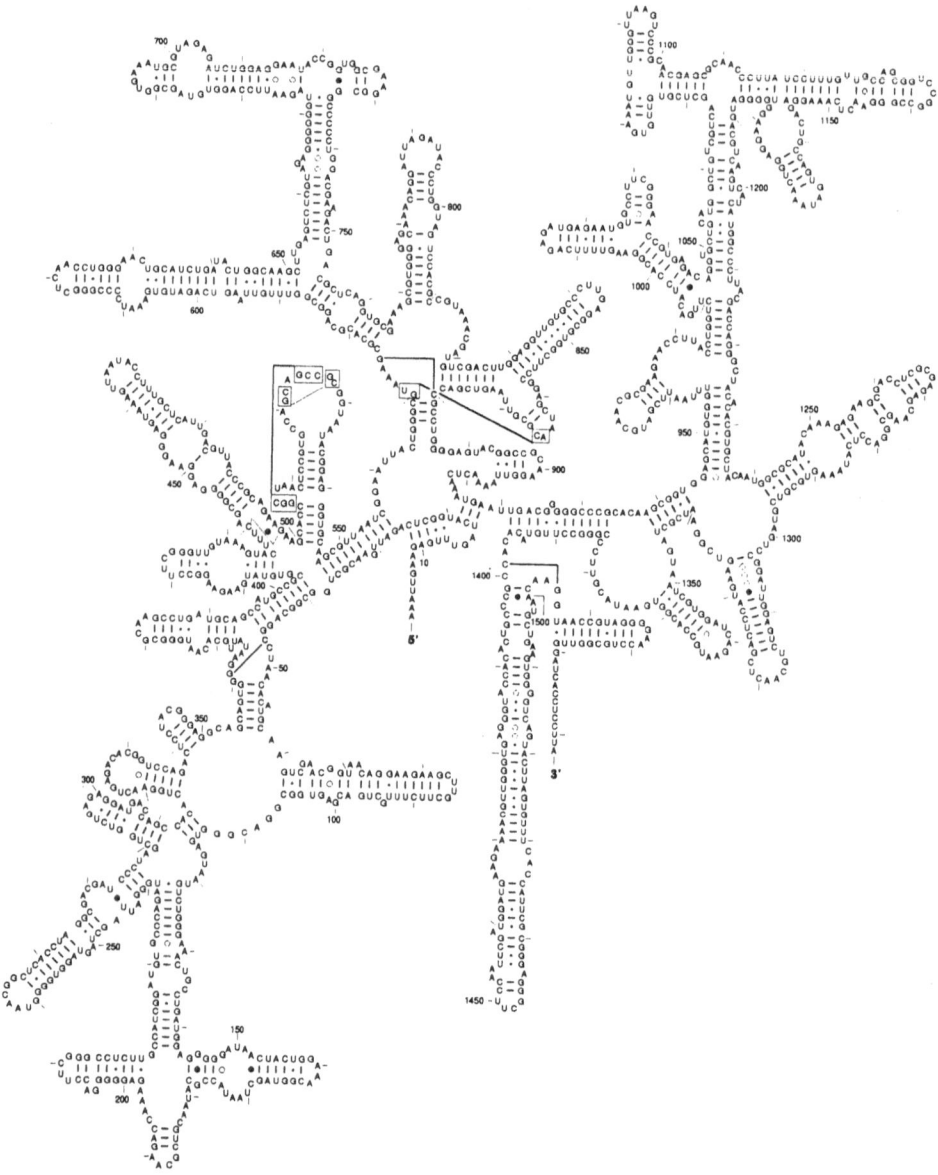

Figure 1. Secondary Structure diagram for *Escherichia coli* 16S rRNA, which is slightly revised from the most previous version (Gutell *et al*. 1993). The primary structure was determined by Brosius *et al*. (1978, 1981). Canonical (C:G, G:C, *etc.*) base pairs are connected by lines, G:U pairs by dots, A:G type pairs by open circles, and other noncanonical pairings with closed circles. Tertiary interactions are connected by thicker (and longer) solid lines. Every 10th position is marked with a tick mark, every 50th is numbered.

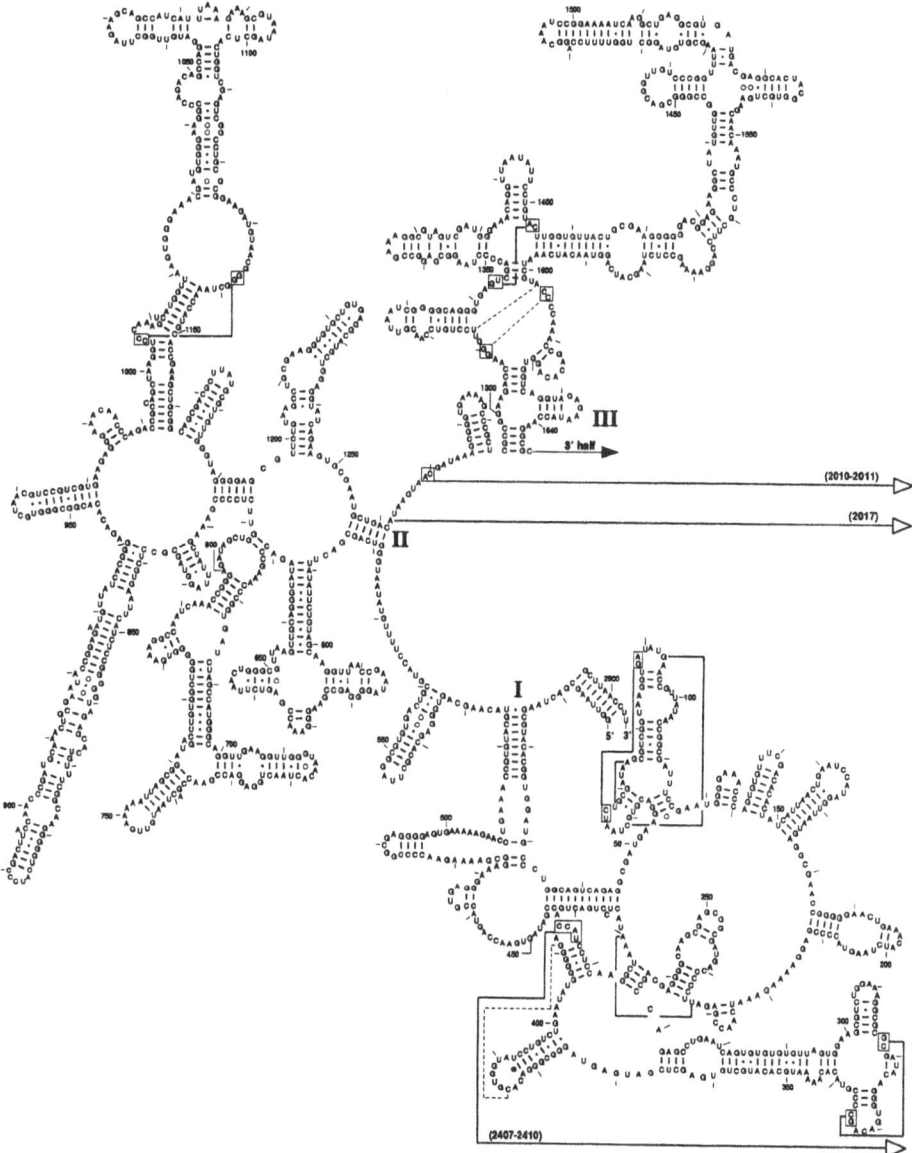

Figure 2. Secondary Structure diagram for *Escherichia coli* 23S rRNA, [revised slightly from the most previous version (Gutell *et al*. 1993)]. Basepairing, tertiary interaction, and nucleotide numbering designations is the same as in Figure 1. The primary structure was determined by Brosius *et al*. (1980). A: 5' half.

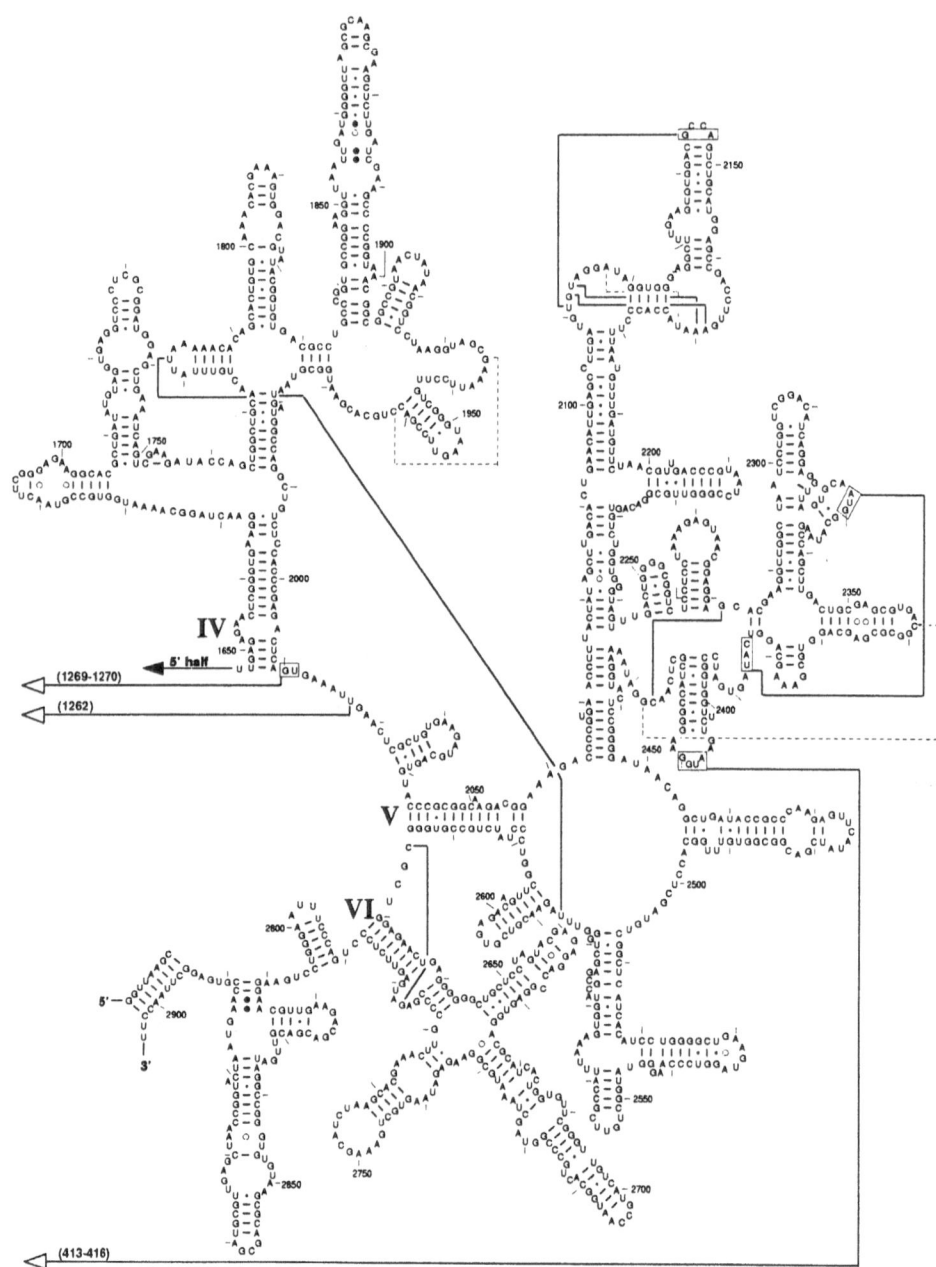

Figure 2B. 3' half of the *Escherichia coli* 23S rRNA structure.

Overall Structural Characteristics: Emerging Principles

Our present covariance algorithm identifies positions with similar patterns of variation, **regardless of:**

the nucleotide pairs generated by associating these two positions,

and the flanking positions' pattern of change, correlations involving these positions, or any structure associated to these positions.

This simple concept has established higher–order structure models for the 16S (and –like) and 23S (and –like) rRNAs and provides a reference point for much of the experimental work on these molecules. This simple concept has also independently elucidated some of the major RNA structure themes. The first and most prominent being Watson–Crick (A–U, G–C) and wobble (G–U) pairing, and the contiguous arrangement of these pairings in an antiparallel orientation (*ie.* a standard secondary structure helix). Outlined below are a few of the other RNA structural features that are beginning to emerge in the 16S and 23S rRNA. [To illustrate these themes, a few (of many) rRNA examples are noted].

Lone–Canonical Pairs: A small number of canonical pairings are not flanked on both sides by other base pairings, and are found in various structural settings. A few examples of such pairings span a simple internal loop or a more complex bifurcation loop. For three cases, all in the 23S rRNA, these single basepair helices are separated by three nucleotides, forming a tri–loop. The majority of these examples exist immediately adjacent to a secondary structure helix, suggesting or consistent with the possibility of an extended coaxial helix.

Non–Canonical Pairs: A small number of correlating positions involve pairings other than A–U, G–C, or G–U. The most prevalent non–canonical pairings and their pairing replacements are: A–G <-> G–A, A–G <-> W–C <-> C–W [W–C: Watson–Crick pairings = A–U, G–C], U–U <-> C–C, A–A <-> G–G, U–G <-> C–A, and U–G <-> C–A <-> W–C. Specific examples of such pairings are discussed elsewhere (Gutell *et al.* 1993).

Tetra Loops: Over 50% of the hairpin loops in *Escherichia coli* 16S rRNA, and 40% in *E.coli* 23S rRNA contain four nucleotides. Of the 256 sequences of size four, these rRNA tetra loops usually contain a small number of sequence motifs, which are: *UUCG*, *GNRA* (where N = any of the four nucleotides, R = A or G), and *CUUG* (Woese *et al.* 1990). Experimental studies have revealed that these loop structures are thermodynamically stable (Tuerk *et al.* 1988, Antao *et al.* 1991) while their three–dimensional structure contains unusual structural properties (Varani *et al.* 1991, Heus and Pardi 1991).

Pseudoknots: This type of structure, defined as a helix crossing an existing secondary structure helix, is found in many RNA types. Probably the most examples of which are found in 16S and 23S rRNA and contain (in the rRNAs) one, two, or three consecutive, antiparallel, and canonical pairings. Given that the vast majority of these structures occur immediately adjacent to the end of a secondary structure helix, it is possible to form a coaxial stack across these two helices. There is a growing amount of experimental evidence that now supports this general class of RNA structure (Pleij 1990, Wyatt *et al.* 1990, Puglisi *et al.* 1990, Powers and Noller 1991).

Parallel Orientation: The majority of all comparatively derived pairings are arranged in an antiparallel orientation; however a few are arranged in parallel. The most exceptional case lies in the 3' half of 23S rRNA, and involves the two internal loops, 2109–2119 and 2162–2173 (Gutell and Woese, 1990, Gutell *et al.* 1993). Moazed and Noller (1989) have mapped the E–site to these two internal loops, suggesting that this unusual structure is associated with ribosomal function.

Coaxial Helix Stacking: See discussion below.

Just A Few More Interactions

More comparative evidence has recently become available for a few additional interactions, some of which have already been mentioned (Gutell *et al.* 1993). These should now be considered more likely given the phylogenetic evidence presented here.

16S rRNA: Interactions between the 1400 and 1500 regions [Table 1]. The previously unpaired regions 1400 and 1500 of 16S rRNA are two of the most highly conserved regions of rRNA, with variation observed only within the mitochondria. Even with a minimal amount of variation within the mitochondrial sequences known in 1985, a small number of coordinated changes were identified, allowing us to tentatively propose a few pairings across these two unpaired strands (C1399/G1504, G1401/C1501, and G1405/C1496) (Gutell *et al.* 1985). These three pairings are arranged in an antiparallel orientation and involve Watson–Crick pairings, however they are not all contiguous with one another. Since 1985 the number of compensatory changes (and the number of phylogenetic events underlying these changes) has increased for these three pairings, and for two new pairing possibilities, C1402/A1500 and C1404/G1497 (see Table 1 and Figure 1). The minimum number of times such coordinated change has occurred at these pairings throughout the evolutionary tree are also noted in Table 1. Of these five pairings, four involve canonical basepairs, while the fifth (1402 and 1500) involves an interchange between C/A and U/G pairs [several examples of this type of basepair replacement in the rRNAs have been noted elsewhere (Gutell *et al.* 1993)]. This irregular helix can be extended with two more pairs, U1406/U1495 and C1407/G1494, both of which remain invariant in the current dataset [The U/U pairing is suggested due to previous comparative findings (Gutell and Woese 1990)]. Recent experimental work suggests that pairings across these two regions are possible (Doring *et al.* 1992; Ofengand, this volume). Given that this region of the rRNA is implicated in rRNA function (reviewed in:Noller *et al.* 1990), these phylogenetically constrained pairings suggested here could well be associated with such function.

Table 1. Comparative tabulation of the seven proposed *Escherichia coli* 16S rRNA pairings, C1399/G1504, G1401/C1501, C1402/A1500, C1404/G1497, G1405/C1496, U1406/U1495, C1407/G1494[*E.coli* nucleotides and numbering]. Variation at these positions, which has only been observed in the *Mitochondrial* 16S–like rRNAs is displayed here.

phylum or subdivision	C1399/G1504 C/G	U/A	U/G	G1401/C1501 G/C	A/U	C1402/A1500 C/A	U/G	C1404/G1497 C/G	U/A	G1405/C1496 G/C	A/U	U1406/U1495 U/U	C1407/G1494 C/G	
MITOCHONDRIA														(a)
Plants	6	-	-	6	-	6	-	6	-	6	-	6	6	
Protists	4	-	1	5	-	5	-	5	-	2	3	5	5	(b)
Animals														
nematodes	-	2	-	-	2	-	2	2	-	-	2	2	2	(c)
All other animals	22	4	2	28	-	28	-	28	-	28	-	28	28	
Fungi/Yeast	3	1	-	-	4	4	-	4	-	3	1	4	4	(d)
Allomyces macrogynus	1	-	-	1	-	-	1	1	-	1	-	1	1	(e)
Suillus sinuspaulianus	1	-	-	1	-	-	1	-	1	1	-	1	1	(f)
Min. # of Phylo. events	3			2		2		1		3		0	0	

(a) Complete mitochondrial alignment (R.Gutell, unpublished collection). (b) Protists, include *Plasmodium, Paramecium, Tetrahymena, Chlamydomonas.* (c) *Caenorhabditis elegans* and *Ascaris suum* (Okimoto *et al.* 1992). (d) *Aspergillus nidulans, Schizosaccharomyces pombe, Saccharomyces cerevisiae, Podospora anserina.* (e) *Allomyces macrogynus:*(unpublished, Franz Lang). (f) *Mushroom:* (unpublished, Tom Bruns).

Table 2. Comparative look at 16S rRNA pairings 521/528 and 522/527. (*E. coli* numbering). The number of indepedent phylogenetic exchanges is displayed at the bottom.

phylum or subdivision	G-521/C-528 G/C	C-522/G-527 C/G	A/U	
(eu)bacteria, Archaea, Eucarya, and plastids	~700	~700	-	(a)
MITOCHONDRIA				
All Mito's except	~60	~56	-	
nematodes	2	-	2	(b)
Suillus sinuspaulianus	1	-	1	(c)
C. reinhardtii	1	-	1	(d)
Min. # of Phylo. events	0	3		

(a) *(eu)Bacterial* and *Archaea* consensus determined from the RDP (Olsen *et al.* 1992) 16S rRNA sequence compilation[Approximately 500 sequences]. Approx. 250 aligned 16S–like rRNA *Eucarya* sequences (R.Gutell, unpublished collection) and approx. 60 aligned 16S–like rRNA Mitochondrial sequences (R.Gutell, unpublished collection).(b) *Caenorhabditis elegans* and *Ascaris suum* (Okimoto *et al.* 1992).(c) Mushroom (unpublished, Tom Bruns). (d) *Chlamydomonas reinhardtii*, Boer and Gray (1988).

16S rRNA: Additional constraint in the 530 loop [Table 2]. Complex and unusual structure is also proposed for another region of the 16S rRNA, a region highly conserved in primary structure and also implicated in ribosomal function (reviewed in: Noller *et al.* 1990). Based on comparative data, positions 505-507 and 524-526 have been implicated in a pseudoknot structure (Woese and Gutell 1989), and now experimentally substantiated (Powers and Noller 1991). Significant comparative evidence now exists for another pair, C–522/G–527 (Gutell *et al.* 1993) [see Table 2 and Figure 1. Note in Table 2 the number of independent phylogenetic events occuring at positions 522 and 527, and the two types of pairing observed, C–G, and A–U]. The invariant positions G–521/C–528 can extend this pairing to a two basepair helix and form another potential coaxial stack onto the 505-7/524-6 pseudoknot structure. Possible coaxial stacking of helices 500-504/541-545 and 511-515/546-550 (see below) only makes this region more complex and intriguing.

23S rRNA: Correlation between positions 2282 and 2427 [Table 3]. Comparative data suggests a new 23S rRNA interaction between positions 2282 and 2427 (see Table 3). This correlation is also consistent with crosslinking experiments (Mitchell *et al.* 1990) which associate positions 2257–2265 with 2427–8. Note for this correlation, there is at least one compensatory exchange event within each of the three phylogenetic domains and mitochondria.

Table 3. Comparative look at the proposed 23S rRNA pair 2282/2427 (*E.coli* numbering).

phylum or subdivision	G-2282/C-2427 G/C	A/U	U/A	C/G	G/U	
(Eu)Bacteria						(a)
All *(eu)bacteria* except:	32	–	–	–	–	
Bacter.,Flavo.,& Cyto.,	–	–	2	–	–	(b)
green sulfur bacteria	–	–	1	–	–	(c)
Campylobacter coli	–	–	1	–	–	(d)
plastids	33	–	–	–	–	(e)
Mitochondria						
protist	6	–	–	–	–	
fungi, yeast	1	2	–	–	2	
Suillus sinuspaulianus	–	1	–	–	–	(f)
plants	4	–	–	–	–	
animals	–	–	–	–	–	(g)
Archaea						
All *Archaea's* except:	17	–	–	–	–	
T.acidophilum	–	–	–	1	–	
Eucarya						
All *Eucarya's* except:	35	–	–	–	–	
Didymium iridis	–	–	–	1	–	
Physarum polycephalum	–	–	–	1	–	
Min. # of Phylo. events	–	1	2	1	–	

(a) Compilation of sequences, structures, organism names, and references can be found in Gutell *et al.* 1992a. (b) *Bacteriodes, flavobacteria and cytophagas and relatives*: *Flavobacterium odoratum* and *Flexibacter flexis*. (c) *Chlorobium limicola*. (d) Unpublished (Trevor Trust) (e) These Chloroplast sequences span a large section of the plant kingdom; The majority of which are unpublished (Monique Turmel and Claude Lemieux). (f) unpublished sequence (Tom Bruns) (g) alignment of sequences in these regions were problematic, thus not included in this discussion.

Coaxial Stacking of helices [Figure 3]. Based on the tRNA crystal structure it is known that two adjacent helices can stack upon one another, forming an extended structure (Kim 1976, Quigley and Rich 1976). It has also been shown experimentally that the two helices in a pseudoknot structure can stack upon one another (Pleij 1990; Puglisi *et al.* 1990). Knowing this, we ask if this RNA structure principle is utilized in the rRNAs, and if so, can comparative analysis be called upon to suggest such conformations? Given the current general and unrestricted rules for coaxial helix stacking, many structures of this type can be proposed in 16S and 23S rRNA. A comparative rationale, proposed by Woese (Woese *et al.* 1983) states that a coaxial helix can be suggested with comparative methods when the two stacking helices vary in length while maintaining the overall length of the extended structure. Two examples of this are shown in Figure 3, one from the 500 region of 16S rRNA (Positions 500 – 545, see 530 loop discussion, and Winker and Woese 1991), and the other is in the alpha–sarcin region of 23S rRNA, at the base of the two helices, 2646–2674 and 2675–2732. In both situations the lengths of the overall structure is maintained within each of the three phylogenetic domains, while the underlying helices vary in length. Given that these two proposed structures reside in close proximity with nucleotides associated with ribosomal function (Noller *et al.* 1990), it is tantalizing to suggest that conformation adjustments in this hinge could be of functional significance. It should also be noted that the 500 region of 16S rRNA, with its pseudoknot structure and the 522–527 pairing can potentially form other coaxial stacking arrangements, although apparently not all can form simultaneously.

16S rRNA: Additional constraint on the 570/866 interaction [Table 4]. A comparatively derived rRNA pseudoknot interaction was proposed in 1985 (Gutell *et al.* 1985), and involved 16S rRNA positions 570 and 866. A different type of correlation involving position 863, with the pair 570/866, can now be observed with a significantly larger collection of sequences(see Table 4). While the 570/866 covariance changes in a strict 1:1 fashion (*ie.* positions 570 and 866 change coordinately), position 863 changes in a less than 1:1 fashion with this pair; When position 863 is a G, the pair 570/866 is a U/A or C/G (Y/R), and when position 863 is a U, C, or A, the pair 570/866 is a G/C or A/U (R/Y). Although the structural interpretation for this constraint is not readily apparent, it is interesting to note a similarity with the observed sequences in tetra loops. The first and last nucleotides in rRNA tetra loops (Woese *et al.* 1990) are constrained, as is the case here, although the details are not identical. The generalized rRNA tetra loop pattern for the first and last nucleotides are: G––A, and (C or U)––G, here the generalized pattern is: G––(A or G), and U––C. Physical studies of two different types of tetra loops reveal an unusual pairing between the first and last nucleotides (Varani *et al.* 1991, Heus and Pardi 1991).

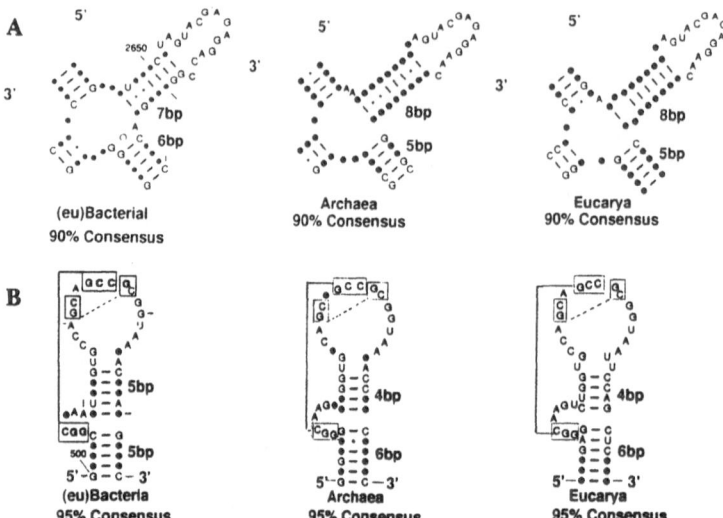

Figure 3. Comparative diagrams of coaxial stacking structures in the 2650 region of 23S rRNA (A.) and the 530 region of 16S rRNA (B.) (*Escherichia coli* numbering). The number of basepairs for each stacking helix is shown for each phylogenetic domain consensus structure; The total length is 13 bp in A., 10 bp in B. Closed circles represent positions with less than the minimum consensus level, nucleotides shown satisfy this consensus minimum.

Table 4. Revisiting the 16S 570:866 pseudoknot pairing. A comparative perspective on positions 570/866 and 863. The U–571 / A–865 pair is invariant in all sequences noted below.

phylum or subdivision	G–570/C–866				U–863				(a)
	G/C	A/U	U/A	C/G	G	U	C	A	
(Eu)Bacteria									
Gram positive bacteria									
low GC group + others	164	–	–	–	–	147	2	15	
high GC group	–	2	–	–	–	1	–	1	
high GC group	76	–	–	–	–	71	2	3	
Purple bacteria	104	–	–	–	–	100	2	2	
Cyanobacteria	1	–	–	–	–	–	1	–	
plastids	17	–	–	–	–	17	–	–	
Flavobact./Bacteriodes	–	–	52	–	52	–	–	–	
Planctomyces	–	–	1	–	1	–	–	–	
Other *(eu)bacteria*	32	–	–	–	2	25	1	4	
Mitochondria									
protists, plants	12	–	–	–	–	12	–	–	
animals	20	–	–	–	–	10	2	8	
fungi, yeast	–	–	3	1	4	–	–	–	
Archaea									
Methanomicrobium	–	–	8	12	20	–	–	–	
Extreme–thermophiles	–	–	1	4	5	–	–	–	
Other *Archaea's*	–	–	–	18	18	–	–	–	
Eucarya									
All	–	–	~251	–	~250	1	–	–	

(a) *Archaea* and *(eu)bacterial* sequences were compiled by the RDP (Olsen *et al.*1992). Mitochondrial and *Eucarya* sequence alignments were compiled by R.Gutell (unpublished).

23S rRNA: Non–canonical and canonical pairings in the 1870 region [Table 5]. The 1858/1884 pairing in 23S rRNA interconverts between A–G and G–A within the *(eu)bacteria* (see Table 5). [Within *(eu)bacterial* 16S rRNA, a similar set of pairings is found at positions 1357 and 1365, while this pair is canonical within the *Archaea* and *Eucarya* (Woese *et al.*1983)]. Two other pairings [1855–6/1886–7] in this vicinity are partially non–canonical within the beta and gamma purple bacteria, while canonical within the rest of the *(eu)bacterial* sequences [with several examples of compensatory change]. The last pair, 1859/1883 is U/U in some *beta* and *gamma purple bacteria*, and *Bacteroides and Flavobacteria*, A/G in other *(eu)bacteria* except *Planctomyces* and *Cyanobacteria* where it is G/C [Within the chloroplast, this pair explores other canonical arrangements].

Table 5. Comparative look at four pairings in the 1852–1890 region of 23S rRNA (*Escherichia coli* numbering). The pairings [U-1855/C-1887, U-1856/U-1886, A-1858/G-1884, and U-1859/U-1883] are compared with analogous pairings in other *(eu)bacterial* sequences, broken down by their phylogenetic classification. Mitochondrial sequences could not be unambiguously aligned in this region, and thus not considered. *Archaea* and *Eucarya* do not have equivalent pairings.

phylum or subdivision	U-1855/C-1887				U-1856/U-1886			A-1858/G-1884			U-1859/U-1883						(a)
	U/C	A/U	G/U	G/C	U/U	A/U	G/C	A/G	G/A	A/A	U/U	A/G	G/C	G/U	A/U	G/G	
(Eu)Bacteria																	
Purple Bacteria																	
gamma subdivision	3	1	-	-	4	-	-	4	-	-	4	-	-	-	-	-	
beta subdivision	1	1	-	1	3	-	-	3	-	-	2	1	-	-	-	-	
alpha subdivision	-	5	-	-	-	4	1	-	5	-	-	5	-	-	-	-	
Gram Positive Bacteria																	
low GC group (+others)	-	-	6	3	-	5	4	-	9	-	-	9	-	-	-	-	
high GC group	-	-	5	-	-	1	4	-	5	-	-	5	-	-	-	-	
Bacteriodes/Flavobact.	32	-	1	1	-	2	-	-	2	-	-	1	-	-	-	1	
Green–sulfur bacteria	-	-	-	1	-	-	1	-	1	-	-	1	-	-	-	-	
Radio–resis. micrococci	-	-	-	1	-	-	1	-	1	-	-	1	-	-	-	-	
Thermotoga maritima	-	-	-	1	-	-	1	-	1	-	-	1	-	-	-	-	
Planctomyces	-	-	-	1	-	-	1	1	-	-	-	-	1	-	-	-	
Spirochetes and relatives	-	-	2	-	-	2	-	-	2	-	-	2	-	-	-	-	
Cyanobacteria	-	-	-	1	-	-	1	1	-	-	-	-	1	-	-	-	
Plastids	-	5	3	26	-	-	34	-	30	4	-	-	27	5	2	-	(b)

(a)Compilation of sequences, structures, organism names, and references can be found in Gutell *et al.*1992.(b) The majority of the *Chlamydomonas* chloroplast 23S rRNA sequences have been determined by Claude Lemieux and Monique Turmel (unpublished).

Closing Thoughts

Embedded within this large and encompassing sequence collection is much detailed structural and phylogenetic information. While the analysis of these sequences has resulted in higher–order structure proposals for the 16S and 23S rRNAs, it is the belief here that much more remains to be elucidated. These sequences are the result of the evolutionary process; and although we don't necessarily understand the underlying RNA structural principles *a priori*, our search identifies constraints on the population of structural and functionally homologous molecules from which we infer structure. These inferred structures and RNA structure principles have been congruent with experimental findings, both in hindsight and concurrent. Thus we are encouraged that proposed structures not yet experimentally substantiated will follow suit. Given the success in the first few stages of this analysis, we have begun developing a more powerful set of comparative algorithms (Gutell *et al.* 1992b). These methods are already revealing weak correlations that were refractory to earlier methods and could well be suggesting additional structural constraints. The continued development of such methods will include a variety of approaches, including a phylogenetic events filter (Winker *et al.* 1990), and detailed analysis of the extent and types of compensatory changes occuring at each pairing. In conclusion, these refined methods and the ever increasing rRNA datasets provide us with the resources to continue this deciphering process of the 16S and 23S rRNA datasets, a process that should reveal more structural constraints and uncover more principles of RNA structure.

ACKNOWLEDGMENTS

I would like to thank Tom Bruns, Franz Lang, Claude Lemieux, Monique Turmel, and Trevor Trust for sharing unpublished sequences, sequences that offered additional comparative support for a few of the interactions discussed here. I would also like to acknowledge Bryn Weiser and Tom Macke for the development of computer code that makes all of this analysis possible. And of course, Carl Woese and Harry Noller for the path they inspired me to take. RRG is an Associate in the Program in Evolutionary Biology of the *Canadian Institute of Advanced Research*. This research has been supported in part by the NIH (GM 48207). I also wish to thank the W.M. Keck Foundation for their generous support of RNA science on the Boulder campus, and *SUN* Microsystems for their timely donation of computer equipment.

REFERENCES

Antao V.P., Lai S.Y. and Tinoco I.Jr. (1991). Nucleic Acids Res. 19:5901–5905.

Branlant C., Krol A., Machatt M.A., Pouyet J., Ebel J.P., Edwards K., and Kossel H. (1981). Nucleic Acids Res. 9:4303–4324.

Brosius J., Palmer M.L., Kennedy, P.J., and Noller H.F. (1978). Proc. Natl. Acad. Sci. USA 75:4801–4805.

Brosius J., Dull T., and Noller H.F. (1980). Proc.Natl.Acad.Sci. USA 77:201–204.

Brosius J., Dull,T.J., Sleeter,D.D. and Noller,H.F. (1981) J. Mol. Biol. 148:107–127.

Boer P.H. and Gray M.W. (1988). Cell 55:399–411.

Chastain M. and Tinoco I. (1991). Prog. Nucleic Acid Res. and Mol. Biol. 41:131–177.

Chiu D.K.Y. and Kolodziejczak T. (1991). CABIOS 7:347–352.

Doring T., Greuer B., and Brimacombe R. (1992). Nucleic Acids Res. 20:1593–1597.

Glotz C., Zwieb C., and Brimacombe R. (1981). Nucleic Acids Res. 9:3287–3306.

Gutell R.R., Weiser B., Woese C.R., and Noller H.F. (1985). Prog. Nucleic Acid Res. and Mol. Biol. 32:155–216.

Gutell R.R., Noller H.F., Woese C.R. (1986). The EMBO Journal 5:1111–1113.

Gutell R.R. and C.R. Woese (1990). Proc. Natl. Acad. Sci. (USA) 87:663–667.

Gutell R.R., Schnare M.N., and Gray M.W. (1992a). Nucleic Acids Res.20:supplement. 2095–2109.

Gutell R.R., Power A., Hertz G.Z., Putz E.J., and Stormo G.D. (1992b). Nucleic Acids Res. 20:5785–5795.

Gutell R.R., Larsen N., and Woese C.R. (1993). In– Ribosomal RNA: Structure, Evolution, Gene Ex pression and Function in Protein Synthesis (Zimmermann,R.A., and Dahlberg,A.E., eds.), CRC Press, Inc. Boca Raton, Florida, *in press*.

Haselman T., Chappelear J.E., and Fox G.E. (1988). Nucleic Acids Res. 16:5673–5684.

Heus H.A. and Pardi A. (1991). Science 253:191–194.

Kim S–H. (1976). Prog. Nucleic Acid Res. and Mol. Biol. 17:181–216.

Mitchell P., Osswald M., Schueler D., and Brimacombe R. (1990). Nucleic Acids Res. 18:4325–4333.

Moazed D. and Noller H.F. (1989). Cell 57:585–597.

Noller H.F., Kop J., Wheaton V., Brosius J., Gutell R.R., Kopylov A.M., Dohme F., Herr W., Stahl D.A., Gupta R., and Woese C.R. (1981). Nucleic Acids Res. 9:6167–6189.

Noller H.F., Moazed D., Stern S., Powers T., Allen P.N., Robertson J.M., Weiser B. and Triman K. (1990). pp.73–92. The RIBOSOME: Structure, Function & Evolution. eds. Hill *et al.* American Society for Microbiology.

Okimoto R., Macfarlane J.L., Clary D.O., and Wolstenholme D.R. (1992). Genetics 130:471–498.

Olsen G.J., (1983). Ph.D. thesis. University of Colorado.

Olsen G.J., Overbeek R., Larsen N., Marsh T.L., McCaughey M.J. Maciukenas M.A., Kuan W–M., Macke T.J. Xing Y. and Woese C.R. (1992). Nucleic Acids Research, 20:supplement 2199–2200.

Pleij C.W.A. (1990). TIBS 15:143–147.

Powers T. and Noller H.F. (1991). The EMBO Journal 10:2203–2214.

Puglisi J.D., Wyatt J.R., and Tinoco I. (1990). J.Mol.Biol. 214:437–453.

Quigley G.J. and Rich A. (1976). Science 194:796–806.

Stiegler R., Carbon P., Zuker M., Ebel J.P. Ehresmann C. (1980). C.R.Acad.Sci (Paris) Ser.D. 291:937–940.

Tuerk C., Gauss P., Thermes C., Groebe D.R., Gayle M., Guild N., Stormo G., D'Aubenton–Carafa Y., Uhlenbeck O.C., Tinoco I., Brody E.N., and Gold L. (1988). Proc.Natl.Acad.Sci.(USA). 85:1364–1368.

Varani G., Cheong C., and Tinoco I. Jr. (1991). Biochemistry 30:3280–3289.

Winker S., Overbeek R., Woese C.R., Olsen G.J., and Pfluger N. (1990). CABIOS 6:365–371.

Winker S., and Woese, C. R. (1991) System. Appl. Microbiol. 14:305–310

Woese C.R. (1980). pp.357–373. In: RIBOSOMES, Structure, Function, and Genetics. University Park Press, Baltimore, Maryland.

Woese C.R., Magrum L.J., Gupta R., Siegel R.B., Stahl D.A., Kop J., Crawford N., Brosius J., Gutell R., Hogan J.J., and Noller H.F. (1980). Nucleic Acids Res. 8:2275–2293.

Woese C.R., Gutell R., Gupta R., Noller H.F. (1983). Microbiology reviews 47:621–669.

Woese C.R. (1987). Microbiological Reviews 51:221–271.

Woese C.R. and Gutell R.R. (1989). Proc. Natl. Acad. Sci. (USA) 86:3119–3122.

Woese, C.R, Winker S., and Gutell R.R. (1990). Proc. Natl. Acad. Sci. (USA) 87:8467–8471.

Wyatt J.R., Puglisi J.D., Tinoco I.Jr. (1990). J.Mol.Biol. 214:455–470.

Zwieb C., Glotz C., Brimacombe R. (1981). Nucleic Acids Res. 9:3621–40.

THE FUNCTIONAL ROLE OF CONSERVED SEQUENCES
OF 16S RIBOSOMAL RNA IN PROTEIN SYNTHESIS

James Ofengand, Andrey Bakin, and Kelvin Nurse

Roche Institute of Molecular Biology
Roche Research Center
Nutley, NJ 07110, USA

INTRODUCTION

A crucial feature of all ribosomes is the conserved nature of its structural organization. This is nowhere more evident than in the secondary structure of the RNA components. Nevertheless, extensive sequence conservation, at least for the small subunit RNA (16S in prokaryotes), is found only at three single-stranded regions (Noller, 1984). One of these sequences (518-533) is in the 5'-domain while the remaining two (1394-1408; 1492-1505) are in the 3'-minor domain (Fig. 1). All three regions have been implicated in tRNA binding and other protein synthesis functions (reviewed in Noller, 1991). For example, C1400 (XL in Fig. 1) which is located in the middle of the 1394-1408 sequence was shown to be at the decoding site of the ribosome (reviewed in Ofengand, *et al.*, 1986; 1988).

Nevertheless, the precise function of these three conserved sequences remains to be deciphered. A step in this direction was taken by Gutell and colleagues (Gutell *et al.*, 1985; Gutell and Woese, 1990) who used phylogenetic analysis to show the possible existence of three tertiary base pairs, 1399/1504, 1401/1501, and 1405/1496. These results led directly to the further proposal that the nearby residues 1404/1497 and 1407/1494 were also base-paired but inasmuch as these residues were constant, their existence could not be verified phylogenetically. All five of the base pairs link two of the three conserved regions to each other (Fig. 2, the 1407/1494 pair is not shown). This set of tertiary interactions is particularly intriguing because, as illustrated in the right panel of Fig. 2, current information about the topographical disposition of the two single-stranded regions places them on either side of the cleft of the subunit, the cleft being that morphological feature which separates the large projection from the head and neck of the particle. Depending on the exact location of the single-stranded segments, tertiary base pair formation may require bridging the cleft in some manner, perhaps only transiently, either by invoking movement of the platform relative to the body, or by looping out the RNA sequences so as to span the cleft. Note, however, that the detailed structure of the 30S subunit, including the cleft region, is still a matter of debate (Vasiliev *et al.*, 1983; Gornicki *et al.*, 1984; Stöffler-Meilicke and Stöffler, 1990; Frank *et al.*, 1991; see also this Volume).

The Translational Apparatus, Edited by K.H. Nierhaus
et al., Plenum Press, New York, 1993

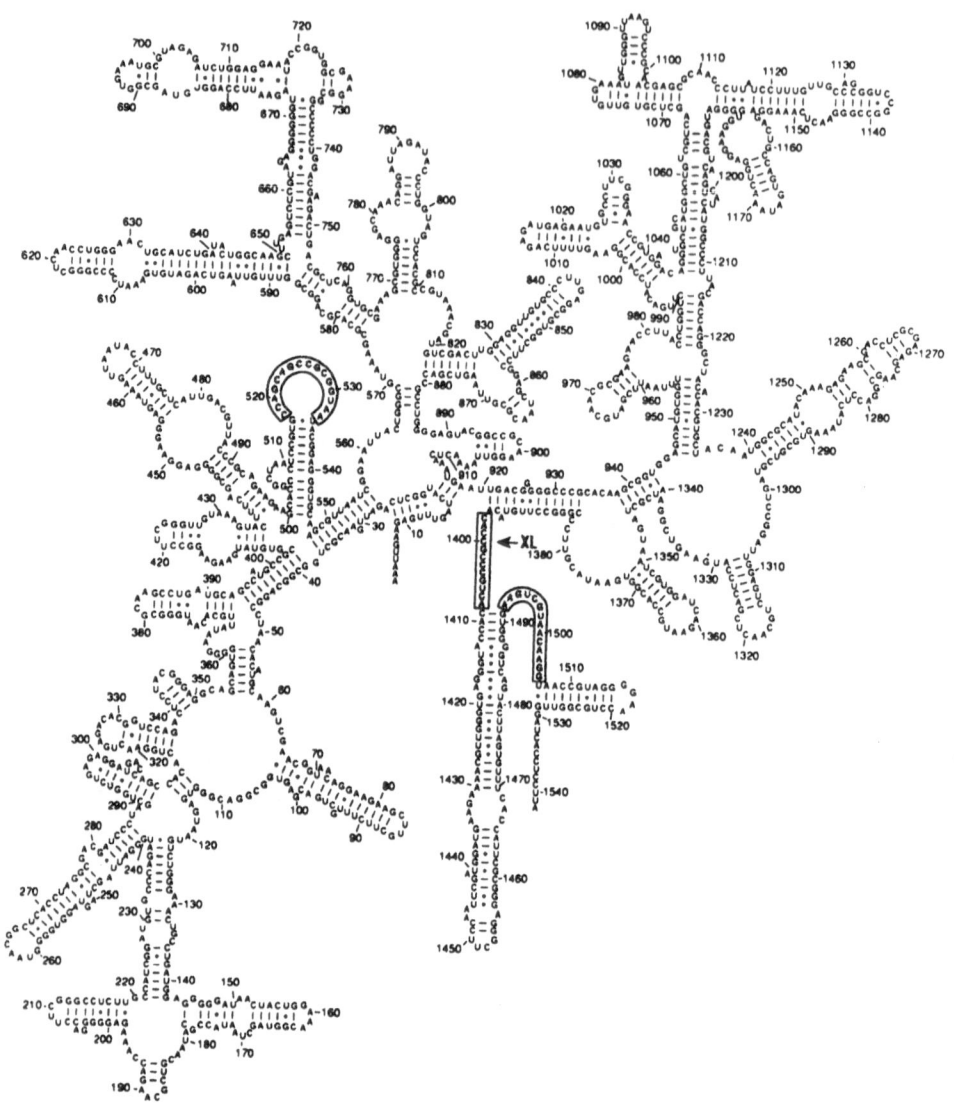

Figure 1. Location of the highly conserved sequences in 16S rRNA. The secondary structure of *E. coli* 16S RNA is according to Stern *et al.* (1989). The three highly conserved sequences are outlined and shaded. XL marks the site of crosslinking to the anticodon of P site-bound tRNA (Ofengand *et al.*, 1986; 1988).

The third conserved sequence, in the 525 region, has been placed distant from the 1400 and 1500 regions in the three-dimensional models proposed by Stern *et al.* (1988) and Brimacombe *et al.* (1988). Thus, the functional effects related to decoding which have been detected in this region have been ascribed to allosteric effects (Noller, 1991). However, recent crosslinking results obtained by Dontsova *et al.* (1992) suggest that the 525 region may in fact be physically near the 1400 and 1500 regions. While this would require a drastic revision of the three-dimensional models of the 30S subunit (see the chapter by R. Brimacombe, this Volume), it would have the satisfying result of bringing the three highly conserved regions of 16S RNA into physical proximity to one another at the site on the 30S subunit where the action is, namely at the decoding site.

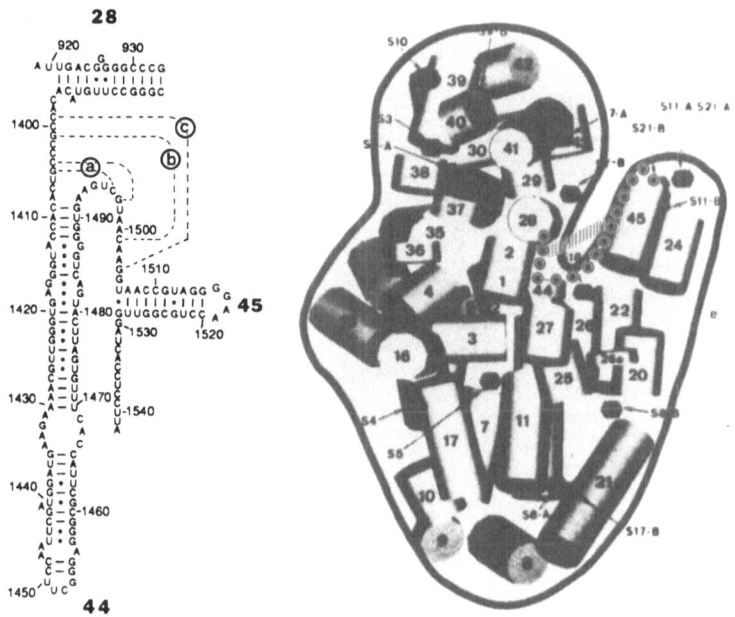

Figure 2. Proposed interaction of the 1400 region conserved segment with the 1500 region segment. Left panel, the 3'-minor domain containing helices 28, 44, and 45 connected by the two single-stranded sequences. Four tertiary interactions, a-c, are shown. Right panel, the helix model of Schüler and Brimacombe (1988) with a superimposed contour outline (Gornicki *et al.*, 1984; Stöffler-Meilicke and Stöffler, 1990) of the 30S subunit. The dots show schematically how the single-stranded sequences connect the helices. The shaded bar symbolizes the proposed tertiary interactions. Reprinted with permission from Cunningham *et al.* (1992b). Copyright 1992 American Chemical Society.

TERTIARY INTERACTION BETWEEN THE CONSERVED 1400 AND 1500 REGIONS

The phylogenetic evidence cited above was derived entirely from mitochondrial ribosomes. So far, all cytoplasmic ribosomes and most of those from chloroplasts are invariant in both the 1397-1407 and 1492-1505 segments, as are some mitochondrial ribosomes. Whether this is due to some variation in the mitochondrial protein synthesis apparatus which allows for minor structural variation at the decoding site, to the greater rate of evolution of mitochondrial ribosomes, or to a combination of both phenomena is not known. The currently available set of variants, all from mitochondria, are listed in Table 1. The table lists not only the three tertiary base pairs mentioned above but also several other potential base pairs. These other potential base pairs, indicated by the dotted lines, have been added because of the fact that organisms exist where the postulated CA pairs, first noted by R. Gutell for the C1402/A1500 pair (pers. communic.; see also chapter in this Volume) are replaced by UA or UG pairs, or where the C, m^3U "pair" is replaced by CA or UA. The data for the table were obtained from the sequence compilations of R. Gutell (see chapter in this Volume), and were kindly supplied by him. The base pairings listed in the table can be best understood by reference to Fig. 3A which shows in detail how the tertiary interactions are proposed to occur. In addition to the sequences listed in Table 1, the figure includes data from two additional sequences (see chapter by R. Gutell, this Volume) which increases the number of occurences of U1402/G1500 from 2 to 4, and the first example of a replacement of C1404/G1497 by UA. It is worth noting that despite the various changes, the relative positions of purines and pyrimidines are never exchanged.

Table 1. Proposed Base Pairing According to Scheme A of Figure 3.

Organism	1398 1505	1399 1504	1400 1502	1401 1501	1402 1500	1403 1498	1405 1496
E. coli	AG	CG	CA	GC	CA	CU	GC
D. pulex		UG					
D. virilis		UA					
D. yakuba		UA					
A, albopictus		UA					AU
S. purpuratus		UA					
P. lividus		UA					
A. suum		UA		AU	UG		AU
C. elegans		UA		AU	UG		AU
P. primaurelia			UA			CA	AU
P. tetraurelia			UA			CA	AU
T. pyriformis						CA	AU
C. reinhardtii		UG	UA				
P. wickerhamii	CG						
A. nidulans				AU		UA	
P. anserina				AU		UA	
S. pombe			UA	AU		UA	
S. cerevisiae		UA		AU		UA	AU
M. edulis		UG					
P. maximus			UA				
M. polymorpha			UA				
P. falciparum						UU	

Table 2. Proposed Base Pairing According to Scheme B of Figure 3.

Organism	1398 1506	1399 1505	1400 1504	1401 1501	1402 1500	1403 1498	1405 1496
E. coli	AU	CG	CG	GC	CA	CU	GC
D. pulex		UG					
D. virilis		UG	CA				
D. yakuba		UG	CA				
A, albopictus		UG	CA				AU
S. purpuratus		UA	CA				
P. lividus		UA	CA				
A. suum		UG	CA	AU	UG		AU
C. elegans		UG	CA	AU	UG		AU
P. primaurelia			UG			CA	AU
P. tetraurelia			UG			CA	AU
T. pyriformis						CA	AU
C. reinhardtii		UG	UG				
P. wickerhamii	CU						
A. nidulans				AU		UA	
P. anserina				AU		UA	
S. pombe			UG	AU		UA	
S. cerevisiae		UG	CA	AU		UA	AU
M. edulis		UG					
P. maximus			UG				
M. polymorpha			UG				
P. falciparum						UU	

Sequence information was obtained through the courtesy of R. Gutell, Univ. of Colorado

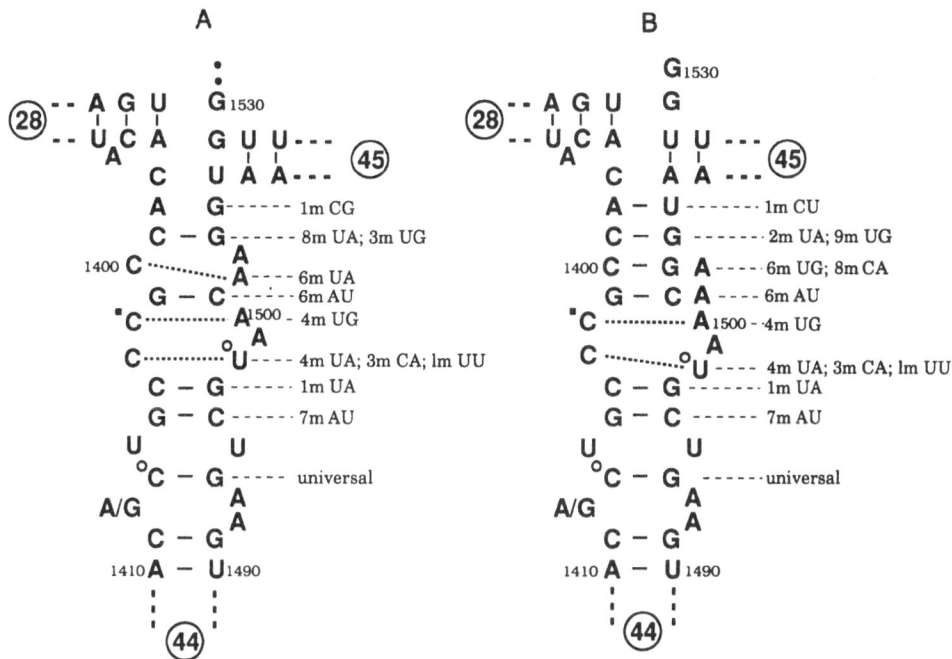

Figure 3. Proposed tertiary interactions between the 1397-1408 and 1492-1505 conserved single-stranded sequences in *E. coli* 16S ribosomal RNA. The sequences shown are invariant in the cytoplasmic ribosomes of all eukaryotes, archaebacteria, eubacteria, in the ribosomes of all but two chloroplast species, and in some mitochondrial ribosomes. The variations indicated are found only in other species of mitochondrial ribosomes. Sequences from trypanosome kinetoplasts have not been considered. The number and type of each variation is indicated, with *m* referring to mitochondrial. The circled numbers refer to the helices of Fig. 2. *C, °C, and °U are modified bases. Panel A, the tertiary interactions proposed by Gutell *et al.* (1985) and Gutell and Woese (1990) are shown as solid lines. The interactions indicated by the dotted lines may also occur. See text for discussion. Panel B, as in Panel A but pairing C1400 with G1504, C1399 with G1505, and A1398 with U1506.

Covariance in base pair substitutions was searched for by manual inspection of Table 1. That is, we looked for cases where the change of a GC pair to AU such as at 1401/1501 was accompanied by the corresponding change of an adjacent pair such as at 1400/1502 or 1402/1500, reasoning that this might be necessary to maintain the same overall stability of the region. However, no such correlation could be established.

An alternate base pairing scheme is shown in Fig. 3B, and the base pairs involved are listed in Table 2. It differs in that C1400 is paired with G1504, creating a bulge on only one side of the helix. This allows pairing of C1399 with G1505 and A1398 with U1506. This scheme appears equally plausible and would result in a potentially more stable upper section of the helix. However, it does not explain why C1400 should be so readily crosslinked to the anticodon of tRNA from both the P (Ofengand *et al.*. 1986; 1988) and A sites (Ciesiolka *et al.*, 1985). At the P site, cyclobutane dimer formation requires the specific availability of the C_5-C_6 double bond of C1400. The A site crosslink involved an aryl nitrene linked by a 23 Å leash to the same anticodon base of tRNA that crosslinked directly to C1400 from the P site. Despite the nonspecific nature of nitrene insertions, this probe only reacted with C1400, suggesting that only C1400 was readily available. The structure in panel A, with a looped out C1400, makes these reactions more plausible.

Covariance in base pair substitutions was looked for with this scheme of base pairings also but as was the case with scheme A no correlation could be established.

FUNCTIONAL EVIDENCE FOR THE TERTIARY BASE PAIRS

Analysis by Mutation

A classical way to test for the presence of base pairs is to reverse them by mutational or other means and then to search for restoration of some measurable parameter. Aside from second order effects due to stacking interactions, a CG pair is expected to be the structural equivalent of a GC pair. Where functional considerations are involved, however, it is possible that CG and GC might be recognized quite differently by a protein or other nucleic acid. Nevertheless, if the functional significance of tertiary base pairs is to be studied, there are few other approaches available. Consequently, we decided several years ago to focus the potential of our "synthetic" ribosome system (Krzyzosiak *et al.*, 1987) on the issue of the functional significance of the tertiary base pairs between the 1400 and 1500 regions.

The 1404/1497 and 1405/1496 Tertiary Interaction

In earlier studies (summarized in Cunningham *et al.*, 1990), we showed that mutating the C residues in the 1404/1497 and 1405/1496 base pairs to G, and the G residues to C, were inhibitory to A and P site binding of tRNA, only 10-30 % of control activity being retained. However, when double mutants were constructed such that each of the base pairs were reversed, activity was considerably restored. The restoration was much greater than the amount calculated if each mutation acted independently, but only reached 40-70 % of wild-type activity. Thus, while functional evidence was obtained for a positive interaction in each pair of mutants, thus verifying the predicted base pairs, it was also clear that reversed base pairs were not as functional as the ones Nature had constructed.

The 1401/1501 Tertiary Interaction

Effect on Function. We first observed that mutation of G1401 to C, U, or A almost completely inactivated the ribosome for all functions despite the fact that mutations at C1400 or C1402 had little or no effect (Cunningham *et al.*, 1992a). Subsequently we showed that mutation of C1501 to G also inactivated all functions of the ribosome. However, creation of the double mutant C1401/G1501 restored activity almost to wild-type levels (Cunningham *et al.*, 1992b). The recovery of activity as a result of making the reciprocal base pair was considerably greater in this case than it was with the 1404/1497 or 1405/1496 pairs.

The interaction appeared to be of the Watson-Crick type, since the pairs C1401:C1501, A1401:C1501, U1401:C1501, and G1401:G1501 were all inactive (Cunningham *et al.*, 1992a; 1992b). Other combinations such as A:U, U:A, G:U, U:G, and C:A remain to be tested. In view of its proximity to C1400, known to be at or near the ribosomal decoding site, formation of the base pair may be essential for codon recognition. This could explain why both the tRNA binding and peptide bond formation assays (both of which are codon-dependent) were inhibited by the single mutants which could not form base pairs. While the complementary base pair restored 30S initiation complex formation (I site), formation of the first peptide bond (fMet-Val) was inhibited. This may explain why the double mutant C1401/G1501 has so far not been found in Nature. Since both A site binding and poly(Phe) synthesis were also not affected, the block would seem to be in the later stages of initiation.

Localization of the initiation defect in the C1401/G1501 mutant. In order to delineate the initiation step which was inhibited, 50S subunits were added to 30S initiation complexes formed with the double mutant, and the ability to form fMet-puromycin was tested. Normally, this requires association with the 50S subunit, hydrolysis of GTP, and release of all 3 initiation factors before the P site bound fMet-tRNA is in a position to react (Hershey, 1987; see Fig. 4). Consequently, fMet-puromycin formation should be a measure of the ability of the mutant 30S to form the 70S initiation complex. When tested in this way, the rate and yield of fMet-puromycin was the same for the wild-type control as for the mutant C1401/G1501, and all of the bound fMet-tRNA reacted (Cunningham *et al.*, 1992b). We conclude that the double mutant is *not* defective in formation of the 70S initiation complex.

What then might be the defect in fMet-Val formation? Both A site binding and peptide bond formation were normal according to our assays. A clue may be found in the state of the A site when the first peptide bond is formed versus that during subsequent

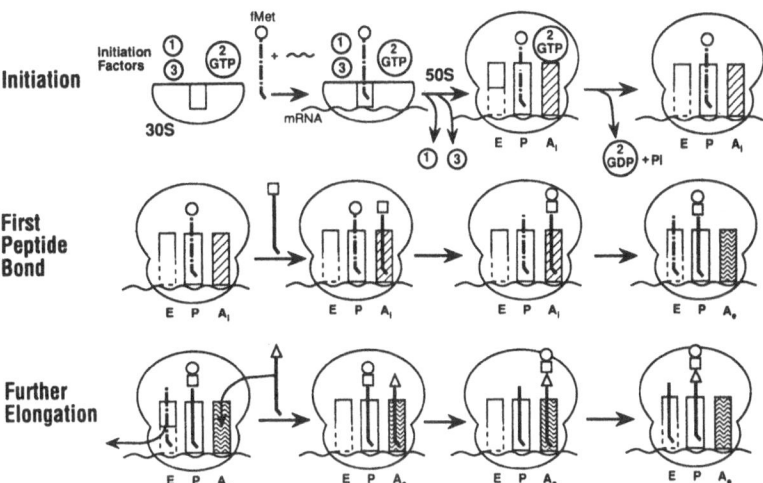

Figure 4. The state of tRNA binding sites during initiation and elongation on the ribosome. The scheme incorporates current knowledge of ribosomal protein synthesis except for the hybrid site concept of Moazed and Noller (1989). The A site is considered to exist in two states, A_i and A_e, as proposed by Nierhaus and co-workers (Hausner *et al.*, 1988; Nierhaus, 1990). Reprinted with permission from Cunningham *et al.* (1992b). Copyright 1992 American Chemical Society.

elongation. As pointed out by Nierhaus (Hausner, *et al.*, 1988; Nierhaus, 1990), the E site is empty when the A site is initially occupied (A_i in Fig. 4), and only is filled during subsequent rounds of elongation (A_e). A_i and A_e sites differ in several respects (Nierhaus, 1990). Thus, the effect of the reversed base pair could be understood in terms of an inability to properly form an A_i site, while having no effect on the A_e site.

Codon recognition by the C1401/G1501 mutant. The effect of this double mutant on the decoding site of the 30S subunit was assessed in two ways (Cunningham *et al.*, 1992b). First, the crosslinking assay was used as a sensitive measure of the stereochemical relationship between the anticodon of tRNA and the decoding site nucleotides of 16S RNA (Ofengand *et al.*, 1986; 1988). Specifically, we asked if the double mutant could form the cmo[5]U34-C1400 crosslink with the same yield, and whether the site of crosslinking was altered. We found that while the yield of crosslinking did not change, the site did. Whereas in our earlier work even insertion of a base between C1400 and G1401 did not change the site of crosslinking if the site was defined as the base 5' to G1401 (Denman *et al.*, 1989), in this case crosslinking was approximately equally distributed between C1399 and C1400. It is as if those ribosomal elements which normally act to precisely define the relationship between C1400 and the 5'-anticodon base of P site bound tRNA were loosened by the double mutation.

This apparent "wobble" at the decoding site suggested that the double mutant might have an increased tendency to misread codons. This possibility was tested *in vitro* by measuring the ability of a ribosome to insert leucine into peptide linkage in response to a polyU message (Ruusala *et al.*, 1982). No evidence for miscoding was found when the double mutant was compared to either isolated 30S or a reconstituted wild-type synthetic control. Despite the lack of intrinsic miscoding, the addition of streptomycin, a known inducer of miscoding (Ruusala *et al.*, 1984) was similarly effective in stimulating miscoding in the control and mutant samples by approximately 20-fold.

The 1399/1504 tertiary interaction

The functional necessity of this tertiary base pair was tested in a similar manner to the others by generating the appropriate single and double mutants. In addition we considered the possibility that C1399 actually pairs with G1505 (Fig. 3B) and constructed those single and double mutants as well. The results were both different and similar to those obtained with the 1401/1501 mutant.

The difference was in the fact that while all of the single mutants retained 40-70% of the wild-type activity for P, A, and I site binding, and for poly(Phe) synthesis (except for G1399 in the I site assay which only retained 20% activity), both double mutants, 1399/1504 and 1399/1505, were almost completely inactive. The fMet-Val formation assay was not done in this series. We conclude with some surprise that sites 1399 and 1504/5, not normally associated with A site function, when combined act to inactivate the A site as well as the P and I sites.

The similarity to the 1401/1501 mutant was that the two sites, 1399 and 1504 or 1505, are also connected, although in this case through a negative rather than a positive functional cooperativity. We conclude that these two regions appear to interact with each other, although it is perhaps premature to conclude that they form standard base pairs. Because both double mutants were so inactive, no further tests such as crosslinking or miscoding were possible.

Mutation of U1498 to G, C, or A

The result of the U to G mutation is to create an additional CG base pair which augments the 1404/1497 and 1405/1496 CG pairs. In Nature there are only a few natural examples of base pairing at these positions and none are CG (Fig. 3). Mutation to C would not be expected to have an effect, whereas mutation to A could create a putative CA pair. When the mutants were made and tested, it was found that mutation to C or A was without effect in any assay while mutation to G affected only formation of the first peptide bond, fMet-Val (summarized in Cunningham *et al.*, 1990). The G1498 mutant, like the C1401/C1501 double mutant, was fully active in 30S initiation complex formation and reacted with puromycin like the wild-type. Thus this mutation, by forming an additional strong base pair near the site of coding, may also prevent the formation of a functional A_i site.

The effect of the G1498 mutation on crosslinking and miscoding was also examined. There was no effect on the rate or yield of crosslinking and no detectable miscoding. Unlike the C1401/G1501 case, the crosslinking site remained strictly at C1400. There was no evidence for any crosslinking to C1399. Thus while the functional assays were very similar to those found for the C1401/G1501 mutant, there are clear structural differences in how the two mutants interact with tRNA at the decoding site.

Mutation of G530 to U

This mutant was first described by Powers and Noller (1990) as a dominant lethal mutation *in vivo*. Because the U530 mutant subunits were found in 70S ribosomes but not in polysomes, Powers and Noller suggested that the U530 mutant was defective in elongation. In view of the protection of this residue by A site bound tRNA (Moazed and Noller, 1990), it was natural to suppose that the elongation defect was in A site binding. In collaboration with M. Santer and U. Santer, Haverford College, we set out to investigate the matter in more detail, being motivated by our general interest in the conserved regions of 16S RNA.

To our surprise, the U530 mutant was as active in A site binding as the G530 wild-type, and was no more sensitive to the concentration of magnesium in the assay than the wild-type. P and I site binding was likewise as good as the control, as was polyPhe synthesis. The only assay to be affected was fMet-Val synthesis, which like the G1498 and C1401/C1501 mutants, was severely affected. As with the other mutants with this defect, I site binding was intact as was the ability to form fMet-puromycin. The rate, yield, and site of crosslinking to C1400 was also unaffected, and no miscoding could be detected. However, the U530 mutant could not be stimulated to miscode by addition of streptomycin.

The same pattern of inhibition of initiation *after* formation of a 70S initiation complex but without a detectable effect on standard A site binding or peptide bond formation has now been seen in three cases, G1401/C1501, G1498, and U530. The first two mutants affect tertiary base pairing in the region known to be at the decoding site, while the coding involvement of the 530 region has up to now been considered to be allosteric. The functional similarity of all three mutants lends some credence to the view (Dontsova *et al.*, 1992) that the 530 region may be physically part of the decoding site also.

ORGANIZATION OF RIBOSOMAL RNA AT THE DECODING SITE

Results of Others

Do other studies either support or contradict the model put forth in Fig. 3? Mutations in this region which were made and studied by others *in vivo*, and other studies which investigated this region have been reviewed in detail in earlier publications (Cunningham *et al.*, 1992a; 1992b). None of the reports disagree with the structural and functional ideas proposed here, except for the paper by Almehdi *et al.* (1991) who reported that deletion of residues 1401-1404 only inhibited the ability to synthesize MS2 coat protein by 50% or less. This startling finding is not in agreement with our work and also contradicts a number of other studies.

A recent crosslinking study (Döring *et al.*, 1992) provides strong support for the model of Fig. 3. In this work, a zero-length UV light-induced crosslink between residues 1402-1403 and 1498-1501 was found. This result is consistent with the postulated base pairs between 1402 and 1500 and between 1403 and 1498 which are proposed in Fig. 3. Three other crosslinks were reported in this work which are also consistent with the model, but the lack of precision in identification of the nucleotides involved make them less useful for testing its correctness.

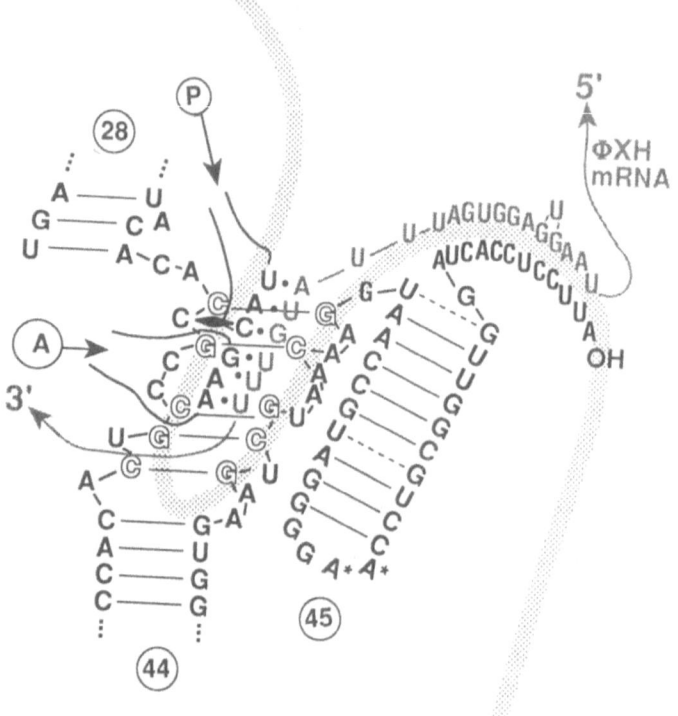

Figure 5. Schematic version of the arrangement of tRNAs, mRNA, and rRNA around the decoding site of the ribosome. The five C-G pairs shown in Fig. 3 are indicated by larger outline type. The anticodon of P and A site bound tRNAs are shown (not to scale), and the cyclobutane dimer crosslink between P site anticodon position 34 and C1400 of the 16S rRNA is indicated by the solid diamond. See text for further description. Reprinted in modified form from Cunningham *et al.* (1992b) with permission. Copyright 1992 American Chemical Society.

A Model for Four RNA Molecules at the Decoding Site

A schematic two-dimensional view of how rRNA, mRNA, and two anticodon arms of tRNA may disport themselves on the ribosome is presented in Fig. 5. In this figure, I have attempted to convey a sense of the stereochemical constraints operative in this region rather than to give an accurate representation of the location of each molecule. The figure shows the cleft and large projection of the *E. coli* 30S subunit (Gornicki *et al.* 1984) in outline form. The approximate location and orientation of helices 28, 44, and 45 of the 3'-end of 16S RNA in the 30S subunit is derived from the three-dimensional models of Brimacombe *et al.* (1988) and Stern *et al.* (1988). The placement of the 3'-end and connecting single-strand of rRNA to the tip of helix 45 is in keeping with the immunoelectron microscopic localization of the 3'-terminus (referenced in Gornicki *et al.*, 1984). The 1400 and 1500 region single-strands are arranged as in Fig. 3A. mRNA, illustrated by ΦXH mRNA, is placed so as to achieve maximal Shine-Dalgarno pairing.

Placement of the initiating AUG is constrained by the need to juxtapose the 5'-anticodon base of P site bound tRNA to C1400 of the rRNA in view of the cyclobutane dimer crosslink (diamond) which can be formed (Ofengand *et al.*, 1986, 1988). This arrangement leaves a distance of only 20Å (two unpaired nucleotides) between the 3'-end of the Shine-Dalgarno sequence and the A of the AUG (Gornicki *et al.*, 1984). The A site tRNA is placed so that its anticodon is adjacent to that of the P site in the direction required by the 5' to 3' direction of mRNA translation. The width of helices 28, 44 and 45 and the tertiary base pairs between the 1400 and 1500 regions are approximately correctly proportioned to the scale indicated by the outline of the ribosomal surface, but the codon-anticodon base pairs are *not* shown in their true size. The constraint on the distance between the initiator AUG and the Shine-Dalgarno region implicit in this diagram provides additional evidence for the juxtaposition of the large projection and the head of the 30S subunit at some stage(s) of translation.

It is rather clear from this diagram that the decoding site of the ribosome must be a "rat's nest" of RNA strands. If one includes the potential physical juxtaposition and possible base-base interaction with the conserved 518-533 region, it becomes clear that study of the mechanism of decoding on the ribosome is not, to paraphrase Noller (1993), a pursuit that is recommended for the faint of heart.

REFERENCES

Almehdi, M., Yoo, Y.S., and Schaup, H.W., 1991, *Nucleic Acids Res.* 19:6895.

Brimacombe, R., Atmadja, J., Stiege, W., and Schüler, D., 1988, *J. Mol. Biol.* 199:115.

Ciesiolka, J., Gornicki, P., and Ofengand, J., 1985, *Biochemistry* 24:4931.

Cunningham, P.R., Weitzmann, C., Nègre, D., Sinning, J.G., Frick, V., Nurse, K., and, Ofengand, J., 1990, *in* "The Ribosome. Structure, Function, and Evolution", W. Hill, A. Dahlberg, R. Garrett, P. Moore, D. Schlessinger, and J. Warner, eds., pp. 243-252, American Society for Microbiology, Washington, DC.

Cunningham, P.R., Nurse, K., Weitzmann, C.J., Nègre, D., and Ofengand, J., 1992a, *Biochemistry* 31:7629.

Cunningham, P.R., Nurse, K., Bakin, A., Weitzmann, C.J., Pflumm, M., and Ofengand, J., 1992b, *Biochemistry* 31:in press.

Denman, R., Weitzmann, C., Cunningham, P.R., Nègre, D., Nurse, K., Colgan, J., Pan, Y.-C., Miedel, M., and Ofengand, J., 1989, *Biochemistry* 28:1002.

Dontsova, O., Dokudovskaya, S., Kopylov, A., Bogdanov, A., Rinke-Appel, J., Jünke, N., and Brimacombe, R., 1992, *EMBO J.* 11:3105.

Döring, T., Greuer, B., and Brimacombe, R., 1992, *Nucleic Acids Res.* 20:1593.

Frank, J., Penczek, P., Grassucci, R., and Srivastava, S., 1991, *J. Cell Biol.* 115:597.

Gornicki, P., Nurse, K., Hellmann, W., Boublik, M., and Ofengand, J., 1984, *J. Biol. Chem.* 259:10493.

Gutell, R.R., Weiser, B., Woese, C.R., and Noller, H.J., 1985, *in* "Nucleic Acid Research and Molecular Biology", W.E. Cohn, and K. Moldave, eds., vol. 32, pp. 155-216, Academic Press, New York.

Gutell, R.R., and Woese, C.R., 1990, *Proc. Natl. Acad. Sci. USA* 87:663.

Hausner, T.P., Geigenmüller, U., and Nierhaus, K.H., 1988, *J. Biol. Chem.* 263:13103.

Hershey, J.W.B., 1987, *in* "*Escherichia coli* and *Salmonella typhimurium*. Cellular and Molecular Biology", F.C. Neidhardt, J.L. Ingraham, K.B. Low, B. Magasanik, M. Schaechter, and H.E. Umbarger, eds., vol. 1., pp. 613-647, American Society for Microbiology, Washington, D.C.

Krzyzosiak, W., Denman, R., Nurse, K., Hellmann, W., Boublik, M., Gehrke, C. W., Agris, P. F., and Ofengand, J., 1987, *Biochemistry* 26:2353.

Moazed, D., and Noller, H.F., 1989, *Nature* 342:142.

Moazed, D. and Noller, H.F., 1990, *J. Mol. Biol.* 211:135.

Nierhaus, K.H., 1990, *Biochemistry* 29:4997.

Noller, H.F., 1991, *Ann. Rev. Biochem.* 60:191.

Noller, H.F., 1984, *Ann. Rev. Biochem.* 53:119.

Noller, H.F., 1993, *FASEB J.* (in press).

Ofengand, J., Ciesiolka, J., and Nurse, K., 1986, *in* "Structure and Dynamics of RNA", P.H. van Knippenberg, and C.W. Hilbers, eds., pp. 273-287, Plenum Publishing, New York.

Ofengand, J., Denman, R., Nègre, D., Krzyzosiak, W., Nurse, K., and Colgan, J., 1988, *in* "Structure and Expression: I. From Proteins to Ribosomes", R.H. Sarma, and M.H. Sarma, eds., vol. 1, 209-228. Adenine Press, Albany NY.

Powers, T., and Noller, H.F., 1990, *Proc. Natl. Acad. Sci. USA* 87:1042.

Ruusala, T., Ehrenberg, M., and Kurland, C.G., 1982, *EMBO J.* 1:741.

Ruusala, T., Andersson, D., Ehrenberg, M., and Kurland, C.G., 1984, *EMBO J.* 3:2575.

Schüler, D., and Brimacombe, R., 1988, *EMBO J.* 7:1509.

Stern, S., Powers, T., Changchien, L.M., and Noller, H.F., 1989, *Science* 44:783.

Stern, S., Weiser, B., and Noller, H.F., 1988, *J. Mol. Biol.* 204:447.

Stöffler-Meilicke, M., and Stöffler, G., 1990, *in* "The Ribosome. Structure, Function, and Evolution" W. Hill, R. Dahlberg, P. Garett, D. Moore, D. Schlessinger, and J. Warner, eds., pp. 123-133, American Society for Microbiology, Washington DC.

Vasiliev, V.D., Selivanova, O.M., Baranov, V.I., and Spirin, A.S., 1983, *FEBS Lett.* 155:167.

CRYSTALLIZATION AND DIFFRACTION STUDIES OF *THERMUS FLAVUS* 5S rRNA AND SYNTHETIC FRAGMENTS OF THE 5S rRNA

V. A. Erdmann[1], S. Lorenz[1], E. Raderschall[1], J. P. Fürste[1], R. Bald[1], M. Zhang[1], Ch. Betzel[2], and K. S. Wilson[2]

[1]Institut für Biochemie, Fachbereich Chemie, Freie Universität Berlin, Thielallee 63, D-1000 Berlin 33, Germany, [2]EMBL c/o DESY, Notkestr. 85, D-2000 Hamburg 52, Germany

INTRODUCTION

Although it has been established that the ribosomal 5S RNA is essential for the biological function of the ribosomes (Nomura and Erdmann, 1970; Erdmann et al., 1971 a, b; Nierhaus and Dohme, 1974; Dohme and Nierhaus, 1976; Hartmann et al., 1988), its precise role in protein synthesis remains to be elucidated. This dilemma is partially due to the fact that its structure in the ribosome and how it interacts with the ribosomal proteins is still unknown. Since the numerous chemical, biochemical and computer aided studies were insufficient in describing the three-dimensional structure of this molecule (Wolters and Erdmann, 1988; Lorenz et al., 1989; Specht et al., 1991; Bald et al., 1992), we are attempting its crystallization. In this communication we will summarize our efforts to crystallize the ribosomal 5S RNAs, their protein complexes and chemically synthesized domains of *Thermus flavus* 5S rRNA.

THE CRYSTALLIZATION OF *THERMUS FLAVUS* 5S rRNA

Based on the experiences gained in other laboratories that not all tRNA molecules are crystallizeable, we set out to isolate as many different 5S rRNA species as possible in order to test their suitabilities for crystallization. This long term study included the 5S rRNAs from the following organisms: *Azobacter vinelandii*, *Bacillus licheniformis*, *Bacillus stearothermophilus*, *Bacillus subtilis*, *Coulobacter* sp., *Escherichia coli*, *Halobacter cutirubrum*, *Micrococcus luteus*, *Proteus vulgaris*, *Pseudomonas fluorescens*, rat liver, *Staphylococcus aureus*, *Thermotoga maritima*, *Thermus aquaticus*, *Thermus flavus*, *Thermus thermophilus*, wheat germ and yeast (Lorenz et al, 1991). From these 18 different 5S rRNA species we found that only a few of them can be crystallized, namely the ones from *E. coli*, *Thermus aquaticus*, *Thermus thermophilus* and *Thermus flavus*, and that of this group the 5S rRNA from *Thermus flavus* was found to yield the best crystals.

As can be seen in Fig. 1 the 5S rRNA from *Thermus flavus* can be separated into two species, which we have designated as the A and B form. Crystallization experiments with the A and B form mixture, the isolated A form and the isolated B form have shown, that only the mixture of the A and the B form and the isolated A form can be crystallized (Fig. 2 and Lorenz et al., 1991). The structural differences between the A and B form are still unknown, for several analyses have yielded no differences between them.

The best crystals so far obtained showed in their analyses by synchroton radiation that they diffract to a resolution of 8 Å. The crystals have the monoclinic space-group *C*2. The unit cell dimensions are a = 190 Å, b =110 Å, c = 138 Å and ß = 117°. From the cell volume we calculated that there are four 5S rRNA molecules per asymmetric unit. Considering a relative molecular mass of 160 000 for the 5S rRNAs per asymmetric unit, we have calculated the packing volume to be 4.3 Å^3/dalton, which is higher than the range usually observed for protein crystals (1.7 to 3.5 Å^3/dalton), but it is in the same order as observed for tRNA crystals. Out of these data we have calculated the water content of the crystals to be approximately 70%.

From the data presented it is evident that the resolution of the *Thermus flavus* 5S rRNA crystals are not high enough to warrant a structural analysis at atomic resolution. We are therefore currently following two approaches to improve the quality of these crystals by further studying the crystallization conditions with the aid of an automatic device, and by engineering the 5S rRNA molecule in such a way that its inherent structure may become more stable.

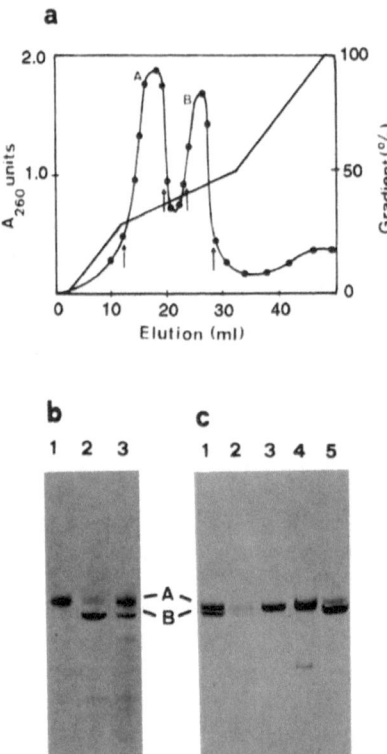

Figure 1. Separation of the two 5S rRNA forms A and B from *Thermus flavus* by (a) FPLC Phenyl-Superose chromatography and (b) and (c) denaturing gel electrophoreses (Lorenz et al, 1991).

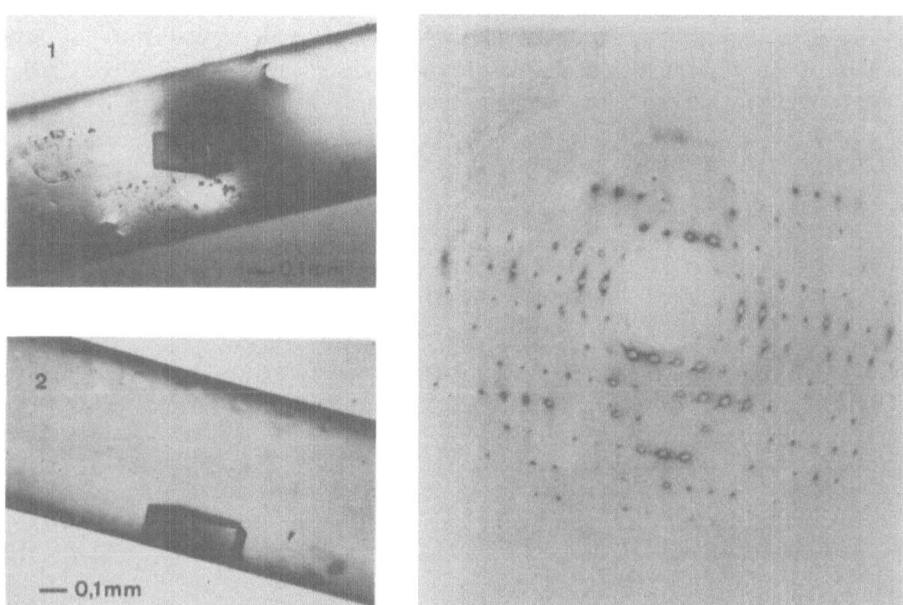

Figure 2. Typical crystals from *Thermus flavus* 5S rRNA A and B form (1) and A form (2). Right panel: crystal diffraction pattern of the A form 5S rRNA crystal from *Th. flavus* (Lorenz et al., 1991).

THE CRYSTALLIZATION OF 5S rRNA-PROTEIN COMPLEXS

Since it has been demonstrated that the *E. coli* 5S rRNA can readily assume different structural conformations, we assume that 5S rRNA-protein complexes may be more stable and thus easier to crystallize. Following our earlier discovery that *E. coli* 5S rRNA interacts with the proteins EL-5, EL-18 and EL-25 and the *B. stearothermopholus* 5S rRNA with the proteins BL-5 and BL-22 (Horne and Erdmann, 1972), we isolated these 5S rRNA-protein complexes for crystallization purposes. The results of such crystallization experiments are shown in Fig. 3. Although both types of 5S rRNA-protein complexes yielded in their optical appearence reasonable crystals, X-ray diffraction studies indicated very low quality with

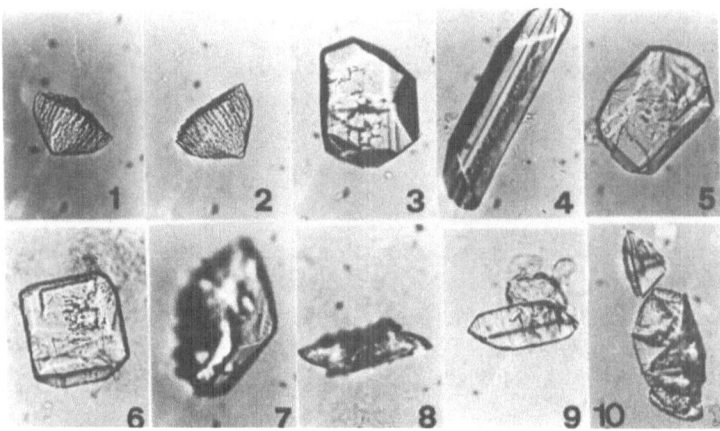

Figure 3. Crystals of 5S rRNA-protein complexes from *E. coli* (1-6) and *B. stearothermophilus* (7-10).

503

respect to their atomic resolution. Another difficulty connected with these crystallization experiments is due to the fact that the crystals required several years to grow. Because of the difficulties with the 5S rRNA-protein complexes from *E. coli* and *B. stearothermophilus*, and our observation that the isolated 5S rRNA from *Thermus flavus* is

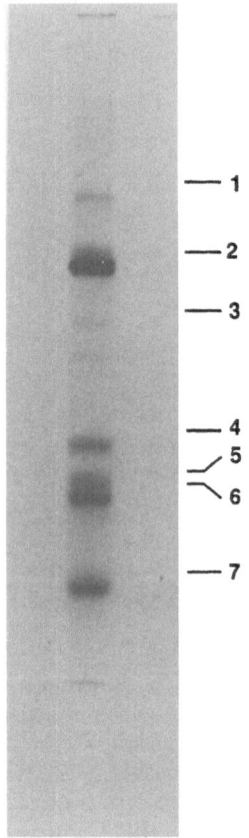

Figure 4.
Electrophoretic separation of *Th. flavus* ribosomal proteins as isolated from a 5S rRNA affinity column. Based upon partial protein sequence data (Schröder, W., et al., unpublished results) the bands were found to correspond to *B. stearothermophilus* L2 (1), an unknown protein (2), *Th. aquaticus* S7 (3), *B. stearothermophilus* L18 (4 and 6), *B. stearothermophilus* S13 (5) and *Th. thermophilus* DNA binding protein II (7).

best suited for crystallization, we have been attempting to isolate and characterize the 5S rRNA-protein complexes from *Th. flavus*. Our current knowledge about the *Th. flavus* 5S rRNA binding proteins are shown in Fig. 4. At this time experiments are on their way in which we are trying to crystallize a reconstituted 5S rRNA-protein complex consisting of *Th. flavus* 5S rRNA and its L18 protein.

THE CRYSTALLIZATION OF 5S rRNA DOMAINS

Since the resolution of the 5S rRNA crystals so far obtained do not allow the determination of their atomic structure, we have initiated a long term project with the purpose to chemically synthesize RNA molecules in amounts large enough to attempt their crystallization. In addition to tert.-butyldimethylsilyl-protected synthons (Ogilvie et al., 1988), we have synthesized ribonucleoside phosphoramidite reagents with triisopropylsilyl- or thexyldimethylsilyl-groups for 2'-hydroxyl protection (Tab. 1 and Bald et al., 1992). Furthermore, as shown in Table 2, we have prepared a number of phosphoramidite synthons for the construction of modified RNA molecules. Several of the modified synthons shown in this table are suitable for cross-linking and fluorescent energy transfer experiments, which we are currently employing in our ribozyme studies and which will be used in structural studies on 5S rRNAs and their protein complexes.

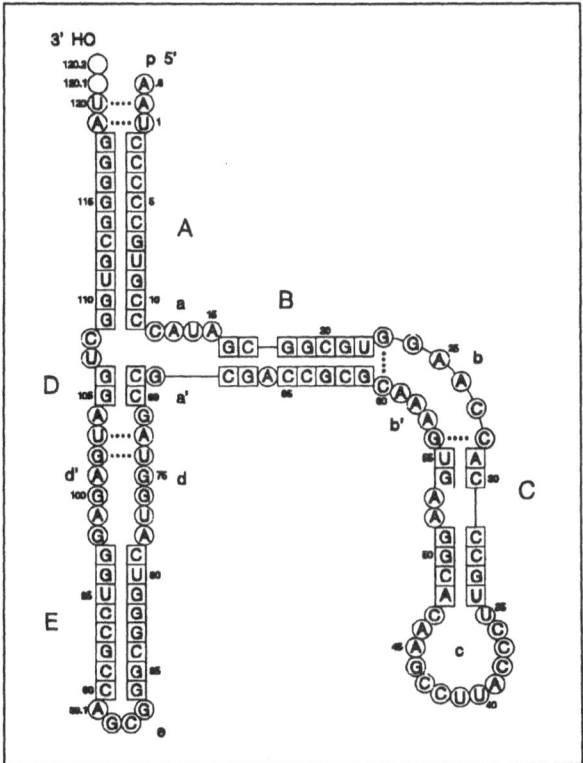

Figure 5. Secondary structure of *Th. flavus* 5S rRNA. The domains A - E are currently synthesized and used in crystallization experiments.

Currently, our main efforts are concentrated on the chemical synthesis of the structural domains A-E of *Th. flavus* 5S rRNA (Fig. 5) to use these molecules in crystallization experiments. Most advanced is the project involving the domain A (helix A) of *Th. flavus* 5S rRNA (Betzel et al., 1992). Crystallization experiments with an automated pipetting device showed that small crystals could be obtained after several days, and that these crystals could be grown to 1 mm length by repeated crystal seeding experiments (Fig. 6).

Table1. Ribonucleoside phosphoramidite reagents for the chemical synthesis of RNA.

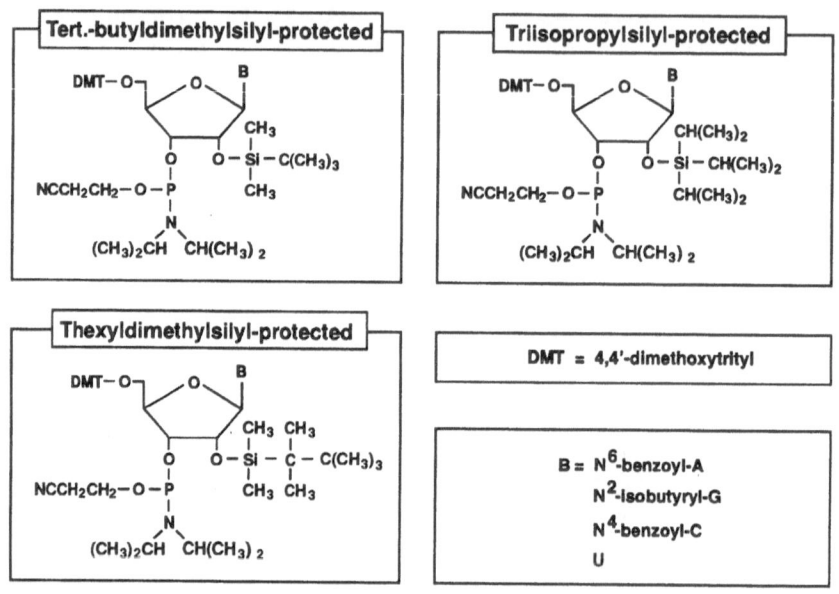

Table 2. Phosphoramidite synthons for the preparation of oligoribonucleotides carrying modifications at specific positions.

N^6-methyladenosine	N^6,N^6-dimethyladenosine	N^1,N^6-ethenoadenosine	2-aminopurineriboside
2-aminoadenosine	5-methylcytidine	5-bromocytidine	isocytidine
5-methyluridine	5-bromouridine	pseudouridine	8-azidoadenosine

2'-O-methylderivatives	2'-deoxyderivatives

Figure 6. Crystal of *Th. flavus* helix A (see Fig. 5). The length of the crystal is 1 mm and its maximal resolution is 2.6 Å.

Such crystals were mounted in a glass capillary and used to collect data up to 3.0 Å on a conventional sealed tube X-ray source with MoK_{α} radiation and a graphite monochromator using a MAR imaging plate detector. The space group of the observed crystal was evaluated to be $P4_3$ or $P4_1$ with unit cell parameters of a and b = 30.10 Å and c = 86.80 Å. One helical structure per asymmertric unit gives a packing parameter V_M of 2.6 Å^3/dalton. The reduced data set contains 1477 reflections and shows a completeness of 94%.

The structure solution was achieved by molecular replacement using the coordinates of the synthetic RNA helix: $[U(UA)_6A]_2$ as a starting model (Dock-Bregeon et al, 1989) and a new rotation and translation function recently written in the program AMORE (Navaza et al, 1992). The final R-value was determined to be 15.8%. Our most recent analysis of the data allowed us to locate 64 solvent water molecules in association with the helix A structure (Betzel et al., 1992).

Current data are collected from a 1 mm helix A crystal by synchroton radiation with the hope that a higher resolution will be obtained. Other *Thermus flavus* 5S rRNA domains have been synthesized and are currently tested in crystallization experiments.

CONCLUSIONS

The data presented in this communication demonstrated that the ribosomal 5S RNA from *Thermus flavus* and its chemically synthesized domain A can be crystallized, and that this progress makes us optimistic to believe that the first non-tRNA structure to be determined by X-ray analysis will be the one of a ribosomal 5S RNA.

ACKNOWLEDGEMENTS

The work was supported by grants of the Deutshe Forschungsgemeinschaft (Gottfried Wilhelm Leibniz - Programm, SFB 9/B5 and SFB 344/C1 and C2), the Fonds der chemischen Industrie e.V. and the Bundesminister für Forschung und Technologie (BMFT).

507

REFERENCES

Bald, R., Brumm, K., Buchholz, B., Fürste, J.P., Hartmann, R.K., Jäschke, A., Kretschmer-Kazemi Far, R., Lorenz, S., Raderschall, E., Schlegl, J., Specht, T., Zhang, M., Cech, D., and Erdmann, V.A., 1992, in:" Structural Tools for the Analysis of Protein-Nucleic Acid Complexes", D. Lilley, H. Heumann and D. Suck, ed., Birkhäuser Verlag, Basel, Switzerland., p.442.

Betzel, C., Lorenz, S., Fürste, J.P., Bald, R., Zhang, M., Raderschall, E., Wilson, K.S., and Erdmann, V.A., 1993, Acta Crystallographica, in press.

Dock-Bregeon, A.C., Chevrier, B., Podjarny, A., Johnson, J., de Bear, J.S., Gough, G.R., Gilham, P.T., and Moras, D., 1989, *J. Mol. Biol.* 209: 450.

Dohme, F., and Nierhaus, K.M., 1976, *J. Mol. Biol.* 107: 585.

Erdmann, V.A., Fahnestock, S., Hogo, K., and Nomara, M., 1971a, *Proc. Nat. Acad. Sci. U.S.A.* 68: 2932.

Erdmann, V.A., Doberer, H.G., and Sprinzl, M., 1971b, *Mol. Gen. Genet.* 114: 89.

Hartmann, R.K., Vogel, D.W., Walker, R.T., and Erdmann, V.A., 1988, *Nucl. Acids Res.* 16: 3511.

Horne, J.R., and Erdmann, V.A., 1972, *Mol. Gen. Genet.* 119: 337.

Lorenz, S., Hartmann, R.K., Schultze, S., Ulbrich, N., and Erdmann, V.A., 1989, *Biochemie* 71: 1185.

Lorenz, S., Betzel, Ch., Raderschall, E., Dauter, Z., Wilson, K. S., and Erdmann, V. A., 1991, *J. Mol. Biol.* 219: 399

Navaza, J., 1992, Proceeding of the Daresbury Study Weekend on Molecular Replacement, Daresbury, England.

Nierhaus, K.M., and Dohme, F., 1974, *Proc. Natl. Acad. Sci. U.S.A.* 71: 4713.

Nomura, M., and Erdmann, V.A., 1970, *Nature (London)* 228: 744.

Ogilvie, K.K., Usmann, N., Nicoghosian, K., and Cedergren, R.G., 1988, *Proc. Nat. Acad. Sci. U.S.A.* 85: 5764

Specht,T., Wolters, J., and Erdmann, V. A., 1991, *Nucl. Acids Res.* 19 (Suppl.): 2189.

Wolter, J., and Erdmann, V.A., 1988, *Nucl. Acids Res.* 16 (Suppl.): r1.

TOWARDS RIBOSOMAL STRUCTURE AT PEPTIDE LEVEL: USE OF CROSSLINKING, ANTIPEPTIDE ANTIBODIES AND LIMITED PROTEOLYSIS

Volker Kruft[1], Oliver Bischof[1], Ulrike Bergmann[2], Elke Herfurth[1], and Brigitte Wittmann-Liebold[1]

[1]Max-Delbrück-Centrum für Molekulare Medizin, Robert-Rössle-Str. 10, O-1115 Berlin-Buch, Germany
[2]Max-Planck-Institut für Molekulare Genetik, Ihnestr. 73, W-1000 Berlin 33, Germany

INTRODUCTION

Although remarkable progress has been achieved in the past decades towards an understanding of the structure and function of the ribosome, the interactions of nucleic acids and proteins involved in protein biosynthesis remain largely unresolved. It has convincingly been demonstrated that ribosomal RNA is essential for ribosomal function (e.g. Schulze and Nierhaus, 1982; Dahlberg, 1989; Noller et al., 1992), yet there can be no doubt that in contemporary ribosomes complexes of RNA and proteins constitute the functional units. Models of the tertiary structure of the 16S RNA have been derived from footprinting and crosslinking experiments (Stern et al., 1988; Brimacombe et al., 1988; Nagano et al., 1988). In addition, functionally important domains and even nucleotides were identified by affinity labelling and site-directed mutagenesis.

In contrast, the models for the distribution of the proteins in the subunits have limited resolution (Capel et al., 1988; Walleczek et al., 1988; May et al., 1992), primarily because the methodologies employed do not lend themselves easily to an analysis at peptide or amino acid level. Therefore, ribosomal models including both, the protein and the nucleic acid components are preliminary due to assumptions about the proteins' quarternary structures (Brimacombe, 1992). Ultimately, X-ray analysis of appropriate ribosomal crystals will provide molecular information of the ribosomal fine structure (Yonath and Wittmann, 1989; von Böhlen et al., 1991). Yet even when data sets at sufficient resolution obtained from native and derivatized ribosomal crystals are available, further biophysical and detailed biochemical information will be required to assign the peptide backbones in the model and to define the peptide moieties of functional centers (Eisenstein et al., 1991).

Knowledge of the primary structures of all proteins from *Escherichia coli* ribosomes (Wittmann-Liebold et al., 1990) and of most of *Bacillus stearothermophilus* (Arndt et al., 1991; Kruft et al., 1991; Herfurth et al., 1991, Herwig et al., 1992a) is the basis for investigations of the higher order structures of those proteins. In this chapter we will discuss crosslinking experiments, the use of antipeptide antibodies and limited proteolysis as means to obtain the molecular information necessary for a refinement of ribosomal topography and investigations of the structure-function relationships in ribonucleoprotein particles.

CROSSLINKING

The analysis of chemical or UV-induced crosslinks in isolated ribosomes as well as affinity-labelling with tRNA, mRNA and antibiotics has been valuable in defining the neighborhoods of ribosomal components. The proposed ribosomal models rely heavily on information derived from such experiments (Brimacombe et al., 1988; Stern et al., 1988; Walleczek et al., 1988) but have limited resolution regarding the protein components. Structural refinements were obtained by the exact determination of nucleotides in RNA-RNA crosslinks and protein-RNA-crosslinks (Brimacombe et al., 1990; Brimacombe, 1991), but only in a few cases have crosslinks induced in intact subunits been analyzed at amino acid level (Ehresmann et al., 1976; Möller et al., 1978; Allen et al., 1979; Maly et al., 1980; Brockmöller and Kamp, 1988; Pohl and Wittmann-Liebold, 1988; Zecherle et al., 1992). During the past few years, our laboratory has carried out crosslink analyses at the amino acid level in order to gain information about the contact sites between neighboring proteins. We did not limit our studies to the *E. coli* ribosome but also included the Gram-positive organism *B. stearothermophilus* and the archaebacterium *Haloarcula marismortui*. Data obtained for ribosomes isolated from different organisms allow an assessment of the conservation of the fine structure of the ribosome between those species.

The identification of crosslinked complexes solely by electrophoretic methods, two-dimensional and diagonal electrophoresis (Sommer and Traut, 1974), often yielded results that were difficult to interpret due to altered electrophoretic mobilities of the proteins. The immunochemical analysis of protein-protein crosslinks avoids this problem but requires a complete set of pure and well-characterized antibodies (Stöffler et al., 1988). However, both methods are limited to the identification of the crosslinked proteins and do not allow peptide analysis. In our studies the application of sensitive methods for the purification and sequencing of proteins and peptides, and in addition the use of mass spectrometry, lead to unambiguous results in the identification of crosslinked peptides and amino acids.

The number and yield of formation of individual crosslinks is strongly dependent on the reactivity and specificity of the crosslinking reagent used. The usefulness of commercially available reagents for protein-protein crosslinking in ribosomes has been reviewed by Kamp (1988) and will not be discussed in detail in this chapter.

One of the main protein pairs formed in 30S ribosomal subunits from both *B. stearothermophilus* and *E. coli* by treatment with diepoxybutane (4Å crosslinking length) was S13-S19 (Brockmöller and Kamp, 1988; Pohl and Wittmann-Liebold, 1988). The proximity of the two proteins in the 30S subunit has also been demonstrated by earlier crosslinking experiments (Traut et al., 1986) and immuno electron microscopy (Stöffler-Meilicke and Stöffler, 1990) but slightly differs from distances measured by neutron scattering (Capel et al., 1988). The analysis of the crosslinked amino acids revealed that the crosslink site is identical in both protein-pairs: for *B. stearothermophilus* Cys-83 (Cys-84 in *E. coli*) of S13 and His-68 of S19 in both organisms are the crosslinked amino acids (Brockmöller and Kamp, 1988; Pohl and Wittmann-Liebold, 1988). The crosslinking sites are located within the most highly conserved domain of the two proteins. Thus, in addition to conservation of the primary sequence the topography of the contact region between the two proteins is conserved in Gram-positive and Gram-negative eubacteria.

Several protein-pairs from 50S subunits were analyzed at molecular detail. Among these is protein pair L3-L19 which was crosslinked by diepoxybutane in *B. stearothermophilus*. The amino acids involved in crosslinking have been identified as His-28 of L3 and each of the three N-terminal amino acids of L19, Met-1, His-2 or His-3 (Herwig et al., 1992b). The crosslink L3-L19 has also been isolated from *E. coli* 50S subunits using dithiobis(succinimidyl propionate) and o-phenylenedimaleimide as crosslinking reagents with a span of 12 Å and 5.2 Å, respectively (Walleczek et al., 1989), but the crosslinked amino acids were not analyzed. However, these crosslinking data established the approximate location of L3 within the large subunit which could not be done by specific antibody binding in immuno electron microscopy (Walleczek et al., 1988).

L23-L29 is one of the major crosslinks induced in the large subunit. The complex has been identified in *E. coli* crosslinked by o-phenylenedimaleimide (Walleczek et al., 1989) and was formed by diepoxybutane in *B. stearothermophilus* (Brockmöller and Kamp, 1986). Recently, the corresponding protein pair was obtained using diepoxybutane and dithiobis(succinimidylpropionate) in the 50S subunit of the *H. marismortui*. The crosslink sites in both protein pairs were determined at amino acid level: For *B. stearothermophilus* Met-1 of L23 was crosslinked to Lys-4 of L29 (Herwig et al., 1992b) whereas in *H. marismortui* Ser-1 of L23 was found to be crosslinked to Lys-57 of L29 (Bergmann and Wittmann-Liebold, submitted). Therefore, the topography of the L23/L29 neighborhood is not fully conserved at amino acid level between eubacteria and archaebacteria; however, in both protein-pairs the N-terminus of L23 is the crosslink site to protein L29.

Our results are interesting with respect to the location of L23 within the 50S subunit. L23 has been placed on the back at the base of the 50S subunit close to L29 by immuno electron microscopy (Stöffler-Meilicke et al., 1983a; Hackl and Stöffler-Meilicke, 1988). Furthermore, L23 and L29 were crosslinked to adjacent sites in 23S RNA (Wower et al., 1981). On the other hand, affinity labelling of L23 by the 3'-end of aminoacylated tRNA (Ofengand et al., 1986) and by A-site bound puromycin (Jaynes et al., 1978; Weitzmann and Cooperman, 1985) argue for surface exposure of L23 at the interface side of the large subunit. These results are supported by crosslinking of L23 to L15, L16, and L27 carried out by Traut and coworkers (1986). Additionally, an elongated structure was measured in solution for isolated protein L23 (Giri et al., 1984). From our results it seems very likely that the N-terminal domain of L23 is in close proximity to L29 while the C-terminal part spans the body of the 50S subunit with a domain exposed close to the peptidyltransferase center, thereby giving experimental evidence to the model proposed by Nagano et al. (1988) in which L23 has a length of about 90 Å running parallel to the channel shielding the nascent peptide chain (Yonath et al., 1987).

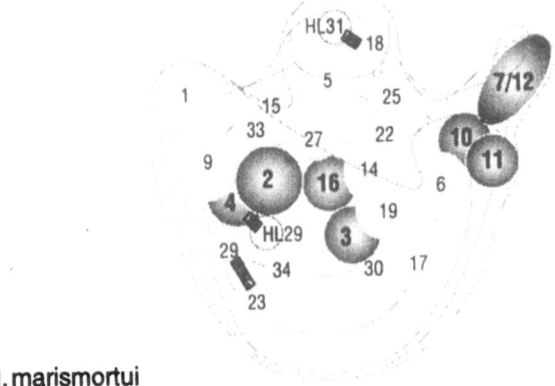

H. marismortui

Figure 1. Location of ribosomal proteins HmaL23, HL29, and HL31 as determined by protein-protein crosslinking. The protein distribution within the 50S subunit is based on the model of Walleczek et al., 1988.

By providing information about protein neighborhoods in intact subunits, protein-protein crosslinking can also be helpful in positioning proteins that cannot be localized by other methods. Proteins L3, L13, L22, L32 and L33 were placed in the protein model of the *E. coli* 50S subunit by crosslinking to proteins whose position was determined by immuno electron microscopy and crosslinking (Walleczek et al., 1988). However, it should be noted that the positions inferred from crosslinking relate to single domains of the protein concerned and do not necessarily coincide with the centres of gravity as determined by neutron scattering (May et al., 1992). Still, the approach should prove valuable for establishing the approximate positions of proteins in halophilic ribosomes since antibody binding in the buffers necessary to maintain a native ribosome structure is not possible.

Assuming the position of HmaL18 in the halophilic 50S subunit to be roughly the same as for the *E. coli* equivalent (Tischendorf et al., 1974), HL31 must be located close to or in the central protuberance (Figure 1) (Bergmann and Wittmann-Liebold, submitted).

Similarly, HL29 was positioned close to the peptidyltransferase center in the neighborhood of protein L4 (Figure 1) (Lotti et al., 1989; Bergmann and Wittmann-Liebold, submitted).

Protein-protein crosslinking has also been applied to eukaryotic ribosomes, such as the 40S and the 60S subunit of rat liver ribosomes (Uchiumi et al., 1981; Uchiumi et al., 1985a, b), as well as to cytoplasmic ribosomes from *Saccharomyces cerevisiae* (Xiang and Lee, 1989). However, due to the incomplete sets of primary structures and the overall high number of proteins in eukaryotic ribosomes that still have to be correlated to the existing 2D-gel patterns no detailed topographical analysis is possible yet. Therefore, more experimental data will be necessary to construct ribosomal models not only of the eubacterial ribosome but also of the archaebacterial and the eukaryotic organelle.

IMMUNOLOGY

Various immunological approaches have been used to investigate the structure and function of the ribosome. Immunochemical analyses combined with immuno eletron microscopy were used to locate ribosomal proteins *in situ* at the surface of ribosomes (Stöffler-Meilicke and Stöffler, 1990; Scheinmann et al., 1992). Most of the antibodies used have been derived from polyclonal antisera which contain heterogenous antibodies directed towards an unknown number of epitopes and are therefore suited to locate the proteins and to reveal the number of surface domains and their extent (Stöffler-Meilicke et al., 1983). Yet the localization of defined domains of ribosomal components is only possible with antibodies directed against defined antigenic epitopes (Winkelmann and Kahan, 1980).

Several methods have recently been applied to achieve the localization of predetermined epitopes on the ribosomal surface:

Antibodies reactive with haptens like fluorescein, coumarine, antibiotics, dinitrophenol groups or modified nucleotides are incubated with ribosomes exposing those haptens as ligands, on reconstituted proteins or RNA-molecules that were site-specifically derivatized, thereby rendering the definition of the antigenic determinants easy. Examples exist for the direct immunological visualization of protein sites (Stöffler-Meilicke et al., 1983b, 1984), rRNA domains (Stöffler-Meilicke et al., 1981; Evstafieva et al., 1985; Lührmann et al., 1981a; Olson and Glitz, 1979), 3´ and 5´ ends of mRNA (Evstafieva et al., 1983), the nascent polypeptide chain (Ryabova et al., 1988; Picking et al., 1992) and various antibiotic binding sites (for example: Lührmann et al., 1981b; Olson et al., 1982; Grant et al., 1983; Stöffler and Stöffler-Meilicke, 1983).

Alternatively, monoclonal antibodies raised against ribosomal proteins or protein fragments are used to locate proteins in the subunits (Breitenreuter et al., 1984; Schwedler-Breitenreuter, 1985; Nag et al., 1987; Nag et al., 1991), but increased resolution can only be achieved after characterization of the antigenic epitopes recognized by the different monoclonal antibodies (Sommer et al., 1985; Olson et al., 1986; Nag et al., 1986; Syu et al., 1990; Olson et al., 1991).

Use of polyclonal antisera raised against isolated peptides is a third alternative to increase the resolution of the immunological approach (Walleczek et al., 1990; Moffat et al., 1991). Since synthetic peptides can induce antibodies specific for the same target sequence in intact proteins (Lerner et al., 1982) we decided to apply this powerful technique to the analysis of the peptide topography of the ribosome (Herfurth, 1992). Synthetic peptides of 10 to 20 amino acids in length were designed according to the predicted antigenicity, turn probability and overall hydrophilicity (Devereux et al., 1984) of the corresponding domain in the primary structure of various ribosomal proteins from *E. coli* and *B. stearothermophilus*.

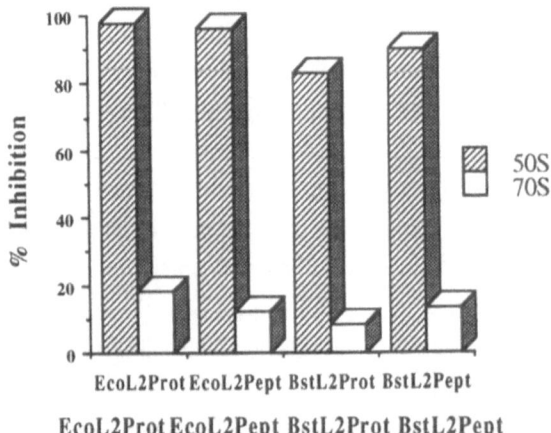

Figure 2. Percent inhibition by preincubation with 50S and 70S subunits of the reaction of antibodies against synthetic C-terminal peptides of protein L2 with purified L2 protein or the peptides.

Purified synthetic peptides were coupled to bovine serum albumin (Green et al., 1982) and the conjugates used to raise antisera. To assess the reactivity of these antibodies with intact ribosomes a solid-phase competition assay was performed (Herfurth and Wittmann-Liebold, submitted): Antibodies were allowed to react with subunits or 70S ribosomes and the amount of unreacted antibody was then measured by reacting it with immobilized protein or peptide. The degree of antigenic reactivity of the ribosomal subunit or the ribosome is inversely proportional to the level of peptide antibody that can still be detected after the first incubation. As an example, results obtained with antibodies directed against the C-terminal 12 and 19 amino acids of protein L2 from *E. coli* and *B. stearothermophilus*, respectively, are shown in Figure 2.

Clearly, the antibodies were reactive with isolated 50S subunits but not with 70S ribosomes, indicating that the C-terminus of both L2 proteins is exposed at the interface of the large subunit. These results concerning the position of L2 in the ribosome as well as the differential exposure of the C-terminal region are in good agreement with other immunological studies (Hackl et al., 1988; Olson et al., 1991) as well as with data obtained by limited proteolysis (see below).

A systematic immunological investigation of the ribosomal surface as decribed above will add valuable data about ribosomal topography and peptides located at functional centers of the ribosome.

LIMITED PROTEOLYSIS

Soon after the ribosome was identified as the principal protein-biosynthetic organelle, proteolytic treatment was used to probe ribosomal structure and the accessibility of the proteins in intact particles (Kaji et al., 1966; Zak et al., 1966; Östner and Hultin, 1968; Chang and Flaks, 1970).

However, a classification of cleavable proteins as external and inaccessible ones as internal is problematic because of assumptions about the structural stability of the ribosome under digestion conditions (Crichton and Wittmann, 1971). Despite the lack of primary sequence information, these early studies showed good agreement with results from other structural investigations, for example modification studies (Craven and Gupta, 1970). Limited proteolysis, although a common method in protein structure analysis (for a review: Wilson, 1991), has only been applied to the ribosome and its subunits to probe structural changes upon factor interaction (Gudkov and Gongadze, 1984; Gudkov and Bubunenko, 1989; Bubunenko et al., 1992) or to determine the length of the nascent polypeptide that is protected by the translating ribosome (Malkin and Rich, 1967; Blobel and Sabatini, 1970; Smith et al., 1978).

We have used limited proteolysis to identify primary cleavage sites, and hence exposed peptide areas, on intact ribosomes and their subunits. The experimental approach minimizes the amount of protease used in order to leave the overall protein composition and thus ribosomal structure intact while generating a small number of fragments barely visible in two-dimensional electrophoresis. The amino acid sequences of those peptides reveal the proteins that are surface exposed as well as the cleavage sites on the proteins. After an initial study on 50S ribosomes (Kruft and Wittmann-Liebold, 1991) we have now extended the approach to include 30S subunits (Kruft, Bischof, and Wittmann-Liebold, submitted).

Figures 3A and 3B schematically summarize the data obtained for 50S and 30S ribosomal subunits. To date, surface-exposed domains were experimentally determined for 14 L- and 13 S-proteins from *E. coli*, as well as 6 and 11 from *B. stearothermophilus* and 8 and 6, respectively, for *H. marismortui*. As expected, for *E. coli* only those proteins are accessible to protease that have previously been mapped on the surface immunologically

(Stöffler-Meilicke and Stöffler, 1990). Furthermore, proteins like L9, L11, and L15 that simultaneously bind more than one IgG molecule in dimeric immunocomplexes (Stöffler-Meilicke et al., 1983a) have extended cleavable domains in intact subunits. In contrast, core proteins (Spillmann et al., 1977) are not accessible or expose only very distinct N- or C-terminal regions.

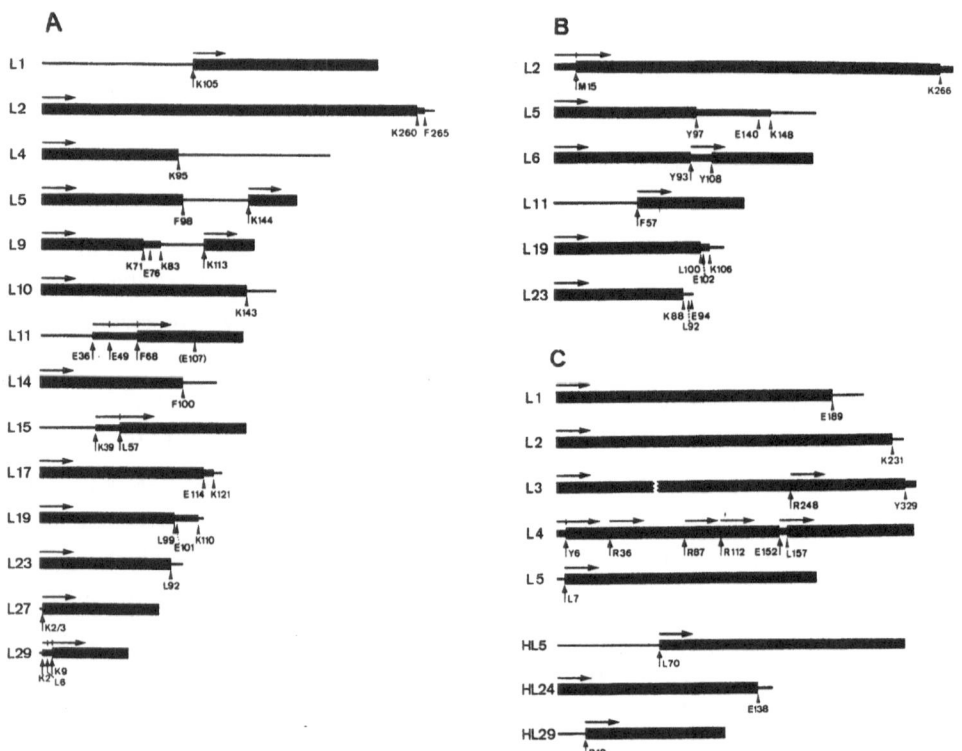

Figure 3A. Surface-exposed peptides on the 50S subunit as determined after limited proteolysis. A: *E. coli*, B: *B. stearothermophilus*, C: *H. marismortui*. Proteins are drawn as bars to relative size. Thick bars indicate peptide areas that stay anchored in the cores, bars of intermediate thickness mark extended exposed domains. Arrows underneath the bars mark directly determined cleavage sites, pointers deduced cleavage sites, arrows above the bars the corresponding peptide sequences. The amino acid at the cleavage site is given in one-letter code and numbered according to the position in the primary structure.

Comparison of the data obtained for *E. coli* and *B. stearothermophilus* allows an assessment of the conservation of the ribosomal surface topography. For most proteins, identical or similar cleavage sites were obtained indicating an overall conservation of the ribosomal fine structure as was also concluded from immunological (Stöffler-Meilicke and Stöffler, 1990) and crosslinking experiments (Brockmöller and Kamp, 1988; Pohl and Wittmann-Liebold, 1988). For several reasons the comparison of the structural data for the eubacteria and *H. marismortui* is complicated: The overall negative charges of the halophilic proteins per se lead to a different distribution of cleavage sites at the ribosomal surface, and

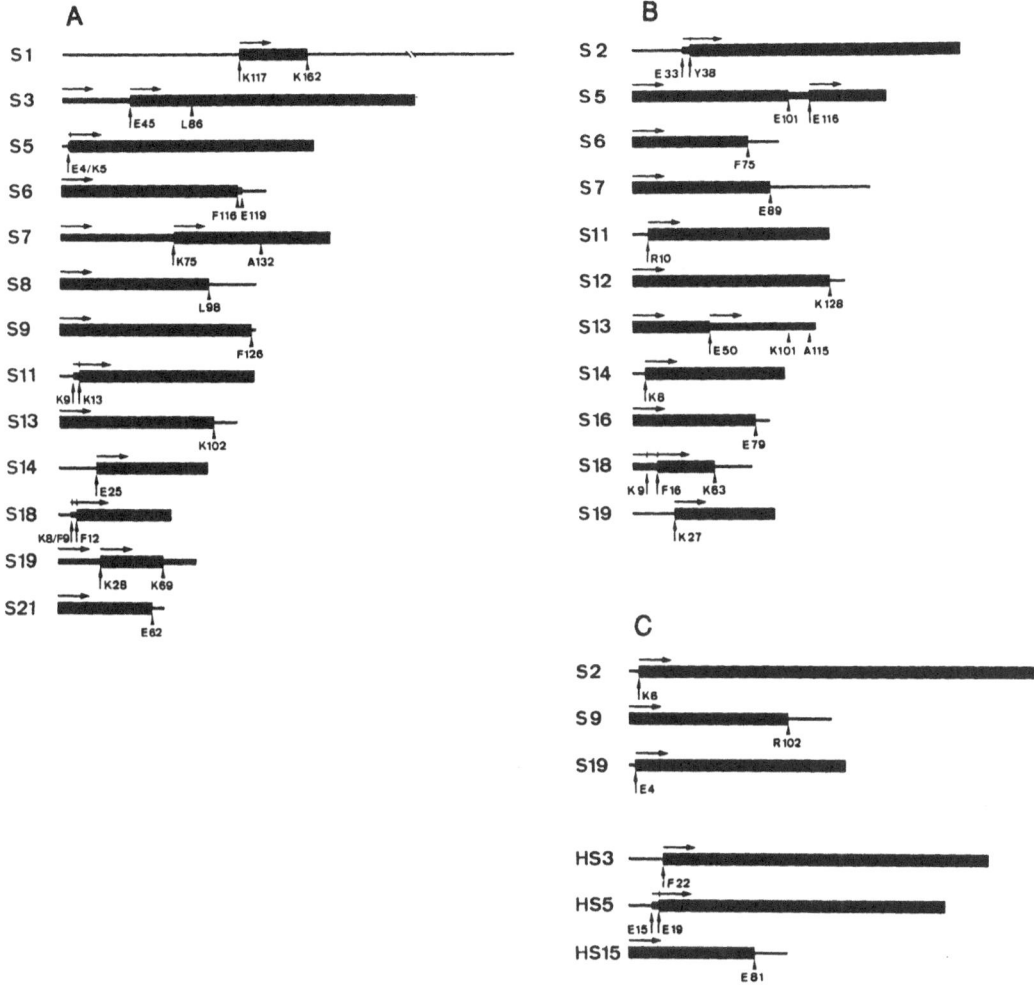

Figure 3B. Surface-exposed peptides on the 30S subunit as determined after limited proteolysis. A: *E. coli*, B: *B. stearothermophilus*, C: *H. marismortui.*. For explanation of symbols, see legend to figure 3A.

the presence of proteins with unresolved primary structures or without counterparts in eubacteria additionally complicates the analysis. However, the data clearly demonstrate that sequence extensions relative to shorter eubacterial and eukaryotic counterparts are exposed on the ribosomal surface: for example, HmaL3, that is cleaved at its C-terminus, bears a C-terminal extension compared EcoL3 and BstL3, and HL5, that is about 90 amino acids longer than the eukaryotic L32 proteins is cleaved at position 70 (Figure 3A)

The excellent correlation of the data presented in this section with immunological data indicates that the sites of proteolytic cleavage and the antigenic determinants are located in the same domain of the target protein. To add experimental proof to this hypothesis tight couple ribosomes were subjected to limited proteolysis. The protection of the C-terminal cleavage sites of interface proteins L2 and L19 by the 30S subunit demonstrates that for a given protein protease cleavage and antibody binding on intact subunits very likely occur at identical suface locations (Kruft and Wittmann-Liebold, 1991).

CONCLUSIONS

Three approaches to increase the resolution of the ribosomal models to peptide level are discussed in this chapter. Antipeptide antibodies and limited proteolysis can efficiently be utilized to identify and position peptides at the ribosomal surface. Data about the surface topography are thereby obtained and verified by two experimentally independent methods (see figures 2 and 3). Analysis of protein crosslinks at the amino acid level introduces further structural constraints for individual proteins. This information, together with results from footprinting and other experiments, will lead to a biochemically derived ribosomal model including both RNA and protein components that will not only facilitate the interpretation of X-ray diffraction data but also help in understanding the molecular basis of ribosomal function.

REFERENCES

Allen, G., Capasso, R., and Gualerzi, C., 1979, *J. Biol. Chem.* 254:9800.
Arndt, E., Scholzen, T., Krömer, W., Hatakeyama, T., and Kimura, M., 1991, *Biochimie* 73:657.
Blobel, G., and Sabatini, D.D., 1970, *J. Cell Biol.* 45:130.
Breitenreuter, G., Lotti, M., Stöffler-Meilicke, M., and Stöffler, G., 1984, *Mol. Gen. Genet.* 197:189.
Brimacombe, R., 1991, *Biochimie* 73:927.
Brimacombe, R., 1992, *Biochimie* 74:319.
Brimacombe, R., Atmadja, J., Stiege, W., and Schüler, D., 1988, *J. Mol. Biol.* 199:115.
Brimacombe, R., Greuer, B., Mitchell, P., Osswald, M., Rinke-Appel, J., Schüler, D., and Stade, K., 1990, *in:* The Ribosome. Structure, Function and Evolution, W.E. Hill, A. Dahlberg, R.A. Garrett, P.M. Moore, D.Schlessinger, and J.R. Warner, eds., American Society for Microbiology, Washington, D.C., 93-106.
Brockmöller, J., and Kamp, R.M., 1986, *Biol. Chem. Hoppe-Seyler* 367:925.
Brockmöller, J., and Kamp, R.M., 1988, *Biochemistry* 27:3372.
Bubunenko, M.G., Kireeva, M.L., and Gudkov, A.T., 1992, *Biochimie* 74:419.
Capel, M.S., Kjelgaard, M., Engelmann, D.M., and Moore, P.B., 1988, *J. Mol. Biol.* 200:65.
Chang, F.N., and Flaks, J.G., 1970, *Proc. Natl. Acad. Sci. USA* 67:1321.
Craven, G.R., and Gupta, V., 1970, *Proc. Natl. Acad. Sci. USA* 67:1329.
Crichton, R.R., and Wittmann, H.G., 1971, *Mol. Gen. Genet.* 114:95.
Dahlberg, A.E., 1989, *Cell* 57:525.
Devereux, J., Haeberli, P., and Smithies, O., 1984, *Nucl. Acid. Res.* 12:387.
Ehresmann, B., Reinbolt, J., Backendorf, C., Tritsch, D., and Ebel, J.P., 1976, *FEBS Lett.* 67:316.
Eisenstein, M., Sharon, R., Berkovitch-Yellin, Z., Gewitz, H.S., Weinstein, S., Pebay-Peyroula, E., Roth, M., and Yonath, A., 1991, *Biochimie* 73: 879.
Evstafieva, A.G., Shatsky, I.N., Bogdanov, A.A., Semenkov, Y.P., and Vasiliev, V.D., 1983, *EMBO J.* 2:799.
Evstafieva, A.G., Shatsky, I.N., Bogdanov, A.A., and Vasiliev, V.D., 1985, *FEBS Lett.* 185:57.
Giri, L., Hill, W.E., Wittmann, H.G., and Wittmann-Liebold, B., 1984, *Adv. Prot. Chem.* 36:1.
Grant, P.G., Olson, H.M., Glitz, D.G., and Cooperman, B.S., 1983, *J. Biol. Chem.* 258:11305.
Green, N., Alexander, H., Olson, A., Alexander, S., Shinnick, T.M., Sutcliffe, J.G., and Lerner, R.A., 1982, *Cell* 28:477.
Gudkov, A.T., and Bubunenko, M.G., 1989, *Biochimie* 71:779.
Gudkov, A.T., and Gongadze, G.M., 1984, *FEBS Lett.* 176:32.
Hackl, W., and Stöffler-Meilicke, M., 1988, *Eur. J. Biochem.* 174:431.
Hackl, W., Stöffler-Meilicke, M., and Stöffler, G., 1988, *FEBS Lett.* 233:119.

Herfurth, E., Hirano, H., and Wittmann-Liebold, B., 1991, *Biol. Chem. Hoppe-Seyler* 372:955.

Herfurth, E., 1992, Ph.D. thesis, Free University of Berlin.

Herwig, S., Kruft, V., and Wittmann-Liebold, B., 1992a, *Eur. J. Biochem.* 207:877.

Herwig, S., Kruft, V., Eckart, K., and Wittmann-Liebold, B., 1992b, *J. Biol.Chem.*, in press.

Jaynes, E.N., Jr., Grant, P.G., Giangrande, G., Wieder, R., and Cooperman, B.S., 1978, *Biochemistry* 17:561.

Kaji, H., Suzuka, I., and Kaji, A., 1966, *J. Mol. Biol.* 18:219.

Kamp, R.M., 1988, *in:* Modern Methods in Protein Chemistry, H. Tschesche, ed., Walter de Gruyter & Co. Verlag, Berlin, New York, 275-298.

Kruft, V., and Wittmann-Liebold, B., 1991, *Biochemistry* 30:11781.

Kruft, V., Kapp, U., and Wittmann-Liebold, B., 1991, *Biochimie* 73:855.

Lerner, R. A., 1982, *Nature* 299:592.

Lotti, M., Noah, M., Stöffler-Meilicke, M., and Stöffler, G., 1989, *Mol. Gen. Genet.* 216:245.

Lührmann, R., Stöffler-Meilicke, M., and Stöffler, G., 1981a, *Mol. Gen. Genet.* 182:369.

Lührmann, R., Bald, R., Stöffler-Meilicke, M., and Stöffler, G., 1981b, *Proc. Natl. Acad. Sci. USA* 78:7276.

Malkin, L.I., and Rich, A., 1967, *J. Mol. Biol.* 26:329.

Maly, P., Rinke, J., Ulmer, E., Zwieb, C., and Brimacombe, R., 1980, *Biochemistry* 19:4179.

May, R.P., Nowotny, V., Nowotny, P., Voss, H., and Nierhaus, K.H., 1992, *EMBO J.* 11:373.

Moffat, J.G., Timms, K.M., Trotman, C.N.A., and Tate, W.P., 1991, *Biochimie* 73:1113.

Möller, K., Zwieb, C., and Brimacombe, R., 1978, *J. Mol. Biol.* 126:489.

Nag, B., Tewari, D.S., Etchison, J.R., Sommer, A., and Traut R.R., 1986, *J. Biol. Chem.* 261:13892.

Nag., B., Tewari, D.S., Sommer, A., Olson, H.M., Glitz, D.G., and Traut, R.R., 1987, *J. Biol. Chem.* 262:9681.

Nag., B., Akella, S.S., Cann, P.A., Tewari, D.S., Glitz, D.G., and Traut, R.R., 1991, *J. Biol. Chem.* 266:22129.

Nagano, K., Harel, M., and Takezawa, M., 1988, *J. Theor. Biol.* 134:199.

Noller, H.F., Hoffarth, V., and Zimniak, L. , 1992, *Science* 256:1416.

Ofengand, J., Ciesiolka, J., Denman, R. and Nurse, K., 1986, *in:* Structure, Function and Genetics of Ribosomes, B. Hardesty, and G. Kramer, eds., Springer-Verlag, Heidelberg, New York, 473-494.

Olson, H.M., and Glitz, D.G., 1979, *Proc. Natl. Acad. Sci. USA* 76:3769.

Olson, H.M., Grant, P.G., Cooperman, B.S., and Glitz, D.G., 1982, *J. Biol. Chem.* 257:2649.

Olson, H.M. , Sommer, A., Tewari, D.S., Traut, R.R., and Glitz, D.G., 1986, *J. Biol. Chem.* 261:6924.

Olson, H.M., Nag, B., Etchison, J.R., Traut, R.R., and Glitz, D.G., 1991, *J. Biol. Chem.* 266:1898.

Östner, U., and Hultin, T., 1968, *Biochim. Biophys. Acta* 154:376.

Picking, W.D., Picking, W.L., Odom, O.W., and Hardesty, B., 1992, *Biochemistry* 31:2368.

Pohl, T., and Wittmann-Liebold, B., 1988, *J. Biol. Chem.* 263:4293.

Ryabova, L.A., Selivanova, O.M., Baranov, V.I., Vasiliev, V.D., and Spirin, A.S., 1988, *FEBS Lett.* 226:255.

Scheinmann, A., Atla, T., Aguinalda, A.M., Kahan, L., Schankweiler, G., and Lake, J.A., 1992, *Biochimie* 74:307.

Schulze, H., and Nierhaus, K.H., 1982, *EMBO J.* 1:609.

Schwedler-Breitenreuter, G.M., Lotti, M., Stöffler-Meilicke, M., and Stöffler, G., 1985, *EMBO J.* 4:2109.

Smith, W.P., Tai, P.-C., and Davis, B.D., 1978, *Proc. Natl. Acad. Sci. USA* 75:5922.

Sommer, A., and Traut, R.R., 1974, *Proc. Natl. Acad. Sci. USA* 71:3946.

Sommer, A., Etchison, J.R., Gavino, G., Zecherle, N., Casiano, C., and Traut, R.R., 1985, *J. Biol. Chem.* 260:6522.

Spillmann, S., Dohme, F., and Nierhaus, K.H., 1977, *J. Mol. Biol.* 115:513.

Stern, S., Weiser, B., and Noller, H.F., 1988, *J. Mol. Biol.* 204:447.

Stöffler, G., and Stöffler-Meilicke, M., 1983, *in:* Modern Methods in Protein Chemistry, B. Tschesche, ed., Walter de Gruyter & Co., Berlin, New York, 409-455.

Stöffler, G., Redl, B., Walleczek, J., and Stöffler-Meilicke, M., 1988, *in:* Methods in Enzymology, Volume 164: Ribosomes, H.F. Noller, and K. Moldave, eds., Academic Press, New York, 64-76.

Stöffler-Meilicke, M., Stöffler, G., Odom, O.W., Zinn, A., Kramer, G., and Hardesty, B., 1981, *Proc. Natl. Acad. Sci. USA* 78:5538.

Stöffler-Meilicke, M., Noah, M., and Stöffler, G., 1983a, *Proc. Nat. Acad. Sci. USA* 80:6780.

Stöffler-Meilicke, M., Epe, B., Steinhäuser, K.G., Woolley, P., and Stöffler, G., 1983b, *FEBS Lett.* 163:94.

Stöffler-Meilicke, M., Epe, B., Wolley, P., Lotti, M., Littlechild, J., and Stöffler, G., 1984, *Mol. Gen. Genet.* 197:8.

Stöffler-Meilicke, M., and Stöffler, G., 1990, *in:* The Ribosome. Structure, Function and Evolution, W.E. Hill, A. Dahlberg, R.A. Garrett, P.M. Moore, D.Schlessinger, and J.R. Warner, eds., American Society for Microbiology, Washington, D.C., 123-133.

Syu, W.-J., Kahan, B., and Kahan, L., 1990, *J. Prot. Chem.* 9:159.

Tischendorf, G.W., Zeichardt, H., and Stöffler, G., 1974, *Mol. Gen. Genet.* 143:187.

Traut, R.R., Tewari, D.S., Sommer, A., Gavino, G.R., Olson, H.M., and Glitz, D.G., 1986, *in:* Structure, Function and Genetics of Ribosomes, B. Hardesty, and G. Kramer, eds., Springer-Verlag, Heidelberg, New York, 286-308.

Uchiumi, T., Terao, K., and Ogata, K., 1981, *J. Biochem.* 90:185.

Uchiumi, T., Kikuchi, M., Terao, K., and Ogata, K., 1985a, *J. Biol. Chem.* 260:5669.

Uchiumi, T., Kikuchi, M., Terao, K., and Ogata, K., 1985b, *J. Biol. Chem.* 260:5675.

von Böhlen, K., Makowski, I., Hansen, H.A.S., Bartels, H., Zaytzev-Bashan, A., Meyer, S., Paulke, C., Franceschi, F., and Yonath, A., 1991, *J. Mol. Biol.* 222:11.

Walleczek, J., Schüler, D., Stöffler-Meilicke, M., Brimacombe, R., and Stöffler, G., 1988, *EMBO J.* 7:3571.

Walleczek, J., Martin, T., Redl, B., Stöffler-Meilicke, M., and Stöffler, G., 1989, *Biochemistry* 28:4099.

Walleczek, J., Albrecht-Ehrlich, R., Stöffler, G., and Stöffler-Meilicke, M., 1990, *J. Biol. Chem.* 265:11338.

Weitzmann, C.J., and Cooperman, B.S., 1985, *Biochemistry* 24:2268.

Wilson, J.E., 1991, *in:* Methods of Biochemical Analysis, Volume 35: Protein Structure Determination, C.H. Suelter, ed., John Wiley & Sons, New York, 207-250.

Winkelmann, D., and Kahan,L., 1980, *in:* Ribosomes. Structure, Function and Genetics, G. Chambliss, G.R. Craven, J. Davies, L. Kahan, and M. Nomura, eds., University Park Press, Baltimore, 255-266.

Wittmann-Liebold, B. Köpke, A., Arndt, E., Krömer, W., Hatakeyama, T., and Wittmann, H. G., 1990, *in:* The Ribosome. Structure, Function and Evolution, W.E. Hill, A. Dahlberg, R.A. Garrett, P.M. Moore, D.Schlessinger, and J.R. Warner, eds., American Society for Microbiology, Washington, D.C., 558-616.

Wower, I., Wower, J., Meinke, M., and Brimacombe, R., 1981, *Nucl. Acids Res.* 9:4285.

Xiang, R.H., and Lee, J.C., 1989, *J. Biol. Chem.* 264:10542.

Yonath, A., Leonard, K.R., and Wittmann, H.G., 1987, *Science* 236:813.

Yonath, A., and Wittmann, H.G., 1989, *Trends Biochem. Sci.* 14:329.

Zak, R., Nair, K.G., and Rabinowitz, M., 1966, *Nature* 210:169.

Zecherle, G.N., Oleinikov, A., and Traut, R.R., 1992, *J. Biol. Chem.* 267:5889.

STRUCTURE AND FUNCTION OF *ESCHERICHIA COLI* RIBOSOMAL PROTEIN L7/L12: EFFECT OF CROSS-LINKS AND DELETIONS

Robert R. Traut, Andrew V. Oleinikov, Evgeny Makarov, George Jokhadze, Bertrand Perroud, and Bruce Wang

Department of Biological Chemistry
School of Medicine
University of California
Davis , CA 95616

INTRODUCTION

Ribosomal protein L7/L12 of *E. coli* is the most extensively investigated representative of the small, four-copy, dimeric acidic proteins that are found in large ribosomal subunits of all organisms. First characterized as the acidic, alanine-rich ribosomal A-protein (Möller and Castleman 1967) and later renamed L7/Ll2 (Kaltschmidt and Wittmann 1970), the protein was studied extensively with respect to both structure and function (Liljas 1982; Möller and Maassen 1986). In eubacteria, eukaryotes and archaea the acidic proteins always exist as a conserved quaternary structural element in which two dimers are integrated into the ribosome through binding to a common anchoring protein (Casiano, Matheson et al., 1990; Liljas 1982; Uchiumi 1987). One or both of the L7/L12 dimers forms a conspicuous morphological feature on the ribosome known in *E. coli* as the L7/L12 stalk (Strycharz, Nomura et al., 1978). The proteins can be simply and selectively removed from and restored to the ribosome with the concomitant loss and regain of activity (Hamel, Koka et al., 1972). In both eubacteria and eukaryotes the proteins are required for the binding of elongation factors, and also initiation and termination factors. This is a major example of ribosome function in which specific proteins play a clearly defined and perhaps dominant role.

L7/L12 DIMERS ARE FOUND IN DIFFERENT LOCATIONS ON THE RIBOSOME

Protein L7/L12 of *E. coli* is composed of two distinct organized structural domains separated by a putative flexible hinge (Leijonmarck, Petterson et al., 1981): an elongated, helical N-terminal domain, residues 1-36, that is responsible for the strong dimer interaction (Gudkov and Behlke 1978a) through a likely coiled-coil interaction of the two α-helices (Möller and

The Translational Apparatus, Edited by K.H. Nierhaus *et al.*, Plenum Press, New York, 1993

Maassen 1986), and a globular C-terminal domain, residues 53-120. The high
resolution crystal structure of the C-terminal domain of the protein has been
determined (Leijonmarck and Liljas, 1987). It is this element that has been
implicated in factor binding, since truncated L7/L12 fragments that lack the C-
terminal domain fail to support protein synthesis even though they bind to the
ribosome (Koteliansky, Domogatsky et al., 1978a; Agthoven, Maassen et al.,
1975), and antibodies to the C-terminal domain inhibit the binding of
elongation factors as well as protein synthesis (Sommer, Etchison et al., 1985).
The two domains can be separated by mild proteolysis of the intact molecule
where they are connected to each other by amino acid residues 37 to 52, a
region that has properties consistent with a flexible hinge structure (Liljas and
Gudkov, 1987). Flexibility of L7/L12 has been demonstrated both by proton
NMR Cowgill, Nichols et al., 1984a; Gudkov, Gongadze et al., 1982) and by
electron microscopy (Verschoor, Frank et al., 1986). Determining more
precisely the ribosomal location of the functionally important C-terminal
domains has been a goal of work in this laboratory. Figure 1 summarizes
present information concerning the location of the C-terminal domain. It
represents two conformations for the L7/L12 dimer, bent (II), and extended (I
and III); and three locations for the C-terminal domain.

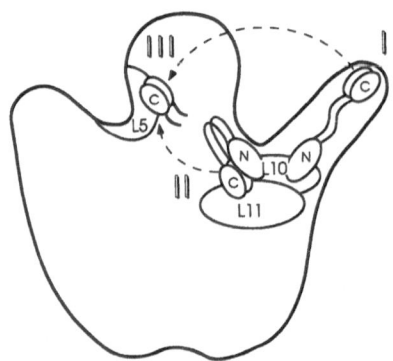

Figure 1. Alternate locations and conformations for L7/L12 in the ribosome.

That L7/L12 comprises the stalk observed on large ribosomal subunits is
well established from electron microscopy and the effects of selective extraction
of this protein. The location of the C-terminal domain in Site I at the tip of the
stalk was inferred from the demonstration that the N-terminal domain is
responsible for binding to L10 and thereby to the ribosome (Koteliansky,
Domogatsky et al., 1978a; Agthoven, Maassen et al., 1975), and shown directly
by immune electron microscopy with monoclonal antibodies to an epitope in
the C-terminal domain of L7/L12 between residues 74 and 120 (Olson, Sommer
et al., 1986). Immune electron microscopy with polyclonal antibodies
suggested that both dimers were present in the stalk (Tokimatsu, Strycharz et
al., 1981); however it was also shown that one dimer per particle was sufficient
to form a visible stalk (Möller, Schrier et al., 1983). The clearly established
location of EF-G on the body of the ribosome near the base of the stalk
(Girshovich, Kurtskhalia et al., 1981) appears incompatible with the static
presence of the functionally required C-terminal domains of both dimers at the
end of the stalk.

Evidence for the location of the C-terminal domain at Site II came first
from immune electron microscopy with the monoclonal antibody mentioned
above which showed a secondary antibody binding site at the periphery of

stalkless subunits in an area from which the stalk might be expected to project (Olson, Sommer et al., 1986); i.e., near the site of EF-G binding. The stalk had been removed from these particles by incubation with a monoclonal antibody to an epitope in the N-terminal domain of L7/L12. Additional evidence that the C-terminal domain can occupy a location on the body of the ribosome near the base of the stalk came from cross-linking between a predetermined location in the C-terminal domain, Cys-89, and Cys-70 of L10. The maximum length of the bridge provided by the cross-linking reagent, 1,4-di[3'-(2'-pyridyldithio)propion-amido]butane, defined a maximum distance between the cysteine residues of 16Å.

Previous studies on the identification of protein-protein cross-links involving L7/L12, using derivatization of ribosomal lysine residues with 2-iminothiolane and oxidative disulfide cross-linking between introduced and/or endogenous SH groups, showed that L7/L12 was able to form cross-links not only with proteins near the base of the stalk such as L10 and L11, but also with proteins near the 5S RNA (L7/L12-L5) and peptidyl transferase domains (L7/L12-L2, L7/L12-L9), regions distant from the stalk and the EF-G binding site (Traut, Lambert et al., 1983; Traut, Tewari et al., 1986). A cross-link between L7/L12 and L5 has also been reported using a different lysine-specific cross-linking and detection method (Redl, Walleczek et al., 1989). The extended length of L7/L12 is estimated to be 100Å, sufficient to permit occupancy of Site III by the C-terminal domain, which contains 11 of the 14 lysine residues in L7/L12, and cross-linking to L5. The Leiden group (Möller and Maassen, 1986) had earlier provided evidence for the location of one L7/L12 dimer on the body of the 50S particle in an extended conformation directed toward the central protuberance. Those results support Site III, and are less consistent with Site II.

Precise correlation of the locations of the C-terminal domains and the functional state of the ribosome remains to be established experimentally. Site II is clearly related to factor binding and the requirement for the C-terminal domain of L7/L12. Possible models and dynamics have been discussed by Möller (1986, 1990) and by Liljas and Gudkov (1987). Any detailed understanding of molecular mechanisms needs to account for the existence of two dimers, transitions between different locations and conformations. Implicit in these transitions is the flexibility of L7/L12. Occupancy of Site II with the dimer in the bent conformation clearly requires flexibility and is presumed to be conferred by the hinge region. A transition between Sites I and III would require a different kind of motion, a "ball-and-socket" binding of L7/L12 to L10 that allows the extended dimer to swivel between the stalk and inward locations. Alternatively, it may be only the transition between Sites II and III that takes place. The mobility of L7/L12 has been demonstrated by ^{1}H-NMR (Cowgill, Nichols et al., 1984b; Gudkov, Gongadze et al., 1982) and contributions by the hinge region documented (Bushuev, Gudkov et al., 1989). Flexibility of the hinge has been proposed to account for the bent conformation and the occupancy of Site II (Zecherle, Oleinikov et al., 1992).

CYSTEINE SUBSTITUTION AND HINGE DELETION VARIANTS OF L7/L12 MADE BY SITE-DIRECTED MUTAGENESIS

Two types of L7/L12 variants have been constructed as summarized in Figure 2: 1. Amino acid substitutions to introduce single cysteine residues at residues 33, 63 and 89. 2. Deletions of 11 and 18 residues in the putative hinge region. In addition, proteins combining one of the deletions with a cysteine substitution were made. The purpose of making the cysteine substitutions is to use the unique sulfhydryl group for the attachment of cross-linking and fluorescent probes and to compare the results with the probes attached in different domains of the protein and on different surfaces of the C-terminal globular domain. The deletions were made in order to test the effect on ribosome binding and activity, and, in combination with the cysteine substitutions, to test the effect on cross-linking and protein dynamics monitored with fluorescent probes. Testing the properties of the L7/L12 variant proteins is facilitated by the ease and specificity with which wild type L7/L12 can be removed from 70S ribosomes or 50S subunits and be replaced by reconstitution with the L7/L12 variant.

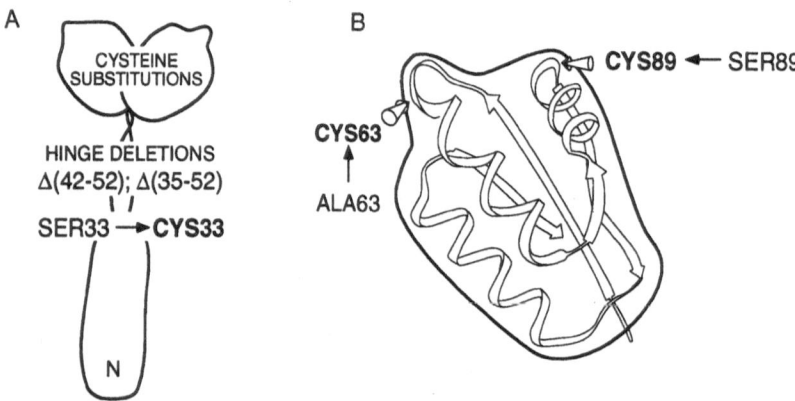

Figure 2. Schematic diagrams showing mutant forms of L7/L12 constructed. A. The dimer of L7/L12 based on the model of Liljas (1982). B. A single C-terminal domain of L7/L12 from a representation of the crystal structure (Liljas et al., 1986).

The proteins were overexpressed in a conditional system indicated in the legend to Figure 3 and purified by methods used previously, all in the absence of urea or other denaturants. The purity of the proteins is shown in Figure 3, which also shows the altered mobility of the deletion variants. The presence of the intended cysteine residues was verified by DNA sequencing of the entire rplL genes, and by modification of the cysteine variants by SH-specific reagents.

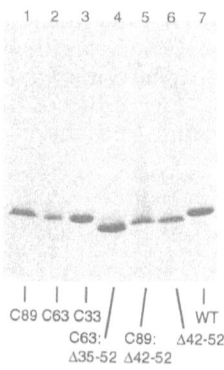

Figure 3. SDS gel electrophoresis (Pharmacia, PhastGel, 20% acrylamide) of purified variant proteins used in biochemical studies. Proteins were expressed and purified as described in Olcinikov et al., 1993.

SITE-SPECIFIC ZERO-LENGTH CROSS-LINKING OF RESIDUE 89 IN THE C-TERMINAL DOMAIN OF L7/L12 TO L10 IN THE RIBOSOME AND IN THE L8 PENTAMERIC COMPLEX

The proximity of the C-terminal domain of L7/L12 to L10 shown previously with cross-linking reagents, and electron microscopy was defined more precisely by using mild oxidation with Cu^{2+}(phenanthroline)$_3$ to form disulfide bonds, zero-length cross-links, between neighboring cysteine residues in ribosomes from a strain in which rplL::Cys89 replaced the gene for wild type L7/L12, or ribosomes reconstituted in vitro with L7/L12Cys-89. The same approach was used with L7/L12Cys-89 in solution and with the L8 pentameric complex, (L7/L12)$_4$.L10, reconstituted with L7/L12Cys-89 and L10, or isolated from ribosomes from the allele replacement strain. Disulfide Cys89-Cys89 cross-links formed in high yield both with L7/L12Cys-89 in solution, in the pentameric complex, and in the ribosome. In addition as shown in Figure 4, L7/L12-Cys89 formed a disulfide cross-link with the single cysteine, Cys-70 in L10. Its formation in the pentameric complex implies that the protein can exist in the bent conformation here as well as in the ribosome. The model of crosslinked L8 complex is represented in Figure 5.

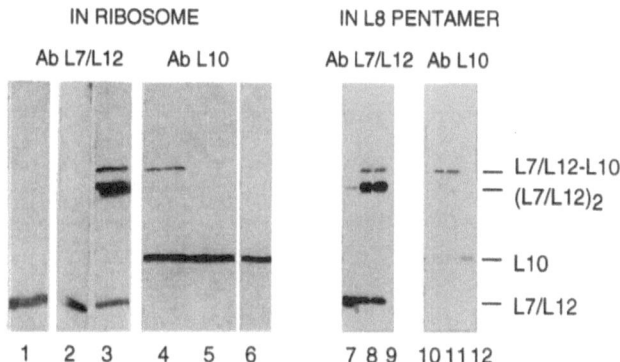

Figure 4. Demonstration of L7/L12-L10 zero length cross-link by SDS gel electrophoresis and western blotting with monoclonal antibodies to L7/L12 or L10. Ribosomes (3mg/ml) from the allele replacement strain NZ4402 bearing rplLC89 as the only gene encoding L7/L12 were oxidized with 500µM Cu^{2+}(phenanthroline)$_3$ for 1h at 0°C as described (Kobashi, 1968; Falke and Koshland, 1987). Oxidation was terminated by the addition of EDTA and iodoacetamide to final concentrations of 50mM and 40mM, respectively. Lanes 1 and 6, reduced NZ4402 ribosomes; Lanes 2 and 5 oxidized wild type ribosomes; Lanes 3 and 4, oxidized NZ4402 ribosomes. The L8 complex was extracted and purified from the NZ4402 ribosomes, or alternatively, reconstituted from pure L7/L12Cys-89 and L10. Oxidation was at 0.1mg/ml with 10µM Cu^{2+}(phenanthroline)$_3$ at room temperature for 15 min. Lanes 7 and 12, L8 reconstituted not oxidized; Lanes 8 and 11, L8 reconstituted and oxidized; Lanes 9 and 10, L8 extracted from ribosomes and oxidized.

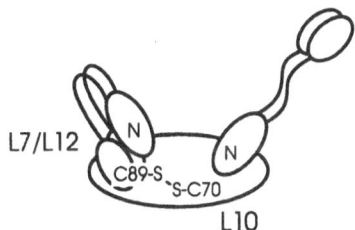

Figure 5. Zero-length cross-link between single cysteines in L10 and L7/L12Cys89.

STRUCTURAL AND FUNCTIONAL EFFECTS OF DELETIONS IN THE L7/L12 HINGE REGION

According to the model shown in Figures 1 and 5, the location of the C-terminal domain of L7/L12 at Site II near the base of the stalk where EF-G has been shown bind (Girshovich, Kurtskhalia et al., 1981) depends on the existence of the flexible hinge region as essential for the bent or closed conformation. In order to test this hypothesis L7/L12 variants that lack eleven (Δ42-52) and eighteen residues (Δ35-52) from the hinge region were prepared, characterized and tested for *in vitro* protein synthesis activity. The purified wild type, deletion proteins and Cys-63-substituted variant proteins were incubated with P0 cores prepared from wild type 70S ribosomes, shown to be specifically and completely lacking wild type L7/L12, in a ratio of 8 copies of mutant L7/L12 per particle. Samples of this mixture were tested for activity in poly(U)-dependent polyphenylalanine synthesis without isolation of the reconstituted particles by centrifugation. Table I shows the results which indicate that neither the longer nor shorter deletion variants restore any activity to P0 cores lacking L7/L12.

Table 1. Activity of ribosomes reconstituted *in vitro* with hinge deletion L7/L12 variant proteins.

Reconstituted particles	Activity ([^{14}C] phe/particle/15 min)
1. P0 cores alone	1.6
2. P0 + L7/L12 wild type	11.2
3. P0 + L7/L12:Cys63	10.8
4. P0 + L7/L12:Δ35-52	2.1
5. P0 + L7/L12:Δ42-52	1.9

Certain structural properties of the deletion variants *in vitro* were examined in order to account for their lack of activity. The possibility that dimer formation was defective was investigated by cross-linking with dimethylsuberimidate. The yield of cross-linked dimers was approximately the same as for wild type proteins and indicate that the deletion proteins were not defective in dimer formation.

Perpendicular gel electrophoresis (Creighton, 1979) across a gradient of urea concentration was used to examine possible effects of the deletions on the compact structure of the C-terminal domain. In order to separate this effect from those on dimerization, an interaction attributable to the N-terminal domain, the proteins were oxidized to prevent dimerization. This treatment does not influence the structure of the C-terminal domain (Gudkov, Khechinashvili et al., 1978). Figure 6A shows the results for full-length and shorter deletion protein. For wild type L7/L12 there is a transition from a form migrating more rapidly at zero or low urea concentration to a form migrating more slowly at a higher concentration of urea. This is interpreted to represent unfolding of a folded domain. There is a sharp and similar transition for the Δ42-52 deletion variant that occurs at a urea concentration (midpoint about 3 M) slightly lower than that for the wild type L7/L12 (midpoint approximately 3.6 M). The result suggests that the deletion protein retains the compact folded structure of the C-terminal domain. Similar results were obtained for the longer deletion.

Similar experiments were carried out with the unoxidized, dimeric proteins and are shown in Figure 6B. The transitions for both proteins are broader and more complex compared to the monomers, making it difficult to distinguish overlapping unfolding of the C-terminal domain and dissociation of the dimer. The deletion protein retains an overall similarity to the wild type. The longer deletion caused a clear broadening and shift toward decreased stability compared to wild type. The results are consistent with the cross-linking results that show that the deletion proteins are dimeric in the absence of urea.

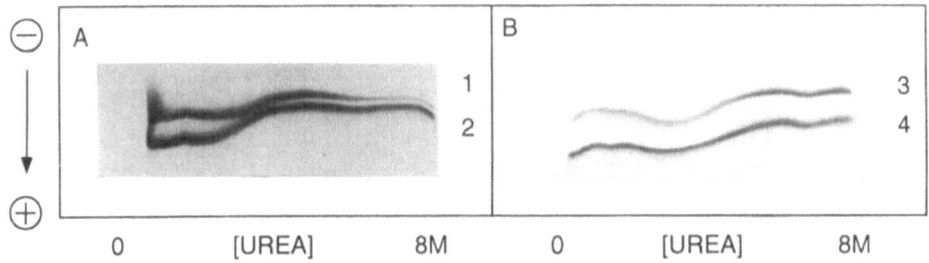

Figure 6. Perpendicular gel electrophoresis of L7/L12 variants. A. Proteins oxidized to produce monomers (Gudkov and Behlke ,1978b). B. Unoxidized dimers. Samples 1 and 3, wild type L7/L12; Lanes 2 and 4, L7/L12Δ42-52. The top protein samples on each gel were added at an arbitrary time interval after starting the electrophoresis of the first protein

The capacity of the deletion protein variants to bind *in vitro* to L7/L12 deficient core particles was quantified using the double-mutant proteins containing a cysteine substitution to which a radiolabel was attached. Proteins L7/L12Δ(35-52)Cys-63 and L7/L12Δ(42-52)Cys-89 were labelled by [^{14}C] iodoacetamide. Protein L7/L12Cys-89 had previously been shown to be fully active in protein synthesis (Zecherle, Oleinikov et al., 1992), and L7/L12Cys-63 was also fully active as shown in Table 1. Modification of these proteins with iodoacetamide had no effect on activity, nor did the much larger group, N-[4-(*p*-azidosalicylamido)butyl]3'-(2'-pyridyldithio) propionamide. Particles were reconstituted by incubation of P0 core particles, prepared from 70S ribosomes or 50S subunits, from which L7/L12 had been quantitatively and selectively removed, with 8 equivalents of the radiolabelled deletion protein variants, as well as the single amino acid substitution variants. The reconstituted particles were purified from unbound L7/L12 by high speed centrifugation through a sucrose cushion. The concentration of each resuspended ribosome solution was determined and samples were counted to determine the bound L7/L12. The results are summarized in Table 2. The single cysteine substituted proteins both gave approximately 4 copies bound per particle, the number expected for two dimers. The shorter deletion L7/L12Δ(42-52) also bound in approximately 4 copies. By contrast, the longer deletion variant, L7/L12Δ(35-52) bound to the extent of only 2.5 copies per particle. The results were qualitatively the same when the particles reconstituted with L7/L12Δ(35-52)Cys-63 were isolated by gel filtration instead of by centrifugation.

Previous studies had shown that fragments containing residues 1 to 26 (Koteliansky, Domogatsky et al., 1978b) and 1 to 55 (Schop and Maassen 1982) did not dimerize. The present results show that residues 35 to 52 are not important in the dimer interaction and imply the importance of residues 1 to 34 in this interaction. The binding of the longer deletion L7/L12Δ(35-52) variant to the ribosome is impaired and the results suggest that residues 35-41 also contribute to the normal interaction with the ribosome.

Table 2. Binding of L7/L12 variants to 70S and 50S P0 core particles determined after high speed centrifugation.

	Ribosomal particle[1]	L7/L12 variant added	Copies of L7/L12 per particle
1.	70S P0	Cys-63	4.0
	70S P0	Cys-89:Δ42-52	3.7
	70S P0	Cys-63:Δ35-52	2.5
2.	70S P0	Cys-63:	4.0
	70S P0	Cys-63:Δ35-52	2.5
3.	50S P0	Cys-63	3.9
	50S P0	Cys-89:Δ42-52	3.7
	50S P0	Cys-63:Δ35-52	2.3
4.	50S P0	Cys-63	3.9
	50S P0	Cys-89:Δ42-52	3.8

[1]P0 particles were shown to lack completely and specifically L7/L12.

Deletions starting from residue 52 and extending toward the N-terminus should have no effect on the highly ordered C-terminal domain, but should, depending upon their length, perturb the hinge region and make the molecule more rigid. The L7/L12Δ 42-52 variant dimerizes, retains the folded conformation of the C-terminal domain, binds in four copies to P0 cores, but is inactive. The results suggest that either the length and/or the flexibility of the hinge are required for the activity of L7/L12 by allowing the C-terminal domain to occupy a location near the base of the L7/L12 stalk for the normal interaction with elongation factors, or allowing movement during the elongation cycle. It is also possible that other functionally relevant movements of L7/L12 require the hinge. Work is in progress to determine the effect of L7/L12Cys-89Δ42-52, reconstituted into ribosomes four copies, on the formation of the cross-link to L10, and on the binding of elongation factors.

OXIDATION OF CYS VARIANTS AND ACTIVITY OF DISULFIDE CROSS-LINKED L7/L12 DIMERS

The three cysteine substitution variants of L7/L12 indicated in Figure 2 were tested for the formation of intramolecular disulfide bonds between the two members of the dimer in solution. The location of residues 63 and 89 in the model based on the high resolution crystallographic structure is shown in Figure 7. Residue 89 is located in the turn between the αB helix and the βB sheet, and residue 63 in the turn between the βA sheet and the αA helix. Neither cysteine substitution had any effect on the activity of reconstituted particles, not even when iodoacetamide or more bulky moieties were attached at these sites. Although the two Cys-89 residues are on the facing surfaces of the two monomers in the crystal structure, the distance across the two-fold axis appears too great to permit efficient intradimer disulfide formation (A. Liljas, personal communication). The distance between the two Cys-63 residues is so great in the crystal structure as to totally preclude intramolecular disulfide bond formation.

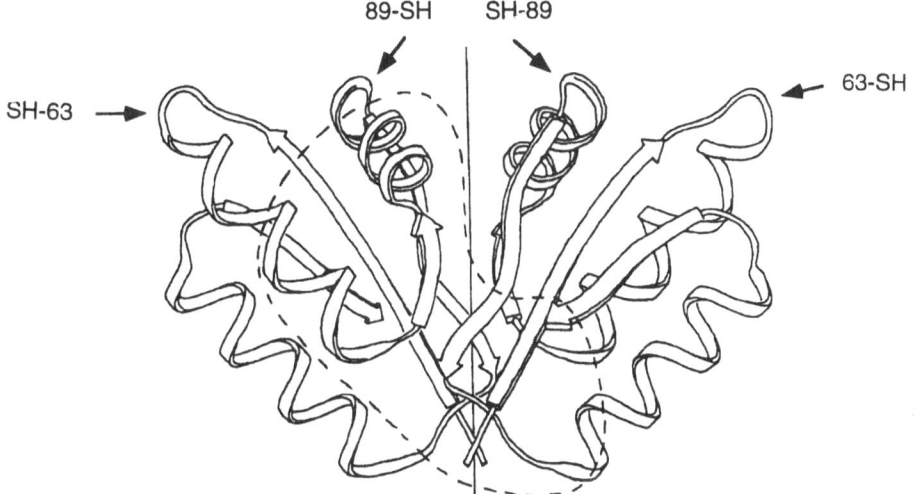

Figure 7. Location of Cys-63 and Cys-89 in the simplified model of the crystallographic structure of the C-terminal domain. The dashed line represents the proposed conserved surface (Liljas, Kirsebom et al., 1986).

Oxidation of the dimers in solution was promoted by Cu^{2+}(phenanthroline)$_3$ and the formation of covalent, disulfide-linked dimers was analyzed by SDS polyacrylamide gel electrophoresis, which gives a clear separation of monomers and dimers. The results in Figure 8 show that all three L7/L12 variants became oxidized almost completely. Analysis by gel filtration of the products of oxidative cross-linking confirmed that the reaction was intramolecular, and not attributable to interaction between two dimers. Dimer formation was abolished by addition of guanidine:HCl. The rate of oxidation catalyzed by Cu^{2+}(phenanthroline)$_3$ was rapid and nearly the same for the three proteins. The rate is dependent on the concentration of Cu^{2+}(phenanthroline)$_3$. In typical experiments approximately 50% of oxidized dimers formed within 30 seconds at 15μM Cu^{2+}(phenanthroline)$_3$ and room temperature.

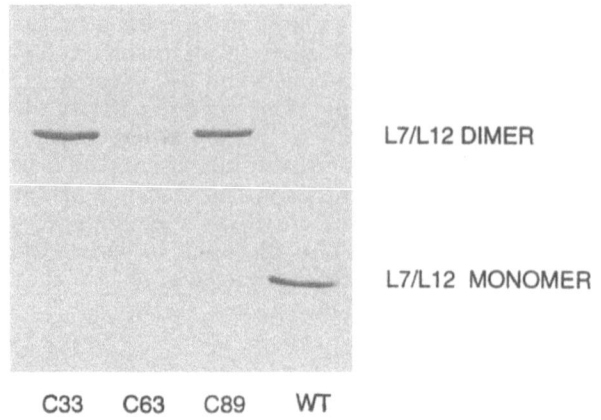

Figure 8. SDS polyacrylamide gel electrophoresis of oxidized cysteine variants. The fully reduced proteins were incubated at a concentration of approximately 0.2 mg/ml in the absence of reducing agents for 10 min at 25°C with 1mM Cu^{2+}/(phenanthroline)$_3$. The oxidation reaction was stopped by addition of EDTA to remove Cu^{2+}.

The three different oxidized L7/L12 dimers were tested for their ability to restore activity to P0 cores depleted of wild type L7/L12. It was necessary to omit the customary reducing agents from the incubation mixture for polyphenylalanine synthesis. The results in Table 3 show that all three reduced dimers are fully active compared to wild type L7/L12. In the oxidized state, both the L7/L12Cys-89 and L7/L12Cys-63 dimers retained full activity. Oxidation of the L7/L12Cys-33 dimer caused a total loss of activity. The possibility that the disulfide bond became reduced during the translation assay was ruled out by SDS gel electrophoresis of total ribosomal protein after the assay, and western blotting with an antibody to L7/L12, as shown in Figure 9. The dimers remain covalently linked.

Table 3. Activity of 70S ribosomes reconstituted from P0 cores with disulfide cross-linked L7/L12 dimers formed by oxidation.[1]

L7/L12 variant added	Expt.	Reduced dimers	Expt.	Oxidized dimers[2]
P0	1	0.4	4	0.7
P0 + WT		10.1		7.0
P0 + C89		10.4		8.6
P0	2	1.5	5	2.3
P0 + WT		10.0		10.3
P0 + C63		11.0		11.7
P0	3	2.3	6	2.8
P0 + WT		14.0		14.1
P0 + C33		14.4		3.2

[1]Results represent Phe/70S particle/15 min. in poly [U]-directed polyphenylalanine synthesis.
[2]Dimers were oxidized before reconstitution and reducing agents were absent from the incubation mixtures.

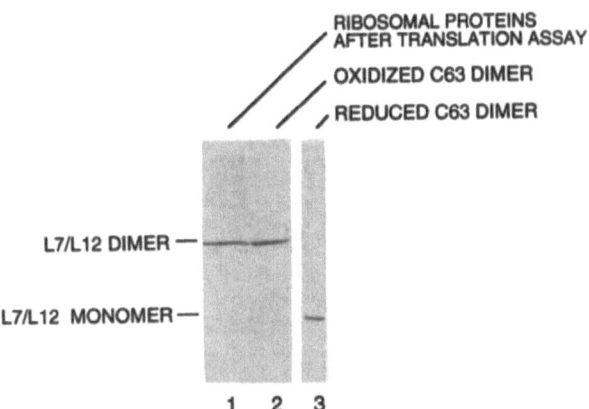

RIBOSOMAL PROTEINS
AFTER TRANSLATION ASSAY

OXIDIZED C63 DIMER

REDUCED C63 DIMER

L7/L12 DIMER —

L7/L12 MONOMER —

1 2 3

Figure 9. Analysis of total translation incubation mixture by western blotting with an antibody to L7/L12 after reconstitution of ribosomes with oxidized L7/L12Cys-63 and *in vitro* protein synthesis for 15 min.

The formation of the zero-length cross-linked dimer between the Cys-33 residues supports a parallel, non-staggered model, and further indicates either that the residues are facing each other, or that residue 33 is in a relatively flexible region. Disulfide cross-linking at position 33 in the N-terminal domain led to almost complete loss of ability to restore activity to core particles. It has been shown that Phe-30 is involved in the interaction with L10 (Gudkov, Khechinashvili et al., 1978) and from NMR experiments it was concluded that Ser-33 is in an organized structural region of the dimer (Bushuev, Gudkov et al., 1989). It is possible that the Cys-33 cross-link perturbs the structure of the region of L7/L12 responsible for the interaction of one or both dimers with the ribosome. Preliminary results indicate that the oxidized Cys-33 dimer fails to bind to the ribosome.

The finding that the rate and extent of intramolecular, disulfide-linked dimer formation for the Cys-63 and Cys-89 variants were comparable and independent of the location of the cysteine substitution was unexpected in relation to the crystal structure of the C-terminal domain dimer. Dimer formation has been demonstrated in the ribosome as well as in solution. The result indicates that there is substantial flexibility of the two C-terminal domains that allows different surfaces to come into proximity to each other. The oxidized dimers must be locked in orientations both disparate from each other and from the dimer interaction surface evident in the crystal structure. On the basis of comparative sequence analysis Liljas proposed the functional importance of a conserved contiguous surface comprised of portions of both C-terminal domains in the crystal structure (Liljas et al.,1986). It now appears that the two C-terminal domains of a dimer can be locked into disparate orientations with retention of full activity, and that there is no obligatory specific functional arrangement or interaction of the two C-terminal domains, nor a requirement for the two globular domains to move independently of one another.

REFERENCES

Agthoven, A., Maassen, J.A., Schrier, P.I., and Möller, W., 1975, Biochem.. Biophys. Res. Commun. 64:1184.
Bushuev, V.N., Gudkov, A.T., Liljas, A., and Sepetov, N.F., 1989, J. Biol. Chem. 264:4498.
Casiano, C., Matheson, A.T., and Traut, R.R., 1990, J. Biol. Chem. 265:18757.

Cowgill, C.A., Nichols, B.G., Kenny, J.W., Butler, P., Bradbury, E.M., and Traut, R.R., 1984a, J. Biol. Chem. 259:15257.

Cowgill, C.A., Nichols, B.G., Kenny, J.W., Butler, P.D., Bradbury, E.M., and Traut, R.R., 1984b, J Biol. Chem. 259:15257.

Creighton, T.E., 1979, J. Mol. Biol. 129:235.

Girshovich, A.S., Kurtskhalia, T.V., Ovchinnikov, Y.A., and Vasiliev, V.D., 1981, FEBS Lett. 130:54.

Gudkov, A.T. and Behlke, J., 1978a, Eur. J. Biochem. 90:309.

Gudkov, A.T. and Behlke, J., 1978b, Eur. J. Biochem. 90:309.

Gudkov, A.T., Gongadze, G.M., Bushuev, V.N., and Okon, M.S., 1982, FEBS Lett. 2:229.

Gudkov, A.T., Khechinashvili, N.N., and Bushuev, V.N., 1978, Eur. J. Biochem. 90:313.

Hamel, E., Koka, M., and Nakamoto, T., 1972, J. Biol. Chem. 10:805.

Kaltschmidt, E. and Wittmann, H.G., 1970, Proc. Natl. Acad. Sci. USA 67:1276.

Koteliansky, V.E., Domogatsky, S.P., and Gudkov, A.T., 1978b, Eur. J. Biochem. 90:319.

Leijonmarck, M. and Liljas, A., 1987, J. Mol. Biol. 195:555.

Leijonmarck, M., Petterson, I., and Liljas, A. 1981. "Structural Studies on the Protein L7/L12 from E. coli Ribosomes", in Structural Aspects of Recognition and Assembly of Biological Macromolecules., M. Balaban, J. L. Sussman, W. Traub , and A. Yonath, eds., Rehovot.

Liljas, A., 1982, Prog. Biophys. Mol. Biol. 40:161.

Liljas, A. and Gudkov, A.T., 1987, Biochimie 69:1043.

Liljas, A., Kirsebom, L.A., and Leijonmark, M. 1986. "Structural Studies of the Factor Binding Domain" in Structure,Function and Genetics of Ribosomes, B. Hardesty and G. Kramer, eds., Springer-Verlag, New York.

Möller, W. and Castleman, H., 1967, Nature 215:1293.

Möller, W. and Maassen, J.A., 1986, "On the Structure, Function, and Dynamics of L7/L12 from Escherichia coil Ribosomes", in Structure,Function and Genetics of Ribosomes, B. Hardesty and G. Kramer, eds., Springer-Verlag, New York.

Möller, W., Schrier, P.I., Maassen, J.A., Zantema, A., and Schop, E.R., H., 1983, J. Mol. Biol. 163:553.

Oleinikov, A.,Perroud, B.,Wang, B., and Traut, R.R., 1993, J. Biol.. Chem.., in press.

Olson, H.M., Sommer, A., Tewari, D., Traut, R.R., and Glitz, D.G., 1986, J. Biol. Chem. 261:6924.

Redl, B., Walleczek, J., Stöffler, M.M., and Stöffler, G., 1989, Eur. J. Biochem. 181:351.

Schop, R.N. and Massen, J., 1982, Eur. J. Biochem. 128:371.

Sommer, A.,Etchison, J.R.,Gavino, G.,Zecherle, N.,Casiano, C., and Traut, R.R., 1985, J. Biol. Chem. 260:6522.

Strycharz, W.A., Nomura, M., and Lake, J.A., 1978, J. Mol. Biol. 126:123.

Tokimatsu, H., Strycharz, W.A., and Dahlberg, A.E., 1981, J. Mol. Biol. 152:397.

Traut, R.R., Lambert, J.M., and Kenny, J.W., 1983, J. Biol. Chem. 258:14592.

Traut, R.R., Tewari, D.S., Sommer, A.,Gavino, G.,Olson, H.M., and Glitz, D.G., 1986, "Protein Topography of Ribosomal Functional Domains: Effects of Monoclonal Antibodies to Different Epitopes in Escherichia coli Protein L7/L12 on Ribosome Function and Structure", in Structure,Function and Genetics of Ribosomes, B. Hardesty and G. Kramer, eds., Springer-Verlag, New York.

Uchiumi, T., Wahba, A. J. and Traut, R. R., 1987, Proc. Natl. Acad. Sci. USA 84:5580.

Verschoor, A., Frank, J., and Boublic, M., 1986, J. Ultrastruc. Research 92:180.

Zecherle, G.N., Oleinikov, A., and Traut, R.R., 1992, J. Biol. Chem. 267:5889.

STRUCTURAL STUDIES ON PROKARYOTIC RIBOSOMAL PROTEINS

V. Ramakrishnan[1], Sue Ellen Gerchman[1], Barbara L. Golden[2],
David W. Hoffman[2], J.H. Kycia[1], Stephanie J. Porter[2]
and Stephen W. White[2]

[1]Biology Department, Brookhaven National Laboratory
Upton, New York, NY 11973, USA
[2]Department of Microbiology, Duke University Medical Center
Durham, North Carolina, NC 27710, USA

INTRODUCTION

Until relatively recently, it was generally accepted that the functional components of the ribosome were the proteins and that the RNA component provided the rigid scaffold for maintaining their correct functional locations and orientations. Consequently, much early research was directed towards determining the locations, functions and structures of the individual ribosomal proteins (Wittmann, 1982). However, with the discovery that isolated RNA can manifest enzymic properties (Cech et al., 1981), the focus has shifted away from the proteins to the ribosomal RNA, and it is now generally accepted that the original roles for these components should be reversed. The highly conserved nature of ribosomal RNA (Noller, 1991) and the recent demonstration that 23S ribosomal RNA stripped of proteins still retains some peptidyl transferase activity (Noller et al., 1992) appear to support this notion. The ribosome is now regarded as an RNA-based organelle which has an absolute requirement for precisely folded RNA molecules in order to function correctly. The proteins' principal role appears to be that of directing the ribosomal RNA molecules to these correctly folded structures during the assembly process, and perhaps modulating these structures during protein synthesis.

Our protein crystallographic work on individual ribosomal proteins started over ten years ago and followed the elucidation of the first structure, the carboxyterminal half of L7/L12 (L12CTF), by Liljas and coworkers (Leijonmarck et al., 1980). We have continued this work for several important reasons. First, very little is known in general about the structural aspects of protein-RNA interactions. Second, many mutations with distinct phenotypes have been found in ribosomal proteins, and these represent an excellent probe of the structural basis of the protein synthetic machinery. Finally, the ribosome is an

The Translational Apparatus, Edited by K.H. Nierhaus
et al., Plenum Press, New York, 1993

ancient and highly conserved organelle, and protein structure can provide clues as to its origins and relationships to other cellular apparatus. Recently, we developed a procedure for cloning the genes for the ribosomal proteins and expressing them in large quantities (Ramakrishnan and Gerchman, 1991). This has considerably helped our structural investigations since a regular supply of milligram quantities of pure protein is no longer a practical limitation. Also, due to the relatively small size of many of the ribosomal proteins, we have been able to initiate high resolution NMR studies in parallel with the X-ray crystallographic work. Here we present our results to date on the structures of 3 proteins, S5 and L6 by X-ray crystallography, and S17 by multi-dimensional NMR. We have restricted our studies to the proteins from the thermophilic organism *Bacillus stearothermophilus* which, being more resistant to thermal denaturation, are more amenable to structural analysis by X-ray and NMR methods.

CLONING AND OVEREXPRESSION OF GENES FOR RIBOSOMAL PROTEINS

A steady supply of milligram amounts of protein is usually required for structural studies. The classical method of obtaining ribosomal proteins from bacteria involves large-scale growths, isolation of ribosomes, step-wise salt elution and ion exchange chromatography. In order to circumvent this tedious and inefficient procedure, we decided to clone and overexpress the genes for these proteins in *Escherichia coli*, using the T7 expression system (Studier et al., 1990). This has the added advantage of being able to choose which proteins to study rather than be limited to those that can merely be purified in sufficient amounts.

Due largely to the work of Wittmann-Liebold and coworkers, ribosomal proteins fall into a peculiar class by today's criteria; in most cases, the protein sequence is known, but not the DNA coding sequence. We used the known amino acid sequence of the protein to construct a degenerate family of primers that complemented the anti-sense strand at the 5' end of the coding sequence, and a similar primer family that complemented the sense strand at the 3' end. These two families of primers were then used to amplify the coding sequence by Polymerase Chain Reaction (PCR) on total genomic *B. stearothermophilus* DNA. In constructing the primer families, we took advantage of the codon usage of *B. stearothermophilus* to reduce the degeneracy; rare codons were not used at all. The optimal annealing temperature (around 50°C) was determined separately for each protein. In addition, restriction sites were put in at both the 5' and 3' ends of the coding sequence. The initiator methionine codon was made part of an *Nde*I site, and the stop codon was followed by a *Bam*HI site.

The coding sequence of each PCR-amplified ribosomal protein gene was inserted into the T7 vector pet-3a. This plasmid contains the T7 promoter, followed by a Shine-Dalgarno sequence from T7 gene 10. This is followed by *Nde*I and *Bam*HI sites for insertion of the target gene. Gene expression is done by introducing this plasmid into *E. coli* strain BL21(DE3), which has the gene for T7 RNA polymerase under *lac* control on the host chromosome. For some of the ribosomal protein genes, we were unable to transform plasmids containing the genes on the pet-3a plasmid into the *E. coli* strain BL21(DE3). This suggests that the *B. stearothermophilus* gene is extremely toxic to *E. coli*. We therefore used a modified T7 expression vector, pet-11a (Dubendorff and Studier, 1991). In this vector, there is also a gene for the *lac* repressor on the plasmid, and a *lac* operator sequence just downstream of the T7 promoter of the gene to be expressed. The combination of a higher level of *lac* repressor concentration, and the presence of the repressor just downstream of the T7 promoter, results in a more tightly regulated system than earlier T7 vectors such as pet-3a (Studier et al., 1990; Dubendorff and Studier, 1991). However, we found that, even in this more tightly regulated vector, the plasmid bearing the *B. stearothermophilus* S5 gene still tends to be unstable. Therefore, we constructed a vector, pet-13a, that was identical except that the gene for kanamycin resistance was used rather than ampicillin resistance. The problems with the use of ampicillin as a selection antibiotic have been discussed (Studier et al., 1990), and it is likely that kanamycin provides more stringent selection pressure. Typically, about 100 mg of protein per liter of culture were obtained by these methods. In addition to ensuring a steady supply of protein, these methods also open up other options that are useful in structural analyses. In particular, we can isotopically label the proteins for NMR studies, and use site-directed mutagenesis techniques for the introduction of cysteines as potential heavy atom sites in X-ray crystallography.

RIBOSOMAL PROTEIN S5

Introduction

In prokaryotes, S5 from the 30S subunit has a molecular weight of 17,500 and contains some 166 amino acids (Wittmann-Liebold and Greuer, 1978). It has also been found in eukaryotes and archaebacteria and, although somewhat larger in these organisms, the protein is highly conserved and is evidently an important ribosomal component(Kimura, 1984; All-Robyn et al., 1990; Scholzen and Arndt, 1991). The ribosomal components that constitute the neighbourhood of S5 within the 30S subunit have been well characterized by a variety of techniques including neutron diffraction (Capel et al., 1987), protein-protein (Lambert et al., 1983) and protein-RNA crosslinking (Osswald et al., 1987; Greuer et al., 1987), and chemical protection of RNA (Stern et al., 1989). Proteins S2, S4 and S8 are adjacent to S5, and S3, S12, S16 and S17 are close by. Although not a primary binder to 16S rRNA, it is clearly adjacent to regions around bases 900, 560 and the 5' terminus, and most likely binds specifically to one or more of these sites during ribosome assembly. Mutations in S5 result in several phenotypes that suggest a role in translational fidelity and translocation. These include ribosome ambiguity or *ram* (Piepersberg et al., 1975a) and resistance to spectinomycin (Piepersberg et al., 1975b). Also, a cold-sensitive, spectinomycin-resistant mutant of S5 has been identified that is defective in initiation (Nomura, 1987). The protein was originally purified from *B. stearothermophilus* ribosomes and this provided sufficient material to grow suitable crystals (Appelt et al., 1983a) and to determine the low resolution (5Å) structure (White et al., 1983). The gene for the protein has now been cloned and overexpressed (Ramakrishnan and Gerchman, 1991), and this furnished the quantities of material necessary to complete the high resolution structure.

Crystallization and Crystallography

The crystals are in the trigonal space group $P3_221$ with cell dimensions a=b=59.3Å, c=109.8Å. They grow from 1.2 M phosphate between pH 7.5 and 8.5, and diffract to a nominal resolution of 2.4Å. The structure was determined by the isomorphous replacement method using a single gold ($KAu(CN)_2$) derivative. Following heavy atom refinement, phase calculation and solvent flattening, the electron density was of sufficient quality to trace the polypeptide backbone. A model was built into the electron density using molecular graphics (Ramakrishnan and White, 1992), and this is currently being refined with X-PLOR.

Description of the Structure

S5 has an α/β structure and the polypeptide backbone is folded into two distinct domains to form a rather flat elongated molecule with approximate dimensions 45Å x 40Å x 25Å (Figure 1). The N-terminal half consists of a disordered N-terminus, three contiguous β-strands (*beta1, beta2* and *beta3*) arranged as an antiparallel β-pleated sheet, followed by a four-turn α-helix (*alpha1*) lying on the β-sheet surface. These elements are connected by two loops (*loop1*, N-terminus -> *beta1; loop2, beta1* -> *beta2*), and two turns (*turn1, beta2* -> *beta3; turn2, beta3* -> *alpha1*). The C-terminal half also contains a three-stranded β-pleated sheet (*beta5, beta6* and *beta7*), but has two α-helices (*alpha2* and *alpha3*) that both lie on the same surface of the β-sheet. These elements are connected by one loop (*loop4, beta6* -> *alpha2*) and three turns (*turn3, beta5* -> *beta6; turn4, alpha2* -> *beta7; turn5, beta7* -> *alpha3*), and the C-terminus is disordered. The domains are linked by a 20 residue stretch that starts in β conformation (*beta4*) and ends as a loop (*loop3*). The two domains associate such that the empty surface of the domain 1 β-sheet packs against the α-helical surface of domain 2. This produces two unusual protein structural features. First, one surface of the second β-sheet is totally exposed. Second, α-helices *alpha2* and *alpha3* are packed between two β-pleated sheets.

There currently exist 13 sequences for S5, and many of the residues that are crucial to its structural integrity are highly conserved and reflect the invariance of the structure. First are the residues in the three distinct hydrophobic cores; those of each domain, and that which cements the domains together. Second, several regions of the polypeptide backbone are unusually close in the structure and are populated with small residues (usually glycines

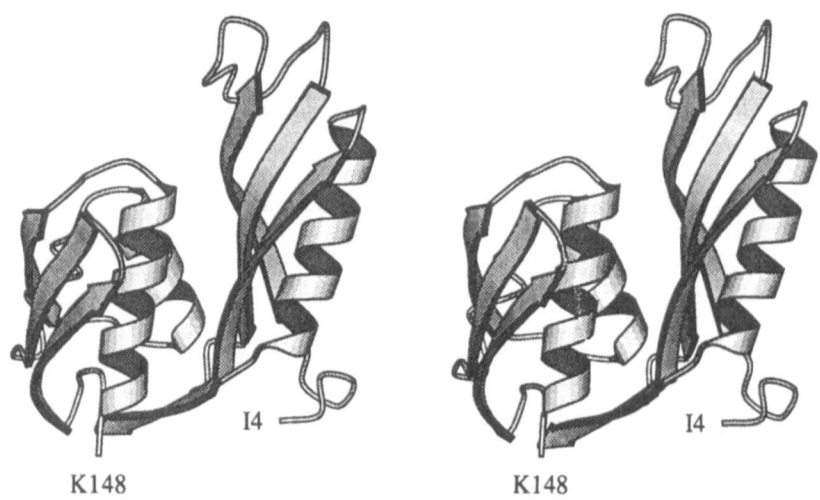

K148   K148

Figure 1. Stereo view of ribosomal protein S5 from *Bacillus stearothermophilus* showing the fold of the polypeptide backbone. The N-terminal domain is to the right and the C-terminal domain to the left. The visible N- and C-terminal residues are indicated. The α-helices are shown as spirals and β-strands as arrows pointing N- to C-terminus. Note the two α-helices sandwiched between the two β-sheets, and the exposed surface of the C-terminal β-sheet.

and alanines). Most notable are residues 47, 49 and 51 of *beta3* and 59, 63 and 67 of *alpha1* where these elements are extremely close. Third, 6 prolines occupy important positions; 3 are at the N-terminus of each α-helix, 2 are in *loop3* connecting the domains, and 1 is close by at the start of *loop4* between *beta6* and *alpha2*. Finally, there appear to be several patches of charged residues on the molecular surface. Many can only be inferred from the structure as determined in the highly ionic crystal environment but several are clear and occur in sensible locations. For example, as is common in many α-helices, *alpha1* has alternating positive and negative residues (E58, R61, K62, E65, D66 and K69) along its exposed surface. Also, E10, E12, R14, K42, R112 and E116 create a patch that spans the two domains and may help to stabilize their association.

Mutations that cause spectinomycin resistance

The antibiotic spectinomycin is thought to disrupt protein synthesis at the level of translocation (Cundliffe, 1990) and its binding site has been located on 16S rRNA (Moazed and Noller, 1987; Sigmund et al., 1984) in the vicinity of base-paired nucleotides G1064-C1192 in helix 34 (Brimacombe, 1991). Resistance to spectiomycin can be introduced into ribosomes by specific mutations in S5 at residues 19, 20 and 21 (Wittmann-Liebold and Greuer, 1978) (*E. Coli* numbering; these correspond to 20, 21 and 22 in *B*.

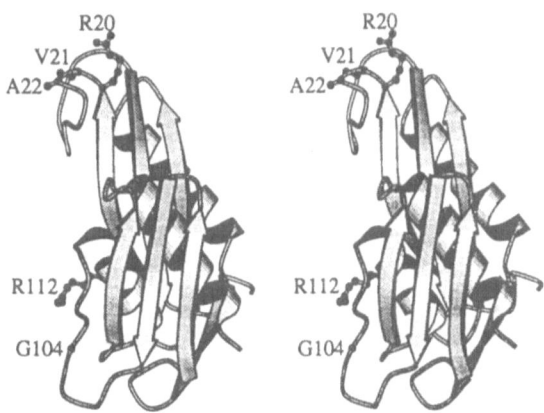

Figure 2. Stereo view of ribosomal protein S5 from *Bacillus stearothermophilus* showing the locations of residues that can be mutated to produce spectinomycin resistance and reversion from streptomycin dependence. The former, arginine 20, valine 21 and alanine 22 are located in a convoluted loop at the top of the molecule that also contains many conserved arginine and lysine residues. The latter, glycine 104 and arginine 112, are adjacent and close to the domain interface.

stearothermophilus). In the 3-dimensional structure, these are located within the convoluted *loop2* that also contains 6 highly conserved lysine and arginine residues (Figure 2). Such residues are often clustered in and around sites of interaction with nucleic acid (Ollis and White, 1987), and the obvious conclusion is that this region of S5 interacts directly with helix 34.

Mutations that affect translational fidelity

S5 is a component of the so-called 'proofreading domain' that also includes S4, S12, the 530 stem-loop or helix 18 (Brimacombe, 1991) and 5' terminus of 16S rRNA (Noller, 1991). These components are adjacent within the 30S subunit (Lambert et al., 1983; Capel et al., 1987; Greuer et al., 1987; Osswald et al., 1987; Stern et al., 1989) and have been shown to control the level of translational fidelity. In the case of S5, mutations at positions 103 and 111 (Wittmann-Liebold and Greuer, 1978) (*E. coli* numbering; 104 and 112 in *B. stearothermophilus*) introduce a ribosome ambiguity or *ram* phenotype (Piepersberg et al., 1975a) which can also relieve the ribosome's dependence on streptomycin (Piepersberg et al., 1975b). The streptomycin binding site is located in the vicinity of nucleotide C912 at the base of the 900 stem-loop or helix 27 (Montandon et al., 1986; Moazed and Noller, 1987; Brimacombe, 1991). Residues 104 (glycine) and 112 (arginine) are highly conserved, adjacent in the molecule (Figure 2) and could mediate the close approach of a region of 16S rRNA. Again, the straightforward conclusion is that this rRNA region is close to the streptomycin binding site.

The S5 structure and models of the 30S subunit

Based on the antibiotic related mutations of S5, we have proposed a rather detailed positioning of the protein within the 30S subunit relative to the ribosomal RNA. How well does this agree with current models (Stern et al., 1988; Brimacombe et al., 1988)? As regards the proofreading domain, the agreement is excellent since, as mentioned earlier, these components are known to be clustered together. The proposed direct interaction with helix 34, however, is more controversial since current models place it some 70Å distant.

Recent results from Brimacombe and coworkers seriously question the validity of the 30S models and impinge directly on S5 (Dontsova et al., 1991). These results not only appear to demonstrate the juxtaposition of S5 and helix 34, but they also support a relocation of the proofreading domain. Classically, this is placed on the 'back' of the 30S subunit some 65Å away from the codon recognition site in the cleft at the subunit interface. It is proposed that communication between these important sites is mediated by changes in rRNA conformation that are able to be propagated over distance (Noller, 1991). Brimacombe proposes that the proofreading components are part of the decoding site environment and that their effects are direct. Certainly, if our interpretation of the S5 spectinomycin mutations is correct, it argues for a re-evaluation of the current 30S models.

RIBOSOMAL PROTEIN L6

Introduction

Ribosomal protein L6 is a highly conserved protein located at the interface of the small and large subunits of the prokaryotic ribosome. Immune Electron Microscopy (IEM) localizes L6 to the same site in the *B. stearothermophilus* and *E. coli* ribosomes (Hackl and Stöffler-Meilicke, 1988). A related and probably homologous eukaryotic protein, labelled L9, has been identified in rat liver ribosomes (Suzuki et al., 1990). In *E. coli*, L6 has been crosslinked to L7/L12, L18, L19 and L31 as well as elongation factor EF-G (Traut et al., 1985). Its location at the interface of the two subunits is confirmed by a crosslink to the small subunit protein S13 (Traut et al., 1985). While not a primary binder of 23S rRNA, L6 has been cross-linked to a site within positions 2473-2481 (Wower et al., 1981). Mutations within L6 produce hyperaccurate ribosomes that in turn increase their resistance to all misreading-inducing aminoglycosides, in particular gentamicin (Kühberger et al., 1979). This is somewhat unusual because these types of mutation usually map to components of the 30S subunit.

Structure Determination

L6 from *B. stearothermophilus* was originally purified and crystallized in 1983 (Appelt et al., 1983b). Although a low resolution structure was published (Appelt et al., 1984), further progress was hindered by the lack of material and difficulty in obtaining good heavy atom derivatives. The project was revitalized when the gene encoding *B. stearothermophilus* L6 was cloned and overexpressed in *E. coli* (Ramakrishnan and Gerchman, 1991). A cysteine residue was engineered into the recombinant protein to facilitate the binding of metals and thereby overcome the problem of generating heavy atom derivatives. The site valine 124 was chosen because the *E. coli* counterpart has a cysteine at the homologous position in the primary sequence. The mutant protein proved to be soluble, and crystallized under the same conditions and with the same space group as the wild type protein. All of the crystallographic data presented here were obtained on the mutant protein. L6 is purified by ion-exchange chromatography on an S-Sepharose Fast Flow column followed by gel-filtration on a G25 column. Crystals are obtained in 1.8 M Na/K phosphate pH 7.6, 3% 1,4-dioxane, 3 mM 2-mercaptoethanol. Large crystals that diffract to better than 2.6Å form in 3-7 days. The space group of these crystals is $P6_122$, with unit cell dimensions of a=b=71.9Å and c=124.9Å.

All data were collected and processed on an Rigaku RAXIS II image plate system. An excellent isomorphous mercury derivative was obtained by soaking the cysteine mutant crystals in p-chloromercuribenzenesulfonate (PCMBS). We also took advantage of the overexpressing vector to incorporate selenomethionine into the protein for use as an additional derivative. In total, four derivatives were used to calculate the map, the two mentioned above and the two platinum compounds originally used for the low resolution structure (Appelt et al., 1984).

The current electron density map of this protein has been calculated at 2.8Å resolution and is of sufficient quality to trace the polypeptide backbone and build a model. Two domains are clearly visible in the electron density map, each with approximate dimensions 15Å x 15Å x 30Å. These are arranged in a head-to-tail fashion to produce an elongated "L"-shaped molecule. Both domains appear to contain extensive regions of β-pleated sheet with at least one α-helix. The positions of the mercury (i.e. cysteine) and selenomethionine atoms provide convenient landmarks and show which domains are in the N- and C-terminal halves of the amino acid sequence. The complete structure of L6 will be available in the near future.

We plan to perform a structure-function analysis of the L6 molecule to locate regions of the molecule that might interact with ribosomal RNA and other ribosomal proteins. To this end, we are currently sequencing the L6 gene from the gentamicin resistant strains described above (kindly provided by Prof. A. Böck). The site of these mutations on the L6 structure may clarify the nature of the L6-gentamicin-rRNA interaction.

RIBOSOMAL PROTEIN S17

Introduction

E. coli ribosomal protein S17 is one of six primary RNA binding proteins in the small subunit. It binds 16S rRNA tightly and footprints a very localized domain of the 16S rRNA centered on the 240-290 stem-loop or helix 11 (Stern et al., 1989; Brimacombe, 1991). Cross-linking studies confirm this binding site and demonstrate a further interaction with residues 629-633 (Kyriatsoulis et al., 1986). The neutron map of the small subunit places this protein towards the bottom of the particle near the subunit interface (Capel et al., 1987). S17 plays an important role in the assembly of the small subunit and, not surprisingly, has been extremely well conserved throughout evolution. The homologous protein in *B. stearothermophilus* that we have investigated shares 75% identity with its *E. coli* counterpart and an homologous protein, termed S11, has been found in the eukaryotic ribosome (Gantt and Thompson, 1990). The role that S17 plays in small subunit assembly is demonstrated *in vivo* by a temperature sensitive mutant of S17 that grows poorly and is defective in assembly at restrictive temperatures (Herzog et al., 1979).

Mutations that confer resistance to neamine, an aminoglycoside antibiotic that causes miscoding (Cundliffe, 1990), map to the gene encoding S17 (Bollen et al., 1975). Ribosomes incorporating the mutant S17 protein exhibit increased translational fidelity as demonstrated by restriction of informational suppressors (Bollen et al., 1975). This S17 mutation complements the *ram* mutations in S4 or S5 that, like neamine, increase the rate of misreading by the ribosome (Topisirovic et al., 1977; Matkovic et al., 1980).

NMR Studies

The gene for S17 was cloned and overexpressed as described above for other *B. stearothermophilus* ribosomal proteins and purified on a S-Sepharose Fast Flow column. Thus far, S17 has not been crystallized despite extensive trials. However, excellent, well resolved two-dimensional NMR spectra can be acquired from it. These measurements are performed at a protein concentration of 4 mM in 20 mM potassium phosphate pH 6.0 and 200 mM NaCl in 93% H_2O : 7% D_2O or 100% D_2O. Assignments were made using two-dimensional Nuclear Overhauser Effect SpectroscopY (NOESY), Double-Quantum-Filtered COrrelated SpectroscopY (DQF-COSY) and TOtal Correlation SpectroscopY (TOCSY) (Bax, 1989). Two-dimensional heteronuclear experiments (Bax, 1989) have also been performed to both confirm these assignments and to extend the analysis. These measurements were made using protein samples that were either uniformly labelled with ^{15}N or selectively labelled with a single α-^{15}N amino acid type.

Main-chain proton assignments of the protein are over 85% complete and these reveal that the majority of the secondary structure is a five-stranded, antiparallel β-pleated sheet with no regions of α-helical structure. Some twenty amide protons are protected from exchange with deuterium in the 100% D_2O samples. All of these residues are found in the assigned β-sheet. Alignment of the sequence of S17 with homologous proteins has revealed the presence of conserved lysines and arginines at the ends of β-strands which may be important in binding RNA. The majority of these residues are located at one end of the β-sheet in turns or loops. An additional feature of the 2D spectra is the presence of doubled peaks which indicate that a second, less populated conformation with a lifetime of at least 100 ms exists in solution. The alternative conformation is not localized to a particular region of the molecule but appears to be a global feature.

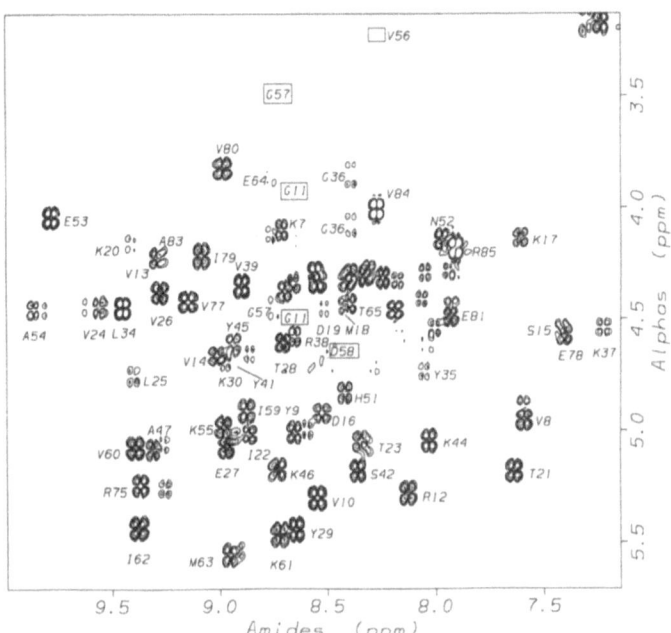

Figure 3. The fingerprint region of the DQF-COSY NMR spectrum of ribosomal protein S17 from *Bacillus stearothermophilus*. The assigned amino acids are labelled in the one letter code. Some peaks are doubled, for example valine 8, threonine 23, valine 24 and alanine 54. These are indicative of a second conformation of the protein with a lifetime of at least 100 ms.

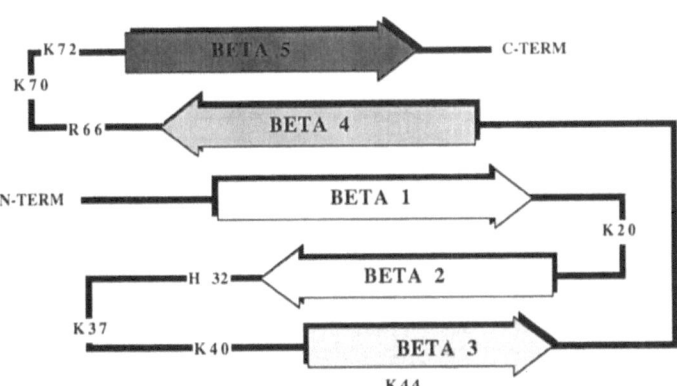

Figure 4. Ribosomal protein S17 forms a five-stranded, anti-parallel β-pleated sheet with a 'Greek key' topology. Each arrow represents a β-strand going from N- to C-termini. Conserved arginine and lysine residues are labelled. These are clustered in the loops between β-strands 2 & 3, and β-strands 4 & 5. Histidine 32, the site of a point mutation that confers resistance to the antibiotic neamine, is also located in the loop between β-strands 2 & 3.

HOMOLOGIES IN RIBOSOMAL PROTEINS

At the time, the structure of the first ribosomal protein L12CTF represented a rather unique protein fold (Leijonmarck et al., 1980). In α/β proteins, α-helices normally cover both surfaces of β-pleated sheets, whereas in L12CTF one surface is completely exposed to the surface. The subsequent structure of L30 proved to be remarkably similar (Wilson et al., 1986). Both molecules contain two α-helices packed on the same side of an anti-parallel, three-stranded β-pleated sheet and, although L12CTF contains an extra α-helix, this is somewhat distant from the body of the molecule and can be regarded as an extension of a loop region of L30. Most notably, the proteins have the same topological arrangements of their secondary structure elements and can be superimposed almost exactly. It was argued that this is good evidence for an evolutionary relationship between these proteins since it is not unreasonable to speculate that they evolved from a smaller group controlling primitive protein synthesis (Leijonmarck et al., 1988).

The structure of S5 considerably complicates this evolutionary scenario. At first sight, both domains with their α-helices packed onto one surface of an anti-parallel three-stranded β-pleated sheet appear to conform to the L12CTF/L30 motif. However, closer inspection reveals that the topological arrangements of α-helices and β-strands in the domains are not only different from each other but also from the L12CTF/L30 motif. It is also intriguing that the RNA recognition motif (RRM) has been found to conform to that of L12CTF/L30 (Nagai et al., 1990; Hoffman et al., 1991). A possible explanation is that this type of structure is particularly suited to protein-RNA interactions, and that the homologies are a result of convergent evolution. There have been several attempts to analyze the primary structures of ribosomal proteins for evidence of evolutionary relatedness (Jue et al., 1980; Wittmann-Liebold et al., 1984) but no clear conclusions have been forthcoming. The structures of L6, S17 and other ribosomal proteins currently under investigation in our laboratories and elsewhere should begin to clarify the evolutionary origins of this ancient organelle.

REFERENCES

All-Robyn, J.A., Brown, N., Otaka, E. and Liebman, S.W., 1990, *Mol. Cell. Biol.* 10:6544.

Appelt, K., White, S.W. and Wilson, K.S., 1983a, *J. Biol. Chem.* 258:13328.

Appelt, K., Dijk, J., White, S.W. and Wilson, K.S., 1983b, *FEBS Lett.* 160:75.

Appelt, K., Tanaka, I., White, S.W. and Wilson, K.S., 1984, *FEBS Lett.* 165:43.

Bax, A., 1989, *Ann. Rev. Biochem.* 58:223.

Bollen, A., Cabezón, T., De Wilde, M., Villarroel, R. and Herzog, A., 1975, *J. Mol. Biol.* 99:795.

Brimacombe, R., 1991, *Biochimie* 73:927.

Brimacombe, R., Atmadja, J., Stiege, W. and Schüler, D., 1988, *J. Mol. Biol.* 199:115.

Capel, M.S., Engleman, D.M., Freeborn, B.R., Kjeldgaard, M., Langer, J.A., Ramakrishnan, V., Schindler, D.G., Schneider, D.K., Schoenborn, B.P., Yabuki, S. and Moore, P.B., 1987, *Science* 238:1403.

Cech, T.R., Zaug, A.J. and Grabowski, P.J., 1981, *Cell* 27:487.

Cundliffe, E., 1990, *in* "The Ribosome, Structure, Function and Evolution" W.E. Hill, P.B. Moore, A. Dahlberg, D. Schlessinger, R.A. Garrett and J.R. Warner, eds., American Society Microbiology, Washington, DC.

Dontsova, O., Kopylov, A. and Brimacombe, R., 1991, *EMBO J.* 10:2613.

Dubendorff, J.W. and Studier, F.W., 1991, *J. Mol. Biol.* 219:45.

Gantt, J.S. and Thompson, M.D., 1990, *J. Biol. Chem.* 265:2763.

Greuer, B., Osswald, M., Brimacombe, R. and Stöffler, G., 1987, *Nucleic Acids Res.* 15:3241.

Hackl, W. and Stöffler-Meilicke, M., 1988, *Eur. J. Bioch⁻m.* 174:431.

Herzog, A., Yaguchi, M., Cabezón, T., Corchuelo, M.-C., Petre, J. and Bollen, A., 1979, *Mol. Gen. Genet.* 171:15.

Hoffman, D.W., Query, C.C., Golden, B.L., White, S.W. and Keene, J.D., 1991, *Proc. Natl. Acad. Sci USA* 88:2495.

Jue, R.A., Woodbury, N.W. and Doolittle, R.F., 1980, *J. Mol. Evol.* 15:129.

Kimura, M., 1984, *J. Biol. Chem.* 259:1051.

Kühberger, R., Piepersberg, W., Petzet, A., Buckel, P. and Böck, A., 1979, *Biochemistry* 18:187.

Kyriatsoulis, A., Maly, P., Greuer, B., Brimacombe, R., Stöffler, G., Frank, R. and Blöcker, H., 1986, *Nucleic Acids Res.* 14:1171.

Lambert, J.M., Boileau, G., Cover, J.A. and Traut, R.R., 1983, *Biochemistry* 22:3913.

Leijonmarck, M., Ericksson, S. and Liljas, A., 1980, *Nature* 286:824.

Leijonmarck, M., Appelt, K., Badger, J., Liljas, A., Wilson, K.S. and White, S.W., 1988, *Proteins* 3:243.

Matkovic, B., Herzog, A., Bollen, A. and Topisirovic, L., 1980, *Mol. Gen. Genet.* 179:135.

Moazed, D. and Noller, H.F., 1987, *Nature* 327:389.

Montandon, P.E., Wagner, R. and Stutz, E., 1986, *EMBO J.* 5:3705.

Nagai, K., Oubridge, C., Jessen, T.H., Li, J. and Evans, P.R., 1990, *Nature* 348:515.

Noller, H.F., 1991, *Ann. Rev. Biochem.* 60:191.

Noller, H.F., Hoffarth, V. and Zimniak, L., 1992, *Science* 256:1416.

Nomura, M., 1987, *Cold Spring Harb. Symp. Quant. Biol.* 52:653.

Ollis, D.L. and White, S.W., 1987, *Chem. Rev.* 87:981.

Osswald, M., Greuer, B., Brimacombe, R., Stöffler, G., Bäumert, H. and Fasold, H., 1987, *Nucleic Acids Res.* 15:3221.

Piepersberg, W., Böck, A. and Wittmann, H.-G., 1975a, *Mol. Gen. Genet.* 140:91.

Piepersberg, W., Böck, A., Yaguchi, M. and Wittmann, H.-G., 1975b, *Mol. Gen. Genet.* 143:43.

Ramakrishnan, V. and Gerchman, S.E., 1991, *J. Biol. Chem.* 266:880.

Ramakrishnan, V. and White, S.W., 1992, *Nature* 358:768.

Scholzen, T. and Arndt, E., 1991, *Mol. Gen. Genet.* 228:70.

Sigmund, C., Ettayebi, M. and Morgan, E.A., 1984, *Nucleic Acids Res.* 12:4653.

Stern, S., Weiser, B. and Noller, H.F., 1988, *J. Mol. Biol.* 204:447.

Stern, S., Powers, T., Changchien, L.-M. and Noller, H.F., 1989, *Science* 244: 783.

Studier, F.W., Rosenberg, A.H., Dunn, J.J. and Dubendorff, J.W., 1990, *Methods Enzymol.* 185:60.

Suzuki, K., Olvera, J. and Wool, I.G., 1990, *Gene* 93:297.

Topisirovic, L., Villarroel, R., De Wilde, M., Herzog, A., Cabezón, T. and Bollen, A., 1977, *Mol. Gen. Genet.* 151:89.

Traut, R.R., Tewari, D.S., Sommer, A., Gavino, G.R., Olson, H.M. and Glitz, D.G., 1985, *in* "Structure, Function and Genetics of Ribosomes" B. Hardesty and G. Kramer, eds., Springer-Verlag, New York.

White, S.W., Appelt, K., Dijk, J. and Wilson, K.S., 1983, *FEBS Lett.* 163:73.

Wilson, K.S., Appelt, K., Badger, J., Tanaka, I. and White, S.W., 1986, *Proc. Natl. Acad. Sci. USA* 83:7251.

Wittmann, H.-G., 1982, *Ann. Rev. Biochem.* 51:155.

Wittmann-Liebold, B. and Greuer, B., 1978, *FEBS Lett.* 95:91.

Wittmann-Liebold, B., Ashman, K. and Dzionara, M., 1984, *Mol. Gen. Genet.* 196:439.

Wower, I., Wower, J., Meinke, M. and Brimacombe, R., 1981, *Nuc. Acids Res.* 9:4285.

MOLECULAR GENETICS OF CHLOROPLAST RIBOSOMES IN *CHLAMYDOMONAS REINHARDTII*

C.R. Hauser,[1] B.L. Randolph-Anderson,[1] T.M. Hohl,[1] E.H. Harris,[1] J.E. Boynton,[1] and N.W. Gillham[2]

Departments of Botany[1] and Zoology[2]
Duke University
Durham, NC 27706 USA

INTRODUCTION

The green alga *Chlamydomonas reinhardtii* is an ideal model in which to study the cooperation of nuclear and chloroplast genes in chloroplast ribosome biogenesis. At least six of the 33 r-proteins identified in the large subunit and 14 of the 31 r-proteins of the small subunit appear to be made in the chloroplast of *C. reinhardtii*, with the remainder being synthesized in the cytoplasm (Schmidt *et al.*, 1983). Antibodies to 16 chloroplast r-proteins from *C. reinhardtii* have been examined for cross reactivity with specific r-proteins from *Anabaena*, *E. coli* and the chloroplast of spinach (Randolph-Anderson *et al.*, 1989; Schmidt *et al.*, 1984). To date 11 chloroplast-encoded r-protein genes have been identified in *C. reinhardtii* (Harris, 1992). The completely sequenced chloroplast genomes of tobacco, rice and the liverwort *Marchantia* have revealed an additional 12 r-protein genes which have not yet been located in *Chlamydomonas* (Shimada and Sugiura, 1991). With a few exceptions, the same set of r-protein genes is found in all chloroplast genomes examined in green algae and land plants (Palmer, 1991). The exceptions include the location of the *rpl5* gene in the chloroplast genome of *Chlamydomonas* and certain other algae, the *rpl21* gene in the chloroplast genome of *Marchantia*, and the *rps16* gene in the chloroplast genome of higher plants but not in *Marchantia*.

A number of mutations affecting chloroplast ribosome structure and function have also been isolated and mapped genetically in *C. reinhardtii* (see Boynton *et al.*, 1992; Harris, 1989 for reviews), but the gene products remain to be established. Recently developed technology for transformation of the chloroplast (see Boynton *et al.*, 1992; Boynton and Gillham 1992 for reviews) and nuclear (Debuchy *et al.*, 1989; Kindle *et al.*, 1989; Mayfield and Kindle, 1990; Diener *et al.*, 1990) genomes of *Chlamydomonas* allows experimental manipulation of genes encoding chloroplast rRNAs and r-proteins to gain a better understanding of the role of particular gene products in the biogenesis of chloroplast ribosomes.

Our research is focused on two aspects of chloroplast ribosome biogenesis in *Chlamydomonas*. First, we are investigating the translational fidelity domain of the small subunit. This is one of two ribosomal domains defined by antibiotic resistance mutations and thus accessible to genetic analysis in *C. reinhardtii*. The other is the peptidyl transferase center of the large subunit. Our aim is to identify and isolate the chloroplast- and nuclear-encoded components of the translational fidelity domain, to compare these components to their cognates in *E. coli*, and to study the coordinate expression of chloroplast and nuclear genes whose products participate in constructing this domain.

The Translational Apparatus, Edited by K.H. Nierhaus
et al., Plenum Press, New York, 1993

Second, we are studying translational control mechanisms which play an important role in the regulation of chloroplast gene expression in *Chlamydomonas* (e.g. Hosler *et al.*, 1989; Liu *et al.*, 1989a; Mayfield, 1990; Rochaix *et al.*, 1991; Rochaix, 1992) and higher plants (Gruissem *et al.*, 1988; Gruissem, 1989). Chloroplast translational control mechanisms can be broadly classified into three groups. Gene specific controls generally involve interactions between nuclear gene products and the untranslated regions flanking the coding sequence of mRNAs for specific chloroplast genes. In *Chlamydomonas* the expression of individual chloroplast genes at the translational level has been found to be under the control of specific nuclear genes (see Rochaix, 1992; Rochaix *et al.*, 1991 for reviews). The same situation obtains in the case of yeast mitochondrial genes (see Costanzo and Fox, 1990; Grivell, 1989 for reviews). Class specific controls discriminate between mRNAs encoding groups of functionally related proteins. For example, in *C. reinhardtii* chloroplast r-protein mRNAs are translated preferentially to mRNAs encoding photosynthetic proteins under conditions of reduced chloroplast protein synthesis (Liu *et al.*, 1989b). General controls apply to the translation of many or all chloroplast mRNAs irrespective of class. For instance, 3' stem-loop structures are important for stability of chloroplast mRNAs (see Gruissem, 1989, and Sugiura, 1991 for reviews). Another general control mechanism operative in *Chlamydomonas* involves the selection of translatable mRNAs from the total mRNA pool. We have found that the proportion of translatable mRNAs for different chloroplast-encoded proteins increases as the mRNA pool is depleted (Hosler *et al.*, 1989).

Our current aim is to understand the molecular basis of class specific mRNA expression that results in preferential synthesis of chloroplast-encoded r-proteins under conditions of reduced chloroplast protein synthesis. Our approach is twofold. First, we are attempting to identify proteins that bind specifically to flanking regions of r-protein mRNAs. Second, we are using gene manipulation *in vitro* and chloroplast transformation to exchange flanking regions of specific r-protein and photosynthesis genes in order to examine the roles played by the 3' and 5' regions in preferential translation of their mRNAs *in vivo*. By combining the two approaches, identification of specific cis acting sequences and trans acting factors responsible for class-specific translational regulation should be possible.

THE TRANSLATIONAL FIDELITY DOMAIN

Putative components of the translational fidelity domain of the small subunit of the chloroplast ribosome are listed in Table 1. Chloroplast encoded molecules include the 16S rRNA, and the S4 and S12 proteins specified by the *rps4* and *rps12* genes respectively. All three genes have been sequenced, and S12 and 16S rRNA are marked by specific antibiotic resistance or dependence mutations. By analogy with *E. coli*, the chloroplast translational fidelity domain should also include the S5 protein. The *rps5* gene encoding this protein is not present in any of the three completely sequenced chloroplast genomes of higher plants (Shimada and Sugiura, 1991), but has been found and sequenced in the plastid (cyanelle) of the alga *Cyanophora paradoxa* (Michalowski *et al.*, 1990). In *Chlamydomonas* the cellular location of the *rps5* gene is unknown. Nuclear genes defined by the *sr-1*, A and *spr-1* mutations may also encode components of the translational fidelity domain (see below).

Mutations at five chloroplast antibiotic resistance loci of *Chlamydomonas* map in a linear array in the 16S rRNA gene (Fig. 1) and result in specific nucleotide substitutions in the rRNA (Harris *et al.*, 1989). Mutations at three of these loci confer streptomycin resistance. The *sr-u-sm3* mutation causes an intermediate level of streptomycin resistance (50-100 μg/ml) and results from a U → G change at the nucleotide equivalent to *E. coli* 13 near the 5' end of the molecule. The *sr-u-2-60* mutation, which confers resistance to >500 μg/ml streptomycin, marks the second locus. The base pair change responsible (A → C) maps in the "530 loop" of the 16S rRNA molecule at a position equivalent to *E. coli* 523. Identical mutations have been found in *Chlamydomonas eugametos* (Gauthier *et al.*, 1988), and *E. coli* (Melançon *et al.*, 1988). A C → U change at 525 confers streptomycin resistance in tobacco (Fromm *et al.*, 1989) and in *E. coli* (Powers and Noller, 1991). Low, high and intermediate level streptomycin resistance mutations at the third locus are the consequence of three different base pair changes in the chloroplast 16S rRNA corresponding to *E. coli* bases 912 (*sr-u-2-23*, C → U, resistant to 50 μg/ml), 914 (*sr-u-sm5*, A → C, resistant to >500 μg/ml), and 915 (*sr-u-sm3a*, A → G, resistant to 100 μg/ml). Mutations in the corresponding region at nucleotide 912 have been reported in *E. coli* (Montandon *et al.*, 1986), tobacco (Etzold *et al.*, 1987), and *Euglena* (Montandon *et al.*, 1985).

Table 1. Translational Fidelity Domain of the Small Subunit of the Chloroplast Ribosome

Component	Gene	Location	Mutant Phenotypes
16S rRNA	16S rDNA	cp	Resistance to streptomycin (3 loci) spectinomycin (1 locus) and neamine/kanamycin (1 locus)
Protein S12	rps12	cp	Streptomycin resistance (sr-u-sm2) and dependence (sd-u-2-24) mutations
Protein S4	rps4	cp	None
?	sr-1	nuc	Streptomycin resistance of small subunit
?	spr-1	nuc	One spectinomycin-resistant mutant with unknown effects on chloroplast ribosomes
Protein S5	rps5	?	None
?	?	cp	45 putative ram mutants with several different ts growth phenotypes
?	A	nuc	Elevates streptomycin resistance of nuclear and chloroplast rRNA and r-protein mutations

Chloroplast mutations at the spectinomycin resistance locus result from three different alterations in a conserved region in the 16S rRNA molecule corresponding to *E. coli* bases 1191-1193. An A → G or C substitution at the equivalent of *E. coli* position 1191 causes high level resistance (100 μg/ml). Mutants with these alterations grow equally well in the presence of spectinomycin whether supplied with CO_2 or acetate as carbon sources. An A → C mutation at 1191 has also been isolated in tobacco (Svab and Maliga, 1991). An *E. coli* mutation at 1192 (C → U or G) results in high level spectinomycin resistance (Sigmund *et al.*, 1984; Makosky and Dahlberg, 1987). In contrast a C → A change at this position confers low level resistance. In tobacco spectinomycin resistance can result from mutation of a nucleotide equivalent to *E. coli* 1064, which normally pairs with 1192 within a stem region in the secondary structure (Fromm *et al.*, 1987), and also from C → U at 1192 (Svab and

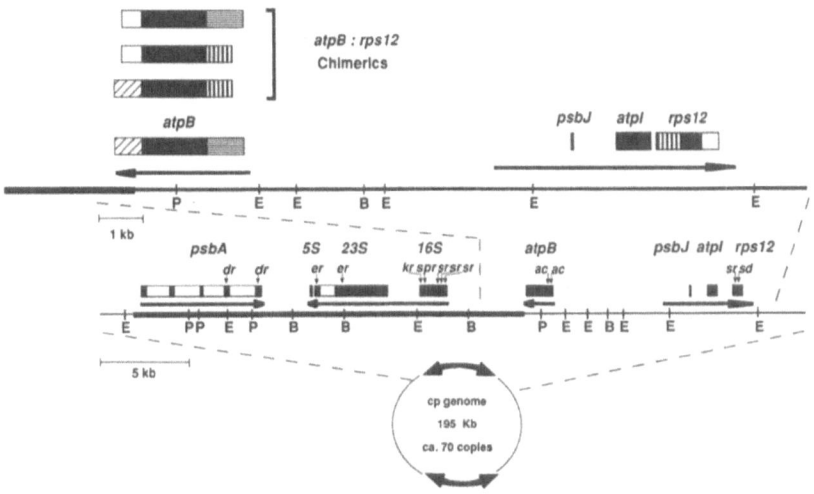

Figure 1. Diagram summarizing the positions of mutations in the *psbA* (herbicide resistant, *dr*); 23S rRNA (erythromycin resistant, *er*); 16S rRNA (kanamycin resistant, *kr*; spectinomycin resistant, *spr*; streptomycin resistant, *sr*); *atpB* (non-photosynthetic, *ac*) and *rps12* (streptomycin resistant, *sr*; streptomycin dependent, *sd*) genes. The *expanded atpB-rps12* region above includes the *atpB* and *rps12* structural genes together with their 5' and 3' UTRs. Chimeric *atpB* constructs involving the 5' and 3' UTRs from *rps12* are shown.

Maliga, 1991). The *C. reinhardtii* mutant *spr-u-1-27-3* results from a G → A change at 1193. This mutant grows well on 100 µg/ml spectinomycin when acetate, but not CO_2, is used as the carbon source since the mutant cannot synthesize chloroplast-encoded photosynthetic proteins efficiently in the presence of the antibiotic (Liu *et al.*, 1989a). Equivalent G → A spectinomycin-resistant mutants have been isolated in tobacco (Fromm *et al.*, 1987). Resistance to neamine and kanamycin in the *C. reinhardtii* chloroplast 16S rRNA is the consequence of an A → G change at a base equivalent to *E. coli* 1408 or of a C → U change at 1409 (Harris *et al.*, 1989). In *E. coli* direct binding of aminoglycoside antibiotics to base 1408 has been demonstrated (Moazed and Noller, 1987), and DeStasio and Dahlberg (1990) have obtained resistance mutations by site-directed mutagenesis of 1409 (C) and 1491 (G) which are paired at the base of a long stem region.

Because the *E. coli* genome has seven rRNA operons, antibiotic resistance mutations must be selected on high copy number plasmids (see Sigmund *et al.*, 1984). In contrast, equivalent mutations can be selected directly in *Chlamydomonas* despite the many copies (ca. 80) of chloroplast genomes per plastid and the duplicate nature of the rRNA operons which are found in the inverted repeat. Newly arising rRNA mutations can copy correct to homozygosity following which somatic (vegetative) segregation of chloroplast genomes yields cells that are homoplasmic for chloroplast genomes carrying the selected antibiotic resistance marker (Boynton *et al.*, 1992). Chloroplast rRNA mutations to antibiotic resistance have proven useful for obtaining rRNA gene transformants (Boynton *et al.*, 1990; Boynton and Gillham, 1992; Newman *et al.*, 1990) and for selecting cotransformants in which a second plasmid contains a chloroplast encoded photosynthetic gene that has been disrupted or altered (Boynton and Gillham, 1992; Newman *et al.*, 1991).

The *rps12* gene encoding r-protein S12 is located not far from the inverted repeat in one of the two large unique sequence regions (Fig. 1). This gene has been sequenced (Liu *et al.*, 1989b) and found to be continuous in *Chlamydomonas* like the corresponding genes of *E. coli* and *Euglena* (see Gillham *et al.*, 1991, for a review). In contrast the *rps12* gene in higher plants is split into three widely separated exons whose mRNAs are assembled by trans splicing (Zaita *et al.*, 1987; Hildebrand *et al.*, 1988). The deduced amino acid sequence of the *Chlamydomonas* S12 protein shows 68% identity with the homologous protein from *E. coli* (Liu *et al.*, 1989b). Furthermore, the *Chlamydomonas rps12* gene is expressed in *E. coli* when introduced on a plasmid and its protein product assembled into functional *E. coli* ribosomes. Mutations to streptomycin resistance *(sr-u-sm2)* and dependence *(sd-u-2-24)* cause amino acid substitutions from Lys42 → Thr and Pro90 → Leu respectively. Similar resistant and dependent phenotypes in *E. coli* result from identical amino acid substitutions in the equivalent positions (Funatsu and Wittmann, 1972; van Acken, 1975).

In *E. coli* ribosomal ambiguity *(ram)* mutations suppress the streptomycin-dependent *(sd)* phenotype of S12 mutants by decreasing hyperactive translational proofreading (Allen and Noller, 1989; Andersson *et al.*, 1986; Kirsebom and Isaksson, 1986). These *ram* mutations result from specific amino acid changes in r-proteins S4 (Gln53 → Leu; van Acken, 1975) and S5 (Gly103 → Arg, Arg111 → Leu; Itoh and Wittmann, 1973) or from a 16S rRNA mutation C1469 → U (Allen and Noller, 1991). In *C. reinhardtii* the chloroplast *rps4* gene has been completely sequenced (X.-Q. Liu, pers. comm.) and contains an Arg (CGT) at a position equivalent to Gln53 which can theoretically be mutated to Leu (CTT).

We are taking two approaches to identify *ram* mutants of *C. reinhardtii*. First, we have isolated 45 UV induced mutations which suppress the streptomycin dependence (sd) phenotype of the *sd-u-2-24* mutant. All suppressed strains are streptomycin sensitive and the suppressed phenotype shows the uniparental pattern of inheritance characteristic of chloroplast genes in *Chlamydomonas* (see Harris, 1989). They fall into three groups with respect to temperature sensitivity for growth. Thirteen strains exhibit wild type growth at 15 and 32°C, 11 strains show reduced growth at 32°C, and 21 show little to no growth at either 15 or 32°C. Thirty-five suppressed strains have been analyzed using PCR sequencing for the predicted *ram* changes in *rps4* and 16S rDNA. No *sd* reversions have been detected in *rps12*, nor have nucleotide substitutions been encountered that would lead to amino acid substitutions at position 42 of S12, the site of the streptomycin resistance mutation *sr-u-sm2*. No nucleotide substitutions in the *rps4* gene leading to the Arg → Leu change expected in *ram* mutations (see above) have been detected. Likewise, no alterations have been found in the 16S rRNA nucleotide equivalent to *E. coli* C1469 which suppresses the sd phenotype of this bacterium. These results demonstrate that we have isolated new suppressors of the sd phenotype. The suppressor mutations may be located either at other sites in the *rps4*, *rps12* or 16S rRNA genes or in other chloroplast genes encoding products that interact with the translational fidelity domain of the chloroplast ribosome. The variety of temperature sensitive phenotypes we see in these suppressed strains suggests that several different nucleotide substitutions are occurring at one or more sites.

We are currently allele testing the suppressor mutations in pairwise crosses of suppressed strains to determine the number of genetic loci involved. We previously used this method to define recombinationally distinct chloroplast loci for antibiotic resistance or photosynthetic function (see Harris, 1989). To identify the genes affected by the suppressor mutations we are attempting to use chloroplast transformation technology to complement the suppressed phenotype. Suppressed *sd* strains are being transformed with plasmids containing the wild type chloroplast 16S rRNA or *rps4* genes and streptomycin-resistant (dependent) transformants selected. The temperature sensitive nature of certain suppressed strains may provide an opportunity to identify other components of the translational fidelity domain since we can select for suppression of the temperature sensitive phenotype. Experiments are also in progress to test directly whether the *ram* alterations seen in the *E. coli rps4* and 16S rRNA genes can suppress the chloroplast *sd* mutant phenotype. The Arg (CGT) → Leu (CTT) alteration at the position equivalent to Gln53 of *E. coli* and the 16S rRNA change from A → C at the base equivalent to *E. coli* U1469 are being made in the respective *Chlamydomonas* chloroplast genes. These altered *rps4* and 16S rRNA genes will be used to attempt transformation of the *sd-u-2-24* strain to streptomycin independence.

Genetic evidence suggests that at least three nuclear gene products function in the translational fidelity domain. The *sr-1* locus mapping on linkage group IX is the best characterized (Harris, 1989). All nuclear mutations conferring resistance to 50-100 µg/ml streptomycin map at the *sr-1* locus and representative *sr-1* mutants have been shown to make the small subunit resistant *in vitro* (Bartlett *et al.*, 1979). A second gene mutation called *A* has little or no phenotypic manifestation itself, but raises the resistance level of both chloroplast and nuclear streptomycin resistant mutants known to affect chloroplast ribosomes to ca. 1 mg/ml (Sager, 1960; Sager and Tsubo, 1961). Mutants in a third nuclear gene, *spr-1*, confer resistance to spectinomycin and map on linkage group XVIII (Harris, 1989). These mutants might be equivalent to mutants in the *E. coli rpsE (rps5)* gene that confer spectinomycin resistance at the level of the EF-G cycle (Bilgin *et al.*, 1990) and affect different amino acids than altered by the *ram* mutants in this gene. Direct effects of the *A* and *spr-1* mutations on *Chlamydomonas* chloroplast ribosomes have not yet been demonstrated.

So far our studies of the translational fidelity domain of the small subunit of the chloroplast ribosome of *C. reinhardtii* have revealed many similarities, but some notable differences, with regard to the *E. coli* small subunit. A unique mutation to streptomycin resistance occurs in *C. reinhardtii* near the 5' end of the 16S rRNA molecular (equivalent to *E. coli* 13). Potential *ram* mutations in *Chlamydomonas* have been isolated that do not affect the classical *ram* sites in the genes encoding S4 or the 16S rRNA in *E. coli*. Possible new genes have been identified (e.g. *sr-1*) which may play a role in translational fidelity but have no equivalents in *E. coli*.

PREFERENTIAL TRANSLATION OF CHLOROPLAST R-PROTEIN MRNAS

Preferential translation of mRNAs encoding chloroplast r-proteins might be expected for two reasons. First, to sustain expression of plastid genes whose products are required for photosynthesis, a functioning chloroplast protein synthesizing system is necessary. Failure to synthesize chloroplast-encoded r-proteins at sufficient rates would lead to the irreversible loss of plastid ribosomes and, therefore, plastid protein synthesis (Walbot and Coe, 1979; Feierabend and Berberich, 1991). Second, the chloroplast protein-synthesizing system may be required to make one or two nonphotosynthetic proteins essential for survival, since genes encoding these components have been retained in plastid genomes of colorless non-photosynthetic algae (e.g. Siemeister and Hachtel, 1990a,b; Siemeister *et al.*, 1990) and plants (de Pamphilis and Palmer, 1988, 1989, 1990; Morden *et al.*, 1991). Several lines of evidence also suggest that some degree of chloroplast ribosome function is essential for survival in *Chlamydomonas* (see Harris, 1989; Gillham *et al.*, 1991; Liu *et al.*, 1989a).

We previously postulated a class-specific translational control mechanism which would ensure synthesis of chloroplast r-proteins in *Chlamydomonas* when chloroplast protein synthesis is reduced (Liu *et al.*, 1989a; Gillham *et al.*, 1991). Mutants deficient in chloroplast protein synthesis exhibit a characteristic syndrome of photosynthetic defects due to a deficiency of chloroplast-encoded photosynthetic proteins, yet accumulate chloroplast ribosomal subunits containing normal complements of chloroplast-encoded r-proteins.

Reduced synthesis of chloroplast-encoded photosynthetic proteins is not paralleled by changes in either their hybridizable or translatable mRNAs.

Any class-specific translational regulatory mechanism should function for all photosynthetic and r-proteins regardless of the organization of the genes encoding them in the chloroplast genome. Most chloroplast r-protein genes in higher plants are organized in two types of gene clusters or operons, whereas a few r-protein genes (e.g. *rps4*) seem to be transcribed monocistronically (Mache, 1990; Subramanian *et al.*, 1990, 1991). The first type of cluster, comprising primarily r-protein genes arranged in the same order that they appear in *E. coli* , is illustrated by the L23 operon, which includes genes found in the *E. coli* S10, *spc* and α operons. However, certain genes which are missing (e.g. *rpl24*, *rps10*) are thought to have been transferred to the nucleus. In *E. coli* certain r-proteins regulate expression of the polycistronic r-protein mRNAs at the translational level by binding to specific mRNA regions (Lindahl and Zengel, 1986). The α, *str*, *spc* and S10 operons are regulated by S4, S7, S8 and L4, respectively. These proteins also bind to rRNA during ribosome biogenesis so that rRNA and mRNA compete for r-protein binding during ribosome assembly (Nomura, 1986). If the same r-proteins are involved in the autoregulation of r-protein operons in chloroplasts, some of them would have to be nuclear gene products (Bourque *et al.*, 1991). Thus, while S4, S7 and S8 are chloroplast gene products, the homologue of L4, if any, is probably coded in the nucleus. This raises the possibility that proteins encoded by nuclear r-protein genes might regulate expression of chloroplast r-protein mRNAs at the translational level. In tobacco a possible binding site for the S7 protein on 16S rRNA and in the intercistronic region between the 3' exon of *rps12* and *rps7* has been identified (Bourque *et al.*, 1991). A similar binding site is found between the homologous genes in *E. coli*. In plant chloroplasts, multiple transcripts have been identified for individual r-protein genes of these clusters, suggesting processing of larger polycistronic transcripts (Subramanian, 1991). In *Chlamydomonas* no single large mRNA has been identified for the L23 r-protein gene cluster (X.-Q. Liu, pers. comm.). In fact many transcripts appear to be monocistronic whereas others are big enough to cover two or more genes. Preferential translation of *rpl2* mRNA from this cluster has already been demonstrated (Liu *et al.*, 1989a). Similarly, the mRNA encoding *Chlamydomonas* r-protein "L-13" also exhibits preferential translation (Liu *et al.*, 1989a). Based on immunological cross-reactivity with *E. coli* L5 (Randolph-Anderson *et al.*, 1989), this protein is probably encoded by the *rpl5* gene which is also part of the L23 gene cluster in *Chlamydomonas* (X.-Q. Liu, pers. comm.).

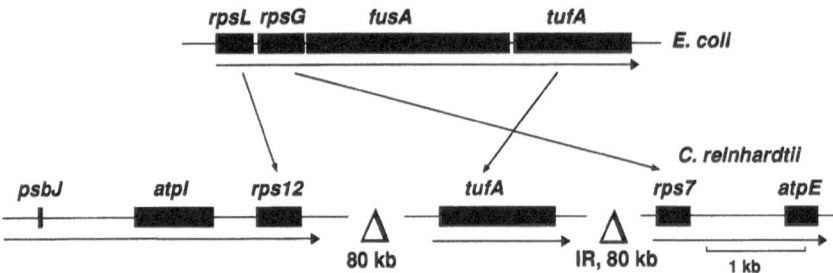

Figure 2. Organization of the *rpsL* (*rps12*), *rpsG* (*rps7*), *fusA* and *tufA* genes in the *E. coli str* operon into three separate transcriptional units in the chloroplast genome of *Chlamydomonas*. The chloroplast *rps7* and *rps12* genes are each cotranscribed with genes encoding photosynthetic proteins (*atpE*, ε subunit of ATP synthase; *atpI*, subunit IV of ATP synthase; *psbJ*, 4 kDa subunit of cytochrome b6/f). The equivalent of the *E. coli fusA* gene is presumably nuclear in *Chlamydomonas*.

The second type of cluster contains a mixture of photosynthetic and r-protein genes which poses unique problems in translational regulation. Examples include the *psaA-psaB-rps14* operon of higher plants (Subramanian *et al.*, 1991) and the *atpE-rps7* and *psbJ-atpI-rps12* operons of *Chlamydomonas* (Fig. 2). In *E. coli* the S12 operon includes the *rpsL* (*rps12*), *rpsG* (*rps7*), *fusA* and *tufA* genes (Fig. 2). In *C. reinhardtii rps7*, *rps12* and *tufA* are all parts of different transcriptional units. Spurious hybridization results (Schmidt *et al.*, 1985) and a sequencing error (Robertson *et al.*, 1990) originally suggested that only the 3' end of the *rps7* gene was cotranscribed with *atpE* while the 5' portion of the gene was adjacent to *rps12* (Gillham *et al.*, 1991). We are using cloned probes and antibodies to determine whether the r-proteins encoded by these mixed operons are synthesized preferentially under conditions of reduced chloroplast protein synthesis.

To identify trans-acting factors which may regulate preferential expression of chloroplast r-protein mRNAs, we have focused on the 5' and 3' untranslated flanking regions (UTRs) of representative photosynthetic and r-protein genes. In *C. reinhardtii* and higher plants, the ubiquitous 3' stem-loop structures found in chloroplast mRNAs seem to be important for their stabilization (Stern and Gruissem, 1987; Stern *et al.*, 1991). Other experiments suggest that some proteins bind to the 3' UTR in a gene-specific fashion (Stern *et al.*, 1989). Trans-acting proteins that bind to 5' UTR sequences are clearly important for translation of mRNAs for specific mitochondrial genes in yeast (Costanzo and Fox, 1990). There is reason to believe the same is true for translation of mRNAs of specific *Chlamydomonas* chloroplast genes as well (e.g. Rochaix *et al.*, 1989).

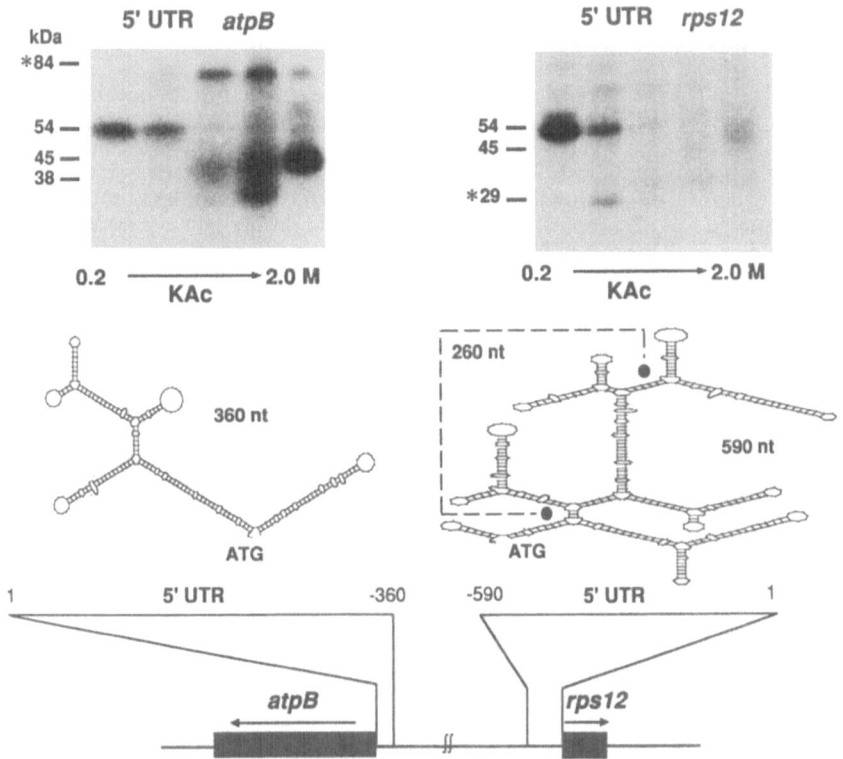

Figure 3. Binding of heparin-Actigel protein column fractions to 5' UTRs of *atpB* and *rps12* mRNAs using UV cross-linking (Danon and Mayfield, 1991). The 5' UTRs were synthesized *in vitro* using T7 polymerase from *atpB* and *rps12* sequences cloned in pGEM vectors. The predicted two-dimensional structures of the *atpB* and *rps12* 5' UTRs were derived from the Zuker-fold program.

We have carried out gel retardation and UV cross linking assays with portions of the 5' UTRs of mRNAs for *rps12* and *atpB*, a monocistronically transcribed photosynthetic protein gene encoding the β subunit of ATP synthase. Both UTRs have several predicted stem-loop structures (Fig. 3). Under normal growth conditions, a protein of 54 kDa binds to the 5' UTRs of both *rps12* and *atpB* mRNAs (Table 2) whereas proteins of 84, 45 and 38 kDa bind to the 5' UTR of the *atpB* mRNA, but not to that of *rps12* mRNA. In protein extracts from cells grown under conditions of reduced chloroplast protein synthesis (growth of the *spr-u-*

Table 2. Proteins crosslinked to the 5' UTRs of *atpB* and *rps12*

	atpB		*rps12*	
	Chloroplast Protein Synthesis		Chloroplast Protein Synthesis	
Protein Species (kDa)	Normal	Reduced	Normal	Reduced
84	yes	yes	no	yes
54	yes	yes	yes	no
45	yes	yes	no	yes
38	yes	no	no	yes

1-27-3 mutant in the presence of antibiotic plus acetate), three of the four proteins (84, 54 and 45 kDa) still crosslink to the 5' UTR of *atpB*, but the 38 kDa protein does not. In contrast, proteins of 84, 45 and 38 kDa are now found crosslinked to the 5' UTR of *rps12*. Our current working hypothesis is that the 38 kDa protein may determine the priority of translation of *rps12* mRNA under conditions of reduced chloroplast protein synthesis, whereas the 45 and 84 kDa proteins may be necessary for differential expression of *atpB* and *rps12* mRNAs under normal conditions. When protein synthesis is reduced, these latter proteins bind to the 5' UTR of *rps12* which may enhance expression of this mRNA.

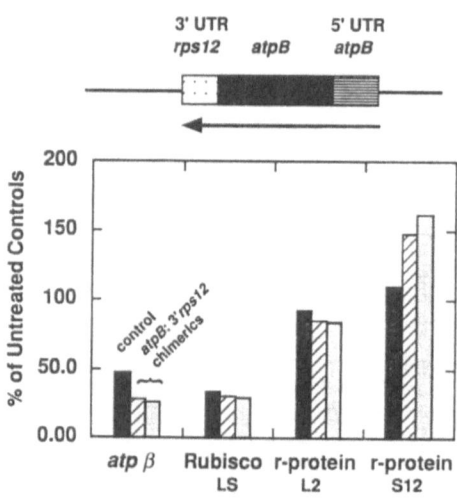

Figure 4. Protein accumulation by transformants in which the *atpB:3'rps12* chimeric has replaced the endogenous *atpB* gene. Each transformant was compared to the untransformed *spr-u-1-27-3* control strain under normal conditions and conditions of reduced chloroplast protein synthesis (acetate medium plus 40 µg/ml spectinomycin). Levels of the β subunit of ATP synthase, RUBISCO LS and r-proteins L1 and S12 for the *spr* control strain (black bars) and two independent *atpB:3' rps12* chimeric transformants (striped, dotted bars) were determined using polyclonal antibodies for the respective proteins and ^{125}I-protein A to probe immunoblots of total cell proteins. Immunoblots were quantified using a Molecular Dynamics Phosphorimager.

To define *in vivo* the cis-acting sequences involved, chimeric *atpB* genes flanked either by the 5' or 3' UTRs of *rps12* have been introduced into an *atpB* deletion mutant possessing the *spr-u-1-27-3* mutation, which shows reduced levels of chloroplast protein synthesis when grown in the presence of spectinomycin (see Liu et al., 1989a). To date we have examined accumulation of photosynthetic and r-proteins under normal conditions and conditions of reduced chloroplast protein synthesis in two homoplasmic transformants that have the 3' UTR of *rps12* fused to *atpB* (Fig. 4 top). In both transformants and the control strain, accumulation of photosynthetic proteins (e.g. the β subunit of the ATP synthase encoded by the *atpB* gene and RUBISCO large subunit encoded by the *rbcL* gene) is affected dramatically whereas r-proteins (e.g. L2 encoded by the *rpl2* gene and S12 encoded by the *rps12* gene) continue to accumulate at normal rates (Fig. 4). These results demonstrate that the 3' UTR of *rps12* is not responsible for preferential translation of r-protein mRNAs under conditions of reduced chloroplast protein synthesis. Both sets of experiments point to the 5' UTRs as the regions responsible for preferential expression of r-protein mRNAs under conditions of reduced chloroplast protein synthesis. Proof of the hypothesis will require demonstration that transformants with the 5' UTR of the *rps12* gene upstream of *atpB* synthesize the β subunit efficiently when chloroplast protein synthesis is reduced. These transformants are now being isolated.

ACKNOWLEDGMENTS

We gratefully acknowledge the technical assistance of Ms. Anita Johnson in the chloroplast transformation experiments and the contributions made to this research by former postdoctoral fellows Xiang-Qin Liu and Dominique Robertson. This work was supported by NIH grant GM-19427 to JEB and NWG and NRSA Fellowship GM-14046 to CRH.

REFERENCES

Allen, P.N., and Noller, H.F., 1989, *J. Mol. Biol.* 208:457.

Allen, P.N., and Noller, H.F., 1991, *Cell* 66:141

Andersson, D.E., Andersson, S.G.E., and Kurland, C.G., 1986, *Biochimie* 68:705.

Bartlett, S.G., Harris, E.H. Grabowy, C.T., Gillham, N.W. and Boynton, J.E., 1979, *Mol. Gen. Genet.* 176:199.

Bilgin, N., Richter, A.A., Ehrenberg, M., Dahlberg, A. and Kurland, C.G., 1990, *EMBO J.* 9: 735.

Bourque, D.P., Elhag, G., Bonham-Smith, P., Thomas, F., McCreery, T., and Glinsmann-Gibson, B., 1991, *in:* "The Translational Apparatus of Photosynthetic Organelles," R. Mache *et al.*, eds., Springer-Verlag, Berlin, p. 85.

Boynton, J.E., Gillham, N.W., Harris, E.H., Newman, S.M., Randolph-Anderson, B.L., Johnson, A.M., and Jones, A.R., 1990, *Current Research in Photosynthesis* III:509.

Boynton, J.E., Gillham, N.W., Newman, S.M. and Harris, E.H., 1992, *Adv. Plant Gene Res.* 6: 3.

Boynton, J.E., and Gillham, N.W., 1992, *Methods in Enzymology* 217:510.

Costanzo, M.C., and Fox, T.D., 1990, *Annu. Rev. Genet.* 24:91.

Danon, A., and Mayfield, S., 1991, *EMBO J.* 10:3993.

Debuchy, R., Purton, S., and Rochaix, J.-D., 1989, *EMBO J.* 8:2803.

de Pamphilis, C.W. and Palmer, J.D., 1988, *in:* "Physiology, Biochemistry and Genetics of Nongreen Plastids", C.D. Boyer *et al.*, eds., American Society for Plant Physiology, Rockville, MD, p. 182.

de Pamphilis, C.W. and Palmer, J.D., 1989, *Nature* 348:337.

De Stasio, E.A. and Dahlberg, A.E., 1990, *J. Mol. Biol.* 212:127.

Diener, D.R., Curry, A.M., Johnson, K.A., Williams, B.D., Lefebvre, P.A., Kindle, K.L. and Rosenbaum, J.L., 1990, *Proc. Natl. Acad. Sci. USA* 87:5739.

Etzold, T., Fritz, C.C., Schell, J., and Schreier, P.H., 1987, *FEBS Lett.* 219:343.

Feierabend, J., and Berberich, Th., 1991, *in:* "The Translational Apparatus of Photosynthetic Organelles," R. Mache *et al.*, eds., Springer-Verlag, Berlin, p. 215.

Fromm, H., Edelman, M., Aviv, D., and Galun, E., 1987, *EMBO J.* 6:3233.

Fromm, H., Galun, E., and Edelman, M., 1989, *Plant Mol. Biol.* 12:499.

Funatsu, G., and Wittmann, H.G., 1972, *J. Mol. Biol.* 68:547.

Gauthier, A., Turmel, M. and Lemieux, C., 1988, *Mol. Gen. Genet.* 214:192.

Gillham, N.W., Harris, E.H., Randolph-Anderson, B.L., Boynton, J.E., Hauser, C.R., McElwain, K.B. and Newman, S.M., 1991, *in:* "The Translational Apparatus of Photosynthetic Organelles," R. Mache *et al.*, eds., Springer-Verlag, Berlin, p. 127.

Grivell, L.A., 1989, *Eur. J. Biochem.* 182:477.

Gruissem, W., 1989, *Cell* 56:161.

Gruissem, W., Barkan, A., Deng, X.-W., and Stern, D., 1988, *Trends in Genetics,* 4:258.

Harris, E.H., 1989, "The *Chlamydomonas* Sourcebook," Academic Press, San Diego.

Harris, E.H., 1992, *in:* "Genetic Maps, Sixth Edition", S. J. O'Brien, ed., Cold Spring Harbor, in press.

Harris, E.H., Burkhart, B.D., Gillham, N.W., and Boynton, J.E., 1989, *Genetics* 123:281.

Hildebrand, M., Hallick, R.B., Passavant, C., and Bourque, D.P., 1988, *Proc. Natl. Acad. Sci. USA* 85:372.

Hosler, J.P., Wurtz, E.A., Harris, E.H., Gillham, N.W., and Boynton, J.E.,1989, *Plant Physiol.* 91:648.

Itoh, T., and Wittmann, H.G., 1973, *Mol. Gen. Genet.* 127:19.

Kindle, K.L., Schnell, R.A., Fernández, E., and Lefebvre, P.A., 1989, *J. Cell Biol.* 109:2589.

Kirsebom, L.A., and Isaksson, L.A., 1986, *Mol. Gen. Genet.* 205:240.

Lindahl, L., and Zengel, J.M., 1986, *Annu. Rev. Genet.* 20:297.

Liu, X.-Q., Hosler, J.P., Boynton, J.E., and Gillham, N.W., 1989a, *Plant Mol. Biol.* 12:385.

Liu, X.-Q., Gillham, N.W., and Boynton, J.E., 1989b, *J. Biol. Chem.* 264:16100.

Mache, R., 1990, *Plant Science* 72:1.

Makosky, P.C., and Dahlberg, A.E., 1987, *Biochimie* 69:885.

Mayfield, S.P., 1990, *Current Opinion Cell Biol.* 2:509.

Mayfield, S.P., and Kindle, K.L., 1990, *Proc. Natl. Acad. Sci. USA* 87:2087.

Melançon, P., Lemieux, C., and Brakier-Gingras, L., 1988, *Nucl. Acids Res.* 16:9631.

Michalowski, C.B., Pfanzage, B., Löffelhardt, W., and Bohnert, H.J., 1990, *Mol. Gen. Genet.* 224:222.

Moazed, D., and Noller, H.F., 1987, *Nature* 327:389.

Montandon, P.-E., Nicolas, P., Schurmann, P., and Stutz, E., 1985, *Nucl. Acids Res.* 13:4299.

Montandon, P.-E., Wagner, R., and Stutz, E., 1986, *EMBO J.* 5:3705.

Morden, C.W., Wolfe, K.H., de Pamphilis, C.W., and Palmer, J.D., 1991, *EMBO J.* 10:3281.

Newman, S.M., Boynton, J.E., Gillham, N.W., Randolph-Anderson, B.L., Johnson, A.M., and Harris, E.H., 1990, *Genetics* 126: 875.

Newman, S.M., Gillham, N.W., Harris, E.H., Johnson, A.M., and Boynton, J.E, 1991, *Mol. Gen. Genet.* 230:65.

Palmer, J.D., 1991, *in:* "The Molecular Biology of Plastids, Cell Culture and Somatic Cell Genetics of Plants, Vol. 7A," L. Bogorad and I.K. Vasil, eds., Academic Press, San Diego, p. 5.

Powers, T., and Noller, H.F., 1991, *EMBO J.* 10:2203.

Randolph-Anderson, B.L., Gillham, N.W., and Boynton, J.E., 1989, *J. Mol. Evol.* 29:68.

Robertson, D., Boynton, J.E., and Gillham, N.W., 1990, *Mol. Gen. Genet.* 221:155.

Rochaix, J.-D., 1992, *Adv. Plant Gene Res.* 6:249.

Rochaix, J.-D., Kuchka, M., Mayfield, S., Schirmer-Rahire, M., Girard-Bascou, J., and Bennoun, P., 1989, *EMBO J.* 8:1013.

Rochaix, J.-D., Goldschmidt-Clermont, M., Choquet, Y., Kuchka, M., and Girard-Bascou, J., 1991, *in:* "Plant Molecular Biology 2", R.G. Herrmann and B. Larkins, eds., Plenum, New York, p. 401.

Sager, R., 1960, *Science* 132:1459.

Sager, R., and Tsubo, Y., 1961, *Z. Vererbungs.* 92:430.

Schmidt, R.J., Richardson, C.B., Gillham, N.W., and Boynton, J.E., 1983, *J. Cell Biol.* 96:1451.

Schmidt, R.J., Myers, A.M., Gillham, N.W., and Boynton, J.E., 1984, *Mol. Biol. Evol.* 1:317.

Schmidt, R.J., Hosler, J.P., Gillham, N.W., and Boynton, J.E., 1985, *in:* "Molecular Biology of the Photosynthetic Apparatus," K.E. Steinback *et al.*, eds., Cold Spring Harbor, New York, p. 417.

Shimada, H., and Sugiura, M., 1991, *Nucl. Acids Res.* 19:983.

Siemeister, G., and Hachtel, W., 1990a, *Plant Mol. Biol.* 14:825.

Siemeister, G., and Hachtel, W., 1990b, *Curr. Genet.* 17:433.

Siemeister, G. Buchholz, C., and Hachtel, W., 1990, *Curr. Genet.* 17:457.

Sigmund, C.D., Ettayebi, M., and Morgan, E.A., 1984, *Nucl. Acids. Res.* 12:4653.

Stern, D.B., and Gruissem, W. 1987, *Cell* 51:1145.

Stern, D.B., Jones, H., and Gruissem, W., 1989, *J. Biol. Chem.* 264:18742.

Stern, D.B., Radwanski, E.R., and Kindle, K.L., 1991, *Plant Cell* 3:285.

Subramanian, A.R., Smooker, P.M., and Giese, K., 1990, *in.:* "The Ribosomes. Structure, Function and Evolution," W.E. Hill *et al.*, eds., American Society for Microbiology, Washington, DC., p. 655.

Subramanian, A.R., Stahl, D, and Prombona, A., 1991, *in:* "The Molecular Biology of Plastids, Cell Culture and Somatic Cell Genetics of Plants, Vol. 7A,", L. Bogorad and I.K. Vasil, eds., Academic Press, San Diego, p. 191.

Sugiura, M., 1991, *in:* "The Molecular Biology of Plastids, Cell Culture and Somatic Cell Genetics of Plants, Vol. 7A,", L. Bogorad and I.K. Vasil, eds., Academic Press, San Diego, p. 125.

van Acken, U., 1975, *Mol. Gen. Genet.* 140:61.

Walbot, V., and Coe, E.H., Jr., 1979, *Proc. Natl. Acad. Sci USA* 76:2760.

Zaita, N., Torazawa, K., Shinozaki, K., and Sugiura, M., 1987, *FEBS Lett.* 210:153.

THE NUCLEAR GENES FOR CHLOROPLAST RIBOSOMAL PROTEINS L11 AND L12 IN HIGHER PLANTS

Jürgen Schmidt, Wolfgang Weglöhner and Alap R. Subramanian

Max-Planck-Institut für Molekulare Genetik
Abteilung Wittmann
Ihnestraße 73, 1000 Berlin 33
Germany

INTRODUCTION

The components of the chloroplast translational apparatus are encoded in two genomes in the plant cell. The rRNA and one-third of the ribosomal proteins (RPs) are encoded in the small circular genome of the chloroplast, while the majority of the RP genes is located in the nucleus. Therefore, the assembly and function of the plastid (chloroplast) ribosome depends on coordinately regulated expression of genes distributed and organized in two cellular compartments.

The sequencing of three complete chloroplast genomes (see Sugiura, 1992) as well as of the complete maize chloroplast RP gene set (e.g. Weglöhner and Subramanian, 1992), their transcriptional analysis (Ohto et al., 1988) and characterization of the gene products (e.g. Schmidt et al., 1992) have led to an insight how the organelle provides its components for the plastid ribosome. More recently, work has also proceeded in characterizing the nuclear RP genes, including identification of genes for chloroplast specific r-proteins (psRPs), and analysis of nuclear RP gene expression. To date nuclear genes for six chloroplast RPs and cDNAs for a total of 16 RPs have been cloned and identified (e.g. Gantt, 1988; Giese and Subramanian, 1989; Zhou and Mache, 1989; Yokoi and Sugiura, 1992; Elhag and Bourque, 1992). All the genes, characterized so far, are single copy genes at the genomic level, unlike the case with eucaryotic cytoplasmic RPs which are often encoded by dispersed multi-gene families.

The Translational Apparatus, Edited by K.H. Nierhaus
et al., Plenum Press, New York, 1993

Endosymbiont theory proposes that the nucleus-located chloroplast RP-genes were originally procaryotic and were transferred to the nuclear genome after the endosymbiotic event. This relocation process led to the acquisition of eucaryotic transcription, mRNA processing and translational signals as well as transit peptides for routing the RPs to chloroplasts. Also the introns, present in these genes show eucaryotic plant intron-exon consensus motifs (Brown, 1986), suggesting that they were also achieved after integration into the nuclear genome. Yet little is known about the regulatory regions or specific transcriptional factors necessary for the precise and coordinate expression of these genes at both intra- and inter-genomic levels.

In this chapter we describe some features of the nuclear genes for chloroplast L11 and L12 and give an overview of all the characterized nuclear genes for chloroplast RPs.

Procaryotic L11 and L12 Proteins and their Gene Organization

Proteins L10, L11 and two dimers of L12 are located in the 100 Å long stalk and its base on the 50S ribosomal subunit (Walleczek et al., 1988) and form an important functional domain of the eubacterial ribosome. This domain is involved in interactions with translational factors and in EF-G-dependent GTP hydrolysis (Möller et al., 1986). It has been shown that L12 (together with its modified form L7) is the only protein present in more than one copy per ribosome (Subramanian, 1975; Hardy, 1975), and that two dimers of L7/L12 are anchored through L10 to the 23S RNA (Beauclerk et al., 1984). The four-copy feature of L12 is also maintained in the chloroplast ribosome (Bartsch et al., 1982). L11 on the other hand is the most heavily methylated RP in *E.coli*, containing an N-terminal trimethylalanine and two internal trimethyllysines (Dognin and Wittmann-Liebold, 1980). For the chloroplast L11 there is evidence for similar modifications at the N-terminus and at the two conserved lysine residues (unpublished results from this lab). Although L11-lacking bacterial mutants have been isolated (e.g. Dabbs, 1991) the conserved features of L11, which are maintained in chloroplasts after more than one billion years of separate evolution, would argue for a significant role for L11 in the overall translational function.

In eubacteria, including the photosynthetic blue-green alga *Synechocystis* (Sibold and Subramanian, 1990; and Schmidt, unpublished), the genes for L11 and L12 are part of a cluster (*rplKAJL*) that encodes L11, L1, L10, and L12. This arrangement for these genes in a cluster is conserved among all eubacteria examined so far and is also maintained in the archaebacterial kingdom. In eubacteria the genes of this cluster are mainly transcribed from two transcription units: the L11-L1 and L10-L12-β-β' operons. Strong promoters in the 5' leader regions of the *L11* and *L10* genes ensure the high level of transcripts. Attenuators, terminators and RNase III-sites within the transcripts modulate the coordinate expression of *L12* and the downstream *rpoBC* genes on transcriptional level (e.g. Climie and Friesen,

1988). Autogenous translational control is provided by the binding of L10 and L1 directly to distinct sites within their mRNAs (e.g. Boughman and Nomura, 1984).

While the expression of these genes in eubacteria is fairly well understood, it is not known how such coordinate expression can be maintained in plants where these genes have been transferred into the nucleus. The relocation of the genes into the host genome could have dispersed them onto different chromosomal locations and subject their expression to the eucaryotic transcriptional - including polyadenylation - and translational control systems. The transcription of eucaryotic genes by polymerase II is typically modulated by the interplay of several sequence specific and non-specific general factors and chromatin components, adapting their expression to developmental stages.

The isolation of all these four genes from the nuclear genome would allow us to examine whether these genes are still clustered in the nuclear genome, how the stoichiometry of their protein products is achieved, and how their expression is coordinately regulated.

RESULTS AND DISCUSSION

RPL11 and *RPL12* Genes

The genomic clones for L11 and L12 were obtained using probes derived from the cDNA clones previously isolated in this laboratory (Figure 1). The mature chloroplast L12 protein has been sequenced (Bartsch et al., 1982) and shown to be the longest member (133 amino acids) of the eubacterial type L12 family. The mature L11 protein has only recently been isolated in pure form (Schmidt and Subramanian, unpublished). The cDNA clones for L11 were isolated using antiserum to a chloroplast RP fraction (Smooker et al., 1991). The cDNA clones for L12 were isolated using a mixed oligonucleotide probe (Giese and Subramanian, 1989). Two spinach cDNA libraries constructed in λgt11 expression vector (Phua et al., 1989; Giese and Subramanian, 1989) served as the source for our chloroplast RP cDNAs. The nucleotide sequences of the two cDNA clones encode cytoplasmic precursors containing sequences of the mature RPs and transit peptides that are cleaved off at the chloroplast envelope. The latter moiety carries the information for targeting the precursor to the chloroplast.

For obtaining the chloroplast RP genes we used an EMBL4 genomic library of *Arabidopsis thaliana* (mouse-ear cress). This small flowering plant has the smallest known plant nuclear genome (< 100 Mbp). Consequently it is likely to contain less repetitive DNA and fewer or smaller introns. Three λ-clones containing the *L11* gene were isolated and a 4.8 kb region from one of them was sequenced. Similarly, screening with the L12 cDNA yielded two recombinant λ-clones, from one of which a 9.1 kb region localizing *rpL12* was sequenced. Screening was also done with an *Arabidopsis* λgt11 cDNA library, using the isolated *L11* and *L12* genes as probes to confirm the in vivo expression of the isolated

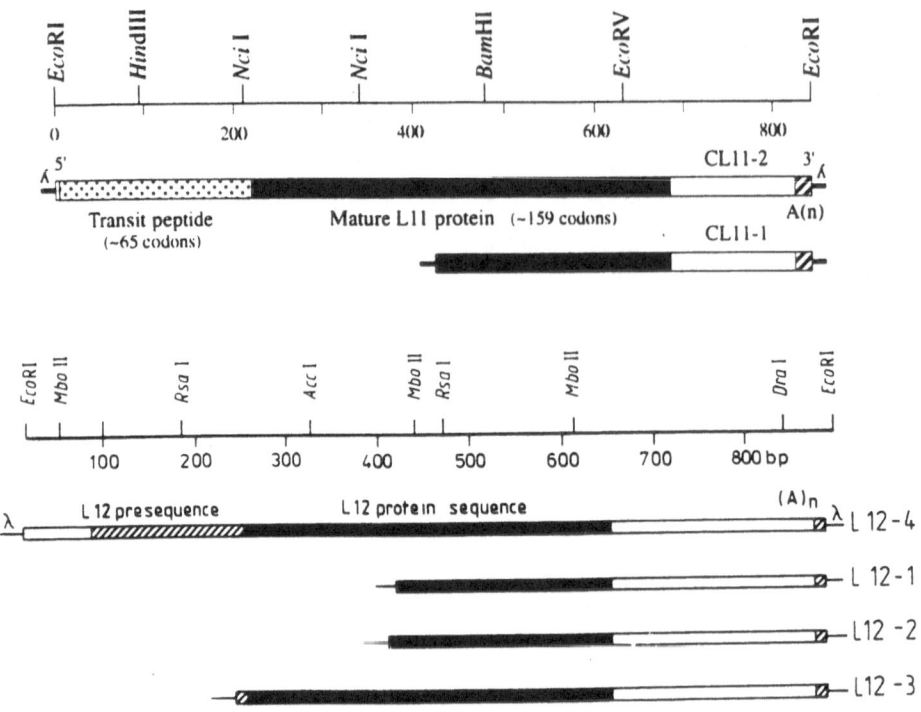

Figure 1. Spinach cDNA clones of chloroplast RPs L11 and L12.

genes. For determining gene copies in the *A.thaliana* genome, DNA was isolated from 9-day old plants (grown from seeds at 16-hour light cycle at 20°C in a growth chamber) by standard procedures. The restricted genomic DNA was subjected to Southern blot analysis using labeled DNA corresponding to the isolated *L11* and *L12* genes.

The *RPL11* Gene. The nucleotide sequence showed that this gene is divided into 4 exons which are interrupted by three relatively small introns (Figure 2). The first exon encodes the transit peptide, an N-terminal extension and the short, basic N-terminal part of the mature protein that is highly conserved among all so far examined eubacteria. The remaining N-terminal and central portion is encoded by exons 2 and 3, which are divided by the second intron. The C-terminal part, which is not well conserved among eubacteria, is expressed by the last exon which contains also the 3' untranslated region of the mRNA. As confirmed by cDNA sequencing, this untranslated region is not seperated by an intron.

Three sequence modules within introns are considered important for the RNA splicing system: i.e. the 5' splice site, internal splice signal and the 3' splice site. While distinct sequences within these motifs are conserved in all eucaryotic lines - fungi, plants and animals - there are some plant specific 5' and 3' splice junction consensus sequences (Hanley and Schuler, 1988) which fit well with the exon/intron boundary sequences within *rpL11*. A polypyrimidine stretch in *rpL11*, which indicates transcription initiation immediately 5' upsteam from the translation start codon, and the absence of a TATA-box are similar to the features found in cytoplasmic RP genes in yeast (e.g. Mager and Planta, 1991) and mammalian cells (Hariharan et al., 1989).

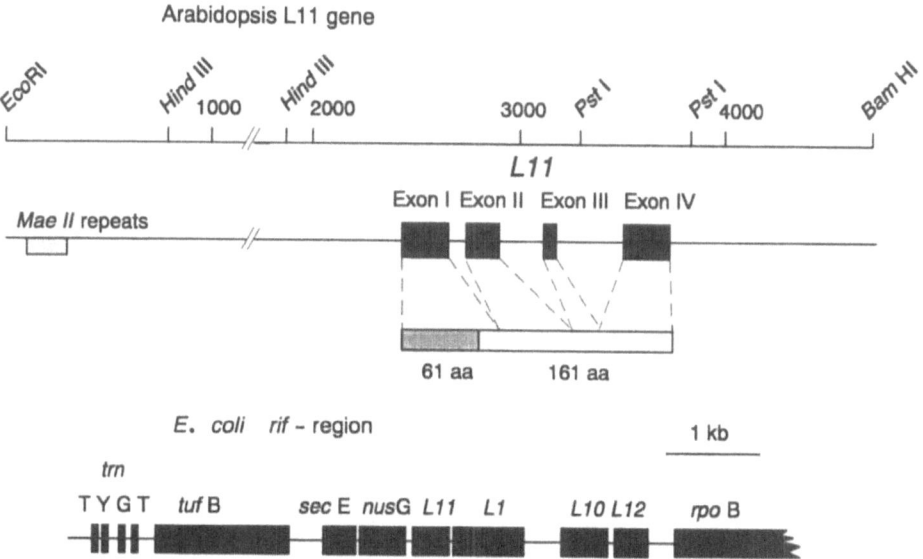

Figure 2. Organization of the nuclear gene for chloroplast RP L11 in *Arabidopsis thaliana*. Shown below is the gene organization of the *rplKAJL* region in *E.coli* genome.

More distant in the 5' upstream region (Figure 2) a highly conserved 7-fold repeat of 30 nucleotides, designated *Mae*II repeats, was found. This distal motif, not so far found in the 5' region of any other genes, could represent a factor binding site that modulates the *L11* gene.

The four exons of the *L11* gene encode a 222 amino acid residues long precursor molecule including a transit peptide. The predicted mature protein (163 amino acid residues) shows identities of 88%, 70% and 54% to the L11 sequences of spinach chloroplast, *Synechocystis* (a cyanobacterium) and *E.coli*, respectively, in the homologous region (Figure 3). It should be mentioned that the three amino acids in *E.coli* L11 (shown underlined in Figure 3) which are methylated are maintained in the chloroplast L11 of *Arabidopsis* (and spinach) and that they occur in conserved contexts.

The *RPL12* Gene. The isolated λ-clones of *RPL12* contained three complete *L12* genes. All three genes lack introns and encode precursors of L12 protein, each consisting of a transit peptide and a mature protein moiety. Two of the genes, *rpL12-A* and *rpL12-C*, code for the same mature protein of 133 amino acids even though there are seven nucleotide substitutions between them. The *rpL12-B* gene is rather different from the other two. The mature protein it encodes has only 74% identity to L12-A/C and it is one residue longer (Figure 4).

```
AthL11   A M A P P K P G G K A K K V V G V I K L A L E A G K A T P A P P V G P A L G S K G V N I M A F C K D
SolL11   • • • • - • • • - • • • • • I • • • • • • • • • • • • • • • • • • • • • • • • • • • • • • • • • • • •
EcoL11             • • • • Q A Y V • • Q V A • • M • N • S • • • • • • • • Q Q • • • • • • E • • • A

AthL11   Y N A R T - A D K A G Y I I P V E I T V F D D K S F T F I L K T P P A S V L L L K A A G V E K G S K
SolL11   • • • • • - • • • • • F V • • • • • • • • • • • • • • • • • • • • • • • • • • • S • • • • • • •
EcoL11   F • • K • D S I E K • L P • • • V • • • Y A • R • • • • V T • • • • • A • • • K • • • • I K S • • G

AthL11   D P Q Q D K V G V I T I D H V R T I A A E K L P D L N C T T I E S A M R I I A G T A A N M G I D I D
SolL11   • • • M E • • • K • • • • Q L • G • • T • • • • • • • • • • • • • • • • • • • • • • • • • • • • • •
EcoL11   K • N K • • • • K • S R A Q L Q E • • Q T • A A • M T G A D • • A M T • S • E • • • R S • • L V V E

AthL11   P P I L E P K K K A V L L        162
SolL11   • • • • • V K • • • E • I F      160
EcoL11   D                               141
```

Figure 3. Comparison between the deduced amino acid sequences of the chloroplast L11 mature proteins of *Arabidopsis thaliana* (Ath), spinach (*Spinacia oleracea*, Sol) and the L11 protein of *E.coli* (Eco). Identical amino acids are marked by asterisks and dashes denote gaps introduced to optimize sequence alignment. The three methylated amino acids in the *E.coli* L11 protein are underlined.

The predicted protein sequence of L12-A/C shows 73%, 47% and 47% identity to the L12 sequences of spinach chloroplast, *Synechocystis* and *E.coli*, respectively. The transit peptides of L12-A, L12-B and L12-C are considerably different in sequence and also in peptide chain length (54-59 residues).

Surprisingly, each of the two *L12-A* and *L12-C* genes is flanked on the 5' side by an identical gene for cytoplasmic tRNA^Pro(UGG). It shows two compensatory nucleotide changes in the acceptor stem but is otherwise identical to the cytoplasmic tRNA^Pro gene of *Phaseolus vulgaris* (Green and Weil, 1989). [Our nucleotide sequence data has been submitted to EMBL/GenBank database with accession number X68046]

Gene Copies of *RPL11* and *L12*. Southern blot analysis of isolated *Arabidopsis* total DNA showed that the haploid genome carries only one copy of the *L11* gene which, from its restriction site pattern, is identified in this work. In the case of the *L12* genes also a similar result was obtained: there is only one cluster of three *L12* genes per haploid genome of *Arabidopsis*.

As mentioned in the Introduction, in *E.coli* L11 and L12 are part of an important functional domain which is essential for ribosome function. Hence it is important to know if the genes identified here are expressed in *Arabidopsis*. Their characterization as single copy (or single cluster for *L12*) genes suggest functional expression, but an experimental

```
AthL12-A/C    A V E A P E K I E K I G S E I S S L T E E A R I L V D Y L Q D K F G V S P L S L A P A A A A V A A
AthL12-B      • • K T • K • • K • • • • • • • • • • • • • • S • • • • • • V • • • • • • • • I • F S • • • • • • L P P
EcoL12            S I T K D Q • I E A V A A M S V M D V V E • I S A M E E • • • • – – – – • A • • V • • • •

AthL12-A/C    P A D – G G A A A V V E E Q T E F D V V I N E V P S S S R I A V I K A V R A L T S L A L K E A K E L
AthL12-B      • L • N • • • T • S • • R • • T • • • • • • D • • R G N • • • • • T • I • • M • • • S • S • S • • •
EcoL12        G P V – E A • – – – – – • • K • • • • • I L K A A G A N – K V • • • • • • • • G A • G • G • • • • • D •

AthL12-A/C    I E G L P K K F K E G I T K D E A E E A K K T L E E A G A K V S I A       133
AthL12-B      • • • F • • • • • • • V • • • • • • • D • T Q • • • P • • • • • • V       134
EcoL12        V • S A • A A L • • • V S • • D • • A L • • A • • • • • E • E V K         120
```

Figure 4. Comparison between the deduced amino acid sequences of the chloroplast L12 mature proteins (L12-A/C and L12-B) of *Arabidopsis thaliana* and the *E.coli* L12. Many of the amino acid substitutions in the L12-B sequence are in the conserved regions of the α-helical and β-strand structures (Liljas et al., 1986) of the *E.coli* L12.

answer to this question was obtained by screening an *Arabidopsis* cDNA library. Probes of both genes yielded several recombinant clones, all of which were purified and sequenced. The sequence of the L11 cDNA showed that it is derived from a polyadenylated mRNA species corresponding to the identified *L11* gene, and the intron/exon borders were exactly as predicted from the sequence/splice site data. In the case of *L12*, the isolated cDNA clones were sequenced to identify, if possible, all the three genes. One of the cDNAs corresponded with *L12-A*, whereas five were products of the *L12-C* gene: none of the sequenced cDNA clones corresponded to the *L12-B* gene.

The *L12-B* Gene: Does it Express a Protein. The unusually divergent sequence of the *L12-B* gene and our failure to detect its cDNA raises the question whether it is expressed at all. There are 92 nucleotide changes in *rpL12-B* when compared to *rpL12-A* (a similar number of changes also between *L12-B* and *L12-C*) and leading to a total of 35 amino acid substitutions. The substitutions occur throughout the whole mature protein, affecting also the highly conserved domains of the eubacterial type L12 sequence.

Crystal data for the *E.coli* L12 C-terminal fragment (Liljas et al., 1986) and alignment of the known eubacterial type L12 proteins have revealed segments of invariant amino acids which are part of a conserved surface. These residues are involved in interaction with translation factors or are part of the potential dimer forming site. Several of the residues in

L12-B do not fit this structure (A. Liljas, personal communication). We infer that *L12-B* is either a silent gene or, if functional, is expressed at a very low level to produce a protein that has a yet unknown function.

Features of the Nuclear Genes for Chloroplast RPs

The nuclear genes for six other chloroplast RPs have recently been reported. They are *S1* and *S22* (*psRP-1*) genes from spinach [Franzetti et al., 1992; and Bisanz-Seyer and Mache, 1992], *L9, L15* and *S17* genes from *A.thaliana* [Thompson et al., 1992] and *L22* gene from pea [Gantt et al., 1991a]. In all these six cases a single gene was obtained. The gene cluster described here for *rpL12* is the only known case of multiple nuclear genes for a chloroplast ribosomal protein. It remains to be seen whether the functional basis for its evolution is to provide the multiple copies of this protein to the ribosome or whether there are additional tissue-specific regulatory aspects.

A translational enhancing feature, i.e. two tandem, in-frame ATG codons, is present in the cDNA for spinach chloroplast L12. Site-specific mutagenesis of this region and transient expression of the resulting constructs in spinach protoplasts has confirmed the enhancing effect (Giese and Subramanian, 1990). All three *Arabidopsis rpL12* genes lack tandem ATG codons. It remains to be seen whether in spinach the *rpL12* gene occurs in a single copy and achieves the 4-fold stoichiometry only through translational enhancement.

Table 1. Features of the characterized genes for nuclear encoded chloroplast r-proteins.

Gene	Start* codon	Number of introns	Lengths of introns [bp]	Stop* codon	Ref.
Sol S1	AACA**ATG**GC	6	84–1350	**TAAA**	a
Ath S17	AAGG**ATG**AT	–	–	**TAGA**	b
Sol S22	AAAG**ATG**GC	4	82–1410	**TAGA**	c
Ath L9	GGCT**ATG**GC	5	83– 334	**TAAA**	b
Ath L11	ATCA**ATG**GC	3	90– 341	**TAAA**	d
Ath L12A	AAAA**ATG**GC	–	–	**TAAG**	d
Ath L12B	TAAA**ATG**GC	–	–	**TAAG**	d
Ath L12C	AACA**ATG**GC	–	–	**TAAG**	d
Ath L15	GACC**ATG**GC	3	71– 292	**TAAA**	b
Psa L22	CGGA**ATG**GC	1	305	**TAGT**	e

* The putative functional contexts of codons included (Lütke et al., 1987; Brown et al., 1990)

Sol: *Spinacia oleracea*, Ath: *Arabidopsis thaliana*, Psa: *Pisum sativum*. [a] Franzetti et al., 1992; [b] Thompson et al., 1992; [c] Bisanz-Seyer and Mache, 1992; [d] this paper; [e] Gantt et al., 1991a.

Table 1 shows some features of the 10 genes including the four genes described here. All of them excepting the genes for L12 and S17 contain varying numbers of introns, the most being six in the *rpS1* gene. The introns in the *Arabidopsis* genes are significantly smaller than those in spinach. As previously mentioned *Arabidopsis* has the smallest nuclear genome size of any known plant, and the choice of this plant genome for our investigation of the chloroplast RP genes appears justified.

The nuclear genes for chloroplast RPs contain certain common sequence motifs (e.g. polypyrimidine stretches) upstream of the coding regions. Trans-acting transcription factors recognizing these motifs could provide a mechanism for the coordinate expression of these genes. Results on these questions from different labs should soon be forthcoming (e.g. Franzetti et al., 1992; Thompson et al., 1992; see also Franzetti et al., this volume).

Conserved sequences in the upstream region of the genes for cytoplasmic ribosomal RPs which are involved in gene expression have been reported in yeast (e.g. Mager and Planta, 1991), and in mouse cells (Hariharan et al., 1989). In plant cells three distinct sets of RP genes are present in the nucleus, i.e. the genes for all cytoplasmic RPs, and those for the majority of the chloroplast and mitochondrial RPs. Sixteen of the plant mitochondrial RPs are encoded in the recently sequenced mito-DNA of a land plant (Takemura et al., 1992). Thus a total of about 200 RP genes belonging to three separate translational systems and requiring coordinate regulation of expression exist in the plant nuclear genome. Although biogenesis of the three types of ribosomes probably takes place at the same developmental stage in leaf growth, it is already known that the regulation of chloroplast RPs is unlinked to that of cytoplasmic RPs (e.g. Gantt et al., 1991b). Further work should reveal the details of the cis-elements and trans-factors that regulate chloroplast RP synthesis and answer the question whether there are some shared common regulatory elements involved in the expression of the RPs of the two organelle systems in plant cells.

REFERENCES

Bartsch, M., Kimura, M, and Subramanian, A.R., 1982, Proc. Natl. Acad. Sci. USA **79**: 6871-6875.
Beauclerk, A.A.D., Cundliffe, E., and Dijk, J, 1984, J. Biol. Chem. **259**: 6559-6563.
Bisanz-Seyer, C., and Mache, R., 1992, Plant Mol. Biol. **18**: 337-344.
Boughman, G., and Nomura, M., 1984, Proc. Natl. Acad. Sci. USA **81**: 5389-5393.
Brown, C.M., Stockwell, P.A., Trotman, C.N., and Tate, W.P., 1990, Nucl. Acids Res. **18**: 6339-6345.
Brown, J.W.S., 1986, Nucl. Acids Res. **14**: 9549-9559.
Climie, S.C., and Friesen, J.D., 1988, J. Mol. Biol. **263**: 15166-15175.
Dabbs, E.R., 1991, Biochimie **73**: 639-646.
Dognin, M.J., and Wittmann-Liebold, B., 1980, Eur. J. Biochem. **112**: 131-151.
Elhag, G.A., and Bourque, D.P., 1992, Biochemistry **31**: 6856-6864.
Franzetti, B., Zhou, D.X., and Mache, R., 1992, Nucl. Acids Res. **20**: 4153-4157.
Gantt, J.S., 1988, Curr. Genet. **14**: 519-528.
Gantt, J.S., Baldauf, S.L., Calie, P.J., Weeden, N.F., and Palmer, J.D., 1991a, EMBO J. **10**: 3073-3078.
Gantt, J.S., Gupta, A., and Thompson, M.D., 1991b, in:"The Translational Apparatus of Photosynthetic Organelles", Mache, R., Stutz, E., and Subramanian, A.R., eds, pp. 207-213, Springer, Berlin.
Giese, K., and Subramanian, A.R., 1989, Biochemistry **28**: 3525-3529.
Giese, K., and Subramanian, A.R., 1990, Biochemistry **29**: 10562-10566.
Green, G.A., and Weil, J.H., 1989, Plant Mol. Biol. **13**: 727-730.
Hanley, B.A., and Schuler, M.A., 1988, Nucl. Acids Res. **16**: 7159-7176.
Hardy, S.J.S., 1975, Mol. Gen. Genet. **140**: 253-274.
Hariharan, N., Kelley, D.E., and Perry, R.P., 1989, Genes & Dev. **3**: 1789-1800.
Liljas, A., Kirsebom, L.A., and Leijonmark, M., 1986, in:"Structure, Function and Genetics of Ribosomes", Hardesty, B., and Kramer, G., eds, pp. 379-390, Springer, Berlin.

Lütke, H.A., Chow, K.C., Mickel, F.S., Moss, K.A., Kern, H.F., and Scheele, G.A., 1987, EMBO J. **6**: 43-48.

Mager, W.H., and Planta, R.J., 1991, Biochim. Biophys. Acta **1050**: 351-360.

Möller, W., and Maassen, J.A., 1986, in:"Structure, Function and Genetics of Ribosomes", Hardesty, B., and Kramer, G., eds, pp. 379-390, Springer, Berlin.

Ohto, C., Torazawa, K., Tanaka, M., Shinozaki, K., and Sugiura, M., 1988, Plant Mol. Biol. **11**: 589-600.

Phua, S.H., Srinivasa, B.R., and Subramanian, A.R., 1989, J. Biol. Chem. **264**: 1968-1971.

Schmidt, J., Herfurth, E., and Subramanian, A.R., 1992, Plant Mol. Biol. **20**: 459-465.

Sibold, C., and Subramanian, A.R., 1990, Biochim. Biophys. Acta **1050**: 61-68.

Smooker, P.M., Schmidt, J., and Subramanian, A.R., 1991, Biochimie **73**: 845-851.

Subramanian, A.R., 1975, J. Mol. Biol. **95**:1-8.

Sugiura, M, 1992, Plant Mol. Biol. **19**: 149-168.

Takemura, M., Oda, K., Yamato, K., Ohta, E., Nakamura, Y., Nozato, N., Akashi, K., and Ohyama, K., 1992, Nucl. Acids Res. **20**: 3199-3205.

Thompson, M.D., Jacks, C.M., Lenvik, T.R., and Gantt, J.S., 1992, Plant Mol. Biol. **18**: 931-944.

Walleczek, J., Schüler, D., Stöffler-Meilike, M., Brimacombe, R., and Stöffler, G., 1988, EMBO J. **7**: 3571-3576.

Weglöhner, W., and Subramanian, A.R., 1992, Plant Mol. Biol., in press.

Yokoi, F., and Sugiura, M., 1992, FEBS Lett. **308**: 258-260.

Zhou, D.-X., and Mache, R., 1989, Mol. Gen. Genet. **219**: 204-208.

THE SPINACH PLASTID RIBOSOME: PROTEIN PROPERTIES AND ASPECTS OF RIBOSOME BIOSYNTHESIS

Régis Mache, Dao-Xiu Zhou, Bruno Franzetti, Thierry Lagrange, Silva Lerbs-Mache and Cordelia Bisanz-Seyer

Laboratoire de Biologie Moléculaire Végétale
CNRS and Université Joseph Fourier
BP 53, F-38041 Grenoble cedex

Plastid ribosomes consist of components which have generally kept a high degree of homology with their prokaryotic ancestors. All types of rRNAs (Kössel, 1991) and most of the ribosomal proteins (for review see Mache, 1991; Subramanian et al., 1991) are highly conserved. However plastid ribosomes have also plastid-specific peculiarities. One is the presence of ribosomal proteins with no eubacterial counterparts (Gantt, 1988; Zhou and Mache, 1989; Johnson et al., 1990). Second, several ribosomal proteins consist of a homologous region which is extended by NH_2- and COOH-terminal regions which have no eubacterial (or eukaryotic) ribosomal counterpart (Zhou et al., 1989; Smooker et al., 1990; Martin et al., 1991). In a few cases higher order structure of ribosomal components has been studied. It also reveals similarity with the prokaryotic models but with differences as well. For example, in collaboration with Ehresman's group, we have studied the higher-order structure of the spinach chloroplast 5S rRNA by using enzymatic and chemical probes (Romby et al., 1989). A Y-shape structure has been proposed with helices II and V not far from coaxiality and with no tertiary interactions between the different domains of the RNA. Several non canonical interactions are present in the loop E region, which are responsible for its intrinsic conformation. Interestingly, for E. coli a different loop E conformation has been proposed (Zhang and Moore, 1989) which is probably due to differences in the nucleotide sequences. Thus, the plastid ribosome, considered at the structural level, has significant differences from the E. coli ribosome.

The localization of the genes coding for the chloroplast r-proteins in two different genetic compartments makes the plastid ribosome an attractive model for the study of gene expression. About one third of these proteins is encoded in the chloroplast genome (Dorne et al., 1984; Sugiura, 1991) and the remaining two thirds are encoded in the nuclear genome. It is therefore interesting to elucidate the mechanisms which control the coordinated synthesis of the ribosomal components in the plastids and in the nucleus. Another important aspect is the control of the level of plastid ribosomes in different plant cells and tissues. It has been known for many years that etiolated leaves grown in the absence of light contain etioplasts which contain a large number of ribosomes. In contrast, root cells, which also develop in the

The Translational Apparatus, Edited by K.H. Nierhaus
et al., Plenum Press, New York, 1993

absence of light, contain a fewer plastid ribosomes. Questions concerning the influence of the external environement, the influence of cell types and of tissues on plastid ribosome accumulation are completely unsolved at present.

With respect to these problems, we will review selected aspects of the structure, function and synthesis of some components of higher plant plastid ribosomes that have been investigated in our laboratory.

STRUCTURE AND FUNCTION OF PLASTID RIBOSOMAL PROTEINS

The Primary Structure of the Ribosomal Protein CS1

We have characterized the ribosomal protein CS1 of spinach chloroplasts (Franzetti *et al.*, 1992). This protein is homologous to the *E. coli* r-protein S1 which has been well characterized at the structural and functional level (Subramanian, 1983; Boni *et al.* 1990). cDNA clones coding for a S1-like ribosomal protein (named CS1) were isolated. They represent 0.04% of the total number of recombinants showing that mRNAs coding for this protein are present in low abundance in young green plantlets. The protein encoded by the cDNA contains a transit peptide which is necessary to transport the cytoplasmically synthesized protein into the chloroplast. The small subunit ribosomal protein which corresponds to the CS1 cDNA clone has been idendified by NH_2-terminal amino acid sequencing. The mature CS1 has a molecular size of 40 kDa. It possesses NH_2- and COOH-terminal extensions besides a central core (residue 51 to 301) homologous to the *E. coli* S1. The homologous core would have a molecular size of about 28 kDa. The conserved region of the CS1 molecule is therefore considerably smaller than the *E. coli* S1 (61 kDa).

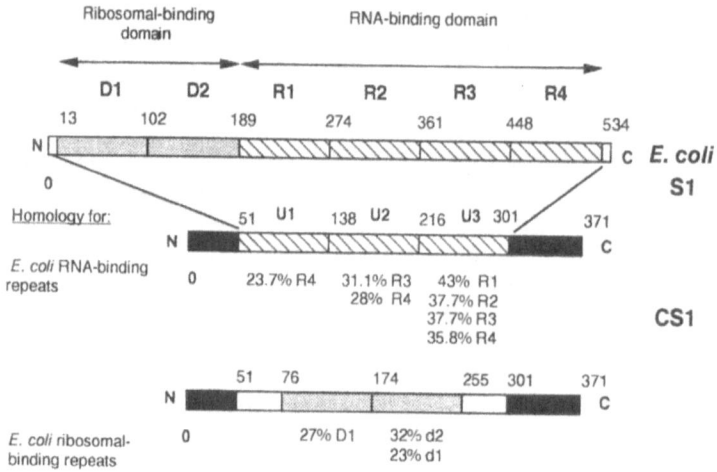

Figure 1. Schematic representation of homologous domains in the chloroplast CS1 and the *E. coli* S1 ribosomal proteins. Striped boxes correspond to repetitive R elements present in the RNA-binding domain of S1 and in their CS1 homologues; grey boxes correspond to the D ribosome-binding domain of S1 and to the homologous regions in S1; black boxes correspond to chloroplast-specific extensions. The CS1 core is divided into 3 units (U1 to U3). Percentage of homology and numbering of amino acid residues are reported. (From Franzetti *et al.*, 1992, with permission).

The comparison of the primary structure of the CS1 protein relative to the evolutionarily distant *E. coli* S1 is particularly interesting. S1 contains six internal repeats with variable degree of homology with each other and which are present in two different domains. The NH_2-terminal domain is responsible for the attachment of the protein to the small ribosomal subunit. The large COOH-terminal domain is involved in the binding of the mRNA to the ribosome (Subramanian, 1983). The comparison of the CS1 primary structure to that of S1 is schematically represented in Figure 1. The CS1 protein has conserved 3 of the repeats which are part of the RNA binding domain. The highest degree of conservation is found for the U3 repeat which has 43% homology to the R1 repeat of S1. The three U repeats in CS1 have the same length as their homologous R repeats in S1 suggesting that these structures are functionally important. The D1 and D2 elements of the S1 ribosome binding domain are not conserved in CS1 as a separate domain in continuity with the R repeat elements. However, by analysing the protein constituents of the chloroplast 30S ribosomal subunit, we found that the ribosomal protein CS1 was present in a stochiometic amount *i. e.* the CS1 is firmly attached to the small ribosomal subunit. Therefore, it should contain a ribosome binding domain although it is not conserved relative to S1. This domain could be located in the homologous core or in the unique peptide extensions.

We have also characterized a tobacco cDNA coding for the CS1 ribosomal protein (unpublished data). The protein is 89% homologous to the spinach protein. This degree of conservation is about 10 % higher than for other nuclear or chloroplast encoded ribosomal proteins known in the two higher plants. It is interesting to note that the U3 element of the spinach CS1, which is the most conserved with respect of the R repeat elements in S1, is almost 100% conserved in tobacco (Figure 2). This observation supports the idea that the U3 element is involved in the RNA binding function. The high degree of conservation of the CS1 ribosomal protein of the two higher plants suggests that this protein has an important function in the plastid translational apparatus. The same correlation holds true for S1 which is highly conserved within the group of Gram-negative bacteria (Schnier and Feist, 1985; Schneier *et al.*, 1988).

A

```
R1  189-LQEGMEVKGIVKNLTDYGAFVDLGGVDGLLHITDMAWKRV         E.coli
U3  216-.GI.SV.L.T.QS.KP....I.I..IN....VSQISHD..          Plastid 1
U3      .GI.SV.T.T.QS.KP....I.I..IN....VSQISHD..          Plastid 2


R1  KHPSEIVNVGDEITVKVLKFDRERTRVSLGLKQLGEDPWVAIAK-272     E.coli
U3  SDIATVLQP..TLK.MI.SH....G....ST.K.EPT.GDM.RN-300     Plastid 1
U3  SDIATVLQP..TLK.MI.SH....G....ST.K.EPT.GDM.RN         Plastid 2
```

B

```
 95-WITLEKAY EDAETVTGVINGKVKGGFTVELDGIRAFLPGSLVDVRPV-141   E. coli
 93-..E.T.IWRS.QNL.K.F.LNS....YA.AIA.YI....K..LRS.K.-140   Mito.
132-.ERCRQLQA.. VV.K.K.V.AN...VVALVE.LRG.V.F.QISSKSS-179   Plastid
```

Figure 2. Most conserved elements in different S1-like ribosomal proteins.(A), alignment of the R1 repeat of *E. coli* S1, of the unit U3 of the CS1 ribosomal protein from spinach chloroplast (plastid 1) and of the unit U3 from tobacco chloroplast (plastid 2). (B), fragments most conserved in S1 of *E. coli*, MS1 of *Marchantia polymorpha* mitochondria (Mito) and CS1 of *Spinacia oleracea* chloroplast (Plastid). Numbers indicate the positions of residues in the amino acid sequence of the mature protein.

Very few S1-like proteins are known at present: besides the three S1-like proteins which have been characterized at the sequence level in Gram-negative bacteria only that of spinach chloroplasts (and tobacco chloroplasts, unpublished data) and that of *Marchantia polymorpha* mitochondria (Oda *et al.,*1992) are known. *Marchantia polymorpha* is a lower plant which diverged from higher plants more than 400 million years ago. With 270 residues the mitochondrial S1 (here named MS1) is even shorter than the chloroplast CS1 (371 residues) with respect to *E. coli* S1 (557 residues). In spite of its intermediate position in the evolutionary scale the mitochondrial MS1 is less conserved than the chloroplast CS1. MS1 contains a 176-residue region which is 31% homologous to the S1 ribosome binding domain. This region is followed by a non homologous 94-residue COOH-terminal extension. A potential RNA binding domain should be located in the homologous region. The most conserved region in the three proteins is shown in Figure 2. It represents a *ca* 50-residue fragment related to the U2 repeat of the chloroplast CS1. It would be interesting to know whether this chloroplast CS1 fragment is involved in the RNA recognition or in the ribosome binding domain.

Poly(A) Binding Properties of the CS1 Ribosomal Protein

In spite of its smaller size the CS1 ribosomal protein possesses a general RNA binding property which is very similar to that of the *E. coli* S1 (Franzetti *et al.,* 1992). We have shown by UV cross-linking that in the chloroplast 30S ribosomal subunit, only the CS1 is in close contact with the mRNA. Interestingly, the small subunit ribosome binds to the sense mRNA but not to the anti-sense mRNA. This observation together with toe-print experiments showed that the 30S ribosomal subunit is positioned on the translation initiation region of the sense mRNA. The CS1 ribosomal protein purified after overproduction in *E. coli* maintains its capacity to bind to the sense mRNA. Moreover, the purified CS1 preferentially binds to poly(A). This is in contrast to the *E. coli* S1 ribosomal protein which has a high affinity for poly(U) (Subramanian, 1983). This specific poly(A) binding property has certainly a biological meaning. The chloroplast genome of higher plants is AT rich and many stretches of poly (A) are present in regions corresponding to translational starts. These stretches are probably important for the control of the translational initiation efficency. Another idea is based on the fact that the CS1 protein is present in a free state in the stroma of chloroplasts (Zhou and Mache, 1989). Thus, it might act as a translational repressor by interacting with the poly(A) stretches present within coding frames.

Function of the CL22 Ribosomal Protein

We have described the functional similarities and particularities of the CS1 ribosomal protein. In a few cases, it has been shown that a chloroplast ribosomal protein can substitute its homologue in the *E. coli* ribosome (Liu *et al.*, 1989; Giese and Subramanian, 1991). However, this "functional" similarity can surely not generalized to all chloroplast ribosomal proteins. For instance, we have shown that the spinach chloroplast CL22 (homologous to the *E. coli* L22) and the CS-L12 (to which the corresponding eubacterial protein, if it exists, has not yet been determined) bind specifically and independently to the chloroplast 5S rRNA (Toukifimpa *et al.,* 1989). The binding sites of these two proteins fit with the proposed tertiary model. In contrast, the ribosomal protein L22 in *E. coli* has no RNA binding activity. Three other ribosomal proteins (L5, L18 and L25), instead of two in the chloroplast ribosome, bind to the 5S rRNA. Interestingly, the CL22 shares with its *E. coli* homologue the property of binding to erythromycin. suggesting that at least one domain of the chloroplast protein occupies the same place in the 50S ribosomal subunit as its bacterial counterpart (Carol *et al.*, submitted). But the chloroplast CL22 protein has another function which is the binding of the 5S rRNA suggesting at the same time another arrangement within the ribosome.

Besides the properties of plastid ribosomal proteins which are homologous to bacteria, it would be interesting to investigate the function of chloroplast-specific ribosomal proteins such as the CS22 protein which has no homologous counterpart in *E. coli*. This 30S ribosomal protein is present in a free state in a relatively large amount in the chloroplast stroma and might have a specific function (Zhou and Mache, 1989).

BIOSYNTHESIS OF PLASTID RIBOSOMES

Differential Accumulation of Ribosomes in Different Plastid Types

Proplastids are present in meristematic tissues and differentiate into different plastid types according to tissues and environmental conditions. Non-green plastids (amyloplasts, leucoplasts, chromoplasts) are derived from proplastids or from the conversion, reversible or not, of fully differentiated chloroplasts. Ribosomes have been detected in almost all types of plastids by biochemical studies or by electron microscopic observations. Etioplasts in dark-grown leaves and chloroplasts in light-grown leaves possess a large amount of ribosomes whilst only a few ribosomes have been detected in proplasts (Bartels and Weier, 1967), during chromoplast (Piechulla *et al.*, 1985) and in amyloplasts differentiation (Aguettaz *et al.*, 1987). Ribosomes are absent in mature leucoplasts (Carde, 1984), in the amyloplasts of vegetative cells of maize pollen (Monéger *et al.*, 1992) and in the fully differentiated chromoplasts of the tomato red fruit (Piechulla *et al.*, 1985), probably because these organelles have reached an irreversible stage of development. The general occurrence of ribosomes in different plastid types is indeed not surprising since a basal level of translation is necessary to express the genes encoded in the plastid genome in the course of plastid differentiation and possibly during the dedifferentiation. The observation of a large difference in the amount of plastid ribosomes according to plastid types raises the problem of the mechanisms involved in the control of such differential synthesis and accumulation of ribosomes. To study this problem we have characterized several genes coding for components of the plastid ribosome.

Organization of Three Nuclear Genes Coding for Chloroplast Ribosomal Proteins

The genes coding for the ribosomal proteins CS1, CS22 and CL21 have been isolated and sequenced. All of these genes are present in one copy per haploid genome (Martin *et al.*, 1991; Bisanz-Seyer and Mache 1992, Franzetti *et al.* 1992a). The three genes contain 4 to 5

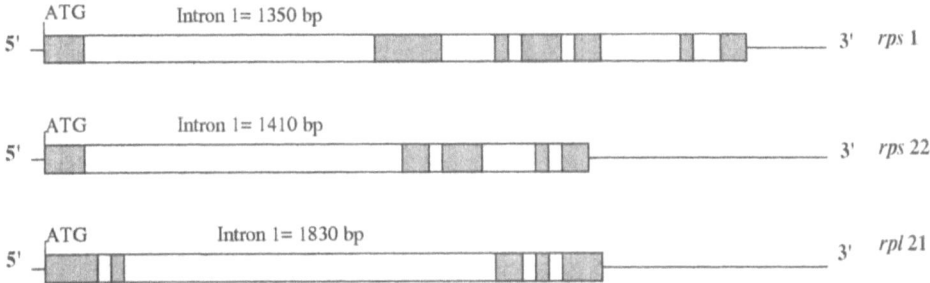

Figure 3. Organization of the nuclear genes coding for the chloroplast ribosomal proteins CS1 (*rps*1), CS22 (*rps*22) and CL21(*rpl*21). Coding sequences are represented by grey boxes and introns by white boxes.

introns (Lagrange *et al.*, Figure 3). The first intron is much larger than most other known plant introns. The significance of this observation is not known at present.

Expression of the Nuclear Genes Coding for the Ribosomal Proteins CS1 and CL21

The difference in the number of ribosomes in relation to plastid types is in accordance with the results obtained by northern analysis of RNA isolated from roots or from the green shoots (leaves and cotyledons) of young plantlets probed with DNA fragments prepared from different cDNA clones. The amount of CL21 mRNA in cotyledon/leaf cells is 7-10 times higher than in root (amyloplast/proplast) cells. The ratio of leaf mRNA *versus* root mRNA is even larger in the case of the *rps*1 mRNA (more than 20 to 1). These results were quantified relatively to a constant amount of cytoplasmic 25S rRNA. They strongly suggest that the level of transcripts is in relation to the level of plastid ribosomes and that regulation operates at the transcriptional level. This assumption is supported by the observation of a positive correlation between the mRNA level and the protein level, at least for some of the nuclear-encoded plastid ribosomal proteins (Bisanz-Seyer *et al.*, 1989). Interestingly, the level of *rps*1 and of *rpl*21 transcripts is the same in dark or light grown cotyledons (Franzetti *et al*, 1992; Lagrange *et al.*, submitted). These results suggest that these genes are regulated in a tissue-specific manner.

The promoter regions of the *rps*1 and *rpl*21 genes have been analysed in detail. Like housekeeping genes, they have no TATA box and possess a polypyrimidine rich sequence close to the transcription initiator. These characteristics are also found in the ribosomal protein genes of animal cells. Surprisingly, two transcription starts are revealed by S1 mapping of the 5' end of transcripts hybridized to corresponding DNA or by reverse transcription of a primer hybridized to mRNA. In both cases, the mRNAs were isolated from cotyledon or leaf tissues. The presence of two independent initiator regions (P1 and P2) has been verified by transient expression in *Arabidopsis* protoplasts of deleted-promoter fragments coupled to the chloramphenicol acetyl transferase gene. Interestingly, only one initiator region (P2) is used in amyloplast/proplasts of root cells. We conclude that a low level constitutive transcriptional activity initiates at P2 in all plastid types and that a large transcriptional activity initiates at P1 in the green tissues. These results verify that the *rps*1 and *rpl*21 genes are regulated at the transcriptional level (Lagrange *et al.*, submitted).

It was interesting to determine the *cis*- and *trans*-acting elements which control the differential expression of the *rps*1 and *rpl*21 genes by usage of two different promoters. The promoter sequence up to about -400 bp of each of the two genes was used for footprinting assays. Nuclear factors which bind to homologous specific *cis*- elements in the two genes were determined (Zhou *et al.*, 1992; Lagrange *et al.*, in preparation). These *cis*-elements are schematically represented in Figure 4. Three different types of *cis*-elements were identified. First, a GT1-like sequence present in the two promoters might be related to the previously characterized GT-1 sites of several light-regulated genes. It is possible that a fine regulation through light exists for these GT-rich sequences yet no significant difference in the accumulation of CS1 or L21 mRNAs was observed in dark-grown (etioplast) *versus* light-grown (chloroplast) cotyledons. The second *cis*-element that is in common for both the *rps*1 and the *rpl*21 genes is the S2F site. The nuclear factor S2F is likely leaf-specific. The function of S2F is presently unknown.

The S1F site (Figure 4) was not previously recognized and was therefore studied in more detail in the *rps*1 gene. The specificity of S1F interaction with a spinach leaf nuclear factor designated S1F was demonstrated by using site-directed mutagenesis. The S1F site is

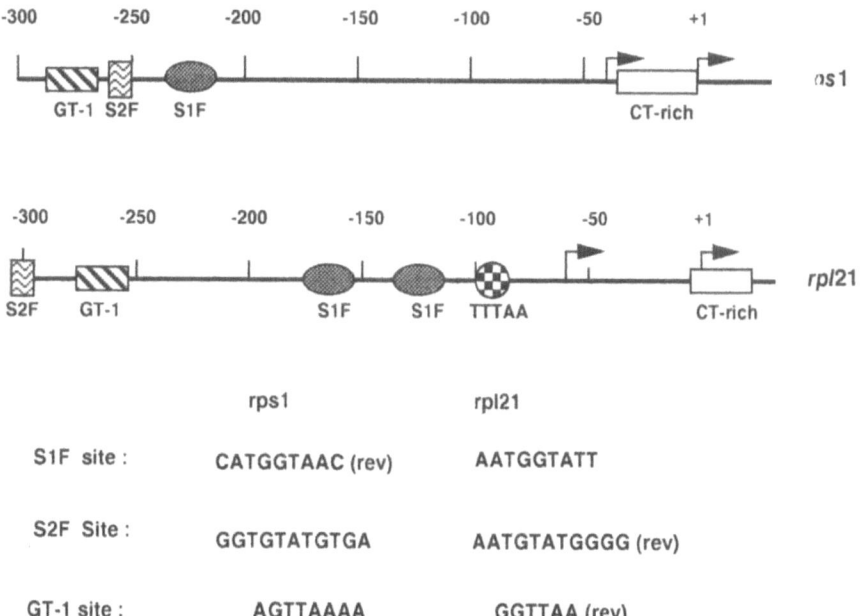

	rps1	rpl21
S1F site :	CATGGTAAC (rev)	AATGGTATT
S2F Site :	GGTGTATGTGA	AATGTATGGGG (rev)
GT-1 site :	AGTTAAAA	GGTTAA (rev)

Figure 4. Organization of the rps1 and of the rpl21 promoters. The two transcription start sites of each gene are indicated by bent arrows. The polypyrimidine stretch in the initiator region is indicated by a white box. The position of the *cis*-elements identified by foot-printing and the protected sequences are reported.

not limited to chloroplast ribosomal protein genes. It is also conserved in the promoter region of several plastid-related genes, such as the *rbc*S gene coding for the small subunit of ribulose bisphosphate carboxylase and the *cab* gene coding for a chlorophyll a/b binding protein (*cab*).

The regulatory function of the S1F site was investigated by transient expression assays using protoplast of soybean suspension cells. It has been shown that the S1F binding site is a negative element down-regulating the transcription of the *rps*1 gene. Interestingly, the soybean suspension cells contain amyloplasts and not chloroplasts. Consequently, the negative function of S1F could be active in root cell amyloplasts and contribute to the low level of expression of the *rps*1 gene in these cells. Experiments are in progress to verify this hypothesis.

The high level of the *rps*1 and *rpl*21 mRNAs in leaf cells relative to root cells is most probably the result of an activation of the promoter activity of each of these genes. The function of the different *cis*-elements observed by foot-printing has to be determined. In particular, it will be worthy to search for the presence of leaf specific factors interacting with sequence-specific elements of the promoter. A first approach has been made by analyzing the promoter regions involved in differential expression of the *rps*1 gene using gel shift assays. It has been found that a *ca* 200 bp DNA fragment upstream of the second initiator region (P2) interacts with nuclear proteins which are only present in leaves (Franzetti *et al.*, 1992).

Differential Transcription in Root Cells and in Leaf Cells of the Chloroplast Encoded 16S rRNA Gene

The different amounts of plastid ribosomes in root and leaf cells results from a differential expression of the nuclear-encoded and of the chloroplast-encoded genes. The effect of light on chloroplast gene expression has been extensively investigated (see Gruissem, 1989) but much less attention has been devoted to the control of plastid differentiation. Deng and Gruissem (1988) have shown that the relative transcriptional activities of ten different genes including two genes required for the translational apparatus (*rrn* and *rpl2*) are very similar in plastids of different plant organs. They concluded that the plastid genome is constitutively transcribed and that most plastid genes are controlled primarily on the post-transcriptional and translational level. In contrast to this conclusion, we found that the spinach 16S rRNA gene expression is regulated primarily on the transcriptional level during plastid development. The activation of 16S rRNA transcription in cotyledon and leaf tissue relative to root tissue is achieved by transcription initiation from a cotyledon/leaf specific promoter (Baeza *et al.*, 1991; Iratni *et al.*, submitted). This promoter is not used in root cells.In root tissue, a basic constitutive expression of the 16S gene occurs. The start of this basic transcript is much further upstream and includes the tRNA^{Val} gene which preceeds the 16S rDNA gene. A schematic representation of the different transcription patterns of the 16S rRNA in root and leaf cells is shown in Figure 5.

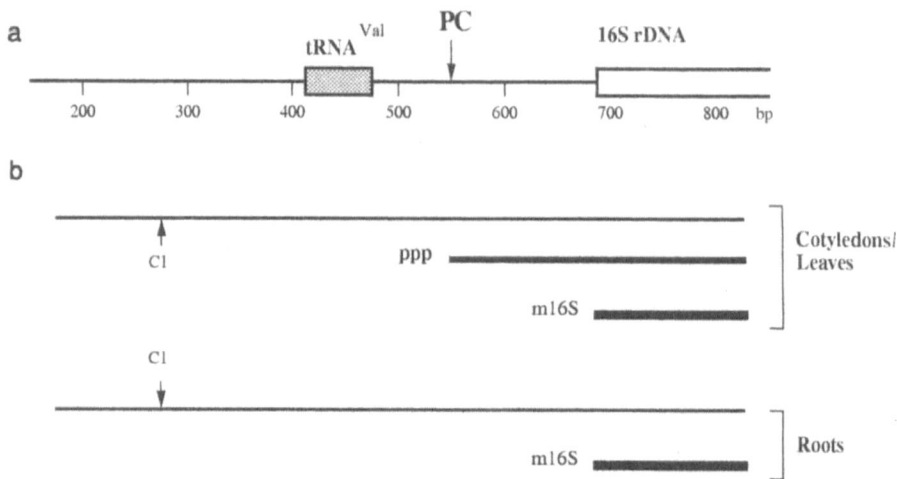

Figure 5. Schematic representation of the different 16S rRNA transcripts in cotyledon/leaf and root cells. The leaf specific promoter (PC) is indicated. C1 represents a processing site of the large transcript which includes the tRNA^{Val}.

CONCLUSION

We have characterized the nuclear genes coding for the ribosomal proteins CS1, CL21 and CS22. The protein CS1 can be distinguished from its *E. coli* homologue by its smaller size, resulting partially from the lost of the region that corresponds to the S1 ribosome binding domain. The CS1 is also characterized by a strong preferential affinity for poly(A) sequence. The results obtained on the expression of the nuclear encoded *rps*1 and *rpl*21 genes and on the transcription of the 16S rRNA genes bring some light on the mechanisms which are responsible for the differential accumulation of plastid ribosomes in root and leaf cells.

REFERENCES

Aguettaz, P., Seyer, P., Pesey, H.,and Lescure, A.-M., 1987, Relations between the plastid dosage and the levels of 16S rRNA and *rbc*L gene transcripts during amyloplast to chloroplast change in mixotrophic spinach cell suspensions. *Plant Mol. Biol.* 8, 169-177.

Baeza, L., Bertrand, A., Mache, R., and Lerbs-Mache, S., 1991, Characterization of a protein binding sequence in the promoter region of the 16S rRNA gene of the spinach chloroplast genome. *Nucleic Acids Res.*, 19, 3577-3581.

Bartels P. G. and Weier T. E., 1967, Particle arrangements in proplastids of *Triticum vulgare* L. seedlings. *J. Cell Biol.*, 33, 243-253

Bisanz-Seyer, C., Li, Y.-F., Seyer, P., and Mache, R., 1989, The components of the 70S ribosome are not accumulated synchronously during the early development of spinach plants. *Plant Mol. Biol.* 12, 201-211.

Bisanz-Seyer, C. and Mache, R., 1992, Organization and expression of the nuclear gene coding for the plastid-specific S22 ribosomal protein from spinach. *Plant Mol. Biol.* 18, 639-651.

Boni, I. V., Isaieva, D. M., Musichenko, M. L., and Tzareva, N. V.,1990, Ribosome-messenger recognition: mRNA target sites for ribosomal protein S1, *Nucleic Acids Res.* 19:155-162

Carde, J.-P., 1984, Leucoplasts: a distinct kind of organelles lacking typical 70S ribosomes and free thylakoids. *Eur. J. Cell Biol.* 34:18-26.

Deng, X.-W. and Gruissem, W., 1988, Constitutive transcription and regulation of gene expression in non-photosynthetic plastids of higher plants. *EMBO J.* , 7:3301-3308

Franzetti, B., Zhou, D.-X., and Mache, R., 1992a, Structure and expression of the nuclear gene coding for the plastid CS1 ribosomal protein from spinach. *Nucl. Acids Res.* 20:4153-4157.

Franzetti, B. Carol, P., and Mache, R., 1992b, Charcterization and RNA-binding properties of a chloroplast S1-like ribosomal protein. J. Biol. Chem. 267, 19075-19081.

Gantt, S., 1988, Nucleotide sequences of cDNAs encoding four complete nuclear-encoded plastid ribosomal proteins. *Curr. Genet.* 14:519-528.

Giese, K. and Subramanian, A.R., 1991, Expression and functional assembly into bacterial ribosomes of a nuclear-encoded chloroplast ribosomal protein with a long NH2-terminal extension. *FEBS Lett.* 288:72-76.

Gruissem, W., 1989, Chloroplast gene expression: how plants turn their plastids on. *Cell* 56:161-170.

Johnson, C. H., Kruft V., and Subramanian, A. R., 1990, Identification of a plastid-specific ribosomal protein in the 30S subunit of chloroplast ribosomess and isolation of the cDNA clone encoding its cytoplasmic precursor. *J. Biol. Chem.* 265:12790-12795.

Kössel, H., 1991, Structure and expression of rRNA genes *in*: "The Translational Apparatus of Plastid Organelles", Mache, R., Stutz, E., and Subramanian, A.R. eds, NATO ASI ser. 51, Springer-Verlag, Berlin, pp. 1-17.

Liu X. Q., Gillham, N.W., and Boynton, J.E., 1989, Chloroplast ribosomal protein gene *rps*12 of *Chlamydomonas reinhardtii*. *J. Biol. Chem.* 264:16100-16108.

Mache, R., 1990, Chloroplast ribosomal proteins and their genes. *Plant Science* 72:1-12.

Martin, W., Lagrange, T., Li, Y.-F., Bisanz-Seyer, C., and Mache, R., 1990, Hypothesis for the evolutionary origin of the chloroplast ribosomal protein L21 of spinach. *Curr. Genet.* 18:553-556.

Monéger, F., Mandaron, P., Niogret, M.-F., Freyssinet, G., and Mache, R., 1992, Expression of chloroplast and mitochondrial geness during microsporogenesis in maize. *Plant Physiol.* 99:396-400.

Oda, K., Yamato K., Ohta E., Nakamura, Y., Takemura, M., Nozato N., Akashi, K., Kanegae T., Ogura, Y., Kohchi T., and Ohyama K., 1992, Gene organization deduced from the complete sequence of liverwort *Marchantia polymorpha* mitochondrial DNA. A primitive form of plant mitochondrial genome. *J. Mol. Biol.* 223:1-7.

Piechulla, B., Chonoles Imlay, K. R. a Gruissem, W., 1985, Plastid gene expression during fruit ripening in tomato. *Plant Mol. Biol.* 5:373-384.

Romby, P., Westhof, E., Toukifimpa, R., Mache, R., Ebel, J.-P., Ehresmann, C., and Ehresmann, B., 1988, High-order structure of chloroplastic 5S ribosomal RNA from spinach. *Biochemistry* 27:4721-4730.

Schnier, J. and Faist, G., 1985, Comparative studies on the structural gene for the ribosomal protein S1 in ten bacterial species. *Mol. Gen. Genet.* 200:476-481.

Schnier, J., Thamm, S., Lurz, R., Hussain, A., Faist, G., and Dobrinski, B. , 1988, Cloning and characterization of a gene from *Rhizobium melilotii 2011* coding for ribosomal protein S1. *Nucleic Acids Res.* 16:3075-3089.

Smooker, P.M., Kruft, V., and Subramanian, A. R., 1990, A ribosomal protein is encoded in the chloroplast DNA in a lower plant but in the nucleus in Angiosperms. *J. Biol. Chem.* 265:16699-16703.

Sugiura, M., 1991, Chloroplast genes coding for ribosomal proteins in land plants, *in*: "The Translational Apparatus of Plastid Organelles", Mache, R., Stutz, E., and Subramanian, A.R.,eds, NATO ASI ser. 51, Springer-Verlag, Berlin, pp. 59-69.

Subramanian, A. R., 1983, Structure and functions of ribosomal protein S1. *Prog. Nucleic Acids Res. Mol. Biol.* 28:101-142.

Subramanian, A. R., Stahl, D., and Prombona, A., 1991, Ribosomal proteins, ribosomes, and translation in plastids, *in*: "The Molecular biology of Plastids", Bogorad, L., and Vasil, I. K., eds, Academic Press, pp. 191-215.

Toukifimpa, R., Romby, P., Rozier, C., Ehresmann, C., Ehresmann, B., and Mache, R., 1989, Characterization and footprint analysis of two 5S rRNA binding proteins from spinach chloroplast ribosomes. *Biochemistry* 28:5840-5846.

Zhou, D-X and Mache, R., 1989, Presence in the stroma of chloroplast of a large pool of one ribosomal protein not structurally related to any *E. coli* ribosomal protein. *Mol. Gen. Genet.* 219:204-208, and *Erratum*,1990, 223:167.

Zhou, D.-X., Quigley, F., Massenet, O. and Mache, R., 1989, Cotranscription of the S10- and spectinomycin-like operons in spinach chloroplasts and identification of three of their gene products. *Mol. Gen. Genet.* 216:439-445.

Zhou, D.-X., Li, Y.-F., Rocipon, M., and Mache, R., 1992, Sequence-specific interaction between S1F, a spinach nuclear factor, and a negative *cis*-element conserved in plastid-related genes. *J. Biol. Chem.* 267 (in press)

STRUCTURE AND FUNCTION OF MAMMALIAN MITOCHONDRIAL RIBOSOMES

Thomas W. O'Brien[1], Nancy D. Denslow[1], Wesley H. Faunce[1],
John C. Anders[1], Jingo Liu[1], and Bonnie J. O'Brien[2]

[1]Department of Biochemistry and Molecular Biology
 PO Box 100245 Health Science Center
College of Medicine
University of Florida
Gainesville, Florida 32610-0245
[2]Interdisciplinary Center for Biotechnology Research
214 Bartram Hall
University of Florida
Gainesville, Florida 32610

INTRODUCTION

The discovery of ribosomes in mitochondria (O'Brien and Kalf, 1967) foreshadowed the surprising diversity of ribosome types in nature. The 55S ribosomes from mammalian mitochondria resemble bacterial ribosomes and eukaryotic cytoplasmic ribosomes in their main functional properties, but in terms of their fine structure and physical chemical properties, they differ unexpectedly from both these kinds of ribosomes, as well as from other kinds of mitochondrial ribosomes (O'Brien, 1977; O'Brien and Matthews, 1976; Kitakawa and Isono, 1991). Also, mammalian mitochondrial ribosomes have an intrinsic GTP binding site in the small subunit, an unprecedented occurrence in translational systems (Denslow et al., 1991).

Several properties of these ribosomes distinguish them from those of prokaryotes and the cytoplasm of eukaryotes. Having about the same mass as bacterial ribosomes, they contain scarcely half as much rRNA and nearly twice as many proteins, differences which markedly affect their sedimentation coefficient and buoyant density (O'Brien et al., 1990). In addition, the mitochondrial r-proteins are distinctive, having no closely related homologues in bacterial or eukaryotic-cytoplasmic ribosomes that have been identified by electrophoretic mobility, immunologic cross-reactivity, or limited protein sequencing (Matthews, et al., 1982; Pietromonaco, et al. 1991; O'Brien, unpublished).

The Translational Apparatus, Edited by K.H. Nierhaus
et al., Plenum Press, New York, 1993

The unusual properties of these ribosomes raise questions about their relation to other kinds of ribosomes, and their large number of proteins raises questions about their functional and structural organization, and also about the identity of individual mitoribosomal proteins that are homologous to proteins in other ribosomes. We are developing the bovine mitochondrial ribosome as a model system for mammalian mitochondrial ribosomes, in general (O'Brien, 1991). The bovine system will be used to address several questions related to the structure, function, biosynthesis and evolution of these interesting ribosomes.

Size of Mammalian Mitochondrial Ribosomes

Despite their lower sedimentation coefficient, 55S mammalian mitoribosomes are actually slightly larger than 70S bacterial (*E. coli*) ribosomes, both on the basis of particle molecular weight (2.8 x 10^6D) (Hamilton and O'Brien, 1974), and physical dimensions (deVries and Vanderkoogh-Schurring, 1973). The 28S small subunit of rat (Grohmann and Stoffler-Meilicke, 1989) and bovine (O'Brien, T. and O'Brien, B., in preparation) mitoribosomes is noticeably larger than 30S *E. coli* small subunits by electron microscopy.

We are developing a three dimensional model of the bovine mitoribosomal small subunit to provide a structural framework for analysis and interpretation of results from structural, functional and affinity labeling studies. We have produced a 3D model of the small subunit based on electron microscopy and the spatial coordinates of conserved stem/loop structures in *E. coli* ribosomes (Stern, et al., 1988). This scaled model provides surface coordinates for the subunit space envelope. Coordinates for the backbone phosphate atoms of mammalian 12S mitoribosomal RNA, as modeled by (Stern, et al., 1988), have been used to model the individual helical stems in bovine 12S mitoribosomal RNA to provide a 3 D working model of the 28S subunit.

Protein Content of Bovine Mitochondrial Ribosomes

It follows that a ribosome somewhat larger than bacterial ribosomes, but containing considerably less RNA (Table 1), should have a correspondingly higher protein content. Indeed, bovine mitochondrial ribosomes contain as many as 52 and 33 proteins in the large and small subunits, respectively (Matthews et al., 1982). These mitochondrial ribosomes thus contain considerably more individual proteins than *E. coli* ribosomes (53) and nearly as many as are in eukaryotic cytoplasmic ribosomes . The molecular weights of the large subunit proteins range from 8.8 x 10^3 to 49 x 10^3, averaging 21.9 x 10^3 (Tables 2 and 3). Thus, the 2-fold greater quantity of total protein found in the mitochondrial ribosomes is due partially to a somewhat larger size of the individual proteins and partially to the larger number of proteins. The sum of the molecular weights of the 52 proteins in the large subunit is 1.14 x 10^6, a value which agrees well with the protein content as calculated from the buoyant density of this subunit (1.10 x 10^6) or from its particle weight (1.11 x 10^6). Similarly, the total of the 33 small subunit proteins, 0.82 x 10^6, is in reasonable agreement with the protein content determined by other methods (Matthews et al.,1982).

It has been pointed out (Gutell et al., 1985) that the major deficiency of RNA in the small subunit of mammalian mitoribosomes, relative to *E. coli* 16S RNA, corresponds to the absence of 17 discrete segments present in bacterial 16S RNA

(Table 1). These missing segments range in size from 5 to 94 nucleotides. How much protein would be required to "replace" these segments? In table 1 is listed the equivalent protein mass required to fill the space of the missing RNA, calculated using the partial specific volumes for RNA and protein (Hamilton and O'Brien, 1974). The aggregate protein equivalent corresponding to the missing RNA amounts to a total mass of 174 kD, which could be contributed by "recruiting" 10 additional proteins of average mass 17 kD. It should be noted that 11 of the more significant RNA deletions correspond to volumes which could be occupied by proteins ranging in size from 7.0-27.6 kD, values well within the range of mitoribosomal proteins (Matthews, et al., 1983). Is it coincidental that the bovine mitoribosomal small subunit contains 33 proteins, 12 more than present in the small subunit of *E. coli* ribosomes?

Table 1.   RNA domains, stems and loops present in *E. coli* 16S rRNA which are absent in bovine mitochondrial small subunit 12S rRNA (Adapted from Gutell, et al. 1985).

"Missing" RNA Region	Nucleotides	Protein Equivalent (kD)
1-5	5	1.5
65-102	38	11.2
135-228	94	27.6
253-264	12	3.5
289-304	16	4.7
315-339	24	7.0
404-436	33	16.5
437-496	60	17.6
588-654	67	19.7
830-864	35	10.3
999-1042	44	12.9
1087-1098	12	3.5
1119-1154	36	10.6
1156-1183	28	8.2
1249-1289	41	12.0
1443-1459	17	5.0
1537-1542	6	1.8

Organization of Proteins in Bovine Mitoribosomes

It might be expected that the arrangement or "packaging" of the components must be fundamentally different in mammalian mitoribosomes, which have a protein: RNA ratio of 2:1, from that which is found in bacterial ribosomes (0.6). Despite their unusual composition, the gross ultrastructural features of bovine mitochondrial ribosomes resemble those of bacterial ribosomes, suggesting that the "extra" proteins in these ribosomes probably have structural roles, possibly functional roles, that are served by RNA in bacterial ribosomes.

Table 2. Properties of proteins in the large subunit of bovine mitochondria ribosomes

	M_r (kD)	Exposure Index	RNA Binding	Core Particle
L1	49.0	39		
L2	46.0	44		x
L3	44.0	43		x
L4	42.0	3		
L5	35.5	12		
L6	35.3	38		
L7	35.6	48	+ +	x
L8	33.6	20	+	x
L9	32.2	---		x
L10	31.0	30		x
L11	31.5	21	+	x
L12	29.4	23		
L13	29.0	7	+ +	x
L14	30.0	24	+ +	x
L15	28.0	13		
L16	27.9	14		
L17	26.4	27		
L18	26.5	25		x
L19	23.5	4		x
L20	21.6	2		
L21	21.1	19	+ +	
L22	20.5	17		
L23	20.2	41		
L24	19.3	9		

M_r (kD)		Exposure Index	RNA Binding	Core Particle
L25	19.2	35	x	
L26	18.8	49	+ +	
L27	18.5	22		
L28	17.5	33	+	x
L29	17.5	26		
L30	17.5	32		
L31	17.3	31		
L32	16.8	34		
L33	16.0	36		x
L34	16.5	28		
L35	15.7	42	+	
L36	16.0	---		
L37	14.5	8		
L38	14.3	15		
L39	14.2	1		
L40	14.0	11	+	
L41	13.0	16		
L42	12.6	6		x
L43	12.9	37		x
L44	13.1	46	+ +	
L45	11.9	---		
L46	11.2	18		
L47	11.0	10		
L48	10.5	5		
L49	11.2	45	+	
L50	9.9	40	+	
L51	8.8	47		
L52	8.8	29		

Molecular weights (M_r) were determined by SDS polyacrylamide gel electrophoresis (Matthews, et al., 1982). The exposure index is the rank order number of proteins organized according to their accessibility to radioiodination in intact subunits, relative to extracted proteins in urea (Denslow and O'Brien, 1984). The most "exposed" protein has an exposure index of 1. The RNA binding properties of individual proteins were assessed using a combination of methods (Piatyszek, et al., 1988). In the last column are indicated proteins predominantly present in 4 M LiCl core particles (Schieber and O'Brien, 1982).

One approach for studying the arrangement of proteins in mitochondrial ribosomes is to determine the surface exposure of individual proteins relative to each other in the subribosomal particles. To assess the relative exposure of proteins in the ribosomal subunits we measured the incorporation of ^{131}I into proteins using a surface specific reaction (Denslow and O'Brien, 1984) and

Table 3. Properties of proteins in the small subunit of bovine mitochondrial ribosomes

	M_r (kD)	Exposure Index		M_r (kD)	Exposure Index
S1	48.0	12	S18	20.5	1
S2	43.0	9	S19	10.0	---
S3	43.0	---	S20	18.8	---
S4	43.0	25	S21	18.6	13
S5	40.8	23	S22	18.0	10
S6	41.5	21	S23	16.9	22
S7	37.4	24	S24	16.0	11
S8	35.6	19	S25	15.3	14
S9	32.8	20	S26	14.0	---
S10	31.5	6	S27	14.0	7
S11	29.0	17	S28	13.0	5
S12	27.5	4	S29	13.1	15
S13	27.6	18	S30	13.0	---
S14	25.6	---	S31	11.9	8
S15	25.4	3	S32	12.5	---
S16	23.0	16	S33	10.0	2
S17	20.8	---			

compared this labeling to the incorporation of ^{125}I into dispersed and denatured r-proteins under conditions in which their labeling is expected to be proportional to their total content of iodinatible residues.

The majority of the mitochondrial r-proteins become significantly radiolabeled in this system indicating that most of these proteins have tyrosine or histidine residues exposed at the surface of the ribosome. This result suggests that the "extra" proteins in these ribosomes are not entirely buried and would seem to preclude a radically different structural organization for these ribosomes.

The individual r-proteins are listed in Tables 2 and 3 with an "exposure index", to indicate their relative exposure in the subunits. The more superficially disposed proteins in the mitoribosomal subunits (smaller index numbers) may be among the last to assemble on mitochondrial ribosomes and, as such, they are in a good position to interact with other macromolecules, including mRNA, tRNA, factors and RNA and proteins of the other subunit. The more buried proteins (larger index numbers) may serve mainly structural roles, perhaps acting as "assembly" proteins since many from this group in the large subunit bind to ribosomal RNA.

RNA Binding Proteins of the Large Subunit of Mammalian Mitochondrial Ribosomes

Attempts to study RNA-protein interactions in these mitochondrial ribosomes are hindered by the large number of proteins and their tendency to aggregate when classical approaches to ribosome assembly are followed. Consequently, we used four alternative approaches (Piatyszek, et al., 1988) to identify RNA binding proteins in the large subunit of bovine mitochondrial ribosomes: (a) binding of radiolabelled rRNA to proteins on Western blots; (b) disassembly of 39S mitoribosomal subunits by treatment with urea, to identify proteins which remain associated with the RNA; (c) binding of proteins to RNA in the presence of urea; and (d) binding of lithium-chloride extracted proteins (Schieber and O'Brien, 1982) to RNA in solution.

Results from these four approaches allowed us to identify a set of six proteins (L7, L13, L14, L21, L26, and L44) which appear to be strong RNA binding proteins. Seven additional proteins (L8, L11, L28, L35, L40, L49, and L50) were identified as secondary RNA binding proteins. RNA binding properties of the proteins in both of these sets were compared with the topographic disposition and susceptibility towards lithium-chloride extraction of the individual proteins. Proteins from the first set are good candidates for early assembly proteins since they have a high affinity for RNA, are generally founding 4 M lithium-chloride "core particles" (Schieber and O'Brien, 1982), and are among the most buried proteins in the large subunit (Table 2).

Functional Roles for Proteins in Mammalian Mitochondrial Ribosomes

Echoes from the RNA world increasingly emphasize the functional involvement of rRNA in translational mechanisms. Nevertheless, it is appropriate to consider possible functional roles for the r-proteins, as well. The large number of proteins in mammalian mitochondrial ribosomes, relative to bacterial ribosomes, raises the possibility that the "extra" proteins may be playing structural (or functional) roles served by RNA in the bacterial ribosome. Indications that some of these proteins may play a more significant role than a simple structural, "space filling", role come from the discovery of a novel GTP binding site on the small subunit of the bovine mitoribosome (O'Brien, et al., 1990; Denslow, et al., 1991). GTP binds in unit stoichiometry and with high affinity ($K_d = 15$ nM) to a site on the small subunit. This binding activity survives high salt washes, indicating that the nucleotide binds to an integral site within the subunit. The GTP binding can be competed by GDP, but not by other nucleotides, suggesting a direct functional role for GTP.

We have used a photoaffinity analogue of GTP to identify the ribosomal components associated with the GTP binding site. Photolysis of the bound analogue results in the specific labelling of a single ribosomal protein, S5 (Anders, et al., 1992).

As isolated by HPLC, S5 is aminoterminally blocked, necessitating cleavage of the protein prior to amino acid sequence analysis. Aminoterminal amino acid sequence analysis of several CNBr cleavage fragments has allowed the construction of a peptide map for S5, and provides information for the design of degenerate oligonucleotide primers used for obtaining cDNA clones for this protein. On the basis of sequence analysis covering more than 50% of the protein, S5 appears to be a unique protein, having no obvious homology to other proteins in the databases, except for consensus elements found in other GTP binding proteins. The amino terminal region of S5 contains a segment (GxxxxGKT) occurring widely in other GTP binding proteins. This element is followed elsewhere in the molecule by a novel version of the consensus sequence NKxD, known to confer guanyl binding specificity in other GTP binding proteins.

This ribosomal protein with a GTP binding site may participate directly in initiation complex formation, further indicating that these ribosomes employ a different mechanism for initiation of protein synthesis from bacterial or eukaryotic cytoplasmic ribosomes which are unable to interact directly with GTP.

Binding of mRNA to Mitochondrial Ribosomes

The mechanism by which mammalian mitoribosomes bind mitochondrial mRNAs is of special interest, since these messages lack the ribosome binding sites commonly found in prokaryotic and eukaryotic mRNAs (Denslow, et al., 1989). In prokaryotic systems, most mRNAs have a leader sequence 5' to the initiation codon containing a stretch of nucleotides that is complementary to a region near the 3' end of the rRNA in the small ribosomal subunit (Shine, J. and Dalgarno, L., 1974). This region of complementarity, which serves as a primary binding site for prokaryotic ribosomes, is absent from mammalian mitochondrial mRNAs. Instead, most of the mitochondrial mRNAs begin immediately with the initiation codon, or have only very short leader sequences of one to eight nucleotides which are not complementary to the 3' end of the small subunit rRNA. While most eukaryotic mRNAs also apparently lack these 5' complementary sequences, they do employ a 5'-terminal modification, the m^7 G(5')ppp cap, as a primary recognition signal. A cap binding protein mediates the binding of capped eukaryotic mRNAs to ribosomes, which then scan the 5'-untranslated mRNA leader for the appropriate initiation codon. Mammalian mitochondrial ribosomes must utilize different mechanisms to initiate protein synthesis since the mitochondrial mRNAs not only lack significant 5' leaders, but also have no caps on their 5' ends.

We have used triplet codons, homopolymers and heteropolymers of various lengths to define properties of the ribosomal template binding site (Denslow, et al., 1989). Presence of the initiation codon is not of itself, adequate to direct the binding of templates to mitoribosomes, since triplet codons, including AUG, do not bind to mitoribosomes under conditions supporting the binding of poly(U). We have determined that, in the absence of initiation factors and initiator tRNA, mitoribosomes do not bind codons or homopolymers (oligo(U)) shorter than about 10 nucleotides. Larger oligomers bind better, with oligomers of 30 nucleotides or greater binding as well as poly(U). It appears that a minimum of 15-30 bases are required for stable interaction between the polynucleotide template and the mitoribosomal RNA binding domain under these conditions. To determine the full extent of the RNA binding domain, we used 5' end-labeled ^{32}P-poly (U,G) in ribosome protection experiments, finding that mitoribosomes are able to protect upwards of 80 bases of bound template from RNase digestion. Thus, while the major binding interaction occurs over the span of about 30 nucleotides, mitoribosomes are able to protect about an 80 nucleotide span of the bound template.

To study the interaction of authentic mitochondrial mRNA with mitochondrial ribosomes, mRNAs isolated from HeLa mitochondria were 5' end-labeled and used in ribosome protection experiments. The mitochondrial mRNA coding for subunit II of the cytochrome oxidase complex (COII) was used in the initial experiments. Under standard conditions supporting the binding of poly(U) and poly(U,G), little mRNA was bound to ribosomes in a manner conferring protection of the 5' terminus from RNase T_1 digestion. The sizes of the major protected fragments under these conditions are 169 and 104 nucleotides long, much longer than the 80 nucleotide span expected from experiments using poly(U,G) as the template. This observation suggested that secondary structure near the 5' end of this mRNA is responsible for the protection of larger fragments, especially since such large fragments exceed the physical dimensions of the small subribosomal particle.

Secondary structure analysis of the COII mRNA by computer RNA folding programs and by single- and double-strand specific RNAses allowed us to derive a model for the first 220 nucleotides of this mRNA (Denslow, et al., 1989). The first guanosines appearing in single-stranded, RNase T_1-susceptible regions occur at positions 104 and 169, which are in fact the main cleavage sites observed in the ribosome protection experiment. The putative secondary structure also predicts that the initiation codon at the 5' terminus of the mRNA is in a stem structure, where it will have to be melted for proper binding of the mRNA and for formation of the initiation complex.

When the above ribosome protection experiment is performed in the presence of a miochondrial extract enriched in mitochondrial initiation factors and initiator tRNA, much larger amounts of protected fragments are obtained, suggesting that some component in the preparation promotes the binding of this mRNA to mitoribosomes. In addition to fragments of 169 and 104 nucleotides, a new protected fragment of 79 nucleotides is produced. The cleavage of the mRNA at position 79 in the presence of the mitochondrial extract implies that the portion of the stem near the 5' end has melted under these conditions. The factor(s) involved in this process appears to be specific for the mitochondrial system, since initiation factors and initiator tRNA from $E. coli$ neither promote binding of this mRNA to mitoribosomes nor lead to the production of a 79-base fragment. Our results suggest that mitochondrial initiation factors are required for the proper recognition and melting of the secondary structure in the 5' terminal regions of mitochondrial mRNAs, as a prerequisite for initiation of protein synthesis in mammalian mitochondria.

This conclusion is consistent with findings of Liao and Spremulli (1989), who studied the binding of a synthetic analogue of bovine COII mRNA to bovine mitochondrial ribosomes. Analysis of the ribosome protected fragments of the bound mRNA showed that the binding did not occur at any specific region of the mRNA, in the absence of added factors.

Interaction of Bacterial IF3 with Bovine Mitoribosomes

The fact that the bacterial initiation factor IF3 binds to mitochondrial ribosomes (Denslow et al., 1988; Spremulli and Kraus, 1987) suggests that a homologous factor will function in mitochondrial systems. Bacterial IF3 binds to the small subunits of bovine mitoribosomes ($K_d = 2.0 \times 10^{-7}$M), with an affinity that is equivalent to that for *E. coli* ribosomes under the same conditions ($K_d = 1.7 \times 10^{-7}$M), suggesting that the ribosomal binding site for IF3 is conserved in mitochondrial ribosomes. This binding site appears to reside on the interfacial aspect of the small mitoribosomal subunit, similar to its location on bacterial ribosomes, where it acts as an anti-association factor. These results suggest that despite the vastly different physical properties of mammalian mitoribosomes, the binding site for IF3 is conserved and has essentially the same molecular features as those involved in the binding of IF3 to bacterial ribosomes. By analogy to the bacterial system, it is expected that the mitochondrial IF3 homologue will play a role in mitoribosomal initiation complex formation, along with the mitochondrial specific initiation factor IF2 (Liao and Spremulli, 1991).

Rapid Evolution of Proteins in Mammalian Mitochondrial Ribosomes

Representative of mammalian mitoribosomes, the bovine mitoribosome contains as many as 85 different proteins (Matthews, et al., 1982), all of which are products of nuclear genes. These proteins are synthesized on cytoplasmic ribosomes (Schieber and O'Brien, 1985) and must therefore be imported by mitochondria for assembly with the mitochondrially encoded rRNA. Sequence analysis of mammalian mitoribosomal RNAs reveals that they are evolving significantly more rapidly than the corresponding RNA in cytoplasmic ribosomes. It is therefore of interest to determine evolutionary rates for the mitoribosomal proteins which, although they are nuclear gene products, must assemble with rapidly evolving RNA.

The proteins in mammalian mitoribosomes appear to be changing more rapidly than those from cytoplasmic ribosomes, giving rise to altered electrophoretic properties (Pietromonaco et al., 1986). In the absence of amino acid sequence information, which is beginning to appear, these differences in mobility are useful in assessing the relatedness of homologous proteins in different species.

From a pairwise analysis of homologous proteins in bovine and rat mitochondrial ribosomes, as well as proteins in bovine and rat cytoplasmic ribosomes (Pietromonaco, et al., 1986), it appears that the mitoribosomal proteins are changing at a rate 13 times higher than that for cytoplasmic r-proteins. Analysis of nucleotide substitution rates for mammalian mitoribosomal RNAs indicates that they are evolving at rates about 23 times that of the more conserved cytoplasmic rRNAs (Pietromonaco, et al., 1986). The result that the mitoribosomal RNA and proteins are changing at comparable rates is especially interesting in view of the fact that they are products of different genomes, where mutational rates are estimated to differ three to ten fold. Comparison of the r-protein and ribosomal RNA rates of evolution suggest that changes are being fixed at comparable rates in both the RNA and protein components of mitochondrial ribosomes, despite the different mutational rates for the RNA and protein. This implies that functional constraints act more or less equally on both kinds of molecules in the ribosome. A similar relationship also holds for cytoplasmic ribosomes in which, despite their overall higher degree of conservation, the RNAs and proteins show essentially the same evolutionary rate (Pietromonaco, et al., 1986). Such concordant evolution of RNA and protein components in two kinds of ribosomes, which are evolving at different rates, implies that the proteins, as well as the RNA, are subjected to effectively similar structural/functional constraints.

It is evident that ribosomes of mammalian mitochondria are evolving more rapidly than their extramitochondrial counterparts in the same cell. This is true for the protein components, found to be changing at a 13 fold higher rate than that for the cytoplasmic-r-proteins as well as for the rRNA. This finding is especially interesting, since both sets of ribosomal proteins, mitochondrial and cytoplasmic, are encoded by nuclear genes (Schieber and O'Brien 1985). The difference in evolutionary rates implies that they are encoded by different sets of genes and that functional constraints operating on ribosomes are relaxed for mitochondrial ribosomes.

REFERENCES

Anders, J. C., Denslow, N.D., and O'Brien, T.W., 1992, submitted for publication.

Denslow, N.D. and O'Brien, T.W., 1984, *J. Biol. Chem.* 259:9867-9873.

Denslow, N.D., LiCata, V.J, Gualerzi, C., and O'Brien, T.W., 1988, *Biochem.* 27:3521-3527.

Denslow, N.D., Michaels, G.S., Montoya, J., Attardi, G., O'Brien, T.W., 1989, *J. Biol. Chem.* 264:8328-8338.

Denslow, N.D., Anders, J.C. and O'Brien, T.W., 1991, *J. Biol. Chem.* 266:9586-9590.

DeVries, H. and Vander Koogh-Schurring, R., 1973, *Biochem. Biophys. Res Commun.* 54:308.

Guttel, R., Weiser, B., Woese, C., and Noller H., 1985, *Prog. Nuc. Acid. Res. Mol. Biol.* 32:155-216.

Grohmann, L., and Stoffler-Meilicke, M., 1989, *Endocytobiosis and Cell Res.* 6:183-192.

Hamilton, M.G. , and O'Brien, T.W., 1974, *Biochem.* 13:5400.

Kitakawa, M. and Isono, K., 1991, *Biochimie* 73:813-825.

Liao, H.X. and Spremulli, L.L., 1989, *J. Biol. Chem.* 264:7518-7522

Liao, H.X. and Spremulli, L.L., 1991, *J. Biol. Chem.* 266:20714-20719.

Matthews, D.E., Hessler, R.A., Denslow, N.D., Edwards, J.S., and O'Brien, T.W., 1982, *J. Biol. Chem.* 257:8788-8794.

O'Brien, T.W. and Kalf, G.F., 1967, *J. Biol. Chem.* 242:2172.

O'Brien, T.W., 1971, *J. Biol. Chem.*, 246:3409-3417.

O'Brien, T.W., and Matthews, D.E. 1976, Handbook of Genetics, Vol. 5, Robert C. King (ed.), Plenum Publishing Co., pp. 535-80.

O'Brien, T.W., 1977, International Cell Biology, B.R. Brinkley and Keith R. Porter (eds.) , The Rockefeller University Press, New York, pp. 245-255.

O'Brien, T.W., Denslow, N.D. , Anders, J.C., and Courtney, B.C., 1990,

    Biochim. Biophys. Acta 1050:174-178.

Piatyszek, M.A., Denslow, N.D., and O'Brien, T.W, 1988, *Nucl. Acid Res.* 16:2565-2583.

Pietromonaco, S.F., Hessler, R.A., and O'Brien, T.W., 1986, *J. Mol. Evol.* 24:110-117.

Pietromonaco, S.F., Denslow, N.D., and O'Brien, T.W. , 1991, *Biochimie* 73:827-836.

Schieber, G.L. and O'Brien, T.W., 1982, *J. Biol. Chem.* 257:8781-8787.

Schieber, G.L. and O'Brien, T.W., 1985, *J. Biol. Chem.* 260:6367-6372.

Shine, J. and Dalgarno, L., 1974, *Proc. Natl. Acad. Sci. USA* 71:1342-1346.

Stern, S., Weiser, B., and Noller, H., 1988, *J. Mol. Biol.* 204:447-481.

Spremulli, L.L and Kraus, B., 1987, *Biochem. Biophys. Res. Commun.* 147:1077-1081.

ESSENTIAL FEATURES OF THE PEPTIDYL TRANSFERASE CENTER IN THE YEAST MITOCHONDRIAL RIBOSOME

Chin Pan, Karen Sirum-Connolly, and Thomas L. Mason

Department of Biochemistry and Molecular Biology and The Graduate Program in Molecular and Cellular Biology
University of Massachusetts
Amherst, MA 01003

INTRODUCTION

The principal function of the mitochondrial translational apparatus in the yeast *Saccharomyces cerevisiae* is to synthesize seven polypeptide subunits of the energy transducing enzyme complexes of the inner mitochondrial membrane. The components of the organellar translation system are specified by a relatively small number of genes in the mitochondrial genome and by a much larger contribution from the nuclear genome. Genes in mtDNA encode the large (21S) and small (15S) rRNAs, a set of 24 tRNAs, an RNA component of mitochondrial RNase P, and the VAR1 protein, an essential component of the mitochondrial small ribosomal subunit (for reviews, see Grivell, 1989; Costanzo and Fox, 1990). The other components, including 70 to 80 ribosomal proteins, are encoded by 120 or more nuclear genes, synthesized in the cytoplasm, and, with a few notable exceptions, targeted for exclusive function in the mitochondrion.

There has been substantial recent progress in the identification and characterization of the nuclear genes required for the synthesis and function of the organellar translational apparatus, most notably gene sequences are now available for over 20 mitochondrial ribosomal proteins (for review, see Kitakawa and Isono, 1991). There are three basic reasons for our interest in these nuclear genes. First, their regulatory properties will provide information about the coordination of nuclear-mitochondrial interactions in the formation of the mitochondrial translational machinery. Second, it is becoming increasingly clear that nuclear-encoded, mRNA-specific translation factors are dominant elements in controlling the expression of mitochondrial genes (Costanzo and Fox, 1990). Thus, a complete understanding of mitochondrial gene expression will require a parallel detailed understanding of the mitochondrial translation system. Finally, the analysis of both conserved and divergent features in the mitochondrial system could lead to a better overall understanding of the translational apparatus, including the prototype system in *E. coli*. In this chapter, we focus on this last area of interest and summarize our recent work on three nuclear genes whose protein products impact on the peptidyl transferase center of the mitochondrial ribosome.

Two of these genes, *MRP7* and *RML16*, encode homologs of *E. coli* ribosomal proteins L27 and L16, respectively, which have been linked by several lines of evidence to the peptidyl transferase center of the ribosome. We have described previously the cloning and initial characterization of *MRP7* (Fearon and Mason, 1988). Here we assess the structure-function relations of the conserved and the nonconserved domains of MRP7p by *in vitro* mutagenesis and comparative sequence analysis. We also describe the identification and preliminary analysis of *RML16* and report that it specifies an essential protein in the mitochondrial ribosome. In addition, we present recent data implicating *PET56* function in an essential ribose methylation at a conserved G nucleotide in the peptidyl transferase center of 21S rRNA.

MAPPING FUNCTIONAL DOMAINS IN MRP7p

The deduced sequence of the 40-kDa MRP7p contains 371 amino acid residues; residues 1 to 27 form a putative mitochondrial targeting presequence, the 85 residues between 28 and 112 align extremely well with the 84 residues of the *E. coli* ribosomal protein L27, and the 259 residues at the carboxy-terminal end show no significant relatedness to other known proteins (Fearon and Mason, 1988). N-terminal sequencing by Grohmann et al. (1991) confirmed that Ala-28 is the amino-terminal residue of mature MRP7p (YmL2 in their nomenclature). The unusual structure of MRP7p led us to propose that the L27-like domain of MRP7p serves a conserved function in the peptidyl transferase center and the nonconserved carboxy-terminal domain provides a function unique to the mitochondrial ribosome. Moreover, the composite nature of this protein suggests that it might have originated from a fusion between smaller genes encoding proteins with separate functions. Our previous gene-disruption analysis showed that MRP7p is an essential component of the ribosome, but this work did not address questions about the relative functional importance of the different domains of the protein.

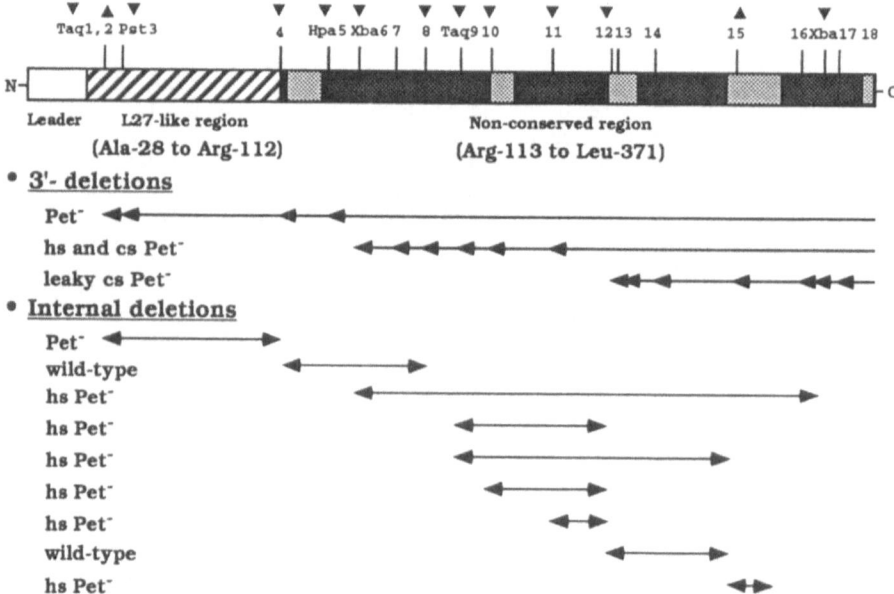

Figure 1. **Summary of mutational analysis of the *MRP7* coding region.** *Bam*HI linker-insertion mutations, 3'-end deletions, and internal in-frame deletions were created *in vitro* as described in the text. The positions of the new *Bam*HI sites are numbered from the 5'-end of the sequence; in-frame, four-codon insertions are indicated by (▼); in-frame, insertion-deletion mutations are indicated by (▲); out-of-frame insertions occurred at positions 7, 13, 14, 16, and 18. The various types of mutant alleles are named according to the following scheme: the original insertion alleles are *mrp7-X*, or *mrp7-ΔX* if an insertion was accompanied by an internal deletion, where X = the site of the insertion; 3'-end deletions, *mrp7-3'ΔX*; in-frame, internal deletions, *mrp7-ΔX-Y*, where X and Y define the 5' and 3' ends of the deleted regions, respectively. The gray shaded boxes in the nonconserved region indicate blocks of five or more amino acids that are identical in the *K. lactis* and *S. cerevisiae* MRP7 proteins (see text and Figure 2).

To further assess the structure-function relationships in the conserved and nonconserved domains of MRP7p, mutant alleles were created by *in vitro* insertion of 12-bp *Bam*HI linkers at six existing restriction sites in the *MRP7* coding region and at additional sites generated randomly by partial digestion with DNase I in the presence of Mn^{2+}. So far, 18 linker insertions in the coding region and numerous others in the flanking sequences have been mapped and sequenced. Existing restriction sites and the new *Bam*HI sites were used to create a set of nested deletions from the 3'-end of the coding region and several in-frame internal deletions. In total, over 45 mutant alleles have been characterized.

Because a functional mitochondrial translation system is required for maintenance of [*rho+*] mtDNA, haploid *mrp7* null mutants convert to [*rho-*] and irreversibly lose the function of the mitochondrial genetic system (Myers et al., 1987). Therefore, a "plasmid shuffle" technique was used to expedite the phenotypic analysis of the new alleles. Briefly, low-copy-number, centromere plasmids carrying the mutated genes were transformed into a *mrp7::LEU2* haploid strain in which *MRP7* function was provided by wild-type *MRP7* on a derivative of YEp24. The mutants were screened for growth phenotypes on nonfermentable carbon sources at 18°, 30° and 37°C after subsequent eviction of the plasmid carrying the wild-type allele.

The results of this mutational analysis are summarized in Fig. 1. Surprisingly, none of the in-frame, four-codon insertions, including two mutations in the conserved N-terminal domain, caused a discernible phenotype, indicating that MRP7p can tolerate a rather high degree of sequence variation. It is noteworthy, however, that none of these simple four-codon insertions happened to fall within conserved sequence blocks (see below). The majority of the deletion mutations produced moderate to severe respiration-deficient phenotypes. Cold-sensitive, respiratory deficiency was associated with even short 3'-end deletions, and a more severe, heat-sensitive phenotype resulted from either 3'-end deletions or in-frame internal deletions that removed a 25-amino-acid sequence specified by codons 233 to 257 (defined by the allele *mrp7-Δ11-12*). Another short deletion-insertion mutation at position 15 (*mrp7-Δ15*) exhibited the curious property of causing a more severe phenotype than much larger 3'-end deletions spanning the same region. Complete loss of function was caused by either long 3'-end deletions that removed 227 amino acids (*mrp7-3'Δ5*) or the entire 259-residue nonconserved region (*mrp7-3'Δ4*) and an internal deletion between insertions 2 and 4 (*mrp7-Δ2-4*) that removed the conserved L27-like sequence.

With the exception of the 3'-end deletions extending upstream beyond the linker insertion at position 6, all of the mutant proteins were detectable by immunoblot analysis of mitochondrial proteins using either a monoclonal antibody to an epitope located between Arg-232 and Ser-257 or a polyclonal antiserum raised against full-length MRP7p (data not shown). When considered together, these results firmly establish an essential role in the ribosome for the conserved N-terminal domain of MRP7p and strongly suggest that the nonconserved C-terminal region is essential as well. Moreover, the deletion alleles *mrp7-Δ11-12* and *mrp7-Δ15* have delineated relatively short sequences that appear to be functionally important elements within the nonconserved region of MRP7p.

If the N-terminal conserved and C-terminal nonconserved regions of MRP7p were derived from separate genes in evolution, then it is conceivable that these sequences could provide full function even when expressed as separate polypeptides, i.e., the proteins encoded by the *mrp7-3'Δ7* and the *mrp7-Δ2-4* alleles might function *in trans*. To test this possibility, we introduced a multi-copy plasmid carrying the *mrp7-Δ2-4* allele into an *mrp7-3'Δ7* heat-sensitive (hs) Pet⁻ strain and tested the transformants for respiratory growth at the restrictive temperature (37°C). Since the resulting strain remained hs Pet⁻, we tentatively conclude that the conserved and nonconserved domains do not provide complementing activities when expressed as separate polypeptides, at least in the present configuration. Interestingly, when cells carrying both mutant alleles were grown at a permissive temperature (30°C) on a nonfermentable carbon source, both polypeptides cosedimented with large ribosomal subunits in sucrose gradients containing 500 mM NH4Cl (data not shown). It appears, therefore, that the C-terminal domain is capable of independent association with the ribosome.

To gain additional information about functionally important sequence elements in MRP7p, we cloned the *MRP7* gene from another budding yeast, *Kluyveromyces lactis*, by colony hybridization using a DNA fragment from the conserved region of *MRP7*. This *K. lactis MRP7* gene complemented the *mrp7* null allele in *S. cerevisiae*. The *K. lactis MRP7* gene was sequenced, and an alignment of the deduced sequences of KIMRP7p and ScMRP7p is shown in Fig. 2. The sequences are 56% identical overall, 88% identical in the L27-like region, and 47% identical in the C-terminal region.

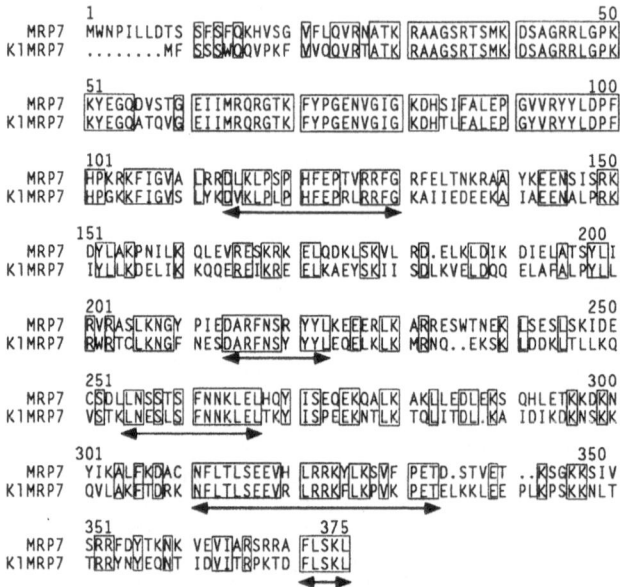

Figure 2. **Alignment of the *K. lactis* and *S. cerevisiae* MRP7p sequences.** The alignment was derived using the BestFit program of the GCG Sequence Analysis Software Package, Version 7.1 (Devereux et al., 1984). Identical amino acids are boxed. Underlined are blocks of five or more amino acids in the C-terminal domain that are identical in the two sequences. The deduced sequence for *S. cerevisiae* MRP7p has been published (Fearon and Mason, 1988); the sequence for *K. lactis* MRP7p was deduced from the DNA sequence of the cloned KI*MRP7* gene (C. Pan, unpublished results).

Inspection of the alignment in the C-terminal region reveals five stretches containing sequence blocks of five or more identical amino acids. These conserved stretches are highlighted in Fig. 1 and underlined in Fig. 2. Surprisingly, the 25-amino-acid segment between Arg-232 and Ser-257 of ScMRP7p that was defined by the *mrp7-Δ11-12* deletion, is not well conserved in KIMRP7p at the level of primary sequence. On the other hand, the sequence removed in the *mrp7-Δ15* allele falls within a conserved region between residues Leu-314 and Asp-333.

IDENTIFICATION OF THE MITOCHONDRIAL HOMOLOG OF L16

The first clue to the identity of the *RML16* gene for the mitochondrial homolog of *E. coli* L16 came from the observation that a complete open reading frame for an L16-like protein was present in the DNA sequence published by Ozier-Kalogeropoulos et al. (1991) for the yeast *URA7* gene and its flanking region (B. Baum, personal communication). This ORF is located upstream, in the opposite orientation, of *URA7* and begins at position 726 and ends at position 193 in the published sequence. The pattern of codon usage in the putative L16 coding region is characteristic for yeast proteins expressed to low levels, including several yeast mitochondrial ribosomal proteins.

```
              1                                                              50
   RML16    .......... ...MPVRTGG SIKGSTLQFG KYGLRLKSEG IRISAQQLKE
    EL16    MLQPKRTKFR KMHKGRNRGL .AQGTDVSFG SFGLKAVGRG .RLTARQIEA
Cyapa-L16   MLSPRRTKFR KQQRGRMKGI STRGNNLVFG DFGLQALEPA .WITSRQIEA
Marpo-L16   MLSPKRTKFR KQHCGNLKGI STRGNVICFG KFPLQALEPS .WITSRQIEA

              51                                                            100
   RML16    ADNAIMRYVR PLNNGHLWRR LCTNVALCIK GNETRMGKGK GGFDHWMVRV
    EL16    ARRAMTRAMK R..QGKIWIR VFPDKPITEK PLAVRMGKGK GNVEYWVALI
Cyapa-L16   SRRAINRYVR R..GGKIWIR IFPDKPVTMR PAETRMGSGK GAPEYWVAIV
Marpo-L16   GRRAITRYAR R..GGKLWIR IFPDKPITIR PAETRMGSGK GSPEYWVAVW

              101                                                          150
   RML16    PTGKILFEIN GDDLHEKVAR EAFRKAGTKL PGVYEFVSLD SLVRVGLHSF
    EL16    QPGKVLYEMD G..VPEELAR EAFKLAAAKL PIKTTFVTKT VM........
Cyapa-L16   KPGRVIFEIN G..VSQEMAK AAFRIATFKL PIKTKFISSR V.........
Marpo-L16   KPGKILYEIS G..VSENIAR AAMKIAAYKM PIRTQFITTS SLNKKQEI..

              151                                      190
   RML16    KNPKDDPVKN FYDENANKPS KKYLNILKSQ EPQYKLFRGR
    EL16    .......... .......... .......... ..........
Cyapa-L16   .......... .......... .......... ..........
Marpo-L16   .......... .......... .......... ..........
```

Figure 3. **Alignment of RML16p with *E. coli* L16 and two other members of the L16 protein family.** The alignment was derived using the PileUp program of the GCG Sequence Analysis Software Package, Version 7.1 (Devereux et al., 1984). Identical amino acids are boxed. The sequence of RML16p was deduced from the DNA sequence published by Ozier-Kalogeropoulos et al. (1991). EL16; *E. coli* L16 (Zurawski and Zurawski, 1985; Brosius and Chen, 1976); Cyapa-L16; *Cyanophora paradoxa* cyanelle L16 (Michalowski et al., 1990); Marpo-L16; *Marcantia polymorpha* chloroplast L16 (Fukuzawa et al., 1988).

With this information, we independently cloned and characterized the *RML16* gene from our laboratory strain 22-2D (Partaledis and Mason, 1988). The following results support the conclusion that *RML16* encodes an essential protein in the large subunit of the yeast mitochondrial ribosome (C. Pan, B. Baum and T. L. Mason, manuscript in preparation):

(i) Antibodies raised against a 167-amino-acid fragment of RML16p expressed in *E. coli* reacted specifically with a 20-kDa polypeptide that was highly enriched in the mitochondrial fraction and cosedimented with the mitochondrial large ribosomal subunit in high-salt sucrose gradients.

(ii) *RML16* was inactivated by replacing the entire coding region with a 3.8-kb DNA fragment containing *URA3*. A linear fragment containing the *rml16::URA3* allele was used to transform a diploid Ura⁻ strain to Ura⁺ prototrophy. Tetrad analysis of spores derived from these diploids showed 2:2 co-segregation of Ura⁺ and Pet⁻ phenotypes, indicating that the disruption of *RML16* caused respiratory deficiency. Southern and Western blot analysis of spores from a representative tetrad confirmed the absence of RML16p in the *rml16::URA3* spores (data not shown). In addition, further genetic analysis demonstrated that the *rml16::URA3* mutants had quantitatively converted to cytoplasmic petites, which is a diagnostic characteristic for a loss-of-function mutation in a gene for an essential ribosomal protein.

(iii) The levels of *RML16* mRNA and protein responded to catabolite repression, and the stable accumulation of RML16p was dependent on the presence of the 21S rRNA. Catabolite repression and rRNA-dependent accumulation are consistent with the properties of several other mitochondrial ribosomal proteins (e.g., MRP13p, Partaledis and Mason, 1988; MRP20p, Fearon and Mason, 1992).

The sequence alignment of RML16p with three other members of the L16 protein family is shown in Fig. 3. The identity between RML16p and EL16 is 36%, which compares with 52% identity between EL16 and the L16 homologs from *C. paradoxa* and *M. polymorpha*, respectively. Since the alignment begins with the highly conserved Gly close to the amino-terminal end of the deduced sequence, RML16p, like at least four other yeast mitochondrial ribosomal proteins (Kitakawa and Isono, 1991), may lack a cleavable presequence.

PET56, which is best known because it is a close neighbor of the divergently transcribed *HIS3* gene, is required for the synthesis or stability of the large subunit of the mitochondrial ribosome. We found that *his3-Δ200* mutants, which are His⁻ and have 10-fold reduced levels of *PET56* transcripts (Struhl, 1985b), were respiration-deficient when grown at 18°C, and when grown at 30°C, these cells contained near normal amounts of the mitochondrial 37S ribosomal subunit, but only 16% of the normal level of the 54S subunit (Figure 4). Moreover, haploid *pet56::URA3* gene disruption mutants converted to *rho-*, which is consistent with an essential role for *PET56* in the synthesis or function of the mitochondrial translation machinery. Immunoblot analysis of ribosomal subunits resolved by centrifugation in sucrose gradients showed that while PET56p is a mitochondrial protein, it is not associated with either the mature 37S or 54S subunits.

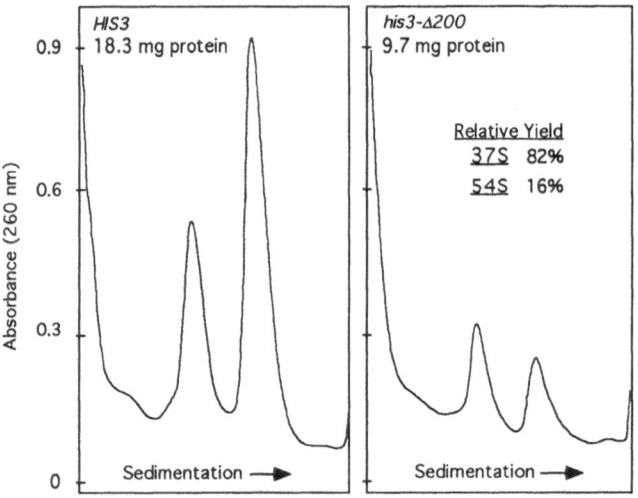

Figure 4. **Sucrose density gradient profiles of mitochondrial ribosomal subunits isolated from wild-type and *his3-Δ200* strains.** Mitochondrial ribosomal subunits from cells grown on galactose-containing media at 30°C were prepared and analyzed as described earlier (Partaledis and Mason, 1988). The relative yield of large and small subunits was determined from the area under each peak and normalized to the amount of protein in the mitochondrial lysate loaded onto each gradient.

A partial sequence of *PET56* was known (Struhl, 1985a), and completion of the sequence showed that the gene contains a long open reading frame capable of specifying a basic (pI = 10.08), 412-amino-acid (46 kDa) polypeptide. The deduced sequence of PET56p, shown in Fig. 5, has 50-55% similarity to two 23S rRNA ribose methylases responsible for thiostrepton resistance in *Streptomyces* (Cundliffe and Thompson, 1981, Bibb et al., 1985 and Li et al., 1990); no other statistically significant matches were found in the current release of the Genbank/EMBL databases. When taken together, these observations raise the exciting possibility that PET56p catalyzes an essential ribose methylation in the mitochondrial 21S rRNA.

As shown in Figure 6, ribose methylation is known to occur at highly conserved sites in several large rRNAs. The three ribose methylations in *E. coli* 23S rRNA all occur within domain V at positions Gm2251, Cm2498, and Um2552 (Nichols and Lane, 1967). *Mycoplasma* 23S rRNA has four ribose methylations including Gm2251, Cm2498, and Gm2553 (Hsuchen and Dubin, 1980). Mitochondrial large rRNAs from *S. carlbergensis* (Klootwijk et al., 1975) and *N. crassa* (Lambowitz and Luck, 1976) each contain two ribose methylations at unknown locations. The mitochondrial large rRNA

Alignment Between Yeast PET56p and *Streptomyces* rRNA Ribose Methylases

```
            1                                                     50
S.actuosus  ..........  ..........  ..........  ..........  ..........
S.azureus   ..........  ..........  ..........  ..........  ..........
PET56       MTSLTNAVFK  RYLAVTPSAH  QALKTRIKKK  SSSFDKFFPQ  QSNSRKKQWE

            51                                                    100
S.actuosus  ..........  ........MT  EPAILINA..  .SDPAVDRII  DVTKHSRASI
S.azureus   ..........  ........MT  ELDTIANP..  .SDPAVDRII  DVTKPSRSNI
PET56       TLNEDKASWF  KRKYAHVHAR  EDDRAADPYG  KKKAHVEKLK  EIKNQAKLNQ

            101                                                   150
S.actuosus  KTI..TLIEDT  EPLMECIRAG  VQFIEVYGSS  G.........  ..........
S.azureus   KTI..TLIEDV  EPLMHSIAAG  VEFIEVYGSD  S.........  ..........
PET56       KSHKSKFQNK  DIALKLMNDN  PIFEYVYGTN  SVYAALLNPS  RNCHSRLLYH

            151                                                   200
S.actuosus  TPLDPALLDL  CRQREIPVRL  IDVSIVNDLF  K.AERKAKUF  GIARVPRP..
S.azureus   SPEPSELLDL  CGRQNIPVRL  IDSSIVNQLF  K.GERKAKTF  GIARVPRP..
PET56       GTIIPSKFLQI  VDELKVTTEL  VDKHRLNLLT  NYGVHNNIAL  ETKRLQPVEI

            201                                                   250
S.actuosus  ARLADIAERG  GDVMV.....  ..........  ..........  ....LDGVK
S.azureus   AREGDIASRR  GDVMV.....  ..........  ..........  ....LDGVK
PET56       AVLGDMDISS  AALSIHELGF  NNENIPHELP  YGTKTDAKKF  PLGLYLDEIT

            251                                                   300
S.actuosus  IVGNIGAIVR  TSLALGAAGI  VLVD...SDL  ATIADRRLLR  ASRGVVFSLP
S.azureus   IVGNIGAIVR  TSLALGASGI  ILVD...SDI  TSIADRRLDR  ASRGVVFSLP
PET56       DPHNIGAIIR  SAYFLGVDFI  VMSRRNCSPL  TPV....VSK  TSSGALELLP

            301                                                   350
S.actuosus  VVLADREEAV  SFLRDNDIAL  MVLDII.....  ........DG  DLGVKDLGDR
S.azureus   VVLSGREEAI  AFIRDSGMDL  MTLKA.....  ........DG  DISVKELGDN
PET56       IFYVDKPLEF  .FTKSQEMGG  WTFIISHLAN  ATSEKYTVGK  TISMHDLNGL

            351                                                   400
S.actuosus  ADRMALVF..  GSEKGG....  ...PSGLFDE  ASAGTVSIPM  LSS...TESL
S.azureus   PDRLALLF..  GSEKGG....  ...PSDLFEE  ASSASVSIPM  MSQ...TESL
PET56       CNEILPVVLVV  GNESQGVRTN  LKMRSDFFEVE  IPFGGIEKGN  RAPEPIVDSL

            401             428
S.actuosus  NVSVSVGIAL  HERSARNFAV  RRAAAQA
S.azureus   NVSVSLGIAL  HERIDRNLAA  NR.....
PET56       NVSVATALLI  DNILTCK...  .......
```

Figure 5. **Alignment between PET56p and *Streptomyces* rRNA ribose methylases.** The alignment was derived as described in the legend to Fig. 3. Identical amino acids and conservative substitutions are boxed. The complete sequence for PET56p was deduced from the partial DNA sequence published by Struhl (1985a) combined with additional unpublished DNA sequence information (K. Sirum-Connolly and T.L. Mason, manuscript in preparation). The bacterial sequences are for proteins specified by the *nsh* gene of *Streptomyces actuosus* (Li et al., 1990), and the *tsr* gene of *Streptomyces azureus* (Bibb et al., 1985). The *Streptomyces* genes encode ribose methylases capable of forming 2'-O-methyladenosine at position A1067 within the GTPase center of the 23S rRNA. This methylation prevents binding of the antibiotic thiostrepton, produced by *Streptomyces*, and results in auto-immunity (Thompson and Cundliffe, 1991).

(17S) from hamster has ribose methylations at three sites corresponding to the universally conserved residues Gm2251, Um2552, and Gm2553 in the *E. coli* 23S rRNA (Dubin and Taylor, 1978; Baer and Dubin, 1981). Significantly, ribose methylations were not detected in the mitochondrial small rRNAs from any of these sources. Cytoplasmic ribosomal RNAs have numerous ribose methylations including those corresponding to Cm2498, Um2552, and Gm2553 in 23S rRNA (Maden, 1990). Despite the high level of conservation of the ribose methylation sites within an important functional center of 23S rRNA, there is no known role for these modifications in ribosome assembly or function.

With a view toward comparing the methylation state of 21S rRNAs from *PET56* and *pet56::URA3* mutant strains, we developed a strategy to probe methylation at positions equivalent to the conserved ribose methylation sites in hamster mitochondrial and *E. coli* large rRNAs. The strategy is based on the fact that the phosphodiester bond 3' to the 2'-O-ribose methylation is totally resistant to alkaline hydrolysis (Maden and Salim, 1974 and references therein). Therefore, primer extension analysis using appropriately positioned [32P]-labeled DNA oligonucleotide primers and partially hydrolyzed template RNA should produce a ladder of extension products with gaps at the positions of the cleavage resistant phosphodiester bonds.

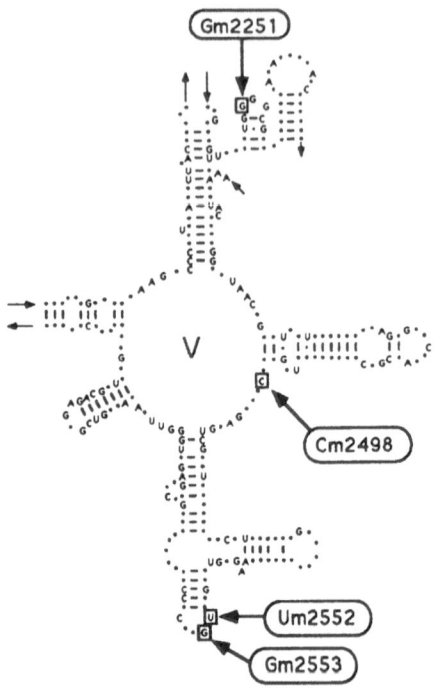

Figure 6. Ribose methylations within domain V of large rRNAs. The secondary structure of the central loop in domain V is shown with universally conserved nucleotides identified by letters (adapted from Egebjerg et al., 1990). The ribose methylations described in the text, Gm2251, Cm2498, Um2552, and Gm2553 are numbered according to the sequence of *E. coli* 23S rRNA (see text for refs.).

This method was applied to RNA isolated from three yeast strains: *PET56* [*rho*⁺]; *PET56* [F11 *rho*⁻] (F11 *rho*⁻ strains retain the mitochondrial gene for 21S rRNA) and *pet56::URA3* [F11 *rho*⁻]. As shown in Figure 7, panel A, primer extension analysis of the RNAs from the two *PET56* strains revealed a gap corresponding to a ribose methylation at position G2270 of the yeast 21S rRNA, which is equivalent to Gm2251 in *E. coli*. This gap is missing in the analysis of an unmodified, *in vitro*-transcribed fragment of 21S rRNA and in the RNA from the *pet56::URA3* [F11 *rho*⁻] strain. These results are consistent with the location of one of the two ribose methylations in 21S rRNA at G2270, and more importantly, the loss of methylation at this site appears to be correlated with a loss of *PET56* function. As shown in Fig. 7, panel B, a gap corresponding to methylation at G2792 was detected with the RNA from both wild-type and mutant strains when primers were used that probed for methylation at U2791 and G2792 (U2552 and G2553 in 23S rRNA). Apparently a second ribose methylation occurs at G2792, and it is not affected by the *pet56::URA3* mutation.

Although our findings implicate PET56p in ribose methylation at G2270, we have not yet demonstrated that this protein is in fact a specific ribose methylase. We are currently trying to establish this linkage by purifying PET56p and characterizing its ability to catalyze ribose methylation of *in vitro*-transcribed RNA substrates. Using reaction conditions developed for a mouse nucleolar 2'-O-methyltransferase by Eichler et al. (1987), we have been able so far to detect ribose methylation activity only in mitochondrial extracts from strains carrying *PET56* on the multicopy plasmid YEp24. Western analysis showed that this strain overproduced PET56p 10 to 15-fold. This methylase activity was G-specific and accepted sense, but not antisense, *in vitro* transcripts containing domains IV and V of 21S rRNA as a substrate. The next step will be to conclusively show that this activity is *PET56* dependent and specific for G2270 in 21S rRNA.

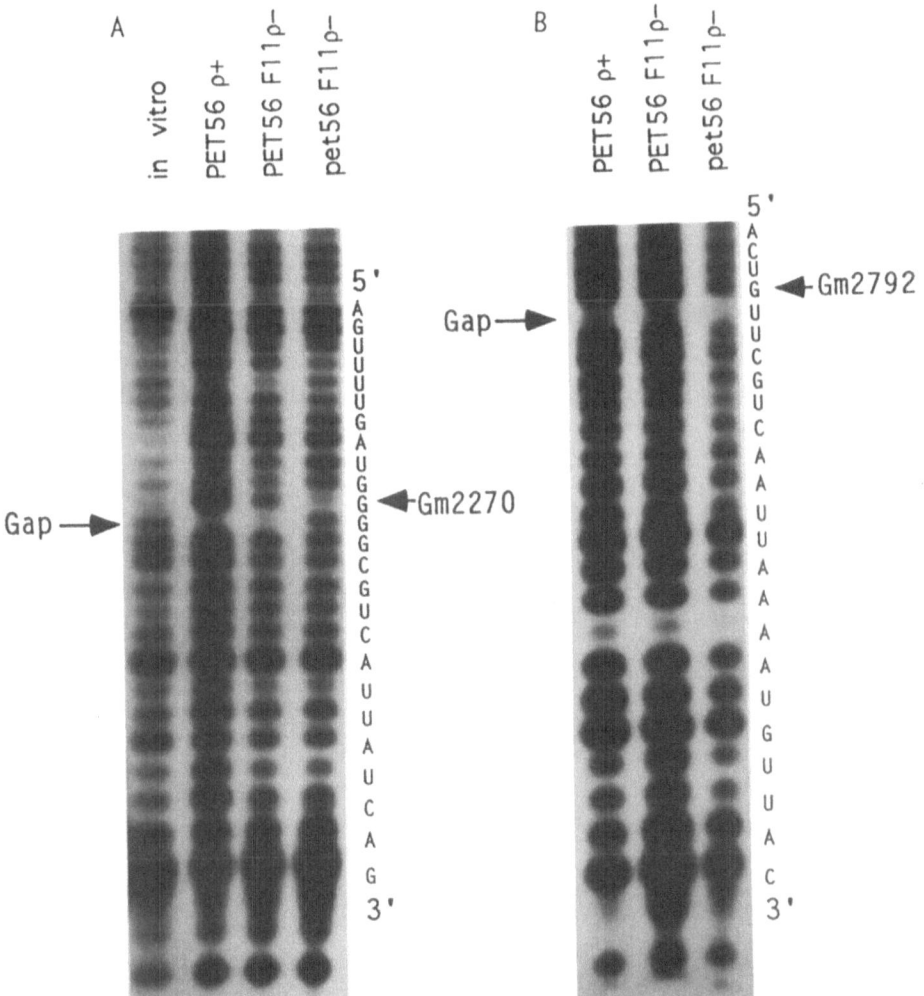

Figure 7. **Primer extension mapping of ribose methylations in 21S rRNA from wild-type and** *pet56::URA3* **strains.** *In vitro* transcripts of domains IV and V of yeast mitochondrial 21S rRNA were generated using T7 RNA polymerase. *In vivo* transcripts were extracted from the mitochondria of *PET56* [*rho+*]; *PET56* [F11 *rho-*] (F11 *rho-* strains retain the mitochondrial gene for 21S rRNA); and *pet56::URA3* [F11 *rho-*] cells. The RNAs were subjected to partial alkaline hydrolysis, and ^{32}P-labeled DNA oligonucleotide primers complementary to either nucleotides 2284-2305 or 2812-2831 of the 21S rRNA were annealed to the hydrolyzed RNA. Reverse transcriptase was used to extend the primers and the extension products were resolved by electrophoresis on an acrylamide gel under denaturing conditions (Kuechler et al., 1988). The nucleotide sequence of the RNA corresponding to the 3' ends of the extension products and the gaps corresponding to Gm2270 and Gm2792 in the wild type strain are indicated. The specificity of the primers was confirmed by RNA sequencing using base specific RNases.

SUMMARY AND FUTURE PROSPECTS

Proteins from the Large Subunit of the Yeast Mitochondrial Ribosome

A systematic effort to isolate and sequence proteins of the large subunit has produced N-terminal sequence data for 26 proteins, and the nuclear genes for several of these proteins have been cloned (Grohmann et al., 1991). Independently, we have identified four nuclear genes for large subunit proteins. Two of these genes, *MRP7* (Fearon and Mason, 1988) and *MRP20* (Fearon and Mason, 1992), encode proteins that match the published partial amino acid sequences for the YmL2 and YmL41 proteins, respectively, whereas *MRP49* (Fearon and Mason, 1992) and *RML16* (this paper; C. Pan, B. Baum, and T. Mason, manuscript in preparation) specify proteins that are not represented in the group of 26 proteins, presumably because they have blocked N-termini or are refractory to purification.

Table 1. Yeast Nuclear Genes for Proteins from the Large Subunit of the Mitochondrial Ribosome.

Gene	kD	Homolog	Essential (Sc)	Essential (Ec)	Reference
1. *MRP7*	40.1	L27	Yes	No/cs[a]	Fearon and Mason, 1988
2. *MRP20*	33.0	L23	Yes	?	Fearon and Mason, 1992
3. *MRP-L9*	27.5	L3	Yes	?	Graack et al., 1992
4. *MRP-L8*	26.8	L17-S13	Yes	?	Kitakawa et al., 1990
5. *RML16*	20.1	L16	Yes	?	Pan et al., in prep.
6. *MRP-L33*	11.0	L30	Yes	No[a]	Kang et al., 1991b
7. *MRP-L4*	34.0	No	?	-	see Grohmann et al., 1991
8. *YMR26*	18.6	No	Yes	-	Kang et al., 1991a
9. *MRP-L20*	20.6	No	Yes	-	Kitakawa et al., 1990
10. *MRP49*	16.0	No	No/cs	-	Fearon and Mason, 1992
11. *MRP-L27*	14.8	No	Yes	-	Graack et al., 1991
12. *MRP-L31*	14.2	No	Yes	-	Graack et al., 1991
13. *YMR44*	11.5	No	?	-	Matsushita et al., 1989
14. *MRP-L13*	21.0	No	?	-	see Grohmann et al., 1991

[a] Based on the phenotype of mutants in *E. coli* that lack these proteins (for refs., see Dabbs, 1991)

An unexpected result is that the majority of the identified large subunit proteins listed in Table 1 do not show sequence relatedness to any other characterized ribosomal protein. With one exception, the mitochondrial large subunit proteins that have been eliminated by gene disruption are essential for mitochondrial protein synthesis, as judged by the inability of these mutants to maintain [rho+] mitochondrial DNA. Two of the essential proteins, MRP7p and MRP-L33p(YmL33), have homologs in *E. coli* that are not essential for viability. At present if is not obvious why mitochondrial ribosomes appear to have a more stringent requirement for individual proteins than bacterial ribosomes.

In this chapter, we have presented new information about two yeast mitochondrial proteins that have bacterial homologs located in the peptidyl transferase center of the ribosome. MRP7p is a particularly intriguing case because its molecular mass is 4.5-fold larger than its homolog EL27. We have shown by mutational analysis that both the L27-like domain and the large C-terminal nonconserved domain are required for MRP7p function. In future work, we hope to confirm, through protein-protein and tRNA-protein crosslinking experiments that MRP7p is in fact located in the peptidyl transferase center of the ribosome. In this regard, it is fortunate that we have been able to identify the yeast mitochondrial homolog of EL16, a close neighbor of L27 in the *E. coli* ribosome (Wallenczek et al., 1988). It will be interesting to determine whether this spatial relationship is conserved in the yeast mitochondrial large subunit.

Does *PET56* Encode an Essential Ribose Methylase?

The data presented in this chapter support the contention that PET56p is required for ribose methylation at the conserved G2270 in the peptidyl transferase center of the 21S rRNA, and the sequence relatedness between PET56p and two *Streptomyces* rRNA methylases suggests that PET56p may be the G2270 methylase. Clearly, purification and *in vitro* characterization of PET56p will be required to establish its direct involvement in rRNA ribose methylation. Our initial efforts in this direction are promising. In particular, we have detected Gm methylase activity in mitochondrial extracts from a PET56p-overproducing strain, and the fact that this activity was detected with *in vitro*-transcribed rRNA as substrate suggests that it does not require a partially assembled ribosome for methylation. This should greatly facilitate the *in vitro* characterization of the enzymatic activity.

Confirmation of an essential ribose methylation at G2270 in 21S rRNA would have important implications for understanding large subunit assembly pathways and the structural and functional elements required in the peptidyl transferase center. It is clear that Gm2270 (Gm2251 in 23S rRNA) occupies a prominent position in a functionally active region of the rRNA. For example, Moazed and Noller (1989) have shown that three nucleotides 3' to Gm2251 are protected by tRNAs in the A and P sites. If this conserved methylation is proven to be essential, then the challenge will be to determine why.

ACKNOWLEDGMENTS

We thank Kathleen Fearon for her assistance in the early stages of this work and Kirill Rosen and Jacek Wower for advice on the analysis of rRNA methylation. We are especially grateful to Bobby Baum for pointing out the presence of the L16 open reading frame in the sequence flanking the *URA7* gene. This research was supported by NSF grant DMB-8719558 to T.L.M.

REFERENCES

Al-Arif, A., and Sporn, M., 1972, *Anal. Biochem.* 48:386.
Baer, R., and Dubin, D.,1981, *Nucleic Acids Res.* 9:323.
Barnard, E., 1969, *Ann. Rev. Biochem.* 33:677.
Bibb, M., Bibb, M., Ward, J., and Cohen, S., 1985, *Mol. Gen. Genet.* 199:26.
Brosius, J., and Chen, R., 1976, *FEBS Lett..* 68:105.
Chelbi-Alix, M., Expert-Bezancon, A., and Hayes, F., 1981, *Eur. J. Biochem.* 115:627.
Costanzo, M., and Fox, T.D., 1990, *Ann. Rev. Genet.* 28:145.
Cundliffe, E. and Thompson, J., 1981, *J. Gen. Microbiol.* 126:185.
Cunningham, P., Richard, R., Weitzmann, C., Nurse, K., and Ofengand, J., 1991, *Biochimie* 73:789.
Dabbs, E., 1991, *Biochimie* 73: 639.
Devereux, J., Haeberli, P., and Smithies, O., 1984, *Nucleic Acids Res.* 12: 387.
Dubin, D., and Taylor, R., 1978, *J. Mol. Biol.* 121:523.
Egebjerg, J., Larsen, N., and Garrett, R., 1990, in *The Ribosome: Structure, Function, and Evolution* (Hill, W.E. et al., eds) pp 168.
Eichler, D., Raber, N., Shumard, C., and Eales, S., 1987, *Biochem.* 26:1639.
Fearon, K. and Mason, T.L., 1988, *Mol. Cell. Biol.* 8:3636.
Fearon, K., and Mason, T.L., 1992, *J. Biol. Chem.* 267:5162.
Fukuzawa, H., Kohchi, T., Sano, T., Shirai, H., Umesono, K., Inokuchi, H., Ozeki, H., and Ohyama, K., 1988, *J Mol. Biol.* 203:333.
Graack, H.-R., Grohmann, L. and Kitakawa, M., 1991, *Biochimie* 73: 837.
Graack, H.-R., Grohmann, L., Kitakawa, M., Schäfer, K.-L., and Kruft, V., 1992, *Eur. J. Biochem.* 206: 373.
Grivell, L.A., 1989, *Eur. J. Biochem..* 182:477.
Grohmann, L., Graack, H.-R., and Kitakawa, M., 1989), *Eur. J. Biochem.* 183:155.
Grohmann, L., Graack, H.-R. V. Kruft, V., Choli, T., Goldschmidt-Reisin, S. and Kitakawa, M., 1991, *FEBS Lett.* 284:51.

Hsuchen, C., and Dubin, D., 1980, *J. Bacteriol.* 144:991.

Kang, W., Matsushita, Y., and Isono, K. , 1991a, *Mol. Gen. Genet.* 225, 474.

Kang, W., Matsushita, Y., Grohmann, L.; Graack, H.-R., Kitakawa, M, and Isono, K., 1991b, *J. Bacteriol.* 173, 4013.

Kitakawa, M., Grohmann, L., Graack, H.-R., and Isono, K, 1990, *Nucleic Acids Res.* 18:1521.

Kitakawa, M., and Isono, K., 1991, *Biochimie* 73:813.

Klootwijk, J., Klein, I., and Grivell, L., 1975, *J. Mol. Biol.* 97:337.

Krzyzosiak, W., Denman, R., Nurse, K., Hellmann, W., Boublik, M., Gehrke, C., Agris, P., and Ofengand, J., 1987, *Biochem.* 26:2353.

Kuechler, E., Steiner, G., and Barta, A., 1988, *Meth. Enz.* 164:361.

Lambowitz, A., and Luck, D., 1976, *J. Biol. Chem.* 251:3081.

Li, Y., Dosch, D., Strohl, W., and Floss, H., 1990, *Gene* 91:9.

Maden, B., 1990, *Progress in Nucleic Acid Res.* 39:241.

Maden, B, and Salim, M., 1974, *J. Mol. Biol.* 88:133.

Matsushita, Y., Kitakawa, M., and Isono, K., 1989, *Mol. Gen. Genet.* 219:119.

Michalowski, C.B., Pfanzagl, B., Loffelhardt,W., and Bohnert, H.J., 1990, *Mol. Gen. Genet .* 224:222.

Myers, A.M., Pape, L.K., and Tzagoloff, A., 1987, *EMBO J.* 4:2087.

Nichols, J., and Lane, B., 1967, *J. Mol. Biol.* 30:477.

Noller, H., 1984, *Ann. Rev. Biochem.* 53:119.

Ozier-Kalogeropoulos, O., Fasiolo, F., Adeline, M.T., Collin, J., and Lacroute, F., 1991, *Mol. Gen. Genet.* 231:7.

Partaledis, J.A., and Mason, T.L., 1988, *Mol. Cell. Biol.* 8:3647.

Struhl, K., 1985a, *Nucleic Acids Res.* 13:8587.

Struhl, K., 1985b, *Proc. Natl. Acad. Sci. U.S.A.* 82:8419.

Thompson, J., and Cundliffe, E., 1991, *Biochimie* 73:1131.

Walleczek, J., Shuler, D., Stoffler-Meilicke, M., Brimacombe, R., and Stoffler, G., 1988, *EMBO J.* 7:3571.

Zurawski, G., and Zurawski, S. M., 1985, *Nucleic Acids Res.* 13:4521.

GENES FOR MITOCHONDRIAL RIBOSOMAL
PROTEINS IN PLANTS

Lutz Grohmann, Axel Brennicke, and Wolfgang Schuster

Institut für Genbiologische Forschung
Ihnestrasse 63
D-W 1000 Berlin 33
Germany

SUMMARY

Plant mitochondrial genomes code for many more proteins than their animal and fungal counterparts. Several ribosomal protein genes have been identified in the mitochondrial genomes of higher and lower plants, but the gene complement varies considerably between species. For example the gene for mitochondrial ribosomal protein S12 is encoded by the mitochondrial genomes of *Arabidopsis*, wheat, maize, and *Petunia*, but is a nuclear gene in *Oenothera*. However, a transcribed fragment of this gene is still present on the *Oenothera* mitochondrial genome, that is lacking the 5'- and 3'-parts, thus representing only a non-functional truncated reading frame.

In mitochondria of higher plants RNA editing changes the information content of almost all protein encoding mRNAs. This process also plays a role in the expression of the mitochondrial ribosomal protein genes, since without RNA editing most of the encoded proteins could not function properly. In the lower plant *Marchantia*, however, RNA editing does not seem to be necessary to correct the mRNA sequence, because in this lower plant the mitochondrial genomic sequences correspond in all instances of RNA editing in higher plants to the respective edited sequences.

Sixteen genes coding for ribosomal proteins have been located on the mitochondrial genome of *Marchantia* (Oda et al., 1992). Fourteen of these are organized in two clusters, similar to the situation in eubacteria. In higher plant mitochondria, however, only rudimentary parts of these cistrons are conserved, most of the ribosomal genes are scattered around in the genome as a result of the high recombinational activity. This process has resulted in new mitochondrial transcription units, differing also between higher plant species. The gene for ribosomal protein L5, for example, is cotranscribed with the downstream *rps14* gene in *Arabidopsis*, while in *Oenothera rpl5* is part of a transcription unit including the gene coding for subunit 3 of the NADH dehydrogenase (*nad3*).

The Translational Apparatus, Edited by K.H. Nierhaus
et al., Plenum Press, New York, 1993

INTRODUCTION

According to the endosymbiotic theory mitochondria are dervived from former eubacterial-like ancestors that invaded the eukaryotic cell. The former presumably independent organisms that evolved to mitochondria brought into the cell their own complete genetic, transcriptional and translational equipment, including their own ribosomes. During evolution many of the genes encoding proteins required for functions of the mitochondrial compartment were transferred to the nucleus that became more and more specialized in storage and retrieval of genetic information. The only genes so far universally found in all mitochondrial genomes are those encoding the ribosomal RNAs and at least part of the necessary set of tRNAs. Only few genes coding for mitochondrial ribosomal proteins (mrp), however, were retained in the mitochondrial genomes. The number of mrp genes still present in mitochondrial genomes varies significantly between different species. While none of the metazoan mtDNA's codes for any mrp gene, a single mrp gene was identified in fungi, and at least four mrp genes are encoded in mitochondria of several protozoa (Pritchard et al., 1990).

In the completely sequenced mitochondrial genome of the lower land plant *Marchantia polymorpha* sixteen genes coding for ribosomal proteins have been located (Takemura et al., 1992). The mitochondrial genome size of this plant is with 186.6 kb more than ten times larger than any animal mtDNA, but still less than a tenth of the largest mitochondrial genomes of higher plants with 2.500 kb (Newton, 1988). These may thus encode a much greater portion of the 60-80 ribosomal proteins generally found associated with mitochondrial ribosomes.

In this contribution we summarize current knowledge of organization and expression of mitochondrial ribosomal protein genes in plant mitochondria with particular emphasis on the differing coding locations of these genes, varying between the nucleus and the mitochondrion.

CONSERVATION OF MRP GENES IN PLANTS

All currently known mrp genes in plant mitochondria have been identified exclusively by their sequence similarity to their bacterial homologues (Grohmann et al., 1989). In the completely sequenced mitochondrial genome of the lower plant *Marchantia* 16 potential ribosomal protein genes were thus found. Additional mrp proteins may be encoded by as yet unidentified mitochondrial reading frames in this moss, however, no striking similarity to other known ribosomal proteins has been noted. In the ten times larger genomes of higher plant mitochondria up to now only a few of the *Marchantia* genes have been identified due to the limited sequence data available for these mtDNAs. Table 1 shows the mrp genes identified to date in monocot as well as dicot plants. In maize mitochondria 4 and in wheat 3 mrp genes have been described. In the dicot plants *Arabidopsis* and *Oenothera* 7 and 8 mrp genes have been analysed respectively.

As expected, primary sequence conservation between the mrp genes is highest among the different higher plant species, ranging between 79.5 and 92.1 % identical amino acids (Table 2). Similarity of the mrp sequences between the moss *Marchantia* and higher plants varies between 49.5 and 80.6 % identical amino acids, while the similarity between the plant and bacterial sequences ranges from 25 to 62.1 %. S12 seems to be one of the best conserved ribosomal proteins with about 60% of the amino acids identical to the *E. coli* sequence, a considerable degree of conservation during evolution. The lower end of sequence similarity with about 25% is observed for the L5 and S3 deduced protein sequences. Other open reading frames to which no function could yet be assigned in either the *Marchantia* or the higher plant mitochondrial genomes may encode further ribosomal proteins of the organelle that can probably only be identified by chemical and physiological studies.

Table 1. Mitochondrial ribosomal protein genes in higher plants.

Organism	Gene	Transcription Unit *	Reference
Monocot plants			
maize	*rps3*		(a)
	rpl16	Ψ*rps19 - rps3 - rpl16*	(a)
	rps12	*nad3 - rps12*	(b)
	rps13	*rps13 - nad1* b/c	(c)
wheat	*rps7*	single	(d)
	rps12	*nad3 - rps12*	(b)
	rps13	not transcribed	(e)
Dicot plants			
Arabidopsis	*rps3*	?	(f)
	rps7	?	(g)
	rps12	*nad3 - rps12*	(h)
	Ψ*rps14*		(i)
	rpl5	*rpl5 -* Ψ*rps14 - cob*	(i)
	rpl16	?	(f)
	Ψ*rps19*	?	(g)
broad bean	*rps14*	Ψ*rpl5 - rps14 - cob*	(j)
Petunia	*rps12*	*nad3 - rps12*	(k)
	rps19		(l)
	rps3		(m)
	rpl16	*rps19 - rps3 - rpl16 - cox2*	(m)
Oenothera	Ψ*rps12*		(n)
	rpl5	*rpl5 - nad3 -* Ψ*rps12*	(g)
	nuc. *rps12*	poly(A)+	(o)
	rps13	*cox1 - rps13 - nad1* b/c	(p)
	Ψ*rps19*		(q)
	rps3		(r)
	rpl16	Ψ*rps19 - rps3 - rpl16*	(r)
	Ψ*rpl2*		(s)
	rps14	Ψ*rpl2 - orf206 - rps14 - cob*	(g)
tobacco	*rpl2*	*orf1 - rpl2 - orf2*	(t)
	rps13	*atp9 - rps13 - nad1* b/c	(c)

* Genes cotranscribed in the respective plant species are connected by dashes.
(a) Hunt and Newton, 1991, (b) Gualberto et al., 1988, Gualberto et al., 1991, (c) Bland et al., 1986, (d) Zhuo and Bonen, 1992,(e) Bonen, 1987, (f) P. Brandt and M. Unseld, unpublished, (g) Schuster et al., 1990a, (h) Heinze et al., unpublished, (i) Brandt et al., manuscript in preparation, (j) Wahleithner and Wolstenholme, 1988, (k) Rasmussen and Hanson, 1989, (l) Conklin and Hanson, 1991, (m) Conklin and Hanson, 1991, (n) Schuster et al., 1990b, (o) Grohmann et al., 1992, (p) Wissinger et al., 1990, Schuster and Brennicke, 1987, (q) Schuster et al., 1991, (r) H. Bock et al., manuscript in preparation, (s) W. Schuster, unpublished, 1992; (t) Vitart et al., 1992.

ORGANIZATION OF THE RIBOSOMAL PROTEIN GENES IN PLANT MITOCHONDRIA

Most mrp genes identified in the mitochondrial genome of the moss *Marchantia* are organized in two major clusters, very similar to the organization of the respective *E.coli str* and S10-*spc-α* ribosomal protein operons (Fig.1; Ceretti et al., 1983; Bedwell et al., 1985). In higher plant mitochondrial genomes on the other hand mrp genes are scattered far apart and are often linked to non-ribosomal protein coding genes. These linkages differ between plant species and thus appear to be in many cases rather fortuitous results of the frequent mitochondrial reorganizations. Some combinations, however, even of non-ribosomal protein and mrp genes, appear to have been conserved throughout the higher plants. For example the *rps12* gene (or at least part of it) is located in all plants analyzed to date downstream of the gene coding for *nad3* (Kim et al., 1991). The *rpl5* gene is located upstream of the *nad3* gene in the *Oenothera* mitochondrial genome, whereas in *Arabidopsis* the *rpl5* gene is encoded adjacent to the *rps14* gene in a completely different environment. The connection between *rpl5* and *rps14* is, however, conserved in the *Marchantia* mitochondrial genome and in bacteria, suggesting this to be the older "original" arrangement (Fig. 1). Downstream of *rps14* lies in all higher plants investigated to date the gene encoding cytochrome *b*.

Rearrangements in the *Oenothera* mitochondrial genome moved virtually all genes of the in bacteria and *Marchantia* conserved ribosomal cistrons to different genomic locations. The *rps19*, *rps3*, *rpl16*, and *rpl5* genes exchanged their location with an open reading frame (*orf206*) probably involved in cytochrome c biogenesis. Another rearrangement translocated the *rpl5* gene upstream of *nad3*. After the deletion of four genes from this ribosomal cluster in *Oenothera* the *rpl2* gene is now linked to the upstream *rps14* gene separated by the inserted *orf206*. The gene arrangement *rps3-rpl16* is the only part of the ribosomal cluster found conserved in all plant mitochondrial genomes investigated so far. In all plant mtDNAs but *Oenothera* these genes overlap analogous to the situation also found in *Marchantia*. The other mrp genes identified in higher plants, as for example the *rps13* and *rps7* genes appear to have also been translocated from the clustered bacterial organization and are now located in different rather distant regions of the mitochondrial genomes (Table 1). Several of the mrp genes comparatively well conserved between *Marchantia* and the respective bacterial sequences, like *rps8*, *rpl6*, and *rps11*, have not yet been identified in any higher plant mitochondrial DNA.

TRANSCRIPTION AND RNA EDITING

Transcription data for the ribosomal cistrons of the lower plant *Marchantia* are not yet available, but several mrp genes of higher plants have been analysed in detail with respect to transcription. As the close physical proximity of many mrp genes with other genes already suggests these adjacent open reading frames are often cotranscribed. In *Oenothera* all mrp genes are part of larger transcription units, e. g. the *rpl5*, *nad3*, and pseudo *rps12* genes are cotranscribed as are the *rpl2*, *orf206*, and *rps14* genes. The *rps13* gene is part of a complex trans-splicing transcript combining three separately transcribed mRNAs, with the *rps13* open reading frame adjacent to an excised intron sequence

It is as yet unclear whether these cotranscribed genes with very different functions share any regulatory connection. The transcription unit in plant mitochondria seems to be a fortuitous result of the many rearrangements in plant mitochondrial genomes that have brought different genes under control of a promoter region which now initiates transcription of the downstream reading frames. These observations suggest that in plant mitochondria no specific promoters are present for the mrp genes.

Table 2. Amino acid sequence identities (in %) among the mitochondrial ribosomal proteins of angiosperms with the corresponding *E. coli* ribosomal proteins are shown. Percent of amino acids identical between *Marchantia* and *E. coli* are indicated in brackets.

Angiosperm (mt)	*E. coli*	*Marchantia* (mt)	*Oenothera* (mt)
S3 (maize)	27 (25.4)	51.4	88.5
S7 (wheat)	36 (35.8)	61.9	-
S12 (wheat/maize)	56.5 (62.1)	80.6	81.6 (nuc.)
S12 (*Oenothera* nuc.)	59.3	79.2	-
S13 (wheat)	38 (38.3)	57.7	79.5
S14 (broad bean)	41.2 (43.3)	69.7	92.1
S14 (*Arabidopsis*)	38	70.7	88.9
S19 (*Petunia*)	37 (37.6)	49.5	90.6
L2 tobacco*	47.5** (44.8)	-	-
L5 (*Arabidopsis*)	25 (28.6)	66.7	88
L16 (maize)	50 (50.4)	71.9	90.3

* identified by the sequence similarity of the first 100 amino acids of ORF1 (Vitart et al., 1992);
** sequence identity in a 60 amino acid overlap with the L2 ribosomal protein from B. stearothermophilus.

The posttranscriptional modifications of mrp mRNAs in plant mitochondria include the exision of introns (e.g. *rps3*) and in all instances investigated to date modification of the primary sequence by the specific RNA editing of higher plant mitochondria. The higher plant mrp genes and other mitochondrial protein coding genes are edited to a similar extent (Covello and Gray, 1989, Gualberto et al., 1989, Hiesel et al., 1989). Numerous cytidines are altered by RNA editing to uridines resulting in a different amino acid sequence from the one predicted by the genomic DNA sequence. In most mrp transcripts RNA editing improves similarity on the amino acid level among plants and also to the respective bacterial encoded polypeptides.

NON-FUNCTIONAL MRP GENES IN PLANT MITOCHONDRIA

Comparative analysis of several mrp genes in different higher plant species surprisingly showed that only some species encode intact mrp genes, while others encode defect copies or have lost the entire genes from the mitochondrial genomes. For example, *rps14* appears to have been deleted completely from the maize mitochondrial genome (Wahleithner and Wolstenholme, 1988), while an intact copy is still present in the mitochondrial genomes of broadbean and *Oenothera*. Other mrp genes have not been lost completely from the mitochondrial genome, but only parts of these genes are still present on the mtDNAs. These remaining gene fragments are the result of the frequent recombinations observed in higher plant mitochondria. For example, recombination via an 11 base pair repeated sequence deleted the 5'-part of the *rps12* gene in *Oenothera*, while the 3'-part of this gene has been lost by a second recombination event with the *nad5* gene.

In other instances mrp genes appear to have been inactivated by genomically encoded stop codons, e.g. the *rps19* gene in *Oenothera*. The RNA editing pattern of this gene has been altered concommitantly and editing has shifted from positions that improve

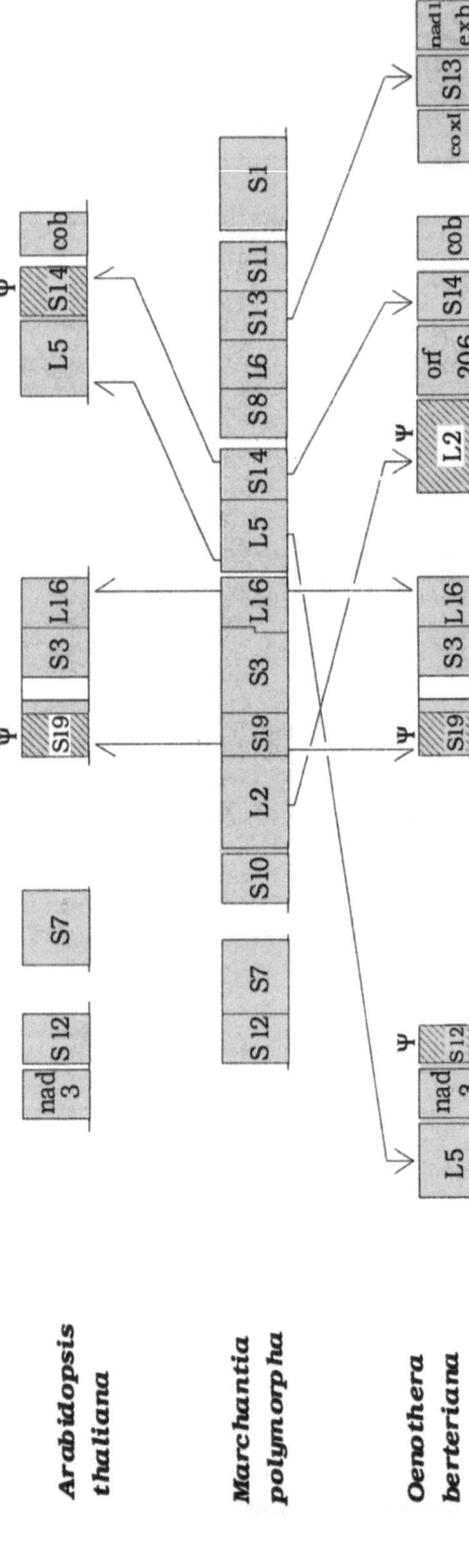

Fig.1. Organization of mitochondrial ribosomal protein genes in *Arabidopsis thaliana*, *Marchantia polymorpha* and *Oenothera berteriana*. The arrangement of the *rps19 - rps3 - rpl16* gene cluster is conserved in both higher plants, the open boxes indicate an intron within the *rps3* open reading frame. In *Arabidopsis* the *rpl5* gene is still linked to the *rps14* gene, whereas in *Oenothera rpl5* is located upstream of the *nad3* gene. The *rps13* gene is not encoded by the mitochondrial genome of *Arabidopsis*. The *rps7* gene in *Oenothera* and *rpl2* in *Arabidopsis* have not yet been identified. Hatched boxes indicate incomplete or disrupted mrp genes.

similarity to apparently senseless editing sites (Schuster and Brennicke, 1991). Transcriptional inactivation is also observed, when apparently intact mrp genes are not transcribed. For example, in wheat mitochondria the *rps*13 gene is a complete open reading frame, but the gene does not seem to be expressed and no transcripts are detectable (Bonen, 1987).

TRANSFER OF MRP GENES TO THE NUCLEUS

Since presumably all mrp genes are required for functional ribosomes in plant mitochondria the protein products of genes lost from or inactivated in the mitochondrial genomes must be encoded somewhere else and have to be imported into the mitochondrion to sustain the required translational apparatus in the organelles. The most likely location of these former mitochondrial genes would of course be the nuclear genome. One example investigated in this respect confirms the transfer of an mrp gene from the mitochondrial genome to the nuclear genome during evolution of the higher plant species. As discussed above, the *Oenothera* mitochondrial genome contains only a fragment of the *rps12* gene (Schuster et al., 1990b), suggesting that the intact gene copy has been transferred to the nucleus. Hybridisations with an intact *rps12* sequence clearly show the complete transcript to be contained in the poly(A)$^+$ fraction of the RNA. Translocation of a mitochondrial gene to the nucleus requires the addition of a mitochondrial target sequence to direct the protein into the mitochondrion. The nuclear copy of the *Oenothera rps12* indeed gained an N-terminal extension to direct the protein to the mitochondrial compartment (Grohmann et al., 1992).

Besides the requirement for such a mitochondrial target sequence, transfer of higher plant mitochondrial genes to the nucleus faces other complications derived by the different modes of transcription. The nuclear gene not only has to acquire correct promoter and regulatory elements for expression, but also has to resemble the edited mitochondrial copy. Since no RNA editing has yet been observed in plant nuclear transcripts, the requirement for editing has to be eliminated during the information transfer to nuclear genome. Indeed the nuclear *rps12* sequence is in all instances identical to the edited versions of the *rps12* mitochondrial mRNAs. This suggests that a gene transfer from the organelle to the nuclear genome would not start at the actual gene, but has to originate from the fully edited mRNA to ensure that the correct sequence is available in the nuclear genome. This mode of transfer of genetic information from the mitochondrial compartment to the nucleus, which requires reverse transcriptase activity at some point, is, of course, not only confined to mrp genes, but also holds true for any other plant mitochondrial protein coding gene (Schuster and Brennicke, 1987). Similar observations have indeed been made for the *cox2* gene, which has been transferred to the nucleus in several legume species (Nugent and Palmer, 1991).

CONCLUSION

Plant mitochondria encode more ribosomal protein genes in their organellar genome than any of the animal or fungal species investigated to date. Nine mrp genes have been identified in higher plant mtDNAs, while the *Marchantia* mitochondrial sequence revealed similarity of 16 open reading frames with ribosomal protein genes in *E. coli*. It is conceivable that some others of the 29 unidentified open reading frames in the *Marchantia* mitochondrial genome may encode additional mrp genes with little similarity to *E. coli* ribosomal protein genes. In yeast for example, only half of the nuclear encoded mrp genes show detectable similarity with eubacterial ribosomal proteins, the others are very distant and may have been derived from different phylogenetic origins. The higher similarity of plant mrp sequences with eubacterial ribosomal proteins may suggest closer phylogenetic origins and different lines of mitochondrial evolution in fungi and higher plants. Such variant phylogenetic origins of mitochondrial ancestors have also been inferred from other observations (Gray, 1989).

The mrp genes of lower and higher plants, although of similar evolutionary origin, differ highly in their organization and even coding capacity between individual species of higher plants. The reason for this surprisingly flexible coding location, i.e. the frequent transfer of mitochondrial mrp genes to the nuclear genome, may be connected with the disruption of the former ancestral mrp cistrons. While two cistrons in the lower plant *Marchantia* are still very similar to the bacterial gene organization, higher plant mitochondrial mrp genes tend to be scattered around the genome to very different environments. Such single mrp genes may allow easier transfer of the organellar gene to the nucleus. More complete analysis of the information content of plant mitochondrial genomes will allow a better estimation of the total number of mrp genes encoded in these organelles and in the nuclear genome respectively.

ACKNOWLEDGEMENTS

We wish to thank Hermann Bock, Petra Brandt, Barbara Heinze and Michael Unseld for information, comments and discussions. This work was supported by grants from the Deutsche Forschungsgemeinschaft and the Bundesministerium für Forschung und Technologie.

REFERENCES

Bedwell, D., Davis, G., Gosink, M., Post, L., Nomura, M., Kesbtler, H., Zengel, J.M.,and Lindahl, L., 1985, Nucleotide sequence of the alpha ribosomal protein operon of *Escherichia coli*, *Nucleic Acids Res.* 13:3891.

Bland, M., Levings III, C.S.,and Matzinger, D.F., 1986, The tobacco mitochondrial ATPase subunit 9 gene is closely linked to an open reading frame for a ribosomal protein, *Mol. Gen. Genet.* 204:8.

Bonen, L., 1987, The mitochondrial S13 ribosomal protein gene is silent in wheat embryos and seedlings, *Nucleic Acids Res.* 15:10393.

Ceretti, D.P., Dean, D., Davis, G.R., Bedwell, D.M., and Nomura, M., 1983, The *spc* ribosomal protein operon of *Escherichia coli*: sequence and cotranscription of the ribosomal protein genes and a protein export gene, *Nucleic Acids Res.* 11:2599.

Conklin, P.L., and Hanson, M.R., 1991, Ribosomal protein S19 is encoded by the mitochondrial genome in *Petunia hybrida*, *Nucleic Acids Res.* 19:2701.

Covello, P.S., and Gray, M.W., 1989, RNA editing in plant mitochondria, *Nature* 341:662.

Gray, M. W., 1989, The evolutionary origins of organelles, *Trends Genet.* 5:294.

Grohmann, L., Graack, H.-R., and Kitakawa, M., 1989, Molecular cloning of the nuclear gene for mitochondrial ribosomal protein YmL31 from Saccharomyces cerevisiae, *Eur. J. Biochem.* 183:155.

Grohmann, L., Brennicke, A., and Schuster, W., 1992, The mitochondrial gene encoding ribosomal protein S12 has been translocated to the nuclear genome in *Oenothera*, *Nucleic Acids Res.* 20:5641.

Gualberto, J.M.,Wintz, H., Weil, J.-H., and Grienenberger, J.-M., 1988, The genes coding for subunit 3 of NADH dehydrogenase and for ribosomal protein S12 are present in the wheat and maize mitochondrial genomes and are co-transcribed, Mol. Gen. Genet. 215:118.

Gualberto, J.M., Lamattina, L., Bonnard, G., Weil, J.-H., and Grienenberger, J.M., 1989, RNA editing in wheat mitochondria results in the conservation of protein sequences, *Nature* 341:660.

Gualberto, J.M., Bonnard, G., Lamattina, L., and Grienenberger, J.M., 1991, Expression of the wheat mitochondrial *nad3*-*rps12* transcription unit: correlation between editing and mRNA maturation, *Plant Cell* 2:1109.

Hiesel, R., Wissinger, B., Schuster, W., and Brennicke, A., 1989, RNA editing in plant mitochondria, Science 246:1632.

Hunt, M.D., and Newton, K.J., 1991, The NCS3 mutation: genetic evidence for the expression of ribosomal protein genes in *Zea mays* mitochondria, *EMBO J.* 10:1045.

Kim, K.-S., Schuster, W., Brennicke, A., and Choi, K.-T., 1991, Korean ginseng mitochondrial DNA encodes an intact *rps12* gene downstream of the *nad3* gene, *Plant Physiol.* 97:1602.

Newton, K.J, 1988, Plant mitochondrial genomes: organization, expression and variation, Ann. Rev. Plant Physiol. Plant Mol. Biol. 39:503.

Nugent, J.M., and Palmer, J.D., 1991, RNA-mediated transfer of the gene *coxII* from the mitochondrion to the nucleus during flowering plant evolution,*Cell* 66:473.

Oda, K., Yamato, K., Ohta, E., Nakamura, Y., Takemura, M., Nozato, N., Akashi,K., Kanegae, T., Ogura, Y., Kohchi, T., and Ohyama, K., 1992, Gene organization deduced from the complete sequence of the liverwort *Marchantia polymorpha* mitochondrial DNA,*J. Mol. Biol.* 223:1.

Pritchard, A. E., Seilhammer, J. J., Mahalingam, R., Saable, C. L., Venuti, S. E., and Cummings, D. J., 1990, Nucleotide sequence of the mitochondrial genome of *Paramecium, Nucleic Acids Res.* 18:173.

Rasmussen J., and Hanson, M.R., 1989, A NADH dehydrogenase subunit gene is co-transcribed with the abnormal *Petunia* mitochondrial gene associated with cytoplasmic male sterility, *Mol. Gen. Genet.* 215:332.

Schuster, W., and Brennicke, A., 1987, Plastid, nuclear and reverse transcriptase sequences in the mitochondrial genome of *Oenothera*: is genetic information transferred between organelles via RNA?, *EMBO J.* 6:2857.

Schuster, W., and Brennicke, A., 1987, Plastid DNA in the mitochondrial genome of *Oenothera*: intra- and interorganellar rearrangements involving part of the plastid ribosomal cistron, *Mol. Gen. Genet.* 210:44.

Schuster, W., Unseld, M., Wissinger, B., and Brennicke, A., 1990a, Ribosomal protein S14 transcripts are edited in *Oenothera* mitochondria, *Nucleic Acids Res.* 18:229.

Schuster, W., Wissinger, B., Unseld, M., and Brennicke, A., 1990b, Transcripts of the NADH-dehydrogenase subunit 3 gene are differentially edited in *Oenothera* mitochondria, *EMBO J.* 9:263.

Schuster, W., and Brennicke, A., 1991, RNA editing makes mistakes in plant mitochondria: editing loses sense in transcripts of a *rps19* pseudogene and in creating stop codons in *cox1* and *rps3* mRNAs of *Oenothera*, *Nucleic Acids Res.* 19:6923.

Takemura, M., Oda, K., Yamato, K., Ohta, E., Nakamura, Y., Nozato, N., Akashi, K., and Ohyama, K., 1992, Gene clusters for ribosomal proteins in the mitochondrial genome of a liverwort, *Marchantia polymorpha, Nucleic Acids Res.* 20:3199.

Wahleithner, J.A.,and Wolstenholme, D.R., 1988, Ribosomal protein S14 genes in broad bean mitochondrial DNA, *Nucleic Acids Res.* 16:6897.

Wissinger, B., Schuster, W., and Brennicke, A., 1990, Species-specific RNA editing patterns in the mitochondrial *rps13* transcripts of *Oenothera* and *Daucus, Mol. Gen. Genet.* 224:389.

Vitart, V., de Paepe, R., Mathieu, C., Chetrit, P., and Vedel, F., 1992, Amplification of substoichiometric recombinant mitochondrial DNA sequences in a nuclear, male sterile mutant regenerated from protoplast culture in *Nicotiana* , *Mol. Gen. Genet.* 233:193.

Zhuo, D., and Bonen, L., 1992, Characterization of the S7 ribosomal protein gene in wheat mitochondria, *Mol. Gen. Genet.*, in press.

EDITING CREATES THE INITIATOR CODON OF THE *rpl2* TRANSCRIPT FROM MAIZE CHLOROPLASTS

Hans Kössel, Brigitte Hoch, Gabor L. Igloi,
Rainer M. Maier, and Stephanie Ruf

Institut für Biologie III
der Universität Freiburg
Schänzle-Str. 1
D-7800 Freiburg
Germany

INTRODUCTION

Sequence analysis of the entire plastomes from tobacco[1], liverwort[2], rice[3], of numerous chloroplast DNA encoded genes from various other higher plants and from algae has shown that the vast majority of chloroplast protein genes have the canonical ATG start codon. In rare cases a GTG codon is used as a start codon. However, as an exception, ACG "start" codons have been reported for the *rpl2* genes from rice[3] and maize[4] chloroplasts as well as for the *psbL* genes from tobacco[1] and spinach[5] chloroplasts. These ACG codons are observed in positions homologous to ATG start codons present in the otherwise highly conserved *rpl2* and *psbL* genes of various plant species. No in-frame ATG or GTG codons upstream or downstream of these genes can be detected which might compensate for the mutated and probably nonfunctional ACG "initiator" codon.

In view of mRNA editing processes originally discovered in the kinetoplast genetic system of trypanosomes[6] and later detected also in nuclear encoded mRNAs[7,8], in mitochondrial mRNAs from higher plants[9-12] and in transcripts of a slime mold[13], we have investigated the possibility that the ACG codon located at the start position of the coding region of the maize chloroplast *rpl2* transcript is converted to a functional AUG start codon by a C-to-U transition caused by a chloroplast-specific editing process. Sequence analysis of cDNAs derived from the maize chloroplast *rpl2* transcript has provided evidence that this C-to-U conversion indeed takes place at the transcript level thereby converting a nontranslatable to a translatable *rpl2* transcript[14]. A similar observation was subsequently made for the initiation codon of the tobacco *psbL* gene[15]. This demonstrates the existence of mRNA editing as a novel step in the expression of chloroplast genes. Identification of editing sites in a number of other chloroplast transcripts in which internal codons are edited by C-to-U transitions, shows that editing is not a rare event in the chloroplast genetic system of higher plants.

The Translational Apparatus, Edited by K.H. Nierhaus
et al., Plenum Press, New York, 1993

RESULTS AND DISCUSSION

The Chloroplast *rpl2* Gene

As depicted in Fig. 1 and in the upper part of Fig. 2, the intron-containing *rpl2* gene is part of the *rpo*A operon which contains a cluster of ribosomal protein genes and the *rpo*A gene coding for the α subunit of chloroplast RNA polymerase at its distal end. The *rpo*A operon which has been analysed in the plastomes of tobacco[1], liverwort[2], rice[3] and maize (unpublished work from this laboratory) extends from the end of the inverted repeat region IR$_B$ into the large single copy region. Among the ribosomal protein genes of the *rpo*A operon the genes *rpl2*, *rpl23* and *rpl16* have been investigated for editing sites by analysing their cDNA sequences. The positions of other chloroplast ribosomal protein genes not contained in the *rpo*A operon, which were included in the cDNA sequence analysis, are also marked in Fig. 1.

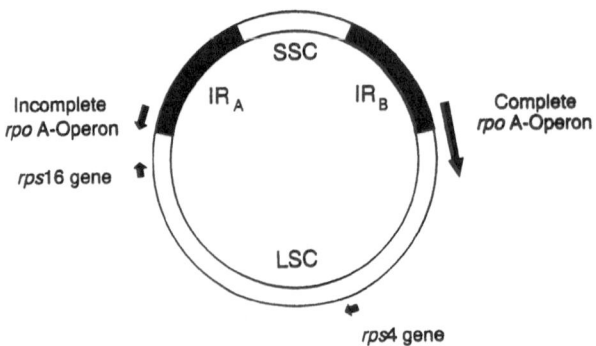

Figure 1. Position of the *rpo*A operon containing the *rpl2*, *rpl23* and *rpl16* genes and of the chloroplast ribosomal protein genes *rps4* and *rps16* on the maize plastome. IR$_A$, IR$_B$, LSC and SSC indicate the inverted repeat regions A/B and the large/small single copy regions of the plastome, respectively. The arrows indicate the orientation of the operon and the direction of transcription of the genes.

As outlined in Fig. 2 for the situation in maize (and the closely related rice[3]), the genes *rpl23*, *rpl2*, *rpl16* and *rpl19* (but only part of *rpl22*) are positioned within the inverted repeat region IR$_B$ and are therefore also present at the end of the inverted repeat region IR$_A$ in the form of a truncated *rpo*A operon which lacks the distal two thirds (including the *rpo*A gene) of the operon. In the lower part of Fig. 2 the first exon and part of the intron and of the second exon sequences of the maize chloroplast *rpl2* gene together with 5'flanking sequences are depicted. The ATG initiator codon present in homologous positions of the tobacco[1] and liverwort[2] *rpl2* genes is substituted by an ACG codon and no in-frame ATG or GTG codon is in the vicinity which might compensate the loss of the initiator codon.

610

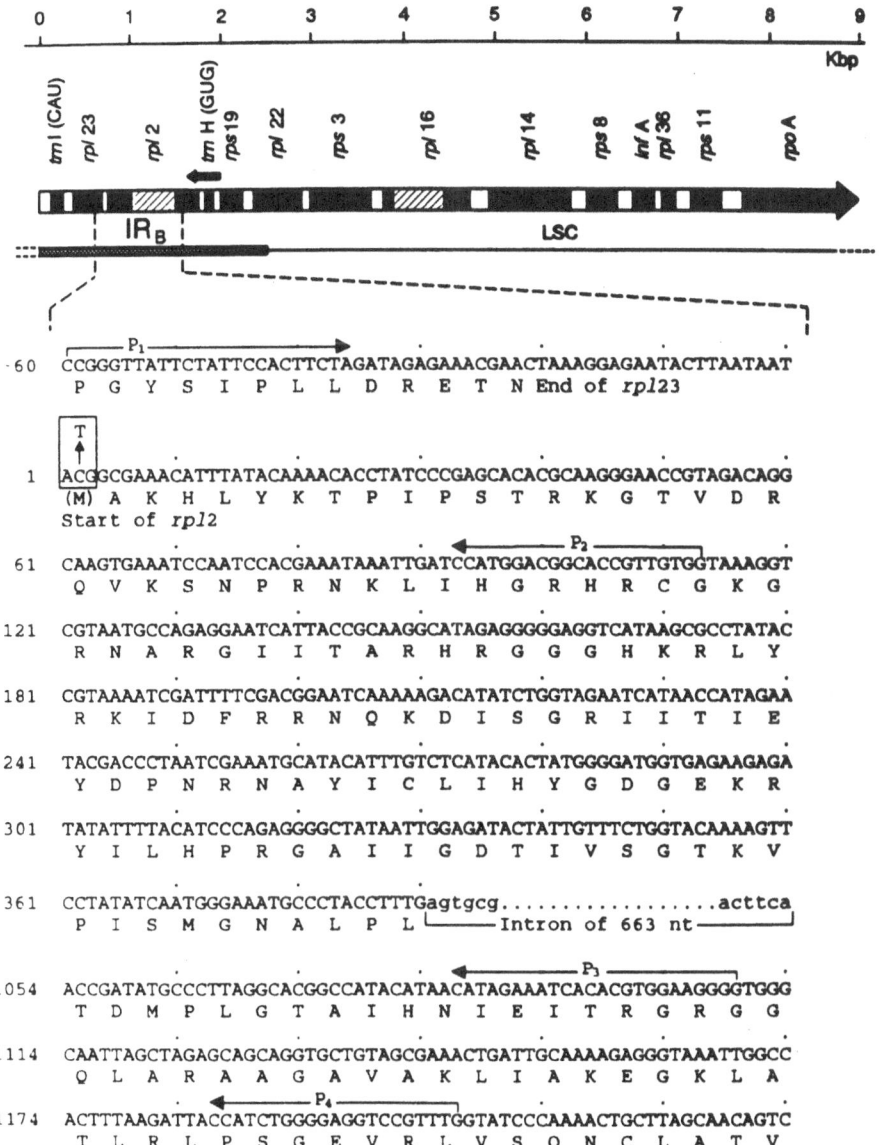

Figure 2. The maize chloroplast *rpo*A operon (upper part) and part of the nucleotide sequence of the *rpl*23 and *rpl*2 region within this operon (lower part). The operon extends from the inverted repeat region (IR$_B$) into the large single copy region (LSC). Intron sequences are marked by striations. In the lower part sequences surrounding the edited initiator codon of the *rpl*2 gene are depicted. The positions and orientations of the primers P$_1$, P$_2$, P$_3$ and P$_4$ used for polymerase chain reactions and sequencing are indicated by arrows. The terminal regions of the intron which comprises 663 base pairs are depicted in lower case letters. The edited start codon of the *rpl*2 gene is marked by framing.

Amplification of *rpl2* Specific Sequences

The positions of the primers used for amplification are indicated in the sequence part of Fig. 2. With the primer pair P_1/P_4 an amplification product of 1267 base pairs is obtained when chloroplast DNA is used as a template (Fig. 3, lane a) which is in accordance with the reported sequence[4]. However, using random primed cDNA as a template, a product of only 604 base pairs is obtained (Fig. 3, lane b) reflecting the loss of the 663 nt intron sequences in the spliced *rpl2* transcript. Correspondingly shorter products of 1170 and 506 base pairs were obtained with the primer pair P_1/P_3 as reported

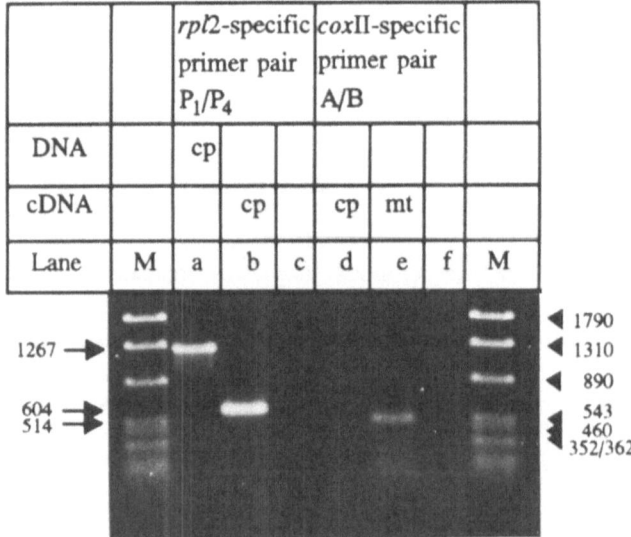

		rpl2-specific primer pair P_1/P_4			coxII-specific primer pair A/B		
DNA		cp					
cDNA			cp		cp	mt	
Lane	M	a	b	c	d	e	f	M

Markers left: 1267, 604, 514

Markers right: 1790, 1310, 890, 543, 460, 352/362

Figure 3. Amplification products obtained by polymerase chain reaction with the primer pair P_1/P_4 specific for the maize chloroplast *rpl2* gene and with the primer combination A/B specific for the mitochondrial *coxII* gene. The templates for the amplifications were either chloroplast DNA (lane a) or cDNA produced by random primed reverse transcription of a chloroplast RNA isolate (lanes b and d) or of isolated mitochondrial RNA (lane e). Control amplifications without DNA were carried out in lanes c and f. The positions and expected sizes in bp of amplification products are indicated by horizontal arrows on the left; the positions of size markers (lanes M) are given by triangles on the right.

earlier[14]. On the other hand, amplification with the primer pair P_1/P_2 which does not comprise the intron sequence (see Fig. 2) leads to identical products of 173 base pairs with either chloroplast DNA or cDNA (data not shown, but see ref. 14), which again is in accordance with the position of the primers in the *rpl2* sequence. Thus, in the case of the amplification products obtained with the primer pair P_1/P_4 (and P_1/P_3)[14] the sizes of the products are indicative of the origin from either the *rpl2* gene or the *rpl2* transcript and mutual contaminations can therefore be excluded.

As an important control, the cDNA preparation had to be tested for contamination with mitochondrial sequences. This control was essential in order to exclude a mitochondrial origin of edited *rpl2* sequences. This might have been the consequence of a gene transfer from the chloroplast to the mitochondrial genome as has been reported in several cases[16,17]. When, however, a primer pair specific for the mitochondrial *cox*II gene[14,18] was used, no amplification products were obtained with maize chloroplast cDNA as template (Fig. 3, lane d), whereas the expected product of 514 base pairs was formed with a maize mitochondrial cDNA as template (Fig. 3, lane e). This clearly excludes a mitochondrial origin of the amplification products obtained with the *rpl2* specific primers and chloroplast cDNA as template.

Sequence Analysis of the *rpl2* Amplification Products

The *rpl2* specific amplification products obtained with the primer pair P₁/P₃ were used directly for sequence analysis with primer P₂ as sequencing primer[14]. As evident from the sequence patterns depicted in Fig. 4 the cDNA shows a G-to-A transition as compared to the genomic DNA sequence which is indicative of a C-to-U transition within the ACG codon at the transcript level. We have been able to confirm this transition of the ACG codon to the AUG initiator codon at the cDNA level with the *rpl2* specific amplification products obtained with the primer pairs P₁/P₂ and P₁/P₄. This has allowed us to conclude that the functional initiator codon of the *rpl2* transcript is created by a C-to-U editing from the ACG codon present in the primary transcript and that this editing event causes the transition from an untranslatable to a translatable *rpl2* transcript. The sequence analysis on both, the genomic and cDNA level was extended using additional primer pairs (not shown in Fig. 2) to the entire *rpl2* gene. The coding sequences thus obtained were in complete agreement with the primary structure reported earlier[4] and thus did not reveal any other editing positions (data not shown).

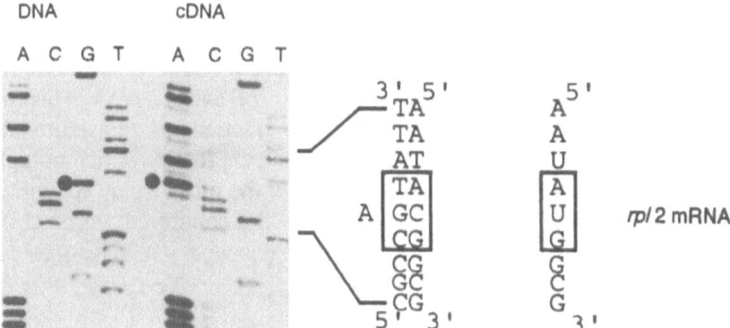

Figure 4. Sequence analysis and comparison of the sequence autoradiograms obtained from *rpl2*-specific amplification products of chloroplast DNA or cDNA. *rpl2* specific amplification products obtained with the primer pair P₁/P₃[14] were isolated and sequenced directly by a modified chain termination method[19] using primer P₂ as sequencing primer. Due to the polarity of this primer the autoradiograms represent sequences complementary to the *rpl2* transcript. The single position differing between the genomic and the cDNA sequence is marked by a dot. The codon affected by the base transition is framed. The sequences around the editing site are written to the right together with the mRNA like DNA strand and the mRNA sequence.

The question arises whether editing and splicing of the *rpl2* transcript occur as independent processing steps or whether they occur in defined order. To address this question a cDNA derived from an unspliced *rpl2* transcript was amplified using an intron-specific 3'-primer (not depicted in Fig. 2) in combination with primer P_1. Sequence analysis of the amplification product as expected showed the presence of the 5'terminal half of the intron sequence thus confirming its origin from an unspliced *rpl2* transcript. However, only a fraction of this product shows the G-to-A transition (C-to-U transition at the mRNA level), leading to the edited initiation codon (data not shown). This leads to the conclusion that splicing need not precede the editing process and vice versa and that the two processing steps occur as independent events during transcript maturation.

Editing of Other Chloroplast Transcripts

The canonical ATG initiator codon is also replaced by an ACG codon in the *rpl2* gene from rice chloroplasts. This suggests that a similar C-to-U editing functions in rice chloroplasts and possibly in other gramminean chloroplasts, which, however, remains to be tested experimentally by cDNA sequence analysis.

The *psbL* genes from tobacco[1] and spinach[5] chloroplasts had also been reported to lack a canonical initiator codon. cDNA analysis of the tobacco *psbL* transcript revealed that the AUG initiator codon again is created by a C-to-U editing from an ACG codon[15]. This shows that editing of chloroplast transcripts is not restricted to monocotyledon plants such as maize and that transcripts also encoded by intronless genes such as *psbL* can be subject to editing.

In view of the two examples in which editing creates initiator codons, the question arose whether chloroplast editing is restricted to the rare cases where non-functional initiator codons are encoded in the respective genes or whether editing also occurs within internal codons of chloroplast mRNAs, as is observed for many plant mitochondrial mRNAs[12]. To answer this question, several of those maize chloroplast genes encoding amino acid substitutions at sites which could be restored by C-to-U editing to otherwise highly conserved positions were selected for sequence analysis of the respective cDNAs. This led us to the detection of several internal C-to-U editing sites in the *ndhA* transcript[20] which was followed by the identification of many additional editing sites contained in the transcripts of the genes *ndhB*, *rpoB*, *petB* and IRF 170[21] and also of the *psbF* gene from spinach (J.Kudla, R.Bock, R.Hagemann and H.Kössel, unpublished). In contrast to this, no deviations between genomic DNA and cDNA sequences could be detected in the maize chloroplast genes *rpoA*[20], *rpoC_1*, *petD*, *rps4*, *rps16*, *rpl16* and *rpl23*. This shows that editing although not necessary for a considerable fraction of transcripts has to be considered as a more general process in chloroplast gene expression which in higher plants may involve perhaps half of the mRNA population. On the other hand, no editing appears to exist in the corresponding liverwort transcripts, as the conserved amino acid residues restored by editing in maize, tobacco and spinach are encoded at the genomic level in all cases observed so far.

REMAINING PROBLEMS

It is tempting to speculate that the creation of an initiator codon by editing provides a mechanism for regulation of gene expression on the translational level. Regulation of the *rpl2* expression in particular can be envisaged to influence the formation and availability of functionally active chloroplast ribosomes and thereby to influence chloroplast gene expression in a more general way. Further work will therefore have to aim at the question of whether editing of the *rpl2* transcript (but also of the other chloroplast transcripts) is constitutive or dependent on the developmental state of the organelle (or of its surrounding cell or tissue). As a first attempt towards this direction the possibility of a differential editing has been tested for the initiator codon of the bell pepper *psbL* transcript[22], which showed that fully edited transcripts exist in both, the chloroplasts of leaves and the chromoplasts of ripe fruit. In contrast to this lack of a differential editing during the chloroplast to chromoplast transition, partial editing could be observed for the same transcript in embryonic and root tissues from spinach, as opposed to fully edited *psbL* transcripts present in green and etiolated leaves (R.Bock, R.Hagemann, H.Kössel, and J.Kudla, unpublished). This may be taken as a first indication that a reduced efficiency of the editing process correlates with certain differential states of plastid development. It remains to be seen whether similar correlations exist in the editing efficiency of other chloroplast transcripts and whether this is only a side effect of plastid development or part of the signal chain leading to plastid development.

As a second major problem the mechanistic steps and components which govern the chloroplast editing process remain to be investigated. While the concept of guide RNAs is well accepted as a mechanism providing the information and the cosubstrate for editing of trypanosome transcripts[6], the question whether guide RNAs are components of other editing systems, in particular of plant organelles[12], is completely open. It remains also to be determined whether the sequences surrounding the individual editing sites act as determinants of the editing process. Mechanistically a deamination reaction would be the simplest explanation of the C-to-U conversion observed for all the chloroplast editing sites so far. However particularly in view of the rare U-to-C transitions observed in the plant mitochondrial editing system[12], which shares several characteristics with the chloroplast editing system[21], base or nucleotide replacements have also to be considered as mechanistic pathways. The answers to these questions must await the development of an *in vitro* editing system which would allow identificaton and characterization of the individual components and of the successive steps underlying the editing process.

A most intriguing third question is <u>why</u> the editing processes exist in the various genetic systems and whether they are evolutionarily old processes - perhaps even vestiges of the RNA world - or whether editing is a recent acquisition evolved as a sophisticated response to very special selective constraints. It is to be hoped that the answer to these questions can be approached by studying the phylogeny of editing which at present can be done at least with respect to the editing sites identified in the various organellar transcripts.

ACKNOWLEDGMENTS

We thank Elfi Schiefermayr and Sigrid Krien for technical assistance. This work was supported by the Deutsche Forschungsgemeinschaft (SFB206) and the Fonds der Chemischen Industrie (H.K.).

REFERENCES

1. K.Shinozaki et al., *EMBO J.* 5:2043 (1986).
2. K.Ohyama et al., *Nature* 322:572 (1986).
3. J. Hiratsuka et al., *Mol. Gen. Genet.* 217:185 (1989).
4. M.Kavousi, K.Giese, I.M.Larrinua, W.E.McLaughlin, and A.R.Subramanian,
      Nucleic Acids Res. 18:4244 (1990).
5. R.G.Herrmann, J.Alt, B.Schiller, W.R.Widger, and W.A.Cramer,
      FEBS Lett. 176:239 (1984).
6. R.Benne, *Trends Genet.* 6:177 (1990).

7. L.M.Powell, S.C.Wallis, R.J.Pease, Y.H.Edwards, T.J.Knott, and J.Scott, *Cell* 50:831 (1987).
8. B.Sommer, M.Köhler, R.Sprengel, and P.H.Seeburg, *Cell* 67:4 (1992).
9. P.S.Covello and M.W.Gray, *Nature* 341:662 (1989).
10. J.M.Gualberto, L.Lamattina, G.Bonnard, J.-H. Weil, and J.-M. Grienenberger, *Nature* 341:660 (1989).
11. R.Hiesel, B.Wissinger, W.Schuster, and A.Brennicke, *Science* 246:1632 (1989).
12. G.Bonnard, J.M.Gualberto, L.Lamattina, and J.M.Grienenberger, *Crit. Rev. Plant Science* 10:503 (1992).
13. R.Mahendran, M.R.Spottswood, and D.L.Miller, *Nature* 349:434 (1991).
14. B.Hoch, R.M. Maier, K.Appel, G.L.Igloi, and H.Kössel, *Nature* 353:178 (1991).
15. J.Kudla, G.L.Igloi, M.Metzlaff, R.Hagemann, and H.Kössel, *EMBO J.* 11:1099 (1992).
16. D.B.Stern, and D.M.Lonsdale, *Nature* 299:698 (1982).
17. D.B.Stern, and J.D.Palmer, *Nucleic Acids Res.* 14:5651 (1986).
18. K.D.Pruitt, and M.R.Hanson, *Curr. Genet.* 19:191 (1991).
19. B.Bachmann, W.Lüke, and G.Hunsmann, *Nucleic Acids Res.* 18:1309 (1990).
20. R.M.Maier, B.Hoch, P.Zeltz, and H.Kössel, *Plant Cell* 4:609 (1992).
21. H.Kössel, B.Hoch, R.M.Maier, G.L.Igloi, J.Kudla, P.Zeltz, R.Freyer, K.Neckermann, and S.Ruf, *in:* "Plant Mitochondria", A.Brennicke and U.Kück, ed., VCH Publishers, Weinheim, (1993), in press.
22. M.Kuntz, B.Camara, J.-H.Weil, and R.Schantz, *Plant.Mol.Biol.*, (1992), in press.

THE PLANT MITOCHONDRIAL TRANSFER RNA POPULATION : A MOSAIC OF SPECIES WITH DIFFERENT GENETIC ORIGINS

André Dietrich[1], Ian Small[2], Thierry Desprez[1], Jean Masson[2], Frédérique Weber[1], Daniel Ramamonjisoa[1], Ginette Souciet[1], Anne Cosset[1], Gaynor Green[1], Pierre Guillemaut[1], Georges Pelletier[2], Jacques-Henry Weil[1] and Laurence Maréchal-Drouard[1]

[1]Institut de Biologie Moléculaire des Plantes du CNRS
12 rue du Général Zimmer
F-67084 Strasbourg-Cedex, France
[2]Centre INRA de Versailles
Route de Saint-Cyr
F-78026 Versailles-Cedex, France

INTRODUCTION

Plant mitochondria still contain a large (from 200 to over 2000 kbp), autonomously replicating genome, although a massive transfer of genes to the nucleus has probably occurred since their endosymbiotic formation (Palmer, 1990). The mitochondrial DNA encodes some of the polypeptides involved in the enzymatic complexes of the respiratory chain, but these organelles depend on the nuclear and cytosolic compartments for most of their proteins, including the aminoacyl-transfer RNA (tRNA) synthetases and the majority of the ribosomal proteins. Plant mitochondria also do not retain all the tRNA genes of the ancestral endosymbiotic genome (called "native" or "genuine" genes). The higher plant mitochondrial tRNA population is therefore of particular complexity. First, a number of chloroplast DNA sequences, some of them comprising tRNA genes, have been integrated into the plant mitochondrial DNA during evolution. Some of these chloroplast-originating tRNA genes are expressed in mitochondria and produce mature, functional "chloroplast-like" species. Second, even including the "chloroplast-like" tRNA genes, plant mitochondria do not contain a complete set of tRNA genes and import some species from the cytosol. The plant mitochondrial tRNA population appears therefore to be a mosaic of species with nuclear, chloroplast and genuine mitochondrial origins.

HIGHER PLANT MITOCHONDRIAL GENOMES CONTAIN tRNA GENES OF CHLOROPLAST ORIGIN

The occurrence of "chloroplast-like" tRNA genes in plant mitochondria is due to the presence of promiscuous chloroplast DNA inserted into the plant mitochondrial genome in the course of evolution (e.g. Stern and Lonsdale, 1982). "Chloroplast-like" tRNA genes have now been found in a number of different plants (for a review see for instance Joyce and Gray, 1989), showing that this phenomenon is general. Although some of them may have inactivating mutations (Dron et al., 1985), most of these genes have remained identical or almost identical to their authentic chloroplast counterparts and are potentially functional.

HIGHER PLANT MITOCHONDRIAL GENOMES DO NOT CONTAIN A COMPLETE SET OF tRNA GENES

Although translation in higher plant mitochondria follows the universal genetic code (Jukes and Ozawa, 1990), the estimated number of expressed tRNA genes (including the "chloroplast-like" tRNA genes) present in the mitochondrial genome of maize and wheat (where the most systematic search was performed) is 16, corresponding to only 12-14 amino acids (Joyce et al., 1988; Sangaré et al., 1990). Genes for alanine, arginine, glycine, leucine, threonine and valine tRNAs seem to be missing from the mitochondrial genome of monocotyledon plants (Table 1).

Table 1. "Native" and "chloroplast-like" tRNAs or tRNA genes in maize, wheat, potato and petunia mitochondria.

Amino acid	Anticodon	Maize(a)	Wheat(b)	Potato(b)	Petunia(a)
"Native"					
Asp	GUC	1	-	+	-
Cys	GCA	-	-	+	1
Gln	UUG	1	+	+	1
Glu	UUC	2	+	+	1
Gly	GCC	-	-	+	1
Ile	CAU(c)	1	+	+	1
Lys	UUU	1	+	+	1
fMet	CAU	1	+	+	2
Phe	GAA	-	-	+	1
Pro	UGG	2	+	+	1
Ser	GCU	1	+	+	1
Ser	UGA	1	+	+	2
Tyr	GUA	1	+	+	1
Val	?	-	-	+	-
Val	?	-	-	+	-
"Chloroplast-like"					
Asn	GUU	2	+	+	1
Cys	GCA	1	+	-	-
His	GUG	1	-	+	1
mMet	CAU	1	+	+	3
Phe	GAA	1	+	-	-
Ser	GGA	-	+	+	1
Trp	CCA	1	+	+	1

a) Number of tRNA genes in the mitochondrial genome. b) tRNAs identified in mitochondria. c) tRNA encoded with a methionine anticodon (CAU) and becoming an isoleucine-specific species after post-transcriptional modification of the C in the anticodon (Weber et al., 1990). Data are taken from Sangaré et al. (1990), Joyce and Gray (1989), Maréchal-Drouard et al. (1990a) and Weber-Lotfi et al. (1992).

The tRNA^{Leu} and tRNA^{Val} genes which are part of the 12 kbp chloroplast DNA insertion in the maize mitochondrial DNA are not expressed (Stern and Lonsdale, 1982). The other sequences in the maize mitochondrial genome which could correspond to some of the missing tRNA genes are actually pseudo-genes or incomplete genes (Dewey et al., 1986). A number of tRNA genes are also absent in the mitochondrial genome of dicotyledon plants, as shown for petunia (Table 1).Several explanations have been envisaged to account for the inability to find in plant mitochondrial genomes the minimal set of tRNAs needed to support protein synthesis. We showed that the tRNAs whose genes are missing in plant mitochondrial genomes are actually nuclear-encoded and imported from the cytosol.

PLANT MITOCHONDRIA CONTAIN NUCLEAR-ENCODED tRNA SPECIES

Evidence for tRNA import from the cytosol arose from the presence, in plant mitochondria, of "cytosolic-like" tRNAs hybridizing to the nuclear genome, besides the "native" and "chloroplast-like" species encoded by the mitochondrial genome. Hybrid selection of labeled total mitochondrial tRNA, followed by two-dimensional polyacrylamide gel electrophoresis of the tRNAs eluted from the hybrids, demonstrated that at least 8 bean mitochondrial tRNAs hybridize to nuclear but not to mitochondrial DNA (Maréchal-Drouard et al., 1988). Among these, there are four tRNAs^{Leu}, three of which have been entirely sequenced. These bean mitochondrial tRNAs^{Leu} are identical to their cytosolic counterparts, except for the presence of a Gm in position 18 in the D loop instead of a G (Green et al., 1987; Maréchal-Drouard and Guillemaut, 1988; Maréchal-Drouard et al., 1988) and they hybridize to nuclear but not to mitochondrial DNA (Maréchal-Drouard et al., 1988). A "cytosolic-like" tRNA^{Leu}, as well as "cytosolic-like" glycine and valine tRNAs were also identified in wheat mitochondria (Joyce and Gray, 1989).

Direct proof for the import of a nuclear-encoded tRNA was recently obtained by *in vivo* experiments (Small et al., 1992). After introduction of the bean nuclear tRNA^{Leu}(C*AA) gene (Green et al., 1987) into potato protoplasts by electroporation, the presence of this foreign bean tRNA could be detected both in the cytosol and inside the mitochondria of regenerated transgenic potato plants (Small et al., 1992).

THE PLANT MITOCHONDRIAL tRNA POPULATION IS A MOSAIC OF SPECIES WITH DIFFERENT GENETIC ORIGINS

We studied in details the genetic origin of the individual tRNAs in the potato mitochondrial tRNA population (Maréchal-Drouard et al., 1990a). Total tRNA extracted from highly purified potato mitochondria was fractionated by two-dimensional polyacrylamide gel electrophoresis and the genetic origin of the different species was determined by hybridization to nuclear or mitochondrial DNA. Out of 31 identified potato mitochondrial tRNAs, only 15 species appeared to be of "native" mitochondrial origin, whereas 5 species turned out to be "chloroplast-like" and 11 species hybridized only to nuclear DNA (Maréchal-Drouard et al., 1990a) (Figure 1).

The "chloroplast-like" and nuclear-encoded mitochondrial tRNA species are functional and do not have "native" counterparts recognizing the same codon. All species are therefore necessary to ensure translation and it is likely that these tRNAs of three different genetic origins really function together during plant mitochondrial protein synthesis. The constraints placed by such a situation on the divergence of mitochondrial-encoded tRNA sequences might explain in part why plant mitochondrial tRNAs have quite "canonical" primary and secondary structures, as compared to their counterparts in metazoa (e.g. Okimoto and Wolstenholme, 1991).

How such a complex situation arose in the course of evolution is not well understood. It has been proposed that some transfers of nucleic acid sequences between different cell compartments during evolution might have been RNA-mediated, involving reverse transcription and genome integration (e.g. Nugent and Palmer, 1991; Schuster and Brennicke, 1987).

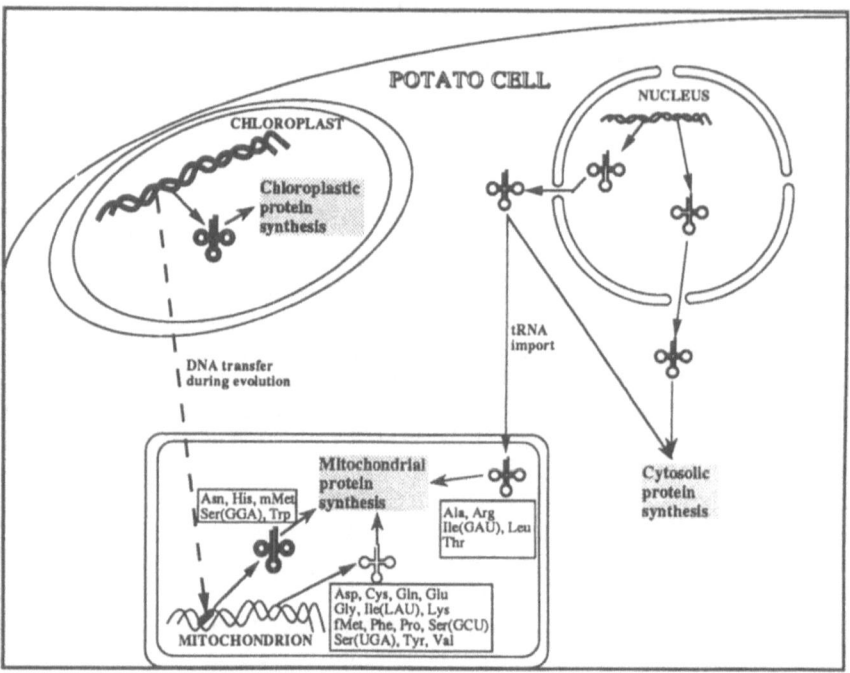

Figure 1. Genetic origin of potato mitochondrial tRNAs. The plant mitochondrial tRNA population comprises species with plastid (due to the presence of chloroplast DNA in the mitochondrial genome), nuclear (due to tRNA import from the cytosol) and genuine mitochondrial origins.

CODON RECOGNITION IN PLANT MITOCHONDRIA

According to the results we obtained for potato (Maréchal-Drouard et al., 1990a), the total number of plant mitochondrial tRNAs should be around 35. Plant mitochondria appear therefore to contain more tRNAs than mitochondria of other organisms (for a review see for instance Maréchal-Drouard et al., 1992), suggesting that in this case the "wobble" mechanism (Crick, 1966) could be sufficient to ensure reading of all 61 codons. However, it appears from partial sequencing that a "two out of three" mechanism (only the first two bases of the codon pair with the anticodon) (Samuelson et al., 1983) is needed in potato mitochondria if the identified tRNA^{Ala}(IGC), tRNA^{Arg}(NCU) and tRNA^{Arg}(ICG) are the only species present to decode all 4 alanine and 6 arginine codons (Maréchal-Drouard et al., 1990a).

"NATIVE" MITOCHONDRIAL-ENCODED tRNAs AND tRNA GENES

We have sequenced several plant mitochondrial "native" tRNAs (bean mitochondrial initiator tRNA^{Met}, tRNA^{Phe}, tRNA^{Pro} and two tRNAs ^{Tyr}, potato mitochondrial tRNA^{Ile}) and "native" tRNA genes (potato tRNA^{Ile} and initiator tRNA^{Met} genes). It appears that, at the level of both the primary sequence and the potential secondary structure, plant mitochondrial "native" tRNAs resemble eubacterial and chloroplast tRNAs (65 to 80% of sequence homology), rather than mitochondrial tRNAs from fungi or animals. For instance, both sequenced bean mitochondrial tRNAs^{Tyr} have a long variable loop (14 nucleotides) as in eubacteria, in chloroplasts and in fungal mitochondria, whereas tRNAs^{Tyr} from archaebacteria, eukaryotic cytosol and mammalian mitochondria have only a short variable loop (4-5 nucleotides) (Sprinzl et al., 1989). Furthermore, "native" mitochondrial tRNAs also appear to contain a low number of post-transcriptionaly modified nucleotides, as in prokaryotes and chloroplasts (e.g. Björk and Kohli, 1990). A tRNA "pseudogene" was also found a hundred nucleotides upstream of the potato initiator tRNA^{Met} gene (Maréchal-Drouard et al., unpublished). It corresponds to 60% of a complete tRNA (from the 5' end to the variable loop) and contains all the invariant and semi-invariant nucleotides. The anticodon would correspond to serine, but this tRNA "pseudogene" shows no significant homology with known tRNAs^{Ser} or tRNA^{Ser} genes.

The potato mitochondrial-encoded "native" tRNA^{Ile} is of special interest, as sequencing showed that this tRNA has a lysidine-like modified nucleotide (L*) in the first position of the anticodon (position 34). The corresponding mitochondrial gene has CAT at the position of the anticodon, which specifies methionine (Weber et al., 1990). The C residue in the anticodon must therefore be post-transcriptionaly modified into L*. In the case of E. coli tRNA^{Ile}(LAU), the single post-transcriptional modification of C into lysidine (L) in the anticodon switches the codon recognition and amino acid specificity of the tRNA from methionine to isoleucine (Muramatsu et al., 1988). The mature potato mitochondrial tRNA^{Ile}(L*AU) shows isoleucine-accepting activity, but no methionine-accepting activity, suggesting that, also in this case, the modification of the C into a lysidine-like residue changes the amino acid specificity of the tRNA, as compared to that of the corresponding gene. From an evolutionary point of view, potato mitochondrial tRNA^{Ile}(L*AU) does not show a high sequence homology with either plant cytosolic, chloroplast and mitochondrial elongator tRNAs^{Met}, or chloroplast and prokaryotic tRNAs^{Ile} containing lysidine or other modified nucleotides at the first position of the anticodon.

"CHLOROPLAST-LIKE" tRNAs AND tRNA GENES IN PLANT MITOCHONDRIA

The isolation from bean mitochondria of a mature "chloroplast-like" tRNA^{Trp} clearly distinguishable from its bean chloroplast counterpart (Maréchal et al., 1985) was the first direct evidence that some of the chloroplast-originating tRNA genes are indeed expressed in mitochondria. Sequencing of the two bean tRNAs^{Trp} also showed that a "chloroplast-like" mitochondrial tRNA can differ from its chloroplast counterpart in the modified nucleotide content (Maréchal et al., 1985), suggesting differences between the mitochondrial and chloroplast post-transcriptional modification systems.

So far, "chloroplast-like" tRNAs or tRNA genes specific for asparagine, cysteine, histidine, methionine (elongator), phenylalanine, serine (GGA) and tryptophane have been shown to be expressed in various plant mitochondria. Among the potato mitochondrial tRNAs that we have identified, tRNA^{Asn}, tRNA^{His}, elongator tRNA^{Met}, tRNA^{Ser}(GGA) and tRNA^{Trp} are of the chloroplast type, but not tRNA^{Cys} and tRNA^{Phe}. A "chloroplast-like" tRNA^{Cys} gene (or pseudogene) seems to be present, but not expressed, in potato mitochondria (Maréchal-Drouard et al., 1990a). As in wheat mitochondria tRNA^{Cys} and tRNA^{Phe} are of chloroplast origin and tRNA^{His} is not (Joyce and Gray, 1989), differences appear between dicotyledon and monocotyledon plants in the distribution and expression of tRNA genes located in promiscuous chloroplast DNA fragments inserted into the mitochondrial genome (Table 1). More data are still required to define whether the same chloroplast tRNA genes have been recruited independently in different plant lineages and to determine at which evolution step(s) transfers of genetic information from chloroplasts to mitochondria have occurred. Detailed studies would in particular indicate whether some of these tRNA genes were transferred before the divergence of monocotyledons and dicotyledons.

The potato "chloroplast-like" mitochondrial tRNA^{His} gene which we have sequenced (Maréchal-Drouard et al., unpublished) is 100% homologous to its sunflower counterpart (Ceci et al., 1989) and shows only two nucleotide differences with those of maize mitochondria (Iams et al., 1985) and tobacco chloroplast (Shinosaki et al., 1986). Flanking sequences are highly homologous in the vicinity of the potato and sunflower mitochondrial tRNA^{His} genes. On the contrary, these sequences are almost unrelated to the flanking sequences of maize mitochondria and tobacco chloroplast tRNA^{His} genes, although in the case of maize the sequences of the mitochondrial and chloroplast tRNA^{His} genes and of their flanking regions are identical.

We also isolated two potato mitochondrial "chloroplast-like" tRNA^{Asn} genes, which we compared to their authentic chloroplast counterparts present on the inverted repeat regions of the potato chloroplast genome (Maréchal-Drouard et al., unpublished). All four tRNA^{Asn} coding sequences are 100% homologous. The chloroplast flanking regions also show 100% homology. By contrast, the flanking regions of the two mitochondrial tRNA^{Asn} genes (tRNA^{Asn}I and II) show very different sequences in the near vicinity. The region downstream of the mitochondrial tRNA^{Asn}I gene is highly homologous to the corresponding chloroplast region. On the other hand, the region upstream of the mitochondrial tRNA^{Asn}II is highly homologous to its chloroplast couterpart. Finally, the sequence of a tomato mitochondrial "chloroplast-like" tRNA^{Asn} gene, which differs by only one nucleotide from its potato mitochondrial and chloroplast counterparts, was reported (Izuchi et al., 1990). The flanking regions of this tomato mitochondrial gene are very different from the corresponding homologous or heterologous chloroplast sequences, but the 5' flanking region of the potato mitochondrial tRNA^{Asn}I gene shows 97% sequence homology with the corresponding tomato mitochondrial region and the 3' flanking region of the potato mitochondrial tRNA^{Asn}II gene shows 95% sequence homology with the corresponding tomato mitochondrial region.

It appears from these data that, although the sequences of the tRNA genes of chloroplast origin expressed in plant mitochondria are highly conserved, comparison of their flanking regions between different plant species suggests that several genetic information transfer events from the chloroplast, followed by different rearrangements of the transferred sequences, have occurred during evolution.

TRANSCRIPTION OF tRNA GENES IN PLANT MITOCHONDRIA

As yet, little is known about the nature or location of promoter sequences for plant mitochondrial "native" or "chloroplast-like" tRNA genes. Joyce et al. (1988) identified a purine-rich motif in a conserved position upstream of wheat mitochondrial tRNA genes and proposed a consensus sequence that might play a role in initiation of transcription. The same motif has subsequently been found upstream of a number of mitochondrial tRNA genes in other plant species, in particular the potato "native" tRNA^{Ile}(L*AU) gene (Weber et al., 1990). However, up to now, there is no direct evidence that this motif plays a role in the expression of plant mitochondrial tRNA genes. On the other hand, none of the plant mitochondrial tRNA genes sequenced so far contains an intron and splicing is therefore not required for maturation of mitochondrial-encoded tRNA precursors.

NUCLEAR-ENCODED tRNAs IN PLANT MITOCHONDRIA

The presence of nuclear-encoded tRNAs which complement the tRNA population encoded by the mitochondrial genome appears to be an important and general phenomenon in plant mitochondria. However, the number and identity of the tRNA genes which are missing in the mitochondrial DNA can differ in various plant lineages, between monocotyledons and dicotyledons but also between higher and lower plants. Whereas we identified 11 nuclear-encoded tRNAs in potato mitochondria, only 3 tRNA genes were found within a sequenced 12.35 kbp region of the 15.8 kbp mitochondrial DNA of the unicellular green alga *Chlamydomonas reinhardtii* and additional information precludes the presence of many more tRNA genes in the remaining 20% of the genome (Boer and Gray, 1988). Liverwort represents the opposite extreme case, as 29 tRNA genes (none of which is "chloroplast-like"), encoding 27 tRNA species, have been located after complete sequencing of the *Marchantia polymorpha* mitochondrial DNA (Oda et al., 1992). Only two undetected tRNA genes are missing to ensure reading of all codons in the liverwort mitochondrial genome.

We found that in potato mitochondria the tRNA^{Ala}, the two tRNAs^{Arg}, a tRNA^{Ile}, the five tRNAs^{Leu} and the two tRNAs^{Thr} are nuclear-encoded (Maréchal-Drouard et al., 1990a). All five potato mitochondrial tRNAs^{Leu} have a Gm$_{18}$, whereas their cytosolic counterparts have a G$_{18}$ (Maréchal-Drouard et al., 1990b and unpublished). The tRNA^{Gly} and the two tRNAs^{Val} identified in potato mitochondria were shown to be coded for by the mitochondrial genome, which indeed contrasts with the situation in a monocotyledon plant like wheat, where "cytosolic-like" mitochondrial tRNA^{Gly} and tRNA^{Val} were characterized (Joyce and Gray, 1989). The case of potato tRNAs^{Ile} underlines another feature of the tRNA import phenomenon, namely that an imported and a mitochondrial-encoded species specific for the same amino acid can both be present in plant mitochondria to ensure decoding of synonymous codons. It is likely that in potato mitochondria the mitochondrial-encoded tRNA^{Ile}(L*AU) (see above) recognizes the AUA codon, as in *Mycoplasma capricolum* and *E. coli* (Andachi et al., 1989; Muramatsu et al., 1988), whereas the nuclear-encoded tRNA^{Ile} reads the AUU and AUC codons.

NUCLEAR GENES CODING FOR MITOCHONDRIAL tRNAs IN PLANTS

We studied a number of bean nuclear tRNA genes coding for species presumably imported into mitochondria (tRNAs^{Leu} and tRNAs^{Thr}) or present only in the cytosol (tRNA^{Pro}) (Green et al., 1986, 1987; Green and Weil, 1989; Ramamonjisoa et al., unpublished). The 5' upstream regions of all these genes are A-T rich (66 to 84%), whereas the 3' downstream regions exhibit oligo(T) stretches which may act as transcription termination sites for RNA polymerase III. Comparison of flanking regions has not revealed characteristic sequences which would be conserved upstream and/or downstream of the nuclear genes coding for tRNAs to be targeted into the mitochondria. Nuclear genes coding for tRNAs present only in the cytosol did not show either specific conserved sequences in their upstream or downstream regions. However, motifs of 5 to 6 nucleotides appear to be conserved in the upstream flanking region, in the near vicinity of the coding sequence, when considering a given gene familly (tRNAs^{Leu}, tRNAs^{Pro}(TGG) or tRNAs^{Thr}). The 5' regions of the two tRNA^{Pro}(AGG) genes studied possess a high homology over 170 nucleotides.

PLANT MITOCHONDRIAL AMINOACYL-tRNA SYNTHETASES

Very little is known about plant mitochondrial aminoacyl-tRNA synthetases, but as in fungi, they are likely to be nuclear-encoded. We found that the bean cytosolic and mitochondrial leucyl-tRNA synthetases cannot be distinguished on the basis of their specificity towards homologous or heterologous tRNAs, chromatographic mobility, molecular weight or immunological specificity (Dietrich, unpublished; Guillemaut et al., 1975; Maréchal-Drouard et al., 1988). The two bean enzymes appear therefore to be very close, if not identical, as are identical (with the exception of the Gm_{18}) the bean (or potato) mitochondrial and cytosolic leucine tRNAs which have been sequenced (see above). On the other hand, microsequencing showed that the bean cytosolic (and presumably mitochondrial) leucyl-tRNA synthetase resembles its fungal cytosolic counterpart, but clearly differs from the fungal mitochondrial leucyl-tRNA synthetase (Dietrich and Metz, unpublished).

Whereas the plant mitochondrial-encoded tRNAs can usually not be aminoacylated in the presence of plant or fungal cytosolic enzymes, the bean or potato mitochondrial tRNAs which hybridize to nuclear DNA could, in all cases tested, be aminoacylated by either the cytosolic or the mitochondrial enzymatic extracts (Green et al., 1987; Maréchal-Drouard et al., 1988; Maréchal-Drouard et al., 1990a and b), which supports both the nuclear origin of these tRNAs and their ability to function in mitochondrial protein synthesis.

MECHANISM OF tRNA IMPORT INTO PLANT MITOCHONDRIA

The mechanism of mitochondrial protein import has now been well documented in fungi, mammals and plants (e.g. Attardi and Schatz, 1988; Chaumont et al., 1990; Hartl et al., 1989; Whelan et al., 1990), but no data are available yet concerning the mechanism of tRNA import into mitochondria (Dietrich et al., 1992). One does not know in particular whether the tRNA is transported as a naked molecule or whether a complex with one or several protein(s) is involved. It is not known either whether the substrate of the import process is the mature cytosolic form of the tRNA or some kind of "precursor", which could be a processing intermediate of the transcript or a molecular species specifically elaborated in the cytosol. Among the proteins which would be involved in mitochondrial tRNA import, the cognate aminoacyl-tRNA synthetase or the tRNA methyl transferase which methylates the ribose at position 18 of the nuclear-encoded imported tRNAs have been proposed as possible candidates. The fact that partial sequence data have shown that some cytosolic counterparts of imported mitochondrial tRNAs also possess a Gm_{18} (Maréchal-Drouard, unpublished) considerably weakens the latter possibility, as well as the hypothesis according to which the Gm_{18} would act as a signal for tRNA import.

It has not yet been possible to set up an *in vitro* system allowing specific and selective tRNA import into isolated plant mitochondria. To study plant mitochondrial tRNA import and check the possible involvement of the aminoacyl-tRNA synthetases in this process, we have developed an *in vivo* system based on the import of heterologous tRNAs into the mitochondria of transgenic plants (Small et al., 1992). Using the bean nuclear tRNA[Leu](C*AA) gene to transform potato, this system brought for the first time direct proof for the import of a nuclear-encoded tRNA into plant mitochondria (see above). Oligonucleotide mutagenesis was then used to create restriction sites in the bean tRNA[Leu](C*AA) gene into which the TCGA sequence was inserted. The insertions were made into the anticodon loop or the variable loop, as it was felt that these regions, which are outside the internal promoter regions, would best tolerate alterations without too greatly affecting expression of the gene. It turned out that the construct containing the TCGA insertion in the anticodon loop was expressed at easily detectable levels in transgenic potato plants and appeared to be correctly processed, whereas modification of the variable loop gave rise to constructs which were expressed at very low or undetectable levels. These *in vivo* expression results correlated with the ability of the corresponding modified tRNAs transcribed *in vitro* to be aminoacylated by purified plant leucyl-tRNA synthetase : the transcript containing the TCGA insertion in the anticodon loop was efficiently aminoacylated, whereas the transcripts containing modified variable regions were poorly aminoacylated, as they were poorly expressed *in vivo*. Analysis of the cytosolic and mitochondrial tRNAs of the potato plants transformed with the bean tRNA^{Leu}(C*AA) gene containing the TCGA insertion in the anticodon loop revealed that the transcript from this modified gene was efficiently imported into mitochondria. It appears therefore possible, in principle at least, to import into mitochondria tRNAs carrying additional inserted sequences and which are presumably non-functional in protein synthesis. Our initial attempts at expressing tRNA genes with larger insertions (up to 113 nucleotides) have been unsuccessful, but with careful attention to the conservation of the structure of the tRNA these problems should be surmountable.

We are developing the same *in vivo* approach to check the possible involvement of the aminoacyl-tRNA synthetases in the import process. In this case, tobacco plants are transformed with tRNA^{Ala} gene constructs (another imported tRNA species) derived from the *Arabidopsis thaliana* genomic sequence and which yield transcripts able or not to interact properly with alanyl-tRNA synthetase.

CONCLUSION

The diverse genetic origin of the mitochondrial tRNA population underlines the complexity of both genetic information transfers between the plant cell compartments and mitochondrial gene divergence during evolution. The differences in the number and/or identity of the "chloroplast-like" and nuclear-encoded mitochondrial tRNA species observed among plants, especially between monocotelydons and dicotyledons, suggest that genetic information transfer has continued after segregation of plant species and that the evolutionary flux is still in progress.

The elucidation of the mechanisms involved in tRNA import may suggest ways of targeting other nucleic acids into mitochondria. As genetic transformation of plant mitochondria is currently impossible, the ability to import RNAs transcribed from genes inserted in the nuclear genome could allow the manipulation of mitochondrial gene expression in ways which are presently unavailable. Our data showing the import into the mitochondria of transgenic potato plants of the bean tRNA^{Leu}(C*AA) containing an insertion in the anticodon loop suggest that naturally imported tRNAs could serve as vectors to carry additional RNA sequences into plant mitochondria.

ACKNOWLEDGMENTS

We wish to thank M. Neuburger and R. Douce (Centre d'Etudes Nucléaires, Grenoble, France) for providing us with highly purified potato mitochondria.

REFERENCES

Andachi, Y., Yamao, Y., Muto, A. and Osawa, S., 1989, *J. Mol. Biol.* 209:37.

Attardi, G. and Schatz, G., 1988, *Annu. Rev. Cell. Biol.* 4:289.

Bartnik, E. and Borsuk, P., 1986, *Nucleic Acids Res.* 14:2407.

Binder, S., Schuster, W., Grienenberger, J.M., Weil, J.H. and Brennicke, A., 1990, *Curr. Genet.* 17:353.

Björk, G.R. and Kohli, J., 1990, *in:*"Chromatography and Modification of Nucleosides", C.N. Gehrke and K.C. Kuo, eds., pp. 813-867, Elsevier, Amsterdam.

Boer, P.H. and Gray, M.W., 1988, *Curr. Genet.* 14:583.

Ceci, L.R., Ambrosini, M., Siculella, L. and Gallerani, R., 1989, *Plant Sci.* 61:219.

Chaumont, F., O'Riordan, V. and Boutry, M., 1990, *J. Biol. Chem.* 265:16856.

Crick, F.H.C., 1966, *J. Mol. Biol.* 19:548.

Dewey, R.E., Levings, C.S.III and Timothy, D.H., 1986, *Cell* 44:439.

Dietrich, A., Weil, J.H. and Maréchal-Drouard, L., 1992, *Annu. Rev. Cell. Biol.* 8:115.

Dron, M., Hartmann, C., Rode, A. and Sevignac, M., 1985, *Nucleic Acids Res.* 13:8603.

Green, A.G., Maréchal, L., Weil, J.H. and Guillemaut, P., 1987, *Plant Mol. Biol.* 10:13.

Green, A.G. and Weil, J.H., 1989, *Plant Mol. Biol.* 13:727.

Green, A.G., Weil, J.H. and Steinmetz, A., 1986, *Plant Mol. Biol.* 7:207.

Guillemaut, P., Steinmetz, A., Burkard, G. and Weil, J.H., 1975, *Biochim. Biophys. Acta* 378:64.

Hartl, F., Pfanner, N., Nicholson, D.W. and Neupert, W., 1989, *Biochim. Biophys. Acta* 988:1.

Iams, K.P., Heckman, J.E. and Sinclair, J.H., 1985, *Plant Mol. Biol.* 4:225.

Izuchi, S., Terachi, T., Sakamoto, M., Mikami, T. and Sugita, M., 1990, *Curr. Genet.* 18:239.

Joyce, P.B.M. and Gray, M.W., 1989, *Nucleic Acids Res.* 17:5461.

Joyce, P.B.M., Spencer, D.F., Bonen, L. and Gray, M.W., 1988, *Plant Mol. Biol.* 10:251.

Jukes, T.H. and Ozawa, S., 1990, *Experientia* 46:1117.

Maréchal-Drouard, L., Dietrich, A. and Weil, J.H., 1992, *in:* "Transfer RNA in Protein synthesis", D.L. Hatfield, B.J. Lee and R.M. Pirtle, eds., pp. 125-140, CRC Press, Boca Raton.

Maréchal-Drouard, L. and Guillemaut, P., 1988, *Nucleic Acids Res.* 16:11812.

Maréchal-Drouard, L., Guillemaut, P., Cosset, A., Arbogast, M., Weber, F., Weil, J.H. and Dietrich, A., 1990a, *Nucleic Acids Res.* 18:3689.

Maréchal-Drouard, L., Neuburger, M., Guillemaut, P., Douce, R., Weil, J.H. and Dietrich, A., 1990b, *FEBS Lett.* 262:170.

Maréchal-Drouard, L., Weil, J.H. and Guillemaut, P., 1988, *Nucleic Acids Res.* 16:4777.

Maréchal, L., Guillemaut, P., Grienenberger, J.M., Jeannin, G. and Weil, J.H., 1985, *Nucleic Acids Res.* 13:4411.

Muramatsu, T., Nishihawa, K., Nemoto, F., Kuchino, Y., Nishinura, S., Miyazawa, T. and Yokoyama, S., 1988, *Nature* 336:179.

Nugent, J.M. and Palmer, J.D., 1991, *Cell* 66:473.

Oda, K., Yamato, K., Ohta, E., Nakamura, Y., Takemura, M., Nozato, N., Akashi, K., Kanegae, T., Ogura, Y., Kohchi, T. and Ohyama, K., 1992, *J. Mol. Biol.* 223:1-7

Okimoto, R. and Wolstenholme, D.R., 1991, *EMBO J.* 9:3405.

Palmer, J.D., 1990, *TIG* 6:115

Samuelson, T., Elias, P., Lustig, F., Axberg, T., Fölsch, G., Akesson, B. and Lagerkvist, U., 1983, *J. Biol. Chem.* 255:4583.

Sangaré, A., Weil, J.H., Grienenberger, J.M., Fauron, C. and Lonsdale, D., 1990, *Mol. Gen. Genet.* 223:224.

Schuster, W. and Brennicke, A., 1987, *EMBO J.* 6:2857.

Shinosaki, K., Ohme, M., Tanaka, M., Wakasuki, T., Hayashida, N., Matsubayashi, T., Zaita, N., Chunwonse, J., Obokata, J., Yamaguchi-Shinozaki, K., Ohto, C., Torazawa, K., Meng, B.Y., Sugita, M., Deno, H., Hamogashira, T., Yamada, K., Kusuda, J., Takaiwa, K., Kato, A. Tohdoh, N., Shimida, H. and Sugiura, M.,1986, *EMBO J.* 5:2043.

Small, I., Maréchal-Drouard, L., Masson, J., Pelletier, G., Cosset, A., Weil, J.H. and Dietrich, A., 1992, *EMBO J.* 11:1291.

Sprinzl, M., Hartmann, T., Weber, J., Blank, J. and Zeidler, R,. 1989, *Nucleic Acids Res.* 17(Suppl), r1.

Stern, D.B. and Lonsdale, D.M., 1982, *Nature* 299:698.

Weber, F., Dietrich, A., Weil, J.H. and Maréchal-Drouard L., 1990, *Nucleic Acids Res.* 18:5027

Weber-Lotfi, F., Maréchal-Drouard, L., Folkerts, O., Hanson, M. and Grienenberger, J.M., 1992, *Plant Mol. Biol.*, in press.

Whelan, J., Knorpp, C. and Glaser, E., 1990, *Plant Mol. Biol.* 14:977.

ASSEMBLY OF SRP FROM SINGLE POLYPEPTIDES AND 7S RNA

Henrich Lütcke and Bernhard Dobberstein

European Molecular Biology Laboratory
Meyerhofstr. 1
W-6900 Heidelberg
Germany

INTRODUCTION

Signal recognition particle (SRP) is a cytoplasmic ribonucleoprotein particle which mediates the targeting of nascent secretory and membrane proteins to the endoplasmic reticulum (ER) by virtue of three activities: (i) The binding to signal sequences once they are exposed on translating ribosomes, (ii) the subsequent slow-down of the elongation until (iii) the binding to the docking protein, the SRP receptor in the membrane of the ER. Upon interacting with the docking protein SRP detaches from the signal sequence and the ribosome. The elongation of the nascent polypeptide resumes and its translocation is initiated. SRP subsequently detaches from the docking protein in a step requiring GTP hydrolysis.

SRP consists of a 7S RNA of 300 nucleotides to which six different polypeptides of 9, 14, 19, 54, 68 and 72 kDa are attached as monomers or heterodimers. The particle has an oblong shape (Andrews et al., 1987) and can be divided structurally and functionally into two parts (Gundelfinger et al., 1983; Siegel and Walter, 1986): One part is crucial for the elongation arrest and consists of the heterodimeric 9/14 kDa protein (SRP9/14) bound to the Alu sequences at the 5' and 3' ends of 7S RNA (Strub and Walter, 1990; cf. Figure). This part of SRP and its function will not be discussed here as they are dealt with in the contribution by Strub et al. (this volume). The second part functions in recognizing and binding signal sequences and targeting to the ER (Siegel and Walter, 1986). It consists of the monomeric 19 kDa (SRP19) and 54 kDa proteins (SRP54) and the heterodimeric 68/72 kDa protein (SRP68/72) which are all attached to the so-called S fragment of 7S RNA (cf. Figure). SRP54 is the signal sequence binding protein of SRP, while SRP68/72 has been implicated in the targeting (see below). In the following we will describe the protein components in the signal recognition/targeting part of SRP and their assembly with 7S RNA.

The Translational Apparatus, Edited by K.H. Nierhaus
et al., Plenum Press, New York, 1993

THE FOUR LARGEST SRP PROTEINS

SRP19

SRP19 is a basic protein (calculated pI = 10.7) (Lingelbach et al., 1988) which directly interacts with 7S RNA. It detaches from the 7S RNA in the presence of EDTA and a high density of positive charges (Walter and Blobel, 1983). *In vitro*-synthesised SRP19 binds to 7S RNA very efficiently (Lingelbach et al., 1988; Römisch et al., 1990). The COOH-terminal 65 amino acids are dispensable for RNA binding (Lütcke et al., 1992a). SRP19 is required for the efficient association of SRP54 with 7S RNA. This was concluded from *in vitro* RNA binding assays with the canine proteins (Walter and Blobel, 1983; Römisch et al., 1990; Zopf et al., 1990). A structutral homologue of SRP19 has been identified in *S.cerevisiae*, SEC65p (Stirling and Hewitt, 1992). In yeast cells containing mutant Sec65p SRP54 was not associated with SRP RNA (Hann et al., 1992). Thus, it appears that also in *S.cerevisiae* the SRP19 homologue is required for the binding of SRP54 to the SRP RNA. The mechanism by which SRP19 facilitates the assembly of SRP54 into SRP is unknown. A direct interaction or cooperation of SRP19 and SRP54 is suggested by the identification of SRP54 as a multicopy suppressor for the mutated, but not for the deleted, Sec65p (Stirling and Hewitt, 1992). It has been postulated that binding of SRP19 to one of the stem loop structures on the 7S RNA provides a binding site for SRP54 on the other stem loop (Römisch et al., 1990).

SRP54

SRP54, like all SRP proteins, is basic (pI = 9.3) and, based on sequence comparison, can be divided into three domains (Römisch et al., 1989; Bernstein et al., 1989): (1) An NH_2-terminal N-domain, (2) a central G-domain with all the consensus elements of GTPases, and (3) a COOH-terminal M-domain, so called because of its richness in methionine residues. A similar G-domain and a less similar N-domain are also present in the α-subunit of the docking protein. The M-domain is unique to SRP54. Proteolytic cleavage of SRP54 resulted in fragments referred to as SRP54N+G (or SRP54G) and SRP54M (Römisch et al., 1990; Zopf et al., 1990).

SRP54 is an RNA binding protein. It specifically binds to the homologues of 7S RNA from E. coli (4.5 S RNA; Ribes et al., 1990) and B. subtilis (scRNA; K. Römisch, thesis, Heidelberg, 1990). Under low salt conditions it also binds to 7S RNA in the absence of SRP19, albeit with reduced efficiency (Poritz et al., 1990; Janiak and Johnson, unpublished). Possible binding sites for SRP54 and for the *E.coli* homologue of SRP54 on 4.5S RNA have been located by mutational analysis at nucleotide positions which are conserved in all 7S RNA homologues (Walter et al., unpublished; Wood et al., 1992).

SRP54 is the signal sequence binding protein of SRP. This was first concluded when selective alkylation of SRP54 abolished the ability of SRP to interact with signal sequences (Siegel and Walter, 1988), and has since been confirmed by cross-linking signal sequences of several different secretory or membrane proteins to SRP54 (Krieg et al., 1986; Kurzchalia et al., 1986; High and Dobberstein, 1991).

SRP54 binds to signal sequences as well as to 7S RNA via its M-domain (SRP54M). **SRP54M** is necessary and sufficient for binding SRP54 to the 7S RNA or the homologous RNAs from *E.coli* and *B.subtilis* (Römisch et al., 1990; Zopf et al., 1990; cf. Figure). The COOH-terminal third of canine SRP54M was found to be dispensable for 7S RNA binding (Lütcke et al., 1992a). In the M-domains of SRP54 and its homologues in *E.coli* (p48 or *ffh*) and in *Mycoplasma mycoides* (Samuelsson, 1992) no known RNA binding consensus motif has been identified. However, the very basic character of SRP54M (pI ~ 10) has been conserved during the evolution.

SRP54M also binds to signal sequences (High and Dobberstein 1991; Zopf et al., 1990). This was shown by analysing the photo-crosslinked complex of SRP with nascent preprolactin. Only SRP54M was found in contact with the signal sequence (High and Dobberstein, 1991). Even free SRP54M of dog (Lütcke et al., 1992a) and *M. mycoides* SRP54 (H. L., B. D. and T. Samuelsson, unpublished observations) interact with signal sequences. This implies that no other SRP component is required for signal sequence binding. However, it does not exclude regulatory functions of other SRP components in the binding or the release of signal sequences.

It has been proposed that three or four amphipathic helices could be formed by the M-domains of *ffh* or the mammalian SRP54, respectively, the hydrophobic faces of which could constitute a signal sequence binding site (Bernstein et al., 1989). Since the hydrophobic faces of the proposed helices are lined by the unbranched methionine side chains the model accounts for the ability of SRP54M to adapt to many different hydrophobic signal sequences. The hydrophilic faces of the helices are predominantly positively charged and have been proposed to function in binding SRP54 to 7S RNA. (High and Dobberstein, 1991). Thus, the intimate contact of the signal sequence and the RNA binding sites may allow the signal sequence binding state of SRP54M to be relayed via the 7S RNA to other regions of SRP.

SRP54N+G is a domain distinct from SRP54M and is released from SRP upon proteolysis with various proteases (Römisch et al., 1990; Zopf et al., 1990). Alkylation of SRP54N+G in intact SRP54 with NEM abolished the binding of a signal sequence to SRP54M. The effect could be reversed by proteolytic removal of the alkylated SRP54N+G (Lütcke et al., 1992a). This suggests that SRP54N+G is close to SRP54M (cf. Figure) and could thus control the binding and/or the release of signal sequences. Another likely function of SRP54N+G is the interaction with the docking protein: SRP54 has been demonstrated to bind and hydrolyse GTP in the presence but not in the absence of docking protein and 7S (or 4.5S) RNA. Binding and hydrolysis of GTP are inhibited in the presence of a signal peptide (Poritz et al., 1990 and P. Walter, unpublished; cf. Walter et al., this volume).

SRP68/72

SRP68 and SRP72 can be released from SRP as a stable heterodimer (SRP68/72). The dimer binds to 7S RNA independently of other SRP proteins (Walter and Blobel, 1983), however with much lower affinity (7±3 nM) than SRP9/14 (<0.1 nM; Janiak et al., 1992).

SRP68/72 can be detached from the 7S RNA with high ionic strength as a heterodimeric complex (Scoulica et al., 1987; Walter and Blobel, 1983). This indicates that the association of SRP68/72 with 7S RNA involves charge interactions, while both of its polypeptide components may be held together by hydrophobic interactions. SRP containing alkylated SRP68/72 still binds signal sequences on nascent polypeptides and arrests elongation but is unable to promote the targeting to the ER. Further, the modified SRP binds to a docking protein matrix with reduced affinity (Siegel and Walter, 1988). Thus, it is likely that SRP68/72 is important for the release of the signal sequence from SRP54 or the subsequent insertion of the signal sequence into the membrane.

SRP68

SRP68 is overall basic (pI = 8.8) and has a calculated M_r of 70.3 kDa (Herz et al., 1990). It binds to 7S RNA independently of SRP72. Amino acid residues involved in RNA binding lie in a very basic region of about 14 kDa (calc. pI = 11), 7 kDa away from the NH_2-terminus (Lütcke et al., 1992b; cf. Figure). A predominance of positive charges is also found in other RNA binding proteins (*e.g.* SRP54M). None of the known RNA binding consensus motifs was found in the 14 kDa region.

Previously, SRP68 had not appeared to bind to 7S RNA when tested by absorption to a DEAE matrix (Herz et al., 1990). It is likely that under these conditions SRP68 and 7S RNA dissociate. Indeed, when the complex of SRP68 with 7S RNA was isolated from a sucrose gradient it dissociated upon absorption to DEAE (H.L. & B.D., unpublished observations). The result suggests an electrostatic interaction of 7S RNA with SRP68.

In vitro synthesised SRP68, SRP72 and 7S RNA form stable complexes that do not dissociate when absorbed to DEAE (H.L. & B.D., unpublished observations). This may indicate that the interaction between SRP68 and 7S RNA is strengthened allosterically by the association with SRP72. However it is also possible that a second RNA binding site is exposed or formed once SRP68 and SRP72 associate.

No association of SRP68 and SRP72 was observed in the absence of 7S RNA. This mode of assembly is different from that found for SRP9 and SRP14 which do not bind to 7S RNA by themselves and form a heterodimer in the absence of 7S RNA (Strub and Walter, 1990).

Binding of SRP72 to SRP68/7S RNA requires a 3 kDa hydrophobic segment near the COOH-terminus of SRP68 (Lütcke et al., 1992b). This may indicate a hydrophobic interaction between SRP68 and SRP72 consistent with the observation that the SRP68/72 heterodimer does not dissociate under non-denaturing conditions.

In vitro assembly studies with SRP68 revealed an unusual intermediate step. SRP68 was found associated with a complex of the size of the small ribosomal subunit when synthesised in the absence of 7S RNA (Lütcke et al., 1992b). The nature of the complex remains to be elucidated. However, as SRP68 was released from the complex upon addition of 7S RNA, the complex may function to maintain SRP68 in a conformation competent for assembly with 7S RNA. Cytoplasmic proteins or protein complexes which mediate oligomeric assembly have been described (Yaffe et al., 1992; Lewis et al., 1992). They are usually termed chaperones. It is conceivable that part of the ribosome functions as such a chaperone.

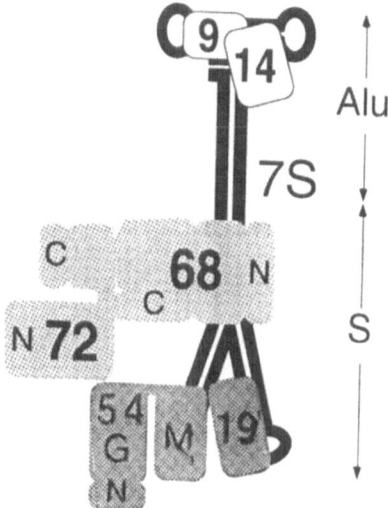

Figure 1. Model of the domains of SRP proteins and their interactions with 7S RNA. 7S RNA (7S) is represented by a thick black line. The Alu region with the attached SRP9/14 heterodimer and the S region of 7S RNA are indicated. Attached to the S region are SRP 19 (19), SRP54 (54) with its N, G and M domains, SRP68 (68), and SRP72 (72), with their amino (N) and carboxyl termini (C). Protease-sensitive sites are indicated by indentations or thin connections between domains.

Like SRP68, SRP72 is overall basic (pI= 10.0) and displays an uneven distribution of its positive charges. These are clustered in the COOH-terminal 18 kDa (calc. pI = 11) (Lütcke et al., 1992b). The remaining NH$_2$-terminal 55 kDa are weakly acidic (calc. pI = 6.7). An NH$_2$-terminal 55 kDa fragment of SRP72 can be released from SRP upon cleavage with elastase (Scoulica et al., 1987; Lütcke et al., 1992b; cf. Figure).

SRP72 remains a freely soluble protein when synthesised in the absence or presence of 7S RNA and only associates with 7S RNA in the presence of SRP68 (Lütcke et al., 1992b). This suggests that it is tethered to SRP only via the interaction with SRP68 (cf. Figure).

A small segment in the COOH-terminal portion of SRP72 is essential for the association with SRP68 (Lütcke et al., 1992b; cf. Figure).

The primary structures of all the SRP proteins are now known, and some functional domains have been identified. However, most of the molecular interactions during the SRP-mediated targeting are yet to be elucidated.

HOW AND WHERE IS SRP ASSEMBLED *IN VIVO* ?

The assembly of ribonucleoprotein particles (RNPs) from their RNA and protein components occurs in the cytoplasm (*e.g.* snRNPs) or in the nucleus (*e.g.* ribosomal subunits). Nothing is known about the steps and cellular sites of assembly of SRP which may occur in either compartment. It is also conceivable that some of the SRP proteins are transported into the nucleus where they bind to 7S RNA, whereas others may only bind, once the partially assembled SRP has reached the cytoplasm. 7S RNA is transcribed by RNA polymerase III (PolIII), and it has been suggested for other PolIII transcripts that their nuclear export is mediated by attached proteins (Zapp, 1992).

Some of the cytoplasmic SRPs are only partially assembled: Reticulocyte lysate was found to contain SRP which bound SRP54, SRP68 and SRP72 but not SRP19 (H.L. and B.D., unpublished observations). Therefore, these proteins may assemble with 7S RNA in the cytoplasm. Cytoplasmic assembly of SRP68 (and SRP72) with 7S RNA is also suggested by the reversible interaction of SRP68 with the cytoplasmic complex mentioned above.

Because of their small size, SRP19 and SRP9/14 could readily enter the nucleus where they could join up with the newly synthesised 7S RNA and thus facilitate or mediate its nuclear export.

SRP54 may be the limiting component in the assembly pathway of SRP, as its synthesis was found severely repressed by two upstream open reading frames (uORFs) present in its mRNA (Römisch et al., 1989). Such uORFs have been described in a number of systems to serve as negative translational control elements, and - as in these cases - removal of the uORFs from the SRP54 mRNA greatly de-repressed the cell-free synthesis of SRP54 (Römisch et al., 1989).

REFERENCES

Andrews, D. W., P. Walter and F. P. Ottensmeyer. (1987). *Embo J.* 6: 3471-3477.

Bernstein, H. D., M. A. Poritz, K. Strub, P. J. Hoben, S. Brenner and P. Walter. (1989). *Nature*. 340: 482-486.

Gundelfinger, E. D., E. Krause, M. Melli and B. Dobberstein. (1983). *Nucleic Acids Res.* 11: 7363-7374.

Hann, B. C., C. J. Stirling and P. Walter. (1992). Nature. 356: 532-533.

Herz, J., N. Flint, K. Stanley, R. Frank and B. Dobberstein. (1990). *FEBS Lett.* 276(1,2): 103-107.

High, S. and B. Dobberstein. (1991). *J. Cell Biol.* 113: 229-233.

Janiak, F., P. Walter and A. E. Johnson. (1992). *Biochemistry*. 31: 5830-5840.

Krieg, U. C., P. Walter and A. E. Johnson. (1986). *Proc. Natl. Acad. Sci. USA.* 83: 8604-8608.

Kurzchalia, T. V., M. Wiedmann, A. S. Girshovich, E. S. Bochkareva, H. Bielka and T. A. Rapoport. (1986). *Nature*. 320: 634-636.

Lewis, V. A., G. M. Hynes, D. Zheng, H. Saibil and K. Willison. (1992). *Nature*. 358: 249-252.

Lingelbach, K., C. Zwieb, J. R. Webb, C. Marshallsay, P. J. Hoben, P. Walter and B. Dobberstein. (1988). *Nucl. Acids Res.* 16: 9431-9442.

Lütcke, H., S. High, K. Römisch, A. J. Ashford and B. Dobberstein. (1992a). *EMBO J.* 11(4): 1543-1551.

Lütcke, H., S. Prehn, A. J. Ashford, R. Frank and B. Dobberstein. (1992b). submitted.

Poritz, M. A., H. D. Bernstein, K. Strub, D. Topf, H. Wilhelm and P. Walter. (1990). *Science*. 250: 1111-1117.

Ribes, V., K. Römisch, A. Giner, B. Dobberstein and D. Tollervey. (1990). *Cell*. 63: 591-600.

Römisch, K., J. Webb, J. Herz, S. Prehn, R. Frank, M. Vingron and B. Dobberstein. (1989). *Nature*. 340: 478-482.

Römisch, K., J. Webb, K. Lingelbach, H. Gausepohl and B. Dobberstein. (1990). *J. Cell Biol.* 111: 1793-1802.

Samuelsson, T. (1992). *Nucl. Acids Res.* 20: 5763-5770.

Scoulica, E., E. Krause, K. Meese and B. Dobberstein. (1987). *Eur. J. Biochem.* 163: 519-528.

Siegel, V. and P. Walter. (1986). *Nature*. 320: 81-84.

Siegel, V. and P. Walter. (1988). *Cell* 52: 39-49.

Stirling, C.J. and E.W. Hewitt. (1992). *Nature* 356: 534-537.

Strub, K. and P. Walter. (1990). *Mol. Cell. Biol.* 10: 777-784.

Walter, P. and G. Blobel. (1983). *Cell*. 34: 525-533.

Wood, H. J. Luirink. and D. Tollervey. (1992). *Nucl. Acids Res*: 20 No.22: in the press.

Yaffe, M. B., G. W. Farr, D. Miklos, A. L. Horwich, M. L. Sternlicht and H. Sternlicht. (1992). *Nature*. 358: 245-248.

Zapp, M. L. (1992). *Semin. Cell Biol.* 3: 289-297.

Zopf, D., H. D. Bernstein, A. E. Johnson and P. Walter. (1990). *EMBO J.* 9: 4511-4517.

THE ALU-DOMAIN OF THE SIGNAL RECOGNITION PARTICLE

Katharina Strub[1], Nicole Wolff[1] and Suzanne Oertle[2]

[1]Département de Biologie Cellulaire
Université de Genève
Sciences III
30, Quai Ernest-Ansermet
CH-1211 Genève 4, Switzerland
[2]Hirnforschungs-Institut der Universität Zürich
A-Forelstr.1
CH-8029 Zürich, Switzerland

INTRODUCTION

A fundamentally important task of all eucaryotic cells is to direct proteins to their final intra- or extracellular location. A cytoplasmic component, the signal recognition particle (SRP), plays a key role in one of these sorting processes. It recognizes secretory, lysosomal and membrane proteins and efficiently targets them to the endoplasmic reticulum (ER). The ER is the first and common subcellular compartment in the localization pathway of these proteins. SRP consists of 6 polypeptides and 1 RNA molecule (Walter and Blobel, 1982) and belongs therefore to a group of cellular components that are called ribonucleoproteins.

A model has been proposed by Walter *et al.* (1984) which describes the different functions of SRP in the sorting pathway of ER-targeted proteins (for review see Walter and Lingappa, 1986). This model summarizes the results obtained in the characterization of SRP using an assay system that reconstitutes the translation-translocation process in vitro (Blobel and Dobberstein, 1975). This system is composed of wheat germ lysate programmed with exogenous mRNAs, of salt-extracted canine microsomal membranes (RM) and of canine SRP (Walter and Blobel, 1980). According to the model, SRP binds specifically to the signal sequence of the nascent polypeptide chain as it emerges from the ribosome. The binding of SRP to the nascent chain-ribosome complex effects an arrest or a delay in the elongation of the nascent chain (Walter and Blobel, 1981; Meyer *et al.*, 1982). The interaction of SRP with the SRP receptor complex (docking protein) targets the SRP-ribosome-nascent chain complex to the ER (Gilmore *et al.*, 1982; Meyer *et al.*, 1982). Protein synthesis resumes at its normal speed and a functional ribosome membrane junction establishes which allows further translation and translocation of the protein across the ER membrane. SRP and SRP receptor complex dissociate from the ribosome and from each other by a GTP-dependent mechanism (Connolly and Gilmore, 1986; Connolly and Gilmore, 1989) and are then free to engage in another targeting round.

In this article we summarize the current understanding of the elongation arrest function of SRP and further characterize the components that play an essential role in mediating the SRP-dependent delay or arrest in the synthesis of ER-targeted polypeptides.

The presence of SRP during the synthesis of preprolactin in wheat germ lysate led to the accumulation of a short peptide of defined length (arrested fragment)(Walter and Blobel, 1981). This observation indicated the existence of the elongation arrest activity of SRP. Similarly, the elongation of nascent immunoglobulin light chains was specifically inhibited when SRP was added to the translation reaction (Meyer *et al.*, 1982). In both cases, the arrest released upon addition of salt-washed membranes. This release was due to the interaction of SRP-ribosome-nascent chain complex with the SRP receptor complex (also called docking protein)(Gilmore *et al.*, 1982; Meyer *et al.*, 1982) The inhibition of full-length presecretory protein synthesis was specific for ER-targeted proteins and depended on the signal recognition function of SRP (Walter, 1981).

Rather than a single arrested fragment, a ladder of arrested fragments accumulated during biosynthesis of several other secretory proteins in the presence of SRP (Lipp *et al.*, 1987). The size of the arrested fragments increased over time resulting in full-length preproteins. These findings suggested that SRP mediated a transient delay in protein synthesis rather than a complete arrest. Likewise, the addition of exogenous canine SRP to the reticulocyte translation system effected a delay in the accumulation of full-length protein (Wolin and Walter, 1989). The distribution of ribosomes along preprolactin mRNA during synthesis of the protein indicated that ribosomes paused naturally at certain sites which were identical in the wheat germ and in the reticulocyte translation systems (Wolin and Walter, 1988). The SRP-mediated delay in biosynthesis of full-length preprolactin was due to enhanced pausing of the ribosomes at these specific sites (Wolin and Walter, 1989). Plant SRP did not effect a complete arrest of secretory protein synthesis in the homologous translation system (Prehn *et al.*, 1987). However, it remains to be examined, whether plant SRP effects a kinetic delay in the synthesis of ER-targeted proteins. For simplicity, the translational control function of SRP will be called elongation arrest activity throughout this article, albeit SRP affects the elongation rate of the nascent polypeptide chain at variable degrees in the different translation systems.

Y. lipolytica contains two functional genes that encode SRP RNA homologues. The simultaneous disruption of both genes is lethal (He *et al.*, 1989) and growth of such a haploid strain becomes thus dependent on an extrachromosomal copy of the 7SL RNA gene. Recently, two independent sets of double mutations in a single extrachromosomal SRP RNA gene copy were identified that resulted in temperature-sensitive growth. The shift to the non-permissive temperature led to a specific decrease in the synthesis of alkaline extracellular protease and not, as might be expected for a translocation-deficient SRP mutant, to the accumulation of precursor protein in the cytoplasm (He *et al.*, 1992; Yaver *et al.*, 1992). Possibly, these mutations inhibit the release of the elongation arrest mediated by SRP. Both double mutations lay within a region that is essential for binding of SRP19 (Zwieb, 1992) and appear to interfere with the assembly of the particle in vivo (He et al., 1992). Further analysis is required to corroborate the current interpretation of these results and to understand the mechanism of the inhibition.

It is important to re-emphasize, that the translocation of ER-targeted proteins across mammalian microsomes, with few exceptions (Schlenstedt *et al.*, 1990), was found to be dependent on ribosomes, SRP and SRP receptor. Some secretory proteins can be translocated until late in synthesis (Ainger and Meyer, 1986) as long as protein synthesis has not been terminated and the nascent chain remains associated with the ribosome (Garcia and Walter, 1988). The translocation without ongoing protein synthesis (Caulfield *et al.*, 1986; Chao *et al.*, 1987; Hansen and Walter, 1988; Mueckler and Lodish, 1986; Perara *et al.*, 1986) and the SRP-dependent targeting of nascent chains to the ER (Siegel and Walter, 1988b) also requires the association of the ribosome with the presecretory polypeptide chain. These findings suggest an important role for the ribosomes in promoting the translocation process and make it clearly distinct from the truly post-translational translocation across yeast microsomes which is not dependent on ribosomes, SRP and SRP receptor complex (Hansen *et al.*, 1986; Rothblatt and Meyer, 1986; Waters and Blobel, 1986).

It is conceivable from these results that the elongation arrest function of SRP ensures the efficiency of the translocation process in vivo by lengthing the time span during which a functional ribosone-membrane junction may be established. This hypothesis is also the conclusion of a proposed mathematical model that takes into account the effects of SRP on translation and translocation of ER-targeted proteins (Rapoport *et al.*, 1987). This mathematical model also predicts that, in vivo, no strong translational inhibition (piling up of ribosomes to the translation initiation site), but rather a local accumulation of ribosomes at the arrest site(s) would be sufficient to reach maximum efficiency for protein translocation.

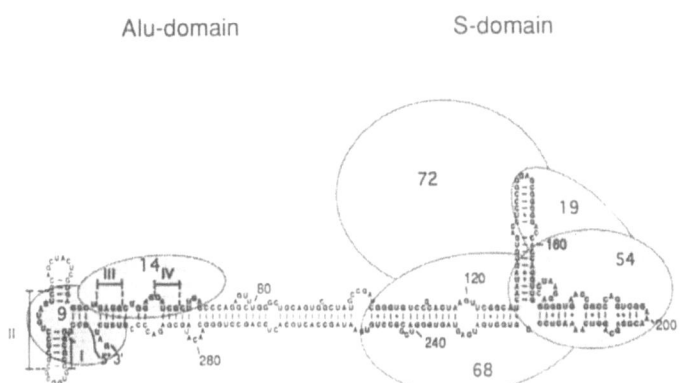

Figure 1. Schematic representation of canine SRP. The proteins are labeled with their apparent molecular weight in kDa (Walter and Blobel, 1980). Regions labeled I, II, III and IV mark contact sites between the RNA and the heterodimer SRP9/14 (Strub *et al.*, 1991).

COMPONENTS OF CANINE SRP THAT MEDIATE ELONGATION ARREST

SRP consists of six polypeptides, named according to their apparent molecular weight in SDS-PAGE (SRP72, SRP68, SRP54, SRP19, SRP14 and SRP9)(Walter and Blobel, 1980), and one RNA molecule of 300 nucleotides (Walter and Blobel, 1982), 7SL RNA or SRP RNA (Ullu *et al.*, 1982). The signal recognition, the targeting and the elongation arrest functions of SRP have been assigned to different components or domains of the particle. The central domain of SRP (S-domain) contains the 4 large polypeptides and the central portion of SRP RNA (Fig. 1, nucleotides 101-250). It can be isolated as a sub-particle, SRP(S), after treating SRP with micrococcal nuclease (Gundelfinger *et al.*, 1983). SRP(S) was found to be essential and sufficient for the signal recognition and targeting activity of SRP (Siegel and Walter, 1986). However, SRP(S) lacks elongation arrest activity. Cross-linking experiments and experiments using biochemically altered SRP preparations assigned the signal recognition function to SRP54 and the targeting function to the heterodimeric protein SRP72/68 (Krieg *et al.*, 1986; Kurzchalia *et al.*, 1986; Siegel and Walter, 1988a).

The remaining components of SRP are contained within the Alu-domain and include 100 nucleotides from the 5' and 50 nucleotides from the 3' end of SRP RNA and the two small polypeptides SRP9 and SRP14 (Fig. 1) which bind to SRP RNA as a heterodimer (see below). The sequences of SRP RNA within this domain are homologous to the mammalian Alu family of repetitive sequences (Ullu *et al.*, 1982). Phylogenetic evidence indicates that SRP is the evolutionary progenitor of the repetitive Alu sequences (Ullu and Tschudi, 1984). Two types of altered SRP particles were identified that had both lost elongation arrest activity: (i) A particle which lacks the heterodimer SRP9/14 but still contains the complete SRP RNA

(Siegel and Walter, 1985) and, as mentioned before, (ii) the sub-particle SRP(S) which lacks RNA and protein moieties of the Alu-domain (Siegel and Walter, 1986). Addition of SRP9/14 to SRP(S) could not restore elongation arrest activity of the sub-particle. These results demonstrated that both the RNA and the proteins in the Alu-domain of SRP, play an essential role in conferring elongation arrest activity to the particle. However, in addition to the components in the Alu-domain, the elongation arrest activity of SRP also depends on the signal recognition function of SRP (Walter and Blobel, 1981; Siegel and Walter, 1988a), a finding that corroborated the specificity of the translational control function of SRP. Furthermore, an SRP particle assembled without SRP68/72, SRP-68/72, also lacks elongation arrest (Siegel and Walter, 1985). The negative phenotype of the SRP-68/72 mutant could simply reflect a defect in the assembly of the particle. Alternatively, SRP68/72 might act as a signal transducer between the two domains.

Both SRP9 and SRP14 Proteins are required for elongation arrest

Murine and canine cDNAs encoding SRP9 and SRP14 proteins were isolated (Strub and Walter, 1989; Strub and Walter, 1990; Strub and Walter, unpublished results). A functional assay confirmed the authenticity of the cDNAs. The proteins derived from the cDNA clones by in vitro transcription followed by the translation of the synthetic mRNAs in wheat germ lysate restored elongation arrest activity to SRP(-9/14)(Strub and Walter, 1990). The primary protein sequence of SRP9 and SRP14 indicated that both proteins are very polar with an overall basic character. The two proteins have no significant similarity to other proteins in the data bank except to SRP19. SRP9 shares a stretch of 6 amino acids sequence identity with SRP19. The conserved motif is located at position 35 to 40 and 90 to 95 in the protein sequences of SRP9 and SRP19, respectively. In addition, some of the amino acids in the flanking regions of the motif are also conserved between the 2 proteins (Strub and Walter, 1990). In SRP14 the core motif is only partially conserved, however, some of the conserved amino acids in the flanking regions are also found in SRP14. Another common feature of SRP14 and of SRP19 is their highly basic C-terminal domain. At the carboxy termini of SRP19 and of SRP14, 7 out of 9 and 9 out of 16 amino acids, respectively, are lysines and arginines. In SRP19 this basic domain is not required for the formation of a SRP RNA-protein complex (Römisch et al., 1990). Possibly, these basic regions could play a role in the interaction with the ribosome.

Despite the overall basic character of SRP9 and SRP14, neither of the two proteins alone bound to SRP RNA. Rather, the protein-SRP RNA complex is exclusively formed with both proteins. Furthermore, the two proteins form a dimer in the absence of SRP RNA. The heterodimer SRP9/14 is therefore most likely an intermediate in the assembly of the RNA-protein complex (Strub and Walter, 1990). It remains to be determined whether the RNA-binding pocket is composed of regions from both proteins or whether dimerization induces the formation of such a pocket in one of the proteins. Sulfhydryl groups in both proteins are specifically shielded from modification in the RNA-protein complex but not in the hetero-dimer (Siegel and Walter, 1988a). This observation rather supports the first model.

The protein-RNA complex is very stable with a $K_d < 0.1$ nM as determined by using fluorescein-labeled RNA in binding experiments. Furthermore, the addition of SRP9/14 to SRP RNA induced an allosteric change in its conformation at the 3' end (Janiak et al., 1992). In similar experiments, the dissociation constant of the SRP68/72-SRP RNA complex was found to be about 10 fold higher and there was no cooperativity in the binding of the two heterodimers to SRP RNA. Thus, the two heterodimers may assemble in a random fashion on SRP RNA. Previous experiments by Walter and Blobel (1983) suggested that binding of SRP proteins to the RNA was cooperative. This interpretation was based on the observation that elongation arrest and translocation efficiency remained constant upon addition of increasing amounts of SRP RNA. Given the results outlined above, such a cooperativity in binding would have to be mediated by SRP19 and/or SRP54.

In summary we can say, that both, SRP14 and SRP9, are required to assemble a functional elongation arrest domain of SRP. It remains to be elucidated whether the functional interaction with the ribosomes involves one or both proteins. Moreover, the heterodimer SRP9/14 appears to belong to aa as yet uncharacterized class of RNA-binding proteins. The functional dissection of the heterodimeric protein will therefore give new insights into protein-protein and protein-RNA interactions.

TOWARDS A SYNTHETIC SRP

Large amounts of SRP proteins and RNA are required to study molecular interacions within SRP and with other cellular components on which SRP exerts its functions, for example the ribosomes. SRP RNA can be produced in large quantities in vitro by phage RNA polymerases and the in vitro synthesized RNA reconstitutes together with canine SRP

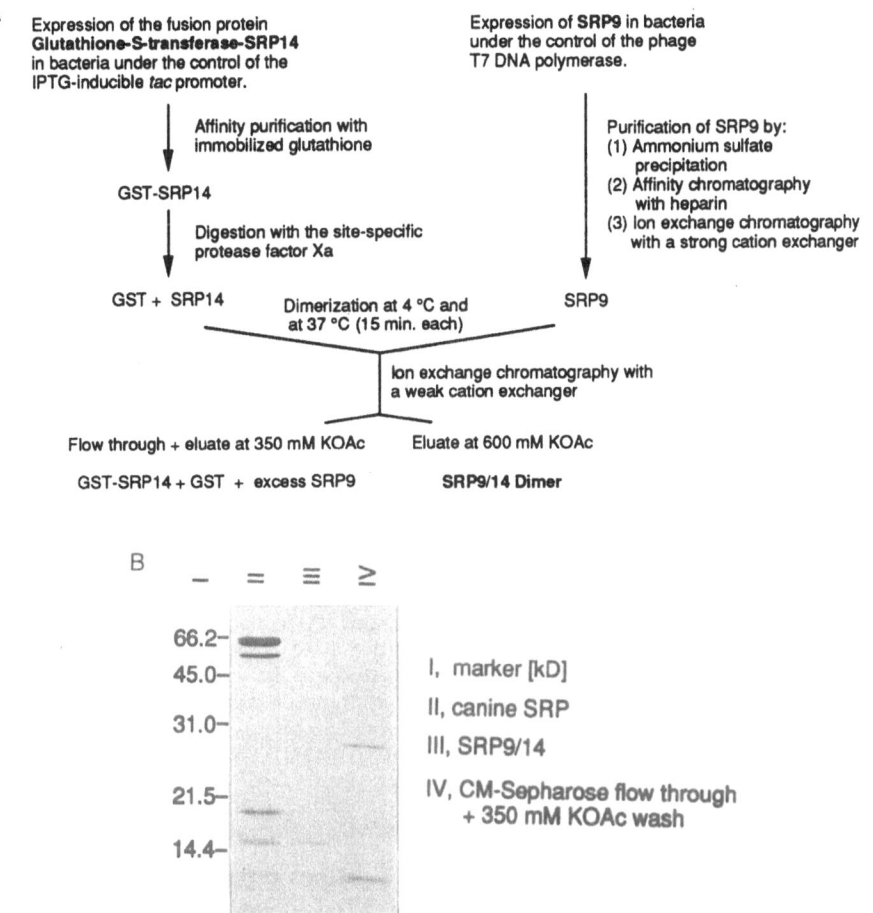

Figure 2. SRP9 and SRP14 produced in *E. coli*. A. Purification scheme for the heterodimer SRP9/14. The glutathione-S-transferase-SRP14 fusion protein was produced and purified as described in Smith and Johnson (1988). The T7 DNA polymerase-dependent expression system is outlined in Studier *et al.* (1990). **B.** Analysis of the various protein fractions on a SDS-17 % polyacrylamide gel. The proteins were visualized by Coomassie blue staining.

proteins into a functional SRP (Strub *et al.*, 1991). Taking advantage of SRP9 and SRP14 cDNAs, we attempted to produce these proteins in bacteria.

We found that SRP9 is readily expressed at high levels in bacteria under the control of the bacteriophage T7 promotor. A purification protocol (Fig. 2A) which includes ammonium sulfate precipitation, heparin affinity and ion exchange chromatography, was subsequently used to purify SRP9 to homogeneity. In contrast, the murine SRP14 was unstable in bacteria when produced under the same experimental conditions. Nevertheless, SRP14 sequences could be expressed at high levels as a glutathione-S-transferase fusion protein (Smith and Johnson, 1988). The fusion protein was purified by affinity chromatography using glutathione covalently linked to agarose beads. The purified fusion protein was cleaved with the sequence-specific protease factor Xa and the products of the cleavage reaction were combined with SRP9. Dimerization of the proteins was allowed to proceed and the heterodimer was purified by cation exchange chromatography on CM-Sepharose. The uncleaved fusion protein, the glutathione-S-transferase and the excess of SRP9 used to drive the dimerization reaction, all eluted in the flow through and wash fractions of the column (Fig. 2B, lane IV). Both proteins migrated as expected in SDS-PAGE (Fig. 2B, lane III) when compared to canine SRP proteins and by taking into account that SRP14 contains 8 additional amino acids at the N-terminus (Fig. 2B, lane II).

The biological activities of the proteins were assayed using wheat germ lysate programmed with exogenous mRNAs and canine microsomes as described in Strub and Walter (1990). A secretory protein and a cytoplasmic protein, preprolactin and cyclin, are both synthesized without SRP (Fig. 3A, lane 1) or in the presence of an incompletely reconstituted SRP lacking SRP9/14 (Fig. 3A, lanes 4 and 5). In contrast, the presence of authentic canine SRP specifically inhibits the biosynthesis of preprolactin as compared to the biosynthesis of cyclin (Fig. 3A, lanes 2 and 3). The lack in elongation arrest activity of SRP-9/14 is rescued by the addition of the recombinant heterodimer SRP9/14 (Fig. 3B, lanes 6 and 7). Likewise, preprolactin is translocated efficiently into mammalian microsomes in the presence of fully reconstituted canine SRP (Fig. 3B, lanes 1 and 2) but not without SRP (Fig. 3B, lane 3) or in the presence of SRP-9/14 (Fig. 3B, lanes 4 and 5). Without the elongation arrest activity of SRP the time window during which preprolactin can be translocated across mammalian microsomes becomes very narrow. Not all the nascent chains will reach the membrane in this short time span and the efficiency of translocation is therefore decreased (Siegel and Walter, 1985). The addition of SRP9 and SRP14 synthesized in bacteria rescued the defect in translocation (Fig. 3B, lane 6 and 7). These results demonstrate that SRP9 and SRP14 proteins produced in E. coli can functionally replace canine SRP proteins.

As mentioned before, it is possible to produce an S-domain sub-particle (SRP(S) by cleaving SRP RNA in the particle with micrococcal nuclease (Gundelfinger et al., 1983; Siegel and Walter, 1986). In contrast, the components in the Alu-domain of SRP dissociate from each other under the same reaction conditions (Siegel and Walter, 1986) and it has therefore not been possible to study the biological activity of an Alu-particle alone. The Alu-portion of SRP RNA synthesized in vitro, comprising the 100 nucleotides at the 5' and the 50 nucleotides at the 3' end, was shown to contain all the structural elements required for SRP9/14 binding (Strub et al., 1991). A synthetic Alu-particle, assembled from the RNA synthesized in vitro and SRP9/14 produced in E. coli, does not inhibit synthesis of proteins in vitro in a concentration range where SRP would completely and specifically arrest elongation of preprolactin (K. Strub, unpublished results). This result may not be surprising, since the elongation arrest activity displayed by SRP is specific for secretory protein synthesis and, thus, dependent on the signal recognition function. The binding of the Alu-domain to the ribosomes may require the cooperative interaction with the S-domain. The lack of elongation arrest activity of the Alu-particle alone could thus be explained by its low affinity for ribosomes. Alternatively, the binding of the S-domain of the particle to the signal sequence could induce a conformational change in the Alu-domain that is required for its function.

Mammalian cDNAs for all SRP proteins have now been isolated (Lingelbach et al., 1988; Bernstein et al., 1989; Römisch et al., 1989; Strub and Walter, 1989; Herz et al., 1990; Strub and Walter, 1990; Lütcke et al., this volume) and it may soon be feasible to assemble biologically active SRP from components that were either produced in vitro or in E. coli. This possibility will greatly facilitate the analysis of the structure-function relationship in SRP. Individual components of SRP can be specifically altered by site-directed mutagenesis of the cDNAs or the SRP RNA gene. These altered components are subsequently used together with unaltered SRP proteins to reconstitute SRP particles. The effect of the alteration on the biological activity of SRP can then be assayed in vitro.

EVOLUTIONARY CONSERVATION OF SRP9/14 BINDING SITES

Complete RNAse-digestion and filter binding experiments demonstrated that the Alu-domain of SRP RNA contains all the structural elements required for binding of the heterodimer SRP9/14. Indeed, in vitro syntesized transcripts, spanning the 60 nucleotides at the 5' end of SRP RNA, bind SRP9/14 efficiently, albeit at lower stringency (at 250mM salt instead of 500 mM salt) than a transcript comprising all the Alu sequences (Strub *et al.*, 1991). These results were consistent with the previous observation that SRP9/14 remained bound to the RNA that is cleaved at position 72 in canine SRP RNA (Gundelfinger *et al.*, 1983).

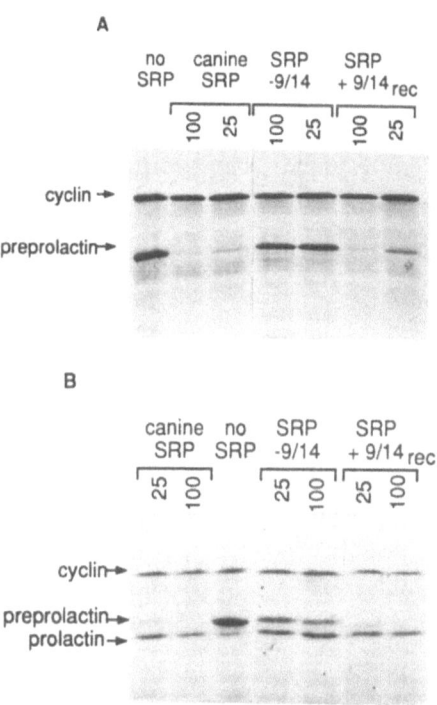

Figure 3. SRP9 and SRP14 produced in *E. coli* can functionally replace canine SRP proteins. A. Elongation arrest activity of SRP samples reconstituted from canine proteins alone or from SRP9/14 produced in *E. coli* (rec) together with canine proteins. Reconstituted SRPs were added at the concentrations indicated above the lanes (nM) to 10 μl wheat germ translation reactions programmed with synthetic preprolactin and cyclin mRNAs. After incubation for 25 min at 26 °C, the proteins were precipitated with trichloroacetic acid and analyzed by SDS-PAGE followed by autoradiography. **B.** Translocation promoting activity of the same SRP samples in the presence of salt-washed canine microsomes. The translation-translocation reactions were incubated for 40 min. at 26 °C before analysis.

Probing with hydroxyl radicals as RNA cleaving reagent (Tullius and Dombrowski, 1986) revealed four specific sites in the Alu domain that are in close contact with the protein (Fig. 1, I, II, III, IV). The region I spans the first half of the stem-loop structure at the 5' end of the RNA. The region II comprises the second half of the first stem-loop structure, the adjacent bulge and two bases of the second stem-loop structure. Regions III and IV are located at the beginning of the central stem structure of SRP RNA and are spaced 11 bases apart. The latter finding suggests that regions III and IV lie on the same face in a putative A-helical structure of the RNA (Strub *et al.*, 1991). The 3' end of SRP RNA, which constitutes the complementary strand of the stem structure in this region, was not protected from hydroxyl radical cleavage in the presence of SRP9/14.

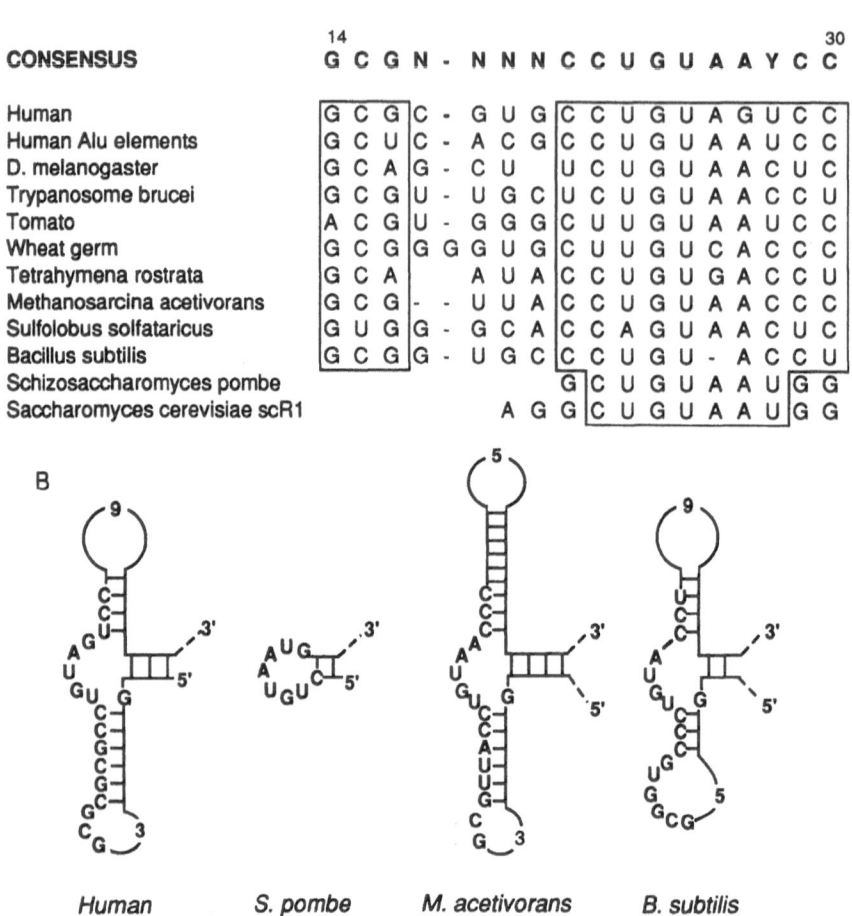

Figure 4. Sequence conservation in the SRP9/14 binding site II. A. The primary sequences in the SRP9/14 binding site II of a representative set of SRP RNAs (for a more detailed sequence comparison see Strub et al. (1991). The complete motif spans nucleotides 14- 30 in human and canine SRP RNA (see Fig. 1). It comprises two highly conserved sequence blocks separated by a region that is base-paired in the secondary structure model of SRP RNA. **B.** The conserved motif shown in the context of the secondary structure model of different SRP RNAs.

Nucleotides 5' of regions I and II were inaccessible for the hydroxyl radical cleaving reagent in the absence of the proteins. According to Celander and Cech (1991), this suggests that these sequences lie on the inside in a putative tertiary structure model of the free SRP RNA. Nevertheless, because of the intrinsic resistance of these regions to the cleavage reagent, it remained uncertain, in this study, whether they are part of the protein binding site. In contrast, Andreazzoli and Gerbi (1991) found that the two G residues 5' of region II, G_{14} and G_{16} in canine SRP RNA, were modified with kethoxal in naked RNA but not in canine SRP, suggesting that these nucleotides are in close proximity of the proteins.

Recently, several components of the mammalian SRP-dependent translocation machinery were found to be conserved throughout evolution. Homologues of SRP54, SRP19 and SRP receptor were identified in yeast and eubacteria (Bernstein *et al.*, 1989; Hann *et al.*, 1989; Römisch *et al.*, 1989; Amaya *et al.*, 1990; Ogg *et al.*; 1992; Samuelsson, 1992; Stirling and Hewitt, 1992) on the basis of primary sequence similarity between the proteins. In contrast, the primary structure of SRP RNA has diverged rapidly and SRP RNA homologues had therefore to be identified on the basis of their overall similarity in secondary structure (for references of SRP RNAs see Larsen and Zwieb, 1991). A structural element in the S-domain of SRP RNA (Poritz et al., 1988; Struck et al., 1988) is particularly well conserved in evolution. This domain most likely serves as a binding site for SRP54 and is therefore involved in the signal recognition function of SRP (Ribes et al., 1990; Römisch et al., 1990; Zopf et al., 1990).

The secondary structure conservation in the Alu-portion of SRP RNA is less striking. Some structural features, like two stem-loop motifs located in the very 5' portion of SRP RNAs, are found conserved in higher eucaryotes and archaebacteria. Their exact secondary structure, such as length of stems and loops, is, however, quite variable. In addition, these stem-loop structures are completely (*S. pombe*, Fig. 4B) or partially missing in yeast SRP RNA homologues. Certain features, like an extended stem structure composed of sequences at the very 5' and 3' ends of the SRP RNA (named terminal helix by Kaine and Merkel (1989)), are even exclusively found in SRP RNA homologues of archaebacteria. The analysis of the SRP9/14 binding sites in SRP RNA revealed, however, that the mostly single-stranded binding site II in the Alu-domain (Fig. 1) is highly conserved in primary structure from eubacteria and archebacteria to yeast and higher eucaryotes (Strub *et al.*, 1991).

The conserved region spans nucleotides 14-30 in canine and human SRP RNA and consists of two highly conserved motifs separated by a stretch of non-conserved bases (Fig. 4A). The stretch of non-conserved sequence is base-paired in the secondary structure model of SRP RNA (Fig. 4B) which may explain why its primary structure has diverged. The region in SRP RNA that is protected from the RNA cleaving reagent in the RNA-protein complex certainly includes the base-paired region and the second highly conserved motif. However, it remains uncertain, for reason outlined before, whether the first conserved sequence motif is also part of the SRP9/14 binding site. The SRP9/14 binding site is also conserved in its location in the overall secondary structure of SRP RNA (Fig. 4B). It is, with few exceptions (for details see Strub *et al.*, 1991), found at the tip of the long central stem in SRP RNA.

In yeast, the second conserved motif partially coincides with another conserved motif, the A box DNA consensus sequence, that has been shown to constitute an important promotor element for the transcription of tRNA, 5S RNA and some viral RNA genes by polymerase III (Geiduschek and Tocchini-Valentini, 1988). However, no evidence has been obtained so far that elements in the putative A box of the SRP RNA gene are involved in the transcription by polymerase III. The analysis of the promotor elements required for the transcription of the mammalian SRP RNA gene by polymerase III has demonstrated that its transcription depends solely on extragenic sequences 5' of the gene and on the internal nucleotides G_{15} and G_{16} (Fig. 4A, second and third nucleotide in the first conserved motif)(Ullu and Weiner, 1985; Kleinert *et al.*, 1988; Bredow *et al.*, 1990). The requirements for the polymerase III transcription of the yeast SRP RNA genes have not been analyzed. The following results, obtained in a recent mutational analysis of the SRP RNA homologue of *S. pombe* , rather support the notion that the primary sequence conservation in this domain is critical for protein binding. The mutation of G_4 (Fig. 4A, the third nucleotide in the second conserved motif) to C_4 results in a conditional phenotype. This nucleotide is flexible in the A box consensus sequence for RNA polymerase III but very highly conserved in the SRP9/14 binding site consensus sequence. In contrast, the mutation of G_9, a nucleotide very highly conserved in the A box consensus, to A_9 did not affect cell growth (Liao *et al.*, 1992).

The high primary sequence conservation in the SRP9/14 binding site II suggests that SRP9 and SRP14 homologues may exist in a vast variety of organisms. If the structural conservation reflects a functional conservation it would indicate, that the elongation arrest function of SRP is also very ancient and important for many organisms.

ACKNOWLEDGMENTS

We thank Didier Picard for reading the manuscript and Mark Poritz for many helpful discussions. This work was supported by a grant from the Swiss National Science Foundation. K.S. is a fellow of the START programme of the Swiss National Science Foundation.

REFERENCES

Ainger, K.J. & Meyer, D.I. (1986) *EMBO J.*, **5**, 951-955.

Amaya, Y., Nakano, A., Ito, K. & Mori, M. (1990) *J. Biochem.*, **107**, 457-463.

Andreazzoli, M. & Gerbi, S.A. (1991) *EMBO J.*, **10**, 767-777.

Bernstein, H.D., Poritz, M.A., Strub, K., Hoben, P.J., Brenner, S. & Walter, P. (1989) *Nature*, **340**, 482-486.

Blobel, G. & Dobberstein, B. (1975) *J. Cell Biol.*, **67**, 835-851.

Bredow, S., Kleinert, H. & Benecke, B.-J. (1990) *Gene*, **86**, 217-225.

Caulfield, M.P., Duong, L.D. & Rosenblatt, M. (1986) *J. Biol. Chem.*, **261**, 10953-10956.

Celander, D.W. & Cech, T.R. (1991) *Science*, **251**, 401-407.

Chao, C.C., Bird, K.P. & Gething, M.J.S., J. (1987) *Mol. Cell. Biol.*, **7**, 3842-3845.

Connolly, T. & Gilmore, R. (1986) *J. Cell Biol.*, **103**, 2253-2261.

Connolly, T. & Gilmore, R. (1989) *Cell*, **57**, 599-610.

Garcia, P.D. & Walter, P. (1988) *J. Cell Biol.*, **106**, 1043-1048.

Geiduschek, P.E. & Tocchini-Valentini, G.P. (1988) *Ann. Rev. Biochem.*, **57**, 873-914.

Gilmore, R., Blobel, G. & Walter, P. (1982) *J. Cell Biol.*, **95**, 463-469.

Gundelfinger, E.D., Krause, E., Melli, M. & Dobberstein, B. (1983) *Nucl. Acids Res.*, **11**, 7363-7374.

Hann, B.C., Poritz, M.A. & Walter, P. (1989) *J. Cell Biol.*, **109**, 3223-3230.

Hansen, W., Garcia, P.D. & Walter, P. (1986) *Cell*, **45**, 397-406.

Hansen, W. & Walter, P. (1988) *J. Cell Biol.*, **106**, 1075-1081.

He, F., Beckerich, J.-M. & Gaillardin, C. (1992) *J. Biol. Chem.*, **267**, 1932-1937.

Herz, J., Flint, N., Stanley, K., Frank, R. & Dobberstein, B. (1990) *Febs Lett.*, **276**, 103-7.

Janiak, F., Walter, P. & Johnson, A.E. (1992) *Biochem.*, **31**, 5830-5840.

Kaine, B.P. & Merkel, V.L. (1989) *J. Bact.*, **171**, 4261-4266.

Kleinert, H., Gladen, A., Geisler, M. & Benecke, B.-J. (1988) *J. Biol. Chem.*, **263**, 11511-11515.

Krieg, U.C., Walter, P. & Johnson, A.E. (1986) *Proc. Natl. Acad. Sci. USA*, **83**, 8604-8608.

Kurzchalia, T.V., Wiedmann, M., Girshovich, A.S., Bochkareva, E.S., Bielka, H. & Rapoport, T.A. (1986) *Nature*, **320**, 634-636.

Larsen, N. & Zwieb, C. (1991) *Nucl. Acids Res.*, **9**, 209-15.

Liao, X., Selinger, D., Althoff, S., Chiang, A., Hamilton, D., Ma, M. & Wise, J.A. (1992) *Nucl. Acids Res.*, **20**, 1607-1615.

Lingelbach, K., Zwieb, C., Webb, J.R., Marshallsay, C., Hoben, P.J., Walter, P. & Dobberstein, B. (1988) *Nucl. Acids Res.*, **16**, 9431-9442.

Lipp, J., Dobberstein, B. & Haeuptle, M.-T. (1987) *J. Biol. Chem.*, **262**, 1680-1684.

Meyer, D.I., Krause, E. & Dobberstein, B. (1982) *Nature*, **297**, 647-650.

Mueckler, M. & Lodish, H.F. (1986) *Nature*, **322**, 549-552.

Ogg, S.C. *et al.* (1992) *Molecular Biology of the Cell*, **3**, 895-911.

Perara, E., E., R.R. & Lingappa, V.R. (1986) *Science*, **232**, 348-352.

Prehn, S., Wiedmann, M., Rapoport, T.A. & Zwieb, C. (1987) *EMBO J.*, **6**, 2093-2097.

Rapoport, T.A., Heinrich, R., Walter, P. & Schulmeister, T. (1987) *J. Mol. Biol.*, **195**, 621-636.

Römisch, K., Webb, J., Herz, J., Prehn, S., Frank, R., Vingron, M. & Dobberstein, B. (1989) *Nature*, **340**, 478-482.

Römisch, K., Webb, J., Lingelbach, K., Gausepohl, H. & Dobberstein, B. (1990) *J. Cell Biol.*, **111**, 1793-1802.

Rothblatt, J.A. & Meyer, D.I. (1986) *Cell*, **44**, 619-628.

Samuelsson, T. (1992) *Nucl. Acids Res.*, **20**, 5763-5770.

Schlenstedt, G., Gudmundsson, G.H., Boman, H.G. & Zimmermann, R. (1990) *J. Biol. Chem.*, **265**, 13960-13968.

Siegel, V. & Walter, P. (1985) *J. Cell Biol.*, **100**, 1913-1921.

Siegel, V. & Walter, P. (1986) *Nature*, **320**, 81-84.

Siegel, V. & Walter, P. (1988a) *Cell*, **52**, 39-49.

Siegel, V. & Walter, P. (1988b) *EMBO J.*, **7**, 1769-1775.

Smith, D.B. & Johnson, K.S. (1988) *Gene*, **67**, 31-40.

Stirling, C.J. & Hewitt, E.W. (1992) *Nature*, **356**, 534-537.

Strub, K. & Walter, P. (1989) *Proc. Natl. Acad. Sci. USA*, **86**, 9747-9751.

Strub, K. & Walter, P. (1990) *Mol. Cell. Biol.*, **10**, 777-784.

Strub, K., Moss, J. & Walter, P. (1991) *Mol. Cell. Biol.*, **11**, 3949-3959.

Studier, F.W., Rosenberg, A.H., Dunn, J.J. & Dubendorff, J.W. (1990) *Methods Enzymol.*, **185**, 60-89.

Tullius, T.D. & Dombrowski, B.A. (1986) *Proc. Natl. Acad. Sci. USA*, **83**, 5469-5473.

Ullu, E., Murphy, S. & Melli, M. (1982) *Cell*, **29**, 195-202.

Ullu, E. & Tschudi, C. (1984) *Nature*, **312**, 171-172.

Ullu, E. & Weiner, A.M. (1985) *Nature*, **318**, 371-374.

Walter, P., Ibrahimi, I., and Blobel G. (1981) *J. Cell Biol.*, **91**, 545-550.

Walter, P. & Blobel, G. (1980) *Proc. Natl. Acad. Sci. USA*, **77**, 7112-7116.

Walter, P. & Blobel, G. (1981) *J. Cell Biol.*, **91**, 557-561.

Walter, P. & Blobel, G. (1982) *Nature*, **99**, 691-698.

Walter, P. & Blobel, G. (1983) *Cell*, **34**, 525-533.

Walter, P., Gilmore, R. & Blobel, G. (1984) *Cell*, **38**, 5-8.

Walter, P. & Lingappa, V.R. (1986) *Ann. Rev. Cell Biol.*, **2**, 499-516.

Waters, G. & Blobel, G. (1986) *J. Cell Biol.*, **102**, 1543-1550.

Wolin, S.L. & Walter, P. (1988) *EMBO J.*, **7**, 3559-3569.

Wolin, S.L. & Walter, P. (1989) *J. Cell Biol.*, **109**, 2617-2622.

Yaver, D.S., Matoba, S. & Ogrydziak, D.M. (1992) *J. Cell Biol.*, **116**, 605-616.

Zwieb, C. (1992) *J. Biol. Chem.*, **267**, 15650-15656.

MOLECULAR MECHANISM OF THE GENETIC CODE VARIATIONS FOUND IN *CANDIDA* SPECIES AND ITS IMPLICATIONS IN EVOLUTION OF THE GENETIC CODE

Kimitsuna Watanabe[1], Takuya Ueda[1], Takashi Yokogawa[2], Tsutomu Suzuki[2], Kazuya Nishikawa[2], Miki Mori[3], Takeshi Ohama[3], Hiroyuki Nakabayashi[3], Takashi Nakase[4], and Syozo Osawa[3]

[1]Department of Industrial Chemistry, Faculty of Engineering, University of Tokyo, Hongo, Bunkyo-ku, Tokyo 113, Japan
[2]Department of Biological Sciences, Faculty of Bioscience and Biotechnology, Tokyo Institute of Technology, Nagatsuta 4259, Midori-ku, Yokohama 227, Japan
[3]Department of Biology, School of Science, Nagoya University, Chikusa-ku, Nagoya 464-01, Japan
[4]Japan Collection of Microorganisms, The Institute of Physical and Chemical Research, Wako, Saitama 351-01, Japan

INTRODUCTION

A deviation from the universal genetic code was first discovered in mammalian mitochondria in 1979. In the 1980s, several unusual genetic codes were also reported in animal organella. However, the genetic code in nuclear systems is less changeable than in organella.

Deviated codons which had been reported in nuclear systems were related to termination codons. However, Kawaguchi et al. (1989) demonstrated that in the asporogenic yeast *Candida cylindracea*, the CUG codon is read as serine instead of leucine. Amongst nuclear codons in bacteria and eukaryotes this is a unique instance in that an assignment for an amino acid codon deviates from the universal code. The codon change from leucine to serine requires at least two base replacements in the anticodon of tRNA.

To infer the evolutionary process of the change of genetic code in *Candida* species, we have sequenced the tRNAs that would have been responsible for the deviated genetic code in *C. cylindracea*; the DNA sequences of their genes have also been determined. The distribution of the deviated codon CUG for serine in yeasts is also demonstrated here. The phylogeny of yeasts has not been established yet. Therefore, 5S rRNA sequences of

Candida species were determined and a phylogenetic tree of these species constructed. Then, the assignment of the codon CUG, serine or leucine, was determined using an *in vitro* translation system with S30 fractions prepared from individual species. It was found that in six *Candida* species, CUG is translated as serine. On the basis of the phylogenetic tree, it is suggested that CUG leucine changed to serine, and then again to leucine as a reversal, during the evolution of yeasts.

ISOLATION OF tRNAs FROM *CANDIDA CYLINDRACEA*

By taking advantage of the fact that serine and leucine tRNAs have longer chain lengths than others, serine and leucine tRNAs of *C. cylindracea* were selectively separated from other tRNAs by gel electrophoresis containing 8M urea. In some experiments, serine isoacceptor tRNAs were charged with serine using partially purified seryl-tRNA synthetase of *C. cylindracea* and isolated by Benzoylated-DEAE-Cellulose column chromatography after chemical modification with N-hydroxysuccinimide-2-naphthoxy-acetate, as described by Andachi et al. (1989) By subsequent RPC 5 column chromatography and gel electrophoresis, five serine and three leucine tRNA species were purified for sequencing.

Amongst these tRNAs thus purified, serine tRNAs for the CUG codon were identified by an *in vitro* translation system using mRNA containing CUG codons, as shown in Fig. 1. After being aminoacylated with [^{3}H]serine with seryl-tRNA synthetase partially purified from *C. cylindracea*, serine tRNAs were employed to an *in vitro* translation system set up according to the method described by Hussain and Leibowitz (1986). Only a single tRNA could incorporate radioactive serine into the acid-insoluble fraction whereas other tRNAs could not. This tRNA was thus a candidate for the molecule responsible for the translation of the codon CUG.

PRIMARY STRUCTURE OF SERINE tRNA FOR CODON CUG

Isolated tRNAs were sequenced by the method of Donis-Keller (1980) and Kuchino et al. (1987). The primary sequence of tRNA translating the in-frame CUG codon as serine in the *in vitro* translation system is presented in Fig 2 a. Since it possesses a CAG anticodon sequence, it was concluded that this tRNA (tRNA^{Ser}CAG) is responsible for translation of the CUG codon (Yokogawa, et al., 1992).

A unique feature observed for tRNA^{Ser}CAG is that unmodified guanosine is located at the position 5'-adjacent to the anticodon. In all tRNAs whose sequences have been reported so far, pyrimidine nuclotides are located in this position. Furthermore the discriminator nucleotide, the fourth nucleotide from the 3' terminus of tRNA, was a uridine nucleotide, while serine tRNAs of bacteria and cytoplasm normally have guanosine at this position.The relationship between these characteristics and the decoding properties of the deviated genetic code in the tRNA is unknown. However, one possible explanation for these unusual structural features is that they play a role as negative determinants to prevent the tRNA from being misacylated by other aminoacyl-tRNA synthetases, e.g., leucyl-tRNA synthetase. If some aminoacyl-tRNA synthetases have affinity to the anticodon sequence, CAG, misacylation would cause incorporation of the wrong amino acid into proteins. The presence of the guanosine residue at position 5'-adjacent to the anticodon may decrease the affinity of other aminoacyl-tRNA synthetases, except seryl-tRNA synthetase, resulting in the prevention of misacylation.

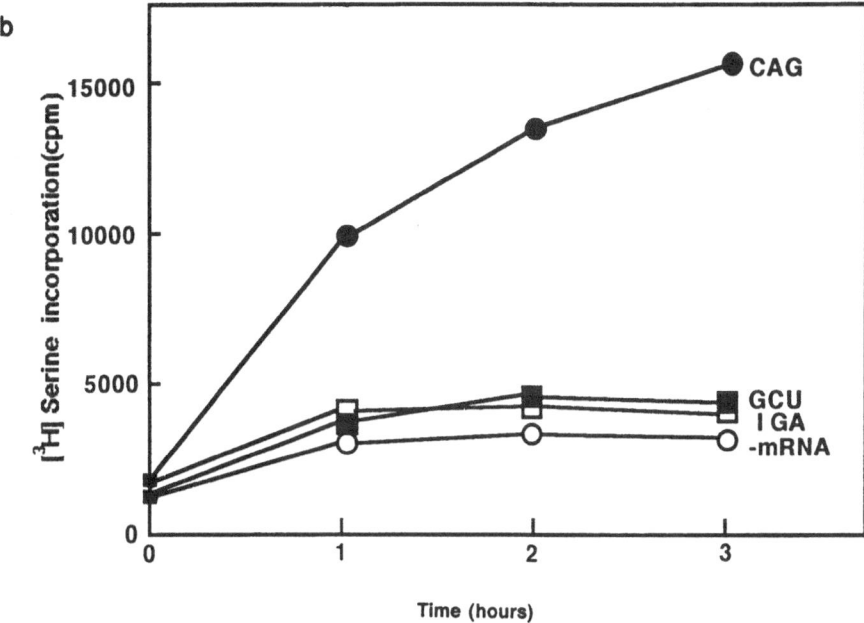

a

| mRNA | (UUU UUU UUU UUU UUG CUG)₂ CUG (UUU UUU UUU UUU UUG CUG)₃ UUU UUU UUU |

Peptide A (Phe Phe Phe Phe Leu Ser)₂ Ser (Phe Phe Phe Phe Leu Ser)₃ Phe Phe Phe

Peptide B (Phe Phe Phe Phe Leu Leu)₂ Leu (Phe Phe Phe Phe Leu Leu)₃ Phe Phe Phe

Fig.1. *In vitro* translation of CUG-containing mRNA with three serine tRNAs. (a) A mRNA containing codons for phenylalanine (UUU) and leucine(UUG) with CUG codons was transcribed by T7 RNA polymerase. Peptides A and B: predicted amino acid sequences following the genetic code for *C. cylindracea* and the universal code, respectively. (b) Three serine tRNAs with anticodons CAG, GCU, and IGA were charged with [³H]serine and employed to an *in vitro* translation system of *C. cylindracea* [From Yokogawa et. al (1992)].

GENE FOR SERINE tRNA HAVING THE ANTICODON CAG

The gene for the tRNA^{Ser}CAG was amplified from genomic DNA of *C. cylindracea* by the polymerase chain reaction using primers synthesized according to the tRNA sequence. The nucleotide sequence was determined directly using these primers. The gene for the tRNA^{Ser}CAG is interrupted by an intron of fifteen nucleotides in the anticodon loop. This shows that the tRNA with the anticodon CAG is created from the precursor molecule by removal of the intron. The splicing should occur at the sites shown by solid arrows in Fig.2 b, in accordance with the specificity of splicing endonuclease in *Saccharomyces cerevisiae*. If C or A at position (α) or (β) in Fig. 2 b is deleted and the intron is removed from the precursor at the sites shown by dotted arrows, the resultant molecule is a tRNA with the anticodon IGA (AGA in DNA) or CGA, each of which translates codons in the UCN serine family box. It is thus likely that serine tRNA for codons UCN and codon CUG originated from same ancestral tRNA gene.

The possibility that this tRNA is of leucine tRNA origin is less likely, because in this pathway the tRNA^{Leu}CAG must undergo many structural changes so as to lose determinants recognizable by leucyl-tRNA synthetase and become able to interact with seryl-tRNA synthetase. As pointed out by Normanly et al. (1992), the conversion of tRNA identity from leucine to serine in *Escherichia coli* without changing the anticodon requires the substitution of eight nucleotides.

Another possible process for the generation of the deviated genetic code, that mutations accumulated in seryl-tRNA synthetase and become able to interact with tRNA^{Leu}CAG in the ancestral organism, was ruled out by aminoacylation experiments using this tRNA with aminoacyl-tRNA synthetases partially purified from *S. cerevisiae* and *C. cylindracea*. Seryl-tRNA synthetase from either organism charged tRNA^{Ser}CAG with [^{14}C]serine, while leucyl-tRNA synthetase of either origin did not (data not shown).

LEUCINE tRNA ISOACCEPTORS

The primary structures of the three major leucine tRNA isoacceptors were also determined, which confirmed the unique feature of the genetic code in *C. cylindracea* (data not shown). Their anticodon sequences indicated that one leucine tRNA having an anticodon CmAA corresponds to a codon UUR, while the other two tRNA molecules have the anticodon IAG, which is able to base-pair with CUU, CUC and CUA in accordance with the wobble hypothesis. This result suggests that the CUA codon is likely to be translated as leucine in *C. cylindracea*. Thus, it is concluded that in *C. cylindracea* CUG belongs to a single-codon family like AUG for methionine and UGG for tryptophan.

ASSIGNMENT OF CODON CUG IN YEAST

In order to determine the assignment of codon CUG, serine or leucine, in yeast,.the translation of a synthetic mRNA having in-frame CUG codons was examined by an *in vitro* translation system using the S30 fractions prepared from various yeast species. In this experiment, the in-frame CUG codon was translated by endogenous tRNAs included in the S30 fraction. With the S30 fraction from six species, *C. albicans, C. cylindracea,C. melibiosica, C. parapsilosis, C.rugosa, and C. zeylanoides,* the codon CUG was translated as serine, while in the other eight species, *Candida azyma, Candida diversa, Candida*

a

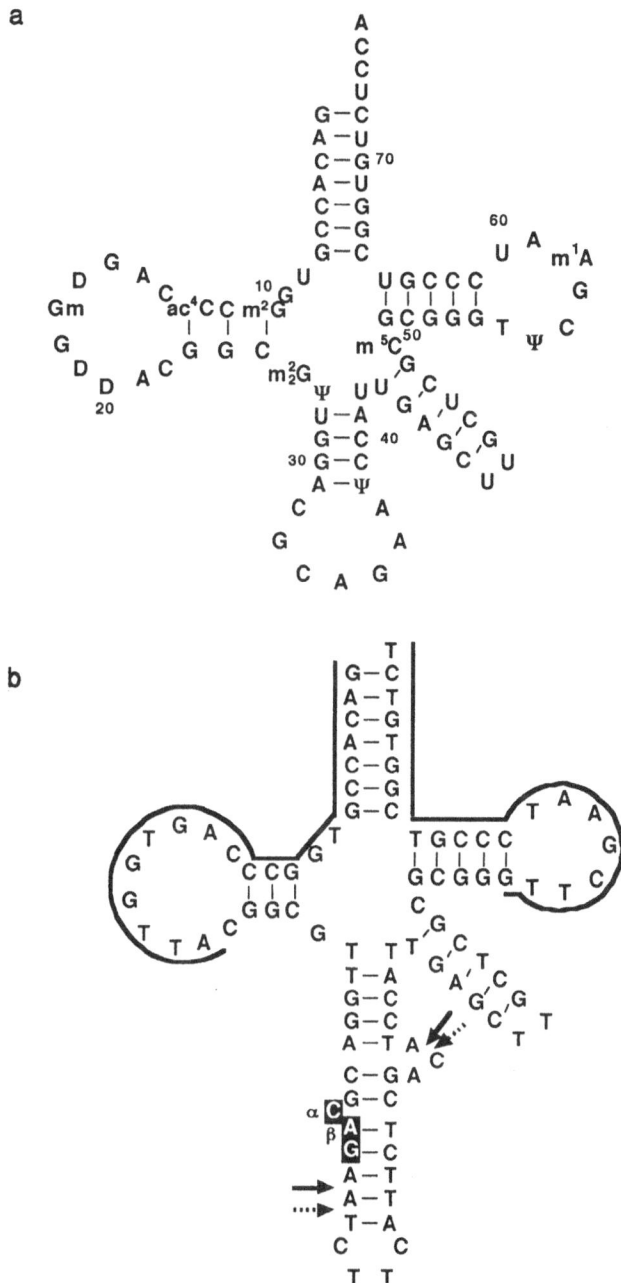

b

Fig. 2. Nucleotide sequences of tRNASerCAG (a) and its gene (b). (a) The sequences of the tRNA were determined as described in text. (b)Two oligonucleotide primers synthesized by referring to the tRNA sequence are indicated by the solid line. The anticodon is shown by a black background. Splicing sites for the tRNA gene and possible splicing sites for the putative ancestral tRNA gene, to which either C or A at position (α) or (β) would have been inserted, are indicated by solid and dotted arrows, respectively [From Yokogawa et. al (1992)].

*magnoliae, Candida rugopelliculosa, Yarrowia lipolytica, Zygoascus hellenicus,
Trichosporon cutaneum* and *Saccharomyces cerevisiae*, CUG was read as leucine. A part of
these results is shown in Fig.3. Thus, this deviated genetic code does not exist solely in *C.
cylindracea* but is distributed widely in yeast species.

Furthermore, to examine whether these yeast species have serine tRNAs decoding the
codon CUG, major serine tRNA species were purified and sequenced. At least four species,
C. zeylanoides, C. albicans*, and C. melibiosica, in which CUG is translated as serine,
contain serine tRNA having the anticodon CAG, as was found in *C. cylindracea* as
described above (Fig. 4) [* Santos et al. reported the same sequence for suppressor tRNA
from *C. albicans,* but charging specificity is not clear (Santos et al. (1991) and Tuite, M. F.,
personal communication)].

Like *C. cylindracea,* these tRNAs have guanosine at the position 5'adjacent to the
anticodon, but they have different discriminator bases. It seems that all tRNA species for the
codon CUG are of the same origin; thus, the appearance of this unusual tRNA would have
occurred once in the ancestral yeast species.

DISTRIBUTION OF THE DEVIATED GENETIC CODE IN YEAST

The classification and phylogeny of yeasts, including *Candida* species, is still in a
chaotic state; a yeast that has been classified into *Candida* is sometimes placed in another
non-*Candida* genus, etc. Therefore, 5S rRNA sequences were determined for the ten
Candida species for which the amino acid assignment of CUG was determined in this study
(*C. azyma, C. cylindracea, C. diversa, C. magnoliae, C. melibiosica, C. parapsilosis, C.
rugopelliculosa, C. rugosa, C. zeylanoides, Y. lipolytica,* and *Z. hellenicus*), and that of
Cryptococcus humicolus (Basidiomycetes) was also determined for comparison. A
phylogenetic tree(Fig. 5) was constructed by comparison of these and the known fungal 5S
rRNA sequences (Wolters and Erdmann., 1988). According to this tree, the *Candida* species
in which the codon CUG is read as serine, appear in one group, suggesting that these
species are derived from one ancestral living organism in which the codon assignment of
CUG changed from leucine to serine. Recently, Barn. et. al constructed a phylogenetic tree
using 18S rRNA sequences of several *Candida* species (1991). In their tree, *Candida*
species in which codon CUG is read as serine are also in one group. Groupings according
to the length of the isoprenoid chain of ubiquinone (Barnett et al., 1990) and the
composition of sugars in cell wall (Gorin and Spencer, 1970) also support the phylogenetic
tree constructed by 5S rRNA sequences.

Another remarkable feature of this tree is that a group of yeasts (Group I; *Zygoascus
hellenicus, C. magnoliae, C. azyma, Yarrowia lipolytica,* and *Schizosaccharomyces
pombe**; CUG = leucine) is separated from Group III yeasts (*C. utilis*, Saccharomyces
cerevisiae, Pichia membranaefaciens*, C. diversa,* and *C. rugopelliculosa*; CUG = Leu) by
six Candida species (Group II; *C. parapsilosis, C. zeylanoides, C. albicans, C. cylindracea,
C. rugosa,* and *C. melibiosica*), in which CUG codes for serine. (The amino acid
assignments for the species marked with asterisks are not known). Since branching of *C.
melibiosica, C. rugosa, C. cylindracea,* and *C. albicans* seems to have occurred at different
points on the dotted stem, as shown in Fig. 5, it must be concluded that the assignment of
codon CUG changed from leucine to serine and then to leucine again as a reversal during the
evolution of yeasts.Thus, it is likely that the amino acid assignment of codon CUG changed
at least twice during the evolution of yeasts. Such a "multiple" code change of one codon
seems not to be restricted to yeasts.

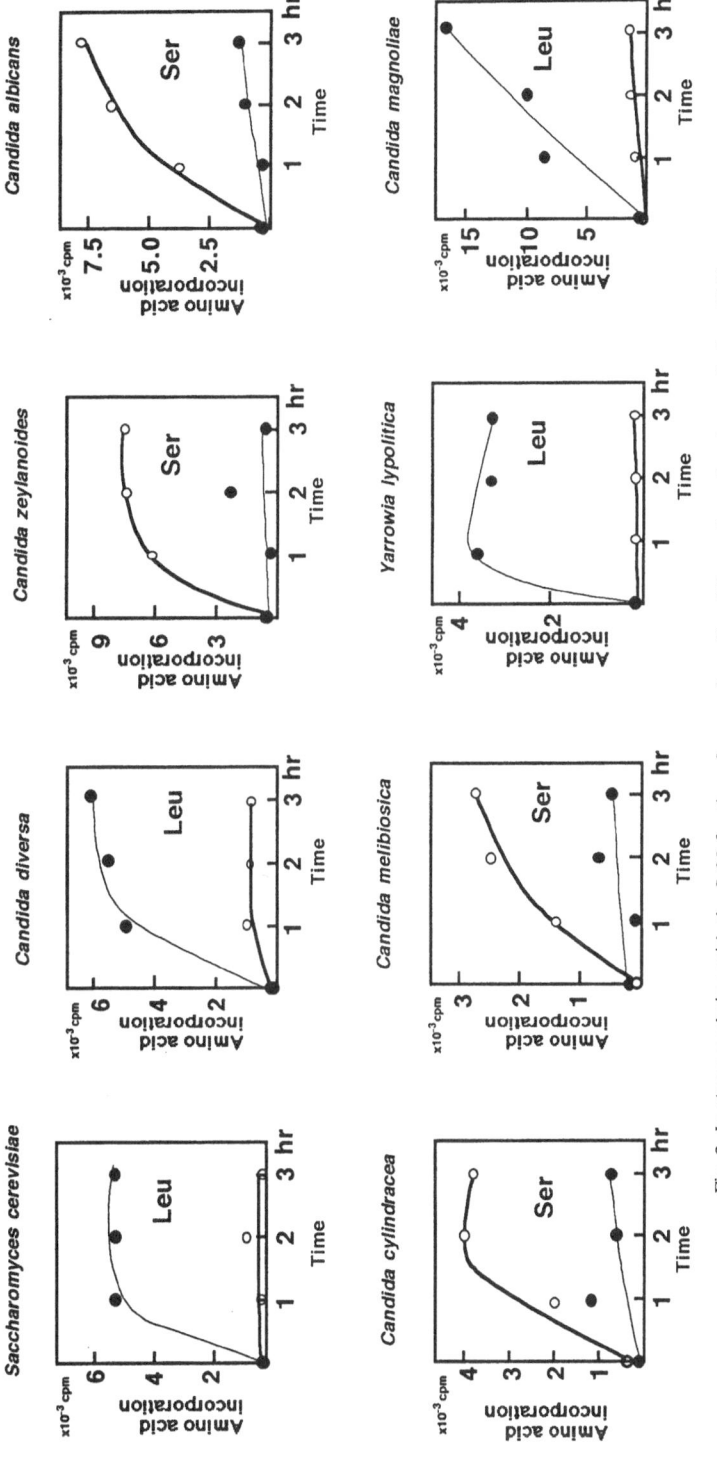

Fig. 3. *In vitro* translation with the S-30 fractions from various *Candida* species using the synthetic mRNA shown in Fig. 1. The reactions were performed in the presence of [³H]serine (o) or [³H]leucine (•).

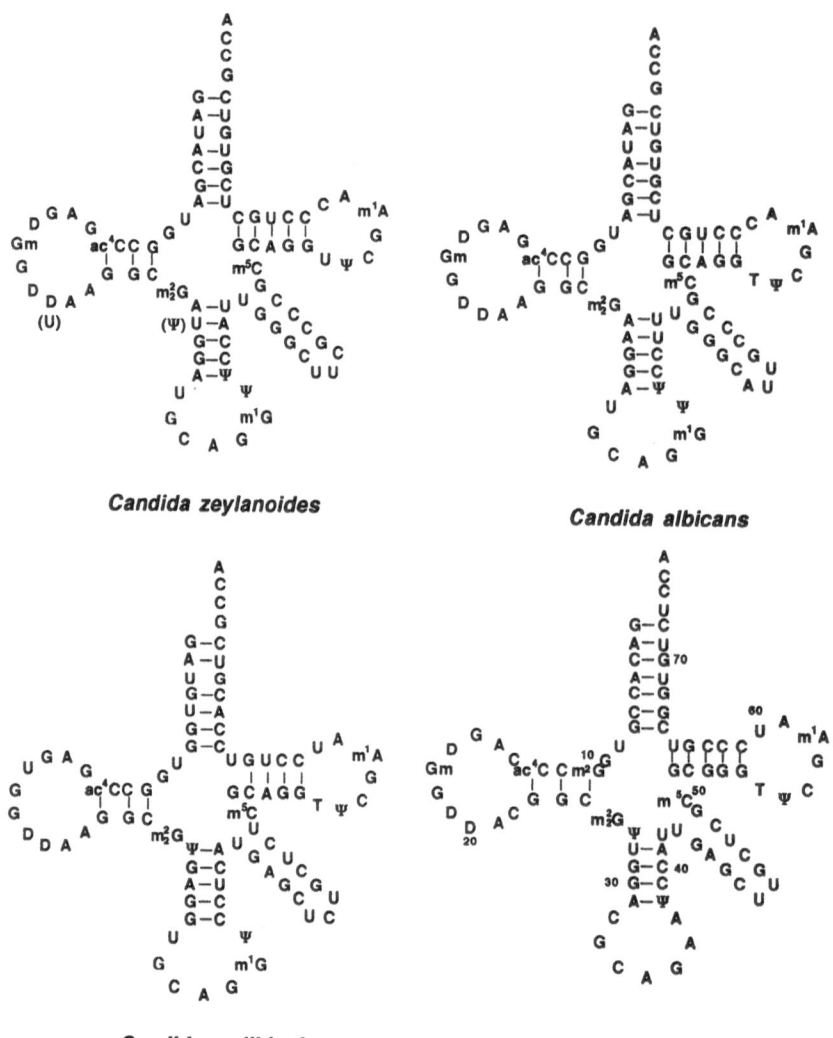

Fig. 4. tRNAs for codon CUG in *Candida* species

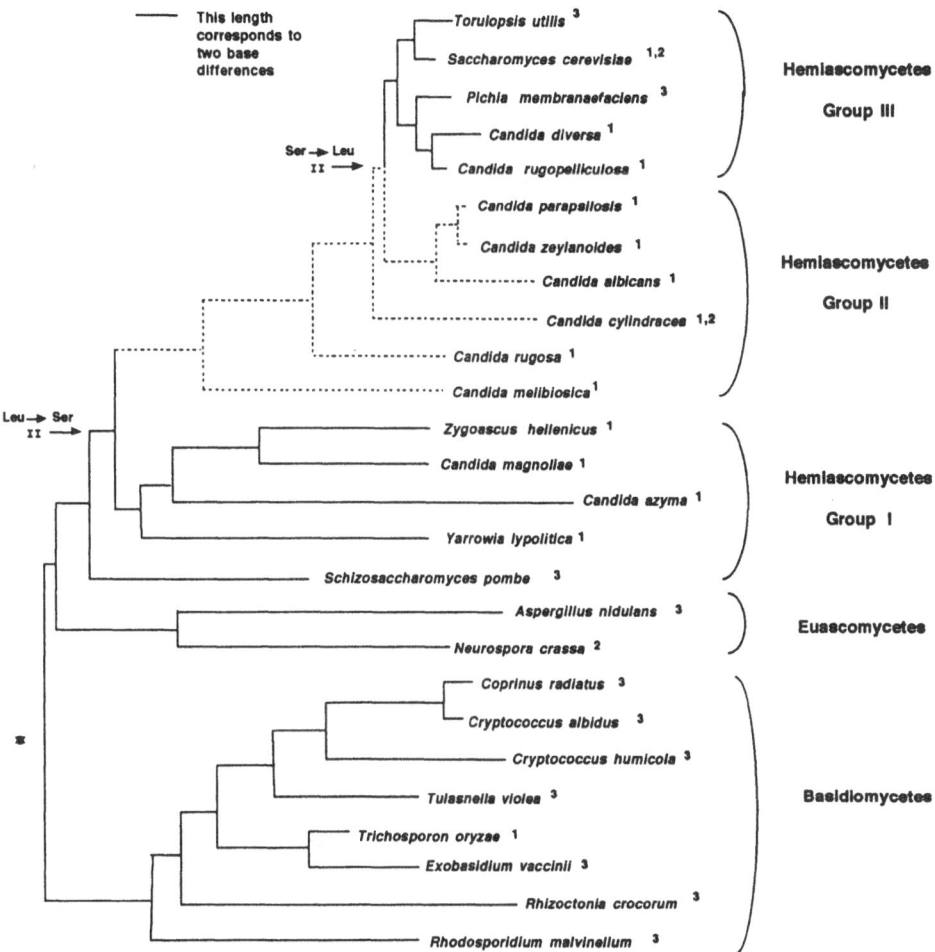

Fig5. Phylogenetic tree of yeasts and related species constructed by 5S rRNA sequences.
1. Determined in this study.
2. The assignment for codon CUG was taken from other studies.
3. The assignment has not been reported.

ACKNOWLEDGEMENT

This work was supported by Grant-in Aid for Scientific Research on Priority Areas from the Ministry of Education, Science and Culture of Japan.

REFERENCES

Andachi, Y., Yamao, F., Muto, A. and Osawa, S., 1989, Codon recognition patterns as deduced from sequences of the complete set of transfer RNA species in *Mycoplasma capricolum*. *J. Mol. Biol*, 209:37.

Barnett, J. A., Payne, R. W. and Yarrow, D., 1990, "Yeasts: Characteristics and Identification", 2nd ed. 1990, Cambridge Univ. Press.Cambridge:1002.

Barns, S. M., Lane, D. J., Sogin, M. L., Bibeau, C. and Weisburg, W. G., 1991,. Evolutionary relationships among pathogenic *Candida* species and relatives, *J. Bacteriol.* 173:2250.

Donis-Keller, H., 1980,. Phy M: an RNase activity specific for U and A residues useful in RNA sequence analysis, *Nucl. Acids Res.* 8:3133.

Gorin, P. A. and Spencer, J. F. T. 1970, Proton magnetic resonance spectroscopy an aid in identification and chemotaxonomy of yeasts, *Advances in Applied Microbiology*, 13: 25.

Hussain, I. & Leibowitz, M. J., 1986, Translation of homologous and heterologous messenger RNAs in a yeast cell-free system. *Gene* 46:13.

Kawaguchi, Y., Honda, H., Taniguchi-Morimura, J. and Iwasaki, S., 1989,. The codon CUG is read as serine in an asporogenic yeast *Candida cylindracea*, *Nature* (London) 341: 164.

Kuchino, Y., Hanyu, N. and Nishimura, S., 1987,. Analysis of modified nucleosides and nucleotide sequence of tRNA, *Methods Enzymol.* 155: 379.

Normanly, J., Ollick, T. and Abelson, J., 1992, Eight base changes are sufficient to convert a leucine-inserting into a serine-inserting tRNA, *Proc. Nat. Acad. Sci. U.S. A.*,89:5680.

Santos, M., Colthuret, D. R., Keith, G., Dirheimer, G. and Tuite, M. F., 1991, Characterization of a novel suppressor tRNA from the human pathogen *Candida albicans*. Abstract of 14th International tRNA workshop, Poznan, Poland, pp21.

Wolters, J. and Erdmann, V. A., 1988,. Compilation of 5S rRNA and 5S rRNA sequences, *Nucl. Acids Res.* 16 (Supplement): r1.

Yokogawa, T., Suzuki, T., Ueda, T.,Mori, M., Ohama, T., Kuchino, Y., Yoshinari, S., Motoki, I., Nishikawa, K., Osawa, S. and Watanabe, K., 1992,. Serine tRNA complementary to the non-universal serine codon CUG in *Candida cylindracea*: Evolutionary implications, *Proc. Nat. Acad. Sci. U.S.A.*89: 7408.

THE AMINOACYL-tRNA SYNTHETASE FAMILY:
AN EVOLUTIONARY VIEW OF THEIR STRUCTURAL ORGANIZATION

Marc Mirande, Myriam Lazard, Pierre Kerjan, Guillaume Bec,
Fabrice Agou, Sophie Quevillon, and Jean-Pierre Waller

Laboratoire d'Enzymologie
Centre National de la Recherche Scientifique
91190 Gif sur Yvette, France

INTRODUCTION

The aminoacyl-tRNA synthetases are a family of ubiquitous enzymes that function at an essential step in the translation of the genetic information. These enzymes are responsible for the accurate esterification of amino acids to the corresponding tRNA species. From an evolutionary point of view, this family of twenty enzymes has long been described as a class of proteins exhibiting similar catalytic functions but a puzzling structural diversity. The early studies revealed a large diversity in quaternary structures and in subunit molecular weights. In *Escherichia coli*, aminoacyl-tRNA synthetases have polypeptide chains ranging from 35 to 108 kDa, and are monomers, dimers or tetramers (Schimmel and Söll, 1979). Whereas these large differences argued against a unifying scheme for their structural organization, it was believed that an extensive relatedness should prevail in the aminoacyl-tRNA synthetase family. In particular, assuming that the primitive system was composed of fewer amino acids and activating enzymes, it was supposed that new synthetases arose through duplications and mutations in the genes of a restricted set of ancestral enzymes. This assumption led Orgel (1968) to ask the pertinent question: *'Is there amino acid sequence homology in the activating enzymes suggesting the course of specialization of these proteins?'*

The primary structures of all the twenty aminoacyl-tRNA synthetases from *Escherichia coli* have now been established. According to the presence of mutually exclusive sets of sequence motifs, synthetases have been partitioned into two distinct groups (Eriani et al., 1990; Burbaum and Schimmel, 1991; Cusack et al., 1991; Steitz, 1991; Moras, 1992) and evolutionary trees have been proposed (Nagel and Doolittle, 1991). The extent of structural relationships between enzymes belonging to the same class has been exemplified when high resolution X-ray structures of several enzymes became available. Synthetases from Class I (specific for amino acids Arg, Cys, Gln, Glu, Ile, Leu, Met, Trp, Tyr, and Val) are characterized by the two sequence motifs HIGH (Schimmel, 1987) and KMSKS (Hountondji et al., 1986) and the catalytic domain of three representatives of this class (glutaminyl-, methionyl- and tyrosyl-tRNA synthetases) is arranged according to a Rossmann nucleotide binding fold (Rould et al., 1989; Brunie et al., 1990; Brick et al.,

The Translational Apparatus, Edited by K.H. Nierhaus
et al., Plenum Press, New York, 1993

Table 1. Quaternary structures and subunit sizes of aminoacyl-tRNA synthetases

Enzyme[a]	Prokaryotes		Lower eukaryotes		Higher eukaryotes[b]	
Cys	α	461	–	–	–	–
Glu	α	471	–	–	CX	1714
Gln	α	550	–	809	CX	–
Arg	α	577	α	–	α, CX	661
Leu	α	860	α	1090	α, CX	–
Ile	α	939	α	1073	α, CX	–
Val	α	951	α	1058	α, CX	1265
Trp	α_2	334	α_2	–	α_2	475
His	α_2	424	α_2	526	α_2	509
Tyr	α_2	424	α_2	–	α_2	–
Ser	α_2	430	α_2	462	α_2	–
Asn	α_2	466	–	–	–	552
Lys	α_2	505	α_2	591	α_2, CX	597
Pro	α_2	572	–	–	CX	1714
Asp	α_2	590	α_2	557	α_2, CX	501
Thr	α_2	642	α_2	734	α_2	724
Met	α_2	677	α	751	CX	–
Ala	α_4	876	α	–	α	967
Gly	$\alpha_2\beta_2$	689+303	α_2	–	α_2	–
Phe	$\alpha_2\beta_2$	795+327	$\alpha_2\beta_2$	595+503	$\alpha_2\beta_2$	–

[a]Synthetases are designated by their amino acid substrates. Molecular sizes (in number of amino acid residues per subunit) are indicated when determined from completed sequences of the cloned genes. For sequences, see Burbaum and Schimmel (1991), Mirande (1991) and for synthetases specific for: Leu from *S. cerevisiae*, Hohmann and Thevelein (1992); Glu and Pro from *Drosophila*, Cerini et al. (1991); Val from human, Hsieh and Campbell (1991); Trp from bovine, Garret et al. (1991); His from human, Raben et al. (1992); Thr from human, Cruzen and Arfin (1991); Ala from *B. mori*, Chang and Dignam (1990); putative Asn from *B. malayi*, Nilsen et al. (1988) and Cusak et al. (1991); Arg and Lys from hamster, Lazard, d'Andrea and Mirande (unpublished).
[b]Aminoacyl-tRNA synthetases that associate within high-molecular-weight complexes are marked CX; quaternary structures of dissociated components are indicated.

1989). By contrast, Class II synthetases (specific for amino acids Asn, Asp, His, Lys, Phe, Pro, Ser, Thr, Ala, and Gly) share quite different conserved motifs (Eriani et al., 1990), and the catalytic domain of seryl- and aspartyl-tRNA synthetases is organized around an antiparallel β-sheet (Cusack et al., 1990; Ruff et al., 1991). The partition of aminoacyl-tRNA synthetases into two distinct structural classes suggests that this family of twenty enzymes has evolved from at least two different ancestral proteins.

The validity of the partitioning of aminoacyl-tRNA synthetases into two classes is also assessed by the finding that functionally homologous synthetases from various procaryotes and eukaryotes belong to the same group. This implies that these two synthetase classes already existed in the ancestral organism, before the emergence of modern bacteria and eukaryotic cells.

Despite extensive structural relationships with their prokaryotic counterparts, eukaryotic aminoacyl-tRNA synthetases are characterized by a more complex structural organization (Mirande, 1991). As a general rule, eukaryotic synthetases have subunit sizes significantly larger than those of the homologous prokaryotic enzymes (Table 1). In addition, ten synthetases from higher eukaryotes form supramolecular assemblies. Two distinct complexes have been described: a multisynthetase complex containing the nine enzymes specific for amino acids Glu, Pro, Ile, Leu, Met, Gln, Lys, Arg and Asp associated with three proteins of 43, 38, and 18 kDa of unknown function (Mirande et al., 1985a; Cirakoglu and Waller, 1985c; Perego and Del Monte, 1986; Godar et al., 1988; Norcum, 1989), and a complex formed by association of valyl-tRNA synthetase with the α, β, γ and δ subunits of elongation factor 1 (Bec et al., 1989; Motorin et al., 1991). The purpose of this paper is to describe in detail the polypeptide chain extensions that distinguish eukaryotic synthetases from their prokaryotic analogs, and to discuss the structural features and the biological functions of these additional protein domains.

FROM PROKARYOTES TO EUKARYOTES:
COMMON AND DISTINCTIVE FEATURES OF THE SYNTHETASES

To date, the primary structure of the five aminoacyl-tRNA synthetases specific for amino acids Val, His, Lys, Asp and Thr have been determined for the enzymes from bacteria, yeast and mammals. A schematic alignment of homologous sequences (Figure 1) points to three noticeable features.

(i) Regions of sequence similarities are generaly shared by the three homologous enzymes, histidyl- and aspartyl-tRNA synthetases from *E. coli* being more distantly related to their eukaryotic counterparts. The larger subunit size of bacterial aspartyl-tRNA synthetase, as compared to the eukaryotic homologous enzymes, is accounted for by a polypeptide chain insertion of about 100 residues located between motifs 2 and 3.

(ii) The extensions that characterize eukaryotic synthetases are located at the amino terminal extremity of their polypeptide chains and display a large variety in size (from 30 amino acid residues for yeast histidyl-tRNA synthetase to 300 residues for human valyl-tRNA synthetase). This rule also applies for most of the other eukaryotic enzymes listed in Table 1, for which primary structures are known.

(iii) Extensions of homologous enzymes from lower and higher eukaryotic cells do not display amino acid sequence similarities. One exception is valyl-tRNA synthetase: the mammalian enzyme possesses a region similar to the extension of the yeast enzyme to which is joined a polypeptide segment of about 200 residues that displays strong similarity with the elongation factor 1 γ-chain from brine shrimp (Hsieh and Campbell, 1991).

A large body of data indicates that the prokaryotic-like segment of eukaryotic aminoacyl-tRNA synthetases is sufficient to sustain the tRNA aminoacylation function. The removal of the extensions by controlled proteolysis of the purified enzymes or by genetic engineering of the structural genes do not affect the catalytic step, as assessed by *in vitro* or *in vivo* approaches. Truncated, yet active forms of eukaryotic enzymes were obtained for glutaminyl- (Ludmerer and Schimmel, 1987), lysyl- (Martinez and Mirande, 1992; Cirakoglu and Waller, 1985a), aspartyl- (Eriani et al., 1991), methionyl- (Walter et al., 1989), and arginyl- (Vellekamp et al., 1985) tRNA synthetases. The resulting modified proteins have molecular sizes very similar to those of their prokaryotic counterparts.

Since the polypeptide chain extensions of eukaryotic synthetases are dispensable for catalysis, it is generally believed that they fulfil a distinct, specific function related to a higher order of cellular organization of the protein biosynthesis machinery in eukaryotic cells (Ryazanov et al., 1987). In that connection, the general polyanion-binding property displayed by aminoacyl-tRNA synthetases from eukaryotes (Alzhanova et al., 1980; Cirakoglu and Waller, 1985b), that is thought to be the reflection of their *in vivo* interaction with subcellular structures, could be ascribed to the lysine-rich domains contributed by their polypeptide chain extensions. With regard to the hypothesis of a spatial organization of the translational machinery, it is conceivable that the formation of multienzyme complexes in higher eukaryotic cells provide a more efficient way to promote cellular compartmentalization of these enzymes (Cirakoglu et al., 1985). Hydrophobic-based interactions, contributed by the amino-terminal extensions of individual synthetases, play a primordial role in holding these enzymes within the complexes (Mirande et al., 1992). As an ultimate step for promoting stable association of synthetase activities, it was found that a component of the multisynthetase complex is a multifunctional aminoacyl-tRNA synthetase composed of distinct domains for glutamyl- and prolyl-tRNA synthetases (Cerini et al., 1991).

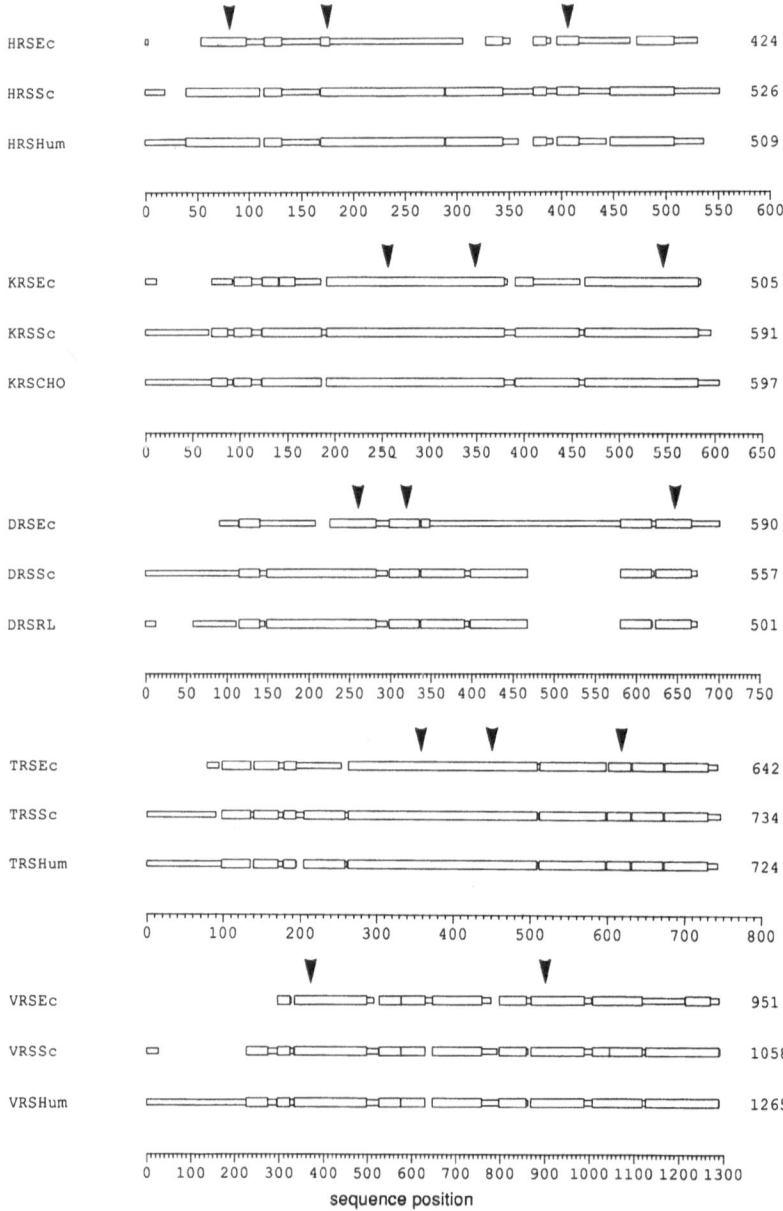

Figure 1. Schematic alignment of homologous aminoacyl-tRNA synthetases. Histidyl- (HRS), lysyl-(KRS), aspartyl- (DRS), threonyl- (TRS) and valyl- (VRS) tRNA synthetases from *E. coli* (Ec), *Saccharomyces cerevisiae* (Sc) and mammalian origin (Human, Hum; Chinese hamster ovary cells, CHO; Rat Liver, RL) were compared by using the MACAW program (Schuler et al., 1991). Larger blocks indicate regions of sequence similarities between two or three enzymes. The location of conserved motifs for Class I and Class II aminoacyl-tRNA synthetases are indicated by arrow heads.

Table 2. Isoelectric points of aminoacyl-tRNA synthetases

	Cys	Glu	Gln	Arg	Leu	Ile	Val	Trp	His	Tyr	Ser	Asn	Lys	Pro	Asp	Thr	Met	Ala	Gly	Phe
Ec	5.79	6.15	6.72	5.62	5.44	6.70	5.47	7.78	6.10	5.97	5.63	5.41	5.35	5.34	5.82	6.46	6.04	6.19	5.33	5.64
Sc		8.58		6.01	6.27	6.53		7.35		6.29		6.35			6.84	7.02	6.88			5.85
Me		7.84		7.33				7.84	6.34	6.00		5.97	6.44	7.84	6.71	6.85		6.21		

Isoelectic points are calculated from the completed sequences of the cloned genes from *E. coli* (Ec), *S. cerevisiae* (Sc) or metazoans (Me). For glycyl- and phenylalanyl-tRNA synthetases, values are for the sum of the α and β subunits. For metazoan aminoacyl-tRNA synthetases, values for the glutamyl- and prolyl-enzymes are given for the multifunctional synthetase, and for the asparaginyl-enzyme for the protein from *B. malayi* that is supposed to be asparaginyl-tRNA synthetase.

THE POLYANION-BINDING DOMAINS: A DISTINCTIVE FEATURE OF EUKARYOTIC SYNTHETASES

At first glance, the polyanion-binding property displayed by eukaryotic aminoacyl-tRNA synthetases is related to the overall amino acid composition of their subunits. As shown in Table 2, the isoelectric points calculated from the sequences of the eukaryotic enzymes are most often significantly higher than those determined for the homologous enzymes from *E. coli*.

As discussed elsewhere (Mirande and Waller, 1988; Mirande, 1991), this increase in pI's is contributed by a higher content in basic residues that are not randomly distributed along the polypeptide chain. The prokaryotic-like domain of the eukaryotic synthetases have pI's very similar to the corresponding *E. coli* enzymes. A clustering of lysine residues is observed within the polypeptide extensions of eukaryotic enzymes. Whereas isoleucyl-tRNA synthetase from yeast is more acidic than its prokaryotic counterpart (Table 2), a cluster of lysine residues is also observed between residues 894 to 918. The finding that the removal of these extensions leads to active enzymes that have lost their ability to bind to negatively-charged carriers (Cirakoglu and Waller, 1985a, 1985b; Lorber et al., 1988) supports the idea according to which the lysine-rich regions confer on eukaryotic synthetases their binding

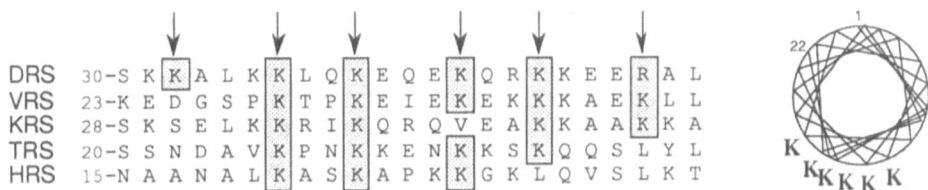

Figure 2. Alignment of the lysine-rich regions from the amino-terminal extensions of aspartyl- (DRS), valyl- (VRS), lysyl- (KRS), threonyl- (TRS) and histidyl- (HRS) tRNA synthetases from *Saccharomyces cerevisiae*. According to the predicted α-helical structure of these polypeptide segments, the lysine residues indicated by arrows should be located on one side of the proposed helices. At right, helical-wheel representation of an idealized α-helix, made of 6 helix turns, containing lysine residues distributed as indicated by the arrows.

properties. Moreover, the finding that valyl- and lysyl-tRNA synthetases from yeast, but not the amino-terminal truncated form of lysyl-tRNA synthetase, are able to interact in a MAP-like fashion with the acidic carboxy-terminal region of α- and β-tubulin (Melki et al., 1991), also argues in favor of a large accessibility of the polycationic domains of synthetases.

661

When the lysine-rich sequences from aspartyl-, valyl-, lysyl-, threonyl- and histidyl-tRNA synthetases are aligned (Figure 2), only few amino acid sequence similarities are recovered, but a similar distribution of lysine residues is observed. These polypeptide segments contain three, four, five or six lysine residues that are regularly spaced four or three residues apart and hence could be viewed as being repeated every turn of an α-helix. Consistently with this model, these lysine-rich regions are predicted to be α-helical (Lorber et al., 1988; Mirande and Waller, 1988). The actual 3D-structures of these domains remain to be established by direct approaches. The crystal structure of the yeast aspartyl-tRNA synthetase/tRNA^{Asp} complex has been solved (Ruff et al., 1991), but the structure of the amino-terminal region, containing the polycationic extension, is not yet known.

Assuming that the lysine-rich segments of eukaryotic aminoacyl-tRNA synthetases do fold into α-helical structures, leading to the segregation of cationic residues on one side of the helix, it is tempting to speculate that this type of structural organization generates high-affinity sites for polyanionic carriers. Although their general occurence in eukaryotic aminoacyl-tRNA synthetases designates these binding domains as important biological interfaces, the expression, in yeast, of alleles encoding truncated forms of glutaminyl-(Ludmerer and Schimmel, 1987), methionyl- (Walter et al., 1989) or lysyl- (Martinez and Mirande, 1992) tRNA synthetases did not provide definitive evidence in the function of these domains. The knowledge of the completed 3D-structure of a eukaryotic synthetase should prove useful to elucidate the role of these extensions.

A HYDROPHOBIC DOMAIN: A SPECIFIC FEATURE OF SYNTHETASES THAT ASSEMBLE INTO COMPLEXES

A characteristic property displayed by aminoacyl-tRNA synthetases that form high molecular weight complexes in higher eukaryotic cells is their ability to bind to hydrophobic carriers in conditions where synthetases from prokaryotes or lower eukaryotes, as well as synthetases from higher eukaryotes that are not associated into complexes, do not bind. Leucyl-, lysyl- and isoleucyl-tRNA synthetases dissociated from the multisynthetase complex (Cirakoglu and Waller, 1985a; Lazard et al., 1985), and valyl-tRNA synthetase isolated from its complex with elongation factor EF-1H (Bec and Waller, 1989) are hydrophobic proteins. The finding that proteolytic conversion of lysyl-tRNA synthetase, a dimer made of identical subunits of 79 kDa, to a fully active dimeric form of 2 x 64 kDa led to the loss of its hydrophobic properties, suggested that terminal extensions, forming hydrophobic domains, are involved in the assembly of individual synthetases into complexes.

The primary structure of aspartyl-tRNA synthetase from rat (Mirande and Waller, 1989) or human (Jacobo-Molina et al., 1989) has been determined via gene cloning and sequencing. The amino acid sequences of the yeast and mammalian enzymes display extensive homology (Figure 1), except for their amino-terminal extensions. Jacobo-Molina et al. (1989) observed that this polypeptide extension may form an amphiphilic helix, leading to the clustering of hydrophobic residues. The deletion of this extension does not affect the catalytic activity of aspartyl-tRNA synthetase, but prevents its association to the other components of the complex, as assessed by *in vivo* approaches (Mirande et al., 1992). Moreover, over-expression of native aspartyl-tRNA synthetase did not result in the association of additional copies of that enzyme within the complex. This implies that the hydrophobic domain of a particular enzyme should trigger site-directed association, through hydropathy complementarity with an adequate partner. As a consequence, the complex association domains of the individual components of the multisynthetase complex should display distinct structural organizations.

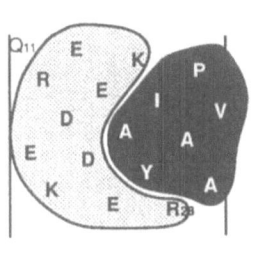 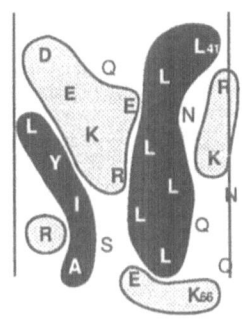 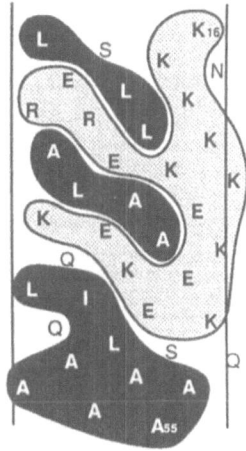

Figure 3. Helical net diagram for the complex association domains of aspartyl- (left), arginyl- (middle) and lysyl- (right) tRNA synthetases of the multisynthetase complex from mammalian origin.

The cDNAs for arginyl- and lysyl-tRNA synthetases from Chinese hamster ovary cells have been cloned and sequenced (M. Lazard, S. d'Andréa, and M. Mirande, unpublished). As shown in Figure 3, the overall organization of the association domains for three components of the multisynthetase complex are quite different. A region of their amino-terminal-extensions, rich in hydrophobic as well as charged amino acid residues, is predicted to be α-helical, but very distinct patterns are observed: a hydrophobic pocket for aspartyl-tRNA synthetase; a leucine alignment for arginyl-tRNA synthetase; a stacking of hydrophobic and hydrophilic blocks for lysyl-tRNA synthetase. Specific interactions between the complex association domains of individual synthetases could be the major element in holding these enzymes in the complex. Alternatively, one of the three proteins of unknown function, of 43, 38 and 18 kDa, that are copurified with the synthetase components of the complex may serve as a template for association of individual enzymes.

Mammalian valyl-tRNA synthetase forms a complex with the α, β, γ and δ subunits of elongation factor 1 (Bec et al., 1989; Motorin et al., 1991). It is worth noting that its amino-terminal extension is homologous to the amino-terminal region of the elongation factor 1γ-chain (Hsieh and Campbell, 1991). Interestingly, this part of the polypeptide chain from EF-1γ has been shown to interact with the EF-1β subunit (Van Damme et al., 1991). This suggests that the homologous region from valyl-tRNA synthetase also participates in the association of that enzyme with the components of EF-1.

A GENE FUSION: AN ULTIMATE STEP ?

The multisynthetase complex, composed of a noncovalent aggregate of proteins that catalyze nonsuccessive steps of a metabolic pathway, is an atypical example of a multienzyme complex. Substrate channeling and active-site coupling, properties inferred for multienzyme complexes, do not apply for the multisynthetase complex. The synthetase components are functionally independent (Mirande et al., 1983). Presumably, this type of structural organization is related to compartmentalization of the protein-synthesizing apparatus in higher eukaryotic cells. The finding that two synthetase activities are carried by one of the polypeptide components of this complex also gives credit to this hypothesis.

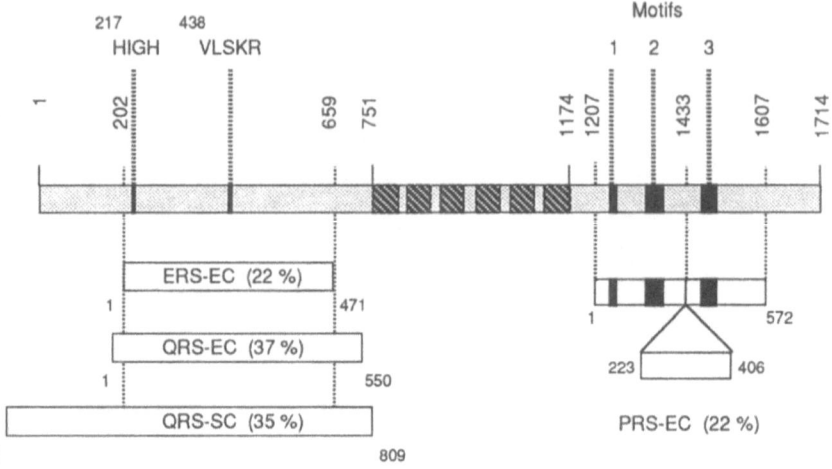

Figure 4. Domain structure of the multifunctional aminoacyl-tRNA synthetase. The large polypeptide encoded by the *Drosophila* cDNA is composed of three distinct domains (Cerini et al., 1991). The amino-terminal domain, from residues 1 to 751, containing the HIGH and VLSKR consensus sequences of class I aminoacyl-tRNA synthetases, specifies glutamyl-tRNA synthetase. It displays convincing homologies with *E. coli* glutamyl-tRNA synthetase (ERS-EC) and glutaminyl-tRNA synthetases from *E. coli* (QRS-EC) and *S. cerevisiae* (QRS-SC). The central domain (residues 751 to 1174) is composed of six tandemly repeated units (hatched boxes). The carboxy-terminal domain displays the three conserved motifs of class II aminoacyl-tRNA synthetases and homologizes with *E. coli* prolyl-tRNA synthetase (PRS-EC, 22% of identity).

The cDNA encoding the largest polypeptide component of the multisynthetase complex from *Drosophila* (Cerini et al., 1991) or human (Fett and Knippers, 1991) was cloned and sequenced. Glutamyl- and prolyl-tRNA synthetases are two discrete and autonomous domains of the same polypeptide (Figure 4). Since the isolated synthetase domains also aminoacylate their cognate tRNA species (Cerini et al., 1991; Kerjan et al., 1992), this naturally occurring chimeric protein most probably arose by a gene fusion event.

The individual synthetases of the complex are held together by noncovalent, hydrophobic interactions. The hydrophobicity-based contacts must be specific enough so that the correct complex forms. Covalent linkage of two components of the multisynthetase complex offers an efficient alternative possibility to generate specific interactions and could aid stability of this multimolecular assembly.

A remarkable feature of the amino acid sequence of this multifunctional protein is the presence of six (three in the human polypeptide) imperfectly repeated sequences, of ≈ 50 amino acids each, that link the two synthetase domains (Figure 4). The composition and the proposed structure for these repeated units suggest that they may serve as a matrix for binding of other proteins (Cerini et al., 1991). They may provide a template for association of the other components of the complex or play a role in anchoring the complex to the cytoskeletal framework of mammalian cells (Dang et al., 1983; Mirande et al., 1985b).

CONCLUSIONS AND PERSPECTIVES

As discussed in this paper, the acquisition of additional structural elements by eukaryotic aminoacyl-tRNA synthetases, contributed by terminal-extensions of their polypeptide chains, and the covalent attachment of two synthetase domains could be the reflection of *in situ* compartmentalization of protein biosynthesis. The cellular organization of protein synthesis could increase the efficiency of this complex process (Ryazanov et al., 1987; Mirande, 1991).

The association of valyl-tRNA synthetase with EF-1 lends credence to this hypothesis. These proteins catalyse two consecutive steps: the aminoacylation of tRNA^{Val} and the formation of the ternary complex EF-1α·GTP·aminoacyl-tRNA, the active form of aminoacylated tRNA for translation. This type of association could be endowed with a channelling function. The amino-terminal region of the multifunctional synthetase, as observed for valyl-tRNA synthetase, exhibits some sequence homology with the amino-terminal region of the elongation-factor 1γ-chain (Fett and Knippers, 1991), although to a lesser extent. This suggests a possible transient interaction between the multisynthetase complex and EF-1 (Sariski and Yang, 1991).

Most certainly, the subcellular organization of protein synthesis results form the sum of many weak and dynamic electrostatic interactions that do not survive cell disruption (Mirande, 1991). Recent data provide some evidence for a channelling model of protein biosynthesis (Negrutskii and Deutscher, 1991, 1992). Aminoacyl-tRNAs are thought to be channelled from synthetases, to the elongation factor and to the ribosome without equilibration with the cellular aqueous phase.

ACKNOWLEDGMENTS

Work from the author's laboratory was supported by Centre National de la Recherche Scientifique (ERS 0029), Association pour la Recherche sur le Cancer and Fondation pour la Recherche Médicale. We thank Dr. Dominique Thomas for his expert help in using the MACAW alignment program.

REFERENCES

Alzhanova, A.T., Fedorov, A.N., Ovchinnikov, L.P., and Spirin, A.S., 1980, Eukaryotic aminoacyl-tRNA synthetases are RNA-binding proteins whereas prokaryotic ones are not, *FEBS Lett.* 120:225.

Bec, G., Kerjan, P., Zha, X.D., and Waller, J.P., 1989, Valyl-tRNA synthetase from rabbit liver: purification as a heterotypic complex in association with elongation factor 1, *J. Biol. Chem.* 264:21131.

Bec, G., and Waller, J.P., 1989, Valyl-tRNA synthetase from rabbit liver: the enzyme derived from the high-Mr complex displays hydrophobic as well as polyanion-binding properties, *J. Biol. Chem.* 264:21138.

Brick, P., Bhat, T.N., and Blow, D.M., 1989, Structure of tyrosyl-tRNA synthetase refined at 2.3 Å resolution, *J. Mol. Biol.* 208:83.

Brunie, S., Zelwer, C., and Risler, J.L., 1990, Crystallographic study at 2.5 Å resolution of the interaction of methionyl-tRNA synthetase from *Escherichia coli* with ATP, *J. Mol. Biol.* 216:411.

Burbaum, J.J., and Schimmel, P., 1991, Structural relationships and the classification of aminoacyl-tRNA synthetases, *J. Biol. Chem.* 266:16965.

Cerini, C., Kerjan, P., Astier, M., Gratecos, D., Mirande, M., and Sémériva, M., 1991, A component of the multisynthetase complex is a multifunctional aminoacyl-tRNA synthetase, *EMBO J.* 10:4267.

Chang, P.K., and Dignam, J.D., 1990, Primary structure of alanyl-tRNA synthetase and the regulation of its mRNA levels in *Bombix mori*, *J. Biol. Chem.* 265:20898.

Cirakoglu, B., and Waller, J.P., 1985a, Leucyl-tRNA and lysyl-tRNA synthetases, derived from the high-Mr complex of sheep liver, are hydrophobic proteins, *Eur. J. Biochem.* 151:101.

Cirakoglu, B., and Waller, J.P., 1985b, Do yeast aminoacyl-tRNA synthetases exist as soluble enzymes within the cytoplasm? *Eur. J. Biochem.* 149:353.

Cirakoglu, B., and Waller, J.P., 1985c, Multiple forms of arginyl- and lysyl-tRNA synthetases in rat liver: a re-evaluation, *Biochim. Biophys. Acta* 829:173.

Cirakoglu, B., Mirande, M., and Waller, J.P., 1985, A model for the structural organization of aminoacyl-tRNA synthetases in mammalian cells, *FEBS Lett.* 183:185.

Cruzen, M.E., and Arfin, S.M., 1991, Nucleotide and deduced amino acid sequence of human threonyl-tRNA synthetase reveals extensive homology to the *Escherichia coli* and yeast enzymes, *J. Biol. Chem.* 266:9919.

Cusack, S., Berthet-Colominas, C., Härtlein, M., Nassar, N., and Leberman, R., 1990, A second class of synthetase structure revealed by X-ray analysis of *Escherichia coli* seryl-tRNA synthetase at 2.5 Å, *Nature* 347:249.

Cusack, S., Härtlein, M., and Leberman, R., 1991, Sequence, structural and evolutionary relationships between class 2 aminoacyl-tRNA synthetases, *Nucleic Acids Res.* 19:3489.

Dang, C.V., Yang, D.C.H., and Pollard, T.D., 1983, Association of methionyl-tRNA synthetase with detergent-insoluble components of the rough endoplasmic reticulum, *J. Cell Biol.* 96:1138.

Eriani, G., Delarue, M., Poch, O., Gangloff, J., and Moras, D., 1990, Partition of tRNA synthetases into two classes based on mutually exclusive sets of sequence motifs, *Nature* 347:203.

Eriani, G., Prevost, D., Kern, D., Vincendon, P., Dirheimer, G., and Gangloff, J., 1991, Cytoplasmic aspartyl-tRNA synthetase from *Saccharomyces cerevisiae*: study of its functional organisation by deletion analysis, *Eur. J. Biochem.* 200:337.

Fett, R., and Knippers, R., 1991, The primary structure of human glutaminyl-tRNA synthetase: a highly conserved core, amino acid repeat regions, and homologies with translation elongation factors, *J. Biol. Chem.* 266:1448.

Garret, M., Pajot, B., Trézéguet, V., Labouesse, J., Merle, M., Gandar, J.C., Benedetto, J.P., Sallafranque, M.L., Alterio, J., Gueguen, M., Sarger, C., Labouesse, B., and Bonnet, J., 1991, A mammalian tryptophanyl-tRNA synthetase shows little homology to prokaryotic synthetases but near identity with mammalian peptide chain release factor, *Biochemistry* 30:7809.

Godar, D.E., Godar, D.E., Garcia, V., Jacobo, A., Aebi, U., and Yang, D.C.H., 1988, Structural organization of the multienzyme complex of mammalian aminoacyl-tRNA synthetases, *Biochemistry* 27:6921.

Hohmann, S., and Thevelein, J.M., 1992, The cell division cycle gene *CDC60* encodes cytosolic leucyl-tRNA synthetase in *Saccharomyces cerevisiae*, *Gene* 120:43.

Hsieh, S.L., and Campbell, R.D., 1991, Evidence that gene G7a in the human major histocompatibility complex encodes valyl-tRNA synthetase, *Biochem. J.* 278:809.

Hountondji, C., Dessen, P., and Blanquet, S., 1986, Sequence similarities among the family of aminoacyl-tRNA synthetases, *Biochimie* 68:1071.

Jacobo-Molina, A., Peterson, R., and Yang, D.C.H., 1989, cDNA sequence, predicted primary structure, and evolving amphiphilic helix of human aspartyl-tRNA synthetase, *J. Biol. Chem.* 264:16608.

Kerjan, P., Triconnet, M., and Waller, J.P., 1992, Mammalian prolyl-tRNA synthetase corresponds to the 150 kDa subunit of the high-Mr aminoacyl-tRNA synthetase complex, *Biochimie* 74:195.

Lazard, M., Mirande, M., and Waller, J.P., 1985, Purification and characterization of the isoleucyl-tRNA synthetase component from the high molecular weight complex of sheep liver: a hydrophobic metalloprotein, *Biochemistry* 24:5099.

Lorber, B., Mejdoub, H., Reinbolt, J., Boulanger, Y., and Giegé, R., 1988, Properties of N-terminal truncated yeast aspartyl-tRNA synthetase and structural characteristics of the cleaved domain, *Eur. J. Biochem.* 174:155.

Ludmerer, S.W., and Schimmel, P., 1987, Construction and analysis of deletions in the amino-terminal extension of glutamine tRNA synthetase of *Saccharomyces cerevisiae*, *J. Biol. Chem.* 262:10807.

Martinez, R., and Mirande, M., 1992, The polyanion-binding domain of cytoplasmic Lys-tRNA synthetase from *Saccharomyces cerevisiae* is not essential for cell viability, *Eur. J. Biochem.* 207:1.

Melki, R., Kerjan, P., Waller, J.P., Carlier, M.F., and Pantaloni, D., 1991, Interaction of microtubule-associated proteins with microtubules: yeast lysyl- and valyl-tRNA synthetases and τ 218-235 synthetic peptide as model systems, *Biochemistry* 30:11536.

Mirande, M., Cirakoglu, B., and Waller, J.P., 1983, Seven mammalian aminoacyl-tRNA synthetases associated within the same complex are functionally independent, *Eur. J. Biochem.* 131:163.

Mirande, M., LeCorre, D., and Waller, J.P., 1985a, A complex from cultured Chinese hamster ovary cells containing nine aminoacyl-tRNA synthetases, *Eur. J. Biochem.* 147:281.

Mirande, M., LeCorre, D., Louvard, D., Reggio, H., Pailliez, J.P., and Waller, J.P., 1985b, Association of an aminoacyl-tRNA synthetase complex and of phenylalanyl-tRNA synthetase with the cytoskeletal framework fraction from mammalian cells, *Exp. Cell Res.* 156:91.

Mirande, M., and Waller, J.P., 1988, The yeast lysyl-tRNA synthetase gene: evidence for general amino acid control of its expression and domain structure of the encoded protein, *J. Biol. Chem.* 263:18443.

Mirande, M., and Waller, J.P., 1989, Molecular cloning and primary structure of cDNA encoding the catalytic domain of rat liver aspartyl-tRNA synthetase, *J. Biol. Chem.* 264:842.

Mirande, M., 1991, Aminoacyl-tRNA synthetase family from prokaryotes and eukaryotes: structural domains and their implications, *Prog. Nucleic Acid Res. Mol. Biol.* 40:95.

Mirande, M., Lazard, M., Martinez, R., and Latreille, M.T., 1992, Engineering mammalian aspartyl-tRNA synthetase to probe structural features mediating its association with the multisynthetase complex, *Eur. J. Biochem.* 203:459.

Moras, D., 1992, Structural and functional relationships between aminoacyl-tRNA synthetases, *Trends Biochem. Sci.* 17:159.

Motorin, Y.A., Wolfson, A.D., Löhr, D., Orlovsky, A.F., and Gladilin, K.L., 1991, Purification and properties of a high-molecular-mass complex between Val-tRNA synthetase and the heavy form of elongation factor 1 from mammalian cells, *Eur. J. Biochem.* 201:325.

Nagel, G.M., and Doolittle, R.F., 1991, Evolution and relatedness in two aminoacyl-tRNA synthetase families, *Proc. Natl. Acad. Sci. USA* 88:8121.

Negrutskii, B.S., and Deutscher, M.P., 1991, Channeling of aminoacyl-tRNA for protein synthesis *in vivo*, *Proc. Natl. Acad. Sci. USA* 88:4991.

Negrutskii, B.S., and Deutscher, M.P., 1992, A sequestered pool of aminoacyl-tRNA in mammalian cells, *Proc. Natl. Acad. Sci. USA* 89:3601.

Nilsen, T.W., Maroney, P.A., Goodwin, R.G., Perrine, K.G., Denker, J.A., Nanduri, J., and Kazura, J.W., 1988, Cloning and characterization of a potentially protective antigen in lymphatic filariasis, *Proc. Natl. Acad. Sci. USA* 85:3604.

Norcum, M.T., 1989, Isolation and electron microscopic characterization of a high molecular mass aminoacyl-tRNA synthetase complex from murine erythroleukemia cells, *J. Biol. Chem.* 264:15043.

Orgel, L.E., 1968, Evolution of the genetic apparatus, *J. Mol. Biol.* 38:381.

Perego, R., and Del Monte, U., 1986, A stable complex from Yoshida hepatoma AH 130 containing nine aminoacyl-tRNA synthetases, *Cell Biol. Int. Rep.* 10:477.

Raben, N., Borriello, F., Amin, J., Horwitz, R., Fraser, D., and Plotz, P., 1992, Human histidyl-tRNA synthetase: recognition of amino acid signature regions in class 2a aminoacyl-tRNA synthetases, *Nucleic Acids Res.* 20:1075.

Rould, M.A., Perona, J.J., Söll, D., and Steitz, T.A., 1989, Structure of *E. coli* glutaminyl-tRNA synthetase complexed with tRNA[Gln] and ATP at 2.8 Å resolution, *Science* 246:1135.

Ruff, M., Krishnaswamy, S., Boeglin, M., Poterszman, A., Mitschler, A., Podjarny, A., Rees, B., Thierry, J.C., and Moras, D., 1991, Class II aminoacyl-tRNA synthetases: crystal structure of yeast aspartyl-tRNA synthetase complexed with tRNA[Asp], *Science* 252:1682.

Ryazanov, A.G., Ovchinnikov, L.P., and Spirin, A.S., 1987, Development of structural organization of protein-synthesizing machinery from prokaryotes to eukaryotes, *Biosystems* 20:275.

Sariski, V., and Yang, D.C.H., 1991, Co-purification of the aminoacyl-tRNA synthetase complex with the elongation factor eEF1, *Biochem. Biophys. Res. Commun.* 177:757.

Schimmel, P., 1987, Aminoacyl-tRNA synthetases: general scheme of structure-function relationships in the polypeptides and recognition of transfer RNAs, *Ann. Rev. Biochem.* 56:125.

Schimmel, P., and Söll, D., 1979, Aminoacyl-tRNA synthetases: general features and recognition of transfer RNAs, *Ann. Rev. Biochem.* 48:601.

Schuler, G.D., Altschul, S.F., and Lipman, D.J., 1991, A workbench for multiple alignment construction and analysis, *Proteins: Struct. Funct. Genet.* 9:180.

Steitz, T.A., 1991, Aminoacyl-tRNA synthetases: structural aspects of evolution and tRNA recogniton, *Curr. Opin. Struct. Biol.* 1:139.

Van Damme, H.T.F., Amons, R., Janssen, G.M.C., and Möller, W., 1991, Mapping the functional domains of the eukaryotic elongation factor 1βγ, *Eur. J. Biochem.* 197:505.

Vellekamp, G., Sihag, R.K., and Deutscher, M.P., 1985, Comparison of the complexed and free forms of rat liver arginyl-tRNA synthetase and origin of the free form, *J. Biol. Chem.* 260:9843.

Walter, P., Weygand-Durasevic, I., Sanni, A., Ebel, J.P., and Fasiolo, F., 1989, Deletion analysis in the amino-terminal extension of methionyl-tRNA synthetase from *Saccharomyces cerevisiae* shows that a small region is important for the activity and stability of the enzyme, *J. Biol. Chem.* 264:17126.

EVOLUTION OF THE EF-Tu FAMILY

William C. Merrick, Jens Cavallius, Terri Goss Kinzy and
Wendy L. Zoll

Department of Biochemistry
School of Medicine
Case Western Reserve University
Cleveland, Ohio 44106
U.S.A.

INTRODUCTION

Two recent developments have allowed for the identification of a "family" of EF-Tu-like proteins. The first is the biochemical characterization of several proteins which bind aminoacyl-tRNAs in a GTP-dependent manner. The second has been the identification of sequence identity elements that are similar to EF-Tu and many of which would allow for a similar three dimensional structure. A key element in this second part has been the emergence of a 2.6 Å resolution crystal structure for Escherichia coli EF-Tu (Kjeldgaard and Nyborg, 1992). This report discusses the evolution of prokaryotic EF-Tu into its eukaryotic counterpart EF-1α as well as the evolution of EF-Tu into proteins with distinct, but related functional properties. While not all of the family members have been characterized biochemically, based upon their similarity to EF-Tu and the presence of the three GTP consensus elements, it is likely that all the family members function by binding aminoacyl-tRNA in a GTP-dependent manner.

EVOLUTION OF EF-1α FROM EF-Tu

The advent of molecular cloning has allowed for the determination of a large number of amino acid sequences for EF-Tu or EF-1α proteins, over 70 at the

The Translational Apparatus, Edited by K.H. Nierhaus
et al., Plenum Press, New York, 1993

writing of this article (Cavallius and Merrick, 1993). This extensive listing has allowed for the establishment of a phylogenic linkage map (Cousineau et al., 1992). While there is a gradual shift in both the peptide length and sequence similarity in the evolution of bacterial EF-Tu to mammalian EF-1α, there also appear to be discrete break points. These occur in the evolution of bacterial EF-Tu (393 amino acids) to archaebacterial EF-Tu (428 amino acids) to plant EF-1α (448 amino acids) to mammalian EF-1α (462 amino acids). This size evolution is accomplished by 11 inserts, 3 inserts and 2 inserts respectively into the EF-Tu sequence with most inserts being 3 to 7 amino acids. Presumedly not to alter the overall structure of the molecule, almost all of the inserted amino acid sequences are into nonstructured portions of the EF-Tu molecule. Where the proteins have been characterized, they usually are 3 to 10% of the soluble protein in a cell and contain one or more trimethyllysine residues (Merrick, 1992).

While one might have expected the similarity of the evolving EF-1α to be rather uniform for the length of the molecule, this is in fact not the case. Indicated in Figure 1 are those regions which are the areas in the EF-1α most similar to EF-Tu. As can be seen, the highest degree of homology is in the GTP-binding domain, domain 1 of the EF-Tu molecule. Part or all of 5 beta strands and 4 helices contain very conserved sequences. Second, the face of the molecule pointing towards the reader is highly conserved while the face pointing inward is conserved for the number of residues, but not amino acid sequence. Biochemical data suggest that the face pointing out binds the aminoacyl-tRNA and the side facing inward is likely to interact with the ribosome (see below). In addition to the conserved structural elements, three consensus sequences are conserved in the loops that recognize the GTP molecule (Dever et al., 1987). This conservation of amino acid sequence (and presumed three dimensional structure) is also observed in the one other GTP-binding protein for which a crystal structure has been determined, the oncogene RAS (Pai et al., 1990). Conservation of sequence elements continues into domain 2 (5 of the 7 beta strands are conserved), but there appears to be considerable divergence in domain 3 where only 2 beta strands of the 6 stranded beta barrel are conserved. Interestingly, the 2 beta strands form direct contacts with the side chains of helix 2 which is the most highly conserved region in domain 1 (73% identical, 91% similar when comparing E. coli EF-Tu with mammalian EF-1α).

MULTIPLE FUNCTIONAL ROLES FOR EF-1α

Because of its abundance there has been speculation that EF-Tu might be linked to other cellular functions, but to date there has been no positive report. In contrast, the list of associations or possible functions of EF-1α are considerable. These include: being part of mRNP complexes (Greenberg and Slobin, 1987) and the valyl-tRNA synthetase complex (Motrin et al., 1988); association with the endoplasmic reticulum (Hayashi et al., 1989) and the mitotic apparatus (Ohta et

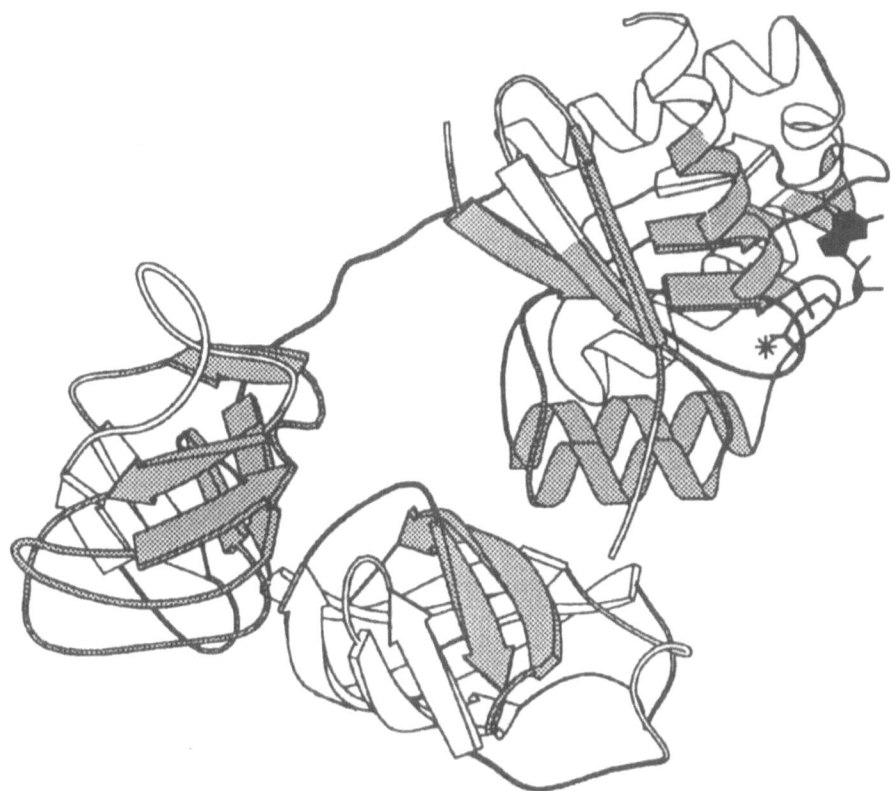

Figure 1.   Comparison of bacterial EF-Tu to mammalian EF-1α

The ribbon diagram for this and subsequent figures was generated from the primary crystallographic data of Kjeldgaard and Nyborg (1992) on the structure of E. coli EF-Tu at 2.6 Å resolution. Shaded areas indicate where regions of EF-1α show a high degree of homology to EF-Tu.

al., 1990); stimulation of enzymatic activities associated with protein turnover (Schwartz , 1992) and phosphotidylinositol phosphorylation (Yang et al., 1993). EF-1α has been shown to bind actin (Yang et al., 1990) and its trimethyllysines residues are probably the most abundant source of free trimethyllysine for the synthesis of carnitine which is required for fatty acid metabolism (McGilvery, 1983). Additional hints of possible EF-1α functions not related to protein synthesis have been the observation of stage or tissue specific expression of the multiple functional genes for EF-1α (Merrick, 1992), the observation that over-expression of EF-1α extends life span (Shepherd et al., 1989) or the most recent finding that constitutive expression of EF-1α predisposes mammalian tissue culture cells to transformation (Taksuka et al., 1992).

While EF-1α might appear to be an excellent candidate for a single cellular protein to sense a variety of different cellular demands, it should be noted that many of the observations on possible EF-1α function not related to its role in protein synthesis elongation may be artifacts. This could easily result from the fact that EF-1α is a very abundant protein (and therefore a likely contaminant of any protein fraction) and that it is a basic protein (and therefore likely to interact non-specifically with nucleic acids). What is required is real proof of function more than association.

THE EF-Tu FAMILY

As noted above, there has been a gradual evolution of bacterial EF-Tu to mammalian EF-1α, proteins which serve the same exact function in translation, the codon specific binding of aminoacyl-tRNA to the A site of the ribosome. Not expected was the now general observation that domain 1, the GTP-binding domain, is conserved in a number of "G-proteins" as characterized by amino acid sequence. This same dichotomy is observed with proteins which make up the EF-Tu family, proteins whose function would appear to be to bind aminoacyl-tRNA in a GTP-dependent manner. Besides EF-Tu and EF-1α, these proteins are Sel B, SUF-12, and eIF-2γ (Cavallius and Merrick, 1993) and Hbs1p (Nelson et al., 1992). Sel B is a bacterial protein which is responsible for the utilization of a UGA codon to specifically insert selenocysteine via a selenocysteyl-tRNA[Ser] (Forchhammer et al., 1989). By sequence comparison, there is an excellent match between Sel B and EF-Tu through domain 1 and 2 , however, the next 420 amino acids in Sel B match poorly with the last 100 amino acids in domain 3 of EF-Tu. One assumes that these last 420 amino acids are responsible for the specificity of Sel B which recognizes only a single aminoacyl-tRNA species and only UGA codons followed on the 3' side by a unique RNA structural element.

The second member of the EF-Tu family is the yeast protein SUF-12 which is highly similar to yeast EF-1α, but contains an amino-terminal 250 amino acid extension (Wilson and Culbertson, 1988). This protein has not been characterized biochemically, but via genetics has been shown to be involved in the control of translational ambiguity and essential for cell viability. Given the near identity of SUF-12 and EF-1α for the carboxy-terminal 440 amino acids, one assumes the unique functional characteristics of SUF-12 reside in the amino-terminal extension.

The third EF-Tu family member is the yeast protein Hbs1p (Nelson et al., 1992). This protein, when over expressed, can suppress the double mutant which knocks out the two heat shock proteins SSB1 and SSB2. These heat shock proteins are associated with polysomes and are suggested to assist in the folding of proteins as they emerge from the ribosome. The authors suggest that Hbs1p may function as a more efficient EF-1α and thus bypass the elongation block that arises in SSB1 and SSB2 deficient strains.

The final member of the EF-Tu family is the γ subunit of eIF-2, a 3 subunit protein which specifically recognizes Met-tRNA$_i$ and none of the other aminoacyl-tRNAs (Merrick, 1992). By amino acid sequence comparison (Hannig et al., 1993), eIF-2γ matches EF-Tu better than EF-1α suggesting an independent evolution of eIF-2γ from EF-Tu. There are two main structural differences between eIF-2γ and EF-Tu which are an amino-terminal extension of 90 amino acids and an insert of 35 amino acids into a part of domain 1 that is analogous to the GAP (GTPase activating protein) binding region in RAS (Pai et al., 1990; Merrick, 1992). A curious feature to this 35 amino acid insert is the presence of 5 cysteine residues and 6 proline residues (Hannig et al., 1993).

eIF-2γ is the largest of the 3 subunits of eIF-2 and in this sense is structurally similar to trimeric "G-proteins" (Bourne et al., 1991). Like G-proteins and EF-Tu, eIF-2 requires an exchange factor (eIF-2B) to facilitate the exchange of bound GDP for GTP. However, at present there appears to be no correlation by amino acid sequence of the α or β subunit of eIF-2 with the corresponding subunits of the trimeric G-proteins, nor has there been any correlation of the sequences of the 5 subunits of eIF-2B with either EF-Ts or the recycling proteins involved in nucleotide exchange with the trimeric G-proteins. Thus it is unclear whether the evolution of eIF-2 and its recycling protein might represent an independent event in spite of formal analogies to EF-Tu and G-proteins.

Figure 2 below presents a pseudo-structural comparison of the EF-Tu family members and highlights areas of extreme conservation of amino acid sequence homology (shaded) and regions which are dissimilar. While it is difficult to predict how the amino-terminal extensions of SUF-12 (250 amino acids), eIF-2γ (90 amino acids) and Hbs1p (70 amino acids) and the carboxy-terminal difference/extension of Sel B (300 amino acids) might effect the folding of the 3 domains similar to EF-Tu, two differences clearly should alter the folding or orientation of domain 1. The first is the insert of 35 amino acids into the turn between the 2nd and 3rd beta strands in eIF-2γ which should alter either the GAP-binding region or the specificity for a particular aminoacyl-tRNA or both relative to EF-Tu. The second is the deletion of 10 to 12 residues from the last helix in domain 1 of Sel B which should cause domain 2 to be closer to domain 1 relative to EF-Tu. Given that there is preliminary data which has been interpreted to indicate that this area of the molecule is near the 3´ end of the aminoacyl-tRNA (Kinzy et al., 1992), it is possible that these two departures from an EF-Tu structural orientation reflect the fact both the eIF-2γ and Sel B are highly specific for a single aminoacyl-tRNA species.

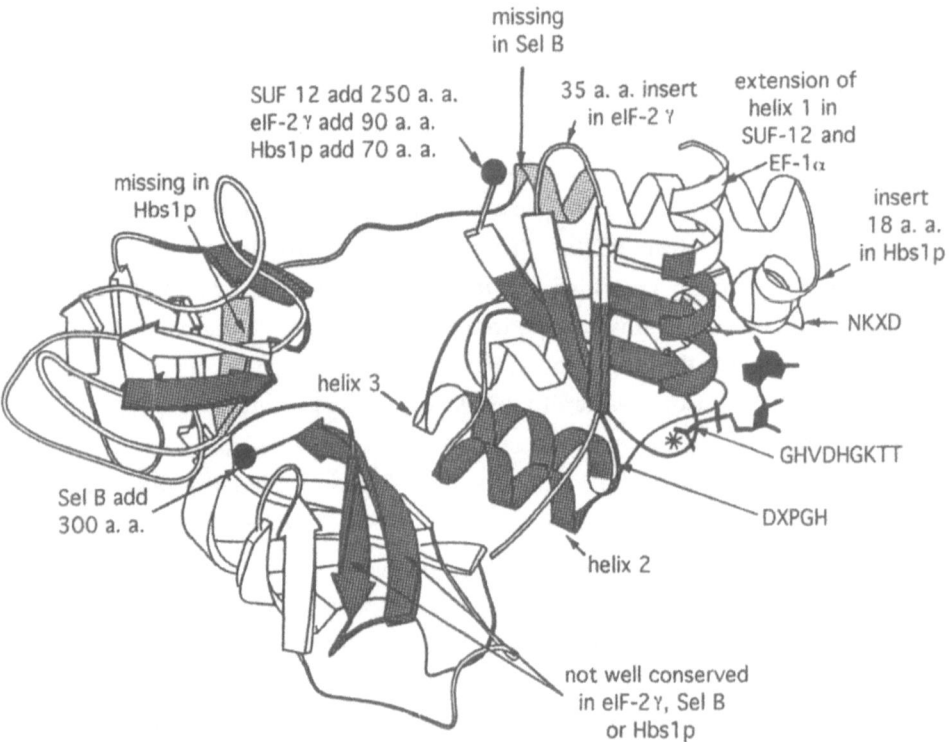

Figure 2.   Comparison of sequence homology of the EF-Tu family members

The darkly shaded regions in the ribbon diagram of EF-Tu indicate where the family members show a high degree of homology to the primary sequence of EF-Tu (see text for description). The lightly shaded regions indicate deletions in Sel B or Hbs1p relative to EF-Tu. "add" indicates an amino- or carboxy- terminal extension of the parental EF-Tu structure and the number following indicates the size of the amino acid extension.

With an understanding of the "apparent" molecular architecture of the members of the EF-Tu family it is appropriate to interpret this as related to function.   Each of the family members binds aminoacyl-tRNA in a GTP-dependent manner (this is assumed for SUF-12 and Hbs1p based upon genetic analyses; Wilson and Culbertson, 1988 and Nelson et al., 1992). EF-Tu and EF-1α bind all aminoacyl-tRNAs except the initiator (f)Met-tRNA₁ while Sel B and eIF-2γ bind only a single aminoacyl-tRNA, selenocysteyl-tRNA and Met-tRNA₁, respectively. Based upon crosslinking and protease protection studies, a model for the binding of aminoacyl-tRNA to EF-1α has been proposed (Kinzy et al., 1992) and a possible orientation of an aminoacyl-tRNA on EF-1α is shown in Figure 3. It should be noted that the orientation of the tRNA on EF-1α is just one of many as there is relatively little data which identifies specific sites in the tRNA in close proximity with specific sites in EF-1α. The one unique exception is His 296 which was crosslinked to the reactive group of N$^\varepsilon$-bromoacetyl-lysyl-tRNA. Assuming, such a model is of some accuracy, it is then possible to guess how eIF-2γ or Sel B might specifically recognize a single tRNA species. The unique 35 amino acid insert in domain 1 or the amino-terminal 90 amino acid

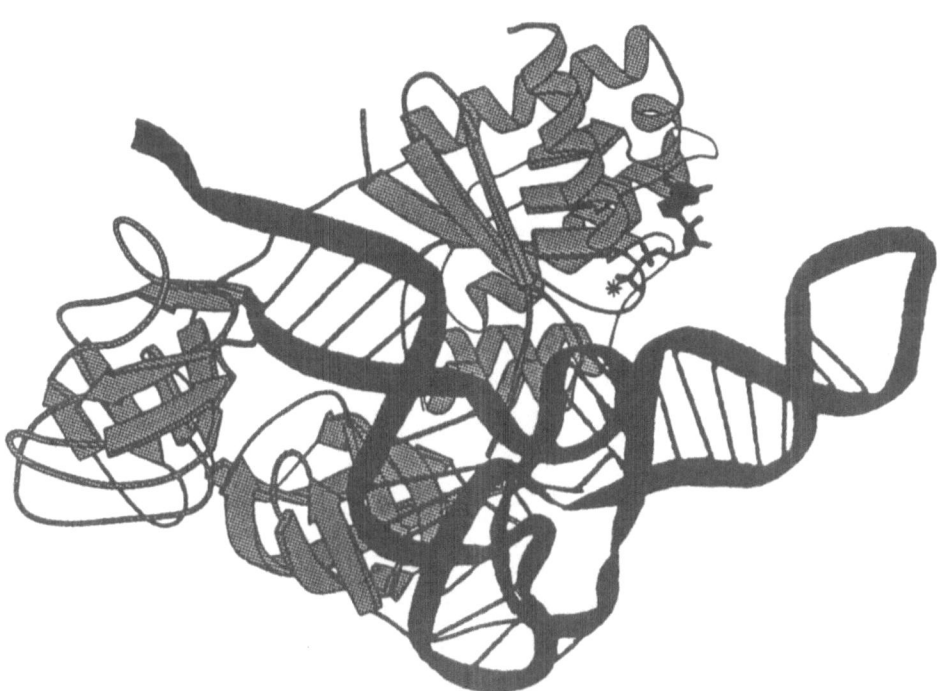

Figure 3.   Model for the aminoacyl-tRNA binding to EF-1α

Using the crystallographic coordinates of Kjeldgaard and Nyborg (1992) and the biochemical data from Kinzy et al. (1992), a tRNA molecule has been positioned on EF-1α. It should be noted that due to the uncertainty in the biochemical data that there are many positionings of the tRNA on the EF-1α structure which would be compatable with the data as is discussed in Kinzy et al. (1992).

extension of eIF-2 could determine amino acid specificity (only allowing for methionine; Wagner et al., 1984) or sequence/structure elements in the amino acid acceptor stem or both. This seems especially likely in that the A_1-U_{72} base pair appears to be a major determinant in eIF-2 recognition of the initiator tRNA (Pawel-Rammington et al., 1992). A second element involved in eIF-2 recognition is the rT_{54}-m^1A_{58} base pair of elongator tRNA species which is A_{54}-m^1A_{58} in the eukaryotic initiator tRNA. In Figure 3, it would appear that rT_{54} would be located in the heel of the tRNA which would be pointed away from the protein. The ability of eIF-2 to sense this modification and base pair may be the result of its other 2 subunits. Consistent with this idea is the observation that forms of eIF-2 which lack the α subunit do not discriminate as well as intact eIF-2 between mammalian Met-tRNA$_i$ and E. coli Met-tRNA$_f$ (Merrick, unpublished observation).

At present, less information is available on how Sel B interacts with selenocysteyl-tRNA. As noted above, the deletion of 11 residues in the last helix in domain 1 (see Figure 2) should cause domains 1 and 2 to be in closer proximity than they are in EF-Tu. This may facilitate an amino acid specific recognition and thus obviate the requirement for an additional sequence element as may be the case for eIF-2. Sel B also contains 300 more amino acids than EF-Tu which could be involved in amino acid or tRNA recognition, but are more likely to be involved in recognition of the RNA sequence/structure required downstream of appropriate UGA codons for correct addition of selenocysteine to the growing polypeptide chain. It might be noted that while seryl-tRNA represents a good choice chemically (seryl-tRNA→ phosphoseryl-tRNA→ selenocysteyl-tRNA), this is the one family of iso-accepting tRNAs for which no base in the anticodon is part of the recognition element by its cognate synthetase. Thus the existence of a complimentary anticodon to UGA would not effect aminocylation.

ACKNOWLEDGEMENTS

The authors are indebted to Drs. Morten Kjeldgaard and Jens Nyborg for providing the structural coordinates of EF-Tu on which Figures 1, 2 and 3 are based and for their exceptional insights into possible structure/function relationships of the domains of EF-Tu. Appreciation is also expressed to Dr. Ernie Hannig for making available the eIF-2γ sequence prior to publication. This work was supported in part by grants from the National Institutes of Health (GM26796,W.C.M. and AM07319,T.G.K.) and from the Northeast Ohio Chapter of the American Heart Association (J.C.)

REFERENCES

H.R. Bourne, D.A. Sanders, and F. McCormick, The GTPase superfamily: conserved structure and molecular mechanism, Nature 349:117 (1991).

J. Cavallius and W.C. Merrick, Eukaryotic translation factors which bind GTP, in: "GTPases in Biology", B. Dickey and L. Birnbaumer, eds. Springer-Verlag, Berlin (in press).

B. Cousineau, C. Cerpa, J. Lefebvre and R. Cedergren, The sequence of the gene encoding elongation factor Tu from Chlamydia trachomatis compared with those from other organisms. Gene (in press).

T. E. Dever, M. T. Glynias and W. C. Merrick, The GTP-binding domain: three consensus sequence elements with distinct spacing, Proc. Natl. Acad. Sci. USA 84: 3826 (1987).

K. Forchhammer, W. Leinfelder, and A. Böck, Identification of a novel translation factor necessary for the incorporation of selenocysteine into protein. Nature 342:453 (1989).

J.R. Greenberg and L.I. Slobin, Eukaryotic elongation factor Tu is present in mRNA-protein complexes, FEBS Lett. 224:54 (1987).

E.A. Hannig, A.M. Cigan, B.A. Freeman and T.G. Kinzy, GCD 11, a negative regulator of GCN 4 expression encodes the γ subunit of eIF-2 in yeast. Mol. Cell. Biol. (in press).

Y. Hayashi, R. Urade, S. Utsumi and M. Kito, Anchoring of peptide elongation factor EF-1α by phoshatidylinositol at the endoplasmic reticulum membrane, J. Biochem. 106:560 (1989).

T.G. Kinzy, J.P. Freeman, A.E. Johnson and W.C. Merrick, A model for the aminoacyl-tRNA binding site of elongation factor 1α, J. Biol. Chem. 267:1623 (1992).

M. Kjeldgaard and J. Nyborg, The refined structure of elongation factor Tu from Escherichia coli, J. Mol. Biol. 223:721 (1992).

R.W. McGilvery. "Biochemistry: A Functional Approach", W.B. Saunders Company, Philadelphia, Pa. (1983).

W.C. Merrick, Mechanism and regulation of eukaryotic protein synthesis, Microbio. Rev. 56:291 (1992).

Y.A. Motrin, A.D. Wolfson, A.F. Orlovsky and K.L. Gladilin, Mammalian valyl-tRNA synthetase forms a complex with the first elongation factor. FEBS Lett. 238:262 (1988).

R. J. Nelson, T Ziegelhoffer, C. Nicolet, M. Werner-Washburne and E. A. Craig, The translation machinery and an 70 kD heat shock protein cooperate in protein synthesis, Cell 71:97 (1992).

K. Ohta, M. Toriyama, M. Miyazaki, H. Murofushi, S. Hosoda, S. Endo and H. Saki, The mitotic apparatus-associated 51 kDa protein from sea urchin eggs is a GTP-binding protein and is immunologically related to yeast polypeptide elongation factor 1α, J. Biol. Chem. 265:3240 (1990).

E.F. Pai, U. Krengel, G.A. Petsko, R.S. Goody, W. Kabasch and A. Wittinghofer, Refined crystal structure of the triphosphate conformation of H-ras p21 at 1.35 Å resolution: implications for the mechanism of GTP hydrolysis, EMBO J. 9:2351 (1990).

U. von Pawel-Rammington, S. Åström and A.S. Byström, Mutational analysis of conserved positions potentially important for initiator tRNA function in Saccharomyces cerevisia, Molec. Cell. Biol. 12:1432 (1992).

Dr. Alan Schwartz, Washington University, St. Louis, Mo., personal communication.

J.C. Shepherd, U. Walldorf, P. Hug and W.J. Gehring, Fruit flies with additional expession of the elongation factor EF-1α live longer, Proc. Natl. Acad. Sci. USA 86:7520 (1989).

T. Wagner, M. Gross and P. Sigler, Isoleucyl-initiator tRNA does not initiate eukaryotic protein synthesis, J. Biol. Chem. 259:4706 (1984)

P.G. Wilson and M.R. Culbertson, SUF 12 suppressor protein of yeast: A fusion protein related to the EF-1 family of elongation factors, J. Mol. Biol. 199:559 (1988).

F. Yang, M. Demma, V. Warren, S. Dharmawardhane and J. Condeelis, Identification of an actin-binding protein from Dictyostelium as elongation factor 1α, Nature 347:494 (1990).

W. Yang, W. Burkhart, J. Cavallius, W. C. Merrick and W. F. Boss, Purification of a phosphatidylinositol4-kinase activator in carrot cells. J. Biol. Chem. (in press).

THE EVOLUTION OF RIBOSOMAL PROTEIN AND RIBOSOMAL RNA OPERONS: CODING SEQUENCES, REGULATORY MECHANISMS AND PROCESSING PATHWAYS

Peter Durovic, Daiqing Liao, Shanthini Mylvaganam
and Patrick P. Dennis*

Canadian Institute for Advanced Research
Program in Evolutionary Biology and
Biochemistry Department
University of British Columbia
2146 Health Sciences Mall
Vancouver, B.C. V6T 1Z3

INTRODUCTION

All extant life forms were derived from a single primordial ancestor that already had a DNA based genetic system and a well established transcription-translation apparatus. Evolutionary diversification driven by natural selection and random processes has produced three easily distinguishable major groups of organisms: eubacteria, archaebacteria, and eucaryotes (or Bacteria, Archaea and Eucarya, respectively; Woese et al, 1990). The translational machinery has been independently refined within each group for efficient and accurate function. Because of this, universal features and components of the ribosome have been an invaluable source of information for deducing the primordial characteristics of the common ancestor and for understanding the origins and relationships of each of the three groups of organisms.

The universal phylogenetic tree based upon small subunit rRNA sequences (Woese et al, 1990) and rooted using the paralogous EF-Tu and Ef-G genes (Iwabe et al, 1989) is depicted in Figure 1. If the general topology of this tree is correct, it would suggest that the eubacteria represent one lineage and that the archaebacteria and eucaryotes are derived from a second lineage. Some of the organisms we have examined and their positions within this tree are indicated. In general, we have studied both prototype species and early branching-slowly evolving species from each group. The early branching-slowly evolving species are important because they are more likely to have retained ancestral features that would be substantially modified or completely lost from later more highly evolved lineages.

Our evolutionary studies have focussed on two separate features of the ribosome. The first is a group of four proteins designated L11, L1, L10 and L12 in *Escherichia coli*. Three of these, L11, L10 and L12, assemble with a stoichiometry of 1:1:4 as a stalk on the large ribosome subunit and function in factor binding and associated GTPase activities during the protein synthesis cycle (Liljas, 1982). The fourth protein, L1,

---

*corresponding author

The Translational Apparatus, Edited by K.H. Nierhaus
et al., Plenum Press, New York, 1993

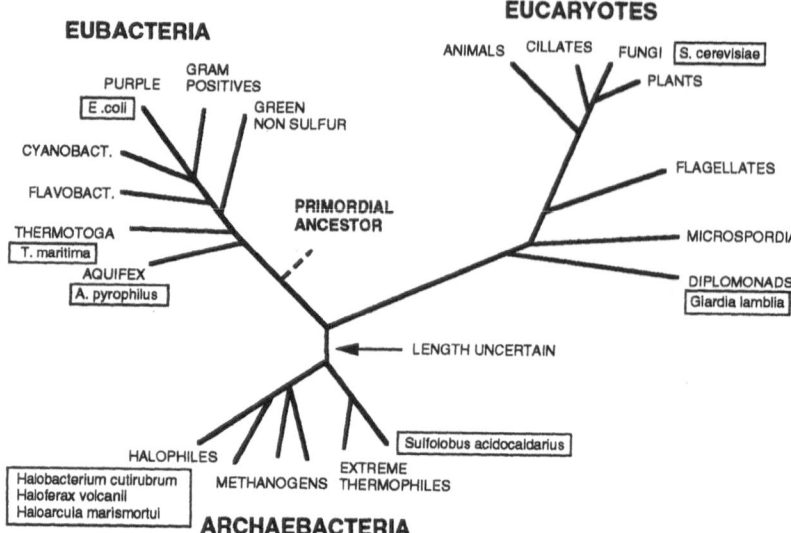

Figure 1. A universal phylogenetic tree. A universal phylogenetic tree based upon small subunit rRNA sequences, and rooted using the paralogous translongation factor genes is presented (adapted from Woese et al, 1990; Iwabe et al, 1989). Organisms that are currently being investigated in our laboratory are named and boxed. The length of the stem at the base of the archaebacterial group is controversial; Woese and colleagues (1990) argue that it is short whereas Rivera and Lake (1992) and colleagues argue that the methanogen-halophile group and the thermophile group branch separately from the lineage that leads to eucaryotes.

stabilizes peptidyl tRNA binding at the P site and indirectly enhances GTPase activity at the factor binding domain (Sander, 1983). In addition to analyzing both gene and protein sequences, we have examined the genomic organization, transcription, and regulation of these genes in a variety of organisms from each of the three groups. The second feature is rRNA gene organization and expression, and rRNA processing and assembly in both halophilic and thermophilic species of archaebacteria.

The L11, L1, L10 and L12 Ribosomal Protein Genes

The genomic organization of the L11, L1, L10 and L12 genes in two eubacteria — *E. coli* and *Thermotoga maritima* — two archaebacteria — *Halobacterium cutirubrum* and *Sulfolobus acidocaldarius*[†] — and the eucaryote *Saccharomyces cerevisiae* is depicted in Figure 2. Comparisons indicate that the eubacterial L11 and L1 proteins are relatively similar in structure and sequence to the respective L11 and L1 proteins from archaebacteria (Liao and Dennis, 1993). A eucaryotic L1 equivalent protein has yet to be identified and the putative eucaryotic L11 equivalent protein, designated "L15" in *S. cerevisiae* and "L12" in *R. rattus*, exhibits at best only marginal sequence and structural similarity to the eubacterial-archaebacterial L11 proteins. In spite of this dissimilarity, the L11 protein from all three groups bind to the same region within large subunit rRNA. The L10 and L12 proteins exhibit the opposite pattern. The eucaryotic L10 and L12 proteins are relatively similar in structure and sequence to the respective L10 and L12 proteins from archaebacteria whereas the eubacterial proteins are dissimilar. This

---

[†] This strain has previously been misidentified as *S. solfataricus*. Here we use the correct species name.

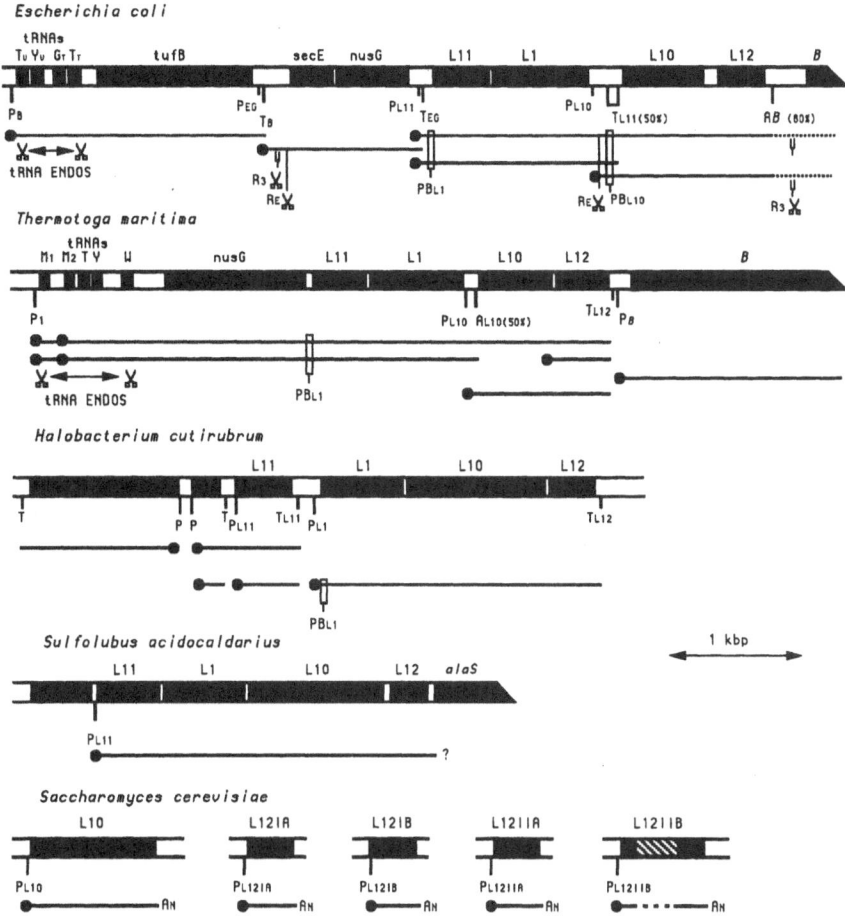

Figure 2. Genomic organization of L11, L1, L10 and L12 genes in eubacteria, archaebacteria and eucaryotes. The genomic region that surrounds the L11, L1, L10 and L12 genes is illustrated for the eubacteria *E. coli* and *T. maritima*, the archaebacteria, *Hb. cutirubrum* and *S. acidocaldarius*, and the eucaryote *S. cerevisiae*. Genes are depicted as solid boxes and identified above. Promoters (P), terminators (T) and attenuators (A) are identified below. Transcripts are depicted as horizontal lines; the filled circles represent 5' triphosphate ends. Endonuclease processing sites are depicted by scissors: R3 is RNaseIII cleavage and RE is RNaseE cleavage. Characterized and putative translational control sites on the mRNAs are boxed: PBL1 and PBL10 are protein binding sites for L1 and L10, respectively. The *S. acidocaldarius* map was determined by C. Ramirez and Matheson (personal communication). The intron in the yeast L12IIB gene is indicated by the cross-hatched area. The gene encoding the putative yeast homolog to L11 ("L15") is not shown; no eucaryotic equivalent to the L1 protein is known.

would imply that the evolution of the L11 and probably also L1 proteins is distinct from the evolution of the L10 and L12 proteins.

Eucaryotic and archaebacterial L10 and L12 proteins exhibit an interesting feature not apparent in their eubacterial counterparts (Shimmin et al, 1989). In any given species of eucaryote or archaebacterium, the L10 and L12 proteins exhibit substantial sequence similarity at their carboxy termini; this similarity presumably is maintained within the species by concerted evolution. Finally, eucaryotes contain two paralogous L12 genes, designated type I and type II that are the result of a duplication occurring either before or very early within the eucaryotic radiation. In yeast there has been a further reduplication to produce types IA, IB, IIA and IIB, all of which exhibit sequence identity to each other and to the single yeast L10 protein at their carboxy termini. We were able to clone all five yeast genes using a single degenerate oligonucleotide complementary to the sequences encoding the hexapeptide DDDMGF, conserved near the carboxy terminus of all five proteins (Newton et al, 1990).

In *E. coli*, expression of these four genes is driven by two promoters, PL11 and PL10, and result in the production of both bicistronic L11-L1 and L10-L12 mRNAs and tetracistronic L11-L1-L10-L12 mRNAs (Downing and Dennis, 1987). The functional integrity of these mRNAs may be related to cleavage by RNaseE at a site within the L1-L10 intergenic space (J. Chow and P. Dennis, unpublished results). Translation of the mRNAs is regulated at two sites. The first, located in the L11 leader, is a sequence and structural mimic of the authentic L1 binding site in 23S rRNA (Yates and Nomura, 1981). In the absence of sufficient rRNA to support ribosome assembly, the protein binds to the mRNA and inhibits further translation of the L11 and L1 cistrons. A second, more complex translational control site occurs in the L1-L10 intergenic space; binding of protein L10 at this site regulates translation of the L10 and L12 cistrons (Fiil et al, 1980; Johnsen et al, 1982).

In the distantly related eubacterium *T. maritima*, the L11, L1, L10 and L12 genes are distally positioned in an operon that contains five leader tRNAs and the *nusG* transcription termination factor gene (Liao and Dennis, 1992). The mRNA clearly retains a sequence and structural mimic of the authentic L1 binding site in 23S rRNA but probably does not contain an L10 translational control site. Instead, a transcriptional attenuator-like structure that retains some sequence and structural similarity to the *E. coli* L10 translational control site is positioned immediately in front of the L10 structural gene. Thus, *E. coli* regulates L10 and L12 production at the level of translation whereas *Thermotoga* apparently regulates at the level of transcript termination. We hope to be able to characterize this putative attenuation control using an in vitro transcription system.

Surprisingly, the genomic organization of the L11, L1, L10 and L12 genes is preserved not only within eubacteria but also within numerous archaebacterial species (Shimmin and Dennis, 1989). In *Hb. cutirubrum* and in other halophiles, the L11 gene is transcribed primarily as a monocistronic mRNA and the L1, L10 and L12 genes as a separate tricistronic mRNA. More surprising was the retention of the L1 translational control site transposed from its L11 proximal position in eubacteria to a position in front of the L1 cistron where it presumably now regulates L1, L10 and L12 translation. In *S. acidocaldarius*, a structural mimic of the L1 rRNA binding site has not been identified in the mRNA. Instead, the four genes are tightly spaced on a single tetracistronic mRNA (Shimmin et al, 1989; Ramirez and Matheson, personal communication).

In eucaryotic organisms, virtually all protein encoding genes are monocistronic. In *S. cerevisiae*, the single L10 and the four L12 genes are well isolated from each other but are coordinately regulated during nutrient up-shifts (Newton et al, 1990). The only unusual feature is the presence of a 301 nucleotide long intron between codons 38 and 39 of the L12IIB gene. This intron does not appear to possess essential regulatory features.

Ribosomal RNA Transcription Units

The rRNA encoding genes in most species are repetitive; copy to copy variation in the nucleotide sequence of these reiterated genes is negligible because of concerted evolutionary processes which maintain intraspecies homogeneity while permitting interspecies variation. Presumably it is important to maintain sequence homogeneity within an organism in order to insure uniformity in molecular interactions involving rRNAs within the ribosome. It is precisely these features, intraspecies sequence homogeneity and interspecies sequence divergence, that have made rRNAs an invaluable resource for elucidating the ancestral relationships between contemporary organisms (Woese, 1987; Woese et al, 1983).

Assembly of ribosomal particles is a complex and highly ordered process that begins with the primary transcripts derived from rRNA encoding genes. In eubacteria the transcripts from rRNA operons are cleaved and processed in a number of sequential steps to produce 16S rRNA, and 23S and 5S rRNAs that ultimately appear in mature

30S and 50S subunits, respectively (Srivastava and Schlessinger, 1990). In *E. coli* there are seven unlinked rRNA operons. The structure of the typical *rrnB* operon is illustrated in Figure 3. The operon, transcribed from two upstream promoters P1 and P2, contains a 16S gene, a spacer tRNA gene, a 23S gene and a 5S gene. There is no distal tRNA in *rrnB*. The 16S and 23S sequences are surrounded by long nearly perfect inverted repeats that are recognized and cleaved by the double strand specific endonuclease RNaseIII (Srivastava and Schlessinger, 1990). The cleavage products contain 5' PO$_4$ and a 3' OH termini. Additional endonuclease activities are required to trim the 16S and 23S precursors at both their 5' and 3' ends. The 5S rRNA is removed from the transcript by RNaseE and the tRNAs are excised by RNaseP and D.

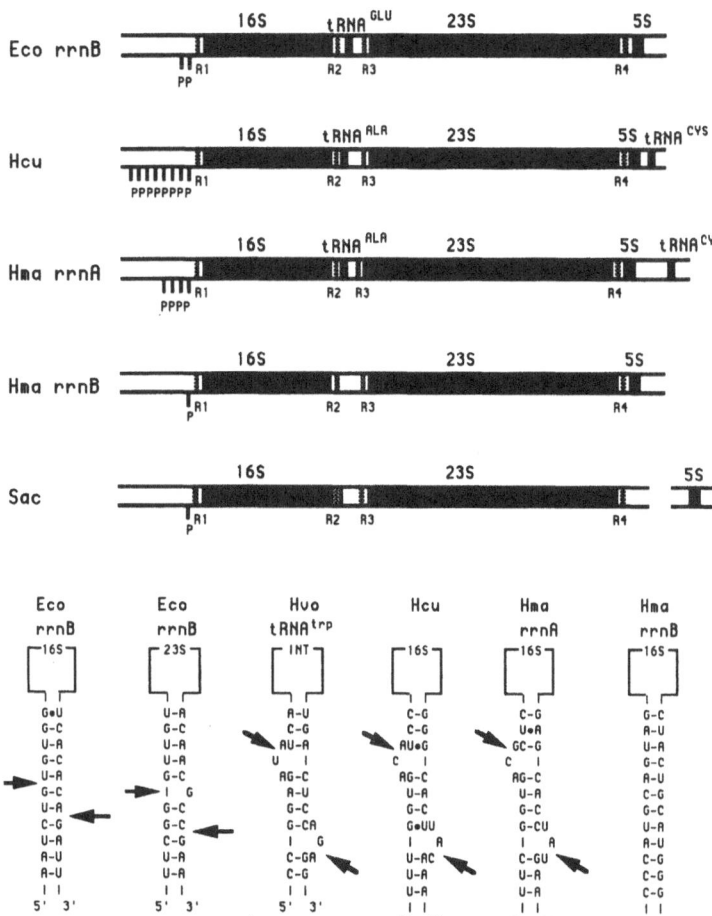

Figure 3. The organization of rRNA genes and the structure of precursor processing sites. Above: The genomic organization of rRNA genes in *E. coli* (Eco *rrnB* operon), *Hb. cutirubrum* (Hcu), *Ha. marismortui* (Hma) and *S. acidocaldarius* (Sac) is depicted. Genes are solid boxes and are labeled above. The inverted repeat sequences surrounding the 16S gene (R1 is complementary to R2) and 23S gene (R3 is complementary to R4) are depicted as shaded boxes. The R2 portion of the 16S inverted repeat in *S. acidocaldarius* is retained within the 16S rRNA found in ribosomal particles. Below: A portion of the helical RNA precursor processing stems is illustrated. The *E. coli* structures are cleaved by RNaseIII at the indicated positions (arrows). The structure of the intron exon boundary in the primary transcript of the tryptophan tRNA is illustrated. The intron is excised by an intron excision endonuclease to liberate the two halves of the tRNA (below the arrows). The same endonuclease appears to be responsible for excision of 16S and 23S precursors in many species of archaebacteria. The structure surrounding the *rrnB* 16S gene of *Ha. marismortui* is apparently not cleaved by this endonuclease.

Halophilic rRNA operons

The structure of rRNA operons, and the in vivo transcripts and processing intermediates have been analyzed in a number of halophilic archaebacteria (Hui and Dennis, 1985; Chant and Dennis, 1986; Mevarech et al, 1989). These organisms contain either one or two rRNA operons which are normally preceded by up to eight tandemly arranged promoters. Halophilic operons superficially resemble those of eubacteria. Typically, they contain a 16S gene, a spacer alanine tRNA gene, a 23S gene, a 5S gene and a distal cysteine tRNA gene. As in eubacterial operons, the 16S and 23S genes are surrounded by long nearly perfect inverted repeat sequences that are used for excision of 16S and 23S precursors from the primary transcript.

The characteristics of the excision endonuclease appear to be fundamentally different from those of eubacterial RNaseIII. In archaebacteria, the substrate for the enzyme is more clearly defined; it consists of two three base loops located on opposite strands of the RNA helix and separated by four base pairs. Cleavage occurs between the second and third base of each loop and the preferred base 3' to the cleavage site is an adenine. Similar structures are present at the intron exon junctions of archaebacterial intron containing tRNA and 23S rRNA sequences (Thompson and Danials, 1988; Thompson et al, 1989). The tRNA introns are usually located adjacent to the anticodon. The tRNA intron excision endonuclease has been partially characterized; it has a rigorous requirement for the two three base loops separated by four base pairs in an extended helix, but not for the adjacent tRNA like structure, and cleaves to produce 5' OH and 3' PO_4 termini. Although we have not tested this activity on an authentic rRNA primary transcript, it efficiently cleaves a short synthetic oligoribonucleotide hairpin that has the sequence and structure of either the 16S or 23S rRNA processing site from *Hb. cutirubrum* (C. Danials and P. Dennis, unpublished results).

Another halophile, *Haloarcula marismortui*, has two rRNA operons in its genome. The *rrnA* operon appears to be a typical halophilic operon whereas the *rrnB* operon is atypical in a number of respects (Myvalganam and Dennis, 1992). For example, it possesses only a single promoter in its 5' flanking region, it lacks that portion of the 5' half of the inverted repeat preceding the 16S gene which is recognized by the excision endonuclease, it lacks an alanine tRNA in the 16S-23S intergenic space and a cysteine tRNA in the distal position, and both the 16S and 23S encoding sequences differ substantially from the *rrnA* operon.

The *rrnA* and *rrnB* 16S gene sequences differ at 74 of 1472 positions. The substitutions are not uniformly distributed but rather are localized within three domains bounded by nucleotide base pairs at positions 56-301, 508-823 and 986-1158 (corresponding to *E. coli* positions 59-353, 570-880 and 1046-1211). Two thirds of the substitutions fall within the 508-823 domain. Regions outside of these domains are identical in sequence except for a single substitution in a tetra loop at position 1216 (*E. coli* position 1258).

Comparisons of the two *Ha. marismortui* 16S sequences with 16S sequences from other halophiles indicates that (i) substitutions (differences between *rrnA* and *rrnB)* occur exclusively at phylogenetically variable positions and never at functionally important or highly conserved positions and (ii) that in their divergence, the unusual *rrnB* gene has accumulated two to three times as many substitutions as the *rrnA* gene. Within the 508-823 central domain, most of the substitutions are compensatory and fall within the two helices that constitute, respectively, the binding sites for proteins S8 and S18. It has been known for some time that proteins S8 and S18 from *E. coli* can form specific complexes with a number of different archaebacterial 16S rRNAs (Thurlow and Zimmerman, 1982). Binding of these proteins requires a conserved rRNA secondary structure but can tolerate nucleotide substitutions at approximately half of the congruent positions.

These anomalies raise the important question — are both operons actively transcribed and do both 16S sequences get assembled into active ribosomes? Using a nuclease S1 protection assay with DNA probes from the two genes, it has been possible to demonstrate that 70S ribosomes contain equal proportions of the two 16S sequences. Our current goals in this area are to fully characterize the processing pathways of the primary transcripts for the *rrnA* and *rrnB* operons and using a genetic approach to attempt to identify hybrid operons produced by the process of concerted evolution within the identical domains of the *rrnA* and *rrnB* rRNA coding regions.

Archaebacteria divide into two sublineages: the methanogen-halophiles (or Euryarchaeota) and the sulfur metabolizing thermophiles (Crenarchaeota or Eocytes) (Woese et al, 1990; Rivera and Lake, 1992). A member of the thermophilic branch *Sulfolobus acidocaldarius* has single copy 16S, 23S and 5S rRNA genes (Olsen and Pace, 1985). The 5S gene is unlinked to the 16S and 23S genes. There is no apparent processing of the 5S transcript, which appears identical to the mature 5S rRNA sequence that is assembled into 50S ribosomal subunits.

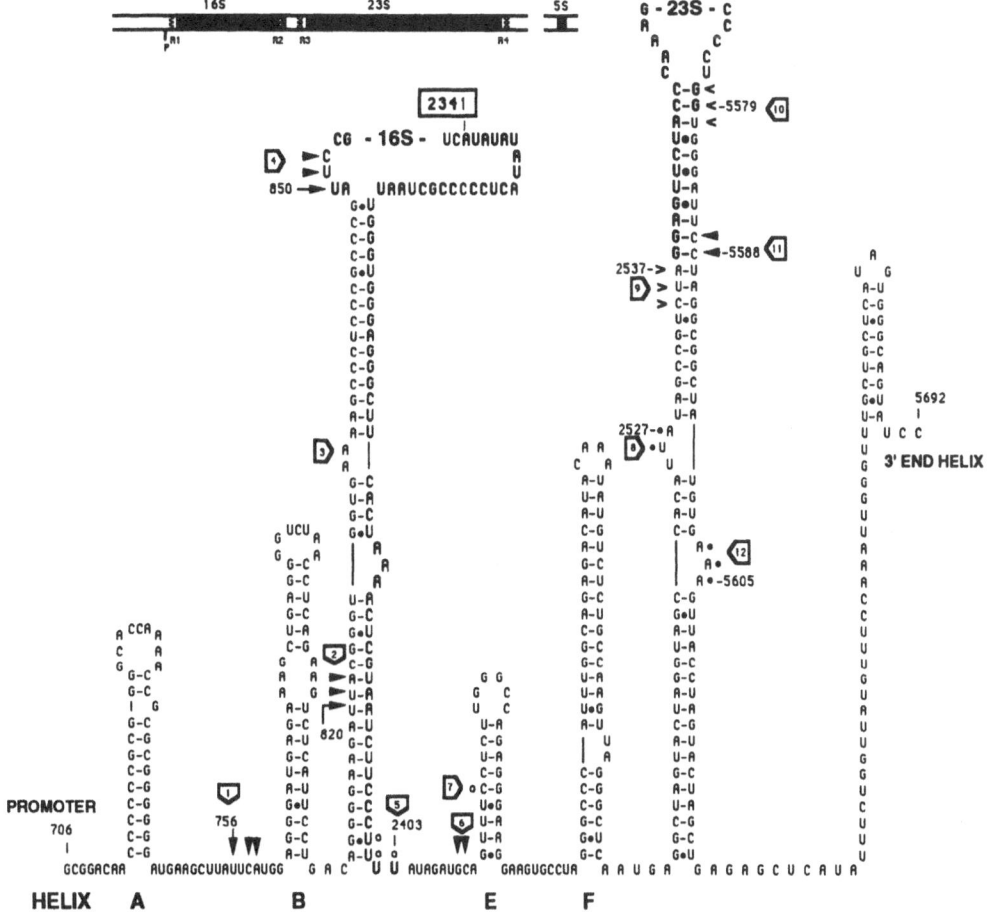

Figure 4. Sequence and structure of the 16S-23S rRNA primary transcript from *S. acidocaldarius*. A region of 7161 nucleotides containing the entire 16S-23S rRNA operon from *S. acidocaldarius* has been sequenced and the primary transcript and major processing intermediates have been characterized by S1 nuclease protection and primer extension analysis. Transcripts are initiated at the G residue at position 706 and extend at least to position 5692. The sequence and secondary structure of the non-coding region of the primary transcript are illustrated. The location of mature 5' and 3' ends of 16S and 23S rRNA are at or near position 850, 2403, 2537, and 5579, respectively. Mature 16S or 23S sequences are in bold. The location of what is generally recognized as the phylogenetically conserved 3' end of 16S RNA is at position 2341 (boxed). The end sites of in vivo processing are indicated as sites 1 through 12; ▶, endonuclease cleavage sites that show sequence similarity to the 5' maturation site on 16S rRNA; ●, endonuclease cleavage site presumably generated by the intron excision endonuclease; O, other end sites of unknown origin; >, 23S maturase sites.

The closely linked single copy 16S and 23S rRNA genes are cotranscribed from a single upstream promoter (Durovic and Dennis, in preparation). Contained within the primary transcript are a 143 nucleotide long 5' leader, a 137 nucleotide long intergenic space, and a 113 nucleotide long 3' trailer. Seven regions of inverted repeat symmetry potentially capable of forming helical secondary structures are located within these noncoding segments (Figure 4). The two largest of these surround, respectively, the 16S and 23S coding sequences and resemble the processing stems found in other eubacterial and archaebacterial rRNA operons. In addition, there are two helices (A and B) in the 5' leader, two (E and F) in the spacer, and one (the 3' end stem) immediately in front of the 3' transcript end.

The processing of the 16S-23S rRNA primary transcript in vivo is complex and unusual. As many as twelve different transcript end sites generated predominantly by endonuclease cleavage events have been detected. Five of these occur within the long nearly perfect inverted repeat structure which surrounds the 23S gene. The two sites designated 9 and 10 are responsible for trimming precursor intermediates to generate the mature 5' and 3' ends of 23S rRNA, respectively. Two other sites, designated 8 and 12, are located on opposite strands of the RNA helix within a region that resembles the substrate for the intron excision endonuclease — two three base loops separated by four base pairs. The last site, designated 11, occurs within the helix at a site opposite the mature 5' end of 23S rRNA. The possible significance of this site is discussed below.

Processing around the 16S coding sequence is more complicated. To begin with, the long helical stem surrounding the 16S sequence contains a region that resembles but definitely is not identical to the substrate for the intron excision endonuclease. The loop on the ascending portion is only two bases in length. A very small amount of product resulting from cleavage at the two base loop is detectable (site 3) but no cleavage is detectable within the three base loop on the opposite descending strand of the helix. Therefore, it seem unlikely that 16S processing utilizes as a first step excision of the precursor by the intron excision endonuclease.

When the mature ends of 16S rRNA were characterized, the 5' end (designated site 4) was located at the expected position whereas the 3' end (designated site 5) was located 62 nucleotides downstream from what is generally recognized as the phylogenetically conserved 3' end of 16S rRNA. The validity of this observation was confirmed by isolating RNA from purified 30S ribosome subunits and carrying out S1 nuclease protection titration assays at RNA:DNA molar ratios varying from 0.3 to 9.0. Even at the lowest submolar RNA concentrations, only the single extended protection product was observed. Thus, the entire descending portion of the 16S inverted repeat is retained at the 3' end of 16S rRNA upon assembly into 30S ribosomal subunits. Three other end sites were also detected in the vicinity of the 16S sequence: the first (site 1) occurs between helix A and B in the 5' leader; the second (site 2) occurs near the base in the ascending stem of the 16S inverted repeat; the third (site 6) occurs immediately preceding helix E in the intergenic space. One additional end, designated site 7, occurs in helix E but is probably not significant.

When the nucleotide sequence in the vicinity of cleavage sites 1, 2, 4 (5' end of mature 16S rRNA), 6, 11 and the 3' end of the primary transcript are compared, a consensus of GAUUCC becomes apparent. Therefore, it is possible that all five of these internal cleavage sites are generated by a single endonuclease activity whose most important role would seem to be to generate the mature 5' end of 16S rRNA. The similarity of this consensus with the most distal 3' end site in the rRNA primary transcript raises the possibility that this site is generated not by transcription termination but rather by processing of an even longer primary transcript. In archaebacteria, poly T track sequences are believed to function as terminators of transcription (Shimmin and Dennis, 1989). It is possible therefore that the primary transcript extends beyond this site but is undetectable because of rapid processing and instability of the distal product. There are two separate penta T tracks in the first 200 nucleotides of sequence beyond the 3' end stem.

The significance of the observed cleavages at positions 1, 2 and 4 (16S 5' maturation) has been demonstrated in vitro using a synthetic RNA leader that extends about 100 nucleotides into 5' end of the mature 16S sequence. Using crude cell extracts, this RNA is cleaved into four fragments (Figure 5). The four fragments extend from the 5' transcript end to site 1, from site 1 to site 2, from site 2 to site 4, and from site 4 to the 3' end of the synthetic transcript. The last fragment is the 5' end of mature 16S rRNA. This experiment is important in demonstrating that (i) the sites detected in vivo are not artifactual, (ii) the activity or activities are endonucleolytic, (iii) processing and maturation at the 5' end of 16S rRNA can occur in the absence of the second half of the precursor processing stem, and (iv) that processing does not require concomitant subunit assembly. We would like to suggest that the 5' leader sequence folds into a large cloverleaf structure which excludes formation of the long inverted repeat surrounding the 16S sequence and by implication, the formation of any structure that might resemble the substrate for the intron excision endonuclease. In this alternate cloverleaf structure, sites 1, 2 and 4 occur within irregularly base paired helical regions. The time course of in vitro processing reveals transient intermediates. One of these is depicted at the top of the autoradiogram in Figure 5. Full characterization of these intermediates may reveal a sequential order for cleavage within the 5' leader.

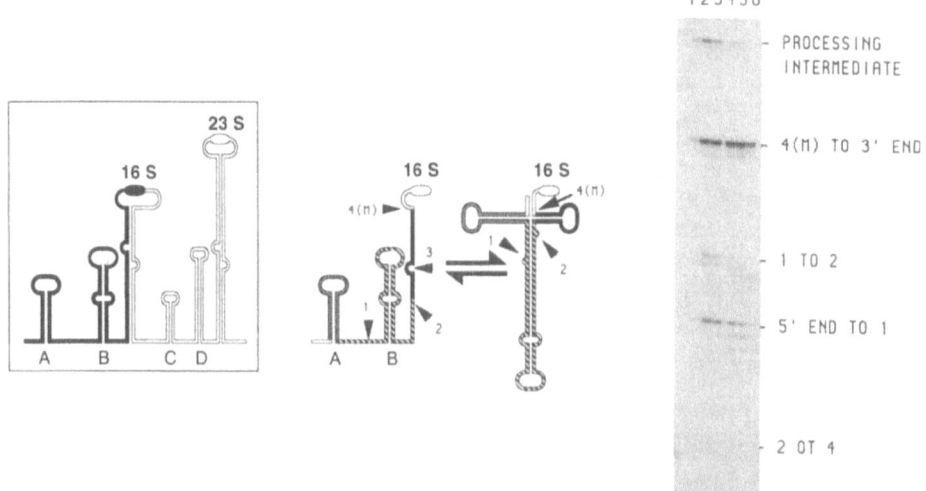

Figure 5. In vitro processing of 5' leader rRNA. The overall structure of the rRNA primary transcript from *S. acidocaldarius* and alternate pattern for 5' leader folding is illustrated on the left. The designations of helical structure and in vivo end sites are as in the legend to Figure 5. A cell extract was incubated with a radioactive leader substrate that contains almost 100 nucleotides of the 16S mature 5' sequence at its 3' end. The mixture was incubated for 0, 2, 4, 8, 15, 30 or 60 min prior to electrophoresis on denaturing polyacrylamide gels. The product bands are identified according to the cleavage sites at their 5' and/or 3' ends.

Our model for processing at the 3' end of 16S rRNA is similar. We suggest that the first step involves cleavage of the precursor at site 6 by the same endonuclease activity required for maturation at the 5' end of 16S rRNA. The leader is then trimmed by 7-8 nucleotides, probably by an exonuclease to generate the mature 3' end of the 16S rRNA that is assembled into 30S subunits. The trailer fragment that is liberated is relatively stable because the excision of precursor 23S rRNA apparently requires that RNA polymerase traverses all the way through the second half of the processing stem located distal to the 23S rRNA gene. Only after formation of the processing helix can the intron excision endonuclease excise precursor 23S rRNA.

In summary, in vivo intermediates in the processing of the 16S-23S rRNA primary transcripts have been identified and characterized by S1 nuclease protection and primer extension analysis. Precursor 23S rRNA is presumably excised from the primary transcript by the intron excision endonuclease. Precursor 16S rRNA excision follows an novel path that probably utilizes alternate folding and one or more different endonucleases. Unexpectedly, this pathway fails to remove approximately 62 nucleotides from the 3' end of 16S rRNA. A cell free processing system is being refined. Preliminary results indicate quite clearly that maturation at the 5' end of 16S rRNA requires neither formation of the processing helix nor concomitant ribosome assembly.

Acknowledgments

This work was supported by a grant from the Medical Research Council of Canada. PPD is a fellow of the Canadian Institute for Advanced Research, Program in Evolutionary Biology.

REFERENCES

Chant, J., and Dennis, P., 1986, *EMBO Journal* 5:1091-1097.

Downing, W., and Dennis, P., 1987, *J. Mol. Biol.* 194:609-620.

El-Baradi, T., de Regt, C., Einerhard, S., Teixido, J., Planta, R., Ballesta, J.P.G., and Rane, H.Q., 1987, *J. Mol. Biol.* 195:909-917.

Fiil, N.P., Friesen, J., Downing, W., and Dennis, P., 1980, *Cell* 19:834-844.

Hui, I., and Dennis, P., 1985, *J. Biol. Chem.* 260:899-906,

Iwabe, N., Kuma, K., Hasigowa, M., Osawa, S., and Miyata, T., 1989, *Proc. Natl. Acad. Sci. USA* 86:9355-9359.

Johnsen, M., Christiansen, T., Fiil, N., and Dennis, P., 1982, *EMBO Journal* 1:999-1004.

Juan-Vidales, F., Sanchez-Madrid, F., Saenz-Robles, M.T., and Ballesta, J.P.G., 1983, *Euro. J. Biochem.* 136:275-281.

Liao, D., and Dennis, P., 1992, *J. Biol. Chem.* 267:22787-22797.

Liao, D., and Dennis, P., 1993, submitted for publication.

Liljas, A., 1982, *Prog. Biophys. Mol. Biol.* 40:161-228.

Mevarech, M., Hirsch-Twizer, S., Goldman, S., Yakobson, E., Eisenberg, H., and Dennis, P., 1989, *J. Bacteriol.* 171:3479-3485.

Myvalganam, S., and Dennis, P., 1992, *Genetics* 130:399-410.

Newton, C., Shimmin, L., Yee, J., and Dennis, P., 1990, *J. Bacteriol.* 172:579-588 and 3535.

Olsen, G.J., Pace, N. R., Nuell, M., Kaire, B.P., Gupta, R. and Woese, C.R., 1985, *J. Mol. Evol.* 22:301-307.

Rivera, M., and Lake, J., 1992, *Sci.* 257:74-76.

Sander, G., 1983, *J. Biol. Chem.* 258:10098-10102.

Shimmin, L., and Dennis, P., 1989, *EMBO Journal* 8:1225-1235.

Shimmin, L., Ramirez, C., Matheson, A., and Dennis, P., 1989, *J. Mol. Evol.* 29:448-462.

Srivastava, A., and Schlessinger, D., 1990, *in*: "The Ribosome: Structure, Function and Evolution", W. Hill, A. Dahlberg, R. Garrett, P. Moore, D. Schlessinger and J. Warner, eds., ASM, Washington, D.C., pp. 417-425.

Thompson, L., and Danials, C., 1988, *J. Biol. Chem.* 263:17951-17957.

Thurlow, D., and Zimmerman, R., 1982, *in*: "Archaebacteria", O. Kandler, ed., Gustav Fracher, Verlag, Stuttgart, pp. 347.

Woese, C., 1987, *Microbiol. Rev.* 51:221-271.

Woese, C., Gutell, R., Gupta, R., and Noller, H., 1983, *Microbiol. Rev.* 46:621-669.

Woese, C., Kandler, O., and Wheeler, M., 1990, *Proc. Natl. Acad. Sci. USA* 87:4576-4579.

Yates, J., and Nomura, M., 1981, *Cell* 24:243-249.

THE RIBOSOMAL PROTEINS: COMPILATION
OF PROTEIN SPECIES EQUIVALENTS
FROM VARIOUS ORGANISMS, BASED
ON EVOLUTIONARY ANALYSES

Eiko Otaka,[1] Tetsuo Hashimoto,[2] Keiko Mizuta,[1] and Katsuyuki Suzuki[1]

[1]Department of Biochemistry and Biophysics
 Research Institute for Nuclear Medicine and Biology
 Hiroshima University
 Kasumi 1-2-3, Minami-ku, Hiroshima, 734 Japan
[2]The Institute of Statistical Mathematics
 4-6-7 Minami-Azabu, Minato-ku, Tokyo, 106 Japan

INTRODUCTION

From the view point of biological evolution, we have been interested in analyzing sequence similarities of ribosomal proteins (r-proteins) from various organisms. According to Doolittle (1981) the ultimate goal of the study of protein evolution is the reconstruction of the past events that gave rise to the vast inventory of proteins in existence today. In order to reach the goal, we believe that studying the r-proteins is a very attractive way.

The ribosome has the central place in the translational apparatus of all organisms. It contains many proteins besides some RNAs. In the well-studied *E. coli,* 54 r-proteins excluding plural copies, 21 in small subunits and 33 in large subunits, are known with still a few believed to exist (Wada and Sako, 1987). Thus the architecture is a highly complicated supramolecule whose properties depend on its constituents. When the protein part is regarded as a gigantic single molecule, that is, a supramolecule, the individual proteins must correspond to partial fragments of the gigantic molecule. It should be understood then that the study of an r-protein with its equivalent proteins from the other ribosome sources is to analyze the evolutionary changes that have occurred locally on a single molecule. It is natural that sequence similarities will differ depending on parts of a molecule.

There comes a time when considerable knowledge of r-protein sequences from various organisms and of the analytical data of their sequence similarities accumulates. To thoroughly understand the r-proteins taking into account the concept of the supramolecule mentioned above, we thought of compiling equivalents among r-proteins from various organisms using a universal code system (UCS, shown below), without having to take into account electrophoretic behavior, but rather basing considerations directly on detailed sequence similarities. Ultimately, after the entire compilation finally comes into being, such a task could come to yield a universal nomenclature for r-proteins throughout all of the various

The Translational Apparatus, Edited by K.H. Nierhaus
et al., Plenum Press, New York, 1993

ribosomes. Considering the accumulation of r-protein sequence data in confusion because of nomenclatures variously devised by each laboratory, such a compilation is a pressing demand. Thus, the compilation would result in bringing light to the perennial nomenclature problem and realize our original purpose. We will introduce here a preliminary compilation.

UNIVERSAL CODE SYSTEM (UCS)

In compiling, since all r-proteins and their sequences from the respective organisms are not yet known, we should have a compilation system adaptable to an ever increasing amount of new sequence data. The following proposals should give such a system, on which this interim report of the concrete compilation will also give a satisfactory explanation.

As equivalents of an r-protein are called r-protein species, the existence of many r-protein species throughout the various ribosomes could be expected. It is estimated that some of the r-protein species will have equivalents in all the organisms through the three primary kingdoms: eukaryotes, metabacteria (members of 'archaea' as proposed by Woese *et al.*, 1990, which we do not accept as our results will show later) and prokaryotes including organelles like chloroplasts or mitochondria (emp, see legend of Fig. 1). In some cases the r-protein species will have equivalents only in both metabacteria and prokaryotes (mp), in both eukaryotes and metabacteria (em), or in both eukaryotes and prokaryotes (ep) while in other cases the r-proteins may be unique to a single kingdom: to prokaryotes (p), to metabacteria (m), and to eukaryotes (e). It is also possible that some r-protein species will only be found in chloroplasts or mitochondria. According to the present results of sequence similarity analyses, the majority of the r-protein species will likely include equivalents from prokaryotes as shown in Fig. 1, that is, as members of groups emp, ep, mp and p. When the numbering system for *E. coli*'s r-protein species is adopted with the prefix UL or US meaning universal code, the majority is acceptably coded, even though grouping is still tentative.

R-protein species in groups em, e, m, and any unique protein species found in mitochondria or chloroplasts are coded by a providing numbering system for them (see Otaka *et al.*, 1993). Their equivalents should be tentatively gathered anyhow and then formally coded after being found, because grouping may depend on further data and must precede coding.

To search for the equivalents of any r-protein, it is most important that sequences should be significantly correlated. As one way, found among various trials for the examination of significant correlation, we have usually adopted a computer program by Lipman and Pearson to pick up sequences which show considerable correlation scores, RELATE and ALIGN to further check the previous selection, and then finally used a computer program that detects weak similarities between distantly related sequences (Otaka *et al.*, 1985, 1993). Recording of the analytical data include references with originally given names of equivalents compiled into US or UL codes, but their sequences in the alignments presented there, have IDs (identification tags) of five letters from scientific terms to clearly show material sources (see legend of Fig. 4). This record, which will be made available as a database, is also under preparation.

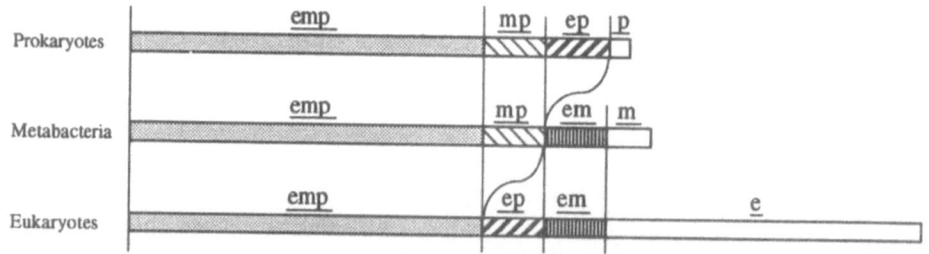

Figure 1. Schematic grouping of r-protein species. e: eukaryotics; m: metabacterials; p: prokaryotics

CODING AND GROUPING OF R-PROTEINS

As would be instantly known, the r-protein codes numbered according to the *E. coli* numbering system are r-proteins US1 to US21 and UL1 to UL36 except for numbers 8 and 26. So-called r-protein L14 of *Bacillus subtilis* and r-protein L14 of *B. stearothermophilus* should be described as r-protein UL14 of BACSU and r-protein UL14 of BACST, respectively. The r-proteins given the same code like above mean they belong to the same r-protein species. Leaving others behind for the present, the r-protein species coded by the *E. coli* numbering system were respectively assigned to one group among emp, ep, mp, or p as shown in Table 1. Table 1 results in showing in which kingdoms the r-protein species exist. As a matter of course, except for group emp, all grouping of coded r-protein species had to be done within the scope of the present information concerning sequences and similarity analyses. It means those groupings cannot help but change depending on further new data, though the code numbers will never change. Table 1 shows that more than half are the r-protein species that exist throughout all kingdoms, emp. Taking account of the fact that the sequence data from eukaryotes and metabacteria are not sufficient, many more r-protein species must be assigned to group emp. It is as surprising as it is clearly revealed that a major part of the various ribosomes from Human to *E. coli* consists of r-proteins derived from a common ancestor, that is, a major part of the prokaryotic or ancestral gigantic molecule as a supramolecule stressed in INTRODUCTION is evolutionarily maintained in all the ribosomes on the earth (see Fig. 1).

Family sequence data of r-proteins with the codes numbered by the *E. coli* numbering system have been registered in great numbers. On the other hand, in any group of other r-proteins there are not many family sequences, although it is known that several r-proteins have been tentatively assigned to groups em or m, but no small number of r-proteins to group e.

Table 1. The present grouping of r-proteins US1 to US21, and UL1 to UL36 except for numbers 8 and 26. e: eukaryotics; m: metabacterials; p: prokaryotics.

US group	emp	ep	mp	p	UL group	emp	ep	mp	p
	US 3	US 6		US 1		UL 2	UL13[2]	UL 1	UL 9
	US 4	US13		US 2		UL 3		UL 4	UL16
	US 5			US16		UL 5			UL17
	US 7			US18		UL 6			UL20
	US 8			US20		UL7/L12			UL21
	US 9			US21		UL10			UL25
	US10					UL11			UL27
	US11					UL14			UL28
	US12					UL15			UL31
	US14					UL18			UL32
	US15					UL19[1]			UL33
	US17					UL22			UL34
	US19					UL23			UL35
total number	13	2	0	6		UL24			UL36
						UL29			
						UL30			
					total number	16	1	2	14

[1,2] Shown from only one prokaryotic, and one eukaryotic (fragment), respectively.

CORRELATION PATTERNS

After detailed analyses of sequence similarities, alignments for members of r-proteins US1 to US21 and UL1 to UL36 in which prokaryotic members are contained (see Table 1) have been entirely constructed, including those for US4 (Mizuta *et al.*, 1991), US5 (Otaka *et al.*, 1993), US6 (Otaka *et al.*, 1986), UL7/L12 (so-called 'A'proteins; Otaka *et al.*, 1989), UL10 (Otaka *et al.*, 1990), UL23 (Raué *et al.*, 1989) and UL30 (Mizuta *et al.*, 1992), having already been presented according to our method. All the patterns of newly treated alignments represent 'one to one correlation' aligning regularly. This is remarkable. That is because, out of the seven alignments cited above, we had met early with three irregular patterns: 'double-transposition' of US6 (eukaryotic US6 must have resulted from the transposition events which occurred at two positions on the prokaryotic US6 gene; Otaka *et al.*, 1986); 'single-transposition' of UL7/L12 (eukaryotic and metabacterial UL7/L12s must have resulted from the transposition event which occurred at the C-terminal one-third of the prokaryotic UL7/L12 gene as shown in Fig. 2; Otaka *et al.*, 1989); and 'fusing of two proteins' of UL10 (eukaryotic and metabacterial UL10s must have resulted from the fusion of the transposed UL7/L12 and the prokaryotic UL10 genes as shown in Fig. 2; Otaka *et al.*, 1990). This encounter confirmed the impression that such irregular cases are not so rare. Such irregular events, however, under the above new presentation, will surely prove to be very unusual, even as further sequence data give rise to new alignments.

As the transposition and gene fusion events are infrequent, we should briefly talk about the details of proteins UL7/L12 and UL10. Being acidic and containing many alanines, protein UL7/L12 is usually called 'A'protein. Eukaryotic and metabacterial UL7/L12s are classified as "transposition type" in contrast to "prototype" for prokaryotes (Otaka *et al.* 1989). The eukaryotic UL7/L12 comprises P1 and P2 line proteins meaning **p**hosphorylated proteins, which are thought to form a tetrameric complex $(P1)_2(P2)_2$. For the pentameric complex as shown in *E. coli* to consist of $(EL7/L12)_4$ and EL10, a third phosphorylation protein P0 is provided, though it is compiled now as eukaryotic UL10. In metabacteria, the transposed UL7/L12 is single in kind. According to our analysis (Otaka *et al.*, 1989), to align all the sequences from the transposition type to the prototype, 11 gaps as deletion/insertion sites of amino acid residues were necessary. Collating them with the crystallographpic data for the C-terminal fragment (CTF) from *E. coli* L7/L12 (Leijonmarck & Liljas, 1987), all the gaps involved in the CTF are located between segments that correspond to structural and functional elements such as α helix, β strand, turning loop or hinge part (Fig. 3A). For this result, we claimed the existence of specific "preservation units" in protein molecules which are conservative in evolutionary phenomenon. In contrast, the transposition event on prototype UL7/L12 had occurred at the center of an α helix element that is involved in a folding domain. The event should have been progressive and drastic in no small measure, being in violation of the general rule concerning the gap-location as mentioned above (Fig. 3A). And by this violation, the ribosomes must have taken long strides together with the complementary molecular coevolution with prototype UL10, explained below (Figs. 2 and 3B). As shown in the right panel of Fig. 2, the N-terminal three-quarters of the eukaryotic and metabacterial sequences align naturally and the first two-thirds of the alignment could involve the prokaryotic UL10. An alignment of the remaining sequences at the C-termini was established, relying on the well-matching sequence similarities between the metabacterial UL7/L12s and their UL10s. Finally, the C-terminal halves of the eukaryotic and metabacterial UL10s corresponded with most of the overall length of the transposed UL7/L12s. Thus the fusion of the prototype UL10 and the transposed UL7/L12 genes have resulted in the formation of the P0 proteins, that is, the eukaryotic and metabacterial UL10. A coupling of this gene fusion and the transposition of prototype UL7/L12 must have given rise to the complementary molecular transformations required for the development toward the mode of higher organisms. As Fig. 3B explains, the C-termini of the fused UL10, homologous with those of the transposed UL7/L12, are considered to form the whole 'CTF' conformation together with the N-termini of the transposed UL7/L12 within a pentameric complex $(UL7/L12)_2(UL7/L12)_2UL10$-like $(P1)_2(P2)_2P0$ which might be involved in given functions

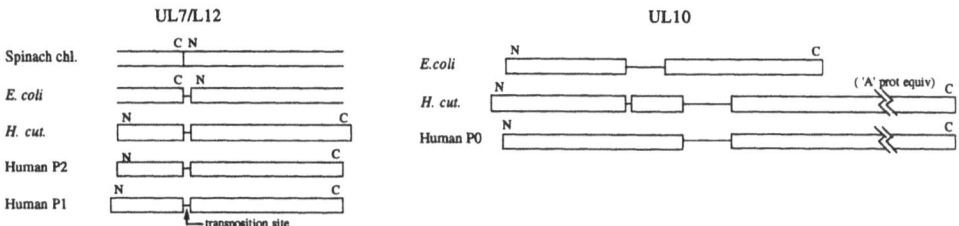

Figure 2. Schematic alignments. Open boxes are meant to correlate with the lower three sequences by transposing. N and C show the amino- and carboxyl-termini, respectively.

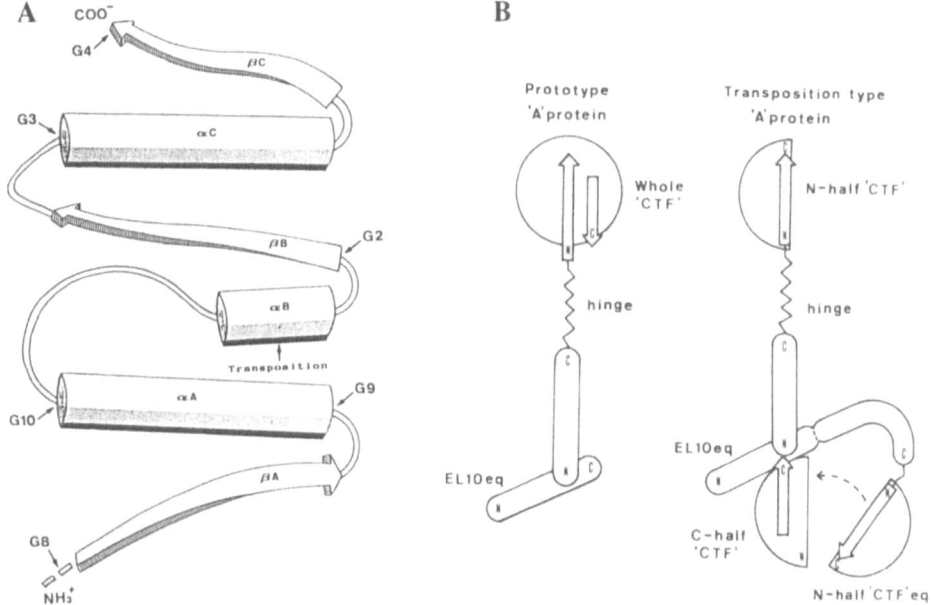

Figure 3. Schematic organization of various elements of the C-terminal fragment (CTF) (A) and possible arrangements of *E. coli* L7/L12 and L10 equivalents before and after the transposition (B). Several gap sites (G) and the transposition site are shown with arrows. See the text for further designation.

as the whole so-called 'CTF'. During the course of ribosomal protein evolution, the greatest event might have been the transposition on the prototype UL7/L12 gene, so that the gene fusion with the prototype UL10 could occur. In addition, it must be brought up here that we could not find such clear traces in any other proteins as proteins UL7/L12 and UL10 showed concerning the complementary molecular coevolution. It is this work concerning the coevolution that has led us to the concept that ribosomal proteins must be regarded as a single gigantic molecule, a supramolecule as mentioned in INTRODUCTION. Moreover development to plural form r-proteins of a prototype r-protein also appears to have been only the one case of UL7/L12 (Otaka *et al.*, 1990).

Depicted in this way, it was revealed that the so-called archaebacteria had already obtained the transposed UL7/L12 and fused UL10 proteins, while their transposed UL7/L12 yet remains a single form like the mode of prokaryotes, whose gene organization also remains prokaryotic. As they had failed to develop toward the mode of higher organisms,.we believe the term **'meta**bacteria' is most suitable for them rather than being termed members of **'archae**a' (Woese *et al.*, 1990).

US2

Figure 4. Alignments. To save space, some representatives out of the family sequences were selected and references were by an irregular format.

BACST:	*Bacillus stearothermophilus*	CANUT:	*Candida utilis*	CRYPHC:	*Cryptomonas phi* chloroplast
CYAPAC:	*Cyanophora paradoxa* cyanelle	DICDI:	*Dictyostelium discoideum*	ECOLI:	*Escherichia coli*
HALCU:	*Halobacterium cutirubrum*	HALMA:	*Halobacterium marismortui*	HUMAN:	Human
MAIZEC:	Maize chloroplast	MARPOC:	*Marchantia polymorpha* chloroplast	METVA:	*Methanococcus vannielii*
MYCCA:	*Mycoplasma capricolum*	NICTAC:	*Nicotiana tabacum* chloroplast	ORYSAC:	*Oryza sativa* (Rice) chloroplast
PISSAC:	*Pisum sativum* (Garden pea) chloroplast	RATNO:	*Rattus norvegicus* (Rat)	SACCA:	*Saccharomyces carlsbergensis*
SACCE:	*Saccharomyces cerevisiae*	SOYBN:	Soybean	SPINAC:	Spinach chloroplast
WHEATC:	Wheat chloroplast				

US2 ECOLI: Wittmann-Liebold B *et al* 1981 FEBS Lett 129:10-16; MARPOC: Ohyama K 1986 submitted to EMBL/GenBank; PISSAC: Hudson GS *et al* 1987 J Mol Biol 196:283-298; NICTAC: Sugiura M 1986 submitted to EMBL/GenBank.

US18 ECOLI: Schnier J *et al* 1986 Mol Gen Genet 204:126-132; BACST: McDougall J *et al* 1989 FEBS Lett 245:253-260; MARPOC: Ohyama K 1986 submitted to EMBL/GenBank; NICTAC: Sugiura M 1989 submitted to EMBL/GenBank; ORYSAC: Sugiura M 1986 submitted to EMBL/GenBank.

UL6 SACCE: Jones DGL *et al* 1991 Nucl Acids Res 19:5785-5785; RATNO: Suzuki K *et al* 1990 Gene 93:297-300; METVA: Auer J *et al* 1989 J Mol Biol 209:21-36; ECOLI: Chen R *et al* 1977 Hoppe-Seyler's Z Physiol Chem 358:531-535; BACST: Kimura M *et al* 1981 FEBS Lett 136:58-64; MYCCA: Ohkubo S *et al* 1987 Mol Gen Genet 210:314-322; CYAPAC: Michalowski CB *et al* 1990 Mol Gen Genet 224:222-231.

US4 DICDI: Steel LF *et al* 1987 Nucl Acids Res 15:10285-10298; SACCE: Mizuta K *et al* 1991 Nucl Acids Res 19:2603-2608; HALCU (fragment):Yaguchi M *et al* 1982 Zbl Bakt Hyg 1 abt orig C3 200-208; ECOLI: Bedwell D *et al* 1985 Nucl Acids Res 13:3891-3903; NICTAC: Shinozaki K *et al* 1986 EMBO J 5:2043-2049; SPINAC: Tahar SB *et al* 1986 Plant Mol Biol 7:63-70; MAIZEC: Subramanian AR *et al* 1983 Nucl Acids Res 1983 11:5277-5287; ORYSAC: Hiratsuka J *et al* 1989 Mol Gen Genet 217:185-194; MARPOC: Ohyama K *et al* 1986 Nature 322:572-574; CRYPHC: Durnford DG *et al* 1990 Nucl Acids Res 18:1903-1903.

US17 RATNO: Tanaka T *et al* 1985 J Biol Chem 260:6329-6333; HUMAN: Lott JB *et al* 1988 Nucl Acids Res 16:1205-1205; SOYBN: Gantt JS *et al* 1990 J Biol Chem 265:2763-2767; METVA: Auer J *et al* 1989 J Mol Biol 209:21-36; HALMA: Kimura J *et al* 1987 J Biol Chem 262:12150-12157; ECOLI: Yaguchi M *et al* FEBS Lett 87:37-40; MYCCA: Ohkubo S *et al* 1987 Mol Gen Genet 210:314-322.

UL30 RATNO: Lin *et al* 1987 J Biol Chem 262:12665-12671; SACCE: Mizuta K *et al* 1992 Nucl Acids Res 20:1011-1016; METVA: Auer J *et al* 1989 J Mol Biol 209:21-36; HALMA: Hatakeyama *et al* 1989 Eur J Biochem 185:685-693; ECOLI: Ritter E *et al* 1975 FEBS Lett 60: 153-155; BACST: Kimura M *et al* 1984 259:1051-1055.

UL23 CANUT: Woudt LP *et al* 1987 Curr Genet 12:193-198; SACCA: Leer RJ *et al* 1984 Nucl Acids Res 12:6685-6700; see Woudt LP *et al* 1987; METVA: Köpke AKE *et al* 1988 FEBS Lett 239:313-318; HALMA: Hatakeyama *et al* 1988 Eur J Biochem 172:703-711; ECOLI: Wittmann-Liebold B *et al* 1979 FEBS Lett 108:69-74; BACST: Kimura M *et al* 1985 Eur J Biochem 150:491-497; WHEATC: Bowman CM *et al* 1988 Curr Genet 14:127-136; MAIZEC: McLaughlin WE *et al* 1988 Nucl Acids Res 16:8183-8183.

UL3 HUMAN and RATNO: Ou J *et al* 1987 Nucl Acids Res 15:8919-8934; SACCE: Schultz LD *et al* 1983 J Bacteriol 155:8-14; HALMA: Arndt E *et al* 1990 J Biol Chem 265:3034-3039; ECOLI: Zurawski G *et al* 1985 Nucl Acids Res 13:4521-4526; MYCCA: Ohkubo S *et al* 1987 Mol Gen Genet 210:314-322; CYAPAC: Michalowski CB *et al* 1990 Mol Gen Genet 224:222-231.

US4

```
                  1         2         3         4         5         6         7         8         9
DICDI    1   MSSNYSKTSHTPRRFFEKERIDAELKVGEFGLRNKNEWRKVQYALAKIRKAARELVLDEKDFKRIFEGSALLRRLHRLGVMEESNRKLDYILANLKVQDFMEERLQTLVFKNGLAKSIHHAARVLIKGRHIRVGKQLVNVPSFLVRVESQKHLGLLAST--6---S
SACCE    1   ...
HALCU    1   MPRAPRTYSKTYSTPKRPYESSRLDAELKLAGEFGLANKKEIYISFQLSKIRRAARDLLTRDEKDFKRLFEGNALIRRLVRVGVLSEDKKKLDYVLALKVEDFLERRLQTQVVYKLGLAKSVHHAARVLITQRHIAVGKQIVNIPSFHVRLDSEKHIDFAPT--6---S
```

```
              1         2         3         4         5         6         7         8         9
BCOLI    1   ARYLGPKLKLSRRBGTDLFLASGVRAIDTKCKIRQAPGQHARKPRLSDYGVQLAEKQKVFRIYGVLERQFRNYTKEAARLKGNFGENIL---------------------ALLEGRLDNVVYRMKFGATRAEBARQLVSHKALMVRNGRVNILASYQVVSFNDVVSIREXAKKQSRVTA
NICTAC   1   ...                                                                      -PRNGSLRNQSRSGCKKSQYIRLEBEOKQKJAFHYGITEKQLLKVRIARIAAKGSTQVLL------------------QLLEWRLDNILFRLGMASTIPAAROLVHHRHILVNGRIVDIPSYRCKFRDIITAKDBQKSRALIQI
SPINAC   1   ...                                                                      -PRNGSLRNQSRSGCKKSQYIRLEBEOKQKJAFHYGITEKQLLKVRIARIAAKGSTQVLL------------------QLLEWRLDNILFRLGMASTIPCGAROLVHNRHILVNGRIVDIPSYRCKFRDIITAKDBQKSRALIQI
MAIZEC   1   ...
ORYSAC   1   ...
MARPOC   1   ...
CRYPHC   1   ...
```

US17

```
              1         2         3         4         5         6         7         8         9         0         1         2         3         4         5
RATNO    1   MADIQTERAYQKQPTIPQNKKRVLLGETGKEKLPRYYKNIGLGFKTPKEAIBGTYIDKKCPFTGNVSIRGRILLSGVVTRMGKGRTIVIHRDTLATYIRKYNRFEKRHKNMSVHLSPCFRDIVQIGDTLVTV-GECRPLSKTVRFHWLAVTKA---AGTKKQFOKF
HUMAN    1   ...                                                                                                                                      -GECRPLSKTVRFHWLAVTKA---AGTKKQFOKF
```

```
                  1         2         3         4         5         6         7         8         9         0
6 DICDI   7   PLAGRFGRRARKMA--KRNSSKGEEEN
SACCE     7   ...
6 BCOLI   7   ...
```

UL30

```
RATNO 1 2 3 4 5 6
MEAVPEKKKKVAAALGTL---10-----KKKKVPAVPETLKKKRRNFAELKVKRLRKKFALKTLRKARRK
MOUSE 1 2 3 4 5 6 7
MEAVPEKKKKVATVPGTLKKKVPAGPKTLKKKVPAVPETLKKKRRNFAELKVKRLRKKFALKTLRKARRK
DICDI 1 2 3 4
MSQKVATKTA-----------------30-----------PVPESILKKRATSQKKAVDVAKLGRAQQLRNSRLNS
SACCE 1 2 3 4
MAAE-----------------28---------KILTPESQLKKSAKQQKTAEQVAAERAARKAANKEKRA

RATNO 7 8 9 0 1 2 3 4 5 6 7 8 9 0 1 2 3 4 5
LIYEKAKHYHKEYRQMYRTEIRMARMARKAGNFYVPAEPKLAFVIRIRGINGVSPKVR-KVLQLLRLRQIFNGTFVKLNKASVNMLRIVEPYIAWGYPNLKSVNELIYKRGYGKINKKRIALTDNSLVARSLGKFGIICMEDLIHEIYTVGKRFKEANNFLWPFKLSSPRGGMK--KKTTHFVEGGDAGNREDQIEHPIKR
MOUSE 8 9 0 1 2 3 4 5 6 7 8 9 0 1 2 3 4 5 6 7
LIYEKAKHYHKEYRQMYRTEIRMARMARKAGNFYVPAEPKLAFVIRIRGINGVSPKVR-KVLQLLRLRQIFNGTFVKLNKASINMLRIVEPYIAWGYPNLKSVNELIYKRGYGKINKKRIALTDNSLIARSLGKFGIICMEDLIHEIYTVGKRFKEANNFLWPFKLSSPRGGMK--KKTTHFVEGGDAGNREDQINRLIRQMN
5 DICDI 6 7 8 9 0 1 2 3 4 5 6 7 8 9 0 1 2 3 4
VYFKRAEKYVSEYHKTEREAIRLNRIAKNSGTFYVPPAAKVAFVIRIRGINGVSPKPR-KVLKLLRLLQLNNGVFVKLNKASINMLKLVEPYVAYGFPNLKSIKELIYKRGHLKIDGQRIPLTSNDMVEKQLGKFGIICVEDIIHEITTCGKHFKQVNNSLWPFKLNCPRGGFN--MKKTPFLQGGDAGNREHLINNLIHRMN
SACCE 5 6 7 8 9 0 1 2 3 4 5 6 7 8 9 0 1 2 3 4
IILERNAAYQKEYETAERNIIQAKRDAKAAGSYYVEAQHKLVFVVRIKGINKIPPKPR-KVLQLLRLTRINSGTFVKVTKATLELLKLIEPYVAYGYPSYSTIRQLVYKRGFGKINKQRVPLSDNAIIEANLGKYGILSIDDLIHEIITVGPHEKQANNFLWPFKLSNPSGGWGVPRKFKHFIQGGSFGNREEFINKLVKSMN
METVA 1 2 3 4 5 6 7 8 9 0 1 2 3 4 5
MAYAVVRVRGSVGVRGDIA-DTMKMLRLHRVNHCVVIPENDHYTGMIKKVKDYVTYGEIDKETLVSLILKRGRL---8----AGNKRLSEELLKELVELPVDALAEKV-LAGEIKLKDTPIKPVFRLHPPRRGYDRGGIKKGFSIGGALGYRAGKINDLLNKMM
HALMA 1 2 3 4 5 6 7 8 9 0 1 2 3 4 5
MHALVQLRGEVNMHTDIQ-DTLEMLNIHHVNHCTLVPETDAYRGMVAKVNDFVAFGEPSQETLETVLATRAEP---7---LEGDADVDDELVAEHTDYDDISGLAFALLSEETTLREQGLSPTLRLHPPRGGHD--GVKHPVKEGGQLGKHDTEGIDLLEAMR
ECOLI 1 2 3 4 5
AKTIKITQTRSAIGRLPKHKATLLGLGLRRIGHTVEREDTPAIRGMINAVSFMVKVEE
BACST 1 2 3 4 5 6
AKKLAITLTRSVIGRPEDQRITVRTLGLRKMHQTVVHNDNPAIRGMINKVAHLVKVKEIEEE
```

UL23

```
CANUT 1 2 3 4 5 6 7 8 9 0 1 2 3 4
MAPSTKAASAKKAVVKGSNGSKALKVRTSTTFRLPKTLKLTRAPKYARKAVPHYQRLDNYKVIVAPIASETAMKKVEDGN-TLVFQVDIKANKHQIKQAVKDLYEVDVLAVNTLIRP-----13------NGTKKAYVRLTADHDALDIANKIGYI
SACCA 1 2 3 4 5 6 7 8 9 0 1 2 3 4
MAPSAKATAAKKAVVKGTNGKKALKVRTSATFRLPKTLKLARAPKYASKAVPHYNRLDSYKVIEQPITSETAMKKVEDGN-ILVFQVSMKANKYQIKKAVKELYEVDVLKVNTLVRP-----13------NGTKKAYVRLTADYDALDIANRIGYI
METVA 1 2 3 4 5 6 7 8
MDAFDVIKTPIVSEKTMKLIEEEN-RLVFYVERKATKEDIKEAIKQLFNAEVAEVNTNITP-----13------KGQKKAYIKLKDEYNAEVAASLPIY
HALMA 1 2 3 4 5 6 7 8
SWDVIKHPHVTEKAMNDMDFQN-KLQFAVDDRASKGEVADAVEEQYDVTVEQVNTQNTM-----13------DGEKKAVVRLSEDDDAQEVASRIGVF
ECOLI 1 2 3 4 5 6 7 8 9 0
MIREERLLKVLRAPHVSEKASTAMEKSN-TIVLKVAKDATKAEIKAAVQKLFEVEVEVVNTLVVKGKVKRHGQRIGRRSDWKKAYVTLKEGQNLDFVGGAE
BACST 1 2 3 4 5 6 7 8 9
MKDPRDIIKRPIITENTMNLIGQKK--YTFEVDVKANKTEVKDAVEKIFGVKVEKVNIMNYKGKFKRVGRYSGYTNRRKKAIVTLTPDSKEIELFEV
MYCCA 1 2 3 4 5 6 7 8 9
MHITEVLKKPVLTEK--SFAGHKDNVYTFLVDKKANKVQIKKTFEEIFEVKVESVRTINYDAKEKRLGKYVGKKPSYKKAIITLKEG-QKLDVLSDL
MARPOC 1 2 3 4 5 6 7 8 9
MNQVKYPVLTEKTIRLLEKNQ--YSFDVNIDSNKTQIKKWIELFFNVKVISVNSHRLPKKKKKIGTTTGYTVRYKRMIIKLQSG-YSIPLFSNK
WHEATC 1 2 3 4 5 6 7 8 9
MDGIKYAVFTEKSLRLLGKNQ--YTFNVESGFTKTEIKHWVELFFGVKVVAVNSHRLPGKGRRIGPILGHTMHYRRMIITLQPG-YSIPLLDREKN
MAIZEC 1 2 3 4 5 6 7 8 9
MDGIKYAVFTEKSLRLLGKNQ--YTFNVESGFTKTEIKHWVELFFGVKVVAVNSHRLPGKGRRMGPILGHTMHYRRMIITLQPG-YSIPLLDRETN
```

Figure 4 (Continued)

UL3

```
HUMAN    1        2        3        4        5
MPGWRLLTQVGAQVLGRLGDGLGAALGFGNRTHIMLFVRGLAGKSGTWWDEHL
RATNO    1        2        3        4        5
MPGWRLLAQGGAQVLGGGAVGGLGAAPGLGSRKNMILFVVRNLHSKSSTWWDEHL
```

```
HUMAN    6        7        8        9        0        1                    2         3        4
SEENVPFIKQLVSDEDKAQOLASKLCPLKDEPWPIHPWEPGSFRVGLIALKLGWWPLWT---10-----KDGQKRHVTLLQVQDCHVLKYTSKEKCNGKWAT-----40
RATNO    6        7        8        9        0        1
SEENVSFVKQLVSDENKAQLTSLLNPLKDEPWPLAHPWERGSSRVGLIALKLGWWPLWT---10-----KDGGKHAVTLLQVQDCHVLKYTPKEDHNGKTAT-----40
SACCE    1        2        3        4        5        6        7        8        9        0        1        2        3
MSHRKYEAPRHGHLGFLPRKRAASIRARVKAFPKDDRSKFVALTSFLGYKAGWTTIVRDLDRPGSKFHKREVEAVTVVDTPPVVVVGVGVETPRGLRSLITVWAEHLSDEVRRFYKNWYKSKKKAAFTKYSAKYAQDGAGIERELARIKKYASVWRVLVHTVQIRKT-----PLAQKKAHLAEIQLAKGGISEKVDWA
HALFA    1        2        3        4        5        6        7
MRPQFSRPRKGSLGFGPRKRSTSETPRFNSWPSDDGQP--GVGQFAGYKAGWTHVVLVNDEPNSPREGWEETVPVTVIETPWRAVALRAYEDTPYGQRPLTEVWTDEPHSELDRTLDV----15-----PEDHDPDAAEEQIRDAHEAGDLGDLRLLTHTVPDA--VPSVPKKKKPDVWETRVGGGSVSRLDHA
         ECOLI    1        2        3        4
         MIGLVGRKVGWMTRIFT----10-----EDGVSIPVTVIEVEANRVTQVKDLANDGYRAIQ
         MYCCA    1        2        3        4
         MKGILGRKVBMTQVFT----10-----NSSQLVPVTVWEVLPMTVLQVKTIDSDGYVAVQ
         CYAPAC   1        2        3
         MSIGILGTKLGWTQIFD----10-----EAGNAIPVTIIQAGRCPITVQIKTTATDGYNAIQ
```

```
HUMAN    9        0        1        2        3        4
DNAAIIKRGTPLYA-AHFRPGQYVDVTAKTIGKGFGQGVMKRWGFKG--QPATHGQTKTHRRRGAVAT----27-----GDIGRVWRGTKMPGRKGNGNIYRTEYGLKVWRINT
RATNO    9        0        1        2        3        4
DNAVIKRGTPLYA-AHFRPGQYVDVTAKTIGKGFGQGVMKRWGFKG--QPASRGQTKTHRRRGAIST----27-----GDIARVWRGTKMPGRKGNGNQNRTVYGLKVWRVNT
SACCE    0        1        2        3        4        5        6        7        8
RERFE--KTVAD-SVFEQNEMIDAIAVTNGHGFEGVTHRWGTK-LPRKTHRGL--RKVACIGACHPAHVMSVARAQGRGTHSRTSINHKIYRVGKGDDEANGATSFDRTKKTITPHGGVHYGEIKNDFIMVKGCIPGNRKRIVTLRKSLYTNTSRKALEEVSLAKVIDTASKFGKGRFQTPAENRAPWGTLKKDL
HALFA    9        0        1        2        3        4        5        6        7        8
LDIVEDGGEAWN-DIFRAGEYADVAGVTNGKGTGFGSPVKRWGVQKRGKRGHUARGGW-RRRIGNLGPWNFSKVRSTVPQQQGTGTHQRTELNKRLIDIGEGDEPT----14-----VDGGFVNYGEVDGPYTLVKGSVPGFDKRLVPFRPAVRPNDQPRLD--PEVRVVSNESNGG
         ECOLI    0        1        2        3        4        5        6        7        8
         BGREFTVGQSISV-ELFADVKKVDVTGTSKGKGFAGTVKRWNFRT--QDATHGNSLLSHRVPGSIGQN--QTPGRVFKGKKMAGQMGNERVVYQSLDVVRVDA----27-----ERNLLVKGAVPGATGSDLIVKPAVKA
         MYCCA    0        1        2        3        4        5        6        7        8
         NMQGYEIGQVINVSDIFVSGEYVDVIGLSKGKGFAGGIKRHNYSR--GPMAHGSGY-HRGIGSKGAI-----INRIFKSKRMPGRMGNAARTIQNLEIIAIDQ----27-----SNRIMLIKGPIPGPKNSFVQIKGNVKGHSSKQAVELLNRNASVQA
         CYAPAC   0        1        2        3        4        5        6        7
         SSDSIEVEKPITV-ELPNDNDIVNIQGYSIGRGFSGVQKRHNFAR--GPMSHGSKN-HRLPGSIGAG--STPGRVYTPGTRAAGRKGRSKITYIRGLKIYKVDS----27-----ERSLLIVKGSVPGKRPGGLLITITQVKKV
```

Figure 4

698

ALIGNMENTS

Some selected alignments (bold faced codes) are displayed in Fig. 4 in the order treated below. As most who have even once participated in aligning find, intimate family sequences are similar in sequence length as well as in sequence similarity, resulting in relative ease of overall aligning and in uniformity of endings at the N- and C-terminal sides. Family sequences from each of the three kingdoms are generally in such a condition. A typical case is shown by the members of **US2**. Some irregularities in sequence length and/or the ending are occasionally encountered in chloroplasts' and in mitochondorials as shown by **US18** and others. Concerning those irregularities and others, especially on mitochondorial ribosomes, we need more information. Upon examination of the sequences from the respective three kingdoms, that is, that concerning group emp, the eukaryotics are not always longer than the metabacterials and/or prokaryotics in sequence length, but sometimes vary as described below, though the eukaryotic ribosomes are bigger than others' in sedimentation coefficients. Many proteins belonging to group e might be the basic cause of the bigger ribosome. Belonging to emp, members of US8 and **UL6** are respectively similar in length and their alignments reveal similar endings at both termini. Members of US14 show a similar alignment pattern to the above, but the eukaryotics and metabacterials have shorter sequences than these other members because of an internal wide deletion site of 45 residues. In contrast, **US4**, members of which have almost the same sequence lengths, is intricate: the prokaryotics are longer than others by about 30 residues at the N-terminal site and the eukaryotics contain an insertion of over 40 residues in the middle portion preceding the region by over 40 residues well-conserved through all sequences (Mizuta *et al.*, 1991). That of **US17** is doorstep-like. Its variation is a case of **UL30**. Following the elongated N-terminal region by over 80 residues of the eukaryotic UL30s, the prokaryotic UL30s start corresponding exactly to the metabacterials in the N-terminal region and are lacking a long equivalent of the metabacterial and eukaryotic C-terminal region (Mizuta *et al.*, 1992). Those of US10, US15, US19, and UL23 correspond to the longer cases at the eukaryotic and/or metabacterial N-terminal sides, a typical alignment of which is shown by **UL23**. According to Raué *et al.* (1989), the elongated sequence of UL23 should result from becoming a eukaryote since it includes nuclear import signals. Metabacterial UL23s do not have such an elongated part but have the same wide deletion site as the eukaryotics do, typically or suggestively indicating the evolutionary place of metabacteria. The next two alignments show an elongated pattern at the eukaryotic C-terminal side. The eukaryotic UL18s have such a pattern plus over 120 residues and the metabacterial UL18 follows them with wide deletion sites. There are three regions well-conserved throughout, where rRNA-binding function might be inferred to arise. The eukaryotic and metabacterial UL10s have the elongated sequences from the fusion with the other r-protein species sequences, EL7/L12 equivalents, as mentioned in the above section (see Figs. 2 and 3B and Otaka *et al.*, 1990).

Generally, even looking only at the alignments mentioned here, it is clear that members from every kingdom, adding the organelles, show grouping changes in deletion/insertion size as well as in sequence length. Taking **UL30** as a typical example, usually metabacterial sequences show such a midway during the changing course that they occasionally act as an intermediate to align prokaryotics and eukaryotics. The changes in internal transposition and in gene fusion are also seen similarly, supporting the claim concerning the term 'metabacteria'. Compared with amino acid conversion on sequences, such changes must be regarded as extremely great events in protein evolution. The grouping changes come out so tightly that even higher eukaryotes, like mammals, maintain in common such great changes as occurred in lower eukaryotes, like the unicellular yeasts. Moreover, those changing sites are not so close. It means the existence of proper sites so that such changes may occur in those molecules, and thus totally back up the claim of specific "preservation units" mentioned in the previous section. We would like to introduce an episode. We thought that one of the very rare cases contrary to this phenomenon was seen in the similarity analysis of **UL3** members. The sequences from higher eukaryotes correlate strongly to those from prokaryotes, but very weakly to that from yeast. The yeast sequence aligns significantly with that from halobacterium, the gene organization of which is similar to that of the prokaryotes (Arndt *et al.*, 1990). The halobacterium sequence aligns only slightly with the prokaryotics. When, considering the positive alignments, a whole alignment is constructed, there are some conserved parts throughout those sequences. The final alignment pattern hardly allows the yeast sequence to be one of the eukaryotic grouping members nor the halobacterium sequence to be the intermediate, as mentioned above. However, it was found that the sequence data of these "higher eukaryotics" have to be regarded as mitochondrials because of the existence of rat cytoplasmic L3 similar to the yeast sequence (Kuwano and Wool, 1992). Although the alignment between *S. cerevisiae* YL35 (88 residues) and rat L37 (111 residues) contains a wide gap internally in spite of being members of the same kingdom (Otaka *et al.*, 1985), having been assigned to group e at present, this case appears also to have misled as did the above because of sequencing error in the rat (personal communication by K. Suzuki). Thus, all the known analytical data resulted in supporting the claims mentioned above. On the other hand, we know the great difference found between mammals and yeasts in analytical patterns of ribosomal proteins by 2D-PAGE (Otaka and Osawa, 1981). The amino acid conversion might be the main cause of such differences. Anyhow it is a pending major problem that no events shown here have yet been numerically evaluated for the evolutionary analysis.

There remain those caches of e, m and em which must be taken up in future, an undertaking now in progress.

SUMMARY

An attempt to systematically compile equivalents of ribosomal protein species from various organisms from the view point of molecular evolution was introduced. Even though we are undertaking the task of compilation, the evolutionary profile of ribosomal proteins is emerging: (1) in no small number there are protein species which exist throughout the three primary kingdoms, (2) almost all the alignments must show 'one to one correlation' to align regularly, that is, irregular correlation as internal-transposition appears to be rarely found, (3) considering the organelles, generally grouping changes of sequences from every kingdom appear as wide insertion/deletions, elongation in length and so on, totally backing up the existence of specific "preservation units" in those molecules as previously stressed (Otaka *et al.*, 1989), and (4) the greatest event during the course of ribosomal protein evolution must be the transposition on UL7/L12, allowing the gene fusion together with UL10.

REFERENCES

Arndt, E., Krömer, W., and Hatakeyama, T., 1990, *J. Biol. Chem.* 256:3034-3039.
Doolittle, R.F., 1981, *Science* 214:149-159.
Kuwano, Y., and Wool, I.G., 1992, *Biochem. Biophys. Res. Comm.* 187:58-64.
Leijonmarck, M., and Liljas, A., 1987, *J. Mol. Biol.* 195:555-580.
Mizuta, K., Hashimoto, T., Suzuki, K., and Otaka, E., 1991, *Nucl. Acids Res.* 19:2603-2608.
Mizuta, K., Hashimoto, T., and Otaka, E., 1992, *Nucl. Acids Res.* 20:1011-1016.
Otaka, E., and Osawa, S., 1981, *Mol. Gen. Genet.* 181:176-182.
Otaka, E., Ooi, T., Kumazaki, T., and Itoh, T., 1985, *J. Mol. Evol.* 22:342-350.
Otaka, E., Ooi, T., Itoh, T., and Kumazaki, T., 1986, *J. Mol. Evol.* 23:337-342.
Otaka, E., Ooi, T., and Suzuki, K., 1989, *Prot. Seq. Data Anal.* 2:395-402.
Otaka, E., Suzuki, K., and Hashimoto, T., 1990, *Prot. Seq. Data Anal.* 3:11-19.
Otaka, E., Hashimoto, T., and Mizuta, K., 1992, *Prot. Seq. Data Anal.* in press.
Raué, H.A., Otaka, E., and Suzuki, K., 1989, *J. Mol. Evol.* 28:418-426.
Wada, A., and Sako, T., 1987, *J. Biochem.* 101:817-820.
Woese, C.R., Kandler, O., and Wheelis, M.L., 1990, *Proc. Natl. Acad. Sci. USA* 87:4576-4579.

EXPRESSION AND ASSEMBLY OF CHLOROPLAST RIBOSOMAL PROTEINS IN *E.COLI*

Wolfgang Weglöhner, Jürgen Schmidt, Klaus Giese, and Alap R. Subramanian

Max-Planck-Institut für Molekulare Genetik
Abteilung Wittmann
Ihnestraße 73, 1000 Berlin 33
Germany

INTRODUCTION

Chloroplast ribosomes have many similarities to eubacterial ribosomes which underlie the endosymbiont theory wherein the origins of chloroplasts are rooted in photosynthetic, oxygen-evolving procaryotes by monophyletic or even polyphyletic events (Bogorad, 1975). The evolutionary time scale for these events go back to over one billion years. Separate evolution for such a long period has however produced remarkable differences between chloroplast and eubacterial ribosomes, especially in ribosomal proteins (RPs). The chloroplast ribosomes contain more proteins than the *E.coli* or *Bacillus* ribosomes. The additional novel proteins (called psRPs, see Subramanian et al., 1991) are generally encoded in the nucleus. At least five psRPs have so far been identified (Gantt, 1988; Zhou et al., 1989; Johnson et al., 1990; unpublished results from our lab). Among the homologous RPs, sequence identity varies over a wide range, i.e between ca 70% and 25%, indicating a considerable disparity in the evolutionary pressure for their sequence conservation. Additionally, many of the chloroplast RPs contain long extensions at their N- and/or C-termini, and often reveal different post-translational processing and modifications (e.g. L2, S16; Schmidt et al., 1992), as compared to their eubacterial homologues. It is an open question whether the psRPs and the extensions confer new properties to the chloroplast ribosome or are simply adaptions to some features of the organelle in its eucaryotic environment.

The Translational Apparatus, Edited by K.H. Nierhaus
et al., Plenum Press, New York, 1993

We have made constructs of three chloroplast RP genes, each with some special features, and expressed them in *E.coli* to examine whether they can compete with the homologous *E.coli* RPs to assemble into the eubacterial ribosome. The results from these experiments are summarized in this chapter. In earlier experiments, Liu et al. (1989) have shown that the protein expressed from the highly conserved organelle gene for chloroplast S12 of *Chlamydomonas* can assemble into *E.coli* ribosomes.

The overexpression of chloroplast RPs in *E.coli* should provide opportunities in other directions as well. Thus milligram amounts of RPs can then be readily purified (instead of struggling with hundreds of kg of plant material). The purified proteins can be used, e.g. for *in vitro* assembly in a partial *E.coli* system since no organelle based systems are available to study the function of individual chloroplast RPs. The functional roles of the procaryotic homologous regions or the extensions can also be studied with the appropriate gene constructs. Additional uses for expressed purified proteins would eventually include crystallization, especially for the psRPs which are unique to chloroplast ribosomes.

RESULTS AND DISCUSSION

Three chloroplast RPs were chosen for the expression experiments. They included the nuclear encoded RPs L11 and L13 whose previously isolated cDNA clones (Smooker et al., 1991; Phua et al., 1989) were used. A representative from the organelle-encoded RPs was S18 with a remarkable heptapeptide repeat in its N-terminal extension (Weglöhner and Subramanian, 1991). Some of the other characteristics of these three proteins are shown in Figure 1.

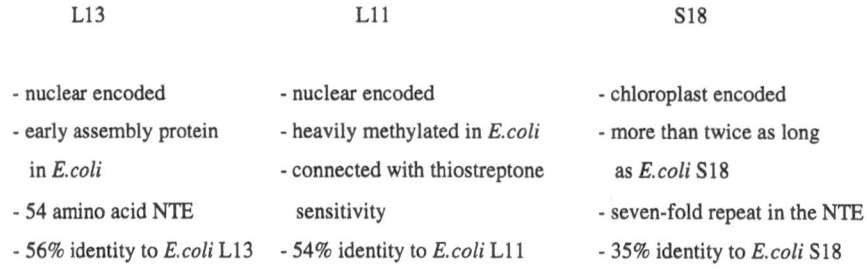

L13	L11	S18
- nuclear encoded	- nuclear encoded	- chloroplast encoded
- early assembly protein in *E.coli*	- heavily methylated in *E.coli*	- more than twice as long as *E.coli* S18
- 54 amino acid NTE	- connected with thiostreptone sensitivity	- seven-fold repeat in the NTE
- 56% identity to *E.coli* L13	- 54% identity to *E.coli* L11	- 35% identity to *E.coli* S18

Figure 1. Characteristics of the three chloroplast RPs that were expressed in *E.coli*.

Gene Constructs. The genes for the three RPs were cloned into the expression vector pJLA502 (Schauder et al., 1987; obtained from Medac). This vector contains the two lambda promoters P_R and P_L under the control of a thermolabile repressor and the transcription terminator region from fd phage. In the case of L13, the mature protein coding region was obtained by polymerase chain reaction (PCR) from its cDNA and was cloned into the initiating ATG and upstream Shine-Dalgarno sequences of the vector, while the L11 clone was made directly from its cDNA (see Figure 2). The S18 construct was cloned with its own 5' region including the SD sequence.

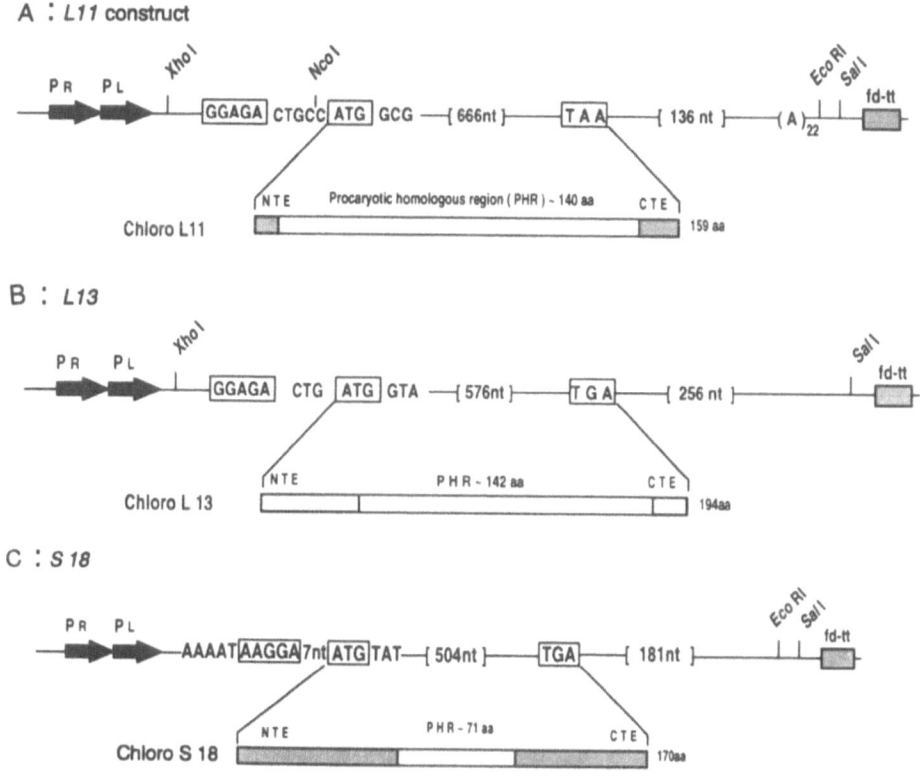

Figure 2. Schematic diagram of the three constructs made for expression of chloroplast RPs L11, L13 and S18 in *E.coli.*

Transformed *E.coli* cells (XL-1 and WK6) containing these constructs were grown at 30°C to early log phase and were then shifted to 42°C to induce expression. Cells were cracked and proteins were analysed by SDS-PAGE after 1, 2, 3, 4, and 14 hrs. For polysome isolation cells were generally harvested three doublings after induction, and they are grown to stationary phase for isolating ribosomes and inclusion bodies.

Expression of Chloroplast L13. Using an antiserum specific to chloroplast L13, two cDNA clones encoding the precursor of chloroplast L13 were isolated and characterized by Phua et al. (1989). The L13 protein has recently been sequenced from the N-terminus to establish the cleavage site in the precursor between the transit peptide and the mature form (Srinivasa and Subramanian, unpublished). The mature chloroplast L13 contains a long N-terminal extension (NTE). Since a long extension might interfere with possible assembly into the ribosome, we made two different contructs, one lacking the NTE (L13") and the other with 43 residues of the NTE (L13'). Both were expressed in *E.coli* WK6 (Giese and Subramanian, 1991).

Interestingly, both L13' and L13" were expressed but neither was overexpressed as superabundant species (see S18 results). The expressed proteins were detected in cell extracts by SDS-gel electrophoresis, Western blot and immunostaining. Ribosomes isolated through a sucrose gradient contained significant amounts of the chloroplast proteins, which in the case of L13" could be detected after 2D-gel electrophoresis by Coomassie blue staining. The spot of L13' comigrated with ribosomal proteins S3/S4 and was therefore detected by immunostaining (Figure 3 A, B and C).

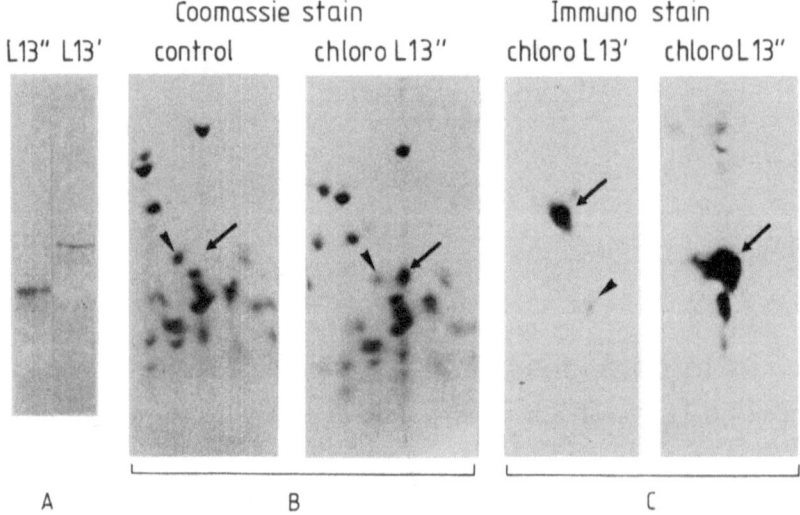

Figure 3. The expression of chloroplast L13' and L13" in *E.coli*. A:detection of chloroplast L13' and L13" in total cell lysate after SDS-PAGE and immuno staining with chloroplast L13-specific antibodies. B and C: 2D-PAGE of TP70 from purified 70S ribosomes. Arrowhead, *E.coli* L13; arrows, expressed proteins.

To test whether ribosomes containg the chloroplast L13 constructs were translationally active, polysomes were isolated and analyzed by immunostaining. The results showed (Figure 4) that polysomes contained similar amounts of L13' or L13" as in the purified 70S ribosomes. The proportion was similar in heavier polysomes as in di- and trisomes. Thus it appears that ribosomes carrying chloroplast L13, with or without the NTE, can translate mRNA. Futher experiments, e.g. analysis of the polysomes for nascent polypeptide chains, are necessary to further confirm this conclusion.

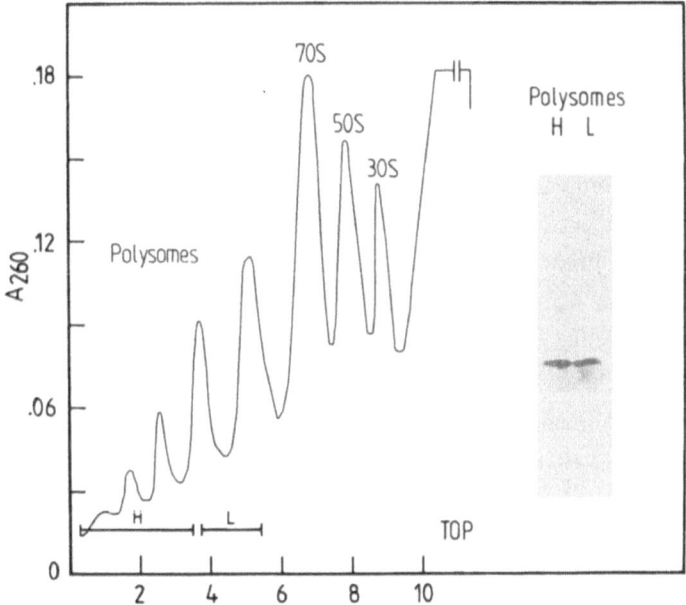

Figure 4. The presence of chloroplast L13' in polysomes from *E.coli* cells expressing the corresponding construct. L, disomes and trisomes; H, heavier polysomes. On the right: detection of L13" in H and L polysomal fractions by immuno staining (similar results were obtained for L13', data not shown).

In further experiments we have found that the NTE in the L13' is readily and specifically removed when the L13' containing ribosomes are incubated with proteases under mild conditions (Giese and Subramanian, 1991). In contrast, the L13" form lacking the NTE was stable to such digestions. It would appear that the NTE must lie exposed on the ribosome surface.

Expression of Chloroplast L11. The construct for chloroplast L11 expression was made with the 836 bp long spinach cDNA previously isolated and cloned into the *Eco*RI-site of pT7T3-19U in our lab (Smooker et al., 1991). The plasmid DNA was digested with *Nco*I and *Eco*RI and the resulting 477 bp fragment, containg the coding sequence for 159 amino acids of the mature L11 protein, the stop codon TAA and 136 bp long 3' downstream region including 22 adenosines of the poly(A) tail, was cloned into the corresponding restriction sites of the pJLA502 expression vector. The gene expression was induced by temperature shift as described earlier.

Total cell extracts, analyzed by SDS-gel electrophoresis and stained by Coomassie blue showed no distinct new bands after temperature shift, but immunostaining of the Western blot showed a band corresponding to chloroplast L11. Two antisera were used for

705

this purpose: 1) antiserum raised against *E.coli* L11 which cross-reacted well with epitopes in chloroplast L11 and 2) antiserum raised against a pooled fraction (pool 44) from spinach chloroplast r-proteins which reacted with both chloroplast and *E.coli* L11 proteins. Both antisera detected the bands of *E.coli* L11 and chloroplast L11 in the cell extracts.

In the next experiment purified 70S ribosomes were isolated by sucrose gradient centrifugation from cells expressing chloroplast L11 (8 hrs) and extracted TP70 was analyzed by 2D-gel electrophoresis and Western blotting. Immunostaining of the blot showed spots of *E.coli* L11 and a new spot corresponding to the calculated Mr (16.9 kDa) and pI (10.3) of chloroplast L11. Thus chloroplast L11 is assembled into *E.coli* ribosomes.

E.coli L11 is the most highly methylated protein in the bacterial ribosome (Dognin and Wittmann-Liebold, 1980). We have evidence for the presence of trimethyllysine in spinach chloroplast L11 and for modification of its N-terminal amino acid (unpublished results from this lab). The sequence contexts of the modified residues of *E.coli* L11 are conserved in the sequence of chloroplast L11 (Smooker et al., 1991).

A further experiment was to complement the thermosensitive, L11 lacking *E.coli* mutant A68 (Dabbs, 1978) with chloroplast L11. We characterized the L11 deletions in mutant A68 by PCR amplification procedure. Two primers corresponding to the 5' and 3' flanking sequences of *E.coli* L11 gene were used for PCR amplification. The PCR product was cloned into pT7T3-19U and its nucleotide sequence was determined. The results showed the insertion of an adenine (A) directly in front of codon 104 in the L11 gene i.e. CGC(Arg)103"A"GCT(Ala)104. The insertion frame-shifts the subsequent codons causing misreading and generates a termination codon after 19 misread codons. Thus the mutant L11 in A68 would be 121 amino acids long (excluding the initiating Met; *E.coli* wt L11 = 141 residues) with a "false" C-terminal region (Figure 5).

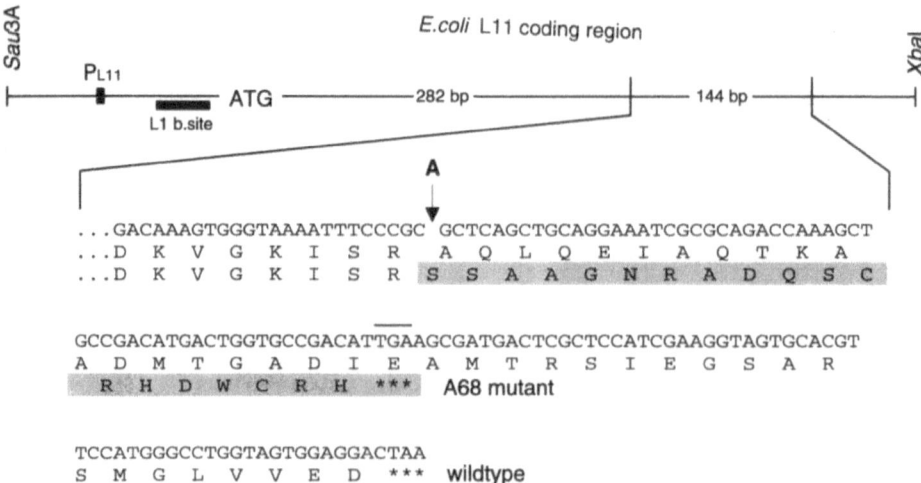

Figure 5. Sequence comparison between the *rp*L11 of wildtype*E.coli* and A68 mutant. The different amino acids in the mutant L11 are shaded. Frame-shift generated termination codon is overlined. Black bars represent promoter- and L1 binding sites.

We have expressed the chloroplast L11 construct in the A68 mutant and further experiments are in progress to determine whether this expression is responsible for complementing temperature sensitivity and thiostreptone resistance in A68.

Expression of Chloroplast S18. Maize chloroplast S18 is an unusual ribosomal protein. It contains 170 amino acid residues as compared to 74 in *E.coli* S18, having both N-terminal and C-terminal extensions of about 50 amino acids each (Weglöhner and Subramanian, 1991). Interestingly, the NTE is made of a repeat motif of seven amino acids with the consensus sequence: SKQPFRK (Figure 6). The number of these repeats in chloroplast S18 appears to depend on the plant family: maize and rye have 7 repeats, rice 6, tobacco 2, a nonphotosynthetic parasite plant (*Epifagus*) one. Liverwort, belonging to the land plant division *Bryophyta* that appeared early on the planet, as well as *E.coli* S18 does not contain this repeat. Whether the repeat has a specific function is unknown, but overexpression and large-scale purification of maize chloroplast S18 (see below) would provide a means to examine this aspect.

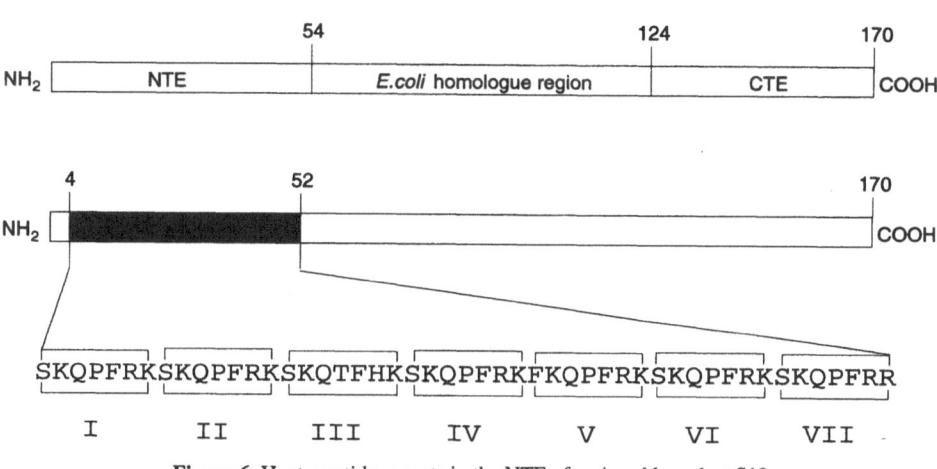

Figure 6. Heptapeptide repeats in the NTE of maize chloroplast S18.

The CTE of maize chloroplast S18 contains no repeats but is very rich in Asn and Gln (17 out of 54 residues). Overall, maize S18 is a very basic protein with a net positive charge of 41 as compared to +14 in *E.coli* S18. In its homologous region there is only 35% amino acid identity with *E.coli* S18. With its extensions, repeat motif and low homology, maize S18 would offer the greatest challenge for assembly into the *E.coli* ribosome.

Thermal induction of S18 expression from the construct shown in Fig. 2 resulted in vast overproduction of this protein. A band corresponding to its Mr (20.6 kDa) dominated the extract, as seen by SDS-gel electrophoresis and Coomassie blue staining (Figure 7).

Analysis of fractionated cell extracts showed that most of the maize S18 was present in the cell debris fraction (10.000 xg pellet) and that it can be isolated from there as inclusion bodies. For large scale purification, a 5-liter culture was grown in a Bench-top fermentor (New Brunswick: Bioflo IIc). The harvested cells (15g) were lysed in a French press and inclusion bodies were isolated by standard procedures (Marston and Hardley, 1990). The latter were solubilized in 6 M guanidine hydrochloride and then dialyzed against 2% acetic acid. This material, which was already >80% pure (Figure 7) could be further purified on a CM-Sepharose column (pH 7.0; 0.5 M to 2.0 M NaCl gradient), and finally by RP-HPLC (Vydac C4, 5μm, pore size 300 Å: gradient of water to 2-propanol in 0.1% TFA). By using this two-column procedure, we can obtain ca. 70 mg pure S18 from one 5L fermentor run.

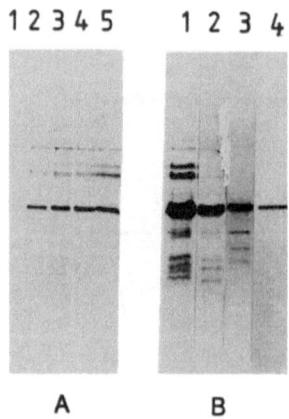

Figure 7. Expression and purification of maize chloroplast S18 expressed in *E.coli*. A: total cell lysate 0, 1, 2, 3, and 4 hours after induction. B: lane 1, cell debris; lane 2, isolated inclusion bodies; lane 3, after ion exchange chromatography; lane 4, after RP-HPLC purification.

The purified protein obtained by RP-HPLC (Figure 8) was also sequenced on an Applied Biosystems sequencer (W.Schröder, FU Berlin). It gave the sequence MYISKQPF (see also Figure 6). Thus the N-terminal processing in *E.coli* removes the formyl group of the initiating formylMet. The retention of the MYI sequence preceding the repeats indicated that all the seven repeats of the maize S18 are maintained in the expressed protein. Further evidence for the maintenance of both NTE and CTE was obtained from the Mr value in SDS gels which corresponded to that (20.6 kDa) calculated for the 170 amino acid residues.

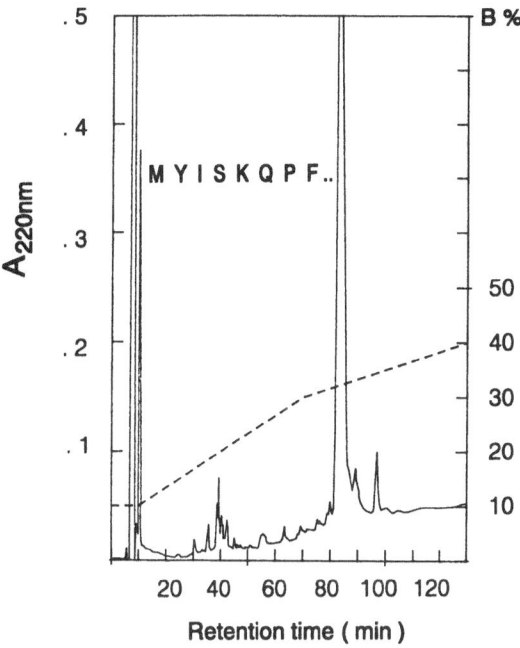

Figure 8. RP-HPLC profile of the expressed maize chloroplast S18 purification and the N-terminal sequence of the purified protein

As in the case of L11 and L13 experiments, purified 70S ribosomes were isolated from S18-expressing cells and were analyzed by 2D-gel electrophoresis. A new spot, corresponding in pI and Mr to maize S18, could be seen by Coomassie blue staining (Figure 9). The identity of this spot was confirmed by coelectrophoresis with purified maize S18.

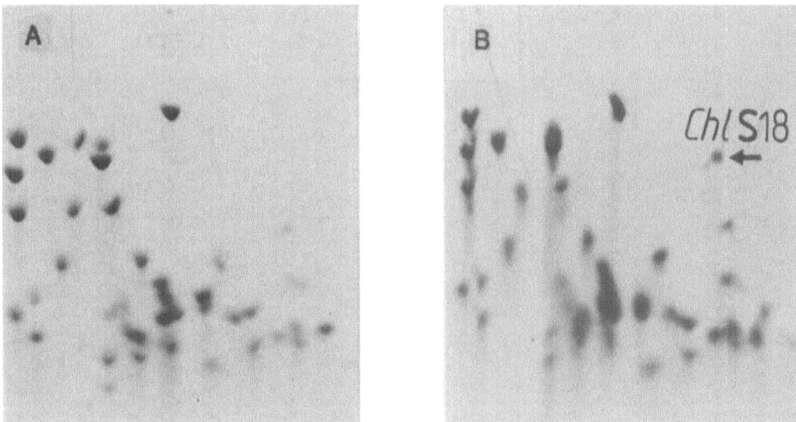

Figure 9. 2D-gel electrophoresis of purified 70S ribosomes from A: control cells with no expressed protein (vector without insert); B: cells·expressing maize chloroplast S18.

Presently we are raising antibodies to purified maize chloroplast S18 to obtain a specific and sensitive reagent that can be used to examine S18 incorporation into polysomes. The antibodies will also useful for experiments with chloroplast lysates to understand the functional significance of the appearance of this varyingly repeated heptapeptide motif in chloroplast ribosomes. The amino acid composition of this motif has similarity to that of nucleotide binding domains in proteins. But computer analysis showed that it is different from the reported (Bandziulis et al., 1989; Corden 1990) conserved motifs found in RNA-binding proteins and in eucaryotic RNA polymerase II.

As in the L13 experiments, we made a construct with the maize S18 gene deleting the coding region for the NTE and attempted its expression. Here we found no overexpression of the protein detectable by Coomassie blue staining. *E.coli* antiserum does not cross-react with chloroplast S18 and hence a sensitive means of detecting low expression (as in the cases of L11 and L13) is presently not available for chloroplast S18.

Aspects of Chloroplast RP Expression in *E.coli*

Expression of chloroplast RPs in *E.coli* raises questions of practical as well as evolutionary importance. In *E.coli* the RPs are among the most abundantly expressed proteins in the cell and they follow a distinct codon usage that avoids many synonymous codons especially in the case of Arg, Gly, and Leu (Andersson and Kurland, 1990). An examination of the codon usage in chloroplast L11, L13 and S18 genes (Table 1) shows that many such rarely used codons of the highly expressed genes in *E.coli* are extensively used in these genes.

Table 1. Extreme codon usage bias in highly expressed *E.coli* genes compared to that in the three expressed genes for chloroplast ribosomal proteins.

Amino acid	Codon	Percent usage in			
		E.coli	Chl S18	Chl L11	Chl L13
Arg	AGG	0	4	0	31
	AGA	0	32	67	25
	CGG	0	0	0	6
	CGA	0	28	33	0
Gly	GGG	1	25	8	33
	GGA	0	25	62	25
Ile	ATA	0	43	6	17
Leu	CTA	0	0	9	0
	TTG	1	8	36	13
	TTA	1	83	0	13
Pro	CCC	1	10	14	8

In our experiments we have observed a difference of 100-fold or more between the expression of L11 and L13 on the one hand and that of S18 on the other. Differences in codon usage (Table 1) does not appear to account for this large effect.

Another aspect is the consequence of the autogenous regulation of *E.coli* RP synthesis. The three RP genes examined here belong to three distinct operons in *E.coli*, i.e. the *rp*L11/L1, *rp*S9/L13 and *rp*S6/S18/L9 operons. In general one of the proteins of the operon acts as the translational repressor of the operon mRNA. In the case of the L11 construct, a potential binding site for L1 is absent. Whether translational repression is the reason for the low expression of the L13 construct has not been verified, but neither has such effects been ruled out.

Evolutionary Significance. As mentioned in the Introduction chloroplasts are assumed to be the products of endosymbiotic evolution that began more than 10^9 years ago. It is evident from the structural data for chloroplast RPs that considerable divergence has taken place during this time, such as extensions, repeats, primary structure differences and novel RPs. The finding that notwithstanding these differences the three RPs we expressed in *E.coli* could compete with homologous eubacterial RPs to assemble into *E.coli* ribosome would therefore appear remarkable. This result would suggest extreme conservation in the tertiary structure of their functional domains and important functional roles for the RPs in the currently functioning translational apparatus.

REFERENCES

Andersson, S.G.E., and Kurland, C.G., 1990, Microbiol. Rev. **54**: 198-210.
Bandziulis, R.J., Swanson, M.S., and Dreyfuss, G., 1989, Genes & Dev. **3**: 431-437.
Bogorad, L., 1975, Science **188**: 891-898.
Corden, J.L., 1990, Trends Biochem. Sci. **15**: 383-387.
Dabbs, E.R., 1978, Mol. Gen. Genet. **165**: 73-78.
Dognin, M.J., and Wittmann-Liebold, B., 1980, Eur. J. Biochem. **112**: 131-151.
Gantt, J.S., 1988, Curr. Genet. **14**: 519-528.
Giese, K., and Subramanian, A.R., 1991, FEBS Lett. **288**: 72-76.
Johnson, C.H., Kruft, V., and Subramanian, A.R., 1990, J. Biol. Chem. **265**: 12790-12795.
Liu, X.-Q., Gillham, N.W., and Boynton, J.E., 1989, J. Biol. Chem. **264**: 16100-16108.
Marston, F.A.O., and Hartley D.L., 1990, Methods Enzymol. **182**: 264-276.
Phua, S.H., Srinivasa, B.R., and Subramanian, A.R., 1989, J. Biol. Chem. **264**: 1968-1971.
Schauder, B., Blöcker, H., Frank, R., and McCarthy, J.E.G., 1987, Gene **52**: 279.283.
Schmidt, J., Herfurth, E., and Subramanian, A.R., 1992, Plant Mol. Biol. **20**: 459-465.
Smooker, P.M., Schmidt, J., and Subramanian, A.R., 1991, Biochimie **73**:845-851.
Subramanian, A.R., Stahl, D., and Prombona, A., 1991, in: "The Molecular Biology of Plastids", Bogorad, L., and Vasil, I.K., eds, pp. 191-215, Academic Press, San Diego.
Weglöhner, W., and Subramanian, A.R., 1991, FEBS Lett. **279**: 193-197.
Zhou, D.-X., and Mache, R., 1989, Mol. Gen. Genet. **219**: 204-208.

THE CELLULAR SLIME MOLD AS A PARADIGM TO STUDY

RIBOSOME POLYMORPHISM IN EUKARYOTES

S. Ramagopal

United States Department of Agriculture
Agricultural Research Service
University of Idaho
Aberdeen, ID 83210-0307

HYPOTHESIS

The eukaryotic organisms undergo complex life cycles which vary in detail depending on the species. Several of the processes of life cycle, especially cell differentiation, organogenesis, and morphogenesis, lead to distinct types of cells and tissues. Many important changes in protein synthesis, e.g., the translation of specific proteins and induction of differential rates, are characteristic features which occur concomitant with the development of specific cell and tissue types. How is this translational complexity accommodated in multicellular eukaryotes? One plausible mechanism how the cells accomplish this is thought to be through the synthesis of unique ribosomes (Sussman, 1970). This hypothesis, therefore, predicts the existence of subpopulations of ribosomes, i.e., a ribosome polymorphism or heterogeneity, in the same organism.

The purpose of this brief review is to describe various studies that support the hypothesis and suggest some directions for future consideration.

Ribosome Polymorphism: A Working Definition

A ribosome population in an organism is considered polymorphic if one or more of the known stable, structural components (rRNA, r-protein) of the mature organelle normally undergoes an alteration (qualitative, quantitative, or chemical) during the life cycle (Ramagopal, 1992a).

POLYMORPHISM IN DIVERSE SPECIES

Polymorphism of ribosome populations has been reported in species ranging from protozoa to humans (Table 1).

Organ-specific r-proteins were first observed by two-dimensional gel electrophoresis in rabbit (Delaunay et al., 1972). Three of the rabbit r-proteins showed qualitative variation: all three were present in caecal appendix, two of which were also present in liver, but none of the three were present in kidney ribosomes. Some differences in the r-proteins of rat, although considered minor, were reported between the ribosomes of skeletal muscle and liver (Sherton and Wool, 1974). It is unclear if changes occur during liver differentiation, because no differences were found between the ribosomes of fetal and adult livers of calf and sheep (Delaunay et al., 1972). However, as shown in fruit fly, it is possible that livers of the correct developmental stage were not

The Translational Apparatus, Edited by K.H. Nierhaus
et al., Plenum Press. New York. 1993

713

Table 1. List of eukaryotes in which ribosome polymorphism has been observed.

Organism	Polymorphism due to		Reference(s)
	rRNA	r-protein	
Plasmodium species	Yes	No	Gunderson *et al* (1987); Waters *et al* (1989)
Dictyostelium species	Yes(?)	Yes	Ozaki *et al* (1984); Ramagopal (1992 a,b)
Polysphondylium species	No	Yes	Ramagopal and Ennis (1984)
Xenopus laevis	Yes	No	Wolffe and Brown (1988)
Drosophila melanogaster	No	Yes	Lambertsson (1975)
Rat	No	Yes	Sherton and Wool (1974)
Rabbit	No	Yes	Delaunay *et al* (1972)
Human	Yes	Yes	Fisher *et al* (1990); Qu *et al* (1991)

studied. In *Drosophila*, distinct ribosome changes were observed in the early (Lambertsson, 1975) but not in the late third instar larvae (Berger, 1974). Lambertsson (1975) investigatednine post embryonic developmental stages and found that 23 of the 74 r-proteins of *D. melanogaster* contributed to both qualitative and quantitative differences in ribosomes. Nine r-proteins were specific to larval ribosomes, and nine different ones were specific to ribosomes of pupae-adult flies.

In addition to r-proteins, rRNA also contributes to ribosome polymorphism. In the malarial parasites, *Plasmodium berghei* and *Plasmodium falcifarum*, two structurally distinct forms of 18S rRNA were found (Gunderson *et al.*, 1987; Waters *et al.*, 1989). They have been termed A and C genes, and the former was expressed in gametocytes and the latter in sporozoites. The major change in transcription coincided with the parasite's asexual stage to the sexual stage. The possibility for the existence of ribosome subpopulations containing different forms of 5S rRNA molecules has also been shown. Two types of 5S rRNA molecules, one specific to oocyte ribosomes and another specific to somatic cell ribosomes, were reported in *Xenopus laevis* (Wolffe and Brown, 1988). Both forms are encoded by multigene families, and their expression was highly linked to oogenesis and embryogenesis.

In humans, ribosome polymorphism appears to emanate from changes in both rRNA and r-protein. Analysis of 28S rRNA has revealed a sequence dimorphism with either A or G at position +60 from the 5' end of the molecule, giving rise to A-form or G-form of rRNA (Qu *et al.*, 1991). Both forms of rRNA coexist in human cells. This A/G variation occurs between the two universally conserved stem structures, and is within the single-stranded, highly conserved motif found in many metazoans. This suggests that the variation may be important for the ribosome function. In human tissues and cell lines, the G-form was the most predominant, but it was not detected in mouse and chimpanzee which expressed the A-form. Structural differences in the ribosomes of human males and females may also be due to one of its r-protein component (Fisher *et al.*, 1990). Two putative genes encoding the human S4 were isolated separately from the sex chromosomes X and Y. Both genes were widely transcribed in human tissues and cells, but the transcript of the S4 gene from the Y chromosome was not found in several tissues (ovary, brain, heart, and liver) of the females, suggesting that the female ribosomes are different from those of the males.

POLYMORPHISM IN CELLULAR SLIME MOLDS

Recently, extensive studies in the family Dictyosteliaceae to which the important cellular slime mold genera *Dictyostelium* and *Polysphondylium* belong, have revealed unique ribosome polymorphisms (see Ramagopal, 1992 a,b for references). Although the cellular slime molds are considered lower eukaryotes, they are endowed with the large 80S cytoplasmic ribosome just as in higher eukaryotic organisms. In one of the best studied species, *Dictyostelium discoideum*, 83 separate r-proteins have been identified and characterized by several criteria. A change in one or more of these r-proteins was found to contribute to distinct subpopulations of ribosomes in vegetative amoebae, cells at different stages of development, and spores. Table 2 summarizes the

Table 2. Proteins involved in creating ribosome polymorphism in *Dictyostelium discoideum.*

Vegetative amoebae	Developing cells and spores	Regulation	Category
S5, A$_1$ L18 E,L,A	E,L,A S5,A$_1$,L18	Presence Absence	Qualitative
S1,S4,S10,S14,S15,S16, S19,S31,S32,S33,A$_2$, A$_3$,L11,L13,L18,L19, L36,L37,L38,L39,L40	D,S6,S24,L3,L4,L7, L10,L42	High relative abundance	Quantitative (stoichiometric)
S6,S10,S13,S30,D,L1, L2,L31 A$_1$	S5,S7,S8,S34,S31, L36 A	Methylation Phosphorylation	Chemical (covalent Modification)

r-proteins that caused the various ribosome polymorphisms in *D. discoideum.* Polymorphism in vegetative amoebae is accomplished during spore germination, a complex developmental process (Figure 1). Polymorphism in developing cells and spores is created when the amoebae attain aggregation competence, and proceed with multicellular development and differentiation.

Studies in other species of Dicytosteliaceae (*D. giganteum, D. mucoroides, D. purpureum, P. pallidum,* and *P. violaceum*) confirm the widespread occurrence of ribosome polymorphism in this group. All these species exhibited complex patterns of r-proteins as in *D. discoideum* and other eukaryotes. Most of the *D. discoideum* r-proteins contributing to qualitative and quantitative changes in ribosomes were also present in all other *Dictyostelium* species. Similarly, ribosome polymorphism was also evident during the life cycle of *Polysphondylium* species. In *P. pallidum,* five r-proteins were specific to vegetative amoebae and six different r-proteins were specific to developing cell (cyst) ribosomes. Spore and amoeba ribosomes in *P. violaceum* differed by the presence of two exclusive r-proteins in each stage.

Other recent studies on the sequences of rRNA suggests the possibility, as described in humans above, that ribosome polymorphism in the cellular slime mold group may also arise from the heterogeneity of rRNA. Unlike many eukaryotes where rDNA occurs as tandem repeats in chromosomes, in *D. discoideum,* rDNA occurs extrachromosomally as a linear 88kb palindromic dimer. There are about 180-200 copies of rRNA genes per haploid genome

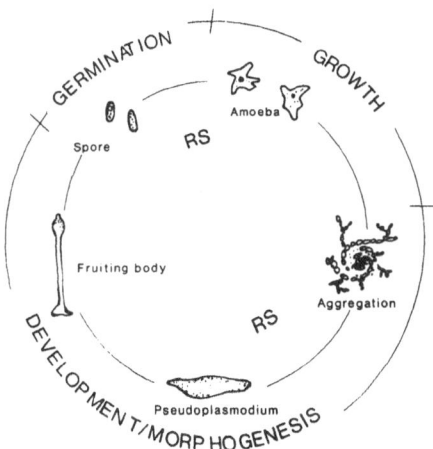

Figure 1. Life cycle of *D. discoideum* showing the developmental stages leading to a major ribosome switch (RS) and polymorphism.

(see Ramagopal, 1992b for references). Genes encoding 5.8S rRNA and 17S rRNA have been independently cloned and sequenced by two groups from the vegetative amoebae (McCarroll *et al.*, 1983; Olsen and Sogin, 1982; Ozaki *et al.*, 1984). Sequence variation at two nucleotide positions was reported in 5.8S rDNA clones. Similarly, variation in nucleotide sequence at nine sites was described between the two 17S rDNA clones. Whether these sequence polymorphisms occur at the conserved or nonconserved regions of the gene has not been determined. The structure of rRNA genes in developing cells has also not been studied.

FUTURE OUTLOOK

It has become more evident from a number of studies described above that the eukaryotic ribosomes are polymorphic and exist in dynamic structures. Polymorphism can arise from one or both the RNA and protein components of the ribosomal particle. However, specific nuances appear to distinguish the phenomenon in various species. Ribosome polymorphism can occur at any phase of the life cycle, but many examples suggest that cell differentiation, development, and morphogenesis play crucial roles. The mechanisms by which ribosome polymorphisms are induced, and the functional properties of the various subpopulations are addressed in another article (Ramagopal, 1992a).

The findings are only a beginning. Application of versatile techniques to each organism will expedite progress. The recently developed gene tagging methods will overcome some of the current limitations, e.g., the lack of a biochemical method to isolate a ribosome subpopulation, or, a classical genetic method to induce nonlethal ribosomal mutations in eukaryotes. Of the many organisms studied so far, *D. discoideum* which belongs to a family with ubiquitous prevalence of ribosome polymorphism, appears to be an excellent system to investigate this subject. With a genome 100 times smaller than that of a mammal, its numerous experimental advantages need no elaboration. Of particular significance is the organism's access to antisense RNA techniques, genomic complementation, homologous recombination, restriction enzyme mediated integration, and transformation technology. These characteristics, as well as amenability to potential approaches (Ramagopal, 1992b), will serve well in elucidating the functional complexities of ribosome polymorphism in *D. discoideum*, and other eukaryotes.

REFERENCES

Berger, E., 1974, The ribosomes of *Drosophila, Molec. gen. Genet.* 128:1.

Delaunay, J., Mathieu, C., and Schapira, G., 1972, Eukaryotic ribosomal proteins, *Eur. J. Biochem.* 31:561.

Fisher, E.M.C., Beer-Romero, P., Brown, L.G., Ridley, A., McNeil, J.A., Lawrence, J.B., Willard, H.F., Bieber, F.R., and Page, D.C., 1990, Homologous ribosomal protein genes on the human X and Y chromosomes, *Cell* 63:1205.

Gunderson, J.H., Sogin, M.L., Wollett, G., Hollingdale, M., de La Cruz, V.F., Waters, A.P., and McCutchan, T.F., 1987, Structurally distinct, stage-specific ribosomes occur in *Plasmodium, Science* 238:933.

Lambertsson, A.G., 1975, The ribosomal proteins of *Drosophila melanogaster, Molec. gen. Genet.* 139:133.

McCarroll, R., Olsen, G.J., Stahl, Y.D., Woese, C.R., and Sogin, M.L., 1983, Nucleotide sequence of the *Dictyostelium discoideum* small-subunit RNA inferred from the gene sequence, *Biochemistry* 22:5858.

Olsen, G.H., and Sogin, M.L., 1982, Nucleotide sequence of *Dictyostelium discoideum* 5.8S rRNA *Biochemistry* 21:2335.

Ozaki, T., Hoshikawa, Y., Iida, Y., and Iwabuchi, M., 1984, Sequence analysis of the transcribed and 5' nontranscribed regions of rRNA gene in *Dictyostelium discoideum, Nucleic Acids Res.* 12:4171.

Qu, L.-H., Nicoloso, M., and Bachellerie, J.P., 1991, A sequence dimorphism in a conserved domain of human 28S rRNA, *Nucleic Acids Res.* 19:1015.

Ramagopal, S., 1992a, Are eukaryotic ribosomes heterogeneous? Affirmations on the horizon, *Biochem. Cell Biol.* 70:269.

Ramagopal, S., 1992b, The *Dictyostelium* ribosome: biochemistry, molecular biology, and developmental regulation, *Biochem. Cell Biol.* (in press)

Ramagopal, S., and Ennis, H.L., 1984, Conservation and variation of ribosomal proteins in several species of the cellular slime molds *Dictyostelium* and *Polysphondylium, Biochim. Biophys. Acta.* 805:300.

Sherton, C.C., and Wool, I.G., 1974, A comparison of the proteins of rat skeletal muscle and liver ribosomes by two-dimensional polyacrylamide gel electrophoresis, *J. Biol. Chem.* 249:2258.

Sussman, M., 1970, Model for quantitative and qualitative control of mRNA translation in eukaryotes, *Nature* 225:1245.

Waters, A.P., Syin, C., and McCutchan, T.F., 1989, Developmental regulation of stage-specific ribosome populations in *Plasmodium, Nature* 342:438.

Wolffe, A.P., and Brown, D.D., 1988, Developmental regulation of two 5S rRNA genes, *Science* 241:1626.

RECONSTITUTION OF A MINIMAL SMALL RIBOSOMAL SUBUNIT

Andrew Scheinman, Anna-Marie Aguinaldo, Agda M. Simpson,
Marian Peris, Gary Shankweiler, Larry Simpson, and James A. Lake

Molecular Biology Institute and Department of Biology
University of California, Los Angeles
Los Angeles, CA 90024

INTRODUCTION

It is appropriate that at a meeting dedicated to H. G. Wittmann we should emphasize comparative studies of the three-dimensional structure of the ribosome since he and his collaborators have made such important contributions to this field. In this paper we present data detailing the first isolation of small mitochondrial ribosomal subunits from the hemoflagellate *Leishmania tarentolae*. Their structure is interesting because these are the smallest ribosomes yet found (their small subunit rRNA sediments at 9S and is only 610 nucleotides long). We also show that particles similar in structure to these small subunits can be reconstituted from *in vitro* transcribed mitochondrial 9S rRNA and *E. coli* proteins.

The 9S and 12S ribosomal RNAs (rRNAs) are the major RNA components of the mitochondrion (kinetoplast) of *Leishmania tarentolae* (Simpson and Simpson, 1978). It has been proposed that the 9S RNA is the small kinetoplastid ribosomal subunit rRNA (de la Cruz *et al.*, 1985a) and the 12S RNA is the large kinetoplastid ribosomal subunit rRNA (de la Cruz *et al.*, 1985b). These assignments were based on the strong similarities between the secondary structures of these RNAs with the consensus small and large subunit rRNA secondary structures and on the occurrence of essential conserved sequences, such as the 520, 720 and 1400 regions (*E. coli* numbering system). However, the existence of kinetoplast ribosomes has not been directly demonstrated.

To test whether 9S RNA can form small subunits, we attempted to reconstitute the small ribosomal subunit (Traub *et al.*, 1971) from heterologous components, *Leishmania tarentolae* 9S RNA and *E. coli* small subunit ribosomal proteins. Heterologous reconstitutions have been previously achieved when the rRNA and ribosomal proteins are from different bacteria (Nomura *et al.*, 1968; Higo *et al.*, 1973; Goldberg and Steitz, 1974; Held *et al.*, 1974). Recently, *in vitro* T7 transcripts of wild type and mutant *E. coli* 16S rRNA have also been reconstituted with *E. coli* ribosomal proteins (Krzyzosiak *et al.*, 1987; Melancon *et al.*, 1987; Scheinman, 1989).

We show in this paper that hybrid subunits can be reconstituted using *in vitro*

The Translational Apparatus, Edited by K.H. Nierhaus
et al., Plenum Press, New York, 1993

transcribed 9S RNA and *E. coli* ribosomal proteins. We also show that a 9S- and 12S-containing kinetoplast-mitochondrial fraction contains ribosomes and ribosomal subunits. The smaller subunit possesses the classical ribosomal structures of the platform, head and base, as do the hybrid reconstituted 9S subunits. This indicates that the 9S rRNA can be assembled *in vivo* and *in vitro* into ribosomal like subunits.

MATERIALS AND METHODS

In Vitro Transcription of 9S RNA

The 9S DNA sequence was amplified in a PCR reaction using plasmid pLt120 (Masuda *et al.*, 1979; de la Cruz *et al.*, 1984) as the template and primers corresponding to positions -28 to +4 (5'-primer) and 575 to 611 (3'-primer) of the 9S DNA and flanking sequence. The 5'-primer also contained a T7 polymerase promoter sequence, and an EcoRI site at its 5'-end. The 3'-primer contained a HindIII site at its 5'-end. PCR conditions used were described elsewhere (Scheinman, 1989). The PCR reaction was run on a 0.8% agarose gel, the appropriate band was excised, and the DNA was recovered by GeneClean. Approximately 1 microgram of this DNA was used for an *in vitro* transcription (Scheinman, 1989), with a final volume of 1 ml.

Reconstitutions

RNA was obtained from transcription reactions by phenol extraction and ethanol precipitation. After resuspension this RNA was combined with the appropriate amount of *E. coli* small ribosomal subunit proteins and reconstituted as described (Scheinman, 1989; Held *et al.*, 1973).

Sucrose Gradient Analyses and Electron Microscopy

Reconstituted subunits were concentrated by ultracentrifugation in a Beckman SW50.1 rotor at 41,000 RPM (157,000 x g) for 3.5 hrs at 4°C, resuspended in 200 microliters of 10 mM Tris-HCl (pH 7.6), 6 mM MgCl$_2$, 50 mM NH$_4$Cl, 1 mM DTT, 0.5 mM EDTA and loaded onto sucrose gradients that were 20-30% sucrose in the same buffer. These gradients were centrifuged for 40 minutes at 50,000 RPM (220,000 x g) at 4°C in a Beckman VTi65 rotor, after which they were fractionated and monitored through an ISCO UA-5 continuous absorbance monitor. Subunits were collected and concentrated by ultracentrifugation, resuspended to a final concentration of 0.2 A$_{260}$ units/ml, negatively stained by the double-layer carbon method (Lake, 1979), and visualized with a Philips 400 electron microscope.

Preparation of L. tarentolae Kinetoplast Small Subunits

L. tarentolae (UC strain) were grown as previously described (Braly *et al.*, 1974). Mid-log-phase cells were harvested by centrifugation and the kinetoplast-mitochondrial fraction was isolated by flotation in Renografin density gradients (Braly *et al.*, 1974; Simpson and Braly, 1970). Kinetoplast ribosomes were isolated by a modification of the method of Spithill *et al.* (1979).

The isolated kinetoplast fraction was resuspended in 125 mM sucrose, 10 mM Tris-HCl (pH 7.9), 1.5 mM MgCl$_2$, and an equal volume of 2X protein synthesis buffer was added (50 mM Tris-HCl (pH 6.7), 20 mM MgCl$_2$, 300 mM KCl, 20 mM KH$_2$PO$_4$, 4 mM ATP, 10 mM ketoglutarate). The resulting solution was incubated for 10 minutes at 25°C, and then 0.01 volume 5 mg/ml puromycin was added. The solution was incubated for three minutes at 25°C, diluted with 4 volumes 250 mM sucrose, 20 mM Tris (pH 7.9), 3 mM MgCl$_2$, and centrifuged at 10,000 x g for 15 minutes. The pellet was resuspended in 200 mM sucrose, 200 mM Tris-HCl (pH 8.0), 35 mM MgCl$_2$, 400 mM KCl, 25 mM EGTA, 2% Triton X-100, 40 mg heparin, the solution was homogenized and then clarified by centrifugation at 10,000 x g for 15 minutes. The solution was then layered on a 15-30% linear sucrose gradient made in the same buffer and centrifuged for 22 hr at 16,000 RPM at 4°C in a Beckman SW27 rotor. RNA was isolated from the gradient fractions by SDS lysis and phenol-chloroform extraction and ethanol precipitation as previously described (Simpson and Simpson, 1978).

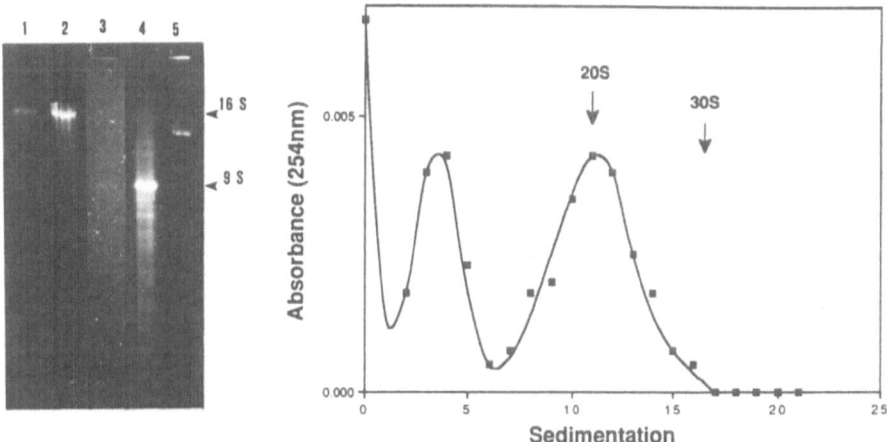

Figure 1. Sucrose gradient sedimentation analysis of reconstituted hybrid small ribosomal subunits (*in vitro* transcribed *L. tarentolae* 9S RNA plus *E. coli* r-proteins) run on a 15-30% gradient. The peak labeled "20S" contains reconstituted particles that, by their migration, rRNA composition and structure, correspond to mitochondrial small subunits. Sedimentation is from left to right and the vertical scale measures absorbance at 254 nm. A 4% PAGE gel is shown at the left. Lanes were loaded with T7 *in vitro* transcribed 16S rRNA (lane 1), native 16S rRNA (lane 2), rRNA extracted from the 20S peak of the adjacent gradient (lane 3 -- contrasted to show a weak band), T7 *in vitro* transcribed 9S rRNA (lane 4), and 9S rDNA used for the transcription (lane 5). RNA was visualized by Ethidium Bromide/UV.

RESULTS AND DISCUSSION

Figure 1 shows the results of a sucrose gradient size fractionation of the reconstituted products. Two peaks are present. The major peak, sedimenting at 20S (labelled 20S), contains ribosomal subunit with a unique morphology. The leading peak consists of unreconstituted components. The position of *E. coli* small subunits in control experiments (labeled "30S") is shown in the figure.

To characterize the 20S particles, the 20S peak was collected from sucrose gradients, concentrated by ultracentrifugation, and either loaded onto polyacrylamide gels for electrophoresis or negatively stained for electron microscopy. Gel electrophoresis (shown at the left in Figure 1) indicated that the 20S peak contained 9S rRNA and did not contain 16S rRNA [possibly introduced with the small subunit proteins]. The left two lanes contain 16S rRNA (*in vitro* and *in vivo* transcribed in lanes 1 and 2, respectively), lane 3 contains RNA extracted from the 20S peak, lane 4 contains *in vitro* transcribed 9S rRNA, and lane 5 contains 9S rDNA. The occurrence of a 9S band in the 20S fraction and the absence of detectible 16S in this peak indicates that the 9S rRNA has reconstituted into the 20S particles, and that the 20S particles are not partially folded 30S subunits.

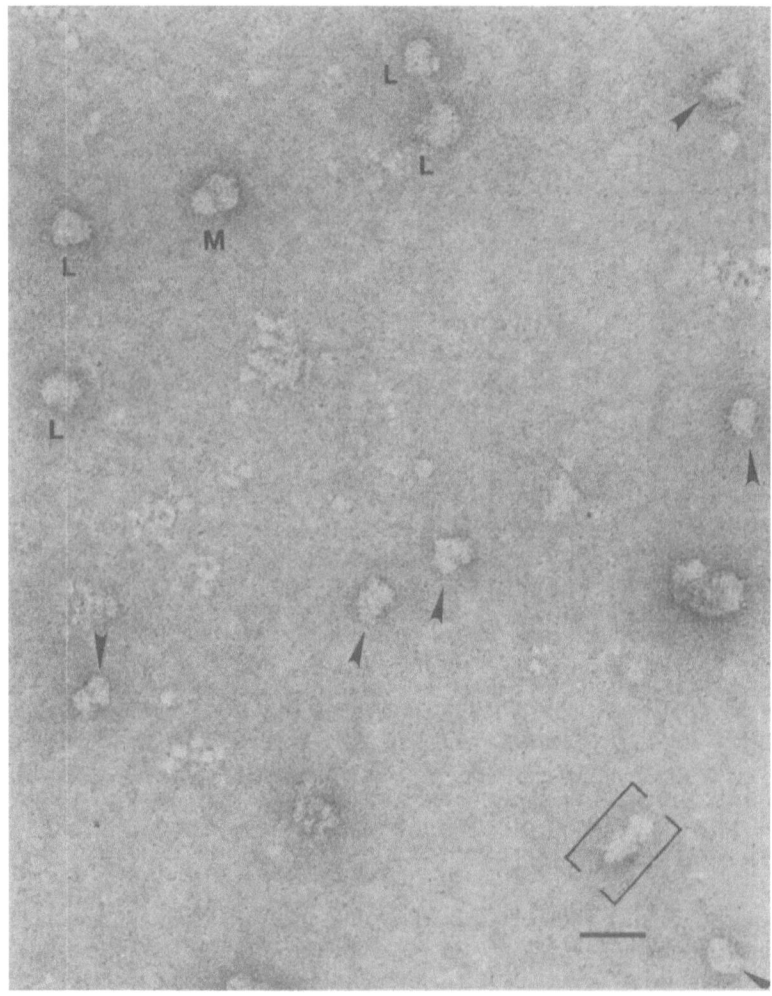

Figure 2. Electron microscopic field of purified mitochondrial small ribosomal subunits. Electron microscopy is described in reference 24. The scale bar represents 25 nanometers. The electron micrograph of *L. tarentolae* ribosomes and subunits was obtained after partial purification on sucrose gradients of the peak shown in Figure 4. Individual small subunits from this preparation are shown in Figure 3B. A single eukaryotic cytoplasmic small subunit is enclosed in brackets and provides a useful size comparison. Mitochondrial ribosomes and large subunits are indicated by 'L' and 'M', respectively. Small subunits are indicated by arrows.

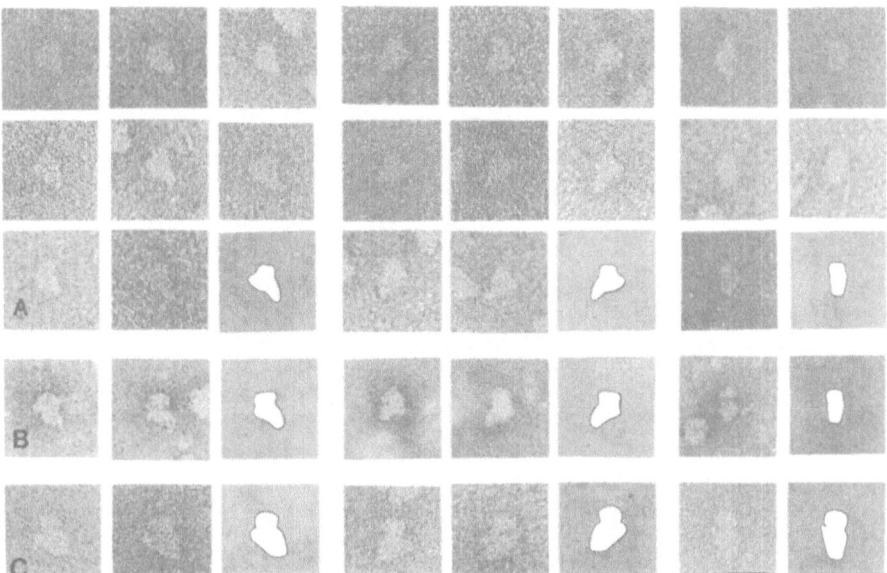

Figure 3. Gallery of representative views of (A) reconstituted hybrid small subunits (*in vitro* transcribed *L. tarentolae* 9S RNA, *E. coli* ribosomal proteins), (B) *L. tarentolae* small subunits, and (C) reconstituted wild-type (*E. coli* 16S rRNA, *E. coli* ribosomal proteins) small subunits. Subunits are shown in the "asymmetric projection" in the first three columns on the left, in the corresponding enantiomorphic projection in the middle three columns, and in the "quasi-symmetric" projection in the two columns on the right (Lake, 1976). Images of reconstituted hybrid- and purified mitochondrial- small subunits and ribosomes were obtained from the fractions in Figures 1 and 4.

Electron microscopy of the 20S peak disclosed ribosomal subunits which, although related to 30S subunits, were clearly smaller and had altered profiles. Figure 3A shows representative projections of these smaller particles, which we tentatively identify as hybrid small subunits. Subunits are shown in the "asymmetric projection" (Lake, 1976) in the first three columns on the left, and in the corresponding enantio-morphic projection in the second three columns. For comparison the corresponding profiles of reconstituted *E. coli* small subunits are shown in Figure 3C. The ribosomal head, base and platform (Lake, 1985) are all present in the hybrid 20S subunits, but both the head and base are smaller than in reconstituted 30S subunits (Fig. 3C). In the "quasi-symmetric" projection of hybrid subunits, shown in the last two columns of Figure 3A, the head and base of the subunits are more similar in size than in the equivalent *E. coli* projections, shown in the last column of Figure 3C.

Ribosomes were also isolated from kinetoplastid mitochondria and the small subunits were compared with the hybrid reconstituted subunits. Sedimentation of broken kinetoplasts on sucrose gradients (Fig 4) yielded several fractions in which 9S sequences were enriched. When fractions across the gradient were collected, run on a polyacryl-amide gel and the gel probed for the 9S sequence in a Northern hybridization, 9S was found to be present mainly in fractions near the 40S peak illustrated in Figure 4 (data not shown). Additional experiments with both 9S and 12S probes indicate the presence of both RNA species in the 40S region (data not shown). These fractions were analyzed by electron microscopy and a representative field is shown in Figure 2. This fraction contains particles tentatively identified by the presence of classical ribosomal structures as monomeric ribosomes (M), large (L) and small ribosomal subunits (marked by arrows) and, rarely (see the RNA gel in Fig. 4), cytoplasmic small ribosomal subunits.

Small subunits from this fraction are shown in Figure 3B and can be identified by the presence of the platform, head and base in the asymmetric projections of the subunit (frames 1, 2, 4 and 5, from the left, of Fig. 3B) and by the head and base in the quasi-symmetric projection of the subunit (frame 7 of Fig. 3B). Additionally, large subunits can be identified by the presence of the central protuberance and the L7/12 stalk (Lake, 1985) (see the two large subunits at the top-center of Fig. 2). The presence of ribosomes and their subunits is in accord with the results of the Northern hybridization that detect both 9S and 12S RNAs. [We assume that the large and small subunits have resulted from the dissociation of monosomes during preparation for microscopy.] The cytoplasmic small subunits are useful size references to compare with mitochondrial subunits, and a contaminating cytoplasmic subunit (enclosed by brackets) included in this field illustrates the small size of the mitochondrial subunits.

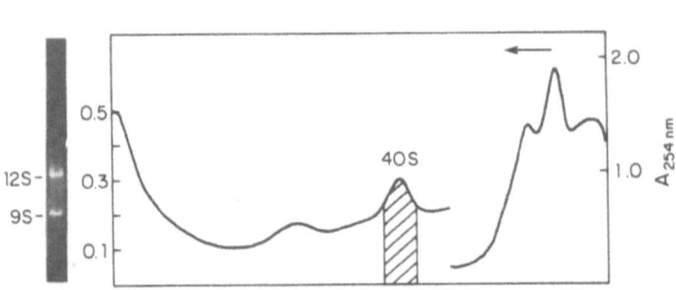

Figure 4. Sucrose gradient sedimentation analysis of partially purified *L. tarentolae* small subunits. On the right side of the figure is the OD profile of mitochondrial ribosome subunits from *L. tarentolae* run on a 15-30% sucrose gradient (the break in the gradient is a change of scale). The hatched peak marked "40S" was collected for electron microscopy (Fig. 2). On the left is a methylmercury agarose gel analysis of the RNA content of the mitochondria preparation used for the gradient. RNA was visualized by Ethidium Bromide/UV. The positions of the 12S and 9S rRNAs are indicated. Note the relative absence of contamination with cytoplasmic rRNAs and their smaller fragments (Campbell *et al.*, 1987) (tRNAs were run off the gel).

The close similarity of the small ribosomal subunits derived from mitochondria and from hybrid reconstitutions corroborates the identity of both. Individual mitochondrial small subunits are shown in Figure 3B for comparison with the hybrid reconstituted subunits (Fig. 3A). Key features present in both the mitochondrial particles and in the reconstituted hybrid particles, but differing from the features of reconstituted *E. coli* subunits (Fig 3C) are (i) the tapered base, and (ii) the reduced head. Both types of particle are more like each other than either is like the *E. coli* small subunit. While these two particles closely resemble each other, they are not identical. This is reasonable since one contains ribosomal proteins from *E. coli* and the others are mitochondrial in origin.

In summary, successful heterologous reconstitution of a small ribosomal subunit using 9S mitochondrial RNA from a kinetoplastid protozoan has established that 9S forms small ribosomal subunits *in vivo* and provides evidence for the presence of a functional translation apparatus in the unusual mitochondrion of these cells.

ACKNOWLEDGEMENTS

We thank Angela Kost for expert electron microscopy, Jerome Arrington, Jr. for photography, and Maria Rivera and Tom Atha for helpful advice. Supported by grants from the NSF and NIH to J.A.L. and from the NIH to L.S.

REFERENCES

P. Braly, L. Simpson and F. Kretzer, Isolation of kinetoplast-mitochondrial complexes from Leishmania tarentolae, *J. Protozool.* 21:782 (1974).

D.A. Campbell, K. Kubo, C.G. Clark and J.C. Boothroyd, Precise identification of cleavage sites involved in the unusual processing of trypanosome ribosomal RNA, *J. Mol. Biol.* 196:113 (1987).

V.F. de la Cruz, N. Neckelmann and L. Simpson, Sequences of six genes and several open reading frames in the kinetoplast maxicircle DNA of Leishmania tarentolae, *J. Biol. Chem.* 259:15136 (1984).

V.F. de la Cruz, J.A. Lake, A.M. Simpson and L. Simpson, A minimal ribosomal RNA: sequence and secondary structure of the 9S kinetoplast ribosomal RNA from Leishmania tarentolae, *Proc. Natl. Acad. Sci. USA.* 82:1401 (1985a).

V.F. de la Cruz, A.M. Simpson, J.A. Lake and L. Simpson, Primary sequence and partial secondary structure of the 12S kinetoplast (mitochondrial) ribosomal RNA from Leishmania tarentolae: conservation of peptidyl-transferase structural elements, *Nucl. Acids Res.* 13:2337 (1985b).

M.L. Goldberg and J.A. Steitz, Cistron specificity of 30S ribosomes heterologously reconstituted with components from Escherichia coli and Bacillus stearothermophilus, *Biochem.* 13:2123 (1974).

W.A. Held, S. Mizushima and M. Nomura, Reconstitution of Escherichia coli 30S Ribosomal Subunits from Purified Molecular Components, *J. Biol. Chem.* 248:5720 (1973).

W.A. Held, W.R. Gette and M. Nomura, Role of 16S ribosomal ribonucleic acid and the 30S ribosomal protein S12 in the initiation of natural messenger ribonucleic acid translation, *Biochem.* 13:2115 (1974).

K. Higo, W. Held, L. Kahan and M. Nomura, Functional correspondence between 30S ribosomal proteins of Escherichia coli and Bacillus stearothermophilus, *Proc. Natl. Acad. Sci. USA.* 70:944 (1973).

W. Krzyzosiak, R. Denman, K. Nurse, W. Hellmann, M. Boublik, C.W. Gehrke, P.F. Agris and J. Ofengand, In vitro synthesis of 16S ribosomal RNA containing single base changes and assembly into a functional 30S ribosome, *Biochem.* 26:2353 (1987).

J.A. Lake, Ribosome structure determined by electron microscopy of Escherichia coli small subunits, large subunits and monomeric ribosomes, *J. Mol. Biol.* 105:131 (1976).

J.A. Lake, Practical aspects of immune electron microscopy, *Meth. Enzymol.* 61:250 (1979).

J.A. Lake, Evolving ribosome structure: domains in archaebacteria, eubacteria, eocytes and eukaryotes, *Ann. Rev. Biochem.* 54:507 (1985).

H. Masuda, L. Simpson, H. Rosenblatt and A.M. Simpson, Restriction map, partial cloning and localization of 9S and 12S kinetoplast RNA genes on the maxicircle component of the kinetoplast DNA of Leishmania tarentolae, *Gene* 6:51 (1979).

P. Melancon, M. Gravel, G. Boileau and L. Brakier-Gingras, Reassembly of active 30S ribosomal subunits with an unmethylated in vitro transcribed 16S rRNA, *Biochem. Cell. Biol.* 65:1022 (1987).

M. Nomura, P. Traub and H. Bechmann, Hybrid 30S ribosomal particles reconstituted from components of different bacterial origins, *Nature.* 219:793 (1968).

A. Scheinman, The construction and analysis of mutant small E. coli ribosomal subunits with inserts in the 16S rRNA, Ph.D. thesis, University of California, Los Angeles (1989).

L. Simpson and P. Braly, Synchronization of Leishmania tarentolae by hydroxyurea, *J. Protozool.* 17:511 (1970).

L. Simpson and A.M. Simpson, Kinetoplast RNA of Leishmania tarentolae, *Cell.* 14:169 (1978).

T.W. Spithill, P. Nagley and A.W. Linnane, Biogenesis of mitochondria 51, *Molec. Gen. Genet.* 173:159 (1979).

P. Traub, S. Mizushima, C.V. Lowry and M. Nomura, Reconstitution of ribosomes from subribosomal components, *Meth. Enzymol.* 20: 391 (1971).

THE BIFUNCTIONAL NATURE OF RIBOSOMAL PROTEINS
AND SPECULATIONS ON THEIR ORIGINS

Ira G. Wool

Department of Biochemistry and Molecular Biology
The University of Chicago
Chicago, IL 60637, U.S.A.

INTRODUCTION

We undertook some years ago to isolate, to characterize, and to determine the structure of eukaryotic (rat) ribosomal proteins and nucleic acids (Wool, 1979). We chose rat ribosomes for reasons that are buried in the history of our research. What was it that led us to this undertaking which was not infrequently derided as an arduous, tedious, time consuming chore lacking excitement or intellectual challenge, indeed, as lacking importance? We made a commitment to the protein and nucleic acid chemistry because of an abiding belief that it was essential for a solution of the structure of the organelle; the structure in turn was needed for a coherent molecular account of the function of ribosomes in protein synthesis. Almost everyone agreed that this last was an enterprise that was to be taken seriously - that it was a formidable intellectual challenge, that it was of great importance, and that it did not lack for excitement. Moreover, no matter how pretentious it sounds, we felt we would be making a contribution to a compilation of data (the primary structure of the seventy to eighty proteins) that could be viewed as a general resource. It is perhaps even more pretentious to say that this prediction has been amply substantiated, that the sequences have been useful to many, including scientists whose research did not touch directly on ribosomes or protein synthesis. From the beginning we suspected that the sequences of amino acids in the proteins might also help in understanding the origins and the evolution of the ribosomal proteins, in unraveling the functions of the proteins, in defining the rules that govern the interaction of the proteins with the rRNAs, and in uncovering the amino acid sequences that direct the proteins to the nucleolus for assembly on nascent rRNA. What we did not, and could not, anticipate was the novel insights that the structure of the ribosomal proteins would provide. Two of these are the subject of this review: The first is the recognition of DNA binding motifs in ribosomal proteins which has led, in turn, to speculations on the origin of the ribosomal proteins. The second is an increasing awareness of the bifunctional nature of ribosomal proteins, i.e. that some have a function apart from the ribosome and from protein synthesis.

The Translational Apparatus, Edited by K.H. Nierhaus
et al., Plenum Press, New York, 1993

ZINC FINGER-LIKE MOTIFS IN RIBOSOMAL PROTEINS

The story begins with the finding that rat ribosomal proteins S27 and S29 have zinc finger-like domains (Chan et al., 1992). This led us to ask what purpose these motifs, most frequently associated with mediating the binding to DNA, serve in ribosomal proteins. In S27 and S29 the domains have the form $-\underline{C}-X_2-\underline{C}-X_{14,15}-\underline{C}-X_2-\underline{C}-$; i.e. they are of the C_2-C_2 variety rather than the more common C_2-H_2 type. The secondary and tertiary structures of zinc fingers of these two types are different, indeed, may not be related (Schwabe and Rhodes, 1991). However, proteins of both types share the potential to coordinate a zinc ion and almost all bind to nucleic acids, most to DNA, a few to RNA (Berg, 1986, 1990; Klug and Rhodes, 1987). For example, transcription factor IIIA (TF IIIA), in which the zinc-finger domain occurs in tandem nine times (Miller et al., 1985), binds to both the internal control region of 5S rDNA and to 5S rRNA, i.e. to both the gene and to its transcript (Brown, 1984).

RS27	37	C	PG	C	YKITTVFSHAQTVVL	C	VG	C
RS29	21	C	RV	C	SNRHGLIRKYGLNM	C	RQ	C
BsuS14	23	C	ER	C	GRPHSVIRKFKL	C	RI	C
McS14	24	C	NH	C	GRPHAVLKKFGI	C	RL	C
MvS14	17	C	KR	C	GRKGPGIIRKYGLDL	C	RQ	C
ScMRP2	78	C	VDS		GHARFVLSDFRL	C	RYQFR	
EcS14	63	C	RQT		GRPHGFLRKFGL		SRIKV	
RL37a	39	C	SF	C	GKTKMKRRAVGIWH	C	GS	C
RL37	18	C	RR	C	GSKAYHLQKST	C	GK	C
HCEP52	20	C	RK	C	YARLHPRAVN	C	RKKK	C
HCEP80	45	C	PSDE	C	GAGVFMASHFDRHY	C	GK	C
BstL32	29	C	PN	C	GEWKLAHRV	C	KA	C
BsuL36	11	C	EK	C	KVIRRKGKVMVI	C	ENPK	H
EcL36	11	C	RN	C	KIVKRDGVIRVI	C	SAEPK	H

Figure 1. Ribosomal proteins with zinc finger-like motifs. The abbreviations for the species and for the ribosomal proteins are: RS27, rat S27; RS29, rat S29; BsuS14, *Bacillus subtilis* S14; McS14, *Mycoplasma capricolum* S14; MvS14, *Methanococcus vannielii* S14; ScMRP2, *Saccharomyces cerevisiae* mitochondrial MRP2; EcS14, *Escherichia coli* S14; RL37a, rat L37a; RL37, rat L37; HCEP 52, human ubiquitin carboxyl-terminal extension protein 52; HCEP80, human ubiquitin carboxyl-terminal extension protein 80; BstL32, *Bacillus stearothermophilus* L32; BsuL36, *Bacillus subtilis* L36; EcL36, *Escherichia coli* L36.

In S27 and S29 the amino acid sequence between the internal cysteinyls, which we shall for convenience refer to as the linker sequence, is dominated by basic and hydrophobic residues (Fig. 1). In S27 2 of the 15 residues are basic and 7 are hydrophobic; in S29 4 of the 14 residues are basic and 5 are hydrophobic. Finally, the linker sequences have aromatic residues - tyrosine and phenylalanine in S27 and tyrosine in S29.

A search of our library of ribosomal protein amino acid sequences revealed that rat ribosomal proteins L37a (Tanaka et al., 1989) and L37 have zinc finger domains of the same form and the linker sequences also have basic and hydrophobic residues (Fig. 1). *B. stearothermophilus* L32, *B. subtilis* L36, and *E. coli* L36 also have C_2-C_2 zinc binding motifs (Fig. 1). Of the archaebacterial and eubacterial S14 family of ribosomal proteins which are related to rat S29 the members from *M. vannielii* (Auer et al., 1989), from *B. subtilis* (Henkin et al., 1989), and from *M. capricolum* (Ohkubo et al., 1989) have the motif; whereas *E. coli* S14 (Yaguchi et al., 1983) has at most a degenerate form of the domain (Fig. 1). In the alignment with the others there is a cysteine in *E. coli* S14 at what would be the initial position; the other three are absent. The hydrophilic and hydrophobic character of the linker sequence is preserved in *E. coli* S14 and 8 of the 12 residues share identity with amino acids at the same positions in *B. subtilis* S14. The yeast nuclear encoded mitochondrial ribosomal protein MRP2 (Meyers et al., 1987) is also related to the S14 family and it too has a degenerate form of the motif; the sequence has only the two cysteinyls that correspond to the first and third sites in the full motif (Fig. 1). Once again the linker sequence is basic (3 of 12 residues), hydrophobic (6 residues), and has two aromatic residues. The spin I choose to give to these findings is that *E. coli* S14 and perhaps yeast mitochondrial MRP2 once had full C_2-C_2 motifs and that parts were lost during divergent evolution; there is, of course, no direct evidence for this bias. Finally, variants of the zinc finger motif are to be found in the human ubiquitin carboxyl-terminal extension proteins CEP52 (Baker and Board, 1991) and CEP80 (Redman and Rechsteiner, 1989) (Fig. 1); the latter is ribosomal protein S27a and the former is likely to be a large ribosomal subunit protein but its identity has not been established.

An approximate consensus amino acid sequence for the linker peptide can be derived. It has the form, -GxBxBxZxBBZxxZ-, where B designates a basic residue, Z a hydrophobic amino acid, and x a position with no consensus. What is most striking is the retention of the initial glycine and of the pattern of basic and hydrophobic residues.

Finally, I note that analysis of rat liver 80S ribosomes and ribosomal 40S and 60S subunits by atomic absorption spectroscopy has indicated the presence of appreciable amounts of Zn (Chan et al., 1992).

THE SIGNIFICANCE OF THE ZINC FINGER MOTIFS IN RIBOSOMAL PROTEINS

There is no evidence that the zinc finger motifs in ribosomal proteins participate directly in binding to RNA, nor for that matter has the possibility been ruled out - there is just no evidence. The zinc finger motifs in ribosomal proteins may serve no purpose; they may be the vestiges of a former function, for example, the binding to DNA (cf. later), that are preserved in ribosomal proteins that have evolved other means for associating specifically with RNA. This is consistent with the preservation in *E. coli* S14 and in yeast MRP2 of only a part of the motif, recognizable but presumably no longer capable of coordinating a metal ion. I would argue further, that it is likely that the same amino acid sequences and/or structures in the *B. subtilis* and *E. coli* S14 proteins are used to bind to 16S rRNA since the binding site is conserved (Stern et al., 1989). If this assumption is correct then the conclusion is that the proteins do not use the cannonical secondary and tertiary structure that is formed by the association of zinc with the motif in binding to rRNA since *E. coli* S14 lacks the capacity to bind the metal and, hence, to form the structure; provided, of course, that coordination of zinc is essential for formation of the characteristic structure. One reconciliation is that the binding of these ribosomal proteins to rRNA employs the side chains of the conserved basic and hydrophobic amino acids in the linker region of the motif rather than the finger structure *per se*.

ON THE ORIGIN OF THE RIBOSOMAL PROTEINS

A question that has nagged at the minds of many who work on ribosomes is where did the proteins come from; what were their origins? It is a tenet of our faith that the *Ur-*ribosome had only RNA. This is a belief that is supported by the mounting evidence that the rRNAs are responsible for the basic biochemistry of protein synthesis: for the binding of mRNA, aminoacyl-tRNA, and the initiation, elongation, and termination factors; for peptide bond formation; and for translocation. The discovery that 23S rRNA may have peptidyl transferase activity (Noller et al., 1992) gives a measure of substance to this recapitulation of the history of the ribosome. The ribosomal proteins are now viewed as a later evolutionary embellishment and are deemed to facilitate the folding and the maintenance of an optimal configuration of the rRNA (Stern et al., 1989), perhaps in this way endowing protein synthesis with speed and accuracy (Wool et al., 1990). This may be too severe a restriction on the role of the proteins in ribosome function, nonetheless, it is likely that RNA preceded the proteins and hence it is pertinent to ask from whence the latter came.

The occasion for the transition during evolution from a ribosome that had only RNA to a RNP machine may have coincided with, or been a response to, the appearance of nucleases which would have put an RNA ribosome at risk. There are at least two possible scenarios for the origin of the ribosomal proteins. They may have been designed specifically for the ribosome, or they may have been coopted from amongst a set of preexistent proteins that already had defined functions. The two possibilities are by no means exclusive nor is it likely that all the proteins were added at one time. If the latter conjecture has substance, the proteins most likely to have been recruited would have been those that already had the capacity to bind to nucleic acids. Considered from this perspective the zinc finger motif in ribosomal proteins might have been used first to bind to DNA or to other RNAs. A possible archetype is TF IIIA which may have evolved to bind to the internal control region of the 5S rRNA gene and only later came to associate with 5S rRNA.

It is relevant here that a potential zinc finger is not the only DNA binding motif to be found in ribosomal proteins. The carboxyl terminus of *E. coli* L12 has a helix-turn-helix motif (Rice and Steitz, 1989) very much like that found in proteins that bind to DNA and that regulate transcription (Steitz, 1990).

A paradigm for the process by which rRNA could have coopted preexisting nucleic acid binding proteins, and indeed the likelihood of its having happened, is exemplified by the *E. coli* basic proteins NS1 and NS2 (Mende et al., 1978). These proteins, which have been likened to eukaryotic histones, bind to single- and to double-stranded DNA and to RNA (Berthold and Geider, 1976). What is significant is that NS1 and NS2 are found with native *E. coli* 30S ribosomal subunits in near stoichiometric amounts but not with native 50S subunits or 70S couples (Suryanarayana and Subramanian, 1978). The binding to 30S subunits is specific and the affinity is strong (Suryanarayana and Subramanian, 1978). There is no evidence, however, that either NS1 or NS2 has a function in protein synthesis: they are not in 70S ribosomes; they are not needed for translation of mRNA; and antibodies specific for NS1 and NS2 do not inhibit protein synthesis (Suryanarayana and Subramanian, 1978). Nonetheless, the observations that the DNA binding proteins NS1 and NS2 can associate specifically with 30S ribosomal subunits presumably by interaction with a particular site on 16S rRNA gives a measure of credence to our proposal concerning the origin of the ribosomal proteins. NS1 and NS2 can be viewed as DNA binding proteins auditioning for a role in ribosome structure by association with native 30S subunits.

That NS1 and NS2 are not required for protein synthesis does not militate against the proposal; indeed, a considerable number of authentic *E. coli* ribosomal proteins are absent from ribosomes in various mutants and the loss is compatible with residual protein synthesis and growth (Dabbs, 1985). Proteins may have been added to ribosomes initially only to coat and protect the RNA. DNA binding proteins would be eminently suitable for the purpose and many bind to RNA. The proteins, either from the beginning, or perhaps only later, tuned the rRNA and thereby optimized the higher order structure and *pari passu* conferred velocity and fidelity on protein synthesis. Still later the proteins may have assumed other functions. NS1 and NS2 may be in the early stages of this evolutionary process.

This raises a related matter: Received wisdom holds that the ribosome has a single function, to catalyze the synthesis of protein. The corollary of this axiom is that any protein not directly or indirectly involved in, or necessary for, protein synthesis is not a ribosomal protein. This reasoning dominated the strategy underlying the identification, the purification, and the characterization of the ribosomal proteins and was responsible for designating as contaminants many proteins more or less tenaciously, and more or less specifically, associated with the particle. The reasoning was heuristic and served those who applied it well. However, the bias inherent in the reasoning may have a defect - it may be wrong. Indeed, there is a prominent exception: *E. coli* ribosomes participate with stringent factor in the synthesis of guanosine tetraphosphate and guanosine pentaphosphate (Cashel and Gallant, 1974; Block and Haseltine, 1974). Although, the synthesis of the guanosine polyphosphates is a by-product of ribosomes idling during protein synthesis it is not necessary for the synthesis of protein and stringent factor is not considered a ribosomal protein. The lesson is that one should be inclined to an open mind concerning the possibility that ribosomes participate in other cellular functions.

THE BIFUNCTIONAL NATURE OF RIBOSOMAL PROTEINS

If rRNA coopted preexisting proteins, and if they retained their original function, then ribosomal proteins would be bi- or even multifunctional. There is evidence for bifunctionality for a number of ribosomal proteins. The ribosomal protein that comes to mind first is *E. coli* S1 which has been known for a long time to be a component of the replicase of some RNA phages (Kamen, 1975). In Qß, for example, replication is catalyzed by a complex containing the product of the viral replicase gene and three host factors; elongation factors Tu and Ts, and ribosomal protein S1. The latter is required for the initiation of replication on the plus strand of the viral RNA.

A second *E. coli* ribosomal protein, S10, participates with λ phage N protein and NusB in antitermination of transcription (Friedman et al., 1981). Apparently NusB binds to N-modified transcription complexes by interacting with *E. coli* ribosomal protein S10 and this interaction is essential for efficient antitermination (Mason et al., 1992).

Replication-inhibiting lesions in *E. coli* DNA increase mutations and induce the multigene SOS response (Woodgate et al., 1989). One of the proteins whose synthesis is increased is UmuC; although its exact function is not certain there is evidence that UmuC forms a complex with UmuD, RecA, and DNA polymerase III holoenzyme and that the complex effects translesion DNA replication with introduction of mutations at the site of damage. The observation that is pertinent here is that *E. coli* ribosomal protein S9 maintains the solubility of UmuC, and presumably its native conformation, during purification (Woodgate et al., 1989). Two possibilities have been suggested to explain the phenomenon (Woodgate et al., 1989): That S9 is necessary for the assembly of the multiprotein complex required for translesion DNA replication just as S10 is needed for the assembly of the λ phage antitermination complex; or that S9 is a "renaturase" or chaperone that specifically, or nonspecifically, stabilizes the conformation of UmuC or facilitates its refolding. Both are functions that would be valuable to the ribosome. S9, for example, might facilitate assembly of protein-protein complexes during ribosome biogenesis; or, if S9 is a chaperone, it might stabilize the unfolded or partially folded nascent peptide. What is not certain is whether the effect of S9 to maintain the solubility of UmuC is physiologic and hence an example of a second function of a ribosomal protein, or merely fortuitous.

There is a most unexpected instance of a separate function for a eukaryotic ribosomal protein. Damage to DNA in mammalian cells occurs all the time and the lesions are ordinarily promptly and efficiently repaired by an ensemble of enzymes (Demple and Linn, 1980; Imlay and Linn 1988). Amongst these enzymes is one that has been designated apurinic/apyrimidinic endonuclease I (API); it catalyzes incision on the 3' side of an apurinic or an apyrimidinic site (Demple and Linn, 1980; Kim and Linn, 1988). Xeroderma pigmentosum is a rare, autosomal recessive disease that is characterized by hypersensitivity to UV irradiation; because of this sensitivity patients with the disease are at great risk of developing sunlight-induced skin cancers (Kraemer et al., 1984). Seven complementation groups for the disease have been defined (Kraemer et al., 1975); group D lacks API activity (Kuhnlein et al., 1976). Linn and colleagues (R. Fellous, J. Kim, A. Admon, D. Pak, J. Stahl, I.G. Wool, and S. Linn, unpublished data) purified the enzyme to homogeneity and determined the sequences of amino acids in a number of proteolytic fragments; the sequences were the same as had been established before for regions of rat ribosomal protein S3 (Chan et al., 1990). This extraordinary, counter intuitive finding raised the possibility that the protein that had been purified contained predominantly S3 but was contaminated with small amounts of API. To test the possibility a rat ribosomal protein S3 cDNA clone was expressed in *E. coli*. The expressed protein had API activity; moreover, the expressed protein reacted with an antiserum raised against rat ribosomal protein S3. Thus S3 appears to be both a ribosomal component and the DNA repair endonuclease denominated API. In conformity with these findings, a cDNA encoding *Drosophila melanogaster* ribosomal protein S3 (the homolog of the rat protein) when expressed in *E. coli* also has API activity (M.R. Kelly, personal communication).

S3 may not be the only ribosomal protein that is involved in DNA repair. A *Drosophila* gene that encodes a separate apurinic/apyrimidinic endonuclease (AP3) was identified using antibody to a related human enzyme (Kelley et al., 1989); the *Drosophila* gene was used in turn to probe a HeLa cell cDNA library (Grabowski et al., 1991). The single human cDNA that was identified encodes a protein that has 66% identity with the *Drosophila* AP3 gene and is identical to human ribosomal protein P0 (Rich and Steitz, 1987) suggesting that this ribosomal protein is also an AP endonuclease. That P0 is induced in human cells by bifunctional alkylating agents that are used in the treatment of cancer and that cause DNA damage supports this deduction (Grabowski et al., 1992).

Finally, damage to DNA by UV irradiation in cells deficient in repair induces the ribosomal protein L7a (Ben-Ishai et al., 1990). No function for L7a in DNA repair has been established, nonetheless, the observation that it is induced by DNA damage and the precedent of S3 and P0 raise a suspicion that it may play a role in the process.

There are findings that can be construed to indicate that ribosomal proteins serve as specific transcription factors during development. In humans there are two transcribed genes for the ribosomal protein S4, one on the X and a second on the Y chromosome (Fisher et al., 1990). What is extraordinary is that the amino acid sequences encoded in the S4X and S4Y alleles differ at 19 of 263 positions; extraordinary since there are seldom more than a few amino acid differences in homologous ribosomal proteins amongst mammalian species and this great a deviation (almost 8%) is more than one would expect to find between human and *Xenopus laevis* (Wool et al., 1990). The S4Y gene is located in the sex-determining region of the Y chromosome and S4X escapes X-inactivation (Fisher et al., 1990). Both genes are transcribed in human cells; mRNAs specific for S4X and S4Y can be demonstrated by northern hybridization and ribosomes from male humans have 90% S4X and 10% S4Y (A. Zinn and D. Page, unpublished data). Thus it is possible that there are male- and female-type ribosomes.

S4Y is in a region of the Y chromosome that has been linked to certain forms of Turner Syndrome (cf. Fisher et al., 1990 for references and discussion). Turner females have gonadal insufficiency, and a variety of anatomic abnormalities including short stature, webbing of the neck, lymphadema, and aortic coarctation. The Turner phenotype obviously is a manifestation of a serious anomaly of development and it is not the only one associated with a mutation, usually a deficiency, of a ribosomal protein. There is a collection of mutations at about 50 separate loci in *Drosophila* that have been designated *Minute* (Lindsley and Grell, 1968). Flies with the *Minute* phenotype have delayed larval development, diminished viability, reduced body size, decreased fertility, thin bristles, and etching of the abdomen. The *Minute* phenotype is also a manifestation of development gone awry. The *Minute* genes that have been identified encode ribosomal proteins (Kongsuwan et al., 1985; Andersson and Lambertsson, 1990).

The explanation given to account for the Turner and *Minute* phenotypes is haploinsufficiency for a ribosomal protein (Fisher et al., 1990); the presumption is that a ribosomal protein insufficiency leads to an insufficiency of competent ribosomes at one or more critical stages of development. One cannot gainsay that this is a logical and economical explanation; a simple means of accounting for complex phenotypes consistant with the small amount of evidence that is available. However, this explanation implies a lack of the regulation of the synthesis of ribosomal proteins, something that is barely credible given a large body of evidence to the contrary (Warner et al., 1985); namely, that their synthesis is precisely adjusted so that equimolar amounts of each are available for packaging with stoichiometric amounts of rRNA. The expectation would be that if the S4Y gene were inactivated there would be an increase in the transcription of the S4X allele or an increase in the translation of the S4X mRNA. Indeed, if the S4X gene accounts for 90% of the protein in ribosomes the adjustment necessary would be small.

There is an alternate explanation: that during development certain ribosomal proteins regulate the expression of genes in a positive or negative manner by binding to DNA or that they participate with the RNA polymerases in transcription of selected genes; S4Y, for example, might serve as a transcription factor for genes that endow the individual with characteristics that we associate with maleness. With respect to this proposal, it may be significant that the degree of repression of the translation of individual ribosomal protein mRNAs in early *Drosophila* embryos varies during development (Al-Atia et al., 1985; Kay and Jacobs-Lorena, 1985). A similar observation, selective repression of the translation of ribosomal protein mRNAs during development, has been made for *Xenopus* as well (Pierandrei-Amaldi et al., 1982). This may reflect a need for specific ribosomal proteins (i.e. the mRNAs whose translation is not repressed) as factors for the regulation of development. This interpretation, of course, may reflect a biased reading of the literature. There most certainly are observations not entirely consistant with my explanation. The effects of combinations of 2 or 3 mutations at *Minute* loci do not increase the severity of the phenotype; this suggests that the loci code for gene products of similar function and favors the interpretation that the phenotype arises from their contribution to ribosome function. It is, however, also possible that the genes encode transcription factors with similar and coordinated functions in development.

There is evidence, albeit circuitous, for the involvement of a ribosomal protein in the import of proteins into mitochondria and as an effector of the v-*fos* oncogene. The viral p55 v-*fos* oncogene and its cellular homolog, c-*fos*, encode nuclear phosphoproteins that can form heterodimeric complexes with the product of the c-*jun* protooncogene (Ransome and Verma, 1990). Heterodimers of Jun with c- or v-Fos are regulators of transcription and constitutive overexpression of either v-*fos* or c-*fos* genes causes cell transformation

(Ransome and Verma, 1990). Revertants of v-*fos*-transformed rat cells were isolated by disruption of a v-*fos* transformation effector gene and a cDNA for this effector protein was isolated and the sequence of nucleotides determined (Kho and Zarbl, 1992). The protein encoded in the cDNA is related to *S. cerevisiae* MFT1 (Garrett et al., 1991). MFT1 participates, in some still to be defined manner, in the import of protein into mitochondria; inactivation of the *MFT*1 gene leads to a failure to transport an ATPase ß-subunit-lacZ fusion protein into the organelle (Garrett et al., 1991). The rat Fos effector protein is a ribosomal protein; the amino acid sequence derived from a rat S3a cDNA (Y.L. Chan, J. Olvera, V. Paz, and I.G. Wool, unpublished data) is identical to that reported for the v-Fos effector protein (Kho and Zarbl, 1992). The MFT1 gene encodes yeast ribosomal protein rp10 (Takakura et al., 1992) the homolog of rat S3a; in an alignment of the amino acid sequences of yeast ribosomal protein rp10 (MFT1) and rat S3a there are 149 identities in 254 possible matches (59%). Thus the proteins that participates in v-Fos mediated transformation and in the import of protein into yeast mitochondria are ribosomal proteins; in both cellular transformation in mammals and in organelle protein import in yeast these homologous ribosomal proteins may be functioning as a chaperone.

The membranes of the nucleus, of the endoplasmic reticulum, and of the Golgi apparatus are dispersed during mitosis and reassembled later (cf. Boman et al., 1992 for references). p27 is a peripheral membrane protein that is associated, in extracts of *Xenopus* eggs, with vesicles formed from the nuclear envelope and from the endoplasmic reticulum (K. Sullivan and K. Wilson, unpublished data). During mitosis p27 in egg extracts is phosphorylated whereas in interphase it is not. The sequence of amino acids in two tryptic peptides from *Xenopus* p27 were determined (K. Sullivan and K. Wilson, unpublished data): one was identical to an amino acid sequence in rat ribosomal protein S8 (residues 129-139) determined earlier by Chan et al., 1987; the sequence of the second was identical at 12 of 13 positions (residues 158-170). Thus the *Xenopus* homolog of rat ribosomal protein S8 is a membrane-associated protein whose phosphorylation and dephosphorylation is correlated temporarily with nuclear envelope disassembly at the beginning of mitosis and reassembly at its conclusion. Phosphorylation of p27/S8 also provides a plausible means for the partition of the protein between the ribosome and intracellular membranes.

These examples suggest that second functions of ribosomal proteins may prove to be the rule rather than the exception and lead us to predict that more will be discovered. In addition, other RNP assemblies, the spliceosome is a prime example, may contain bifunctional proteins of which some may be ribosomal proteins. Indeed, we know that the prediction has substance. Human B lymphocytes infected with Epstein-Barr virus synthesize two small nuclear RNAs called EBER 1 and 2 for Epstein-Barr encoded RNAs (Toczyski and Steitz, 1991). The EBERs bind proteins to form nuclear RNPs; both bind the La antigen and EBER 1 is also associated with a second protein designated EAP (EBER associated protein) (Toczyski and Steitz, 1991). EAP is ribosomal protein L22 (D.P.W. Toczyski, J.A. Steitz, and I.G. Wool, unpublished data).

Is there a precedent for the proposal we make for the origins of the ribosomal proteins and for their bifunctionality? Perhaps there is a paradigm. The crystallins are structural proteins of the lens that contribute to the refraction of light. Some crystallins have no other known function; others, especially species-specific crystallins, are identical to enzymes found in lesser amounts in other tissues (cf. Piatigorsky and Wistow, 1991 for references and discussion). Both of the α-crystallins, αA and αB, are related to heat shock proteins; the ß-and γ-crystallins of vertebrates are distantly related to the dormancy proteins of microorganisms which are also induced by stress. The most abundant mammalian-specific crystallin is identical to aldehyde dehydrogenase. In various species the crystallins include: lactate dehydrogenase-B, argininosuccinate lyase, and α-enolase (birds and reptiles); NADPH-dependent reductases (frogs); and glutathione S-transferases (cephalopods) (Piatigorsky and Wistow, 1991). The enzyme crystallins are bifunctional, just as some ribosomal proteins appear to be; bifunctionality seems to have been acquired by modification of gene expression, a phenomenon called gene sharing (Piatigorsky and Wistow, 1991). The presumption is that a transcriptional or posttranscriptional modification has allowed recruitment of the enzyme to the lens. Later there may have been evolutionary pressure for changes in the structure of the enzyme that improves its function as a crystallin but are neutral with regard to its catalytic activity; or there may have been gene duplication with separation of the functions. There are examples consistant with both modes of evolution (Piatigorsky and Wistow, 1991). The crystallin enzymes are present in lens in amounts that exceed reasonable catalytic need. Indeed, in some cases the enzymes no longer have catalytic activity because of posttranslational modification or because of alteration of the amino acid sequence in the case of gene duplication and separate evolution. Thus, lens crystallins share features with ribosomal proteins. These include one or more separate functions and the presumption of a means (most likely posttranslational modification) to effect partition of the protein to separate sites.

Ribosomal proteins from species that span the three kingdoms have zinc finger motifs; moreover, ribosomes and ribosomal subunits appear to contain Zn (Chan et al., 1992). The identification of zinc finger motifs leads me to suggest that at least some of the ribosomal proteins were recruited from amongst a set of molecules that had already acquired the amino acid sequences necessary for binding to nucleic acids. Consistent with this suggestion concerning the evolutionary origins of the ribosomal proteins are the observations that some have a second function that usually involves association with a nucleic acid. I recognize that the primordial sequence of events may have been the other way around from what I have postulated; that, for example, the *E. coli* ribosomal protein S1 was adopted by small RNA phages to increase the efficiency of their replicase and that the mammalian ribosomal proteins S3 and P0 were adopted as endonucleases for DNA repair.

ACKNOWLEDGEMENTS

I am grateful to my colleagues Drs. Yuen-Ling Chan, Anton Glück, and Katsuyuki Suzuki for advice and for helpful discussions. The research was aided by a grant (GM21769) from the National Institutes of Health.

REFERENCES

Al-Atia, G.R., Fruscolini, P., and Jacobs-Lorena, M., 1985, Translational regulation of mRNAs for ribosomal proteins during early *Drosophila* development, *Biochemistry* 24:5798.

Andersson, S., and Lambertsson, A., 1990, Characterization of a novel *Minute*-locus in *Drosophila melanogaster*: a putative ribosomal protein gene, *Heredity* 65:51.

Auer, J., Spicker, G., and Böck, A., 1989, Organization and structure of the *Methanococcus* transcriptional unit homologous to the *Escherichia coli* "Spectinomycin Operon". Implications for the evolutionary relationship of 70S and 80S ribosomes, *J. Mol. Biol.* 209:21.

Baker, R.T., and Board, P.G., 1991, The human ubiquitin-52 amino acid fusion protein gene shares several structural features with mammalian ribosomal protein genes, *Nucleic Acids Res.* 19:1035.

Ben-Ishai, R., Scharf, R., Sharon, R., and Kapten, I., 1990, A human cellular sequence implicated in *trk* oncogene activation is DNA damage inducible, *Proc. Natl. Acad. Sci. U.S.A.* 87:6039.

Berg, J.M., 1986, Potential metal-binding domains in nucleic acid binding proteins, *Science* 232:485.

Berg, J.M., 1990, Zinc finger domains: Hypotheses and current knowledge, *Annu. Rev. Biophys. Biophys. Chem.* 19:405.

Berthold, V., and Geider, K., 1976, Interaction of DNA with DNA-binding proteins. The characterization of protein HD from *Escherichia coli* and its nucleic acid complexes, *Eur. J. Biochem.* 71:443.

Block, R., and Haseltine, W.A., 1974, *In Vitro* synthesis of ppGpp and pppGpp, *in:* Ribosomes, Nomura, M., Tissières, A., and Lengyel, P., eds., Cold Spring Harbor Laboratory, Cold Spring Harbor.

Bowman, A.L., Delannoy, M.R., and Wilson, K.L., 1992, GTP hydrolysis is required for vesicle fusion during nuclear envelope assembly *in vitro*, *J. Cell. Biol.* 116:281.

Brown, D.D., 1984, The role of stable complexes that repress and activate eucaryotic genes, *Cell* 37:359.

Cashel, M., and Gallant, J., 1974, Cellular regulation of guanosine tetraphosphate and guanosine pentaphosphate, *in*: Ribosomes, Nomura, M., Tissières, A., and Lengyel, P., eds., Cold Spring Harbor Laboratory, Cold Spring Harbor.

Chan, Y.L., Devi, K.R.G., Olvera, J., and Wool, I.G., 1990, The primary structure of rat ribosomal protein S3, *Arch. Biochem. Biophys.* 283:546.

Chan, Y.L., Olvera, J., and Wool, I.G., 1987, The primary structure of rat ribosomal protein S8, *Nucleic Acids Res.* 15:9451.

Chan, Y.L., Suzuki, K., Olvera, J., and Wool, I.G., 1992, Zinc finger-like motifs in rat ribosomal proteins S27 and S29, submitted.

Dabbs, E.R., 1985, Mutant studies on the prokaryotic ribosome, *in*: Structure, Function, and Genetics of Ribosomes, Hardesty, B., and Kramer, G., eds, Springer-Verlag, New York.

Demple, B., and Linn, S., 1980, DNA N-glycosylases and UV repair, *Nature* 287:203.

Fisher, E.M.C., Beer-Romero, P., Brown, L.G., Ridley, A., McNeil, J.A., Lawrence, L.G., Willard, H.F., Bieber, F.R., and Page, D.C., 1990, Homologous ribosomal protein genes on the human X and Y chromosomes: Escape from X inactivation and possible implications for Turner syndrome, *Cell* 63:1205.

Friedman, D.I., Schauer, A.T., Baumann, M.R., Baron, L.S., and Adhya, S.L., 1981, Evidence that ribosomal protein S10 participates in control of transcription termination, *Proc. Natl. Acad. Sci. U.S.A.* 78:1115.

Garrett, J.M., Singh, K.K., Vonder Harr, R.A., and Emr, S.D., 1991, Mitochondrial protein import: isolation and characterization of the *Saccharomyces cerevisiae MFT1* gene, *Mol. Gen. Genet.* 225:483.

Grabowski, D.T., Deutsch, W.A., Derda, D., and Kelley, M.R., 1991, Drosophila AP3, a presumptive DNA repair protein, is homologous to human ribosomal associated protein PO, *Nucleic Acids Res.* 19:4297.

Grabowski, D.T., Pieper, R.O., Futscher, B.W., Deutsch, W.A., Erickson, L.C., and Kelley, M.R., 1992, Expression of ribosomal phosphoprotein P0 is induced by antitumor agents and increased in Mer⁻ human tumor cell lines, *Carcinogenesis* 13:259.

Henkin, T.M., Moon, S.H., Mattheakis, L.C., and Nomura, M., 1989, Cloning and analysis of the *spc* ribosomal protein operon of *Bacillus subtilis*: comparison with the *spc* operon of *Escherichia coli*, *Nucleic Acids Res.* 17:7469.

Imlay, J.A., and Linn, S., 1988, DNA damage and oxygen radical toxicity, *Science* 240:1302.

Kamen, R.I., 1975, Structure and function of the Qβ RNA replicase, *in*: RNA Phages, Zinder, N.D., ed., Cold Spring Harbor Laboratory, Cold Spring Harbor.

Kay, M.A., and Jacobs-Lorena, M., 1987, Selective translational regulation of ribosomal protein gene expression during early development of *Drosophila melanogaster*, *Mol. Cell. Biol.* 5:3583.

Kelley, M.R., Venugopal, S., Harless, J., and Deutsch, W.A., 1989, Antibody to a human DNA repair protein allows for cloning of a *Drosophila* cDNA that encodes an apurinic endonuclease, *Mol. Cell. Biol.* 9:965.

Kho, C.J., and Zarbl, H., 1992, *Fte-1*, a v-*fos* transformation effector gene, encodes the mammalian homologue of a yeast gene involved in protein import into mitochondria, *Proc. Natl. Acad. Sci. U.S.A.* 89:2200.

Kim, J., and Linn, S., 1988, The mechanisms of action of *E. coli* endonuclease III and T4 UV endonuclease (endonuclease V) at AP sites, *Nucleic Acid Res.* 16:1135.

Klug, A., and Rhodes, D., 1987, 'Zinc fingers': a novel protein motif for nucleic acid recognition, *Trends Biochem. Sci.* 12:464.

Kongsuwan, K., Quiang, Y., Vincent, A., Frisardi, M.C., Rosbash, M., Lengyel, J.A., and Merriam, J., 1985, A *Drosophila Minute* gene encodes a ribosomal protein, *Nature* 317:555.

Kraemer, K.H., de Weerd-Kastelein, E.A., Robbins, J.H., Keijzer, W., Barrett, S.F., Petinga, R.A., and Bootsma, D., 1975, Five complementation groups in xeroderma pigmentosum, *Mutation Res.* 33:327.

Kraemer, K.H., Lee, M.M., and Scotto, J., 1984, DNA repair protects against cutaneous and internal neoplasia: evidence from xeroderma pigmentosum, *Carcinogenesis* 5:511.

Kuhnlein, U., Penhoet, E.E., and Linn, S., 1976, An altered apurinic DNA endonuclease activity in group A and group D xeroderma pigmentosum fibroblasts, *Proc. Natl. Acad. Sci. U.S.A.* 73:1169.

Lindsley, D.L., and Grell, E.H., 1986, Genetic variations of *Drosophila melanogaster*, *Carnegie Inst. Washington Publ.*, 627.

Mason, S.W., Li, J., and Greenblatt, J., 1992, Direct interaction between two *Escherichia coli* transcription antitermination factors, NusB and ribosomal protein S10, *J. Mol. Biol.* 223:55.

Mende, L., Timm, B., and Subramanian, A.R., 1978, Primary structures of two homologous ribosome-associated DNA-binding proteins of *Escherichia coli*, *FEBS Lett.* 96:395.

Miller, J., McLachlan, A.D., and Klug, A., 1985, Repetitive zinc-binding domains in the protein transcription factor IIIA from *Xenopus* oocytes, *EMBO J.* 4:1609.

Myers, A.M., Crivellone, M.D., and Tzagoloff, A., 1987, Assembly of the mitochondrial membrane system: *MRP1* and *MRP2*, two yeast nuclear genes coding for mitochondrial ribosomal proteins, *J. Biol. Chem.* 262:3388.

Noller, H.F., Hoffarth, V., and Zimniak, L., 1992, Unusual resistance of peptidyl transferase to protein extraction procedures, *Science*, 256:1416.

Ohkubo, S., Muto, A., Kawauchi, Y., Yamao, F., and Osawa, S., 1987, The ribosomal protein gene cluster of *Mycoplasma capricolum*, *Mol. Gen. Genet.* 210:314.

Piatigorsky, J., and Wistow, G., 1991, The recruitment of crystallins: new functions precede gene duplication, *Science* 252:1078.

Pierandrei-Amaldi, P., Campioni, N., Beccari, E., Bozzoni, I., and Amaldi, F., 1982, Expression of ribosomal-protein genes in *Xenopus laevis* development, *Cell* 30:163.

Ransone, L.J., and Verma, I.M., 1990, Nuclear proto-oncogenes *Fos* and *Jun*, *Annu. Rev. Cell Biol.* 6:539.

Redman, K.L., and Rechsteiner, M., 1989, Identification of the long ubiquitin extension as ribosomal protein S27a, *Nature* 338:438.

Rice, P.A., and Steitz, T.A., 1989, Ribosomal protein L7/L12 has a helix-turn-helix motif similar to that found in DNA-binding regulatory proteins, *Nucleic Acid Res.* 17:3757.

Rich, B.E., and Steitz, J.A., 1987, Human acidic ribosomal phosphoproteins P0, P1, and P2: analysis of cDNA clones, *in vitro* synthesis, and assembly, *Mol. Cell. Biol.* 7:4065.

Schwabe, J.W.R., and Rhodes, D., 1991, Beyond zinc fingers: steroid hormone receptors have a novel structural motif for DNA recognition, *Trends Biochem. Sci.* 16:291.

Steitz, T.A., 1990, Structural studies of protein-nucleic acid interaction: the sources of sequence-specific binding, *Quart. Rev. Biophys.* 23:205.

Stern, S., Powers, T., Changchien, L.M., and Noller, H.F., 1989, RNA-protein interactions in 30S ribosomal subunits: folding and function of 16S rRNA, *Science* 244:783.

Suryanarayana, T., and Subramanian, A.R., 1978, Specific association of two homologous DNA-binding proteins to the native 30S ribosomal subunits of *Escherichia coli*, *Biochim. Biophys. Acta* 520:342.

Tanaka, T., Aoyama, Y., Chan, Y.L., and Wool, I.G., 1989, The primary structure of rat ribosomal protein L37a, *Eur. J. Biochem.* 183:15.

Takakura, H., Tsunasawa, S., Miyagi, M., and Warner, J.R., 1992, NH_2-terminal acetylation of ribosomal proteins of *Saccharomyces cerevisiae*, *J. Biol. Chem.* 267:5442.

Toczyski, D.P.W., and Steitz, J.A., 1991, EAP, a highly conserved cellular protein associated with Epstein-Barr virus small RNAs (EBERs), *EMBO J.* 10:459.

Warner, J.R., Elion, E.A., Dabeva, M.D., and Schwindeger, W.F., 1985, The ribosomal genes of yeast and their regulation, *in*: Structure, Function, and Genetics of Ribosomes, Hardesty, B., and Kramer, G., eds., Springer-Verlag, New York.

Woodgate, R., Rajagopalan, M., Lu, C., and Echols, H., 1989, UmuC mutagenesis protein of *Escherichia coli*: Purification and interaction with UmuD and UmuD', *Proc. Natl. Acad. Sci. U.S.A.* 86:7301.

Wool, I.G., 1979, The structure and function of eukaryotic ribosomes, *Ann. Rev. Biochem.*, 48:719.

Wool, I.G., Endo, Y., Chan, Y.L., and Glück, A., 1990, Structure, function, and evolution of mammalian ribosomes, *in*: The Ribosome: Structure, Function, and Evolution, Hill, W.E., Dahlberg, A., Garrett, R.A., Moore, P.B., Schlessinger, D., and Warner, J.R., eds., Amer. Soc. Microbiol., Washington, D.C..

Yaguchi, M., Roy, C., Reithmeier, R.A.F., Wittmann-Liebold, B., and Wittmann, H.G., 1983, The primary structure of protein S14 from the small ribosomal subunit of *Escherichia coli*, *FEBS Lett.* 154:21.

INDEX

Ribosomal proteins *(cont'd)*
L9, 417, 523
L10, 522, 680
L10(L12), 343
L11, 343, 679
  binding site, 344
  mutant sequence, 706
L12, *see also* L7/L12, 680, 681, 730
  paralogous genes, type I and type II, 681
  structural homologies, 542
L18, 412, 416
L22 (human), 734
L23, 471
  HmaL23 location, 512
  location, 511
  N-terminal domain, 512
L24
  assembly initiation, 176
  assembly mutant, 176
  binding to 23S rRNA, 176
L27, 460, 462
L30, 114
  structural homologies, 542
L33, 460
L37/L37a (rat), 728
mitochondrial gene organization, 602, 604
mitochondrial L16, 590
mitochondrial MRP7 (L27), 588
mitochondrial nuclear genes, 596
mitochondrial ribosomal proteins, 601
mitochondrial RML16, 588
mitochondrial S5 (GTP binding), 582
mitochondrial transcription, editing, 602
mitochondrias, 603
mitochondria: transfer of genes to the nucleus,
    605
NS1, 730
NS2, 730
nuclear genes of chloroplast proteins
  organization, 569
origin of the ~, 730
overexpression, 534
p27 *Xenopus*, 734
paralogous L12 genes, 681
plastid CL21, 569
plastid CL22, 568
plastid CS1, 566
plastid CS22, 569
plastid *rpl*2, 609
plastid S1, 567
P0 (human), 732
post-translational modifications, 174
protein 59, 113
psRPs, 701
rabbit S6, 415
S1, 473, 731
S1 to S21, 691
S2
  crosslinking to S5, 535
S3
  crosslinking to S5, 535
S3 (rat), 732
S3a (rat), 734

Ribosomal proteins *(cont'd)*
S4, 116, 376, 473
  crosslinking to S5, 535
S4X and S4Y (human), 732
S5, 116, 439, 534
  crosslinking to r-proteins, 535
  position on 30S, 538
  structural homologies, 542
  structure, 535
  translational fidelity, 538
S7, 460, 462, 473
  crosslinks to mRNA, 426
S8, 684
  crosslinking to S5, 535
S8 (rat), 734
S9, 731
S10, 114, 731
  attenuation of transcription, 135
  regulation of S10 operon, 131
S10 leader, 131
S10 operon, 131
  repression of translation, 136
  secondary structure of the leader, 132
S11, 460
S12, 376, 473
  and ribosome assembly, 185
  crosslinking to S5, 535
  hyperaccuracy, 386
  hyperaccurate ribosomes, 389
  streptomycin resistance, 389
  thermo-sensitive mutant, 185
S13, 113, 461, 462
S14 (rat), 729
S16
  crosslinking to S5, 535
S17, 534
  crosslinking to S5, 535
  fidelity, 540
  in assembly, 540
  NMR spectra, 540
  resistance to neamine, 540
  structure, 541
S18, 473, 684
  crosslinks to mRNA, 426
S19, 461, 462
S21, 473
  crosslinks to mRNA, 426
S27 (rat), 728
S28, 114
S29 (rat), 728
yeast L10
  phosphorylation, 114
yeast L32, 111, 113
  role of splicing, 113
zinc-finger like motifs, 728
Ribosomal RNA genes in yeast
  transcription, 101
Ribosomal surface topography, 516
Ribosome
  assembly (*see also* rRNA processing), 173
    impaired assembly of 50S, 174
    ~ mutants, 174
    suppressors of assembly mutants, 174
  crystals of ~, parameters, 399